Clinical Kinesiology and Biomechanics

Gordon J. Alderink • Blake M. Ashby

# Clinical Kinesiology and Biomechanics

## A Problem-Based Learning Approach

Gordon J. Alderink
Coopersville, MI, USA

Blake M. Ashby
Grand Valley State University
Grand Rapids, MI, USA

ISBN 978-3-031-25324-9 ISBN 978-3-031-25322-5 (eBook)
https://doi.org/10.1007/978-3-031-25322-5



This Springer imprint is published by the registered company Springer Nature Switzerland AG
The registered company address is: Gewerbestrasse 11, 6330 Cham, Switzerland

*From Gordon J. Alderink: "To those naturally curious who are self-directed and aspire to life-long learning and conscientization."*

*From Blake M. Ashby: "To my wife, Meagan, for her endless support, love, and encouragement."*

# Preface

The intention of this text is to provide a template for teaching kinesiology and biomechanics to a clinical audience using a hybrid problem-based learning (PBL) format. Problem-based learning originated at McMaster University in the early 1970s when the new medical school designed a revolutionary curriculum using medical problems as the basis for learning. Since that time, PBL has been accepted as a legitimate pedagogical approach, supported by many theories of adult learning, in medical schools, programs of allied health professions, e.g., physical and occupational therapy, as well as in undergraduate disciplines. Furthermore, PBL education outcomes, e.g., board examinations for medical practitioners, graduation success, etc., rival those of the more traditional, teacher-centered pedagogical models.

Let me digress. Within the first 5 years of my teaching career as a Physical Therapy (PT) faculty member at Grand Valley State University, I received formal recognition by the GVSU Alumni Association for teaching excellence. I was recognized by former students for being knowledgeable in my areas of expertise, ability to deliver the information effectively, enthusiasm, and passion for student learning. I used the traditional lecture/discussion teaching method, mostly imitating professors from my own experiences as both an undergraduate and graduate student. By the way, most junior faculty who are trained in disciplines other than education are forced to do this because we lack formal training in educational philosophy and pedagogy. Being recognized by my students as an outstanding teacher was gratifying and I assumed that they had the same passion for learning as I had for teaching. So I was confused, as time went on, by some students' seeming inattention, i.e., lack of engagement, with what I was presenting. I never confronted students, but those episodes forced me to confront myself, and rethink my classroom approach. I concluded that, although part of the problem was with those particular students, I needed to do something different. First, reflect on what I was doing, and second seek some assistance. I concluded that I had deficits in my knowledge and understanding of education in general and decided to engage in my own learning in that discipline. I contacted a colleague in our School of Education and asked for help. My journey of life-long learning about learning began. Fortunately, our university had an office of faculty development that offered workshops each Fall, so that is where I started. From those workshops, I began to collect resources on educational theory and began to read, and along with that, dialogue with some new-found colleagues in our College of Education. Sometime later I attended a workshop sponsored by the Physiotherapy Program at McMaster University. Over the course of 2 days, I was introduced to the theoretical construct of PBL and observed and conversed with both students and faculty. Every student and student group I encountered was always fully engaged with the material they were learning (getting on their own), and their peers and faculty. I was convinced that PBL was the way to teach, and, thus, began my transition to learning and practicing different ways of using the method in my classes. I was confronted by several major constraints: (1) curriculum, since the PT curriculum design at my University was traditional, and (2) student resistance since PBL required them to not only re-think what teaching excellence was, but also their own effort, and (3) faculty and administration resistance. These constraints remain today, but that is another story.

We offer a two-semester sequence of courses for first-year PT students: Clinical Kinesiology and Biomechanics I taught in the fall, and Clinical Kinesiology and Biomechanics II taught in the winter. I used to teach both courses, but now only teach the second course. The fall course is taught using a traditional educational pedagogy, with a focus on functional anatomy (our students take gross human anatomy concurrently) and the osteo- and arthrokinematics for the trunk, pelvis, and extremities. Since our curriculum constrained how I could modify biomechanics, I could not use the authentic PBL model, but developed a modified-PBL format for the winter course, using clinical vignettes and an anatomical regional approach.

Words matter. We chose to use both kinesiology and biomechanics in the course titles for the following reasons: (1) the words provided some degree of continuity, (2) kinesiology suggests emphases, in the first course, on something of vital importance to physical therapists, that is, obtaining a broad understanding of human movement, and (3) biomechanics suggests that the second course would utilize first principles of kinematics, but emphasize the application of classical mechanics with the fields of biology and physiology. There are several excellent texts on kinesiology and biomechanics for health care practitioners in medicine, physical therapy, and occupational therapy. To name one, I have used Neumann's *Kinesiology of the Musculoskeletal System, Foundations for Rehabilitation*, an encyclopedia of kinesiology. I do not intend to offer a competing text of kinesiology. What my text intends to do is illustrate how one can apply abstract biomechanical principles to explain different aspects of common clinical problems.

In Chap. 1, we detail the theoretical framework of both teacher-centered and student-centered learning. Contrasting these theories, we attempt to demonstrate the advantages of the PBL model. Details on the roles of the tutor and students will be provided, as well as how groups are set up and operationalized.

The clinical problems posed in this text, although common and appear straightforward, are ambiguous and complex, so when we use the word "solve," we do not mean to look for and find a single solution. Most of the solutions presented for each regional problem entail a list of related hypotheses that suggest a level of clarification and understanding. Often, we pose additional questions for the reader to pursue on their own, hopefully using similar approaches to those presented in the text. When the analysis of the problem is complete, the hypotheses developed could certainly lead to a discussion of how one might create an intervention, based on the same biomechanical concepts used to explain the problem. However, we leave that exercise for the reader.

The following chapters take a regional approach:

In Chap. 2, we provide basic biomechanical definitions and render a broad overview of the field, discussing how we generally approach using biomechanics to solve and understand neuro-muscular problems.

Chapter 3: Shoulder Girdle Complex – A 57-year-old woman presents with a medical diagnosis of left shoulder impingement. She has multiple co-morbidities. The primary objective is to, given her medical history, clarify the medical diagnosis from a mechanical perspective. We pose a statics problem at the glenohumeral joint to simulate a clinical testing position for the purpose of examining joint bone-on-bone forces.

Chapter 4: Elbow – An adolescent baseball player has significant medial elbow pain and can no longer perform at a functional level. When working with athletes, it is absolutely critical to have a profound understanding of the biomechanics of the primary movement patterns used in each specific sport. For this case, having that understanding also provides the background for detailing the mechanism(s) of injury, which aids in explaining the player's signs and symptoms. This young man had a reproduction of one aspect of his elbow pain when he underwent resisted muscle testing. We use a statics problem related to the manual muscle testing positions for the elbow flexors and extensors to provide insight into his pain.

Chapter 5: Wrist – A young male adult carpenter falls and fractures his right (dominant hand) scaphoid, but presents with residual pain and dysfunction as he begins his rehabilitation. This case demands a detailed exploration of the injury mechanism, deleterious mechanical effects of immobilization, and mechanical explanation of his present pain generators.

Chapter 6: Hand – Surgical repair of a lacerated flexor pollicis longus tendon, and the rehabilitation protocol that follows affords the student an opportunity to apply mechanical principles related to aspects of the normal tendon function, the inflammatory cycle, including healing constraints, treatment decisions, and the basis for the rehabilitation protocol.

Chapter 7: Cervical Spine – A lateral whiplash injury is not common, but includes many of the same features of a flexion/extension injury including the general mechanism(s) of injury, the number of soft tissues that could sustain a strain injury, and the often-chronic nature of continued signs and symptoms. The reader can follow a qualitative analysis of biomechanical trauma, using this detailed mechanical analysis to help clarify the medical diagnosis.

Chapter 8: Thoracic Spine and Rib Cage – This middle-aged male, long-time smoker apparently sustained acute strain injuries related to splitting wood. This case will afford the student to examine the mechanical dysfunctions related to chronic obstructive pulmonary disease and provide another chance to perform a qualitative analysis of biomechanical trauma.

Chapter 9: Lumbo-Pelvic Complex – A 35-year-old man presents with chronic low back pain likely related to his football playing days and lumbosacral instability. This case forces the student to examine the types of external forces and moments experienced by middle linebacker football players that may have contributed to this chronic condition. Because this individual does a lot of repetitive lifting and works in semi-flexed postures, simulating those circumstances and solving associated static forces provides insight into the role his abnormal lumbosacral angle plays in the etiology of his injury, as well as his current pain status.

All aspects of observational and instrumented gait analysis will be incorporated in the final three chapters. Students have an opportunity to explore the use of electromyography, three-dimensional kinematics and kinetics, and inverse dynamics to compare and contrast normal and pathological gait.

Chapter 10: Hip – One of the most common lower extremity pathologies described in the literature is hip degenerative joint disease (DJD). Students explore mechanical hypotheses related to the development of hip DJD and examine the mechanical problems in a pre-surgical presentation of a clinically mild case of advanced disease. Instrumented gait measures are introduced and the reader is presented with a systematic algorithm for performing an observational and three-dimensional gait analysis. Along with knowledge gained about how stress is distributed at the hip joint, a static analysis of the bone-on-bone forces there assists us in understanding the etiology of his DJD.

Chapter 11: Knee – A young lady sustained multiple lateral knee meniscus injuries, which ultimately resulted in two failed meniscal allograph procedures. Although the tibial-femoral joint is described as a hinge joint, implying a relatively simple mechanical system, this case affords the student to peel off layers of anatomy and biomechanics to gain a more profound mechanical understanding. Gait deviations/pathologies identified using instrumented gait analysis detail the results of this chronic condition, not its cause.

Chapter 12: Ankle/Foot – An elite high school cross-country and track runner was referred to a clinical motion analysis lab for a biomechanical explanation for recurrent stress fractures. This case allows students to begin to appreciate the complexity of the foot/ankle anatomy and biomechanics as they attempt to look for the relationship between anatomical, historical, and physical examination, and instrumented gait analysis features to help answer the question: why?

Each chapter includes the clinical vignette and essential anatomical, kinesiological, biomechanical, and neurological information that one might gather in order to understand the clinical problem from a biomechanical perspective. Perhaps the most important feature of each chapter will be the integration of general principles with the particularity of each case to make conclusions and draw generalizations that could be used in future cases. The uniqueness of this text is the integration and explication of clinical cases from the application of the most commonly used biomechanical principles. The approach taken in this text may appear to be no different from the typical deterministic or reductionist approaches that are typically used to explain complex processes. The reader will notice that there is some redundancy across the text. This is by design because we believe that repetition, within the context of difference, can help seal memory.

Let's explain. The human body, as a system, is both complicated and complex. Many use the words complicated and complex interchangeably. However, these words, which represent a specific concept, should not be conflated because they do not represent the same kind of system. Complicated systems are made up of many parts, or subsystems and modules, that must function effectively in order for the system to work properly. But the parts of a complicated system function independently. That is, one subsystem does not depend on any of the other subsystems that make up the whole. In the reductionist approach to the analysis of a complicated system, the system would be deconstructed, i.e., divided into its parts, where each part could be examined in isolation, and finally reconstructed, but with a fuller understanding of the system. Human-engineered mechanical devices, or systems, are complicated, i.e., they have many moving parts that need to work efficiently, and a reductionist approach was used to engineer and build them. Complex systems, like the human body, cannot be analyzed or understood using reductionism, although most of the time that is what biomechanists and clinicians do because, believe it or not, it makes things simpler, and in many cases that is where we often have to begin a process.

What makes complex and complicated systems different? Complex systems have parts that work interdependently, so the effectiveness and efficiency of the system depend on the integration and cooperation of multiple systems, subsystems, modules within subsystems, etc. Researchers in physics, biology, cognitive and behavior psychology, and more recently in motor control have been using informational, mathematical, statistical, and other measurement tools for the past several decades to explore the complexity of the natural world.

Complex systems theory (an aspect of General Systems Theory), also known as dynamical systems theory, is multifaceted and has application to biomechanics, and the application of biomechanics in clinical practice. This text is not the place to provide details about dynamical systems theory, but let me list some of its elementary concepts, and how I see the relevance of the theory to the cases in this text. First, the theory offers two major challenges for the understanding of dynamic patterns of complex systems, explaining how: (1) a pattern(s) is constructed from a very large number of material components – called the complexity of substance, and (2) not just one pattern, but many patterns are produced to accommodate different circumstances – called pattern complexity. Here are some elementary concepts of complex systems:

- Have many inter-related, inter-acting parts (degrees of freedom or DoF) that operate in a coordinated, non-linear fashion.
- Have DoF that characterize emerging patterns, which are created by the coordination of the inter-related parts, called order parameters, which, in turn, influence the behavior of parts.
- Self-organize: patterns arise spontaneously as the result of interacting components so that the motion of the whole, for example, is not only greater but different than the sum of the motion of the parts.
- Are dissipative and far from equilibrium, where many DoF are suppressed or controlled, with only a few contributing to system behavior at any one time.
- Their loss of stability occurs near non-equilibrium phase transitions, which can give rise to new or different patterns and/or switching between patterns (i.e., system fluctuations, flexibility, and adaptability).

Learning biomechanics in the context of clinical cases using a PBL approach, I believe, induces us to practice thinking about, and applying, complexity theory to the practice of learning. For each case, we will identify key facts provided in the vignette that serve as clues about which systems we need to study to clarify the problem from a biomechanical perspective. For example, in the first case, we learn that our patient had sustained a proximal humeral fracture several years prior to her referral to us. One of the questions we should address is, "what might the relationship be between this prior injury and the current problem?" Subsequently, we decide to learn more about the type and magnitude of forces/loads that cause fractures, and since the fracture occurred close to the joint, how the fracture-related load might have affected glenohumeral joint integrity. In this process, we not only begin to integrate history taking and understanding of key facts but our exploration and findings also lead us to integrate an understanding of possible mechanisms of injury with anatomy and mechanics. In essence, I anticipate we will integrate reductionist and complex thinking and problem-solving. Let's have some fun and see how it goes.

Typically, each problem vignette should be guided by behavioral objectives. When I taught Clinical Kinesiology and Biomechanics II, I provided general behavioral objectives for the course and required students to develop specific objectives for each case. To help students in that exercise, I provided them with references on writing objectives, e.g., Bloom's taxonomy, and wrote objectives for the first case, i.e., shoulder girdle, as a guideline for their own work. I will provide the same information for the reader as an appendix.

Several additional appendices in this text will include physical conversion factors, trigonometric, geometric, and calculus review, anthropometric measurement data, concepts of forces and moments, principles of deformable body mechanics, a general static equilibrium problem-solving approach, and a review of biomechanical instrumentation, e.g., electromyography, motion capture, etc.

Coopersville, MI, USA Gordon J. Alderink

## References

Kelso JAS (1995) Dynamic patterns, the self-organization of brain and behavior. The MIT Press, Cambridge, MA

Neumann DA (2017) Kinesiology of the musculoskeletal system, foundations for rehabilitation, 3rd edn. Elsevier, St. Louis

Tranquillo J (2019) An introduction to complex systems, making sense of a changing world. Springer, Cham

von Bertalanffy L (1969) General systems theory, foundations, development, applications, 2nd edn. George Braziller, Inc., New York

# Acknowledgments

I want to take this opportunity to acknowledge a number of individuals who, directly and indirectly, have helped shape me as a person, teacher-learner, and biomechanist, and made this text possible. I apologize to those who I may have inadvertently overlooked.

I begin by thanking my wife Sally, who has supported, without hesitation, my professional endeavors for 44 years. She has, indeed, been steadfast. Sally also helped me grow as an individual. For example, oftentimes, in her wisdom and using humor, she often asked me (tongue in cheek) "Who do you think you are?", in "her" way, which always forced me to take a step back and reflect a bit. Moreover, during my years as a full-time educator and part-time doctoral student, Sally, and my daughters, Jenny and Liz, were left to their own devices more times than I can count. So, in addition to Sally, I want to thank Jenny and Liz for allowing me the time I needed to teach, study and do biomechanics. I also want to acknowledge my parents, Marian and Gordon, who were models of unconditional love, self-direction, and life-long learning. I have never forgotten what my mother would say to me just about every day as I left the house for school or to go play ball, "Do the best that you can today." It seems to me that this mantra was good advice and a guide for how to conduct one's life. It has served me well.

I also want to acknowledge Blake Ashby, my co-author, colleague, mentor, and friend. This project would not have achieved the level of excellence I was seeking without our conversations, and his readings, questions, criticisms, and writings. I also acknowledge the skill, creativity, attention to detail, and desire for excellence provided by Emily Hromi, this text's primary illustrator. I am indebted to her ability to communicate, assist in organizing the project's needs and coordinate the efforts of illustrators Lindsey Behrend, Michelle Padley, and Cecilia Smith. Blake and Emily glimpsed my vision for this book and took it to another level. Thanks also to Springer, Michael McCabe (the project editor) for the opportunity and Cynthya Pushparaj for her technical guidance.

As we make our way in the world, we are afforded innumerable opportunities and choices. I want to highlight a few of the opportunities afforded to me that likely led to this book project. As a physical therapy student, my first course in kinesiology was taught by Dr. James Youdas. Although I had to remediate my final examination, my enthusiasm for the discipline was set by Dr. Youdas, eventually influencing my decision to pursue a doctoral degree in engineering mechanics/biomechanics. Early in my practice as a physical therapist, I began a master's degree in exercise physiology at the University of Michigan and had the opportunity to study muscle physiology under the mentorship of Tim White. After working with Dr. White, and his cadre of graduate students, over a 2-year period, I was convinced that an academic career was what I wanted. I was fortunate to be able to join the Physical Therapy faculty at Grand Valley State University in 1984. I want to recognize Drs. Herman Triezenberg and Elizabeth Mostrom, who saw something in an untested, gullible, but enthusiastic physical therapy clinician and offered me the opportunity to teach physical therapy. Drs. Triezenberg and Mostrom set standards of excellence and expectations for faculty and students at Grand Valley and were instrumental in my development as an academician. Early in my tenure at Grand Valley I designed and began teaching a course in clinical biomechanics. Preparation for that course cemented my desire to pursue doctoral work in biomechanics. Somehow (I don't recall how) I was encouraged to meet with Dr. Robert Soutas-Little, the director of a clinical master's degree in biomechanics offered by the College of Osteopathic Medicine at Michigan State University. To make a long story shorter, Robert outlined for me how I could eventually obtain a doctoral degree in engineering mechanics, with an emphasis in biomechanics at Michigan State University. I began my training in biomechanics by initiating a second master's degree at Michigan State. I want to thank Robert for taking an interest in me, for serving as my first mentor, for his insistence on a commitment to learning, and for showing me a "road not taken." Working with Robert and other graduate students in a motion analysis laboratory at Michigan State University provided many opportunities for learning and to begin networking with biomechanists nationally, and set the stage for my later work in clinical instrumented gait analysis at Mary Free Bed Rehabilitation Hospital. In fact, two of my graduate school classmates, Brock Horsley and David Marchinda helped me establish and manage the motion analysis laboratory at Mary Free Bed for several years. I also want to acknowledge several other individuals who helped me along the way during my graduate years at Michigan State University: Bob Hubbard (primary dissertation advisor), Gary Cloud, Clarence Nicodemus, Tammy Reid-Bush, Phil Greenman, Yasin Dhaher, and Krisanne Chapin. I want particularly to recognize the late Dr. Hubbard who shared my desire to help bridge engineering and clinical practice and consistently reminded me to create a balance between scientific rigor, clinical relevance, and life. I now recall that one of my biomechanics instructors at Michigan State asked me why I wanted to pursue a doctoral degree in engineering when I already had a fulfilling profession, i.e., physical therapy. I replied that I did not intend to leave physical therapy but wanted to use an engineering degree to assist in my work and that I was sure that by obtaining a doctoral degree that I could make a difference. I can see now that this book project can be used to help bridge engineering and physical therapy and that it could make a difference for some aspiring students and clinicians.

My thinking and teaching have been influenced by many other individuals, including my physical therapy faculty colleagues in the Department of Physical Therapy and Athletic Training at Grand Valley State University: Todd Sander, Dan Vaughn, Brianna Chesser, Meri Goehring, Mary Green, Cathy Harro, Barb Hoogenboom, Lisa Kenyon, Bonnie Kinne, Yunju Lee, Karen Ozga, Jon Rose, Mike Shoemaker, Corey Sobeck, and Laurie Stickler. Two other faculty at Grand Valley who taught anatomy to our physical therapy students, Tim Strickler and Brian Curry, also need to be recognized. These faculty colleagues were all committed to excellence in teaching, practice, and research and were wonderful inspirations and models for me and our students. In particular, I want to acknowledge my former department chair, Dr. John Peck. When I began experimenting with the use of a hybrid problem-based teaching-learning approach in Clinical Kinesiology and Biomechanics II, John was encouraging and supportive. This became more difficult in years when students complained to him about my method of teaching. When we met to discuss student concerns, he asked pertinent critical questions, then listened as I addressed each concern with hints of compromise in some cases, and rebuttal and defense of my methods in other instances. In the end, John never wavered in his support of what I was trying to accomplish. I wish also to acknowledge other faculty colleagues in Grand Valley's School of Engineering, Dr. Samhita Rhodes, Dr. John Farris, and Dr. Wendy Reffeor, and the Department of Statistics, Dr. David Zeitler. Finally, I would be remiss if I did not acknowledge the professional and personal mentoring I gathered from Stephen Glass, Tonya Parker, Peter Loubert, Amy Lenz, and Jennifer Yentes.

The biomechanics community is relatively small but has been a rich source of knowledge, skill, and professional development, all of which have provided a rich resource for this book project. In particular, my membership in two biomechanics societies: the Gait and Clinical Movement Analysis Society (GCMAS) and the American Society of Biomechanics (ASB) has provided a network of information and collegiality. I was a charter member of the GCMAS, and years later was appointed to the Commission for Motion Laboratory Accreditation (CMLA). I want to acknowledge the present and former Board members of the CMLA: Robert Kay, James McCarthy, Aloysia Schwabe, Wendy Pierce, Joseph Krzak, Audrey Zucker-Levin, John Henley, Jason Long, James Carollo, Jean Stout, Sylvia Ounpuu, Kristen Pierz, Jon

Davids, Wayne Stuberg, Katherine Alter, James Richards, Dennis Matthews, Freeman Miller, Tom Novacheck, and Juan Garbalosa. With these professional colleagues, I have gained clinical biomechanics insight, improved my critical thinking skills, and had opportunities to lead.

Over the years, I have also had the occasion to develop professional and personal relationships with equipment and software vendors and want to recognize the engineers and biomechanists from Vicon, Advanced Mechanical Technology, Inc. (AMTI), Motion Lab Systems, and C-Motion, entities I have worked with closely over many years, and these special people: Tom Kepple, Scott Selbie, Katie Bradley, Edmund Cramp, Cindy Samaan, Gary Blanchard, Jeff Ovadya, Nev Pires, John Porter, Felix Tsui, Katlin Nolte, Mike Braman, Korey Herber, and Mike Kocourek.

I can't help myself…but I am now going to "name-drop." Over the past 25 years during participation at the annual conferences of the GCMAS and ASB, I have had the opportunity to hear presentations from, and later meet and engage with, legacy bio- and clinical scientists, David Winter, Jacquelin Perry, David Sutherland, James Gage, Kenton Kaufman, Stephen Piazza, and Scott Delp, to name a few. These icons of basic and clinical biomechanics were great teachers and researchers who have lived professional and personal lives of integrity, made great contributions to the discipline, and served as models of excellence who have influenced me in many ways. I have been fortunate to have benefitted from reading their work and interacting with them in small ways.

Finally, I want to recognize the many undergraduate and graduate students who I have interacted with in the biomechanics laboratory and classroom. I am especially appreciative of those students who appeared to understand, in real time, what I was attempting to do with problem-based learning, and who then fully engaged with the material, the process, and me, their teacher-learner. The dialectic of student enthusiasm and resistance to my teaching philosophy provided a constant source of energy that fueled my desire to try and get it right.

This project was partially funded by Grand Valley State University through two sources: the College of Health Professions Supplemental Professional Development Fund and the University's Center for Scholarly and Creative Excellence.

Gordon J. Alderink

Writing acknowledgments is always tricky. There are too many people to count who have influenced my personal and professional development in meaningful and positive ways, so I apologize to those who are left out. First and foremost, I acknowledge my wife, Meagan. She is the best thing to ever happen to me and has supported me unreservedly for the past 15-plus years. Along with her, I thank my children: Tiffany, Addie, Kaiya, Jace, and Lexi for their willingness to share me with this book. (They don't know this yet, but I think they have more than earned a trip to Disney World later this year.) I cannot neglect to mention my wonderful parents, Wayne and Janice Ashby, who instilled in me sincere faith and the value of diligently seeking learning and truth.

Then there is my colleague, collaborator, and friend, Gordy Alderink. I know of no one who loves and embraces learning more than Gordy. Only a special person would pursue a Ph.D. in engineering without an undergraduate degree in engineering while working full-time as a physical therapy professor. Biomechanics is by its nature an interdisciplinary field and Gordy's training and expertise in both physical therapy and engineering mechanics have allowed him to contribute in unique and consequential ways by "bridging the gap" throughout his career. This book was his vision throughout. I thank him for pulling me into this project as I have learned so much through the experience. He was on the search committee that brought me to Grand Valley State University (GVSU) and became my first collaborator when I arrived. Since then, I have benefited greatly from his mentoring and assistance as we have worked together on numerous combined projects over the years.

My first interest in biomechanics, although I did not know the word at the time, was ignited by my high school AP physics teacher and cross-country coach, the late Mike Guarino. I loved how he used physics principles to explain how we could optimize our running styles and inspired me to be a better runner and scholar than I thought I could be. I studied mechanical engineering as an undergraduate student at Utah State University. I truly enjoyed my studies, but I realized that my passion was not necessarily in designing machines made of metal and plastic. As I pursued my graduate studies at Stanford University, I confirmed that my passion lay in applying mechanical engineering principles to the human body. I am grateful to my first graduate adviser, Dr. Jean Heegaard, who nurtured my interest in biomechanics and willingly invested countless hours tutoring me in the difficult concepts of applying advanced dynamics and optimization principles to modeling the human body. I also thank my second graduate adviser, Dr. Scott Delp, who picked up where Jean left off, patiently mentoring and guiding me through the completion of my Ph.D. As a student at Stanford, I benefited and learned from my associations with numerous other brilliant biomechanists including Sylvia Blemker, Saryn Goldberg, Jill Higginson, Kate Saul, Rob Siston, May Liu, Jay Henderson, Clay Anderson, Felix Zajac, Wendy Murray, Deanna Asakawa, Darryl Thelen, Allison Arnold, Rick Neptune, Thor Besier, Tom Andriacchi, Gene Alexander, Matt Kaplan, Rich Bragg, and Ajit Chaudhari.

I acknowledge all of my friends and colleagues from GVSU in the School of Engineering, Department of Physical Therapy and Athletic Training, Department of Occupational Science and Therapy, Department of Movement Science, and Department of Statistics. There are too many to name, and I know I would leave someone out, so I regretfully refrain from listing you individually. Please know that I recognize and am truly grateful for your significant contributions to my personal and career journey.

Finally, I express appreciation to my students, both those who have taken my classes and those who have conducted research with me for their graduate studies including Kundan Joshi, Lauren Hickox, Arif Ahmed Sohel, Nathan Vliestra, Andrew Vander Moren, Justin Bjorum, Jennifer Edwards, and Austin Filush. Teaching is what I love to do, and it is because of them that I love to do it.

Blake M. Ashby

# Contents

# Reviewers and Content Consultants

**Jeanine Beasley, EdD, OTRL, CHT, FAOTA** Occupational Science and Therapy Department, Coordinator of the OST Hybrid Program, Grand Valley State University, Grand Rapids, MI, USA

**Scott D. Burgess, MD** Orthopaedic Associates of Michigan, Grand Rapids, MI, USA

**Krisanne Chapin, Ph.D.** Department of Movement Science, Grand Valley State University, Allendale, MI, USA

**Barbara Hoogenboom, PT, EdD, SCS, AT (retired)** Department of Physical Therapy & Athletic Training, Grand Valley State University, Grand Rapids, MI, USA

**Joseph J. Krzak, PT, Ph.D., PCS** Physical Therapy Program, Midwestern University, Downers Grove, IL, USA

**Sylvia Ounpuu, MSc** Center for Motion Analysis, Connecticut Children's Medical Center, Farmington, CT, USA

**Michael Shoemaker, PT, DPT, Ph.D., GCS** Department of Physical Therapy & Athletic Training, Grand Valley State University, Grand Rapids, MI, USA

**Corey Sobeck, PT, DScPT, OCS, OMPT** Department of Physical Therapy & Athletic Training, Grand Valley State University, Grand Rapids, MI, USA

**James W. Youdas, PT, MS** Mayo Clinic College of Medicine and Science, Mayo Clinic School of Health Sciences, Program in Physical Therapy, Department of Physical Medicine and Rehabilitation, Mayo Clinic, Rochester, MN, USA

# 1 Introduction

## 1.1 Learning in the Profession

The development of the professional requires integrating knowledge and skill in a particular discipline. Thus, a professional can rightly claim extraordinary expertise in matters of great importance. In a sense, the process of becoming a professional and the special knowledge one accrues contribute to a community of practitioners who distinguish themselves from other individuals concerning the holding of special rights and privileges. We suggest that professionals share the tradition of a "calling," i.e., vocation, as they operate in particular kinds of institutional settings, with practices that are structured specifically to a discipline, including specific preferences and norms, as well as ethical and moral behaviors.

The acquisition and practice of theories and techniques derived from systematic endeavors directed toward solving instrumental problems lead to professional competence. Typically, two types of situations are encountered in practice: the routine and the unfamiliar. Common concerns can be addressed by the professional's knowing-in-action, i.e., applying knowledge and skill intuitively, that is, the sorts of know-how we reveal in our intelligent action – publicly observable, physical performances – where knowing is the action that is revealed by our spontaneous, skill execution of the performance, yet are characteristically unable to make it verbally explicit.

In the unknown, however, the problem is complex and ill-structured (i.e., initially not apparent), and the situation may be uncertain, unstable, and infused with value conflicts. Thus, there seems to be no evident fit between the characteristics of the problem and the available repertoire of theories, techniques, and skills. In the unfamiliar, professionals still draw on their intuition but experience a surprise that leads them to rethink (reflection-in-action) their knowing-in-action in ways that go beyond available rules, facts, theories, and operations. This reflection (or metacognition) may occur after the fact (reflecting-on-action) or "in the moment." Reflection-in-action has a critical function, questioning the assumptional structure of knowing-in-action, thus leading the professional to restructure some of their previous strategies of action, ideas of phenomena, and ways of framing the problem. Thus, they invent on-the-spot experiments induced by curiosity and inquiry and test revised or new sets of facts and "understandings."

Peter Vaill recognized that a professional's environment was complex, uncertain, unstable, ambiguous, and constantly changing. That is, their practice would be full of surprises and novel problems (i.e., issues not anticipated or imagined), as well as messy, ill-structured events, which would recur (yet, perhaps not in the same way). As a result, Vaill proposed that the natural world would require professionals to employ their curiosity and participate in ongoing, imaginative, and creative learning processes. Most would agree with Vaill. The question is how to best prepare.

## 1.2 Institutional Learning

Traditionally, the "how" is accomplished through institutional learning, which is primarily transactional, based on a philosophy of goal-directedness. Learning is assumed to depend on the learner's desire to achieve a specific goal, which may not originate with the learner. Yet under the institutional learning model, learning, defined as the path toward possessing new knowledge or some skill, is promised to inevitably lead to a secure, well-paying job. Under this model, the volume of material, i.e., new facts learned, and the efficiency and speed with which this occurs, is a primary concern. Some of the hallmarks of institutional learning include (1) learning is answer oriented; (2) getting the correct answer results from the manipulation of tools or data or ideas the "right" way, i.e., according to the teacher; (3) competitiveness leaves the learner "alone"; (4) the learner enters the system in a state of inferiority, tentativeness, and dependence on those in authority, with suppression of here-and-now feelings in favor of anticipation of future comfort and success; and (5) the learner experiences a diminished sense of self, with feelings of inauthenticity, which lingers until



G. J. Alderink, B. M. Ashby, *Clinical Kinesiology and Biomechanics*, https://doi.org/10.1007/978-3-031-25322-5_1

learning is complete. Despite the aforementioned characteristics, institutional learning "works" as some of the brightest and most energetic learners achieve "success" under the criteria and assumptions of institutional learning. Students realize their success partly by orienting their attention toward those who possess expert knowledge, by not challenging the system, and may even exhibit creativity, but always within the goals and objectives defined by the instructor and institution.

On the other hand, these individuals often sacrifice their innate curiosity and lack confidence about their performance in new learning situations. Another outcome of institutional learning is the development of well-educated individuals, schooled in the way of knowing that treat what they encounter in the world (ideas, people, position, etc.) as objects to be dissected and manipulated, with a form of learning that gives them power over their world. Success may lead to an arrogance that sees knowledge as an arbitrary process, subject only to the rules of whatever educational or "cultural game" one happens to be playing at the time. With competition as the driving force in institutional learning, the individual perceives win-lose or lose-win situations as the only possibilities. Vaill suggested that ultimately, training the professional in this way often results in individuals who experience a sense of weariness, withdrawal, and cynicism. Vaill's perspective may be an over-generalization of the present educational system in the United States (and, perhaps elsewhere), but, notwithstanding, we encourage readers to use Vaill's ideas as a mirror to reconsider their perspective on teaching and learning.

We have hinted at how the teacher and student might function within the institutional learning environment, but now let's examine the advantages and disadvantages of the teacher-centered philosophy, as practiced in institutional learning. Within the teacher-centered learning model, the teacher is solely responsible for what the student will learn. The teacher decides the sequence, pacing, and information/skills students should know. This process is efficient for both teacher and student, creating a relieved and content student who is satisfied that the teacher has total control. Finally, teaching excellence is recognized by the administration and students if the instructor ensures that the customer, i.e., the student, is satisfied and follows the model that is assumed to be correct. Concurrent with the teacher-centered model is the emphasis on learning by *subject*, putting different disciplines in silos and teaching them separately, as if there were no relationship between them. So, not only is the power structure hierarchical but so also is the content. The teacher and learner assume that best learning occurs under hierarchical organization models, which leads to building layers of knowledge and skill from basic to advanced. Students attempt to learn new facts and essential concepts in some depth, with the teacher assuming that the learner will figure out how to apply concepts to future tasks, i.e., solving problems.

What we have described may sound familiar, and as noted earlier, most of us have succeeded under this system, so one might ask, "what's wrong with the system if it works?" A criticism of the teacher- and subject-centered model is not necessarily with the end product but the means. Using a radical or fundamental critique of the Brazilian educational system, Freire characterized the institutional learning defined by Vaill as *banking education*. Although Freire's motivation was to enlighten indigenous peoples to assist in their struggle to make social change and overcome oppression, his characterization of the teacher-centered model is worth examining because it may provide us with additional insight. Freire posited that any analysis of the teacher-student relationship reveals its fundamentally contradictory and narrative character; narration (with the teacher as narrator) leads the students to memorize the narrated content mechanically. The "teacher as narrator" model turns students into containers, into receptacles to be filled by the teacher. The more completely the teacher fills the receptacles, the better the teacher, and the more meekly the receptacles permit themselves to be filled, the better and more successful the student. Thus, education becomes the act of depositing. *Banking education* maintains and even stimulates the teacher-student contradiction through the following attitudes and practices; thus the teacher:

- Teaches and the students are taught
- Knows everything, and the students know nothing
- Talks and the students meekly listen
- Disciplines and the students are disciplined
- Chooses and enforces while the students comply
- Acts and the students have the illusion of acting through the action of the teacher
- Selects the program content, and the students (who were not consulted) adapt to it
- Confuses the authority of knowledge with their professional authority, which they set in opposition to the freedom (and power) of the students
- Is the *subject* of the learning process, while the students are mere objects

In the banking concept, knowledge is a gift bestowed by those who consider themselves knowledgeable, which implicitly projects ignorance onto others, alienating and dehumanizing them, and negates education and knowledge as a process of inquiry fused by curiosity as a dialogical relationship between teacher-student and student-teacher.

## 1.3 Problem-Based Learning

Vaill's proposed alternative learning model, *learning as a way of being*, suggests a more holistic approach to teaching and learning. Inherent in his model is the ideal of lifelong learning, motivated primarily by curiosity, encompassed by the intrapersonal and interpersonal; that is, my learning (as a teacher) as a way of being will exist in relation to your learning (as a student) as a way of being. Similarly, Freire suggested that a teacher-student/student-teacher relationship would be realized in his proposed alternative to banking education, what he called *problem-posing education.* Although problem-posing education and problem-based learning are not synonymous, it will be helpful to examine Freire's problem-posing model because its potential importance in our context transcends politico-socio-cultural circumstances. For example, dialogic action by both the teacher and student is key to problem-posing education. The primary assumption of dialogic action is that human life can hold meaning through authentic social interaction and communication. In other words, the teacher's thinking is authenticated only by the authenticity of the students' thinking. One of the hypothetical outcomes of this learning model is the humanization of the learning process. In abandoning the educational goal of deposit-making in favor of the posing of problems, Freire proposed that the essence of learning, i.e., consciousness, is something that rejects communiqués in favor of communication and metacognition. Freire defined metacognition as *consciousness of consciousness*, so that instead of using education for transferals of information, a la institutional learning or banking education, there is a response to authentic curiosity and the fusing of inquiry and acts of cognition. The teacher no longer merely teaches but is also taught by being in dialogue with the students, who, while being taught, also teach; they become jointly responsible. The students – no longer docile listeners – are now critical co-investigators in dialogue with the teacher. The role of the problem-posing teacher is to create, together with the students, an environment under which knowledge will be constructed, deconstructed, reconstructed, and used. Under problem-posing education, there is a constant unveiling of reality. And, as students are increasingly challenged with complex problems relating to themselves in the world and with the world, they will respond to that challenge in increasingly skilled and creative ways. Over time, students will see challenging problems and related processes as liberating and claim the corresponding power and responsibility to anticipate future challenges.

Problem-based learning is a method by which the learner is presented with a problem or situation in order to identify learning needs. PBL facilitates the learning that results from working toward an understanding or resolution of a problem or situation. It has two primary educational objectives: (1) acquisition of an *integrated* body of knowledge and (2) development or application of critical thought and problem-solving skills. Self-directed (i.e., student-directed) learning is the hallmark of PBL. Encouraging self-direction allows students to identify their learning objectives and needs, induces students to develop strategies to meet those needs, and assumes an interest in evaluating their progress toward achieving the claimed goals. The work is accomplished within a small group setting based on the well-accepted assumption that the best learning occurs in social situations; thus, PBL involves the development of intra- as well as inter-dependent study. That is, intra-dependent learning (and self-directed), which occurs within a person or "closed group," can be transcended by learning that occurs outside oneself, not only within the PBL group but also with the entire class of peers.

PBL is student-centered rather than teacher-centered. In this model, the teacher may have considerable responsibility at the beginning of a learning module. But eventually, the student will assume most responsibility for their learning with facilitation, i.e., guidance, by the teacher/tutor. Students' active acquisition of information and skills is expected, although the teacher may be more responsible for initially setting up learning objectives, learning resources, and evaluation methods. The advantages of using a student-centered model are that students are, in a sense, "forced" to learn how to learn and students become active participants, i.e., reigniting their inherent curiosity, raising questions, finding resources, etc., rather than acting as docile recipients. Under a student-centered model, motivation and rewards can become more intrinsic, in comparison to the external motivating factors students responded to under the institutional learning or banking models of their prior learning experiences. Furthermore, students actively learn how to assess and evaluate their work and the work of their peers. The primary disadvantages of a student-centered approach to learning affect both the teacher and students. For example, teachers have to relinquish some control and plan for more individualized oversight. Whereas students may initially feel less secure, their maturity level may not be appropriately matched to the intent and responsibility required for self-direction.

The theoretical framework for PBL is attributed to many ideas associated with John Dewey, one of which was learning by doing. Learning by doing was based on the premise that the construction/acquisition of new knowledge requires activation of prior knowledge. Knowledge is remembered best in the context in which it is learned. The problem scenarios are used in tutorials, clinical skills laboratories, and inquiry seminar courses. The problem, e.g., medically related, promotes the exploration of (at least for this text) the underlying anatomical, mechanical, physiological, developmental, and clinical determinants of health and disease. Problems require integrating previous learning with the acquisition of new knowledge, skills, and behaviors. The problems or situations are designed to provide a context that resembles the natural world's complexity and professional circumstances as much as possible. Some of the advantages of this method (over a teacher-centered model) include (1) information, concepts, and skills are learned in the context of real problems; (2) recall is enhanced; (3) students use problems as a stimulus to their curiosity to focus on the study of multiple disciplines/subjects, actively integrating information into a system that can be applied to present and future problems; (4) critical thinking, critical appraisal, and problem-solving are enhanced; (5) students become active and self-directed learners; (6) students gain a more broad and positive perspective on the relevance of basic science information; (7) motivation and rewards are inverted from external to internal; (8) students develop a sense of freedom, power, and responsibility; (9) learning is more fun, and students regain/refashion their inherent human curiosity; (10) lifelong learning is promoted; (11) students develop skills in self-assessment and self-evaluation; (12) students develop reflective and metacognitive thinking; and (13) students participate in a dialogic, collegial relationship with faculty tutors.

As with any model, PBL has limitations. Since PBL originated in medical school curricula, one major disadvantage is the perception that clinical concepts override the importance of basic sciences. Other somewhat contentious issues relate to prioritizing problem-solving processes over the learning of content, inefficiency of the process (at least in the early stages of using PBL), insecurity concerning passing certifying examinations, and using evaluation methods that are considered "soft." These disadvantages are notable and important to keep in mind, i.e., continually reflecting with students on how things are going. On the other hand, in our experience, the primary challenge to transitioning from a teacher- to a problem-based-centered pedagogical learning model is overcoming the inertia of preconceived notions of what constitutes teaching excellence and understanding the importance of the development of clear communication between the teacher (or tutor) and student.

Under the PBL model, the tutor's knowledge and skills, personal attributes, and responsibilities are extensive and critical to the success of the educational endeavor. Although we know that tutors do not necessarily need to be content experts, they do need sufficient training in skills related to facilitative teaching, such as (1) demonstrating to students how to ask open-ended, stimulating questions, and challenge students when appropriate, i.e., withholding challenges to see if students first challenge each other; (2) summarizing consequences of student conclusions, as needed; (3) using questioning to induce students to seek additional external information that may be required; (4) helping identify additional resources; and (5) while acknowledging one's participating as a teacher-student, avoiding dominating the presentation of information. Additionally, tutors must promote individual and group problem-solving and critical thinking and encourage efficient group dynamics by, for example, modeling professional behaviors and productive ways of giving feedback. Moreover, tutors need to assist individual group members in their learning by fostering the use of study plans and exploring multiple study methods, and in student assessment evaluation, including assisting students in defining personal objectives and participating in self-, peer-, and tutor-evaluation.

Supported by decades of experience using PBL, the most critical elements contributing to the success of the PBL process include skilled facilitation, student engagement, and, ultimately, the humanization of learning. A satisfying career/vocation is the "conscientization" of the teacher and student. *Conscientization*, according to Freire, is the immediate goal of transformational learning that is expected under problem-posing (and problem-based) learning. Building on the work of Freire, Allman expressed the *conscientization* process in this way: "the term 'conscientization' expresses the inseparable unity between critically acting to transform relations (i.e., between teacher and student) and the critical transformation of consciousness." Allman goes on to claim that "it is only within the experience of transforming relations and the experience of the transformation that our critical consciousness can fully develop."

What are Freire's and Allman's intentions? To understand their motivation, we need to go back and restate Freire's primary critique of banking education, which was the inherent contradictory relationship between the teacher and student. Freire claimed that the teacher-student relationship was a unity of opposites, an antagonistic contradiction that needed to be changed (i.e., ablated or dialectically resolved). In other words, teachers and students constitute different groups in which teaching and learning processes had become separated or dichotomized. One consequence of this is that the "act of creating new knowledge" can become separated from the "act of acquiring extant knowledge." These separations are antagonistic because they limit both groups', i.e., teacher's and learner's, creative and learning potential. Allman suggested that Freire's problem-posing education conceived teaching and learning as two internally related processes within each person, which explains his use of the phrase "teacher-learners and learner-teachers." Practically, let's imagine this model as one where:

1. Teachers do not cease being teachers but cease being the exclusive or only teacher in the learning group; they need to relinquish the system's hierarchically established power structure, i.e., authoritarianism, but not their

authority; and they must have plans, projects, and goals and an overall intent with which they work, including their learning.

2. Learner-teachers do not cease being primary learners but join together with teachers in a mutual process, a unity of teaching and learning.

This transformational process sounds reasonably straightforward but is constrained by the inherent bias of both teachers and learners who, most likely, developed under an institutional learning model, i.e., banking education. Teachers (i.e., tutors), on their own, cannot transform the relationship. They can initiate the process by challenging learners to consider the limitations of the traditional model. Still, only when learners accept the challenge does the collective struggle to transform the relationship begin. Initiating the process (by the teacher) and taking the challenge (by the learner) requires a new mode of communication and maturing dialogic skills, i.e., clear speaking/articulation, open listening, and commitment to stay engaged, which brings its own set of challenges.

Yet, according to Allman, at least one other major constraint challenges teachers and learners in the transformation process, and that is their respective relationship to knowledge. What does Allman mean by this? First, let's note that in the socioeconomic system that most of us live in, liberal capitalism, facts, and knowledge are perceived as a "thing," a commodity, and static possession. Often social relationships (including the teacher-learner relationship) are merely transactional ones and often feel like things. For example, we often hear university administrators refer to students as the customer, another commodity version. Therefore, because of how knowledge is generally perceived by most of us, if we possess it, it positively affects who we are, our status, and our self-esteem. Still, there is an equal and opposite effect on seeing and defining ourselves if we do not own knowledge. In problem-posing education, Freire anticipated that teachers and learners must also transform how they relate to knowledge, i.e., "being or relating differently is inextricably bound up with knowing differently." This transformed relation to knowledge involves constantly reflecting on what we know and how we come to know things and continually testing its adequacy as a tool for future learning and informing our actions. Knowledge, therefore, cannot be perceived as static but only as mediation or a tool between people and the world in which they live and work. According to Freire, knowledge becomes an object to which we direct critical thought and action, i.e., it must be constantly tested and questioned. In other words, accessing new knowledge is a means by which we begin learning. It becomes a springboard for the creation of new knowledge and a deeper understanding of our world; thus, knowledge cannot, or must not, be an end in itself.

## 1.4 Summary

This concludes our formal theoretical discussion of the teacher- and student-centered educational models. We have provided the advantages and limitations of both models. Furthermore, we showed striking similarities between two revolutionary teaching methodologies, i.e., problem-posing and problem-based models. For the problem-based model, we defined the unique role of the tutor. In contrast, for Freire's problem-posing model, we highlighted the unique challenges teachers and students face when they attempt to initiate a radically new learning process and anticipate transformational changes in their thinking. There are many details about setting up a problem-based curriculum and learning groups, which we leave to the reader to explore.

## Bibliography


Allman P (1999) Revolutionary social transformation, democratic hopes, political possibilities and critical education. Bergin & Garvey, Westport

Barrows HS (1988) The tutorial process. Southern Illinois University School of Medicine, Springfield

Barrows HS (1994) Problem-based learning applied to medical education. Southern Illinois University School of Medicine, Springfield

Barrows HS, Tamblyn RM (1980) Problem-based learning, an approach to medical education. Springer Publishing Company, New York

Barrows HS, Wee Keng Neo L (2010) Principles & practice of a PBL, revised edn. Southern Illinois University School of Medicine, Springfield

Dewey J (1966) Democracy and education. Free Press, New York. (originally published in 1916)

Eyler JR (2018) How humans learn, the effective science and stories behind effective college teaching. West Virginia University Press, Morgantown

Freire P (2000) Pedagogy of the oppressed. The Continuum International Publishing Group Inc., New York

Vaill PB (1996) Learning as a way of being, strategies for survival in a world of permanent white water. Jossey-Bass Publishers, San Francisco

# Biomechanics: Overview, Terminology, and Concepts

# 2

## 2.1 Biomechanics

Engineering mechanics involves the analysis and design of mechanical systems. Biomechanics is a multidisciplinary science concerned with the application of classical mechanical principles to the study of biological structures. Our interest here is the application to the human body at rest and in motion. Mechanics is typically divided into three general categories: rigid body, deformable body, and fluid mechanics (Fig. 2.1).

In rigid body mechanics, we assume that the body being studied does not deform. Although this is not strictly true, if it is known that the deformations are small under external loading conditions, they are ignored in an effort to simplify biomechanical models. Statics is the study of forces and moments on rigid bodies at rest or under conditions of constant velocity. In the cases presented in this text, you will see several applications of static analyses, which present reasonably good approximations of the effect of external forces and moments on selected internal structures.

Dynamics deals with bodies in motion. In studying the human body in motion, whether under experimental or simulation conditions, we typically use rigid body and link-segment modeling, with the following simplifying assumptions:

1. Segments have fixed masses.
2. Each segment's center of mass (COM) remains fixed, with respect to the body, during the movement.
3. Joints are considered to be hinge or ball-and-socket.
4. The mass moment of inertia of each segment about its mass center (or about either proximal or distal joints) is constant during the movement.
5. Segment length remains constant during the movement (i.e., the distance between joint centers remains constant).

Rigid body mechanics and link-segment modeling, which requires a full kinematic description, accurate anthropometric measures (i.e., segment masses, COMs, joint centers, and moments of inertia), and measured external forces, allow us to determine net intersegment forces (which includes contributions from bone-on-bone contact and muscle forces) and net joint (i.e., muscle) moments using a process called inverse dynamics. Inverse dynamics analyses assist the biomechanist and/or clinical scientist in the design of medical equipment, surgical decision-making, rehabilitation decisions, and describing normal and pathological movement.

Kinematics is a branch of dynamics that analyzes the position and time-dependent aspects of motion, without regard to the forces and moments causing the motion. Osteokinematics refers to the movement, e.g., rotation, of one bone relative to its adjacent partner, usually described relative to the three cardinal planes of the body, i.e., sagittal, frontal, and transverse (or horizontal). The rotation of one bone relative to its neighbor takes place about the joint axis of rotation (passing through the joint center), which is located perpendicular to the plane in which the motion is taking place (Fig. 2.2). Note that most of the time, osteokinematic or physiological motions are under voluntary control. Joint degrees of freedom (DoF) often refer only to the number of independent coordinates required to describe the motion, e.g., three DoF at the ball-and-socket glenohumeral joint include the clinically defined movements of flexion/extension, abduction/adduction, and long-axis rotation, i.e., internal/external rotation. However, theoretically, each joint has six DoF, with three rotations about the cardinal axes and a translation along each axis.

Arthrokinematics describes the movements that occur between the articular surfaces of joints, where the direction and magnitude of motion are primarily dictated by periarticular constraints, i.e., the joint capsule, ligaments, and fibrocartilage, as well as by the shape of joint articular surfaces, i.e., convex, concave, concavo-convex (saddle), or flat. Typical arthrokinematic movements include roll (sometimes referred to as "rock"), slide (glide or translation), and spin (Fig. 2.3). Arthrokinematic, or accessory, movements typically are not under voluntary control but are considered essential to the normal pain-free movements required for



G. J. Alderink, B. M. Ashby, *Clinical Kinesiology and Biomechanics*, https://doi.org/10.1007/978-3-031-25322-5_2

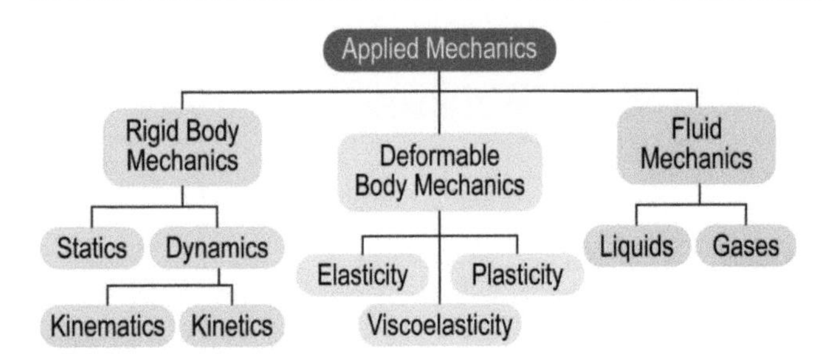


**Fig. 2.1** Classification of mechanics

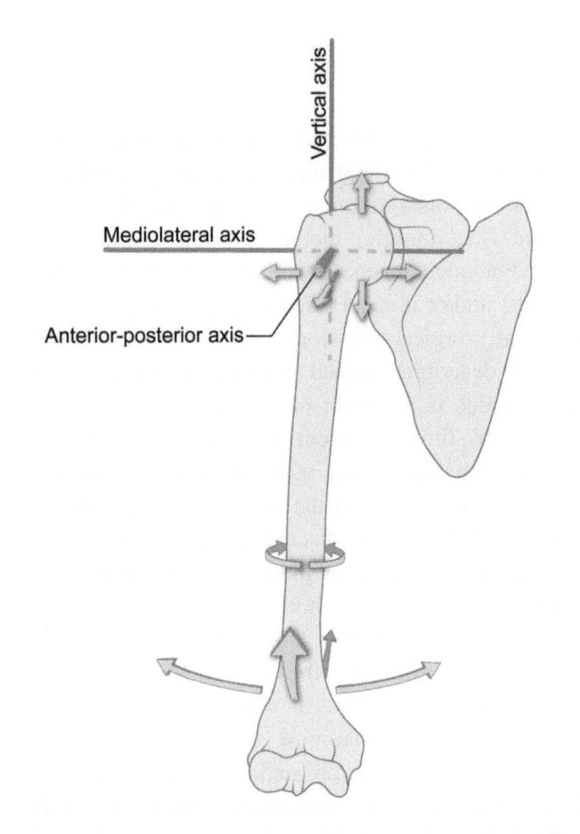


**Fig. 2.2** The right glenohumeral joint and its three orthogonal axes of rotation and associated translations illustrate all six degrees of freedom (DoF)

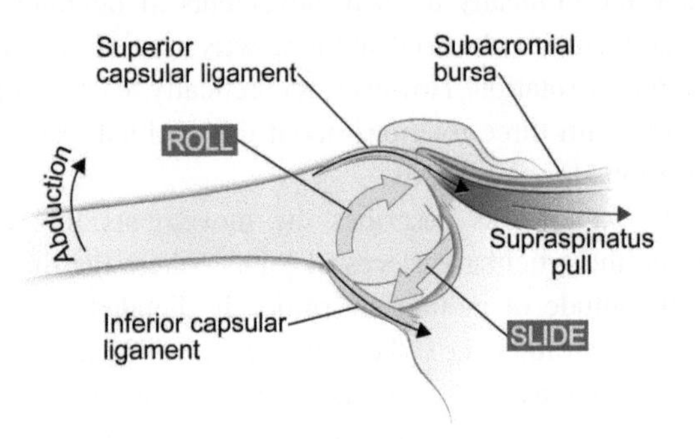


**Fig. 2.3** Arthrokinematics of glenohumeral joint abduction, i.e., superior movement or roll of the humerus. Note that the glenoid fossa, of a relatively fixed scapula, is concave and the head of the humerus is convex necessitating, arthrokinematically, a glide opposite the direction of roll, i.e., an inferior glide of the head of the humerus as the humerus rolls superiorly

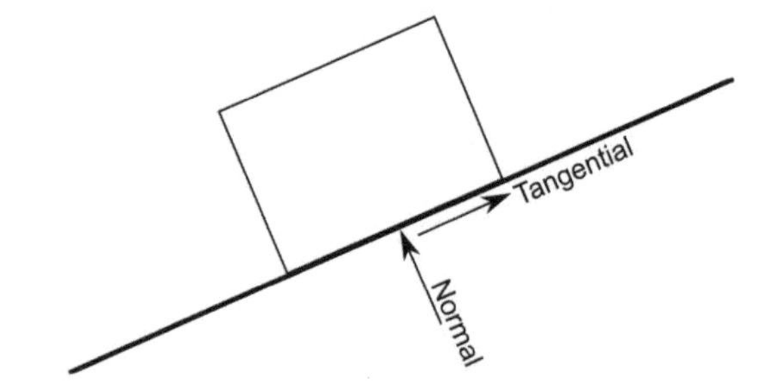


**Fig. 2.4** Example of forces applied normal (perpendicular) and tangential (parallel) to the surfaces in contact

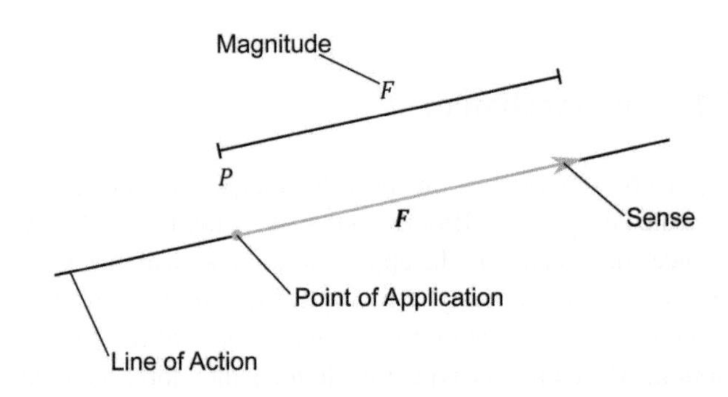


**Fig. 2.5** Representation of a force vector, where a force vector $\boldsymbol{F}$ is applied to a body at point $P$

activities of daily living. The cases studied in the forthcoming chapters describe the details of joint osteo- and arthrokinematics under normal and pathological conditions.

Kinetics is related to kinematics, taking into account forces, moments, and inertias, e.g., masses. Force, sometimes called a load, can be thought of as a mechanical disturbance that typically results in a movement. In simple terms, a force is a push or a pull. Forces can be external, e.g., gravity and kicking a ball, or internal, e.g., forces inside the body that resist deformation or bind tissues and muscle action. The same force can be considered either external or internal, depending on the context. For example, when considering the entire upper extremity, the muscle forces and joint reaction forces at the elbow are considered to be internal forces to the upper extremity. However, when considering just the forearm, the muscle forces and joint reaction forces at the elbow are considered to be external forces to the forearm. There are several force systems, e.g., collinear, parallel, and concurrent. Forces acting between two bodies are often divided into components normal (perpendicular) and tangential (parallel) to the surfaces in contact (Fig. 2.4). For example, frictional forces between two bodies act in the tangential direction. Forces are also categorized as "tensile" if the force tends to elongate the body or "compressive" if the force tends to shorten the body. In setting up free-body diagrams (FBD) for solving static equilibrium problems, a force is depicted as a vector. The force vector is affixed to the body showing its point of application, magnitude, and sense of direction (or orientation) (Fig. 2.5).

When a force is applied to a body segment and is not balanced by opposing forces, that segment translates (i.e., moves in a linear direction), rotates, and/or deforms. When a force is applied to a body segment at some distance from the mass center (i.e., the center of mass (COM) of that segment), it tends to rotate that segment, depending on the presence of other forces or constraints. The force acting at some distance from a point is referred to as a moment of force or torque, and if that moment is not balanced, a rotation is induced. In engineering mechanics, torque is technically defined as a pure moment that induces a rotation, whereas a moment is related to a bending effect. However, in biomechanics the words torque and moment are often used interchangeably. The magnitude of the moment of a force about a point is defined as the magnitude of the force times the length of the shortest distance (i.e., in a planar coordinate system, the perpendicular distance) between the point and the line of action of the force. A *couple* is a special case where two equal parallel forces, equal in magnitude but opposite in direction, determine a *couple-moment* equal to the magnitude of one of the forces times the distance between them. Moments, described as free vectors because it makes no difference where on the body the moment is applied, are depicted with a magnitude and direction (i.e., in a planar coordinate system perpendicular to the plane in which the force is applied) (Fig. 2.6).

In statics and dynamics, we examine the effect of external forces and moments on body segment position, under the assumption that the segments do not deform. For deformable body mechanics, we extend our study of external loads relative to internal structures, e.g., bone, tendon, etc. The field of material science involves an examination of the constituent parts, e.g., molecular structure, bone, ligament, etc., of the body segments of interest. Deformable body mechanics is subdivided into three parts: elastic, plastic, and viscoelastic. The elastic response of a body is one that recovers its original form once the deforming force is removed, e.g., like a spring (Fig. 2.7). On the other hand, plastic deformation is permanent. For example, if a large pulling force is applied to a small spring, the spring does not return to its resting position but stays longer, unable to recover its original shape. Obviously, depending on the magnitude and duration of the external force, any material body may undergo an elastic, plastic, or elastoplastic deformation.

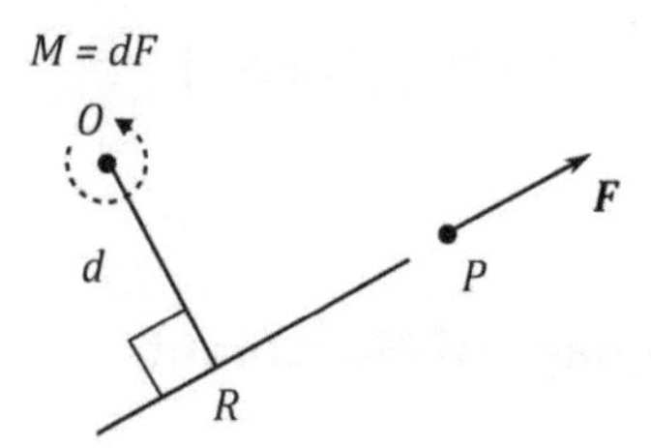


**Fig. 2.6** Point $O$ is a distance $d$ from the line of action of a force, $\boldsymbol{F}$. The moment, $M$, produced by this force acts in a counter-clockwise direction as shown with a magnitude equal to $d$ times $F$

Viscoelasticity is a complex phenomenon that is found in all biological tissues. Because of their material constituents, i.e., water, cells, collagen, etc., biological tissues have both solid and fluid properties. A feature of a solid is that if a force (or load) is applied, it deforms to some degree, as noted previously, but its deformation does not increase over time if the load remains. On the other hand, when a fluid body is placed under a load over time, it deforms continuously (i.e., flows) as long as the load remains. Viscosity is a property of flow and is measured as the resistance to flow. Viscoelastic tissues, then, have both elastic, i.e., solid body, and viscous, i.e., fluid body, properties, which makes their response to external loads remarkably adaptable. The concept of viscoelasticity is quite abstract when first encountered, but each case in this text provides an opportunity to re-examine the general phenomenon of viscoelasticity and the unique viscoelastic response of different body tissues in the context of normal and pathological conditions (a more detailed presentation on viscoelasticity is presented in Appendix H).

Fluid mechanics is the third category of applied mechanics and includes the mechanics of fluids and gases. In this text, we work to achieve a reasonably robust conceptual understanding of fluid mechanics by examining two fundamental types of articular cartilage lubrication: boundary and fluid film. We compare the fluid mechanics of articular cartilage under normal and degenerative conditions. The mechan-

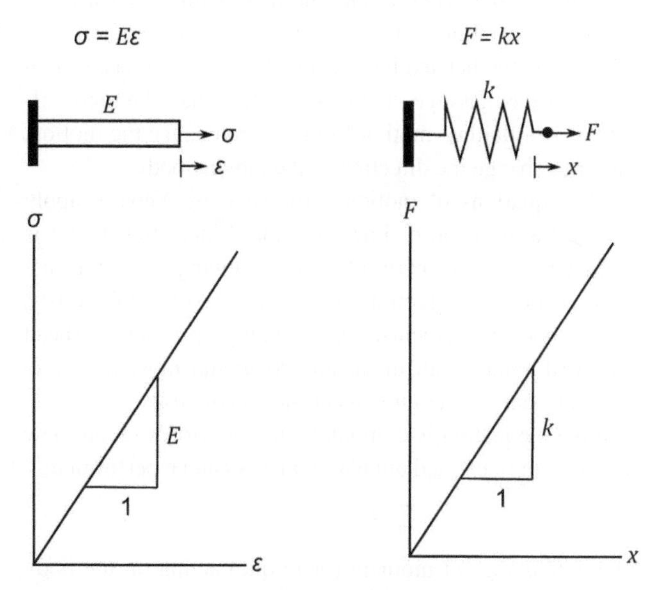


**Fig. 2.7** Illustration of the similarity between the mechanical behavior of a linear spring and an elastic solid. An elastic material deforms, stores energy, and recovers deformations in a manner similar to a spring. The elastic modulus $E$ for a linearly elastic material relates stresses ($\sigma$) and strains ($\varepsilon$), whereas the spring constant, $k$, for a linear spring relates applied forces and corresponding deformations. Both $E$ and $k$ are measures of material, or body segment, stiffness

ics of the gases most essential to life, oxygen and carbon dioxide transport during normal breathing, are examined relative to normal and pathological thoracic cage kinematics and chronic obstructive pulmonary disease.

## 2.2 Newton's Laws

Newton's three laws of motion provide the cornerstone for static and dynamic analyses. Newton's first law of motion, i.e., the law of inertia, states that a body remains at rest or moves with a constant velocity in a straight line if the net force acting on the body is zero. The first law is summarized as:

$$\sum \boldsymbol{F} = 0$$

Note: bolding the letter that represents a force or a moment implies that the letter represents a vector.

Newton's second law of motion, i.e., the law of acceleration, states that a non-zero net (or resultant) force that acts on a body accelerates that body in the same direction as the net force:

$$\sum \boldsymbol{F} = m\boldsymbol{a}$$

Known as the equations of motion, where $\boldsymbol{F}$ represents the net force acting on a body, $m$ is the mass of the body, and $\boldsymbol{a}$ is its acceleration. Note that we need to distinguish the mass, which is a quantitative measure of inertia, from weight force, i.e., mass times the force of gravity, which is a force with magnitude equal to the weight. Inertia is defined as resistance to change in motion of a rigid body. The magnitude of a body's mass is dictated by how much "matter" it has, so that the larger the mass, i.e., the more matter it has, the more difficult it is to initiate motion from rest, to change the motion, and/or to change the direction of the moving body.

The equations of motion formulated by Newton apply only to translational or linear motion. Linear translation is characterized as all parts of a body moving the same distance, in the same direction, and at the same time. Obviously, normal body movements, e.g., reaching up to an overhead cupboard, entail both linear and rotational movements, so dynamic analyses require an additional equation.

Euler's equation is combined with Newton's second law so that both linear and angular analyses can be performed:

$$\sum M = I\alpha$$

where $M$ is the net moment (or torque) acting on the body segment, $I$ is the mass moment of inertia of the body segment, and $\alpha$ represents body's angular acceleration. The mass moment of inertia measures how the mass is distributed throughout the body and represents the body's resistance to changes in its rotational (or angular) velocity. Note that this form of Euler's equation is only valid for bodies undergoing planar, i.e., two-dimensional, motion (see Appendix J for Euler's equation for three-dimensional motion). With pure planar rotation, a body segment rotates in a clockwise or counterclockwise direction, where all points on the body follow the same circular path in the same direction through the same angle at the same time. With general motion, a body can both translate and rotate at the same time.

Newton's third law implies that forces belong to interactions. It is usually stated as: "to every action, there is an equal and opposite reaction." However, in keeping with the idea of body-to-body interaction, we shall claim that the interaction between any two bodies is represented by forces that act equally on each body but in opposite directions. For example, from body A to body B and from B to A the forces are:

$$\boldsymbol{F}(A\,on\,B) = -\boldsymbol{F}(B\,on\,A).$$

## 2.3 Dimensions, Systems of Units, and Conversion of Units

Dimension can refer to space, e.g., two-dimensional, as well as denote the nature of quantities. For example, a meter is a unit of measurement, but its general dimension is length. Primary dimensions typically used in biomechanics include

length ($L$), time ($T$), and mass ($M$). Secondary dimensions are derived from primary dimensions, for example, area = length × length = $L^2$, velocity = position/time = $L/T$, and acceleration = velocity/time = $L/T^2$.

Why is this important? For solving biomechanics problems, setting up and solving equations involves knowing the appropriate dimensions and related units. Adhering to consistent dimensional and unit conventions allows us to validate not only the magnitudes of solutions but also the appropriate units. In this text, we typically use the SI system for units of length, time, and mass as a meter (m), second (s), and kilogram (kg), respectively. Other units can be derived from the units for these primary dimensions. For example, the unit for force in the SI system is the Newton (N), which can be derived from Newton's second law ($F = ma$). One Newton is equal to the amount of force required to accelerate a 1-kg mass at 1 m/s$^2$:

$$1\,\mathrm{N} = (1\,\mathrm{kg})(1\,\mathrm{m/s^2})$$

## 2.4 Language of Movement

Readers that do not have a fundamental knowledge of terminology commonly used in kinesiology and biomechanics, anatomical planes and directions, and joint motions, can consult Appendix B for a brief overview.

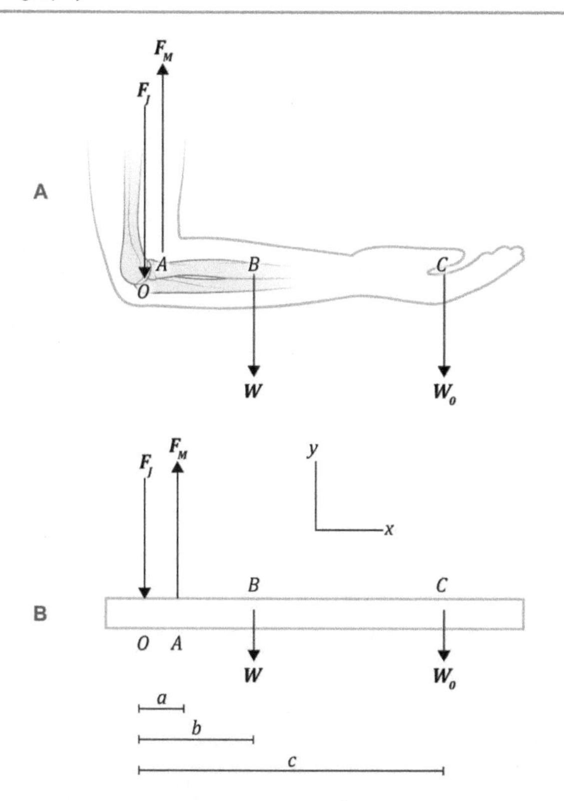


**Fig. 2.8** Illustration of (**A**) forces acting on the forearm and a corresponding (**B**) free body diagram. Note: $\boldsymbol{F_M}$ = muscle force vector; $\boldsymbol{F_J}$ = joint reaction force vector; $\boldsymbol{W}$ = force vector of forearm weight; $\boldsymbol{W_O}$ = force vector of the hand and the object's weight

## 2.5 Modeling and Mathematics

For the cases presented in this text, modeling is restricted to two dimensions for the purpose of solving static equilibrium problems. This requires some background and basic skills in geometry, algebra, trigonometry, vectors as well as vector composition and resolution, and skill in the creation of free-body diagrams (FBD) (see Appendices A, C, D, E, F, G, and I).

An FBD is constructed to help identify the forces and moments acting on individual parts of the system being modeled. The parts of the system can be isolated from the environment making it easier to identify and accurately portray all known and unknown external (to that body) force and moment vectors, including their points of application and orientations (Fig. 2.8). Sometimes we have to assume the sense of forces and moments. Our assumptions may not be correct, but if we have properly set up the model with regard to coordinate axes and operational definitions for "sign" assignment for force and moment directions, the correct orientation and magnitude of the unknowns naturally follow in the solution.

## 2.6 Conclusion

The intention of this chapter was to present a broad foundation of biomechanics. In addition to your previous experiences, we hope that this framework of biomechanics allows you to begin the process of learning more biomechanics in the context of examining and "solving" the ensuing clinical vignettes.

## Bibliography


Neumann DA (2017) Getting started. In: Neumann DA (ed) Kinesiology of the musculoskeletal system, foundations for rehabilitation, 3rd edn. Elsevier, St. Louis

Nordin M, Frankel VH (2012) Basic biomechanics of the musculoskeletal system, 4th edn. Lippincott Williams & Wilkins, Baltimore

Özkaya N, Leger D, Goldsheyder D, Nordin M (2017) Fundamentals of biomechanics, equilibrium, motion and deformation, 4th edn. Springer International Publishing, Cham

Winter DA (2009) Biomechanics and motor control of human movement, 4th edn. Wiley, Hoboken

# Part I
# Getting Started

Prior to getting into the details of the shoulder case, we want to present a brief overview of some logistical information for the formation and operation of learning groups (see Appendix L). Since we know that best learning occurs in social settings, we typically form learning groups comprised of four to six individuals. Before each learning session begins, the group selects a reader and a scribe; these assignments should change each time the group meets. The reader's tasks include a presentation of the case, an overview of previous meetings, and updating progress on case discussions. The reader is also responsible for assisting in the monitoring/guiding and supervision of group discussions, attending to the agreed-upon agenda for the learning session, and monitoring/encouraging all members of the group to engage with the process. The scribe's job is to take notes during the learning session. These notes are not detailed trails of information presented by group members, but an outline of the meeting's topics and a list of the learning issues (i.e., topics that are insufficiently known or completely new) identified during the session. A few minutes before the end of the session, the scribe reviews the meeting's accomplishments, and the group decides on learning issue assignments. All group members are responsible for contributing to discussions by asking questions and researching and presenting information on their assigned learning issues. The tutor has an important role on several levels, but most importantly works to ensure that the teacher-learner/learner-teacher dialogue is collegial, involves all group members, and remains primary. As we present each case, we will have already selected what we think are the most appropriate learning issues to be the subject of the chapter. We want to remind the reader that the naming of learning issues is typically generated by the student's innate curiosity. Therefore, look at the learning issues as answers to your own questions about the case, what you are curious about, and what you want to learn.

Prior to delving into the details of the shoulder case, we want to present a brief overview of some logistical information for the formation and operation of learning groups (see Appendix I). Since we know that most learning occurs in social settings, we typically form learning groups composed of four to six individuals. Before each learning session begins, the group selects a leader and a scribe; these assignments should change each time the group meets. The leader's tasks include presenting orienting questions, an overview of previous [illegible] and ongoing progress on each learning issue. The leader is also responsible for assisting in the monitoring, guiding and supervising of group discussions, adhering to the agreed-upon agenda for the learning session, and monitoring/encouraging all members of the group to engage with the process. The scribe's role is to take notes during the learning session. These notes should detail [illegible] of information presented by group members, a list of the meeting's agenda and a list of the learning issues (i.e., topics that are insufficiently known or completely new) identified during the session. A few minutes before the end of the session, the scribe reviews the meeting's accomplishments, and the group identifies the learning issue assignments. All group members are responsible for contributing to discussions by asking questions and researching and presenting information about the assigned learning issues. The tutor has an important role as a facilitator [illegible], to ensure that the learner-guided discussion is helpful, engages all group members, and remains primary. As we present each case, we will have already selected what we think are the most appropriate learning issues to be the subject of the chapter. We want to remind the reader that the naming of learning issues is typically generated by the student's innate curiosity. Therefore, look at the learning issues as answers to your own questions about the case, what you are curious about, and what you want to learn.

# 3 Shoulder: Impingement Syndrome

## 3.1 Introduction

The primary purpose of this case is to clarify the medical diagnosis, impingement syndrome, from a mechanical point of view. In addition, this patient had signs and symptoms that were likely related to her comorbidities and other pathomechanical issues, which also need to be explored in some detail. Although the diagnosis, as presented, is one of the most common problems involving the shoulder girdle, the presentation of this particular case is far from simple because of the pre-existing medical and skeletal issues that characterize this patient.

This is one of many *complex* cases involving individuals in their middle age where we must be aware of the primary problem in the context of comorbidities. Some think that the words *complex* and *complicated* are synonymous. Complicated systems have many parts that make them appear to be complex; however, having many parts, i.e., elements, that are unrelated does not fit the definition of complex. A *complex system* is one that consists of many elements that are inter-related or dynamically coupled; thus, studying the parts of a complex system in a reductive manner, i.e., deconstructing the system and then putting it back together as if "the whole is the sum of its parts," cannot account for the dynamic coupling of all of the elements. In this case, we have at least two elements, i.e., subsystems really, the thoracic spine and cage and the shoulder girdle, which are both complex in their own way and that are dynamic in themselves and have elements that interact. For example, the individual, in this case, had a pre-existing injury to the thoracic spine that she adapted to, which in the process may have contributed to the impairment of her shoulder. Thus, we need to consider the tenets of complexity, that is, complex systems are systems whose states change as a result of interactions and whose interactions change concomitantly as a result of states. That is, complex systems are adaptive and co-evolutionary, are prone to collapse, yet are resilient. The reader is reminded of the notion of complexity throughout this text.

## 3.2 Case Presentation

Mrs. Buckler is a 57-year-old postmenopausal housewife with a 5-year history of osteoporosis. She was referred to you by a physician with a diagnosis of impingement syndrome of the left shoulder. She slipped and fell on the ice outside of the grocery store approximately 8 weeks ago. As she was falling, she reached out with her left arm (since she was carrying a bag in her right arm) to break her fall. Past medical history included a fractured left proximal humerus at age 45. Present symptoms included sharp, intermittent pain with activities of daily living (ADLs) that involved overhead arm movements. She complained of stiffness in the morning and a dull ache in the left shoulder at the end of the day. She also complained of intermittent paresthesias in the ring and little fingers of both hands, which she has had for about 4 years. Radiographic findings revealed moderate to severe glenohumeral joint osteoarthritis, with joint space loss and humeral osteophytes; the humeral head was relatively centered on the glenoid on the axillary view (Fig. 3.1). There was also evidence of stable compression fractures of the vertebral bodies of T3, T4, and L1, as well as old T6 and T11 transverse process fractures (Fig. 3.2); note T represents thoracic and L represents lumbar.

Mrs. Buckler presented with an ecto-mesomorphic body type. On inspection, you noted a moderate forward head, a slight increase in cervical lordosis, and a mild Dowager's hump. The scapular position was observed with the patient standing, revealing a mildly abducted scapula bilaterally, along with a posterior prominence of the inferior angle of the left scapular. In observing her gait, you noticed nothing unusual.

Active range of motion (ROM) testing of the cervical (i.e., abbreviated C) spine was within normal limits in all planes, except for the mild limitation of extension and side-bending bilaterally; with cervical extension testing, it appeared that motion "hinged" at the C6/7 junction. Active-assisted extension ROM testing of the upper thoracic spine,


G. J. Alderink, B. M. Ashby, *Clinical Kinesiology and Biomechanics*, https://doi.org/10.1007/978-3-031-25322-5_3

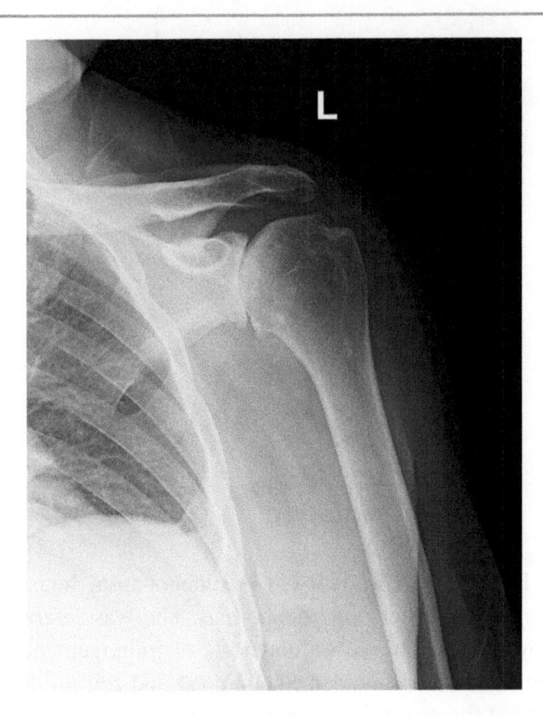


**Fig. 3.1** Abnormal radiographic findings of the left glenohumeral joint indicative of joint osteoarthritis. Note relative loss in joint space and a humeral osteophyte (lower margin)

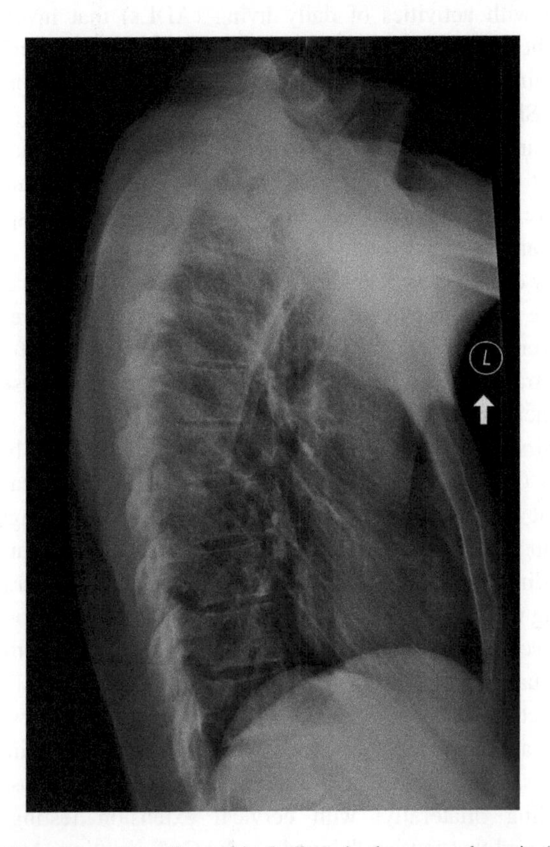


**Fig. 3.2** Abnormal radiographic findings in the upper thoracic (i.e., stable fractures at T3 and T4) and lumbar (L1) regions indicative of fractures related to osteoporosis

with the patient sitting, was moderately limited; there was a mild symmetrical limitation in sidebending and rotation. Neurological screening, i.e., dermatome and upper extremity reflex tests, was negative. The compression/distraction test of the cervical spine was unremarkable. Active ROM testing of shoulder motion revealed a painful arc on the left side. Observation of scapula-humeral rhythm during active ROM revealed a dyskinesis of the scapula bilaterally, worse on the left side. Goniometric measures of active ROM were:

| | *Left* | *Right* | *Normal* |
|---|---|---|---|
| Abduction | 165° | 170° | 180° |
| Flexion | 160° | 170° | 180° |
| External rotation (ER) | 75° | 90° | 90° |
| Internal rotation (IR) | 55° | 60° | 60° |

It was noted that with overpressure at the completion of active ROM, the end feel for both ER and IR was firmer, with less "spring" than on the right side. Joint play testing revealed moderate loss of mobility in the acromioclavicular joint bilaterally and moderate loss of joint play, i.e., inferior and anterior/posterior glide, in the left glenohumeral joint, with mild loss of play in the same directions in the right glenohumeral joint. Note: joint play testing is a passive manual mobility test for passive translatory, i.e., glides, movements at the joint surfaces.

Resisted strength, or manual muscle (MMT), testing revealed:

| | *Left* | *Right* |
|---|---|---|
| Abductors | Weak/painless | Strong/painless |
| External rotators | Weak/painless | Strong/painless |
| Internal rotators | Strong/painless | Strong/painless |
| Adductors | Strong/painless | Strong/painless |
| Serratus anterior | Weak/painless | Weak/painless |
| Middle/inferior trap | Weak/painless | Weak/painless |

Palpation testing of bony and soft tissue structures of the cervical and thoracic spine, shoulder girdle, and upper arm were unremarkable.

### 3.2.1 Key Facts, Learning Issues, and Hypotheses: Shoulder Impingement Syndrome

The first thing we need to do is identify key facts from the medical and physical examination and justify why these facts may be important relative to our overall task. Discussion of key facts contributes to the development of hypotheses, both particular and general, and learning issues. There may be some differences in how individuals classify or prioritize key facts, but we begin with her medical diagnosis of impingement syndrome (IS). This appears to be Mrs. Buckler's primary problem. It interferes (because of pain) with her activities of daily living. On the other hand, the problem did not seem to be getting worse, and she did not have any functional limitations. Our first learning issue needs to address the following questions: (1) what is impingement syndrome?, (2) what are the general characteristics of IS?, (3) what are the pathomechanical characteristics of IS?, and (4) how might the fall be related to her IS? Our purpose in this text is to address each relevant learning issue as it arises relative to key facts.

**Timeout** In a typical PBL session, learning issues are initially identified and listed; additional learning issues may surface as the investigation of the problem proceeds. Then, near the end of the session, learning issues are delegated to team members who will, in the interim between sessions, find resources to address what needs to be known about each learning issue. Typically, learning issues are not addressed at the time they are identified, but presented at subsequent meetings. In this text, however, at times we present extended detail on learning issues that are immediately relevant or present extended coverage of an important topic in a later section.

It is easy to get trapped by a perceived desire/need to determine cause and effect. In our experience, even with common problems like shoulder impingement, clinical problems are complex, and we often find ourselves with the "chicken or the egg" dilemma: what came first? So, rather than get caught in a vicious cycle, we examine key facts and subsequent hypotheses in terms of how they might be related. Since our task is to clarify Mrs. Buckler's medical diagnosis and other signs and symptoms from a mechanical perspective, that is how we address impingement syndrome. Syndromes are generally defined as a set of signs and symptoms. Impingement syndrome is associated with repeated and potentially damaging compression or pinching of several tissues below the subacromial space: supraspinatus tendon, the tendon of the long head of the biceps, the superior glenohumeral capsule, and the subacromial bursa (Fig. 3.3).

These tissues are compressed when the arm is raised overhead, so hallmark signs of impingement include intermittent pain, pain with overhead activities, and a painful arc. A painful arc typically refers to pain during only a part of the range of motion; in the case of glenohumeral impingement, this occurs in the midrange, i.e., approximately 70–120° of arm elevation. Mrs. Buckler presented with physical examination findings and radiographic findings that provided clues about possible pathomechanics of the shoulder girdle complex that may contribute to impingement: intermittent pain, painful arc, symptom reproduction with overhead activities, reduced subacromial space, degenerative changes in the glenohumeral joint (see Fig. 3.1), abnormal cervical and thoracic posture, reduced glenohumeral joint mobility, and muscle

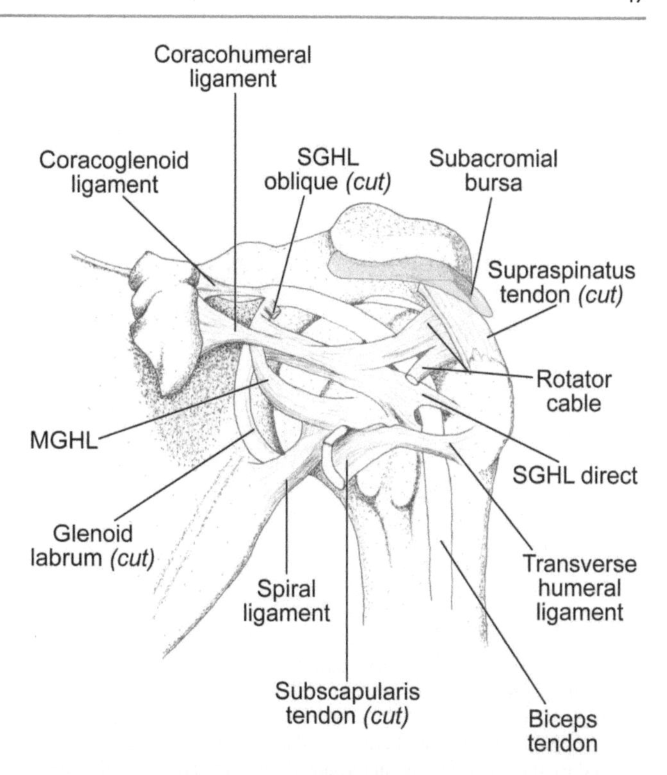


**Fig. 3.3** A portion of the soft tissue anatomy of the glenohumeral joint showing some of the deep tissues that may be subjected to impingement. Note: SGHL = superior glenohumeral ligament; MGHL = middle glenohumeral ligament

weakness. Before we can make more definitive links between her findings and the impingement, we need to review the normal kinematic and kinetic relationships in the upper quarter.

## 3.3 Upper Quarter Kinematics and Kinetics

### 3.3.1 The Diarthrodial Joint: Osteo- and Arthrokinematics

Prior to detailing the osteo- and arthrokinematics of the joints comprising the upper quarter, we need to examine a few fundamental kinematic concepts and the characteristics of diarthrodial (synovial) joints. Kinematics is the study of motion (position, velocity, and acceleration) without regard to the forces and moments that produce the motion. One of the characteristics of kinematic modeling is the use of a Cartesian coordinate system that defines the cardinal planes and axes (Fig. 3.4).

Osteokinematics is defined as a rotation or movement of a bone (or rigid segment) relative to another bone, in one, or combination, of the three cardinal planes. In Fig. 3.4, it is apparent that the origin of the coordinate system is located in the approximate location of the body's center of mass. However, this is one of several locations for body-referenced

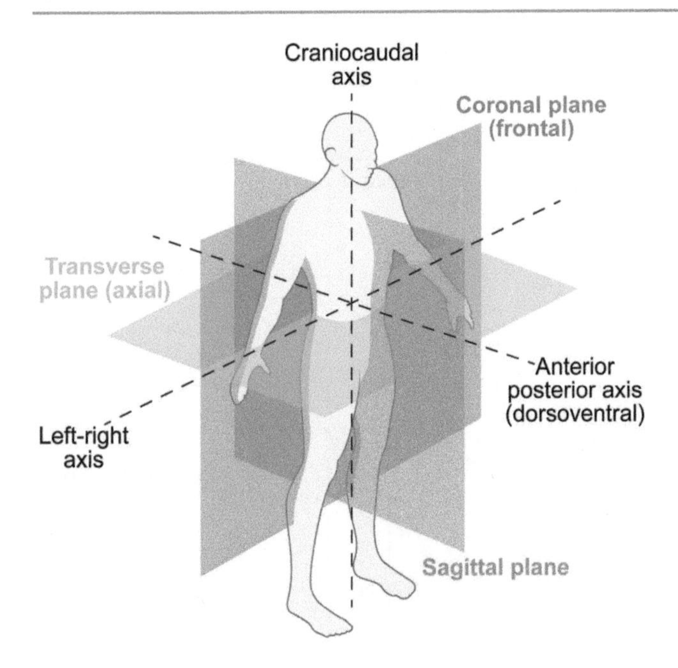


**Fig. 3.4** Cartesian coordinate system consisting of craniocaudal, anteroposterior, and left-right axes. The coronal or frontal plane is defined by the left-right and craniocaudal axes. The transverse plane is defined by the left-right and anteroposterior axes. The sagittal plane is defined by the anteroposterior and craniocaudal axes

coordinates systems. We can also imagine a coordinate system representing an individual body segment with an origin located at the segment's center of mass, or at an approximate joint center. So, imagining that a coordinate system with its origin at the geometric center of the head of the humerus, we can describe the osteokinematic motion of the humerus, relative to the glenoid fossa of the scapula, about the $x$-axis, as occurring in the sagittal plane, using the clinical terms flexion/extension; motion about the $y$-axis in the frontal plane as abduction/adduction; and motion about the $z$-axis in the transverse (or horizontal) plane as internal/external rotation. In classical terms, the shoulder joint, typically labeled as a ball-and-socket joint, has been classified as a joint with three degrees of freedom (DoF), that is, it requires three rotations about three separate axes to describe its movement. In this text, we do not refer to joint motion based on the classical mechanical analogies, e.g., ball-and-socket, hinge, etc., but assume that all of the synovial joints we study move with six DoF, comprised of three rotations and three translations. See Fig. 3.5 for a representative model of the potential six DoF of movements for the segments that comprise the shoulder girdle; note that the potential translations along each axis are not shown.

Moving the arms overhead through a full range of motion, in a pain-free fashion, involves the dynamic coupling and coordination of multiple segments, including the thoracic cage (vertebra and ribs), scapula, clavicle, and humerus. The thoracic cage is its own complex system that includes intervertebral, costovertebral, and costo-sternal joints. The system that is involved with elevating the arm is typically thought to include the scapula-thoracic articulation, acromioclavicular, sternoclavicular, and glenohumeral joints. Although the cervical spine segments and head segment are

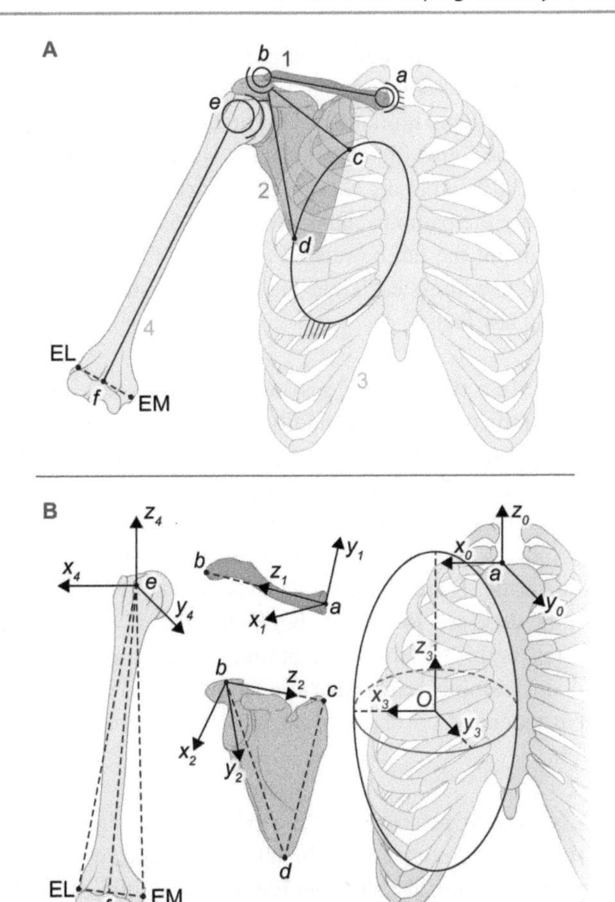


**Fig. 3.5** Components of a model of the shoulder. (**A**) schematic of shoulder girdle segments: (1) clavicle, (2) scapula, (3) thorax, (4) humerus, and (**B**) shoulder girdle segments with attached reference frames. Note that for each segment, rotations about and translations along each axis, i.e., $x_i$, $y_i$, $z_i$, where $i = 1, 2, 3, \ldots$, can induce six DoF of movement of the thorax and scapular segments, as well as at the sternoclavicular (SC), acromioclavicular (AC), and glenohumeral (GH) joints. In this model, the SC, AC, and GH joints are denoted as $a$, $b$, and $c$, respectively, and are treated as ideal three-DoF ball-and-socket joints. The scapula achieves two-point contact with the thoracic ellipsoid surface through its medio-lateral edges $c$ and $d$; the long axis of the humerus represents component 4, that is, the direction of connecting the center of the GH joint $e$ and the center of the elbow $f$, where the elbow joint $f$ is defined as the midpoint of the EL (lateral humeral condyle) and EM (medial humeral condyle). Note: $a$ is the origin of the global coordinate system $S_0 = \{a - x_0, y_0, z_0\}$. The SC joint at point $a$ and axes $x_0$, $y_0$, and $z_0$ are, respectively, parallel to three intersections of the human anatomical coronal, sagittal, and transverse planes: for the clavicular system $S_1 = \{a - x_1, y_1, z_1\}$; for the scapular system $S_2 = \{b - x_2, y_2, z_3\}$, whose origin is located at the AC joint at $b$; for the thorax $S_3$ parallel to $S_0$; and for the humeral system $S_4 = \{e - x_4, y_4, z_4\}$ whose origin is located at the center $e$ of the glenoid fossa, where the $y_4$ axis is perpendicular to the plane determined by point $e$, EL, and EM; the $x_4$ axis is determined by the right-hand rule (see Appendix G for a definition of the right-hand rule). (Modified with permission from Zhibin et al. (2018))

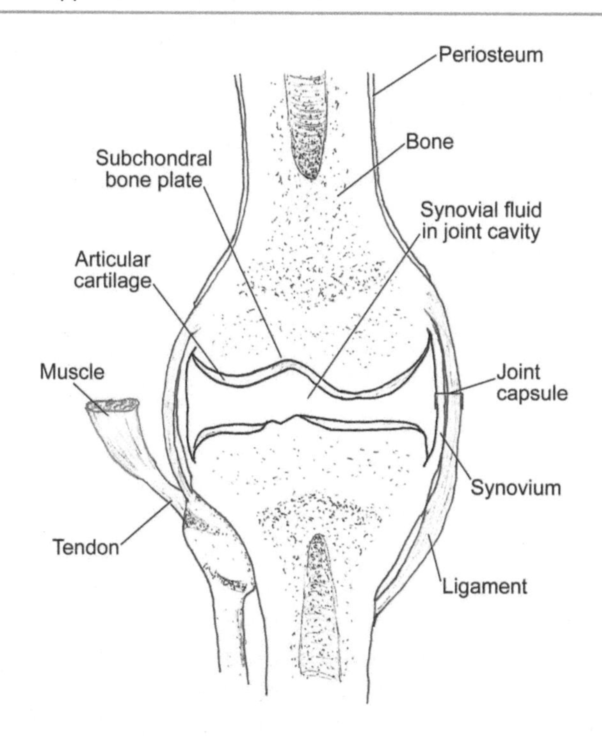


**Fig. 3.6** Characteristic elements of a normal diarthrodial (synovial) joint

linked to the thorax, for the purpose of this presentation, we assume that their role in overhead arm movements is minimal.

Since most of the joints involved with overhead movements are diarthrodial in nature, let's first examine the general characteristics of those types of joints (Fig. 3.6). A diarthrodial or synovial joint is an articulation between two bones that allows small, moderate, and large movements. Notice that the metaphysis of a long bone is made up largely of spongy (trabecular) bone and a subchondral bone plate adjacent to the articular cartilage. The ends of the articulating bones are covered with articular (hyaline) cartilage, and the joint is enclosed by connective tissue referred to as the joint capsule. The external layer of the capsule is composed of dense connective tissue that serves to contain joint contents and assists in joint stability. The internal layer of the capsule consists of a synovial membrane that manufactures synovial fluid. Some joints have synovial plicae (i.e., redundant folds of synovial membrane tissue), which increases the synovial surface area, and may contribute to less tension on the synovial membrane, thus promoting full joint motion. The synovial fluid and its lubricating proteins coat the articular surfaces of the joint, thereby reducing friction between joint surfaces and providing nutrients to the articular cartilage. Ligaments connect the bones of a joint. They are composed of dense connective tissue that can be sheet-like capsular thickenings (i.e., capsular ligaments) or more distinct cord-like extracapsular structures. The capsular thickenings provide overall joint stability, whereas extracapsular ligaments provide joint constraint in particular planes, e.g., the medial collateral ligament of the knee is the primary constraint for excessive tibial abduction or valgus and may also assist in the production of joint moments. Unlike the articular cartilage, the joint capsule has a rich blood supply, as well as nociceptors for pain and mechano-receptors for proprioception. Discussion on the mechanics of the various tissues, e.g., articular cartilage, etc., making up a synovial joint is presented in more detail in subsequent sections of this chapter.

Before we move forward, let's briefly examine the sensory apparatus associated with the synovial joint that helps to control the body's mechanical systems. These sensory structures or joint mechanoreceptors can be found in the skin, periarticular tissues, and muscles crossing synovial joints. Proprioception is the ability of the nervous system to detect the static and dynamic position of a joint in three-dimensional space. Because this is a sort of kinematic measurement system, these joint receptors sense touch/pressure; joint positional change, i.e., applied tensile loading; and the rate or velocity, at which these changes take place. Proprioception is essential to normal, pain-free movement, and injury to this system, related to trauma or joint degeneration, impairs function. See Table 3.1 for a description of the five common joint mechanoreceptors.

Let's now frame the assumptions around the concept of arthrokinematics. Arthrokinematics is defined as the motions, i.e., roll, slide, and spin, that occur within, or between, the articular surfaces of a joint. These motions are not under voluntary control and are often referred to as accessory motions (or joint play). MacConaill and Basmajian pioneered the concepts relating joint surface morphology (or shape) and the distinct non-voluntary movements between the articular surfaces. Based on their anatomical work, they concluded that there were no flat (or "plane") articular surfaces, but all joint surfaces had either an ovoid or sellar shape. The ovoid surface was everywhere convex or concave in all directions, whereas the sellar (or saddle) joint was one in which a section of the surface in one direction was convex and a section cut at right angles to the first direction had a concave shape (Fig. 3.7).

It is important to note that when describing arthrokinematics motions, we must clearly state our frame of reference. There are two choices: (1) describe joint surface motions based on how the proximal segment (or bone) moves relative to the distal segment or (2) how the distal segment moves relative to the proximal segment. For our purpose at the moment, let's assume the distal segment is moving relative to the proximal segment. Based on anatomical and empirical evidence, MacConaill proposed the following hypotheses with regard to arthrokinematics: (1) if the distal segment joint surface is ovoid concave, the concave surface can both spin and slide (also called glide) on the ovoid convex surface

**Table 3.1** Periarticular joint receptors

| | Ruffini endings | Pacinian corpuscles | Golgi tendon organs | Free nerve endings | Muscle spindles |
|---|---|---|---|---|---|
| Classification | Type I | Type II | Type III | Type IV | Ia/II |
| Location | Superficial layers of the fibrous joint capsule | Deep layers of the fibrous joint capsule and fat pad | Intrinsic and extrinsic ligaments | Fibrous capsule, ligaments, articular fat pads, and blood vessel walls | Intrafusal muscle fibers |
| Activation threshold | Low | Low | High | High | Low |
| Adaptation | Slow | Rapid | Slow | – | Rapid |
| Range of activation | Rest | Initiation and cessation of movement | End of the range of movement | Abnormal chemical and mechanical irritation | Initiation and cessation of movement |
| Function | Feedback regarding static joint position and joint acceleration; sensitive to tensile forces | Feedback regarding joint acceleration; sensitive to compression forces | Active at extremes of joint motion; feedback regarding tissue deformation | Signal presence of noxious, chemical, mechanical, and inflammatory stimuli | Feedback regarding velocity and magnitude of length change |

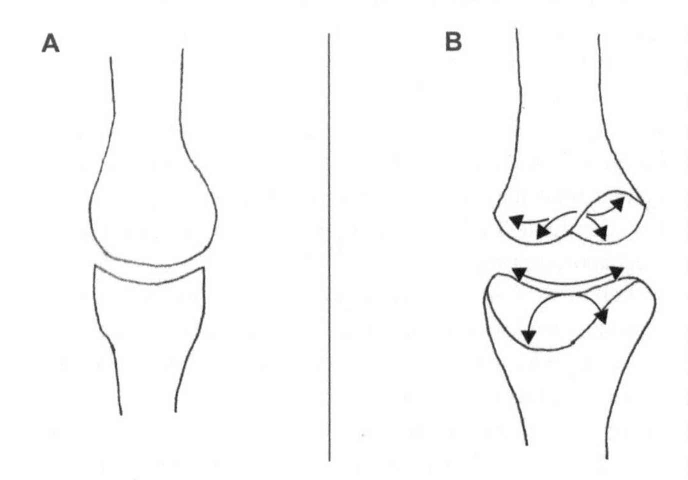


**Fig. 3.7** (**A**) Ovoid convex on concave and (**B**) sellar joint surfaces. Note that a sellar joint surface is characterized by a convexo-concavo shape, i.e., convexity from one observational perspective and concavity from a perspective that is orthogonal to the other

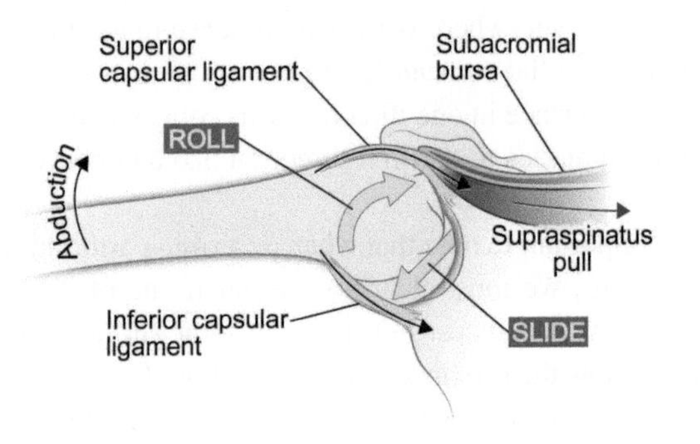


**Fig. 3.8** Osteokinematic elevation, i.e., abduction, sees the head of the humerus roll upward (superiorly) but glide inferior at the joint surface. In this case, the distal segment joint surface, i.e., the head of the humerus, is ovoid convex rolling relative to the proximal ovoid concave segment joint surface, i.e., the glenoid fossa of the scapula. Note the proximity of the under surface of the acromion process and the subacromial burse, supraspinatus tendon, and superior joint ligament (i.e., capsule)

of the proximal segment joint surface; (2) if the distal segment joint surface is ovoid concave and rolling (also called rocking), relative to the proximal segment which is ovoid convex, it slides in the same direction, except at the very beginning and the very end of the movement; (3) convex surfaces moving on concave surfaces can both slide and spin, but rolling is the principal movement of a convex surface; and (4) the direction of the slide that accompanies the roll when a convex surface moves relative to the concave surface is opposite to that of the roll (Fig. 3.8). According to MacConaill, the effect of the roll/slide combination is important for the economical use of articular cartilage, therefore the long-term functional and pain-free range of motion needed for daily activities.

MacConaill notes that rolling does not take place when the joint is close-packed. Note that a close-packed position occurs near or at the end of a joint's range of motion and is characterized by (1) partnered joint surfaces that are maximally congruent; (2) tautness of periarticular structures, e.g., ligaments and joint capsules, meaning they are more constraining; (3) maximal joint stability; and (4) minimal to no accessory motion.

The third type of joint surface movement between curved surfaces has also been identified. This movement is referred to as a *spin* and is characterized by a rotation around a mechanical axis and a "single" point of contact of an articular surface rotating relative to a single point of its articulating partner (Fig. 3.9). An example of the complexity of joint arthrokinematics that combine a roll/slide and spin mechanics can be seen at the tibiofemoral joint during flexion and extension (Fig. 3.10). Imagine an individual sitting in a chair. From your perspective facing the individual you ask that person to straighten, i.e., extend, one of their knees. During knee extension, as the tibia moves anteriorly, there is an anterior roll/glide of the concave proximal tibia relative to the convex femoral condyles. However, because the medial femoral

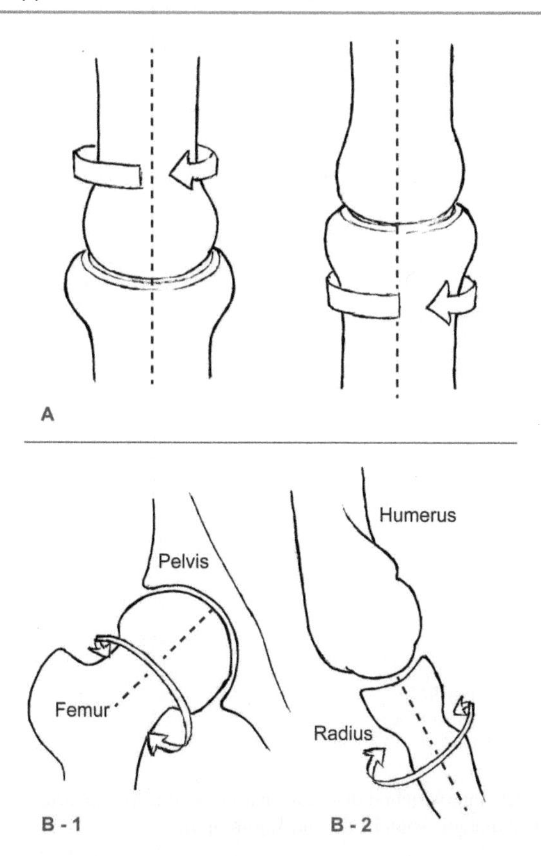


**Fig. 3.9** (**A**) Joint spin can occur when an ovoid convex surface moves on an ovoid concave surface or vice versa; (**B**) examples of spin: (1) spin of the head of the femur relative to the acetabulum during hip flexion and (2) spin of the proximal radial head relative to the capitellum of the humerus during forearm pronation/supination

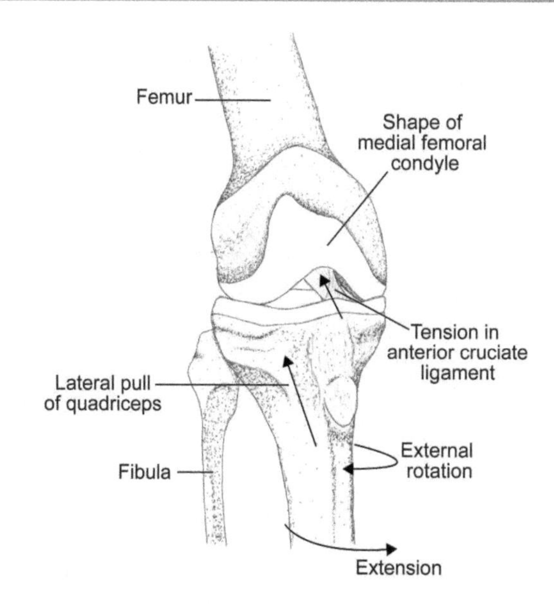


**Fig. 3.10** During terminal knee extension, with the tibial moving relative to the femur, there is a lateral spin of the tibia. This conjunct rotation has been referred to as the screw-home mechanism since it induces the tibiofemoral joint into a close-packed, and more stable, position. It is notable that this conjunct rotation creates increased tension in the anterior and posterior cruciate, and medial and lateral collateral, ligaments

condyle has a larger anteroposterior diameter than the lateral femoral condyle, there is also a conjunct external rotation, i.e., lateral spin, of the tibia during the last 20–30° of knee extension. As an exercise, we leave it to the reader to describe the mechanics when initiating knee flexion from an extended knee when the tibia is moving relative to the femur. A related exercise is to imagine knee arthrokinematics when the femur moves relative to the tibia with flexion initiation and terminal knee extension during the loading response and terminal swing subphases of the gait cycle, respectively.

In the preceding discussion, we presented a theoretical model of joint surface motion that simplifies a complex function. When biomechanists create models, we do so under assumptions that allow us to reductively study complex phenomena. Clinicians often use those models and apply them in their practice and, in the process, collect empirical evidence that may corroborate or refute the model. The clinical model which prescribes joint arthrokinematics based on the convex-concave and concave-convex rule has been used for many years in practice and seems to be mostly correct. However, there is some experimental evidence that challenges the clinical rule, suggesting that more research is needed. On the other hand, we should be aware that laboratory-controlled conditions cannot mimic clinical reality. Perhaps as technology improves, we will be able to measure these very small joint surface excursions with better accuracy and precision both in and outside of the laboratory. In the meantime, however, anatomists and biomechanists presently agree that joint surface motion and joint stability are related to the following: (1) the shape of articular surfaces; (2) the variability of joint surface geometry, which make joint partners more or less congruent, i.e., similar or dissimilar; (3) joints with more congruent surfaces which are more stable; (4) joint surface curvature, and its magnitude, affecting joint mobility and stability; (5) joint surfaces with similar radii of curvature that are more congruent; (6) the amount of curvature of articulating surfaces and their congruence influencing the direction and amount of roll/glide; (7) joint partners that are biconvex, e.g., atlanto-axial joint, which tend to allow freer movements; and (8) intra- and extracapsular ligamentous designs that provide variable degrees of joint constraint and mobility. In spite of the experimental evidence that challenges the convex-concave and concave-convex rules, we repeatedly refer back to these claims as we explore other regional clinical problems.

#### 3.3.1.1 Spinal Osteo- and Arthrokinematics

Now let's examine the anatomical morphology of a typical thoracic vertebral unit: superior vertebral segment, intervening intervertebral disc, inferior vertebral segment, and adjoining ribs (Fig. 3.11). Examining this morphology is

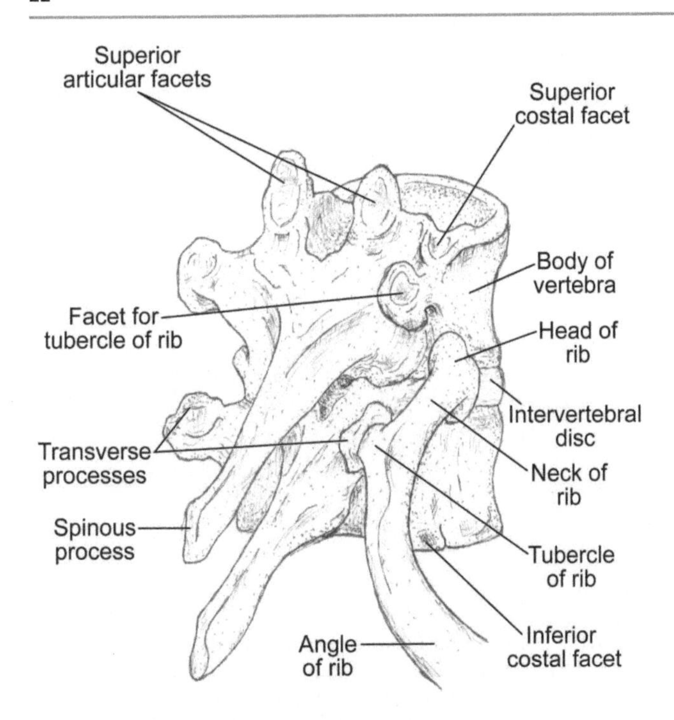


**Fig. 3.11** Typical thoracic spinal unit: superior segment, an intervening intervertebral disc, inferior segment, and adjoining typical rib

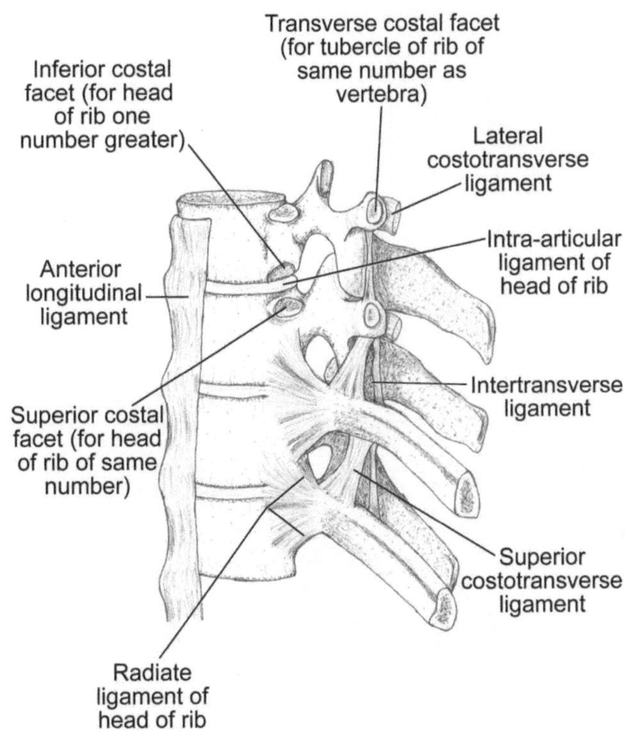


**Fig. 3.12** Intervertebral disc, capsular and extracapsular constraints of a typical thoracic spinal unit, and adjoining ribs

important because it informs our understanding of the joint arthrokinematics associated with spinal movements of flexion, extension, sidebending, and rotation.

The second through the ninth thoracic vertebrae demonstrates similar features, like that shown in Fig. 3.11. For any one segment, the superior articular facets face posteriorly, while the inferior articular facets face anteriorly. Once articulated, these flat-shaped facets form the synovial apophyseal joints, which are vertical yet pitched slightly forward and oriented in the frontal plane. The intervertebral articulation is a cartilaginous, not a synovial, joint. The proximal end of a typical rib is ovoid convex and articulates with the concave demi-facets of adjoining vertebral bodies, termed the costovertebral (or costocorporeal) joint. Just lateral to the rib head, the ovoid convex rib tubercle articulates with its concave partner on the vertebral transverse process, forming the costotransverse joint. Both the costovertebral and costotransverse articulations are synovial joints. The thoracic spinal unit and rib pairs are constrained by the outer annulus of the intervertebral disc, strong joint capsules, and extracapsular ligaments (Fig. 3.12), limiting motion and ensuring thoracic spine and cage stability.

Thoracic spine motion is constrained by the rib articulations anteriorly as well as posteriorly (Fig. 3.13). The manubrium and sternum make a fibrocartilaginous (synarthrosis) joint where they fuse; early in life, this junction is filled with a partial cartilaginous disc, which later ossifies. The sternocostal joints are comprised of two sections: costochondral junctions which represent the interface of the end of the rib

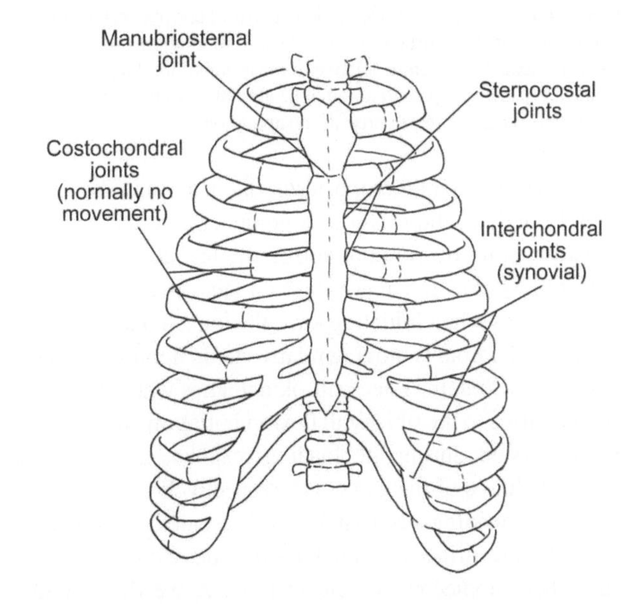


**Fig. 3.13** The anterior aspect of the thoracic cage with four additional sets of articulations, i.e., manubriosternal, sternocostal (costochondral and chondrosternal), and interchondral

and fibrocartilage and chondrosternal junctions between the cartilage and slightly concave sternal costal facets. The first chondrosternal junction is a synarthrosis, but ribs 2–7 form synovial joints with the sternum, where small accessory movements are permitted. Fibrocartilaginous discs are frequently present in the lower sternocostal joints. Synovial

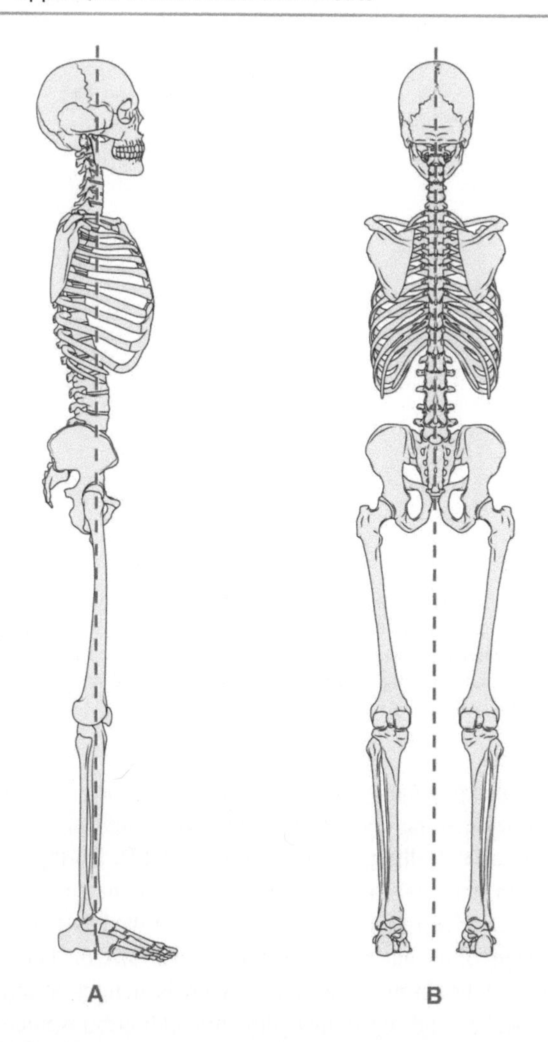


**Fig. 3.14** Ideal standing posture (**A**) sagittal view, (**B**) posterior frontal view. Note the plumb line in (**A**) bisecting the external auditory meatus, the upper cervical vertebral bodies, the acromion process, through the center of the thorax, lumbar vertebral bodies, just behind the axis of the hip joint, just in front of the axis of the knee joint, and just in front of the lateral malleolus. The plumb line in (**B**) runs through the center of the cranium, all vertebral bodies, and the midline of the sacrum

lined interchondral joints link ribs 5–10. These joints are constrained by inner- and extracapsular ligaments. Ribs 11 and 12 are "floating," i.e., not attached anteriorly, and are not shown.

Finally, before we review the osteo- and arthrokinematics of the thoracic cage, we need to look at the normal posture of the upper half of the body. When we observe the ideal standing posture (Fig. 3.14), and focus on the sagittal view of the thoracic cage, we notice that the curve in the upper to middle regions, i.e., T1–T7, is convex posteriorly, also called kyphosis. This mild kyphosis is normal but holds the vertebral segments in this region in a relatively flexed position. Because of this position, the thoracic spine must extend in order to allow/promote full shoulder elevation bilaterally. With a unilateral elevation of the arm, thoracic cage movement gets more complicated because vertebral sidebending and rota-

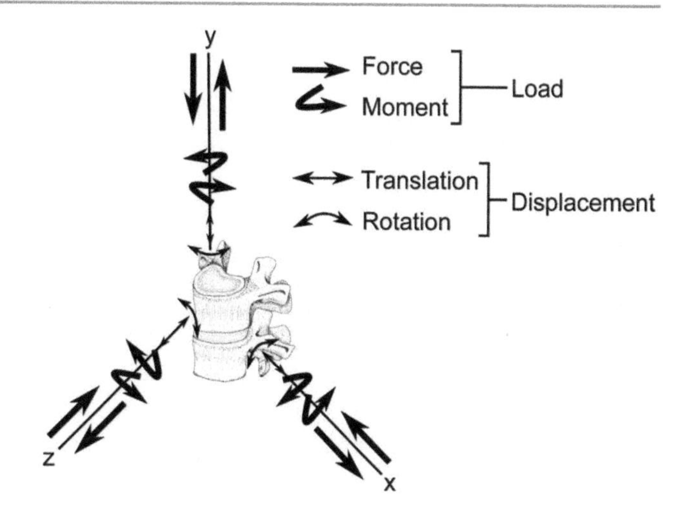


**Fig. 3.15** The spinal functional unit includes a superior vertebra, the intervening intervertebral disc (IVD), and inferior vertebrae. Note a Cartesian coordinate system with its origin located within the center of the IVD and the six DoF of movement allowed in the spinal unit. (Modified with permission from White and Panjabi (1990))

tion, and rib rotation, are coupled with thoracic extension. For example, during abduction of the left upper extremity, the thorax sidebends to the right and concomitantly rotates (either in the same or opposite direction). Let's illustrate these ideas a bit more.

However, before getting into the details of thoracic spine/cage osteo- and arthrokinematics, we need to be aware of the general convention for describing spinal functional unit movements (Fig. 3.15). Internal forces and moments both produce and control complex spinal complex movements. Reductively, if we fix a coordinate system to the center of the vertebral body, we can describe the six DoF of spinal segmental movement as three rotations about, and three translations along, the coordinate axes, as illustrated in Fig. 3.15. By accepted convention, vertebral segmental movement is described by how the superior segment moves relative to the one below it. During clinical motion testing, we may ask the individual to bend in a cardinal plane, e.g., flexion/extension, but functionally, the spinal segments never move in pure cardinal planes but move in coupled patterns. For example, during flexion, the superior segment rotates, for example, about the $x$-axis and translates along the $z$-axis.

**Label warning** We strongly encourage the reader to have a plastic model of the spine that can be examined and moved while attempting to follow the osteo- and arthrokinematic instructions in the forthcoming sections.

Let's examine the combination of movements in the thoracic spine/cage when someone elevates their arm through a full range of motion. The clinical term extension is a concise way to describe spine backward bending or a posterior

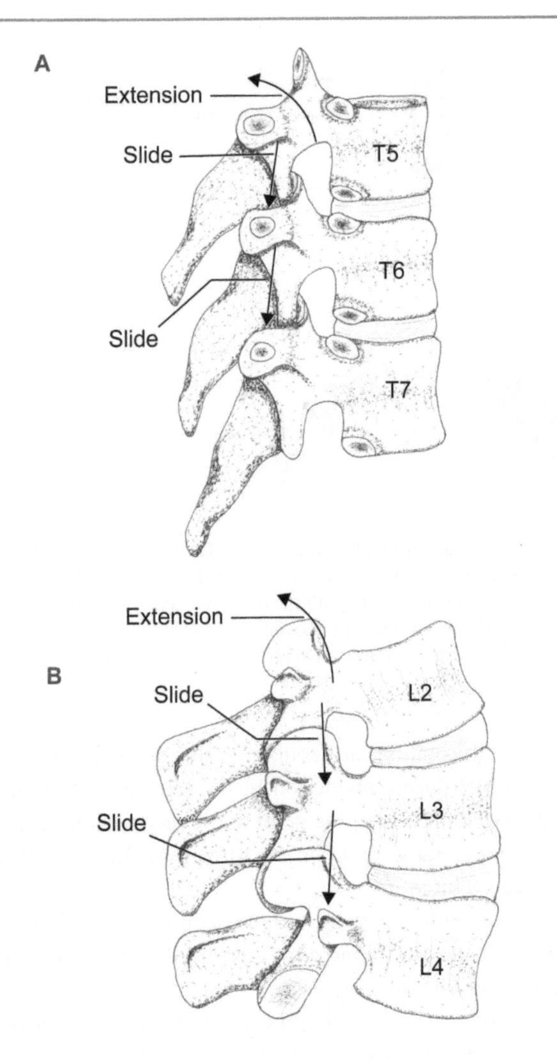


**Fig. 3.16** Osteo- and arthrokinematics of (**A**) thoracic and (**B**) lumbar spinal units during extension. As the vertebral body rolls (i.e., tilts or rocks) backward, there is a posterior translation of the superior vertebral segment relative to the inferior vertebral segment. Simultaneously, the inferior articular facet of the superior vertebra will glide slightly posteroinferiorly relative to the superior articular facet of the inferior vertebra

movement of the spine/cage (Fig. 3.16). During this movement, the superior vertebral body rotates (or rocks), and glides, posteriorly. With vertebral extension the posterior aspect of the intervertebral disc is placed under compression loads, which induces shear and tensile stresses within the outer annular fibers anteriorly; these outer annular fibers provide a constraint, not unlike joint ligaments. Concomitantly, the right and left inferior facets of the superior segment glide posterior/inferior relative to the superior facets of the inferior segment. To avoid deviation to the right or left during extension, there must be symmetrical posterior/inferior gliding at both facet joints. The gliding that takes place at the intervertebral and facet articulations does not follow the convex/concave rule since these surfaces do not have ovoid shapes, but are more planar, i.e., flat.

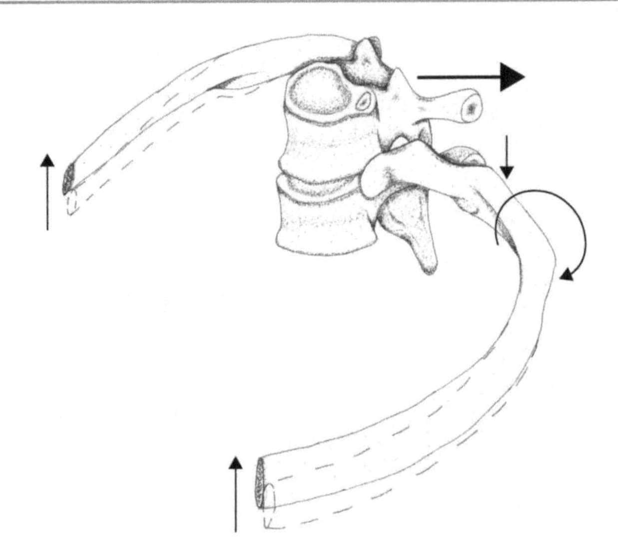

**Fig. 3.17** Typical rib osteo- and arthrokinematic movements during thoracic extension. As the spine extends, elevation of the typical ribs anteriorly induce a posterior rotation of the body of the ribs that result in gliding and spinning actions at the costocorporeal and costotransverse joints, respectively

As noted previously, rib attachments posteriorly and anteriorly provide some constraint to these movements; however, the ribs are not idling, i.e., passive (Fig. 3.17). During spinal extension, the ribs rotate posteriorly (osteokinematic movement) around a paracoronal axis, also described as a superior movement of the anterior aspect of the rib. Since the head and tubercle of the rib are ovoid convex, there is an inferior tangential glide, i.e., glide opposite in direction of bone movement, at the costocorporeal joint, and a posterior pivot/spin at the costotransverse joint. Anteriorly, at the costochondral fibrocartilaginous junction, there may be a small amount of posterior roll/glide, but most likely a tensile loading results in elastic deformation of the cartilage, whereas at the sternocostal (or chondrosternal) joints, where the sternal articulation is slightly concave, there may be a slight posterior pivot glide.

What happens when only one arm is elevated? The thoracic spine segments still extends slightly, but now, since the movement is asymmetrical, spinal segment and rib movements are more complex. With an elevation of the left arm, in addition to the slight segmental extension, involved spinal segments bend slightly to the right, i.e., right side (or lateral) bending, and rotate slightly to the right as well. Concomitant sidebending and axial rotation, called coupled motion, in the spine always take place. Note that when we describe coupled motion, we describe the direction the vertebral body is moving. There is some disagreement among clinicians about the direction these coupled motions take in the thoracic spine, but for now, we assume that when the thoracic spine is generally in a neutral posture, sidebending and rotation movements couple in the opposite direction, but when the spine is flexed or extended (as in this case), sidebending and rotation couple in the same direction. Getting back to our example of elevation of the left arm, with right sidebending, we see right rotation and right glide of the superior vertebral segment. There is an anterior/superior glide of the left inferior facet of the superior segment and a posterior/inferior glide of the right inferior facet. Since the superior segment is sidebent and rotated right, the ribs on the left are distracted and rotate anteriorly, whereas the ribs on the right approximate and rotate posteriorly. With the left ribs anteriorly rotated, there is a superior tangential glide at the left costocorporeal joints and an anterior spin at the costotransverse joints; concomitantly, with the right ribs posteriorly rotated, we see inferior tangential glides at the right costocorporeal joints and a posterior spin at the costotransverse joints. Anteriorly, at the costochondral junction bilaterally, both tensile and torsional elastic deformation loads are mitigated by fibrocartilage, while slight anterior and posterior pivot glides likely take place at left and right costochondral joints, respectively.

Of course, the thoracic spine is not moving in isolation, so we now need to examine the anatomy and mechanics of the cervical spine. Our review does not include the atypical first and second cervical segments, not because they are not important, but because the small movements that occur there were normal in Mrs. Buckler's situation and they are not critical to shoulder elevation. The typical cervical spinal unit is similar to the thoracic unit previously described (Fig. 3.18). Let's first note the cervical-thoracic junction, i.e., C7/T1. From the anterolateral view of this region, we see that the first thoracic vertebral segment, i.e., T1, has a full costal facet (ovoid concave) that articulates with the head of the first rib. Similar to typical ribs, the first rib's tubercle articulates with the transverse process of T1. The first rib has the typical costochondral and chondrosternal articulations, but with the manubrium of the sternum, where it sits very close to the manubrial facet that accepts the proximal end of the clavicle. From this perspective, we might anticipate that thoraco-cervical/first rib and/or proximal manubrial pathomechanics could impair full, pain-free shoulder elevation.

As noted previously, the osteo- and arthrokinematic description of the typical cervical spine includes the motion of the superior vertebrae relative to the one below it. In general, major differences between the morphology of the typical cervical and thoracic segments influences the kinematics. Similar to the thoracic segment, the cervical facets are flat, but with facet joint surfaces oriented midway between the frontal (coronal) and horizontal planes. The inferior facets of the superior segment face antero-inferior, whereas the superior facets of the inferior segment face postero-superior. Another notable difference in the cervical spine is that relative to the height of the vertebral body, the intervertebral discs have a greater height (or thickness) compared to discs in the thoracic spine. Both the orientation of typical cervical facets and greater disc height, as well as the absence of rib attachments, contribute to the greater overall mobility in the cervical spine. Another unique feature of the cervical vertebral morphology is the uncinate processes (Fig. 3.19), which

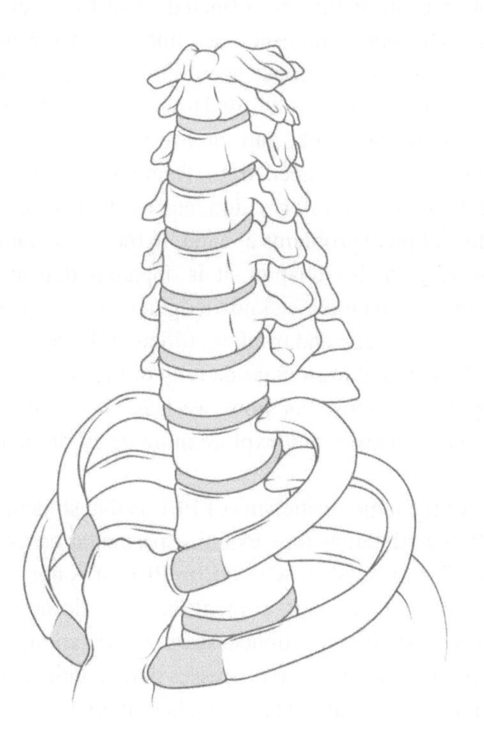

**Fig. 3.18** Anterolateral view of the thoraco-cervical region, first rib, and typical cervical spinal unit, i.e., superior vertebrae, intervening disc, and inferior vertebrae

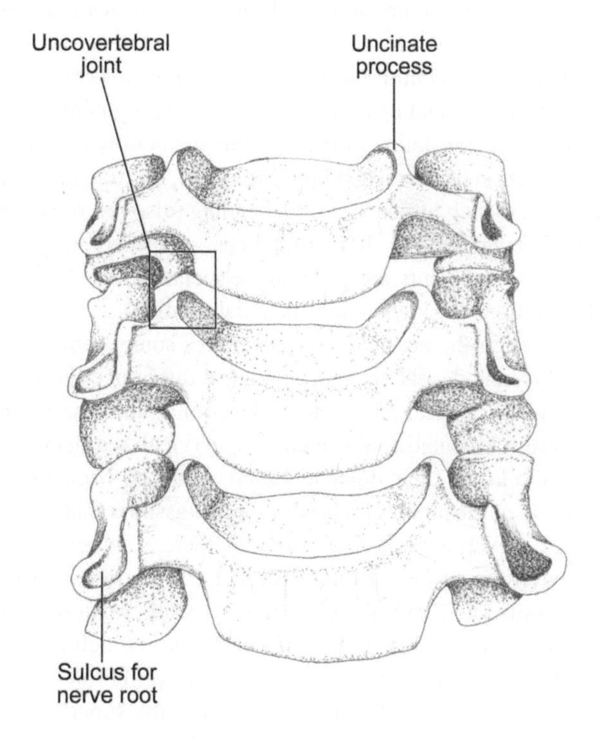


**Fig. 3.19** Uncinate processes which form the uncovertebral joints as part of a typical cervical spinal unit

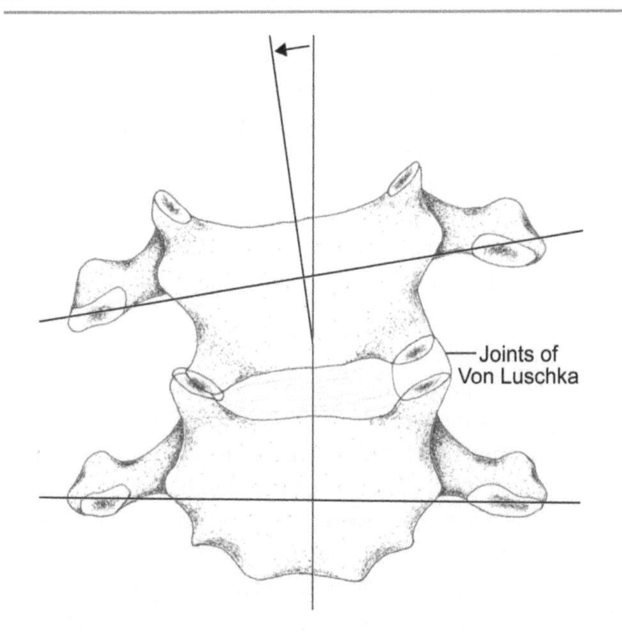


**Fig. 3.20** Anterior view of a typical cervical functional unit illustrating the proposed uncovertebral joint, i.e., joint of Von Luschka, mechanics during cervical sidebending to the left

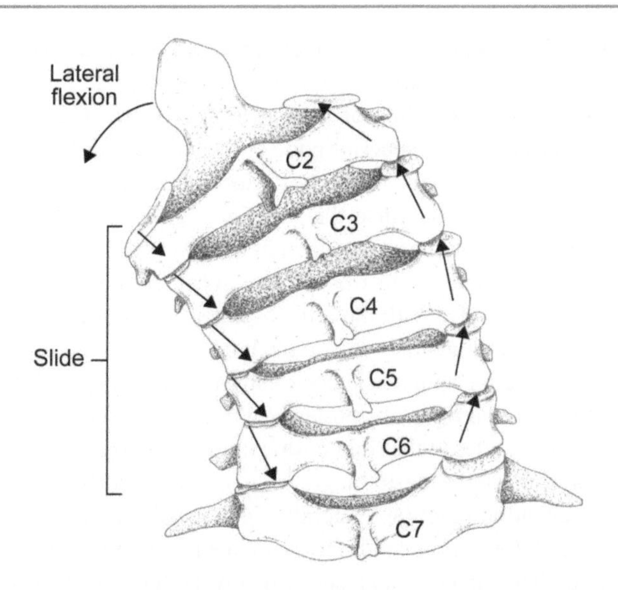


**Fig. 3.21** Left lateral flexion (i.e., bending) coupled with axial rotation in the same direction in the cervical spine. Note the posteroinferior direction of sliding of the inferior facet of the superior segment on the concave side and concomitant antero-superior direction of sliding of the inferior facet of the superior segment on the convex side

form uncovertebral joints (also referred to as the joints of Von Luschka) on the lateral borders of the vertebral bodies. The uncovertebral joints appear to be synovial joints that likely play a mechanical role in providing frontal plane stability during lateral bending. Given the increased cervical spine disc height, which may increase translation of the cervical segment, the uncovertebral joints may provide a bony constraint to excessive lateral translation, in essence helping to maintain the central positioning of the intervertebral disc. During lateral bending of the typical cervical segments, there is a relative translation of the superior segment in the opposite direction of lateral bending (Fig. 3.20), i.e., translation to the right during left cervical sidebending (also referred to as lateral bending). It has also been proposed that the uncovertebral joints are generally associated with the coupled motions of sidebending and axial rotation.

Although the cervical spine sacrifices some stability for mobility, and despite the morphological differences between the typical cervical and thoracic functional unit, the osteo- and arthrokinematic movements of cervical segments are very similar to their thoracic cohort. Let's go back to our example of elevation of the left arm and assume that during the full elevation of the left extremity, there is a slight cervical extension, with concomitant left lateral flexion and rotation. Associated with these rotations is slight posterior and left lateral translation of the vertebral bodies, which induces elastic deformation of the intervertebral disc. Arthrokinematically, the inferior facets of the superior segment along the left articular pillar glide posteroinferiorly, while the corresponding facets along the right articular pillar glide antero-superiorly (Fig. 3.21).

**Timeout** We acknowledge the dynamical and complex nature of the many DoF that the central nervous system must control to achieve full, coordinated, pain-free elevation of the arm. Of course, this includes muscles of the neck and trunk, which provide proper positioning of the segments, as well as trunk (or core) stability. However, one of the tenants of PBL is that we seek, and choose to use, only the most relevant information needed to frame key facts, provide needed fundamental information, and explain/solve the most important clinical problems at hand. In the particular case we are studying in this chapter, it is apparent that a general, rather than a detailed, acknowledgment of the functional anatomy of the neck and trunk is sufficient. In later cases, we revisit the detailed muscular anatomy of the neck, trunk, and lumbopelvic regions. For now, we leave it to the reader to follow their curiosity and explore more detail on their own.

One of the major criticisms of PBL is that students do not get sufficient detail as they examine particular cases. This is not true. For those of us who utilize PBL, it is not necessary to overload students with content that they do not immediately need. In lieu of unnecessary information, students develop skills in important processes that include, but are not limited to, (1) critically examining key findings; (2) learning how to ask their own probing questions that are needed to understand, and perhaps reframe, the nature of a problem(s);

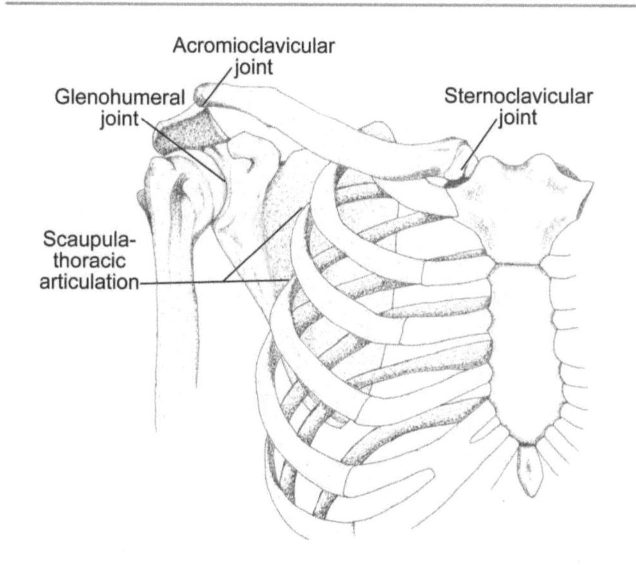


**Fig. 3.22** Joints of the right shoulder girdle, including the scapula-thoracic articulation

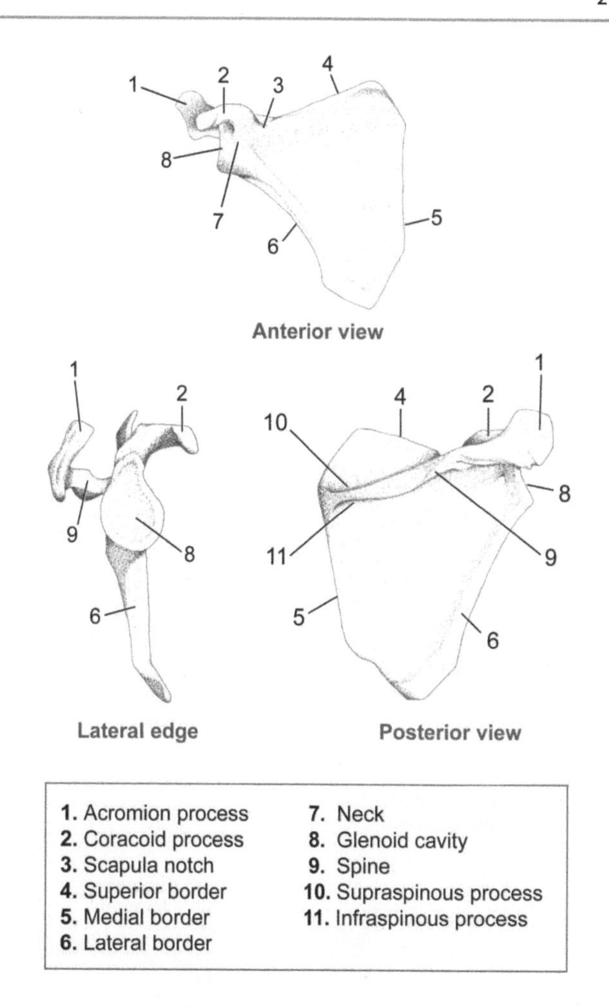


**Fig. 3.23** Anatomical landmarks of the scapula

(3) identifying the information that is needed to master an understanding the problem (even going down "blind alleys" is a learning process germane to problem-solving); (4) seeking and critically examining relevant resources, which includes making decisions about what information is not relevant to the problem; and (5) constructing new "knowledge" in the context of their problem-solving.

#### 3.3.1.2 Shoulder Girdle Osteo- and Arthrokinematics and Kinetics

Let's next examine the functional anatomy, arthrology, kinematics, and kinetics of the shoulder girdle complex including the (1) scapula-thoracic articulation, (2) glenohumeral joint, (3) acromioclavicular joint and subacromial space, and (4) sternoclavicular joint (Fig. 3.22). The scapula is a triangular-shaped bone with three angles, superior, lateral, and inferior, and three borders, medial (vertebral), lateral (axially), and superior (Fig. 3.23). The features of the scapula most germane to our problem include the flat anterior end of the acromion process (clavicular facet) and the ovoid concave glenoid cavity (or fossa). The glenoid fossa is inclined slightly upward, relative to a horizontal axis, which may afford some inherent stability at the glenohumeral (GH) joint. At rest, the scapula is nestled on the posterolateral wall of the thoracic cage, with the glenoid fossa angled slightly anterior to the frontal plane. This orientation of the scapula and its articulating partner, the humerus, is referred to as the scapular plane. Abduction (or elevation) of the arm in the scapular plane may be the least constraining movement related to scapula-humeral functional motions.

The ovoid convex humeral head, comprising about half of a full sphere, articulates with the glenoid fossa of the scapula. The head faces medio-superiorly and forms an approximate 135° angle of inclination with the long axis of the humeral shaft. Relative to a horizontal axis at the elbow joint, the adult humeral head is retroverted (i.e., rotated posteriorly) by approximately 30°. It has been suggested that the upward inclination of the glenoid fossa and retroverted humeral head are evolutionary compensations for the inherent bony instability at the GH joint secondary to the relative incongruence of the concave/convex fit between the two bones (Fig. 3.24).

The inherent instability at the GH secondary to insufficient bony congruence is carefully guarded by the joint capsule and GH capsular ligaments, coracohumeral ligament, rotator cuff muscles, long head of the biceps brachii, and glenoid labrum (see Fig. 3.2). The glenoid labrum is a fibrocartilaginous structure that deepens the glenoid cavity. This arrangement increases the surface area of contact between the head of the humerus and the glenoid fossa. The labrum also serves as an anchoring site for the joint capsule and other related structures, e.g., the long head of the biceps, which has important mechanical implications.

The synovial acromioclavicular (AC) joint is the articulation between the acromion of the scapula and the lateral end

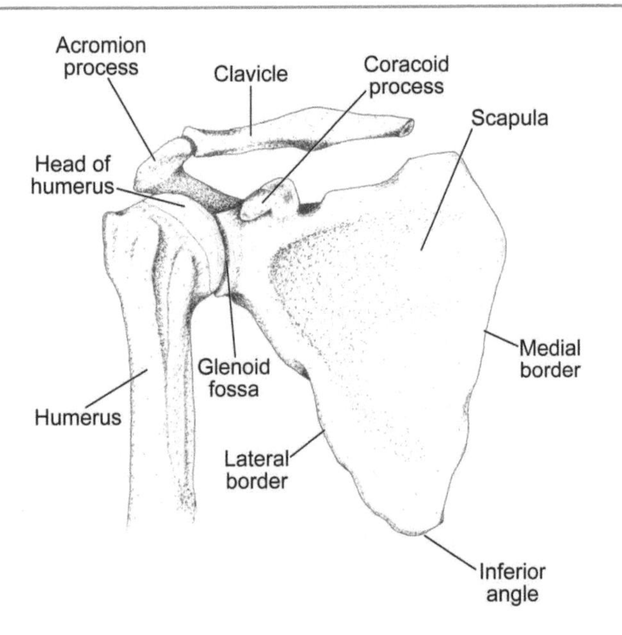


**Fig. 3.24** Anterior perspective demonstrating humeral head angle of inclination and orientation relative to the glenoid fossa of the scapula. Note that the humeral head is also slightly retroverted (not shown)

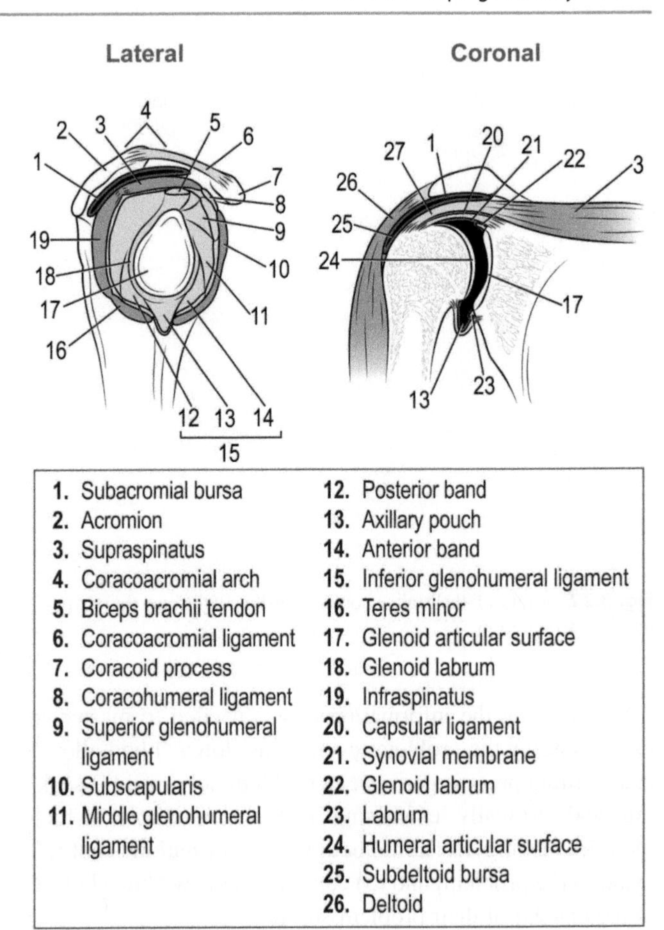


**Fig. 3.26** Lateral and coronal view of subacromial space and related structures

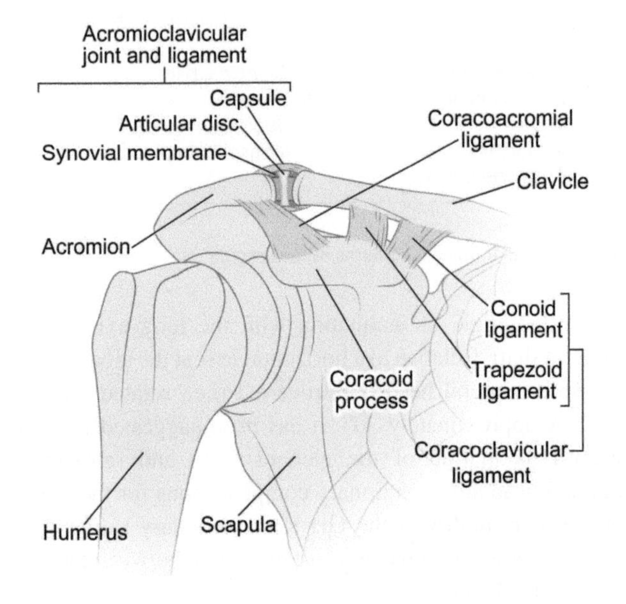


**Fig. 3.25** Acromioclavicular joint and supporting ligaments

of the clavicle (Fig. 3.25). The paired joint surfaces are relatively flat, lined with a layer of fibrocartilaginous tissue. The AC joint may be separated by an articular disc, which enhances joint stability and may also increase the joint surface contact area. Because of the lack of bony congruency at this joint, stability is mostly dependent on periarticular soft tissues, i.e., joint capsule, superior and inferior acromioclavicular ligaments, coracoclavicular ligaments (conoid, trapezoid), and the deltoid and upper trapezius muscles. Note the space between the undersurface of the acromion process and the head of the humerus. In the healthy adult, this space, referred to as acromio-humeral distance (AHD), approximates 1 cm with the arm at the side. However, the ADH fluctuates throughout the range of active humeral abduction, from approximately 8–10 mm at 20° to 3 mm between 60° and 100° and 4–5 mm at full abduction. These fluctuations have mechanical implications. Moreover, with the changes in periarticular soft tissues, selective muscle weakness or abnormal control, and impaired glenohumeral arthrokinematics associated with degenerative joint disease, the tissues housed in this space can become irritated, as in the case of Mrs. Buckler (Fig. 3.26).

The AC joint is the first of two links that connect the appendicular skeleton with the axial skeleton, and the sternoclavicular (SC) is the second. This joint has a more complex articulation involving the medial end of the clavicle, its facet partner on the sternum, and the superior border of the cartilage of the first rib (Fig. 3.27). The design of the SC joint paradoxically contributes to a considerable range of motion during arm elevation, as well as significant soft tissue constraint at the manubrium of the sternum.

Sternoclavicular joint constraints are comprised of extensive periarticular connective tissues, an intraarticular disc,

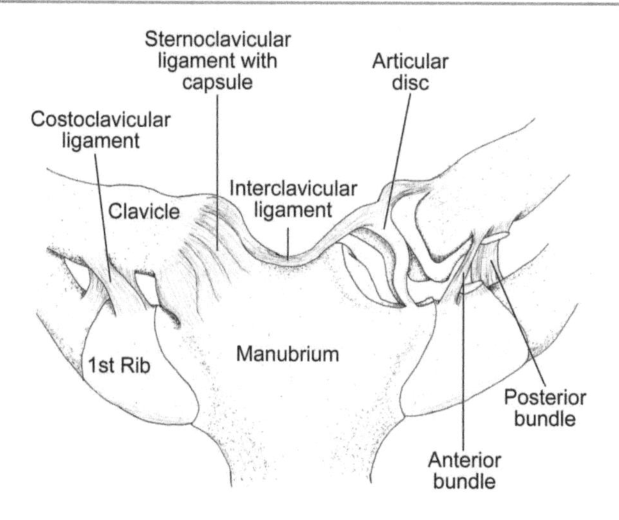


**Fig. 3.27** Sternoclavicular joints and supporting periarticular and deep (articular disc) structures. Note the lateral section of the anterior bundle of costoclavicular ligament is removed to show the posterior bundle

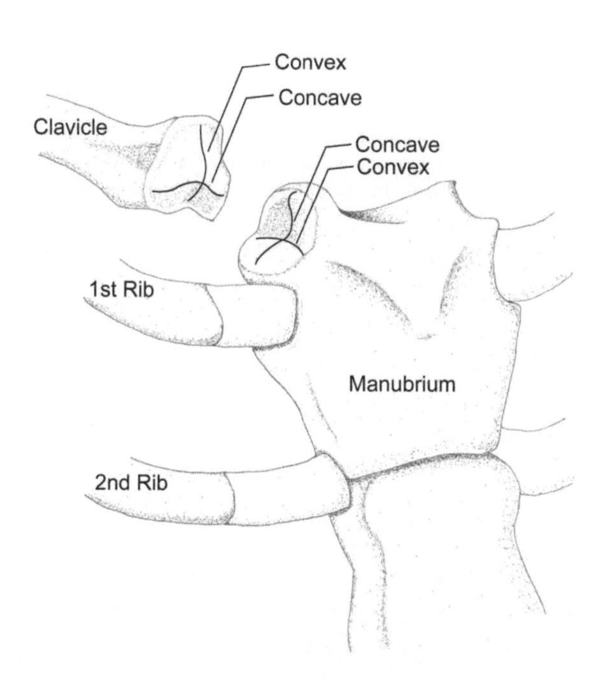


**Fig. 3.28** Saddle-shaped convex-concave (frontal plane) and concave-convex (horizontal plane) partners of the sternoclavicular joint

saddle-shaped joint surfaces, and several muscles (sternocleidomastoid, sternothyroid, sternohyoid, and subclavius). The medial end of the clavicle is ovoid convex along its longitudinal diameter and slightly concave along its transverse diameter. The sternal facet is reciprocally shaped (Fig. 3.28). As a result of bony morphology, capsular and extracapsular constraints and the presence of an articular disc at the SC joint arthrokinematic movements vary depending on the plane of movement.

The functional activities that reproduce Mrs. Buckler's symptoms are those that require her to elevate her arm. When the arm is elevated during functional movements, it rarely follows cardinal plane movements, i.e., flexion, abduction, or even abduction in the scapular plane. However, regardless, both functional and cardinal plane elevations can be characterized by similar kinematic sequences, involving movement of the scapula relative to the posterolateral wall of the thorax, which inevitably involves the AC and SC joints, and movement of the humerus relative to the scapula. We describe specific kinematic and arthrokinematic features of general arm elevation but will leave it to the reader to further explore specific kinematic and arthrokinematic nuances associated with cardinal plane motions of the scapula and arm.

Let's examine the directional movements of the scapula. Scapular movement is unique because the scapula resides over deeper layers of muscle that separate it from the ribs and by the fact that we only describe its osteokinematics. It is notable that scapular stability is reliant on muscle and its dynamic coupling to the axial skeleton via the AC and SC joints. During the elevation of the arm, the scapula (relative to the thorax) elevates, upwardly rotates, externally rotates, and tilts posteriorly. A scapulohumeral rhythm (Fig. 3.29) has been described, which defines the relatively consistent relationship between humeral elevation (approximately 120°) and scapular upward rotation (approximately 60°), resulting in a full arc of 180° arm elevation. Upward rotation of the scapula is particularly important for three reasons: (1) it projects the glenoid fossa anterolateral and upward, which optimizes matching of the concave glenoid and convex humeral head, and promotes the full upward and lateral reach of the arm; (2) preserves optimal length-tension for deltoid and supraspinatus action; and (3) maximizes the volume of the subacromial space, i.e., space between the undersurface of the acromion process and the greater tubercle of the humeral head. In general, the scapulohumeral rhythm depends on the precise control of muscle synergies. Actions of the upper, middle, and lower trapezius and serratus anterior produce upward rotation of the scapula (Fig. 3.30). The scapular posterior tilt and external rotations are conjunct (accessory) motions that occur near the end range of elevation; they are likely related to complex actions of the lower and middle trapezius, and serratus anterior, and perhaps partially dictated by the undersurface of the scapula and shape of the thorax (Fig. 3.31). The three-dimensional rotations of the scapula during humeral elevation are the result of complex force couples, e.g., equal and opposite forces of the upper and lower trapezius, and serratus anterior, to produce the upward rotation (see Appendix G for more detailed mechanical details of force couples).

Movements at the AC joint during humeral elevation can be characterized by either separate movements about the Cartesian axis system or as a movement about a helical axis; that is, we imagine that the scapular rotations (just described above) dictate a movement about and along a helical axis at the AC joint as the clavicle appears to "go along for the ride"

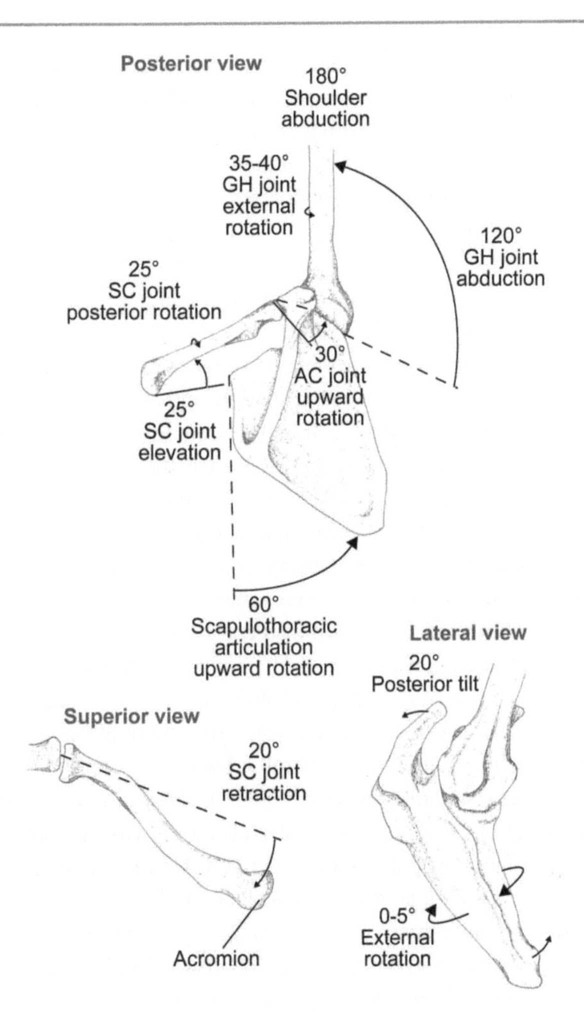


**Fig. 3.29** Posterior, superior, and lateral views of the right shoulder complex illustrating the scapulohumeral rhythm associated with 180° of arm elevation. Note that in order to achieve full pain-free shoulder elevation, the accessory motions of scapular posterior tilt and external rotation, which appear to drive sternoclavicular retraction, is paramount. GH = glenohumeral; SC = sternoclavicular; AC = acromioclavicular

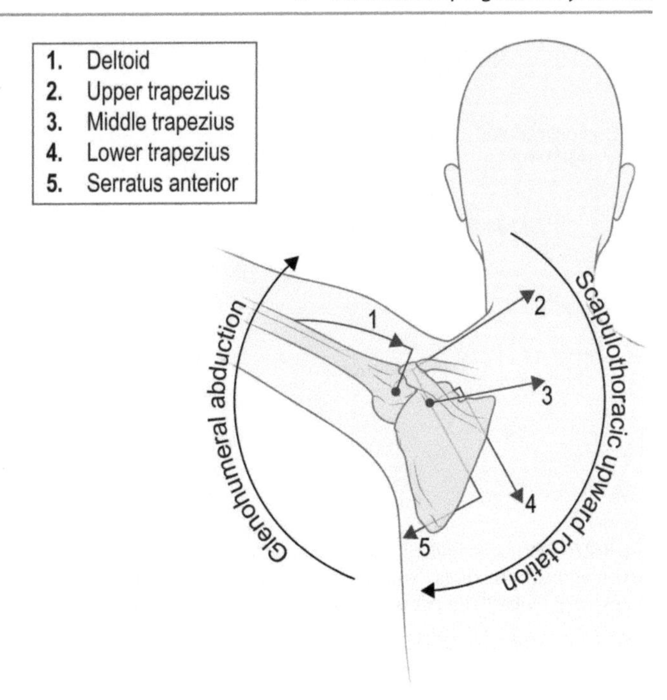


**Fig. 3.30** Synergistic action of the deltoid, three parts of the trapezius, and serratus anterior to upwardly rotate the scapula during the elevation of the arm, e.g., glenohumeral abduction. Note the approximate centers of rotation (red dot) of the glenohumeral joint and scapula-thoracic articulation

(Fig. 3.32). It is likely that a helical type of movement is what really takes place at the AC joint, yet because a helical axis description defies clinically meaningful terms, it is rarely used. In fact, there are no clinical terms for rotations at the AC joint, i.e., rotation of the scapular acromion relative to the distal clavicle, or rotation of the distal clavicle relative to the acromion. Instead, the classical descriptions of AC movements are actually descriptions of how the scapula moves in space, i.e., relative to some spatial reference frame: elevation/depression, upward/downward rotation, and protraction/retraction (Fig. 3.32).

Because the respective joint surfaces at the AC joint are planar, the arthrokinematics do not follow, what is termed, the convex-concave rule, but instead consist of small rolling and gliding adjustments that have a helical nature.

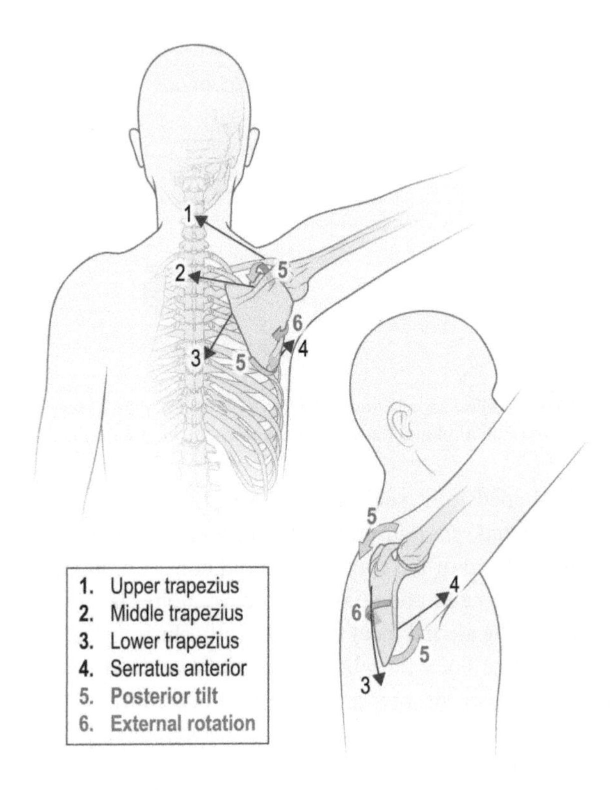


**Fig. 3.31** The muscular mechanism for scapular upward rotation, external rotation, and posterior tilt during shoulder girdle elevation

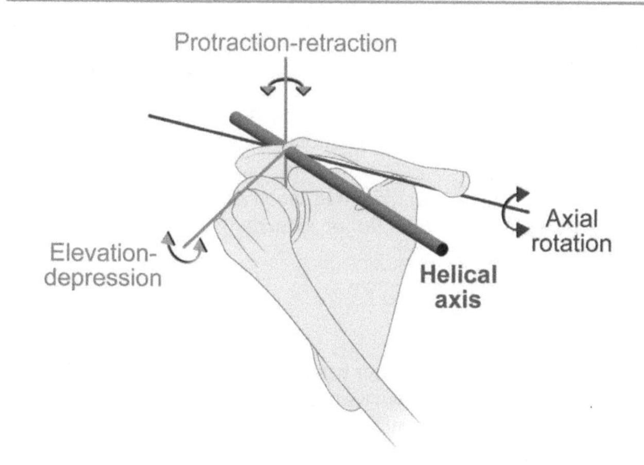


**Fig. 3.32** Cartesian coordinate axes located at the approximate AC joint center are most commonly used to describe movements (rotations and translations) at the AC joint as elevation-depression about/along an anteroposterior axis, anterior-posterior rotation about/along a medio-lateral axis, and protraction-retraction about/along a vertical axis. We could also depict the six-degree-of-freedom of movement at the AC joint as a translation and rotation about a helical axis. As described by Sahara et al. (2006), it appears that during glenohumeral abduction the helical (also called screw) axis is oriented as a combination of the anteroposterior and medio-lateral Cartesian axes

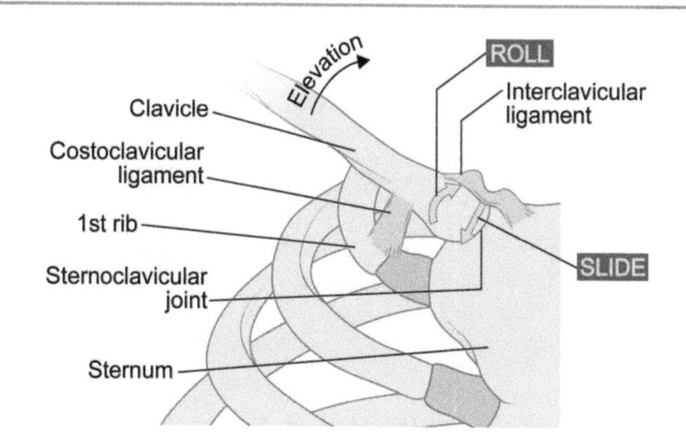


**Fig. 3.34** Arthrokinematics, i.e., superior roll and inferior glide, at the SC joint during elevation of the arm. Note that concomitant tautness of the costoclavicular ligament may contribute to the inferior glide of the clavicular head

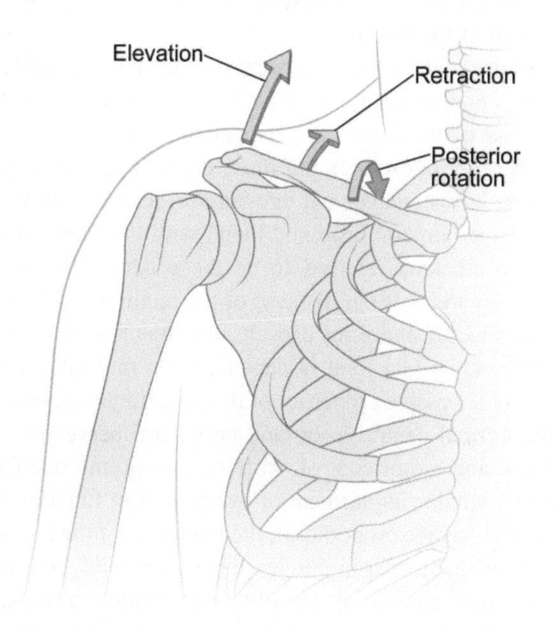


**Fig. 3.33** Elevation of the humerus and concomitant elevation, retraction and posterior rotation of the clavicle at the SC joint

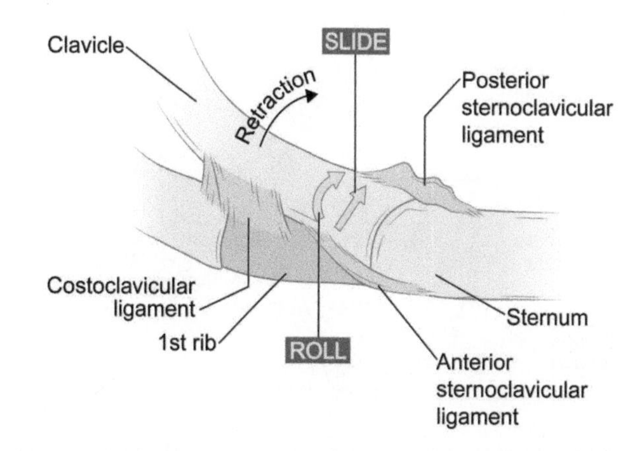


**Fig. 3.35** Arthrokinematics, i.e., posterior roll and glide, at the SC joint with clavicular retraction. Note the relative increase in tension in the anterior sternoclavicular ligament, with concomitant reduced tension in the posterior sternoclavicular ligament

As the arm is elevated (i.e., osteokinematic movement), scapular rotation produces concomitant elevation of the clavicle at the SC joint. Furthermore, it is hypothesized that scapular rotation and slight retraction create tension in the coracoclavicular ligament that induces a posterior rotation of the clavicle about its long axis, i.e., a posterior spin at the SC joint (Fig. 3.33). Assuming elevation in the scapular plane, i.e., primarily a frontal plane rotation, the arthrokinematic movements at the SC joint follow the convex-concave rule; that is, since, in the frontal plane the distal end of the clavicle is convex, and its reciprocal manubrial partner is concave, there is a superior roll and inferior glide (Fig. 3.34). As noted in the figure, we see an example, e.g., coracoclavicular ligament, of how the passive properties of ligaments can contribute to joint moments (i.e., called passive elastic moments) and translations. Since there is a mild degree of retraction (related to the posterior tilt of the scapula) during elevation in the scapular plane, we need also to consider a more complicated roll/glide sequencing (Fig. 3.35), where there is a posterior roll/glide.

With retraction, then, the osteokinematic motion involves posterior movement of the lateral end of the clavicle, and since the medial end of the clavicle is concave, the roll and glide occur in the same direction as the osteokinematic motion. Although retraction is clearly secondary during arm elevation, it is important to be cognizant of the three-dimensional complexities of motions at the joint surface.

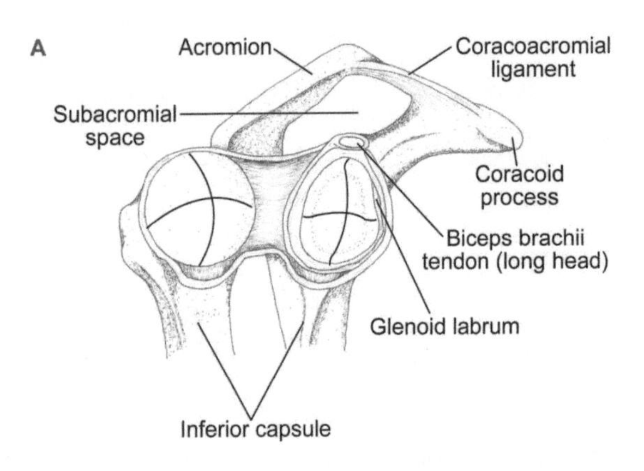


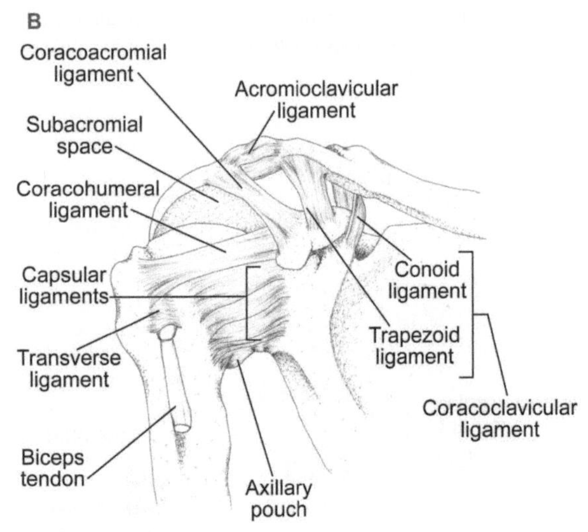


**Fig. 3.36** (**A**) Mismatched convex and concave ovoid diameters and (**B**) extracapsular ligaments of the glenohumeral joint

It is notable that the glenohumeral (GH) joint favors mobility over stability. The large rotational mobility at the GH joint in all three planes is due to (1) an ovoid convex humeral head with a larger longitudinal and transverse diameter than its concave partner (Fig. 3.36A) and (2) the fact that the internal and external capsular ligaments (Table 3.2), despite their redundancy, are inherently lax (Figs. 3.3 and 3.36B). Despite their relative laxity, the internal and external capsular ligaments can still restrain excessive translation of the humeral head along the three cardinal axes. Additionally, because of the tension developed in these ligaments during various movements of the arm, and the fact that the line of application of these passive forces is eccentrically located relative to the geometric center of rotation (approximate center of the humeral head), these ligaments also contribute to GH moments, e.g., humeral external rotation.

Despite reduced joint surface congruence, some bony stability is provided by the slight upward tilt of the glenoid fossa (Fig. 3.23), which may be increased dynamically secondary to upward rotation of the scapula during GH elevation. Additional passive stability is also provided by the glenoid labrum, a fibrocartilaginous ring that significantly deepens the glenoid cavity, but also has significant collagenous attachments to the GH joint capsule, supraspinatus, and long head of the biceps.

**Table 3.2** Function of glenohumeral capsular ligaments

| Ligaments | Motions checked |
|---|---|
| Superior glenohumeral ligament | External rotation; inferior and anterior translation of the humeral head |
| Middle glenohumeral ligament | Anterior translation of the humeral head (45°–90° of abduction); external rotation |
| Inferior glenohumeral ligament: | |
| Anterior band | Anterior band: 90° of abduction and full external rotation; anterior translation of the humeral head |
| Posterior band | Posterior band: 90° of abduction and full internal rotation |
| Axially pouch | Axially pouch: 90° of abduction, combined with anteroposterior and inferior translations |
| Coracohumeral ligament | Inferior translation of the humeral head; external rotation |

Because of the inherent looseness of the passive joint constraints, i.e., GH ligaments, the rotator cuff muscles, and the long head of the biceps are critical for providing dynamic stability. The four rotator cuff muscles, subscapularis, supraspinatus, infraspinatus, and teres minor, form a cuff around most of the humeral head and blend with the joint capsule, save for the area referred to as the rotator cuff interval (Figs. 3.26 and 3.36B). Because of the relationship between the rotator cuff and joint capsule, the dynamic mechanical stability of the GH joint is dependent on multiple factors: viscoelastic passive properties of the periarticular connective tissue; normal central, root, and peripheral nerve function; strength; and motor control, with the loss of any one factor leading to major impairment of overhead arm function. The passive and active properties of the long head of the bicipital-tendon unit are just as important in both the static, i.e., arm at side position, and during glenohumeral abduction (Figs. 3.3 and 3.36B). For example, during the acceleration phase of pitching with the abducted shoulder rapidly internally rotating, and the elbow rapidly extending, the long head of the biceps checks rapid elbow extension while simultaneously providing a compression force that helps to centralize and stabilize the humeral head into the glenoid fossa.

Beginning with the arm at the side of the body, glenohumeral abduction can also be described as a movement of the humerus in a superior direction. With a superior movement of the humerus, the ovoid convex humeral head rolls superiorly and glides inferiorly, relative to the ovoid concave glenoid fossa (Fig. 3.37). The inferior glide occurs close to

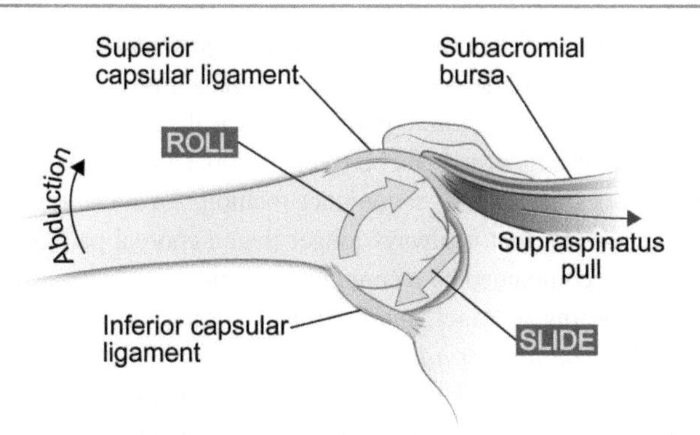


**Fig. 3.37** Glenohumeral arthrokinematics during shoulder abduction. Note the supraspinatus portion of the rotator cuff and its relationship to the superior capsular ligament. Both of these structures, as well as the subacromial bursa, can be subjected to compression from the overhanging acromial process if the inferior glide is inadequate or if there is tightness of the inferior GH capsule (i.e., inferior capsular ligament)

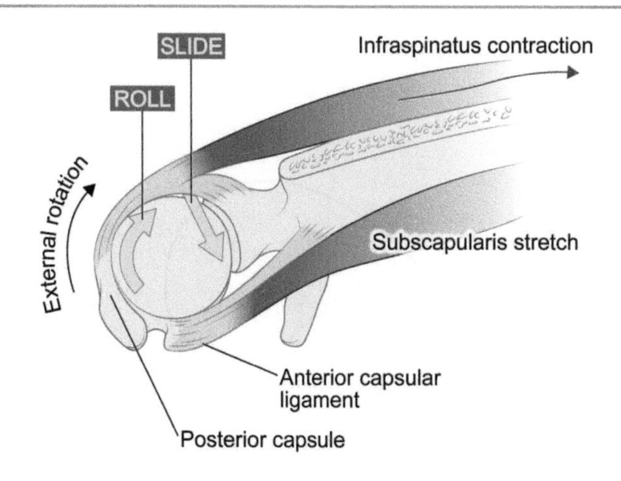


**Fig. 3.39** Superior view of GH arthrokinematics related to the conjunct external rotation that takes place during shoulder abduction

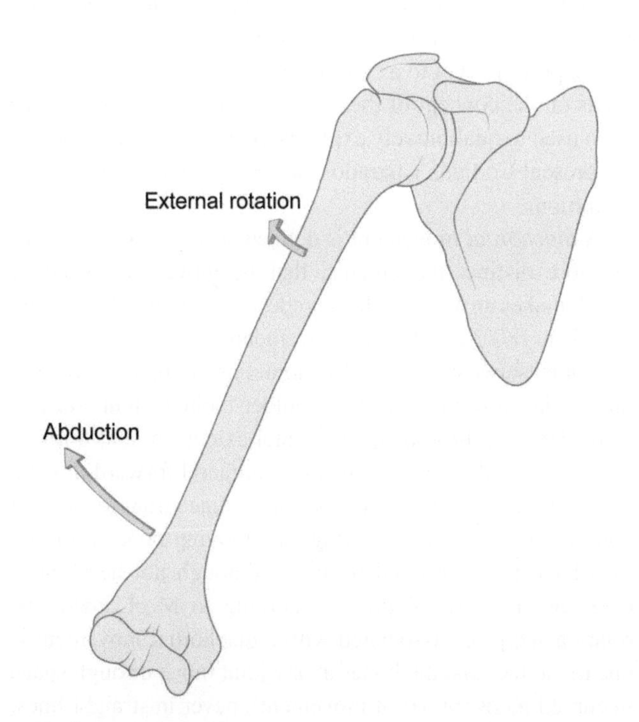


**Fig. 3.38** Overhead movement of the arm, e.g., abduction, is a complex movement that combines the osteokinematic motions of abduction and a conjunct external rotation

the longitudinal diameter of the glenoid (see Fig. 3.36A), suggesting that during abduction, the head of the humerus is kept in a nearly centric position relative to the glenoid fossa.

But there's more. It turns out that during abduction, there is a conjunct, i.e., involuntary or automatic, external rotation of the humerus (Fig. 3.38). This external rotation is controlled primarily by rotator cuff muscle action and assisted by the passive elastic properties of the GH capsule. The magnitude and timing of this external rotation is a critical element of a dynamic coupling of simultaneous rotations that allow for free movement of the humeral head and greater tubercle of the humerus under the coracoacromial arch. Osteokinematically, external rotation of the humerus can also be described as a posterior roll, so that arthrokinematically, we expect according to the convex-concave rule, an anterior glide of the humeral head relative to the glenoid (Fig. 3.39). This coupled roll-glide takes place along the transverse diameters of the humeral head and glenoid fossa, again, keeping the large convex humeral head in a relatively centric position throughout the range of motion.

Before we examine the muscular controls involved with glenohumeral movements, we want to digress and discuss, in some detail, the general idea of conjunct rotation and its application specifically as it relates to Codman's paradox. In his presentation of osteokinematics, MacConaill distinguished two ways by which bones move: they spin (or rotate) or they swing. MacConaill defined the physiological movements we call flexion, extension, adduction, and abduction as *swings*. In a *swing*, one joint surface slides upon the other, whereas in a rotation, e.g., internal or external rotation, one joint surface turns about an axis that is normal or "perpendicular" to a joint surface. Recall that earlier in this chapter, we defined a *spin* as a rotation about the mechanical axis of the bone. MacConaill further distinguished swings as *pure* and *impure*: a *pure* swing (or *cardinal swing*) is a movement of the bone without any accompanying spin whereas an *impure* swing is when the swing also undergoes a *spin* (also referred to as an *arcuate swing*). Let's introduce a few more operational definitions specific to MacConaill's kinesiological theories.

It is notable that any point located on a bone moving at a joint moves in a curved line. This is so since most joint surfaces are ovoid-shaped, i.e., convex, concave, or saddle-shaped. According to MacConaill, when analyzing a swing, we first must consider a point on the mechanical axis as it traces a line along the joint surface. Therefore, the mechanical

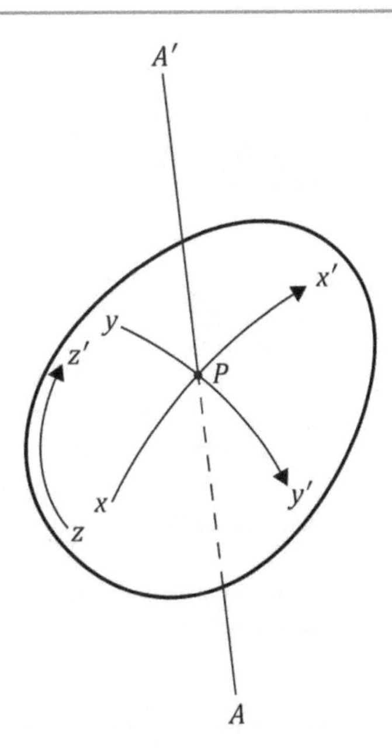


**Fig. 3.40** Let P be the point on the mechanical axis (*AA*′) of a bone moving along the joint surface. Then *xx*′, *yy*′, and *zz*′ represent three possible paths of P on the ovoid of motion. (Modified with permission from MacConaill and Basmajian (1977))

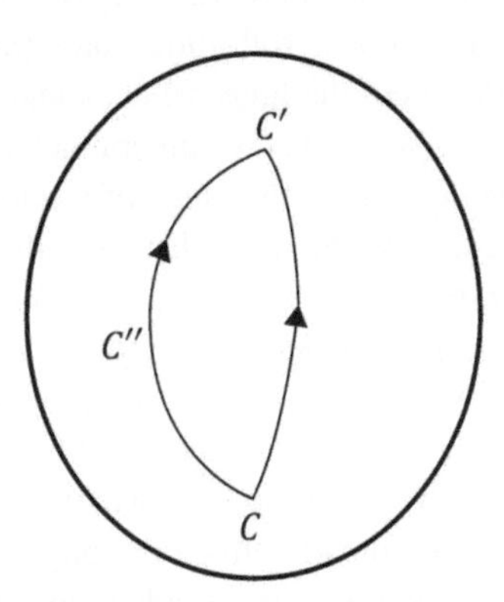


**Fig. 3.41** Chordal (*CC*′) and arcuate (*CC*″*C*′) paths between the same two points. The arcuate path is longer than the chordal path, which is defined as the shortest path between the first and last points (marked by the mechanical axis as it traces its path along a joint surface). (Modified with permission from MacConaill and Basmajian (1977))

axis of all swings traces a line on a joint surface that is curved, with some lines of greater curvature than others, depending on the joint. MacConaill referred to these traced lines as the *ovoid of movement* (Fig. 3.40). Because joint surfaces are not flat, it can be said that this line corresponds to a meridian of longitude on a sphere (or to a straight line in the case of a joint surface that is flatter). For *pure swings*, this point moves from its initial to its final position along the shortest possible curved line, with this line defined as a *chord* between the same points. With an *impure swing*, the movement is always along some curved line other than the shortest, defined as the *arc* between two points on the surface (Fig. 3.41).

In summary, the concept of the *ovoid of movement* is key to describing and understanding osteokinematic movements. With *cardinal* swings, a *chordal* path is traced as the shortest path for any point in a moving bone between its initial and final position, and it does not include a conjunct rotation. An *arcuate* swing creates an *arc* that is always longer than a chordal path and is always accompanied by a conjunct rotation.

According to MacConaill, spins (or rotations) are manifested as *conjunct* or *adjunct*. Generally, the sense of either a spin or an arcuate swing is clockwise (CW) or counterclockwise (CCW). A conjunct rotation occurs when the movement of a bone along a path or a set of successive paths *necessarily* involves a CW or CCW rotation. It is termed a conjunct rotation because it is always conjoined with the arcuate swing, although the effect of this rotation may be increased, diminished, nullified, or reversed by an adjunct rotation (also termed an *antispin*). The adjunct rotation is any other rotation of a bone when it is adjoined either to no rotation at all (i.e., a pure spin) or to a conjunct rotation. Going forward we focus on the concept of the conjunct rotation and how it has been used to qualitatively explain Codman's paradox. Before we present Codman's paradox, we need to make a few more definitions.

A *diachordal* movement is defined as a succession of two or more distinct movements, that is, movements each of which makes an angle with its predecessor greater than 0° and less than 180°. Diachordal movements occur only at joints with at least two degrees of freedom. For example, a two-step diachordal movement at the shoulder includes humeral forward flexion followed by horizontal extension, and a three-step diachordal movement includes humeral forward flexion, horizontal extension, and adduction. A diachordal movement that is *closed* is one that brings the moving bone from and then back to its original position, although not necessarily back into its original pose. According to MacConaill, the rotation sequence associated with a diachordal movement is due to the fact that the bones at any joint move through space in curved paths (ovoid of movement), never in straight lines, secondary to the morphology of the joint surfaces. If the path of motion of a bone is along one chord, an accessory rotation does not take place. However, if the path of motion moves first along one chord and then along a different chord, a conjunct rotation naturally occurs simply because of the geometry of ovoid surfaces.

A closed, three-sided figure traced on any joint surface is called a *triangle* if and only if all three sides are chords (Fig. 3.42A). On an ovoid surface, the sum of the three (interior) angles of a triangle is always greater than two right angles, i.e., 90°, a key geometrical fact about the kinematics of bones. For example, during the two-step diachordal movement as the bone follows a triangular path, a conjunct rotation takes place during the second stage of the sequence. The accessory rotation during the second stage of the sequence is called a conjunct rotation because it is conjoined with the

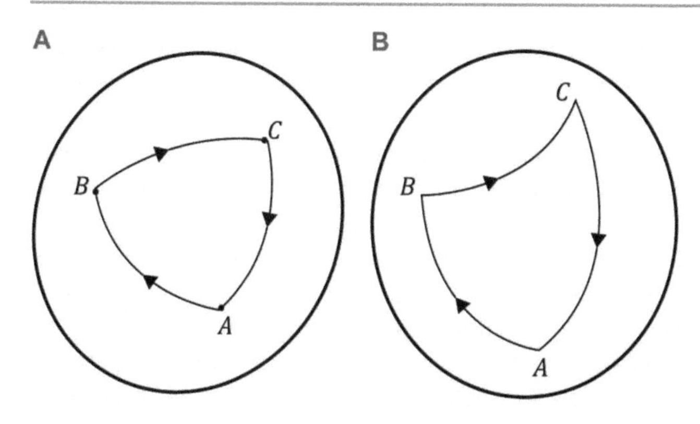


**Fig. 3.42** Ovoid of movements along a triangular pathway AB-BC-CA (on the left) and a trigonal pathway, where AB and CA are chords and BC is an arc (on the right). (Modified with permission from MacConaill and Basmajian (1977))

movement of the bone along the second side of the triangle ABC. It is "as if" this chord functioned like an arc; that is, the rotation necessarily involved a spin (rotation) of the bone. In the second case, that is, if one, or more, of the sides of the movement path consists of arcs, not chords, the pathway of the bone at the joint surface ascribes a *trigone* (Fig. 3.42B). Despite the differences between the triangle and trigone, as in the case of the triangle, a trigonal sequence AB-BC-CA also induces a conjunct rotation.

In his 1934 book on the shoulder, Codman describes a specific pattern of movement at the shoulder joint that appears paradoxical. The movement pattern involved two or three sequential glenohumeral rotations that did not involve any rotations about the longitudinal axis yet appeared to result from an axial rotation. MacConaill, and later Kapandji, describe this phenomenon. Kapandji summarizes one rendition of the paradox as:

- In the position of reference, the upper limb hangs down vertically alongside the trunk, with the thumb facing anteriorly and the palm facing medially.
- The limb is abducted 180°.
- From the fully abducted position with the palm facing laterally, the limb is extended 180° in the sagittal plane.
- It is now back in the original position alongside the body, but now the palm faces laterally and the thumb is pointing posteriorly (Fig. 3.43A).

Another version (Fig. 3.43B) of the paradox involves three sequential arm swings that begin in the same initial position but is initiated by glenohumeral flexion to 90°, which is followed by a 90° horizontal extension and then bringing the arm back to the side by adducting the humerus. The paradox in the form of a question is: How can this sequence of movements result in different initial and final positions when no known axial rotation of the arm was initiated?

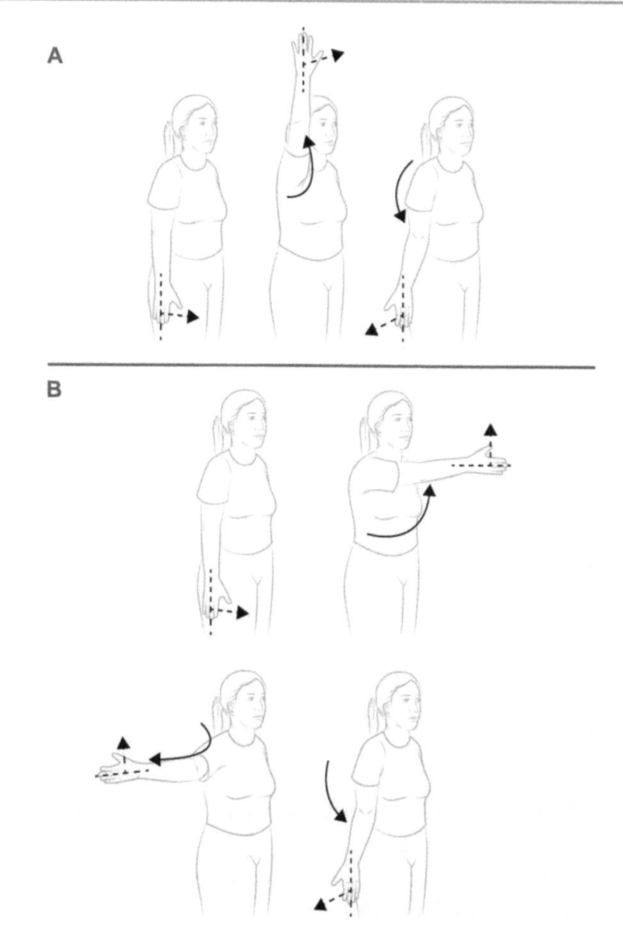


**Fig. 3.43** Kapandji's depiction of Codman's paradox with the sequence of movements from (**A**) abduction to extension and (**B**) flexion to horizontal extension to adduction

Let's now examine MacConaill's mini-experiment, a different version of the paradox, that demonstrates Codman's observations (Fig. 3.44): (1) take and hold an object like a baton and let the right upper limb hang by the side of the body with the object pointing forward, i.e., the object lies in the parasagittal plane, and with the forearm semi-pronated which is to be maintained throughout the movement steps; (2) swing the upper limb upward and forward, i.e., forward flexion, until it is in the horizontal plane, which makes the object point upward; (3) swing the arm backward in the horizontal plane, i.e., horizontal extension, placing the object in the scapular plane; notice that the object is in the same position as if the limb had simply been abducted in the scapular plane and then rotated laterally through a right angle. This is obvious since, in order to bring the object back into its original position (1), the limb would have to be adducted and medially rotated 90°; (4) having completed steps (1)–(3), adduct the limb to the side while maintaining the lateral rotation of the glenohumeral joint (5); repeat the cycle of operations from (1) to (3) with the initial condition of a laterally

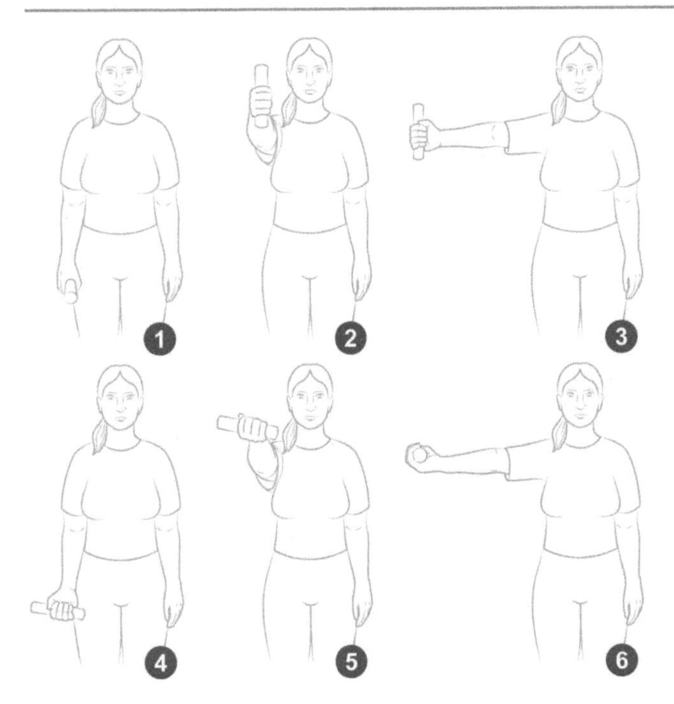


**Fig. 3.44** MacConnaill's version of Codman's paradox

rotated shoulder; and (6) verify that the cycle can be repeated just once, i.e., a stage is reached when full flexion of the glenohumeral joint becomes impossible unless the limb is rotated medially with flexion.

Again we ask: How does some sort of axial rotation about a longitudinal axis of the arm occur during two or three sequential rotations (MacConaill would say *arcuate* swings) that did not involve any rotations about a longitudinal axis, i.e., Codman's paradox? Codman's paradox has been the subject of several inquiries over the years with little consensus on an anatomical or mathematical explanation. According to Cheng, the lack of consensus may be related to the fact that the problem has not been well defined and that conventional kinematic analytical methods may not have been suitable for the problem. Cheng framed Codman's paradox with three characteristics: (1) the paradox involved a closed-loop motion that consisted of three sequential rotations of the long axis of the arm in a reference coordinate system; (2) the first rotation was about an axis that was perpendicular to the initial position of the long axis of the arm, with the second rotation taken about an axis that coincided with the initial position of the long axis, and the third rotation was used to move the long axis back to its initial position; and (3) the questioned axial rotation angle was defined as a rotation about the long axis.

The closed-loop long axis rotation is referred to as Codman's rotations, that is, there are three sequential rotations: the first referred to as *elevation*, the second referred to as *swing* that is usually described as a rotation about a vertical axis, and the third termed *descending* that enables the long axis to return to its initial position. Note: the *swing* was also generalized to involve rotation about the initial position of the long axis at a non-neutral position. By applying a two-step rotation analytical method in the spherical rotation coordinate system, and running several different simulations, Cheng suggested a general law that appeared to address Codman's paradox: "when the long axis of the arm performs a closed-loop motion by three sequential rotations, defined as Codman's rotation, it produces an equivalent axial rotation angle about the long axis…the equivalent axial rotation angle equals the angle of the swing – the second rotation in the three sequential long axis rotations."

Tondu, applying a robotic analytical approach to simulate and interpret Codman's paradox, corroborated Cheng's thesis and offered a revision of Cheng's proposed law of motion: "assuming the arm initially set in a neutral attitude, i.e., with zero joint angles, Codman's paradox involves a closed-loop, eventually repeated, motion of the arm's long axis that consists of a sequence of abduction-adduction and flexion-extension relative movements, mixed with eventual swings about a fixed vertical axis – relative means that successive movements are performed with respect to the given axis." Tondu suggested that considering the determination of a possible conjunct rotation may be necessary, in addition to accepting the ability of the upper limb to "sum up voluntary and involuntary arm rotations," may best reveal Codman's paradox. In fact, Politti et al. claimed that Codman's paradox was not a paradox. They demonstrated, using both Cartesian and polar coordinate systems, that the phenomenon was a mechanical property that can be mathematically described by the equivalence between the matricial product of three orthogonal rotation matrices applied to a position vector and the matricial product of a single rotation matrix applied to the same vector. With polar coordinates, it was shown that the total shift vector clearly was equivalent to a single net rotation about the longitudinal axis of the arm. Perhaps what appears to be missing in the more complex mathematical modeling is the clinical correlate. Several authors have suggested that MacConaill's conjunct rotation concept may likely be a key to providing a qualitative answer to the Codman conundrum.

MacConaill's attempt to derive a general law to explain Codman's paradox was expressed this way: "it can be shown, both theoretically and experimentally, that the amount of conjunct rotation is directly proportional to the amount of backward swing" (likely a swing around a vertical axis). He proposed a "Law of the Conservation of Axial Rotation" in which he suggested that "the effect of any axial rotation is conserved unless means be taken either to prevent this or undo it…plays the same part in mechanics of joints as the 'Law of the Conservation of Energy' does in mechanics generally." MacConaill's explanation for Codman's paradox came in the form of his description of diachordal movements previously presented. In a follow-up to his explanation of Codman's paradox, MacConaill also ventured a theoretical approximation of the magnitude of the conjunct rotation: "the magnitude of conjunct rotation is equal to the difference between two right angles and the sum of the angles of the triangle (or trigone) formed by the two 'legs' of the diachordal path and the 'direct' path between the first and last positions of the bony point considered" or, more practically, the angle between the plane of the "direct" swing and the plane of the first stage of the diachordal path, approximately 90°. In some corners, biomechanists have distinguished differences between an apparent and true axial rotation associated with Codman's paradox, suggesting that an induced long axis rotation (an apparent rotation) may not occur about a physical axis, but simply because of changes in latitude/longitude. Although it appears that MacConaill's concept of a conjunct rotation is accepted as a factor that may explain Codman's paradox, no one to date has quantified the magnitude of the "true" axial rotation that takes place during glenohumeral abduction.

The reader might wonder why we took this "fork." In retrospect, this digression appears to have provided insight into the MacConaill-formulated concept of conjunct rotation and how he used it to explain Codman's paradox. This understanding is useful to us as we develop mechanical hypotheses related to Mrs. Buckler's impingement syndrome. How so? Given the importance of conjunct rotations in joints with at least two degrees of freedom, it is possible that in Mrs. Buckler's case there may be some disruption in the many subtle conjunct rotations that must occur at the glenohumeral and sternoclavicular joints, as well as at the scapula-thoracic articulation and the joints of the thoracic cage that could be contributing to her painful shoulder when she attempts to raise her arm overhead.

Let's move on. The dynamics, i.e., kinetics, of shoulder abduction is an example of one of the body's most complex force couples that combines the actions of a powerful prime mover, deltoid, with the actions of the four fine-tuner rotator cuff muscles (Fig. 3.45). The central control of the deltoid and rotator cuff muscle mechanism manages to provide large GH moments for lifting and throwing activities and at the same time maintain humeral head stability. More specifically, supraspinatus action contributes to the upward roll of the humeral head, which adds to joint stability by compressing the head into the glenoid cavity. At the same time, it serves as a buffer between the acromion process and the superior aspect of the glenohumeral joint capsule by restricting the superior glide of the humeral head. Augmenting supraspinatus function, the combined force line of actions of the infraspinatus, teres minor, and subscapularis appear to pull the humeral head inferiorly. These mechanisms are important in preventing the impingement of soft tissues between the acromion process and the head

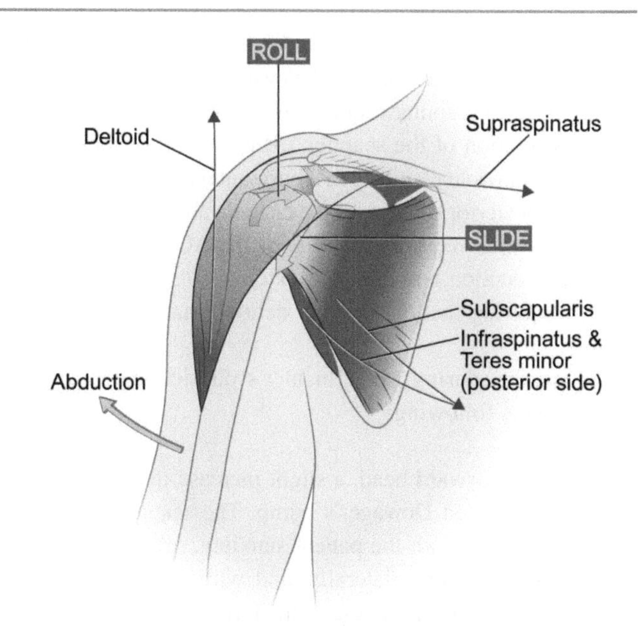


**Fig. 3.45** Deltoid rotator cuff force couple action that contributes to shoulder abduction

of the humerus. Concomitantly, the infraspinatus and teres minor also contribute to the posterior roll/anterior glide of the humeral head, which is an additional mechanism that helps to clear the greater tubercle of the humerus from under the coracoacromial arch. We should note that the deltoid has sufficient ability to abduct the humerus against reasonably large loads without the rotator cuff, e.g., in instances of large rotator cuff insufficiency. However, the action of the deltoid alone would result in a superior glide of the humeral head and large compression of the tissues under the acromial hood, a scenario leading to an impingement syndrome and further shoulder impairment.

#### 3.3.1.3 Summary: Relating Key Facts from Mrs. Buckler's Physical Examination, Learning Issues, and Hypotheses Regarding the Diagnosis of Impingement Syndrome

We draw on Neumann's synopsis of key kinematic absolutes associated with shoulder abduction. However, restricting the principles to the shoulder girdle does not account for the importance of the entire upper quarter. Let this brief summary allow us to see how we might begin to explain/understand part of Mrs. Buckler's problem. In order to achieve repeatable, pain-free, full overhead arm activities, the following normal mechanisms must be in place: (1) cervical and upper thoracic extension and associated posterior rotation of the ribs; (2) scapulohumeral rhythm consisting of 120° humeral elevation and 60° of scapular upward rotation that contribute to a total of 180° of shoulder abduction; (3) 60° of upward scapular rotation produced by the synergistic actions/force couple produced by the upper, middle, and lower trapezius, and serratus anterior, which produces a helical rotation at the AC joint, and elevation at the SC joint; (4) during upward scapular rotation, associated posterior tilt and external rotation of the scapula; (5) retraction of the lateral end of the clavicle resulting in a horizontal rotation at the SC joint; (6) posterior rotation of the clavicle about its longitudinal axis, which requires mobility at the AC and SC joints; and (7) abduction and conjunct external rotation at the GH joint, requiring action of the deltoid/rotator cuff force couple.

Recall that during Mrs. Buckler's physical examination we found the following:

- Moderate forward head, a slight increase in cervical lordosis, and mild Dowager's hump. The scapular position was observed with the patient standing, revealing mildly abducted scapula bilaterally, and with posterior prominence of the inferior angle of the left scapular.
- Active range of motion (ROM) testing of the cervical spine was within normal limits in all planes, except for the mild limitation of extension, and sidebending bilaterally; with cervical extension testing, it appeared that motion hinged at the C6–7 junction.
- Active-assisted extension ROM testing of the upper thoracic spine, with the patient sitting, was moderately limited.
- Active assisted ROM testing revealed a painful arc on the left side.
- Goniometric measures of active ROM were (normal values provided previously):

| | *Left* | *Right* |
|---|---|---|
| Abduction | 165° | 170° |
| Flexion | 160° | 170° |
| External rotation | 75° | 90° |
| Internal rotation | 55° | 60° |

- Note: Observation of scapular mobility during shoulder elevation revealed mild "dyskinesis" bilaterally. Note: Passive joint play, i.e., accessory motion, testing revealed moderate loss of mobility in acromion-clavicular joint bilaterally and moderate loss of joint play, i.e., inferior and anterior/posterior glide, at the left glenohumeral joint, with mild loss of play in the same directions at the right glenohumeral joint.
- Manual muscle testing (MMT) revealed:

| | *Left* | *Right* |
|---|---|---|
| Abductors | Weak/painless | Strong/painless |
| External rotators | Weak/painless | Strong/painless |
| Internal rotators | Strong/painless | Strong/painless |
| Adductors | Strong/painless | Strong/painless |
| Serratus anterior | Weak/painless | Weak/painless |
| Middle/inferior trap | Weak/painless | Weak/painless |

Mrs. Buckler's radiographic findings suggested GH degenerative joint disease, which likely predisposed the development of shoulder impingement. Accordingly, we cannot conclude that the fall on an outstretched hand directly caused her impingement. We know that joint degeneration is associated with (1) damaged articular cartilage and synovial membrane that can impair joint mobility; (2) changes in intrinsic capsular ligaments that reduce joint mobility; (3) arthrogenic pain, related to intra-articular swelling, and bone-on-bone forces, which can result in neurologic inhibition of muscle actions that cross the affected joint; (4) muscle atrophy from disuse or altered use; and (5) reduced joint space related to AC degeneration, resulting in bone remodeling, i.e., osteophytes, which can interfere with joint motion. It is clear that Mrs. Buckler's probable GH joint degeneration is likely also related to her physical examination findings of abnormal scapular posture; reduced shoulder range of motion, including joint accessory movements (reduced joint play); scapular dyskinesis, i.e., abnormal scapula-humeral rhythm; and muscle weakness. Possible AC joint degeneration and reduced subacromial space predispose the soft tissues housed beneath the subacromial arch to intermittent (or constant at times) and repetitive compression, ultimately leading to an inflammatory response, painful condition, and pain with overhead activities. The combination of reduced capsular extensibility and abnormal strength and/or motor control of key scapular and rotator cuff muscles also make it likely that soft tissues in the subacromial space are compressed. This is a good time for the reader to re-examine the previous extensive presentation on relevant upper quarter anatomy and joint kinematics and kinetics to ensure that this summary makes sense.

## 3.4 Key Facts, Learning Issues, and Hypotheses: History of Osteoporosis and Radiographic Evidence of Old Compression Fractures in the Mid-thoracic Spine

### 3.4.1 Osteoporosis

Mrs. Buckler had a 5-year history of osteoporosis and radiographic evidence of upper thoracic compression fractures. Let's briefly examine this disease and its implications with regard to bone health. Bone is a remarkable and complex structure that has two major functions. First, it provides the endoskeleton that is crucial for the support of the body, acts as a rigid system of levers that transfers forces from muscle, and provides protection for vital organs. Bone continually adapts and changes its structure in order to manage these tasks. The second function of bone is to store minerals, particularly calcium ($CA^{++}$), needed to maintain mineral homeostasis in the

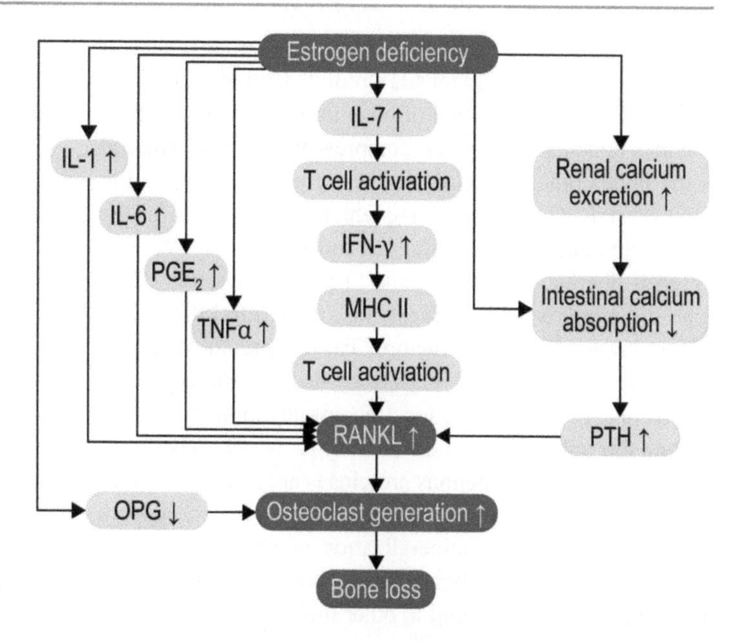


**Fig. 3.46** Model of the effects of estrogen deficiency on bone loss. (Modified with permission from Sipos et al. (2009))

body. Osteoporosis is a disease that upends the body's normal mineral homeostasis. Osteoporosis is defined as a skeletal disorder characterized by compromised bone strength, leading to an increased risk of fracture. Bone strength is influenced by bone quality and density, and there is a clear correlation between each standard deviation (SD) decrease in bone mineral density (BMD) and the risk of fracture. Osteopenia is defined as a BMD between 1 and 2.5 SD below the mean value for young adults, whereas osteoporosis is defined as a BMD more than 2.5 SD below the adult mean value. In menopausal and postmenopausal women, osteoporosis is caused primarily by estrogen deficiency, although other factors, e.g., genetics, lifestyle, nutrition, and medications, may also contribute. Although menopausal osteoporosis mainly affects trabecular bone, later phases of the disorder can also affect cortical bone.

Bone is a very dynamic tissue and is constantly adapting and remodeling, which is critical for repairing microdamage, adapting the skeleton to mechanical loading, and maintaining calcium and phosphorous homeostasis. Bone remodeling is primarily controlled by the coupled action of bone-resorbing cells (osteoclasts) and bone-forming cells (osteoblasts). Osteoblasts, as well as related cell types, i.e., adipocytes, chondrocytes, fibroblasts, and myoblasts, differentiate from mesenchymal stem cells. They synthesize and secrete unmineralized collagenous bone matrix (osteoid). Osteoblasts also appear to participate in the calcification of bone and seem to regulate the flux of calcium and phosphate in and out of bone. Osteoclasts, on the other hand, are derived from the hematopoietic mononuclear cell group. Two cytokines, which are mainly produced by bone marrow stromal cells and osteoblasts, are essential for osteoclastogenesis: macrophage colony-stimulating factor (M-CSF) and receptor activator of nuclear factor-κB ligand (RANKL). RANK is the receptor for RANKL and is expressed on mononuclear osteoclast precursors. Osteoprotegerin (OPG), which is also produced by stromal cells and osteoblasts, is a natural decoy receptor for RANKL and is an antagonist to the osteoclastogenic action of RANKL. Estrogen deficiency in postmenopausal women leads to an upregulation of RANKL on bone marrow cells, which is an important determinant of increased bone resorption, whereas estrogen stimulates OPG production in osteoblasts, exerting anti-resorptive effects on bone (Fig. 3.46). The effects of extraskeletal estrogen deficiency is primarily based on increased renal calcium excretion and decreased intestinal calcium resorption. It has also been shown that estrogen deficiency parallels a continuous decrease in serum parathyroid hormone (PTH) levels and increases the sensitivity of bone to PTH. Hyperparathyroidism is compensatory for net calcium losses in the aging body, while estrogen also seems to have a direct depressive action on the parathyroid gland. Vitamin D deficiency is also responsible for inadequate intestinal calcium absorption associated with aging, while impaired metabolism of vitamin D to its active form and a decrease in intestinal vitamin D receptors can accentuate osteoclastic generation.

### 3.4.2 Bone Structure and Mechanical Properties

#### 3.4.2.1 Structure

Information on the pathogenesis of osteoporosis and its effect on bone are necessary, but not sufficient. Therefore, we need to examine the structural and mechanical properties of bone. Let's look at its structure first. Bone is composed of collagen, water, hydroxyapatite mineral, and small amounts of proteoglycans and noncollagenous proteins. Collagen is a structural protein that organizes itself into strong fibers. It gives bone flexibility and tensile strength. Collagen also provides loci for the nucleation of bone mineral crystals, which provides bone its rigidity and compressive strength. Mineral in the bone comes primarily in the form of hydroxyapatite (HA) crystals, $Ca_{10}(PO_4)_6(OH)_2$, which make up 60% of the dry weight of bone. HA crystals are found primarily between collagen fibers and have ceramic-like properties, providing bone the property of brittleness, i.e., tolerating little deformation before failure or fracture. The ground substance of bone consists of proteoglycans, e.g., decorin and biglycan, which appear to control the location or rate of mineralization in bone through their calcium-binding properties. The function of all the noncollagenous proteins is not clear, but one of these proteins, osteocalcin, secreted by osteoblasts, appears to be important in the mineralization of new bone. Water makes up about 25% of bone content, some of which is free and some of which is bound to other molecules.

Immature bone is referred to as woven. In woven bone, the collagen fibers are arranged and distributed more randomly, giving, more or less, strength to the bone in all directions. Woven bone is weaker than mature bone, but is more flexible, providing more resilience that may accommodate the many traumas of childhood, i.e., tumbles and falls. As bone matures, osteoclastic activity creates "tunnels" in the bone, as osteoblasts line the tunnels with type I collagen. Hydroxyapatite crystals are deposited in the spaces of the hierarchical collagenous framework, thus mineralizing the bone. Osteocytes, i.e., bone cells, appear to control bone mineralization, as well as extracellular calcium and phosphorous. The interplay of the specialized bone cells creates Haversian canals, which are lined with layers of bone (i.e., lamellae) that become oriented primarily along the load-bearing directions of bone, e.g., the longitudinal axis of the femur (Fig. 3.47). The Haversian canals (also called osteons), representing the structural units of bone, also serve as passages for blood vessels and nerves. In maturing bone, the primary osteons eventually are replaced in more mature bone by secondary osteons. Volkmann's canals run perpendicular to the secondary osteons, also serving as passages for vessels.

Macroscopically, there are two major types of bone tissue, compact (or cortical) and cancellous (or spongy; trabecular). Cortical bone is denser than cancellous bone. It forms the bone's outer shell and is of variable thickness. The external periosteal surface or periosteum is a specialized outer membrane, which is layered, highly vascularized, and innervated. The outer layer of the periosteum consists of dense connective tissue and related cells, i.e., collagen and fibroblasts, which are relatively impermeable and under constant tension. The inner surface of the medullary canal, referred to as the endosteum, lacks an extensive connective tissue structure and is cell-lined. The diaphyseal region of a long bone is the mid-shaft, and the epiphyseal regions, at the end of the bone, contain the growth plates. The flared regions in between the diaphysis and epiphysis are referred to as the metaphysis. Cancellous bone has a porosity of 75–90% and is found in cuboidal (e.g., vertebral bodies), flat, and the end of long bones. The pores in cancellous bone are connected and filled with marrow. The bone matrix in cancellous bone is in the form of plates or struts, called trabeculae; thus, cancellous bone is often also called trabecular bone. The arrangement/orientation of the trabeculae is variable and influenced by the orientation of the external forces and moments imposed on the bone (Fig. 3.47).

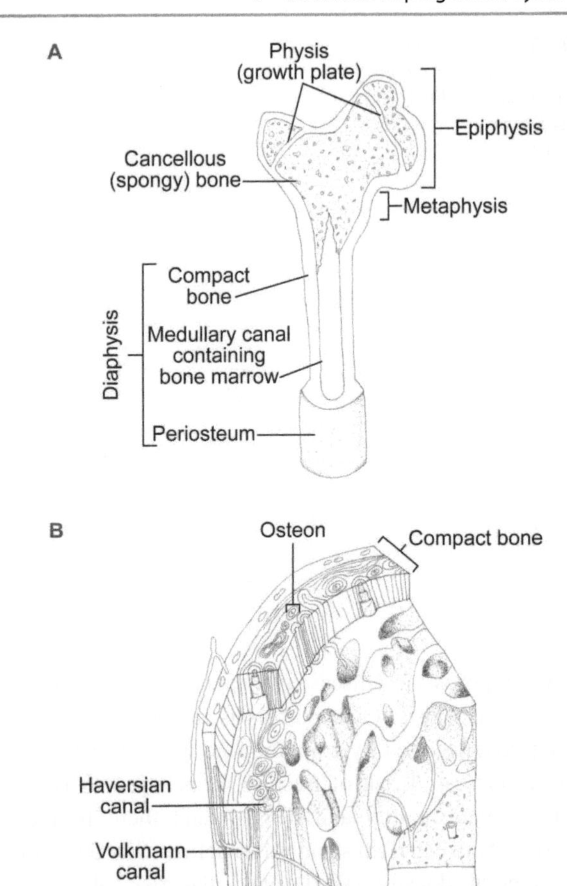


**Fig. 3.47** A human long bone showing (**A**) longitudinal section illustrating bone regions, and compact and cancellous bone, and (**B**) cross section of the typical bone structure inside the diaphysis

Before we examine the mechanical properties of bone and its relevance to the case we are studying, we need to briefly discuss the concepts related to bone adaptation and remodeling. Progress has been made in understanding the cellular mechanisms that accomplish absorption and deposition of bone, but the details of physiological mechanisms have not been clearly identified, and much work remains. We provide a framework for what is presently known. It has been suggested that the mechanosensing processes of an osteocyte enable it to sense the presence of, and respond to, external physical loads. Tissue sensitivity is a general property of a connected set of cells that is accomplished by the intracellular processes of mechanoreception and mechanotransduction. Mechanoreception is the process that transmits the content of an extracellular mechanical stimulus to a receptor cell, while mechanotransduction refers to the process that transforms the mechanical stimuli content into an intracellular signal. There are many such processes of intercellular transmission of transduced signals at the tissue, organ, and organismal structural levels.

Osteoprogenitor cells, osteoblasts, bone-lining cells, and osteoclasts are all located on the cellular interface (IC), which is contained on the surfaces of the tubular cavities of bone, the osteonal canals, Volkmann's canals, as well as the endosteum and periosteum (Fig. 3.47). The osteoblasts on the IC and osteocytes, buried in the bone matrix, are interconnected by the cell processes of the osteocytes, forming a connected cellular network (CNN). The interconnectivity of the CNN is provided by the touching cell processes of neighboring bone cells and their gap junctions, i.e., a channel connecting two cells. The CNN represents the hard wiring that connects all osteoblasts to neighboring osteoblasts and to osteocytes located perpendicular to the IC. The stimulus for bone adaptation/remodeling is related to its strain history, and a corresponding internal threshold, that is employed by the bone to sense its mechanical load environment. The CNN is the site of intracellular stimulus reception, signal transduction, and intercellular signal transmission. Research suggests that stimulus reception occurs in the osteocyte and that the CNN transmits the signal to the IC, where the osteoblasts directly regulate bone deposition and maintenance and indirectly regulate osteoclastic resorption. Several possible mechanisms for the osteocytic processes include stretch and voltage-activated ion channels, cytomatrix sensation-transduction processes, cyto-sensation by fluid shear stresses, cyto-sensation by streaming potentials, and exogenous electric field strength. The leading hypothesis is that an intracellular potential is generated and is responsible for the trigger transducing signals from osteocytes to neighboring cells in the CNN and on to osteoblasts in the IC. In summary, three key concepts explain bone's ability to adapt to changing mechanical loads: (1) bone structure optimizes its strength relative to its density; (2) trabeculae align with principal stress directions, and (3) 1 and 2 are accomplished by a self-regulating system of cells that respond to a mechanical stimulus. These principles have conventionally been referred to as Wolff's law.

#### 3.4.2.2 Mechanical Properties: Fracture Mechanics

Bone has a nonhomogeneous structure; therefore, it is characterized by its anisotropic material properties. Because of

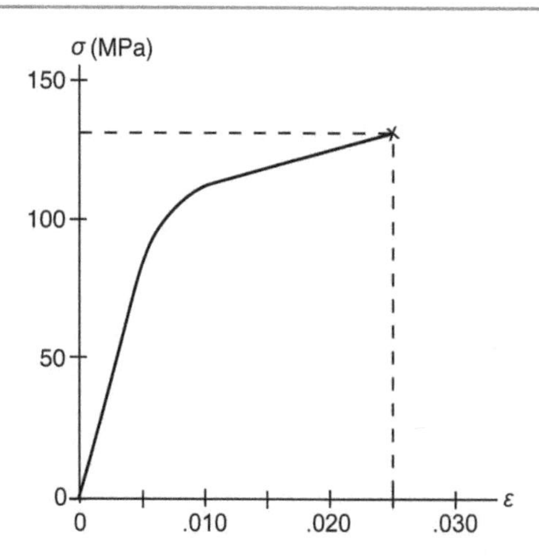


**Fig. 3.48** Example of a stress-strain diagram for human cortical bone loaded longitudinally under tension (strain rate, $\dot{\varepsilon} = 0.05\ s^{-1}$). Note: stress ($\sigma$) expressed as MPa ($10^6\ N/m^2$); $\varepsilon$ = strain, i.e., percent elongation

these properties, the mechanical response of bone is dependent on the direction and magnitude of external loads. Note also that bone possesses viscoelastic material properties so that its response to applied loads is rate-dependent (see Appendix H for more details on the general features of deformable body mechanics).

Bone is a complex structure with different mechanical responses to loads under compression, tension, shear, torsion, and bending moments. Typically, laboratory experiments are used to test bone specimens or whole bones under a variety of conditions. For example, a tensile stress-strain ($\sigma - \varepsilon$) diagram for cortical bone, as seen in Fig. 3.48, provides a good illustration of the type of information provided by laboratory experiments. The $\sigma - \varepsilon$ curve has three regions: (1) a linearly elastic region, where the $\sigma - \varepsilon$ is nearly a straight line and where the slope (elastic modulus) represents the bone's stiffness; (2) an intermediate region characterized by the nonlinear elastoplastic material behavior, and where yielding takes place; and (3) a region where the bone demonstrates a linear plastic, i.e., permanent, response. The cortical bone in this experiment would have fractured, i.e., failed, when the tensile stress reached approximately 128 MPa, and a tensile strain of about 2.6%. It is notable that cortical bone is stiffer than cancellous bone, withstanding greater stress, but less strain before failure.

The elastic moduli and strength values for bone are dependent on test conditions, including the rate of loading and the orientation of the bone with respect to the direction of loading. In Fig. 3.49, we see an example of the viscoelastic nature of bone tissue that was tested under different strain rates ($\dot{\varepsilon}$). With more rapid loading rates (high $\dot{\varepsilon}$), bone demonstrates greater stiffness, ultimate strength, and energy absorbed (area under the $\sigma - \varepsilon$ curve) before failure. It is interesting that during normal activities of daily living, bone

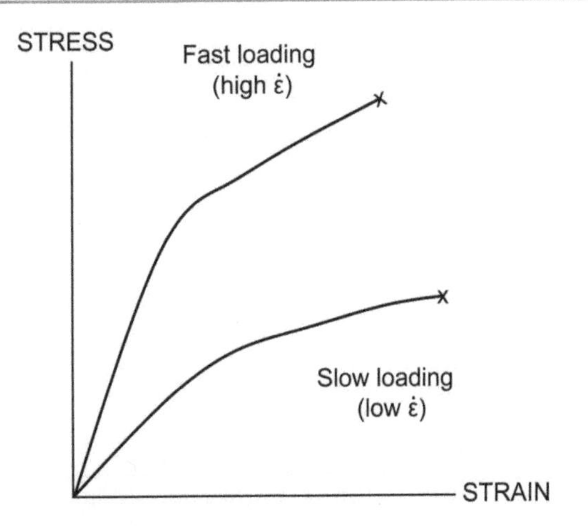


**Fig. 3.49** Illustration of the effect of different strain rates ( $\dot{\varepsilon}$ ) on the relation between stress and strain, e.g., a faster loading rate induces increased stiffness within the bone

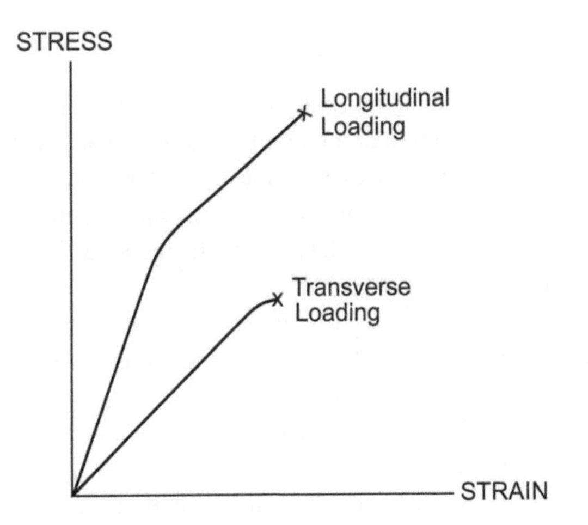


**Fig. 3.50** Example of the different stress-strain responses of bone tissue depending on how it is loaded. Generally, bone is stiffer, and its ultimate failure point is greater when it is loaded longitudinally, i.e., placed under tension, as opposed to transverse loading

is subjected to a strain rate of about 0.01 $s^{-1}$. At very high strain rates (>1 $s^{-1}$), representing impact loading, the bone becomes more brittle. The loading rate is clinically meaningful because it influences the fracture pattern and amount of tissue damage at fracture. At a high loading rate, the greater energy stored cannot dissipate rapidly enough, resulting in the comminution of the bone and extensive soft tissue damage.

The anisotropic behavior of bone is demonstrated when a bone is loaded under different orientation configurations (Fig. 3.50). In general, when a cortical bone is loaded longitudinally, its stress-strain behavior exhibits greater elastic modulus (i.e., it is stiffer) and ultimate strength when compared to transverse loading. Further, we see that with transverse loading, bone shows less yielding before failure, i.e., demonstrating more brittle behavior. Tables 3.3 and 3.4 compare ultimate strength values for adult femoral cortical bone under various loading conditions. Note that cortical bone strength is greatest under compression loading in a longitudinal direction (direction of osteon orientation secondary to large axial loading) and lowest with tensile testing in the transverse direction (direction perpendicular to longitudinal). We also see that the elastic modulus of cortical bone is greater when it is loaded longitudinally (i.e., axially); that is, it is stiffer. The different loading modes, e.g., compression, tension, shear, torsion, and bending, can be tested separately under controlled experimental conditions. But laboratory methods can only provide a glimpse into the complexity of how bones are loaded during daily activities. For example, in vivo measurement of strains on the anteromedial surface of an adult human tibia during jogging has been demonstrated to experience a combination of tensile, compressive, and shear stresses, with a large tensile peak at toe-off. Moreover, with an increase in running speed, the external forces changed in distribution and magnitude. This kind of information is useful because it provides insight with regard to running injuries such as plantar fasciitis, anterior shank compartment syndrome, and stress fractures.

**Table 3.3** Ultimate strength for human femoral cortical bone

| Loading modes | Ultimate strength to failure (MPa) |
|---|---|
| *Longitudinal* | |
| Compression | 133 |
| Tension | 93 |
| Shear | 68 |
| *Transverse* | |
| Compression | 51 |
| Tension | 122 |

Note: 1 MPa = $10^6$ Pa

**Table 3.4** Ultimate elastic and shear moduli for human femoral cortical bone

| Loading modes | Ultimate strength to failure (GPa) |
|---|---|
| *Elastic moduli, E* | |
| Longitudinal | 17.0 |
| Transverse | 11.5 |
| *Shear modulus, G* | 3.3 |

Note: 1 GPa = $10^9$ Pa

Up to this point, our discussion has referred to the material properties of cortical bone. Although the chemical composition of cortical and cancellous bone is similar, their material properties are distinguished by differences in density, i.e., where density is defined as the mass of bone present in a unit volume. It has been shown that cancellous bone is approximately 25% as dense as cortical bone. The material properties of cancellous bone, as with cortical bone, depend on the degree of porosity (which can differ from bone to bone), as well as the mode and rate of loading. In general, we see (Fig. 3.51) that cancellous bone is less stiff (on average

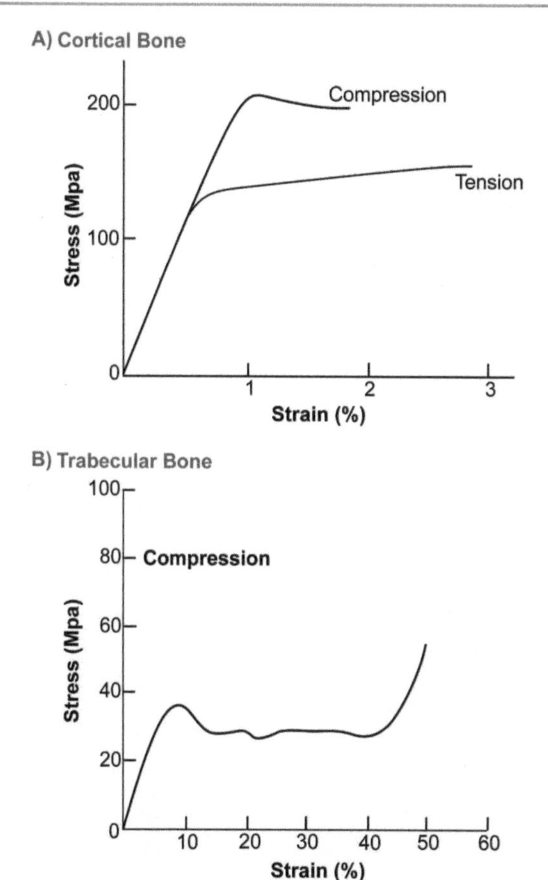


**Fig. 3.51** Differences in stress-strain response between (**A**) cortical and (**B**) cancellous bone under compression loading

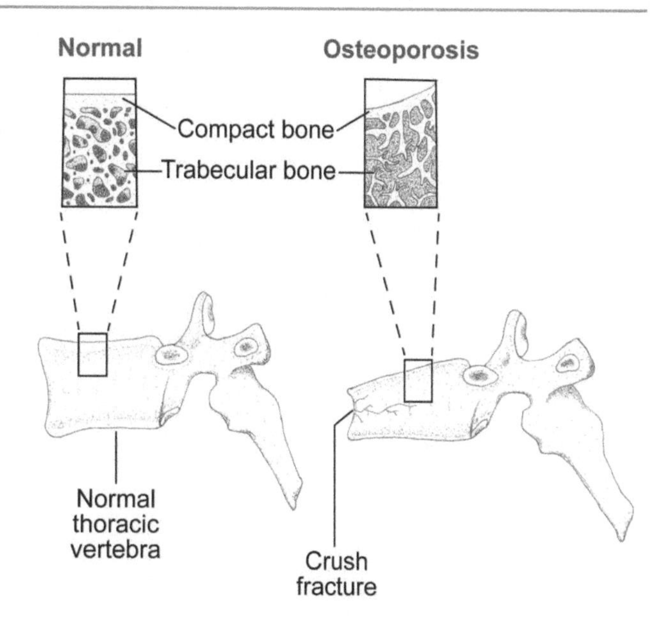


**Fig. 3.52** Matrix of trabeculae in normal and osteoporotic bone in a typical thoracic vertebrae

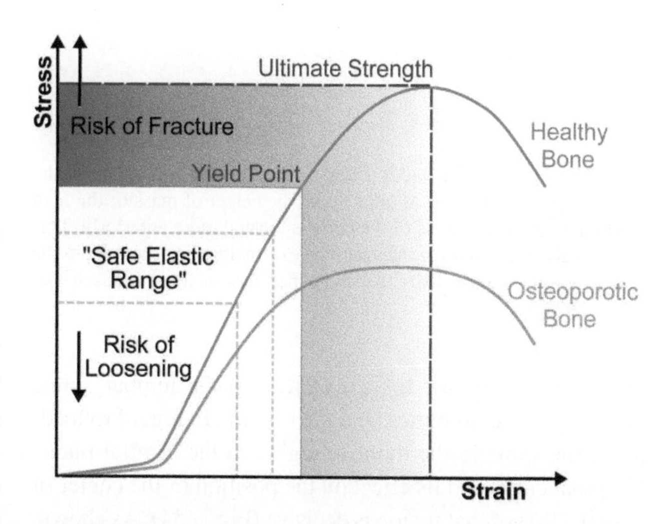


**Fig. 3.53** Demonstration of notable reductions in the stiffness, strength, and stress at failure in osteoporotic bone under an external flexion moment. The internal stresses are controlled primarily by the posterior spinal muscles, e.g., erector spinae, posterior annulus fibrosis, and other posterior spinal and secondarily by the constraints imposed by the costovertebral joints and the rib cage, particularly the sternocostal joints anteriorly. The magnitude of the external flexion moment is, of course, dictated by both the superincumbent body weight and length of the moment arm, i.e., the horizontal distance from the center of mass (COM) to the joint center. This external flexion moment, thus, can change depending on whether the posture is relatively fixed or dynamically changing

5–10%) but more ductile (up to five times greater) sustaining up to 50% strain before yielding and with a larger capacity for energy storage (calculated area under the stress-strain curve) than cortical bone. In Fig. 3.51, the stress-strain curve shows an initial linear elastic region up to a strain of about 5%, with material yielding at something less than 10% strain, indicating initial trabecular fracture. After yielding, the region that follows demonstrates a plateauing, indicating near-constant stress until fracture, exhibiting ductile material behavior. Not shown, typically cancellous bone fractures abruptly under tensile forces, therefore demonstrating brittle material behavior.

As noted previously, osteoporosis reduces the density and, thus, the integrity of both cortical and cancellous bone. Since cancellous bone is less porous than cortical bone, we assume that osteoporosis adversely affects cancellous bone to a larger degree (Fig. 3.52). In fact, the material properties of osteoporotic cancellous bone are significantly impaired under compression loading conditions (Fig. 3.53).

As a result, individuals with osteoporosis are predisposed to bone fractures, even under normal loading conditions. For example, since both trunk flexion movements, and an increased thoracic kyphotic posture, significantly increase the external compression loads on both the vertebral bodies and intervertebral disc, compression fractures are common in the thoracic spine. Let's briefly digress and examine why this is so. In the upright position, the primary function of the thoracic spine is the transmission of loads from the upper

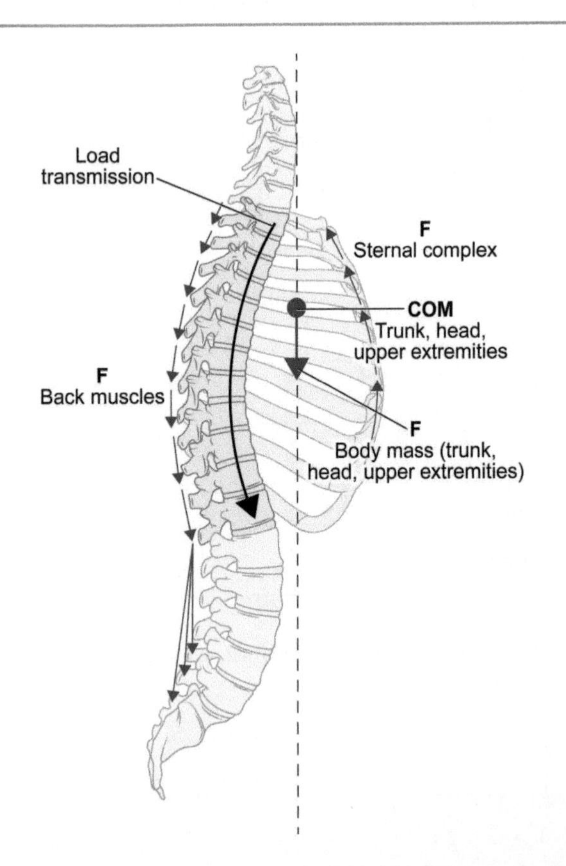


**Fig. 3.54** Schematic of forces acting on the thoracic spine. Due to the normal kyphosis (flexion) of this region, the center of mass of the load imposed by the upper third of the body is located more anteriorly. This downwardly directed force increases the external flexion moment in the normal upright stance. (Modified with permission from Liebsch and Wilke (2022))

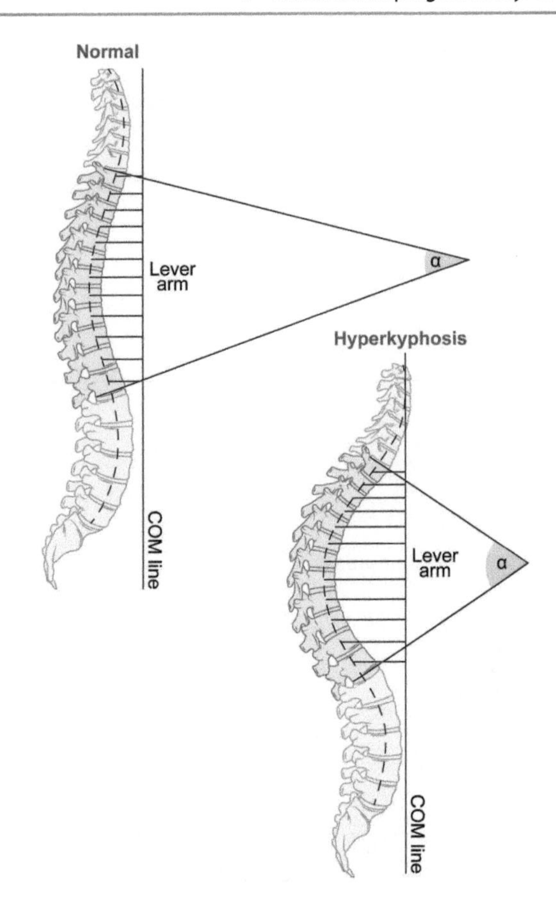


**Fig. 3.55** Schematic demonstrating the effect of the thoracic kyphosis angle on the magnitude of the moment arms to the center of mass (COM) line with a normal (left) and increased kyphosis (right). (Modified with permission from Liebsch and Wilke (2022))

third of the body and the extremities to the lumbar spine, sacrum, lower extremities, and the ground. In regard to loading of the spine in the thoracic region in the sagittal plane, the spinal curve and its effect on the position of the center of mass (COM) of that region is decisive (Fig. 3.54). As shown, the normal thoracic curve, or kyphosis, puts the spine in a flexed posture. This essentially shifts the downwardly directed forces from the superincumbent weight anteriorly. Mechanically, this process induces shear and tensile stresses on the posterior elements, e.g., facets, and compression stresses on the anterior aspects of the vertebral bodies and intervertebral discs (IVD), and anterior structures, e.g., sternocostal joint complex. Interestingly, this complex system has adapted to the upright posture and the nature of the forces in that the vertebral bodies and IVDs normally have a mild wedge-shaped appearance, i.e., the anterior aspect of the vertebrae and IVDs demonstrate reduced heights and thicknesses, respectively, compared to the posterior aspect. This downwardly directed force creates an external flexion moment, which is controlled by the posterior spinal muscles, primarily erector spinae, posterior annulus (which is very thick posteriorly), and other posterior ligaments of the spine. Secondarily, the constraints imposed by the costovertebral joints and, in general, by the rib cage, particularly the sternocostal joints anteriorly. The magnitude of the external flexion moment is, of course, dictated by both the superincumbent body weight and length of the moment arm, i.e., the perpendicular distance from the action line of the force (located at the center of mass (COM)) to the approximate joint center of the external force. This external flexion moment, thus, can change depending on whether the posture is relatively fixed or dynamically changing.

The external moment arms depend on the initial kyphotic angulation. Over time, individuals develop postures that increase the thoracic kyphosis secondary to work habits, development of scoliosis, angulation of the lumbar curve, or trauma. In Mrs. Buckler's case, her kyphosis did increase over time, which, of course, increased the magnitude of the external moment arm relative to the thoracic COM (Fig. 3.55).

Although not evident, the altered moment arm also redistributes the superincumbent weight, which then adds to the increased flexion moment. The larger flexion moment increases the tensile loads on the posterior soft tissues and the compression load on the vertebral bodies and IVDs. Note

**Table 3.5** Stages of bone healing

| | Stage 1<br>Inflammation | Stage 2<br>Soft callus | Stage 3<br>Hard callus | Stage 4<br>Remodeling |
|---|---|---|---|---|
| Time | *Before week 1* | *Weeks 1 and 2* | *Weeks 2–4* | *Weeks 4–8* |
| Predominant cell type | Inflammatory cells, platelets, macrophages | Chondrocytes, fibroblast, mesenchymal progenitors | Osteoblasts, "chondroclasts" | Osteoclast, osteoblast |
| Matrix | Hematoma, granulation tissue | Extracellular matrix proteins (collagen II, collagen X) | Mineralized bone matrix, collagen I, woven bone | Lamellar bone, cortical and trabecular formation |
| Fracture healing processes | Reorganization migration of mesenchymal stromal cells | Endochondral ossification, matrix mineralization | Vascular invasion, replacement of cartilage by bone | Bone and matrix degradation, new bone formation |

that the compression on the anterior aspect of the vertebral segments was already greater under normal conditions and this asymmetry is worsened with hyperkyphosis. Complex loading of the anterior elements of the thoracic cage, e.g., sternocostal joints, increases joint and periarticular stresses, but the inherent stability of this region seems capable of adapting. Mrs. Buckler, then, was predisposed to fracture injuries to her thoracic spine for two reasons: her posture and osteoporosis.

Embedded in the preceding discussion on the stress-strain properties of cortical and cancellous bone under different modes of loading, we learned that bone fails, i.e., fractures, when the applied stress exceeds a critical threshold. There is a range of how bone fractures are initiated, but for our work in this text, we classify fractures into two broad categories: macro-traumatic and micro-traumatic. As noted earlier, during activities of daily living, bone is subjected to a combination of different types of external loads. Thus, fracture of bone may be the result of a combination of loading modes, e.g., compression + shear + torsion. Fracture due to macro-trauma occurs when the external stress suddenly exceeds the yield point and progresses through the linear plastic region to bone failure. Recall that the stress-strain magnitude of the critical threshold, i.e., yield point, is variable depending on the geometry of the bone, mode of loading, and rate of loading. When the yield point is exceeded, cracks in the osteons are initiated and then rapidly progress until the structure can no longer sustain the externally applied stress, resulting in a fracture. In most cases, fractures from macro-trauma result from an external force impulse, i.e., high magnitude force applied to the bone over a short period of time, i.e., a high rate of loading. Although we learned that bone stiffness increases under higher rates of loading due to its viscoelastic nature, large impulsive loads simply overwhelm bone's inherent properties. In general, fractures related to failure as a result of compression loads are stable, whereas fractures initiated by tension, shear, or torsion may be more catastrophic.

Bone can also sustain a fracture following micro-trauma to its structure under two common conditions: (1) application of repeated loads that approach and mildly exceed the yield point, followed by insufficient time for recovery/rest from the outside forces, with repetition of this pattern, and (2) application of normal loads related to activities of daily living or postures sustained over long periods of time to the bone that is not normal, e.g., osteoporotic. These kinds of fractures have been defined as fatigue or stress fractures.

#### Fracture Healing: Remodeling

Let's briefly describe the process of fracture healing and bone remodeling. Fracture healing begins immediately after the fracture site is reduced and stabilized and initiated by the inflammatory process that may be completed within a week (Table 3.5). The second stage of recovery takes place over the next 2 weeks when a proliferation of small blood vessels and progenitor cells creates a fibrocartilaginous soft callus. In weeks two through four, osteoblastic activity is increased, and the soft callus mineralizes, with the replacement of cartilage by bone. Finally, weeks four to eight are characterized by more normal osteoblastic/osteoclastic activity leading to the differentiation of cortical and cancellous bone. Of course, there are many factors that dictate the staging and timing of this process, such as the severity of the fracture, how long the fracture site was immobilized, and the nutritional and preexisting medical status of the individual. Generally, by 8–10 weeks, movement constraints are gradually removed, and individuals begin to slowly resume activities of daily living. In addition, many individuals may begin rehabilitation programs that target particular joint and functional impairments. The key to successful rehabilitation is to provide the necessary mechanical stimuli, e.g., controlled external forces and moments, that trigger normal bone adaptation and remodeling processes (as described previously). Although rehabilitation may be completed over the course of 4–6 weeks, bone adaptation and remodeling continue for several weeks/months as the individual resumes all normal activities.

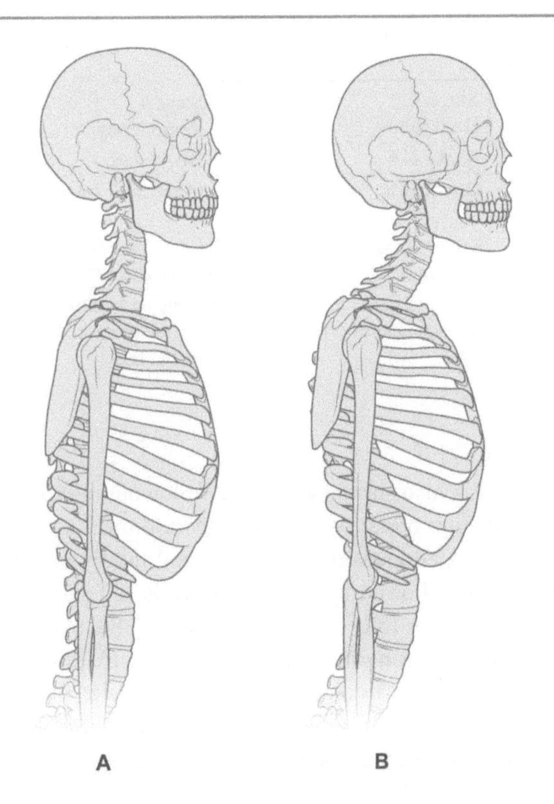


Fig. 3.56 The sagittal perspective of (A) ideal and (B) abnormal postures of the cervical and upper thoracic regions. Note in (B) a moderate increased thoracic kyphosis with a mild Dowager's hump

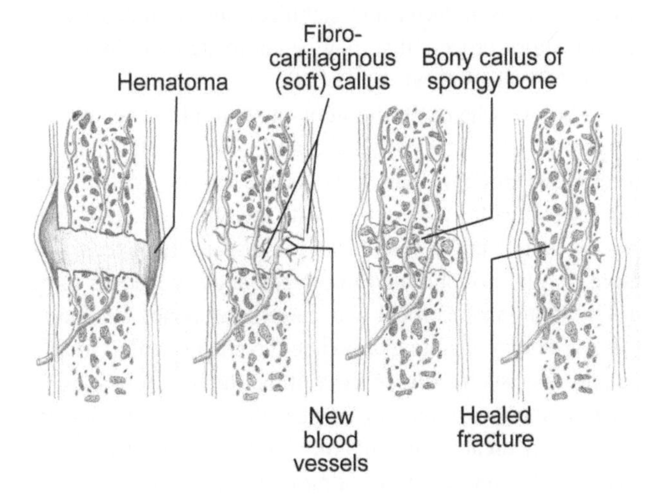


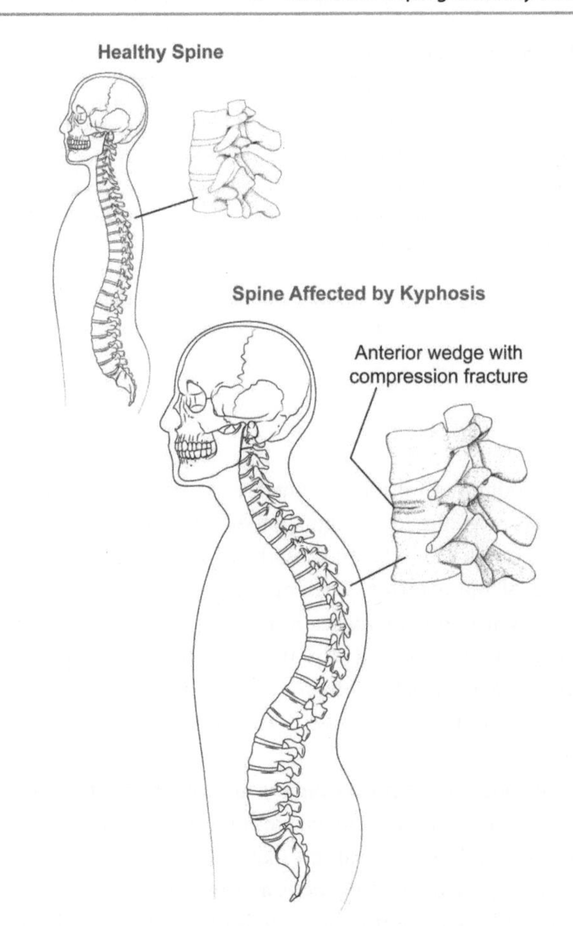


Fig. 3.57 A mid-thoracic compression fracture is likely due to moderately increased thoracic kyphosis

### 3.4.2.3 Summary: Relating Key Facts from Mrs. Buckler's History, Physical Examination, Learning Issues on Osteoporosis, Bone and Fracture Mechanics, and Hypotheses Regarding the Diagnosis of Impingement Syndrome

Mrs. Buckler had a 5-year history of osteoporosis and radiographic evidence of old compression fractures. On physical examination, we noted that she had a moderate increase in thoracic kyphosis and Dowager's hump (Fig. 3.56), likely conditions induced by long-term postural habits. The difference between an ideal posture (Fig. 3.56A) and the abnormal posture (Fig. 3.56B) suggests that an increased thoracic kyphosis, i.e., flexed posture, creates external compression loads on the anterior aspect of the vertebral bodies, as well as the intervertebral disc. Note that the magnitude of these loads is not abnormally large, nor are the forces impulsive. But Mrs. Buckler's postural condition can lead to the typical mechanism of injury leading to fatigue, i.e., stress, fractures to one or more thoracic vertebrae. This is so because the loads are sustained over long periods of time, i.e., in most of Mrs. Buckler's waking hours, she maintains an abnormal flexed posture. We can see that the fracture results in a collapse of the vertebral body anteriorly, creating a "wedged" appearance (Fig. 3.57).

This process is gradual, and we can imagine that it is painful. Let's assume that Mrs. Buckler sought health-care management for her painful condition. Thoracic stress fractures are inherently stable because of the bony rib cage and are treated conservatively. Conservative care generally includes oral anti-inflammatory medication, and/or other pain medication, and modification of aggravating activities. Since the primary cause of the fractures is related to sustained flexion postures, Mrs. Buckler would likely have been prescribed a thoracic corset that prevented a flexed posture and excessive thoracic movements. The course of treatment would, as noted above, span approximately 8–10 weeks to allow the fracture sites to heal and for bone remodeling to occur. Since Mrs. Buckler had osteoporosis, which actually predisposed her to stress fractures, it is possible that bone healing and remodeling could have been delayed to some degree (or may not have occurred at all). Mrs. Buckler may or may not have participated in a formal rehabilitation program; many individuals are not referred by their primary care physician to receive rehabilitation.

Regardless of how Mrs. Buckler's stress fractures were treated, when we examined Mrs. Buckler, we found a moderate increase in thoracic kyphosis that was now "fixed" in flexion; in other words, permanent plastic changes in the vertebral bodies, intervertebral discs, ribs, facet capsules, and perhaps muscles, as well as spinal ligaments, had constrained her ability to extend and rotate the spine in a normal fashion as part of activities of daily living. Furthermore, this posture likely changed the resting position of the scapula and other important skeletal elements associated with overhead arm movements. Thus, it is predictable that this combination of changes was related to the scapula-humeral dyskinesis noted in Mrs. Buckler's upper limb movement screen and contributed to her impingement syndrome.

## 3.5 Key Facts, Learning Issues, and Hypotheses: History of Prior Left Proximal Humeral Fracture at Age 45 Years, and Relationship to Present Glenohumeral Degenerative Joint Disease

There was radiographic evidence of degenerative joint disease (DJD) at the glenohumeral joint, with likely narrowing of GH joint space, decreased subacromial space, and sclerosis of the humeral head and distal clavicle. What is DJD? Why do these findings suggest DJD? How might DJD be related to the history of the left proximal humeral fracture? What might these findings have to do with subacromial impingement?

### 3.5.1 Degenerative Joint Disease

The lay term for degenerative joint disease (DJD or osteoarthritis/arthrosis) is arthritis. The expression arthritis implies that there is an inflammatory component to this type of joint disease, which is only partly true. We know that inflammation is a normal biological response to injury, whether it is related to trauma or some systemic disease process, but it is not indefinite. That is, inflammation runs its course, which is followed by other physiological processes that attempt to replace damaged tissue, followed by remodeling and adaptation. Since DJD appears, on the surface, to involve both bone (i.e., osteo) and joint (i.e., arthro) related structures, we need to examine these issues separately. We have already examined, in some depth, the structure, and mechanics, of bone. Let's now focus on the mechanical nature of the diarthrodial or synovial joint (Fig. 3.6) and start with articular cartilage. We return to a discussion about osteoarthritis after examining the mechanics of selected periarticular structures of the synovial joint.

### 3.5.2 Articular Cartilage Structure and Mechanical Properties

#### 3.5.2.1 Structure

Articular cartilage (AC) is a thin layer of hyaline cartilage (typically 2–4 mm thick) that covers the articulating surface of bones that make up a synovial joint (also referred to as arthrodial, diarthrodial, or investing cartilage). Some have described AC as an isolated tissue because it is devoid of blood vessels, lymph channels, and nerves. These deficiencies have significant implications with regard to the maintenance of AC health and its reaction to trauma. For example, with a significant strain injury to AC, the absence of nociceptors results in the lack of pain sensation provided to the central nervous system, as well as deficient sensory feedback related to the inflammatory cycle and tissue healing/remodeling. AC has two primary functions: (1) distribute joint loads (we also refer to these as "bone-on-bone" forces) over the joint surface area of contact, which decreases contacting surface stresses, and (2) direct relative movement between opposing joint surfaces, i.e., arthrokinematics, with minimal friction and wear. Articular cartilage is a soft tissue comprised of a proteoglycan matrix (15–40% of the dry weight) reinforced with collagen (40–70% of the dry weight) and is highly hydrated (ranging from 60% to 85% water).

Cartilage cells or chondrocytes vary in size, morphology, and arrangement, are sparsely distributed in articular cartilage, and account for less than 10% of the total AC volume. Despite their sparse distribution, chondrocytes are the engine of AC, as they drive the manufacture, secretion, organization, and maintenance of the organic component of the extracellular matrix (ECM) or matrix (Fig. 3.58). Unlike osteocytes, chondrocytes do not have processes that facilitate communication, and there is no direct cell-cell contact between them. To date, it is not known how interchondrocyte signal transduction and cell coordination work. The ECM immediately surrounding the chondrocytes is referred to as the pericellular matrix and is surrounded by a capsule. The capsule, pericellular matrix, and chondrocytes constitute the "chondron,"

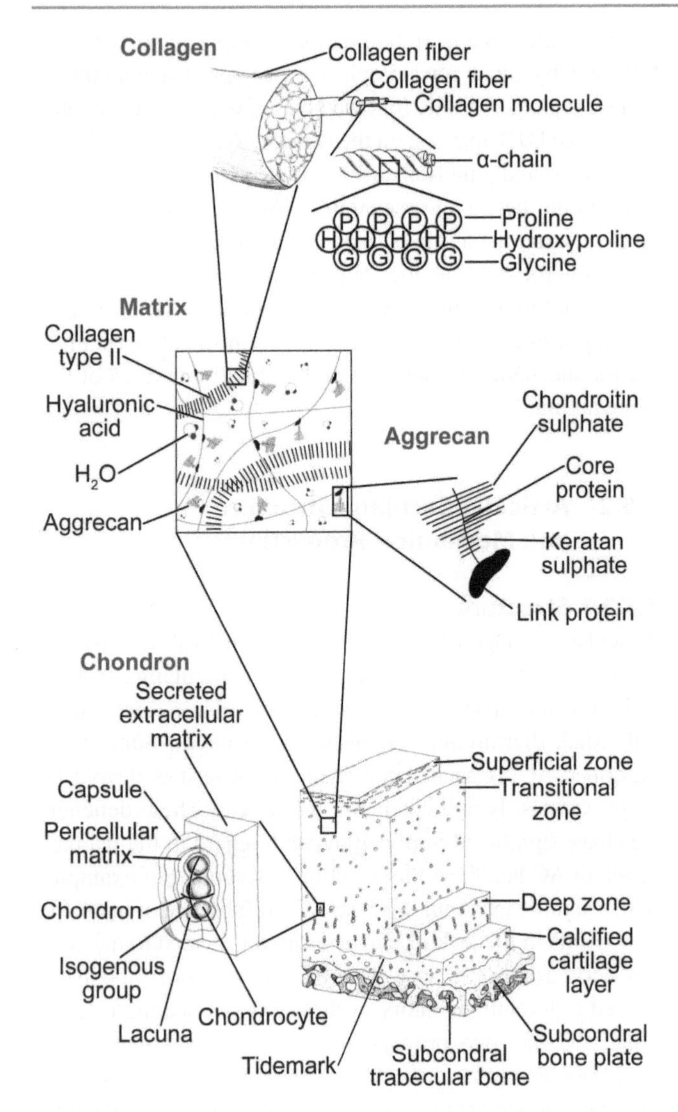


**Fig. 3.58** Schematic of zonal arrangement of chondrocytes, illustrating the changes in size and orientation with depth. Included are constituents that constitute articular cartilage matrix, e.g., collagen, aggrecan, and water

**Table 3.6** Types of collagen in articular cartilage and their proposed roles

| Collagen type | Function |
|---|---|
| II | Structural; mechanical; a primary constituent of AC |
| VI | Pericellular adhesion molecule |
| IX | Fibril association; stabilizes type II |
| X | Hypertrophic zone of growth plate; role in calcification |
| XI | The core of type II; controls fibril growth |

the primary functional and metabolic unit of cartilage that functions to protect the integrity of the chondrocyte, and its pericellular environment, during joint loading. The organic matrix is composed of a dense network of fine collagen fibrils (primarily type II collagen) that are enmeshed in a concentrated solution of proteoglycans (PGs). Collagen and PGs are dispersed differently throughout the AC depth, with more collagen toward the surface and more PGs in the deeper layers of the matrix. Collagen fibrils and PGs are the structural components that support the internal mechanical stresses that result from loads being applied to the joint cartilage.

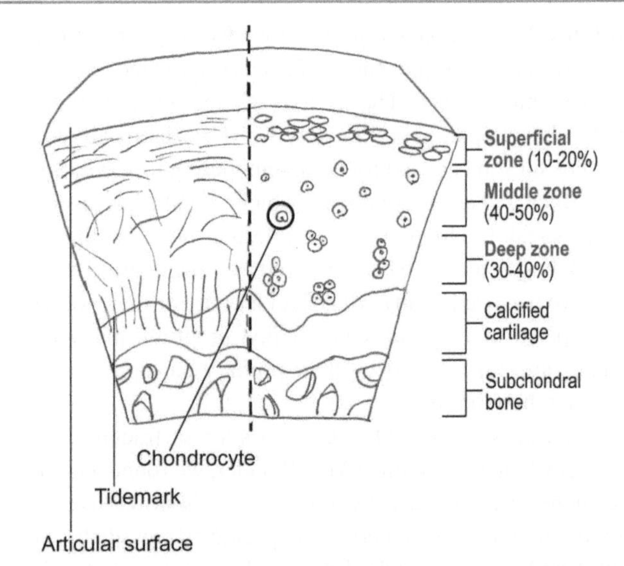


**Fig. 3.59** Relationship between chondrocytes and type II collagen dispersion in the superficial, middle, and deep zonal network of articular cartilage

Let's look at collagen in a bit more detail. The organization of collagen in articular cartilage provides the tissue its infrastructure and strength, i.e., resistance to tensile loads. Type II collagen is distinguished from the type I collagen found in bone, ligaments, and tendons, because it has thinner fibrils, allowing it to be maximally dispersed throughout the cartilage tissue. Although type II collagen is predominant, other types of collagen have been identified (Table 3.6). It is notable that collagen dispersion is not homogeneous: (1) in the superficial tangential zone, collagen is expressed as densely packed fibers randomly woven in planes parallel to the articular surface (Fig. 3.59), which protects deeper layers from shear stresses; (2) in the middle zone, there is more dispersion between the randomly oriented collagen fibers, which provides a front line of resistance to compressive forces; and (3) in the deep zone, collagen fibers are closely aligned in parallel fashion, oriented radially, providing, along with the highest concentration of proteoglycans, the greatest resistance to compressive forces.

The collagen in the deep zone crosses the tidemark, i.e., the interface between AC and the calcified cartilage beneath it, to enter the calcified cartilage, which forms an interlocking basement system that anchors the cartilage to the underlying bone. It appears that the inhomogeneous zonal variation in collagen content, and the collagen-PG concentration, has mechanical implications. For example, large tensile stresses within the articular surfaces and at the edges of joint contact

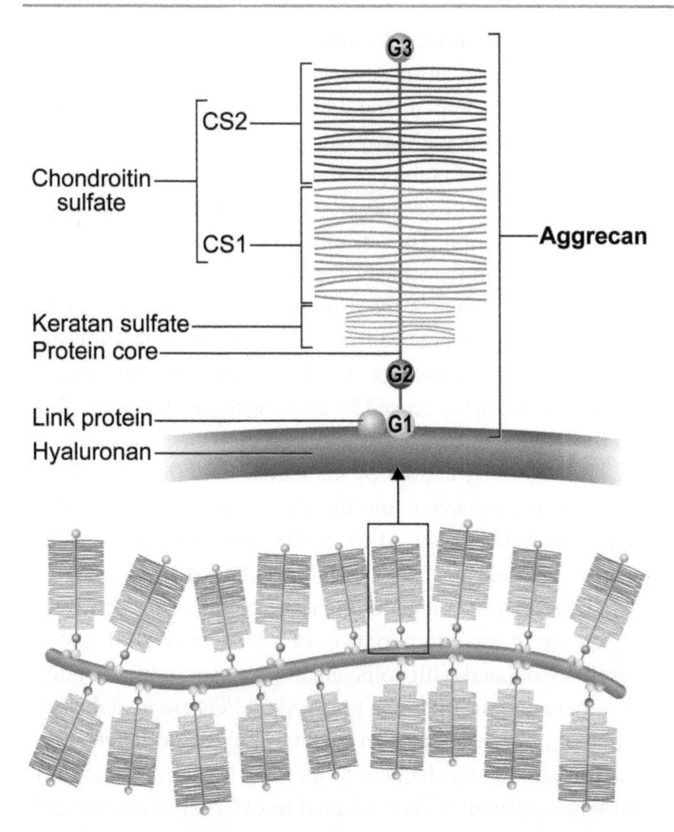


**Fig. 3.60** Schematic of aggrecan, composed of GAGs bound by a protein core. Aggrecan is linked by hyaluronan to form the large proteoglycan macromolecule

areas are resisted primarily by tangentially oriented collagen fibers. Additionally, the layering of surface collagen appears to facilitate interstitial fluid flow, resist shear stresses, and optimize frictional properties of AC, whereas the compressive and hydrostatic stresses found in the deeper layers of cartilage are resisted by the incompressibility of water, which is held chemically by the hydrophilic aggrecan molecules. More on that later. Like bone, the material properties of AC differ, i.e., anisotropic, with the direction in loading. This property is likely related to the varying collagen arrangements in zonal regions, variations in collagen cross-linking, as well as variations in collagen-PG interactions.

Proteoglycans are large protein-polysaccharide molecules composed of a protein core to which one or more glycosaminoglycans (GAGs) are attached (Fig. 3.60). The molecules are composed of central hyaluronic acid chains, i.e., hyaluronan, which are attached to core proteins by link proteins. The core proteins form the attachment points for glycosaminoglycans (GAGs), which find themselves linked to repeating disaccharide units, chondroitin 4-sulfate, chondroitin 6-sulfate, and keratin sulfate, all of which carry a negative charge. These molecules are covalently bonded to the core protein to form a proteoglycan monomer which, when linked to hyaluronan, becomes the proteoglycan aggregate known as aggrecan. This PG aggregation appears to promote immobilization of the PGs within the fine collagen framework, contributing to the structural stability and rigidity of the ECM.

The extracellular matrix consists of several distinct regions based on proximity to chondrocytes, composition, and collagen fibril diameter and organization. The pericellular matrix completely surrounds the chondrocyte and contains mainly proteoglycans. This region likely plays a functional role in initiating signal transduction within cartilage during load-bearing. The territorial matrix (TM) surrounds the pericellular matrix, is composed primarily of fine collagen fibrils, and appears to protect chondrocytes against mechanical stresses. The TM may contribute to the resiliency of the articular cartilage framework. The interterritorial region contributes most to the biomechanical properties of AC. It is characterized by randomly oriented bundles of large collagen fibrils, arranged parallel to the surface of the superficial zone, obliquely in the middle zone, and perpendicular to the joint surface in the deep zone. Proteoglycans are abundant in the interterritorial region.

Water is the most abundant component of articular cartilage and is concentrated near the articular surface. The water here contains many free mobile cations, e.g., $Na^+$, $K^+$, and $Ca^{2+}$, that influence the biochemical and mechanical behavior of AC. Since there are no blood vessels in AC, the interstitial fluid of AC is essential for promoting the movement of nutrients and waste between chondrocytes and joint synovial fluid. Equally important is the interaction between water and the collagen-PG structure in controlling osmotic pressure. For example, when loaded under compression, approximately 70% of the water may be moved. This interstitial fluid movement is largely responsible for the mechanical behavior and joint lubrication of AC. Let's examine the mechanical relationship between water, collagen, and the PG aggregate.

### 3.5.2.2 Mechanical Properties

The mechanical behavior of AC is highly dependent on its porous structure and proteoglycan content. The matrix combination of collagen and proteoglycan molecules constitutes about 30% of the tissue; the rest is essentially water. Since proteoglycans are hydrophilic, AC is essentially homeostatic in a hypersaturated state due to the physiochemical and biochemical interactions within and between PGs and collagen. As such, AC can be treated as a biphasic material consisting of two incompressible, inhomogeneous, and distinct phases: an interstitial fluid phase and a porous-permeable phase. The challenge faced by AC during joint loading and movements is in the distribution of very large internal stresses, e.g., compression, tension, and shear, that vary in magnitude, but can be as large as ten times body weight. By exploring the material properties of AC, we gain an understanding of how AC responds to large physiological loads under normal and pathological conditions.

Like most biological tissues, articular cartilage demonstrates viscoelastic material properties (see Appendix H). However, because AC is very porous and hypersaturated, its response to various loading rate conditions is much different than what we see in the bone, for example. Recall, that when a bone is loaded at a higher rate, it becomes stiffer and, therefore, better able to resist deformation. Articular cartilage, on the other hand, is subjected, during the course of a day, to loading rates that vary widely, resulting in elastic moduli that vary widely. It has been claimed that the function of AC is to absorb shock. However, because it is much less stiff than cortical bone, even under high loading rates, cartilage cannot serve to reduce impulsive forces in joints. It cannot do this because it is not as thick as the bone and does not have the capacity to absorb energy, i.e., computed area under the elastic region of a stress-strain curve, like subchondral bone. Rather, AC provides a relatively friction-free lubricating surface, where high internal stresses are optimally distributed using a self-renewing system. So, how does that system work?

Articular cartilage acts like a combination of a viscous fluid (i.e., dashpot) and an elastic solid (i.e., spring). Under experimental conditions, when a viscoelastic solid is placed under a constant load, it responds with a rapid initial deformation followed by a slow (time-dependent) progressive deformation (or *creep*), until equilibrium is reached. Conversely, when a viscoelastic solid is instantaneously strained, i.e., deformed, we see a rapid, high initial stress followed by a slow (time-dependent), progressive decrease in stress, referred to as *stress relaxation*. With AC, the creep and stress-relaxation phenomenon act simultaneously under loading conditions.

The key to understanding how AC distributes stresses under external loads is found in the relationship between the proteoglycan aggregates and the collagen network. The aggrecans attached to the hyaluronic acid backbone of the PG macromolecule are negatively charged related to the chemical structure of the sulfated GAGs that compose them. These charges cause the aggrecans to repel each other and branch out into something that has the appearance of a bottlebrush (Fig. 3.60). This negatively charged body intrinsically attracts counter-ions to maintain electroneutrality, thus making the aggrecan hydrophilic; that is, it attracts the cations contained within the interstitial fluid, i.e., water, of the AC. With the "bristles" of the bottlebrush strongly bonded to water, a large swelling pressure is created, which is balanced by tension developed in the collagen network. This swelling pressure creates a significant prestress even in the absence of external loads.

When a joint surface is subjected to an external compression load, there is an instantaneous deformation caused primarily by a change in the PG molecular aggregate. This external stress concomitantly causes an increase in the internal pressure in the matrix to exceed the swelling pressure, forcing water to flow out of the tissue, which reduces internal stress, i.e., stress relaxation. The outward flow of fluid continues, i.e., creeps, while simultaneously, the PG concentration increases, which in turn increases the osmotic swelling pressure or the charge-to-charge repulsive force until they are in equilibrium with the external stress. For AC, the compressive viscoelastic behavior is primarily caused by the flow of the interstitial fluid and the frictional drag associated with this flow. The aspect of AC viscoelasticity caused by interstitial fluid flow is referred to as biphasic behavior, and the aspect caused by macromolecular motion is referred to as the flow-dependent, or the intrinsic viscoelastic, behavior of the collagen-PG matrix.

In summary, it is apparent that the collagen and PG interactions are of great importance. PGs have been shown to be closely associated with collagen and may serve as a bonding agent between the collagen fibrils. And PGs are also thought to play an important role in maintaining the collagen infrastructure and its mechanical properties. Moreover, it is clear that when normal AC is subjected to external loads, the collagen-PG solid matrix and interstitial fluid are coupled dynamically to uniquely reduce its permeability to protect the ECM. We primarily examine the material properties of AC under compression loads, which are predominant during activities of daily living. However, external tensile loads also elicit AC viscoelastic behaviors, anisotropically, i.e., in regions where the loads are aligned with the collagen framework. Likewise, external shear stresses on AC are mitigated, but primarily by its collagen content or collagen-PG interaction, rather than its biphasic viscoelastic behavior.

#### 3.5.2.3 Lubrication

Human articular cartilage is subjected to mechanical loads imposed during normal daily activities that vary in magnitude, rate of loading, and areas of concentration in the lower and upper limbs. For example, consider the variety of loads imposed on the hip or knee joints while sleeping, sitting in a chair, transferring from sitting to standing, walking, jumping, or hopping on one leg. It is really quite remarkable that under these normal circumstances, joint cartilage sustains little wear. In addition to the mechanical behavior which manages this variety of loads, sophisticated lubrication processes also offer protection to joint cartilage. In general, synovial joint lubrication involves load-driven transport processes, i.e., imposition of external loads, that move interstitial fluid out of the articular cartilage. This movement changes osmotic pressures causing the cartilage to imbibe fresh synovial fluid containing nutrients to maintain chon-

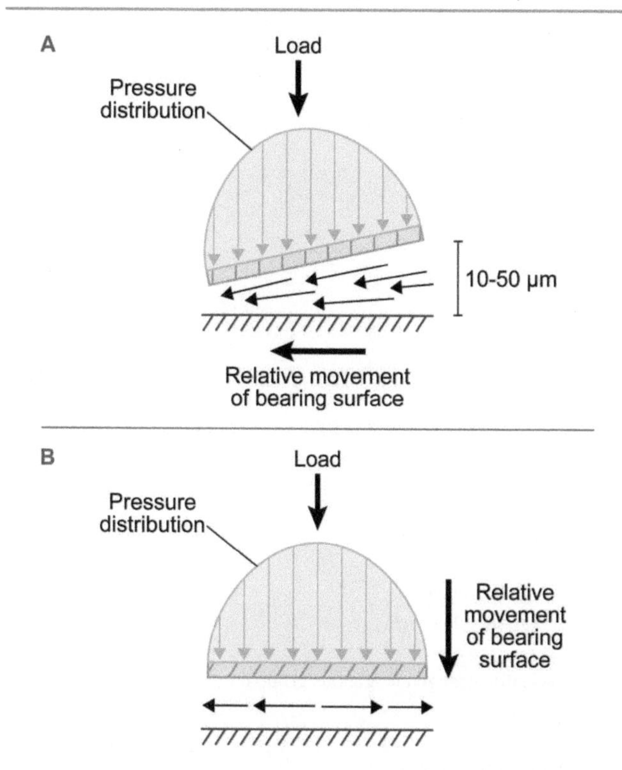


**Fig. 3.61** (**A**) Sliding or hydrodynamic lubrication in which non-parallel rigid surfaces move relative to one another, forming a wedge of fluid between the surfaces. A lifting pressure between the surfaces, proportional to the viscosity of the fluid, is generated. (**B**) Squeeze-film lubrication in which parallel rigid surfaces are moved together, squeezing the fluid film from between the surfaces. As in (**A**), a lifting pressure between the surfaces is created, which is proportional to the viscosity of the fluid. (Modified with permission from Ateshian and Mow (2005))

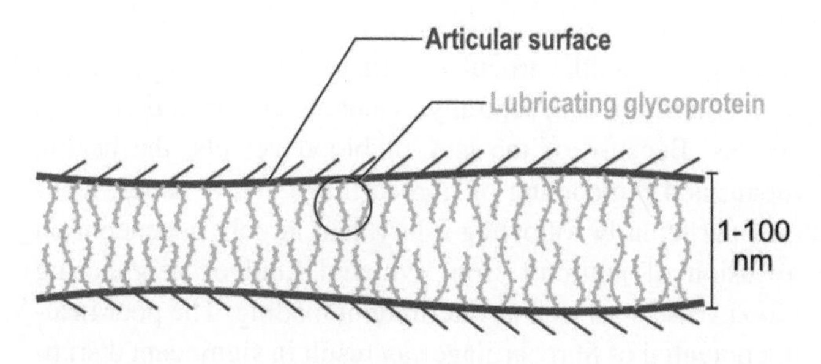


**Fig. 3.62** Boundary lubrication of articular cartilage where the internal stress on the articulating surfaces is mitigated by the glycoprotein lubricin. (Modified with permission from Ateshian and Mow (2005))

drocyte health. A related benefit provided by the lubrication processes to joint cartilage is the maintenance of a very low coefficient of friction, permitting human joint bearings decades of smooth rolling and gliding.

Several types of lubrication processes, i.e., borrowed theories from engineering, have been described, but only the two primary types, fluid film, and boundary lubrication are briefly reviewed here. With fluid film lubrication, a thin film separates the two load-bearing surfaces that are sliding relative to one another, with no surface-to-surface contact

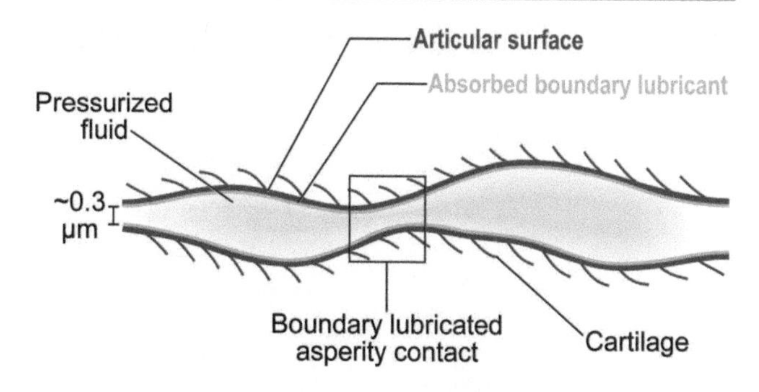


**Fig. 3.63** Mixed lubrication in articular cartilage. Boundary lubrication is likely to predominate when the thickness of the fluid-film is on the same order as the roughness of the articular surface, whereas fluid-film lubrication takes place with more widely spaced surfaces. (Modified with permission from Ateshian and Mow (2005)).

(Fig. 3.61). The pressure created, which is proportional to the viscosity of the fluid, in the fluid film by the relative sliding of the two bones balances loads transverse to the joint surfaces. By engineering standards, this mode of lubrication generally requires high relative sliding speeds between the two solid surfaces to create a lubricant film that would manage significant joint loads.

There is no fluid film in boundary lubrication. Boundary lubrication describes a situation in which the lubricant, i.e., bearing material, is between joint surfaces moving at relatively slow speeds while the joint cartilage is mitigating high loads. This process appears to involve a monolayer of lubricant molecules adsorbed on each bearing surface, which prevents direct surface-to-surface contact and eliminates most surface wear. It appears that in diarthrodial joints, a glycoprotein lubricin is the major component of the synovial fluid that is responsible for the boundary lubrication. This uniquely makes boundary lubrication independent of the viscosity of a lubricant or the stiffness of the tissues (Fig. 3.62).

Since joint surfaces are not perfectly congruent, articular cartilage is not perfectly smooth. Thus, in synovial joints, it is hypothesized that situations often occur where a mixed-mode of lubrication is operating; that is, the surface load is mitigated by both the fluid-film pressure in areas of non-contact and by boundary lubrication in the areas of asperity, i.e., unevenness of surface or roughness, contact. In this case, most of the joint friction, which is still very low, is generated in the boundary lubricated areas, while most of the load is carried by the fluid film (Fig. 3.63).

In summary, it appears that the predominant mode of lubrication depends on the applied loads and the relative velocity between the articulating surfaces. Boundary lubrication and absorption of lubricin seem to be most important under severe loading conditions, i.e., contact with high magnitude loads, low relative speeds, and long duration. Conversely, fluid-film lubrication seems to be most prevalent

**Table 3.7** Features comparing hyaline (articular) and fibrocartilage

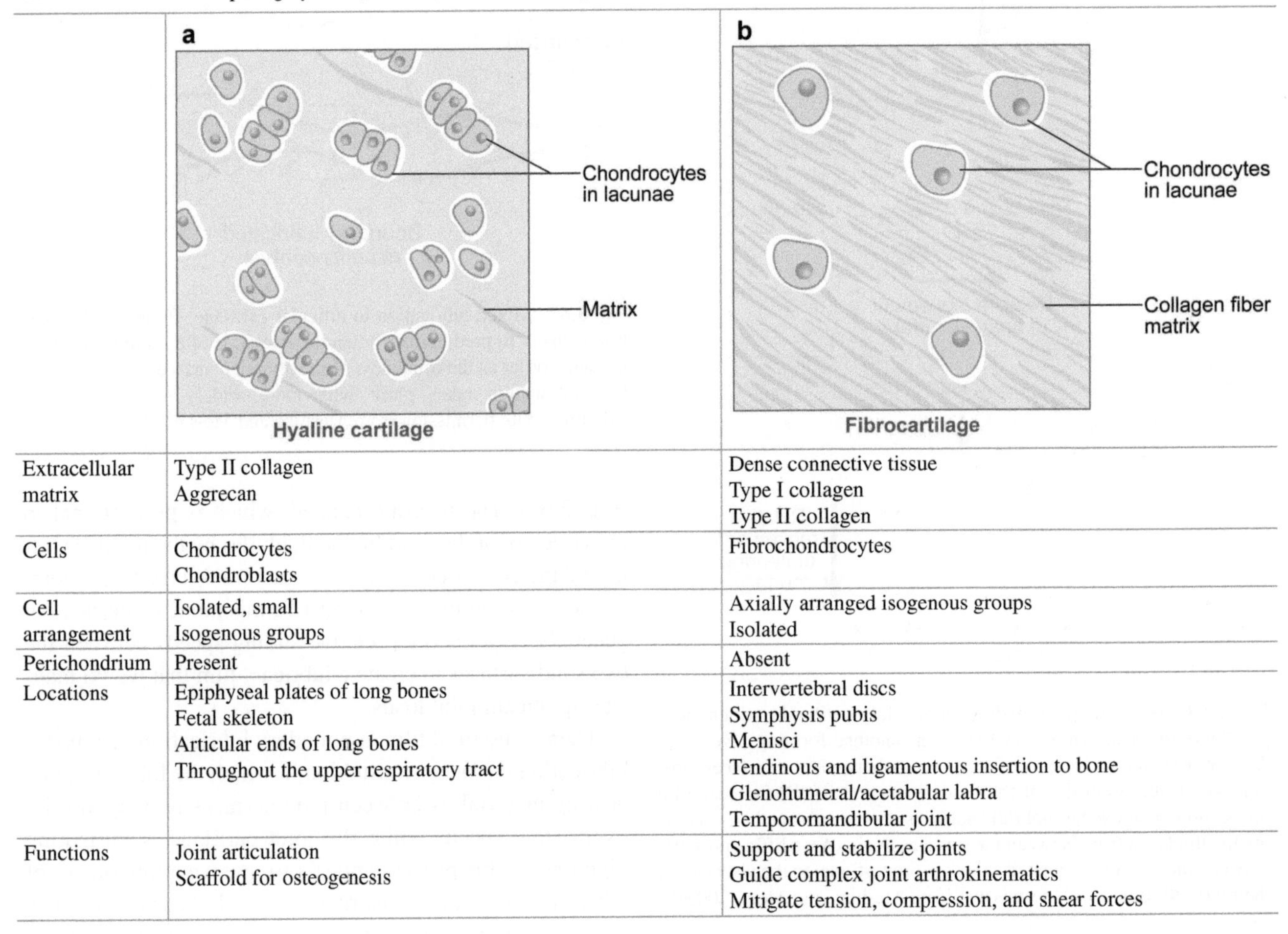


| | Hyaline cartilage | Fibrocartilage |
|---|---|---|
| Extracellular matrix | Type II collagen<br>Aggrecan | Dense connective tissue<br>Type I collagen<br>Type II collagen |
| Cells | Chondrocytes<br>Chondroblasts | Fibrochondrocytes |
| Cell arrangement | Isolated, small<br>Isogenous groups | Axially arranged isogenous groups<br>Isolated |
| Perichondrium | Present | Absent |
| Locations | Epiphyseal plates of long bones<br>Fetal skeleton<br>Articular ends of long bones<br>Throughout the upper respiratory tract | Intervertebral discs<br>Symphysis pubis<br>Menisci<br>Tendinous and ligamentous insertion to bone<br>Glenohumeral/acetabular labra<br>Temporomandibular joint |
| Functions | Joint articulation<br>Scaffold for osteogenesis | Support and stabilize joints<br>Guide complex joint arthrokinematics<br>Mitigate tension, compression, and shear forces |

under loading conditions of smaller magnitude and/or oscillating magnitudes and when the articulating surfaces are moving at relatively faster speeds.

### 3.5.3 Fibrocartilage Structure and Mechanical Properties

Fibrocartilage is a composite between dense connective tissue and articular cartilage (Table 3.7). Although it has some cells similar in nature to chondrocytes, i.e., called fibrocytes, the cells are arranged in short rows of three or four and separated by dense bundles of type I collagen, which forms a three-dimensional framework. According to the adage, "function dictates structure," we suggest that fibrocartilage's multidimensional cartilaginous framework is different for each region, i.e., glenoid labrum, intervertebral disc, etc., because of the fact that the external stresses imposed on it are unique to the joints where it has developed. The presence of moderate amounts of proteoglycan aggregates gives fibrocartilage hyaline-like physiochemical properties, providing this tissue with material properties very similar to articular cartilage. Also, like articular cartilage, fibrocartilage lacks a perichondrium, and, centrally, is aneural and devoid of blood vessels. Because of the lack of blood vessels, the health, repair, and remodeling of fibrocartilage are, at times, tenuous, particularly following injury, and largely dependent on diffusion of nutrients from synovial fluid or neighboring blood vessels, as well as intermittent loading. The poor healing potential of fibrocartilage can result in significant disruption of its material properties under deforming external loads, as well as interference with normal joint arthrokinematics. Because no nociceptors are present in fibrocartilage, internal derangement of this tissue cannot be sensed by the central nervous system. Likewise, the absence of mechanoreceptors in fibrocartilage reduces central proprioceptive input, although this is partially compensated by sensory fibers located peripherally where fibrocartilage interfaces with tendinous or ligamentous insertions.

Because fibrocartilage has a structure very similar to articular cartilage, its mechanical properties closely mirror those of AC. However, fibrocartilage does not appear to have zonal arrangements of, and varying concentrations of, collagen and PGs, as seen in AC, thus making this tissue

**Table 3.8** Approximate distribution of collagen, proteoglycans, and water in articular and fibrocartilage

| Tissue | Collagen (% dry weight) | Proteoglycans (% dry weight) | Water (% dry weight) |
|---|---|---|---|
| Articular cartilage | 50–75 | 15–30 | 58–78 |
| Menisci | 75–80 | 2–6 | ~70 |
| IVD (NP) | 15–25 | ~50 | 70–90 |
| IVD (AF) | 50–70 | 10–20 | 60–70 |
| Tendon | ~23–30 | ~7 | ~70 |
| Ligament | ~23 | ~7 | ~70 |

Reprinted with permission from Mow et al. (1990)
*IVD* intervertebral disc, *NP* nucleus pulposis, *AF* annulus fibrosis

more uniform in how it manages external loads. For example, fibrocartilage is much more densely packed with collagen, which would appear to make it better able to resist tensile loads. On the other hand, the complex inter-relationship between collagen, proteoglycans, and water in fibrocartilage suggests that its swelling and viscoelastic properties perfectly match the loading rates and magnitudes experienced, depending on its location and functional role, e.g., intervertebral disc, labra, etc. (Table 3.8). In Table 3.8 the notable differences in the proportion of collagen, proteoglycans, and water from the tissues made up primarily of fibrocartilage, e.g., menisci, suggest there are nuances in how these tissues mitigate external loads.

### 3.5.4 Hypotheses Regarding the Development of Osteoarthritis: The Possible Relationship Between Mrs. Buckler's Previous Fracture, Radiographic Joint Degenerative Changes, and Impingement Syndrome

We frequently borrow and apply theoretical concepts from engineering in our reductive studies of the complexity of the human body. In our study of how joints degenerate, we take the engineering definition of wear, i.e., mechanical action that removes material from solid surfaces, as our starting point. There are two components to wear: interfacial and fatigue. Interfacial wear results from the direct interaction of bearing surfaces without lubricant film. This type of wear can result in adhesions or abrasions. Adhesive wear is the result of contact between bearing surfaces, which produce fragments that adhere to one another and eventually loosen secondary to relative surface sliding. Abrasive wear is the result of a harder material, or loose particles, abrading the softer bearing partner. Although conceptually, interfacial wear seems a plausible starting point in our study of joint cartilaginous injury (which appears to precipitate joint degeneration), it has been shown that synovial joint lubrication robustly prevents direct surface-to-surface contact between the asperities (i.e., microscopic surface roughness) of the opposing cartilaginous surfaces. On the other hand, adhesion and abrasion may take place on joint surfaces that are impaired in some alternative way. Once the joint surface sustains microdamage and/or decreases in mass (perhaps secondary to early stages of degeneration), it becomes softer and more permeable. The increased permeability is likely related to impaired collagen-PG coupling in the ECM, which can also initiate biochemical events that change lubricating processes leading to abnormal surface-to-surface contact and interfacial wear.

Fatigue wear of load-bearing surfaces, not unlike fatigue fractures, is an accumulation of microdamage within the bearing material, i.e., ECM, as a result of repetitive stresses. This damage can be the result of either repeated application of high external loads, which mildly or markedly exceeds the yield point of the material, over a short period of time, or with excessively repeated, or sustained, low loads over an extended period of time. In the latter case, the loads do not exceed the yield point, but only approach it over thousands of cycles, which eventually leads to a type of plastic deformation. It is notable that fatigue wear can take place in an environment of well-lubricated surface bearings. These scenarios very likely take place in synovial joints that experience cyclical loading that varies in its rate, magnitude, and location. Normal joint arthrokinematics, i.e., roll and glide, theoretically optimize joint surface stress mitigation. However, if there were some impairment in joint mobility, and because joint surfaces are not perfectly congruent or smooth, specific regions of that surface may be more susceptible to injurious deformations. Thus, repetitive joint movement can cause stressing of the solid matrix, with repeated exudation and imbibition of the tissue interstitial fluid, which ultimately may give rise to two possible mechanisms of fatigue wear: (1) disruption of collagen-PG dynamics in the ECM and (2) PG "washout." It is hypothesized that this fatigue wear is likely what initiates a cascade of events that results in degenerative joint disease.

In the first case, i.e., damage to the collagen-PG solid matrix, from high load repetitive stresses over a relatively short time span or normal load repetitive stresses over a long time period, can lead to disruption of the collagen infrastructure (i.e., from tensile stresses at, or exceeding, the yield point), the PG macromolecules, and/or the dynamics of PG-collagen complex. These types of disruptions lower the stiffness of articular cartilage, making it more susceptible to highly intermittent and sustained loads, even normal day-to-day stresses. In the second instance, the large and repetitive loss of interstitial fluid, followed by the imbibition of water, may cause damaged PGs to "wash out" of the ECM, also contributing to the loss of tissue stiffness. Concomitantly, cartilage tissue becomes more permeable, which impairs the stress-shielding mechanism of interstitial fluid-load support.

These two hypothetical mechanisms initiate a vicious cycle of hyaline cartilage degeneration.

A third possibility that may explain how cartilage sustains a third-degree strain that would accelerate degenerative joint changes has been offered. If an impulsive load, i.e., rapid load application of high magnitude, is applied to a synovial joint, it is hypothesized that the resultant damage, without sufficient time to repair, may initiate joint degeneration. Recall that with normal physiological loading and joint lubrication, the viscoelastic properties of articular cartilage mitigate large compression loads with the fluid redistribution, i.e., collagen-PG-water dynamics, within the articular cartilage over time. Typically, within 2–5 s, cartilage tissue undergoes progressive creep and stress relaxation (internal stress may decrease by as much as 60%) in the compacted region. Even with higher rates of loading, the viscoelastic nature, i.e., increased stiffness, of AC can adapt sufficiently to escape injury. However, with impact loading, e.g., repeated jumping in basketball, repeated lumbar extension with blocking in football, a sudden fall onto an outstretched arm, etc., the load is applied so quickly that there is insufficient time for the internal mechanism of fluid redistribution to relieve the highly stressed region of a joint surface. A series of impulsive loads, without sufficient recovery time, may be enough to damage the collagen-PG matrix and initiate a degenerative cascade.

Because articular cartilage is aneural (no nociception), the microdamage that initiates a degenerative cascade of events, i.e., a vicious cycle of micro-injury, may not be detected for weeks, months, or years. Furthermore, articular cartilage has a very limited capacity for healing, repair, and regeneration because of an insufficient blood supply. These same caveats are true for the intrinsic ECM of fibrocartilaginous structures, such as the glenoid labrum, acetabular labrum, knee menisci, or intervertebral disc. With this information in hand, we speculate about Mrs. Buckler's previous proximal humeral fracture and how it might be related to the radiographic findings of her left shoulder. Recall that she did not have radiographic abnormalities of her uninvolved extremity.

Although Mrs. Buckler could not recall a mechanism of injury related to the left proximal humeral fracture, we hypothesize that, however she might have fallen or how the arm might have been struck, a large impulsive load was responsible for the fracture. Furthermore, we suspect that structures in the vicinity of the proximal humerus, i.e., left glenohumeral, acromioclavicular, and sternoclavicular joints, likely also sustained impact loading. It is not unreasonable to assume that the degree of strain to the periarticular joint structures was enough to initiate (1) changes in the intrinsic molecular and microscopic structure of the collagen-PG matrix, leading to increased cartilage permeability, and (2) changes in the mechanical properties, i.e., decreased stiffness, of the articular and fibrocartilage tissues, both factors associated with degenerative joint disease. It is known that these two factors result in a loss of mass (i.e., cartilage thickness) due to excessive exudation of interstitial fluid. Additionally, these alterations lead to the calcification of cartilage and the extension of the tidemark zone into subchondral bone, which is typically seen as decreased joint space between two articulating bones on standard radiographs, as reported in Mrs. Buckler's case (Fig. 3.64). Moreover it has also been shown that other periarticular structures, e.g., joint capsule, etc., change over time as a result of the articular cartilage impairment. For example, a slight loss of the normal roll and glide can lead to impairments in the material properties of the joint capsule.

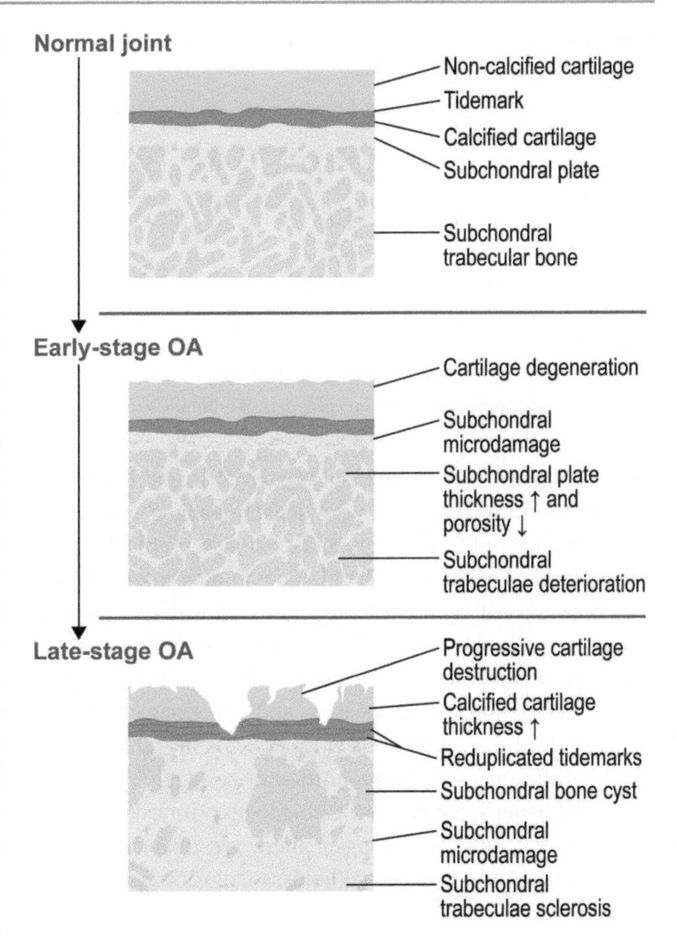


**Fig. 3.64** Changes in articular cartilage and subchondral bone related to advancing osteoarthritis

Because of the inability of damaged cartilage to mitigate/distribute normal repetitive compression, tension, and shear stresses, the subchondral bone is subjected to stresses that AC is no longer able to distribute and mitigate. Our previous examination of bone structure and material properties demonstrates that bone structure is very dynamic and responds to increases in stress by stimulating osteoblastic activity. Therefore, evidence of increased density of the subchondral

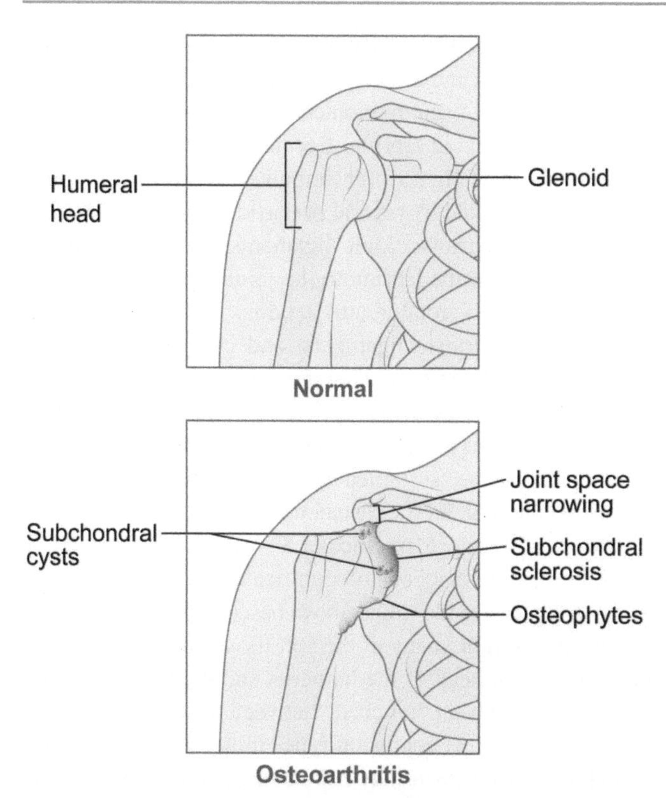


**Fig. 3.65** Schematic of osteoarthritis changes at the glenohumeral joint demonstrating changes in subchondral bone, narrowing of the subacromial space, osteophyte formation at the joint margins

bone of the humeral head, as was likely in Mrs. Buckler's case, suggests that this represents a thickening of the bone and is a normal adaptation to increased stresses, likely related to cartilage degeneration. Another bony adaptation that often takes place as compensation for reduced AC capacity is the development of osteophytes at the joint margins and entheseal sites (attachment sites of dense connective tissue to bone, e.g., joint capsule, etc.). This new bone is added by endochondral ossification, similar to the cellular mechanisms of skeletal growth and development, and stimulated by mechanical loading. Some have suggested that osteophyte formation may serve to provide additional joint stabilization. We hypothesize that Mrs. Buckler may have osteophytes that have contributed to the narrowing of the subacromial space found on her radiograph, similar to what is shown in Fig. 3.65.

Changes to other periarticular structures, as a result of osteoarthritis, include thickening (hypertrophy) of the synovial membrane and joint capsule, decreased stiffness of joint ligaments and fibrocartilaginous tissues, and alterations in motor control, i.e., neuromuscular activity, secondary to adaptations in joint mechanoreceptors, muscle spindles, and arthrogenic pain. There is evidence of some of these changes, as indicated by Mrs. Buckler's physical examination. For example, the goniometric measures of active ROM showed:

| | *Left* | *Right* |
|---|---|---|
| Abduction | 165° | 170° |
| Flexion | 160° | 170° |
| External rotation | 75° | 90° |
| Internal rotation | 55° | 60° |

Additionally, with overpressure at the completion of active ROM, the end feels for both ER and IR were firmer, with less spring than on the right side. Joint mobility testing revealed moderate loss of mobility in the acromion-clavicular joint bilaterally and moderate loss of joint play, i.e., inferior, anterior, and posterior glide, at the left glenohumeral joint, with mild loss of play in the same directions at the right glenohumeral joint. These findings are suggestive of GH and acromioclavicular joint capsular changes consistent with thickening and increased stiffness of the capsule, signs indicative of joint degeneration.

Loss of normal arthrokinematics (normal roll/glide) during overhead arm activities do not allow joint stresses to be distributed over the joint surface in a way to optimally mitigate large bone-on-bone forces. What we cannot discern is whether the impaired joint mobility has been present for some time, perhaps beginning after her proximal humeral fracture, or developed as a result of her recent fall. What we do know is that altered joint mobility can contribute to the cascade of events leading to the progression of degenerative joint disease. Earlier we suggest that Mrs. Buckler's abnormal arthrokinematics and muscle weakness likely contributed to her impingement syndrome. Now, we are suggesting that the physical impairments may have been pre-existing and related to the development of degenerative changes in her glenohumeral joint and other joints in the shoulder girdle complex.

## 3.6 Key Facts, Learning Issues, and Hypotheses Relating to the Mechanism of Injury, i.e., Fall on an Outstretched Hand, and Impingement Syndrome

Mrs. Buckler attempted to break her fall on an outstretched hand. This is a common mechanism of injury often resulting in a distal radius or scaphoid fracture, ulnar coronoid fracture, and/or third-degree strain of the ulnar collateral ligament at the elbow, dislocation of the glenohumeral joint, or separation of the acromioclavicular joint. Given her age and endocrine status, it is surprising that Mrs. Buckler escaped these injuries. In this section, we focus our inquiry on the injury mechanism in the shoulder girdle region.

**Table 3.9** Classification of strain injuries

| Severity | Pathology | Signs and symptoms |
|---|---|---|
| Grade I (mild) | Mild stretch, i.e., strain just past the yield point of the tissue; no instability | No hemorrhage, minimal swelling, point tenderness, and negative stress test |
| Grade II (moderate) | Moderate stretch, i.e., strained into the plastic region of the tissue; mild to moderate hypermobility or instability | Hemorrhage, localized swelling, moderate/ marked point tenderness, and positive stress test |
| Grade III (severe) | Reached failure point of the tissue; significant hypermobility or instability | Significant hemorrhage; diffuse swelling, extreme point tenderness, and positive stress test |

Note: yield point = the point on the stress/strain curve at which the material starts to deform plastically; plastic region = the region of a stress-strain curve where the material deforms permanently; failure point = point on the stress/strain curve where at which the failure (e.g., fracture) of the material takes place

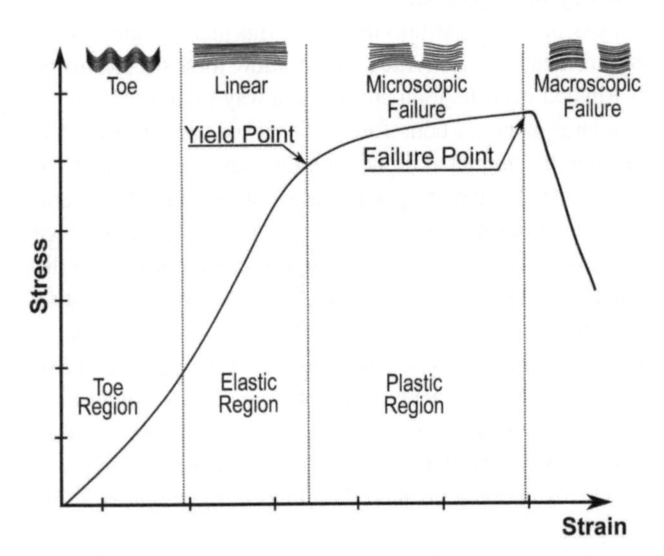


**Fig. 3.66** Typical stress/strain diagram for ligaments and tendons. The toe region represents the "uncrimping" of the collagenous framework under low stress. In the elastic region, we see a linear relationship between stress and strain, with increasing resistance of the tissue under greater stress. Microdamage, i.e., mild strain injury, at the yield point begins the region of a plastic, i.e., permanent, strain injury to the tissue. At the failure point, there is a rapid fall in the stress the tissue can resist related to significant deformation of the tissue

### 3.6.1 External Forces and Moments at the Glenohumeral and Acromioclavicular Joints and Their Mitigation

Falling on an outstretched hand likely created a large impulsive force, projected cranially through the long axis of the forearm and humerus. This external axial force likely pushed the humerus in a superior direction, causing compression and shear loads between the head of the humerus, the glenoid fossa and labrum, and the acromion process. The compression force would have been distributed/mitigated by the GH joint cartilage and glenoid labrum. Because there was a superior shear at the GH joint, a high impulsive tensile load would have been mitigated by several periarticular soft tissues: GH intrinsic and extrinsic joint ligaments, fibrocartilaginous attachments from the labrum to the joint capsule, long head of biceps brachii, rotator cuff tendons, acromioclavicular intrinsic and extrinsic ligaments, and coracoacromial and coracohumeral ligaments. Therefore, it is possible that one, or more, of the bony and soft tissues designed to attenuate the external forces and moments related to the mechanism of injury, may have sustained a first-degree strain injury (Table 3.9 and Fig. 3.66). We can be confident in this hypothesis because we saw Mrs. Buckler 8 weeks following the fall and there was no evidence of significant injury and/or instability to the tissues identified above based on physical examination. In addition, many of the soft tissues listed above that lie between the head of the humerus and the acromion process could have been "pinched" between the two bones, perhaps sustaining a contusion-type injury. As noted, the "pinching" mechanism may have caused a first- or second-degree contusion that likely resolved within the 8-week time frame. On the other hand, it is also likely that a strain injury to one or more tissues could have exacerbated previous impairments, e.g., degenerative conditions at the GH or acromioclavicular joints, and contributed to the many factors already identified as part of Mrs. Buckler's impingement syndrome.

## 3.7 Estimation of Glenohumeral Joint Bone-on-Bone Forces Associated with Rehabilitation and Activities of Daily Living

Given Mrs. Buckler's impaired strength of several muscles involved with overhead activities, the physical therapist would likely create an intervention program of progressive resistive exercises for those muscles, i.e., serratus anterior, middle, and lower trapezius, and GH abductors and external rotations. Because she has signs and symptoms of degenerative joint disease at the GH, we may need to reconsider how to approach muscle strength impairments. Let's estimate the GH bone-on-bone forces with the arm abducted 90°.

As part of the physical examination, we examined the active range of motion of the GH joint of flexion/extension, abduction/adduction, and internal/external rotation. These tests are usually performed with the patient sitting with the elbow fully extended. Of course, during abduction in the frontal plane, the joint passes through 90° abduction. Let's virtually stop her at 90° abduction for a static equilibrium

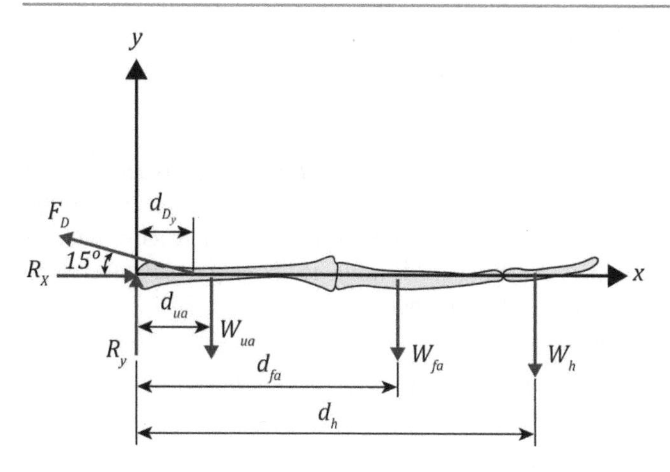


**Fig. 3.67** Free body diagram for solving for glenohumeral joint contact forces

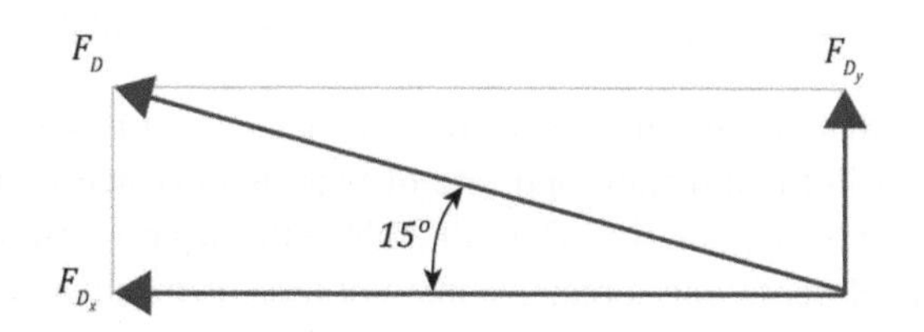


**Fig. 3.68** Resolving $F_D$ into components

determination of the left deltoid muscle force and related GH joint bone-on-bone forces.

Static equilibrium assumptions:

1. Static analysis is a good estimate of dynamic loads.
2. Rigid body mechanics.
3. The deltoid is the only muscle that is active; there is no antagonistic muscle action.
4. Only planar forces, i.e., $x - y$ (frontal) plane.

The solution begins with reporting information that is known:

Body mass $m$ = 66 kg (body weight $W = mg$ = (66 kg) (9.81 m/s$^2$) = 647.5 N)
Upper arm weight $W_{ua} = 0.028\ W = 18.13$ N.
Forearm weight $W_{fa} = 0.016\ W = 10.36$ N
Hand weight $W_h = 0.006\ W = 3.88$ N

Note: see Appendix A for anthropometric body measures.

We assume that at 90° GH abduction with the arm fully extended, the deltoid ($F_D$) makes an angle of 15° relative to the long axis of the humerus. The distance from the GH joint to the insertion of the deltoid on the humerus is $d_{D_y}$ = 0.12 m. The moment arms (i.e., the perpendicular distance from the line of action of the force vector to the center of rotation of the GH joint) for the weights of the humerus, forearm, and hand are $d_{ua}$ = 0.13 m, $d_{fa}$ = 0.40 m, and $d_h$ = 0.57 m, respectively.

We need to define the "signs" of forces and moments. Therefore, forces that point up/right are positive, and counterclockwise moments are positive. The free-body diagram is a simplified sketch that depicts the relationship between the segments of interest and the environment. We establish a segmental body-fixed coordinate system attached to the humerus with the origin located at the GH joint center. All relevant, known and unknown, external and internal forces (e.g., gravity, muscle, friction, ground reaction, bone-on-bone or joint reaction, etc.) acting on the system are shown. Forces are represented by vectors, with their point of application, line of action, and orientation (direction) (Fig. 3.67).

Note: the distance $d_{D_y}$ from the insertion of the deltoid to the joint center of the GH joint is not the moment arm of $F_D$. Therefore, we can do one of two things: (1) resolve $F_D$ into its $x$ and $y$ components and use this distance as the moment arm of $F_{D_y}$ or (2) determine the moment arm of $F_D$.

1. Let's first resolve $F_D$ into its $x$ and $y$ components. We use the graphical method of vector addition and create a right triangle (Fig. 3.68). If we choose the triangle to make a parallelogram, $F_{D_y}$ , its opposite link is equal; the same is true for $F_{D_x}$ and its opposite link.

   Knowing that $F_D$ makes a 15° angle relative to the $x$-axis, $F_{Dx} = F_D \cos 15°$ and $F_{D_y} = F_D \sin 15°$ .

   We can describe $F_{D_y}$ as the rotatory component of $F_D$ (i.e., the part of $F_D$ creates a counterclockwise rotation, i.e., abduction, or controls humeral adduction). The moment arm for $F_{D_y}$ is $d_{D_y} = 0.12\,\text{m}$ . Also, the line of action for $F_{D_x}$ passes through the GH joint center, so this force component has no moment arm and therefore does not contribute to the moment. The $F_{D_x}$ force acts to translate the head of the humerus toward the glenoid fossa, which induces an external compression force at the joint surface, i.e., a direct bone-on-bone force. The internal stress due to this force is mitigated by articular cartilage and the subchondral, i.e., trabecular, bone.

   The moment about the shoulder joint for the deltoid force using this first approach is then $F_{D_y} d_{D_y} = F_D d_{D_y} \sin 15°$ .
2. Since the moment arm of a force is the shortest perpendicular distance from the action line of the force to point about which the moment is taken, we can also create a right triangle using the 15° angle and the distance $d_{D_y} = 0.12\,\text{m}$ (distance from the insertion of the deltoid to the GH joint center, point C in Fig. 3.69). In this case, the moment arm $d_D$ of $F_D$ is the side opposite of the hypotenuse, i.e., $d_{D_y}$ , so that $d_D = d_{D_y} \sin 15° = 0.031\,\text{m}$ . In comparison to the moment arms of the external forces, i.e., weight of the upper arm, forearm, and hand, the magnitude of the deltoid moment arm is significantly less. In general, the mechanical disadvantage of human muscles results in large joint bone-on-bone forces when muscles either produce or control joint movements.

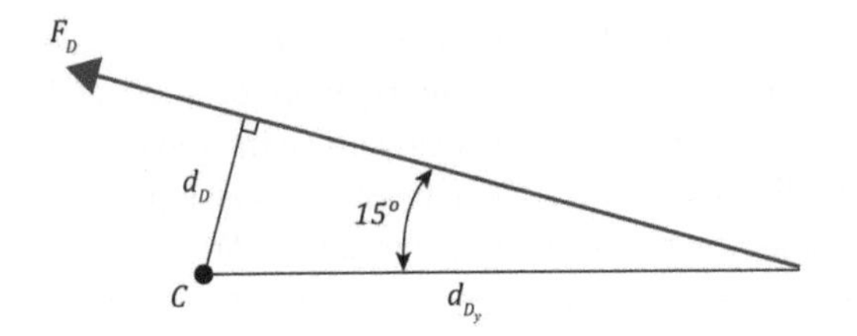


**Fig. 3.69** Determining the moment arm of $F_D$

The moment about the shoulder joint for the deltoid force using this second approach is then $F_D d_D = F_D d_{D_y} \sin 15°$, which is the same as what we found for the first approach.

Finally, let's write the static equilibrium equations. The equilibrium conditions are $\sum F = 0$ ($\sum F_x = 0$ and $\sum F_y = 0$) and $\sum M = 0$. Let's first determine the muscle force, $F_D$, using the moment equation. Recall that a moment = force times the shortest perpendicular distance from the action line of the force to the point of reference and that we defined a counterclockwise moment as positive. Therefore, summing the moments about the shoulder joint (S) (using the first approach from above):

$$\sum M_s = -W_{ua} d_{ua} - W_{fa} d_{fa} - W_h d_h + F_{D_y} d_{D_y} = 0$$

Note the signs for the moments due to the segment weights are all negative because they induce a clockwise rotation about the shoulder and the sign for the moment due to the $y$-component of the muscle force is positive because it induces a counterclockwise rotation about the shoulder.

Solving for $F_{D_y}$:

$$F_{D_y} = \frac{W_{ua} d_{ua} + W_{fa} d_{fa} + W_h d_h}{d_D}$$

$$F_{D_y} = \frac{(18.13\,\text{N})(0.13\,\text{m}) + (10.36\,\text{N})(0.40\,\text{m}) + (3.88\,\text{N})(0.57\,\text{m})}{0.12\,\text{m}}$$

$$F_{D_y} = 72.62\,\text{Nm}$$

The total muscle force $F_D$ and $x$-component of the force $F_{D_x}$ are then calculated as:

$$F_D = \frac{F_{D_y}}{\sin 15°} = 280.6\,\text{N}$$

$$F_{D_x} = F_D \cos 15° = 271.0\,\text{N}$$

As noted, $F_{D_y}$ represents the rotatory component of $F_D$, and $F_{D_x}$ represents the compression component.

Now, let's determine the unknown joint reaction, or bone-on-bone forces, using force equilibrium equations. Recall that force vectors oriented up/right are positive:

$$\sum F_x = -F_{D_x} + R_x = 0$$

Solving for $R_x$:

$$R_x = F_{D_x} = 271.0\,\text{N}$$

Note that because the result of $R_x$ is positive, it is acting in the same direction drawn in the free body diagram (i.e., to the right or in the $+x$-direciton). $R_x$ represents a reactive internal compressive force that is mitigated largely by articular cartilage and subchondral bone.

Next,

$$\sum F_y = F_{D_y} - W_{ua} - W_{fa} - W_{ha} + R_y = 0$$

Solve for $R_y$:

$$R_y = -F_{D_y} + W_{ua} + W_{fa} + W_{ha}$$
$$R_y = -72.62\,\text{N} + 18.13\,\text{N} + 10.36\,\text{N} + 3.88\,\text{N} = -40.25\,\text{N}$$

Note that because the result for $R_y$ is negative, it is actually acting in the direction opposite of what is indicated on the free body diagram (i.e., $R_y$ is 40.25 N in the negative $y$ direction). $R_y$ represents a reactive internal force perpendicular to the joint surface that is largely mitigated by articular cartilage.

To get the resultant reaction force, $R$, we realize that $R_x$ and $R_y$ are sides of a right triangle and $R$ is the hypotenuse, so we can use the Pythagorean theorem:

$$R = \sqrt{R_x^2 + R_y^2} = 274.0\,\text{N}$$

First, notice that the resultant bone-on-bone force is dominated by the $F_{D_x}$ component of $F_D$. This is so because there were no external forces in this particular situation and the muscle force is large compared to the segmental weights. Second, if we look at the magnitude of $R$ relative to Mrs. Buckler's body weight (i.e., $R/W \times 100$), we see that $R$ is 42% of the body weight.

We can use the inverse tangent to determine the orientation (angle relative to the $x$-axis) of $R$. Let $\theta$ be the angle we wish to determine: $\theta = \tan^{-1}(R_y/R_x) = -8.4°$. This suggests that $R$ is almost parallel to $F_{D_x}$, largely because the $R_y$ component is small compared to $R_x$.

The resultant joint force, $R$, is quite large considering Mrs. Buckler is sustaining only the weight of her arm. We can imagine that if we, as part of a progressive resistance program to strengthen any of her weak muscles, e.g., ask her to hold a 5-kg mass in her hand with the shoulder at 90° abduction, the calculated GH joint bone-on-bone would substantially increase (i.e., to 1.8 times her body weight). Given the degenerative status at her GH joint, we should be concerned about these very large GH bone-on-bone forces that cannot be mitigated well by her degenerated AC. The physical therapist may need to consider more creative ways to strengthen her shoulder girdle muscles and perhaps improve their control, in order to partly resolve her impingement symptoms.

Within the limitations of this statics problem, we make several conclusions. First, it is clear that muscle forces make the primary contribution to the large joint bone-on-bone forces. This is always true because external loads have a mechanical advantage, i.e., larger moment arm, over balancing muscle forces. A more detailed analysis of muscle mechanics later in this text provides additional insight into the potential for force development in the muscle-tendon unit that corroborates our first conclusion. Second, joint forces are complex due to two-dimensional muscle force vectors, producing loading modes that are also multidimensional, i.e., compression, shear, etc. Third, the impaired mechanical response of the articular cartilage secondary to degenerative joint disease is not likely able to distribute large joint loads, which can exacerbate degenerative processes, adding to both joint dysfunction and pain.

## 3.8 Chapter Summary

Mrs. Buckler reported a medical diagnosis of impingement syndrome. The syndrome, as with many other syndromes, includes a cluster of signs and symptoms that needed to be clarified. In this case, we explore several issues/perspectives, including the mechanism of the injury, i.e., fall, medical history, radiographic findings, and physical examination results. In the context of Mrs. Buckler's history, we review the relevant anatomy and biomechanics of the cervical and thoracic spine, shoulder girdle, and particulars of synovial joints. In addition, we examine the structure and material properties of bone and periarticular structures, e.g., articular cartilage, etc. We simulate a joint loading situation to provide some insight into an aspect of potential treatment interventions. Our problem-solving process uses a frequently employed hypothetical-deductive or reductive approach, which reveals that the common orthopedic problem we were presented with was actually quite complex. We conclude that resolution of this problem requires consideration of multiple, in some ways competing, parts that demand a creative intervention regiment.

**Reflection** At the conclusion of each case we study, we want you to think about your thinking. That is, how did you take the given, i.e., your previous knowledge or experiences, or discovered information and use it to reach a conclusion? We want you to consider that you will be using one or more of the following different approaches to thinking and decision-making: *deduction*, *induction*, and *abduction*. What exactly are these forms of thinking? Considering their roots might help. All three words are based on the Latin *ducere*, meaning "to lead." The prefix *de*- means "from"; therefore deduction derives *from* generally accepted statements or facts. The prefix *in*- means "to" or "toward," so induction leads you *to* a generalization or abstraction. The prefix *ab*- means "away"; thus, given some information, you take *away* the best explanation in abduction.

Most of the time we use deductive thinking. However, in many instances, we do not have enough known facts to make definitive conclusions. Moreover, we often make conclusions based on insufficient information or previous ill-conceived notions or assumptions. One of our expectations is that with these cases, you will practice inductive and abductive thinking. With regard to inductive thinking, think about it this way. Each case is presenting you with some particular information regarding an individual. In the process of studying each case, you will examine known information related to anatomy, physiology, and mechanics and combine it with the given information in an attempt to clarify the clinical problem. When you are finished with that process, how then can you take some of the specific information and generalize or abstract it to future similar (or perhaps dissimilar) cases? For example, in the statics problem for which we were interested in solving for the glenohumeral/glenoid fossa joint forces when Mrs. Buckler was asked to abduct her shoulder, we find that the joint bone-on-bone forces are largely related to the force of the deltoid muscle. As we explore this solution closer, we discover that the reasons for this were, perhaps, related to the length-tension of the muscle-tendon, but more likely related to the poor mechanical advantage of the muscle-tendon compared to the mechanical advantage of the external forces. It is important for us to derive an estimate of the glenohumeral joint forces because of her radiographic diagnosis of degenerative joint disease (DJD) and our discovery of the implications of DJD with regard to a joint's inability to mitigate large joint forces. Inductive thinking takes the set of facts from this particular case and suggests the following generalization that can be applied to any case involving an individual who has joint degenerative disease: "any test I perform as part of my examination or any action/movement that I ask of the patient that involves large muscle forces will create large forces at the joint those muscles cross; as a result, I will need to consider how these large joint forces may be manifested by the patient." This is just one of many inductions that can be made in Mrs. Buckler's case. Now you try to generate an additional list of generalizations related to this case.

## Bibliography


Argatov I, Mishuris G (2015) Contact mechanics of articular cartilage layers, asymptomatic models. Springer, Cham

Ateshian GA, Mow VC (2005) Friction, lubrication, and wear of articular cartilage and diarthrodial joints. In: Mow VC, Huskies R (eds) Basic orthopaedic biomechanics and mechano-biology, 3rd edn. Lippincott Williams & Wilkins, Philadelphia

Boissonnault WG (1999) Prevalence of comorbid conditions, surgeries, and medication use in a physical therapy outpatient population: a multicentered study. J Orthop Sports Phys Ther 29:506–525

Cheng PL (2006a) Simulation of Codman's paradox reveals a general law of motion. J Biomech 39:1201–1207. https://doi.org/10.1016/j.jbiomech.2005.03.017

Cheng PL (2006b) Response to Dr. Stepan and Dr. Otabal: a mathematical note on the simulation of Codman's paradox. J Biomech 39:3082–3084. https://doi.org/10.1016/j.jbiomech.2006.09.013

Cheng PL, Nicol AC, Paul JP (2000) Determination of axial rotation angles of limb segments, a new method. J Biomech 33:837–843. https://doi.org/10.1016/S0021-9290(00)00032-4

Codman A (1934) The shoulder. Robert E. Kreiger Publishing Company, Malabar

Coronado RA, Alappattu MJ, Hart DL et al (2011) Total number and severity of comorbidities do not differ based on anatomical region of musculoskeletal pain. J Orthop Sports Phys Ther 41(7):477–485. https://doi.org/10.2519/jospt.2011.3686

Cowin SC, Doty SB (2007) Tissue mechanics. Springer, New York

Fox AJS, Bedi A, Rodeo SA (2009) The basic science of articular cartilage: structure, composition, and function. Sports Health 1(6):461–468. https://doi.org/10.1177/1941738109350438

Hearney RP (1998) Pathophysiology of osteoporosis. Endocrinol Metab Clin North Am 27(2):255–265

Hung CT, Ateshian GA (2022) Biomechanics of articular cartilage. In: Nordin M, Frankel VH (eds) Basic biomechanics of the musculoskeletal system, 5th edn. Wolters Kluwer, Philadelphia

Jeon KW (2004) Biology of fibrocartilage cells. Int Rev Cytol 233:1–45

Kapandji AI (2018) The shoulder. In: Kapandji AI (ed) The physiology of the joints, Vol 1, The upper limb, 7th edn. Handspring Publishing, Edinburgh

Kibler WB, Sciascia A (2010) Current concepts: scapular dyskinesis. Br J Sports Med 44:300–305. https://doi.org/10.1136/bjsm.2009.058834

Kibler WB, Ludewig PM, McClure PW et al (2013) Clinical implications of scapular dyskinesis in shoulder injury: the 2013 consensus statement from the "Scapular Summit". Br J Sports Med 47(14):877–885. https://doi.org/10.1136/bjsports-2013-092425

LeVeau B (1977) Williams and Lissner: biomechanics of human motion. W.B. Saunders Company, Philadelphia

Lewis JS, Green A, Wright C (2005) Subacromial impingement syndrome: the role of posture and muscle imbalance. J Shoulder Elb Surg 14(4):385–392. https://doi.org/10.1016/j.jse.2004.08.007

Liebsch C, Wilke H-J (2018) Basic biomechanics of the thoracic spine and rib cage. In: Galbusera F, Wilke H-J (eds) Basic biomechanics of the spine, basic concepts, spinal disorders and treatments. Elsevier, Academic Press, London, pp 35–50. https://doi.org/10.1016/B978-0-12-812851-0.00003-3

Liebsch C, Wilke H-J (2022) Biomechanics of the thoracic spine and rib cage. In: Nordin M, Frankel VH (eds) Basic biomechanics of the musculoskeletal system, 5th edn. Wolters Kluwer, Philadelphia

Loeser RF, Goldring SR, Scanzello CR et al (2012) Osteoarthritis: a disease of the joint as an organ. Arthritis Rheum 64(6):1697–1707. https://doi.org/10.1002/art.34453

MacConaill MA (1948) The movements of bones and joints. I. Fundamental principles with particular reference to rotational movement. J Bone Joint Surg 30B(2):322–326

MacConaill MA, Basmajian JV (1977) Muscle and movements, a basis for human kinesiology, 2nd edn. Robert E. Kreiger Publishing Co., Inc., Huntington

Martin RB, Burr DB, Sharkey NA (1998) Skeletal tissue mechanics. Springer, New York

Mow VC, How JS, Owens JM et al (1990) Biphasic and quasilinear viscoelastic theories for hydrated soft tissues. In: Mow VC, Ratcliffe A, Woo SL-Y (eds) Biomechanics of diarthrodial joints, vol 1. Springer-Verlag, New York

Neumann DA (2017) Chapter 5 – Shoulder complex; Chapter 9 – Axial skeleton: osteology and arthrology. In: Neumann DA (ed) Kinesiology of the musculoskeletal system, foundations for rehabilitation, 3rd edn. Elsevier, St. Louis

Nordin M, Frankel VH (2022) Basic biomechanics of the musculoskeletal system, 5th edn. Wolters Kluwer, Philadelphia

Oatis CA (2017) Kinesiology, 3rd edn. Wolters Kluwer, Philadelphia

Özkaya N, Leger D, Goldsheyder D, Nordin M (2017) Fundamentals of biomechanics, equilibrium, motion and deformation, 4th edn. Springer, Cham

Politti JC, Goroso G, Valentinuzzi M et al (1998) Codman's paradox of the arm rotations: mathematic validation. Med Eng Phys 20:257–260. https://doi.org/10.1016/S1350-4533(98)00020-4

Sahara W, Sugamoto K, Murai M et al (2006) 3D kinematic analysis of the acromioclavicular joint during arm abduction using vertically open MRI. J Orthop Res 24(9):1823–1831. https://doi.org/10.1002/jor.20208

Sipos W, Pietschmann P, Rauner M et al (2009) Pathophysiology of osteoporosis. Wien Med Wochenschr 159(9–10):230–234. https://doi.org/10.1007/s10354-009-0647-y

Thurner S, Hanel R, Klimek P (2018) Introduction to the theory of complex systems. Oxford University Press, Oxford

Tondu B (2018) A robotic model for Codman's paradox simulation and interpretation. C R Mecanique 346:855–867. https://doi.org/10.1016/j.crme.2018.04.016

White AA, Panjabi MM (1990) Physical properties and functional biomechanics of the spine. In: White AA, Panjabi MM (eds) Clinical biomechanics of the spine, 2nd edn. J.B. Lippincott Company, Philadelphia

Wolf SI, Fradet L, Rettig O (2009) Conjunct rotation: Codman's paradox revisited. Med Biol Eng Comput 47:551–556. https://doi.org/10.1007/s11517-009-0484-6

Zhibin S, Tianyu M, Chao N et al (2018) A new skeleton model and the motion rhythm analysis for human shoulder complex oriented to rehabilitation robotics. Appl Bionics Biomech 2018:2719631. https://doi.org/10.1155/2018/2719631

# Elbow-Forearm: Elbow Pain

# 4

## 4.1 Introduction

For this case we examine the mechanism of injury related to pitching a baseball, delineating the external forces and moments that may explain the injury to the elbow. Additionally, we explain the pathomechanics of each problem identified and the potential functional impact of each. Finally, we solve a statics equilibrium problem that aids in the understanding of the role muscles play in contributing to joint bone-on-bone forces and their potential contribution to joint pain during resisted muscle testing.

When working with athletes, it is essential to have a profound understanding of the biomechanics (kinematics, kinetics, and muscle activity) of the primary movement patterns that they perform on a daily basis. After we review the medical history and physical findings and examine the relevant anatomy of the elbow, we study the details of the mechanism of injury: throwing/pitching. To get insight into why particular tissues are predisposed to injury related to the pitching act, we extend our study of the structure and material properties of several additional biological tissues.

## 4.2 Case Presentation

Brad had become the star pitcher for his little league baseball team by the time he was 10 years old. He often played outfield on days he was not pitching. He stopped pitching at age 13 because of pain in his left elbow. Brad continued playing baseball, but only as an outfielder, which may have extended his playing days. However, by age 16 his elbow pain was so severe he could no longer perform and finally consulted a physician.

Brad had an athletic, mesomorphic body habitus, with well-defined upper extremity muscles, bilaterally (height = 182.9 cm; mass = 82 kg; body mass index (BMI) = 24.5 kg/m$^2$). Physical examination of the right shoulder and elbow/forearm and right and left wrist/hand function found they were normal; however, left glenohumeral external rotation was greater than the right by 20°, with concomitant reduced internal rotation by 12°. He demonstrated full active and passive left elbow flexion range of motion but had a 10–15° restriction of active and passive elbow extension. There was a mild restriction in forearm pronation and supination range of motion. The elbow carrying angle on the left appeared slightly greater than on the right. Brad's left lateral and medial elbow pain was constant. Some of his symptoms were reproduced with valgus stress testing of the left humeroulnar joint. This stress test also produced mild laxity with an elastic end feel and mild deep pain on the lateral side of his elbow. The varus stress test of his left humeroulnar joint was normal. Palpation of the left upper arm revealed point tenderness along the medial and lateral joint lines, over the medial epicondyle of the humerus and the ulnar groove, and at the radial head. Upper extremity sensory neurological tests, i.e., dermatomal and peripheral nerve regions, and reflex tests were normal, although he reported intermittent prickly feelings around the hypothenar region and little and ring fingers of this left hand. Resisted manual muscle testing demonstrated normal strength and general elbow pain with testing of the elbow and wrist flexors and extensors, as well as forearm supinators and pronators. Radiographs showed a large area of erosion of articular cartilage of the capitellum, enlargement of the radial head and humeral shaft, and closed displaced avulsion fracture of the medial epicondylar epiphysis (Figs. 4.1 and 4.2).

### 4.2.1 Key Facts, Learning Issues, and Hypotheses: Left Elbow Pain

Brad had restricted left elbow and forearm range of motion, increased carrying angle, reproduction of symptoms and laxity with valgus stress testing, and radiographic evidence of

**Supplementary Information** The online version contains supplementary material available at https://doi.org/10.1007/978-3-031-25322-5_4.


G. J. Alderink, B. M. Ashby, *Clinical Kinesiology and Biomechanics*, https://doi.org/10.1007/978-3-031-25322-5_4

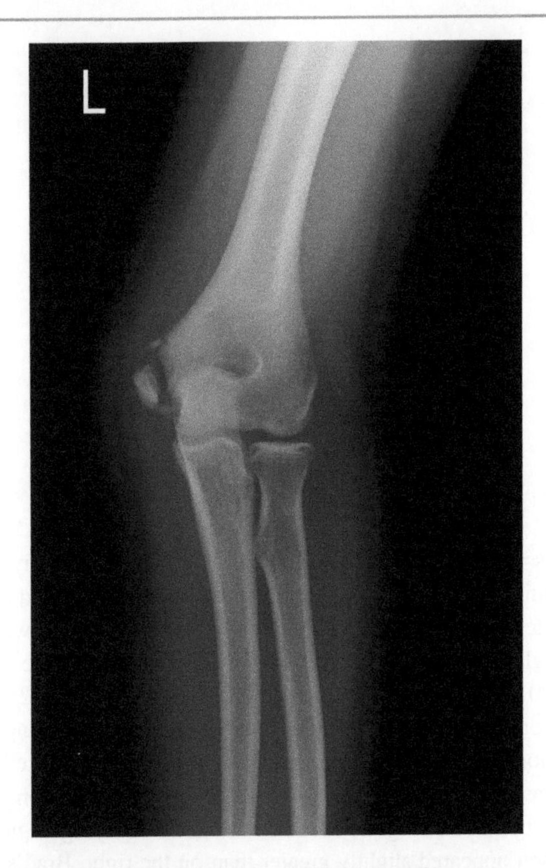


**Fig. 4.1** Medial epicondylar growth plate abnormality

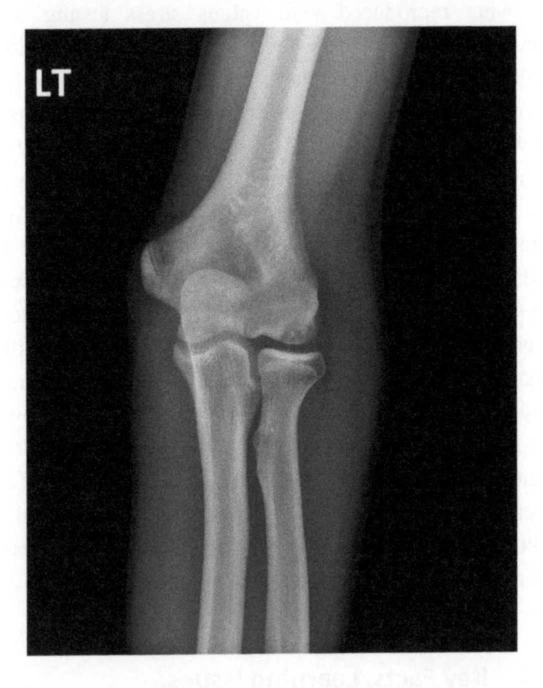


**Fig. 4.2** Osteochondritis dissecans of humeral capitellum

radio-capitellar osteochondritis dissecans and medial epicondylar injury. Let's review the anatomy and osteo- and arthrokinematics of this region in an attempt to clarify these findings.

#### 4.2.1.1 Elbow and Forearm Structure

The elbow and forearm complex include the humeroulnar, humeroradial, and proximal and distal radioulnar joints (Fig. 4.3). Together these joints contribute to the motions of flexion, extension, pronation, and supination; all necessarily coupled to allow effective completion of the tasks of daily living, e.g., reaching, pulling, opening/closing doors, and throwing, to name a few. Let's look at some of the details of the elbow anatomy that are most relevant to this case. The lateral and medial epicondyles of the distal humerus serve as attachment sites for periarticular extrinsic ligaments and forearm muscles. In developing athletes, the epicondyles are the sites of long bone growth and are not fully fused, which is relevant to the study of Brad's elbow pain. Let's digress to briefly look at the growth centers around the elbow (Fig. 4.4). For example, note that, for boys, the medial epicondyle apophysis appears between the ages of 7–9 years, but the growth plate does not fully fuse until approximately age 17 years. *Brad's radiographs showed abnormal changes in his medial epicondylar growth plate, which may be contributing to his elbow pain.* Parents and coaches of little league baseballers need to be mindful of the potential external stressors placed on all of these developing ossification centers, particularly the medial epicondylar growth region. We revisit this topic in more detail later.

Let's get back to our general discussion of key landmarks of the elbow-forearm complex (Fig. 4.3). Laterally, the capitulum of the distal humerus has an ovoid convex shape that articulates with the concave fovea of the radial head. Compared to the humeroulnar joint, the humeroradial bony anatomy provides minimal sagittal plane stability to the elbow. Paradoxically, however, it serves as a good buttress to excessive ulnar abduction, due to the natural angulation of the proximal radius, the depth of the radial fossa, and the linking of lateral and medial extrinsic ligaments. Medially, the sellar-shaped trochlea of the humerus is largely congruent with its ulnar partner, the reciprocally shaped trochlear notch. Anteriorly, the coronoid process of the ulna matches with the humeral coronoid fossa, while posteriorly, the olecranon process of the ulnar fits snugly into the olecranon fossa of the humerus. These anterior and posterior bony configurations/articulations on the medial aspect of the humeroulnar joint provide approximately 60% of the humeroulnar joint stability, with the coronoid acting as a primary boney constraint to excessive movement, particularly at end-range elbow flexion.

The proximal radioulnar joint is formed by the small convex radial head and concave radial notch, located lateral to the trochlear notch. As we follow the radius distally, we see that it widens, and notice that its concave end articulates with the proximal row of the medial carpals, forming the radiocarpal joint. Distally, the convex ulnar head articulates with the reciprocally shaped ulnar notch on the distal radius. The

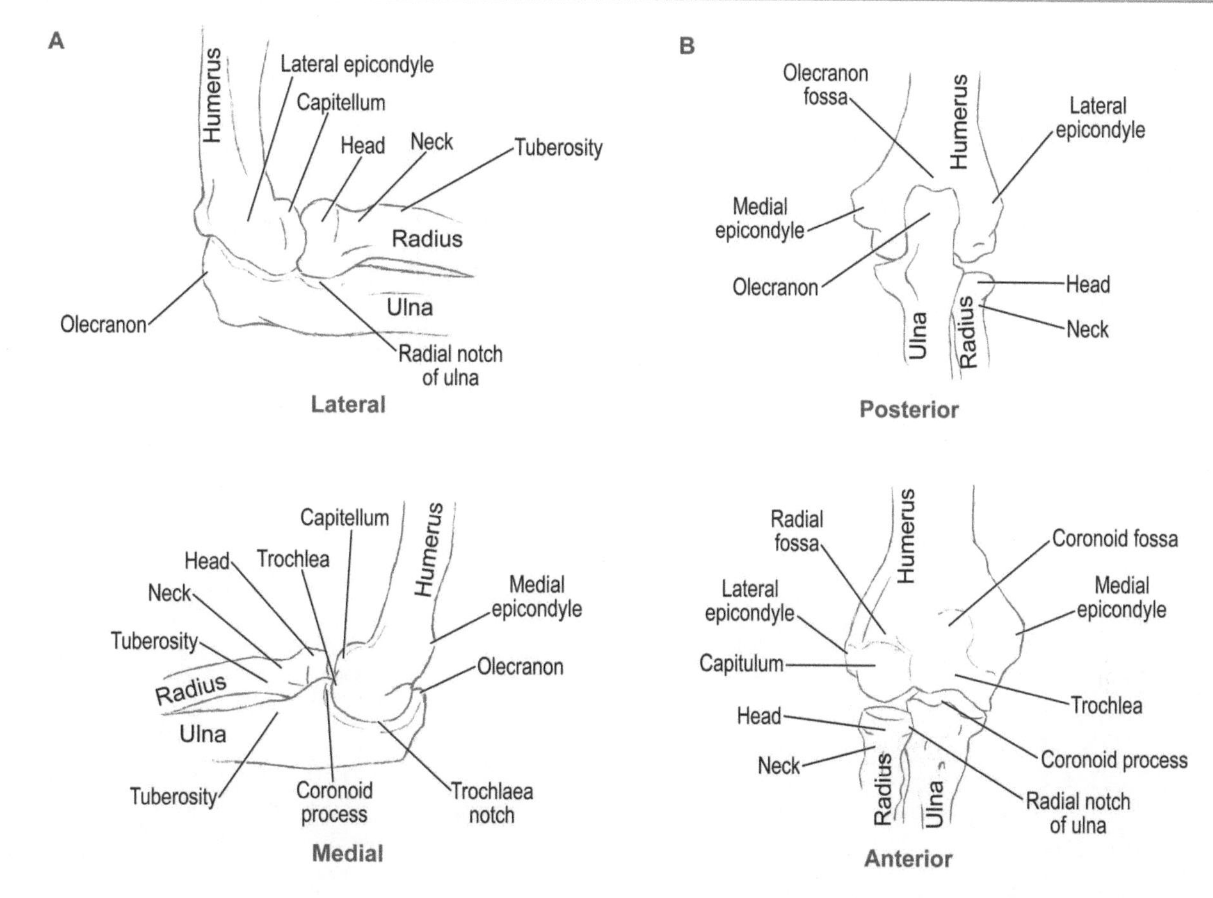


**Fig. 4.3** (**A**) Lateral and medial and (**B**) posterior and anterior views of key bony landmarks of the proximal elbow-forearm complex

articular disc (also named the triangular fibrocartilage), located laterally at the wrist joint, connects the distal radius, ulna, and triquetrum and serves as a primary stabilizer for the ulnar side of the distal forearm and ulnar side of the wrist. The anterior and posterior edges of the articular disc are continuous with the anterior (palmer) and posterior (dorsal) aspects of the radioulnar joint capsular ligaments, with additional links to the ulnar collateral ligament (Fig. 4.5).

Intrinsic ligaments, i.e., joint capsule, enclose the humeroulnar, humeroradial, and proximal radioulnar joints and are lined by a synovial membrane; the thin capsule is reinforced anteriorly (not shown) by dense fibrous connective tissue (Fig. 4.6). The extrinsic ligaments medially include the medial collateral ligament (MCL) complex, which provides additional multiplanar support to the joint capsule. The MCL is primarily responsible for constraining abduction, i.e., valgus, and lateral translation movements of the ulna relative to the distal humerus. The MCL has three bands: anterior, posterior, and transverse. Both the anterior and posterior bands resist ulnar abduction; however, the anterior bands are the most robust and provide this constraint throughout the entire flexion/extension range of motion, whereas the posterior band becomes most taut in extreme flexion. The transverse band may be a vestigial part of the joint capsule. Since it does not cross the elbow joint line, it cannot directly provide any constraint to excessive movements.

Besides the bony architecture, the MCL provides major constraints to forces that abduct the humeroulnar joint during sport-related activities that involve rapid and forceful overarm motions like baseball throwing/pitching. *Given that Brad's signs and symptoms are related to his participation in baseball, it is likely that he has injured his MCL.* We examine the material properties of ligaments and the biomechanics of pitching that contribute to MCL strain in an upcoming section.

The extrinsic ligaments on the lateral side of the humeroradial joint include the radial collateral ligament, which merges with the annular ligament, and the lateral collateral ligament (LCL). Both ligaments constrain excessive adduction, i.e., varus, movement of the radius, and ulna relative to the humerus throughout full elbow flexion/extension range of motion. Because the LCL has fibers that merge with the anterior band of the MCL, these ligaments appear to provide complex dynamic coupling constraints to excessive frontal plane movements of the elbow. Distally, the LCL also provides resistance to excessive external rotation of the radius. Table 4.1 provides a summary of the articular and soft tissue restraining forces during the primary, i.e., flexion and extension,

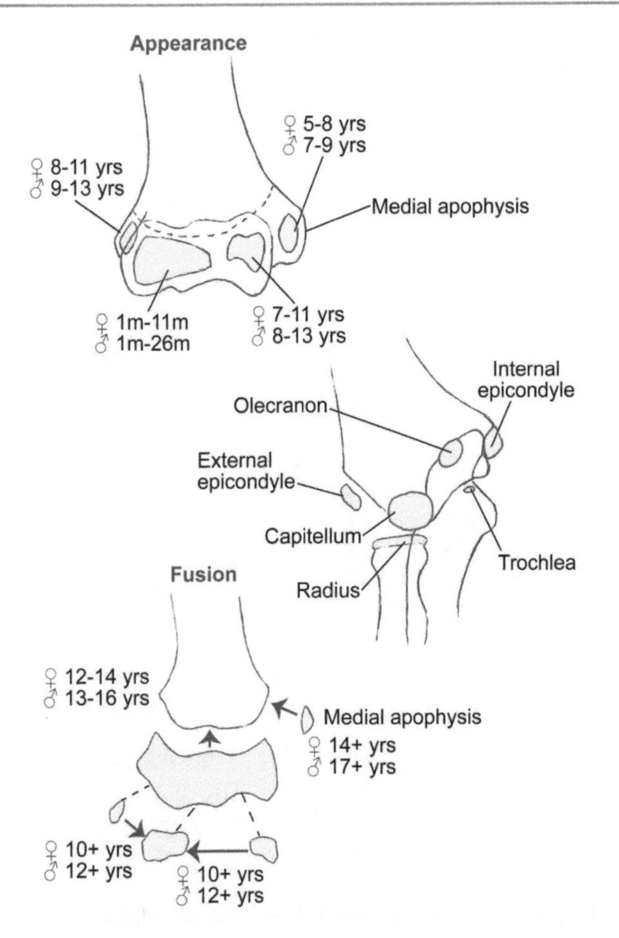


**Fig. 4.4** Ages at which key elbow bony landmarks appear and fuse. Note that the medial epicondylar apophysis does not typically fuse in males until age 17 years

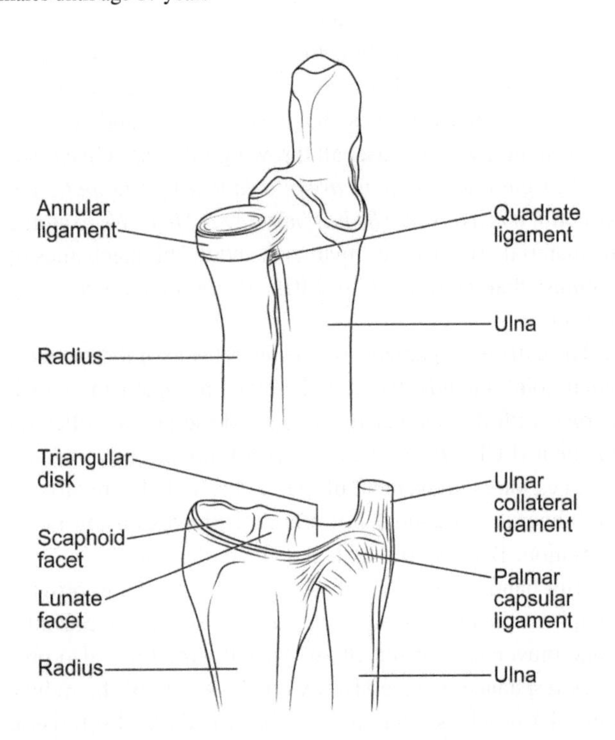


**Fig. 4.5** Proximal and distal bony and ligamentous anatomy of the radioulnar joints

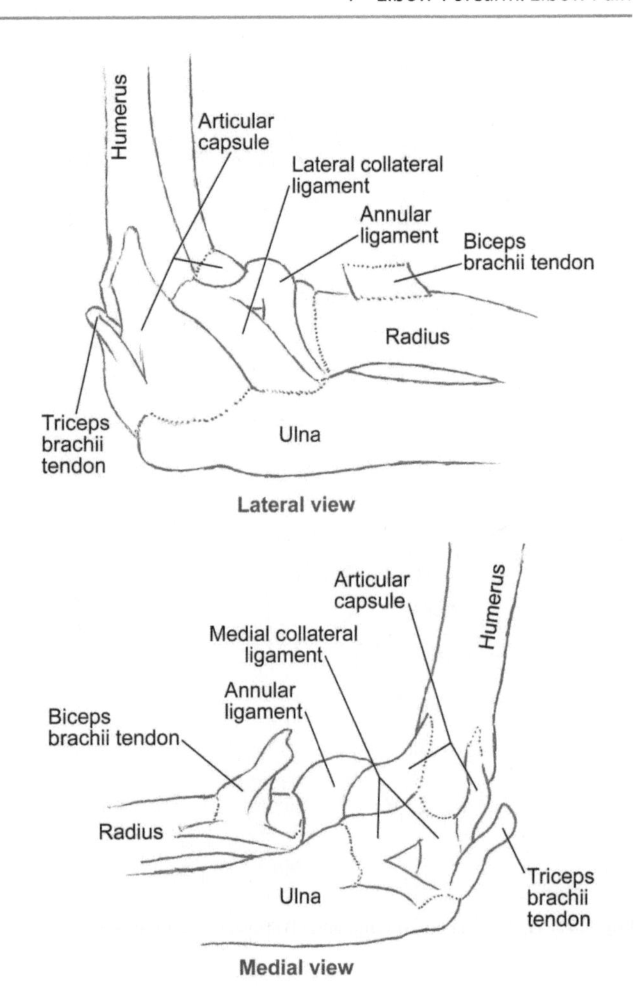


**Fig. 4.6** Lateral and medial views of the periarticular stabilizing soft tissues surrounding the humeroulnar, humeroradial, and proximal radioulnar joints

and secondary, e.g., valgus or abduction, etc., movements at the elbow joint.

In addition to an annular ligament that surrounds the head of the radius snugging it against the radial notch of the ulna, there is a thin quadrate ligament that originates just below the radial notch inserted on the neck of the radius. The quadrate ligament is most tense at the extreme of forearm supination (Fig. 4.5).

The shafts of the radius and ulna are bound by a syndesmosis or interosseous membrane (Fig. 4.7). The primary fibers, comprising the central band, are oriented obliquely from the radius to the ulna. This thick collagenous band has a tensile stiffness similar to some of the strongest ligaments in the lower extremity, e.g., the patellar ligament. The interosseous membrane serves as an attachment site for several extrinsic muscles of the hand but serves primarily as a conduit to transmit large external compression forces generated at the distal radio-carpal joint and radius, e.g., "catching" a fall on an outstretched hand with wrist extended, across to the ulna and proximally to the humeroulnar joint.

**Table 4.1** Percent contribution of constraint to excessive distractive or rotational displacement during elbow flexion and extension

| Movement | Constraint | Distraction | Varus | Valgus |
|---|---|---|---|---|
| Extension | MCL | 12 | – | 31 |
| | LCL | 10 | 14 | – |
| | Capsule | 70 | 32 | 38 |
| | Articulation | – | 55 | 31 |
| Flexion | MCL | 78 | – | 54 |
| | LCL | 10 | 9 | – |
| | Capsule | 8 | 13 | 10 |
| | Articulation | – | 75 | 33 |

Modified with permission from Rosenthal et al. (2022)
*MCL* medial collateral ligament complex, *LCL* lateral collateral ligament complex

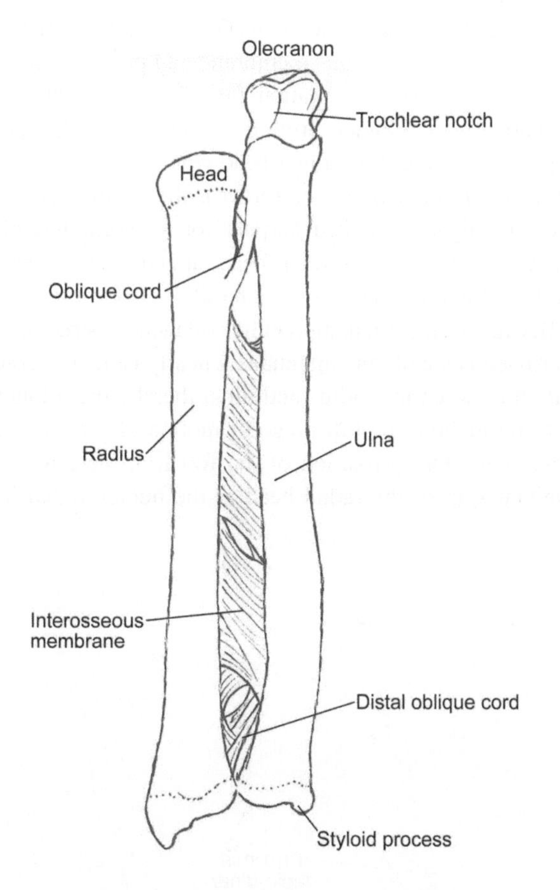


**Fig. 4.7** Interosseous membrane showing central, oblique, and distal oblique cords

#### 4.2.1.2 Osteo- and Arthrokinematics

The elbow is often described as a hinge or modified hinge joint, with one degree of freedom. Although the primary motions at this joint are in the sagittal plane, i.e., flexion and extension, movements in the transverse and frontal plane are critical to normal, pain-free function. One clue to the complexity of the kinematics of the elbow joint is evident when we examine the pose of the elbow joint complex from a frontal perspective when an individual is standing in an anatomical position, i.e., elbow extended with the forearm supinated (Fig. 4.8). Note the slightly abducted ulna relative to the long axis of the humerus. This pose is related to the asymmetrical shape of the humeral trochlea, i.e., the medial lip resting slightly more distal, which creates an oblique axis of rotation that pierces the humeral epicondyles, infero-medial to supero-lateral. This valgus angle approximates 5–15° in adults and is termed the carrying angle, i.e., cubitus valgus. Evidence for the six degrees of freedom of movement that is available at the elbow joint is demonstrated when, as the elbow moves from extension to flexion, the ulna adducts (not shown); that is, the carrying angle is no longer apparent when the elbow is held in a flexed position. Excessive cubitus valgus may result from trauma, such as a fractured medial epicondylar growth plate. *Brad demonstrated a mild increase in the carrying angle of his throwing arm, which may be related to the throwing stress-related injuries to his MCL or medial epicondylar growth plate.*

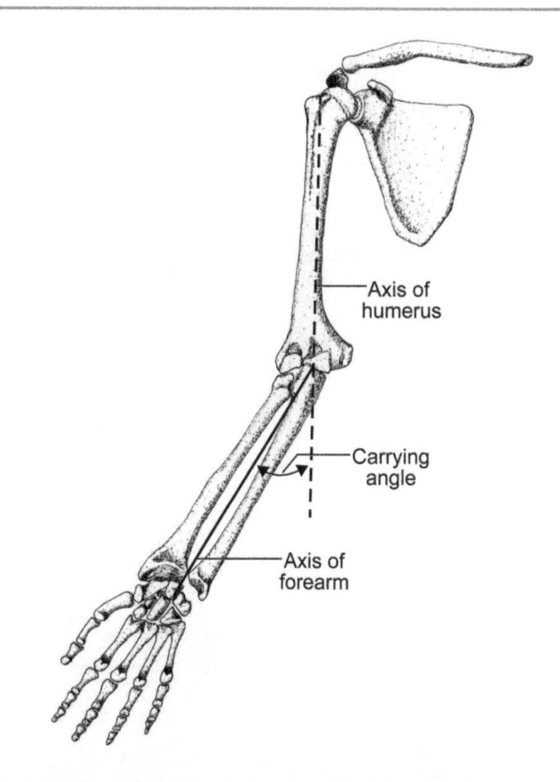


**Fig. 4.8** The carrying angle (also called cubitus valgus) is evident when the elbow is extended. This angle is measured as the medial angle between the respective long axes of the humerus and forearm

Flexion and extension are the primary motions of the elbow. These movements are essential, obviously, for pushing and pulling, lifting, reaching, and throwing activities. Considering the anatomical position as our reference, flexion is the osteokinematic, i.e., movement of the ulna in an antero-superior direction, movement of the ulna and radius relative to the humerus; elbow extension can be described as a posteroinferior movement of the ulna and radius. Arthrokinematically during flexion, the ulna rolls and glides antero-superiorly relative to the convex humeral trochlea. Note that in the sagittal plane, the sellar-shaped ulnar trochlear notch is concave (Fig. 4.9).

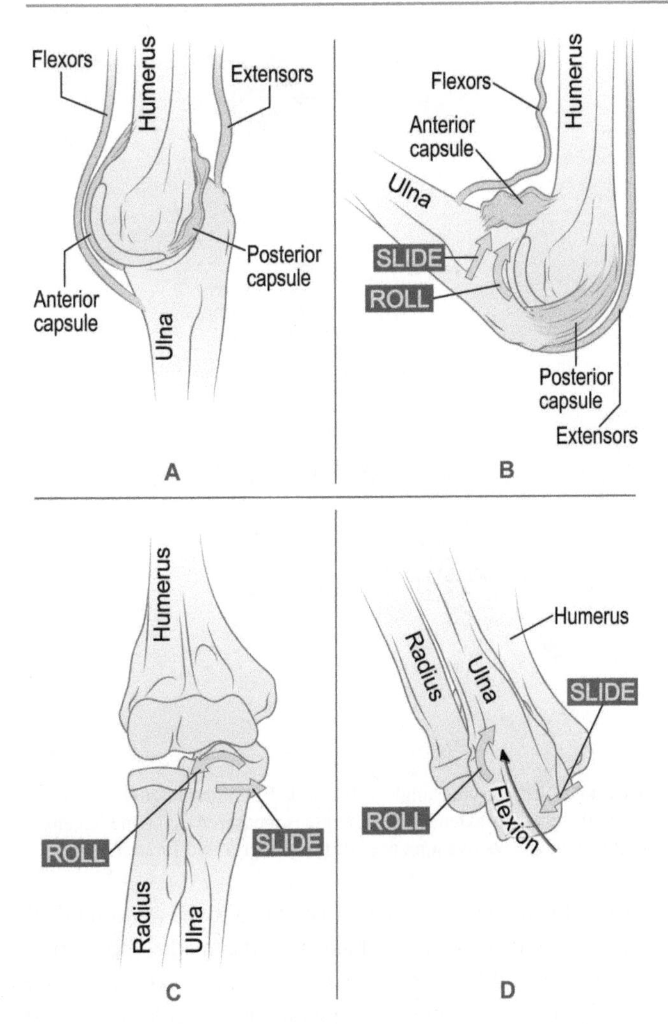


**Fig. 4.9** Sagittal view of (**A**) extended humeroulnar joint, (**B**) roll/glide action associated with flexion at the humeroulnar joint, (**C**) frontal view; when the elbow extends, it is notable that the ulna abducts, with its corresponding roll/glide, and (**D**) roll/glide associated with adduction of the ulna during flexion

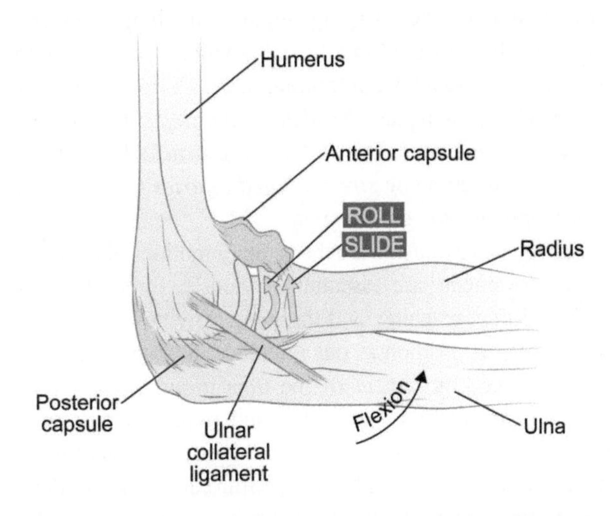


**Fig. 4.10** Sagittal lateral view of the humeroradial joint during elbow flexion illustrating the direction of roll/glide. Note the relative reduced tension in the anterior capsule during flexion

Concomitantly during elbow flexion, the concave radial head also rolls and glides antero-superiorly relative to the capitulum of the humerus (Fig. 4.10). Since the ulna adducts as the elbow flexes, there is necessarily an accessory joint motion between the ulna and humerus in the frontal plane. Recall, that in the frontal plane, the trochlea notch of the ulna is convex, while its humeral partner is concave. Therefore, during the osteokinematic medial movement of the adducting ulna, the trochlea glides in the opposite direction, i.e., lateral, relative to the humerus (Fig. 4.9). Since the radius is going along for the ride, i.e., following adduction of the ulna, it moves medially, while at the joint surface, the radial head rolls/glides medially.

As noted previously, the radius and ulna are kinematically linked by the interosseous membrane and proximal and distal joints. The axis of rotation for forearm pronation and supination runs obliquely from the head of the radius proximally through the distal ulnar head (Fig. 4.11). During pronation (and supination, not shown), the radius, triangular disc, and ligament-linked carpals rotate circumferentially around a relatively fixed ulnar head; at end range pronation, the distal ulna translates dorso-laterally.

Because of the orientation of its collagenous structure, the interosseous membrane maintains a nearly isometric configuration between the radius and ulna, thereby providing stability throughout the full range of motion of pronation and supination. During rotation of the forearm, there is a concomitant spin of the radial head at the humeroradial joint.

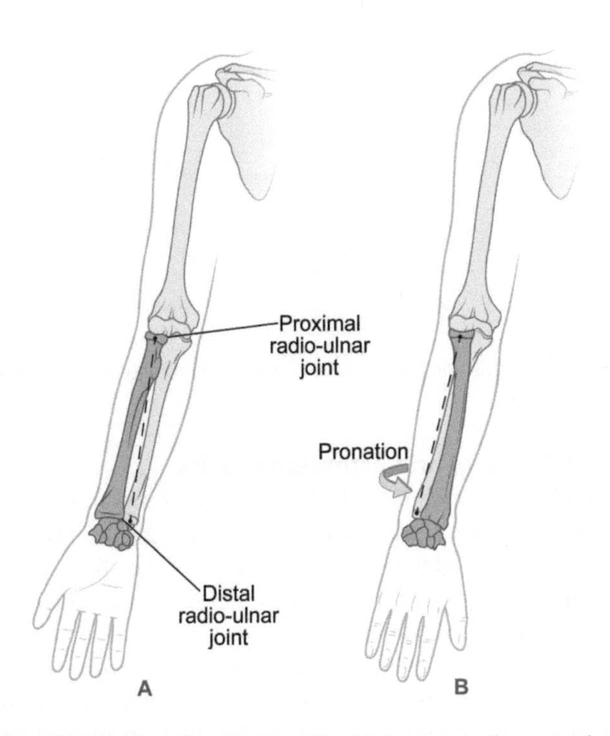


**Fig. 4.11** Position of supination (**A**) with the arm in the anatomical position and (**B**) forearm pronation, i.e., palm facing posterior (dorsal). Note the dashed line which represents the approximate axis of rotation for forearm pronation/supination

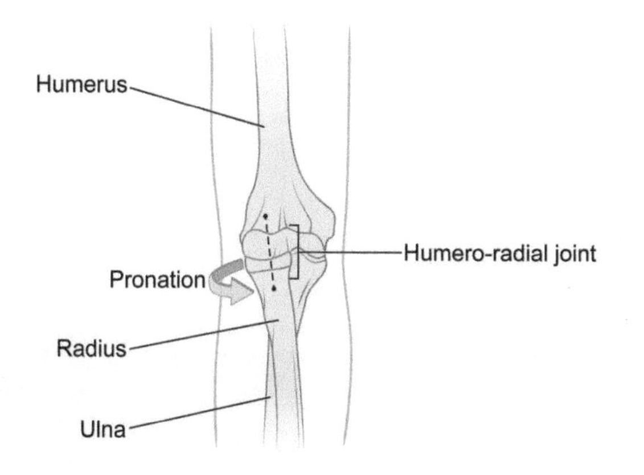


**Fig. 4.12** Frontal view of the humeroradial joint during pronation demonstrating a spin of the radius relative to the humerus about a longitudinal axis oriented nearly parallel to the proximal-distal axis from the radial to ulnar heads

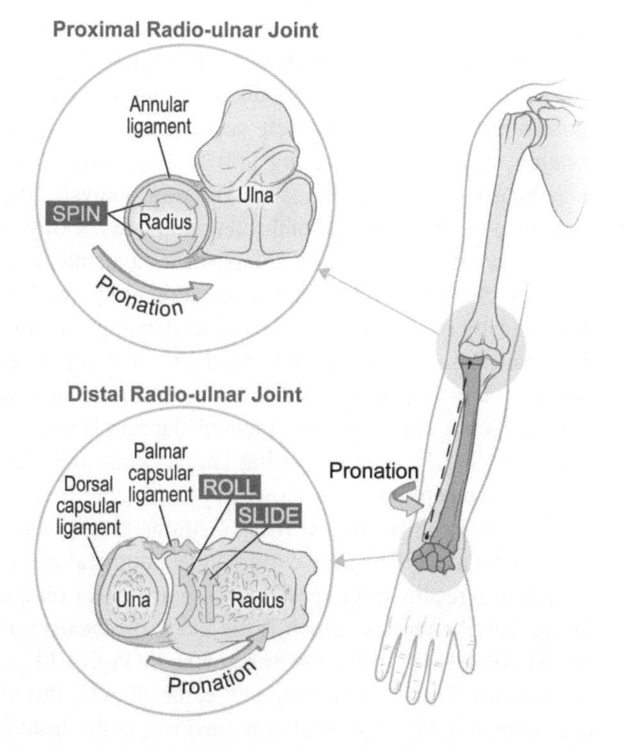


**Fig. 4.13** Arthrokinematics at the proximal and distal radioulnar joints during pronation. At the proximal joint, the radius spins relative to its ulnar partner, while at the distal joint the radius rolls and glides in the same direction as the moving radius

Spinning at a joint surface occurs when the longitudinal axis of the moving bone is orthogonal to its articular partner (Fig. 4.12). During both pronation and supination, there is no accessory glide at the proximal radioulnar joint because of the strong annular fibro-ligamentous constraint. However, distally the concave radius roll/glides in the same direction as the rotating radius (Fig. 4.13). Accessory joint motions at the proximal and distal radioulnar joints are similar during both pronation and supination movements.

Given Brad's mild reduction in elbow extension, forearm pronation, and supination, it is likely that he had impaired arthrokinematics at the humeroulnar, humeroradial, and proximal radioulnar joints. These impairments could be related to repeated first-/second-degree strains to both the intrinsic and extrinsic ligamentous complexes. Additionally, osteochondritis dissecans, and perhaps lateral joint degenerative changes, have likely caused secondary changes, i.e., thickening and related changes in material properties of the elbow joint capsule.

Understanding musculoskeletal clinical problems must always begin with, and return to, the structure and function of the anatomical region of interest. In this section, we focus our study on the bony architecture and the bone and ligamentous constraints to excessive movements of the humeroulnar, humeroradial, proximal radioulnar, and distal radioulnar joints. We also examine the osteo- and arthrokinematics of the same joints, which adds to the foundation for our understanding of injury mechanisms. Based on Brad's history, we hypothesized that his pain was associated with strain injuries to bone, cartilaginous, and soft tissue structures related to pitching/throwing. Although we acknowledge the complex joint dynamics that take place during the act of pitching, which includes the lower extremities, shoulder girdle complex, trunk, and pelvis, we take a reductive approach here to focus on the elbow/forearm complex to somewhat simplify our problem-solving approach. Let's now examine the mechanics of pitching.

## 4.3 Mechanism of Injury: Pitching

Approximately 17–58% of baseball players experience elbow injuries, with pitchers being the most susceptible. Injuries sustained by pitchers tend to be more severe; that is, approximately 73% of injuries in high schoolers requiring surgery were sustained by pitchers. In the past decade, there has been a surge in the percentage of youth baseball pitchers who have sustained serious injuries to their throwing arm. These injuries have been associated with the following risk factors: (1) unsafe participation in practice and games, including insufficient recovery time; (2) suboptimal physical characteristics and inadequate attention to developmental biomechanics (understanding of the impact of external forces and moments on immature skeletal structures); (3) improper pitching techniques, which occur during training; and (4) poor pitching mechanics. With regard to improper pitching mechanics, then, it is important to thoroughly examine, and understand, all aspects of pitching biomechanics as part of the problem-solving approach to clarify Brad's elbow pain.

Motor control experts, kinesiologists, and biomechanists intuitively divide complex movement patterns into phases

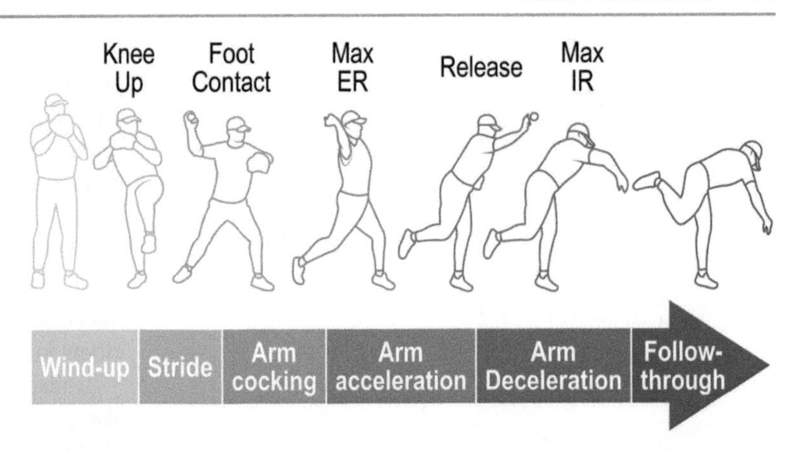


**Fig. 4.14** Phases of throwing for a right-handed pitcher. In reference to the shoulder, Max = maximum, ER = external rotation, IR = internal rotation. Not evident is the extreme valgus (or abducted) position of the humeral-ulnar joint during the arm cocking phase

and operationally define each phase. This practice facilitates both description and analysis. Therefore, let's briefly examine the six phases of pitching: windup, stride, arm cocking, acceleration, deceleration, and follow-through for a right-handed pitcher (Fig. 4.14).

1. The wind-up initiates the generation of whole-body momentum, which is ultimately essential for imparting maximum velocity on the pitched ball. The forces and moments imparted to the throwing arm are negligible during wind-up, but control of the body's center of mass (COM), i.e., total body balance, is critical for everything that follows. Therefore, the synergistic actions of the trunk, pelvis, and upper and lower extremity muscles, and the coordination and timing of the body segment positioning, i.e., dynamic coupling of joint motion, is important in order to optimize pitching mechanics and performance. The wind-up phase is completed when the left knee reaches its maximum height. At that point in time, the right knee and hip are flexed with the pitcher's body weight centered over the left hind- and mid-foot. Control of the body's COM is critical. Loss of this control creates compensations in the kinetic chain that induces abnormal joint forces in the upper extremity joints.
2. The stride begins at the end of the wind-up as the lead leg falls and begins to move toward the target. At the same time, the timing and coordination of arm separation contribute to dynamic balance and places the glove-side arm in a position to keep the trunk closed (prevent premature rotation) and the throwing arm in an early position of strength, i.e., optimally elevated and rotated to maximize muscle length-tension and moment arms. Often during this time period, the young pitcher may lose his balance, as the COM must be controlled both anteroposteriorly and medio-laterally, which requires mature motor control and significant core, i.e., trunk, and lower extremity strength. This phase ends when the lead foot contacts the ground.
3. Cocking of the arm occurs in two subphases: early and late. The early cocking phase is reached just when the stride ends, as the glenohumeral joint horizontally abducts and externally rotates. Late cocking is reached at maximum glenohumeral joint external rotation, i.e., 170–190°. This external rotation is produced primarily by the momentum created by the left pelvic and upper torso rotation, causing the throwing arm to "lag behind." In fact, based on three-dimensional motion analysis, the extreme glenohumeral external rotation in late cocking is only apparent and is really a composite of dynamic coupling of glenohumeral rotation, scapulothoracic retraction, and thoracic spine extension/sidebending/rotation. The late phase of cocking has been linked to shoulder injuries including subacromial impingement, posterior impingement, and superior glenoid labrum anterior-posterior (SLAP) lesions. During late cocking, the long head of the biceps has been shown to provide anterior glenohumeral stability and provide restraint to excessive external rotation. Thus, the extreme glenohumeral external rotation results in increased tension and shear on the biceps-labral complex, which is the likely mechanism of the SLAP lesion. From the schematic in Fig. 4.14, it appears that during the cocking phase, the elbow is flexed approximately 90°, but what is not evident is the frontal plane position, i.e., abduction or valgus, of the humeroulnar joint. We examine the kinematics and kinetics at the elbow during the cocking phase in more detail shortly.
4. Arm acceleration is the very short time period from maximum glenohumeral external rotation to ball release. During this phase, the glenohumeral joint horizontally adducts and internally rotates at a rate of approximately 6000–7000°/s, primarily due to the concentric action of the pectoralis major, teres major, and latissimus dorsi, and forward momentum of the body related to pelvic and upper torso rotation. This phase also sees the elbow extend at a rate of approximately 2000°/s, with concomitant forearm pronation, just prior to ball release. It has

been demonstrated that forearm pronation occurs with all types of thrown pitches, i.e., fastball, curveball, etc.

5. Arm deceleration lasts only a few hundredths of a second, spanning the time from the instant of ball release to maximum glenohumeral internal rotation. The hallmark of this phase is the eccentric action of several posterior muscles of the scapula-humeral complex, e.g., scapular adductors, posterior deltoid, and rotator cuff muscles, etc., which induce very large tensile stresses on the active muscle elements and their tendons. Large tensile forces are also encountered by glenohumeral, humeroulnar, and humeroradial joint capsules.
6. The follow-through phase, a continuation of deceleration, begins at maximum shoulder internal rotation, ending once the pitcher regains a balanced position.

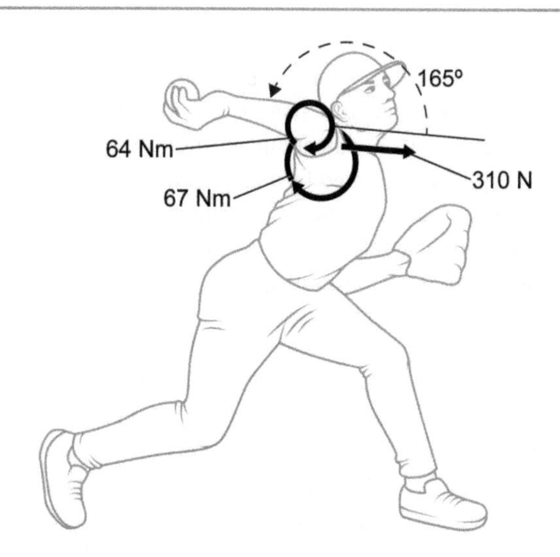


**Fig. 4.15** Glenohumeral and ulno-humeral forces and moments associated with late cocking phase from 26 highly skilled adult pitchers. Shortly before maximum external rotation was achieved, the arms were externally rotated 165°, and the elbow was flexed 95°. There were 67 N-m internal rotation and 64 N-m varus torques at the shoulder and elbow, respectively, and a 310 N anterior force at the shoulder. (Modified with permission from Fleisig et al. (1995))

Of the six phases described above, arm cocking, acceleration, and deceleration are associated with the greatest joint forces and moments at the shoulder and elbow. I briefly described common shoulder injuries related to the acceleration phase and made inferences to strain injuries to the muscle-tendon units of other muscles during the deceleration phase. However, for this case, we focus on the musculoskeletal tissues most susceptible at the elbow, which are typically strained during the cocking phase. First, let's go back and discuss some root concepts related to the pitching act.

Research on pitching has certainly been motivated by human curiosity, but it has mostly been driven by the need to use an understanding of the mechanics of pitching to enhance performance and reduce throwing-related injuries to both little and major leaguers. Atwater was one of the first researchers to propose that the velocity imparted to a pitched baseball is the result of a sequential summation of the acceleration of kinetic links from the legs (distal) to the pelvis/trunk/shoulder girdle (root or proximal), and then to the arm and hand (distal). These movement patterns typically involve each segment initially lagging behind its preceding segment and then accelerating to greater angular velocities while the preceding segment lags behind. This "whip-like" summation of angular velocities requires exquisite neuromotor control, coordinated timing of segment movements, and core balance. In the decades since Atwater's work, many have corroborated her findings using three-dimensional kinematic analyses and inverse dynamics to demonstrate the magnitude of joint forces and moments (Fig. 4.15), and have proposed hypotheses regarding mechanisms that explain how ball velocity is generated, discuss the causes of common injuries, and develop programs to prevent injury and enhance performance. We have learned much about the pitching act, but the inverse dynamic method is reductive (i.e., assumes that "the whole is the sum of its parts") and cannot resolve the individual musculotendon and ligament forces related to the muscular and non-muscular intersegmental torques that increase (or decrease) the angular velocities of coupled segments to produce desired results. On the other hand, more recent research on pitching, using induced acceleration and velocity analysis, has shed more light on the complexity of pitching, which could enhance our ability to improve training methods and, perhaps, prevent injury. Examining the details of induced acceleration analysis and its application to pitching is beyond the scope of this text. Our purpose for introducing the more advanced dynamic analysis methods was to caution the reader about the limitations of inverse dynamics, despite the fact that this review relies primarily on inverse dynamics research of pitching.

Let's focus on the series of events that we speculate have contributed to Brad's elbow injuries. During the arm cocking and acceleration phases, as noted above, the forces and moments produced by the distal segments (ankle to knee to hip joints) produce rapid sequential rotations of the pelvis, upper torso, and shoulder girdle, i.e., root proximal segments, which results in distal segments, i.e., arm, elbow, and wrist, lagging behind; both intersegmental muscular and non-muscular torques contribute to the temporal sequencing relationship between proximal and distal segment rotations. It is notable that although the proximal segments rotate earlier, there is temporal overlap between movements of the proximal and distal segments; that is, the distal segments start to rotate while the proximal segments are approaching large angular velocities. These actions result in the effective transfer of momentum to the distal segments. The lag also contributes to the deformation, i.e., elongation or stretching, of muscle-tendon units that cross the joints, which is stored as potential (or strain) energy that is ultimately used to produce effective stretch-shortening cycles. Although highly efficient (from an energy standpoint), the distal joint lag places several joints in vulnerable positions; that is, proximal and distal joints are forced beyond normal ranges of motion that stress joint constraints, i.e., intrinsic and extrinsic ligaments. The angular velocities at the glenohumeral and elbow joints are large, which results in loads that stress joint structures to a greater extent.

In late cocking, as noted previously, shoulder external rotation angles can exceed 170°. The excessive rotations, and associated large torsional stresses on the humerus, likely contribute to injuries of the biceps-labral complex and posterior rotator cuff. They have also been associated with changes in the morphology of the proximal humerus, e.g., increased retroversion, growth plate injury to the humeral physis, as well as the humeral shaft, e.g., increased circumference/diameter. Obviously, injury to the humeral physis would likely be a painful condition and could lead to future pathomechanics. On the other hand, the changes in humeral torsion and size are normal adaptations, i.e., consistent with Wolff's law, to the chronic overloads associated with the repetitive acts of pitching.

It is notable that in Brad's case, there was radiographic evidence of adaptive humeral shaft changes to his throwing arm. Additionally, the increased glenohumeral external rotation and decreased internal rotation range of motion in Brad's throwing are consistent with increased humeral retroversion, as has been reported in the literature (see the review by Whiteley).

Excessive shoulder external rotation also results in excessive abduction of the ulna relative to the humerus (also often referred to as valgus positioning), which creates large tensile stresses on the medial elbow structures and concomitant compression forces on lateral joint structures. Excessive ulnar abduction creates large internal ulno-humeral varus moments. In this instance, we use the phrase ulno-humeral because it more accurately describes the moment produced at the humeroulnar joint by the distal on the proximal segment. (Note: the elbow moment has also been referred to as a valgus overload or torque, which refers to the moment produced at the humeroulnar joint by external forces). The magnitude of the elbow varus moment has been reported to range from 64 to 120 Nm; most estimates for adult collegiate and professional pitchers approximate 60–75 Nm, high school pitchers approximate 40–50 Nm, and youth pitchers approximate 20–30 Nm. The differences in these moment estimations are likely related to measurement methods, the pitcher's body weight, and strength and pitching skill. Regardless of the factors that distinguish pitchers at various levels, in all instances, the largest contribution (approximately 54% of the total) to the varus moment is provided by the medial (ulnar) collateral ligament (UCL) complex, particularly the anterior band. It is notable that adult cadaveric mechanical testing demonstrated that the UCL withstands a varus moment of 32.1 (±9.6) Nm before ultimate failure, suggesting that the UCL may undergo mild to moderate plastic deformation during the pitching act with every pitch; however, remember that the UCL varus restraint is supplemented by bony and muscle-tendon constraints.

Fortunately, controlling excessive ulnar abduction is not just reliant on the integrity of the UCL. As described previously, some inherent stability of the elbow is provided by bone morphology. For example, the elbow is particularly stable in extension due to (1) the snug fit of the olecranon and coronoid processes in their respective fossae and (2) the natural inclination of the proximal radius, size, and depth of the radial fossa and the compressive forces produced by muscle activation. However, since the elbow is flexed approximately 90° during the cocking phase when the external valgus moment is maximized, ulno-humeral mediolateral joint stability is provided by the reciprocal and highly congruent sellar shapes of the ulnar trochlea notch and trochlea of the humerus, which may not offer much additional support to the UCL.

The final source of medial elbow stability during the cocking and acceleration phases is provided by the muscle-tendon units of elbow and forearm muscles that cross the humeroulnar and humeroradial joints. Recall that during the acceleration phase of pitching, the abducted shoulder internally rotates as the elbow extends. The long head of the biceps has been shown to provide anterior shoulder stability during late cocking, but also appears to control rapid elbow extension during acceleration and forearm pronation just prior to ball release. Thus, as a result of the directions of the large angular velocities during the acceleration phase of pitching, the biceps muscle acts eccentrically at the glenohumeral, humeroulnar, and proximal radioulnar joints. Although the long head of the biceps cannot provide a significant varus moment, its role in controlling extension moments is significant. It is notable that the estimated 300 N compression/shear forces at the elbow joint, somewhat attributed to the eccentric action of the biceps muscle, can provide aid in joint stability but likely also may contribute to strain injury to the articular cartilage at the surface of the humeroradial joint. Surprisingly, despite its small moment arm to do so, the triceps has also been shown to provide some dynamic constraint to external valgus moments at the elbow at the time of peak valgus loading in late cocking/early acceleration. Finally, several researchers have demonstrated the important contribution to the varus moment by the flexor-pronator muscles: brachialis, pronator teres, flexor carpi radialis, flexor carpi ulnaris, and the flexor digitorum superficialis (FDS). Of these, the FDS appears to provide the largest constraining varus moment. Based on experimental simulations, Buffi et al. concluded that without muscle-tendon actuation, the magnitude of the external valgus load on the UCL during pitching would be approximately 330% of its reported failure load. These findings have major implications with regard to sources of possible strain injury, as well as insight into performance training and injury prevention.

Large linear and angular accelerations of the upper extremity and ball in pitching result in large elbow valgus moments, which in turn create excessive tensile stresses on the medial elbow periarticular structures and a combination of compression and shear stresses at both the posteromedial and posterolateral joint compartments of the elbow. Besides the injury to the UCL, other pitching-related injuries include medial epicondylitis, ulnar neuropathy, stress fracture, and damage to growth physes, e.g., medial epicondylar growth plate. Lateral compression contributes to osteochondral defects and includes osteochondrosis of the capitulum, osteochondritis dissecans (OCD) of the capitulum, OCD and enlargement and angulation of the radial head, and Panner's disease. Osteochondrosis is an umbrella term affecting the immature skeleton, whereas OCD refers to inflammation of the osteochondral articular surface. Many have grouped OCD and Panner's disease into the same category, although it appears that these conditions are a continuum of disordered endochondral ossification, with overall outcome dependent on age and severity of the lesion (Fig. 4.16).

OCD is characterized by reduced elbow extension and lateral elbow pain, similar to Brad's presentation. Although Brad did not report this, young athletes often complain of locking and catching at, or within, the elbow during use, signaling the presence of loose bodies, i.e., fragments of cartilage and bone, within the joint. Identification of loose bodies often helps to distinguish OCD from Panner's disease. Brad's OCD may be classified as type II, that is, as open/unstable, characterized as cartilage fracture, with collapse or partial displacement of subchondral bone.

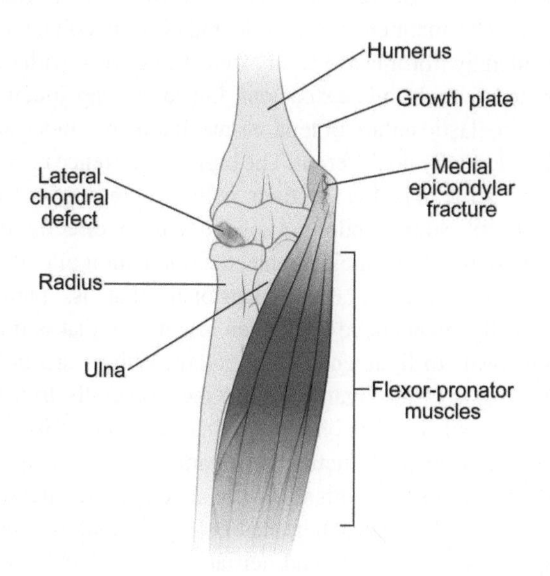


**Fig. 4.16** Medial tensile and lateral compression forces during late cocking/early acceleration can result in medial epicondylar growth plate disturbances and chondral defects laterally

As a result of our examination of pitching mechanics, in general, and the likely mechanism of injury to elbow structures associated with the cocking and acceleration phases, several biological tissues have been identified as potential pain generators, i.e., UCL, medial epicondyle, medial epicondylar growth plate, flexor-pronator muscle-tendon actuators, lateral elbow articular cartilage, and the ulnar nerve. Extensive exploration of bone and joint articular cartilage structure and material properties is presented in the previous case. However, this case has introduced problems suggesting that we also need to examine, in more detail, the structure and material properties of muscle, tendon, ligament, and the peripheral nerve.

## 4.4 Tendons and Ligaments: Structure and Material Properties

### 4.4.1 Structure

Tendons and extrinsic and intrinsic (joint capsules) ligaments are deep connective tissues that, although mechanically passive, play an important role in joint mobility and stability. Tendons attach muscle to the bone and transmit its tension to the bone to produce joint motion, provide dynamic stability, and contribute to the maintenance of posture. Tendons have smaller cross-sectional areas than muscles, making them less bulky, and are thus more easily accommodated by the joints they cross. Despite the smaller cross-sectional areas, tendons can transmit large tensile loads to their insertions. Compared to muscle, tendons are stiffer, have greater tensile strength, and are able to endure larger stresses with small deformations. Ligaments connect bone partners that comprise the joint; enhance stability, i.e., prevent excessive motion; and, in many cases, can also guide joint motion (i.e., produce passive elastic joint moments). Both tendons and ligaments play important roles in motor control due to the associated neural structures, e.g., mechanoreceptors, etc., that provide proprioceptive feedback to the central nervous system (refer to Chap. 3 for details on types of joint mechanoreceptors).

Tendons transmit large tensile loads generated by muscle actions, or passive elongation, to their bony insertions. Tendons are composed of dense connective tissue. Their structure consists of an extracellular matrix (ECM) dominated by a parallel arrangement of collagen fibers, interspersed with few, but metabolically active, fibroblastic cells called tenocytes. Tenocytes, lying in longitudinal rows parallel to the collagen fibrils, are subjected to the same tensile loads. Tenocytes occupy approximately 20% of the total tissue volume, while the ECM makes up the remaining 80%. Though few in number, tenocytes play a critical role in tendon metabolism, i.e., production and degradation of the ECM. As important is the tenocyte's role in responding to mechanical stimuli, i.e., tensile loads on the tendon, which signals (called mechanotransduction) collagen production. The ECM is comprised of water (55–70% of total ECM volume) and solids: collagen (60–80% of solid volume), proteoglycans, elastin, and other proteins. Similar to articular cartilage, the complex collagen-proteoglycan-water relationship contributes to the bi-phasic viscoelastic material properties of the tendon, making it an ideal structure to mitigate and transmit large, and sometimes impulsive, tensile loads.

In the previous chapter, we briefly discuss the role collagen plays in the structure and function of bone and articular cartilage. Because collagen is a common constituent in all biological tissue and appears to also play a predominant role in the tendon and ligament, it seems prudent to digress and detail the structure and maintenance of collagen. The collagen framework in tendons and ligaments is dominated by type I fibers. These are characterized by their capacity to transmit and mitigate large tensile loads. As previously noted, collagen is synthesized by tenocytes in a complex process that contributes to the integrity of the collagen molecule

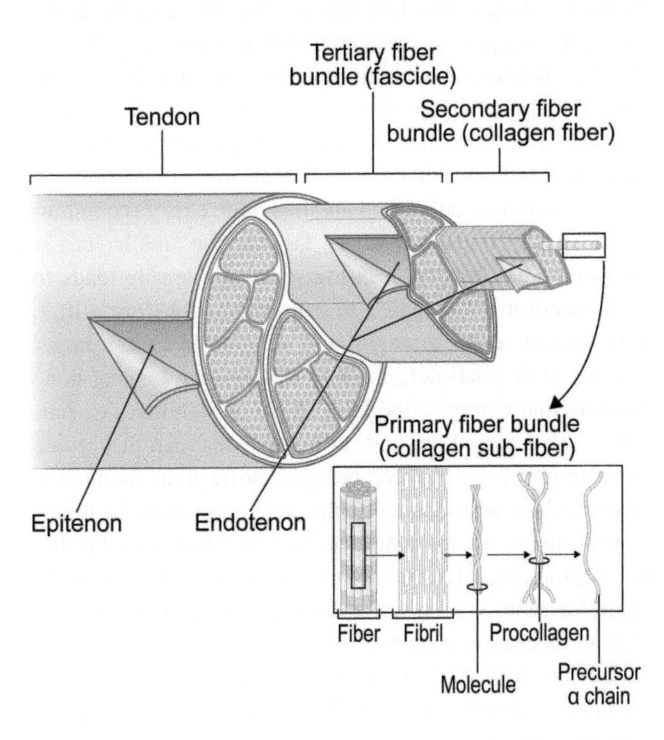


**Fig. 4.17** Schematic representation of collagen precursor alpha chains, procollagen, tropocollagen, collagen fibrils, fibers, and bundles in tendons (and similarly in ligaments). Collagen molecules and triple helices are synthesized and secreted by fibroblasts, e.g., tenocytes, aggregate in the extracellular matrix in a parallel arrangement to form microfibrils and fibrils, which then become arranged into fibers and densely packed fiber bundles

(Fig. 4.17). It turns out that tendons, ligaments, and bones, which are dominated by collagen type I, have similar synthesis and degradation processes.

The synthesis and degradation of collagen in tendons begin at the membrane of the tenocytes (fibroblasts in ligaments). Integrin molecules serve as a vital link between the cytoskeleton and the ECM (Fig. 4.18), serving as sensors that respond to external mechanical loads, i.e., tension, and transferring this sensitivity from outside to inside the cell. The integrin-induced mechanotransduction, along with the presence of several different growth factors, interleukins, and prostaglandins appear to be the main regulators of collagen. The most crucial regulator, mitogen-activated kinase (MAPK), induces signaling to the cell nucleus, which mediates gene expression and the activation of protein synthesis to initiate procollagen synthesis intracellularly and eventually extracellular production of collagen molecules and fibrils. A feature of collagen fibrils is their cross-linking, which provides the stiffness needed to function under tensile loading. Aggregation of fibrils forms collagen fibers, then fiber bundles, which become aligned in the direction of the mechanical load; tenocytes, elongated in rows between collagen bundles, have multiple extensions that stretch extensively within the ECM, thus allowing for multidimensional communication via gap junctions. An interesting note: in an unloaded state, fibril segments have a characteristic "wavy" form referred to as *crimp*, whose biomechanical function is to straighten during the initial stage of tensile loading.

The composition and structure of ligaments are similar to tendons, with a couple of notable differences. Instead of tenocytes, fibroblasts in ligaments are responsible for controlling the metabolism of the ECM. Like tendons, fibroblastic activity is sensitive to mechanical loading, i.e., primarily tension. The maintenance and degradation of collagen are controlled by fibroblasts via an extensive network with other cells and cytoplasmic extensions linked by gap junctions. The viscoelastic nature of tendons and ligaments under loading is essentially no different. The biggest difference between the two structures lies in their collagen fiber orientation (Fig. 4.19). Since tendons transmit tension directly from muscle to their bony attachments, the orientation of collagen bundles is primarily one-dimensional, that is, parallel, whereas ligaments have a collagen orientation that is multidimensional. In ligaments, most collagen fibers are in line with the axis of the ligament since they generally transmit/mitigate tensile loads in one predominant direction. However, because a primary function of ligaments is to constrain excessive motion in joints that have six degrees of freedom, a portion of their substance has collagen bundles aligned along other, secondary and tertiary, axes of movement. Generally, the cross-sectional area of collagen bundles along the secondary axes is less, meaning they bear smaller tensile loads. The orientation and thickness of collagen fiber bun-

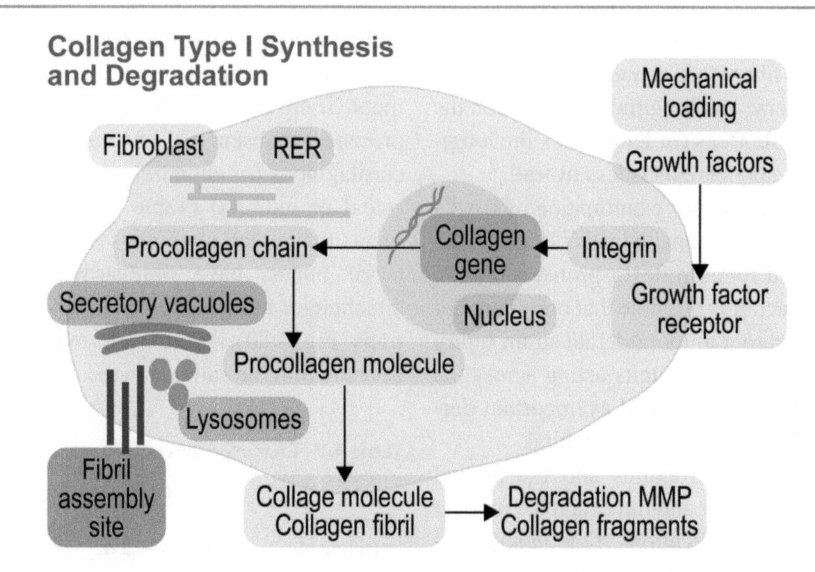


**Fig. 4.18** Schematic showing the process of mechanotransduction as a result of tensile loads within the muscle-tendon complex. In the presence of the mechanical loading and growth factors, a tenocyte responds in a series of events involving signaling integrin, mitogen-activated kinase (MAPK), and the cell nucleus to trigger a cascade of events producing procollagen and collagen fibrils, the primary constituent of the tendon that resists tensile loads. Note: RER = rough endoplasmic reticulum, MMP = matrix metalloproteinases. (Modified with permission from Kjaer (2003))

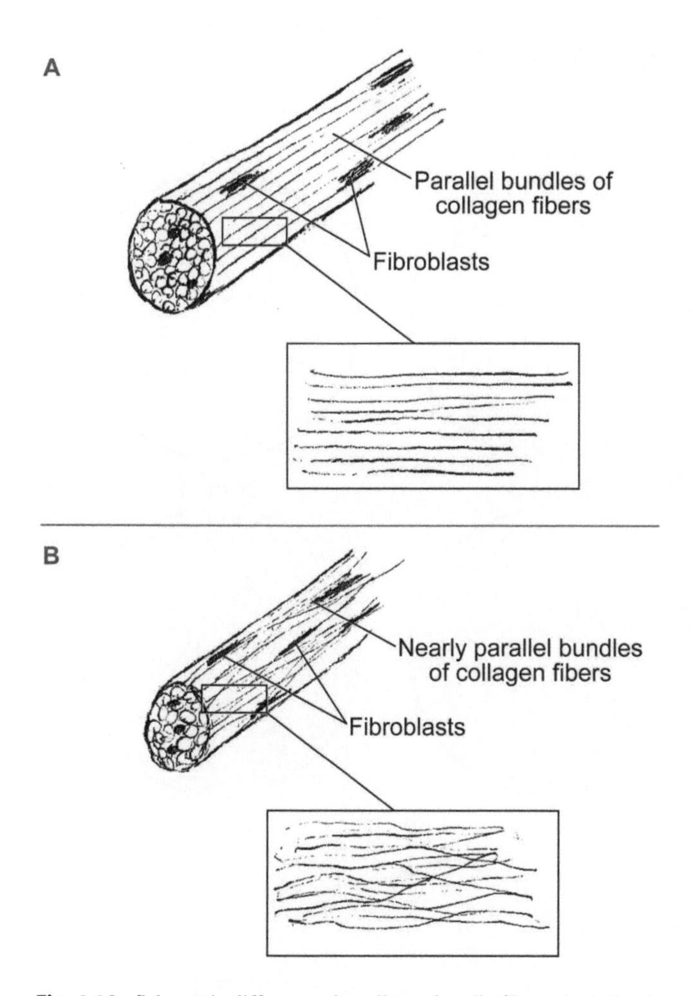


**Fig. 4.19** Schematic differences in collagen bundle fiber orientation in (**A**) tendon and (**B**) ligament. (Modified with permission from Snell (1984))

dles vary among ligaments and are related to specific joint functions.

We know that the internal structure of a tendon, i.e., parallel arrangement of collagen bundles, is well-suited to, and largely dictated by, its function, that is, the transmission of large tensile forces from muscle actions to their bony levers. However, given the differences in the architecture of muscle and tendon, what kind of interface between the contractile machinery of muscle and tendon is required to manage the task of force transmission between them? The interface or transitional structure, known as the aponeurosis, consists of a flat sheet of collagen fibers that narrows and thickens to form a tendon. This sheet, therefore, consists of another parallel arrangement of collagen formed along the lines of stress. Preceding the transitional zone, individual muscle fibers attach to these tendinous extensions at both ends of the muscle, where at each junction the force developed by muscle action is transferred to the ECM of the tendon. The details of the aponeurotic apparatus are as follows: (1) the thickness of the aponeurosis changes along its length, such that it becomes thicker closer to the free tendon, where the junction site is characterized by extensive membrane folding, which increases the surface area available to mediate large forces; (2) the angles between the muscle fibers and tendon collagen are close to zero; therefore, (3) the combination of (1) and (2) enhances junctional strength. Details on human aponeurosis mechanical behavior in vivo are scarce, but it is clear that the tendon and aponeurosis have different functional roles during force transmission. It has been suggested that the muscle force may influence the aponeurosis behavior along its length such that the passive tensile strain of the aponeurosis exceeds that of the free tendon. On the other hand, because the muscle fibers that connect to the aponeurosis can exert both transverse and longitudinal forces on the aponeurosis as the muscle shortens (or lengthens), the longitudinal deformation of the aponeurosis is less overall.

This information is relevant to our examination of Brad's circumstance because of the large tensile loads placed on the long head of the biceps and forearm flexor-pronator muscle group during the pitching act. It is possible that one source of pain generation was related to a mild/moderate strain injury to one or more of the muscle-tendon units acting across the elbow/forearm complex that we identified as important during the cocking and acceleration phases of pitching.

Tendon and ligament outer structures and transitional attachments to the bone are similar and important to consider because there are mechanical implications. Tendon and ligaments each have distinct loose areolar connective tissue coverings that are part of transitional zones. Tendons are associated with three layers or coverings: (1) paratenon is the tissue between the tendon and its sheath; (2) epitenon (or epitendineum) is the synovial sheath around the tendon; and (3) endotenon is the fibrous sheath around each collagen fiber in a tendon. In ligaments, the tissue is referred to as the epiligament or fascia. The term enthesis is used to denote the junction between a tendon or ligament and its bony attachment. There are two different types of entheses related to the presence or absence of fibrocartilage at attachment sites: (1) cartilaginous (or chondral or direct) and (2) fibrous (or periosteal or indirect) (Table 4.2). Usually, tendons and ligaments that attach to epiphyses of long bones, or to carpal or tarsal bones, have fibrocartilaginous entheses, whereas attachments to metaphyses or diaphysis have fibrous connections (Fig. 4.20). At the tendo-osseous junction, the collagen fibers within the endotenon mesh into the bone as perforating

**Table 4.2** Characteristics of entheses

| | Fibrous entheses | Fibrocartilaginous entheses |
|---|---|---|
| Common attachment | Metaphyses and diaphysis of long bones | Epiphyses and apophyses |
| Composition | Perforating mineralized collagen fibers | Four distinct zones |
| Angle of insertion | Insertion angles change slightly during movements | Prone to overuse injuries as the insertion angle changes are greater |

Apostoiakos et al. (2014)

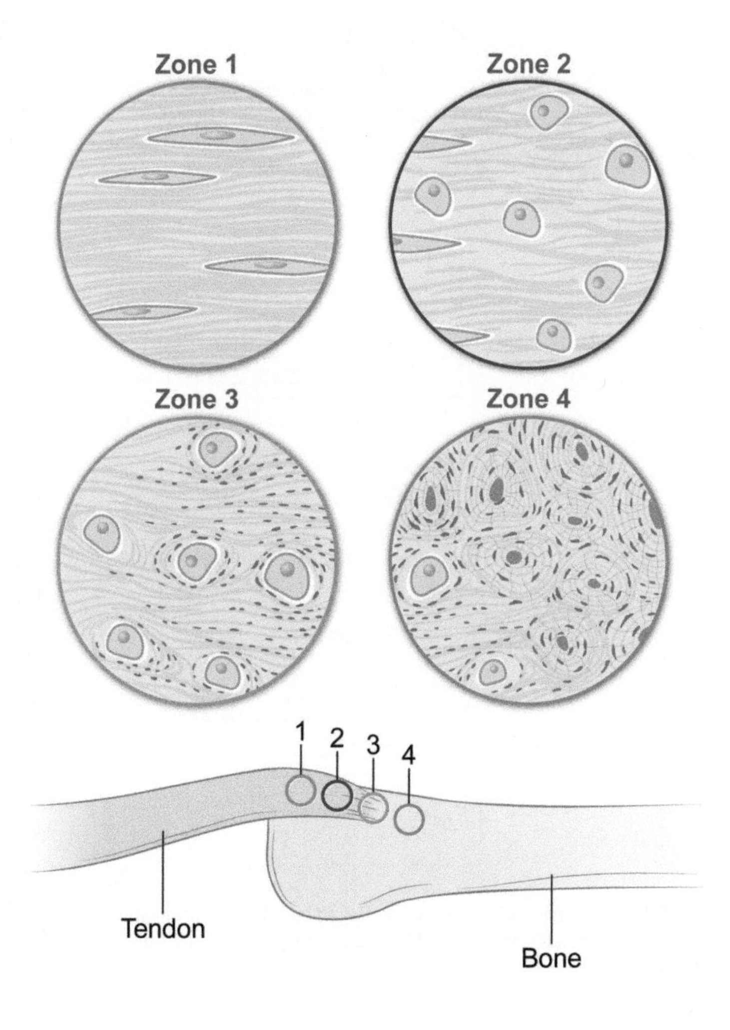


**Fig. 4.20** Schematic illustration of tendon-bone transitions zones

Sharpey's fibers, becoming continuous with bone periosteum.

In the direct attachment arrangement, the structure of the tendon and ligamentous attachments to the bone can be delineated over four zones (Fig. 4.20). The end of the tendon (or ligament) proper (zone 1) is a pure dense fibrous connective tissue composed of pure tendon and highly populated by fibroblasts. The mechanical properties of this zone are similar to those of a mid-substance tendon, with a composition consisting primarily of linearly arranged type I collagen, as well as some type III collagen, elastin, and proteoglycans housed within the ground substance. In zone 2, there is collagenous meshing with fibrocartilage, i.e., uncalcified fibrocartilage. This area is an avascularized region of uncalcified, or unmineralized, fibrocartilage consisting of fibrochondrocytes and proteoglycan aggrecan and types I, II, and III collagen. Because fibrocartilaginous insertions are more complex than those of fibrous entheses, the uncalcified fibrocartilage zone functions as a damper, dissipating stress generated by bending collagen fibers during joint movements, making this region more susceptible to chronic overuse. A tidemark serves to separate the uncalcified and calcified fibrocartilaginous zones. Calcified fibrocartilage makes up zone 3. It consists of mineralized fibrocartilage populated by fibrochondrocytes, primarily type II collagen as well as aggrecan, and smaller densities of types I and X collagen. This zone represents the junction of the tendon to the bone, creating a boundary with the subchondral bone. This area is highly irregular, which is functionally important as it provides mechanical stability to the enthesis. Cortical bone (zone 4) consists of osteoclasts, osteocytes, and osteoblasts that reside in a matrix of type I collagen and carbonated apatite mineral. The gradual increase in the stiffness of the tissue, i.e., tendon or ligament, as it merges into the bone suggests that stress concentrations differ from zone to zone; that is, susceptibility to strain injuries to the tendon (or ligaments) also differs by zonal region.

In younger individuals, the substance of a tendon or ligament is often more resistant to failure than the transitional zones. However, in Brad's case, it appears that a moderate failure of the UCL occurred in its mid-substance. On the other hand, the tension created at the bony insertion of the UCL during the cocking and acceleration phases was also transmitted to the medial epicondylar growth plate, which appears to have been injured. More details of an injury exposure to the medial epicondylar growth plate are forthcoming.

### 4.4.2 Material Properties

Tendons can withstand large tensile stresses created by muscle actions because collagen bundles are aligned along those lines of action. Yet, tendons have the flexibility to serve a pulley-like function as they angulate around bone surfaces and deflect under the retinacula in order to create a more effective muscle action and/or redirect muscle forces. Ligaments, as compared to tendons, often contain a greater proportion of elastin that accounts for greater extensibility yet are noted for their ability to mitigate large tensile forces when they act to constrain excessive joint motion and, at the same time, perhaps contribute to net joint moments. The complexity of joint motion is reflected in the multidimensional structure of collagen bundles in ligaments. Both tendons and ligaments are viscoelastic structures and their material, i.e., mechanical, characteristics depend upon the properties of the collagen bundles, as well as the arrangement and proportion of collagen, elastin, and the collagen-proteoglycan-water framework.

The mechanical properties of tendons and ligaments have typically been evaluated by mounting or fixing specimens, e.g., bone-ligament-bone, in a material testing machine programmed to apply external tensile loads of varying combinations of magnitudes and rates. Other tissue testing methods include the use of buckle transducer instrumentation at insertion sites, robotic/universal force-moment sensor testing, kinematic linkage measurements, magnetic resonance and elastographic imaging, and finite element modeling. Typical tensile tests of a tendon-ligament structure are usually performed along the collagen bundle fiber axis as this is the predominant loading direction in vivo to obtain load-deformation (or force-elongation) curves and stress-strain diagrams (see Appendix H). A load-deformation curve provides information on the tensile capacity of a tendon-ligament structure after loading it to failure (Fig. 4.21). Load-deformation

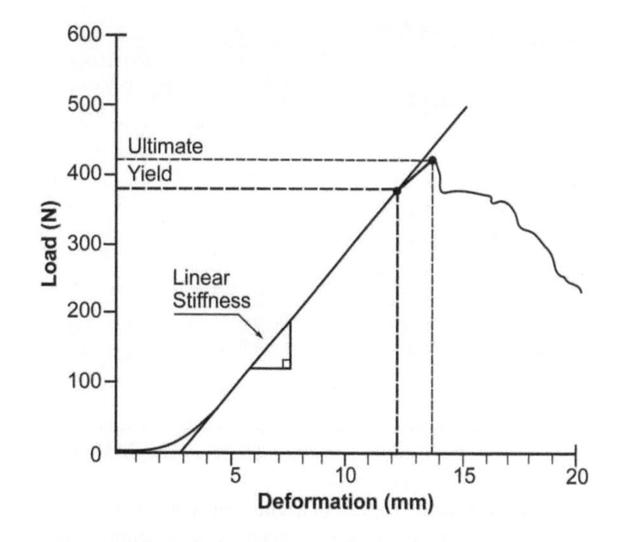


**Fig. 4.21** Typical load-deformation curve of a tendon-ligament structure after loading to failure. The x-axis is the deformation (or change in length) of the structure as a result of loading, measured in mm. The y-axis is the tensile load (or force) applied to the tissue in Newtons (N)

curves have several regions that characterize the behavior of the tissue. The first region, represented by the upward concave part of the curve, is called the "toe" region. This region represents a change in the wavy pattern or crimp of the resting, i.e., relaxed/unloaded, collagen fibers. In the toe region, the tendon-ligament deforms easily under a small load, the collagen fibers lose their wavy appearance, and sliding occurs between fibers and fascicles. As the load increases in magnitude, the tissue deforms further, as shown by a rapid increase in the slope of the curve. This region is referred to as the linear or elastic region, whose slope represents the stiffness (N/mm) of the tendon-ligament structure, and the endpoint is demarcated by the yield point. The ultimate load (N) is the highest load placed on the tissue before failure. The failure point signals the beginning of plastic, or permanent, changes in the deformed tissue and its ultimate deformation or elongation (mm). The ultimate elongation essentially represents a complete failure of the load-supporting ability of the tendon or ligament. The area under the entire curve is the energy absorbed (N-mm) at failure, which represents the maximum energy stored by the tendon-ligament structure.

Because the tendon-ligament structure is anisotropic, i.e., inhomogeneity of size, shape, length, and cross-sectional area, different tissues demonstrate different load-deformation characteristics. Therefore, it is advantageous to evaluate tendon-ligament structural material properties using a stress-strain diagram (Fig. 4.22). Stress can be thought of as the normalized load, and strain can be thought of as the normalized deformation. In a stress-strain diagram, stress is equal to the load divided by the cross-sectional area of the tissue, and strain is the deformation of the tissue as a percentage of the original length of the tendon-ligament structure. From a stress-strain curve, a modulus is determined from the slope of the linear portion of the curve between two limits of strain. Young's modulus, i.e., the modulus of elasticity, represents a proportional relationship between stress and strain ($E = \sigma/\varepsilon$) or the stiffness of the tissue. Tendons generally demonstrate a greater modulus (stiffer) than ligaments likely due to the highly organized, i.e., parallel, collagen bundle arrangement. Tendons function at strains of up to approximately 4% (near their yield point) and rupture at strains near 10%, whereas ligaments deform elastically up to approximately 20% strain. The proportion of elastic proteins in ligaments is important for the deformations that ligaments undergo because they allow the tissue to store energy, which can be used to restore the material to its original shape and size after the stress is removed, and/or contribute to the production of a joint moment. Some of the energy in tendon-ligament structures is lost as part of the loading cycle and is referred to as hysteresis (Fig. 4.23). The hysteresis loop is one feature of the viscoelastic nature of the tendon-ligament structure that allows it to use stored energy that results from applied loads.

Physiologic loading in tendon-ligament structures typically produces 3–5% strains, well within the elastic region of the stress-strain curve. Activities that generally produce strains in the range of 4–8% for tendons and 10–15% for ligaments lead to the development of microdamage to the tendon-ligament structure, which is particularly problematic under conditions of repetitive use without adequate recovery time between episodes of use.

Repetitive use, or overuse, without adequate recovery is the most common extrinsic factor leading to the failure of the UCL in baseball players at all levels. It is likely that Brad's problems began at an early age because, as a player with perhaps more advanced skills than his peers, he pitched and

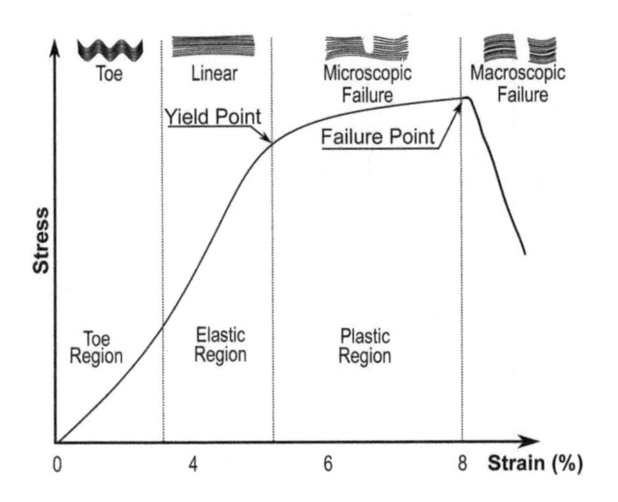


**Fig. 4.22** A stress-strain diagram of a tendon-ligament structure under tensile loading. The x-axis is the percentage of deformation expressed as strain ($\varepsilon$), where the y-axis is the load per unit of area (N/mm$^2$ or MPa) expressed as stress ($\sigma$). The tensile strength (N/mm$^2$) is the maximum stress achieved, the ultimate strain is the strain at failure, and the strain energy density is the total area under the stress-strain curve

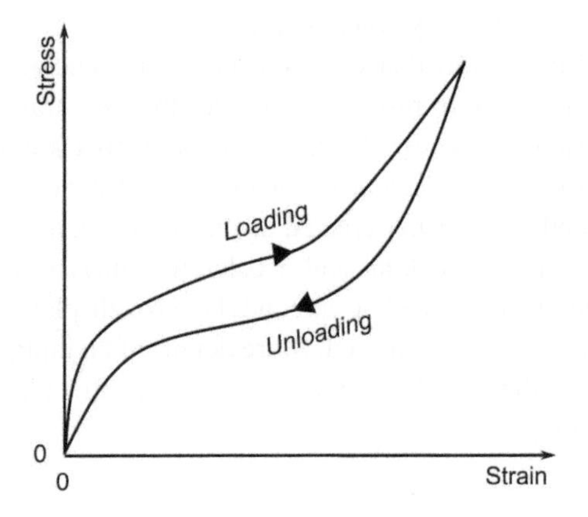


**Fig. 4.23** Schematic showing loading (top) and unloading (bottom) curves in the elastic region of a stress-strain diagram for a ligament. The nonlinear curves form a hysteresis loop, where the area between the curves represents the energy dissipated in the tissue as a result of it being stretched

played a position that required repetitive and long throws and likely practice/played with inadequate recovery periods.

Previously, we noted that all biological tissue exhibited viscoelastic material properties (see Appendix H and Chap. 3). Therefore, the tendon/ligament complexes also display these same rate- and magnitude-dependent behaviors when loaded. Recall that this behavior is largely controlled by the collagen-proteoglycan-water framework. When loaded at different rates, the tendon/ligament structure demonstrates different degrees of stiffness. For example, at higher rates of loading, the linear portion of the stress-strain curve becomes steeper, indicating greater stiffness of the tissue. Although the material property variations depend on the specific structure/function of a particular tissue, generally, with greater strain rates, the tendon/ligament structure deforms less and requires more load to rupture, consequently, storing less energy (i.e., energy is lost) (Fig. 4.24). These features would appear to provide some inherent protection to the tendon/ligament structure during sporting activities, like pitching, which includes rapid joint angular velocities, large muscle forces, and related impact loads on other joint structures.

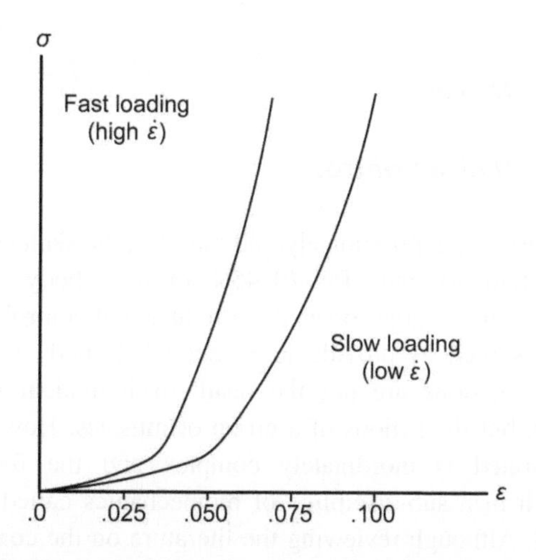


**Fig. 4.24** Illustration of the effect of the rate of loading on the stress-strain curves for tendons. It is notable that faster rate of loading induces a greater stiffness response (slope of the elastic region) by the tendon or ligament

Two additional features of viscoelasticity include the creep and stress-relaxation phenomenon (Fig. 4.25). Stress relaxation has been demonstrated in vitro when the loading of a specimen is terminated well below the yield point of the load-deformation curve. If the specimen is maintained at a constant length over an extended period of time, the internal stress decreases, at first rapidly, until it reaches some sort of equilibrium, i.e., it no longer changes. The creep phenomenon (Fig. 4.26) takes place when loading a specimen is halted safely below the yield point of the load-deformation curve and the amount of load remains constant over an extended period of time. There is a nearly instantaneous deformation upon loading, which over time decreases slowly. A rudimentary understanding of the stress relaxation and creep responses would suggest that they have the potential to protect the tendon/ligament structure from strain injuries. However, since the experimental conditions for demonstrating these phenomena call for controlling the magnitude, i.e., keeping it in the elastic region of the stress-strain curve, and rate of loading, creep and stress relaxation are typically not evident during most activities of daily living. They certainly do not come into play during the act of pitching because of the rapid rate of loading, as well as the repetitive nature of the activity under training or game conditions.

Experimental cyclic loading of the tendon/ligament structure displaces the stress-strain curve to the right along with the elongation (strain) axis (Fig. 4.27). During this type of testing, rate- and magnitude-controlled loads, typically only in the linear region of the stress-strain curve, are applied and released over specific time periods. The shifting of the stress-strain curve to the right suggests the presence of nonelastic, i.e., plastic or permanent, deformations of the tissue that progressively increases with each cycle. We also see that as repetitive loading progresses, the tissue demonstrates a decrease in elastic stiffness as a result of the molecular displacement. These findings are evident even when loading the tissue within physiological ranges. This phenomenon suggests that if loading occurs cyclically (repetitively), in the range of 10–15% strain for a ligament (4–8% for a tendon), microdamage to the collagenous framework (and proteoglycan-water network) could

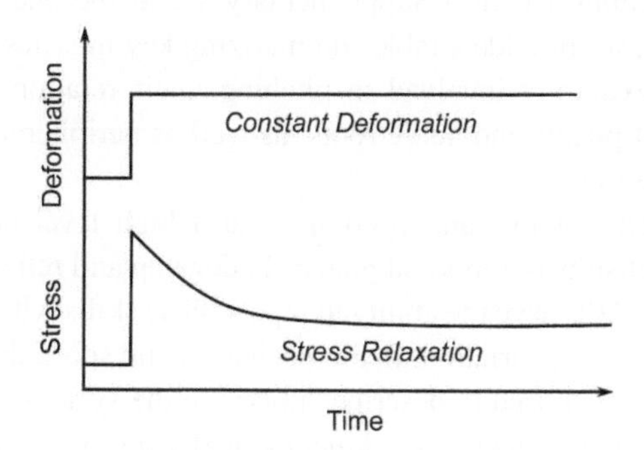


**Fig. 4.25** Stress-relaxation phenomenon

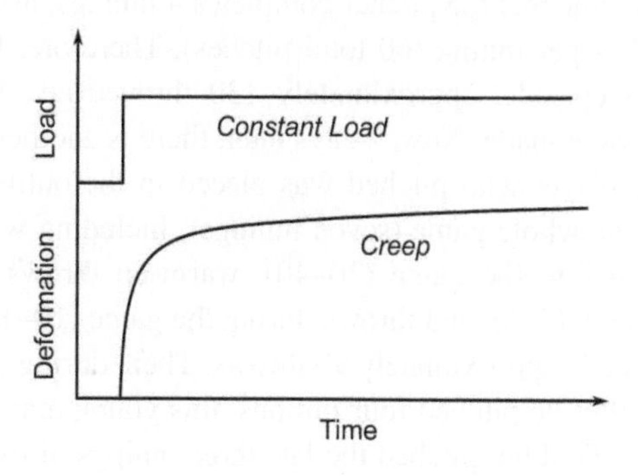


**Fig. 4.26** Creep response phenomenon

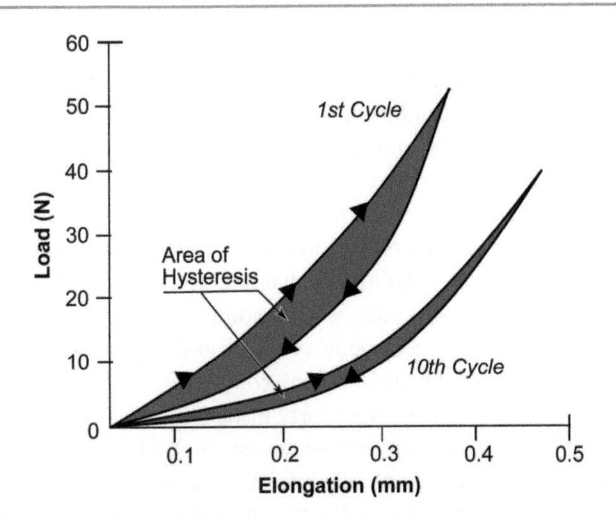


**Fig. 4.27** Typical cyclic loading and unloading of a tendon/ligament structure, where the y-axis depicts the load applied and the x-axis deformation within the tissue. The area within the curves represent energy stored/lost (i.e., hysteresis) in the tissue. Note that with repetitive loading, the total energy lost (i.e., dissipated) is reduced

ensue. Microdamage can result in mild pain or muscle soreness, but does not initially impair the performance, and repairs if the tissue is allowed sufficient time to recover and before another pattern of repetitive loading takes place. Let's see how that might play out for pitchers in general and Brad in particular.

Recall that the results of cyclic loading, just described, occur under controlled experimental conditions. Let's contrast those conditions to the ones imposed by the pitcher preparing (warming up) for and then performing in a competitive game. Under ideal conditions, a pitcher's warm-up consists of general calisthenics, stretching of key muscle groups of the lower extremities and throwing arm, and progressive light to moderate running intensity for a total of about 15 minutes. This is followed by progressive throwing intensity, first over short-to-long distances on flat ground, then finishing on the pitching mound, for a total of 12–15 minutes, completing approximately a total of 50–70 throws. Note: this sequence of supervised warm-up activities rarely occurs with young players (little league through high school). Let's assume that the pitcher completes 4 innings, averaging 15 pitches per inning (60 total pitches). Therefore, for this pitching episode, approximately 130 throws, i.e., loading cycles, were made. Now, 2 days later, there is another game, and the player who pitched was placed in the outfield and played the whole game (seven innings). Including warm-up throws before the game (30–40), warm-up throws before each inning (3–5), and throws during the game (10–15), this player made approximately 90 throws. Then, during a game, 5 days after he pitched four innings, this young man started in the outfield but pitched the last three innings of the game in relief of the starting pitcher (115 total throws). What we have just described is an example of a cumulative, i.e., cyclic loading, exposure to high strain rate loading (a combination of high rate- and magnitude-loading over many repetitions). In other words, the mechanical tolerance of the exposed tissues, i.e., muscle, tendon, ligaments, etc., is exceeded, likely resulting in micro-trauma. Just in a 1-week series of events, we can imagine repetitive loading resulting in plastic deformation of one or more tissues, where the micro-trauma could end up exceeding normal healing and repair processes. It is obvious that this scenario, if continued, would likely lead to conditions of chronic inflammation, susceptibility to more serious injury, and ultimately failure of the weakened tissue.

It is likely that given Brad's advanced baseball skills at his age, he was put in situations similar to what was just described. We know that injuries to the throwing arm in younger players result from a combination of many factors, two of which are often discussed: improper pitching mechanics and overuse. In most cases, overuse, without adequate recovery, is the major factor contributing to shoulder and elbow injuries in little league pitchers. The mechanics of cycling load may explain the overuse phenomenon Brad experienced.

## 4.5 Muscle

### 4.5.1 Motor Control

Comprised of approximately 430 muscles, the skeletal muscle system accounts for 40–45% of total body weight. Muscles create large forces that produce and control movement, as well as provide joint and whole-body stability. These functions are not the result of individual muscle actions, but the actions of a group of muscles. How this is orchestrated is inordinately complex and the focus of research in a sub-discipline of biomechanics called motor control. Although reviewing the literature on the control of muscles cannot be a topic covered in detail in this text, we want to briefly review one overarching perspective on the control of the muscles and movements. This digression is relevant to this case because of the large role muscle plays in the pitching act. In a Supplementary 4.1 at the end of the chapter, we provide a table summarizing key muscles of the upper extremity involved in pitching, their relation to the brachial plexus and nerve roots, as well as peripheral nerve information.

Athletes who want to compete at a high level need to accomplish two important goals: (1) develop and refine their skills and (2) have the ability to repeat those skills with minimal error, i.e., variation. In biomechanics and sport, the term error is often used to describe "noise" in the system or variability in the kinematics, kinetics, and patterns of muscle activation that result in suboptimal performance. However, in the past several decades, those in motor control have sought to elucidate and expand the work of Bernstein whose expression "repetition without repetition" was used to describe consecutive attempts at solving a motor task. In other words, each trial of a movement and set of muscle actions, involved unique, nonrepetitive neural and motor patterns. Thus, researchers in motor control today view variability as a window into the central nervous system's (CNS) organization that produces voluntary movements. One obvious origin of motor variability is motor redundancy, that is, having more muscles available to produce an action than is "needed." This poses a problem with regard to how multiple muscles are recruited (or not) and controlled.

This problem was famously posed as "Bernstein's problem" or the problem of "elimination of redundant degrees of freedom (DoF)" as the central issue of motor control. What does this mean? At many levels of analysis of the system for the production of voluntary movements, there are many more elements that can contribute to the performance than are absolutely necessary to solve the motor task. Thus, since a motor task does not appear to prescribe a single, particular motor pattern, the CNS is confronted with the problem of how to select a particular way of solving each particular problem, i.e., how to move to accomplish a task. Biomechanics might ask, how does the CNS decide how to produce such and such joint kinematics and kinetics related to a particular action, e.g., pitching a baseball?

Since Bernstein's original work, several motor control theories have been developed, e.g., dynamical systems, neurophysiological approach, neuromechanical task-level control, uncontrolled manifold approach, optimal feedback control, and adaptive model control. All of these theories have been motivated by Bernstein's intuition that for a complex system to become controllable, the number of elements subject to individual control, i.e., the degrees of freedom, must be less than the sum of those available. Let's frame this problem around a definition of coordination as "the organization of the different elements of a complex body or activity so as to enable them to work together effectively" and "a cooperative effort resulting in an effective relationship" (Shorter Oxford English Dictionary, 2007). In Bernstein's hierarchical neural model, movements are controlled via the organization and relationship among various elements, which can assure the stability/repeatability of motor performance (Table 4.3). A structural unit within the hierarchy exists and is organized in a task-specific manner such that if an element introduces an error into the common output, other elements change their contribution in order to minimize the original error; this happens at the subcortical level. If that happens as it should, no central control is required, based on the principle of minimal interaction. Systems that demonstrate error compensation among elements are called synergies. Using synergy as an organizing principle, it appears that the DoF problem is not solved by examining how elements, i.e., specific muscles, are selected, but by examining how all of the DoF participate (principle of abundance) in all tasks, assuring both flexibility and stability of the performance.

**Table 4.3** Bernstein's hierarchical levels

| | |
|---|---|
| Level of tonus | Red nucleus, tonic reflexes, musculoskeletal, and fascial systems |
| Level of synergies | The spinal cord, brainstem nuclei, thalamus, and cerebellum |
| Level of space | The striatum, primary motor, and sensory cortices |
| Level of action | Premotor cortex and supplementary cortex |

Bernstein's model was built bottom-up, that is, with the older or more primitive levels of control, e.g., basal ganglia, forming the base of the hierarchy. However, functionally the model operates top-down, with the upper two levels advantaging the functional capabilities of the lower levels so as to decrease the upper level's involvement. In his model, any movement requires at least two levels, leading and background, requiring a complex relationship where the leading refers to the level that controls any given movement and the background level refers to the level, or levels, that provide the necessary support (Table 4.3).

What follows is a brief description of the primary themes of each level, as offered in a review by Profeta and Turvey. The first level, the level of tonus, establishes the communication between the neural and muscular systems. Residing in the spinal cord, this level's responsibility is to prepare the movement structure to respond adequately to commands, i.e., influences, instructions, and constraints, coming from upper levels of movement by altering the facilitation of sensory and motor cells (Fig. 4.28). Level one is never a leading, but only always a background, level. Therefore, the form that a movement takes in solving a particular movement or performance is never defined at the level of tonus. It has been suggested that the traditional neural-centered view of tonus

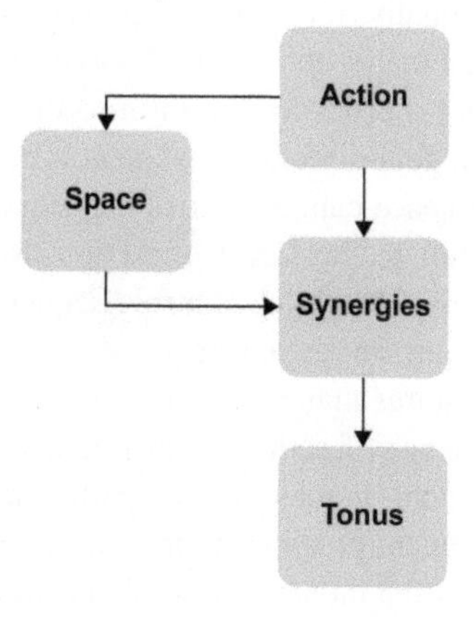


**Fig. 4.28** Representation of Bernstein's levels of construction of movements. Arrows indicate the direction of dominance between levels (Profeta & Turvey, 2018)

ignores contributions from static connective tissue in the maintenance of overall basal tension. Therefore, taking account of all tissue properties, e.g., elasticity, viscosity, and contractility, tonus has been redefined as a mechanical property of the muscular-connective tissue-skeletal system. This recent redefinition of tonus has significant mechanical implications and is explored in detail in a forthcoming section. As a result, it is clear that without the level of tonus, the movement apparatus would not be prepared to respond adequately to upper-level commands.

The level of synergies (or level of muscular-articular links), also a background level, resides in the middle brain. Its role is to constrain the DoF of the motor system, guarantee the coordination of movements, and correct the details of movements. In shaping the coherence of synergistic action of muscles, level two receives proprioceptive input from the entire body. The level of synergies does not receive afferent input from the visual or auditory systems, so it is not capable of assisting in the anticipation of, or compensation for, nonmechanical environmental perturbations.

The level of space can lead to purposeful movements. It exhibits what Bernstein defined as dexterity, that is, "the ability to find a motor solution for any external situation…to adequately solve any emerging motor problem correctly, quickly, rationally and resourcefully." This level refers to how a goal-directed movement is performed in the environment. The level receives optical, acoustical, and haptic input that is used to generate a space field, that is, a perception of the environment based on the interweaving of different sensory sources and previous experiences. The space field performs a dual function: (1) objective perception of the relation between the body and external space and (2) the ability to use external space and supports classes of movements that entail displacing the body or segments of the body from one location to another and classes of movements that involve object manipulation. The two classes of movement problems are solved by the extrapyramidal and pyramidal neural subsystems. Additionally, the level of space controls anticipatory behaviors. Finally, the level of space does not inhibit how a movement is produced but informs the level of synergies that a given goal-directed movement is possible.

The level of space cannot control the sequence of movements; that is left to the Level of Action. Residing in the frontal cortex, this 4th level controls sequences of movements by creating an organization that accommodates a problem that requires many steps to solve. The order of elements in a sequence of actions is of primary importance to the solution of goal-directed action, particularly since each individual movement provides an intermediate response to a specific aspect of the movement problem. Because there are many possible solutions, each with its own unique and intrinsic organization, to an action-oriented movement, flexibility and adaptability at the Level of Action is required.

Within the context of this interpretation of Bernstein's theory of movement, dynamical systems theory challenges the notion that muscle synergies, i.e., patterned and skilled movements, are invariant. In fact, variability (some have called this "noise") in movement systems has been related to the sensorimotor equivalence that arises from the abundance of motor system degrees of freedom. Thus, variability of performance is viewed as more functional, since a consistent outcome, i.e., of a skilled movement, can be achieved by different patterns of joint relations owing to the dynamics of the joint biomechanical DoF. In dynamical systems, spontaneous pattern formation between component parts (or elements of a system) has been found to emerge through processes of self-organization. This self-organization is manifested as transitions between different organizational states emerging due to internal and external constraints, i.e., boundaries or features which form to limit, or constrain in some way, biological actions.

At the behavioral level, Bernstein's suggested solution to the redundant DoF problem was that humans eliminated a portion of the DoF by "fixing" joint articulations to restrict segmental rotation and translation. Thus, movement systems cope with redundant DoF by introducing temporarily strong, rigid couplings between multiple DoF resulting in more controllable single "virtual" DoF complexes termed "coordinative structures." Coordinative structures constrain the inter-connections between the elements of the movement system. On the one hand, coordinative structures reduce the dimensionality, i.e., complexity, of the movement system by allowing humans to exploit the inherent connectedness of the anatomical/biomechanical systems. On the other hand, an essential feature of coordinative structures is that if one of the elements introduces an error into the common output, other elements automatically vary their contribution to movement organization in an effort to minimize the original error. A key feature of dynamical systems is the flexibility to adapt to environmental conditions (external and internal) by fine-tuning coordinative structures based on information, e.g., muscle spindles, mechanoreceptors, etc.

In summary, this review has focused on Bernstein's theory of the construction of the movement, which provides a framework that involves the interaction and coordination of several subsystems in the control of human movement. Let it be said, however, that this review is incomplete, and the interested reader should further explore more recent theories of motor control, e.g., dynamical systems theory, uncontrolled manifold hypothesis, and generalized motor program theory. This exercise provides a perspective and framework on the complex control of the musculoskeletal system involved in the pitching act, and a perfect context to study the microscopic anatomy of muscle, including its neural control (see Supplementary 4.1 at the end of this chapter for details on some of the neural anatomy associated

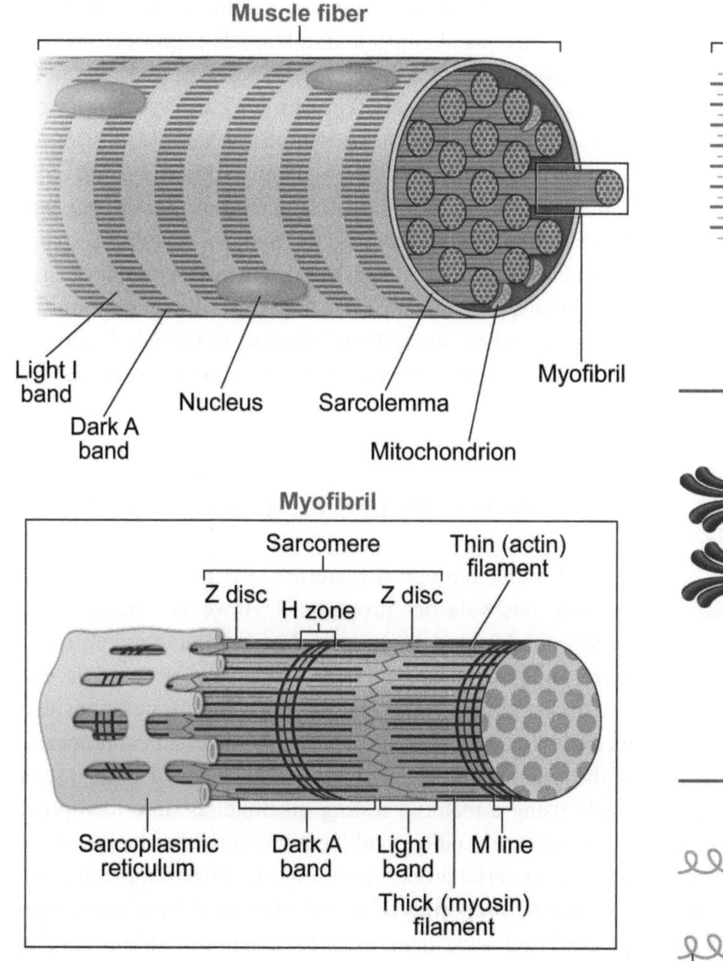


**Fig. 4.29** Schematic illustrating the hierarchical structure of the muscle fiber, including its functional unit, i.e., sarcomere, and myofibrils

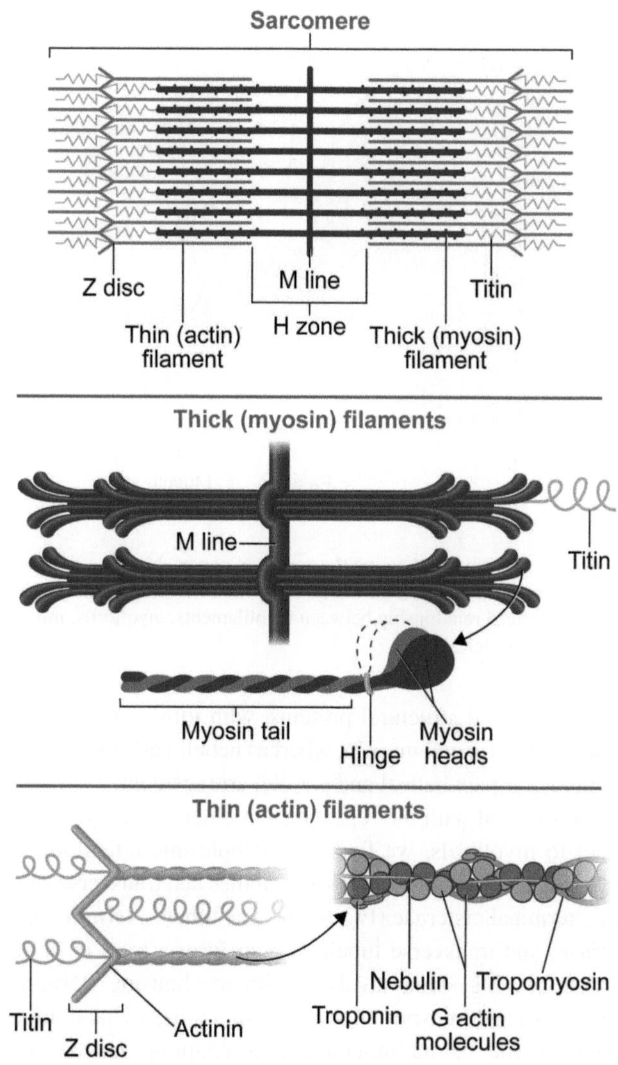


**Fig. 4.30** Schematic of arrangement of thick (myosin) and thin (actin) filaments in relation to the cytoskeletal proteins, titin, and nebulin

with the parts of the upper extremity most important in the pitching act).

## 4.5.2 Structure

What follows is an overview of the microstructure of the muscle-tendon actuator, not a review of muscular anatomy and physiology. This is so, not because the gross anatomy, i.e., muscle origin and insertion, and the excitation-contraction coupling mechanism are not important, but for purposes of focus. We strongly encourage the interested reader to seek out, and use, additional resources to enrich your knowledge of the anatomy and physiology of skeletal muscle.

Skeletal muscle is composed of multinucleated and neurally innervated cells called muscle fibers. Muscle fibers range in length from 1 to 30 cm (Fig. 4.29). Fibers are comprised of many myofibrils, which are surrounded by a plasma membrane, the sarcolemma. The sarcolemma is connected by vinculin- and dystrophin-rich costameres with the sarcomere Z lines, which are part of the extramyofibrillar cytoskeleton. Myofibrils, lying parallel within the sarcoplasm of the muscle fiber and extending throughout its length, vary in number from a few, up to thousands, depending on the diameter of the muscle fiber, and muscle type, i.e., fast or slow twitch. Each myofibril is comprised of repeating sarcomere units, which are composed of thin (actin), thick (myosin), elastic (titan), and inelastic (nebulin) filaments. Actin and myosin are the contractile proteins and the active components that contribute to muscle actions. The titan and nebulin filaments, referred to as structural proteins, play an important role in the generation and transmission of force by (1) generating passive tension when stretched, (2) providing internal and external support and alignment of muscle fibers, and (3) helping to transfer active forces throughout the muscle (Fig. 4.30). Note the intimate relationship between the

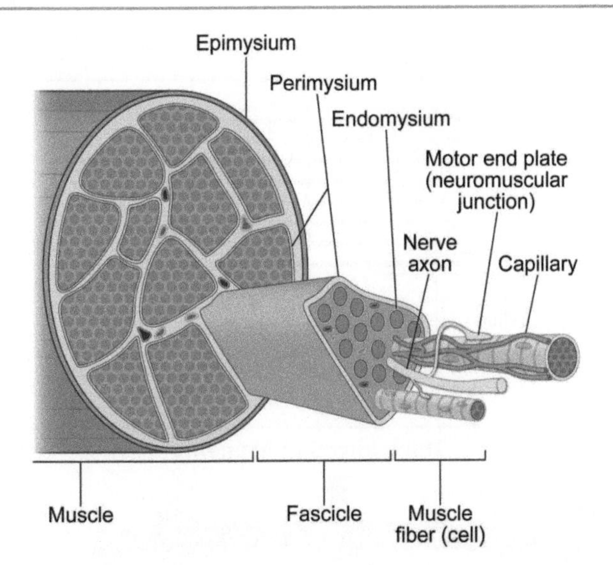


**Fig. 4.31** Schematic showing the course of nerves, blood vessels, and attached motor endplates on the sarcolemma of a muscle fiber within the hierarchical relationship between myofilaments, myofibrils, muscle fibers, and fascicles

contractile and structural proteins, with titin more in series with both actin and myosin, whereas nebulin appears to have a more complex helical and parallel arrangement.

Associated with the repeating sarcomeres, and lying parallel to myofibrils, we find the sarcoplasmic reticulum, an organized network of small sacrotubules and transverse sacs, i.e., terminal cisternae (Fig. 4.31). The tubules, terminal cisternae, and transverse tubule system form a triad of structures, i.e., T-system, involved with the housing (terminal cisternae) and transportation of calcium ions, a vital component of the excitation-contraction coupling mechanism. Excitation-contraction is initiated by an action potential in the central nervous system and propagated via nerve axons to motor endplates, i.e., neuromuscular junction, of each muscle fiber. An endplate potential is then generated that depolarizes the sarcolemma (muscle membrane) with an inward spread of the action potential along the T-system, generating a muscle fiber action potential over the membrane surface, ultimately leading to the final stages of the contraction mechanism within the sarcomere.

In addition to the structural proteins within the sarcomere apparatus, there is a complex system of deep extracellular connective tissues, composed of collagen and elastin. These connective tissues provide structural support, as well as contribute to the non-contractile viscoelastic properties of muscle. The endomysium, immediately external to the sarcolemma, surrounds individual muscle fibers and marks the location of capillaries, the source of metabolic exchange. This thin tissue is composed of a dense meshwork of collagen fibers that has lateral connections from muscle fibers, thereby serving as a conduit of force from the contractile elements. The endomysium also has fascial connections to the perimysium. The perimysium lies within the epimysium and organizes the whole muscle into bundles of muscle fibers or fascicles. The space between fascicles serves as a conduit for nerves and blood vessels. The epimysium is a tough fascial structure that surrounds the entire surface of the muscle belly, giving it form. The epimysium and perimysium collagen framework at the ends of muscles are continuous with those in the tendon, forming the aponeurosis that connects the muscle to a tendon (Fig. 4.31). The extracellular connective tissues are interwoven, thus interconnecting muscle fibers, providing strength and support to the whole muscle, as well as acting as a parallel elastic component to the contractile apparatus.

### 4.5.3 Material Properties: Hill Muscle Model

Like all other biological materials, the muscle-tendon unit exhibits viscoelastic properties. However, because the muscle-tendon unit is composed of both active, i.e., sarcomere, and passive, i.e., aponeurosis, tendon, deep fascia, etc., structures, its material properties are a complex composite of its individual constituents. That is, the mechanical properties of the tendon and the aponeurosis can be tested in vitro separately using a material testing machine, as described previously, and are usually combined, in a sort-of reductive sense, with the material properties of muscle. For example, the passive tensile properties of muscle fascicles have been tested in vitro and have demonstrated maximum strains ranging from 60% to 115%, far exceeding maximum strains of a tendon, i.e., approximately 10%. Combining these bits of information provide some sort of reading as to how the passive properties of the tendon, aponeuroses, and muscle fascicles might interact. However, generally, the mechanical properties of the contractile proteins have been modeled, e.g., the Hill model, in isolation from the passive material properties of tendons. But even in this instance, researchers have attempted to "sum" the material properties of muscle and structural connective tissue to provide an overall explanation for how muscle works. What is unique about muscle, as compared to a tendon, is that it converts signals from the nervous system into force, which, in turn, produces bodily movements, i.e., joint rotations; and there is a neural feedback loop among multiple sensors, e.g., skin, joint, muscle, and tendons, between the periphery and the CNS that contributes to both facilitation and control (Fig. 4.32; Table 4.4). Therefore, what makes muscle more complex is the interrelationship between its constituent parts, i.e., active and structural components, and its ability to respond to an infinite number of stimuli and environmental conditions. A. V. Hill developed a simple, but powerful, conceptual model of muscle function. Although many advances in the understanding of muscle function have occurred since Hill's original work,

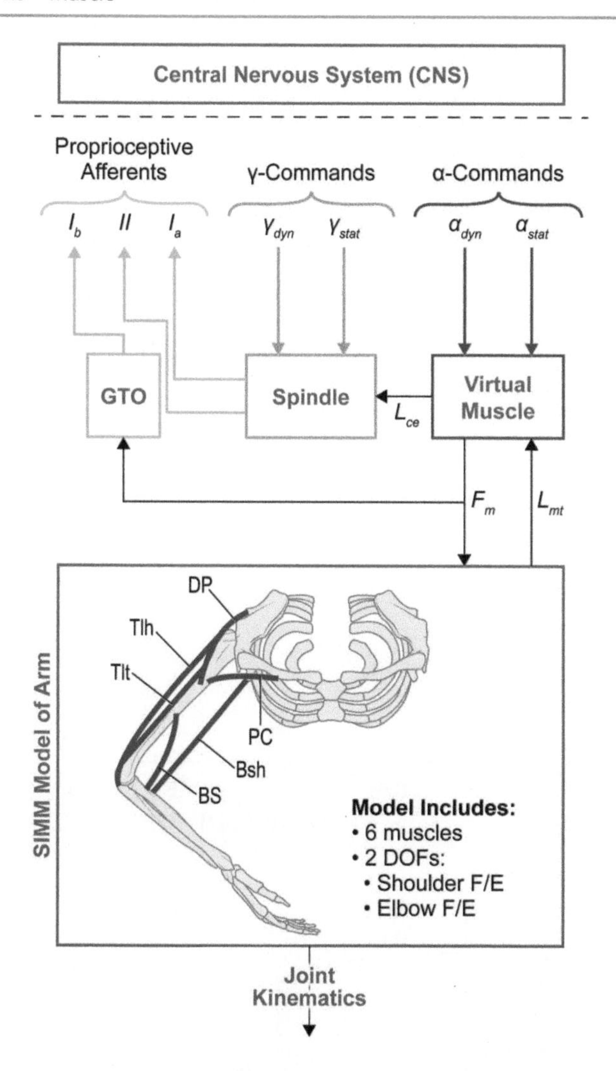


**Fig. 4.32** Schematic of integrated neuromuscular sensorimotor systems model using MATLAB's SIMULINK (MathWorks®, Inc.). This model has two DoF in the horizontal plane (shoulder flexion/extension, elbow flexion/extension), driven by six muscles: PC = clavicular portion of the pectoralis major, DP = posterior deltoid, BS = brachialis, Tlt = lateral head of triceps brachii, Bsh = short head of biceps brachii, Tlh = long head of triceps brachii. Neural commands from motor programs in the central nervous system (CNS) are routed to the muscle via alpha ($\alpha$) motor neurons to manage both static (stat) and dynamic (dyn) conditions. The force ($F_m$) produced by a muscle-tendon actuator is applied to a bone to initiate a movement, i.e., joint kinematics. The amount of force produced is influenced by the length ($L_{mt}$) in contractile elements. The change in instantaneous muscle fascicle length ($L_{ce}$) is sensed by the spindle, which provides input into (via *Ia* and *II* fibers), and receives from, the CNS via the gamma ($\gamma$) motor system. Force magnitude ($F_m$), i.e., contraction forces, is also sensed by the *Ib* fibers of the Golgi tendon organ (GTO), providing additional sensory information to the CNS. (Modified with permission from Nan and He (2012))

his model is still considered the standard for describing the fundamentals of muscle mechanics.

The original Hill model (Fig. 4.33) consists of a contractile component (CC), series elastic component (SEC), and parallel elastic component (PEC). Let's agree that, in general, models serve to simplify complex structural functions, which inherently contains limitations. The Hill model does just that, representing muscle behavior rather than structure, although we can see how individual model components may coincide with particular muscle structures, e.g., the CC ≈ sarcomere unit. Let's first consider the CC. The contractile component represents the "active" elements that convert neural signals into force production, whose magnitude depends on four relationships: (1) stimulation-activation (SA), (2) force-activation (FA), (3) force-velocity (FV), and (4) force-length (FL). Stimulation-activation has to do with how the neural signal relates to the contractile apparatus; in model parlay, stimulation is the input, and activation is the output. Briefly, alpha motor neuron input from the central nervous system triggers motor unit action potentials (MUAP) that are carried to the transverse tubule system, which stimulates the release of calcium ions into individual sarcomeres, i.e., stimulation. The actin-myosin complex responds to the calcium influx by changing its resting state to an activated state in which cross-bridge formation is capable of producing force, i.e., activation. Stimulation-activation takes the system to a state of force potential, that is, an activated state in which force can be produced, rather than an actual force level. Force-activation is a conceptual relation in that it is assumed there is a linear relationship between activation and actual force (e.g., measured in Newtons); therefore, 20% activation equals 20% force. An underlying question regarding both the SA and FA relationships is how much activation and force is produced for a given stimulation. The answer to this question is complex, beyond the scope of our review for the elbow case but can be sensed by reviewing Fig. 4.32.

The force-velocity relationship may be one of the most important properties of the CC in regard to the magnitude of force production (Fig. 4.34). Although Fig. 4.34 was not drawn from Hill's work, it represents the essence of his results. Hill's research, building on earlier contributions that reported on a relationship between muscle work/energy and speed of movement, examined the force-velocity relationship related to isometric and concentric muscle actions in an animal model in the first half of the twentieth century. Hill's famous equation for a rectangular hyperbola:

$$(P+a)(v+b)=(P_0+a)b$$

demonstrated the force-velocity relationship, where $P$ and $v$ represent the CC force and velocity at a given instant in time, respectively, and $P_0$ represents the force level the CC would attain at that instant if it were isometric. Hill conceived the values $a$ and $b$ to be muscular dynamic constants, which represented energy liberation. Hill noted that the dynamic constants were both muscle- and species-specific and dictated the shape of the rectangular hyperbola and its intercepts on the force ($P_0$) and velocity axes. For example, it has been

**Table 4.4** Sensory receptors involved with central nervous system control of skeletal muscle

| Group | Sensory receptor | Function | Stimulus of receptor |
|---|---|---|---|
| Ia | Muscle spindle (primary) | Increases excitability of agonist; decreases the excitability of antagonist | Rate of muscle stretch |
| II | Golgi tendon organ (GTO) | Decreases excitability of agonist; increases excitability of antagonist | Muscle-tendon force |
| II | Muscle spindle (secondary) | Increases excitability of agonist; decreases the excitability of antagonist | Muscle stretch |
| III | Mechanoreceptor | Increases cardiovascular and ventilatory output; inhibits central motor drive | Change of intra-muscular pressure |
| IV | Metaboreceptor | As above | Change in muscular metabolism |

Note: Roman numerals represent the classification of the nerve fiber associated with a particular receptor

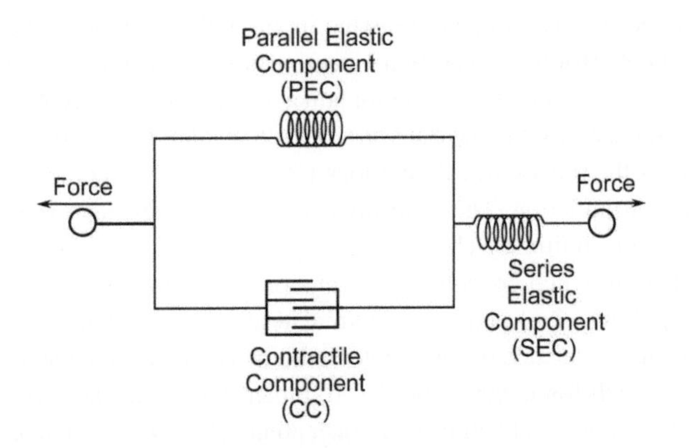


**Fig. 4.33** The three-component Hill muscle model composed of a contractile (CC), series elastic (SEC), and parallel elastic (PEC) components

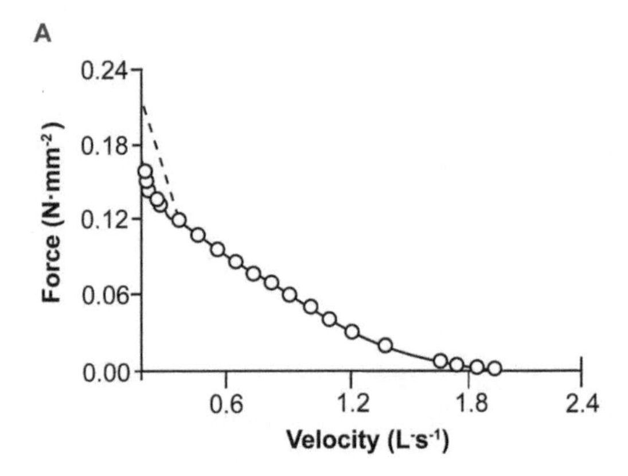


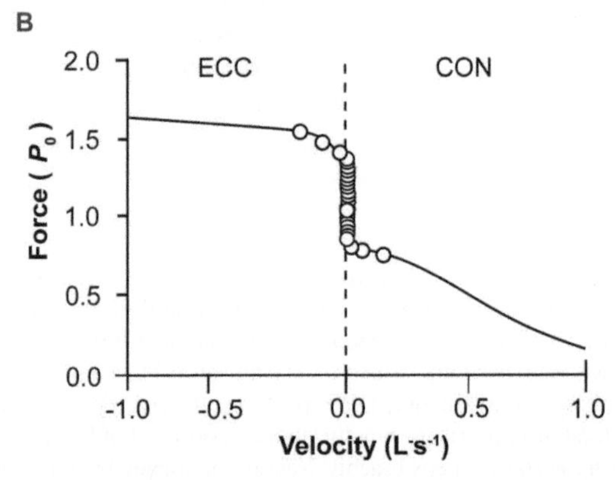


**Fig. 4.34** Force-velocity relationship during (**A**) isometric (i.e., zero velocity) and concentric muscles actions and (**B**) transition from concentric to eccentric muscle actions in a single muscle fiber from frog tibialis anterior muscle. Note in (**A**) N·mm$^{-2}$ = newton per millimeter squared; in (**B**) $P_o$ = maximal isometric force; L·s$^{-1}$ = length per second (Modified with permission from Alcazar et al. (2019))

shown that muscles with predominantly fast-twitch muscle fibers have roughly 2.5 times greater maximal shortening velocity, which means that a specific muscle would have a different and uniquely shaped rectangular hyperbola representing the force-velocity relationship. The data in Fig. 4.34 were derived by Edman from a single muscle fiber from the tibialis anterior of a frog. Examination of graphs A and B shows that isometric (velocity = 0) force magnitude is greater than during concentric, i.e., muscle shortening, action. This can be explained by two factors: (1) under isometric conditions, a maximal number of attached cross-bridges exist within a given sarcomere at any instant in time, and (2) more motor units are recruited in the time it takes to reach maximum tension. Furthermore, we can see that as the velocity of muscle shortening increases, force magnitude decreases. Winter provided two possible explanations for this phenomenon: (1) loss of force production or tension as the cross-bridges in the CC break and then reform in a shortened condition, that is, at higher velocities, the number of cross-bridges at any given time is less than when the muscle is acting more slowly, and (2) fluid viscosity in the CC, SEC, and PEC, where viscosity results in friction that requires an internal force to overcome, thereby reducing tendon force. In Fig. 4.34B, we see that there is a direct relationship between increased muscle force and increased velocity of muscle lengthening, i.e., eccentric muscle action. It is hypothesized that the increase in muscle tension, when it is acting eccentrically, may be related to the idea that within the cross-bridges, the force required to break the links is greater than that required to hold it at its isometric length. Additionally, it is thought that viscous friction in the fluid surrounding the muscle fibers must be overcome, adding to the requirement for more force.

The fourth relationship related to CC includes force-length (FL), which refers to the dependence of maximum

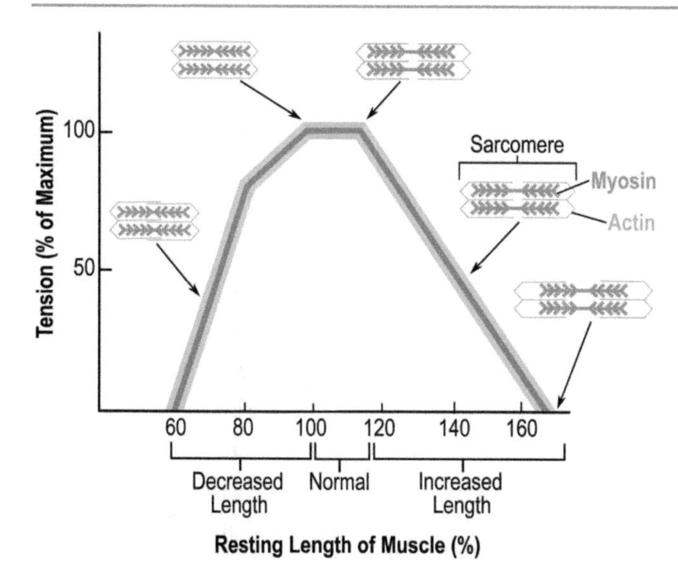


**Fig. 4.35** The CC relationship between force and length when the muscle is maximally activated

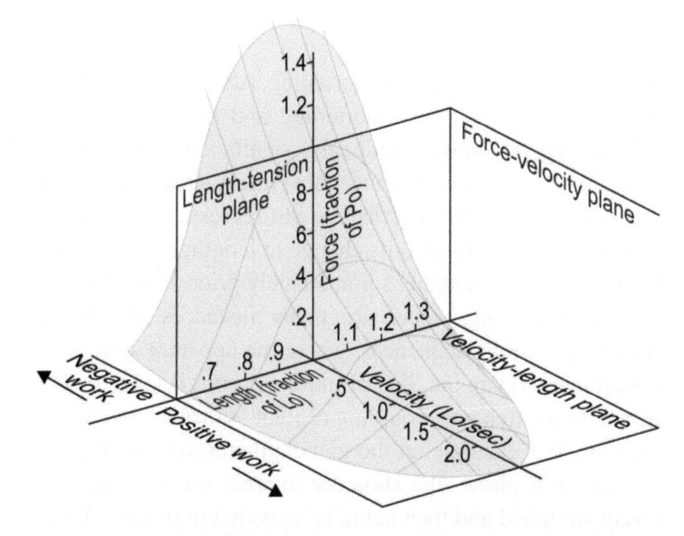


**Fig. 4.36** A three-dimensional plot showing the changes in CC tension, during maximum activation, as a function of muscle length and shortening or lengthening velocity. Note: (1) negative work corresponds with eccentric muscle action and positive work corresponds with concentric muscle action, where the arrows represent increasing velocity; (2) $L_o$ = resting muscle length; (3) $P_o$ = maximum isometric tension; and (4) length-tension is synonymous with force-length. (Modified with permission from Winter (2009b))

isometric force production on CC length (Fig. 4.35). Building on the work of Ramsey and Street, and the pre-established sliding-filament theory of muscular contraction, Gordon et al. isolated sarcomeres from striated frog muscle and showed that isometric force production was greatest at intermediate CC lengths and declined as the CC was either shortened or lengthened. An implication of these findings suggests that the maximum force at zero velocity, i.e., isometric, must then be a function of CC length. Referring to Fig. 4.35, we see that when the muscle is at its optimal length, i.e., 100–120% of resting length, there are a maximum number of cross-bridges between the filaments, making maximum muscle tension possible. To complete the picture, we see that as the muscle's length increases, the filaments become increasingly separated, the number of cross-bridges decreases, and tension decreases, and when the muscle shortens, there is overlapping of cross-bridges, resulting in interference and a reduction of tension.

Since the four CC properties (SA, FA, FV, and FL) do not operate in isolation, let's briefly examine the underlying dynamics of this complex system (Fig. 4.36). Remember that muscle force is not due just to the CC properties but is also modulated by motor unit firing rate and recruitment (see the electromyography section of Appendix J). In Fig. 4.36, we see the functional relationship between force, length, and velocity at maximal activation. Notice that the force-generating capacity of a muscle at a particular length and velocity is a product of corresponding values on separate force-length and force-velocity curves. For example, a muscle acting at a rapid velocity at its shortened length produces relatively lower force levels, even under maximum activation levels, whereas a muscle acting under near isometric conditions hypothetically produces significantly greater active force. The hypothetical control of the interactions among these subsystems is complex and can be reviewed by examining the model shown in Fig. 4.32.

Let's take a brief timeout to link our discussion of the force-velocity relationship in skeletal muscle and Brad's elbow pain. Recall that during the cocking and acceleration phases of pitching, the ulno-humero joint is placed in extreme abduction, i.e., valgus, and the shoulder moves from maximal external rotation followed by very rapid internal rotation, respectively. During the cocking phase, in addition to the large tensile stresses imposed on the UCL, the flexor-pronator muscles develop large forces due to their rapid eccentric actions, as they attempt to control the ulno-humero valgus moment. The large tensile forces in the flexor-pronator muscles could lead to strain injuries to the muscle fascicles, tendons, or both. As well, those forces impose large tensile strains on the medial epiphyseal growth plate.

The elbow rapidly extends while the abducted shoulder internally rotates during the acceleration phase. We know that, as a two-joint muscle, the tension in the long head of the biceps stabilizes, i.e., compresses, the abducted glenohumeral joint, while it controls the extensor moment at the elbow. The elbow extends at a rate of approximately 2000°/s which means that the long head of the biceps is undergoing an eccentric action at a very rapid rate. The force generated by the long head of the biceps generously stabilizes the glenohumeral joint, which has been shown to contribute to strain injuries to the antero-superior glenoid labrum (not Brad's problem). At the other end of the segment, the large

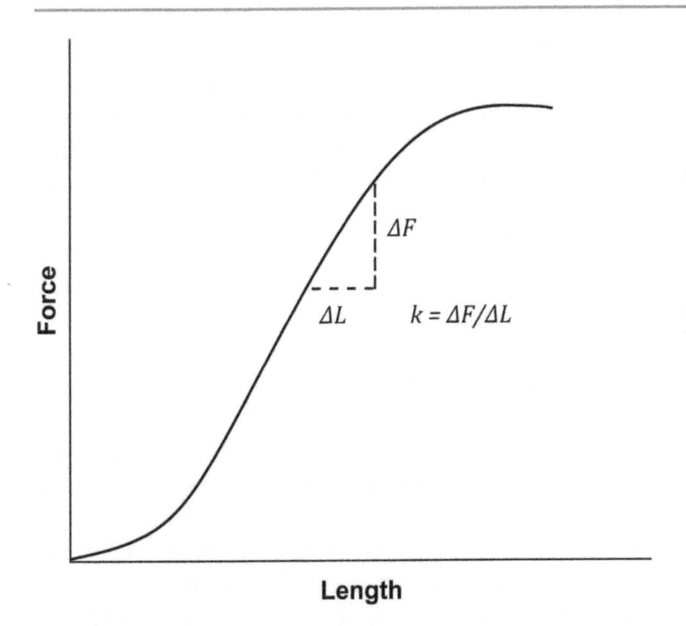


Fig. 4.37 The SEC relationship between force ($\Delta F$) and length ($\Delta l$), where $k$ represents the spring constant

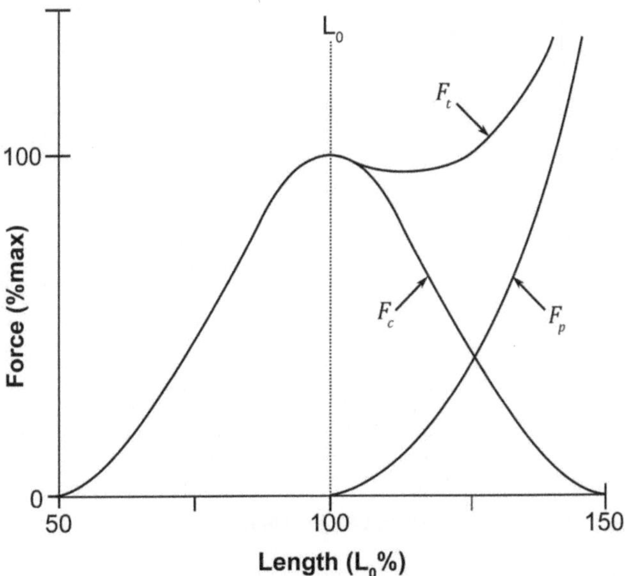


Fig. 4.38 Contractile element (CC) producing maximum force ($F_C$) along with tension from PEC ($F_P$), so that total tendon tension is $F_T = F_C + F_P$

biceps force likely also produces large compressive loads on the humeroradial joint, which may have contributed to Brad's osteochondritis dissecans.

Let's now examine the role of the connective tissues involved with the muscle-tendon actuator. The passive components of the force-length relationship are not neurally activated, so are not considered in Fig. 4.36; however, they do make important contributions to total force production in healthy muscles. All dense regular connective tissues in series with CC are referred to as the series elastic component (SEC). Tissues comprising the SEC include tendon, aponeuroses, and connective tissues within the sarcomere, i.e., titin, nebulin, etc. The SEC exhibits biphasic material properties, i.e., viscoelastic, as all other biological tissues; that is, under higher rates of loading, their elastic character changes, i.e., they become stiffer. Generally, during dynamic situations, i.e., non-isometric, the nonlinear characteristics of the SEC due to viscous damping influence the time course of the muscle tension. However, this has been shown to have a relatively small influence on overall behavior in muscle modeling, so typically only the elastic properties of the SEC are considered. Therefore, SEC elastic behavior (Fig. 4.37) is modeled as a spring, with constant $k$, calculated as the change in applied force divided by the resulting change in length of the tissue ($k = \Delta F/\Delta l$). Whether the material displays linear or nonlinear characteristics, the spring stiffness is simply the slope of the line of force versus displacement. According to Winter, during isometric muscle action, the SEC is placed under tension and is stretched only a finite amount. Because, under isometric conditions, the overall length of the muscle remains constant, stretching of the SEC would only occur if there was equal shortening of the CC itself, described as internal shortening. Although interesting, the interplay of internal shortening and SEC lengthening does not appear to be functionally significant. On the other hand, during high-performance movements such as throwing and jumping, it is thought that the SEC is responsible for the storage of energy (due to its change in length; see Appendix H) as the muscle lengthens immediately prior to rapid shortening. In pitching, there are many instances of muscle stretch-shortening sequences, where the apparent storage of potential energy in the SEC of muscle groups is used to contribute to the dynamic coupling of joint velocities and accelerations. For example, at the glenohumeral joint during the late cocking phase, the shoulder internal rotators are passively stretched and then neurally activated to initiate shoulder internal rotation that eventually reaches an angular velocity as high as 7000°/s.

Inactive muscles display viscoelastic behavior even if the contractile component has not been activated. If a resting muscle is loaded, it resists and is deformed, i.e., stretched (lengthened). The series elastic component plays only a small role in this response because the contractile component has not been activated. Instead, the elastic response is produced by structures that are in parallel, i.e., parallel elastic component, to the contractile component. The parallel elastic component is identified with the hierarchical layers of deep fascia associated with muscle, i.e., endo-, epi-, and perimysium. Similar to the series elastic component, the force-deformation relationship of the parallel elastic component is nonlinear in nature, showing increased stiffness as the muscle lengthens under higher rates and magnitudes of loading (Fig. 4.38). Under laboratory isometric testing conditions, the measured force response is the sum of the active force and passive force associated with the isometric length of the muscle. At shorter muscle lengths (parallel elastic component not stretched), total tension comes entirely from the contractile component. As the muscle acts under conditions when it is lengthened, the parallel elastic component is stretched, and its force response is added to the active contractile component's response. The actual shape of the combined force-length response depends on (1) the percentage of excitation, (2) the contractile component length at which the parallel elastic component first generates force in relation to the optimal length ($L_O$), and (3) the stiffness of the parallel elastic component, which is related to the rate and magnitude of loading.

Although the connective tissue components, i.e., endomysium, perimysium, and epimysium, of skeletal muscle comprise only about 10% of the total mass, their role in force transmission throughout the muscle is complex. The other roles of connective tissue include lateral transmission among muscle fibers, energy storage that enhances function, buffering length changes that may prevent injury, aligning sarcomeres within and between muscle fibers, and altering the operating range by allowing sarcomere shortening. One or more of these properties can be altered by injury or disease.

Our examination of muscle force production would be incomplete unless we also discussed the contributions from other factors, such as muscle morphology, architecture, and motor unit recruitment. As compared to tendons, muscles have lower tensile strengths. Thus, the relatively decreased ultimate strength of muscle requires it to have larger cross-sectional areas in order to transmit large forces without sustaining plastic strain injuries. Muscle morphology (i.e., its basic shape) provides some clues about the variety of ways our muscular system has evolved to generate large forces under a variety of circumstances (Fig. 4.39).

The two most common shapes are fusiform, e.g., biceps brachii, where the fibers run parallel to one another and to the connecting aponeurosis and tendon, and pennate. Pennate muscle fibers approach the central tendon obliquely. Most muscles in the human body are considered pennate and show a variety of patterns, depending on the number of similarly angled sets of fibers that attach to the tendon, for example, unipennate (e.g., extensor digitorum longus), bipennate (e.g., rectus femoris), or multipennate (e.g., gastrocnemius, deltoid). Generally, what all pennate muscles have in common is a larger number of fibers within a given cross-sectional area. Therefore, it is apparent that the interaction of the pennation angle and physiological cross-sectional area plays a significant role in the magnitude of tension that is transmitted from the muscle and tendon to the skeletal levers. It is obvious how cross-sectional area contributes to maximal force production, i.e., the larger the muscle is in multiple dimensions, the greater its potential to generate more tension. How pennation angle influences muscle force production is not so obvious, so let's look closer at this phenomenon.

Pennation angle refers to the orientation muscle fibers make relative to the central tendon. A fusiform muscle has a pennation angle of zero degrees. In that case, all of the force generated by that muscle is transmitted to the tendon and bony lever. A fusiform muscle generally has a relatively smaller cross-sectional area and thus have a reduced potential for generating large magnitudes of force. If, however, the

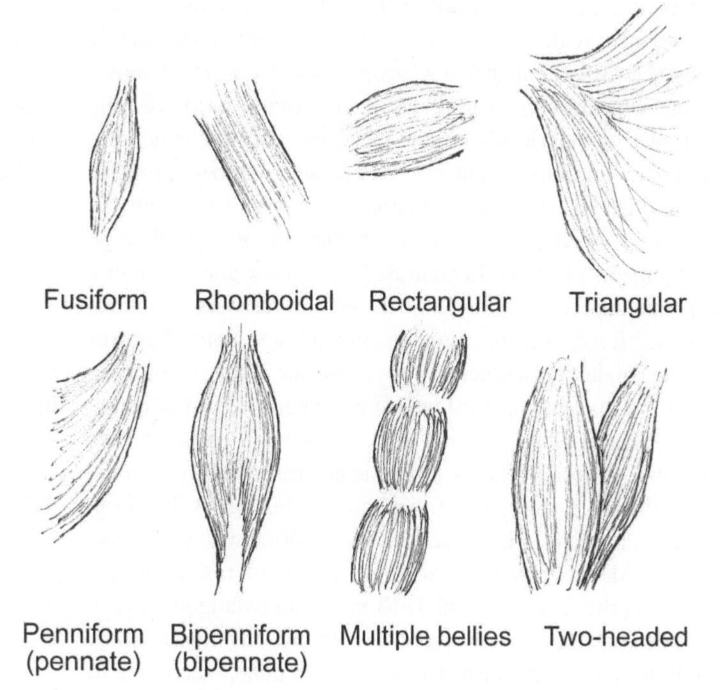


**Fig. 4.39** A variety of muscle morphologies have evolved in the human body, likely related to its specific function

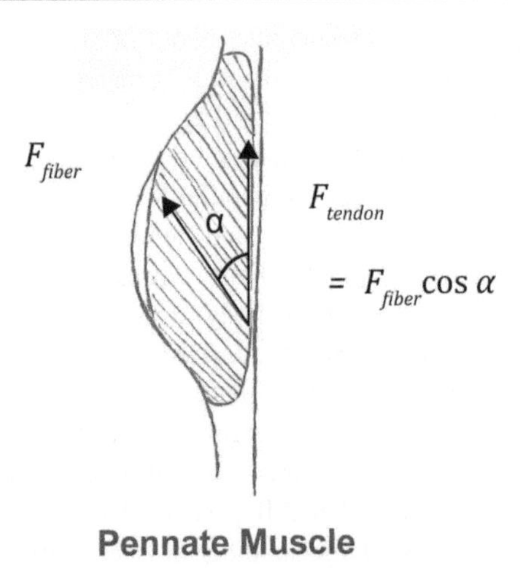


**Fig. 4.40** Muscle with a unipennate angle of α relative to the central tendon. The proportion of the total muscle force transmitted to the central tendon ($F_t$) is equal to the force of muscle fibers ($F_f$) times the cos $\alpha$

pennation angle is 30° only 87% (i.e., $= \cos 30° = 0.87 = 87\%$) of the total muscle force is transmitted through the tendon (Fig. 4.40). If only 87% of the force is transmitted to the tendon in a unipennate muscle, what advantage does pennation offer? It turns out that orienting muscle fibers obliquely to the central tendon allows for an increased number of muscle fibers in a given area. Thus, the reduced transfer of force from the pennate muscle to the central tendon is small compared to the increased total physiological cross-sectional area of a muscle.

Motor unit recruitment is the third major component that must be considered when examining contributions to total muscle force production. See Appendix J (section on electromyography) for a detailed examination of neural activation involving the motor unit. Recall that all of the muscle fibers associated with each motor unit share similar contractile characteristics and that any given muscle possesses motor units with different variations in innervation ratio, i.e., the number of muscle fibers innervated by each motor neuron. Small motor units, i.e., those with a smaller number of muscle fibers, are associated with synergistic actions of muscles that generate smaller forces used for fine motor control, e.g., movements of digits. In contrast, large motor units contain a greater number of muscle fibers and are associated with a greater force potential used to produce and control movements of larger segments over a greater range of motion. Motor units are classified into three categories, fast fatigable (FF), fast fatigable-resistant (FR), and slow (S), based on the number of muscle fibers (size), isometric twitch responses, and physiological/histochemical profile (Fig. 4.41). The slow motor unit type is characterized by its slow-twitch response and oxidative potential, i.e., the ability to resist fatigue, whereas the fast glycolytic (FF) motor units fatigue quickly but exhibit a fast-twitch response. Although the muscle fibers within a motor unit are recruited as "all or none," according to the size principle, the small motor units are typically recruited prior to large motor units. The orderly recruitment of motor units allows for a smooth, controlled increase in force development. In special circumstances, e.g., reaction to environmental conditions, however, large motor units, with simultaneous rapid large muscle force development, are preferentially recruited. Recruiting larger motor units also occurs routinely for high-performance activities such as pitching a baseball or performing a triple jump. Thus, we again see the importance of a central nervous system that can manage the extremes of muscle performance related to activities ranging from posture control to impulsive movements.

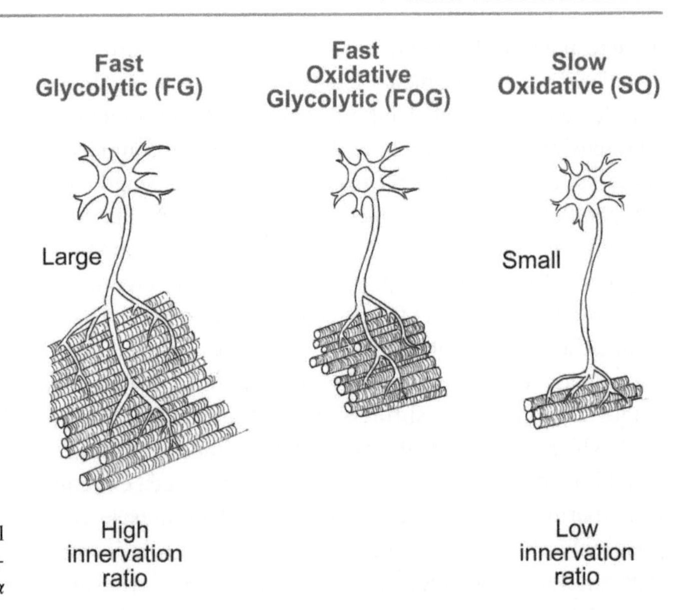


**Fig. 4.41** Classification of motor unit types based on innervation ratio (number of of muscle fibers innervated per axon)

Summary: In examining the structure and mechanical properties of the muscle-tendon actuator, we learn that the system is materially and neurally complex. With this understanding, we gain more detailed insight into the mechanics of the cocking and acceleration phases of pitching related to the generation of large muscle and joint forces that could explain Brad's signs and symptoms. Let's move forward and examine two additional problems identified in Brad's history and radiological examination: (1) premature closure of the medial epicondylar epiphysis and (2) intermittent prickly feelings in the hypothenar region and little and ring fingers of this left hand.

## 4.6 Peripheral Nerves: Structure and Material Properties

The intermittent altered sensation in the hypothenar region of Brad's left hand, projecting into the ring and little fingers, suggests impairment of the ulnar nerve.

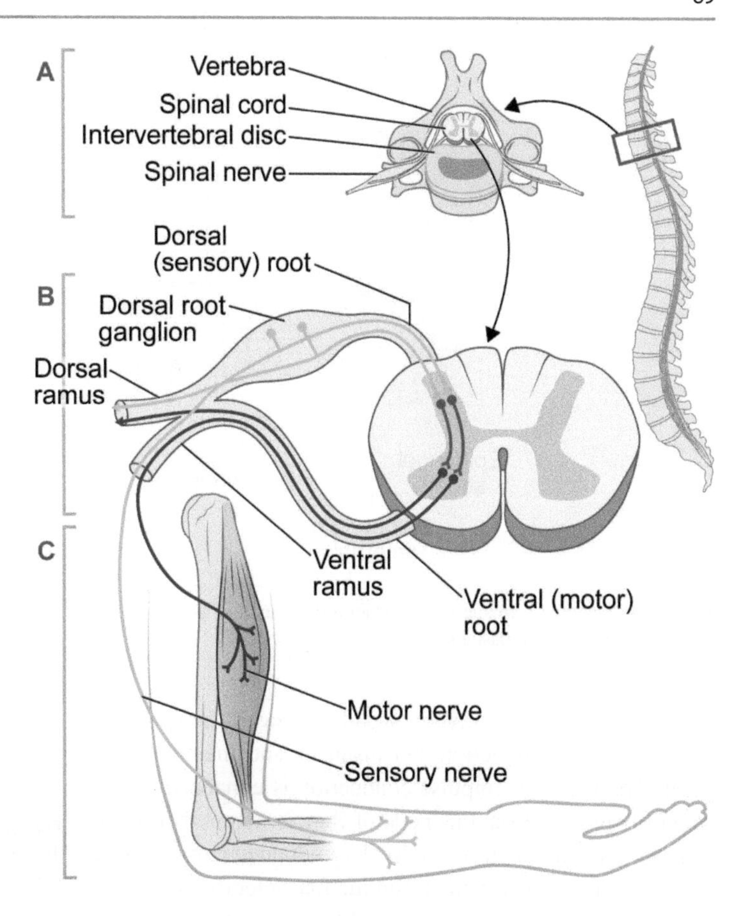


**Fig. 4.42** Schematic cross section drawing of the vertebral column, spinal cord, and spinal nerves. (**A**) Relationship between a cervical vertebra, intervertebral disc, spinal cord, and exiting spinal nerve at the intervertebral foramen. (**B**) Ventral (anterior) and dorsal (posterior) horns and sensory and motor roots. (**C**) Ventral (motor) and dorsal (sensory) root ganglia, ventral and dorsal rami, with peripheral motor and sensory nerves to/from effector end-organ, e.g., muscle

## 4.6.1 Structure of Peripheral Nerves

Let's begin our examination of the potential ulnar nerve injury to Brad with an overview of the nervous system. The nervous system has three broad responsibilities: (1) sense changes in body systems, (2) interpret the changes, and (3) initiate actions to respond to changes in the environment, e.g., muscle action to produce or control movement. The nervous system is comprised of two subsystems, the central (CNS) and peripheral (PNS) nervous systems (which includes the autonomic or involuntary subsystem). This review focuses on the peripheral nervous system. The PNS consists of peripheral nerve processes that extend from the spinal cord to organs and tissues that provide input into the CNS from sensory organs, e.g., receptors in the skin, joints, muscles, fascia, tendons, and viscera. These same peripheral nerve processes are also used to provide output from the CNS to the end effectors, e.g., muscles (Fig. 4.42). The anterior (ventral) and posterior (dorsal) roots within the spinal cord unite to form the spinal nerves at the intervertebral foramen. The posterior roots are responsible for conducting sensory information to and from the periphery, whereas the anterior roots contain mainly fibers that innervate motor neurons (i.e., the transmission of signals from the CNS to muscles). Just distal to the intervertebral foramen, the spinal nerves divide into dorsal and ventral rami. The dorsal rami innervate the muscles and skin of the head, neck, and back. In addition to innervating the ventral and lateral aspects of the head, neck, and back, the ventral rami, via nerve plexuses in the cervical, e.g., brachial plexus, and lumbopelvic region, innervate the muscles of the upper and lower extremities.

Examination of the images in Fig. 4.42 reminds us that the nervous system is a complex system that is a composite of inter-related connective tissues, blood vessels, and nerve fibers.

The peripheral nerves are no less complex. Let's first examine the anatomy of nerve fibers (Fig. 4.43).

Nerve fibers include the axon emanating from a nerve cell body along with its myelin sheath and Schwann cells (Figs. 4.42 and 4.43). Nerve fibers from sensory organs transmit impulses from the skin, muscle, and periarticular joint structures to the CNS. The CNS responds through nerve fibers connected to motor neurons in the spinal cord to activate muscle contractions.

Nerve fibers also serve as an anatomic connection between central nerve cells and end organs. An axonal two-way transport system is designed to maintain this connection by transporting various proteins synthesized in central neurons to the peripheral structures.

Most peripheral nerve axons are surrounded by multilayered, segmented coverings known as myelin sheaths. These sheaths are produced by Schwann cells arranged along and

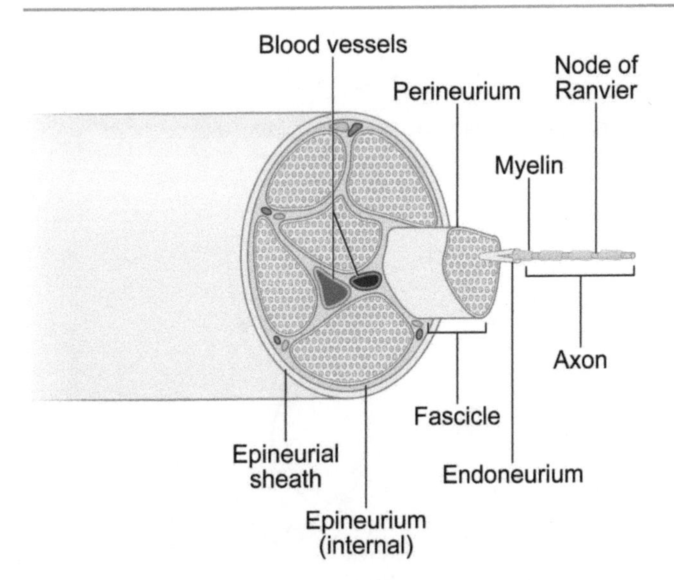


**Fig. 4.43** Schematic image of the structural features of a nerve fiber and related connective tissue structures

around the axon. The myelin sheath increases the speed of nerve impulse conduction and insulates and maintains the axon. The speed of impulse conduction is enhanced as the impulse "jumps" from one node of Ranvier to another. Both sensory and motor nerves have large-diameter myelinated fibers, which enhance their conduction velocity, a feature that characterizes the importance of these two types of nerve fibers. Nerve fibers are organized into functional subunits or fascicles, which are then bundled to form the nerve itself.

Just as we saw with muscle, there is a complex succession of deep fascial layers that surround nerve fibers (Fig. 4.43). These facias function to protect nerve fibers' continuity, which is essential to guard against excessive tensile and compressive loads. The outermost layer, the epineurium, surrounds the fascicle bundles. This, more loose, connective tissue serves as a cushion protecting fascicles from trauma and helps to maintain the vascular system and supply of oxygen. Where nerves lie close to a bone, e.g., the ulnar nerve (Fig. 4.44), a functional adaptation of the epineurium makes it more abundant, i.e., it is thickened.

The perineurium is a lamellar sheath covering each fascicle. This sheath provides a strong barrier to large compression forces, with the ability to mitigate an internal pressure of approximately 1000 mm of mercury before it fails. Its barrier function also chemically isolates nerve fibers from their surroundings, preserving the essential internal environment for nerve fibers. The innermost layer, endoneurium, is composed of fibroblasts and collagen, providing some structural support for nerve axons. The interstitial fluid pressure, essential to nerve functioning, slightly exceeds that found in subcutaneous and muscle tissue. However, even a slight elevation of this pressure due to edema following trauma negatively impacts nerve microcirculation.

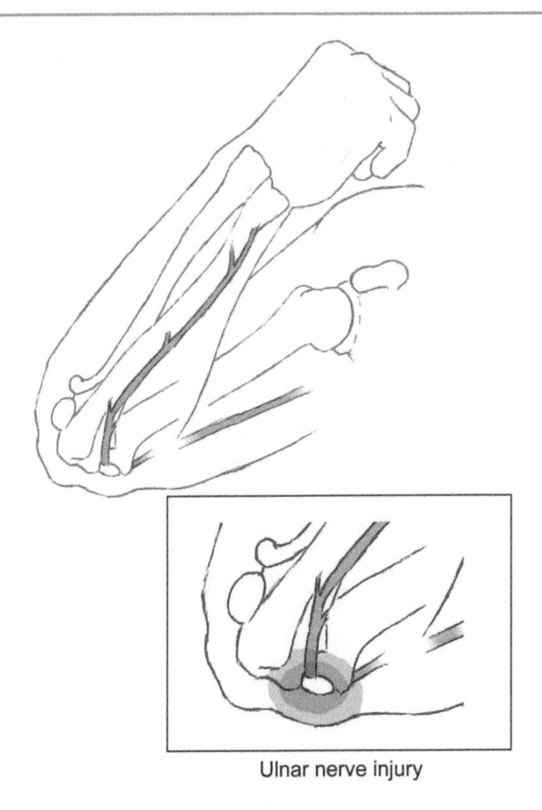


**Fig. 4.44** Anatomical location of the ulnar nerve as it passes behind the medial epicondyle into the cubital tunnel. As seen (inset highlighted area), this is a common site of ulnar nerve entrapment and irritation

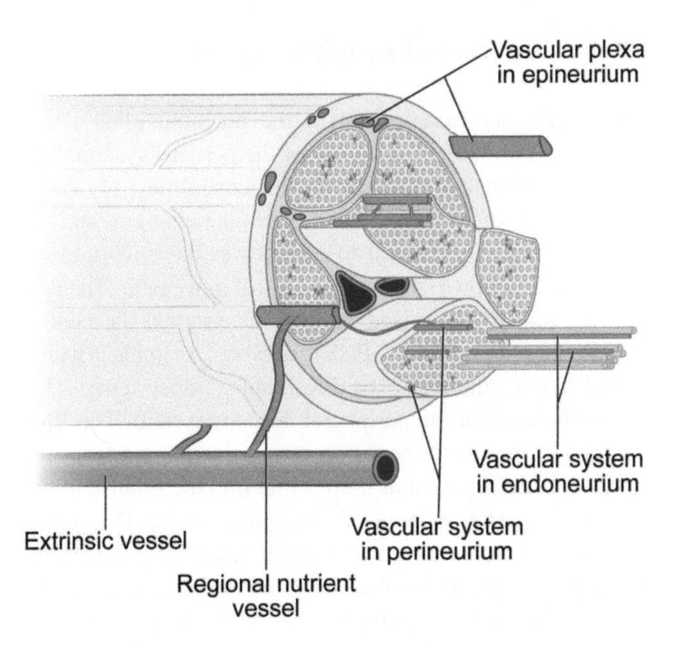


**Fig. 4.45** Schematic of a vascular network external to and within the epineurium, perineurium, and endoneurium fascial layers of a typical peripheral nerve

Because nerve impulse propagation and axonal transport require a large oxygen supply, peripheral nerves are highly vascularized. The vascular network extends through all layers of connective tissue (Fig. 4.45). Blood vessels are located externally and segmentally along the course of the nerve and then divide into ascending and descending, regional, branches. These vessels run longitudinally and anastomose with other vessels in the perineurium and endoneurium. Each fascicle houses longitudinally oriented capillary plexuses, which run oblique courses throughout the perineurium. Increases in intra-fascicular pressure can easily close these capillary networks like valves, which is often associated with external compression forces, leading to nerve ischemia. On the other hand, the extensive network of vessel anastomoses affords a fairly wide margin of protection to nerve fibers.

### 4.6.2 Material Properties: Tension and Compression

Under normal physiological conditions, e.g., posture and body movement, nerves are exposed to various mechanical stresses. Tensile stresses are applied parallel, i.e., longitudinally, and compression stresses are applied perpendicular to the length of a nerve. When joint motion causes elongation of the nerve bed, the nerve glides along the nerve bed but is also stretched, resulting in tensile stress of the nerve. Before we provide a clinical picture of how the ulnar nerve may become impaired, let's first examine the general material properties of peripheral nerves under tensile and compressive loading conditions.

Relying primarily on the use of animal models, research on the material properties of peripheral nerves under tensile and compressive loading is decades old. Overall, nerves have considerable tensile strength, while compression has been found to increase pressure within the connective tissue linings and may be more detrimental. With either loading condition, it appears that disturbances in intraneural blood flow, axonal transport, and nerve function occur well before plastic changes occur in the tissues. Like all biological structures, nerves are anisotropic (i.e., composite structures with unique mechanical properties at each layer of connective tissue) and demonstrate viscoelastic properties (Fig. 4.46). Under tensile loading conditions, the load-elongation curve and stress-strain diagrams are similar to typical plots for tendons and ligaments examined previously in this chapter (also see Appendix H). A typical rabbit tibial nerve has an extended toe region, suggesting that nerve elongation is significant under minimal stress during the initial stages of loading. At 6–8-mm elongation, the nerve has strained approximately 15–20%. From 8 to 16 mm of elongation (up to ~35% strain), the nerve becomes increasingly stiffer, as noted by the steepness of the linear portion of the two loading curves. It appears that failure of a peripheral nerve occurs at about 35–40% strain. With experimental tests of strain to failure in a stress-strain diagram representing human cadaver sections of tibial and peroneal nerves, we see somewhat different responses (Fig. 4.47). In human nerves, the toe region is less extended with an almost immediate linear elastic response. Complete failure appears to occur sooner in the tibial nerve specimens

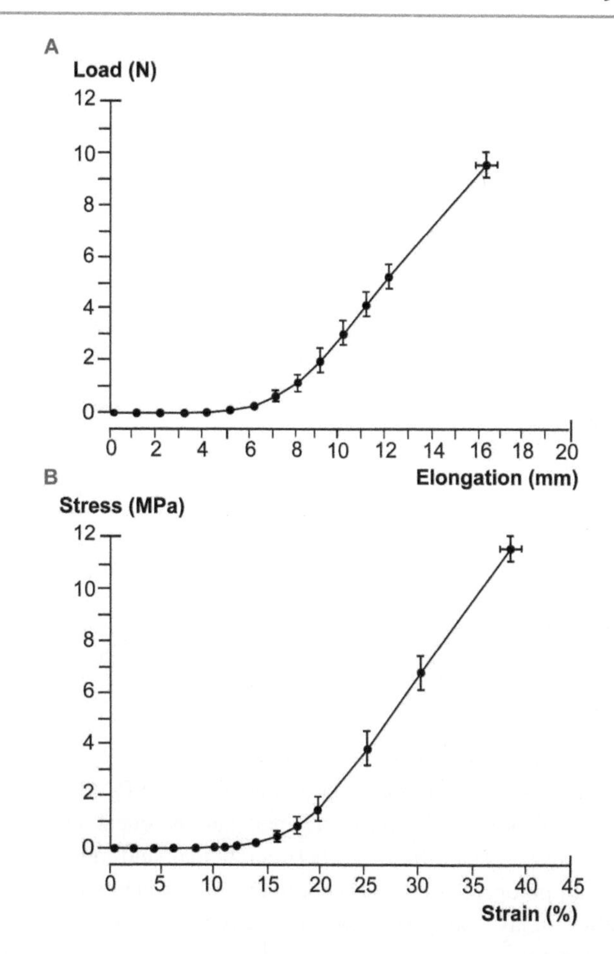


**Fig. 4.46** Typical (**A**) load-elongation (deformation) and (**B**) stress-strain curves of a rabbit tibial nerve. (Modified with permission from Kwan et al. (1992))

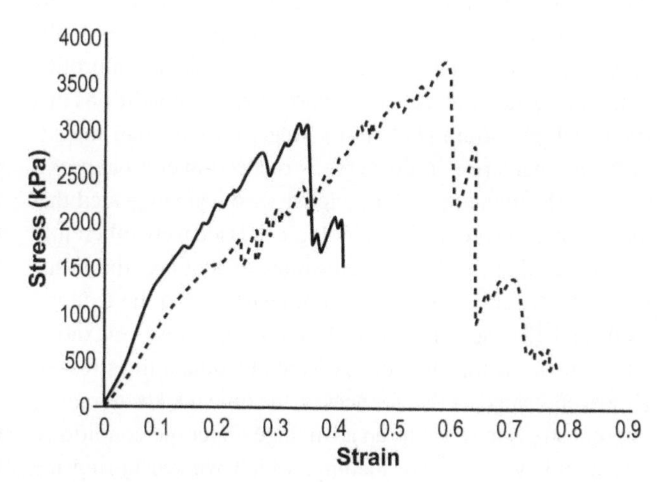


**Fig. 4.47** Stress-strain curves for harvested human tibial and peroneal sections tested to failure under tensile loading. Solid = tibial; dashed = peroneal. (Modified with permission from Kerns et al. (2019))

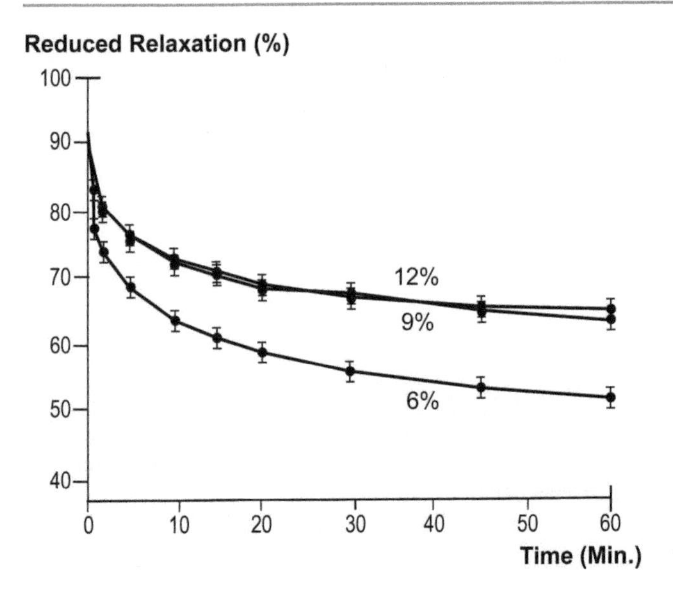


**Fig. 4.48** Stress-relaxation experiment using a rabbit tibial nerve. Note the difference in the magnitude of stress relaxation with less elongation. (Modified with permission from Kwan et al. (1992))

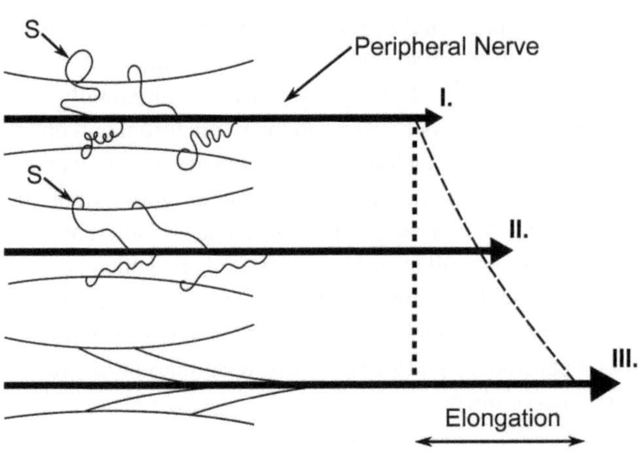


**Fig. 4.49** Schematic of a peripheral nerve bed and accompanying vessels at different stages of tensile loading: I. The small blood vessels (S) are normally crimped or coiled to allow for normal physiologic excursions, i.e., gliding of the nerve; II. Under increasing tension/elongation, these vessels become stretched impairing blood flow; and III. The cross-sectional area of the nerve is reduced (not shown) with increased tensile loading, further impairing intraneuronal blood flow. (Modified with permission from Lundborg and Rydevik (1973))

(35% strain) compared to peroneal nerves (~60% strain). Ultimate stress to failure approximated 3000 kPa (3 MPa) and 3500 kPa (3.5 MPa) for tibial and peroneal nerves, respectively. The differences between human and rabbit tibial and peroneal nerves in the ultimate stress (and strain) at failure may be explained by differences in testing procedures.

The biphasic nature of biological tissue includes the viscoelastic phenomena of creep and stress relaxation (see Appendix H for a review of the mechanics of deformable bodies). In a stress-relaxation test of a rabbit tibial nerve, the magnitude of elongation affects the stress-relaxation profile, with more relaxation occurring under loading conditions that produced less strain (Fig. 4.48). Regardless of load magnitude, internal stress in the nerve is reduced over an approximately 20-minute period. Previously, we had suggested that the viscoelastic properties of biological tissue were inherently protective; that is, (1) at greater rates of loading, the tissue becomes stiffer, i.e., offering more resistance to the external load, and (2) the properties of creep and stress relaxation allow the tissue time to reduce potentially damaging stresses. However, it appears that for nerves, the time it takes to reduce internal stress may not be an advantage under the conditions, i.e., impulsive repetitive loading, which we see during the pitching act. Thus, it appears that the creep and stress-relaxation phenomena may provide protection for nerves only under conditions where they may be under prolonged tensile loads, e.g., static postures.

Strain, i.e., deformation or elongation, of nerve results in a reduction in its cross-sectional area. This action produces increased pressure in the endoneurial compartment. It is possible that the outer connective tissue constraining this inner pressurized neural core contributes significantly to the resistance a nerve may provide under tensile loading. In fact, upon elongation, the increased intra-endoneurial pressure resists the transverse contraction and contributes to the elastic stiffness of the nerve. However, with increasing tensile load, structural separation occurs in the core-sheath interface, followed by damage to the axons and connective tissues of the epi- and perineurium. In their tests of rabbit peripheral nerves, Rydevik et al.'s histological analysis revealed significant disruption of the perineurium along the length of the nerve and diffuse damage to axons and blood vessels within the endoneurium (Fig. 4.49), without visible evidence of damage to the outer fascia coverings. Rydevik's group also noted that complete cessation of all blood flow in the nerve usually occurred at approximately 15% strain. They conclude that the perineurium provides primary resistance to tensile loading, whereas the nerve fibers located within the fascicles make small, if any, contributions to resistance.

Nerves are also routinely exposed to static and dynamic compressive stresses. Extrinsic and intrinsic compression stress typically occurs by approximation of the nerve to adjacent tissues, such as muscle, tendon, or bone, or by pressure increases in the extra-neural environment. A number of different methods have been used experimentally to show that mild compression of the nerve induces structural and functional changes. For example, local prolonged, i.e., 4–6 hours, compression on a nerve at 30 mm mercury (Hg) has been shown to result in impairment of blood flow and changes in axonal transport. Furthermore, prolonged compression induces intraneural edema, which may contribute to scar formation. Moreover, direct compression on a nerve at 80 mm

Hg has been shown to result in complete cessation of intraneural blood flow so that the nerve becomes completely ischemic. Fortunately, compression relief restores normal nerve function within 2 hours. It has been shown that compression of a nerve segment causes displacement of its internal contents in longitudinal and transverse directions, noting that damage to axons and myelin is greatest at the edges of the compressed zone. All of these mechanical effects of compression on nerves must be considered in light of its magnitude and duration, as well as how the pressure is applied.

Let's now briefly apply what is known about the material properties of peripheral nerves to the potential effects on the ulnar nerve related to pitching. In general, the direction and magnitude of nerve excursions, i.e., both elongation and gliding, are dependent upon the anatomical relationship between the nerve and the joint axis of rotation. For example, when a nerve bed is elongated, the nerve is placed under increased tensile stress, and the nerve glides toward the moving joint. Conversely, if nerve bed tension is relieved during joint motion, the nerve realigns by gliding away from the moving joint. It has been demonstrated that when the shoulder is held in 90° abduction/external rotation, with the elbow flexed approximately 90°, the ulnar nerve glides distally, i.e., toward the hand, between 5 and 14 mm; however, it is unknown how ulno-humero abduction affects the ulnar nerve bed. Concomitantly, the tension in the ulnar nerve is maximized, which increases compression of the nerve in the cubital tunnel. Recall that during the cocking phase of throwing, the glenohumeral joint is abducted 90°, externally rotated 170–190°, the elbow is flexed approximately 90° with maximal abduction at the humeroulnar joint, and the wrist is extended. Note that, in general, the ultimate magnitude of ulnar nerve stress depends on the angular velocity at the shoulder, elbow, and wrist, the time spent in each position, and the repetitive nature of the movement. In pitching, the shoulder and elbow angular velocities during the cocking phase are much lower compared to the very high angular velocities recorded during the acceleration phase, which approach 7000°/s at the shoulder and 2000°/s at the elbow. Although the positions of the shoulder and elbow during the cocking phase of pitching place the ulnar nerve "on notice," the time of the stress exposure to the ulnar nerve is finite. On the other hand, the frequency with which the ulnar nerve is stressed, under both tension and compression loads, in both practice and game situations is high; and often the recovery time after performances is limited, which does not allow time for damaged tissues to heal. Furthermore, repetitive stress, i.e., loading and unloading, likely leads to plastic changes in the nerve over time (see Fig. 4.27) and may create a chronic inflammatory cycle leading to scar formation and localized edema, both of which limit the normal excursion of the nerve during joint movements and contribute to chronic compression of the nerve in the cubital tunnel.

Another examination of the several tissues that can develop strain injuries, inflammatory reactions, and scar formation is worthwhile (Fig. 4.50). Proximal to the medial epicondyle, the ulnar nerve is securely wrapped by the arcade of Struthers and medial intermuscular septum, which likely is not significantly stressed during the pitching act. As the ulnar nerve approaches the cubital tunnel, it is fairly exposed, but then takes a sharp turn distally and is fixed securely within the tunnel by Osbornes' ligament/fascia superficially and the deeper ulnar collateral ligament. It is in this region where the ulnar nerve is most loaded under tension and compression. We know that the UCL is placed under large tensile loads with the humeroulnar joint flexed and abducted during the cocking phase, and the figure makes it clear that the Osborne tissues could also be placed under significant tensile loading. Finally, it is notable that the ulnar nerve is embedded within the flexor-pronator muscle group after it exits the cubital tunnel, which is another potential site for ulnar nerve entrapment and irritation. Thus, the ulnar nerve is likely very susceptible to tensile and compressive strain injury during the late cocking phase of pitching.

Stresses induced on the ulnar nerve change slightly during the acceleration phase of pitching. During acceleration, the shoulder remains abducted at 90° as it rapidly internally rotates and horizontally adducts. Although ulno-humero abduction does not change significantly during the acceleration phase, the elbow extends nearly fully, as the forearm pronates and the wrist flexes. The ulnar nerve bed has been shown to shorten, reducing tensile strain, when the elbow extends with an abducted shoulder. Thus, it appears that tensile strain on the ulnar nerve may be reduced during a very brief time during the acceleration phase of pitching. What does not change, of course, are the large shoulder and elbow angular velocities, which induce impulsive loads on the ulnar nerve, and the repetitive nature of the pitching act.

It is likely that a portion of Brad's medial elbow pain and abnormal sensation in the ulnar aspect of his hand is related to an ulnar nerve injury induced by the large and repetitive tensile and compressive stresses incurred during the pitching act.

## 4.7 Medial Epicondylar Growth Plate

Brad had tenderness over the medial humeral epicondyle. Of course, this is the site of attachment for multiple structures that might have sustained strain injuries, e.g., flexor-pronator muscle group and UCL. Point tenderness in this region could also be referred from an impaired ulnar nerve and/or the

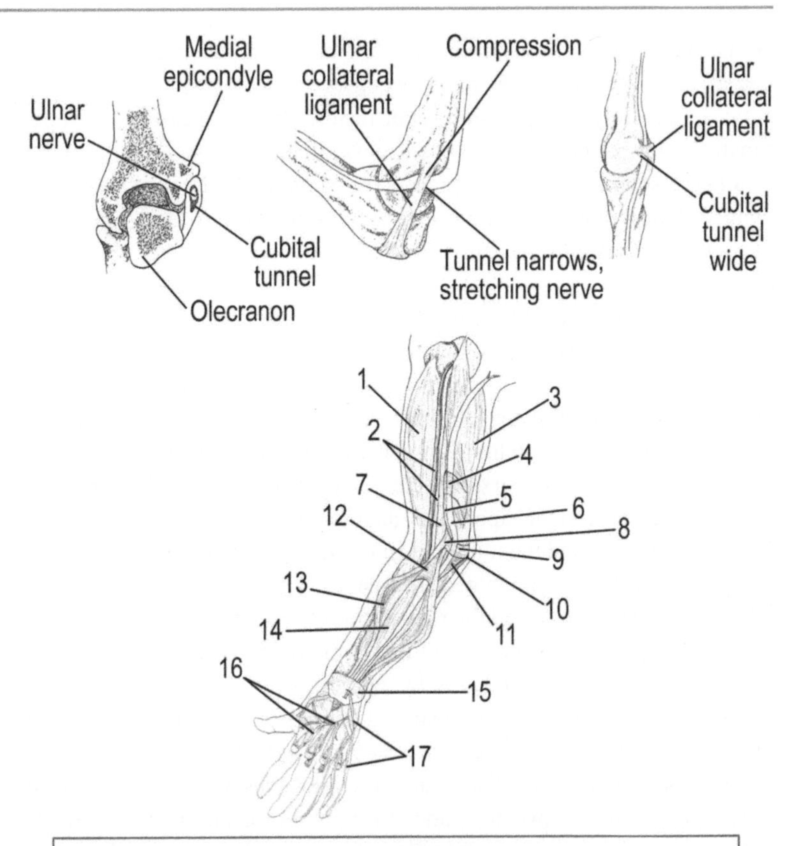


**Fig. 4.50** Bony, muscular, and soft tissue anatomical relationships to the ulnar nerve as it maneuvers around the medial epicondyle and meanders through the cubital tunnel

medial epicondylar growth plate (MEGP). Based on his radiographs, we know that the MEGP had been damaged. This anatomical region is likely another source of Brad's elbow pain, so we need to closely examine its structure and the material properties of its constituent parts.

### 4.7.1 Structure

For boys, the MEGP appears between the ages of 7–9 years, but it does not mature, i.e., fuse, until approximately 17 years of age (Fig. 4.4). Although most of our understanding of growth plate mechanobiology has centered on the epiphyseal regions of long bones (Fig. 4.51), the features of that research can be generalized to apophyseal growth plates in other regions as well, e.g., MEGP. The medial apophysis growth plate is an avascular and aneural region of chondrocytes embedded in a rich extracellular matrix, which is the site that induces longitudinal bone growth. The matrix of the growth plate is similar to that of articular cartilage (which we discuss in Chap. 3, Sect. 3.5.2); however, it consists primarily of type II collagen and aggrecan, and its role and regulation is entirely different. Sections of the growth plate reveal ordered columns of chondrocytes that are smaller and flatter at the epiphyseal end and larger and rounder toward the metaphysis. The chondrocytes are clearly differentiated into distinct zones: reserve, proliferative, and hypertrophic. Growth plates, e.g., epiphyses of long bones, are considered monop-

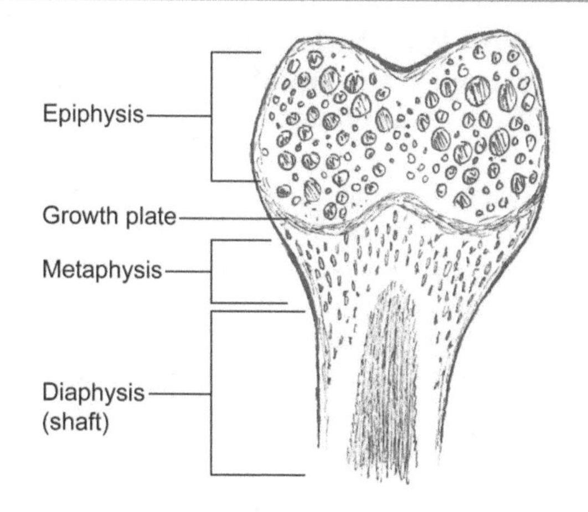


**Fig. 4.51** The location of the growth plate in a long bone relative to the metaphysis and diaphysis

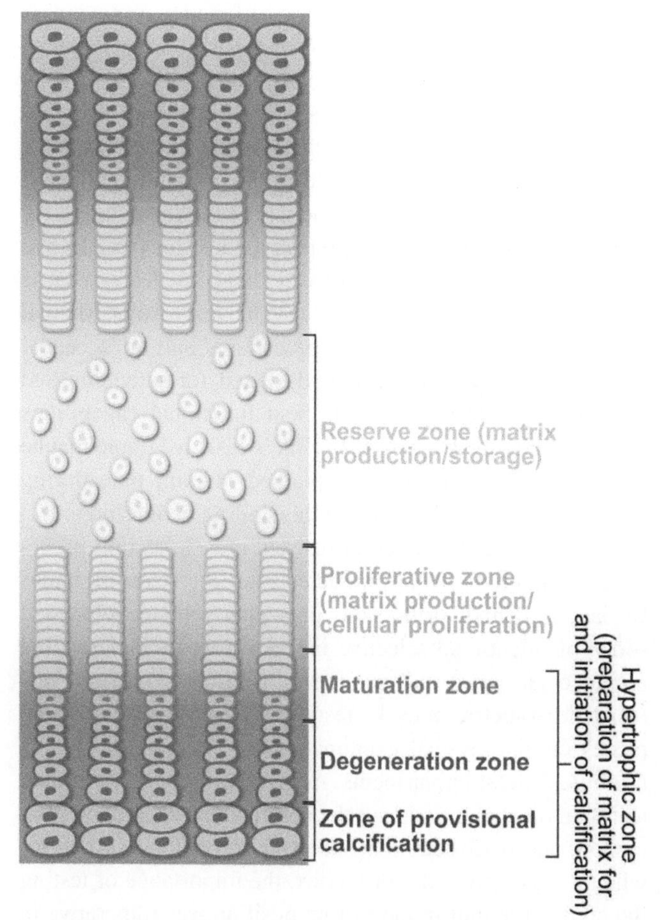


**Fig. 4.52** Schematic epiphyseal growth plate showing the proposed bipolar growth arrangement of the reserve, proliferative, and hypertrophic zones

olar (Fig. 4.52), whereas other growth cartilages such as apophyses, e.g., medial epicondylar apophysis, are bipolar. Bipolar growth cartilages have reserve chondrocytes in the middle and proliferative and hypertrophic chondrocytes on both ends. Longitudinal growth results from the complex dynamics between cell division (proliferation), hypertrophy (cell enlargement), extracellular matrix synthesis, and controlled degradation (which occurs in both the proliferative and hypertrophic zones). Eventually, chondrocytes undergo apoptosis (invasion by blood vessels) and calcification in the zone of provisional calcification.

The regulation of bone growth is complex and poorly understood. It involves many inter-related factors, including (1) genetic control; (2) systemic hormonal levels, local action of growth factors and cytokines, nutritional status, and blood supply; (3) transport of nutrients and signaling molecules; (4) autocrinal and periclinal factors, i.e., secretion of a substance by a cell that acts on the same or different cell to produce an action; and (5) mechanical loading of soft tissue. Growth and growth rate regulation may follow normal maturation, be altered by pathology or mediated by mechanical loading, and is achieved by a combination of (1) change in the rate of proliferation, (2) a change in the number of chondrocytes undergoing proliferation, and (3) a different height of hypertrophic cells in the growth direction. The accomplishments of these actions assume that all of the cells that are produced in the proliferative zone complete their differentiation into mature hypertrophic cells and that matrix synthesis and degradation are well balanced, i.e., the matrix between the cells in the growth direction is negligible. It is known that the level of activity has little influence on the eventual size and stature of an individual, suggesting that mechanical factors have a limited effect on the longitudinal growth and growth rate of immature bone plates. On the other hand, Wolff's law relates the adaptation of bone to its mechanical environment; that is, bone apposition is stimulated by intermittent increased stress, and bone resorption is induced following reduced stress. The apparent paradox between bone and growth plate response to mechanical stimuli may be explained by examining the details of growth plate behavior under different mechanical loading conditions.

### 4.7.2 Material Properties

Similar in nature to all biological tissues, the growth plate is anisotropic and demonstrates heterogeneous responses to external loads depending on the region tested, i.e., resting, proliferative, or hypertrophic zone, and the magnitude and rate of loading. For example, the hypertrophic region is less stiff, with greater compressive strains demonstrated in regions overlapping the reserve and hypertrophic zones. With tensile tests, research demonstrated lower ultimate strain in the reserve zone, likely related to the random orientation of the collagen framework in that region. Generally, under controlled tensile and compressive experimental testing conditions, growth plate material properties mirror the viscoelastic behavior seen in articular cartilage.

Recall, bone remodeling is governed by Wolff's law, which relates bone apposition to an increase in intermittently applied loads, i.e., compression, tension, shear and torsional, and bone resorption to a decrease or absence of intermittent loading. Longitudinal bone growth, however, appears to be controlled by the Hueter-Volkmann law (HVL). The transduction of mechanical signals to growth plate chondrocyte transition zones is likely induced by cellular stresses (and strains), as well as the state of the surrounding ECM. The relatively thin growth plate, embedded within the stiffer bone tissue, experiences primarily compression due to large muscular forces associated with normal physiologic loading. The Hueter-Volkmann law relates the reduction of growth rate and growth to increased sustained compression on a growth plate. Therefore, although the HVL does not seem to account for loading rate and static versus cyclic loading, the HVL defines a relationship that is different from that described by Wolff's law, which relates primarily to alterations in the internal structure of bone (particularly trabeculae) resulting from several different types of external loads. The causes of the altered growth rate under static compression appear to result from a complex interaction of changes in the chondrocyte zones. Sustained compression leads to a reduction in the cellular shape, number, and proportion of proliferating chondrocytes, a thinning in both the proliferative and hypertrophic zones, and degradation of the ECM (type II and X collagens). Although chondrocyte proliferative activity is not halted, it is significantly reduced, as is the number, height, and volume of hypertrophic chondrocytes. Conversely, a sustained reduction of compression and/or tensile loading results in the overall thickening of the growth plate. Although there appears to be a disruption of the chondrocyte columns under tensile loading, enhanced proliferation and hypertrophic thickening predominate due to an increase in the number, height, and volume of chondrocytes. Therefore, while it is unclear exactly how the growth plate responds to dynamic intermittent compression and tension, compression appears to be related to decreased bone growth, and cyclic tension appears to accelerate growth, likely resulting in early closure or fusion.

We know from our review of anatomy and study of the phases of pitching that the two major soft tissue structures that attach to the medial humeral epicondyle, i.e., ulnar collateral ligament and flexor-pronator muscle group, experience large, impulsive, intermittent, but repetitive tensile loads during the cocking and acceleration phases. We, therefore, speculate that the tensile loads, over time, significantly altered the mechano-transduction of signals to the medial apophysis and accelerated the injury to the growth plate. Since the growth plate was avulsed, we can be confident that this is a likely source of some of Brad's symptoms, and we can be sure of altered structure and function at this site. The long-term effects of this avulsion fracture of the medial growth plate on elbow/forearm function are not known.

## 4.8 Elbow Pain During Manual Muscle Testing: Static Equilibrium Analysis Offers an Explanation

Physical examination of the musculoskeletal system usually includes tests of muscle strength. These tests could use instruments such as handheld dynamometers or the more sophisticated, and perhaps more valid and reliable, tool such as an isokinetic dynamometer device. For expediency, Brad's strength was tested using a manual technique. Resisted muscle testing using manual techniques (MMT) involves the following steps, in this order: (1) ask the individual to move the joints of interest through as large a range of motion as possible in order to determine the ability and willingness to move; (2) place the segments of interest in a specific test position, which assumes that the muscle(s) (it is not possible to test muscles in isolation because of functional movement synergies) are tested in their "resting position," in order to perform an isometric (i.e., no joint movement) test; (3) ask an explanation of what is required of the individual; (4) apply increasing manual resistance to the distal aspect of the appropriate segment attempting to "match" the maximum effort of the individual; (5) ask the individual to hold the position for a count of 5 seconds; and (6) repeat the test for three to five repetitions to give the individual a chance to learn what is required and attempt to get valid/reliable test results. Manual testing has clear standardized procedures and an accepted scoring/rating system, but it is designed for screening and does not provide information about functional capabilities and is marginally subjective. In Brad's case, we found that he had normal strength, but complained of joint pain during several of the tests. How might this finding be interpreted?

The physical examination resisted muscle testing procedures outlined above are modeled after the work of James Cyriax, a British orthopedic physician who developed a series of intentional selective tissue testing procedures that included an active and passive joint range of motion and resisted isometric muscle testing. These were important pieces of his physical examination arsenal for diagnosing musculoskeletal impairments. According to Cyriax, resisted testing indicated the state of the contractile unit, i.e., muscle, tendon, and periosteal attachment, by provoking symptoms when it was activated. For Cyriax, the importance of testing the contractile unit in the resting position was imperative in order not to strain periarticular inert structures, e.g., ligament, capsule, bursa, fascia, dura mater, and nerve. Thus, if the correct testing position is found, activation of the muscle(s) primarily stresses (i.e., under tension) the contractile unit, implicating it as the cause of dysfunction if the response is painful, weak, or both. A normal grade is given for a test response of strong and painless, indicating that the contractile unit is intact. If pain is elicited only with strong repeated muscle actions, intermittent claudication must be considered. A strong and painful response suggests a minor, i.e., first-degree, strain injury to the muscle(s), tendon, and/or periosteal attachments. A weak and painful response is consistent with a second- or third-degree muscle-tendon strain or bone fracture, whereas a weak and painless response suggests a complete failure/rupture of the muscle-tendon or nerve palsy. Based on Cyriax's hypotheses regarding symptom provocation with resisted testing, intuition suggests that Brad sustained minor strain injuries to the elbow flexors and extensors, forearm supinators and pronators, and wrist flexors and extensors. It is possible that multiple muscles and tendons were strained secondary to actions related to pitching, but it is highly unlikely for a couple of reasons: (1) multiple muscle strains like this are not common in pitchers at any level and (2) strains to the forearm supinators and wrist extensors related to pitching have not been reported in the sports literature. So, let's look for an alternative explanation.

Given Brad's radiological examination that noted osteochondritis dissecans, a type of advanced degenerative joint condition, we hypothesized, based on our findings in the shoulder case, that the pain he experienced during resisted testing could be due to large bone-on-bone forces at the humeroulnar and humeroradial joints. Setting up a static equilibrium situation mimicking resisted testing of the elbow flexors and extensors may provide us with additional information regarding his elbow pain. We always start with stating the problem assumptions:

1. No friction at joint surfaces.
2. Biceps and triceps are the only muscles acting during the resisted testing. Synergists are not active, and there is no co-action of antagonists.
3. No passive elastic internal flexor and extensor moments are produced by non-contractile elements, i.e., series elastic component, parallel elastic component, etc.
4. Rigid body mechanics, i.e., there is no deformation of tissue.
5. All forces are assumed to lie in the plane perpendicular to the flexion-extension axis.

The solution begins with presenting information that is known and consistent with the anthropometric values in Appendix A.

1. Body mass, 82 kg → body weight, $W$ = (82 kg)(9.81 m/s$^2$) = 804.4 N.
2. Forearm weight, $W_{\text{fa}} = 0.016\ W = 12.9$ N
3. Weight of hand, $W_{\text{h}} = 0.006\ W = 4.8$ N
4. Flexion resistance, $F_{\text{flex}} = 30$ lb (4.448 N/lb) = 133.4 N
5. Extension resistance, $F_{\text{ext}} = 25$ lb (4.448 N/lb) = 111.2 N

The free-body diagram is a simplified sketch that depicts the relationship between the segments of interest and the environment (Fig. 4.53). First, establish a segmental body-fixed coordinate system at the elbow joint. All relevant, known and unknown, external and internal forces (i.e., gravity, muscle, friction, ground reaction, bone-on-bone or joint reaction, etc.) acting on the system are displayed. Forces are represented by vectors, with their points of application, lines of action, and orientations (directions) (see also Supplementary 4.2 for an additional statics problem exercise).

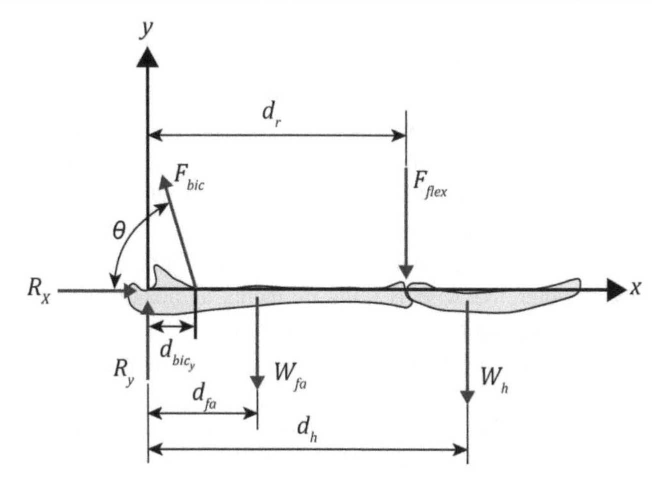


**Fig. 4.53** Free body diagram used to set up static solution for elbow flexor resisted strength test

The biceps force ($F_{\text{bic}}$) makes an angle of $\theta = 73°$ relative to the negative $x$-axis; the insertion of the biceps is $d_{\text{bic}y} = 0.05\text{m}$ from the elbow joint, but this distance is not the moment arm of $F_{\text{bic}}$. The distance $d_{\text{bic}_y}$ is the moment arm of the $y$-component, $F_{\text{bic}_y}$. The moment arms of the vectors representing the weight of the forearm ($W_{\text{fa}}$), MMT test resistance ($F_{\text{flex}}$), and weight of the hand ($W_{\text{h}}$) are 0.115, 0.267, and 0.337 m, respectively. $R_x$ and $R_y$ represent the components of the unknown bone-on-bone, i.e., joint reaction, forces.

Let's illustrate how the moment arms of $W_{\text{fa}}$, $F_{\text{flex}}$, and $W_{\text{h}}$ were determined. Using information from Appendix A, we first need to estimate the lengths of the forearm and hand segments. Using Fig. A3, these lengths are determined relative to Brad's height (1.83 m): (1) length of the forearm = (0.146)(1.83 m) = 0.267 m; (2) length of the hand = (0.108)(1.83 m) = 0.198 m. Next, we need to estimate the location of the center of mass for the forearm and hand segments. From Table A1, we see that the COM distance from the elbow for the forearm is 0.430 times the segment length, which results in $d_{\text{fa}} = (0.430)(0.267\ \text{m}) = 0.115$ m. The COM distance from the wrist for the hand is 0.506 times the segment length. Here is where we must be careful: the segment length calculated above for the hand using Fig. A3 is the distance from the wrist to the fingertips. From Table A1, the segment length is defined as the distance from the wrist to knuckle II of the middle finger. We estimate the reference hand length as 70% of the distance from the wrist to the fingertip: (0.70)(0.506)(0.198 m) = 0.070 m. The total moment arm for the weight of the hand from the elbow is then $d_h$ = 0.267 m + 0.070 m = 0.337 m. We note that the manual resistance to test the elbow flexor strength was applied at the distal forearm, i.e., the wrist joint, which is $d_r$ = 0.267 m from the elbow joint center.

We define forces that point up and to the right as positive and counterclockwise (CCW) moments as positive. Then, we write the static equilibrium equations.

Let's solve for the muscle forces using the moment equilibrium equation. First, we use the equilibrium moment equation with respect to the elbow joint to solve for the *y*-component of the biceps force vector:

$$\sum M = -W_{fa}d_{fa} - F_{flex}d_r - W_h d_h + F_{bic_y} d_{bic_y} = 0$$

Solving for $F_{bic_y}$:

$$F_{bic_y} = \frac{W_{fa}d_{fa} + F_{flex}d_r + W_h d_h}{d_{bic_y}}$$

$$F_{bic_y} = \frac{(12.9\text{N})(0.115\text{m}) + (133.4\text{N})(0.267\text{m}) + (4.8\text{N})(0.337\text{m})}{0.05\text{m}}$$

$$F_{bic_y} = 775\text{N}(\uparrow, \text{positive force directed upward}).$$

Now, we solve for the resultant biceps force, $F_{bic}$, and the force component of $F_{bic}$ in the *x*-direction, $F_{bic_x}$:

$$F_{bic} = F_{bic_y} / \sin\theta° = (775\text{N}) / \sin 73° = 811\text{N}$$

$$F_{bic_x} = -F_{bic}\cos\theta = -(811\text{N})\cos 73° = -237\text{N}.$$

The negative sign indicates that the *x*-component of the biceps force is directed to the left. Based on our assumption that the biceps was the only elbow flexor activated, its force component in the *x*-direction, i.e., 811 N (or about 182 lb), would not likely have induced a strain injury that may have contributed to the discomfort Brad experienced during the muscle test. Yet, if we qualitatively account for the brachialis activation during the flexion test, its smaller moment arm would have required a larger muscle force resulting in larger joint contact forces than if the biceps acted alone.

Next, we set up the equations to solve for the bone-on-bone forces at the joint surface. We know that the sum of all of the forces must be equal to zero, $\sum F = 0$. Let's first solve for the reaction force vector in the *x*-direction:

$$\sum F_x = F_{bic_x} + R_x = -237\text{N} + R_x = 0$$

$$R_x = 237\text{N}(\rightarrow, \text{positive force directed to the right}).$$

Note that the magnitude of this force is equal to 0.29 BW. Next, we solve for the reaction force vector in the *y*-direction:

$$\sum F_y = F_{bic_y} - W_{fa} - F_{ext} - W_h + R_y = 0$$

Solving for $R_y$:

$$R_y = -F_{bic_y} + W_{fa} + F_{ext} + W_h = -624\text{N}$$

The negative sign for $R_y$ indicates that its sense is opposite to what is indicated in the free body diagram, e.g., the direction of $R_y$ is down. The magnitude of the *y*-component of the reaction force approximates 0.78 BW.

Finally, we compute the resultant (*R*) joint bone-on-bone force:

$$R = \sqrt{R_x^2 + R_y^2} = 668\text{N}(\sim 0.83\text{BW})$$

*R* acts at an angle of $\theta = \tan^{-1}(R_y/R_x)$ = − 69.2° (or 69.2° below the positive *x*-axis) (Fig. 4.53).

Because of the large concavity, i.e., relatively large contact surface area, of the proximal ulna, the bone-on-bone forces generated in the *x*- and *y*-directions are likely well-distributed. On the other hand, the humeroradial joint has a simpler morphology with a shallow concavity and smaller area of surface contact, compared to the humeroulnar joint. This analysis does not address how the elbow joint reaction forces are distributed between the humeroradial and humeroulnar joints. However, the smaller contact area of the humeroradial joint could lead to problematic stresses in some cases. Based on this information, it is likely that the large compression stress caused by resisted muscle testing was not well distributed/mitigated by the degenerated lateral joint resulting in pain from bone nociception.

Let's repeat our analysis for resisted testing of the elbow extensors (Fig. 4.54):

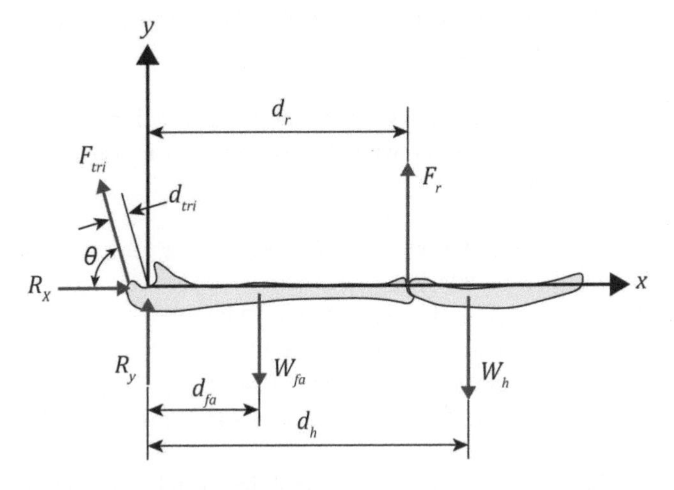


**Fig. 4.54** Free body diagram for static solution to elbow extensor resisted strength test

The triceps force ($F_{tri}$) makes an angle of $\theta = 74°$ relative to the negative $x$-axis. In this case, the moment arm of $F_{tri}$ is $d_{tri} = 0.02$ m. The moment arms of the vectors representing the weight of the forearm $W_{fa}$, MMT test resistance ($F_{ext}$), and weight of the hand ($W_h$) are the same as they were for the elbow flexor resistance example. $R_x$ and $R_y$ represent components of the unknown bone-on-bone, i.e., joint reaction, forces. We do not know the exact "signs" of these reaction components, but we learn the direction of forces after we complete the computations.

Let's solve for the muscle forces using the moment equilibrium equation:

$$\sum M = -W_{fa}d_{fa} + F_{ext}d_r - W_h d_h - F_{tri}d_{tri} = 0$$

Solving for $F_{tri}$:

$$F_{tri} = \frac{-W_{fa}d_{fa} + F_{ext}d_r - W_h d_h}{d_{tri}}$$

$$F_{tri} = \frac{-(12.9\text{N})(0.115\text{m}) + (133.4\text{N})(0.267\text{m}) - (4.8\text{N})(0.337\text{m})}{0.02\text{m}}$$

$$F_{tri} = 1330\text{N}(\uparrow, \text{positive force directed upward}).$$

Now, we solve for $F_{tri_x}$ and $F_{tri_y}$:

$$F_{tri_x} = -F_{tri}\cos 80° = -367\text{N}$$

$$F_{tri_y} = F_{tri}\sin 80° = 1279\text{N}.$$

The negative sign for $F_{tri_x}$ indicates that the $x$-component of the triceps force is directed to the left, while the positive sign for $F_{tri_y}$ indicates the $y$-component is directed upward.

Next, we set up the equations to solve for the bone-on-bone forces at the joint surface. We know that the sum of all of the forces must be equal to zero, $\sum F = 0$. Let's first solve for the reaction force vector in the $x$-direction:

$$\sum F_x = F_{tri_x} + R_x = -367\text{N} + R_x = 0$$

$$R_x = 367\text{N}(\rightarrow, \text{positive force directed to the right}).$$

We note that the magnitude of the force in this direction is quite small relative to body weight (~0.25 BW). Next, we solve for the reaction force vector in the $y$-direction:

$$\sum F_y = F_{tri_y} - W_{fa} + F_{ext} - W_h + R_y = 0.$$

Solving for $R_y$:

$$R_y = -F_{tri_y} + W_{fa} - F_{ext} + W_h = -1372\text{N}$$

The negative sign for $R_y$ indicates that its sense is opposite to what is indicated in the free body diagram, e.g., the direction of $R_y$ is down. The magnitude of this force is quite large (~1.71 BW).

Finally, we compute the resultant ($R$) joint bone force:

$$R = \sqrt{R_x^2 + R_y^2} = 1420\text{N}(\sim 1.77\text{BW}),$$

where $R$ acts at an angle of $\theta = \tan^{-1}(R_y/R_x) = -75.0°$ (or 75.0° below the positive $x$-axis).

We see similar complex internal stresses associated with strength testing of the elbow extensors, but the magnitudes of the $x$- and $y$-components of the triceps muscle are much larger despite the fact that the MMT test was performed with gravity, i.e., the weight of the forearm and hand, assisting the triceps in producing the internal extensor moment. The triceps muscle had to generate more force than the biceps because the length of its moment arm was about 2.5 times less than that of the biceps.

Based on the solutions to our statics problem, we can make two conclusions: (1) it is clear that the internal muscle forces that are generated across a joint to balance external loads make the primary contribution to large joint bone-on-bone forces. This is always true because external loads have a mechanical advantage, i.e., larger moment arm, over resisting muscle forces, and (2) impaired mechanical properties of the articular cartilage at the humeroradial joint secondary to osteochondritis dissecans are not likely able to distribute large joint loads (particularly with elbow flexion resisted tests), which can only exacerbate degenerative processes, adding to both joint dysfunction and pain.

## 4.9 Chapter Summary

Brad, a former little league baseball star, experienced a significant reduction in his pitching/throwing performance and was reduced to an average player in his teen years due to marked pain in his elbow. His history is not unlike thousands of youngsters who develop overuse injuries to their throwing arms. Brad presented with radiologically diagnosed osteochondritis dissecans and avulsion fracture of the medical epicondylar growth plate and clinical presentation of moderate strain injury to the ulnar collateral ligament and ulnar nerve. His pain was likely related to a complex interplay of these multiple injuries. Our study includes detailed examination of the anatomy of the elbow and forearm and osteo- and arthrokinematics of the humeroulnar, humeroradial, and proximal and distal radioulnar joints. Additionally, we study the structure and material properties of ligaments, tendons, muscle-tendon actuators, peripheral nerves, and growth plates. The solutions to a static equilibrium problem of resisted muscle testing of the elbow flexors and extensors suggest that one contribution to Brad's elbow pain likely originated from the advanced degenerative-like changes in humeroradial joint. The sequence of our analysis for this problem is typical of the approach used in problem-based learning; that is, a reductive, yet a somewhat nonlinear, approach that attempts to account for the complexity and interrelationship among biological systems and subsystems.

What can we abstract from this case? There is at least one thing we abstract because we think we can use this information in future cases. Cyriax suggested that a strong and painful response resulting from resisted muscle-tendon testing was that the structure had likely sustained a minor (i.e., first degree) strain. In this case, we obtain a strong and painful responses with the resisted testing of muscles (muscle synergists) that were not likely strained related to the pitching mechanism. We hypothesize that an alternative explanation for the reproduction of elbow pain with resisted muscle testing is related to the degenerative-like joint condition, i.e., osteochondritis dissecans, of the lateral elbow compartment. The abstract notion we carry forward is that although we think we can isolate the contractile properties of muscle with resisted muscle tests, they are provocative to multiple articular and periarticular structures.

Now, you make your list.

## Bibliography

Aguinaldo A, Escamilla R (2019) Segmental power analysis of sequential body motion and elbow valgus loading during baseball pitching. Comparison between professional and high school baseball players. Orthopaedic Journal of Sports Medicine 7(2):2325967119827924. https://doi.org/10.1177/2325967119827924

Alcazar J, Csapo R, Ara I et al (2019) On the shape of the force-velocity relationship in skeletal muscles: The linear, hyperbolic, and the double-hyperbolic. Frontiers in Physiology 10:769. https://doi.org/10.3389/fphys.2019.00769

Apostoiakos J, Durant TJS, Dwyer CR et al (2014) The enthesis: A review of the tendon-to-bone insertion. Muscles, Ligaments and Tendons Journal 4(3):333–342. https://doi.org/10.11138/MLTJ/2014.4.3.333

Atwater AE (1979) Biomechanics of overarm throwing movements and of throwing injuries. Exercise and Sport Sciences Reviews 1:217–258

Binder-Markey BI, Sychowski D, Lieber RL (2021) Systematic review of skeletal muscle passive mechanics experimental methodology. Journal of Biomechanics 129:110839. https://doi.org/10.1016/j.jbiomech.2021.110839

Bojsen-Møller J, Magnusson SP (2019) Mechanical properties, physiological behavior, and function of aponeurosis and tendon. Journal of Applied Physiology 126:1800–1807. https://doi.org/10.1152/japplphysiol.00671.2018

Bramson MTK, Van Houten SK, Corr DT (2021) Mechanobiology in tendon, ligament and skeletal muscle tissue engineering. Journal of Biomechanical Engineering 143:070801–070801. https://doi.org/10.1115/1.4050035

Bruton M, O'Dwyer N (2018) Synergies in coordination: A comprehensive overview of neural, computational, and behavioral approaches. Journal of Neurophysiology 120:2761–2774. https://doi.org/10.1152/jn.00052.2018

Buffi JH, Werner K, Kepple T et al (2014) Computing muscle, ligament, and osseous contributions to the elbow varus moment during baseball pitching. Annals of Biomedical Engineering 43(2):404–415. https://doi.org/10.1007/s10439-014-1144-z

Cinque ME, Schickendantz M, Frangiamore S (2020) Review of anatomy of the medial ulnar collateral ligament complex of the elbow. Current Reviews in Musculoskeletal Medicine 13:96–102. https://doi.org/10.1007/s12178-020-09609-z

Clauser CE, McConville JT, Young JW (1969) Weight, volume and center of mass of segments of the human body. Wright-Patterson Air Force Base, Ohio (see LeVeau below)

Cyriax J (1982) Textbook of orthopaedic medicine. Volume I. Diagnosis of soft tissue lesions. Balliere Tindall, London, UK

David K, Glazier P, Araújo D et al (2003) Movement systems as dynamical systems, the functional role of variability and its implication for sports medicine. Sports Medicine 33(4):245–260. https://doi.org/10.2165/00007256-200333040-00001

Emons J, Chagin AS, Sävendahl L et al (2011) Mechanisms of growth plate maturation and epiphyseal fusion. Hormone Research in Pædiatrics 75:383–391. https://doi.org/10.1159/000327788

Fleisig GS, Andrews JR, Dillman CJ (1995) Kinetics of baseball pitching with implications about injury mechanisms. The American Journal of Sports Medicine 23(2):233–239

Frangiamore SJ, Bigart K, Nagle T et al (2018a) Biomechanical analysis of elbow medial ulnar collateral ligament tear location and its effect on rotational stability. Journal of Shoulder and Elbow Surgery 27:2068–2076. https://doi.org/10.1016/j.jse.2018.05.020

Frangiamore S, Moashe G, Kruckeberg BM et al (2018b) Qualitative and quantitative analysis of the dynamic and static stabilizers of the medial elbow: An anatomic study. The American Journal of Sports Medicine 46:687–694. https://doi.org/10.1177/0363546517743749

Gordon AM, Huxley AF, Julian FJ (1966) The variation in isometric tension with sarcomere length in vertebrate muscle fibres. The Journal of Physiology 184:170–192

Hill AV (1970) First and last experiments in muscle mechanics. Cambridge University Press, London

Hoang PD, Herbert RD, Todd G et al (2007) Passive mechanical properties of human gastrocnemius muscle-tendon units, muscle fascicles and tendons *in vivo*. The Journal of Experimental Biology 210:4159–4168. https://doi.org/10.1242/jeb.002204

Hunter SK, Senefeld JW, Neumann DA (2017) Muscle: The primary stabilizer and mover of the skeletal system. In: Neumann DA (ed) Kinesiology of the musculoskeletal system, foundations for rehabilitation, 3rd edn. Elsevier, St. Louis

Hurd WJ, Kaufmann KR, Murthy NS (2011) Relationship between the medial elbow adduction moment during pitching and ulnar collateral ligament appearance during magnetic resonance imaging evaluation. The American Journal of Sports Medicine 39(6):1233–1237. https://doi.org/10.1177/0363546510396319

Ikezu M, Edama M, Inai T et al (2021) The effects of differences in the morphologies of the ulnar collateral ligament and common tendon of the flexor-pronator muscles on elbow valgus braking function: A stimulation study. International Journal of Environmental Research and Public Health 18:1986. https://doi.org/10.3390/ijerph18041986

Jackson TJ, Jarrell SE, Adamson GJ et al (2016) Biomechanical differences of the anterior and posterior bands of the ulnar collateral ligament of the elbow. Knee Surgery, Sports Traumatology, Arthroscopy 24:2319–2323. https://doi.org/10.1007/s00167-014-3482-7

Kerns J, Piponov H, Helder C et al (2019) Mechanical properties of the human tibial and peroneal nerves following stretch with histological correlations. The Anatomical Record 302:2030–2039. https://doi.org/10.1002/ar.24250

Kjaer M (2003) Role of extracellular matrix in adaptation of tendon and skeletal muscle to mechanical loading. Physiological Reviews 84:649–698. https://doi.org/10.1152/physrev.00031.2003

Kobayasbi K, Burton KJ, Rodner C et al (2004) Lateral compression injuries in the pediatric elbow: Panner's disease and osteochondritis dissecans of the capitellum. The Journal of the American Academy of Orthopaedic Surgeons 12:246–254

Kwan MK, Wall EJ, Massie J et al (1992) Strain, stress and stretch of peripheral nerve rabbit experiments in vitro and in vivo. Acta Orthopaedica Scandinavica 63:267–272. https://doi.org/10.3109/17453679209154780

Kwansa AL, Freeman JW (2015) Ligament tissue engineering. In: Nukavarapu SP, Freeman JW, Laurencin CT (eds) Regenerative engineering of musculoskeletal tissues and interfaces. Elsevier, Ltd. https://doi.org/10.1016/C2014-0-02826-2

Latash ML, Schotz JP, Schöner G (2002) Motor control strategies: Revealed in the structure of motor variability. Exercise and Sport Sciences Reviews 30(1):26–31

LeVeau B (1977) Williams and Lissner: Biomechanics of human motion. W. B Saunders Company, Philadelphia, PA

Lis A, de Castro C, Nordin M (2022) Biomechanics of tendons and ligaments. In: Nordin M, Frankel VH (eds) Basic biomechanics of the musculoskeletal system, 5th edn. Wolters Kluwer, Philadelphia

Lundborg G, Hansson HA (1980) Nerve regeneration through preformed pseudosynovial tubes. A preliminary report of a new experimental model for studying the regeneration and reorganization capacity of peripheral nerve tissue. The Journal of Hand Surgery 5(1):35–38. https://doi.org/10.1016/s0363-5023(80)80041-4

Lundborg G, Rydevik B (1973) Effects of stretching the tibial nerve of the rabbit: A preliminary study of the intraneural circulation and the barrier function of the perineurium. Journal of Bone and Joint Surgery 55B(2):390–401

Millesi H, Zöch G, Reihsner R (1995) Mechanical properties of peripheral nerves. Clinical Orthopaedics and Related Research 314:76–83

Momma D, Funakoshi T, Endo K et al (2018) Alteration in stress distribution patterns through the elbow joint in professional and college baseball pitchers: Using computed tomography osteoabsorptiometry. Journal of Orthopaedic Science 23:948–952. https://doi.org/10.1016/j.jos.2018.06.006

Morrey BF, An KN (2005) Stability of the elbow: Osseous constraints. Journal of Shoulder and Elbow Surgery 14:174S–178S

Murray WM, Buchanan TS, Delp SL (2002) Scaling of peak moment arms of elbow muscles with upper extremity bone dimensions. Journal of Biomechanics 35:19–26

Naito K, Takagi H, Yamanda N et al (2014) Intersegmental dynamics of 3D upper arm and forearm longitudinal axis rotations during baseball pitching. Human Movement Science 38:116–132. https://doi.org/10.1016/j.humov.2014.08.010

Nan N, He X (2012) Fusimotor control of spindle sensitivity regulates central and peripheral coding of joint angles. Frontiers in Computational Neuroscience 6(Article 66):1–13. https://doi.org/10.3389/fncom.2012.00066

Neumann DA (2017) Elbow and forearm. In: Neumann DA (ed) Kinesiology of the musculoskeletal system, foundations for rehabilitation, 3rd edn. Elsevier, St. Louis

Oyama S (2012) Baseball pitching kinematics, joint loads, and injury prevention. Journal of Sport and Health Science 1:80–91. https://doi.org/10.1016/j.jshs.2012.06.004

Özkaya N, Leger D, Goldsheyder D, Nordin M (2017) Fundamentals of biomechanics, equilibrium, motion and deformation, 4th edn. Springer, Cham, Switzerland

Park MC, Ahmad CS (2004) Dynamic contributions of the flexor-pronator mass to elbow valgus stability. The Journal of Bone and Joint Surgery. American Volume 86:2268–2274

Pomianowski S, O'Driscoll SW, Neal PG et al (2001) The effect of forearm rotation on laxity and stability of the elbow. Clinical biomechanics 16:401–407

Profeta VLS, Turvey MT (2018) Bernstein's levels of movement construction: A contemporary perspective. Human Movement Science 57:111–133. https://doi.org/10.1016/j.humov.2017.11.013

Robertson DGE, Caldwell GE, Hamill J et al (2014) Research methods in biomechanics, 2nd edn. Human Kinetics, Champaign

Rosenthal Y, Virk MS, Zuckerman JD (2022) Biomechanics of the elbow. In: Nordin M, Frankel VH (eds) Basic biomechanics of the musculoskeletal system, 5th edn. Wolters Kluwer, Philadelphia

Rydevik BL, Kwan MK, Myers RR et al (1990) An in vitro mechanical and histological study of acute stretching on rabbit tibial nerve. Journal of Orthopaedic Research 8:694–701

Sabick MB, Torry MR, Lawton RL et al (2004) Valgus torque in youth baseball players: A biomechanical study. Journal of Shoulder and Elbow Surgery 13:349–355

Scarborough DM, Bassett AJ, Mayer LW et al (2020) Kinematic sequence patterns in the overhead baseball pitch. Sports Biomechanics 19(5):569–586. https://doi.org/10.1080/14763141.2018.1503321

Seroyer ST, Nho SJ, Back BR et al (2010) The kinetic chain in overhand pitching: Its potential role for performance enhancement and injury prevention. Sports Health 2(2):135–146. https://doi.org/10.1177/1941738110362656

Shorter Oxford English Dictionary, 6th edn (2007). Oxford University Press, Inc., New York

Snell RS (1984) Clinical and functional histology for medical students. Little, Brown, and Company, Boston, MA

Topp KS, Boyd BS (2006) Structure and biomechanics of peripheral nerves: Nerve responses to physical stresses and implications for physical therapy practice. Physical Therapy 86(1):92–109

Trinick J (1992) Understanding the functions of titin and nebulin. FEBS Letters 307(1):44–48

Tskhovrebova L, Trinick J (2010) Roles of titin in structure and elasticity of the sarcomere. Journal of Biomedicine & Biotechnology 2010:1. https://doi.org/10.1155/2010/612482

Uchida TK, Delp S (2020) Biomechanics of movement, the science of sports, robotics, and rehabilitation. The MIT Press, Cambridge

Villemure I, Stokes IAF (2009) Growth plate mechanics and mechanobiology: A survey of present understanding. Journal of Biomechanics 42(12):1793–1803. https://doi.org/10.1016/j.jbiomech.2009.05.021

Werner SL, Fleisig GS, Dillman CJ et al (1993) Biomechanics of the elbow during baseball pitching. The Journal of Orthopaedic and Sports Physical Therapy 17(6):274–278

Whitley R (2007) Baseball throwing mechanics as they relate to pathology and performance – A review. Journal of Sports Science and Medicine 6:1–20

Wilke DR (1950) The relation between force and velocity in human muscle. The Journal of Physiology 110:249–280

Winter DA (2009a) Muscle mechanics. In: Winter DA (ed) Biomechanics and motor control of human movement, 4th edn. Wiley, Hoboken, NJ

Winter DA (2009b) Kinesiological electromyography. In: Winter DA (ed) Biomechanics and motor control of human movement, 4th edn. Wiley, Hoboken, NJ

# 5 Wrist: Fractured Scaphoid

## 5.1 Introduction

Entire books have been written on the complex anatomy and biomechanics of the wrist, as well as the functional limitations imposed after particular common fractures, e.g., scaphoid, of which there are many classifications. In this chapter, we conduct an extensive review of wrist joint anatomy, including the salient morphological features of the carpal bones, related passive soft tissue constraints, and relevant extrinsic and intrinsic muscles. The remaining tasks related to this case are to (1) review the important function of the scaphoid in wrist/hand function; (2) describe the mechanism of injury, including structures exposed to external forces and moments and their response; and (3) generate hypotheses with regard to potential symptom generators, as well as the functional limitations imposed by our patient's present pathology.

## 5.2 Case Presentation

Mr. Cobb (Jim), a 28-year-old carpenter, tripped and fell, sustaining a transverse (i.e., waist) fracture of the middle third of the right scaphoid.

Jim is right-handed and was working in his shop when he *f*ell *o*n his *o*ut*s*tretched *h*and (FOOSH). He experienced immediate right wrist/hand pain, and after some time of self-triaging the extent of this injury (he noticed swelling at the base of his thumb, but no deformity), Jim realized he would be unable to continue his work.

An emergency room initial examination and radiograph revealed no fracture, and Jim was sent home with a forearm splint with the wrist immobilized, but the fingers and thumb free to move. He was instructed to control swelling by elevating his arm when not in use, applying a cold pack, and using over-the-counter anti-inflammatory medication, as needed. He was also told to schedule a visit with his primary care physician if his condition did not improve over a 7- to 10-day period. When his symptoms did not change significantly at 2 weeks, Jim visited his doctor who ordered another imaging study, which revealed a non-displaced scaphoid waist fracture line. A short-arm cast was applied with the wrist in slight extension and the thumb free to move to the metacarpal phalangeal joint. Jim reported for physical therapy evaluation and treatment approximately 14 weeks after he had been immobilized and about 2 weeks after cast removal. A repeat radiograph at the time the cast was removed (16 weeks after the injury) revealed a malunion of the scaphoid. He was referred to physical therapy because of difficulty with residual pain and inability to perform his job and other activities of daily living (ADL). A decision on how to handle the malunion would be made following a 2- to 3-week attempt at conservative rehabilitation (Fig. 5.1).

On observation, he had no evidence of residual deformity, but notable atrophy of right hand and forearm musculature and mild swelling around the anatomical snuffbox. Active wrist range of motion (AROM) testing revealed:

| | Right | Left | Normal |
|---|---|---|---|
| Flexion | 55° | 70° | 70–85° |
| Extension | 40° (pain[a]) | 55° | 60–75° |
| Radial deviation | 10° (pain[a]) | 15° | 15–20° |
| Ulnar deviation | 15° | 30° | 35–40° |

[a]Pain occurred during active movements throughout the range of motion

Passive range of motion testing of the right wrist was mildly increased (compared to AROM) with mild pain; the end feel was empty in all directions. Joint mobility testing of the radiocarpal, midcarpal, and intercarpal joints was mildly decreased in all directions, with muscle guarding noted particularly with the palmar mobility test at the radiocarpal joint. Resisted muscle testing revealed normal strength of the major muscle

**Supplementary Information** The online version contains supplementary material available at https://doi.org/10.1007/978-3-031-25322-5_5.



G. J. Alderink, B. M. Ashby, *Clinical Kinesiology and Biomechanics*, https://doi.org/10.1007/978-3-031-25322-5_5

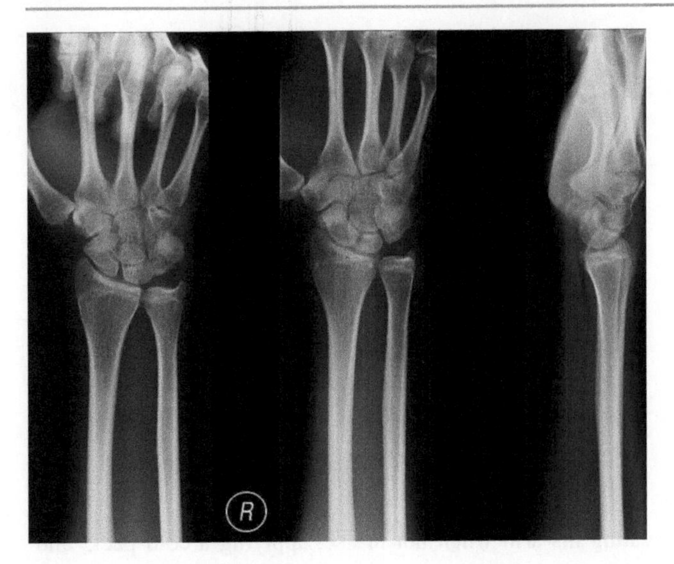


**Fig. 5.1** Left to right, posteroanterior, relatively oblique, and lateral radiographs that show signs of healing of the scaphoid but in an apparent humpback posture

groups at the shoulder and elbow, but Jim was weak and painful with resisted testing of the right wrist extensors and strong and painful with resisted wrist flexors. He had normal strength of the left wrist extensors and flexors. Handheld dynamometry grip strength testing revealed right 30 kg and left 50 kg. Vascular and neurological screening tests, i.e., reflexes and sensation, were normal. Palpation revealed mild tenderness over the anatomical snuffbox and moderate point tenderness over the dorsal and palmar aspect of the lunate on the right side.

## 5.3 Key Facts, Learning Issues, and Hypotheses

A brief review of the epidemiological literature indicates that fractures of the scaphoid account for 50–80% of all carpal bone fractures in young and active individuals. Although scaphoid fractures are the most common carpal bone fractures, up to 25% of them are missed with plain radiographs during the acute stage of fracture because of the anatomical complexity of the scaphoid. This fracture is known to compromise the blood supply to the bone, which can lead to slow (delayed union) or no healing (non-union). Non-union occurs in approximately 5–10% of non-displaced scaphoid fractures. Another common complication associated with a scaphoid fracture, whether it is treated with or without internal fixation, is malunion, described as a humpback deformity.

Examination of this case takes us down several paths. We need to review fracture mechanics, which is also presented in Chap. 3. Since, in this case, "healing" had taken a bit of a tortuous, although not uncommon, route, we look in more detail at the normal vascular supply to the radial side of the wrist and hand and the normal inflammatory cycle and healing stages of biological tissues.

It is clear from the physical examination that Jim has, in addition to continued wrist/hand pain, significant impairments of range of motion of the right wrist and weakness of several muscles that function across the wrist joint. He also demonstrated generalized forearm muscle atrophy, which is likely related to the length of time he was immobilized, and perhaps arthrogenic pain (with neurological inhibition of normal muscle action). Knowing that the scaphoid is a keystone for wrist/hand function, we examine the anatomy and osteo- and arthrokinematics of the wrist; as well, we speculate how a humpback deformity of the scaphoid may contribute to abnormal wrist biomechanics. We also need to explore the complex of passive and active stabilizing soft tissues around the wrist joint. Based on the mechanism of injury, i.e., forced wrist hyperextension, and muscle guarding during palmar joint mobility testing of the radiocarpal joint, it is possible Jim sustained a significant strain injury to one or more palmar extrinsic ligaments. Because he was immobilized for an extended period of time and continues to be unable to participate in normal daily activities using his right hand, we also study the effects of immobilization on biological tissues, e.g., bone, articular cartilage, ligaments, tendons, and muscle.

## 5.4 Mechanism of Injury: Impulse-Momentum Approach to Estimate Forces on the Scaphoid During a FOOSH

Falling on an outstretched hand is the most common mechanism of injury to the distal radius and scaphoid. In Jim's case, apparently, the scaphoid was the weak link in the right upper extremity because it was the structure that failed. Recall that in the presentation on the mechanical properties of bone, we observe from the study of the stress/strain diagram (Fig. 5.2) that with externally applied loads, bone deforms (strains) in a linear fashion until the yield point is reached. At the yield point, the tissue is subjected to microdamage, which is exceeded as the external load continues to be applied. Tissue deformation after the yield point is permanent until eventually the tissue can no longer sustain internal stresses and the tissue fails; in the case of bone, ultimate strain or failure is the fracture. What was it about the hyperextension of the wrist that resulted in a fracture to the scaphoid?

As noted in the range of motion test data for Jim, normal active wrist extension ranges from 60° to 75°. Normally, with passive overpressure in the direction of extension, an additional 10° of extension can be achieved. Note: we typically apply overpressure to a joint in multiple directions but

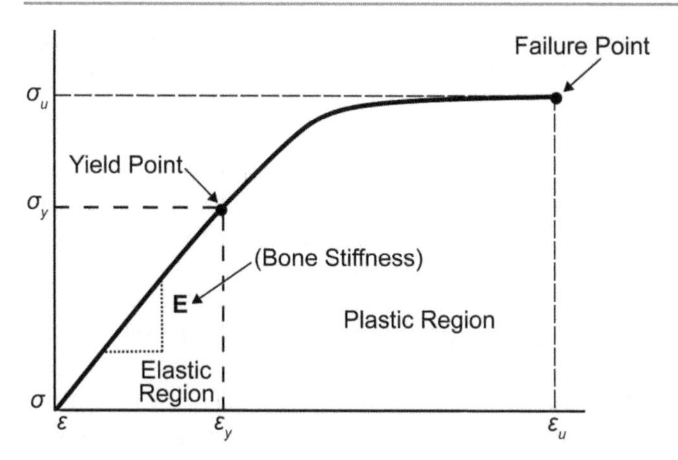


**Fig. 5.2** Schematic of typical stress/strain diagram of compression loading of bone to failure. $\sigma$ = stress (load); $\sigma_u$ = ultimate stress (maximum load); $\sigma_y$ = yield stress (yield load); $\varepsilon$ = strain (deformation); $\varepsilon_y$ = yield strain; $\varepsilon_u$ = ultimate strain; $E$ = elastic (or Young's) modulus

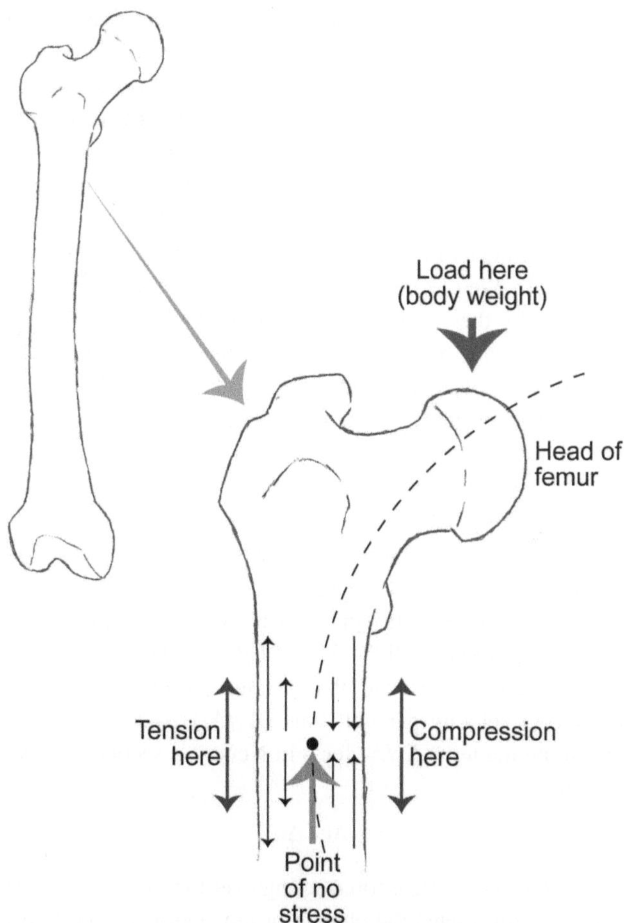


**Fig. 5.3** Schematic illustrating a bending moment produced in the upper diaphysis of the femur. The moment is roughly equivalent to the force of the body weight applied to the head of the femur times a moment arm approximately equal to the length of the femoral neck. This produces a bend in the femoral shaft that imposes compression and tension stresses on the concave and convex sides of the bend, respectively

are particularly interested in testing joints in their close-packed position, which, for the wrist, is hyperextension. Recall (from Chap. 3) that the close-packed position is hypothetically the position where the involved bones that make up the joint are maximally congruent (i.e., fit together the best), and the intrinsic and extrinsic periarticular connective tissues are maximally taut. In other words, the close-packed position is thought to be a position of maximum joint stability, so that with overpressure testing there is a point where a passively applied force yields no further joint movement. When Jim tripped, however, his body accelerated to the floor so that the force imposed on the outstretched hand when it contacted the floor was very large, forcing his wrist into about 95–100° of extension. We define the forces imposed by the FOOSH mechanism on the various tissues of the wrist, i.e., bones and soft tissues, as *impulsive* because of the force magnitude and the high rate of loading. It has been reported (see Neumann 2017) that when the wrist joint is loaded under compression, the lateral column, i.e., scaphoid-distal radius, absorbs 80% of the applied load. About 20% of the load is shouldered by the medial column, i.e., ulna-menisco-triquetral complex, depending on the frontal plane position of the wrist at impact and the radioulnar interosseous membrane. Therefore, when the wrist was forcibly close-packed, a complex combination of impulsive loads was imposed on the distal radius and scaphoid. Because of the unique morphology of the scaphoid, however, large compression loads were likely combined with a bending moment about the waist region of the bone. Recall, with bending moments, the bone is placed under tension on the convex side of the bend and compression on the concave side of the bend (Fig. 5.3). Since it has been shown that cortical bone is weaker under tensile loading, it is possible that the scaphoid failed under the tensile load associated with a bending moment.

But there is more. Because of the maximum tautness that was placed on the periarticular soft tissues when the radiocarpal joint was placed in its close-packed position, it is possible that significant strain injuries may have been imposed on several ligaments, including the primary constraints to palmar glide of the scaphoid and lunate bones, e.g., scapholunate ligaments, etc. Finally, we have to consider the possibility of strain injuries imposed on flexor muscle-tendon actuators related to their high rate of reflexive eccentric actions when the wrist was forced into hyperextension.

Let's first look in particular at the effect of external forces on bones. In previous chapters, we used static equilibrium problems to provide insight into the bone-on-bone forces produced at joint surfaces as a result of muscle actions. In this case, the injury mechanism provides an opportunity to use a dynamics *impulse-momentum* approach to clarify the injury. We are interested in estimating the compressive force absorbed by the scaphoid when falling on the hand. We assume, in this case, that when Jim fell, the load applied to the hand was concentrated on the radial half of the palm with the wrist in approximately 95–100° extension. In a simulated fracture experiment, Weber and Chao (1978) determined that the scaphoid fractured under a force ranging from 460 to 960 lb secondary to compression and bending moment at the distal pole. We can compare this force to the ultimate force tolerated by the scaphoid based on previous biomechanical studies that may be reported in the literature, or alternatively, we may need to extrapolate from studies on adult cortical bone. Note: failure is actually determined by the ultimate stress, not the ultimate force; in order to determine fracture stress, we would need to know the cross-sectional area of the scaphoid along with the force acting on it.

An impulse-momentum approach is particularly useful when a force acts on an object over a period of time. Using an impulse-momentum approach is essential when a collision is involved (like the collision between the hand and the floor or ground). Recall that average acceleration is the change in velocity per unit time ($\Delta v/\Delta t$). Therefore, if we substitute the term $\Delta v/\Delta t$ for $a$ in Newton's second law, we obtain:

$$F = m\Delta v / \Delta t$$

where $F$ is the average force acting over the time interval $\Delta t$ and $m\Delta v$ represents the change in momentum of the body. This equation may be manipulated so that:

$$F\Delta t = m\Delta v$$

or

$$F\Delta t = mv_2 - mv_1.$$

The product of force and change in time is called the *impulse*. A greater force, a force applied over a longer time, or a combination of both increases the magnitude of the change in momentum. The change in the velocity vector points in the same direction as the resultant force. Thus, depending upon the direction of the force, the velocity may be increased or retarded. The impulse involved to stop a moving object corresponds directly to the change of momentum of the object. The force and time of the impulse are inversely related. A longer time taken to stop a moving object requires less force by allowing an increased time for the momentum to change. Conversely, an increased force reduces the duration necessary to stop the object.

Given Jim's body mass of 100 kg, with a center of mass 1.0 m above the ground, let's estimate the force on his outstretched hand after he tripped. Assume that the center of mass of the body falls freely for 0.6 m; i.e., distance ($d$). The velocity at the instant the hand strikes the floor may be calculated using constant acceleration kinematic principles. The constant acceleration kinematics equation that uses velocity, acceleration, and displacement is:

$$v^2 = v_0^2 + 2a\left(x - x_0\right)$$

Solving for the velocity $v$ and assuming the initial velocity $v_0 = 0$, the acceleration $a = g = 9.81$ m/s$^2$, and the change in displacement $x - x_0 = d = 0.6$ m:

$$v = \sqrt{2gd} = \sqrt{2\left(9.81\,\text{m}/\text{s}^2\right)\left(0.6\,\text{m}\right)} = 3.43\,\text{m}/\text{s}.$$

Conservatively assuming an impact duration of $\Delta t = 0.1$ s, with a velocity at the end of the impact of zero, along with Jim's mass of 100 kg, we can now use the impulse-momentum equation to find the impulse and average impact force:

$$F\Delta t = m\Delta v$$

$$F\Delta t = \left(100\,\text{kg}\right)\left(3.43\,\text{m}/\text{s}\right)$$

$$F\Delta t = 343\,\text{Ns}$$

With an impulse of 343 Ns acting over $\Delta t = 0.1$ s, the average impact force is found by:

$$F = \left(343\,\text{Ns}\right)/\left(0.1\,\text{s}\right) = 3431\,\text{N}.$$

Thus, with Jim's fall on the outstretched hand, an average force of approximately 3431 N or 771 lb was induced on his extended wrist. This was the average force; the peak force would have been as much as two times higher or about 1543 lb. Now, not all of this force is applied to the radioscaphoid joint since it has been shown that approximately 20% of the compressive load is transferred to the radiolunate and ulno-menisco-triquetral joints and triangular cartilage. If that is the case, a peak compressive force of approximately 1234 lb was imposed on Jim's radio-scaphoid joint. Based on the work of Weber and Chao, we know that this force magnitude could certainly result in a waist fracture of the scaphoid.

## 5.5 Wrist Structure and Function

### 5.5.1 Structure: Osteology, Ligaments

The wrist is composed of the distal radius, ulna (and related triangular cartilage), and a proximal and distal row of carpals that forms the radiocarpal and midcarpal joints. The two rows of bones consist of eight carpal bones that form a functional bridge connecting the forearm and hand. The primary motions of the wrist occur in the sagittal (allowing physiological flexion/extension) and frontal planes (allowing physiological radial and ulnar deviation). However, we remind the

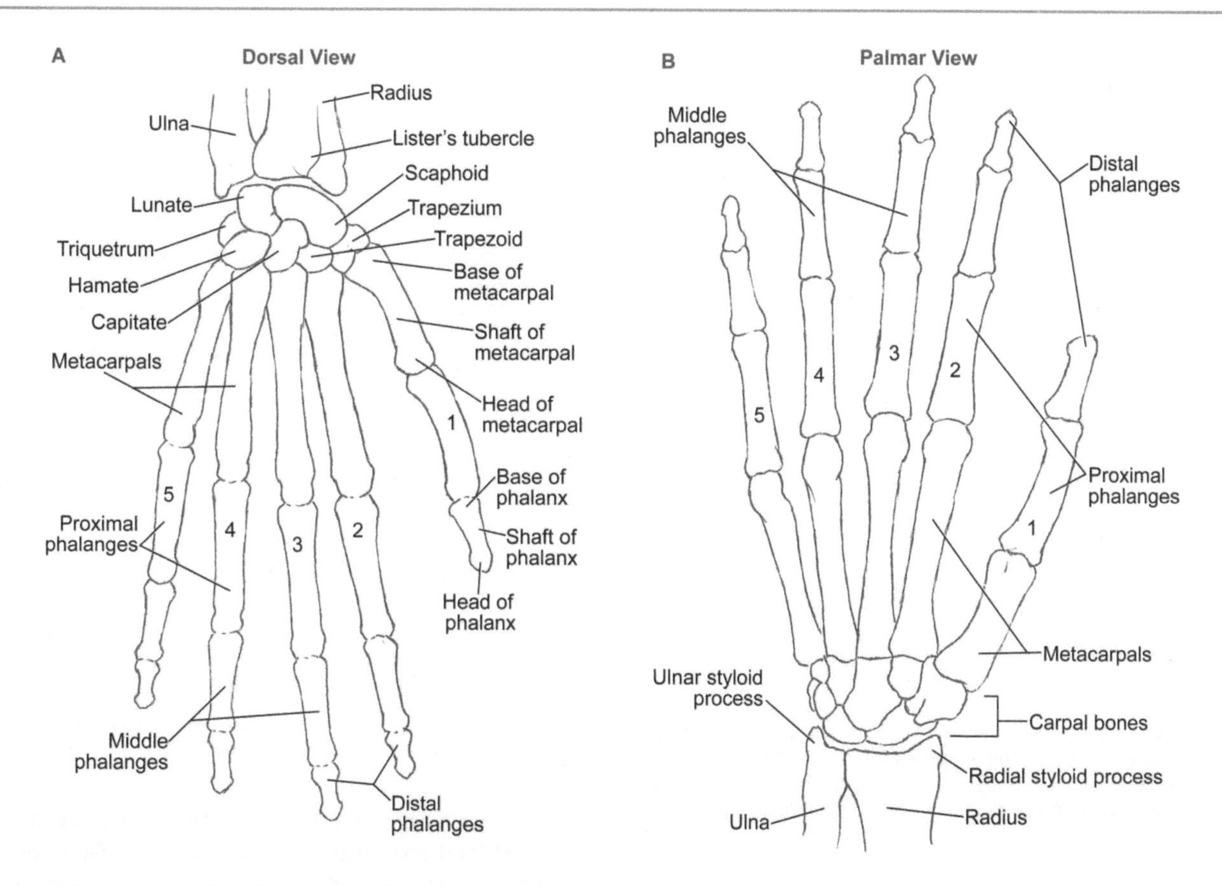


**Fig. 5.4** (A) Dorsal and (B) palmar (ventral) perspectives of the bones that comprise the right wrist

reader that the wrist complex, like all joints in the human body, has the potential to move with six degrees of freedom, which makes it possible to position the wrist and hand for performing a myriad of functional tasks. As we can see in Jim's case, the issues he presents to us as a result of the fractured scaphoid and residual pathomechanics have significantly affected the use of his hand for work activities as well as routine daily tasks. A detailed study of the functional anatomy of the wrist may provide us with additional insight into Jim's problems.

The volar and dorsal view of the distal radius and ulna, proximal and distal carpals, and metacarpals reveal their unique configurations that have structural and functional importance (Fig. 5.4). For example, notice the several grooves and tubercles dorsally that serve to separate and guide distal tendons to the wrist and hand. The volar and dorsal surfaces of the distal radius and ulna serve as attachments for the proximal wrist joint capsule and associated extrinsic ligaments. We see that the conjoined distal surfaces of the ulna and radius form ovoid concave surfaces that articulate with the proximal row of carpals, which have reciprocally-shaped geometries.

A closer look at the distal radius and its angulating tilt in both frontal and sagittal views is interesting (Fig. 5.5). In the frontal plane, the distal radius tilts toward the ulna,

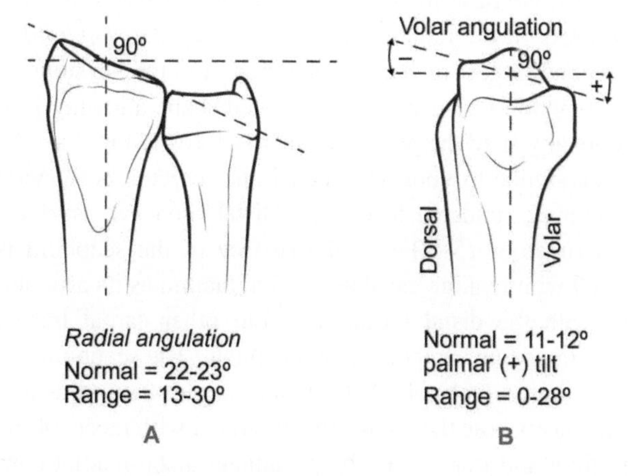


**Fig. 5.5** Schematic illustrating distal radius angulations in the (A) frontal and (B) sagittal planes

which kinematically limits radial deviation and facilitates ulnar deviation of the wrist. A sagittal view of the distal radius reveals a slight palmar tilt that predisposes the wrist to a slightly greater flexion, than extension, range of motion.

The proximal row of carpals includes the scaphoid (formally navicular), lunate, triquetrum, and pisiform. In the distal row, we see the trapezium, trapezoid, capitate, and hamate. Each carpal has a unique geometry, all essential for

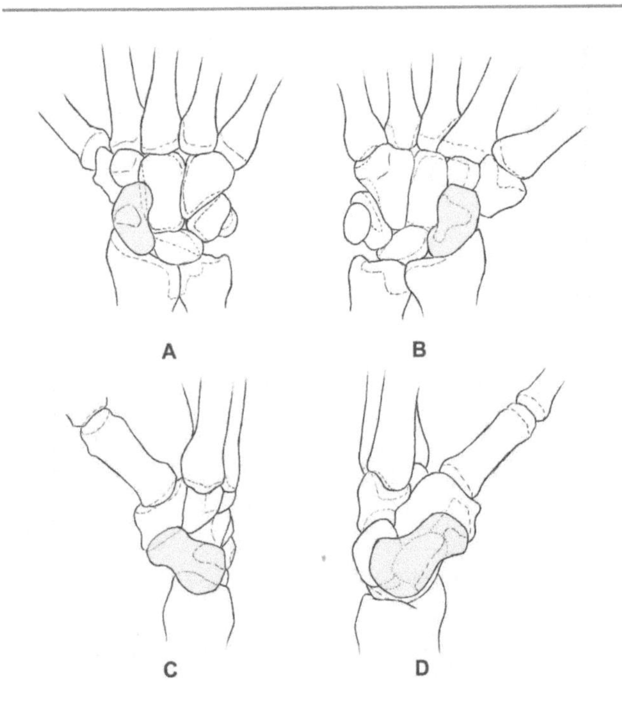


**Fig. 5.6** Schematic of radiocarpal and midcarpal joints highlighting the scaphoid showing its complex geometry from multiple views: (**A**) dorsal frontal, (**B**) palmar frontal, (**C**) radial sagittal, and (**D**) ulnar sagittal. It is notable that the extent of articular cartilage encasing the scaphoid is significant

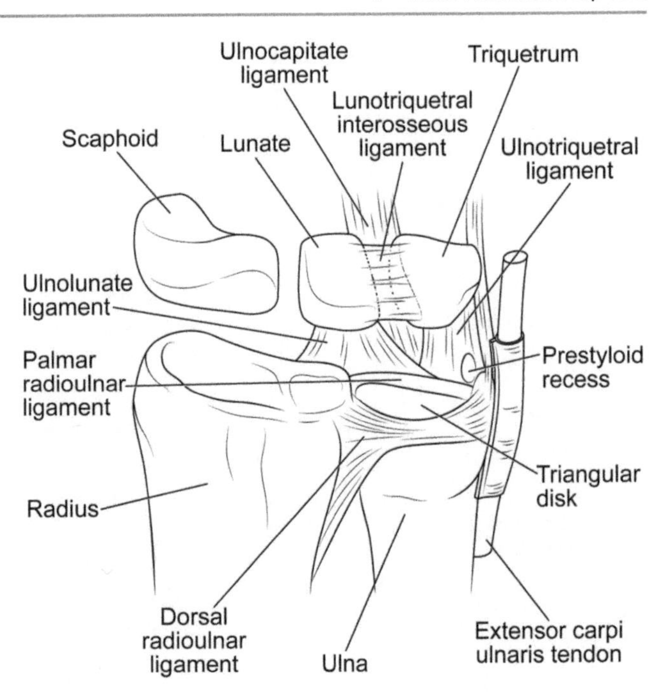


**Fig. 5.7** Schematic of the ulno-mensico-triquetral, triangular disc complex and its associated ligamentous tissues

the inherent stability and mobility of radiocarpal and midcarpal joints. We describe in detail the salient features of each. Positioned on the radial side of the carpus, the *scaphoid* is the second largest of the carpal bones. It is the keystone to wrist function because of its elongated shape allowing it to functionally span the proximal and distal rows (Fig. 5.6). Its shape is similar to a boat (from the Latin *scaphos*, boat), with the convex undersurface and distal concave surface. Approximately 75–80% of the surface of the scaphoid is covered with hyaline cartilage, which facilitates its articulations with the distal radius and four other carpal bones: lunate, trapezium, trapezoid, and capitate. The scaphoid has a convex proximal pole that articulates with the radius and distal convex pole that is mostly congruent with facets of the trapezium and trapezoid. The prominent distal-medial concave facet of the scaphoid fits neatly with the head of the capitate, whereas a smaller medial semi-lunar facet articulates with the lateral aspect of the lunate. There are two primary locations for attachments of ligaments on the scaphoid: (1) near the small facet on the scaphoid's medial side that articulates with the lunate, which is reinforced by the scapholunate ligament, and (2) on the proximal volar surface of the scaphoid tubercle that serves as an attachment site for the transverse carpal ligament. These areas are also important because they are primary locations at which the scaphoid receives its vascular supply.

The *lunate* (from the Latin *luna*, moon) is a wedge- or crescent-shaped bone that lies in the central column of the proximal row. It has five articulations: distal radius, the proximal pole of the scaphoid, the distal radial aspect with the triquetrum, the distal central aspect with the capitate, and the distal hamate. Despite these many bony contacts, the lunate is the least stable carpal because of its shape and poor dynamic stabilization. In addition, its ligamentous connection to the very stable capitate is insufficient. The lunate's proximal surface is convex, whereas its distal surface is concave that matches two articulations with the convexities of the capitate and apex of the hamate. Both the volar and dorsal surfaces of the lunate are sites of vascular ingrowth.

The triangular *triquetrum* forms the ulnar border of the carpus, along with its riding partner, the *pisiform*. Despite the fact that the pisiform is embedded within the tendon of flexor carpi ulnaris and serves as the attachment for the abductor digiti minimi and several ligaments, it functions as a sesamoid with no significant role in wrist mechanics. The triquetrum is part of the important ulno-meniscal-triquetral or triangular fibrocartilage complex (TFCC) (Fig. 5.7). Although the medial radiocarpal column was not directly affected by Jim's scaphoid fracture, the concept of dynamic coupling suggests that we consider the role that the TFCC plays in wrist mechanics because its structures likely adapt to the pathomechanics located in the lateral column. The complexity of the TFCC is attested by its constituents: (1) triangular disc, (2) distal radioulnar capsular ligaments, (3) ulnotriquetral and ulnolunate ligaments, (4) ulnar collateral ligament, and (5) fascial sheath that surrounds the extensor carpi ulnaris tendon. The triangular fibrocartilage forms the structural framework for the entire complex. From Chap. 3, we know that fibrocartilage, in general, performs the important dual functions of enlarging articular surfaces and serving as attachment sites for periarticular ligamentous constraints. The TFCC satisfies these general characteristics in that its proximal surface accepts a portion of the distal ulna at the distal radioulnar joint, while its distal surface articulates with the convex parts of the lunate and triquetrum at the radiocarpal joint. Accordingly, its primary specific function is to provide stability to the distal radius and ulna while promoting forearm pronation and supination. Secondarily, the TFCC reinforces ligamentous constraints ulnarly and provides a fibrocartilaginous layer between the ulnar and triquetral joint surfaces that can distribute/mitigate large compression forces.

The *capitate* (from the Latin root *caput* meaning head) is the largest carpal bone and occupies a central location within the wrist, thus serving as the center of rotation for all wrist motions. The proximal convex head articulates with the concavities of the scaphoid and lunate. Sitting tightly between the hamate and trapezoid, its stability is secured by intracarpal ligaments. Relatively rigid articulations distally between the capitate and third metacarpal (and possibly 2nd and 4th metacarpals) dictate stout longitudinal stability of the central column, of which it is a keystone.

The *trapezium* has an asymmetrical shape to suit the unique functional role of the thumb for grasping. Its proximal convex surface articulates with the reciprocally shaped scaphoid, while its distal articular geometry features a saddle-shaped surface. Recall that the characteristic of a saddle joint is that each articular surface is concave in one direction and convex in the other. In the palmar-to-dorsal direction, the trapezium is concave, whereas in its transverse diameter, the articular surface is convex in a medial-to-lateral direction (the contours of the base of the first metacarpal's articulating surface are reciprocally shaped). A tubercle on the palmar surface of the trapeziums serves as an attachment for the transverse carpal ligament (Fig. 5.8).

There are two carpals that play essential, but small, roles in wrist biomechanics: the *trapezoid* and *hamate*. The relatively small trapezoid fits snugly between the trapezium and capitate. Its proximal surface is slightly concave relative to the convexity of the scaphoid, with a distal and immobile articulation with the base of the second metacarpal, making the second ray a very stable central aspect of the carpus. The hamate is wedged between the lunate proximally, triquetrum medially, and capitate laterally and articulates with the bases of the fourth and fifth metacarpal bases. The articulation with the metacarpals importantly contributes to the "cupping" of the hand. The hook of the hamate serves as an additional medial attachment for the transverse carpal ligament.

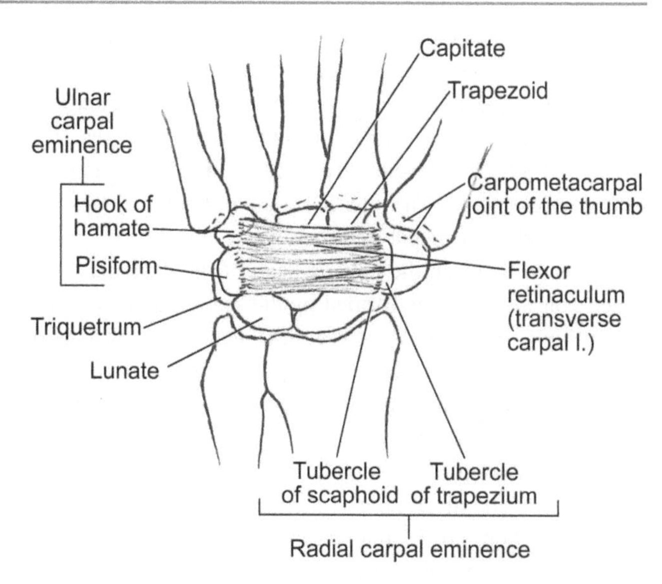


**Fig. 5.8** Schematic of the volar aspect of the wrist with the overarching transverse carpal ligament, its points of attachment, and relationship to proximal and distal carpals

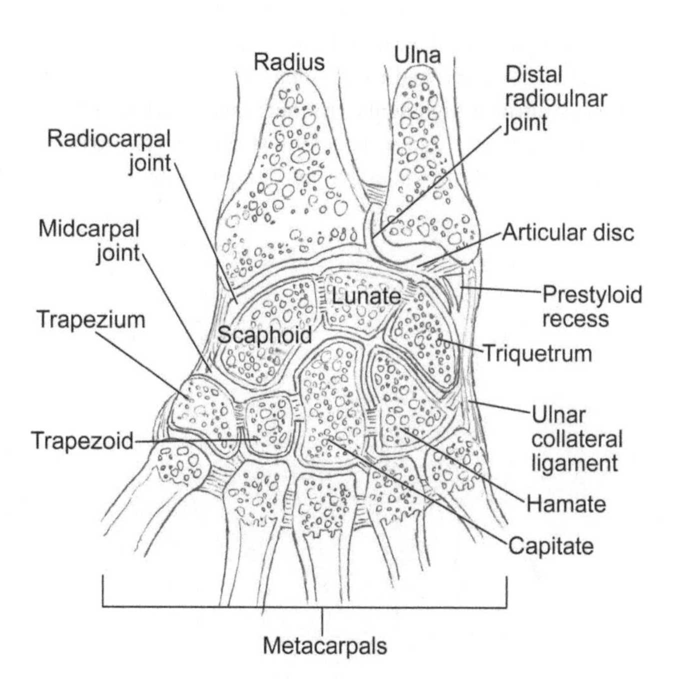


**Fig. 5.9** Cross section schematic of radiocarpal, midcarpal, and intercarpal joints

Joint stability is provided by a relationship between dynamic (muscle-tendon actuators) and static stabilizers. See the Supplementary 5.1 that provides a summary of the volar and dorsal muscles that cross the wrist and provide dynamic stabilization. What follows below is a more focused review of the static stabilizers of the wrist: first, a brief summary of the organization of the joints of the wrist complex (Fig. 5.9). The wrist consists of two primary articulations: radiocarpal and midcarpal. The radiocarpal joint includes articulation between concave surfaces of the distal radius, ulna and triangular disc, and convex surfaces of the scaphoid and lunate. The midcarpal joint includes the articulations between the proximal and distal rows of carpal bones. The midcarpal joint is conveniently divided into medial and lateral compartments. The medial compartment includes the capitate and hamate but is functionally dominated by the relationship between the convex head of the capitate; combined concave surfaces formed by the scaphoid, lunate, and triquetrum; and the base of the third metacarpal, where, in essence, the head of the capitate and its articulations function like a ball-and-socket joint during physiological movements of the hand in the sagittal and frontal planes. The lateral compartment, although less mobile, but no less important, is composed of the distal pole of the scaphoid, trapezium, and trapezoid. There are 13 intercarpal joints whose small, six-degree-of-freedom movements are importantly dictated by the unique geometries of each carpal, intrinsic, and extrinsic ligaments, and the wrist/hand muscle-tendons.

Like all diarthrodial joints, static stabilization of the wrist complex is ensured by intrinsic, e.g., joint capsule, and extrinsic ligaments. Extrinsic ligaments have their proximal attachments on the radius or ulna and attach to carpals, whereas intrinsic ligaments originate and insert into bones within the carpus. Ligaments are most notable for providing passive constraint to excessive movement of the bones that make up articulations, but remember that they also (1) store potential energy when elastically strained that contributes to joint movement; (2) contain mechanoreceptors that convey joint position and rate of position change to the CNS, which provides feedback to joint effectors, i.e., muscles; and (3) provide information via a direct feedback loop to joint-related muscle-tendon actuators that are activated to provide dynamic support.

Given what we know about the complex interplay among affective structures located in the skin, ligaments, joint capsules, muscles, and tendons which activate and control muscle-tendon actuators to optimize function, intuition suggests that Jim's fracture and possible strain injury to one or more palmar radiocarpal ligaments have likely disrupted normal neurological sensory-motor function and mechanical stability.

Ligaments within the fibrous joint capsule of the wrist joint are typically named by the bones of their attachments. The extrinsic carpal ligaments are divided into three main groups: palmar radiocarpal, palmar ulnocarpal, and dorsal radiocarpal. There are no formally named ligaments on the dorsal ulnar aspect since this region is occupied by the TFCC (Fig. 5.10). Four ligaments comprise the palmar radiocarpal group: radioscaphoid (RSL), radioscaphocapitate (RSCL), and the long and short radiolunate (LRLL). The RSCL is particularly important for scaphoid stability and serves as a fulcrum for

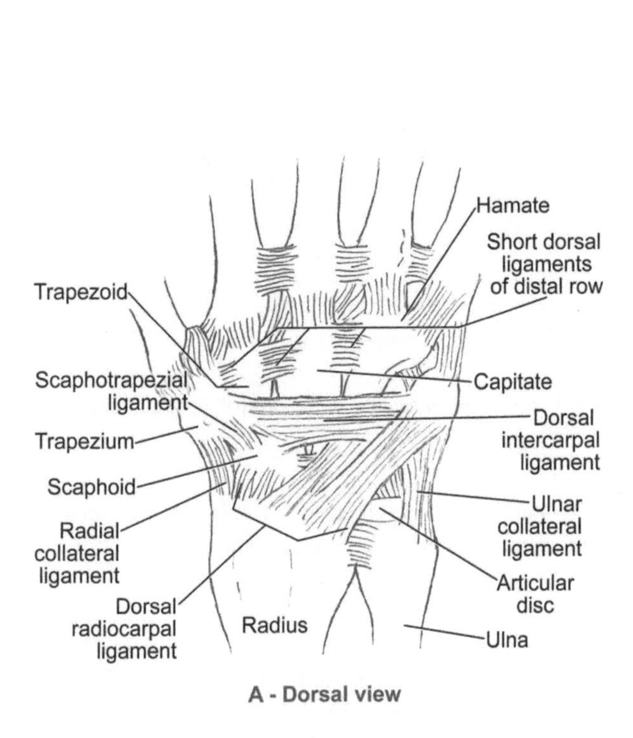


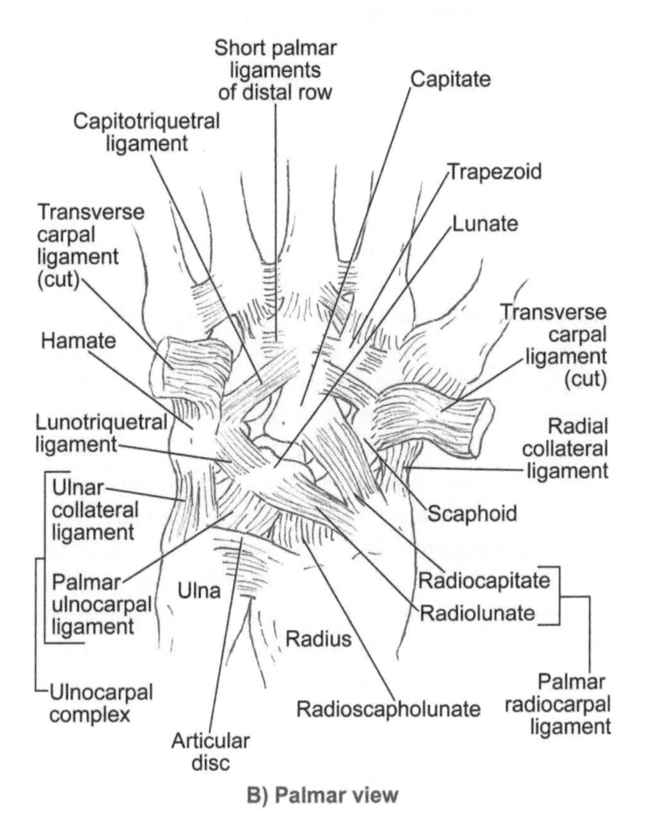


**Fig. 5.10** Primary (**A**) dorsal and (**B**) palmar ligaments of the wrist. Note the grouping of ligaments associated with the ulnocarpal (TFCC) complex

scaphoid rotation. It is notable that the interval between the RSCL and LRLL, known as the *space of Poirier*, is an area of weakness that often allows lunate subluxation secondary to tensile strain injury from forced wrist hyperextension.

The palmar ulnocarpal group of ligaments, organized into superficial and deep structures, includes the ulnar capitate (UCL), ulnar triquetral (UTL), and ulnar lunate (ULL) (all deep to the palmar ulnocarpal ligament in Fig. 5.10b). The ulnar collateral ligament represents a thickening of the medial wrist joint capsule, which, with the UTL and flexor and extensor carpi ulnaris muscles, reinforce the ulnar side of the wrist.

The dorsal radiocarpal ligament, also referred to as the dorsal radial triquetral ligament (DRTL), is a broad thin ligament with numerus attachments onto the triquetrum, lunate, and scaphoid. This ligament reinforces the radiocarpal joint and contributes to the arthrokinematics of the proximal carpus. Fibers that attach to the lunate may help control excessive palmar (volar) glide produced by forces exaggerating wrist extension. This thin dorsal ligament appears to be rich in sensory mechanoreceptors suggesting an important role in wrist proprioception.

Controlled arthrokinematics and joint stability of every intercarpal joint are provided by one, or more, intrinsic ligaments. These ligaments have been classified according to their relative length: short, intermediate, or long. The short ligaments are primarily restricted to the distal row of carpals, providing firm constraint, thus allowing this row of bones to move as a functional unit. Of the four major intermediate ligaments, lunotriquetral, scapholunate, scaphotrapezial, and scaphotrapezoidal, the dorsal, palmar, and proximal parts of the scapholunate appear to be most crucial for scaphoid stability. The long palmar intercarpal ligament consists of two legs, the lateral which runs obliquely from the capitate to the distal pole of the scaphoid and the medial which runs obliquely from the capitate to the triquetrum. This pair of ligaments, giving an inverted-V appearance, is essential for normal wrist arthrokinematics. The dorsal intercarpal ligament connects the trapezium, scaphoid, triquetrum, and lunate, which notably is responsible for controlling excessive dorsal glide of the scaphoid and lunate and, like the dorsal radiocarpal ligament, richly packed with mechanoreceptors.

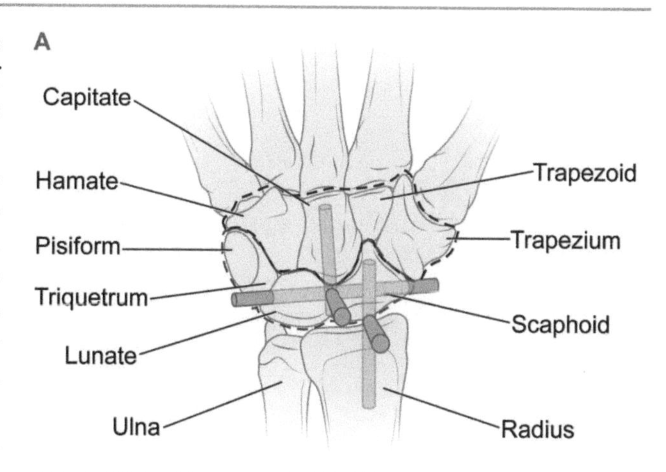


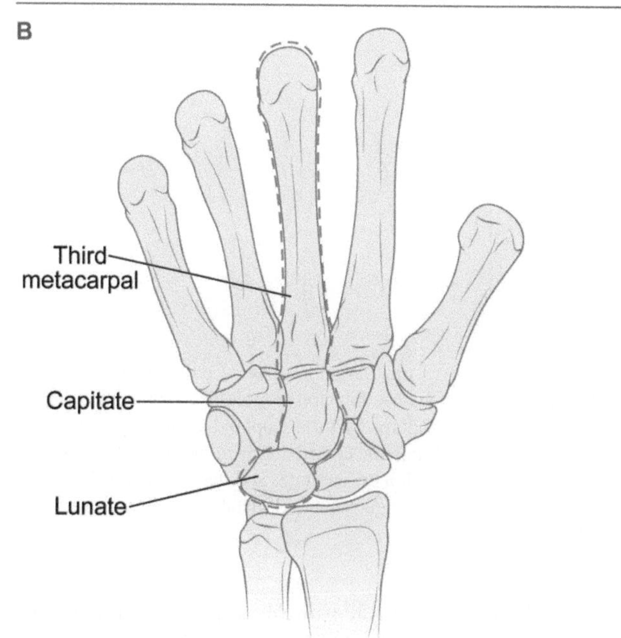


**Fig. 5.11** Schematic showing biplanar axes within the wrist complex located approximately at (**A**) radiocarpal and midcarpal joints, with (**B**) the capitate as a key carpal

### 5.5.2 Kinematics: Osteo- and Arthrokinematics

#### 5.5.2.1 Osteokinematics

Throughout this text, we emphasize the potential for six degrees of freedom of movement in all human joints. In our examination of the wrist complex to this point, it would not be difficult to accept that premise. Historically, however, the wrist has been classified as an ellipsoid (or condyloid) joint with two degrees of freedom: flexion/extension and ulnar/radial deviation. Ex vivo cadaver studies and computational methods, with the aid of computed tomography and magnetic resonance imaging scanning, have provided empirical evidence for a relatively simple model of wrist osteokinematics with Cartesian-like biplanar axes representing rotations in the sagittal and frontal planes (Fig. 5.11). As well, others have provided a detailed description of three-dimensional intercarpal movements. Three general theories have been proposed to explain wrist osteokinematics: row, column, and ring. With the row theory, the carpal bones work as two functional units, with the distal and proximal carpal rows moving as separate units (Fig. 5.11A). Full wrist flexion/extension occurs as a result of the coordination of rotations at both the radiocarpal and midcarpal joints. Ulnar and radial deviation range of motion appears to be the result of combined rota-

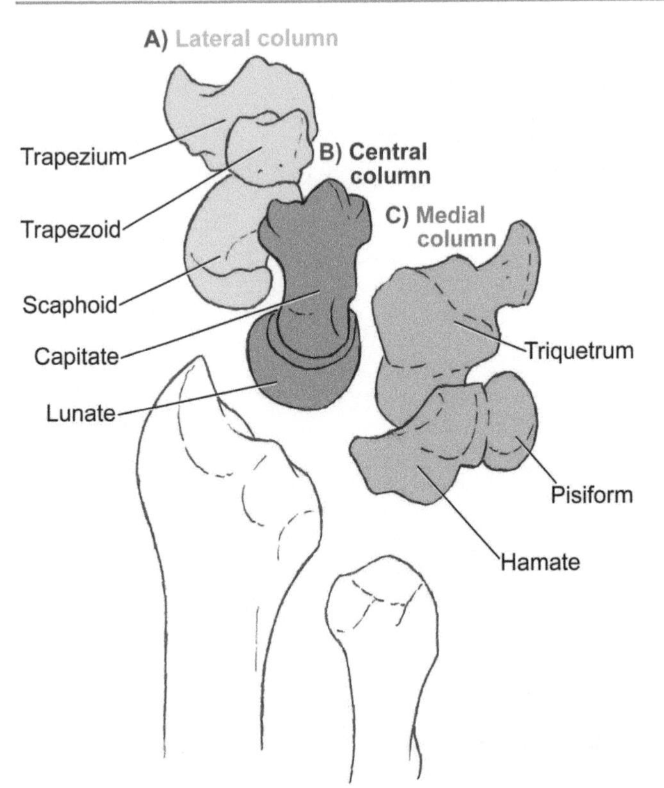


**Fig. 5.12** Schematic of one variation of wrist column theory, where (**A**) the lateral column, whose function is mobility, includes the scaphoid, trapezoid, and trapezium, with the scaphoid as the key carpal; (**B**) the central column with notable contributions to flexion/extension provided by mobility of the lunate, capitate, and hamate; and (**C**) medial column comprised of the triquetrum and distal carpal row whose primary function is rotation

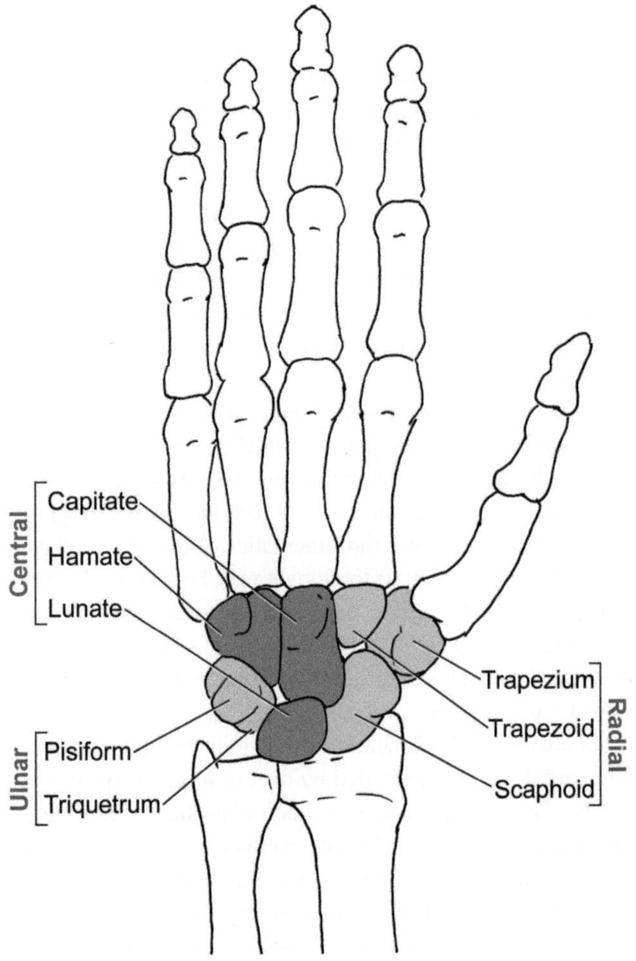


**Fig. 5.13** Schematic of central column (red) theory showing linkages between distal radius, lunate, capitate, and third metacarpal. Note the ulnar column (orange) is comprised of the triquetrum and pisiform and the radial column (green) is comprised of the scaphoid, trapezium, and trapezoid

tions, or pivoting, of each row of carpals. Although this model seems to capture fundamental osteokinematic features of the wrist, it does not appear to provide insight into the complex interplay of carpal movements.

Perhaps, the more accepted model suggests that the wrist complex functions as three columns, each with its own unique function. Let's note at the outset that biomechanists and research clinicians have proposed several variations of column theory without reaching a consensus about which model best explains wrist function. We present a few of the variations. One central column model has three components, the first composed of the lunate, hamate, and capitate, which primarily contribute to flexion/extension. In this version, the triquetrum/pisiform is associated with pronation/supination and makes up the medial column (note: the pisiform was later removed from the model because of its primary role as a sesamoid). The lateral column is formed by the scaphoid, trapezium, and trapezoid, primarily, though, for load mitigation rather than wrist mobility. Another modification of the three-column model added that (1) the distal row of carpals interacts with the lunate to facilitate flexion and extension and (2) the triquetrum and scaphoid control radial and ulnar deviation through rotation about the central column (Fig. 5.12).

Another version of the column theory model can be described in this way: assuming that the wrist is a double-joint system, with movements occurring at the radiocarpal and midcarpal joints simultaneously, carpal kinematics are characterized by a central column formed by linkages between the distal radius, lunate, capitate, and third metacarpal (Fig. 5.13). The radiocarpal joint in this representation of the central column includes exclusive articulation of the radius and lunate. The mid-portion of the central column forms a part of the midcarpal joint that includes the lunate-capitate articulation, whose central function makes a primary contribution to wrist flexion/extension. Finally, the carpometacarpal joint formed by the capitate and third metacarpal base provides relative rigid stability to the central column as a whole.

The final theory modeling wrist motion and carpal stability is similar to row theory and is referred to as the ring model. In this model, the carpal bones are linked in an oval ring, with the key links connecting the scaphoid, lunate, triquetrum, and distal row of carpals. With this model, the lunate is tapped as the key proximal row carpal. As such, the lunate's spatial position and contribution to wrist stability are influenced by key ligamentous connections to the capitate, scaphoid, and triquetrum (Fig. 5.10). Having several, somewhat disparate, theories that attempt to explain wrist kinematics and stability may be more confusing than helpful. On the other hand, being presented with this variety of models attests to the complexity of a system that clinical biomechanists continue to study. It has been suggested that wide inter-individual variability in types of wrists contributes most to disputes about a one-size-fits-all modeling approach. Despite the many alternative writs joint models, most researchers agree that wrist mobility and stability are influenced by the morphology of individual carpals, the constraint system (static and dynamic), and intercarpal dynamic coupling. Certainly, more research is needed.

#### 5.5.2.2 Arthrokinematics

Accurate and precise descriptions of wrist complex arthrokinematics are hampered by the desire to consider all eight carpals, each with unique morphologies and a propensity to move within multiple planes simultaneously. As with all modeling, we use a more simplified model and describe wrist arthrokinematics in terms of how the central column moves. During wrist extension at the radiocarpal joint, the convex lunate rolls dorsally and glides ventrally, relative to the concave surface of the distal radius. At the midcarpal joint, the convex head of the capitate also rolls dorsally and glides ventrally relative to the concave facet of the lunate (Fig. 5.14). The combined movements at the radiocarpal and midcarpal joints contribute relatively equally to total physiological wrist extension. Recall that end-range wrist extension is the close-packed position of the wrist complex, which hypothetically places all of the carpals, relative to one another, in maximally congruent positions. Furthermore, in general, the palmar static and dynamic soft tissues are placed under tension when the wrist is placed in its close-packed position. In particular, the ventral glide of the inherently least stable lunate places increased tensile stress on the palmar radiocarpal ligaments (Fig. 5.10).

During flexion, joint movement between the lunate and distal radius and capitate and lunate occur in the reverse directions, that is, the lunate rolls ventrally and glides dorsally. Concomitantly, at end-range wrist flexion, the dorsal soft tissues become taut and ventral structures slacken. We identify a significant problem when wrist arthrokinematics is simplified when using only central column theory, that is, not accounting for the key role the scaphoid plays in functionally

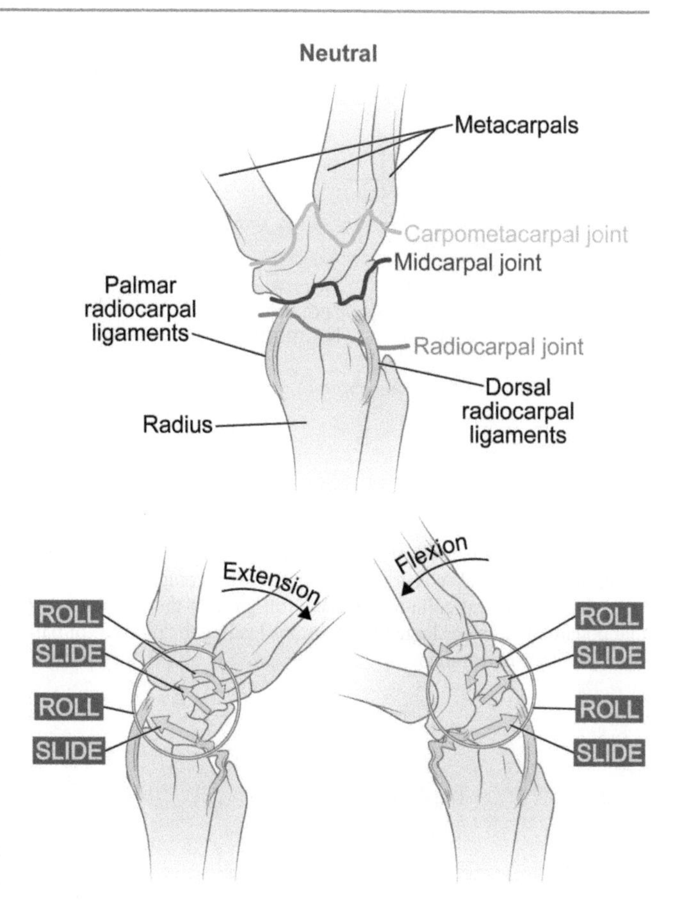


**Fig. 5.14** Use of a central column model of wrist kinematics to illustrate the arthrokinematics during wrist extension and wrist flexion. Notice the change in tension in the dorsal and ventral radiocarpal ligaments at the extreme of motion, e.g., at end range wrist extension, the dorsal ligament is slack, as the ventral spring becomes taut. We see a reverse response of the ligaments with end-range wrist flexion

spanning the proximal and distal carpals. Although the unusual shape of the scaphoid likely contributes to its more complex contribution to wrist arthrokinematics, we assume that during wrist flexion, its roll/glide action at the radiocarpal joint is similar to that of the lunate.

Let's take a brief timeout to examine the connection between abnormal kinematics and Jim's problem. We can be certain that Jim's scaphoid fracture and subsequent malunion are contributing to abnormal wrist arthrokinematics and Jim's continued pain and dysfunction. Additionally, based on the mechanism of injury, i.e., forced wrist hyperextension, and Jim's physical examination findings, e.g., increased muscle guarding with palmar mobility testing of the proximal row of carpals, it is likely that one or more of the palmar radiocarpal ligaments have sustained second-degree strain injuries, leading to ligamentous laxity. Ligamentous laxity has likely induced excessive joint mobility of the carpals. The combination of scaphoid malunion and ligamentous laxity appear to be major determinants of Jim's problems and complaints and a common clinical entity, similar to that described by Neumann; that is, Jim's situation may suggest a

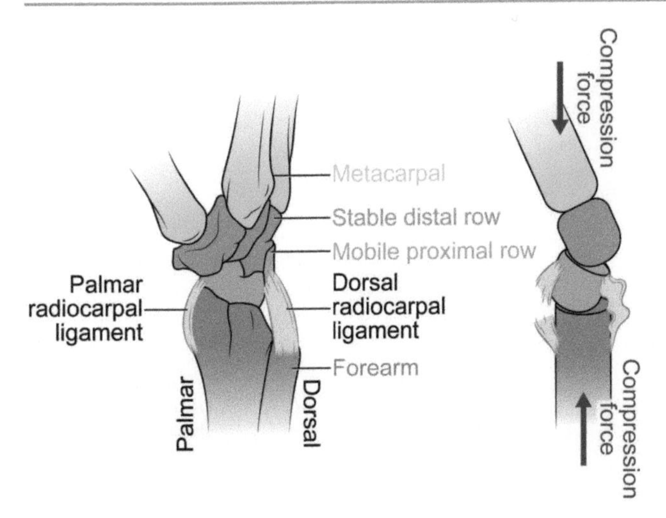


**Fig. 5.15** A schematic illustrating a proposed general mechanism for disruption of carpal constraints secondary to compression forces imposed on a central column. Note the roll/glide of the proximal row that potentially induces large tensile strains leading to "rupture" of palmar soft tissue constraints

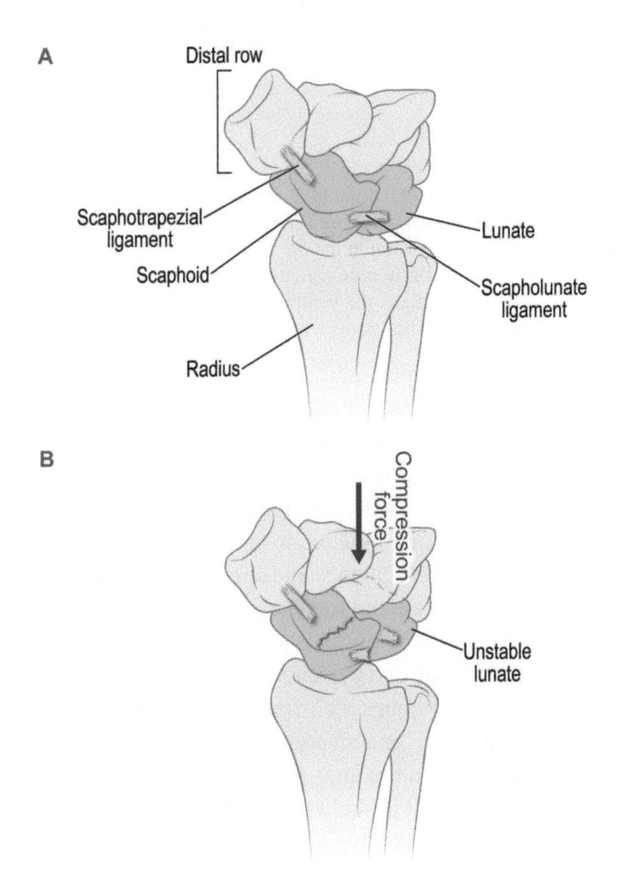


**Fig. 5.16** Schematic of (**A**) normal and (**B**) abnormal scaphoid/lunate configuration with the wrist in a neutral position. Under normal conditions, the scaphoid, and associated key ligaments, spanning both radiocarpal and midcarpal joints contribute to mobility and stability. With dissociation of the scapholunate relationship, wrist mechanics is impaired

potential collapse of carpal bones from a combination of internal compression forces induced by muscle action and external compression forces (Fig. 5.15). The injuries sustained by Jim have likely resulted in scapholunate dissociation, malalignment of the scaphoid and lunate, and pathomechanics of wrist movements (Fig. 5.16). This condition, known as dorsal intercalated segment instability, (DISI) induces a painful and unstable wrist and poor foundation for hand function. Static instability is related to chronic excessive tensile stresses resulting in plastic length changes of extrinsic and intrinsic ligaments and reduction in dynamic stability due to pathological changes in muscle-tendon actuator length-tension and moments. Long term, DISI is associated with changes in bone geometry, chronic inflammation, and advanced degenerative joint disease.

Let's complete our examination of wrist arthrokinematics. In the frontal plane, unlike in the sagittal plane, it appears that a more row-centric theory of kinematics is applied. We note coordinated simultaneous movements of convex-shaped carpals relative to their reciprocal joint partners at the radiocarpal and midcarpal joints. During ulnar deviation at the radiocarpal joint, the scaphoid, lunate, and triquetrum roll in the ulnar direction and glide in the radial direction. Simultaneously, at the midcarpal joint, the capitate rolls ulnarly and glides radially in the large concavity created by neighboring hamate, lunate, and scaphoid bones. The arthrokinematics associated with radial deviation occur in a reverse fashion, but with a more limited range of motion, as noted earlier due to distal radial morphology (Fig. 5.17). The reduction in the distance moved during radial deviation is explained by the constraint imposed by the radial styloid process. The close association of the lunate and scaphoid during ulnar and radial deviation is dependent on the integrity of the palmar radiocarpal ligaments, e.g., the scapholunate ligament.

There are several additional accessory joint rotations in the sagittal and transverse planes that must necessarily be available to foster normal ulnar and radial deviation. For example, during radial deviation, the proximal row flexes slightly; conversely, a slight extension of the proximal row is apparent during ulnar deviation. This pattern of movement is notable by the altered appearance of the scaphoid, which looks shortened at end range radial deviation (as depicted in Fig. 5.17). The "shortened" appearance of the scaphoid occurs because it flexes approximately 15° beyond neutral during radial deviation. This out-of-plane rotation of the scaphoid is a kinematic mechanism that likely promotes additional rolling of the scaphoid before it impacts the radial styloid process. This mechanism has not been fully explained but provides another instance of the complex dynamic coupling of related joints, likely produced and controlled by the coordinated action of muscle-tendon actuators, and extrinsic and intrinsic ligaments.

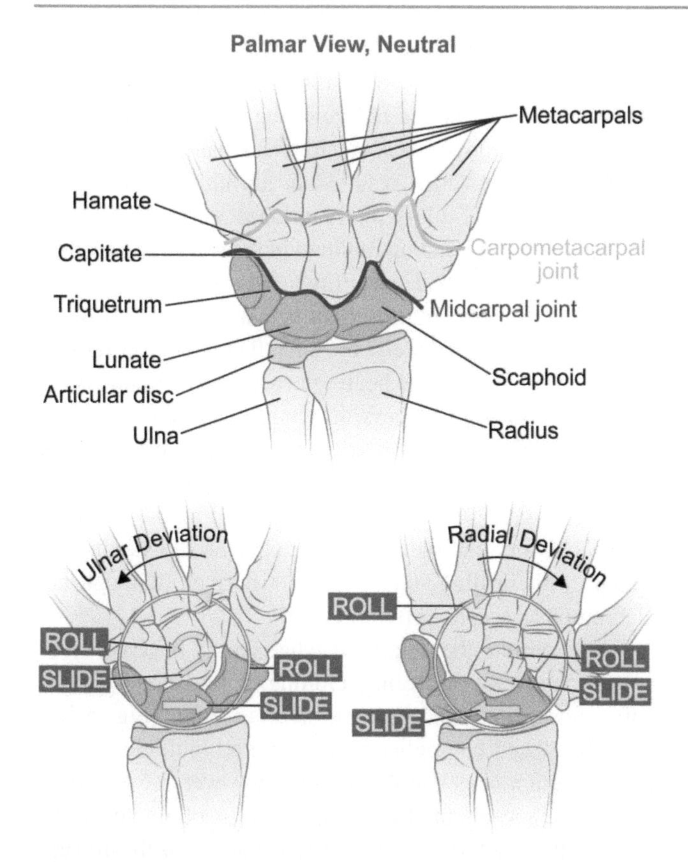


**Fig. 5.17** Schematic demonstrating position and change in position of carpals during ulna and radial deviation. In general, we see that the direction of carpal glides occur in the opposite direction of the rotation, e.g., ulnar deviation induces radial glides

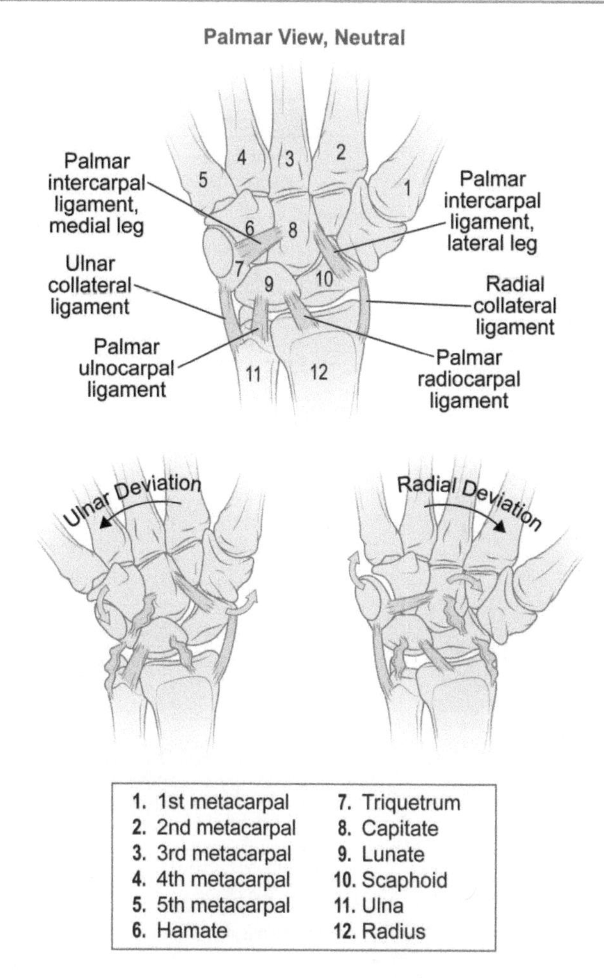


**Fig. 5.18** Schematic of proposed actions of the double-V system of ligaments during ulnar and radial deviation of the wrist. In general, intercarpal ligaments on the ulnar side of the wrist become slackened, with concomitant increasing tension on intercarpal ligaments on the radial side during ulnar deviation and vice versa during radial deviation

One possible explanation for the kinematics related to ulnar and radial deviation described above is offered by our understanding of functional ligamentous anatomy. During ulnar and radial deviation, it is hypothesized that coupled mobility and stability are provided by the double-V system of ligaments that cross the radiocarpal and midcarpal joints (Fig. 5.18). Four ligaments are identified as inverted Vs: a distal pair and a proximal pair. The distal inverted V is comprised of the medial (capitate to triquetrum) and lateral (capitate to scaphoid) legs of the palmar intercarpal ligament (Fig. 5.10); the proximal inverted V includes the lunate attachments of the palmar ulnocarpal and radiocarpal ligaments. In the neutral wrist, all four ligaments are placed under some resting tension. Then, with ulnar deviation, increased tension in the capitoscaphoid ligament appears to facilitate radial glide of the proximal carpals, while the increased tension in the palmar ulnocarpal ligament appears to control excessive radial glide of the proximal carpals. We see an opposite diagonal of ligamentous tensions (capitotriquetral and radiocarpal) during radial deviation. Tension in the radial and ulnar collateral ligaments during ulnar and radial deviation, respectively, likely assists the controlling mechanism of the double-V system.

## 5.6 Inflammatory Cycle and Stages of Healing: Vascular Bed of the Scaphoid

### 5.6.1 Inflammatory Cycle and Stages of Healing

Although this is a text on clinical biomechanics, we are compelled to examine two important physiological aspects of biology: the vascular supply of the scaphoid and the inflammatory cycle and stages of healing. It is known that a major determinant of abnormal or failed scaphoid fracture healing is related to a disruption in the vascular supply to the scaphoid. Because a normal vascular supply is an important aspect of the inflammatory cycle and stages of healing following fracture, let's review those processes first.

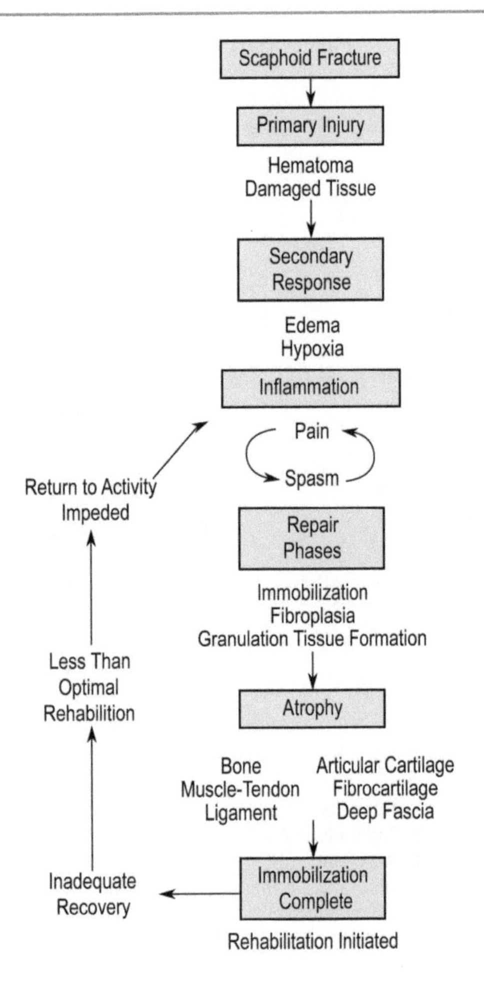


**Fig. 5.19** Injury recovery sequence that Jim likely experienced

A bone fractures when the internal stresses induced by external loads are exceeded, resulting in plastic deformation and, ultimately, complete failure of the tissue. Recall that in this case, Jim fell on his outstretched hand (i.e., macrotraumatic event), which resulted in an injury to his wrist. That mechanism of injury likely induced an impulsive compressive force and bending moment in the waist region of the scaphoid that exceeded the ultimate stress resulting in the fracture. A fracture of this nature, classified as an acute injury, is met immediately by an inflammatory response, which begins a two-stage healing process: repair and regeneration. The healing process, i.e., injury cycle, is complex, involves several steps, and is unique to the mechanism of injury, tissue damaged, immediate physiological environment, and health status of the individual (Fig. 5.19). Based on Jim's history and radiographic and physical examination evidence, his healing states were not normal, although not uncommon as scaphoid fractures go.

Let's examine the details of what should have occurred during each phase of healing: clotting, inflammatory, proliferative/fibroplasia, and maturation/remodeling (Fig. 5.20). This review can be applied to any injured biological structure; at certain stages in this review, however, we highlight what is unique about the normal healing of fractures. As result of the fracture, blood and lymphatic vessels are ruptured as well, and the cellular and plasma components of blood and lymph enter the wound.

The blood and lymph that are forced out (extravasation) cause swelling, and the vascular cellular components

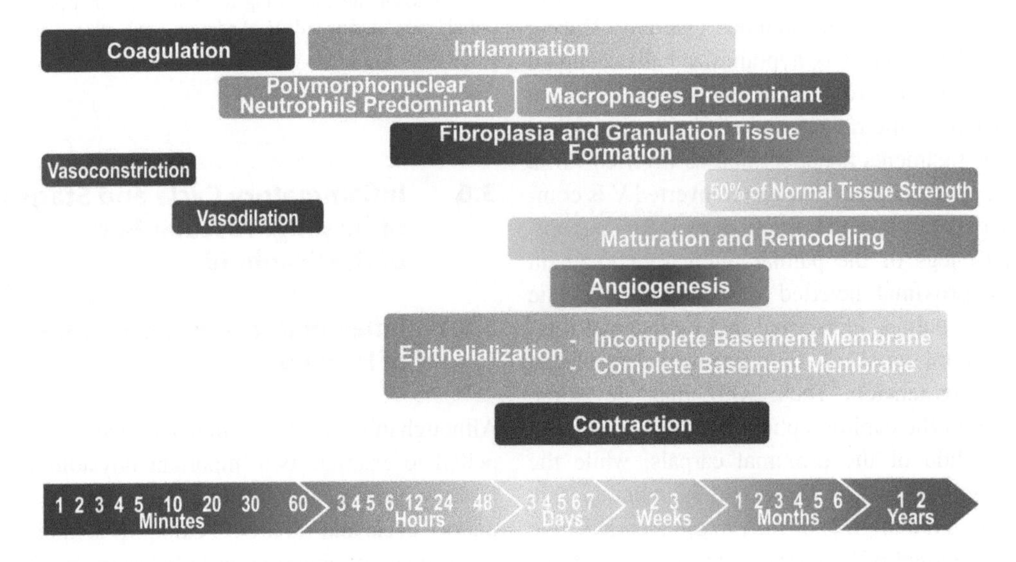


**Fig. 5.20** An overview of the healing process. Although we conveniently divide the process into four cycles, in reality, repair and regeneration are a series of overlapping phases

(i.e., platelets, red blood cells, and other blood-borne cells) die. Swelling and vasospasm (vasoconstriction) lead to anoxia (no oxygen secondary to hypoxic injury) due to stasis of blood and the environment outside the blood vessels. The pool of blood eventually congeals or coagulates forming a clot. This platelet-like plug acts like a low-tensile-strength "glue" to hold the wound edges together and localize the injury. Eventually, the clot is transformed into granulation tissue and an acellular collagenous scar. The damaged tissue releases, and the clot contains, chemoattractants that invite vasoactive chemicals and increase vascular permeability, which allows further extravasation of plasma products into the injury site. Because approximately 80% of the surface of the scaphoid is covered by articular cartilage (i.e., the bone is part of several synovial joints within the wrist), damage to the synovial membrane adds to the fluid accumulation, which may interfere with the clotting mechanism. These combined actions are the main causes of the soft tissue inflammatory response that includes redness, swelling, heat, pain, muscle spasm, and impaired mobility. The clotting mechanism may take up to 48 h to be completed, as was likely in Jim's case.

Inflammation is not considered a negative series of events, but a necessary part of the healing stages, yet something that needs to be monitored and controlled (i.e., anti-inflammatory medication, immobilization, elevation, application of cold, etc.). Healing can only commence effectively once inflammation begins to subside. Therefore, within just a few hours post-injury, several localized factors in the damaged area are activated, which induce several actions, including vasodilatation, increased vessel permeability, increased nociceptive activity, and the attraction of leukocytes, i.e., migratory white blood cells. Leukocytes and similar functioning lymphocytes, monocytes (macrophages), and granulocytes act as phagocytes and clean the wound of dead tissue and disable toxic proteins seen with cellular injury. Macrophages are also important because of their association with growth factors, cytokines, and other chemicals that modulate the inflammatory response. As wound debris is removed, there is a shift from the inflammatory to the proliferative phase, in which fibroblasts (for repair) and other specialized cells, e.g., myoblasts and tenoblasts (for regeneration), migrate into the area. As the original clot dissociates, it is replaced with an extracellular matrix called granulation tissue. Granulation tissue is comprised of capillaries, myofibroblasts, macrophages, fibrin matrix, endothelial cells, and a nutritive matrix. Normal functioning endothelia cells are critical because they are involved with the production of new capillary buds. Note, however, that the continued presence of prostaglandins is paradoxically involved with some ongoing inflammation as well as repair. In fact, the inflammatory process largely overlaps the proliferative phase. With normal healing, inflammatory effects in the proliferative phase become less prominent, and a normal repair and regenerative response ensue, particularly if the time period of immobilization can be minimized.

Following a fracture, the repair is not accomplished by existing osteoblasts and chondroblasts but depends on the creation of a new network of cells specific to the task stimulated by stem cells. In this process, once it is set in motion, each succeeding step is predicated on the steps before it, so that if an error occurs or if there is a complication, the process is derailed, in a sense, with no recourse to start over. It is possible that non-unions or delayed unions result from a failure in the proper sequencing of steps in this early phase of repair.

We speculate that two factors played a role in disrupting the normal early process of reacting to fracture in Jim's case: disruption of normal vascular supply to the scaphoid and prolonged immobilization.

Let's consider the importance of the inflammatory, and early proliferative, phase with regard to the role of a well-constituted vascular system. We can assume that with the fracture of any bone, small and medium lymphatic and blood vessels are damaged. Thus, in the inflammatory and early proliferative phases, in addition to wound stabilization, the vascular and lymphatic tissues are rapidly repairing and reconstituting their function. These activities would be paramount for both removing waste material and providing necessary nutrients for anabolic processes. If the vascular system is compromised or the inflammatory process is persistent, the time course of repair and regeneration is slowed, and healing is generally poor. As well, the side effects of inflammation, i.e., swelling, pain, impaired muscle activity, etc., become chronic leading to many other potential problems, including muscle atrophy, articular cartilage degeneration, and joint adhesions, to name a few. And, of course, one or more of these unintended consequences in combination would likely induce, as in Jim's case, a compromised and/or poor outcome.

Tissue repair and regeneration occur in the proliferative/fibroplasia phase of healing. In the early stages, repair predominates as granulation tissue replaces the original tissue with a scar in an attempt to provide some degree of barrier protection and mechanical strength. The clinical priority is to protect, i.e., immobilize, the injured structures/area while maintaining some function in surrounding tissues/joints. The period of scar formation is known as fibroplasia and lasts approximately 4–6 weeks or longer in some cases. At the same time, regenerative processes begin to replace the damaged tissue, i.e., scar, with the same type of tissue having a nearly similar function as the original tissue. The growth of endothelial capillary buds into capillaries improves the vascular bed, reconstitutes and increases the blood supply, and allows the wound to heal aerobically. As capillaries proliferate, fibroblasts accumulate at the wound site. These cells synthesize an extracellular matrix that contains collagen,

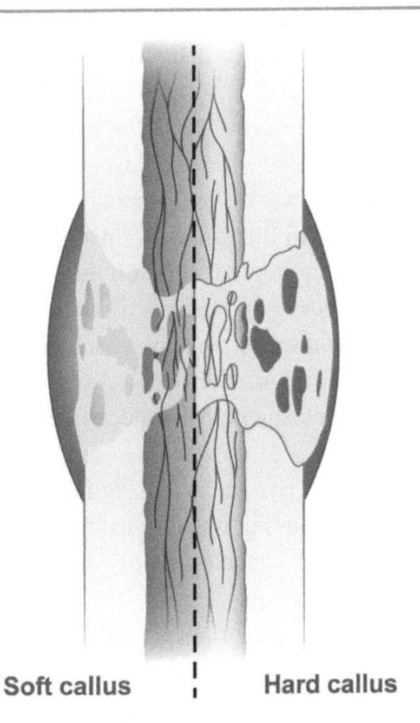


**Fig. 5.21** Schematic of two steps: soft callus and hard callus in the reparative phase of fracture healing

elastin, ground substance (proteoglycans), glycosaminoglycans, and fluid, all constituents of the more mature substance of normal tissue. The production of the extracellular matrix and proliferation of collagen improves the tensile strength of the regenerating tissue and requires controlled external tensile loads as part of the healing and repair process. Beginning in the middle stages of this phase, mobilization and controlled movement gradually reduce the size of the unorganized scar and produce a reconstituted tissue with better structural organization, stiffness, and strength.

The reparative phase following the fracture is highlighted by periosteal and medullary calluses formed by osteoblasts from the periosteum and marrow. The stem cells for medullary osteoblasts also serve as progenitors for fibroblasts, chondrocytes, and some marrow cells. The production of the mesenchymal cells is associated with great vascular proliferation, which enhances woven bone formation during this phase. Inadequate vascularization, insufficient immobilization, or unusual stresses at the fracture site during this time period can induce chondroblast rather than osteoblast formation. If this minor error occurs, it is usually corrected as the chondrogenic regions calcify and are replaced by lamellar bone. The reparative phase is also dependent upon the application of controlled multidimensional forces and is complete with transformation of a provisional or soft callus into one that is bony. The hard callus is formed through endochondral ossification and direct bone formation; the woven bone then replaces the soft callus with a hard callus around the bone fragments. The hard callus is less rigid than normal bone, but because of its increased cross-sectional area, its strength is functionally equivalent to cortical bone (Fig. 5.21).

With maturation, the number of fibroblasts diminishes, while the density and organization of collagen increases, which increases the stiffness of the new tissue. Normally, the sequence of repair attempts to minimize scar formation. However, the amount and extent of scarring depend on such factors as the site of injury, the tissue injured, the nature of the injury, the degree of strain injury, the patient's genetics and general health, the intervention provided, and the presence or lack of complications. The maturation phase is a long-term process, ranging from 12 to 18 months. For some tissues, the strength of the maturing structure regains only approximately 70–80% of its original strength by 12 months. This implies that individuals are at some risk if they return to the activity that produced the injury prior to 12 months. At the end of the maturation phase, the new tissue has been remodeled as a result of rehabilitation, i.e., the use of controlled compressive, tensile, and torsional loads, and a gradual and safe return to normal daily activities. In some cases, the new tissue's structure and strength are no different than the original, whereas in other cases the original structure is replaced with an organized scar. The scar has a similar, but not identical, structure that allows the individual to return to most of the activities they enjoyed before the injury, but that tissue is never completely normal in appearance or function.

As opposed to the extent of soft tissue repair at 12 months, repaired bone (not a scar) is typically as strong as the intact bone was because of its increased mass. Modeling and remodeling of the fracture site over time gradually restore the original contour, internal structure, mass, and mechanical efficiency of the healed bone. If the fracture heals with incorrect angulation, e.g., malunion of scaphoid, this may be partially corrected by remodeling induced by participation in normal daily activities. Generally, individuals with scaphoid malunion who are treated conservatively have similar functional outcomes compared to individuals who were treated surgically.

### 5.6.2 Vascular Anatomy of the Scaphoid

We have alluded to the vascular challenges trauma induces when the scaphoid is fractured. This challenge is not unique to the individual but is inherent to the vascular anatomy of this region. The scaphoid receives the majority of its blood supply from dorsal vessels at or just distal to the waist area (Fig. 5.22). Important vascular branches of the radial artery enter the scaphoid through foramina along its dorsal ridge and perfuse the proximal pole in a retrograde fashion. The dorsal vessels supply approximately 70–80% of the bone, including the entire proximal pole. The second group of vessels arise from palmar and superficial palmar branches of the radial artery and enter the distal end of the bone. The palmar vessels supply 20–30% of the distal pole and the scaphoid tubercle. Jim sustained a fracture in the waist region, which is commonly associated with damage to selected dorsal branches of the radial artery (Fig. 5.23). Because of how the vessels enter the bone foraminae in different regions of the scaphoid, waist fractures (proximal pole) leave a small fragment of the scaphoid with insufficient or no blood supply. It has been shown that as the fracture site nears the proximal pole, the percentage of normal bone union decreases dramatically.

## 5.7 Effects of Immobilization

Intervention immediately (within 7–10 days) following fracture involves managing inflammation and pain and stabilizing bone fragments. In theory, with proper triage and appropriate intervention, successful tissue repair and regeneration and return to normal function can be assured. In general, two approaches have been used to stabilize the fracture site: surgical (internal) fixation or external immobilization (plaster or fiberglass cast). There has been considerable debate on the value of operative versus nonoperative treatment following a scaphoid fracture. Overall, functional outcomes do not differ when comparing operative to nonoperative approaches, although the patients treated with surgical fixation usually return to previous levels of functioning sooner. A nonoperative treatment approach offers a few choices about how best to immobilize the wrist and forearm (and elbow in some cases). When planning to use cast immobilization following scaphoid fracture, the following factors are considered: type of fracture; presence of other potential injuries, e.g., ligament strain; degree of instability; and how best to position the wrist and forearm (and elbow) to maximize stability and minimize internal stresses to the fractured fragments and related soft tissue. For non-displaced fractures, two common options about how to position the relevant joints to be immobilized have been considered:

1. Scaphoid cast – with the thumb metacarpal phalangeal joint immobilized and distal phalanx free, wrist neutral in the sagittal plane or flexed approximately 10° with slight radial deviation and elbow free to move
2. Colles' cast – with the thumb free to the metacarpal phalangeal joint, wrist extended 10° to 20° without deviation radially or ulnarly, and elbow free to move

The immobilization timeframe generally follows well-known healing patterns of biological tissue, in the case of bone, approximately 6–8 weeks. However, based on the natural history of scaphoid fracture recovery, and the known complications related to vascular damage to vessels supplying the scaphoid, it is typical to immobilize these fractures for 12–16 weeks.

Recall that the treatment of choice for Jim was conservative; he was immobilized for 12 weeks with a Colles' cast. The rationale for this choice was based on the fact that his fracture was non-displaced and that placing him in a Colles'

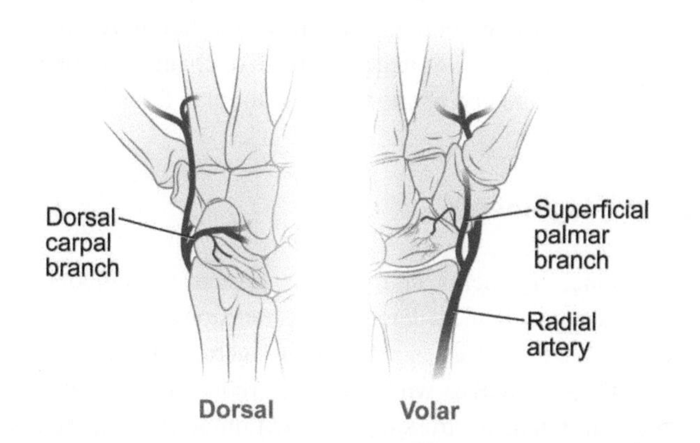


**Fig. 5.22** Branches of the radial artery perfuse the scaphoid via dorsal and volar branches (Amadio and Moran 2005)

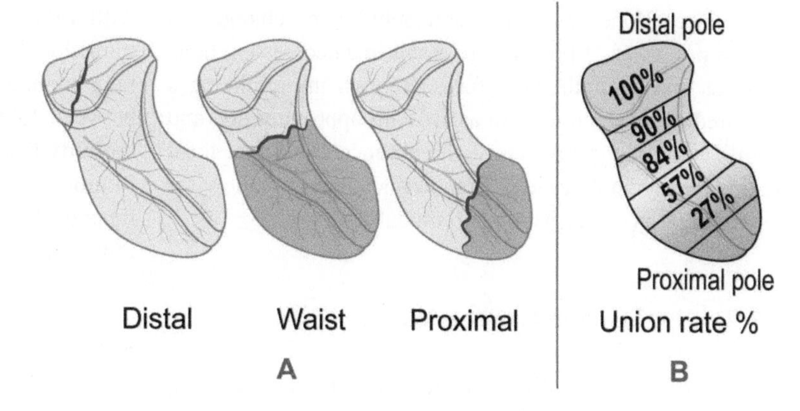


**Fig. 5.23** Schematic showing (**A**) vascular branching and fractures in the distal, waist, and proximal regions of the scaphoid and (**B**) progressive reduction in union rate percentage when the fracture line moves from the distal to the proximal pole

cast would put his wrist and thumb in more functional positions; that is, he would be able to perform simple, low-force grasping tasks using his fingers and thumb, and he would be free to move his elbow. Theoretically, moving his elbow and fingers to do simple, involving low-force, daily tasks would allow him to use the muscles of his arm, forearm, and digits that would create minimal internal stresses in the bones and soft tissues of the forearm and wrist that could prevent significant muscle atrophy and enhance the repair and regeneration processes. It was assumed that these internal stresses would not increase instability or interfere with normal healing and repair processes and that Jim would be compliant, i.e., not attempt activities that would require power grips.

Obviously, the treatment plan was not successful as noted by the radiographic evidence of malunion, Jim's symptoms, and inability to perform his normal activities. It is possible that the major determinants of this failure are related to things we have already discussed, i.e., abnormal wrist biomechanics related to the scaphoid malunion, impaired vascular system, and radiolunate ligament strain resulting in lunate hypermobility/instability. However, it is possible that the known deleterious effects of immobilization on biological tissues may have exacerbated the more obvious problems.

### 5.7.1 Bone

Wolff's law claims that the remodeling of bone is influenced and modulated by mechanical stresses. Generally, loads on the skeleton are imposed by muscle activity and gravity. Conversely, disuse or inactivity, i.e., reduced or absent loads, has significant deleterious effects on the skeleton. For example, on one extreme, it has been shown that bed rest induces a bone mass decrease of approximately 1% per week, which demonstrates the rapidity with which the skeleton responds to the reduction in mechanical stimuli. Thus, when a bone is not subjected to the usual mechanical stresses and strains, as happens with cast immobilization following fracture, resorption of the periosteal and subperiosteal bone, decreased collagen and mineral content, and reduction in bone density take place. Bone resorption and subsequent changes in related processes of bone health result in a decrease in bone stiffness and strength. Therefore, although immobilizing a fractured bone is required to assist with approximation and stabilization of the fragments, the absence of normal local strains for moderate to long time periods may not allow for the resumption of normal bone remodeling.

### 5.7.2 Hyaline and Fibrocartilage

Because of its inherently poor blood supply, the health of both hyaline (articular) and fibro-cartilage is highly dependent on intermittent compression, shear, and tensile stresses, which are created by muscle-tendon actuators that produce and control joint movements. It has been shown that with the loss of normal joint motion and reduction in these daily stresses, articular cartilage exhibits degenerative-like changes as early as the first 4 weeks of immobilization. The degenerative changes include thinning, softening, and swelling, decreased ability to imbibe water because of an altered extracellular matrix, breakdown of collagen, reduction of proteoglycan content, increased permeability, and alteration in joint lubrication. All of these effects reduce the ability of articular cartilage to distribute/mitigate joint loads once immobilization has been discontinued. Although articular cartilage is devoid of nociception, joint surface degeneration leads to a cascade of events affecting other periarticular structures, e.g., joint capsules, ligaments, and subchondral bone, which can be sources of pain generation. Arthrogenic pain has been shown to inhibit normal muscle activity, which usually results in muscle atrophic changes. Recall that because all of the carpals articulate with multiple structures, much of their surface area is articular cartilage, and even microdamage (which would not be sensed) to the articular cartilage of several carpals can lead to an accumulation of impairments within the carpus that could exacerbate the primary problem. For example, the ulno-menisco-triquetral complex, although not damaged during the fall, may experience impairment due to degenerative-like changes in the triangular cartilage and abnormal strain to related ligaments.

### 5.7.3 Ligaments

Like bone, ligaments and tendons appear to remodel in response to mechanical demands (i.e., tension). They become stiffer and stronger when subjected to increased stresses and less stiff and weaker when stress is reduced or absent. A reduction in tensile stress induced by immobilization results in changes to the joint capsules (intrinsic ligament) and ligaments leading to production of immature collagen, decreased cross-linking within the collagenous framework, disorganization of collagen, and reduced cross-sectional area of the tissues, all leading to reduced tensile and directional strength within their mid-substance, as well as at the ligament/bone transitional zones. Additionally, the proliferation of fibrofatty connective tissue within the joint spaces and adhesions between folds of the synovium is reported to reduce joint mobility (referred to as increased stiffness in some circles). Intuition suggests that because of the abundant number of ligaments, external and internal to the carpus, multiple impairments are likely related to prolonged immobilization.

### 5.7.4 Muscle-Tendon Actuators

Muscle and tendon are linked because they work functionally as a unit. However, they have distinct biological structures

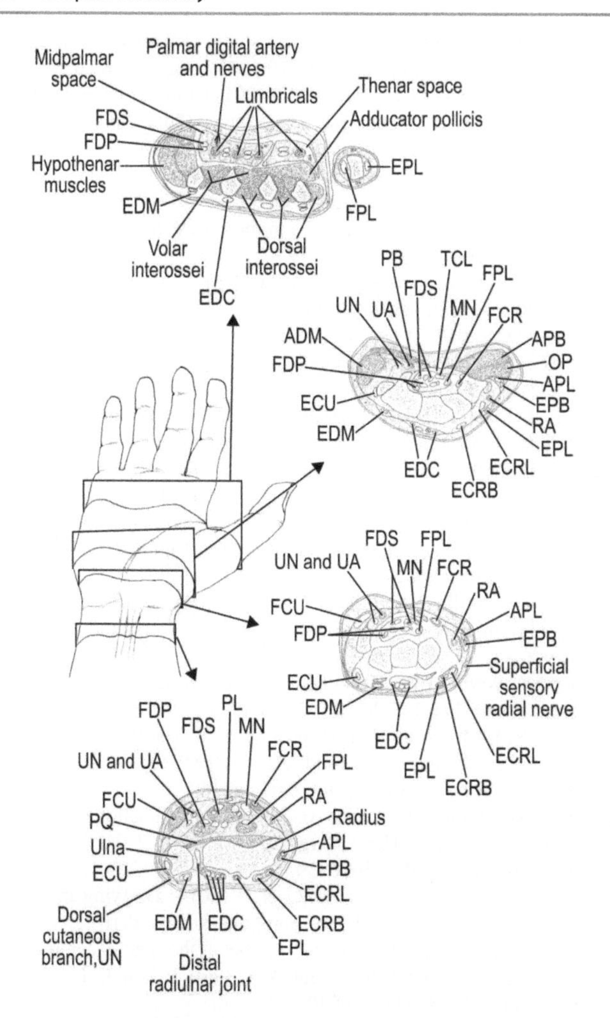


**Fig. 5.24** Cross-sectional cuts at different levels of the forearm, wrist, and hand demonstrate the location of bones, volar and dorsal tendons, and vessels. FDS = flexor digitorum superficialis, FDP = flexor digitorum profundus, EDM = extensor digiti minimi, EDC = extensor digitorum communis, FPL = flexor pollicis longus, EPL = extensor pollicis longus, FCR = flexor carpi radialis, APB = abductor pollicis brevis, ADM = abductor digiti minimi, APL = abductor pollicis longus, EPB = extensor pollicis brevis, ECU = extensor carpi ulnaris, ECRB = extensor carpi radials brevis, ECRL = extensor carpi radials longus, FCU = flexor carpi ulnaris, PL = palmaris longus, PQ = pronator quadratus, UN = ulnar nerve, UA = ulnar artery, RA = radial artery, MN = median nerve, TCL = transverse carpal ligament, P = pisiform, Tq = triquetrum, L = lunate, S = scaphoid, H = hamate, C = capitate, Td = trapezoid, Tm = trapezium; numbers 1–5 represent metacarpals 1–5

that dictate their mechanical properties, as noted previously. For practical purposes, the effects of immobilization on tendons are identical to that of ligaments. In one sense, muscle is very similar to ligament and tendon in that it remodels within the context of the presence or lack of tensile stress. Muscle remodels, i.e., lose their cross-sectional dimension or atrophies, in response to inactivity and immobilization. The loss of cross-sectional area and strength has been shown to be significant as early as 4 weeks of immobilization. The early reductions in muscle size and strength suggest changes in both muscle protein and neurological influence. Similar to what has been reported about ligaments and tendons, the reduction in muscle protein with immobilization or disuse is related to a decrease in tensile loading. For example, it has been demonstrated that immobilization which places muscles in shortened positions induces greater atrophy, due to loss of sarcomeres, compared to muscles immobilized in lengthened positions. Because Jim was immobilized with his wrist extended, we speculate that there was likely greater atrophy of the wrist and digital extensors: (1) these muscle-tendon actuators are placed in statically shortened positions, and (2) the muscles on the volar aspect of the wrist are placed in lengthened positions, and (3) activation of the digital flexors for grasping tasks likely spare those muscle-tendon units from atrophic changes (Fig. 5.24). The plastic changes muscles experience as a result of immobilization may also affect both the functional muscle moment arms (particularly in Jim's case if there is carpal hypermobility or instability) and muscle length-tension.

In summary, Jim sustained a common fracture secondary to a FOOSH mechanism. Immobilization is commonly the treatment of choice to stabilize the fracture site and attempt to create an environment in which repair and regeneration can occur. However, immobilization induces its own set of limitations which may impair the stages of healing by creating deficits in the mechanical properties of bone, articular and fibrocartilage, ligaments, and the muscle-tendon actuators.

## 5.8 Chapter Summary

A 28-year-old carpenter slipped and, in catching his fall with his outstretched right hand, fractured the scaphoid, which was diagnosed 2 weeks after the incident. He was treated with closed reduction and Colles' cast immobilization. His cast, which allowed grasping movements of all digits, was removed after 12 weeks. At the time of cast removal, a repeat radiograph revealed a malunion of the scaphoid. When Jim reported for rehabilitation, he described a decreased ability to perform routine daily and work activities due to pain and weakness. To clarify this case, our study examines the mechanism of injury, regional anatomy, physiology and biomechanics, and stages of healing. We conclude that Jim primarily had a kinematic problem that was hampered by continued pain: paradoxically both hypomobility and hypermobility and secondarily impaired motor function, i.e., weakness.

What becomes clear from the particulars of this case was the complexity of the anatomy and biomechanics of the wrist complex. Our use of abducted thinking leads to speculation, based on the best available evidence, that Jim's problems were most likely related to malunion and pathomechanics of wrist during activities of daily living, exacerbated by lunate hypermobility. We suggest that the deleterious effects of prolonged immobilization on multiple biological tissues may have factored into Jim's wrist problems. We suggest two take-aways (generalizations) from this case: (1) specific to the particulars, balancing the need for immobilization following fracture and assertive rehabilitation and (2) the dynamics of carpal mechanics that are yet to be determined so we are reminded that the "whole is not the sum of its parts."

## Bibliography


Alnaeem H, Aldekhayel S, Kanevsky J et al (2016) A systematic review and meta-analysis examining the differences between nonsurgical management and percutaneous fixation of minimally and nondisplaced scaphoid fractures. J Hand Surg Am 41(12):1135–1144. https://doi.org/10.1016/j.jhsa.2016.08.023

Alshryda S, Shah A, Odak S et al (2012) Acute fractures of the scaphoid bone: systematic review and meta-analysis. Surgeon 10:218–229. https://doi.org/10.1016/j.surge.2012.03.004

Amadio PC, Moran SL (2005) Fractures of carpal bones. In: Green DP, Pederson WC, Hotchkiss RN, Wolfe SW (eds) Green's operative hand surgery, vol 1, 5th edn. Elsevier, Philadelphia

Beer FP, Johnson ER Jr, Cornell PJ (2013) Vector mechanics for engineers: dynamics. McGraw-Hill Companies, Inc., New York

Berdia S, Wolff SW (2001) Effects of scaphoid fractures on the biomechanics of the wrist. Hand Clin 17(4):533–540

Burgess RC (1987) The effect of a simulated scaphoid malunion on wrist motion. J Hand Surg Am 12A(2 Pt 1):774–776

Chambers SB (2019) The impact of scaphoid malunion on wrist kinematics & kinetics: a biomechanical investigation. Dissertation, The University of Western Ontario

Freedman DM, Botte MJ, Gelberman RH (2001) Vascularity of the carpus. Clin Orthop Relat Res 383:47–59

Furey MJ, White NJ, Dhaliwal GS (2019) Scapholunate ligament injury and the effect of scaphoid lengthening. J Wrist Surg 9:76–80

Gillette BP, Amadio PC, Kakar S (2017) Long-term outcomes of scaphoid malunion. Hand 12(1):26–30. https://doi.org/10.1177/1558947716643295

Lee C-H, Lee K-H, Lee B-G et al (2015) Clinical outcome of scaphoid malunion as a result of scaphoid fracture nonunion surgical treatment: a 5-year minimum follow-up study. Orthop Traumatol Surg Res 101:359–363. https://doi.org/10.1016/j.otsr.2014.09.026

Levangie PK, Norkin CC, Lewek MD (2019) Joint structure & function, a comprehensive analysis. F. A. Davis, Philadelphia

LeVeau B (1977) Williams and Lissner: biomechanics of human motion. W. B Saunders Company, Philadelphia

Magee DJ, Zachazewski JE, Quillen WS (eds) (2007) Scientific foundations and principles of practice in musculoskeletal rehabilitation, vol 2. Saunders Elsevier, St. Louis

Martin RB, Burr DB, Sharkey NA et al (2015) Skeletal tissue mechanics, 2nd edn. Springer, New York

Neumann DA (2017) Wrist. In: Neumann DA (ed) Kinesiology of the musculoskeletal system, foundations for rehabilitation, 3rd edn. Elsevier, St. Louis

Nigg BM, MacIntosh BR, Mester J (eds) (2000) Biomechanics and biology of movement. Human Kinetics, Champaign

Nordin M, Frankel VH (2022) Basic biomechanics of the musculoskeletal system, 5th edn. Wolters Kluwer, Philadelphia

Oatis CA (2017) Kinesiology, the mechanics and pathomechanics of human movement, 3rd edn. Wolters Kluwer, Lippincott Williams & Wilkins, Philadelphia

Ökaya N, Nordin M (1991) Fundamentals of biomechanics; equilibrium, motion, and deformation. Van Norstrand Reinhold, New York

Palmer AK, Werner FW (1984) Biomechanics of the distal radioulnar joint. Clin Orthop 187:26–35

Patterson RM, Moritomo H, Yamaguchi S et al (2003) Scaphoid anatomy and mechanics: update and review. Oper Tech Orthop 13(1):2–10. https://doi.org/10.1053/otor.2003.36316

Rainbow MJ, Wolff AL, Crisco JJ et al (2016) Functional kinematics of the wrist. J Hand Surg Eur Vol 41(1):7–21. https://doi.org/10.1177/1753193415616939

Sandow MJ, Fisher TJ, Howard CQ et al (2014) Unifying model of carpal mechanics based on computationally derived isometric constraints and rules-based motion – the stable central column theory. J Hand Surg Eur Vol 39(4):353–363. https://doi.org/10.1177/1753193413505407

Schuind F, Cooney WP, Linscheid RL et al (1995) Force and pressure transmission through the normal wrist, a theoretical two-dimensional study in the posteroanterior plane. J Biomech 28(5):587–601

Tözeren A (2000) Human body dynamics, classical mechanics and human movement. Springer, New York

Weber ER, Chao EY (1978) An experimental approach to the mechanism of scaphoid waist fractures. J Hand Surg Am 3(2):142–148

# Hand: Lacerated Flexor Pollicis Longus

# 6

## 6.1 Introduction

The hand serves equally important roles as a sensory and an effector organ. Its functions include performing delicate prehensile tasks, exploring small objects through manipulations between the thumb and fingers, and employing powerful grasp patterns. Use of the hand also includes supporting and stabilizing objects to free the contralateral extremity for other tasks. Its anatomical and biomechanical complexity is formulated by the presence of 27 bones and joints and 34 extrinsic and intrinsic muscle-tendon actuators. A major injury, e.g., fracture or tendon damage, affecting any part of the hand can lead to significantly impaired function. All parts of the hand make unique contributions to human activities, but the function of the thumb is particularly crucial for use of the hand because of its ability to contribute to both fine and gross motor activities. We do not study the entirety of the hand in this case but focus on the particulars of the thumb region. Recall that in problem-based learning, it is assumed that learning everything is not as crucial as learning a process and the specifics of a particular problem and then abstracting what has been learned to future similar problems.

## 6.2 Case Presentation

Mr. Jack Bishop (Jack) was a right-handed, 54-year-old high school mathematics teacher. Approximately 8 weeks ago, he lacerated his left flexor pollicis longus (FPL) tendon just proximal to the distal interphalangeal joint.

Jack sustained the cut to his left thumb while recycling plastics, tin cans, and cardboard into a dumpster. He believes his thumb was cut by broken glass hidden in a garbage bag he pushed to make room for the material he was recycling. He was fortunate to have a towel in his car and immediately wrapped his thumb to control bleeding and apply pressure. He was able to drive home without difficulty. Fortunately, when he got home, his daughter, who was visiting for the weekend and was an emergency room (ER) nurse, helped clean and tape the wound. She told Jack he would likely need the wound stitched and to go to urgent care for further evaluation.

At the urgent care center, the attending physician (a fourth-year orthopedic surgical resident) cleaned and debrided the wound close to the interphalangeal (IP) and examined the extent of skin and underlying tissue injury. He suspected a tendon laceration, searched for the tendon, and performed a resisted muscle test of the flexor pollicis longus (FPL). The resisted test of the FPL demonstrated a strong/mild pain response. The physician was surprised because a lacerated tendon should have tested weak and painful. Regardless, he proceeded with debridement and finally located the tendon, noted the damage, and estimated that it had sustained an 80% laceration. He recommended that Jack see a hand surgeon as soon as possible.

The urgent care physician sutured and bandaged the wound. Then, he placed the thumb in approximately 20° flexion at both the metacarpal phalangeal (MCP) and IP joints and applied a makeshift semi-rigid splint to restrict both flexion and extension movements of the thumb. Movements of the thumb carpometacarpal (CMC), wrist, and elbow were not restricted.

Jack was examined by a hand surgeon (who had received communication from the urgent care physician) the day following the injury. He confirmed the need for a surgical repair for the left thumb FPL tendon and radial digital nerve lacerations.

The tendon repair took place 3 days following the injury, with details of the repair provided below:

A Brunner incision was utilized extending the volar laceration proximally and distally. The incision was taken sharply through

**Supplementary Information** The online version contains supplementary material available at https://doi.org/10.1007/978-3-031-25322-5_6.


G. J. Alderink, B. M. Ashby, *Clinical Kinesiology and Biomechanics*, https://doi.org/10.1007/978-3-031-25322-5_6

the skin, the subcutaneous tissue was dissected using tenotomy scissors, and bipolar electrocautery was used for hemostasis. Full-thickness skin flaps were elevated off the flexor tendon sheath. The patient was noted to have a laceration just proximal to the IP joint, i.e., Zone II, through the tendon sheath with 90% plus percent laceration of the FPL tendon. The tendon sheath was then excised proximally and distally for exposure, taking care to preserve the oblique pulley. The ends of the tendon were then sharply debrided and freshened up, and a core suture repair was performed using 3-0 looped Supramid suture and a modified four-strand Tsuge repair. Epitendinous sutures were then placed using a 6-0 Prolene suture using an over-and-over stitch circumferentially around the tendon. Next, the thumb was taken through a passive range of motion. There was no impingement of the oblique pulley noted and no gapping. Next, attention was turned to the digital nerve repair.

We know (see Chap. 4) that the structure of a tendon, comprised of a parallel and dense arrangement of collagen fibrils, enables it to mitigate/transmit large tensile stresses from the muscle to its boney attachment. The parallel arrangement is not an accident, but a function of the tensile forces generated in the muscle that is attached to the tendon. It has been demonstrated that tendons exhibit a rate-dependent resistance, i.e., stiffness, to tensile loading and do not fail until about 8–10% strain, i.e., elongation. A primary goal of a tendon repair following laceration (or failure) is to approximate the two ends assuming that this will facilitate the healing and regeneration of the collagenous framework and restore the integrity of the tendon's tensile strength. The modified Tsuge technique, similar to what was used for Jack's repair, is one of many surgical procedures used for tendon reconstruction (Fig. 6.1). In the study by Laban et al., the ultimate strength of the four-strand Tsuge was compared to a six-strand version. Following the repair of fresh frozen human cadaver flexor pollicis longus tendons, the forces at failure, i.e., rupture of the sutures, and at initial gapping were recorded using an Instron materials testing device. Results showed that maximum forces at failure for the four-strand and six-strand repairs were 48.4 (±10.7) and 64.2 (±11.0) Newtons (N), respectively. The maximum forces at which initial gapping was observed were 40.7 (±12.3) and 56.1 (±9.7) N for the respective techniques. In each instance, it was evident that the six-strand Tsuge was stronger. However, since previous research reported by Labana et al. had demonstrated that the forces in finger flexor tendons during unrestricted active range of motion exercise did not exceed 35 N, Labana and co-workers concluded that the four-strand repair would adequately protect a repaired FPL tendon during the early stages of rehabilitation. Of note, the strength of a flexor tendon is proportional to the number of strands crossing the repair site. The additional suture material is assumed to increase the bulk of the repair, thus increasing resistance to the gliding of the tendon within its sheath. The selection of four- versus six-strand repair is based on tendon morphology and the anticipated postoperative, rehabilitative compliance of the patient.

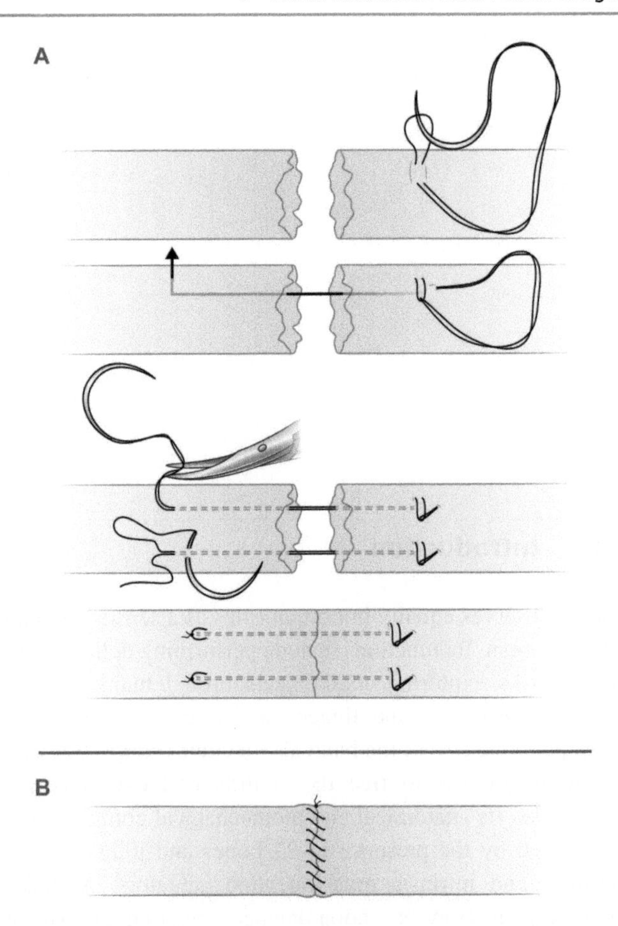


**Fig. 6.1** (**A**) Modified four-strand Tsuge repair of a lacerated tendon. Note the attempt to align the parallel collagenous arrangement by approximating the tendon ends, (**B**) over-and-over epitendinous circumference stitch. (Modified with permission from Labana et al. (2001))

The ends of the radial nerve were dissected off the surrounding soft tissues using tenotomy scissors and were debrided using micro-scissors under loupe magnification. Proximal neuroma and distal glioma were excised sharply, and a primary repair was performed in a tension-free fashion using 8-0 nylon sutures under loupe magnification. Four interrupted simple sutures were placed in the epineurium obtaining a tension-free repair.

The wounds were then thoroughly irrigated and closed using 5-0 nylon sutures. A digital nerve block was performed with 1% lidocaine and 0.5% bupivacaine. Sterile dressings were then applied followed by a thumb spica splint with the wrist flexed. Finally, the tourniquet was released, and the patient was reversed from anesthesia and taken to the recovery room in stable condition.

In conversation with the surgeon following surgery, Jack learned that in the time from the injury to surgery, the tendon had sustained additional separation at the laceration site, i.e., the tendon was noted to be ~90% lacerated at the time of

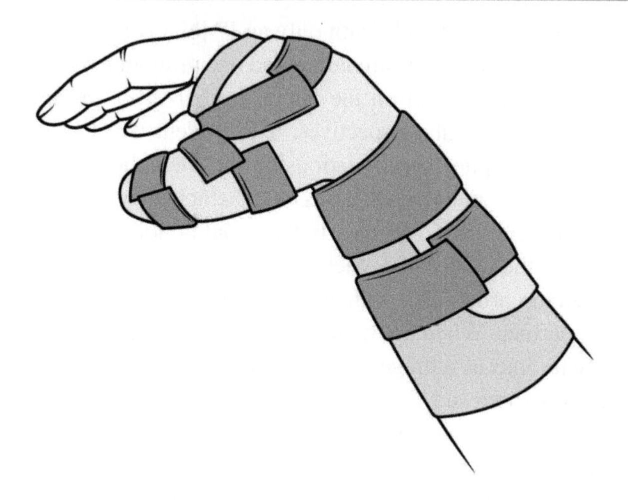

**Fig. 6.2** Dorsal blocking splint that blocks extension of the wrist and all joints of the thumb

**Fig. 6.3** Boutonniere deformity of a right thumb. Note the prominent appearance of the tendon of the flexor pollicis longus (arrow) because it is no longer tethered by one or more of the pulley systems

surgery. Recall that the attending physician at urgent care immobilized the thumb IP and MCP joints to prevent extension movements that could have placed tension on the damaged tendon, but did not restrict CMC or wrist motion, nor did he provide Jack with use precautions of his left arm and hand. It turns out that since Jack did not experience any significant pain, he used his left arm within the constraints of the immobilized thumb; that is, he performed functional tasks that included motions of the left wrist and thumb CMC wrist extension, which likely placed the partially lacerated FPL tendon under too much tension. Jack was fortunate that the FPL tendon did not completely separate for that would have led to a more complex surgical exploration and reconstruction, which would likely have complicated his rehabilitation.

Jack returned to the surgeon's office approximately 10 days post-surgery when the stitches were removed, and he was referred to occupational therapy to begin a tendon repair rehabilitation protocol (modified Duran protocol). Following the occupational therapy evaluation, Jack was fit with a dorsal blocking orthosis (DBS) (Fig. 6.2). The orthosis placed the wrist in 20° of flexion, the thumb CMC in 20° of palmar abduction, 20° of MCP flexion, and slight (5°) of thumb IP flexion.

The postoperative program for the first 4 weeks included passive range of motion (PROM) for thumb flexion, but extension only within the confines of the DBS; Jack was instructed to avoid full extension of the tendon. Gentle active range of motion (AROM) and a "place and hold flexion" exercise were initiated by the fifth postoperative week. Putty exercises were initiated postoperative week eight with low resistance putty. Jack appeared to have been fully compliant with his rehabilitation program without incident until just recently. At a follow-up visit, Jack reported having felt a sharp pain and reduced strength while doing his putty exercises. Observation revealed a Boutonniere deformity of the left thumb. Examination of the active and passive range of motion of the thumb IP and MCP joints were found to be within functional limits and were not painful. Resisted strength testing of the FPL produced moderate weakness without pain, compared to the contralateral side.

## 6.3 Key Facts, Learning Issues, and Hypotheses

The injury, physical examination, surgical, and early rehabilitation history are straightforward. Jack sustained a partial, but marked, laceration of the left FPL tendon and to the radial digital (sensory) nerve of the thumb (a branch of the median nerve), and they were repaired without complication. Jack was making progress with his rehabilitation program when he experienced a possible reinjury to the FPL tendon. This would be a reasonable hypothesis because cardinal signs of a poster-operative rupture include experiencing a pop, and sudden loss of tendon function with no pain or strength with an examination, similar to what Jack reported. On the other hand, Jack was able to complete an active ROM, within his available range at the time, without pain, suggesting that the tendon was working. Observation revealed a Boutonniere deformity (Fig. 6.3) of the left thumb suggesting a different mechanism of injury perhaps involving volar subluxation of the EPL due to a disruption of the extensor pollicis brevis and/or extensor hood of the thumb, collateral ligaments, or the tendon pulley system. If the tendon had ruptured, Jack would have to consider a repeat reconstruction. Our other hypotheses suggest the need for additional testing, imaging, or surgical exploration, in order to make a definitive diagnosis and develop a new intervention plan.

For this case we (1) review the normal structure and function of the portions of the hand most relevant to this case, (2) explore the anatomical and biomechanical rationale of tendon surgical repair, (3) review the injury cycle and healing constraints, (4) define principles of rehabilitation for tendon repairs, and (5) develop hypotheses (biomechanical) related regarding the development of a Boutonniere deformity. This case provides the reader with an opportunity to examine the physiological and biomechanical basis for a commonly used tendon repair rehabilitation protocol, i.e., the modified Duran. Although as noted above, the anatomy, biomechanics, and function of the hand are very complex, we maintain that our focused (to the exclusion of looking at all detailed aspects of the hand) study plan on the form and function of the thumb, in the context of this case, gives the reader a requisite foundation to build from. As we do in the cases reported in previous chapters, let's examine the structure and function of the hand, in general, but with a focus on the thumb.

## 6.4 Hand Structure and Function

### 6.4.1 Structure: Osteology, Ligaments

The wrist carpus includes eight carpal bones, with the distal row articulating with the metacarpus. A ray includes the metacarpal and its related phalanges. Articulations in the hand include the carpometacarpal (CMC), metacarpal phalangeal (MCP), and interphalangeal (IP) joints. Each finger has a proximal (PIP) and distal (DIP) interphalangeal joint, except the thumb, with only an IP joint (Fig. 6.4). The metacarpals and digits are numbered one through five beginning with the thumb on the lateral (radial side). From the palmar and dorsal perspectives, it is evident that there are complex morphological relationships between the distal carpals and the metacarpal bases with which they articulate, although the first CMC clearly has a saddle joint configuration. On the other hand, each metacarpal head with its convex shape articulates with its shallow concave phalangeal base partner. Whereas tradition has classified the interphalangeal joints as a hinge, Neumann suggests that those joints (both proximal and distal) have more of a "tongue-in-groove" shape that favors primarily the sagittal plane movements. Finally, it is obvious that, in the anatomical position (Fig. 6.4A), the first metacarpal is oriented approximately 45° in palmar abduction relative to the other metacarpals. This orientation has clear and significant implications with regard to the thumb's unique ability to move across the palm, i.e., opposition, and interact with each digit for a variety of prehensile actions.

The ability to grasp, hold, and manipulate a multitude of diversely shaped objects is a testament to the dynamic kinematic and kinetic coupling of the carpus and metacarpus. The unique shape of the hand at rest is sculpted by three arches, one longitudinal and two transverse (Fig. 6.5). The proximal, relatively constrained, transverse arch is formed by the distal row of carpals as the carpal tunnel. The carpal tunnel, housing all of the wrist and digital, i.e., finger, flexors, and median nerve, is covered by the transverse carpal ligament. The keystone of the proximal transverse arch is the capitate, which is reinforced by its carpal neighbors and strong carpal ligaments. The distal transverse arch passes

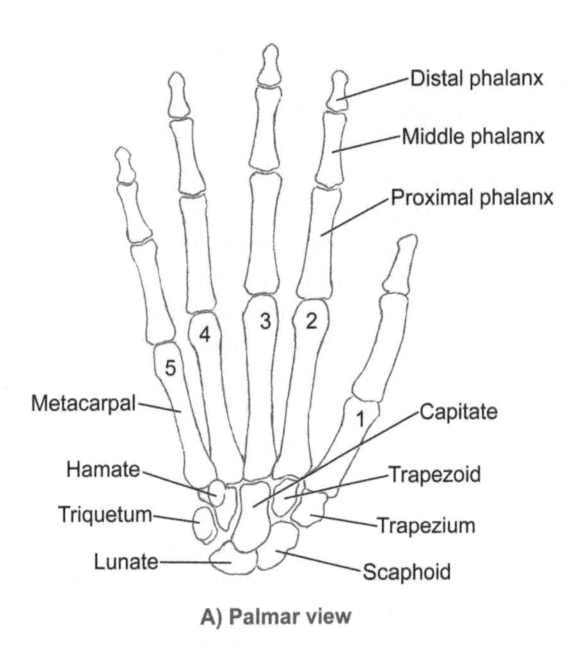


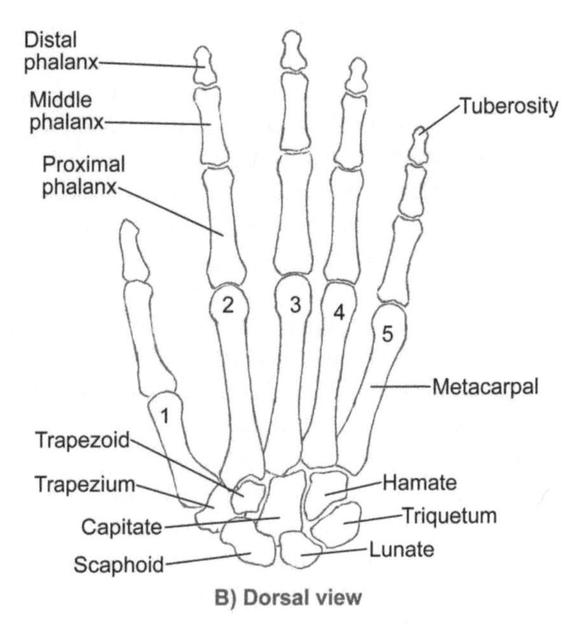


**Fig. 6.4** (**A**) Palmar (ventral or anterior) and (**B**) dorsal (posterior) views of the bones of the wrist and hand

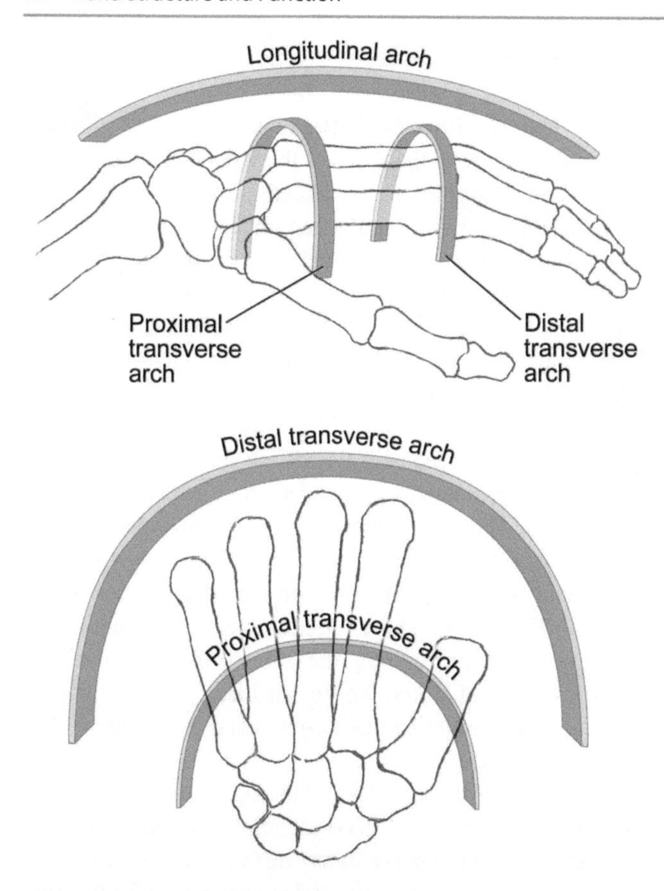


**Fig. 6.5** Schematic of the proximal and distal transverse and longitudinal arches of the hand, an integrated system of arches fundamental to biomechanical stability for hand function

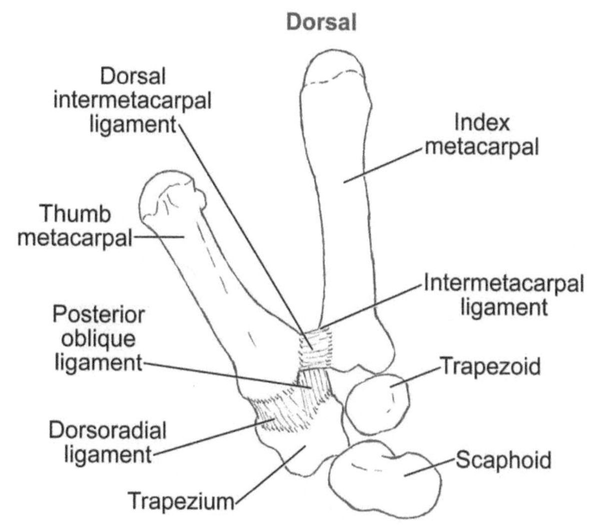


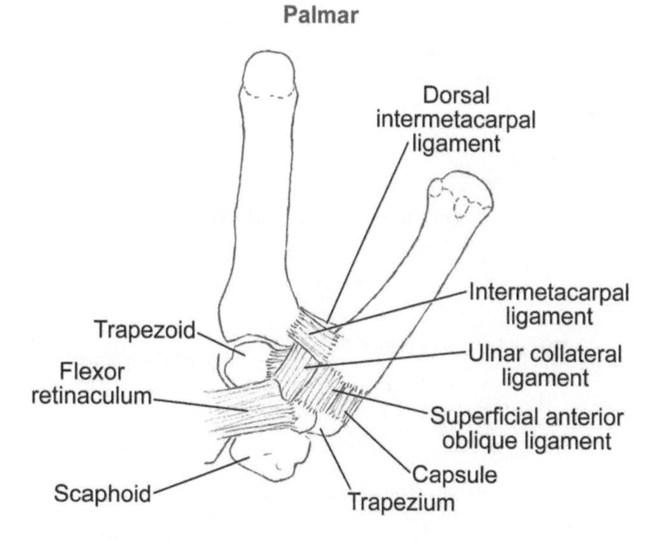


**Fig. 6.6** Dosal and palmar views of important CMC ligaments of the thumb

through the MCP joints. Compared to the proximal arch, this arch has considerable mobility due to the relatively freely moving first, fourth, and fifth metacarpals. The distal arch's signature is its ability to "rotate" around the relatively immobile central column, i.e., second and third metacarpals (the keystone of the distal arch) and capitate. The longitudinal arch follows the shape of the central column and related digits, with the proximal end anchored by the second and third CMC joints and the distal end defined by the phalanges. The longitudinal arch shares its keystone with the distal transverse arch. As noted at the outset of this section, all three arches are mechanically linked, and failure of any one arch leads to significant functional impairments within the hand.

The CMC, MCP, and IP joints of the thumb are stabilized by their capsules and multiple dorsal, palmar, and collateral ligaments. The CMC joint's bony configuration is inherently unstable in its resting position, related to the anatomical fact that the midsagittal (dorsal-volar) diameter of the thumb metacarpal base and the articulating trapezium are mismatched by approximately 34%. Thus, the CMC is primarily reliant on several ligaments for its stability (Fig. 6.6). The superficial anterior oblique ligament is thin, slack when the thumb is opposed, flexed, and abducted, but taut in full extension. The ulnar collateral ligament constrains abduction and extension, which is complemented by the constraint to opposition, flexion, and abduction provided by the intermetacarpal ligament. The dorsal radial (also referred to as the radial collateral) and posterior oblique ligaments are the strongest CMC ligaments, densely populated with sensory fibers and taut in opposition, flexion, and abduction; these are considered the primary constraints to thumb opposition. Thus, although the CMC joint is mildly incongruous and lax in the resting position without rotation, congruity is increased with the screw home torque rotation phase of opposition, induced by the combination of intrinsic muscle action and the dorsal ligamentous complex.

The collateral ligaments of the MCP and IP joints are similar in the thumb and digits (Fig. 6.7). The MCP joints are comprised of a relatively larger convex metacarpal head that

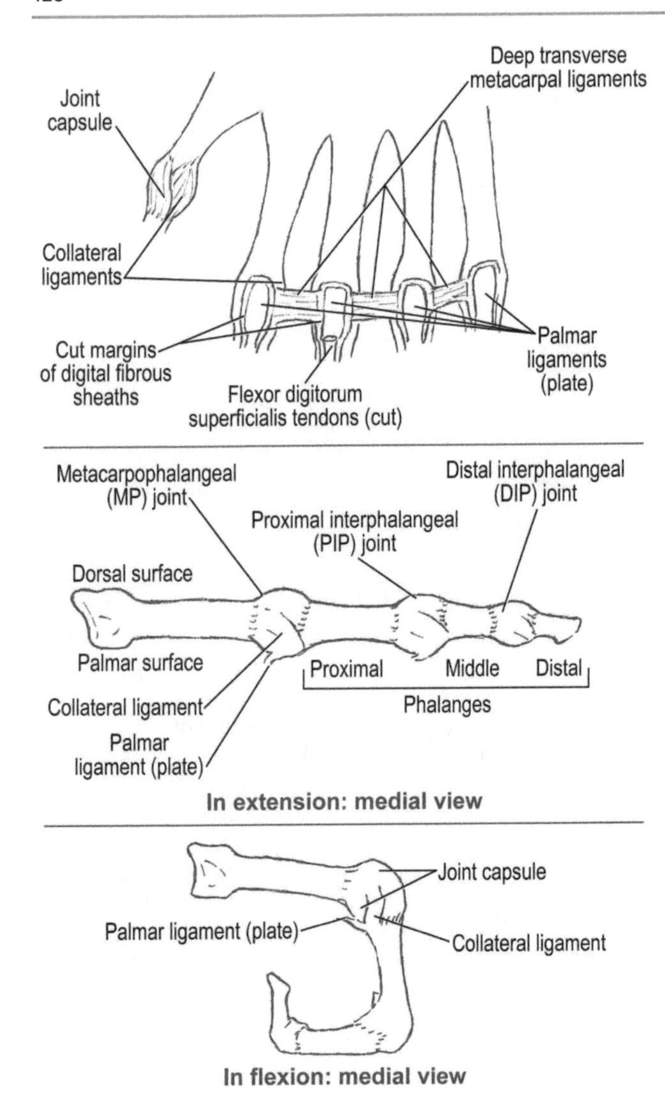


**Fig. 6.7** Collateral ligament system of the digits and thumb

articulates with a shallower concave phalangeal base. Recall that the MCP joints form the keystone of the distal transverse arch so the stability of the MCP joints is foundational for the integrity of digital and overall hand function. Each MCP joint capsule is reinforced by collateral and palmar ligaments. The collateral ligaments are drawn taut in flexion, particularly for digits two to five, which reduces accessory joint motion and enhances stability. This arrangement makes sense in that the largest joint loads are applied to these joints when the hand is working in flexed postures. The palmar ligaments, referred to as palmar (or volar) plates, consist of dense, thick fibrocartilage. These plates serve as stout attachment sites for the fibrous digital sheaths, which form tunnels or pulleys for the flexor tendons. The volar plates reinforce the MCP joints and serve as a primary constraint to hyperextension. The deep transverse metacarpal ligaments attach to the volar plates, loosely securing the second through the fifth metacarpals.

### 6.4.2 Function

#### 6.4.2.1 Muscle-Tendon Actuators

Dynamic coupling of the joints of the wrist and hand to initiate, and adjust, during functional activities is the result of an exquisite interplay between a central controller; effectors, i.e., muscle-tendon actuators; and sensory organs to accommodate a range of motor actions, from precision to power grips (Fig. 6.8A, B). In Fig. 6.8A, we show a schematic of the general central and peripheral elements of a movement of the upper extremity. Now let's imagine the model in Fig. 6.8B representing the actions at the thumb carpometacarpal (CMC), metacarpophalangeal (MCP), and interphalangeal (IP) joints, represented by $\psi_i$ and their orientation, $\theta_i$, controlled by the abductor pollicis longus, extensor pollicis longus, flexor pollicis longus, adductor pollicis, flexor pollicis brevis, and opponens pollicis, represented by forces $F_i$ (as shown in Fig. 6.8B). Thus, for example, if we think about opposing and repositioning the thumb for grasping and releasing a cup of water, we get an idea of how a biomechanical model might be diagrammed and the elements that compose the model.

Innervation of the muscles and the skin of the hand is supplied by three peripheral nerves: radial, median, and ulnar (Fig. 6.9). The radial nerve innervates the extrinsic extensor muscles of the digits and wrist and is responsible for the sensation on the dorsal radial aspect of the wrist and hand. The median nerve innervates most of the extrinsic flexors of the digits and then, after it enters the carpal tunnel, innervates most (with exception of the deep head of the flexor pollicis brevis and adductor pollicis) of the intrinsic thenar muscles, i.e., muscles of the thumb, and two lateral lumbricals. The sensory aspect of the median nerve covers the lateral palm, including the palmar region of the lateral $3\frac{1}{2}$ digits. The ulnar nerve innervates the medial half of the flexor digitorum profundus, intrinsic hypothenar muscles, and medial two lumbricals. The deep motor branch of the ulnar nerve innervates the remainder of intrinsic muscles, i.e., all palmar and dorsal interossei and the adductor pollicis. The ulnar nerve covers the sensation of the ulnar border of both the dorsal and volar aspect of the hand and ulnar $1\frac{1}{2}$ digits. The joints of the hand receive sensory nerve fibers from dorsal nerve roots at the spinal cord level: C6 for sensation from the thumb and index finger, C7 for sensation from the middle finger, and C8 for sensation from the ring and small fingers.

The three major flexors of the digits and thumb include the extrinsic flexor digitorum superficialis, flexor digitorum profundus, and flexor pollicis longus (see the Supplementary 6.1 for a listing of all extrinsic and intrinsic muscles of the hand and detailed actions of the thumb muscles) (Fig. 6.10). Notice that the tendon of the flexor digitorum superficialis splits, at the level of the proximal phalanx, to make room for the flexor digitorum profundus and its distal digital

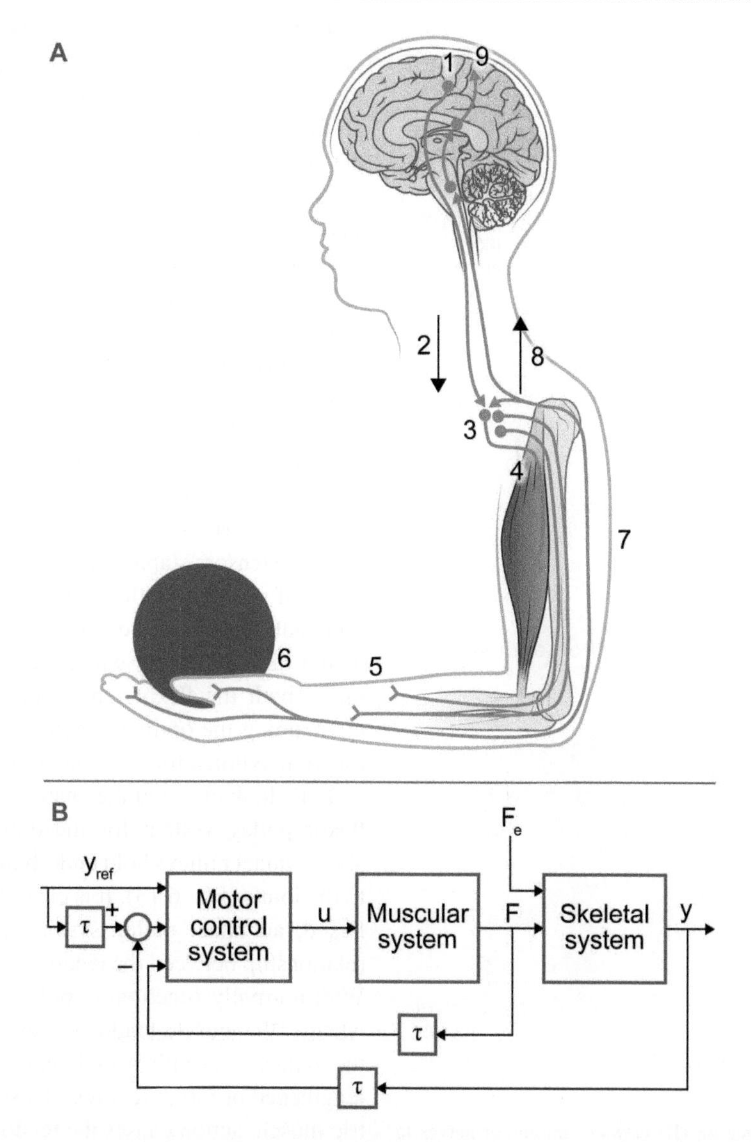


**Fig. 6.8** (**A**) Schematic diagram of central and peripheral nervous system components involved in the control of voluntary movements. A voluntary action is initiated in selected brain regions, e.g., precentral cortex. (1) Cortical motor neurons project along the spinal cord (2) and synapse at various levels. Peripheral motor neurons have cell bodies in the spinal cord (3) and axons along with the spinal roots and peripheral nerves (4) that terminate in neuromuscular synapses in various limb muscles (5). Sensory feedback arises from a variety of receptors, e.g., cutaneous receptors in the fingertips, mechanoreceptors in joint capsules, etc. (6). Sensory information is transmitted centrally along afferent nerve fibers (7) that project directly or indirectly onto the motor neurons (3) to produce spinal reflexes. Sensory cells also send ascending branches along the spinal cord (8) that, via relay neurons, project to various sensory areas of the cortex, in particular the postcentral cortex (9). Connections among various areas of the cortex and subcortical structures (brainstem, cerebellum, basal ganglia) participate in motor control loops. The control of voluntary movements includes sensory feedback from cutaneous afferents, proprioceptive afferents from muscles and joints, and visual, auditory, and vestibular sensory feedback. (From Hoffer et al. (1996).) (**B**) A block diagram of the neuromuscular system. The motor control system generates neural input signals $\boldsymbol{u}$ based on a reference trajectory, $y_{ref} = [y_1, \dot{y}_1, \ddot{y}_1, y_2, \dot{y}_2, \ddot{y}_2, y_3, \dot{y}_3, \ddot{y}_3]$, representing the thumb CMC, MCP, and IP joints, efferent position and velocity signals $y = [q_1, \dot{q}_1, q_2, \dot{q}_2, q_3, \dot{q}_3]$, and muscle forces $\boldsymbol{F} = [F_1, F_2, F_3, F_4, F_5, F_6]$, i.e., a synergy (see text for a listing of the six thumb muscles). Afferent signals, e.g., from the skin, muscle (spindles), tendons (Golgi tendon organ), and mechanoreceptors (joint capsule and ligaments), are available after a time delay, $\tau$. The musculoskeletal system is induced to move by the muscles forces $\boldsymbol{F}$ and external forces $\boldsymbol{F}_e$. Note: the symbols $\dot{\psi}_i$ and $y_i$ represent the kinematic variables angular velocity and acceleration, respectively. (Modified with permission from Stroeve (1998))

attachment. The superficialis, profundus, and flexor pollicis longus have the potential to flex each joint that it crosses, although the palmar interossei and lumbricals are primary flexors of the MCP joints.

Superficial, and beginning proximal, to the synovial sheaths, we find the palmar aponeurosis (Fig. 6.11). Because of their position at the proximal margin of the digital flexor tendon sheaths, it appears that the transverse fascicular fibers

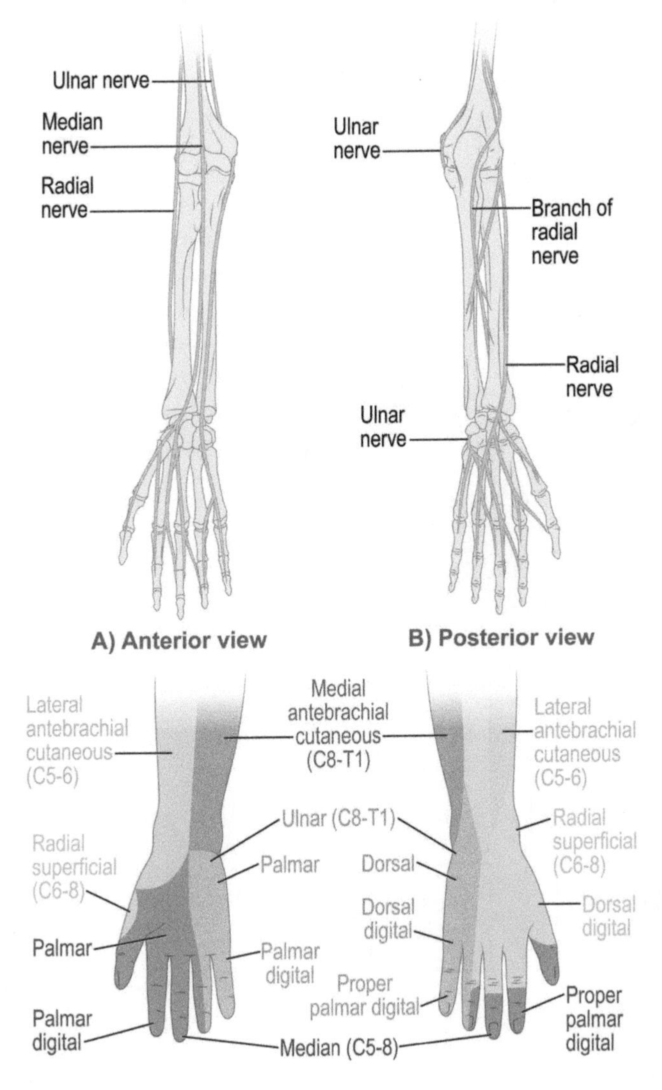


**Fig. 6.9** Anterior (**A**) and posterior (**B**) views of the motor nerves to the wrist and hand and the sensory nerve distribution to the distal forearm, wrist, and hand. Note anterior is also referred to as volar and posterior is also referred to as dorsal

and paratendinous bands of this apparatus form a tunnel under which the flexor tendons pass. Deep into the palmar aponeurosis, we note the ulnar and radial synovial sheaths that surround the tendons of the flexor digitorum superficialis and profundus and flexor pollicis longus, respectively (Fig. 6.10). In addition, the tendons of the long finger and thumb flexors travel through fibro-osseous tunnels called fibrous digital sheaths, which are anchored to the phalanges and volar (palmar) plates. Flexor pulleys, e.g. A1, A2, etc., are bands of tissue that lie deep to the digital sheaths and hold the tendons relatively close to the MCP and IP joint centers (Fig. 6.10). It has been suggested that the palmar aponeurosis forms a pulley in conjunction with the first and second annual pulleys of the digital flexor mechanism; this mechanism does not appear to influence the thumb pulley system. Finally, the digital synovial sheaths, lying deep to the pulley systems, serve both nutritional and lubrication functions, i.e., reduce friction between the superficialis and profundus tendons for digits two to five, along the pullies of the digits enclosing the tendons.

A closer look at the pulley system of the thumb (Fig. 6.12) shows four pulleys: A1, Av, oblique, and A2. The location of the oblique and $A_2$ pulleys is notable, i.e., close to the IP joint, since these are located in the vicinity of where Jack's FPL was lacerated.

As mentioned, for the hand to function efficiently, the extrinsic and intrinsic muscles must work synergistically, but this is also true of muscles that are considered antagonists. For example, the long digital flexors are capable of providing the power needed to grasp, hold, and move heavy objects but only if the wrist is held in approximately 20° extension and digital extensors stabilize the respective joints. Moreover, by virtue of positioning the wrist in extension and providing IP joint stabilization, the extensors optimize the length-tension of the digital flexors, while the flexor pulley system optimizes both the flexor tendon moment arms and assists in maintaining the optimal length of the muscle-tendon actuators to maximize force production.

Let's look closer at the anatomy and biomechanics of the flexor pulley system for the digits and thumb. As noted above, intact pulleys hold underlying tendons relatively close to the joints (Fig. 6.13). It is clear that when the digit is fully flexed, an intact pulley system maintains a near-constant relationship between the tendon and the joints that it crosses. With normally functioning pulleys, these restraints operate whether the muscle-tendon forces are small or large. Without the pulleys or with a deficient pulley system, i.e., overlengthened or torn, the force of a strong concentric or eccentric muscle action causes the tendon to be pulled away from the joint centers. This pulling away of the tendon is referred to as "bowstringing" (Fig. 6.14). It is obvious that if one or more pulleys are not secured, tensioning of the tendon results in an approximation of the end of the tendons that are secured, producing this bowstringing appearance. Apart from inducing a cosmetic aberration (Fig. 6.3), bowstringing results in significant pathomechanics of the fingers and thumb during activities that require large flexion moments.

Under normal conditions, there is evidence that the moment arms for the finger flexors do not change with changes in joint orientation; that is, there is a linear relationship between tendon excursion and joint angle change. Neumann discusses how an overstretched or torn pulley alters the moment arm of the flexor tendon (Fig. 6.15). He suggests that the primary function of the intact pulley system is to maintain near-constant moment arm lengths of the extrinsic flexor tendons, i.e., flexor digitorum superficialis and profundus and flexor pollicis longus. Bowstringing of a tendon crossing a joint increases its moment arm, which

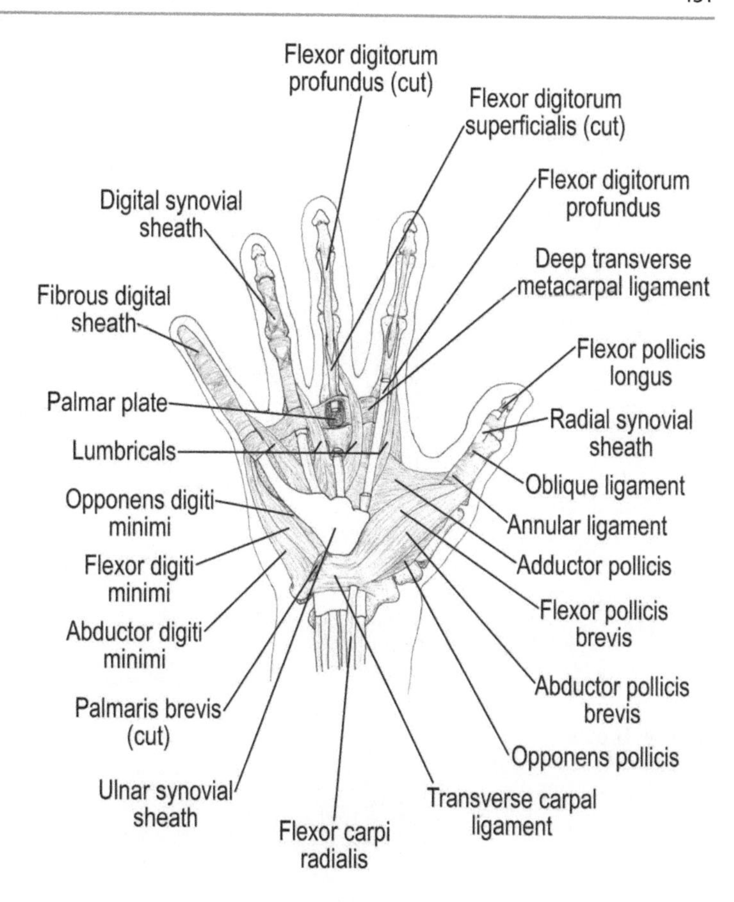


**Fig. 6.10** Palmar layered view of the hand shows the relationship among many periarticular connective tissues and extrinsic and intrinsic muscles

could increase the mechanical advantage of the muscle at that joint. That is, given that the muscle force is sustained, i.e., not increased or decreased, an increase in its moment arm produces a larger muscle moment (torque) at the joint the muscle-tendon actuator crosses (see Appendix C for a more detailed discussion of a model rope-pulley system). This would seem to be an advantage; however, bowstringing also induces a suboptimal change in the muscle-tendon length-tension, thus perhaps offsetting the increased moment arm. Moreover, the increased moment arm reduces the angular rotation of the joint per linear distance of the muscle action. Thus, we see, for example (Fig. 6.15), that rupture of the A2 and A3 pulleys results in a significant increase in the moment arm across the proximal interphalangeal PIP joint, but an approximate 57° reduction in PIP flexion. Besides, then, the ungainly cosmetic appearance of bowstringing (Fig. 6.3), a physiological reduction in full IP flexion that results from a pulley rupture, induces a significant loss in a strong functional grip.

At his 8-week rehabilitation visit, Jack reported thumb pain related to performing his resisted exercises. A Boutonniere deformity was demonstrably obvious (similar in appearance to the photo in Fig. 6.3). As noted previously, it is possible that the pain he reported could be related to a strain injury to the repaired FPL tendon, since it is known that at 6–8 weeks, the regenerating tendon is very weak. However, a strain injury would not produce a Boutonniere deformity. Several events may explain his Boutonniere deformity: (1) third-degree strain injury, i.e., rupture, to the collateral ligaments and other tissues with attachments to the volar plate and adjacent bones, since we know that the pulley system is linked to the collaterals via the volar plate (see Fig. 6.7); (2) rupture of one or more pulleys, e.g., Av, A1, A2, and oblique (see Fig. 6.12); or (3) failure of the extensor hood, i.e., central slip, of the extensor pollicis brevis with volar migration of the lateral bands. Rupture of flexor pulleys is very common in rock climbers, but thumb flexor pulley ruptures are rare. There are only a small number of case reports of thumb flexor pulley ruptures which include (1) an acute pulley rupture related to opening a jar lid, (2) post-repeated corticosteroid injections into the thumb for symptoms of tenosynovitis (it has been shown that strong corticosteroids can damage the biochemical bonds of the collagen framework of connective tissue), and (3) an acute pulley rupture while changing a car tire. In all of these cases, it was determined that the bowstringing was related to rupture of the A1, A2, and oblique pulleys. Since Jack's laceration was just proximal to the IP joint, it is possible that the Av, A2, and/or oblique pulleys were "nicked" as part of the original injury, and not seen by the surgeon at the time of

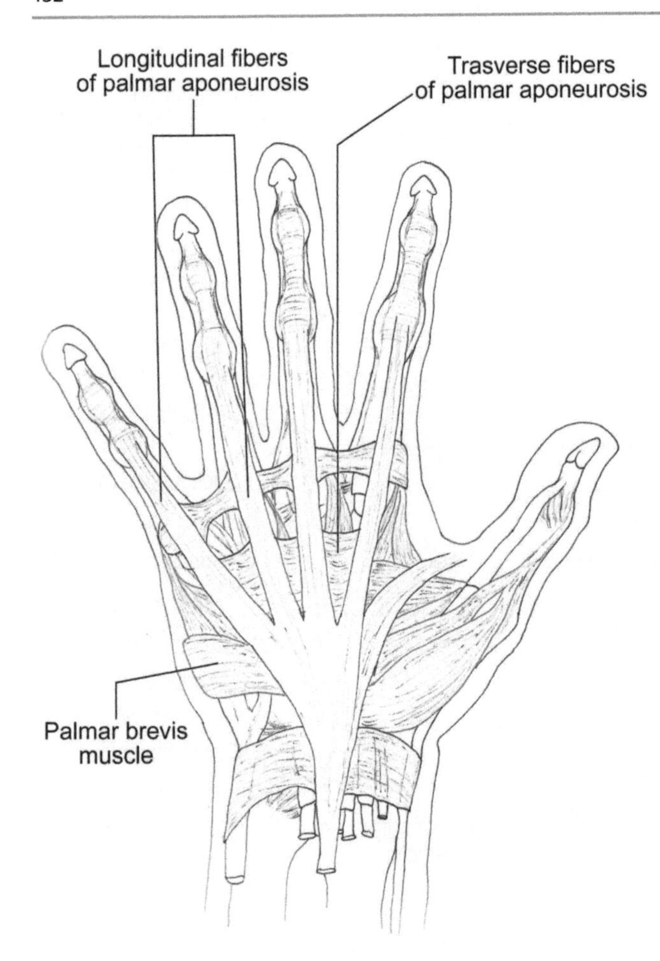


**Fig. 6.11** Schematic of palmar aponeurosis system including its longitudinal and transverse components

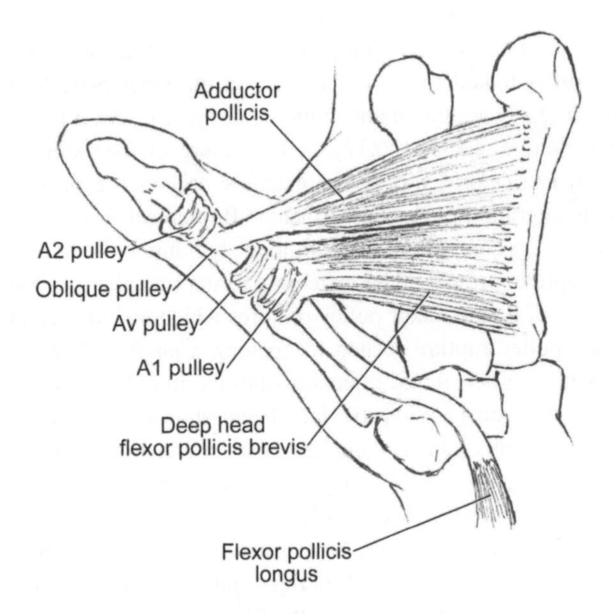


**Fig. 6.12** Schematic illustrating detailed pulley system of the thumb. Note the relationship between the A2 and oblique pulleys and the IP joint

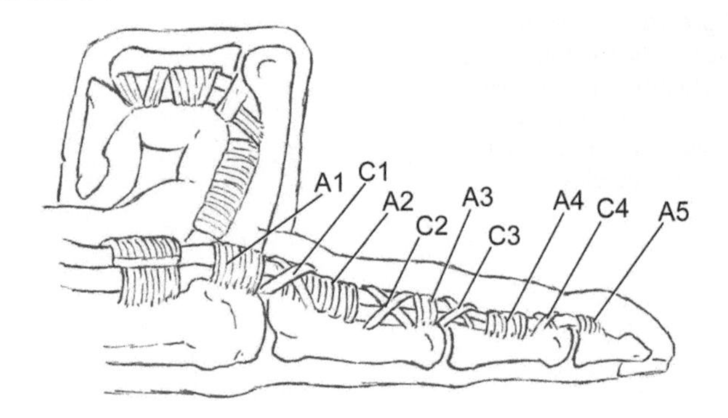


**Fig. 6.13** Schematic of the complex pulley system of the index finger in an extended and fully flexed position. We can see that as the finger flexes, the pulley complex maintains the approximation of the tendon and phalanges

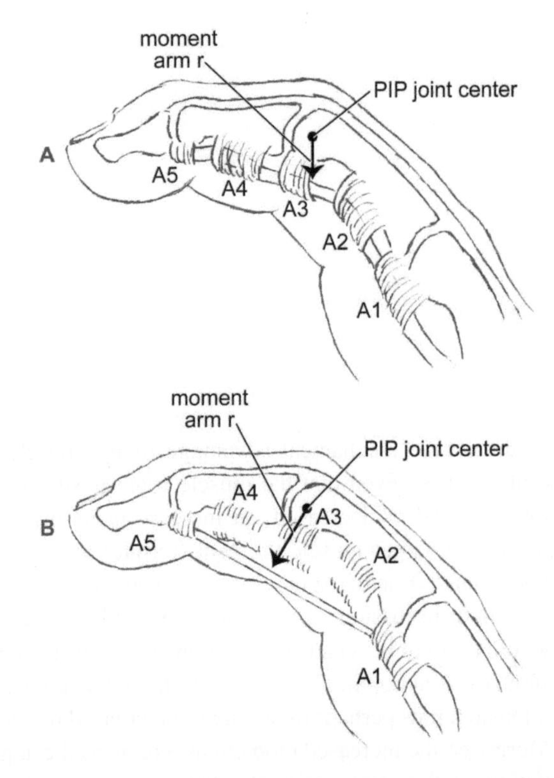


**Fig. 6.14** Schematic of (**A**) normal and (**B**) "bowstringing" tendon that results from rupture of the A2, A3, and A4 pulleys of a digital flexor. Note in both (**A**) and (**B**) the moment arm, *r*; in (**B**), the moment arm of the digital flexor tendon at the PIP joint is increased due to pulley ruptures

repair. Then, secondary to highly repetitive passive, followed by active flexion/extension range of motion exercises, the "bowstringing" force and friction between the FPL tendon and pulleys may have induced a fatigue injury that resulted in ultimate failure. Recall that experimental cyclic loading of a tendon-ligament structure displaces the stress-strain curve to the right along the elongation (strain) axis (Fig. 6.16). During this type of testing, rate- and magnitude-controlled loads, typically occurring only in the linear region of the

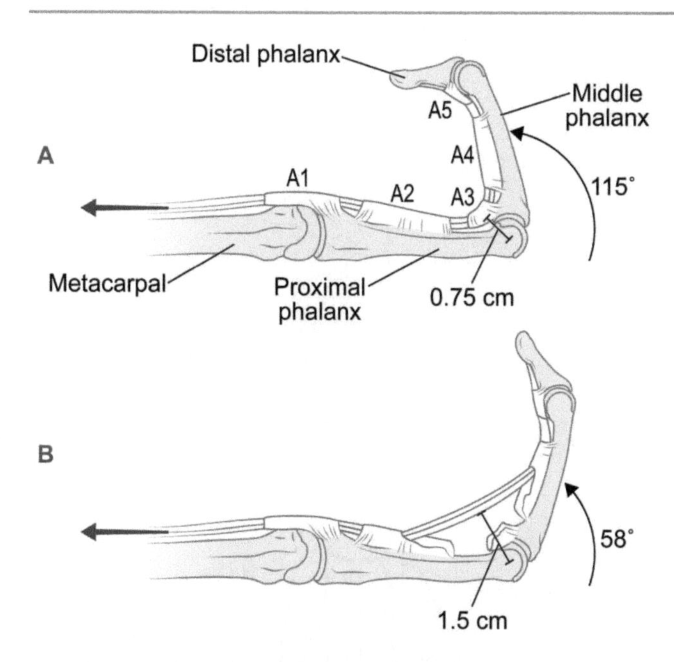


**Fig. 6.15** Schematic illustrating (**A**) normal action of a flexor pulley and (**B**) theoretical altered mechanics following rupture of A2 and A3 pulleys

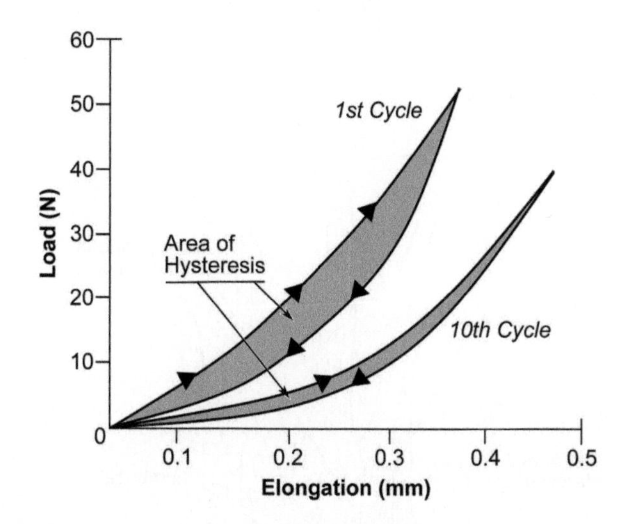


**Fig. 6.16** Typical cyclic loading and unloading of a tendon-ligament structure, where the *y*-axis depicts the load applied and the *x*-axis deformation within the tissue. The area within the curves represent energy stored/lost (i.e., hysteresis) in the tissue. Note the reduction in energy stored/lost with repeated loading, i.e., cycle #10

stress-strain curve, are applied and released over specific time periods. A shifting of the stress-strain curve to the right suggests the presence of a nonelastic, i.e., plastic or permanent, deformation of the tissue that progressively increases with each cycle. We also see that as repetitive loading progresses, the tissue demonstrates a decrease in elastic stiffness. These findings are evident even when loading the tissue within physiological ranges. Our intuition suggests that if loading occurs cyclically (repetitively), in the range of 5–8% strain, microdamage to the collagenous framework (and proteoglycan-water network) could have contributed to the rupture of one or more pulleys in Jack's thumb.

Let's briefly examine the superficial and deep dorsal muscles of the fingers and thumb, including the extensor mechanism. The primary extrinsic extensors include the extensor digitorum (ED) (also a primary wrist extensor), extensor pollicis longus (EPL), extensor indicis (EI), and extensor digiti minimi (EDM) (Fig. 6.17).

The digital extensor tendons cross the wrist within synovial-lined compartments deep to the extensor retinaculum (Fig. 6.18). Distal to the extensor retinaculum, the digital tendons lie dorsal to the MCP joints before they connect with their distal attachments. Intertendinous connections, juncturae tendinae, in the proximity of the MCP joints, connect the tendons of the extensor digitorum, providing direction to the tendons and some transverse stability. It is notable that the extensor tendons do not have a well-defined pulley system but are integrated into a complex connective tissue system called the extensor expansion (also extensor mechanism or hood). The extensor expansion is comprised of the extrinsic digital extensors, and two sets of intrinsic muscles, e.g., lumbricals and interossei. Acting in isolation, the ED acts to hyperextend the MCP joints but cannot extend the proximal (PIP) and distal (DIP) joints. With their insertion into the extensor hood, the interossei, and lumbricals, via the oblique and transverse fibers (Fig. 6.18), assist the ED in extending the two distal IP joints. Our review of the entire digital extensor apparatus was cursory for the simple reason that Jack's injury did not involve these joints and tissues. We leave it to the curious reader to use the resources provided at the end of this chapter to further explore the intricacies of hand anatomy and biomechanics.

We do, however, need to explore in some more detail the extrinsic and intrinsic muscles of the thumb. The extrinsic muscles of the thumb include the extensor pollicis longus (EPL), extensor pollicis brevis (EPB), and abductor pollicis longus (APL). Tendons from both the EPL and EPB are part of the extensor mechanism of the thumb, but only the EPL can extend the CMC, MCP, and IP joints of the thumb. Actions of the EPL also include slight lateral rotation (an accessory conjunct rotation) and adduction, components of thumb repositioning. The abductor pollicis longus is more of a thumb extensor but also abducts the thumb at the CMC joint (Fig. 6.18).

We would be remiss if we did not briefly describe three key muscles of the thenar eminence: flexor pollicis brevis, abductor pollicis brevis, opponens pollicis, and first dorsal interosseous (Fig. 6.10). As a unit, these muscles may be most important because of their role in providing dynamic CMC stability, but they also contribute to thumb opposition, motion about the CMC joint that includes abduction, flexion, and medial rotation (another accessory conjunct rotation).

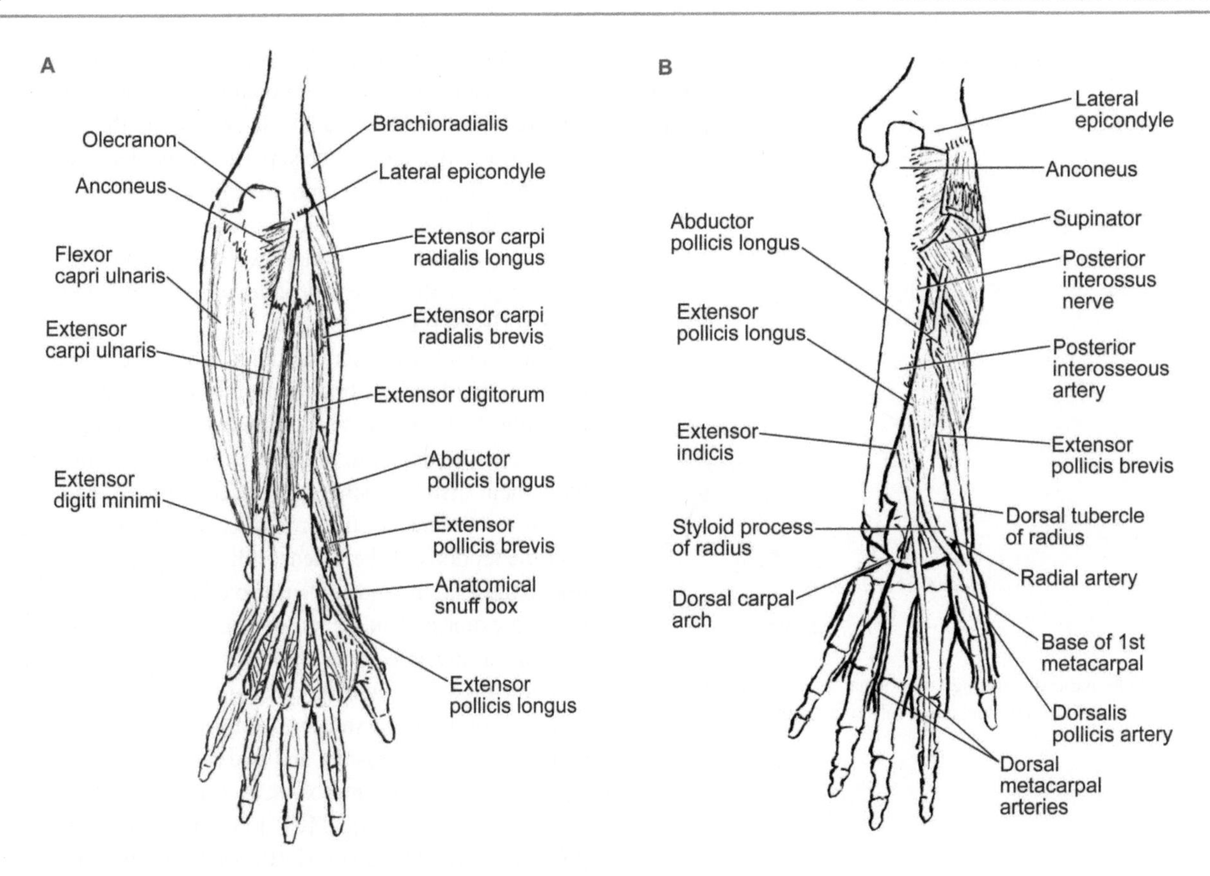


**Fig. 6.17** Dorsal view of the (**A**) superficial and (**B**) deep wrist, finger, and thumb extensors

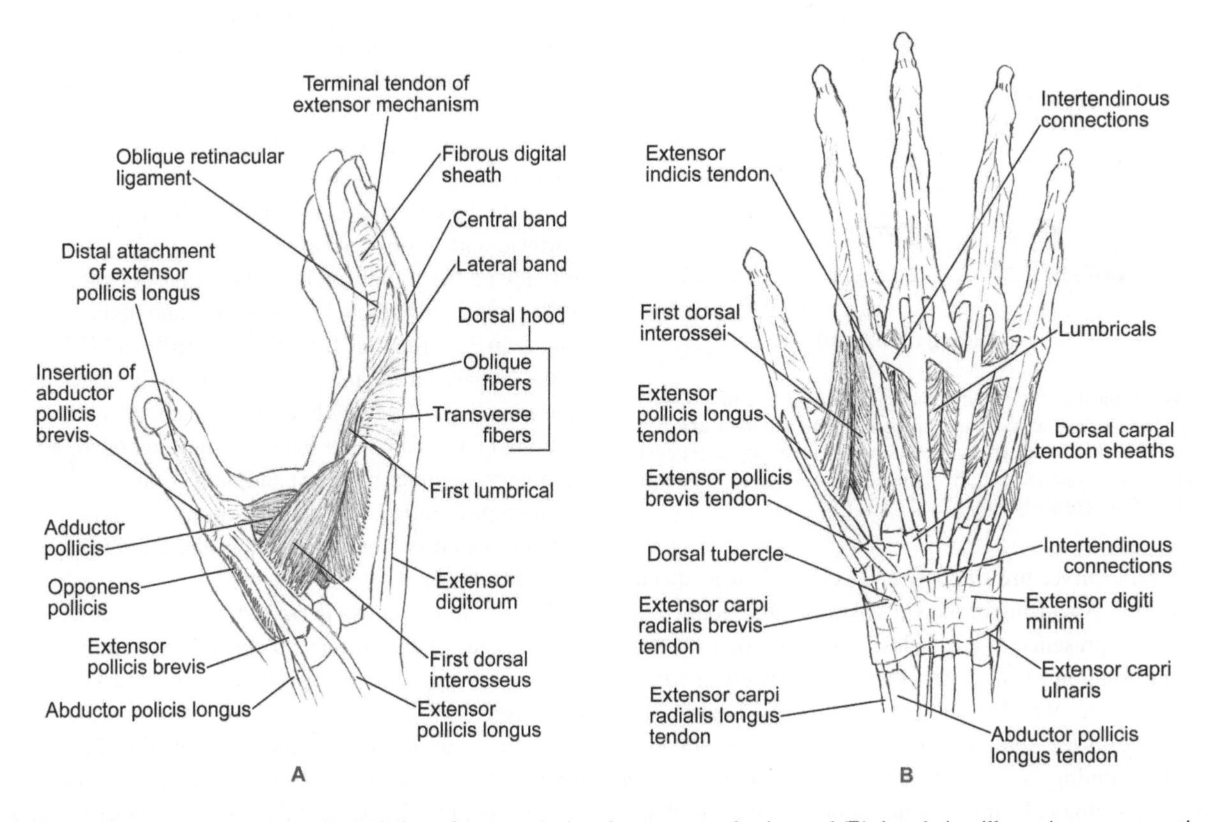


**Fig. 6.18** (**A**) Lateral view depicting a detailed view of the complexity of extensor mechanism and (**B**) dorsal view illustrating extensor retinaculum, carpal tendon sheaths, and intertendinous connections of thumb and index finger

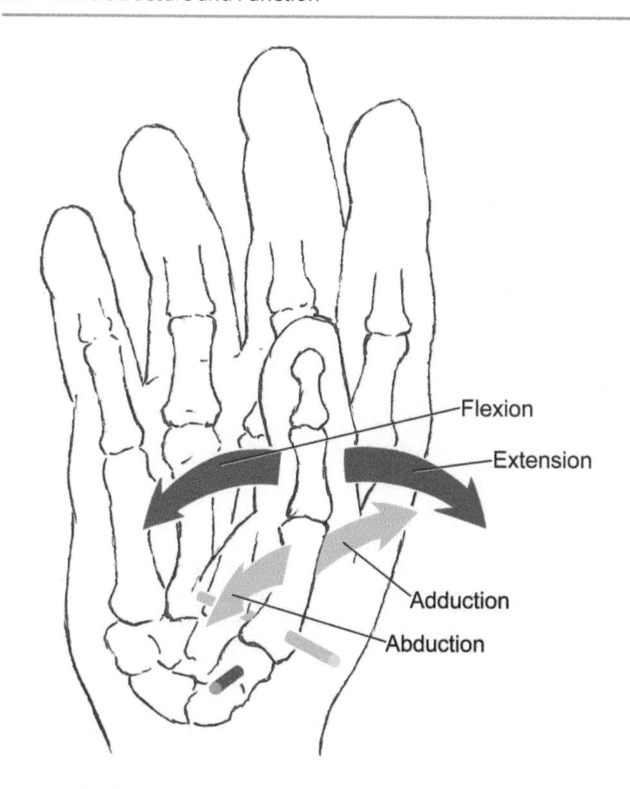


**Fig. 6.19** Sagittal and frontal plane movements of the thumb. The green and and blue cylinders indicate the axes of rotation for abduction/adduction and flexion/extension, respectively

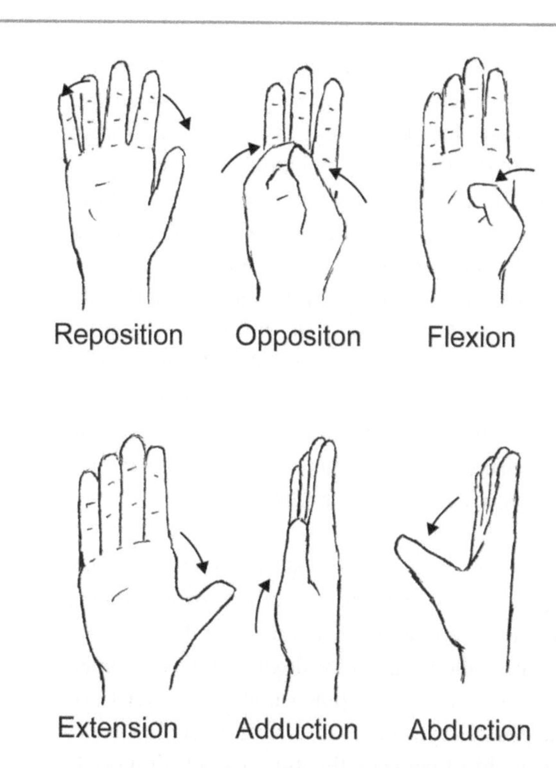


**Fig. 6.20** Illustration of osteokinematic movements of the thumb

Opposition is vitally important for all grasping actions of the thumb and fingers (note: there is a similar opposition movement at the CMC in the hypothenar eminence; see Fig. 6.10). The adductor pollicis is one of the strongest muscles of the body based on cross-sectional area and is responsible for the strong lateral and pulp pinch between the thumb and index and long fingers. Additionally, we see that the oblique pulley originates at the insertion of the adductor pollicis, perhaps making this muscle an important partner in the thumb flexor pulley system (Fig. 6.12). Bringing attention to these four intrinsic muscles may be important because of their location relative to the flexor pollicis longus tendon and, thus, their potential impairment secondary to the FPL tendon injury sustained by Jack.

#### 6.4.2.2 Kinematics: Osteokinematics

The clinical description of joint movement is always dependent on some reference system or starting position. In most cases, we assume that the reference position for describing movements of the digits and thumb is the anatomical position: elbow extended, forearm supinated, and wrist in neutral. To describe the physiological movements of the digits, we use the standard cardinal plane convention with flexion/extension occurring in the sagittal plane, abduction/adduction occurring in the frontal plane, and long-axis rotation occurring in the horizontal or transverse plane. With the middle finger as the reference digit, its frontal plane movement is described as radial or ulnar deviation.

Since the thumb is oriented medially relative to the anatomical position's frontal plane, i.e., or relative to the fingers, we must apply our standard clinical terms differently. Flexion and extension are frontal plane movements of the thumb, with the volar surface of the thumb moving across the palm during flexion; extension is a return to the anatomical position. Flexion/extension occurs in all three thumb joints: the CMC, MCP, and IP joints. Abduction is a movement away from the palm in an anterior direction, i.e., a sagittal plane movement, with adduction a return to the anatomical position; note that this movement primarily takes place at the CMC joint, usually with the MCP and IP joints extended (Fig. 6.19).

Opposition of the thumb is the movement that brings the tip of the thumb to meet the tip of any other finger; reposition is a return to the anatomical position. Opposition is a type of helical rotation that could be described as a combination of flexion and adduction involving the CMC, MCP, and IP joints, with reposition the return of the thumb to the anatomical position (Fig. 6.20). Of course, during functional activities, our digits do not actually move in the cardinal planes, yet it is useful because the clinical range of motion measurements are based on these definitions.

Being able to manipulate objects involves intricate and highly coordinated movements, i.e., combinations of physiological rotations of all digits in multiple planes, of the fingers and thumb. Some activities of manipulation are very repetitive, e.g., scratching, typing, etc., which are managed by a central controller and system of sensors and effectors (see Fig. 6.8A, B). Other manipulations, e.g., writing, drawing, etc., are more continuous and variable and perhaps more complex, requiring movement synergies and frequent coordinated adjustments of position and force.

Prehension is the ability of the fingers and thumb to work together to pick up, grasp, and hold while simultaneously manipulating an object. Generally, prehension includes grasping or pinching for power or precision; power requires a moderate to a large force, but less specificity, with precision requiring less force but accuracy of movement. There are five basic categories of prehension:

1. The power grip requires large forces and involves the digital flexors, lumbricals and interossei (MCP flexion), and the thumb flexor and adductor muscles. There is co-action of the long finger flexors and wrist extensors, to maintain the wrist in approximately 20° wrist extension.
2. The power key pinch, i.e., lateral pinch, involves stabilizing an object between the thumb and lateral border of the index finger. This position involves the action of the adductor pollicis and the first dorsal interosseous muscles.
3. A precision grip generally requires less force and is used to complete tasks for accuracy. Precision grips always involve the thumb and two (i.e., three-point pinch or three-jaw chuck) or more digits. These grips require the actions of digital flexors and thumb abductors.
4. Two types of precision pinches involve just the thumb and index finger when more fine control is required: tip-to-tip or pad-to-pad. Tip-to-tip is used for smaller objects which need to be manipulated accurately, e.g., threading a needle, whereas pad-to-pad pinches are used when a larger surface area of contact is required, e.g., holding and writing with a pencil.
5. The hook grip only requires the use of finger flexors, e.g., carrying a briefcase or holding a bucket of water. A hook grip is usually only a static position and is held as long as needed to accomplish the task.

The hand is vitally important because of its ability to do a variety of specialized tasks related to support, manipulation, and prehension. For example, the non-dominant hand is used to support, i.e., brace or stabilize, an object, freeing up the dominant hand to perform a task. Jack was fortunate, in a sense, because his non-dominant thumb was injured. So, although he would be limited in using his left hand to forcefully grasp and stabilize objects in performing functional activities related to home maintenance or repair, his job-related activities were likely less affected. Jack was free to use the digits of his non-dominant hand as long as he did not compromise the repair and recovery of his left thumb. Without the use of his left thumb, he could not perform most other functional activities that required both hands, with the exception of the hook grip.

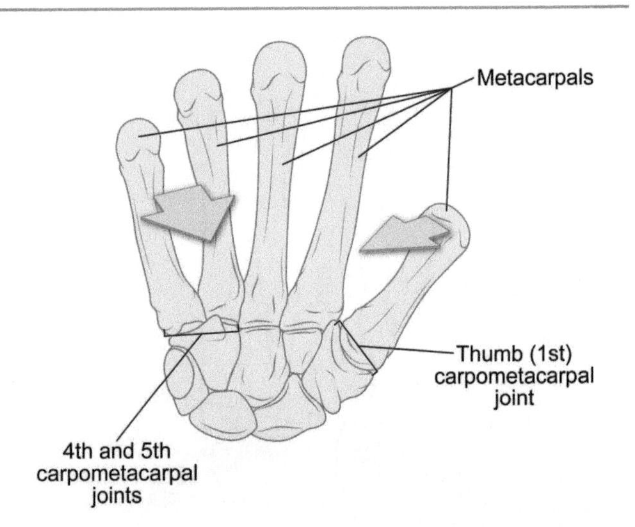


**Fig. 6.21** Schematic palmar view of the carpometacarpal joints of digits 1–5. Note the opposition of the thenar and hypothenar eminences that contribute to a type of cupping of the radial and ulnar aspects of the hand when cylindrical objects need to be held

#### 6.4.2.3 Kinematics: Arthrokinematics

The carpometacarpal joints of the hand are formed by the distal row of carpals and bases of the five metacarpal bones (Fig. 6.21). The first, fourth, and fifth CMC joints are quite mobile. The coupled flexion and internal rotation produced at these joints are primarily responsible for making it possible, i.e., opposition of thenar and hypothenar eminences, to grasp cylindrical and spherical objects of different sizes and shapes. Because of the irregular and varied shapes of the articulating surfaces of the proximal carpals and their metacarpal partners, we cannot describe typical patterns of roll-and-glide arthrokinematics. However, remember that each joint in the CMC complex has six degrees of freedom of movement, suggesting that immobilization of these joints has the potential to induce joint restriction. The second and third metacarpals in articulation with the trapezoid, capitate, and hamate exhibit more constrained motion and, as a result, provide necessary stability to the central column.

The CMC joint of the thumb is comprised of the base of the first metacarpal and trapezium. This articulation has the well-defined saddle morphology, where each respective articular surface is convex in one plane and concave in the other, sometimes also referred to as a double saddle joint, where both surfaces are saddle-shaped, one atop the other, upside down, and rotated 90°. At the first CMC, the articular surface of the metacarpal base is convex in the longitudinal direction in a palmar-to-dorsal direction and concave in the transverse diameter in a medial-lateral direction. Its

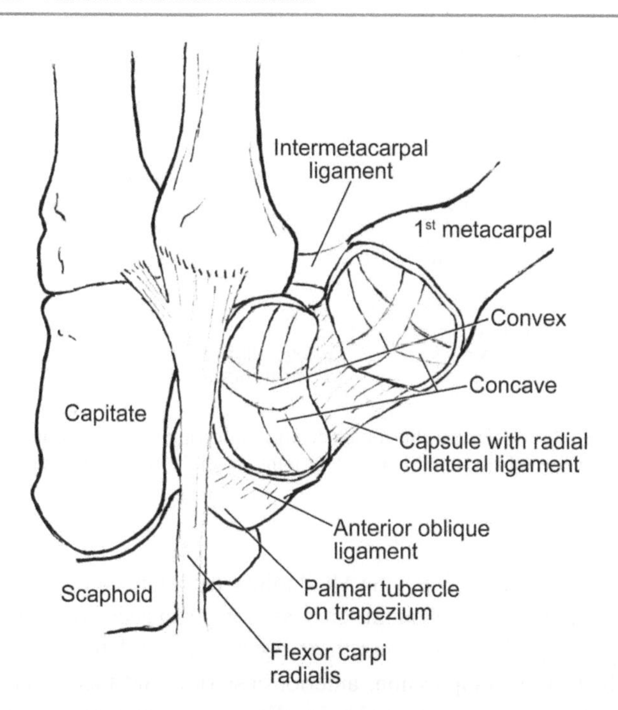


**Fig. 6.22** Palmar view of an opened carpometacarpal joint of the thumb illustrating the saddle configuration of this joint. Note the supporting periarticular ligaments that contribute to joint arthrokinematics and stability

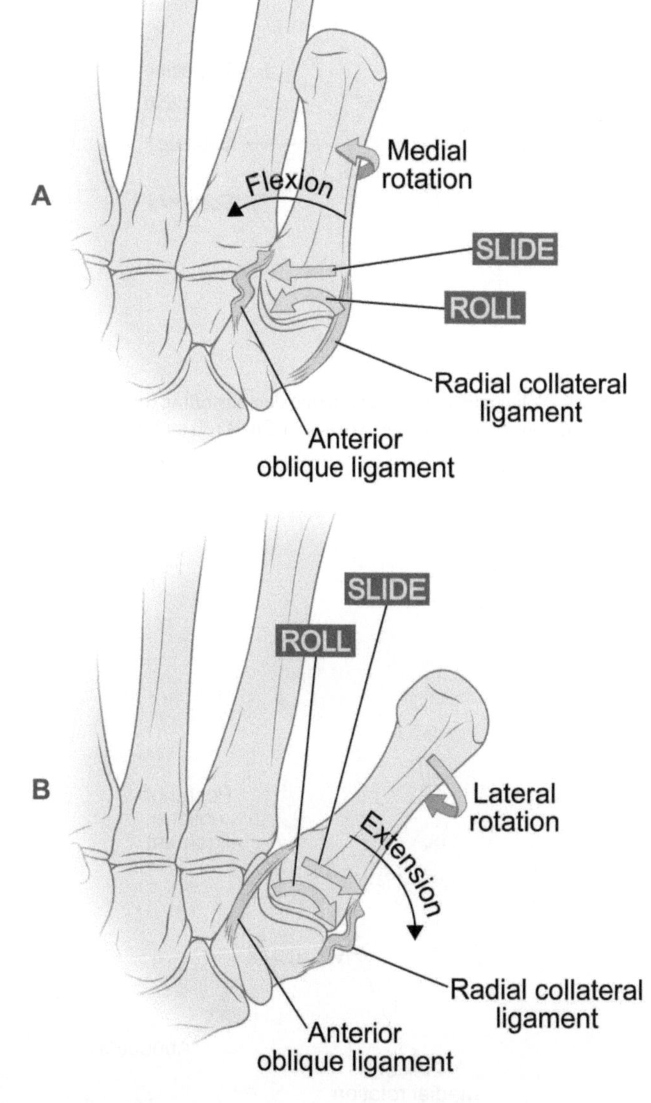


**Fig. 6.23** Schematic of carpometacarpal joint motion of the thumb during (**A**) flexion and (**B**) extension. In the sagittal plane, the concave surface of the proximal metacarpal rolls and glides in the same direction as the osteokinematic movement of the bone

articulating partner, i.e., trapezium, has a reciprocal shape, which contributes to the good congruency of the joint surfaces of the first CMC joint (Fig. 6.22). A saddle joint typically provides two degrees of freedom, i.e., rotations in two dimensions, that is, flexion and extension in the frontal plane, and the sagittal plane movements of abduction and adduction. For thumb opposition, these rotations occur simultaneously (flexion + internal rotation) in a kind of helical fashion.

For the physiological movements of flexion/extension, with the thumb's initial position neutral and the hand in the anatomical position, flexion can be described as an ulnar (medial) roll and glide, as the concave articular surface of the metacarpal moves relative to the trapezium's convex surface. There is a conjunct, i.e., non-voluntary or accessory, medial rotation of the metacarpal, which might be analogous to a spin; sometimes this rotation is referred to as pronation (Fig. 6.23). With extension, the metacarpal laterally rotates slightly as it rolls-glides radially.

The reference position of the thumb, i.e., zero position, is as if the thumb is resting on the volar surface of the second digit. When the metacarpal moves anterior, i.e., away from, relative to the plane of the hand that is in the anatomical position, it abducts. In this plane, the proximal end of the metacarpal is convex, and the trapezium is concave. Therefore, the rule of arthrokinematics prescribes that the metacarpal rolls anteriorly, i.e., in a palmar direction, and glides posteriorly or dorsally. When the thumb returns to its starting position the motions are reversed at the metacarpal/trapezium interface, i.e., dorsal roll and palmar glide (Fig. 6.24).

As noted for thumb abduction, we call the starting position of the thumb for opposition to be with the thumb resting on the volar surface of the second digit. Opposition at the CMC joint allows tip-to-tip prehension between the thumb and little finger (see Fig. 6.20). This involves abduction, then flexion, and medial rotation at the CMC joint. As you can imagine, the arthrokinematics is a composite of what we describe for the separate movements of flexion and abduction (Fig. 6.25). According to Neumann, the movement of opposition, initiated by the opponens pollicis, is augmented

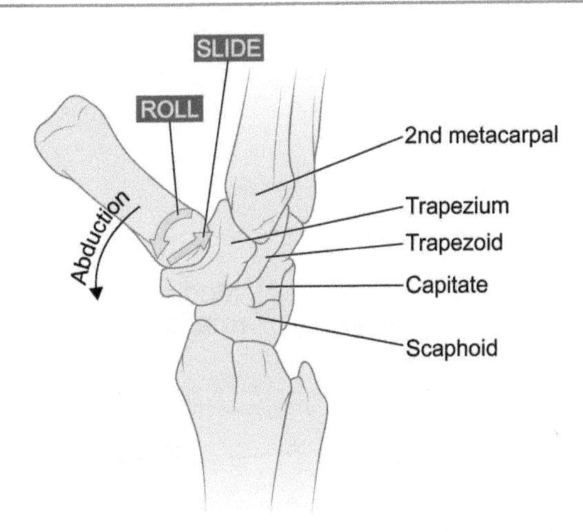


**Fig. 6.24** Schematic of CMC abduction demonstrating the arthrokinematics motions of anterior (i.e., volar or palmar) roll and posterior (i.e., dorsal) glide

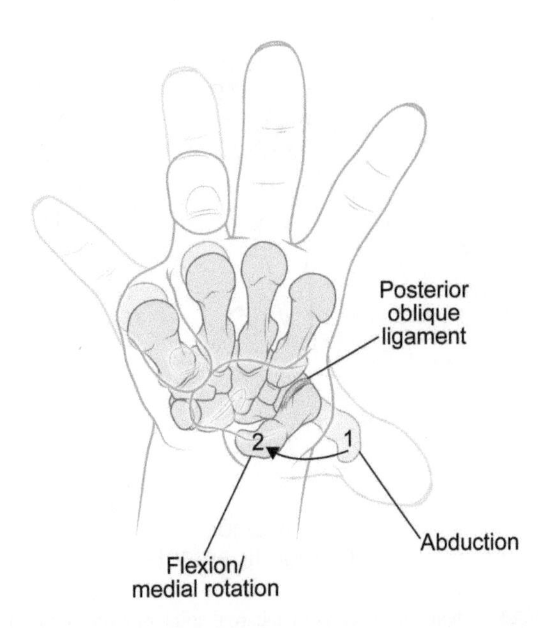


**Fig. 6.25** Opposition at the CMC joint of the thumb involves combined abduction, flexion, and medial rotation of the metacarpal relative to the trapezium. Arthrokinematically, there is a volar roll/palmer glide (abduction) followed by an ulnar roll/glide with concomitant medial rotation (or spin)

by increased tension in the posterior oblique and radial collateral ligaments (Fig. 6.6). At the end of the range of motion, the CMC reaches a close-packed position. It is notable that accessory internal rotation at the MCP and IP joints and trapezium (relative to the scaphoid and trapezoid) add to total thumb opposition motion.

We have already described the morphology and constraints of the MCP and IP joints of all five digits. Tradition has defined motion at the MCP and IP joints with two and one degrees of freedom, respectively: flexion/extension and abduction/adduction at the MCP joints and flexion/extension

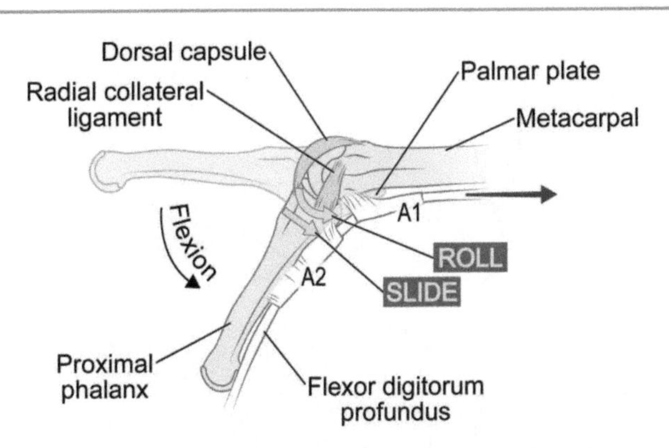


**Fig. 6.26** Schematic illustration demonstrating joint surface motion of volar roll/glide at the MCP and IP joints of digits two to five during flexion

at the IP joints. When the MCP joints are extended for digits two to five, the collaterals are less taut, so an accessory rotation, e.g., long axis, and three accessory translations, e.g., distraction/compression, anterior/posterior, and medio- lateral, are available. As defined earlier, accessory motions cannot be induced voluntarily, but are essential degrees of freedom of movement for full pain-free mobility of the joint. The PIP and IP joints of digits two to five, however, are close-packed when the joints are extended, leaving little room for additional mobility (although it does exist); the collateral ligaments are most taut at the PIP and IP joints when those joints are extended. For digits two to five, the morphology of the MCP and IP joints are similar, that is, the proximal ends of the bones, i.e., heads, are convex, and the distal ends of the bones, bases, are concave. Therefore, during flexion of the MCP and IP joints, there is a volar roll/glide at the joint surface, which is facilitated by the tautness of the dorsal joint capsule (Fig. 6.26). Extension at these joints sees a dorsal roll/glide at the joint surface.

Voluntary abduction/adduction is only available at the MCP joints of digits two to five. With abduction of the second digit, for example, the metacarpal rolls/glides in the radial direction; for the fourth and fifth digits, abduction induces an ulnar roll/glide (Fig. 6.27); adduction produces a roll/glide in the reverse directions. Morphologies and arthrokinematics at the thumb MCP and IP joints are similar to their respective joints of digits two to five, except that abduction/adduction at the thumb MCP is accessory only; the additional accessory motions listed previously for the MCP joints of digits two to five are also available and important. Flexion/extension occurs about an axis in an anterior/posterior direction and involves osteokinematic movement of the phalanx in the frontal plane, i.e., ulnar and radial directions, respectively. Since the base of the moving bone is concave, there is an ulnar roll/glide during flexion and radial roll/glide during extension (Fig. 6.28).

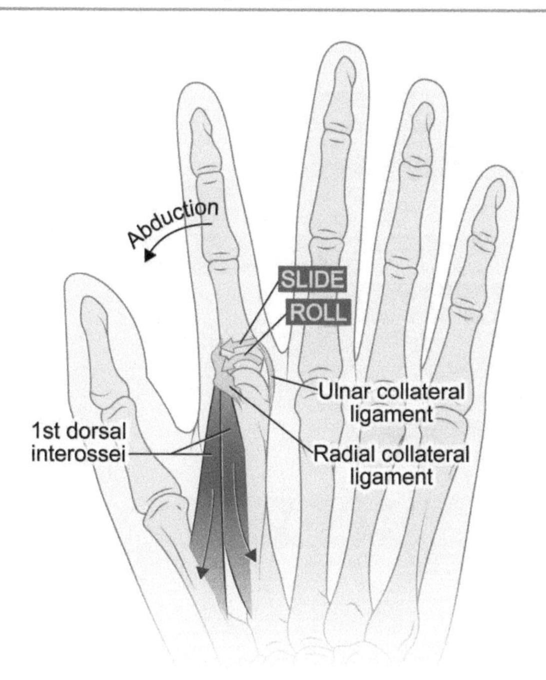


**Fig. 6.27** Abduction at the second MCP joint is initiated by the first dorsal interossei with a demonstration of the radial roll/glide at the joint surface of the concave base of the proximal phalanx and convex head of the metacarpal

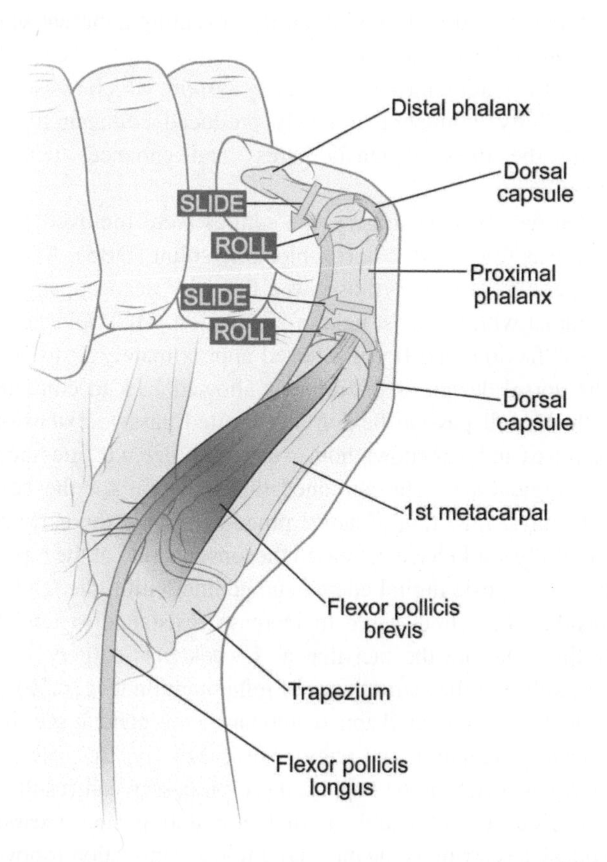


**Fig. 6.28** Schematic of the initiation of MCP and IP flexion by the flexor pollicis longus and brevis, also illustrating the ulnar roll/glide at the respective joints

Since this case involved a lacerated tendon and its repair, you may wonder why we spend significant time examining the structure and kinematics of all of the digits. It is important to do so because we need to appreciate the similarities and differences between digits two to five and the thumb, as well as the overall complexity of the hand. We also need to understand how those structures and their functions may be altered as a result of the repair of the lacerated tendon and related periods of immobilization and reduced use during the rehabilitation phases. Although our main focus in rehabilitation would be the repaired FPL tendon, we need to be cognizant of the fact that reduced use of other structures could lead to alterations of normal joint motion and disuse atrophy of select muscles. Additionally, with the injury occurring in close proximity to periarticular structures of the thumb IP joint, it is likely that normal proprioceptive feedback from both joint mechanoreceptors and muscle spindles was also altered.

## 6.5 Rehabilitation: How Physiology and Mechanics Are Linked, Stages of Healing: Modified Duran Program

In Chap. 5 we examine the details of the inflammatory and healing stages following fracture of the scaphoid. We are ready to use that information again. Doing so is a great example of abstracting or generalizing previous learning experiences to new and different ones, which we think is unique to the PBL pedagogy, an example of the importance of some repetition. Let's briefly review the cycle of healing and regeneration of biological tissues in general, with emphasis on tendon injury recovery (Fig. 6.29). To link physiology with mechanics, we examine the stages of healing and stages of the modified Duran protocol (MDP) in tandem. The MDP is commonly used following the surgical repair and reconstruction of flexor tendons of the hand and is based on the theory of early passive mobilization. What is the theory of early mobilization? Recall that we had previously touted the phrase, "function dictates structure," which means that external, and the reactive internal, forces that are imposed on bones, ligaments, tendons, etc. are critical for the maintenance and enhancement of the function of those structures; for bone, this principle was referred to as Wolff's law. The phrase "specific adaptation to imposed demand" (SAID) is an analogy to Wolff's law and informs the idea of passive mobilization as the basis for the MDP. Since we know that the maintenance and enhancement of tendon performance are related to its ability to mitigate and transfer tension, the application of controlled tension is the mechanical principle that frames the MDP. Let's look at the details of Jack's rehabilitation program.

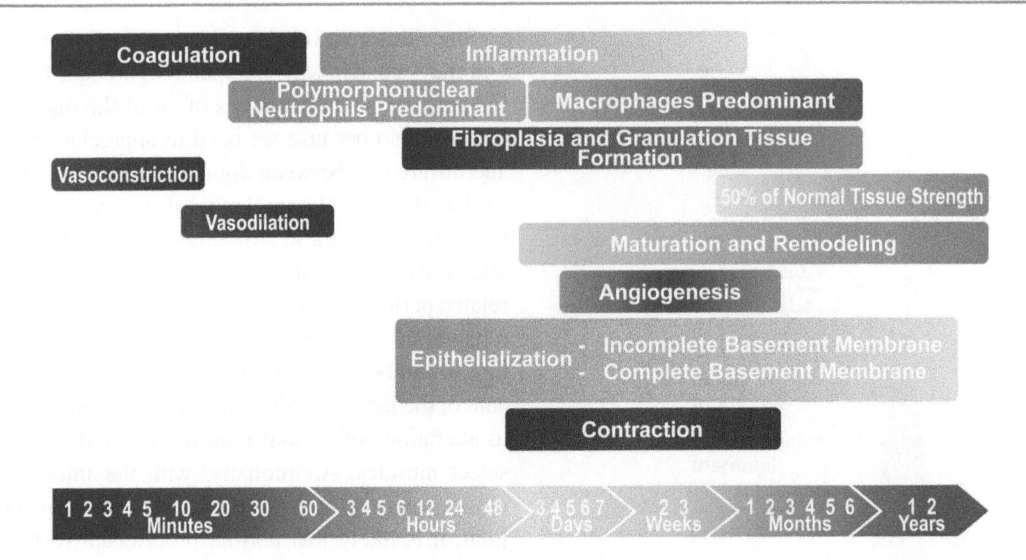


**Fig. 6.29** Stages of inflammation, healing, and maturation following a tendon injury

Immediate post-repair, but prior to closing the wound, the surgeon performed a final check of the repair by performing a passive range of motion test of the thumb CMC, MCP, and IP joints. This was to ensure that no appreciative gapping occurred at the repair site and that there was smooth gliding, i.e., no "catching," of the tendon within its sheaths and pulleys. As noted in the surgical report, following skin closure, the thumb was placed in a soft dressing and spica splint, with the thumb (CMC, MCP, and IP joints) and wrist secured in a flexed position. The purpose of securing these joints in flexion was to decrease the maximum power potential of the repaired flexor muscle-tendon unit; that is, flexing the wrist and thumb places the extrinsic flexor at a mechanical disadvantage if accidentally activated. The splinting also reduces forced passive tensile forces on the repaired tendon.

During the first 10 days following surgery, Jack was instructed to wear the splint 24 hours but could remove it once a day to clean the wound. After the splint was replaced, but before it was secured, full passive thumb flexion and then extension range of motion were performed within the constraints of the splint. As we can see (Fig. 6.29), during the first 7–10 days, several normal physiological inflammatory and immunological processes take place: polymorphonuclear neutrophil and macrophage activity to clear wound debris, formation of fibroplasia and granulation tissue, and epithelialization. The formation of scar within the tendon and surrounding area is normal during the later phases of healing and important to provide stability to the wound. The unwanted side effect of scaring, i.e., restrictive adhesions, can be prevented with early and frequent passive range of motion exercise; Jack was instructed to perform the passive exercises for 25 repetitions, three to four times throughout the day. Making sure that a full passive range of motion toward extension was achieved within the constraints of the splint, full passive flexion to the distal palmar crease places very small tensile loads on the tendon and instructs the tendon to glide distally from the repair site in both directions. It is hypothesized that in addition to preventing adherent scar formation, flexion and extension movements foster free gliding of the tendon through the pulley system, which assists in the orderly alignment of newly produced collagen fibrils along the lines of tensile stress and enhances tendon nutrition.

On day 10 post-surgery, the sutures were removed, and Jack was fitted with a dorsal blocking splint (DBS). A low-temperature plastic orthosis was fitted by an occupational therapist with the wrist held in 20° flexion, MCP joint fixed at 20° flexion, and IP joint flexed approximately 5° to 10°. The dorsal design of the orthosis allowed Jack to continue with the full passive flexion and limited passive extension exercises and was shown how to self-mobilize, i.e., massage, the surgical scar. He continued this regiment for the next 7–10 days. Initiation of active range of motion too early or forcefully could have increased the tensile stress at the repair due to extrinsic digital edema and edema within the tendon sheath, which both serve to increase resistance to tendon gliding. Despite the fact that at 1-week post-surgery Jack was well past the acute stages of inflammation (Fig. 6.29), it is clear that the wound area would have continued to see significant granulation and angiogenesis activity. Because the hand has a rich blood supply, these processes can result in prodigious external and internal granulation sites (scars). Adhesive scarring is the most common complication following tendon repair surgery. Jack was encouraged to use regular skin lotion as a lubricant and perform cross-scar massage, as tolerated, for 3–5 minutes, three times daily. It is believed that passive massage facilitates a more regular arrangement or alignment of collagen fibril formation, and it has been shown to reduce the severity of the adhesions of the skin (external) and also internal scarring.

Beginning the third postoperative week, active flexion and extension exercises were added to complement the passive exercises, still within the constraints of the DBS; the frequency and duration of the exercises remained the same. Active range of motion obviously calls for the neurological activation of the muscle-tendon unit, which obviously increases the tensile load of the repaired tendon. The intensity of muscle activation during active range of motion is low, assuring that the repair would not be exposed to damaging tensile forces. During this subacute phase, i.e., 3 weeks, there is a high rate of new collagen synthesis, but the modeling and maturation phases are just beginning (Fig. 6.29). It is hypothesized that the minimal muscle activation enhances the active distribution of synovial fluid for tendon nutrition, as well as tendon regeneration and modeling.

By week 4, Jack still experienced significant joint stiffness, but post-surgical pain had disappeared likely contributing to renewed confidence in the use of his thumb. He was cautioned that, despite the fact that surgical pain was no longer a deterrent, his tendon was very weak and susceptible to reinjury if he was not careful. Between weeks 4 and 5, exercise frequency and duration were maintained but were conducted outside the constraints of the DBS. Active flexion and extension of the wrist and fingers, closing and opening a composite fist, and a "place and hold" flexion exercise (i.e., passively placing the IP and MCP in end-range flexion and performing a gentle isometric hold for 3–5 seconds) were added to the program. Jack was encouraged to take his wrist and fingers through a full active range of motion in both directions, adding mild over-pressure only at the end range flexion (passive extension was not allowed). During this time period of rehabilitation, as the granulation, angiogenesis, and epithelialization processes decline, modeling and maturation of regenerated collagen can progress, aided by controlled increases in tensile loading of the repaired tendon. Scar massage continued to be a frequent daily routine. Jack was instructed to continue to don the DBS when he was not exercising.

Twenty-four-hour donning of the DBS was discontinued at the sixth week, although Jack was encouraged to wear it while sleeping and when involved with activities that could accidentally hyperextend his left thumb. He continued the active flexion and extension exercises and massage routine. Blocking (to IP joint only) and passive extension for both the MCP and IP joints were added. A blocking exercise at the IP joint entailed supporting the MCP in full extension, i.e., holding the metacarpal with the free hand while activating the flexor pollicis longus, thus inducing moderately strong active IP flexion. At this point in the modeling and maturation phase of healing, the repaired tendon would have regained approximately 50% of normal tensile strength so that passive and active stretching was safe to perform (Fig. 6.29). At 6 weeks, Jack's range of motion had improved to the point where he was much less conscious of having had the repair. He was again warned about the danger of damaging the repair by over-exercising or using his left hand for routine daily activities. Up to this point, Jack had been very compliant and appeared to understand his situation.

Although Jack's thumb had been immobilized, the early passive, and later active, mobilization that is a feature of the MDP meant that most of the deleterious effects of immobilization we studied previously were not an issue. The most important feature of the MDP, i.e., specific adaptation to imposed demand, for flexor tendon injuries protected and maintained the function, to a certain degree, of bone, joint hyaline and fibrocartilage, and muscle. The DBS blocked thumb function during the very important early phases of the inflammation, granulation, and epithelialization periods, but the digits were free to move which theoretically provided a neurological stimulus to the inactive FPL and other related thumb muscles. Therefore, while there would be some atrophy of thumb musculature, it was not as significant as what we observed in our previous wrist case (Chap. 5). Furthermore, the passive and active ROM exercises performed by Jack were augmented by joint mobilization, based on arthrokinematics principles, performed by his occupational therapist.

At week 8, in addition to maintaining active ROM and gentle passive stretching exercises for the MCP and IP joints of the thumb, submaximal resistive exercises were introduced, e.g., light squeezing of therapeutic putty, a light hand exerciser, and light free weights for the wrist, forearm, and upper arm. These exercises were to be performed for 1–3 sets of 25 repetitions once daily. For the most part, Jack felt like he was back to normal and wanted to resume all of his activities. However, he was reminded to refrain from lifting and moving heavy objects and performing maximum isometric wrist and finger/thumb flexion exercises because the tendon was still very weak. Resistive exercises are introduced at this time to increase the intensity of tensile loading of the maturing tendon (Fig. 6.29), but low-intensity, high-repetition exercises would need to continue for up to 12 weeks. Jack was told that at about 12–14 weeks, it would be safe for him to resume all activities of daily living, but that full range of motion and strength would not likely be achieved for approximately 1 year.

Unfortunately, soon after he began increasing his exercise workload, Jack had an incident that produced a painful response and a Boutonniere deformity. As we hypothesized previously, it is likely that he sustained a third-degree strain to one or more of his flexor pulleys, i.e., $A_2$ or oblique. We might wonder why a pulley could have failed before his repaired tendon, since, based on the timeline of healing, a repaired tendon has been shown to have only about 50–60% normal strength at 8 weeks. Although we cannot know exactly why this happened, we can conclude that the weakest link, i.e., one or more of the flexor pulleys, is what failed. And we can also be sure that it failed under tensile loading because we know that when the tendon tension increases, the normal "bowstring" effect that it creates increased tension in the pulleys as well.

## 6.6 Chapter Summary

Mr. Bishop was a 54-year-old mathematics teacher who lacerated the tendon of his left flexor pollicis longus while recycling plastics and cardboard. Fortunately, he was seen by an orthopedic resident in an urgent care center who had prior experience with similar injuries, triaged his injury accurately, and facilitated a referral to a hand surgeon. Jack's tendon and digital nerve repairs, which occurred within 3 days of his injury, were completed without complication. Jack's rehabilitation, based on the early passive-mobilization model of the modified Duran protocol, began 1–2 days following his repair and progressed well. However, in the eighth week of the program, Jack experienced some pain in the process of performing resistive muscle exercise and reported to therapy with a Boutonniere deformity in the left thumb. We hypothesize that the tendon repair was spared and that it was likely Jack sustained a third-degree tension strain injury to one or more of his thumb flexor pulleys.

Although the injury, in this case, was reasonably straightforward, its presentation provided the opportunity to examine the complexity of the anatomy, biomechanics, and function of the hand while focusing on the thumb. We limit a detailed study to the thumb because a problem-based inquiry theorizes that a broad perspective, while important, is secondary to the details of a particular problem. For this case, we explore a portion of the anatomical and biomechanical detail of the other digits because it is necessary to examine the similarities and differences between the digits and thumb. As well, although the extensor side of the thumb was not involved, it is important to study because its function may have been impaired by the FPL injury, and also a Boutonniere deformity most often occurs due to disruption of the extensor mechanism. This case forces us to draw on our experiences with previous cases, such as the structure and biomechanics of tendon, stages of healing, and arthrokinematics principles. Exploring the biomechanical principles used by the orthopedic surgeon in his selection of best practices for the repair and rehabilitation protocol demonstrates the breadth of the application of biomechanics. In the elbow case (Chap. 4), it is important for us to study the details of the mechanics of pitching because it provides insight into Jack's injuries. Drawing on the details of the inflammatory and repair cycle explored in the wrist case (Chap. 5) is an example of the importance of the development of the skill in abstracting or generalizing and application to other problems. In this case, examining, i.e., applying, the details of the stages of healing following a tendon repair in tandem with the rehabilitation protocol demonstrates its physiological/biomechanical rationale.

## Bibliography

Amadio PC, Lin G-T, An K-N (1989) Anatomy and pathomechanics of the flexor pulley system. J Hand Ther 2(2):138–141

Bayat A, Shaaban H, Giakas G et al (2002) The pulley system of the thumb: anatomic and biomechanical study. J Hand Surg Am 27:628–635. https://doi.org/10.1053/jhsu.2002.34008

Boyer MI, Gelberman RH, Burns ME et al (2001) Intrasynovial flexor tendon repair, an experimental study comparing low and high levels of in vivo force during rehabilitation in canines. J Bone Joint Surg Am 83(6):891–899

Doyle JR (1990) Anatomy and function of the palmar aponeurosis pulley. J Hand Surg Am 15:78–82

Doyle JR (2001) Palmar and digital flexor tendon pulleys. Clin Orthop Relat Res 383:84–96

Doyle JF, Blythe WF (1977) Anatomy of the flexor tendon sheath and pulleys of the thumb. J Hand Surg 2(2):149–151

Edmunds JO (2011) Current concepts of the anatomy of the thumb trapeziometacarpal joint. J Hand Surg Am 36:170–182. https://doi.org/10.1016/j.jhsa.2010.10.029

Fazilleau F, Cheval D, Richou J et al (2014) Reconstruction of closed rupture of thumb flexor tendon pulleys with a single free palmaris longus tendon graft: a case report and review of the literature. Chir Main 33:51–54. https://doi.org/10.1016/j.main.2013.11.009

Francis-Pester FW, Thomas R, Sforzin D et al (2021) The moment arms and leverage of the human finger muscles. J Biomech 116:110180. https://doi.org/10.1016/j.jbiomech.2020.110180

Hoffer JA, Stein RB, Haugland MK et al (1996) Neural signals for command control and feedback in functional neuromuscular stimulation: a review. J Rehabil Res Dev 33(2):145–157

Imaeda T, An K-N, Cooney WP (1992) Functional anatomy and biomechanics of the thumb. Hand Clin 8(1):9–15

Kin-Ho L, Pak-Cheong H (2012) A rare case of thumb boutonniere deformity: flexor pulley rupture. J Orthop Trauma Rehabil 16:75–77. https://doi.org/10.1016/j.jotr.2012.09.008

Kline SC, Beach V, Moore JR (1992) The transverse carpal ligament. An important component of the digital flexor pulley system. J Bone Joint Surg Am 74(10):1478–1485

Komatsu I, Lubahn JD (2018) Anatomy and biomechanics of the thumb carpometacarpal joint. Oper Tech Orthop 28:1–5. https://doi.org/10.1053/j.oto.2017.12.002

Kosiyatrakul A, Jitprapaikulsarn S, Durand S et al (2009) Closed flexor pulley rupture of the thumb: case report and review of literature. Hand Surg 14:139–142

Labana N, Messer T, Lautenschlager E et al (2001) A biomechanical analysis of the modified Tsuge technique for repair of flexor tendon lacerations. J Hand Surg Br 26(4):297–300

Levangie PK, Norkin CC, Lewek MD (2019) Joint structure & function, a comprehensive analysis, 6th edn. F. A. Davis Company, Philadelphia

Leversedge FJ (2008) Anatomy and pathomechanics of the thumb. Hand Clin 24:219–229. https://doi.org/10.1016/j.hcl.2008.03.010

Lin G-T, Amadio PC, An K-N et al (1989) Biomechanical analysis of finger flexor pulley reconstruction. J Hand Surg Br 14:278–282
Lin G-T, Cooney WP, Amidio PC et al (1990) Mechanical properties of human pulleys. J Hand Surg Br 15:429–434
Manske PR, Lesker PA (1983) Palmar aponeurosis pulley. J Hand Surg Am 8(3):259–263
Moor BK, Nagy L, Snedeker JG et al (2009) Friction between finger flexor tendons and the pulley system in the crimp grip position. Clin Biomech 24:20–25. https://doi.org/10.1016/j.clinbiomech.2008.10.002
Neumann DA (2017) Hand. In: Neumann DA (ed) Kinesiology of the musculoskeletal system, foundations for rehabilitation, 3rd edn. Elsevier, St. Louis
Nordin M, Frankel VH (2022) Basic biomechanics of the musculoskeletal system, 5th edn. Wolters Kluwer, Philadelphia
O'Brien VH, Giveans MR (2013) Effects of dynamic stability approach in conservative intervention of the carpometacarpal joint of the thumb: a retrospective study. J Hand Ther 26:44–52. https://doi.org/10.1016/j.jht.2012.10.005
Phillips C, Mass D (1996) Mechanical analysis of the palmar aponeurosis pulley in human cadavers. J Hand Surg Am 21:240–244
Renner C, Corella F, Fischer N (2015) Biomechanical evaluation of 4-strand flexor tendon repair techniques, including a combine Kessler-Tsuge approach. J Hand Surg Am 40(2):229–235. https://doi.org/10.1016/j.jhsa.2014.10.055
Schiven TM, Osterman AL, Fedorczyk J et al (eds) (2011) Rehabilitation of the hand and upper extremity, 6th edn. Mosby
Schubert MF, Shah VS, Craig CL et al (2012) Varied anatomy of the thumb pulley system: implications for successful trigger thumb release. J Hand Surg Am 37:2278–2285. https://doi.org/10.1016/j.jhsa.2012.08.005
Schuind F, Garcia-Elias M, Cooney WP et al (1992) Flexor tendon forces: in vivo measurements. J Hand Surg Am 17:291–298
Spies CK, Heuvens J, Langer MF et al (2021) Viscoelastic properties of the human A2 finger pulley. Arch Orthop Trauma Surg 141:1073–1080. https://doi.org/10.1007/s00402-021-03781-8
Stroeve S (1998) Neuromuscular control model of the arm including feedback and feedforward components. Acta Psychol 100:117–131
Tsuge K, Ikuta Y, Matsuishi Y (1977) Repair of flexor tendons by intratendinous tendon suture. J Hand Surg Am 2(6):436–440
Vigouroux L, Quaine F, Labarre-Vila A et al (2006) Estimation of finger muscle tendon tensions and pulley forces during specific sport-climbing grip techniques. J Biomech 39:2583–2592. https://doi.org/10.1016/j.biomech.2005.08.027
Wilson SM, Roulot E, Le Viet D (2005) Close rupture of the thumb flexor tendon pulleys. J Hand Surg Br 30(6):621–623
Zafonte B, Rendulic D, Szabo RM (2014) Flexor pulley system: anatomy, injury, and management. J Hand Surg Am 39:2525–2532. https://doi.org/10.1016/j.jhsa.2014.06.005

# 7 Cervical Spine: Lateral Whiplash, Neck Pain, and Nerve Palsy

## 7.1 Introduction

The axial skeleton includes the cranium, vertebral column, ribs, and sternum. In the case for this chapter, a motor vehicle accident imposed large impulsive external forces leading to both macro- and micro-trauma to different regions of the axial and appendicular (i.e., extremities, clavicle, scapula, and pelvis) skeleton. When these regions are traumatized, we often see both neurological and musculoskeletal impairments because of the close relationship between the neural elements, e.g., spinal cord, nerve roots, plexuses, and peripheral nerves, and skeletal, muscular, and periarticular soft tissues. For this case we (1) review the neural, skeletal, muscular, and ligamentous anatomy of the cervical spine; (2) describe the mechanism of injury (MOI), including related external and internal forces and moments, and its potential effect on all relevant bony and soft tissues of the cervical spine and related upper quarter; (3) describe the structure and material properties of the intervertebral disc; (4) explain the biomechanics of relevant special tests used as part of the clinical physical examination; and (5) briefly examine the etiology/biomechanics of whiplash-related concussion.

## 7.2 Case Presentation

Sandra Arendsen was a 21-year-old collegiate basketball player involved in a motor vehicle accident (MVA) 1 month prior to reporting to our office for evaluation and treatment. She was a passenger in a car (which was not moving) when it was hit broadside (referred to as a T-bone or far side impact) on the driver's side by a decelerating, i.e., braking, car, which had been traveling approximately 25 miles per hour. In anticipation of the impending impact, Sandra reached for, and grabbed, the edge of the seat cushion with her right hand. She recalled her body being tossed to the left and then, on the rebound, hitting the passenger-side door frame with her shoulders and head.

Both the driver and Sandra were transported to an emergency room (ER) for examination. Sandra reported that she had not lost consciousness. Based on her symptoms of headache, dizziness, and mild ringing in her ears, she was diagnosed with a mild traumatic brain injury, i.e., concussion. Sandra reported some soreness in her neck and right arm muscles but was cleared of any significant bone or soft tissue injuries by the ER medical staff. She was told she had sustained a whiplash injury and instructed to take Tylenol, as needed, for neck soreness and headache and to see her primary care physician (or team athletic trainer and physician) for follow-up.

Sandra's concussion symptoms resolved within 1 week of the accident. However, since she had some persistent signs and symptoms on the right side of her neck and arm, her primary care physician referred her for physical therapy. When we saw her, she complained of intermittent right-sided neck and right arm pain (5 on a 10-point visual analog pain scale). The neck pain was aggravated during the day when she was active, i.e., studying, basketball practice, etc., but was reduced when she was sitting still or lying down. Her sleep was not disturbed, and she usually felt better in the morning. She described having a "catch" in her neck with turning and/or looking up. Sandra also described a moderate decrease in performance in basketball, shooting, and passing, secondary to a general "fatigue" or "heavy" feeling, and a general deep ache in her right arm.

With active range of motion testing of the neck, she was found to be mildly limited in cervical extension, with a deviation of her head slightly to the left when she extended. She complained of pain on the right side of her neck during cervical extension. She also had mild limitation and pain, i.e., that "catch" feeling, with right sidebending and axial rotation (these motions reproduced her symptoms to some degree). Passive mobility tests of the atlantooccipital and atlantoaxial joints were within normal limits. Sandra had a reproduction of neck pain with cervical compression and reproduction of right arm symptoms with distraction. The brachial plexus


G. J. Alderink, B. M. Ashby, *Clinical Kinesiology and Biomechanics*, https://doi.org/10.1007/978-3-031-25322-5_7

tension test (sometimes referred to as the upper limb tension test) was positive on the right side. Craniovertebral stress tests were negative. Resisted muscle testing of key myotomes and muscles of the cervical and bilateral shoulder, elbow, wrist, and hand joints was performed to check for nerve and muscle integrity. Sandra demonstrated a weak (fair+) and painless response only with resisted testing of the right shoulder internal rotators.

## 7.3 Key Facts, Learning Issues, and Hypotheses

Anyone involved in a motor vehicle accident has large, impulsive external forces imposed on multiple body segments. In this case, Sandra recalled that her body first moved to the left, then back toward the right; hitting the passenger-side door somewhat stopped the side-to-side movements of her head and body. She likely experienced a whiplash mechanism of injury to her cervical and upper thoracic cage. Whiplash injuries are most often reported as flexion-extension injuries because they are associated with car accidents where the victim's car is rear-ended by another car. Sandra's whiplash was atypical because it involved repeated left- and right-sidebending of the cervical and upper thoracic cage regions. She had signs and symptoms that suggested an injury to one or more structures in her neck, e.g., ligaments, joint capsules, intervertebral disc, and muscle, but she reported no symptoms in the thoracic cage, e.g., vertebral or rib contusion, fractures, etc. We need to make a detailed study of the injury mechanism to help sort out Sandra's neck pain.

Sandra reported attempting to protect herself from injury by grabbing onto the seat cushion with her right hand. Now she has complaints of weakness and fatigue in her right arm, especially impacting her basketball performance. Resisted muscle testing demonstrated objective weakness of her right shoulder internal rotator muscles, yet the resisted tests did not reproduce the symptoms in her neck. The combined mechanisms of rapidly repeated sidebending of her neck and perhaps a traction force on her right arm suggest the potential for both neurological and musculoskeletal injury. Therefore, we need to study the important neurological structures, e.g., the spinal cord, etc., that are associated with the cervical spine and upper extremities.

Sandra sustained a mild concussion but reported no ongoing concussion-related symptoms at her first physical therapy visit. Despite this, we cannot rule out some sort of residual neurological and/or musculoskeletal-related injury since it has been shown that mild central nervous system impairments, e.g., control of standing balance, can persist for several weeks following even mild concussions. Therefore, additional research is needed to gain a more detailed understanding of whiplash-related traumatic head injuries.

Because of the impulsive nature of the external forces and moments associated with whiplash, we need to consider significant injuries to ligaments that stabilize the upper cervical spine. Performing stress tests of key craniovertebral ligaments are one of the first things we did in our physical examination. Fortunately, these tests were negative, suggesting that Sandra's symptoms were not likely related to serious injury to those ligaments. However, it is not uncommon to get false-negative results (i.e., the test is negative, but one or more structures were actually damaged) with any ligamentous stress test, so we should not completely ignore the upper cervical region as we move forward. We thoroughly examine the anatomy and biomechanics of the upper cervical spine.

## 7.4 Mechanism of Injury (MOI): Revisit Newton-Euler and Related Neuromusculoskeletal Structure and Function

Recall that Newton's first law states that a body remains at rest or moves at a constant linear (and/or angular) velocity until an external force (or torque) is applied to that body (see Chap. 2 for a review of Newton's laws). Some refer to this law as the *law of inertia* since inertia is the resistance a body offers to changes in its motion. The inertia of a body or body segment, e.g., head and neck, can be represented by its mass. The center of mass of single body segment is the weighted average position of all of the particles in the segment. The center of mass (COM) of a system of body segments is the weighted average of the COM of each segment. When we take the forces of gravity into consideration, the location of the COM corresponds to its center of gravity (COG), where the COG is the point about which the effects of gravity are balanced. For example, we can imagine that with Sandra seated in the car (Fig. 7.1), prior to contact made by the oncoming car, the COM of the head and torso was located approximately 1/3 of the distance from the shoulders to the hips below the shoulders. We can imagine that prior to the impact Sandra's COM was located within the body and remained relatively stationary within the body, with only small variations caused by subtle dynamic changes, e.g., adjusting her seated posture within the car seat. However, it is likely that upon impact a significant change was induced in the spatial location of Sandra's body segments moving the location of her COM, perhaps even outside of the body. Just as the central nervous system (CNS) provides control of the body's center of mass when it moves outside of the body during walking (Fig. 7.2), the CNS would have been reflexively activated in response to changes in the location of Sandra's

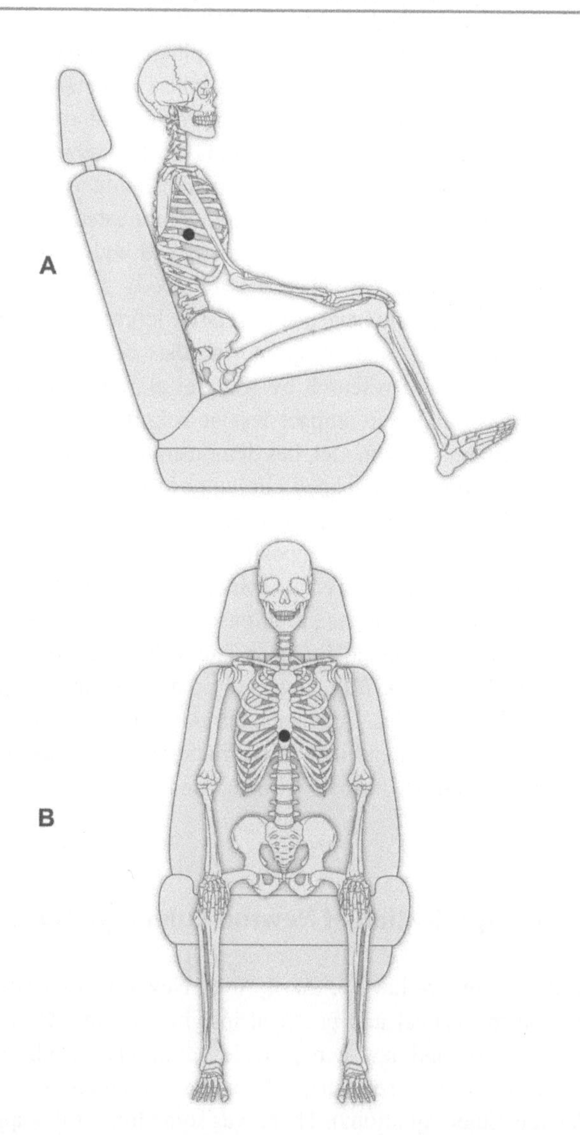


**Fig. 7.1** Schematic of skeleton assuming a typical posture while seated in a car. Note that the approximate center of mass, i.e., red dot, of the head and torso is located mid-thorax at the level of the xiphoid process of the sternum midway between the right and left glenohumeral joints

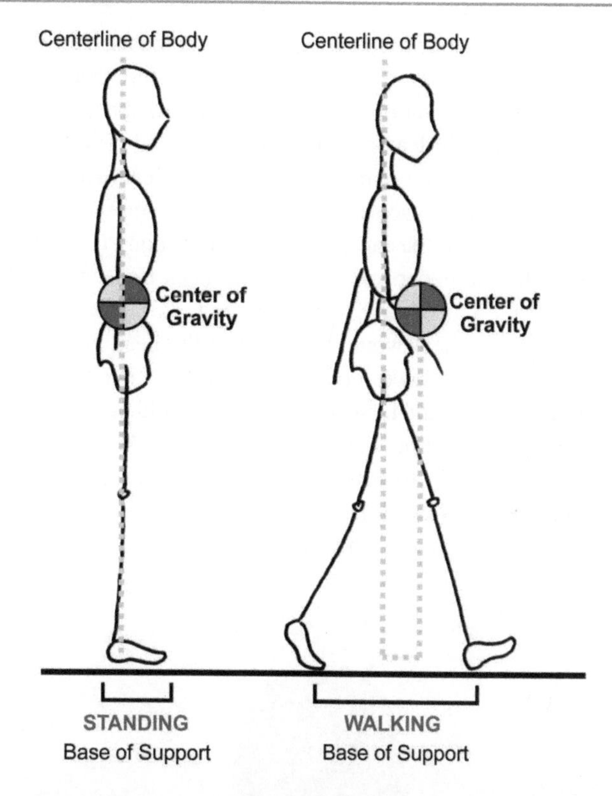


**Fig. 7.2** Schematic illustrating relocation of the body's center of mass from within the pelvis to a place outside of the body when an individual moves from a static standing posture to initiation of walking

COM as a result of her car being impacted from the side. The movements experienced by Sandra during the impact involved changes in both linear and angular velocities which were resisted by her mass and *mass moment of inertia*, respectively. The mass moment of inertia is a quantity that represents the resistance to changes in angular velocity. The mass moment of inertia depends on both a body's mass and how that mass is distributed with respect to an *axis of rotation*.

The mechanism of injury (MOI) in this case was described as *whiplash*. Most commonly associated with rear-end car collisions, whiplash refers to a back-and-forth movement of the head and neck taking place in the sagittal plane, i.e., flexion-and-extension. However, in this instance, the back-and-forth movement occurred in the frontal and horizontal planes. Research on side-impact motor vehicle accident mechanisms has not been as extensive as the research on the rear- or frontal-impact MVAs. However, what we know is that side-impact MOIs put occupants at higher risk of more severe injuries. The major reason for this is related to the extent of "crumple zones" in the front or rear of a car compared to the side of a car. There is a greater amount of material in the front and rear of a car such that when a frontal or rear impact is experienced deformation of the car lengthens the time of impact, which reduces the peak forces and accelerations. For side impacts, the "crumped zones" are minimal or non-existent so that impact times are shorter, resulting in larger forces and accelerations for the same speed of impact. Another reason may be related to the fact that the seatbelt is less effective, allowing the upper torso to slip out of the shoulder belt.

Because the car Sandra occupied was hit broadside, the impact force likely crushed the driver's side door and pushed the car in the same direction as the velocity of the impacting car, i.e., to the right. Since Sandra was a passenger and belted, her body was, in essence, part of the car, yet Newton's first law would have her stay where she was as her car moved to the right; thus, as the car moved to the right, it would seem like Sandra's body moved to the left relative to the car. However, as noted above, because the seat belt was likely ineffective in restraining the upper body, Sandra moved into the seat belt and toward the driver-side door; that is, her body was minimally restrained by the seat belt, and there was likely excessive rotation of the torso toward the point of impact, essentially leaving the head, neck, and torso unrestrained. Because the upper torso, head, and cervical spine are quite mobile, these segments would have moved to the left with respect to the car. Thus, as Sandra's body translated to the right, the upper torso, head, and neck translated and rotated to the left relative to the car (Fig. 7.3). The body's rightward movement would tend to pull the head and neck along, then as the lower body's rightward motion stopped, the head and neck and torso subsequently would have moved to the right again. If her body, and head and neck, moved far enough to the right to strike the passenger door, as Sandra recalled, then this could have pushed the head and neck to the left again, i.e., a rebound effect.

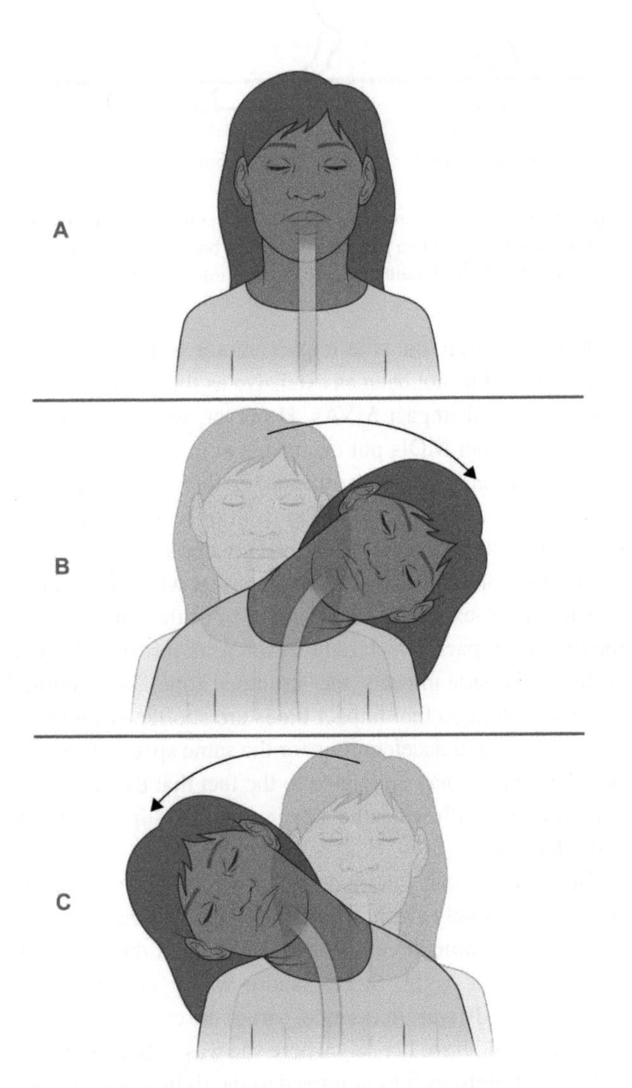


**Fig. 7.3** Schematic of (**A**) neutral cervical spine position prior to impact, (**B**) body pushed to the right inducing left cervical translation/sidebending, and (**C**) rebound body movement to the left inducing right cervical translation/sidebending

Recall, from our discussion of spinal kinematics (see Chap. 3), that coupled motion is an inherent feature of the spinal column, e.g., rotation in the frontal and transverse planes is coupled with translation in the same plane. Therefore, in the cervical spine left translation is coupled with left sidebending. But it is more complicated than that because sidebending is also always coupled with axial rotation and vice versa. So, when Sandra's body was pulled to the right, there would have been a relative left translation and left sidebending and rotation in the cervical spine. Interestingly, prior research by Kang et al., who simulated side MVAs (i.e., sled impact test at 8 km/h) involving a cadaveric head-neck model, has demonstrated that the cervical spine forms an S-shaped curvature in the frontal plane immediately following impact. This S-shaped curvature consists of lateral bending in opposing directions at the cranio-cervical and lower cervical spine. In Kang's research, cervical spine motions, i.e., rotations and shears, were shown to exceed physiological limits in lateral bending at C1/C2 and C3/C4 through C7/T1, suggesting a mechanism of injury to facet joint capsules, ligaments, intra-articular menisci, and including vertebral endplate microfractures, as well as intervertebral annulus fibrosis strains, and nuclear pulposis cracking.

## 7.4.1 Application of Newton-Euler Equations

To this point, we have primarily used Newton's first law to provide theoretical and practical insight into the MOI, but Newton's second law can provide additional insight into Sandra's injuries (see Chap. 2 for the presentation of the Newton-Euler equations). There was some hint of the application of Newton's second law in our description of the cervical sidebending that was induced by a linear force, so let's look a bit closer at the details of Newton's statements. Newton's second law states that the linear acceleration of a body is directly proportional to the force causing it, is along the line of application of the force, and inversely proportional to the mass of the body. In equation form:

$$\sum \boldsymbol{F} = m\boldsymbol{a}$$

where $\sum \boldsymbol{F}$ designates the vector sum (or net) of the forces acting on a body (or body segments), with $m$ and $\boldsymbol{a}$, the mass and acceleration of the body, respectively. Newton-Euler equations also demonstrate that a moment (torque) causes an angular acceleration of a body. For planar problems, this angular acceleration is proportional to the moment causing it, takes place in the same rotatory direction in which the moment acts, and is inversely proportional to the mass moment of inertia of the body:

$$\sum M = I\alpha$$

where $\sum M$ represents the net moment acting to rotate a body (or body segments), with $I$ and $\alpha$ the body's mass moment of inertia and angular acceleration, respectively. Within the musculoskeletal system, the primary moment producers are the muscle-tendon actuators; note that ligaments, joint capsules, and other fascial structures that cross joints can also contribute to net joint moments. Therefore, segmental angular accelerations are proportional to the net moment but inversely proportional to the mass moment of inertia of the segment. A way to think about these relationships is that, given a constant moment of force, the segment with a smaller mass moment of inertia experiences a greater angular acceleration than a segment with a larger mass moment of inertia. Thus, during the impact, the force that pushed Sandra's body to the right was primarily acting on her lower torso (e.g., from the center console and the seat bottom and seat back). These forces would have had a moment arm relative to an "axis of rotation" associated with the cervical spine, resulting in a moment that would induce a left sidebending/rotation of her cervical spine. Furthermore, it is likely that the associated angular acceleration of the head/neck was large, given its relatively small mass moment of inertia (compared to the entire body). The large angular velocities/accelerations imposed on Sandra's neck are associated with an additional relationship that can be derived from Newton's second law. This relationship is the *impulse-momentum relationship*, a concept previously discussed in Chap. 5.

Let's briefly review the derivation of linear and angular impulse momenta. First, let's restate Newton's second law:

$$\sum \boldsymbol{F} = m\boldsymbol{a}$$

Since Newton's second law only describes the application of linear forces on a body, we also restate Euler's equation, which relates applied net moment and angular acceleration for planar bodies:

$$\sum M = I\alpha$$

Now, since average linear acceleration is the rate of change of velocity ($\Delta v/\Delta t$), we can make a substitution in the first equation, so that:

$$\boldsymbol{F}_{ave} = m\Delta \boldsymbol{v} / \Delta t$$

Multiplying both sides by $\Delta t$:

$$\boldsymbol{F}_{ave}\Delta t = m\Delta \boldsymbol{v}$$

where $\boldsymbol{F}_{ave}\Delta t$ expresses a *linear impulse*, and $m\Delta \boldsymbol{v}$ a change in linear momentum. With motor vehicle accidents, we typically see impact forces, often large, applied over a short time period. Note that the braking force which changes the momentum of the impacting car is relatively small. However, the linear momentum of the car that is struck changes very rapidly, which means there is less time for the potentially damaging impact force to be mitigated. We should also recognize that the impact force also changes the momentum of the impacting car. In fact, the change in its momentum is equal and opposite to the change in momentum of the struck car, but in the opposite direction, consistent with Newton's third law (see Chap. 2).

When Sandra's rebounding body and head impacted the door frame, an external impulsive force was imposed on Sandra's head, which had to be mitigated/distributed by the cranium, and its contents. We have referred to impulse loading previously when we studied the material properties of biological tissue. We noted that because biological tissues, e.g., bone, tendon, etc., have viscoelastic properties, higher rates of loading cause the tissues to become stiffer. We also noted, however, that this property has limits depending on the magnitude of external loads and the rate of loading. In Sandra's case, the external compression force imposed on her head likely resulted in a minor contusion of the bone and contributed to her mild head injury.

We now derive the rotation analog to linear impulse-momentum. Since average angular acceleration is the rate of change of angular velocity ($\Delta\omega/\Delta t$), we substitute this expression for the angular acceleration in Euler's equation:

$$M_{ave} = I\Delta\omega / \Delta t$$

multiplying both sides by $\Delta t$:

$$T_{ave}\Delta t = I\Delta\omega$$

where the *angular impulse*, $T_{ave}\Delta t$, is equal to the change in the angular momentum. As discussed above, we saw that the vehicle collision produced a force acting on Sandra's body that, acting with its moment arm, resulted in cervical left sidebending/rotation, which we can now describe as an external angular impulse, an impulse that had to be mitigated over a very short time frame. This impulse would have been distributed among several musculoskeletal tissues, e.g., joint capsules, ligaments, muscle, tendon, etc., in the cervical and shoulder girdle region.

Since we have shown, using Newton-Euler formulations, that injuries associated with motor vehicle accidents are related to the large external forces and moments that must be mitigated/distributed by biological tissues (each with different material properties), we have serendipitously encountered Newton's third law. The third law of motion claims that for every action there is an equal and opposite reaction. Typically, this law is applied to the actions of one rigid body on another. For example, during gait when the foot makes, and maintains, contact with the ground during the stance phase, there is an equal and opposite ground reaction force in the opposite direction; and this reaction force changes in magnitude, direction, and point of application throughout stance. In deformable body mechanics (see Appendix H), we know that external forces (or pressures) imposed on biological tissue result in internal forces (or stresses), that in a sense, are reaction stresses, where the reaction stresses change as the external forces change. Furthermore, both compressive and tensile stresses can be identified. For example, in the first phase of Sandra's whiplash when the cervical spine rapidly side-bent and rotated to the left, a large external compression stress was applied to the cervical facets and annulus of the intervertebral disc (IVD) on the left side that was mitigated by articular cartilage, subchondral bone, and the nucleus of the IVD. Concomitantly, cervical left-sidebending induces high tensile forces, secondary to shear forces, on the right outer annulus of the IVD, facet joint capsules, cervical muscles, and lateral ligaments, just to list a few.

### 7.4.2 Response to External Forces and Moments Induced by the MOI

#### 7.4.2.1 Typical Cervical Segments

To this point in our presentation, we provide a biomechanical framework, with a couple of examples, for describing the forces and moments associated with Sandra's motor vehicle accident. Let's now outline hypothetical mechanisms that may explain Sandra's symptoms, building on what we have presented thus far. The precipitating external force was obviously initiated when the offending car impacted the stationary car. When considering both cars, i.e., the striking car and the impacted car, momentum was conserved when this large external impulsive force pushed Sandra's car and her body to the right until the impacted car (and its contents) eventually stopped moving due to friction between the vehicle's tires and the ground. Inside the car, the force pushed Sandra's body to the right, which resulted in her head and neck moving to the left relative to her body. We explained this counterintuitive movement of the head and neck using concepts related to the law of inertia. When Sandra's body changed its direction of movement and her neck reached its end range of motion directed to the left, the head and neck rapidly moved to the right inducing a right translation and rotation, oscillating like a spring and damper, until eventually her body and head/neck stopped moving. The initial reversal of movement of Sandra's head/neck to the right resulted in her head hitting the door frame, as she recalled when reciting her history. Finally, recall that in anticipation of the two cars making contact, Sandra grasped the car seat cushion with her right hand, a natural reaction in the attempt to prevent her body from moving excessively to the left; in reality, the forces of impact likely exceeded her ability to restrain herself. However, by anchoring her right hand, arm, and shoulder girdle, large impulsive external forces were imposed on the right side of her body, likely inducing large internal tensile stresses to her wrist, elbow, shoulder joints, and the right side of the cervical spine, and to associated soft tissues.

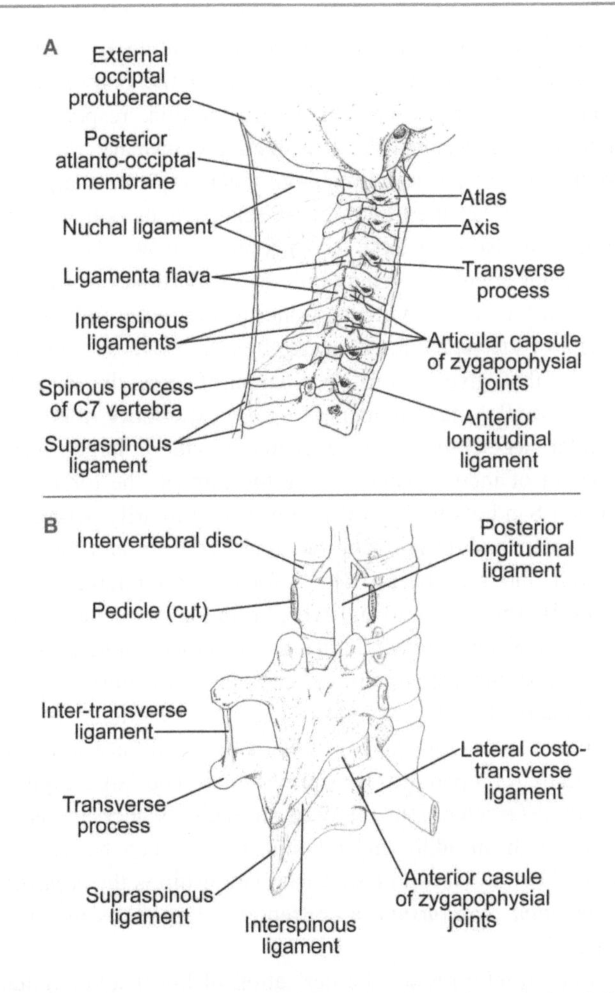


**Fig. 7.4** Schematic illustrating key ligaments from (**A**) sagittal and (**B**) posterior-oblique views that are likely affected in a lateral whiplash neck injury: annulus of the intervertebral disc, articular joint capsules, and intertransverse ligament

Previously, we noted in passing that when the cervical spine sidebends/rotates to the left, large impulsive compression, shear, and tensile forces induce internal stresses on the facet and uncovertebral joints, i.e., articular cartilage, intervertebral disc (internal annulus) and endplate, and subchondral bone. On the contralateral side, tensile forces cause corresponding stress within the outer fibers of the annulus of the intervertebral disc, joint capsules, and related periarticular soft tissues, e.g., intertransverse ligament, etc. (Fig. 7.4). Prior work by Panjabi et al. and Chen et al. have provided some evidence that corroborates our hypotheses regarding tissue injuries showing ligamentous hemorrhage and laceration in the intervertebral discs, ligamentum apicis dentis (apical ligament of dens), ligamentum flavum, and the anterior longitudinal, posterior longitudinal, and capsular ligaments, primarily at the middle and lower cervical spine (C4/C5 through C7/T1).

#### 7.4.2.2 Upper Cervical Spine

The anatomy of the upper cervical spine (sometimes referred to as craniovertebral joints) is uniquely different from the lower cervical spine, i.e., C2 to C7). The atlantooccipital (OA) joints are comprised of the articulation between the convex occipital condyles (C0) and concave facets of the atlas, i.e., first cervical vertebrae (C1). Unlike the two-joint OA articulations, the atlantoaxial (AA or C1/C2) articulations involve three joints. At the AA joint, the articulations are formed by the flat or slightly concave inferior facets of C1 and slightly convex superior facets of the axis (C2). The third joint, or median joint, is comprised of articulation between the dens of the axis with the anterior arch of the atlas, and the transverse ligament. On the front end, an anterior facet of the axis forms a synovial relationship with the posterior border of the anterior arch of C1. Posteriorly, the dens articulates with a fibrocartilage-lined transverse ligament. There are no intervertebral discs between the C0/C1 and C1/C2 articulations (Fig. 7.5). Therefore, in the craniovertebral joints the impulsive compression, shear, and tensile forces generated by the MOI were borne primarily by articular cartilage of the synovial joints and subchondral bone, except in the case of the median joint of the AA complex where shear/tension may be mitigated by the transverse ligament. We can be confident that craniovertebral joint stability is primarily dependent on the intrinsic and extrinsic ligaments and muscles.

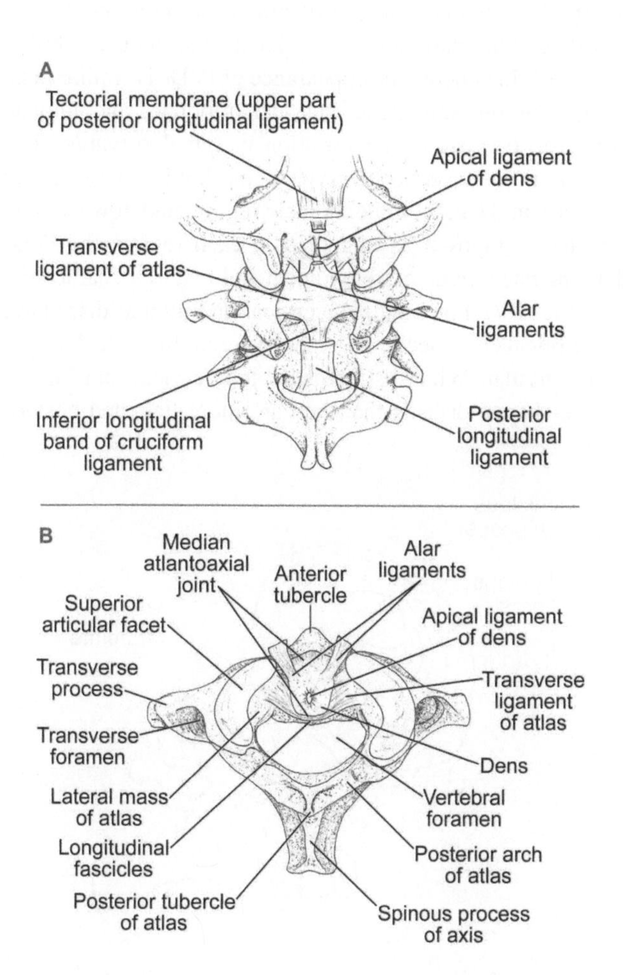


**Fig. 7.5** (**A**) Posterior and (**B**) superior views of craniovertebral joints, atlantooccipital and atlantoaxial, and important extrinsic ligaments: apical, tectorial membrane, transverse/cruciform, and alar

Let's briefly describe how key craniovertebral ligaments likely functioned under the stress of this whiplash mechanism. The apical ligament is believed to be a vestigial remnant of the notochord, is absent in approximately 20% of the population, and makes no appreciable contribution to craniovertebral stability. As a result of Sandra's rapid, and extreme, side-to-side movement of the head, large internal stresses were produced within the apical ligament, possibly resulting in a mild strain injury. Normally, the tectorial membrane (a continuation of the posterior longitudinal ligament) provides resistance primarily to excessive forward flexion, but because of its width may control multi-dimensional movements as well. As noted with the apical ligament, the tectorial membrane also may have sustained a mild strain injury. These minor strains could have contributed to Sandra's head and/or upper cervical symptoms but were unlikely to result in craniovertebral hypermobility.

The key stabilizing structures, in this case, are the transverse (and longitudinal cruciform appendages) and alar ligaments. The transverse ligament contributes to AA rotational integrity and restrains excessive anterior translation of C1 on C2, while the cruciform appendages also prevent posterior migration of the dens onto the spinal cord. Because of the impulsive shear stress associated with the lateral movements of the head, these ligaments may have generated large internal tensile stresses. The alar ligaments likely experienced very large internal tensile stress because of their role in constraining excessive sidebending and rotation of the head. The intrinsic ligaments of the craniovertebral joints, i.e., atlanto-occipital and atlantoaxial ligaments or joint capsules, also likely developed large internal tensile stress as a result of the multi-dimensional movements of the head in this whiplash accident.

#### 7.4.2.3 Intervertebral Disc Complex

Before we consider the role muscle played in this MOI, we discuss two additional intra- and periarticular joint structures: (1) the intervertebral disc of the interbody joints and (2) intra-articular soft tissues within and around the facet (apophyseal) joints, and the uncovertebral joints, that may be a source of Sandra's pain symptoms. Two distinct accessory structures associated with the facet joints include subscapular fat pads and fibro-adipose meniscoids. Fat pads have been described as fillers of small crevices between the joint capsules and underlying synovial membranes and may be extra-capsular if they extend outside the joint through the crevices. Meniscoids, found at the edges of facet joints, range from thickenings of connective tissue along the internal surface of the joint capsule to synovial folds that surround small fat pads, collagen fibers, and blood vessels. Larger meniscoids can extend into the facet joint itself. The exact role of these intra-articular structures has not been described, but it is likely that these joint spacers help to distribute/mitigate joint compression forces, thus providing support to articular cartilage and subchondral bone. It has been shown that these tissues are richly innervated, which also suggests a prominent role in noci- and proprioception. The impulsive joint loads experienced by the cervical spine, in this case, may have strained these protective structures, and since they contain nociceptors, they could be a source of pain. Furthermore, scarred or displaced meniscoids could have contributed to the pathomechanics of the apophyseal joints that were noted in the physical examination of Sandra.

Recall that with cervical spine range of motion testing, Sandra demonstrated mild limitation of right sidebending and axial rotation and complained of a "catch" somewhere in her neck when it turned. Additionally, when she extended her neck, there was a mild limitation in extension with mild deviation to the left, and reproduction of symptoms. These findings could be the result of mild strain injuries to facet joint capsules, intertransverse ligaments, subscapular fat pads, or the meniscoid structures.

Because of what we know about the nature of whiplash MOIs, we routinely check the integrity of the craniovertebral ligaments as part of triaging for potential macrotrauma to bone and soft tissue. In the emergency room, radiographs (or computed tomography or magnetic resonance imaging) are taken with the head placed in simulated stress positions so that the radiologist can visualize potential abnormal positioning of the dens relative to the anterior arch of C1. The distance between the dens and arch of C1 can be measured and compared to normal values. A dens that appears to be more posterior is an indication of a possible third-degree strain of the alar ligament.

On the examination table, we manually stress key ligaments within the head and neck complex. The craniovertebral ligament stress tests that were used in Sandra's examination included the (1) Sharp-Purser test, (2) alar ligament test, and (3) tectorial membrane test. As described in the elbow case, ligament stress tests place affected joints in a position where the practitioner can perform passive maneuvers to apply external tensile forces to a specific structure. The Sharp-Purser test is used to stress the transverse and cruciform ligaments by applying a flexion moment to the occiput while stabilizing the axis. A flexion moment places tension on the transverse ligament as a result of inducing posterior displacement of the dens relative to the anterior arch of C1; posterior translation of the dens occurs as C1 translates anteriorly during flexion. The alar ligaments are stressed by applying a lateral bending moment to the head, relative to C2. Since sidebending and axial rotation are automatically coupled in the spine, this maneuver checks the ability of the alar ligaments to check both sidebending and rotation of the head relative to C2. Testing of the tectorial membrane can be accomplished with the patient seated or lying supine (on their back). For this test, we manually stabilize C2 and apply a distractive, i.e., traction, force to the head and C1. Unfortunately, this maneuver also applies external tensile loads to the tectorial membrane, annulus of the disc, and facet capsules, so a limitation of this test is that our traction force cannot isolate the tectorial membrane. The reader is encouraged to consult other resources for illustrations of these manual techniques. None of these tests resulted in excessive movements of the head, i.e., instability, nor reproduced symptoms when they were performed on Sandra. The clinical findings related to the stress tests performed in this case suggested that if the MOI induced strains in one or more of these ligaments, they were minor; and at weeks post-injury, if they had been strained, they likely completed the normal inflammatory and healing phases.

The intervertebral disc (IVD) of what some refer to as the interbody joints is a fibrocartilaginous structure comprised of two distinct parts: nucleus pulposis and annulus fibrosis (Fig. 7.6). In general, the appearance of IVDs is similar in all regions of the spine, i.e., cervical, thoracic, and lumbar. However, because of the variation in spinal curvature, i.e., lordotic in the case of the cervical and lumbar regions, and kyphotic in the case of the thoracic region, and how external loading is imposed and handled in each region, the IVDs demonstrate unique morphological and material characteristics (Fig. 7.7). For example, cervical and lumbar discs have an elliptical cross-sectional shape, whereas thoracic discs are more circular. What distinguishes the cervical and lumbar ellipse-shaped discs is that they are more flattened posteri-

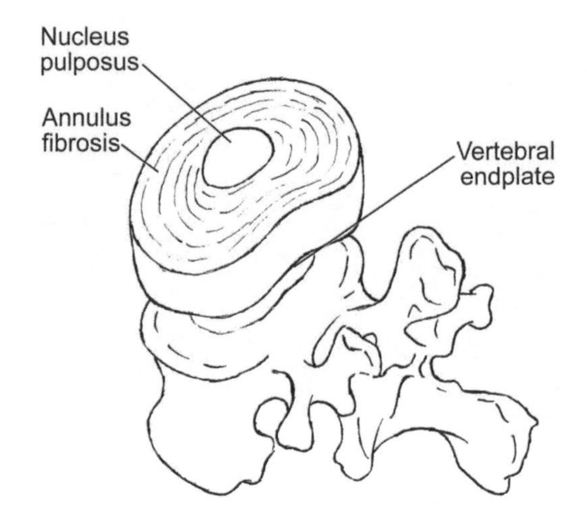


**Fig. 7.6** Schematic of supero-posterior view showing relationship between the intervertebral disc, vertebral endplates, and vertebral body of a typical spinal unit

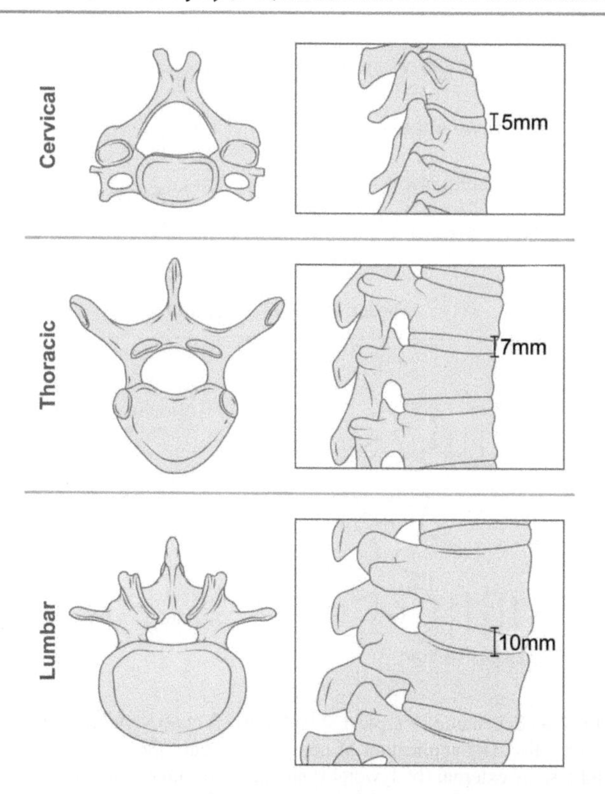


**Fig. 7.7** Superior and sagittal views of regional differences in shape, cross-sectional area, and disc height in cervical, thoracic, and lumbar regions of the spine

orly, which appears to make them stiffer under applied flexor moments; that is because the posterior annular fibers have a reduced length, they provide resistance, i.e., stiffness, earlier in vertebral flexion than vertebral extension. We also see regional differences in disc height (or thickness). The spinal units with thicker discs demonstrate greater interbody joint mobility. The relation between disc thickness and intervertebral mobility suggests that lumbar segments could express greater mobility than cervical and thoracic segments. However, in fact, cervical vertebral segments have greater mobility than lumbar segments in all planes for two reasons: (1) cervical disc thickness relative to vertebral body height is greater than in the lumbar and thoracic regions and (2) facet architecture and ligamentous constraints are more favorable.

The functional spinal unit, i.e., two vertebrae, intervening disc, and related vertebral endplates, is part of the spinal complex that provides redundant degrees of freedom that promotes an incredible range of mobility. Yet, at the same time, the spinal unit provides a system that can control the very large, repetitive, and sometimes impulsive, forces imposed by gravity and muscle actions. The IVD is key to the systems' ability to provide a shock-absorbing system that can protect vertebral bodies and endplates, and the adjoining apophyseal and uncovertebral joints. The nucleus and annulus have unique, yet similar, properties that contribute to the shock-absorbing function of the IVD. Let's examine these characteristics in detail. The nucleus pulposis has a gel-like consistency and is located in the mid-to-posterior part of the disc. In the first two decades of life, the nucleus consists of 70–90% water, which gradually recedes with age. It has been reported that gradual, or sometimes acute, dehydration of the nucleus has been associated and may be exacerbated by disease or trauma. The high-water content of the nucleus of the IVD allows it to function as a modified hydraulic system that efficiently distributes large external compression loads, including those of an impulsive nature. Recall that fibrocartilage has an extracellular matrix similar to that of articular cartilage (which was discussed in detail in Chap. 3). In our previous discussion of articular and fibrocartilage, we explored the normally complex relationship that exists between the collagenous framework (primarily type II collagen) that provides strength and the dense hydrophilic proteoglycan aggregate and chondrocytes and fibrocytes that are responsible for the synthesis and regulation of the proteins and proteoglycans within the matrix. It is this same type of relationship that allows the IVD to provide an effective viscoelastic response to external loads.

The annulus fibrosis consists of concentric layers (i.e., rings) of collagen fibers that surround the nucleus. The annulus has a similar extracellular matrix to that of the nucleus, with a much larger density of collagen (50–60% dry weight versus only 15–20% in the nucleus), but less water content. Elastin fibers, embedded in parallel to collagen, are more abundant in the annulus and provide the circumferential elasticity required for this closed-hydraulic system. Similar to the nucleus, the inner annular fibers consist primarily of type II collagen and are devoid of blood vessels and sensory innervation. The outer rings of the annulus contain both type I and type II collagen, similar to a typical ligament, make strong bonds with the anterior and posterior longitudinal ligaments, rims of the vertebral bodies and endplates, and uncovertebral joints. Thus, a second, but related, function of the IVD complex is spine stability; that is, the annulus functions like any other ligament by providing resistance to excessive intervertebral rotations and translations. The outer annulus is also richly innervated.

If we look at the orientation of the annular rings (Fig. 7.8), we can see why this part of the IVD is well-suited for its primary task. In all regions of the spine, collagen fibers of the annulus are arranged in very precise geometric patterns. In the cervical region, collagen rings are generally oriented about 65 degrees from the vertical (exact orientation not shown in the figure), with adjacent layers oriented in opposite directions. This particular geometry may be a good example of the adage "function dictates structure" in that it is likely that this arrangement is related to the multi-dimensional movements of the spine and related compressive, tensile, and shear forces induced by the activities of daily living. On the

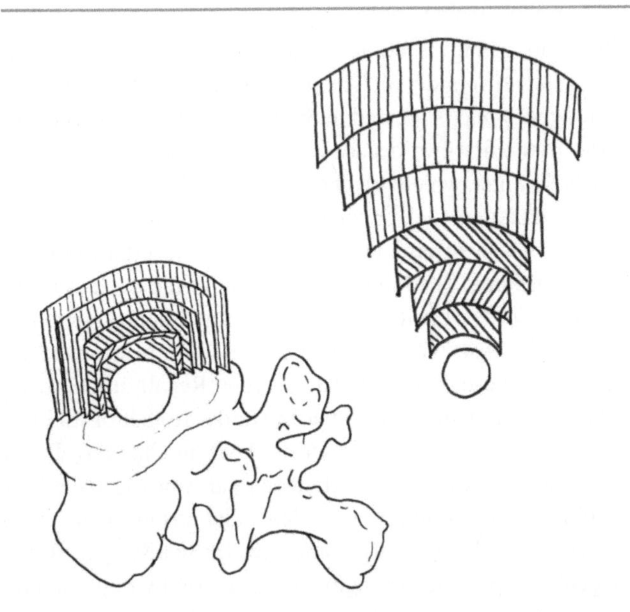

**Fig. 7.8** Schematic showing the relationship of the inner annulus to the nucleus pulposis and geometric pattern of inner and outer annular concentric rings

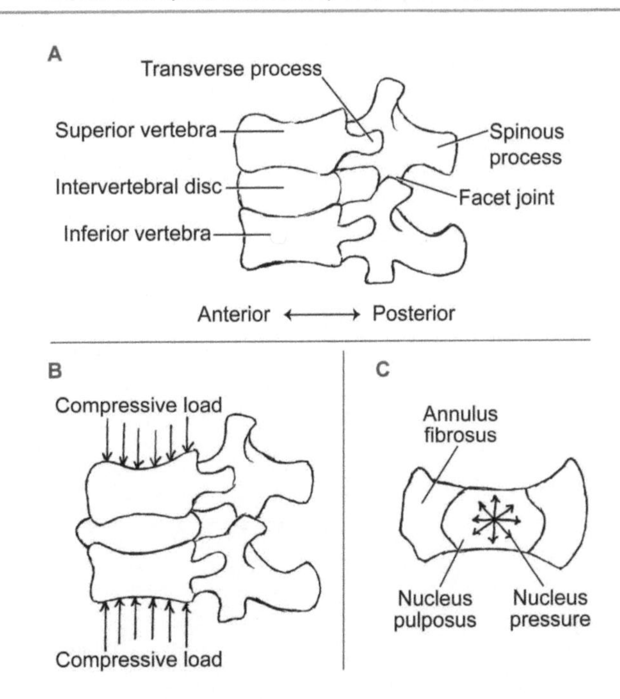


**Fig. 7.9** Schematic of typical spinal unit: (**A**) two vertebra and intervening disc, (**B**) application of compression onto the vertebral bodies and disc by external (bodyweight) and internal (muscle actions) forces, and (**C**) even distribution of compression forces raises the internal hydrostatic pressure in the nucleus which induces uniform distribution of internal tensile stress throughout the inner annular framework. Note: internal apophyseal and interbody compression forces induced by muscle action are not shown

other hand, because the collagen fibers of the annulus are oriented only 25 degrees from the horizontal, large external torsion loads may not be well mitigated. This fault may explain why a combination of flexion and rotation moments imposed on the spine, particularly the larger combination of moments in the lumbar region, has been implicated as a mechanism for disc failure, i.e., rupture. Two final notes: (1) not all annular rings in the cervical disc completely encircle the nucleus. Those rings that do not encircle the nucleus appear to fuse with adjacent rings, a phenomenon particularly evident in the posterior-lateral region of the disc, which suggests that this may be an area of weakness and susceptibility to significant strain with repetitive flexion movements, and (2) the annular rings are nearly non-existent near the uncovertebral joints, which perhaps allows greater freedom of movement to these joints but is an area less able to support the hydraulic system of the IVD.

The unique fibrocartilaginous design and intrinsic viscoelastic nature of the IVD enhance its ability to distribute and mitigate complex external and internal loads. Remember (see Chap. 3) that the mechanical properties of fibrocartilage mimic those of articular cartilage. In a healthy disc, external compression, in combination with internal forces produced by muscle actions, induces large internal compression stress (or hydrostatic pressure) within the water-saturated nucleus (Fig. 7.9). Under controlled laboratory conditions, it has been shown that compression forces imposed uniformly on a spinal unit increase the hydrostatic pressure within the nucleus, which, in turn, induces large tensile forces on the inner annular fibers. Inner annular fiber tension helps to control the radial expansion of the nucleus, while its elastic properties store energy that restores the normal conformation of the annulus following the removal of the external force. At the same time, the increased nuclear pressure is distributed evenly to the vertebral endplates (not shown in Fig. 7.9) and vertebral bodies (subchondral bone). Like all viscoelastic materials, the nucleus and annulus under load demonstrate the properties of creep and stress relaxation (see Appendix H or Chap. 3 for a more complete review of these phenomena). The nucleus and annulus deform (i.e., undergo creep) under compression and tension, respectively, until some equilibrium is reached with regard to the changes in hydrostatic and osmotic pressures, i.e., movement of water in and out of the nuclear matrix. Simultaneously, internal stresses are reduced (i.e., stress relaxation) over some time period secondary to transient material changes in both the nucleus and inner annulus. It is notable that with the impulsive multi-dimensional loading that Sandra's cervical spine experienced (similar to that illustrated in Fig. 7.10), there would have been insufficient time for the IVD's viscous nature to play out.

Under controlled experimental loading conditions, when external forces are removed, the elastic properties of the vertebral bodies and endplates, and annular fibers find a new equilibrium point until the next loading cycle is imposed. The unique viscoelastic nature of each structure, e.g., bone,

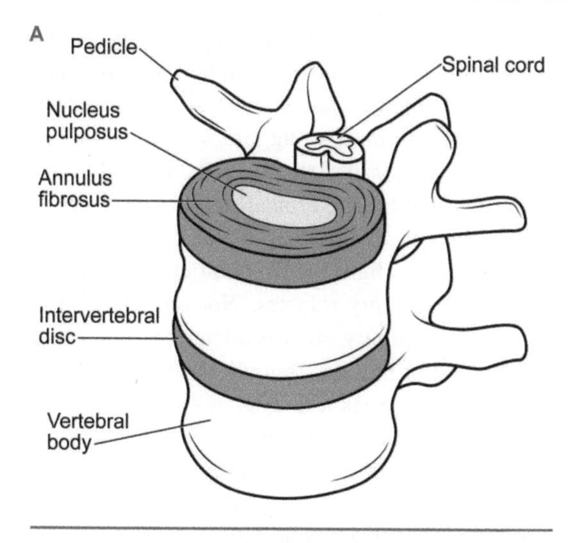


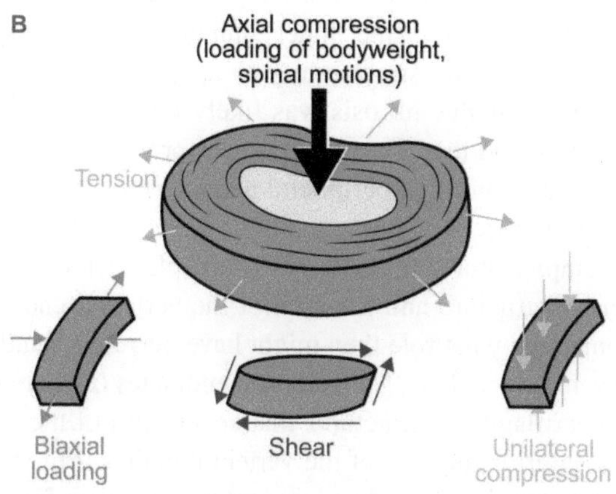


**Fig. 7.10** Schematic illustrating (**A**) functional spinal unit and intervertebral disc and (**B**) distribution of multi-dimensional internal disc stresses that are routinely distributed by the IVD during activities of daily living

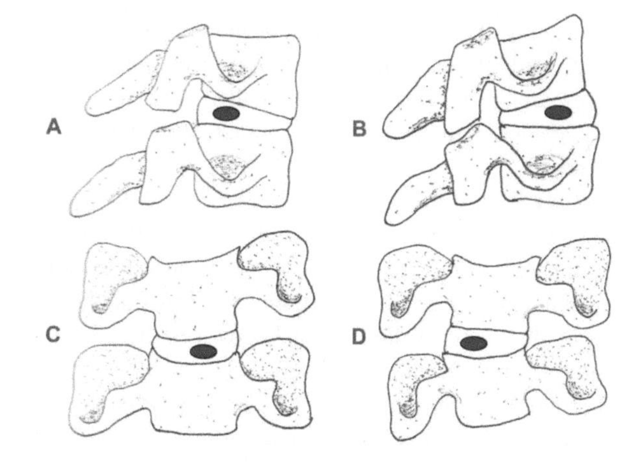


**Fig. 7.11** Schematic illustrating the migration of the nucleus pulposis during spinal (**A**) flexion, (**B**) extension, (**C**) left sidebending, and (**D**) right sidebending. Note the migration of the nucleus pulposis in the direction opposite of the movement, e.g., with flexion a posterior migration

etc., allows the spine to manage both sustained loads and/or repetitive loads (within reason). Recall, also, that viscoelastic structures manage large external impulsive loads (similar to the high rate of loading Sandra experienced in her accident) by becoming stiffer, i.e., that is, offering greater resistance in the elastic loading phases. While the IVD may be considered the primary structure controlling large hydrostatic forces, its relationship to the vertebral endplates and subchondral region of the vertebral bodies cannot be ignored.

Before we explore the structural and functional relationship between the IVD and vertebral endplates, we need to briefly discuss the theoretical kinematics of the nucleus within the IVD. Up to this point, we employ the principles of deformable body mechanics to characterize how the IVD responds to external loads. However, during spinal movements, e.g., flexion, extension, sidebending, and rotation, there are characteristic migration, i.e., kinematic, patterns of the nucleus pulposis (NP) as well (Fig. 7.11). Most research examining NP mobility has focused on the lumbar region, likely because of the greater prevalence of discogenic pathology in that region. However, the few studies that have examined cervical NP mechanics have reported behavior similar to that found in the lumbar spine. As with most research that has potential clinical applications, the early laboratory, i.e., using cadaveric specimens, and in vivo testing, i.e., using young healthy males and females, examined the inter-IVD mobility of normally hydrated, and intact, discs. In some cases, studies examined both normal and abnormal, i.e., degenerative, discs to compare NP mobility patterns. We only describe the movement patterns observed under normal loading conditions. It has been hypothesized that because a normally hydrated disc's behavior is characterized by a hydrostatic mechanism, it deforms toward the area of least load. Therefore, since the compression load is greater on the concave side of the IVD, the NP migrates (or deforms) toward the convexity, which creates a bulge of the IVD. For example, during active cervical flexion, the force of gravity and action of the neck flexors increases IVD compression anteriorly, which typically results in a posterolateral migration of the NP, i.e., the disc appears to have a bulge in its posterolateral corner; the reverse sequence takes place with active cervical extension. Remember that sidebending and rotation are always coupled, and in the same direction for the subcranial vertebrae, so that left sidebending/rotation induces increased compression on the left side of the spinal curve and migration of the NP toward the right, and vice versa. These are the general rules that have emerged from previous research, and the ones we assume were operational as part of the MOI in Sandra's vehicle accident. However, consider a couple of caveats: (1) some research has suggested that a distinct movement of the NP may not be anything more than the deformation of the NP length; this is entirely possible given its fibrocartilaginous character; (2)

in vivo testing showed moderate inter-individual differences in the magnitude and direction of NP migration, and (3) these research results assumed a normally hydrated IVD and cannot be applied to discs with moderate to severe degenerative changes.

We now have a near-complete picture of the IVD complex. To this point, we have focused on the structure and material properties of the IVD, but there is one more aspect of its behavior we need to examine. It turns out that the height or thickness of the IVD that separates the bodies of the vertebrae creates an opening referred to as the intervertebral foramen. In a normal spine, the diameter of this foraminal opening provides adequate space for the spinal nerve roots to emerge after they branch off the spinal cord (Fig. 7.12). Examination of the figure suggests that a strain injury to one or more of the following, apophyseal joint capsules, the annular fibers of the IVD, as well as other spinal ligaments (Fig. 7.4), could impair, i.e., reduce, the diameter of the intervertebral foramen and compress a spinal nerve root. It is possible that the tensile strains imposed on the apophyseal joint capsules, related to the MOI in this case, and resulting plastic deformation and inflammatory response could have altered normal facet joint kinematics affecting the intervertebral diameter. Or the large sidebending moment produced as part of the whiplash MOI and induced tension imposed on the IVD could also have induced a strain injury that deformed the IVD to an extent that its height was reduced, thus compromising a spinal nerve root's exit through the foraminal opening.

We know, based on Sandra's history and physical examination, i.e., resisted muscle tests, that she had a painless weakness of her right shoulder internal rotator muscles. A painless weakness of muscle suggests a nerve palsy, i.e., spinal cord, nerve root, or peripheral nerve injury. Fortunately, Sandra did not exhibit the typical nerve root injury symptoms, i.e., altered dermatomal sensation, or signs, i.e., reduced sensation to light touch, pinprick, and pressure, or reduced upper extremity reflexes. Nor did she exhibit any signs of spinal cord injury, such as bilateral changes in sensation, reflexes and/or muscle strength. Furthermore, the demonstrated shoulder weakness was not associated with a particular myotome, i.e., nerve root. We can safely assume that given Sandra's age and no previous history of cervical spine injury prior to the motor vehicle accident, her IVDs were normal, i.e., NP was normally hydrated and had no radial fissures within the inner annulus. Therefore, we hypothesize that any strain injury to apophyseal joint capsules and/or annulus fibrosis was likely minor, i.e., a first-degree strain and that a nerve palsy was probably related to tensile strain injury to a peripheral nerve or related structure located elsewhere.

To complete our study of the IVD complex, let's examine the normal structure and function of the vertebral endplate, and conjecture what role they might have played in Sandra's motor vehicle accident. The vertebral endplates (EP) are thin hyaline cartilaginous structures that cover most of the superior and inferior surfaces of the vertebral bodies (Fig. 7.13); we note that during early development the endplates function as growth plates for the vertebrae. In adults, the EP occupies

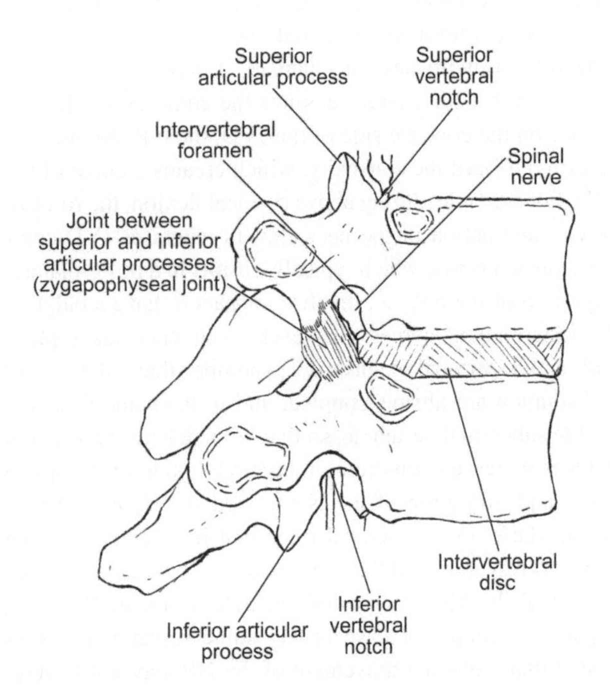


**Fig. 7.12** Schematic illustrating intervertebral foramen, exiting spinal nerve root, and their relationship to the apophyseal and interbody joints

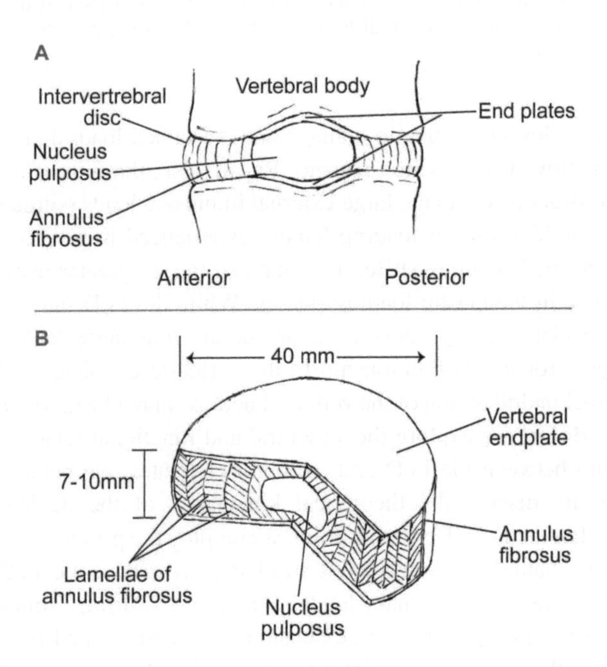


**Fig. 7.13** Schematic (**A**) cross-section of gross morphological relationship within the spinal unit and (**B**) vertical cut detailing nucleus, annular lamellae, and vertebral endplate

approximately 5% of the height of each intervertebral space. The surfaces of the endplates in contact with the annulus have fibrocartilaginous characteristics and are strongly bound to the annulus via collagen fibers. It is notable that the cartilaginous connection between the endplate and annulus forms a stronger bond than the calcified cartilaginous interface of the endplates and subchondral bone of the vertebral body. As an aside, controlled experimental tests, using intact, i.e., normal, cadaveric functional spinal units, have shown that the imposition of large or repetitive compression loads to failure induces endplate fracture before IVD rupture, with the fracture occurring at the EP/bone interface. It is notable that the EP fractures produced in a laboratory setting where pure axial forces are applied are not similar to those that occur commonly in practice. And, in Sandra's case, it is unlikely that the sidebending/rotational torsional stress placed on the functional spinal unit induced axial compression loads large enough to produce an EP fracture.

Endplate fractures perforate the endplate, allowing the extracellular matrix of the NP to "leak," which would obviously impair the closed hydraulic system of the IVD. Chronically, the perforation also allows the invagination of small blood vessels related to the inflammatory process to encroach the outer and inner annular fibers, which initiates a degenerative process (Fig. 7.14). The long-term sequelae of an EP fracture induce an early degenerative disc disease process. We do not anticipate that at Sandra's age any sort of degenerative process had been active. Certainly, a fractured EP and/or vertebral subchondral bone secondary to the MOI could be a source of Sandra's symptoms because of their rich sensory innervation. On the other hand, it's more likely that if an EP strain injury occurred, it was minor since radiographic images in the emergency room did not demonstrate any evidence of fracture sites or Schmorl nodes (note: radiographs may not be able to detect endplate fracture as well as magnetic resonance imaging since the endplate is cartilaginous).

The vertebral endplates are a source of nutrients, provided via blood vessels, and nociceptive innervation only for the outer annulus (Figs. 7.13, 7.14, and 7.15). The vertebral bodies are highly vascularized by vertebral capillaries that provide the principal nutritive support for the endplates, and the matrix and cells of the NP. A nutrient artery enters each vertebra via a bony tunnel in the posterior cortex of the vertebral body, the basivertebral foramen (BVF). Vessels emerge from the BVF and then branch in ascending and descending trajectories to supply the central endplate. As depicted in Fig. 7.15, essential nutrients, e.g., glucose, oxygen, etc., diffuse a significant distance via an osmotic pressure gradient to reach the NP and inner annular fibers. This process is paramount to maintaining the metabolism of the NP; that is, it provides nourishment for metabolic activities within the extracellular matrix to maintain the proteoglycan aggregate, and chondrocyte and fibrocyte activities. Sensory nerves, which vary in size and shape, are strongly associated with the blood vessels. The endplate nerves arise from the basivertebral nerve trunk, which originates from the sinuvertebral nerves via the sympathetic trunk (Figs. 7.16 and 7.17). Generally, the nerve distribution has a large point of entry at the BVF and a greater density along the posterior border of the vertebral body (including the endplate). The endplate may also receive small blood vessels and nerve invagination

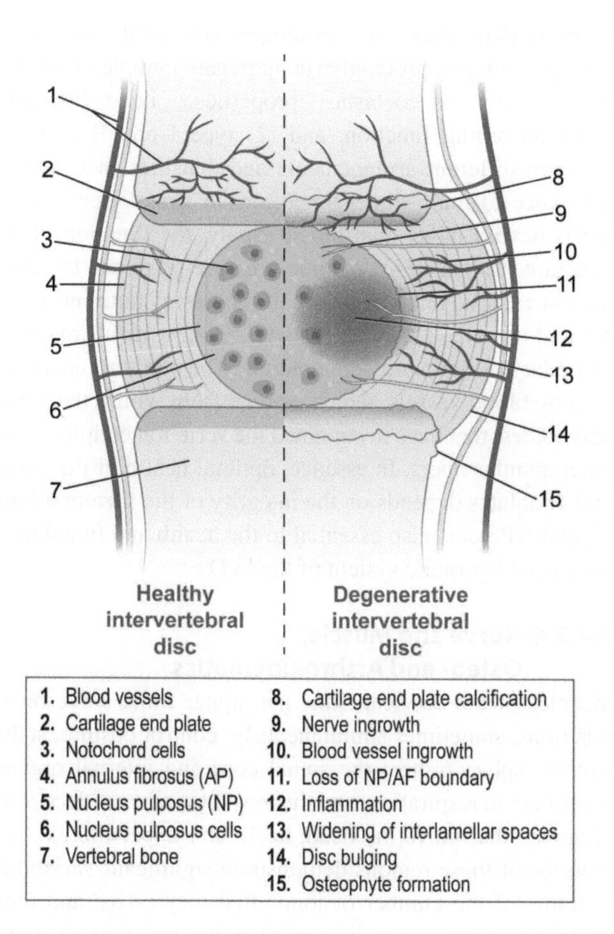


**Fig. 7.14** Schematic illustrating the blood vessel and nerve supply to vertebral endplate and outer annulus under normal, i.e., healthy, and degenerative conditions

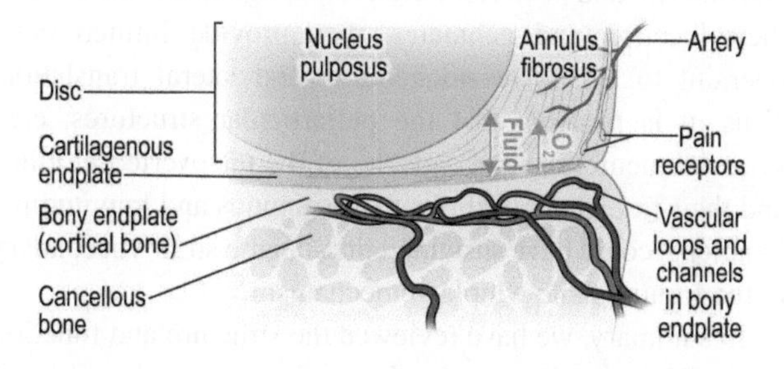


**Fig. 7.15** Schematic illustrating invagination of small blood vessels and nerves of outer annulus and endplate rim and relationship of vertebral subchondral bone, endplate, and intervertebral disc

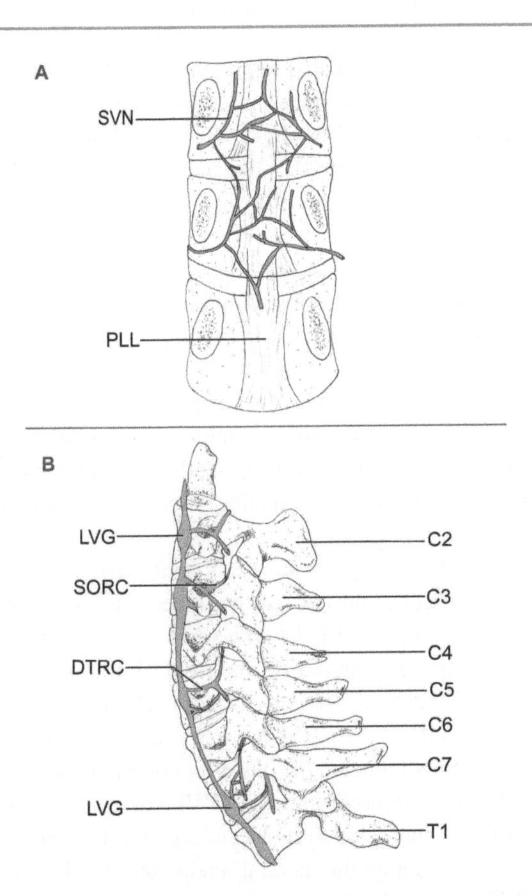


**Fig. 7.16** (**A**) Posterior and (**B**) lateral views of nerve branches that contribute to the innervation of the outer annulus fibrosis and vertebral endplate borders. Note: SVN = sino-vertebral nerve, PLL = posterior longitudinal ligament, LVG = latero-vertebral ganglia, DTRC = deep transverse rami communicans, SORC = superficial oblique rami communicans

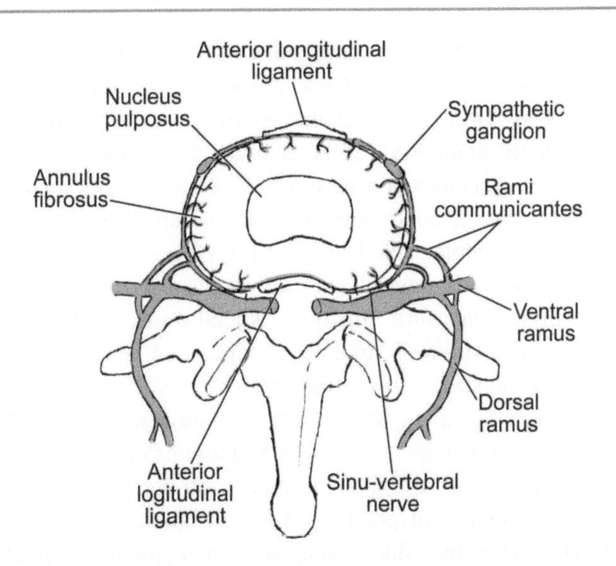


**Fig. 7.17** Schematic showing the relationship between a recurrent meningeal (sinuvertebral) nerve that branches off the extreme proximal aspect of the ventral ramus. In addition to innervating the outer annulus and endplate borders, the sinuvertebral nerves provide sensory and sympathetic input to the meninges of the spinal cord and connective tissues, e.g., posterior longitudinal ligament, associated with the intervertebral joints

indirectly via the outer annular fibers, which originate from the metaphysical arteries found along the lateral vertebral column and nerve branches from the gray matter communications (separate from the sinuvertebral nerves).

A final note about an often-forgotten joint, the uncovertebral joint. We discussed the anatomy and function of this joint in Chap. 3, so we do not reiterate that in toto. However, be reminded that the uncovertebral joint is a synovial joint with peripheral connections to the outer annulus of the IVD and anterior and posterior longitudinal ligaments. Because of their location and geometry, they provide limited bony restraint to excessive sidebending and lateral translation. Thus, it is possible that the periarticular structures, e.g., synovial membrane and capsule, of the uncovertebral joints and their fascial connections to the annulus and longitudinal ligaments could have sustained first-degree strains secondary to the frontal plane whiplash mechanism.

In summary, we have reviewed the structure and function of the IVD complex, i.e., nucleus pulposis, annulus fibrosis, and vertebral endplates. This is a complex that relies on an interdependent relationship between its parts, and in which all parts share common, yet unique, structural characteristics: (1) collagen-proteoglycan aggregate complex that provides their viscoelastic properties, i.e., hydraulic shock-absorbing function, and (2) type I and II collagen fibers in different arrangements and densities that provide resistance to multi-dimensional tensile stress. Because the NP is devoid of its own blood supply, the function of the structure itself, i.e., hydrostatic pressure produced by intermittent external loads, provides the necessary nutrients to the NP and its cells, e.g., chondrocytes, etc., by diffusion and a secondary osmotic pressure gradient. The nutrients are delivered by blood vessels, which emerge from within the vertebral bodies, that have invaginated the vertebral endplates and outer annular fibers. In essence, optimal health of the vertebral endplates depends on the integrity of the vertebral bodies and NP yet is also essential to the health and function of the closed hydraulic system of the IVD.

#### 7.4.2.4 Nerve and Muscle; Osteo- and Arthrokinematics

Muscles of the cervical spine and upper trunk serve many functions, sometimes simultaneously: control posture, stabilize the spine, protect the spinal cord and internal organs, contribute to respiration, and the most obvious, produce joint moments that move the head, neck, and upper thorax. The muscles of these regions demonstrate significant variability in terms of the number of joints that they cross, and their resting length, shape, fiber orientation, and cross-sectional area, all of which reflect a variety of functional demands. In this section, we examine the regional location and actions of the muscles that were most likely involved in the MOI in this case. With this exercise, we generate other hypotheses regarding possible pain generators. In general, with a whiplash effect on the cervical spine, muscle eccentric actions are activated involuntarily through spinal reflexes in an effort to stabilize the joint(s) they cross. We look at the muscles in the subcranial regions, but first present a brief review of the general neuroanatomy of the cervical spine and related regions.

Each spinal nerve root is comprised of a ventral and dorsal nerve root (Fig. 7.18). The ventral roots contain efferent, i.e., outgoing, signals to muscle and autonomic organs. The dorsal roots receive signals from the periphery via the dorsal root ganglion. The sensory signals from peripheral structures, e.g., skin, muscle, etc., are then transmitted to the spinal cord, and central controllers for processing and response. Near the intervertebral foramen, the dorsal and ventral roots merge to form the spinal nerve root, which houses mixed, i.e., information from both sensory and motor nerve fibers, signals. Upon exit through the foraminal opening, the spinal nerve root divides into ventral and dorsal rami. Nerves from the ventral ramus generally innervate joints, muscles, and skin of the antero-lateral regions of the neck, trunk, and extremities, whereas nerves from the dorsal ramus innervate the same structures in the posterior neck and trunk (Fig. 7.19).

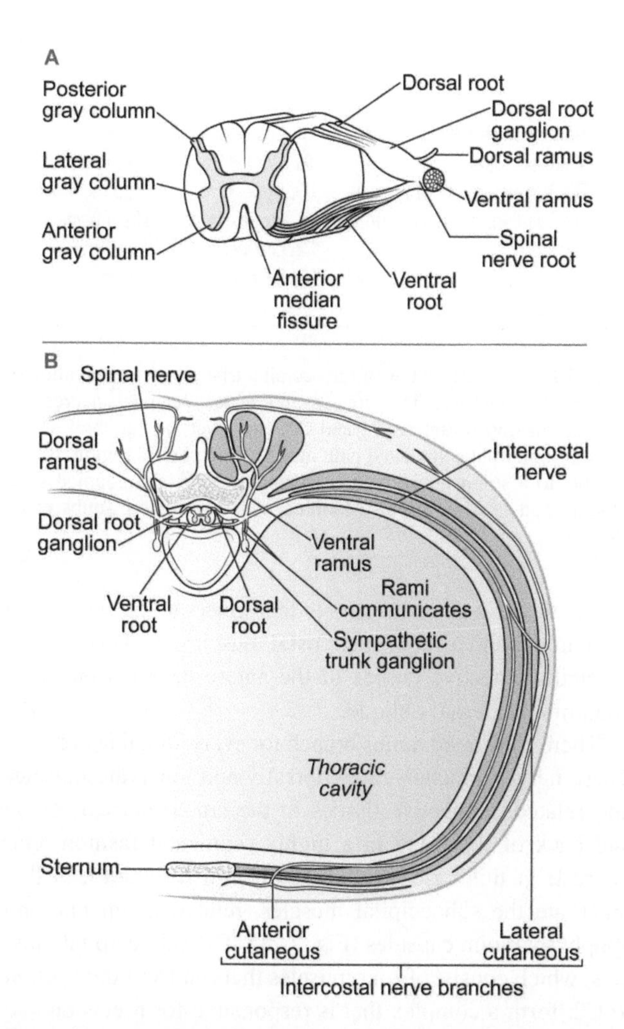


**Fig. 7.18** (**A**) Cross-section of the spinal cord and spinal nerve root. Ventral and dorsal nerve roots originate from the anterior and posterior gray matter, respectively, and then join to form the dorsal root ganglion and spinal nerve root. The spinal nerve root splits into a large ventral and much smaller dorsal ramus. This figure was reproduced from Gray's Anatomy 20th US edition, which can be found in the public domain: https://en.wikipedia.org/wiki/Grey_column#/media/File:Gray675.png. (**B**) Cross-section of a typical thoracic segment illustrating branching of the dorsal ramus to form nerves to innervate structures posteriorly and the ventral ramus and associated nerves to innervate antero-lateral structures

The ventral rami of a spinal nerve root either continue as an individual nerve or form a plexus. A plexus is a complex formation of ventral rami that eventually forms peripheral nerves, e.g., radial, median, etc. Of the four major plexuses, e.g., cervical (C1 to C4) (Fig. 7.19), brachial (C5 to T1) (Fig. 7.20), lumbar (T12 to L4), and sacral (L4 to S4), for this case we are most interested in the cervical and brachial plexuses. Note: the initials C, T, and L represent the cervical, thoracic, and lumbar vertebral segments. Most of the nerves that are formed by the cervical plexus innervate structures associated with the axial skeleton, while most nerves that emerge from the brachial plexus innervate structures in the upper extremities.

Recall that Sandra, in anticipation of being struck broadside, grasped the edge of the car seat with her right hand. By doing so she anchored or constrained her right hand, arm, shoulder girdle, and torso. This may have limited some body motion, but could not prevent excessive movement of her head and neck. Thus, when the car was struck pushing Sandra's lower body to the right, her head and neck sidebent

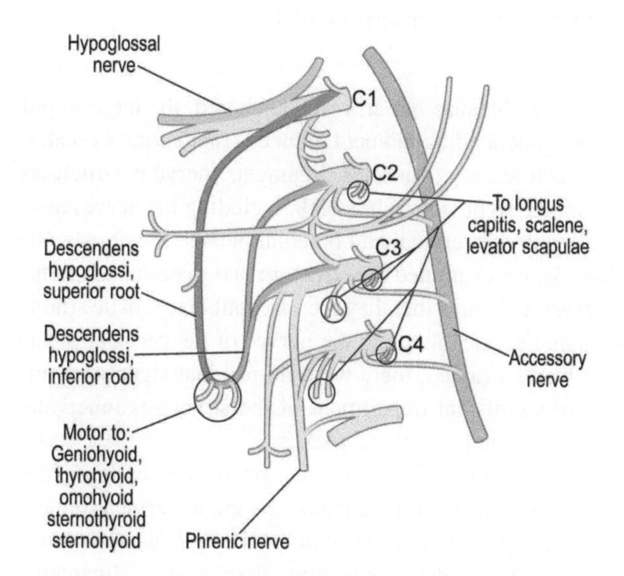


**Fig. 7.19** Schematic of cervical plexus illustrating specific nerve roots that innervate particular muscles or combine to form particular peripheral nerves. For example, the C1, C2, and C3 nerve roots are combined into peripheral nerves to provide motor innervation to the hyoid muscles

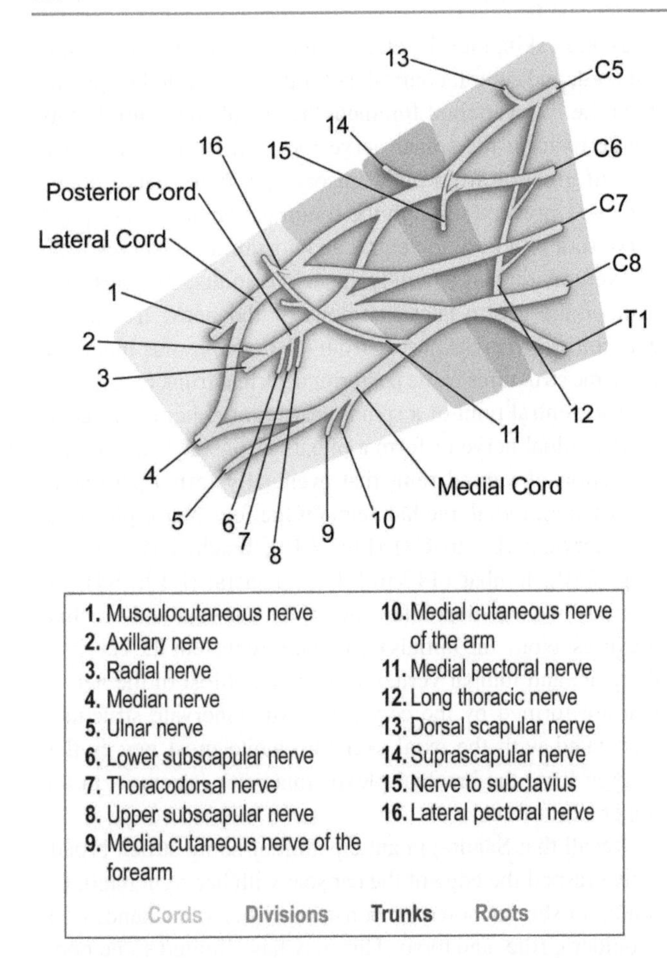


**Fig. 7.20** Schematic of brachial plexus and its organization into roots, trunks, divisions, and cords, and their relationship to the peripheral nerves of the upper extremity. For example, the long thoracic nerve is formed directly from the C5 to C7 nerve roots, and the medial pectoral nerve is a branch off the posterior cord (C5 to C8)

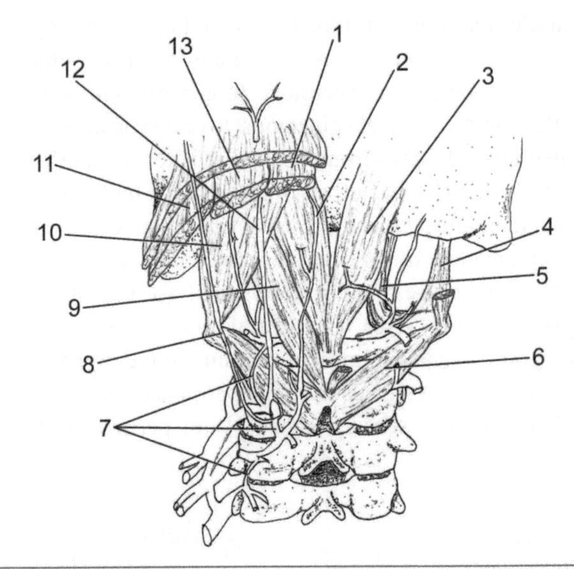


| | |
|---|---|
| **1.** Semispinalis capitis muscle | **8.** Lesser occipital nerve |
| **2.** Third occipital nerve | **9.** Rectus capitis major muscle |
| **3.** Rectus capitis minor muscle | **10.** Superior oblique capitis muscle |
| **4.** Rectus capitis lateralis muscle | **11.** Longissimus capitis muscle |
| **5.** Vertebral artery | **12.** Greater occitptial nerve |
| **6.** Inferior oblique capitis muscle | **13.** Splenius capitis muscle |
| **7.** Cruveilhier plexus | |

**Fig. 7.21** A posterior view of the occipital triangle and innervation of suboccipital muscles. The greater and lesser occipital nerves are branches of dorsal rami at C1 and C2. Note that the vertebral artery, with its somewhat convoluted path in the craniocervical region, is susceptible to a strain injury with cervical extension, and coupled sidebending and axial rotation. (Modified with permission Tubbs et al. (2011))

to the left. Because her arm was anchored, the large impulsive left sidebending moment induced tensile forces on all of the structures, e.g., muscles, ligaments, nervous structures, etc., on the right side of her neck, including the nerves associated with the cervical and brachial plexus. Previously (see Chap. 3), we examined the structure and material properties of nerves and know that they are susceptible to tensile strains. In Sandra's case, although the nerves of the cervical plexus were likely strained, there were no residual signs or symptoms of significant impairment of the structures innervated by nerves from the cervical plexus.

There are many individual peripheral nerves branching off ventral rami within the trunk and craniocervical regions that are not part of the cervical or brachial plexus. As a result, muscles and related connective tissues, e.g., ligaments, superficial and deep fascia, etc., which cross multiple vertebral segments receive innervation across multiple levels of the spine. For example, in Fig. 7.18B notice an intercostal nerve branching off a ventral ramus in the thoracic region. There are 12 intercostal nerves that innervate an intercostal dermatome and a set of intercostal muscles, as well as muscle and connective tissues of the antero-lateral trunk, e.g., abdominal external oblique.

There is a dorsal ramus branch for every spinal nerve root. These nerve branches travel dorsally and innervate muscles and related connective tissues in the craniocervical region and back of the trunk in a highly segmental fashion. Our interest, in this case, is the dorsal rami (C1 and C2) that innervate the suboccipital muscles, related ligaments, and apophyseal joint capsules (Fig. 7.21). The suboccipital muscles, which consist of four muscles that run from the occiput to C2, form a complex that is responsible for precision and monitoring movements of the occiput, atlas, and axis, and provide dynamic stabilization of the craniovertebral spine. As well, these muscles house a relatively large density of muscle spindles that help control the posturing of the head to optimize special sensory functions, i.e., visual, vestibular, auditory, and olfactory systems. The primary and secondary actions of the one-joint antero-lateral and suboccipital muscles are described in Table 7.1. Two additional muscles that are grouped with the muscles of the posterior craniovertebral

**Table 7.1** Summary of actions of the suboccipital muscles at the atlantooccipital and atlantoaxial joints

| Atlantooccipital Joint | | | | | Atlantoaxial Joint | |
|---|---|---|---|---|---|---|
| Muscles | Flex. | Ext. | Lat. flex. | Flex. | Ext. | Axial rot. |
| Rect. cap. anterior | XX | – | X | – | – | – |
| Rect. cap. lateralis | – | – | XX | – | – | – |
| Rect. cap. post. major | – | XXX | XX | – | XXX | XX (ipsilat.) |
| Rect. cap. post. minor | – | XX | X | – | – | – |
| Obl. cap. inferior | – | – | – | – | XX | XXX (ipsilat.) |
| Obl. cap. superior | – | XXX | XXX | – | – | – |

Modified with permission from Neumann (2017)
*Flex.* flexion, *Ext.* extension, *Lat. flex.* lateral flexion (sidebending), *Axial rot.* axial rotation, *Rect. cap.* rectus capitis, *Obl. cap.* obliquus capitis, *post.* posterior, *ipsilat.* ipsilateral (i.e., same side). The letters X, XX, and XXX represent a minimal, moderate, and maximum contribution to the movement, respectively. Note: the rectus capitis lateralis is not considered one of the suboccipital muscles that make up the suboccipital triangle

**Table 7.2** Actions for the joints of the craniocervical region

| Joint | Sagittal plane (°) (Flexion/extension) | Horizontal plane (°) Axial rotation | Frontal plane (°) Sidebending (LF) |
|---|---|---|---|
| Atlantooccipital joint | Flexion: 5<br>Extension: 10 | Negligible | ~5 degrees (to one side) |
| Atlantoaxial joint | Flexion: 5<br>Extension: 10 | 35–40 (to one side) | Negligible |

Modified with permission from Neumann (2017)
*LF* lateral flexion

region include the splenius capitis and splenius cervicis. Both muscles, when acting bilaterally, are strong extensors of the head and neck. Acting unilaterally, the splenii induce ipsilateral sidebending and axial rotation.

Before we examine additional muscles that induce cervical and thoracic vertebral segment movements, we need to digress briefly to discuss the osteo- and arthrokinematics of the craniovertebral joints. In Chap. 3 we discuss the kinematics of the typical cervical and thoracic segments but do not describe craniovertebral kinematics because it is not relevant to that case, but now it is. Historically, the atlantooccipital (AO) and atlantoaxial (AA) joints are noted to have two degrees of freedom (Table 7.2). However, as with all synovial joints in the human body, the craniovertebral joints have the potential for six degrees of freedom, even though the translations and rotations in the third plane and about the third axis, respectively, are small. We make a point of reminding the reader about this because often the impairments in mobility that are elicited as part of a physical examination are related to subtle loss of the small tertiary rotations and translations out of the plane.

Osteokinematically, flexion at the AO joint finds the occiput rotating (tipping) forward (anterior) relative to the atlas. Thus, arthrokinematically, the convex occipital condyles roll anterior and glide posterior relative to the concave condyles of the atlas (Fig. 7.22). Since the inferior facets of C1 are slightly concave and superior facets of C2 are convex, the atlas (C1) rolls and glides anteriorly during flexion of the head. The anterior tipping and sliding of C1 are constrained by the transverse ligament (Fig. 7.5). Extension at the AO and AA joints follows in reverse fashion (Fig. 7.23). Atlantoaxial extension is limited primarily by the confrontation of the anterior arch of C1 with the dens. It is notable that we can also voluntarily protract (translate forward) and retract (translate backward) the cervical spine. During protraction, the craniocervical joints extend and the mid- and lower-cervical segments flex, with the reverse taking place with retraction.

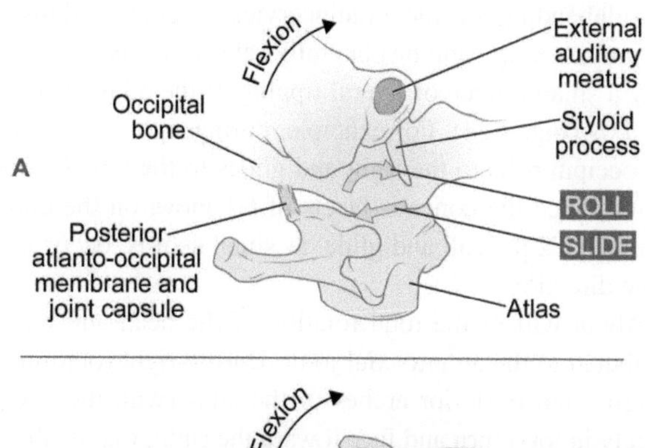


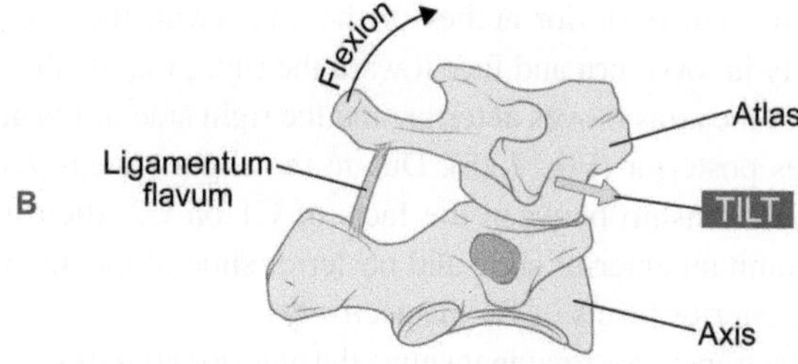


**Fig. 7.22** Osteo- and arthrokinematics during flexion at the (**A**) atlantooccipital and (**B**) atlantoaxial joints. Excessive movement is controlled by posterior intrinsic and extrinsic ligaments

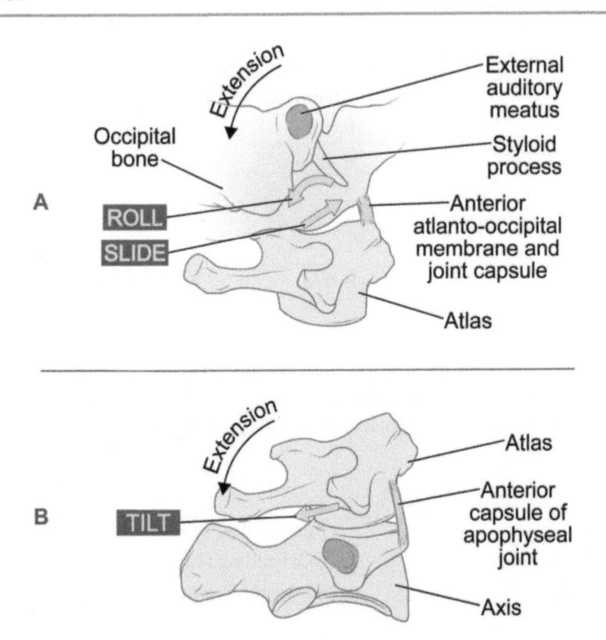


**Fig. 7.23** Osteo- and arthrokinematics during extension at the (**A**) atlantooccipital and (**B**) atlantoaxial joints. Excessive extension is controlled by anterior intrinsic ligaments

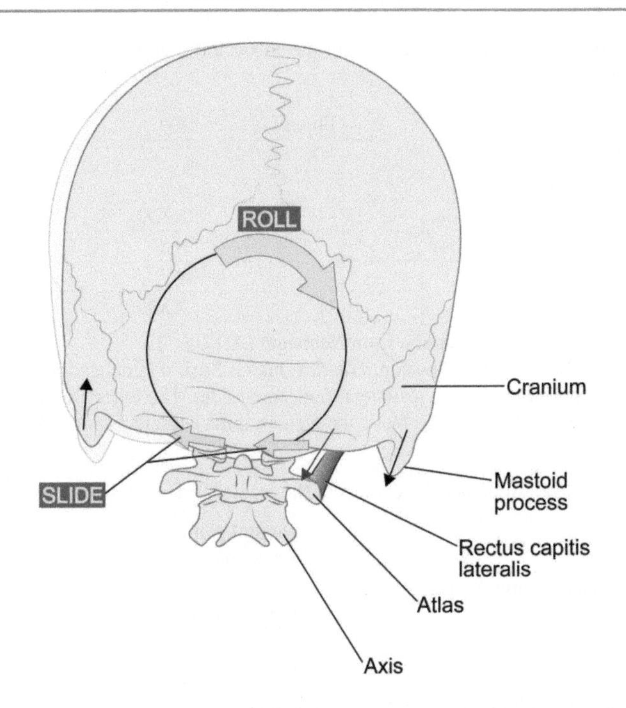


**Fig. 7.24** Osteo- and arthrokinematics of right sidebending at the atlantooccipital and atlantoaxial joints. The rectus capitis lateralis contributes to this action

Sidebending of the craniocervical segments finds the occiput rotating (tipping) laterally relative to the atlas, with only a small degree of lateral tipping of the atlas relative to the axis (Fig. 7.24). For example, during right sidebending, the occiput rolls to the right and glides to the left. As noted above, since the concave facets of C1 move on the convex facets of C2 the roll and glide, as small as it is, occur in the same direction

About half of the total rotation of the head and neck is attributed to the atlantoaxial joint. During right rotation, the anterior and posterior arches of the atlas (with the occiput largely in tow) turn and face toward the right; that is, the left side of the atlas moves anterior, and the right side of the atlas moves posterior (Fig. 7.25). Due to the slight concave/convex relationship between the facts of C1 on C2, there is a concomitant anterior slide and posterior slide of the anterior and posterior facets of C1, respectively.

The transverse ligament snugs the anterior arch of C1 and the dens but does not constrain excessive rotation. Extreme rotation is checked primarily by the alar ligament, facet capsules, and suboccipital muscles.

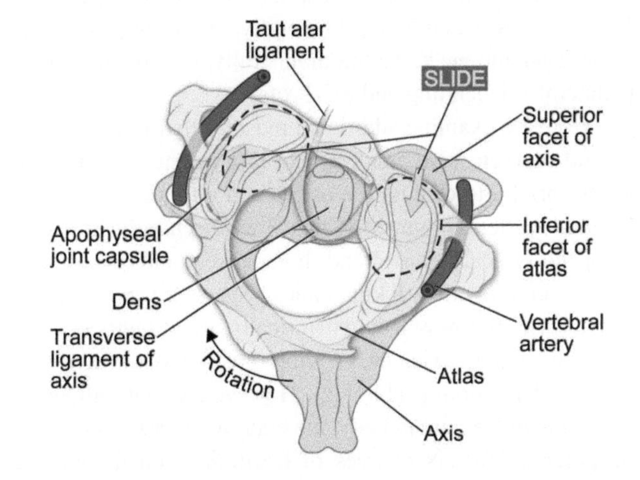


**Fig. 7.25** Osteo- and arthrokinematics of craniocervical rotation. Note the midpoint of the dens as the approximate joint center of rotation. Axial rotation (and its couple sidebending) creates a torsional strain of the vertebral artery in this region of the neck

As with all vertebral segments, coupled movements also occur in the craniocervical complex whenever either sidebending or axial rotation is introduced. Recall (see Chap. 3) that with the cervical segments from C2 to C7, sidebending and axial rotation are always coupled in the same direction, regardless of which motion is introduced first. However, because of the unique nature of joint morphology, ligamentous constraints, e.g., alar ligament, and suboccipital muscle activity, contralateral coupling of sidebending and axial rotation is the signature of the craniocervical complex. It is believed that the primary purpose of the unique coupling patterns of the craniocervical complex is to maintain and optimize the function of the visual, auditory (i.e., vestibular), and olfactory systems. For example, when someone turns their head to the right to the end of the range of motion, i.e., right axial rotation, a concomitant right sidebending of the typical cervical segments occurs, with a subtle left sidebend at the atlantooccipital joint. This left AO sidebend, induced by the left rectus capitis lateralis and tautness in the left alar ligament, is a mechanism that appears to augment the maintenance of a level horizontal visual field. Conversely, right sidebending and rotation of the lower cervical segments induce action of the left obliquus capitis inferior to producing left axial rotation of the subcranial cervical segments.

Recall that a moment is determined by a force and the distance over which that force is acting relative to a joint center. Also, the primary contributor to the net internal joint moment (or torque) is the force produced by the muscle-tendon actuator, where secondary contributions come from periarticular structures, e.g., intrinsic and extrinsic ligaments. For the extremity and spinal joints net internal joint moments are typically defined relative to a Cartesian coordinate system defined by the sagittal, frontal, and transverse (or horizontal) planes. For example, when the brachialis is activated, it flexes the elbow about a joint center located between the medial and lateral humeral epicondyles. The action of the brachialis is not complex because its fibers are unipennate, and it is acting on a "simple" hinge joint. On the other hand, the multiple actions of the pectoralis major at the glenohumeral joint are more complex because of the multipennate orientation of its muscle fibers and multi-dimensional movements at the glenohumeral joint (Fig. 7.26). What the brachialis and pectoralis major have in common is that they cross a joint with a well-defined joint center. In contrast to both the brachialis and pectoralis major, most muscles of the neck and thorax act across multiple bones and joint axes; and joint centers have not been well-defined. So, for example, although the rectus capitis lateralis runs from the base of the occiput to the transverse process of C1 (Fig. 7.21), the "joint" that it crosses, i.e., the atlantooccipital joint, does not have a well-defined joint center. Because of the simplicity of the orientation of the rectus capitis lateralis, however, it can only induce a lateral bending moment at the OA joint. The rectus capitis posterior major, on the other hand, crossing both the OA and atlantoaxial (AA) joints makes contributions to net joint moments in both the sagittal and frontal planes across two different joints, about multiple axes. And, to add to the complexity, because of the inherent dynamic motion coupling in the spine, when the rectus capitis posterior major shortens, it induces both a sidebending and axial rotation moment simultaneously. Such is the complexity of muscle action in the craniovertebral region (Table 7.1), as well as with typical cervical and thoracic vertebral mobile segments. We soon explore the location, innervations, and actions of muscles that act on the sub-cranial, typical cervical, and upper thoracic segments.

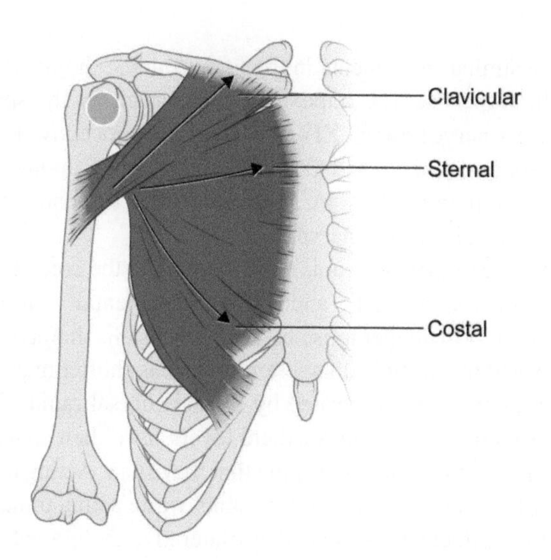


**Fig. 7.26** Schematic of the pectoralis major, illustrating the orientation of its clavicular, sternal, and costal divisions. Imagine its potential multiple actions as it acts on the geometric center of the convex humeral head (marked with a blue circle)

With the MOI in Sandra's case, we speculate that the following muscles would have been activated eccentrically, in response to the rapid external left sidebending/axial rotation moment: rectus capitis anterior, rectus capitis lateralis, rectus capitis posterior major and minor, and obliquus capitis superior. Because of the back-and-forth movement of the head following the collision, the suboccipital muscles would have been activated first on the right side, then on the left side. In our exploration of muscle mechanics in Chap. 4, we explored the unique biomechanical response of the muscle-tendon actuator to passive loads. Recall, as a viscoelastic structure muscle has two mechanisms that are activated when it is loaded at a high rate: (1) muscle spindles respond to both a change and rate of change in muscle length (i.e., when the muscle is stretched), which facilitates a signal to the spinal cord to induce a resistive action by the same muscle, i.e., similar to the deep tendon reflex response; and (2) an inherent increase in stiffness, i.e., passive resistance, of the tissue (muscle fibers, as well as the parallel and series elastic elements) in the elastic region of a stress-strain diagram. We assume that these two mechanisms, operating normally during the incident, provided Sandra with reasonable protection. However, these responses can be overcome if the magnitude and rate of external loads exceed the yield point of the tissues resulting in a strain injury. It is common for individuals who have experienced whiplash to complain of muscle soreness for several days, sometimes weeks, following an accident. This soreness is likely related to first- or mild second-degree strains of multiple muscle-tendon actuators and associated fascial supporting structures. If Sandra sustained minor injuries, we can be confident that she has now recovered from these since resisted isometric testing of her cervical muscles was unremarkable. On the other hand, the complex neurological response to the rapid lengthening actions of cervical muscles during the whiplash may have induced abnormal muscle spindle activations that continue to over-facilitate cervical muscles resulting in joint mobility restrictions.

**Table 7.3** Muscles of the axial skeleton

| Anatomic region | Location | Muscles |
|---|---|---|
| Muscles of the trunk | Posterior trunk | *Superficial*<br>Trapezius, latissimus dorsi, rhomboids, levator scapulae, serratus anterior<br>*Intermediate*<br>Serratus posterior superior<br>Serratus posterior inferior<br>*Deep (3 groups)*<br>1. Erector spinae (spinalis, longissimus, iliocostalis)<br>2. Transversospinal (semispinalis, multifidi, rotators)<br>3. Short segmental (interspinalis, intertransversarius) |
| | Antero-lateral trunk | Rectus abdominis<br>Obliquus internus abdominis<br>Obliquus externus abdominis<br>Transversus abdominis |
| | Additional muscles | Iliopsoas<br>Quadratus lumborum |
| Muscles of the craniovertebral region | Antero-lateral craniovertebral region | Sternocleidomastoid<br>Scalenus (anterior, medius, and posterior)<br>Longus colli<br>Longus capitis<br>Rectus capitis anterior<br>Rectus capitis lateralis |
| | Posterior craniovertebral region | *Superficial*<br>Splenius cervicis<br>Splenius capitis<br>*Deep (suboccipital)*<br>Rectus capitis posterior major<br>Rectus capitis posterior minor<br>Obliquus capitis superior<br>Obliquus capitis inferior |

Modified with permission from Neumann (2017)

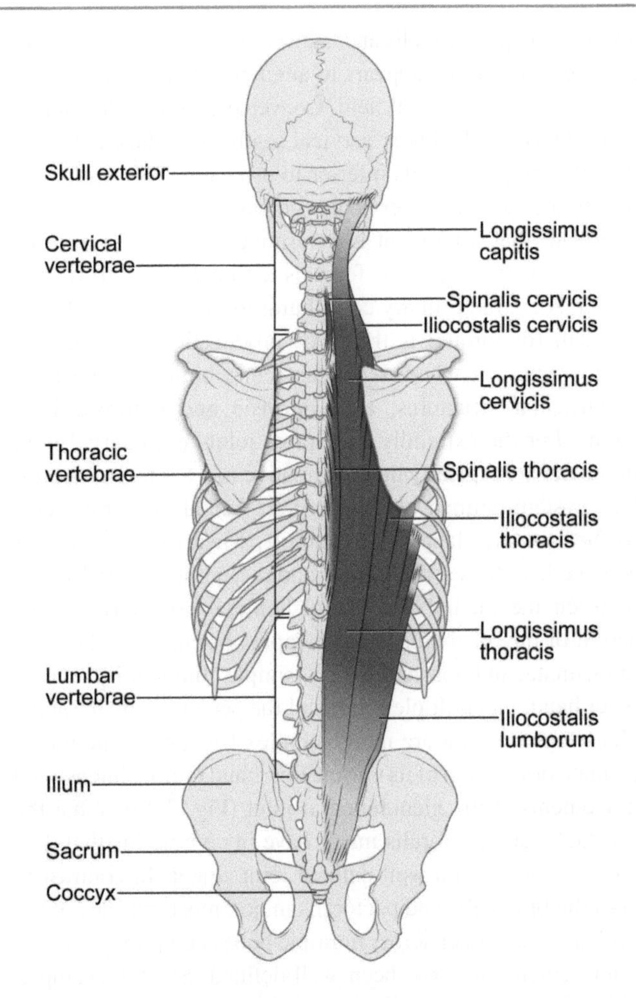


**Fig. 7.27** Each muscle has three components named by the region they span. For spinalis, capitis, cervicis, and thoracis groupings; for longissimus, capitis, cervicis and thoracis groupings; and for iliocostalis, cervicis, thoracis, and lumborum groupings

Let's now briefly review the organization of the muscles of the spine (Table 7.3). In the interest of this case, we describe the muscles that directly affect the head and cervical and thoracic vertebral segments. The superficial muscles of the posterior trunk that can mobilize the cervical spine when their distal attachments are fixed include the trapezius, levator scapulae, and rhomboids. Acting bilaterally, these three muscles extend the cervical spine. Unilateral action of the trapezius sidebends the neck ipsilaterally and rotates it contralaterally, whereas unilateral action of the levator scapulae sidebends and rotates the neck toward the same side. Although not listed in Table 7.3, the rhomboid muscles, with insertions on the spinous processes of thoracic segments, induce similar movements in the upper/middle thorax as the middle trapezius. The trapezius is innervated by the spinal accessory nerve (cranial XI), with some proprioceptive fibers from ventral rami C3 and C4, while the levator scapulae (C3 to C5) and rhomboids (C4/C5) are innervated by ventral rami via the dorsal scapular nerve.

The deep posterior trunk muscles include the erector spinae, transversospinal and short segmental muscles (Fig. 7.27). The longer muscles, e.g., iliocostalis thoracis, are innervated by multiple dorsal rami, whereas shorter muscles, e.g., rotatores, are innervated by a single dorsal ramus. The most superficial and intermediate layers have their muscles named by the region of the spine that they span. Acting bilaterally, these muscles extend the spine, while acting unilaterally they effectively sidebend ipsilaterally; the cranial and cervical components of the longissimus and iliocostalis appear to have effective moment arms to induce axial rotation ipsilaterally.

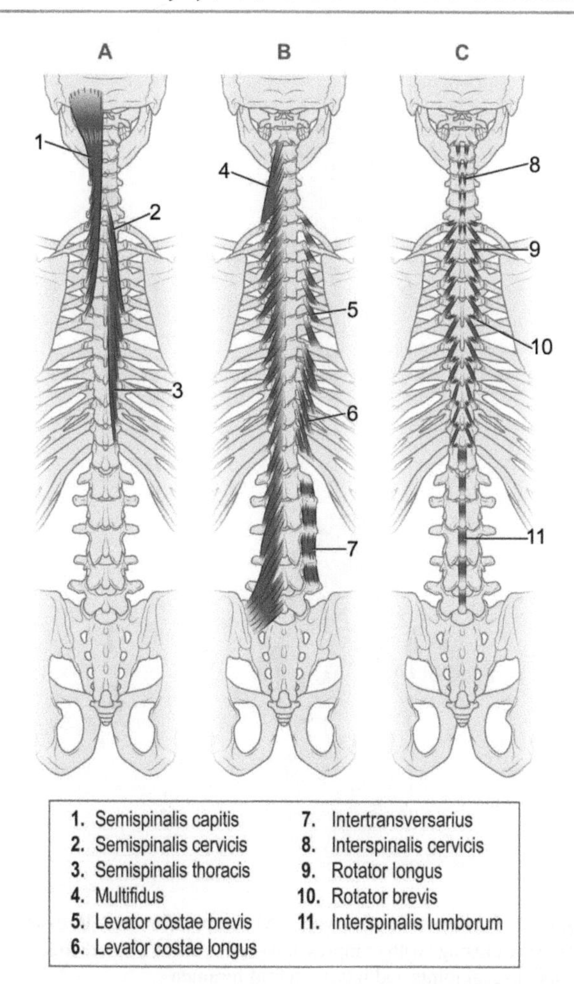


**Fig. 7.28** Posterior view illustrating (**A**) more superficial semispinalis muscles, (**B**) multifidi, and (**C**) interspinalis and intertransversarius muscles

Working from superficial to deep, the muscle fibers of deep muscle groups become shorter and are oriented in a more angular fashion. For example, the semispinalis muscles may span six to eight vertebral segments, whereas the multifidi span two to four segments, with rotatores usually spanning one or two segments (Fig. 7.28).

Like other posterior muscles, when these muscles act bilaterally, they extend the axial skeleton. With the exception of the semispinalis capitis, the transversospinal muscles contribute to ipsilateral sidebending and contralateral rotation, but generally do not have effective moment arms for either movement. The short segmental group of muscles is notable for their density of muscle spindles suggesting a predominant role for these muscles in providing sensory information to the CNS for fine-tuning stability and control, particularly in the craniovertebral region. Unilateral action of the intertransversarius muscles makes segmental contributions to ipsilateral sidebending.

Muscles that make up the antero-lateral group are innervated primarily by nerves from the ventral rami of the cervical plexus, except the sternocleidomastoid which is innervated by cranial nerve XI (Fig. 7.29). The sternocleidomastoid (SCM) is a strong flexor of the mid and lower cervical spine, and also contributes to extension of the head and suboccipital joints. If the longus colli and rectus capitis anterior co-act during SCM activation, craniovertebral extension is inhibited. Acting unilaterally, the SCM sidebends to the same side and induces contralateral rotation. When the cervical spine is stabilized, the scalene muscles act as secondary inspiratory muscles by elevating the first two ribs. The scaleni often become primary inspiratory muscles, as evidenced by their hypertrophy, in individuals with chronic obstructive pulmonary disease whose diaphragm has become biomechanically ineffective. With the ribs fixed, the scaleni, acting bilaterally, are ineffective flexors of the neck. When acting unilaterally the scalene muscles sidebend the neck to the same side. There is some disagreement about the role of the scaleni in cervical rotation. Some claim they rotate the neck ipsilaterally; others claim they rotate the neck contralaterally. What is clear, however, is that if the neck is rotated to the left, the right-sided scaleni return the head and neck to a neutral posture, i.e., rotate the cervical spine to the right. The longus colli lie deep to the trachea and esophagus. Their primary function is to flex and stabilize the cervical spine in its neutral posture, and to prevent hyperextension, yet they have also been defined as weak lateral flexors. The two remaining antero-lateral muscles, the rectus capitis anterior and lateralis, act only on the atlantooccipital joint where they flex and laterally flex the head, respectively.

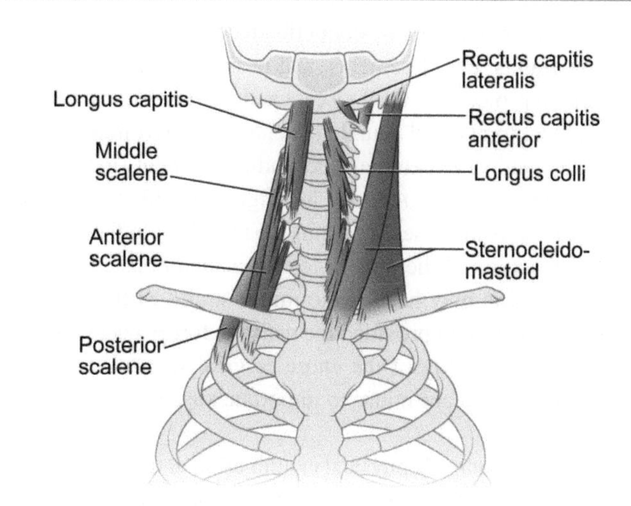


**Fig. 7.29** Anterior view of antero-lateral muscles that control the head and neck

This completes our study of the neuromuscular systems of the craniocervical and subcranial cervical spinal regions. What we see is a complex redundancy of muscle function, layered from superficial to deep, posteriorly and anteriorly. Because of the extensive mobility of the cervical spine, which serves our functional needs, some stability is sacrificed.

However, the muscular system effectively augments the passive restraints, e.g., ligaments, and provides dynamic stability, which protects cervical viscera and vessels, intervertebral discs, facet joints, and neural elements. The MOI in Sandra's case involved large external sidebending moments in the cervical segments both to the left and right. This likely induced eccentric activation, secondary to the muscle spindle mechanism, of several of the muscle-tendon actuators listed above. This activation may have led to minor strains of these muscles, but the system provided the resistive forces necessary to enhance the stability of the entire cervical spine, and likely prevented significant injury to important vascular and neural tissue.

Based on Sandra's physical examination data, we surmise that her symptoms did not originate from the subcranial region of the cervical spine. However, her symptoms did appear to be localized to the mid-cervical spine. With an understanding of the osteo- and arthrokinematics of the typical cervical segments, we conclude that the reproduction of her neck pain during extension was likely related to the inability of the apophyseal joints to glide inferiorly (i.e., close) in one or more of the cervical segments. Also, because Sandra's head deviated to the left during cervical extension, we hypothesize that it is likely that impaired facet closure involved facets on the right side of her neck. This hypothesis is supported by two additional physical examination tests: (1) reproduction of her feelings of a "catch" on the right side of her neck along with a mild reduction of right sidebending and axial rotation, and (2) reproduction of her neck pain with a cervical compression test. We provoked her "catch" when we asked her to perform active right sidebending and axial rotation. We know (see Chap. 3) that with right sidebending and axial rotation, the left-sided facets must glide superiorly (i.e., open) and the right-sided facets must glide inferiorly (i.e., close). Therefore, a reduction in the right sidebending/axial rotation range of motion strongly suggests a right-sided problem. Typically, the application of an axial compression load through the head and cervical vertebral segments (usually with the head and neck in a neutral posture) compresses the IVD and reduces the diameter of the intervertebral foramen (Fig. 7.30). But it is also the case that with cervical compression the apophyseal joints are forced to glide inferiorly, which in Sandra's case, likely provoked her symptoms.

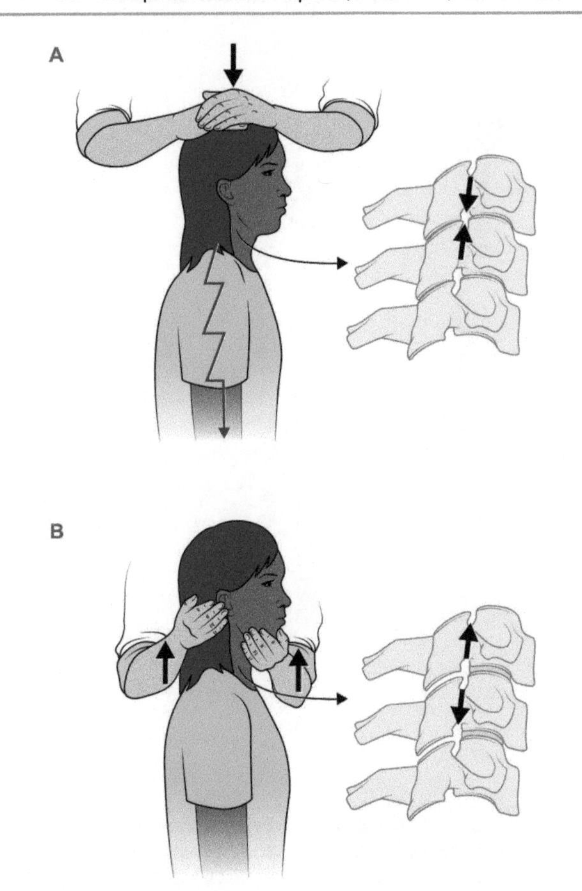


**Fig. 7.30** Performance of cervical (**A**) compression and (**B**) distraction tests. Note closing (with compression) and opening (with distraction) of the apophyseal joints and intervertebral foramen

So, we know the approximate location of Sandra's cervical impairment, and our physical examination tests have allowed us to form hypotheses regarding a likely mechanical fault, yet we cannot provide an explanation for an exact etiology. However, we can generate a list of potential sources for the mechanical problem we have identified. In fact, the entire treatise of this chapter can be combed for a list of structures that may have sustained a strain injury related to her cervical problem. For example, one or more muscles and their tendons on the right side of the neck were exposed to large impulsive tension forces that could have induced strain injuries, or in some way, altered the neuro-control, i.e., muscle spindle facilitation, of one or more one-joint muscles that are now restricting facet closure on the right side of the neck. We leave it to the reader to complete their list of hypothetical etiologies.

#### 7.4.2.5 Nerve Palsy; Brachial Plexus

We have one more problem to examine: Sandra's right arm weakness. Previously, we concluded that it was unlikely that Sandra sustained strain injuries to one or more muscles based on the fact that resisted testing was painless. We know that her weakness was not painful and concluded that it must be related to some type of nerve palsy, yet not related to the spinal cord itself. We also know that cervical distraction and the brachial plexus tension test reproduced her right arm symptoms; that is, external tensile loads provocated symptoms to some degree. Recall that in anticipation of the collision Sandra reached and grasped the car seat with her right hand. We have already decided that that action provided a weak constraint to body movement, but also likely induced large tensile forces to multiple tissues on the right side of her body. Let's look at this a bit closer.

**Table 7.4** Actions and innervations of the muscles that internally rotate the glenohumeral joint

| Muscles | Actions | Innervations | Origin within brachial plexus |
|---|---|---|---|
| Subscapularis | Internal rotation | Upper and lower subscapular nerves (C5, C6, C7) | Posterior cord |
| Latissimus dorsi | Internal rotation, extension, adduction | Thoracodorsal (middle subscapular) (C6, C7, C8) | Posterior cord |
| Pectoralis major | Internal rotation, adduction, flexion | Lateral and medial pectoral (C5, C6, C7, C8, T1) | Medial and lateral cords |
| Teres major | Internal rotation, extension | Lower subscapular (C5, C6, C7) | Posterior cord |
| Anterior deltoid | Internal rotation, flexion | Axially (C5, C6) | Posterior cord |

It is possible that Sandra sustained a severe first- or mild second-degree strain of a nerve root and associated connective tissues, the IVD, or a region of her brachial plexus and related peripheral nerve. Recall that with the physical examination and resisted muscle testing, there was a weak and a painless response to the test of the glenohumeral internal rotator muscles, as a group. There are a number of muscles that contribute to an internal rotation moment at the glenohumeral joint, including the latissimus dorsi, pectoralis major, subscapularis, teres major, and anterior deltoid (Table 7.4). To sort out where the source of Sandra's weakness might be, let's examine what these muscles have in common. All five muscles share the C6 nerve root, four share the C5 and C6 nerve roots, three share the C5, C6, and C7 nerve roots, and two muscles share the C6 and C7 nerve roots. Since all share the C6 nerve root, it is reasonable to hypothesize that a nerve root had sustained either a tension or compression injury. Tension on the nerve root alone was created when the head and neck were rapidly thrust into left sidebending, but that mechanism would have affected all of the nerve roots that make up the cervical and brachial plexus. The same cervical spine movements would have concomitantly increased compression on the left side of the IVD between segments C5 and C6, inducing internal displacement of the nucleus to the right resulting in tensile strains of the internal and external annulus on the right side. If this insult resulted in plastic changes and internal derangement of the IVD, a disc bulge in the posterolateral right corner of the IVD could be compressing the C6 nerve root. We examined these hypotheses previously and concluded that they were less likely because Sandra did not demonstrate any symptoms and signs of nerve root irritation secondary to tensile or compression phenomenon, e.g., altered sensation, specific motor loss of C6 myotomes, and altered deep tendon reflexes of the biceps or brachioradialis. Moreover, an isolated disc rupture resulting in a nerve root injury, without concomitant vertebral endplate fractures, is not a common occurrence.

Let's look closer at the origins of the peripheral nerves within the brachial plexus (Fig. 7.20 and Table 7.4). Four of the five muscles that contribute to a glenohumeral internal rotation moment have their peripheral nerves branch off the posterior cord of the brachial plexus. But it is not likely that the position Sandra placed her shoulder (adducted), elbow (extended), and wrist (extended) at the time of the incident could have induced strain injuries to all of these muscles' peripheral nerves simultaneously. Therefore, the most reasonable hypothesis that could explain the etiology of Sandra's nerve palsy is that the posterior cord sustained a moderate first- or mild second-degree strain injury. In other words, the MOI-created impulsive tension forces to the right side of Sandra's neck and shoulder girdle secondary to an induced left sidebending moment in the cervical spine that exceeded a yield point resulting in plastic deformation of the posterior cord of the brachial plexus. It would not be unreasonable to hypothesize a strain injury to both the medial and lateral cords as well, but there was no evidence of impaired muscle function in muscles innervated by any of the peripheral nerves that branch off those cords, e.g., median, ulnar, radial, etc.

It is obvious that the subscapularis is the only muscle that induces a pure internal rotation moment. The other four muscles produce shoulder moments in other planes, and the body can be positioned differently to test those shoulder moments. Therefore, with resisted muscle testing of the glenohumeral internal rotators, it is not feasible to isolate the internal rotatory component of any of the five muscles, so generally, all five muscles are tested as a group. For testing, the individual can be placed in a variety of positions, but to make the internal rotator muscles resist with the arm in a position where most of the muscles are working against gravity, we place the individual supine holding the arm in a neutrally rotated position (Fig. 7.31). The individual is asked to hold this position for three to five seconds while the operator applies a force orthogonal to the distal volar aspect of the forearm in the direction of external rotation; we test the muscles isometrically. After testing instructions are given, a few submaximal holds are practiced making sure the individual understands the nature of the test. For testing, individuals are asked to hold against maximum resistance provided by the operator, and the testing is repeated at least three to five times. Note: performing multiple repetitions (up to 15 or more) of resisted isometric tests can be useful to assess the fatiguability of the muscles; it has been established that a weakened muscle secondary to nerve palsy fatigues more quickly. In Sandra's case, fatigue testing was not performed.

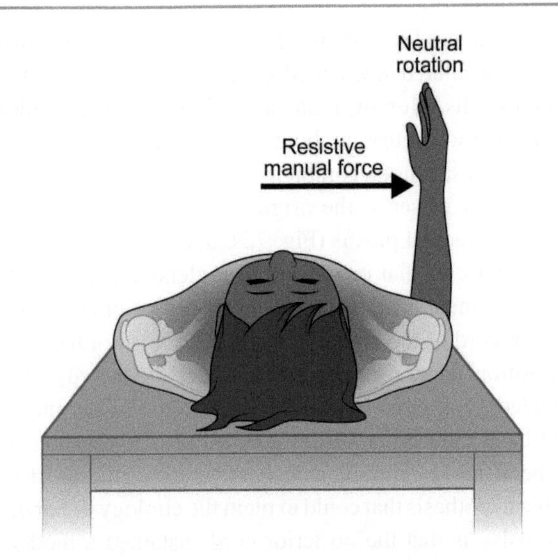


**Fig. 7.31** Schematic of positioning for performing an isometric resistive test for the glenohumeral internal rotators as a group

Although manual resistive testing of these muscles was necessary and important, the information we received was not sufficient. Based on an understanding of the neuroanatomy of the cervical and brachial plexus, and evaluation of the MOI, a special test, i.e., brachial plexus tension test, was utilized to verify the brachial plexus/posterior cord strain hypothesis. First, some background information. There is an extensive complex of connective tissues that protect the neural components of the nervous system that allows neural impulses to be transmitted while the body is placed in a variety of postures and exposed to multi-dimensional forces and moments. Therefore, the importance of the dynamic role of the nervous system needs to be considered. Let's examine what protects the nervous system from external forces, besides the obvious, i.e., bone. The spinal cord dura mater, with its layered and axially directed collagen fibers, possesses great stiffness and tensile strength, along with some elasticity. Collagen fibers of the pia and arachnoid are arranged in a lattice pattern that allows both lengthening and shortening under stresses that are controlled by their inherent viscoelastic properties (Fig. 7.32). Alternative collagen arrangements in peripheral nerves appear to provide the perineurium, endoneurium, and epineurium (see Chap. 4 for more details of peripheral nerve structure and their material properties) with a framework to also resist external tensile and compressive forces. Tension in peripheral nerves is transmitted to the spinal dura mater via the perineurium, which is continuous with the dural sleeves and to epidural tissues via the epineurium. In addition to the protections afforded by the meninges, spinal cord tracts and peripheral nerve fasciculi are wavy and folded, which allow elongation when loaded in the toe region of the stress-strain curve. Moreover, denticulate ligaments (not shown), linking the pia

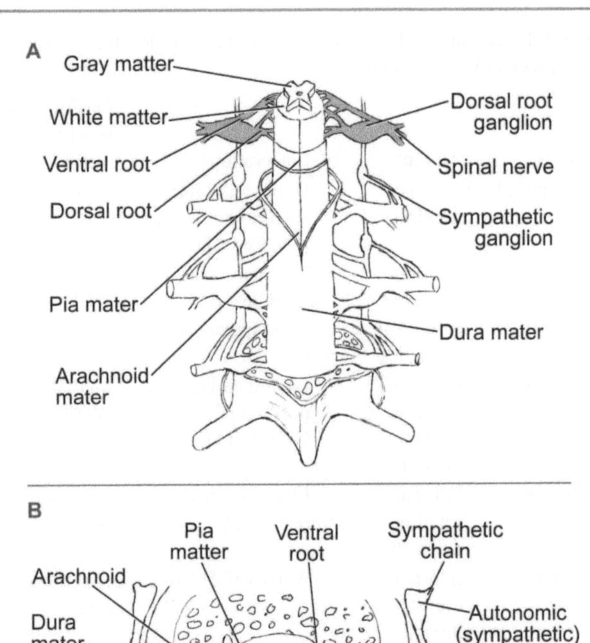


**Fig. 7.32** Schematic of (**A**) posterior and (**B**) cross-sectional views of the spinal cord dura and other protective connective tissue layers

and dura mater, assist with the dissipation and mitigation of tensile loads, as well as suspend the cord centrally in the spinal canal. Additional protection via the cerebrospinal fluid, along with an extensive network of flow reversible veins in the spinal canal, cushions the spinal cord during movement, while also serving a nutritive role.

The nervous system is attached to surrounding tissue via ligamentous and layered fascial structures that impart both restraint and mobility. For example, within the spinal canal, dural ligaments from the anterior dura and the intervertebral disc to the posterior longitudinal ligament, and the dorsal plicae from the posterior dura to the ligamentum flavum, allow movement as well as provide constraints. Peripherally, there is considerable mobility of nerves allowed by epineural attachments that demonstrate a different degree of stiffness depending on the area of the body.

Body movements and postures, as well as environmental conditions, induce tensile, shear, torsional, and compressive stresses to the nervous system. Therefore, the nervous system must adapt, and/or provide limits, to a variety of ranges, speeds, and combinations of movements imposed on it. Adaptation and control by the nervous system occur under two circumstances that are related in complex ways: (1) development of tension or compression within the system and (2) movement relative to the mechanical interface, defined as the most anatomically adjacent tissue to the nervous system that can move independently of the system. For example, the ligamentum flavum provides a mechanical interface to the posterior dura mater, with these two tissues interacting in a unique way during spinal flexion. Moreover, neural elements also have movement relationships with connective tissues, e.g., movement of the spinal cord within the dura mater or movement of myelin sheaths and Schwann cells in relation to the endoneurium. In the upper extremities, for example, the median nerve may slide as much as 2 cm during movements of the wrist and neck in relation to interfacing tissue in the upper arm and/or within the brachial plexus.

In addition to relative movements between the interface, there are also areas referred to as "tension points" where the nerve/interface movement remains constant. It is likely that tension points occur normally in nerves within the plexus, or at the shoulder, elbow, or wrist during arm and neck movement combinations. It is possible to develop new tension points in the system secondary to a macro- or micro-traumatic injury where the inflammatory cycle creates a scar and/or adhesions. During various movements, it is at the tension points where nervous, and related connective, tissues deform as they provide resistance to external loads. It is notable that the nervous system has mobility of its own, independent of the interfacing tissues. Finally, it is important to note that nervous system movement in one part of the body creates tension and movement in the nervous system in other more remote body parts.

Clinically, there are vulnerable sites, e.g., osseous tunnels like the intervertebral foramen, within the body where lesions that affect the elasticity and movement of the nervous system begin. And certainly, trauma to soft tissues, e.g., muscle, etc., and nerves and their related connective tissues can create interface sites with altered material properties that exhibit reduced extensibility and mobility. As noted previously, most of the connective tissues of the nervous system are innervated via sinuvertebral nerves and local autonomic axonal branching. If nerves and neuromeningeal tissues are sensitized by mechanical trauma and/or chemical irritants, i.e., secondary to the inflammatory cycle, even a small internal or external stimulus (non-noxious or noxious) can create a pain response. For example, if a tension point exists at an intervertebral foramen that may be compressing a nerve root, spinal extension could readily irritate the tension point and elicit pain or abnormal sensory feelings. Moreover, inflammation of the nerve roots, plexus, and/or peripheral nerves results in intraneural edema which directly impairs nerve function, i.e., reduces axonal transport, and compromises the blood supply and venous drainage. In addition, ischemia can

**Table 7.5** Common symptoms resulting from mechanical disorders of the nervous system

| Possible local origins | Possible remote origins |
|---|---|
| Hypoxic, damaged, regenerating, or immature axons<br>Connective tissue irritation (e.g., dura mater, epineurium, other attachments)<br>Mechanical interface irritation | Extra-segmental referral from the dura mater and nerve root sleeves<br>Mechanical interface referral (e.g., disc, zygapophyseal joint)<br>Referral from the autonomic nervous system |

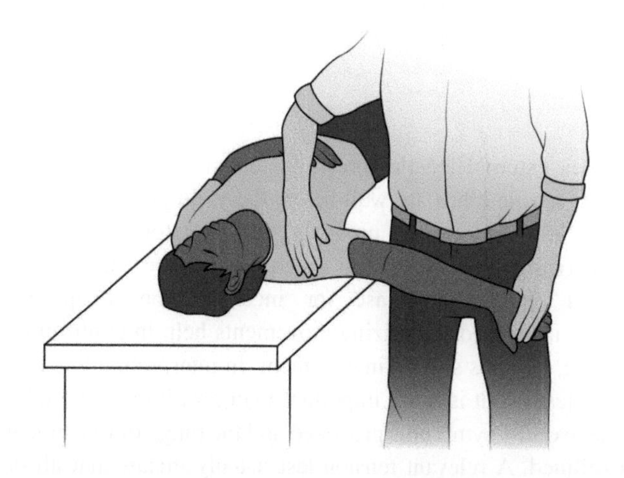

**Fig. 7.33** A general brachial plexus tension test that is often used to test for possible disorders of the nervous system originating at the nerve root, but that may peripheralize through the brachial plexus to peripheral nerves, e.g., median radial, and ulnar. Note: there are several additional tests similar to the general test that can also be used to test more specific peripheral nerves. We leave it to the reader to explore further

damage the dorsal root ganglion, which then can become a source of pain (Table 7.5).

Based on the theoretical mechanics of the two movement characteristics of nervous system tissue described above, the brachial plexus tension test, one of several neural tension tests available, was used as part of Sandra's physical examination (Fig. 7.33). It was chosen based on the posture assumed by Sandra in the car prior to the collision, as well as on the cervical spine movements after impact. How is it performed? The examiner begins the test with the patient's shoulder adducted. The right hand of the examiner is then placed on the superior aspect of the shoulder girdle and depresses it, maintaining that position as the shoulder is abducted and extended. As the operator performs these movements, the patient is asked if symptoms are reproduced (Table 7.6). If not, the operator continues the shoulder movement, as well as adds sensitizing movements, e.g., wrist and finger extension, cervical sidebending away or toward (not shown in the figure), etc., in an attempt to intentionally provoke symptoms. A variety of sensitizing movements can be used to test the mobility of nervous and related connective tissue structures. For example, adding a sensitizing movement to an end-range of movement, e.g., left cervical sidebending,

**Table 7.6** Possible signs and symptoms reproduced by the brachial plexus tension test

| Extraneural | Intraneural |
|---|---|
| Catch or twinges of pain<br>Short duration of symptoms<br>"Lines" of pain<br>Pain through the range of motion and feelings of resistance with tension tests<br>Symptoms provoked by a tension test, then eased by tension at the "other end," e.g., cervical sidebending or wrist flexion easing arm pain | Persistence symptoms<br>Increased duration of symptoms "blocks of pain"<br>Pain at end of the range of motion and feelings of resistance with tension tests (tension applied from "both ends") |

places external tensile forces on tension points; note that these tensile loads are well below the yield point of the tissues. If symptoms are provoked, the operator reverses the sensitizing movements and ascertains whether those positional changes decrease (or increase) the symptoms. Sensitizing or desensitizing movements help in confirming, or not, nervous system involvement. In interpreting this tension test, what is more important than positivity is the relevance of the symptoms provoked and the range of movement exhibited. A relevant tension test usually means that all or part of the patient's symptoms has been reproduced. It can also mean that symptoms produced are different from what is known to be normal, and in the case of limb testing, different from the contralateral limb. There may also be abnormal resistance to the movement when compared with the contralateral limb. Finally, a relevant tension test does not necessarily indicate that there is a mechanical disorder of the nervous system. The tension test could be placing a force on a surrounding symptomatic structure or that part of the nervous system may be irritated and symptom provocative, but the mechanics are normal.

Sandra had a positive brachial plexus tension test with signs and symptoms of both extraneural and intraneural involvement. Based on the MOI, painless weakness of the glenohumeral internal rotators, and positive brachial tension test, our preliminary hypothesis of a significant tension injury to the posterior cord of the brachial plexus was confirmed.

## 7.5 Concussion

Sandra was one of the approximately 6 of every 1000 Americans who are affected every year by a traumatic brain injury (TBI). Typically, a diagnosis of TBI severity is made based on the Glasgow Coma Scale, and the common clinical symptoms that include headache, blurred vision, confusion, dizziness, memory problems, general feelings of malaise, and sleep difficulties. Sandra was diagnosed with a mild TBI (mTBI), or concussion, based on her report of symptoms. Intuition would suggest a likely possibility of a concussion since she reported hitting the right side of her head against the door frame. In addition, Sandra's description of what happened when the car was stuck is consistent with what we would refer to as a whiplash injury. However, it has been demonstrated that the clinical presentation of a whiplash injury and concussion have considerable overlap, given that both diagnoses are based on presenting signs and symptoms and a history of trauma. Furthermore, the biomechanical distinction between whiplash injury and concussion can be equally unclear since car crashes are frequently associated with both a whiplash mechanism and head impact. Additionally, we know that head impacts in other circumstances, e.g., contact and collision sports, generate reaction forces in the neck, which can also create whiplash-like movements. Much research using cadaver specimens, test dummies, animal models, human volunteers (with only low-speed events), and musculoskeletal modeling have examined mechanisms of injury to the head and neck related to both motor vehicle accidents and football, a sport with a high prevalence of reported head injuries. It turns out that although the specifics of the MOIs described by past research related to car crashes and sports are different, the biomechanics of injury to the brain is nearly identical.

### 7.5.1 The Derivation and Application of Velocity and Acceleration Relative to Concussion

To this point, we have examined the biomechanics of the whiplash and the likely response of the neuromusculoskeletal system. From this point forward, we study only the biomechanics of head injury and resultant neural and vascular tissue damage. We draw on research from crash studies and sports-related (primarily football) head injuries. It is clear from research on head injuries associated with MVA and sport that brain damage is induced largely by acceleration-deceleration forces more so than impact alone. Microscopic examination of brain tissue from primates exposed to a non-impact experimental model revealed that shear strain is the likely cause of plastic deformation of axons. These early studies were instrumental in documenting cerebral injury from an apparently mild non-impact head injury. Thus, trauma to the brain results from rapid changes in the head's velocity or change in the vector speed over time, i.e., acceleration (also deceleration, which is a "negative" acceleration). Revisiting fundamental Newtonian formulas related to linear velocity and acceleration is useful as we consider how they have been used to study the relationship between velocity and acceleration, and traumatic brain injury.

We restrict our derivations and analysis to the principles of rectilinear (or straight-line) motion. First, for any analysis of the dynamics of motion, we need to know the initial conditions, i.e., the starting point of the body, and whether the body is moving or not. *Position* is used to indicate where something is, whereas *displacement* indicates the difference between the initial and final positions. A body's *average velocity* (speed) from time $t_1$ to time $t_2$ is determined by simply dividing the distance traveled $\Delta x$ by the elapsed time $\Delta t$:

$$v_{\text{ave}} = \frac{\Delta x}{\Delta t}$$

However, the *instantaneous velocity*, or the velocity at a particular instant of time $t_1$, may actually be of more interest. We can find this by taking the limit as $\Delta t$ goes to zero (see Appendix D):

$$v(t_1) = \lim_{\Delta t \to 0} \frac{x(t_2) - x(t_1)}{t_2 - t_1} = \lim_{\Delta t \to 0} \frac{x(t_1 + \Delta t) - x(t_1)}{(t_1 + \Delta t) - t_1}$$
$$= \lim_{\Delta t \to 0} \frac{\Delta x}{\Delta t} = \frac{dx}{dt} = \dot{x}$$

Then, to determine the acceleration, the time rate of change of velocity, we proceed in a similar fashion:

$$a(t_1) = \lim_{\Delta t \to 0} \frac{v(t_1 + \Delta t) - v(t_1)}{\Delta t} = \lim_{\Delta t \to 0} \frac{\Delta v}{\Delta t} = \frac{dv}{dt} = \frac{d^2 x}{dt^2} = \ddot{x}$$

In kinematics we often measure the acceleration of the body (or body segments) directly, but also want to determine its velocity and position over time. This is managed by reversing the process, progressing from acceleration ($\ddot{x}$) to velocity ($\ddot{x}$), and then to position ($x$). This is accomplished analytically or digitally using a computer with a process called integration. Therefore, to go from acceleration to velocity, we perform the following integration:

$$v(t_2) = v(t_1) + \int_{t_1}^{t_2} a \, dt$$

and to go from velocity to position we proceed in a similar fashion:

$$x(t_2) = x(t_1) + \int_{t_1}^{t_2} v \, dt$$

Let's look at three special cases involving linear acceleration. The motivation for looking at these one-dimensional examples is to imagine how they might be utilized to understand the more complex velocities and accelerations affecting the head during a motor vehicle accident, where the head and neck may sustain an impact and/or is part of a repetitive motion MOI.

In the first case, we assume that acceleration $a$ is a constant. In this instance, the velocity change is linear with time and can be expressed as:

$$v(t_2) = v(t_1) + a \int_{t_2}^{t_1} dt = v(t_1) + a(t_2 - t_1)$$

Now, let $t_1 = 0$, denote $v(t_1)$ as $v_0$, and replace $t_2$ by the general $t$, we get:

$$v(t) = v_0 + at$$

Integrating again yield's the body's position as a function of time:

$$x(t) = x_0 + v_0 t + \frac{1}{2} a t^2$$

where $x_0$ is the position at $t = 0$. What does this equation mean practically? The position of a body at some time $t$ depends on where the body was initially ($x_0$), on what its initial velocity was ($v_0 t$), and on what its acceleration is ($\frac{1}{2} a t^2$). Thus, if the body had no acceleration (or deceleration) and no initial velocity, then it would remain at some initial position $x_0$ (*Note: Sandra's situation just prior to her car being struck*). If there was an initial velocity but still no acceleration, velocity remains constant. If velocity was constant, the body's distance from the starting position would increase linearly with time ($x_0 + v_0 t$).

Now let's look a bit deeper into how position, velocity, and acceleration are related. Recall, acceleration is the time rate of change of the velocity:

$$a = \frac{dv}{dt}$$

and velocity is the time rate of change of position:

$$v = \frac{dx}{dt}$$

Now, we solve both equations for $dt$ which gives:

$$a = \frac{dv}{dt} \Rightarrow dt = \frac{dv}{a}$$

$$v = \frac{dx}{dt} \Rightarrow dt = \frac{dx}{v}$$

We then equate the two expressions for $dt$ to get:

$$\frac{dv}{a} = \frac{dx}{v}$$

Rearranging:

$$v \, dv = a \, dx$$

We see that both the right and left sides of the equation $v\,dv = a\,dx$ can be integrated. Assuming constant acceleration, $a$ can be pulled out of the integral on the right-hand side of the equation:

$$\int_{v_2}^{v_1} v\,dv = a\int_{x_2}^{x_1} dx$$

Completing the integration results in:

$$\frac{v_2^2}{2} - \frac{v_1^2}{2} = a\left(x_2 - x_1\right)$$

which can alternatively be expressed as:

$$v_2^2 = v_1^2 + 2a\left(x_2 - x_1\right)$$

Expressing the displacement $x_2 - x_1$ with the variable $s$ and rearranging the expression to solve for the acceleration $a$:

$$a = \frac{v_2^2 - v_1^2}{2s}$$

### 7.5.2 Newton Returns

The use of $g$ allows for the expression of results in terms of multiples of acceleration due to gravity. At sea level, one $g$ is equivalent to 9.81 m/s$^2$. Providing the results in terms of the number of $g$'s allows us to use an alternative expression for acceleration when we want to apply a sports model to facilitate insight into other trauma-related head injuries.

For example, in a football collision, e.g., a defensive player tackles a runner, the final velocity $v_2$ is assumed to be zero since the runner is brought to a halt. In that case, the magnitude of the acceleration would be:

$$a = \frac{v_1^2}{2s}$$

Then, using data from game film where velocities (directional speeds) and stopping distances can be calculated, consider the following scenario: a running back has a running speed of 3.7 m/s when his head is brought to a stop within a distance of 0.15 m (both of which are realistic). With this information and plugging into the above equation for $a$, the following deceleration can be determined:

$$a = \frac{\left(3.7\,\mathrm{m/s}\right)^2}{2\left(0.15\,\mathrm{m}\right)} = 45.6\,\mathrm{m/s^2} = 4.65\left(9.81\,\mathrm{m/s^2}\right) = 4.65\,g$$

Thus, we see that the decrease in the player's velocity over time, i.e., deceleration, is 4.65 $g$, or more than four times the normal acceleration due to gravity, i.e., 1 $g$.

Further, note that the force required to accelerate the player's mass ($m$), with an acceleration of magnitude $a$ (either acceleration or deceleration), can be determined by:

$$F = ma$$

We can also say that if $a$ is nothing more than the acceleration of gravity, for example, a person falling to the ground with no other forces involved, Newton's law can also be expressed as:

$$F = mg$$

Then, if a person experiences an acceleration (or deceleration) of 10 $g$, the force needed to accelerate any element of mass, e.g., the brain, is $F = m(10\,g)$. To claim that a force 10 times the force of what someone would experience by just falling would induce plastic changes in brain tissue depends on many factors. However, it has been suggested that accelerations below 70 $g$ to 78 $g$ related to football-related incidents and motor vehicle crashes do not result in irreparable brain injury.

Let's now substitute the value for acceleration magnitude determined earlier ($a = v_1^2 / 2s$) into Newton's equation ($F = ma$), so that:

$$F = \frac{mv_1^2}{2s}$$

This equation suggests that if several different collisions all with the same initial speed occurs, then the smaller the stopping distance ($s$), the larger the resulting force on the brain. In motor vehicle accidents, as we noted earlier, there are two conditions that are important to consider: the head impacting an external surface within the car (*in Sandra's case, the head hitting the door frame*), and the head and neck moving at a high-rate back-and-forth, which produces an impact of brain tissue with the inside of the skull. Previously, we referred to the external forces and moments imposed on Sandra's body as impulsive since large forces were applied over a very short time period. This hypothesis appears to be corroborated by the above equation, which suggests that large forces ensue under conditions where the velocities are high and stopping distances are small; note, however that the stopping distance of the body is not likely be the same as the stopping distance of the head. In applying the above equation, we need to be careful to how it is applied. For example, if we are interested in applying the equation to what is happening to the head, $m$ refers to the mass of the head and the force $F$ is the sum of all forces acting external to the head. In this case, the only external force to the head is the internal force of the neck acting on the head. If, on the other hand, we are interested in forces acting on the brain, then $m$ is the mass of the brain, and the forces are the internal forces between the skull and the brain.

Let's briefly summarize what we have discussed to this point. We derive equations that described rectilinear motion and examined the relationship between position, velocity, and acceleration. In that process we derive an equation showing the relationship between initial and final velocity and acceleration, assuming that acceleration was constant. Then, we, with some rearrangement of the equations, apply our findings to estimate, and compare, the magnitude of accelerations during a fall, or as a result of a collision in football. We then briefly extrapolate the results from an example collision and speculate how this process could be applied to provide insight in brain injury induced by a motor vehicle accident. We need to remember that the process we develop is based on a very simple one-dimensional problem, but that the characteristics of sport collisions and motor vehicle accidents are multi-factorial and multi-dimensional.

Before we go into more detail on whiplash and concussion related injuries, we need to briefly explore a bit further two other acceleration conditions that provide insight into the relationship between position, velocity, and acceleration. To this point in our discussion, we only examine this relationship when acceleration is constant; now let's examine the case when acceleration depends on position. This case occurs when, for instance, the body is attached to a spring. We need to digress just a bit for this example.

Recall, that in our study of deformable body mechanics, we claim that all biological tissues, e.g., bone, ligament, tendon, muscle, articular and fibrocartilage, nervous tissue, vessels, and the brain, were viscoelastic; as well, regions of the body that are dynamically linked, e.g., the head and cervical spine, also behave as viscous spring-like solids. With this reminder as a backdrop, we can imagine that the whiplash MOI in this case approximated the actions of a mass-spring-damper system, i.e., a mass attached to a spring and damper.

In this example, the forces acting on the body, and subsequently the body's acceleration, depend on the body's position because the spring force is position-dependent, i.e., as the spring stretches, the force it exerts changes. So, let's rewrite an expression we derive earlier, this way:

$$v\,dv = a(x)\,dx$$

where we see that acceleration is dependent on position, $x$. As before, we can integrate both sides directly:

$$\int_{v_2}^{v_1} v\,dv = \int_{x_2}^{x_1} a(x)\,dx$$

Since the integral of $v\,dv$ is $v^2/2$, we have:

$$\frac{1}{2}\left(v_2^2 - v_1^2\right) = \int_{x_2}^{x_1} a(x)\,dx$$

$$v_2^2 = v_1^2 + 2\int_{x_2}^{x_1} a(x)\,dx$$

This solution suggests that because of the acceleration (deceleration), the velocity of the body (or body segments) is going to change in some way (in our case, in a complex way). We encourage you to use your imagination as you further consider specific implications of this equation.

Finally, what happens when acceleration depends on velocity? One application of this derivation is realized when the body experiences a drag force that depends on velocity. Consider that perhaps the drag force could be the shearing that takes place within brain tissue, which resulted from the linear and angular velocity, and acceleration conditions imposed on the head and neck as a result of the whiplash. Again, we start with the acceleration/velocity relationships:

$$a(v) = \frac{dv}{dt} \Rightarrow dt = \frac{dv}{a(v)}$$

$$v = \frac{dx}{dt} \Rightarrow dt = \frac{dx}{v}$$

which implies:

$$\frac{dx}{v} = \frac{dv}{a(v)} \Rightarrow dx = \frac{v\,dv}{a(v)}$$

and then integrating both sides which results in:

$$x_2 = x_1 + \int_{v_2}^{v_1} \frac{v}{a(v)}\,dv$$

Although very abstract, this solution demonstrates that acceleration has a relatively straightforward relationship with velocity, and the body's new position can be determined by completing the integration. Practically, in the context of Sandra's accident, this result suggests that there likely was a very complex coupling of velocity and acceleration while her head and neck moved back and forth in the frontal and transverse planes, which would not only have determined the final position of these segments but would also have influenced what occurred inside the skull. And remember that in this case, tissue stresses and segmental movements were multi-dimensional, not linear. We are now ready to look a little closer at the mechanics of mTBI.

### 7.5.3 Mechanics of Brain Insult

Mild traumatic brain injuries are caused by impact conditions that induce the acceleration (and deceleration) of brain tissue resulting in focal and/or diffuse damage. Typically, focal injuries are related to pure translational movements and linear velocities. Peak linear accelerations recorded from concussed football players have been reported to range from 60 to 144 *g*, which are much higher than the 10 *g* we present earlier in one of our derivations. Focal lesions consist of contusions, lacerations, hematomas (extra- or intradural), and tentorial/tonsillar herniation (i.e., protrusion of brain tissue into the tentorial notch secondary to increased intracranial pressure). Focal damage may also occur as coup or contrecoup (opposite to the site of impact) (Fig. 7.34). Sandra's concussion was likely related to two different, but related, mechanisms. During the initial impact, her lower body was rapidly accelerated to the right, which caused her head, neck, and upper torso to lag behind and sidebend/rotate to the left with respect to the car. Following the initial impact, the friction between the car's tires and the ground slowed the car to a stop. This would have caused forces acting to the left on Sandra's lower body, resulting in a moment that in turn caused her head, neck, and upper torso to sidebend/rotate back to the right. During this movement, there would have been mild, likely non-injurious, internal forces between the skull and the left side of her brain. Her upper body continued its rightward rotation until the right side of her head struck the passenger side door, resulting in the first focal coup on the right side of Sandra's skull and brain. The impact force acting on the right side of her head would have accelerated her head back to the left. The motion of the brain would have lagged behind the skull and continued moving to the right as the skull was being accelerated to the left. This would have resulted in stresses to the opposite (or left) side of her brain potentially causing some damage to the cerebral tissue there. Thus, Sandra not only likely sustained a mild contusion of the right side of her skull and brain, but a second contrecoup injury on the left side of her brain. It has been shown that coup and contrecoup contusions create cavitation bubbles within the brain due to high negative pressures resulting in local tissue damage. When the impact impulse (i.e., hitting her head on the door frame) was directed to the side of her head, a reactive translational acceleration would have been induced. However, it is likely that Sandra's head was impacted obliquely, which would have induced a more complex loading pattern, resulting in rotational accelerations as well. In addition, the whiplash itself, i.e., repeated right and left sidebending/rotation of the head, cranio-cervical and lower cervical spine, could have created large internal impulsive shear stresses.

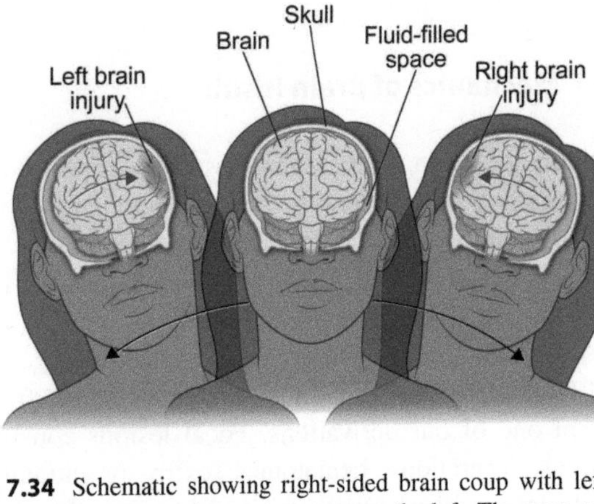

Impulsive loading of the head and neck can result in large angular accelerations and rotations of the head and neck, which appear to be the principal mechanism of diffuse brain injury. Rotational accelerations approximating 5000 rad/s$^2$, recorded in American football-related activities, appear to be related to concussion-related signs and symptoms. Similar magnitudes of accelerations have been identified with side-impact motor vehicle accidents, i.e., lateral loading, and TBI. Ivancic's review suggests that the primary reason for this may be related to the lower mass moment of inertia of the head in the frontal plane, which produces lower mechanical impedance to the rapid angular velocities and accelerations. Since the brain is not rigidly attached to the inside of the skull, rotational accelerations of the brain produce widespread mechanical strain on nerve fibers (Fig. 7.35), other brain tissue and cerebral blood vessels, leading to local and diffuse inflammation, and related swelling that induces increased intracranial pressure.

Diffuse axonal injury (DAI) (Fig. 7.36), now widely recognized in the medical and biomechanical literature on concussion, is found within the cerebral hemispheres, corpus callosum, thalamus, and basal ganglia making intraneural connectivity and communication dysfunctional in most



**Fig. 7.34** Schematic showing right-sided brain coup with left sidebending with a concomitant contrecoup on the left. The coup and contrecoup are repeated with right sidebending

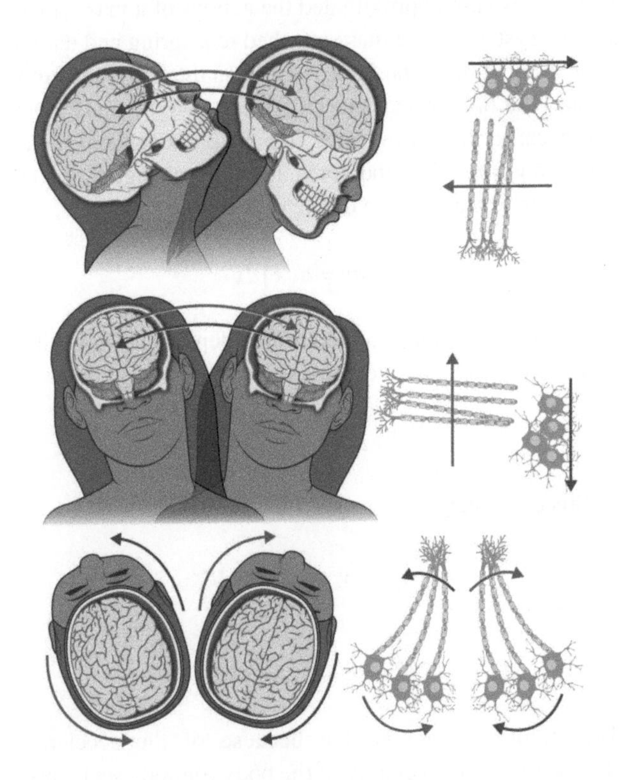

**Fig. 7.35** Schematic illustrating multi-dimensional rotational movements of the brain within the cranium during a whiplash mechanism. Note the directional shear and tensile strains on the neuronal axons with the potential to cause plastic deformation of those important neurological extensions

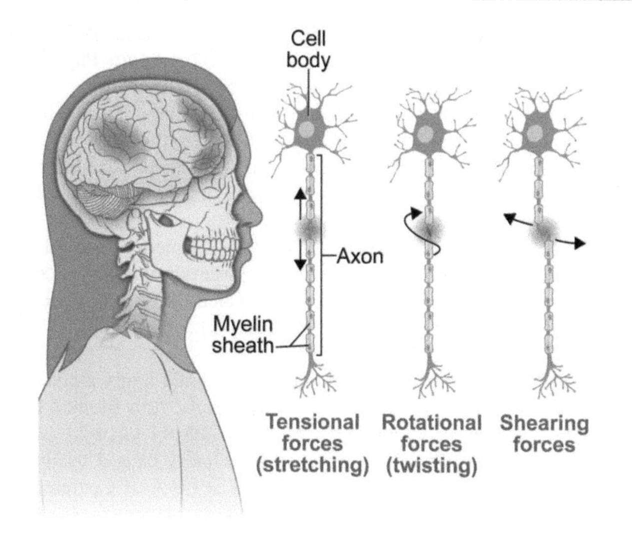


**Fig. 7.36** Schematic illustrating potential diffuse nature of axonal injury as a result of whiplash and concussion mechanisms of injury. Tension, torsional, and shearing forces secondary to brain torsions

cases, and non-functional in severe cases. Likewise, capillary damage results in deoxygenation of local and diffuse regions of the brain leading to additional neuronal compromise and death. Functional outcomes of DAI include impaired ability to learn, memory loss, motor deficits, and long-term post-traumatic amnesia.

Although Sandra initially (in the emergency room) presented with a diagnosis of mild concussion, it was fortunate that the signs and symptoms of her head injury appeared to subside over the course of 1 to 2 weeks following the car crash. Based on Sandra's report to the physical therapist at the time of her first visit, we concluded that the strain injury to her brain was no more than first-degree. On the other hand, damage to nervous tissue takes a longer time to heal than a strain injury to muscle fibers. Thus, Sandra may have subclinical minor impairment of brain and/or neuromuscular function, e.g., dizziness, loss of balance, etc., of which she is not aware.

## 7.6 Chapter Summary

Sandra was a collegiate athlete who sustained a mild concussion and whiplash-related injury to her cervical spine as a result of a side-impact (i.e., T-bone) car collision. When she reported for a physical therapy evaluation about 1-month post-injury, Sandra's concussion symptoms had apparently dissipated, but she was bothered by right-sided neck pain, mild loss of cervical range of motion, and impairments in performing basketball skills due to right arm symptoms. Developing hypotheses to explain Sandra's continued problems involve a thorough study of the anatomy and biomechanics of the cranio- and typical-cervical spine, including intervertebral disc and dissection of the likely mechanisms of injury related to the physics of the whiplash. We also take the opportunity to revisit Newton's laws as they might have pertained to the trauma biomechanics to the head and neck regions. Finally, this case forces resourcing many of the topics discussed in previous chapters, e.g., osteo- and arthrokinematics of the cervical spine, mechanics of muscle under inertial eccentric loading, etc.

We are not able to make many definitive clinical biomechanical conclusions in this case, which is typical of whiplash-related incidents. In fact, the literature suggests that in the majority of cases like this there is a paucity of physical and imaging evidence of tissue injury. Of course, this poses a real problem to the clinical scientist. What we try to illustrate in this case, however, is the importance of using a profound understanding of the functional anatomy and biomechanics of the regions of interest in relation to clinical findings, as well as the ability to use the laws of physics to reconstruct the details of trauma-related biomechanics to develop a list of possible problems. Once a list of hypotheses is created, we look for possible relationships among the systems involved and re-order the hypotheses from least- to most-likely. Throughout the processing and presentation of the details of this case, we take several "time outs" to list the most likely problems related to Sandra's symptoms. For example, we hypothesize that Sandra's cervical symptoms were related to minor strain injuries to many possible structures including facet joint capsules, intertransverse ligaments, outer annular fibers of the IVD, and muscle (or possible muscle spindle impairment). However, we cannot be definitive in our conclusion if one, two, or several of these structures were involved. We also determine that Sandra had a nerve palsy that resulted in impaired right shoulder internal rotation strength. We do not think that the palsy was due to a spinal cord or nerve root injury but was more likely related to a moderate strain injury to the posterior cord of the brachial plexus. The next two cases involve impairments in the thoracic and lumbar regions of the spine with their own unique history of trauma biomechanics. Therefore, we encourage the reader to imagine what may be generalized or abstracted from this case and applied to future cases.

## Bibliography

Barth JT, Freeman JR, Broshek DK et al (2001) Acceleration-deceleration sport-related concussion: the gravity of it all. J Athl Train 36(3):253–256

Bian K, Mao H (2020) Mechanisms and variances of rotation-induced brain injury: a parametric investigation between head kinematics and brain strain. Biomech Model Mechanobiol 19:2323–2341. https://doi.org/10.1007/s10237-020-01341-4

Broglio SP, Schnebel B, Sosnoff JJ et al (2010) Biomechanical properties of concussions in high school football. Med Sci

Sports Exerc 42(11):2064–2071. https://doi.org/10.1249/MSS.0b013e3181dd9156
Butler DS (1989) Adverse mechanical tension in the nervous system: a model for assessment and treatment. Aust J Physiother 35(4):227–238
Chen H-B, Yang KH, Zheng-guo W (2009) Biomechanics of whiplash injury. Chin J Traumatol 12(5):305–314
Elkin BS, Elliot JM, Siegmund GP (2016) Whiplash injury or concussion: a possible biomechanical explanation for concussion symptoms in some individuals following a rear-end collision. J Orthop Sports Phys Ther 46(10):874–885. https://doi.org/10.2519/jospt.2016.7049
Elmaazi A, Morse CI, Lewis S et al (2019) The acute response of the nucleus pulposis of the cervical intervertebral disc to three supine postures in an asymptomatic population. Musculoskelet Sci Pract 44:102038. https://doi.org/10.1016/j.msksp.2019.07.002
El Sayed T, Mota A, Fraternali F et al (2008) Biomechanics of traumatic brain injury. Comput Methods Appl Mech Eng 197:4692–4701. https://doi.org/10.1016/j.cma.2008.06.006
Elvey RL (1986) Treatment of arm pain associated with abnormal brachial plexus tension. Aust J Physiother 32(4):225–230
Fields AJ, Liebenberg EC, Lotz JC (2014) Innervation of pathologies in the lumbar vertebral end plate and intervertebral disc. Spine J 14:513–521. https://doi.org/10.1016/j.spine.2013.06.075
Fazey PG, Song S, Mønsås Å et al (2006) An MRI investigation of intervertebral disc deformation in response to torsion. Clin Biomech 21:538–542. https://doi.org/10.1016/j.clinbiomech.2005.12.008
Fazey PJ, Song S, Price RI et al (2013) Nucleus pulposis deformation in response to rotation at L1-2 and L4-5. Clin Biomech 28:586–589. https://doi.org/10.1016/j.clinbiomech.2013.03.009
Fazey PJ, Takasaki H, Singer KP (2010) Nucleus pulposis deformation in response to lumbar spine lateral flexion: an *In vivo* MRI investigation. Eur Spine J 19:1115–1120. https://doi.org/10.1007/s00586-010-1339-4
Fennell AJ, Jones AP, Hukins DWL (1996) Migration of the nucleus pulposis within the intervertebral disc during flexion and extension of the spine. Spine 21(23):2753–2757
Gennarelli TA (1981) Acceleration induced head injury in the monkey. I. The model, its mechanical and physiological correlates. Acta Neuropathol Suppl 7:23–25
Gil C, Pecq P (2021) How similar are whiplash and mild traumatic brain injury? A systematic review. Neurochirurgie 67:238–243. https://doi.org/10.1016/j.neuchi.2021.01.016
Grauer JN, Panjabi MM, Cholewicki J et al (1997) Whiplash produces an s-shaped curvature of the neck with hyperextension at lower levels. Spine 22(21):2489–2494
Groh AMR, Fournier DE, Battié MC et al (2021) Innervation of the human intervertebral disc: a scoping review. Pain Med 22(6):1281–1304. https://doi.org/10.1093/pm/pnab070
Guo R, Zhou C, Wang C et al (2021) In vivo primary and coupled segment motions of the healthy female head-neck complex during dynamic head axial rotation. J Biomech 123:110513. https://doi.org/10.1016/j.j.biomech.2021.110513
Guskiewicz KM, Mihalik JP (2011) Biomechanics of sport concussion: quest for the elusive injury threshold. Exerc Sport Sci Rev 39(1):4–11
Hartvigsen J, Boyle E, Cassidy D et al (2014) Mild traumatic brain injury after motor vehicle collisions: what are the symptoms and who treats them? A population-based 1-year inception cohort study. Arch Phys Med Rehabil 95(3 Suppl 2):S286–S294. https://doi.org/10.1016/j.apmr.2013.07.029
Ishii T, Mukai Y, Hosono N et al (2004) Kinematics of the upper cervical spine in rotation, *In vivo* three-dimensional analysis. Spine 29:E139–E144
Ivancic PC, Ito S, Tominaga Y et al (2008) Whiplash causes increased laxity of cervical capsular ligament. Clin Biomech 23:159–165
Ivancic PC (2016) Mechanisms and mitigation of head and spinal injuries due to motor vehicle accidents. J Orthop Sports Phys Ther 46(10):826–833. https://doi.org/10.2519/jospt.2016.6716
Jang SH, Kwon YH (2018) A review of traumatic axonal injury following whiplash injury as demonstrated by diffusion tensor tractography. Front Neurol 9:57. https://doi.org/10.3389/fneur.2018.00057
Jull GA (2000) Deep cervical flexor muscle function in whiplash. J Musculoskelet Pain 8(1/2):143–154
Kang Y-B, Jung D-Y, Tanaka M et al (2005) Numerical analysis of three-dimensional cervical behaviors in posterior-oblique car collisions using 3-D human whole body finite element model. JSME Int J Ser C Mech Syst Mach Elem Manuf 48(4):598–606. https://doi.org/10.1299/jsmec.48.598
Kimpara H, Iwamoto M (2012) Mild traumatic brain injury predictors based on angular accelerations during impacts. Ann Biomed Eng 40(1):114–126. https://doi.org/10.1007/s10439-011-0414-2
King AI, Yang KH, Zhang L (2003) Is head injury caused by linear or angular acceleration?. Paper presented at IRCOBI Conference, Lisbon, Portugal, September, 2003
Kleiven S (2013) Why most traumatic brain injuries are not caused by linear acceleration but skull fractures. Front Bioeng Biotechnol 1:15. https://doi.org/10.3389/fbioe.2013.00015
Kumar S, Ferrari R, Narayan Y (2004a) Cervical muscle response to whiplash-type anterolateral impacts. Eur Spine J 13:398–407. https://doi.org/10.1007/s00586-004-0700-x
Kumar S, Ferrari R, Narayan Y (2004b) Electromyographic and kinematic exploration of whiplash-type neck perturbations in left lateral collisions. Spine 29:650–659
Lorente AI, Hidalgo-Garcia C, Fanlo-Mazas P et al (2022) In vitro upper cervical spine kinematics: rotation with combined movements and its variation after alar ligament transaction. J Biomech 130:110872. https://doi.org/10.1016/j.jbiomech.2021.110872
Lloyd JD (2017) Biomechanical evaluation of motorcycle helmets: protection against head and brain injuries. J Forensic Biomed 8:137. https://doi.org/10.4172/2090-2697.1000137
McMorran JG, Gregory DE (2021) The influence of axial compression on the cellular and mechanical function of spinal tissues; emphasis on the nucleus pulposis and annulus fibrosus: a review. J Biomech Eng 143:050802–050801. https://doi.org/10.1115/1.4049749
Mustafy T, El-Rich M, Mesfar W et al (2014) Investigation of impact loading rate effects on the ligamentous cervical spinal load-partitioning using finite element model of functional spinal unit C2-C3. J Biomech 47:2891–2903. https://doi.org/10.1016/j.j.biomech.2014.07.016
Nazari J, Pope MH, Graveling RA (2012) Reality about migration of the nucleus pulposis within the intervertebral disc with changing postures. Clin Biomech 27:213–217. https://doi.org/10.1016/j.clinbiomech.2011.09.011
Neumann DA (2017) Chapter 9 - Axial skeleton: Osteology and arthrology, Chapter 10 - Axial skeleton: muscle and joint interactions. In: Neumann DA (ed) Kinesiology of the musculoskeletal system, foundations for rehabilitation, 3rd edn. Elsevier, St. Louis
Nordon M, Frankel VH (2022) Basic biomechanics of the musculoskeletal system, 5th edn. Wolters Kluwer, Philadelphia
Panjabi MM, Cholewicki J, Nibu K et al (1998) Mechanism of whiplash injury. Clin Biomech 13:239–249
Panjabi MM, Ivancic PC, Maak TG et al (2006) Multiplanar cervical spine injury due to head-turned rear impact. Spine 31(4):420–429
Pooni JS, Hukins DWL, Harris PF et al (1986) Comparison of the structure of human intervertebral discs in the cervical, thoracic and lumbar regions of the spine. Surg Radiol Anat 8:175–182
Siegmund GP, Davis MD, Quinn KP et al (2008) Head-turned postures increase the risk of cervical facet capsule injury during whiplash. Spine 33:1643–1649
Skrzypiec DM, Pollintine P, Przybyla A et al (2007) The internal mechanical properties of cervical intervertebral discs as revealed

by stress profilometry. Eur Spine J 16:1701–1709. https://doi.org/10.1007/s00586-007-0458-z

Silva A, Manso A, Andrade R et al (2014) Quantitative in vivo longitudinal nerve excursion and strain response to joint movement: a systematic literature review. Clin Biomech 29:839–847. https://doi.org/10.1016/j.clinbiomech.2014.07.006

Slobounov SM, Sebastianelli WJ (eds) (2014) Concussions in athletics, from brain to behavior. Springer, New York

Soutas-Little RW, Inman DJ, Balint DS (2009) Engineering mechanics, computational addition, SI edn. CENAGE Learning, Toronto

Stemper BD, Pintar FA (2014) Biomechanics of concussion. Prog Neurol Surg 28:14–37. https://doi.org/10.1159/000358748

Takasaki H (2015) Comparable effect of simulated sidebending and side gliding positions on the direction and magnitude of lumbar disc hydration shift: *In vivo* MRI mechanistic study. J Man Manip Ther 23(2):101–108. https://doi.org/10.1179/2042618613Y.0000000059

Tonge BH (2010) Dynamics, analysis and design of systems in motion, 2nd edn. Wiley, Hoboken

Tubbs RS, Mortazavi MM, Loukas M et al (2011) Cruveilhier plexus: an anatomical study and a potential cause of failed treatments for occipital neuralgia and muscular and facet denervation procedures. J Neurosurg 115:929–933. https://doi.org/10.3171/2011.5.jns102058

Winkelstein BA, Nightingale RW, Richardson WJ et al (2000) The cervical face capsule and its role in whiplash injury. Spine 25(10):1238–1246

Zhou C, Li G, Wang C et al (2021) *In vivo* intervertebral kinematics and disc deformation of the human cervical spine during walking. Med Eng Phys 87:63–72. https://doi.org/10.1016/j.medengphy.2020.11.010

Zhou C, Wang H, Wang C et al (2020) Intervertebral range of motion characteristics of normal cervical spine segments (C0-T1) during *In vivo* neck motions. J Biomech 98:109418. https://doi.org/10.1016/j.biomech.2019.109418

Zhou S-W, Guo L-X, Zhang S-Q et al (2010) Study on cervical spine injuries in vehicle side impact. Open Mech Eng J 4:29–38

# Thoracic Spine and Rib Cage Pain with a Comorbidity of Chronic Obstructive Pulmonary Disease

8

## 8.1 Introduction

This case involves the thoracic spine/cage region. This region is uniquely complex for a couple of reasons: proximity of the ribs and autonomic ganglia (Fig. 8.1), and the dynamic coupling of as many as 11 joints for a typical thoracic vertebral unit. Close examination of the potential mechanism of injury and related self-imposed forces and moments is critical to the development of hypotheses regarding the biological responses and to the understanding of this individual's kinematic impairments. A thorough review of normal thoracic cage anatomy and thoracic spine, rib, and respiratory mechanics is necessary to describe the pathomechanics related to hypotheses regarding pain generators. Exploration of the pathophysiology and pathomechanics of respiration related to chronic obstructive pulmonary disease (COPD) is conducted in parallel with the problem-solving process for identifying the source(s) of this individual's symptoms. Although outside of our biomechanical interests, this case forces us to broaden our perspective about the very common role of comorbidities in individuals, particularly those in the middle ages, who present with musculoskeletal impairments.

## 8.2 Case Presentation

Scott McDonald, a 55-year-old (height, 177.8 cm; mass, 90 kg; body mass index (BMI), 28.5 kg/m$^2$), right-handed male with a 40 pack-year (one pack/day times 40 years) smoking history reported to our clinic with complaints of right mid/lower thoracic pain of 3 days duration. Scott claimed that 3 days ago he was splitting wood and woke up the following morning with stiffness and pain in his back. His pain worsened when he turned to the right and took a deep breath, coughed, and sneezed. Scott was diagnosed with COPD, secondary to cigarette smoking, approximately 10 years ago. Otherwise, within the constraints of COPD, Scott was working full-time (bank accountant) and was healthy and relatively active. Scott reported no prior history of neck or back pain.

Inspection revealed a mildly overweight (based on BMI) man with a moderate barrel chest (Fig. 8.2). As we walked to the examination room alongside Scott, we observed two things about his breathing pattern: (1) his rate of breathing appeared to be rapid and shallow (i.e., dyspnea), and (2) he appeared to exhale by sort of "blowing" air out. When we asked him about the shortness of breath he stated, "I have that all the time. I think it is related to my lung disease."

Posture examination revealed a moderately decreased lumbar lordosis, with mildly increased cervical lordosis, thoracic kyphosis, and a moderate forward head (Fig. 8.3A). In the frontal posterior view, with the patient standing, we noted a mild c-curve in the mid-thoracic (T5 to T8) spine (convex right) (Fig. 8.3B). When Scott was asked if he was aware of the scoliotic curve in the mid-back region, he denied any knowledge of it.

At the end range of the standing forward flexion test, palpation of the transverse processes of T5 to T8 revealed a posterior prominence on the left side. Trunk extension was not tested in standing. With the testing of trunk hyperextension when Scott was placed in the prone position, palpation of the transverse processes revealed no asymmetry between the right and left sides, i.e., the spine was straight.

Trunk sidebending (tested in standing) and rotation (tested while Scott was seated) were limited bilaterally but were more limited toward the right compared to motion testing toward the left. With over-pressure at the end of the range of motion of right rotation, Scott's symptoms were somewhat reproduced and a muscle guarding end-feel was notable.

He was tender to palpation over the spinous processes from T7 to T10 and over the rib angles of the same thoracic

**Supplementary Information** The online version contains supplementary material available at https://doi.org/10.1007/978-3-031-25322-5_8.


G. J. Alderink, B. M. Ashby, *Clinical Kinesiology and Biomechanics*, https://doi.org/10.1007/978-3-031-25322-5_8

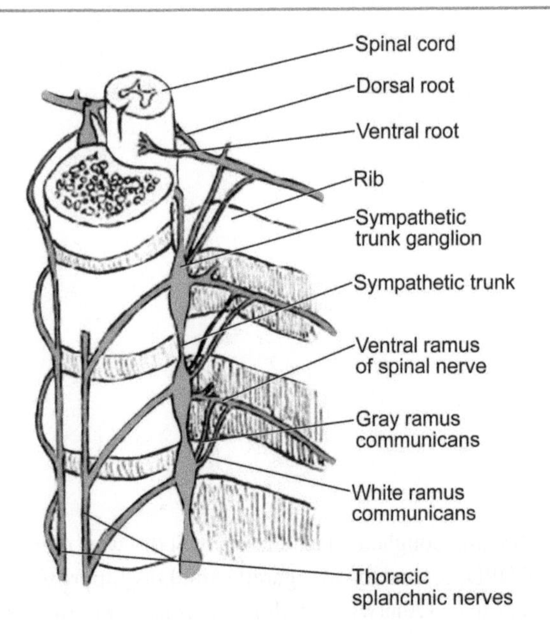


**Fig. 8.1** Relationship between sympathetic trunk, sympathetic trunk ganglion, and ribs

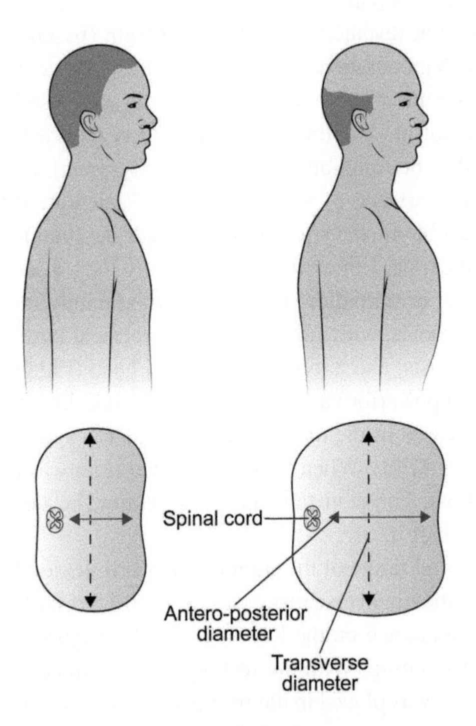


**Fig. 8.2** A comparison of normal and barrel chest, which is characterized by increased medio-lateral and antero-posterior diameter of the thoracic cage associated with chronic obstructive pulmonary disease

segments and costochondral junction of the corresponding ribs on the right side. On testing of gross respiratory function, we found mildly decreased rib cage movement with inhalation and exhalation, bilaterally, but more so on the right side.

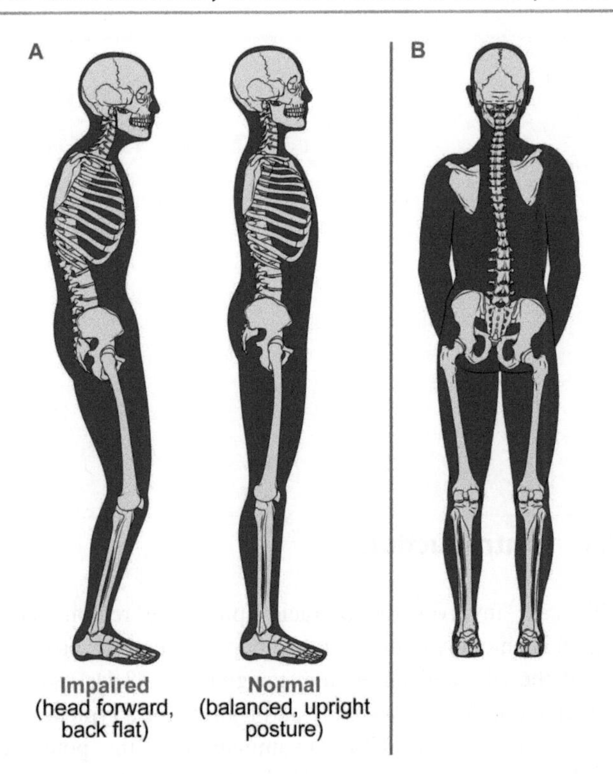


**Fig. 8.3** (**A**) Sagittal view of impaired vs. normal spine curvature and (**B**) right C-curve in the mid/lower thoracic region

Pulses and blood pressure were normal. Auscultation examination of his lungs found that breath sounds were decreased throughout all lung fields.

## 8.3 Key Facts, Learning Issues, and Hypotheses

We saw Scott just 3 days following the activity, i.e., splitting wood, which likely was the root cause of his symptoms. Although altered spine and rib mechanics due to his abnormal posture and barrel chest might have predisposed Scott to a movement-related injury (i.e., this is a hypothesis), it is not likely that COPD was the cause of his complaints. The most common manifestation of Scott's COPD, besides his respiration pattern, was his barrel chest, which is part of the natural history of COPD. We need to examine the etiology of the barrel chest and explore the possible relationship between the barrel chest and changes in his rib cage, posture, and thoracic spine mobility.

Since Scott was right-handed, we asked him to demonstrate how he handled the ax to split wood. In other words, we performed a qualitative motion analysis of the act of splitting wood (potential mechanism of injury or MOI) in order to develop hypotheses regarding the forces and moments related to that activity. As we see in other cases, the

**Fig. 8.4** Prone lying hyperextension, i.e., prone press-up, is used to assess the ability of the thoracic segments to extend or bend backward

details of the MOI, as best as we can reconstruct them, are critical as we advance in the problem-solving process. We discuss the hypothetical MOI a bit later.

Scott demonstrated abnormal static postures and movement patterns. His spine was side-bent (Fig. 8.2) in standing, i.e., scoliosis, and when he was fully flexed, but was straight when he hyperextended his back from the spine-neutral prone-lying position (Fig. 8.4). How can it be that he had the appearance of scoliosis in one posture and not in another? We attempt to explain this apparent contradictory finding later in our discourse. Furthermore, Scott demonstrated mild loss of right sidebending and axial rotation, compared to those same movements to the left. Both the static postures and movement impairments suggest osteo- and arthrokinematics problems in the thoracic spine/cage. Furthermore, the reduction in the range of motion of rib movements found during the examination, and reproduction of symptoms with deep breathing, coughing, and sneezing suggested alterations in the kinematics of the ribs. We need to thoroughly review the normal kinematics of thoracic vertebrae and ribs during cardinal plane movements, as well as with inhalation and exhalation. (Note: in Chap. 3 we present the kinematics of the thoracic spine and ribs, comparing the osteopathic and Nordic models of spinal coupling. In this case, we assume an osteopathic model of vertebral motion.) And we have to consider how COPD may have altered normal thoracic spine and rib kinematics prior to his recent injury.

Finally, we need to study normal respiratory mechanics, which has to include the anatomy and physiology of the lungs, as well as the deformable body mechanics of the lung tissues. After we examine the disease mechanisms related to cigarette smoking, we can explain the pathomechanics of COPD and speculate about the relationship between his premorbid condition, the MOI, and Scott's present condition.

## 8.4 Mechanics of Respiration

### 8.4.1 Thoracic Cage Structure

A comprehensive review of the anatomy of the thoracic spine and rib cage is presented in Chap. 3, but let's take another

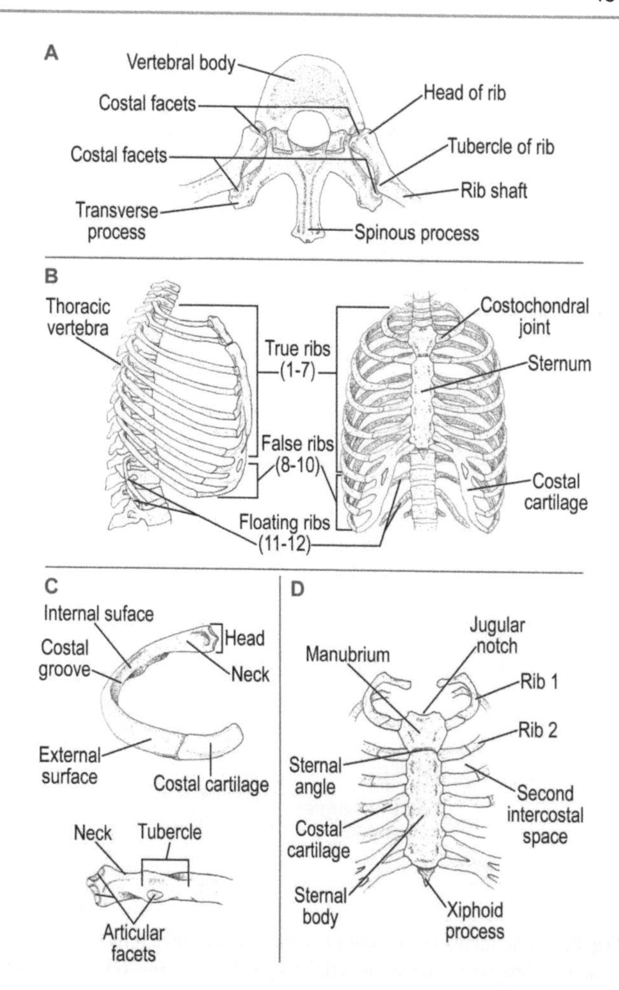


**Fig. 8.5** (**A**) Typical thoracic vertebral segment with attached paired ribs, (**B**) sagittal and frontal perspectives of the thoracic cage, (**C**) morphology of a typical rib with costal cartilage, and (**D**) merging of bony rib with costal cartilage and manubrial and sternal attachments

brief look (Fig. 8.5). Anteriorly we see the large weight-bearing vertebra body, which is joined to the posterior elements, i.e., transverse and spinous processed, by the pedicles. In the typical thoracic functional unit (superior vertebrae, intervening disc, and inferior vertebrae), the convex head of the typical rib articulates with shallow concave demi-facets, and the intervertebral disc, to form the costocorporeal joint. Laterally, a slightly convex tubercle of the rib mates with the concave facet on the transverse process of the corresponding thoracic vertebra to form the costotransverse joint (Fig. 8.5A, C). The synovial costovertebral joints (costocorporeal and costotransverse) anchor the ribs posteriorly yet allow for coupled movements that accommodate multi-dimensional spine and rib cage movements. The costovertebral joints are stabilized by robust intrinsic (capsular) and extrinsic ligaments (Fig. 8.6).

Anteriorly, the ends of ribs are covered by hyaline cartilage which articulates with the sternum's slightly concave facets. For ribs 1 to 7, there are direct attachments to the sternum, but for ribs 8 to 10, the attachments to the sternum

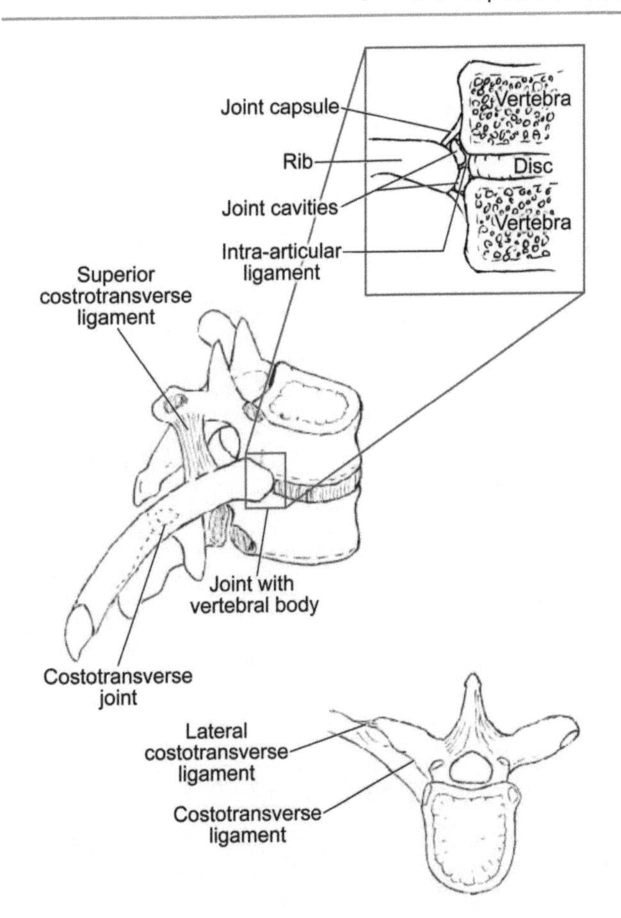


Fig. 8.6 Ligaments (including perhaps fascia connections to the outer annulus fibrosis) of the costovertebral joints that constrain excessive movements

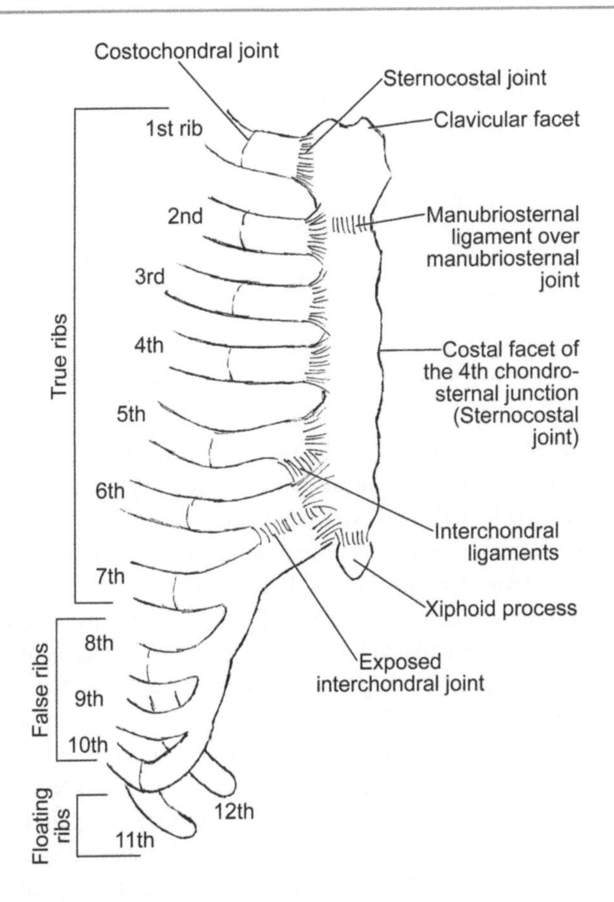


Fig. 8.7 Interchondral joints and associated ligaments

are indirect via a merging of the cartilage of an inferior rib into the one above it (Fig. 8.5B, D). These two "articulations" (i.e., anterior end of bony rib and hyaline cartilage and chondral cartilage to the sternum) are generally referred to as sternocostal joints, but, from a functional perspective, more accurately referred to as costochondral and chondrosternal junctions. The first chondrosternal junction is a synarthrosis, which affords little mobility, whereas the chondrosternal junctions for ribs 2 to 7 have diarthrodial characteristics that permit more movement. Each of these joints is stabilized by strong radiate ligaments. For ribs 5 to 10 the inter-rib cartilages contain small synovial-lined interchondral joints that are reinforced by ligaments (Fig. 8.7). Ribs 11 and 12 are not attached to the chondral cartilage and are referred to as floating ribs.

Vertebrae in each region of the spine share similar characteristics (as noted above) but can be contrasted by the size of their bodies, height of their intervertebral discs, and orientation of the apophyseal (or facet) joints (Fig. 8.8). The height (or thickness) of the IVD in the thoracic region is generally decreased compared to the IVDs in the cervical and lumbar spine, and the nucleus pulposis is relatively small. Because of this, the interbody rotations are generally reduced

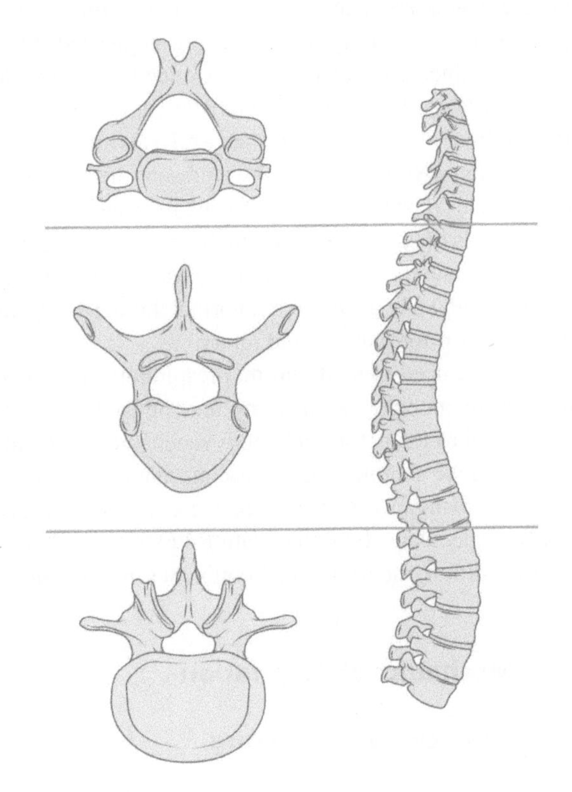

Fig. 8.8 Comparison of the shapes and sizes of cervical (top), thoracic (middle), and lumbar (bottom) vertebra

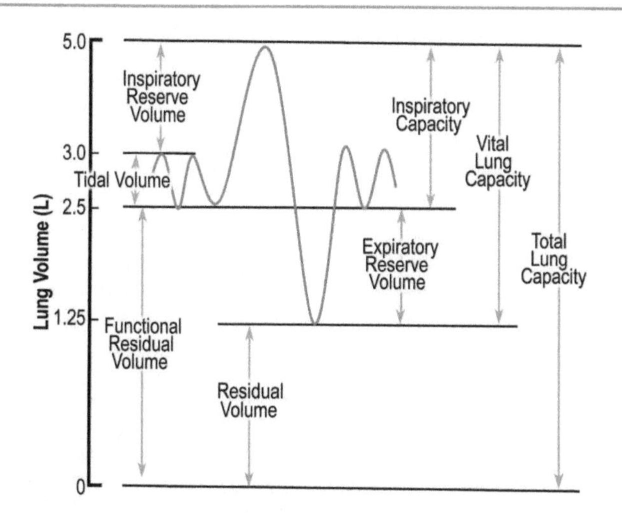


**Fig. 8.9** Lung volumes and capacities for a normal adult

between thoracic segments, in contrast to cervical and lumbar segments. The annulus fibrosis is thicker and stiffer, particularly posteriorly, compared to the IVDs of the other regions, suggesting that the IVD is paramount to thoracic vertebral stability. The superior and inferior articular facets of the thoracic segments are oriented more vertically (approximately 60° relative to the horizontal), with a slight forward pitch, which limits flexion more than extension, and rotation more than sidebending. The movements of thoracic vertebral segments (see Chap. 3 for details of vertebral and rib osteo- and arthrokinematics) are relatively free in three dimensions but constrained by the coupled movements of the costovertebral joints posteriorly and sternocostal articulations anteriorly. Thus, the thoracic cage is the most mechanically stable region of the spine (with the exception of the sacroiliac joints), which allows it to ably distribute/mitigate axial compression and shear stresses, thus providing protection to the spinal cord and internal viscera.

## 8.4.2 Function: Ventilation Mechanics, Kinematics, and Muscle Function

### 8.4.2.1 Ventilation

Ventilation is the mechanical process of breathing in (inhalation or inspiration) and out (exhalation or expiration). Of course, the purpose of ventilation, i.e., respiration, is to promote the exchange of oxygen and carbon dioxide in the parenchyma of the lungs so that our body organs and cells receive the oxygen needed to maintain normal metabolism. Normal quiet inhalation (tidal volume) is an active process driven primarily by the action of the diaphragm, whereas exhalation is primarily a passive process (Fig. 8.9). Total lung capacity (vital capacity + residual volume) approximates five liters of air, whereas the vital capacity is the maximum amount of air that can be exhaled after a maximum inhalation

**Table 8.1** Respiratory volumes and capacities

| Volumes and capacities | Definition |
|---|---|
| Tidal volume (TV) | Air displaced between normal inspiration and normal expiration |
| Inspiratory reserve volume (IRV) | Normal inspiration is prolonged by a forced inspiration |
| Inspiratory capacity (IC) | IRV + TV |
| Expiratory reserve volume (ERV) | Normal expiration is prolonged to the maximum by a forced expiration |
| Vital capacity (VC) | IRV + TV + ERV |
| Residual volume (RV) | The air that cannot be expelled and is still present in the lungs and bronchi after a complete expiration |
| Functional reserve capacity (FRC) | RV + expiratory volume |
| Total lung capacity (TLC) | VC + RV |

Note: With an increased respiratory rate during exercise, the TV increases to some maximum, and the ERV markedly increases which induces a reduction in the IRV

(Table 8.1). Residual volume is what air is left in the lungs following full exhalation. Forced inhalation and exhalation take place under more stressful conditions, e.g., following trauma, exercise, etc., which results in larger volumes of air exchange and requires muscle action during both phases.

Ventilation creates changes in the intrathoracic volume and pressures with characteristics that are dictated by Boyle's law: given a fixed temperature and mass, the volume and pressure of a gas, e.g., oxygen and carbon dioxide, are inversely proportional. In other words, during inhalation the internal lung volume increases, which results in a decrease in the intrathoracic pressure and a drawing in of air; that is, lung expansion reduces alveolar pressure below atmospheric pressure creating a pressure gradient that allows air to move into the lung parenchyma for gas exchange to take place. Conversely, relaxation of inspiratory muscles, during quiet breathing, combined with the elastic recoil of the thorax wall structures and lung parenchyma, i.e., alveoli and related connective tissues, reduces intrathoracic volume and creates a pressure gradient that permits air to flow out of the lungs. Forced exhalation requires the action of the expiratory muscles, primarily the abdominal muscles.

A pressure-volume curve is used to provide information about how the lung deforms during breathing. It describes the mechanical behavior of the chest wall (i.e., thoracic cage) and lungs. Thus, the elasticity of the chest wall and the lung and its ability to expand and stretch, i.e., compliance (or distensibility), is quantified by the slope of the pressure-volume curve (Fig. 8.10). The chest wall is a complex system of interacting parts, e.g., muscle, ligaments, fascia, etc., that have viscoelastic characteristics. Thus, during inhalation, these structures are loaded, i.e., by pressure changes, in their

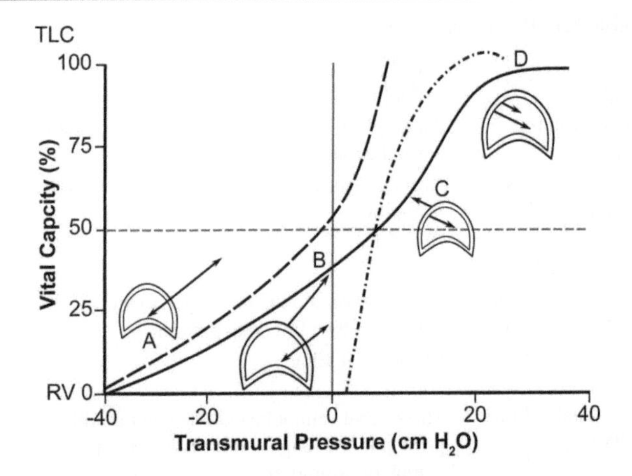


**Fig. 8.10** Transmural pressure-volume curve illustrating the compliance of the chest wall and lung throughout the vital capacity. At any given lung volume, the elastic recoil of the lungs (as evidenced by positive transmural pressures) is always toward lung collapse with both positive and negative transmural pressures. Elastic recoil of the chest may vary, however, based on lung volume. (**A**) At lower lung volumes, the chest recoils toward expansion/rib elevation, (**B**) at intermediate lung volumes the elastic recoil of lung tissue is counter-balanced by chest wall recoil/expansion, (**C**) elastic recoil of lung tissue is not counter-balanced by chest wall recoil/expansion, and (**D**) at high lung volumes, elastic recoil of both the chest and lung recoils inward/rib depression. Note: TLC = total lung capacity, RV = residual volume; FRC = functional reserve capacity; long, dashed line = chest wall; solid line = chest wall + lung; short, dashed line = lung

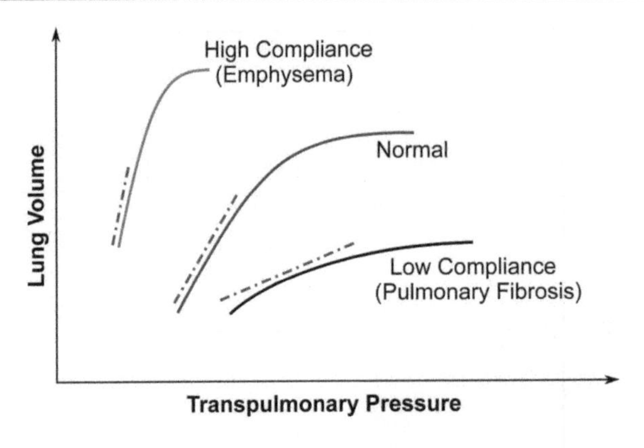


**Fig. 8.11** Graphical representation comparing normal lung compliance to greater compliance, i.e., less stiff, typically associated with chronic obstructive pulmonary disease, and lesser compliance, i.e., stiffer, associated with pulmonary fibrosis

elastic region to a greater or lesser extent, depending on the depth and rate of breathing. Concomitantly, the system internal to the lung, e.g., trachea, bronchus, bronchioles, alveoli, etc., likewise responds to the pressure viscoelastically in a characteristic manner unique to their structure. This response is referred to as lung compliance ($C_L$) with units of mL/cm $H_2O$ to reflect the change in lung volume ($\Delta V$) as a result of a change in the pressure ($\Delta P$) within the lung:

$$C_L = \frac{\Delta V}{\Delta P}.$$

In essence, lung compliance represents the slope of the lung volume vs. pressure curve. Compliance is the relationship between the volume of air and the wall pressure needed to displace it, and, in general, the compliance of the lungs is greater than that of the chest wall. In normal exhalation the chest wall and lungs regain their position of equilibrium, which is like a spring that returns to its resting state; that is, the intra-alveolar pressure and atmospheric pressure return to a state of semi-equilibrium. During forced inspiration, which is comparable to stretching a spring, a negative pressure relative to the atmospheric pressure develops within the thorax. As a result, air enters the trachea, but the elasticity of the thorax again tends to bring it back to its original position. It is notable that the functional residual capacity corresponds to the point where the pressure exerted by the elasticity of the chest wall and that exerted by the elasticity of the lungs toward further exhalation are equal and opposite.

Unlike a typical stress-strain diagram, greater compliance is defined by a thorax/lung that is more easily distended, i.e., less stiff, and is reflected with a steeper pressure-volume curve (or steeper slope), whereas lesser compliance implies a stiffer, or less easily distended, thorax/lung system (Fig. 8.11). Notice that with emphysema (advanced stage of COPD), the system compliance is significantly increased, i.e., steeper slope (dotted line in the figure), compared to the normal lung. In the interpretation of the lung compliance in Fig. 8.11, we find that at low lung volumes the lung distends easily, but at high lung volumes, large changes in pressure only produce small changes in lung volume. This is because at high lung volumes the alveoli and airways have been maximally distended. Typically, to account for the variations in volume, specific compliance (compliance divided by the lung volume) is normalized to the functional residual capacity (FRC). We draw on these concepts when we examine the pathomechanics of the thoracic cage associated with the barrel chest.

#### 8.4.2.2 Kinematics

Normally during the inhalation phase of respiration, the osteo- and arthrokinematics at the apophyseal, intervertebral, costovertebral, sternocostal, and interchondral joints are very similar to what happens when the thoracic spine is actively extended. Osteokinematically, the vertebrae rotate and translate posteriorly, while the arthrokinematics at the costocorporeal and costotransverse joints during posterior rotation of the ribs is dictated by the convex-concave rule. However, rotations of the ribs during inhalation do not strictly occur in the sagittal plane, are regionally dependent, and are dictated by other mechanical and soft tissue factors (Fig. 8.12). Kapandji has noted that an approximate joint line that is drawn to pierce the centers of the costocorporeal and

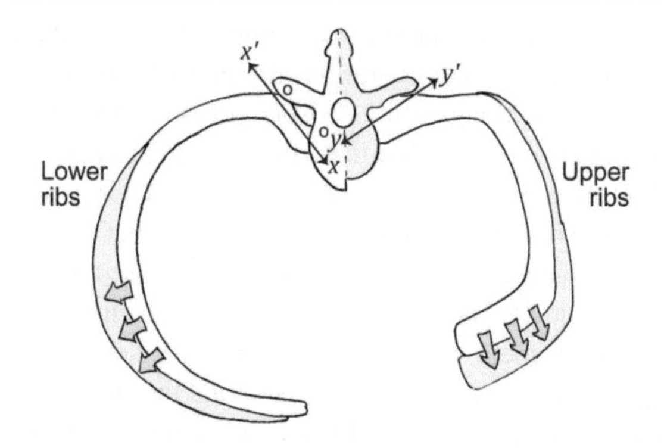


**Fig. 8.12** Schematic of lower ribs with the *xx'* axis piercing the joint centers of the costocorporeal and costotransverse joints, and upper ribs with the *yy'* axis that lies in more of the frontal plane. The thicker arrows depict the direction of rib movement and represent an increase in the transverse (left) and antero-posterior (right) diameter of the rib cage

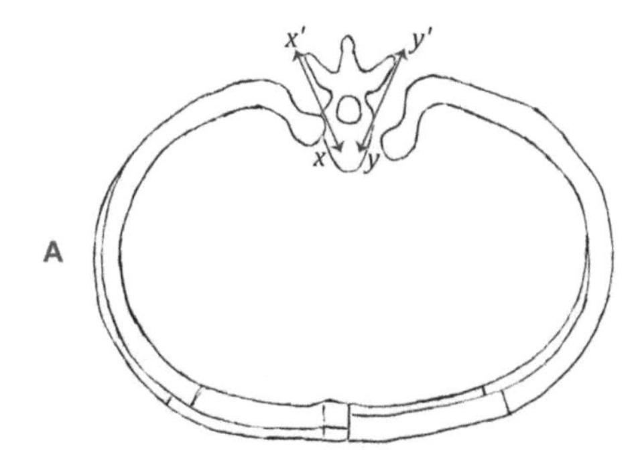


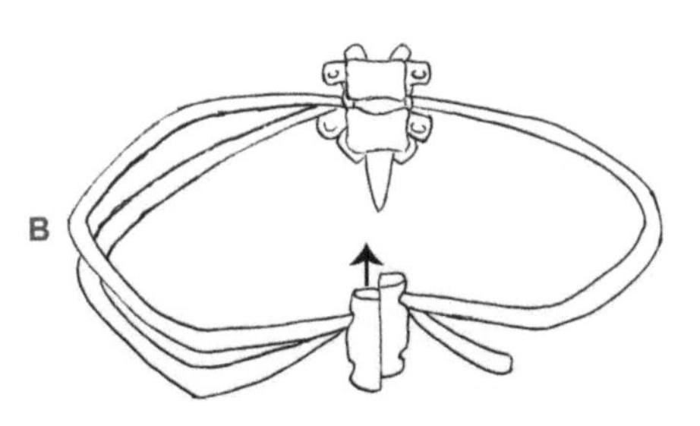


**Fig. 8.13** (**A**) Superior and (**B**) anterior views of a schematic representation of movements of the ribs and costal cartilage during inhalation. It is apparent that there is lateral movement of the ribs away from the axis of symmetry of the body, while the sternum rises (black arrow), making the costal cartilages become more horizontal. The angular movement of the costal cartilage relative to the sternum occurs at the sternocostal joints; at the same time, there is another rotation around the axis of the cartilage at the costochondral joints. Note the obliquity of the axes *xx'* and *yy'* arranged by the relationships between the costovertebral and costotransverse joints

costotransverse articulations creates an oblique axis. That axis lies in more of a frontal plane for a subset of the upper ribs, and in more of a sagittal plane for a subset of the lower ribs. Thus, during inhalation, the lower ribs that rotate about their sagittal plane axes are elevated, increasing the medio-lateral (i.e., transverse) diameter of the thorax, which is often referred to as a *bucket-handle* movement. Concomitantly, the elevating upper ribs increase the antero-posterior diameter as they rotate about their frontal plane oblique axis, often referred to as a *pump-handle* movement. The costovertebral joint axes for the middle region of the ribs run obliquely at 45°; thus inhalation simultaneously increases both the medio-lateral and antero-posterior diameters of the thoracic cage.

If inhalation is associated with thoracic extension, intuition suggests that exhalation must be associated with thoracic flexion. That is, with diaphragmatic relaxation and the elastic recoil of the connective tissues external and internal to the lungs, the vertebral bodies rotate and translate anteriorly, as the ribs depress, i.e., rotate anteriorly, and return to their resting position; that is, the ribs are depressed, as the transverse and antero-posterior diameters of the thorax decrease.

Remember that the rib cage is a closed system with both posterior and anterior regions of mobility and constraint. Let's now examine the relative movements anteriorly, i.e., the ribs, costal cartilages, and sternum. As noted, during inhalation there are small lateral and anterior movements of the ribs away from the symmetry of the body. At the same time, the sternum moves cranially, and the costal cartilage becomes more horizontal relative to the sternocostal joints (Fig. 8.13). Since there are orientation differences in the costovertebral joint axes between the upper and lower ribs, there are also slight differences in the orientation and magnitude of the changes in the costal cartilages and movements at the sternocostal joints, e.g., the costal cartilage of the upper ribs tends to assume a more horizontal position (inferior view in Fig. 8.13). But something else happens to the costal cartilages as well. Because of the rib-fibrocartilage interface that precludes a rotational movement, and the relative immobility at the sternocostal articulation, the costal cartilages become twisted, i.e., undergo torsions, during inhalation. Recall that fibrocartilage has an extra-cellular matrix supported by a collagenous framework that deforms under compressive and tensile loads. Thus, it is not surprising that during inhalation, and given the constraints posteriorly and anteriorly, the ribs are induced to both rotate about their costovertebral axes and simultaneously undergo a long-axis rotation. Because of the viscoelastic nature of the ribs, this long axis rotation induces a slight twist within the bone, and simultaneously induces a torsion in the costal cartilages as well as a change in the angle at the costochondral junction (Fig. 8.14). These induced torsional elastic strains, i.e., deformations, within the ribs and costal cartilages allow these structures to store potential energy. After relaxation of the inspiratory muscles, this stored energy is released allowing the costal cartilage to assist in returning the ribs to their resting position. Of course,

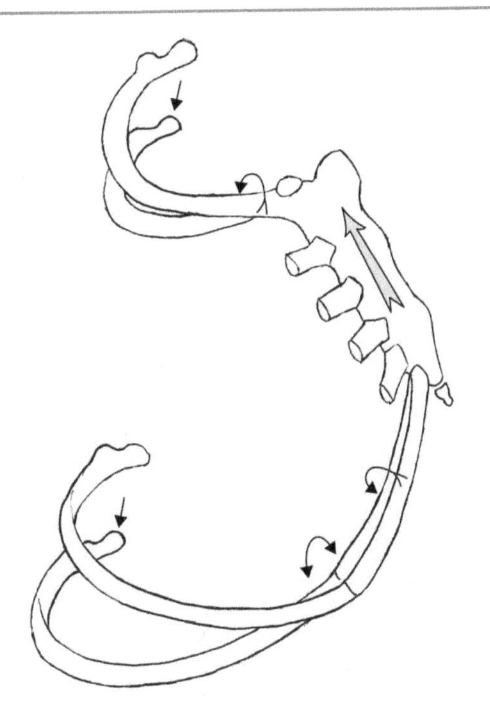

**Fig. 8.14** During inhalation, the sternum rises with concomitant rib elevation and posterior rotation, as well as torsion of the costal cartilage about the rib's long axis

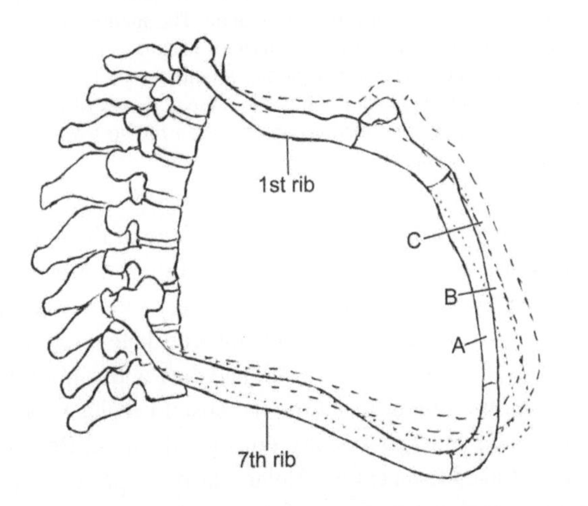


**Fig. 8.15** Schematic demonstrating deformation of the thorax, relative to a stable spinal column, in the sagittal plane during inhalation. This deformation occurs secondary to elevation of the upper and lower ribs, as well as upward movement of the sternum, made possible by rib torsion and sternocostal rotations

these movements are minimal during breathing at rest and become more prominent when larger volumes of air are needed for activities such as exercise.

A final word about changes in the geometry of the thorax that takes place primarily in the sagittal plane during inspiration: we assume that the thoracic cage is moderately constrained posteriorly by the thoracic vertebrae and costovertebral joints, and anteriorly by the sternocostal joints and sternum. Let's consider the following structures as key participants in ventilation mechanics that contribute to changing the cage configuration: the first rib, sternum, and the seventh rib along with its costal cartilages. During inhalation the freely moveable first rib, at its joint of the costal head, is elevated, increasing the anterior diameter of the thorax and raising the sternum (Fig. 8.15); note that the anteroposterior diameter of the upper thorax is increased more than that of the lower thorax. Because there is potentially more anterior movement in the upper ribs than the lower ribs, the sternum does not stay parallel to itself and the angle it forms with the vertical decreases slightly (movement from A to B to C in Fig. 8.15). Simultaneously, as the seventh rib elevates, the angle it forms relative to the sternum increases, which is augmented by the intra-costal cartilage torsion. We can conclude that normal ventilatory mechanics is quite complex requiring the dynamic coupling of joints and joint configurations spanning multiple degrees of freedom. Let's next look at the motors behind these movements.

#### 8.4.2.3 Muscle Actions

In order to maintain partial pressure diffusion gradients of oxygen and carbon dioxide for gas exchange to occur, it is necessary to move air in and out of the lungs. This is achieved by the respiratory pump: respiratory centers, nerves that transmit signals from those centers, the respiratory muscles that are the pressure-generating structures, and the rib cage and abdomen. These comprise the elements of the complex system that function in a highly coordinated fashion by which ventilation goes unnoticed with minimal energy cost. Central control is located in the upper medulla, which integrates input from the periphery and other parts of the nervous system. The output of central control is modulated by mechanical (i.e., musculoskeletal system), cortical, and sensory input, and by the state of oxygenation ($PaO_2$), $CO_2$ concentration ($PaCO_2$), and acid-base (pH) status of circulating arterial blood. The output from the central nervous system (CNS) is distributed by conducting nerves to the respiratory muscles, which move and deform the rib cage and abdomen, and generate intrathoracic pressures, which displaces volume, allowing air to move in and out. The relationship between "drive" and inspiratory pressure or volume is referred to as "coupling," a process that is complex, involves many interacting elements, and occurs with minimal effort, i.e., it is extremely efficient from a metabolic perspective.

There are many muscles involved in the mechanics of ventilation, and they are innervated by a wide array of motor neurons that range from the spinal accessory nerve (cranial nerve XI), providing innervation to the sternocleidomastoid, to the lumbar nerve roots L2/L3, which innervate abdominal muscles. The respiratory cycle is regulated by a complex series of centrally organized neurons that can be voluntarily overridden by the cortex. Some of the respiratory muscles are described as primary and others as secondary. Because many of the secondary muscles may also be involved with the control of posture, movement, and stability of the trunk and craniocervical regions, and indirectly with the upper and lower extremities, impairments of these muscles can have wide-ranging effects. We make this claim not because we examine all, or even a few, of the direct and indirect actions of these muscles, but to raise your awareness of the complexity of the respiratory system (and its relationship to the musculoskeletal system). In general, muscles that can increase the intra-thoracic volume contribute to inhalation, and those that can decrease the intra-thoracic volume contribute to exhalation.

Of the three primary muscles of inspiration (diaphragm, scaleni, and intercostales), the diaphragm is most important because it contributes 60–80% of the work of ventilation during quiet breathing (Fig. 8.16). Although strenuous exercise may be the stimulus for the recruitment of several secondary inspiratory muscles, the diaphragm, if used properly, remains the predominant driver of ventilatory work.

The diaphragm is a thin, dome-shaped muscle that separates the thoracic cavity from the abdominal cavity and is made up of three parts: (1) the costal part attached to the upper margins of the lower six ribs and costal cartilages and the tips of the eleventh and twelfth ribs, (2) a relatively small and variable sternal part that arises from the posterior surface of the xiphoid process, and (3) the thick crural part attached to the bodies of the upper three lumbar vertebrae via the left and right crus. Some of the vertically aligned crural fibers blend with the anterior longitudinal ligament, with other fibers intermixing with proximal fascial attachments of the psoas major. These three sets of attachments converge to form the central tendon at the upper dome of the diaphragm.

The diaphragm's primary role in inhalation stems from the fact that when it is activated it increases the intra-thoracic volume in three directions: vertical, medio-lateral, and antero-posterior, thus, drastically reducing the intra-thoracic pressure. When the diaphragm is activated, the lowering of the central tendon, which increases the vertical diameter of the thorax, is quickly checked by tension in the mediastinal contents and mass of the abdominal viscera. At this juncture, the diaphragmatic muscle fibers continue to shorten, but with the central tendon now a fixed point, elevating the lower ribs, which increases the medio-lateral diameter of the lower thorax (Fig. 8.17). Simultaneously, with help of the sternum, i.e., it moves anteriorly (Fig. 8.15), the diaphragm assists in elevating the upper ribs, thus increasing the antero-posterior diameter of the thorax.

There is no other single inspiratory muscle that can accomplish what the diaphragm does. Under normal conditions, its length-tension and the unique coupling of its moment arms, along with the predominance of slow oxidative muscle fibers, the diaphragm is one of the most efficient muscles in the human body. But there is more. In addition to its primary role in inhalation, the diaphragm appears to aid other core muscles, e.g., abdominals, erector spinae, etc., in stabilizing the lumbar spine. Its ability to stabilize the lumbar region is accomplished through its extensive deep fascial connections, and by increasing intra-abdominal pressure when its actions are coupled with those of the abdominal and pelvic floor muscles.

Other primary and secondary (or accessory) muscles of inspiration are recruited when the magnitude and rate of

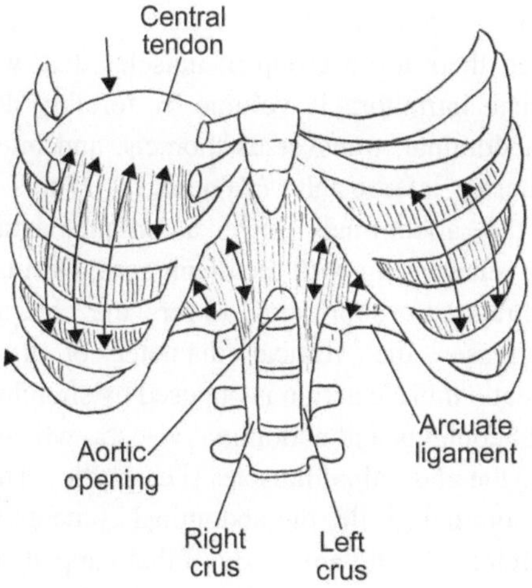

**Fig. 8.16** The action of the diaphragm muscle during inhalation. After initial shortening (downward arrows) which lowers and flattens the dome of the diaphragm, obstruction by the abdominal viscera provides resistance that inhibits further lowering. At this point, the stabilized dome and continued shortening of muscle fibers elevate the lower ribs (upward arrows)

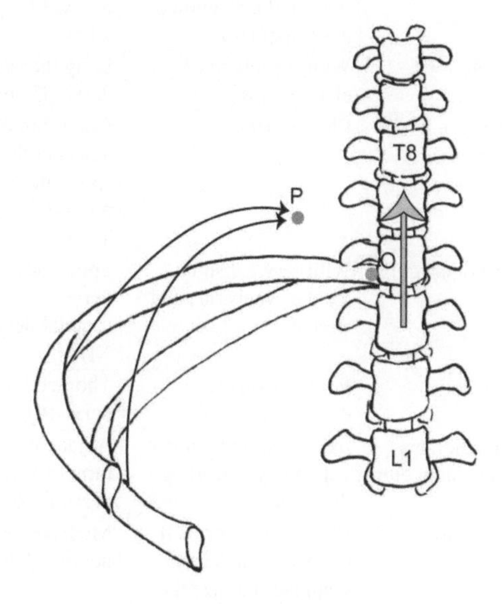


**Fig. 8.17** With the central tendon (green arrow) fixed (at point P), shortening of diaphragmatic muscle fibers elevates the lower ribs (black arrows) inducing rotation of the ribs about point O

**Table 8.2** Primary and secondary muscles of inhalation

| Muscle | Actions | Innervation |
|---|---|---|
| Diaphragm | *Primary*: the dome of the diaphragm lowers increasing the vertical diameter of the thorax<br>*Secondary*: central tendon is fixed allowing continued action of muscle fibers to elevate lower ribs, which increases the transverse diameter<br>*Tertiary*: contribution to anterior movement of the sternum assists in elevation of upper ribs, which increases antero-posterior diameter | Phrenic nerve (C3 to C5) |
| Scaleni | With the cervical spine, fixed action of all three heads elevate the ribs and sternum | Ventral rami of spinal nerve roots (C3 to C7) |
| Intercostales | The parasternal fibers of the interni and the intercostales externi elevate the ribs; the intercostales stabilize intercostal spaces during inhalation to prevent an inward collapse of the thoracic wall | Intercostal nerves (T2 to T12) |
| Serratus posterior superior | Elevates upper ribs | Intercostal nerves (T2 to T5) |
| Serratus posterior inferior | Stabilizes lower ribs for initial activation of the diaphragm | Intercostal nerves (T9 to T12) |
| Serratus anterior | With scapula fixed elevates ribs | Long thoracic nerve (C5 to C7) |
| Levator costarum (longus and brevis) | Elevates ribs | Dorsi ram of adjacent thoracic spinal nerve roots (C7 to T11) |
| Sternocleidomastoid | With cervical spine fixed elevates ribs and sternum | Spinal accessory nerve (cranial nerve XI) |
| Latissimus dorsi | With arms fixed elevates lower ribs | Thoracodorsal nerve (C6 to C8) |
| Iliocostalis thoracis and cervicis of erector spinae | Increases intrathoracic volume by extending the spine | Adjacent dorsal rami of spinal nerve roots |
| Pectoralis minor | Elevates upper ribs if trapezius and levator scapulae stabilize the scapula | Med. pectoral nerve (C8 to T1) |
| Pectoralis major | Elevates middle ribs and sternum if upper arm is fixed at 90° flexion or abduction | Med. pectoral nerve (C8 to T1) |
| Quadratus lumborum | Stabilizes lower ribs for activation of the diaphragm during early force inhalation | Ventral rami of spinal nerve roots (T12 to L3) |

Note: The secondary muscles are primarily recruited when force inhalation is required. *Med.* medial

breathing are such that a greater effort is warranted, e.g., exercise (Table 8.2). In other instances, these muscles are activated to compensate for injury, fatigue, or diseases. For example, in individuals with COPD, several accessory muscles have to be recruited, e.g., scaleni, etc., as the disease progresses and the efficiency of the diaphragm decreases. Evidence of overuse of the accessory muscles of inspiration secondary to COPD comes from observation of excessive scaleni and sternocleidomastoid tendon tension, for example, during resting inhalation; their hypertrophy from overuse is also obvious.

As noted previously, normal quiet exhalation is a passive process. Activation of the primary muscles of inhalation increases the intra-thoracic volume which, in turn, loads (stresses) bone, cartilage, ligaments, fascia, and lung parenchyma under tension. As these tissues deform (strain) under load, they store strain energy. With the relaxation of the inspiratory muscles, the loaded tissues release the strain energy, as they return to their original form, and in the process, expel used gases. So, although inhalation is a process that requires energy, i.e., ATP, normal quiet exhalation is an energy-restoring process.

However, there are a group of muscles that, when activated, reduce intrathoracic volume by forcibly depressing the ribs: abdominal, transversus thoracis, and intercostales interni (internal intercostals) (Table 8.3). Of this group of muscles, the abdominals have a unique antagonistic-synergistic relationship with the diaphragm. During inspiration, diaphragmatic shortening lowers the central tendon which increases the vertical diameter of the thorax. Diaphragmatic muscle action is opposed by stretching of the mediastinal contents and abdominal viscera, which are held in place by the abdominal muscles (i.e., girdle). Thus, without the abdominal girdle, the abdominal contents would be displaced inferiorly and anteriorly. If that happened the central tendon would no longer be able to provide the anchor, or fixed point, for maintenance of optimal muscle moment arms, length-tension, and continued shortening of the diaphragmatic muscle fibers to elevate the lower ribs. Conversely, during exhalation, relaxation of the diaphragm

**Table 8.3** Muscles of forced exhalation

| Muscle | Actions | Innervation |
|---|---|---|
| Abdominal muscles: Rectus abdominis Obliquus externus abdominis Obliquus internus abdominis Transversus abdominis | Rib depression with trunk flexion Compress abdominal wall and contents increase intra-abdominal pressure → relaxed diaphragm is pushed upward which decreases the intrathoracic volume | Intercostal nerves (T7 to L1) |
| Transversus thoracic | Depresses and pulls ribs inward, i.e., constricts thorax | Adjacent intercostal nerves |
| Intercostales interni (interosseous fibers) | Depresses ribs | Intercostal nerves (T2 to T12) |

Note: See the Supplementary 8.1 of this chapter for images of muscles of inhalation and exhalation

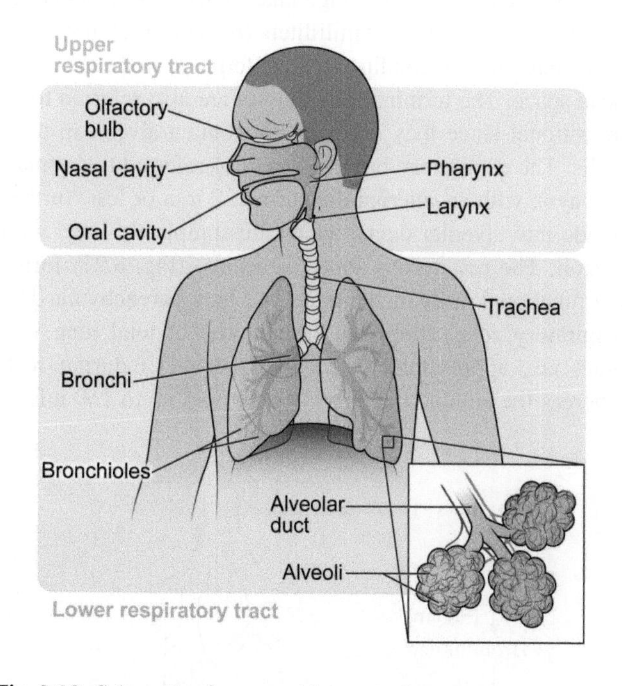


**Fig. 8.18** Schematic of upper and lower respiratory tracts

allows the resting tonus of the abdominal muscles to depress the ribs around the thoracic outlet, thereby decreasing the transverse and antero-posterior diameters of the thorax. The resulting mild increase in intra-abdominal pressure concomitantly pushes the viscera upwards raising the central tendon (restoring the proper length-tension to the diaphragm muscle fibers) and decreasing the vertical diameter of the thorax.

As muscles that can force exhalation to a lesser or greater extent, the abdominals are important for other natural respiratory actions like coughing, sneezing, or performing a Valsalva maneuver (used in lifting, defecation, and childbirth). In individuals with advanced COPD, the abdominal force is used to a lesser degree in what is termed "pursed-lipped" breathing. Pursed-lipped breathing is a controlled forced exhalation by blowing air out with the lips pursed, i.e., lips held in a somewhat closed or pressed position. This method of breathing seems to help slow the rate of breathing, and internally, may control the collapse of the small airways through positive airway pressure-mediating splinting of the airways.

Scott exhibited evidence of dyspnea and shallow breathing as well as purse-lipped exhalation. These observations suggest that Scott's COPD had advanced significantly. More on that later.

## 8.5 Chronic Obstructive Pulmonary Disease

### 8.5.1 Normal Respiratory Anatomy and Physiology

The primary function of the respiratory system is gas exchange. Oxygen from the external environment is transferred into the bloodstream while carbon dioxide is expelled into the outside air. Oxygen first enters the nose and/or mouth during inhalation, allowing the air to pass through the larynx and the trachea. The trachea splits into two bronchi, which then bifurcates into two smaller branches forming bronchial tubes. These tubes form a multitude of pathways within the lung, terminating with connections to tiny sacs called alveoli. The exchange of gases takes place at the alveoli, where oxygen ($O_2$) diffuses into the lung capillaries in exchange for carbon dioxide ($CO_2$). Exhalation begins after the gas exchange, and the air containing $CO_2$ begins the return journey through the bronchial pathways and back out into the external environment through the nose or mouth. Inhalation primarily through the nose is important because the secondary functions of the nasal structures include filtering, warming, and humidifying the inhaled air.

The respiratory system can be separated into regions based on function or anatomy (Fig. 8.18). The conducting zone (nose to bronchioles) consists of organs that form a path to conduct the inhaled air into the deep lung regions. The respiratory zone (alveolar duct to alveoli) consists of the alveoli and the small airways that open into them where the gas exchange takes place. Anatomically, the respiratory system can be divided into the upper and lower respiratory tract. The upper respiratory tract includes the organs located outside the thorax area (i.e., nose, pharynx, and larynx), whereas the lower respiratory tract includes the organs located almost entirely within it (i.e., trachea, bronchi, bronchiole, alveolar duct, and alveoli).

The tracheobronchial tree is the structure of the trachea, bronchi, and bronchioles that form the upper part of the lung airways. The trachea splits into the right and left main bronchi, which further branches out into the progressively smaller airways. The bronchial generation is normally referred to by number, indicating the number of divisions from the trachea, which is assigned as generation 0 or 1. The trachea is a hollow tube connecting the cricoid cartilage in the larynx to the primary bronchi of the lungs. Anteriorly it is made up of c-shaped cartilaginous rings and posteriorly by a flat band of muscle and connective tissue called the posterior tracheal membrane, which closes the c-shaped rings. There are 16 to 20 tracheal rings that prevent it from collapsing. More distally along subsequent bronchi the cartilage support diminishes.

The trachea divides into the main bronchi at the carina, with the right bronchus wider, shorter, and more vertical than the left bronchus (Fig. 8.19). The right bronchus bifurcates posteriorly and inferiorly into the right upper and lobe bronchus, and intermediate (middle) bronchus. The left bronchus passes infero-laterally at a greater angle from the vertical axis than the right bronchus, and divides into two, left upper lobe bronchus and left lower lobe bronchus. The lobar bronchi are further divided into segmental bronchi (tertiary bronchi), which supply the bronchopulmonary segments of each lobe. There are 10 bronchopulmonary segments in each lung. More proximal bronchi continually divide into smaller and smaller bronchi up to about 24 generations of divisions from the main bronchi (Fig. 8.20). With each succeeding generation, and as bronchi become smaller (i.e., reduced diameter), there are some notable structural changes: (1) cartilage rings disappear, so that by generations 12 to 15 the airways, now called bronchioles, lose all cartilaginous support; (2) in the shape of epithelial cells, with a loss of cilia and mucous producing cells; and (3) the presence of smooth muscle in the walls of small airways provide support and neural control of air passage diameter. The generations that run from the trachea to terminal bronchioles are referred to as the conducting zone because no gas exchange takes place. These passages, which contain about 150 milliliters (mL) of air at any given time under normal conditions, are referred to as the anatomic dead space. The terminal bronchioles are also referred to as transitional since they occasionally contain alveoli in their walls. The respiratory bronchioles (also referred to as small airways), with an internal diameter of 2 mm or less, further divide into alveolar ducts, which are completely lined with alveoli. The respiratory zone (or acinus) (Fig. 8.21) forms the functional units of the lung, i.e., lung parenchyma. The respiratory zone comprises the majority of total lung volume, i.e., approximately 2.5 to 3 liters (L) during rest, whereas the conducting airways only hold up to 150 mL of

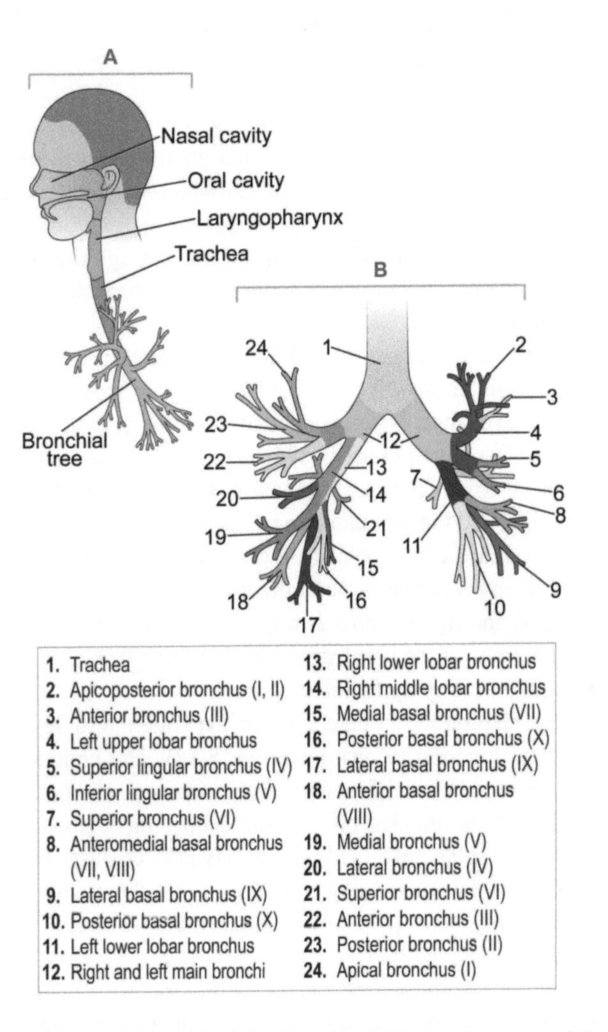


**Fig. 8.19** Schematic of (**A**) left lateral and (**B**) anterior views of the upper and lower respiratory tracts and tracheobronchial airway showing subdivisions in the first generations and where the branches lead into segments of the lung, i.e., the bronchopulmonary segments

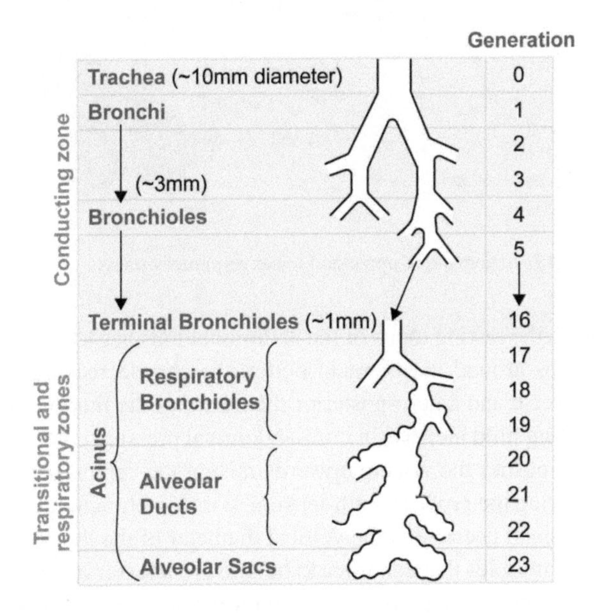


**Fig. 8.20** Schematic of airway generations in the human lung

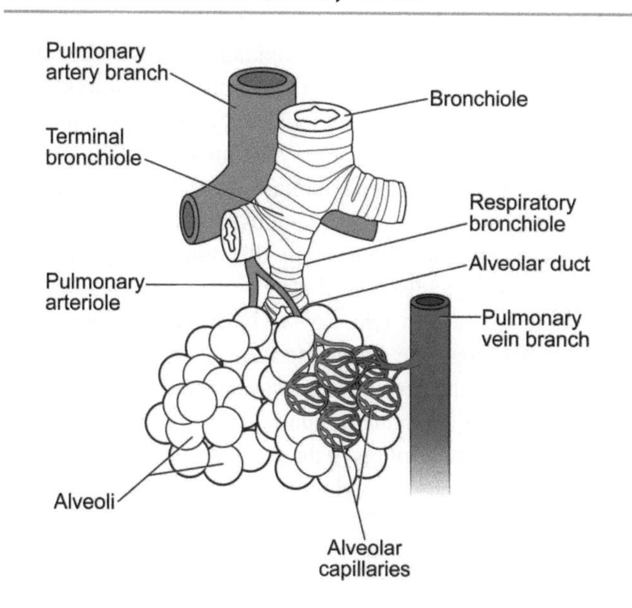


**Fig. 8.21** Schematic of acinus unit showing the relationship among airway passages, blood vessels, and alveoli

gas. The alveolar ducts are short tubes that are supported by a matrix of collagen and elastin fibers. During inhalation, the alveolar wall elastin provides the compliance needed when the walls deform, i.e., are stretched, storing strain energy that is released during exhalation. Collagen performs two functions: it deforms under loading and can store energy that is used during the exhalation phase, and it provides structural strength to alveolar walls. The distal end of the alveolar duct opens into the alveolar sac, which is comprised of an atrium and alveoli, the termination of all airway passages in the respiratory system. Because gas exchange takes place in the acinus, it is surrounded by a network of capillaries. It is estimated that each adult lung has about 300 million alveoli, with a surface area for gas exchange totaling 70–80 square meters ($m^2$).

The alveoli are optimally suited for gas exchange by having a thin (approximately 1–2 μm) moist surface and a very large total surface area. An important factor related to alveolar mechanics relates to the surfactant that is produced by type II alveolar cells. Surfactant is vital for controlling the inherent surface tension that would constrain the stretching of the alveolar walls during inhalation. It is notable that surfactant is most effective at the low volumes that alveoli normally operate under. The surfaces of the alveoli are also covered with a network of capillaries. Oxygen is passed from the alveoli into the surrounding capillaries that contain oxygen-deprived, carbon dioxide-rich blood. Gas exchange takes place where the oxygen is dissolved in the water lining of the alveoli before it diffuses into the blood, while carbon dioxide diffuses from the pulmonary arterioles (capillaries) and into the alveoli where it is expelled from the air passages during exhalation.

The exchange of $O_2$ and $CO_2$ occurs through diffusion, which is the net movement of gas molecules from a region of higher pressure to one of lower pressure (see Sect. 8.4.2.1 on mechanics of ventilation). For example, the partial pressure of oxygen in the inspired air within the alveolar spaces in the lung is greater than the partial pressure of oxygen in the blood which enables oxygen to diffuse into the red blood cells. The diffusion process is defined by Fick's law of diffusion: the diffusion of a gas across a boundary is directly related to its surface area ($A$), the diffusion constant of the specific gas ($D$) and the partial pressure difference of the gas on each side of the boundary ($P_1 - P_2$), and inversely proportional to the boundary thickness ($T$):

$$Diff \propto \frac{AD(P_1 - P_2)}{T}$$

Because the exchange of gases takes place at a molecular level, these molecules move randomly in the direction of the partial pressure gradient and are temperature-dependent. The process occurs until some sort of equilibrium is reached. Because of the increased cross-sectional area of the respiratory zones, the velocity of airflow is sharply reduced as the airstream moves peripherally through the airways yet assists in gas exchange across the 1–2 μm alveolus-capillary interface.

### 8.5.2 Pathophysiology and Altered Respiratory Mechanics with COPD

Chronic respiratory diseases are a major global issue, with chronic obstructive pulmonary disease ranked as the fourth-leading cause of death in the United States. As a result, there are significant healthcare costs and a considerable economic-social burden, as well as a staggering loss of quality of life for those affected. Chronic obstructive pulmonary disease (COPD) is characterized by poorly reversible airflow obstruction and abnormal inflammatory response in the lungs. In Scott's case, the inflammatory response represents the innate and adaptive immune responses to long-term exposure to cigarette smoke. All cigarette smokers have some chronic inflammation in their lungs, but those who develop COPD have an enhanced or abnormal response, which results in mucous hypersecretion (chronic bronchitis), destruction of lung parenchyma (emphysema), and disruption of normal repair and defense mechanisms that perpetuate small airway inflammation and induce excessive fibrosis (bronchiolitis). Emphysema is a histopathologic diagnosis defined as an abnormal permanent enlargement of the air spaces distal to terminal bronchioles, accompanied by destruction of their walls, without fibrosis. In most individuals with more advanced COPD, significant emphysematous changes are present; however, the disease does not affect all portions of the lungs to the same degree.

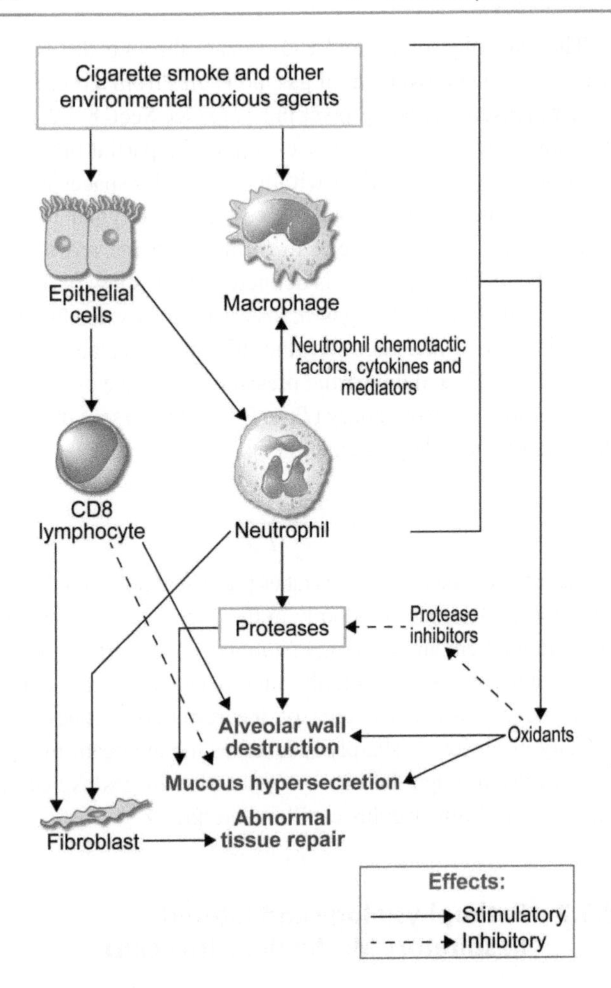


**Fig. 8.22** Schematic illustrating inflammatory mechanisms in COPD

Smokers' lungs are characterized by chronic inflammation. This normally protective response is heightened in smokers because of the nature of the inhaled toxins. For example, increased numbers of neutrophils, macrophages, and T lymphocytes, as well as inflammatory mediators, e.g., leukotriene $B_4$, chemotactic factors, pro-inflammatory cytokines, and growth factors are released (Fig. 8.22).

The release of growth factors is, directly and indirectly, responsible for the activation of fibroblasts that induce repair processes, and unwanted bronchiolar fibrosis. It has also been recognized that the interaction between fibroblasts and inflammatory cells plays a key role in fibrotic remodeling. A combination of abnormal repair and/or excessive fibrosis results in the thickening of the small airways that contribute to obstruction of airflow and increased airway resistance. The chronic inflammatory state of the lungs is also associated with increased production (or activity) of proteases and inactivation (reduced production) of antiproteases, leading to a disequilibrium that results in the destruction of normal tissue, e.g., alveolar sacs. Protease action can also directly contribute to hypersecretion of mucous.

Cigarette smoke increases the oxidative stress in the lungs which creates an imbalance of oxidants and antioxidants, which induces changes in epithelial and other cells that stimulate chemotactic factors, e.g., macrophage inflammatory proteins. Protease factored damage to airway epithelium induces mucosal ulcers and squamous- and goblet-cell metaplasia. A significant increase in the number of mucous-secreting goblet cells contributes to the development of airflow obstruction in two ways: (1) excess production of mucous that alters the surface tension of the airway lining fluid, rendering the peripheral airways unstable and facilitating closure, and (2) inducing luminal occlusion through the formation of mucous plugs in the small airways. Ciliary dysfunction also occurs secondary to squamous metaplasia of epithelial cells and results in a mucociliary escalator (i.e., clearance or transport) and difficulty in the expectoration of mucous.

The cross-sectional area of peripheral airway smooth muscle is increased in smokers with COPD. Let's digress briefly here to examine the normal role of smooth muscle in airway caliber. In a healthy lung, the force generated by the airway smooth muscle is a primary determinate of airway caliber. The force magnitude of airway smooth muscle depends on its contractile properties and total mass. The length of the airway smooth muscle is regulated by the balance between its force, which tends to shorten it, and the magnitude of opposing forces. For a given linear length of airway smooth muscle, the airway caliber is determined by the thickness of the airway internal wall layer. Airways and lung parenchyma are interdependent systems. Thus, when smooth muscle is activated the elastic elements of the surrounding parenchyma are stretched, generating a force that represents an external load against which the airway smooth muscle has to work. The external load on the contracting airway smooth muscle approximates the lung elastic recoil plus the pressure necessary to overcome the tension generated when the outer wall decreases. Thus, the efficacy of this force in maintaining airway caliber is expected to depend on lung volume, lung elastic recoil, and how well this is transmitted to the airway wall.

With COPD there is a strong linear relationship between increased small airway resistance to flow and the size or thickness, i.e., cross-sectional area, of the smooth muscle. The degree of airflow limitation is related to airway narrowing due to wall thickening, which is likely the result of smooth muscle hyperplasia and hypertrophy; hyperplasia due to the activity of inflammatory mediators, cytokines, and growth factors; and hypertrophy as a normal response to increased airflow resistance. The major functional consequence of the increase in smooth muscle mass is that, in airways with thickened walls, the same degree of smooth muscle shortening may cause considerably greater luminal narrowing than in normal airways.

Let's examine the physics of airflow dynamics and resistance. Airflow through the respiratory system is driven by pressure differences as soon as air enters the nose (or mouth) and continues to flow through the respiratory zone. During inspiration, the glottis at the larynx opens allowing gas to flow from the conducting zones through to the respiratory zones. Generally, gas flow in the airway can be defined as laminar or turbulent. Laminar air flows experience low resistance and are orderly and characterized by smooth streamlines. With laminar flow the viscosity of the fluid plays a significant role in holding the fluid together in its layered formation; thus, if a fluid particle was tracked, its path could be predicted. Turbulent flows, on the other hand, are characterized by eddies and random fluctuations, irregular and chaotic motion, and greater levels of vortices; turbulent flow is known to increase where the airways branch or diverge. In straight tubular or pipe flows, this flow regime is defined by the Reynolds number:

$$Re = \frac{\rho DV}{\mu}$$

where $\rho$ is the density, $D$ is the hydraulic diameter, $V$ is the average velocity, and $\mu$ is the dynamic viscosity. For a given fluid with constant density and viscosity, increases in flow velocity contribute to flow turbulence, which tends to be greater in the larger airway passages, e.g., trachea, etc. However, normally in the small airways, and even bronchioles, the small diameters essentially damp out the inertial effects i.e., velocity, and contribute toward a laminar flow (Fig. 8.23).

Airflow resistance ($R$) in the respiratory airways is a concept that describes the opposition to air flow, from its inhalation point to the alveoli, caused by frictional forces. It has units of cm $H_2O$ $s/L$ (where $L$ = liters) and is defined as the ratio of the driving pressure ($\Delta P$) given in cm $H_2O$ to the flow rate ($\dot{V}$) given in $L/s$:

$$R = \frac{\Delta P}{\dot{V}}$$

For laminar flow (only), the flow rate is estimated through Poiseuille's law, given by:

$$R = \frac{8\mu L}{\pi r^4}$$

where $r$ is the radius and $L$, in this case, is the airway length. Setting the two previous equations equal to each other and solving for $\dot{V}$:

$$\dot{V} = \frac{\Delta P \pi r^4}{8\mu L}$$

We can see that the resistance is inversely proportional to the fourth power of the radius and that a change in airway size has a greater effect in comparison to other variables, e.g., $\Delta P$, etc. Thus, under normal conditions, the cross-sectional area within the tracheobronchial tree is by far the most important determinant of airway resistance. The cross-sectional area of any given airway is determined by the balance of opposing forces: those tending to narrow the airway lumen (primarily the force generated by airway smooth muscle) and those tending to enlarge it (the indirect effects of pleural pressure in the intrapulmonary conducting airways or the direct effect of the attached lung parenchyma in the terminal bronchioles). Thus, airway resistance decreases as lung volume increases because the airways distend as the lungs inflate, and wider airways offer less resistance.

During inhalation, the positive pressure within the alveoli and small airways cause their diameter to increase, leading to less resistance to flow; the opposite is true for exhalation since airways narrow secondary to reduced pressure. Smaller airways, such as bronchioles and alveolar ducts, see greater flow resistance than larger airways, e.g., trachea. Thus, the rate of flow in these areas is very low, but because of the branching, there are many smaller airways in parallel, which reduces the total resistance to airflow. Normally, airflow in smaller airways is laminar, but with COPD there is greater resistance to flow secondary to airway wall thickening, increased presence of mucous, and abnormal smooth muscle control (secondary to impaired sympathetic nervous system function), all of which can create an environment for more turbulent flow. An increase in the turbulent flow in the small airways with COPD may significantly alter gas exchange. Intuition suggests that greater inspiratory effort by the diaphragm and accessory muscles can create a pressure difference, i.e., increasing alveolar $pO_2$/decreasing alveolar $pCO_2$ across the respiratory membrane, to overmatch the

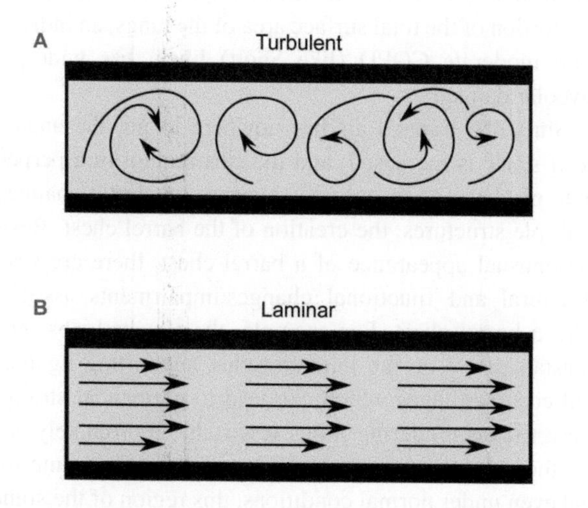


**Fig. 8.23** Schematic diagrams showing (**A**) turbulent flow where the air is not flowing in parallel layers, and direction, velocity, and pressure with the flow is chaotic and (**B**) laminar flow in which air moves through the airway in parallel layers with no disruption between layers, and the central layers flowing with the greatest velocity

resistance, but these actions are actually counter-productive. Moreover, forced exhalation induces alveolar collapse.

Moderately involved and advanced cases of COPD see major changes in the structure and function of the alveoli and related connective tissue. Emphysema, as a histopathologic finding, is defined as a permanent "destructive" enlargement of air spaces distal to the terminal bronchiole including alveolar wall fibrosis. There are two classifications of emphysema: centriacinar (or centrilobular) and panacinar (or panlobular). Centriacinar emphysema is characterized by focal destruction restricted to respiratory bronchioles and the central portions of the acinar, surrounded by areas of grossly normal lung parenchyma. This form is usually most severe in the upper lobes of the lung. Panacinar emphysema is characterized by the destruction of the alveolar walls in a fairly uniform manner, and typically involves the lower lobes rather than the upper lobes. The varieties of emphysema seem to have distinct mechanical properties and peripheral airway involvement. Lung compliance is greater in panacinar than in centriacinar emphysema, whereas the extent of peripheral airway inflammation is greater in the centriacinar form. Thus, airflow limitation in centriacinar emphysema appears to be most related to peripheral airway inflammation, whereas in panacinar emphysema airflow limitation is primarily related to the loss of elastic recoil.

Recall, normally lung parenchyma compliance (i.e., a function of the elastin/collagen matrix) is paramount to the spring-like response that returns the alveoli to their resting position during the exhalation phase of respiration. Damaged/repaired alveolar walls have increased alveolar compliance that significantly impairs a normal elastic response and is the primary factor that contributes to air-trapping and hyperinflation (Fig. 8.11). And, as noted previously, the imbalance between proteases and antiproteases leads directly to alveolar cell wall damage, whose repair may be non-functional or non-existent. It is notable, and a bit paradoxical, that although we know that damaged alveoli are more compliant, their repair also makes their walls stiffer, i.e., resistant to stretch, which may, in some way, impair their ability to distend during inhalation. At the same time, the combination of inflammation, fibrosis, and smooth muscle hypertrophy effects on airway wall thickness facilitates the uncoupling between airways and the parenchyma and contributes to airway closure. Since, the cross-sectional area of the respiratory zones is very large, isolated and small regions of acinus destruction can be tolerated in the early stages of COPD, i.e., an individual may not have any or just mild symptoms. However, with extensive alveolar wall destruction and impairment in advanced COPD, significant gas exchange abnormalities are present, leading to hypoxemia (decrease in the partial pressure of oxygen, i.e., $PaO_2$) and hypercapnia (increase in the partial pressure of carbon dioxide, i.e., $PaCO_2$), and other systemic problems.

We do not have a definitive staging of Scott's COPD but hypothesized that he was moderately involved since he demonstrated a barrel chest, was short of breath when walking even a short distance, and had adapted his control of expiration with the use of pursed-lip breathing. Yet, despite his symptoms he seemed to be functioning reasonably well; for example, he apparently was well enough to split wood. There are many more facts about the pathophysiology and systemic effects of COPD we could discuss, but we leave it to the curious reader to explore this on their own.

After reviewing the etiology and pathophysiology of COPD, we are left with a couple of questions. So, what? What does Scott's COPD have to do with his injury, in particular, and musculoskeletal biomechanics in general? Our general response to these pertinent questions is that examining the details of COPD offers an opportunity to see how clinical biomechanics is applied to our biology seemingly unrelated to the musculoskeletal system. And yet, in doing so it is apparent that there is a relationship between the systemic problem, i.e., COPD, and how it is manifested in the musculoskeletal system. For example, we now know that the main site of airflow obstruction occurs in the small airways. This results from an exaggerated and sustained inflammatory response to the toxins from cigarette smoking, and subsequent narrowing (i.e., airway remodeling) from the fibroblastic activity and smooth muscle hyperplasia and hypertrophy. This obstruction progressively traps air during exhalation, resulting in hyperinflation at rest. Without intervention, i.e., cessation of smoking, these damaging changes take place over a long period of time, which lead to plastic changes in multiple systems, e.g., small airways and related connective tissue, as well as structures in the chest wall, e.g., vertebra, ribs, and associated periarticular soft tissues. Since small airway and alveolar sacs comprise an overwhelming proportion of the total surface area of the lungs, an individual with moderate COPD (like Scott) likely has widespread alveolar damage.

Since the trapped air has nowhere to go, the anatomic dead space is increased, and the ribs remain in a perpetual state of elevation, eventually, because of plastic changes in multiple structures, the creation of the barrel chest. Besides the unusual appearance of a barrel chest, there are several structural and functional changes/impairments associated with a barrel chest. For example, the elevated ribs induce constant stress on the joint capsules, supporting ligaments, and costocartilages which can lead to permanent strains of these tissues rendering joints less stable. Fortunately, since the thoracic cage is constrained posteriorly and anteriorly, and even under normal conditions, this region of the spine is inherently stable. However, even small alterations in the mobility of the costovertebral joints can change the positioning of the ribs relative to the sympathetic ganglia resulting in possible autonomic nervous system irritation, with possible impairment. Furthermore, since the ribs are at or near their end range in this constant state of elevation, an individual's inspiratory potential is significantly reduced, impairing the ability to adapt to situations that require a greater inspiratory effort. In those cases, the individual may resort to a greater rate of breathing that has a much higher energy cost and cannot be sustained. Of course, in the case of COPD, the individual has significant difficulty with greater inspiratory effort because of both small airway obstruction and destruction of the alveolar walls. In addition, altered rib mechanics could affect vertebral segment mechanics, not only limiting the vertebral range of motion in all three dimensions but also inducing secondary effects on joint mechanoreceptors, as well as the muscle spindles of all muscles crossing the affected joints. It is not difficult to imagine the myriad of unintended musculoskeletal dysfunctions that could arise related to a barrel chest. For example, since most spinal rotation that is required for some functional activities, e.g., splitting wood, swinging a golf club, etc., is provided by the thoracic spine, many activities are impaired; reduced mobility in the thoracic spine may not only increase the risk to that region with activities that require large movements but may also lead to greater mobility demands on the lumbar spine putting that region at risk for injury as well.

Although we cannot claim that COPD directly caused Scott's present symptoms, it is not unreasonable to hypothesize that he was predisposed to a variety of musculoskeletal injuries, e.g., to the spine, rib cage, or even peripheral joints, because of his pre-morbid condition.

Air trapping and the resulting lung hyperinflation also impact the muscular system. As the prime motor for inhalation, the diaphragm is ideally suited to a job that requires mechanical efficiency and a low energy cost. However, when a sustained increased residual volume, secondary to airway obstruction and air trapping, has increased to a critical point, the diaphragm's function is affected in three critical ways: (1) loss of optimal operating length-tension, (2) altered/reduced effective moment arms, and (3) reduced viscoelastic response of its contractile and non-contractile elements (note: this is also an issue with the secondary inspiratory muscles). This occurs secondary to lung hyperinflation that causes deformation of all structures in the thoracic cage, but, in particular, flattens the diaphragmatic dome by pushing the muscle inferiorly. As a result, the one muscle that can increase the intra-thoracic volume in multi-dimensions most effectively is rendered, not useless, but much less effective. Of course, the obvious problem with a dysfunctional diaphragm is the inability to maintain efficient tidal volume breathing, but, most dramatically, the ability to adapt to situations that require a greater inspiratory effort is significantly impaired. The central nervous system has an amazing ability to adapt, and thus recruits accessory muscles of inspiration. These muscles, however, do not have the primary function of contributing to inhalation, cannot increase the intra-thoracic volume in multiple dimensions, and typically do not have the muscle fiber type, i.e., slow-oxidative, make-up to sustain the system as efficiently. We know that muscle has the capacity to adapt to applied demand so the accessory muscles that have fewer motor units comprised of slow-oxidative fibers see some conversion of fast-glycolytic to fast-oxidative. However, typically these adaptions do not meet the functional demands. So, the energy cost of breathing increases dramatically. In later stages of the disease progression (we are not sure Scott is there yet), the energy cost of breathing is such that the individual is limited to essential activities of daily living only so that they develop generalized disuse muscle atrophy and become severely deconditioned. Consider, also, that since many accessory muscles of inhalation are called to second duty, i.e., breathing, their primary functions may be impaired affecting the musculoskeletal system in a variety of ways that have negative long-term effects.

We also know that moderately advanced COPD is associated with changes in small airway smooth muscle, i.e., hyperplasia and hypertrophy. These changes mostly contribute to airway obstruction because of increased mass, but it is likely that there are alterations in central nervous system control that affect airway diameter. For example, the changes in the costovertebral joint mechanics may have effects on sympathetic ganglia regionally, as already noted. Abnormal stimulation or inhibition of the sympathetic trunks affects control of all systemic functions in the affected regions, including airway smooth muscle. In addition, the chronic inflammatory state associated with COPD has an effect on receptors in smooth muscle that can impact their function.

Without early cessation of smoking, one can readily see that the disease progression we have described ends with a dysfunctional system of inter-related parts that has positive feedback loops that perpetuate dysfunctional actions of many elements in the thoracic cage complex.

## 8.6 Mechanism of Injury; Injury Hypotheses

We conclude that Scott's pre-morbid condition, i.e., COPD, did not play a direct role in his present symptoms. On the other hand, we hypothesize that his barrel chest, which is a permanent and progressive condition, with associated plastic changes in the vertebral and costovertebral joints, i.e., reduction in normal ranges of motion, and related soft tissues, may have predisposed Scott to the nature of his injury(s). Therefore, it seems plausible that the forces and moments associated with splitting wood resulted in minor strain injuries, to one or more joints, which are related to Scott's symptoms. So, let's reconstruct and perform a qualitative biomechanical analysis of the wood-splitting act. First, we

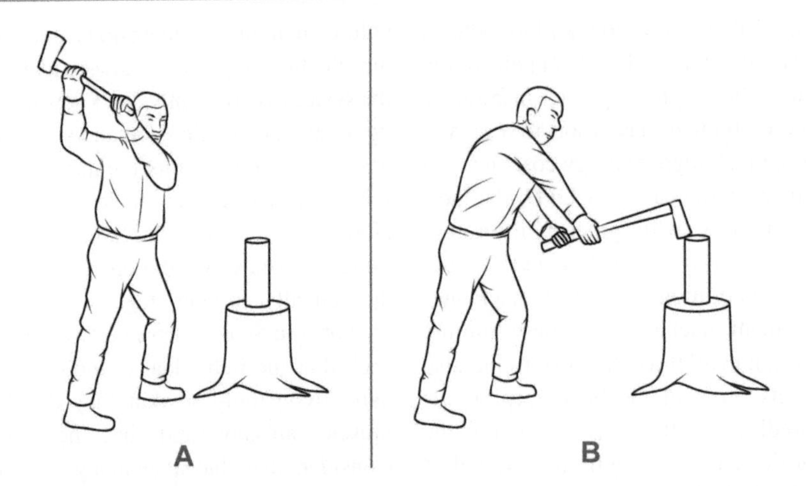


**Fig. 8.24** Likely movements for splitting wood for a right-handed dominant individual, where (**A**) raising the ax overhead occurs in phase I and (**B**) moving the ax from phase I to striking the wood occurs in phase II

need to establish a couple of assumptions. Assumption #1: since Scott is right-handed, we claim that he split wood as shown in Fig. 8.24. Similar to how we analyze the pitching act (Chap. 4), or how any complex movement would be analyzed, we divide the act of splitting wood into two phases: (1) wind-up or preparatory and (2) acceleration and contact. Although we acknowledge that this activity involves the dynamic coupling of joints progressing from the lower extremities, through the pelvis, thoracic cage, and upper extremities, we focus our analysis on the thoracic spine and rib cage. Thus, we can see that in raising the ax, both shoulders are elevated, which induces thoracic cage extension, and right thoracic cage rotation coupled with left side-bending. We define the initiation of the second phase as deceleration of the wind-up movements, then the acceleration of the arms and thoracic cage in the opposite direction: flexion, coupled with left sidebending and rotation, ending with the contact of the ax head on a piece of wood. Note: striking the wood clearly imposes other forms of external impulsive loads on the body, but to simplify our analysis we ignore them and focus on the preparatory and acceleration phases of the wood-splitting actions.

Our second assumption is related to normal sidebending/rotation coupling movement patterns that have been described for the thoracic region of the spine. Recall that we discussed thoracic spine coupling in Chap. 3, but it is important that we review vertebral coupled motion theories before we state the second assumption. Coupled motion refers to the simultaneous rotations and translations that always occur within a spinal unit during normal movements of the spine. Often coupled rotations are described as those which occur about a Cartesian axis system fixed in the center of the intervertebral disc of the interbody joints. By convention, we describe the movement, e.g., flexion, extension, etc., of the

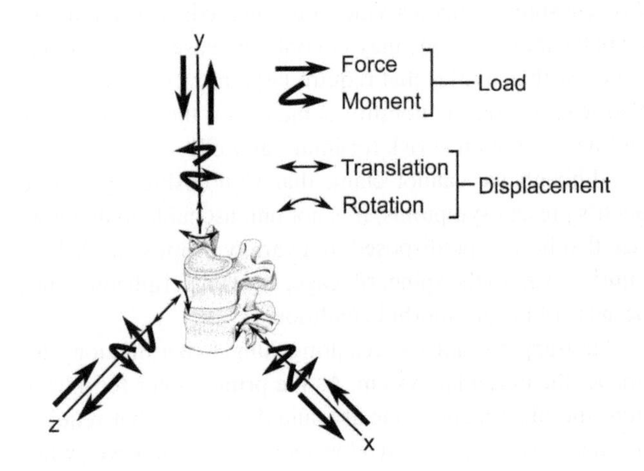


**Fig. 8.25** Functional spinal unit = superior vertebrae, intervertebral disc and inferior vertebrae. The motion of the superior vertebrae relative to the one below it occurs when an external force is applied. If the force is applied at some distance from the joint center, it induces both a translation and rotation of the superior vertebra. Any spinal movement includes coupled movements, i.e., translation and simultaneous rotation, about a helical-type axis. (Modified with permission of White and Panjabi (1990))

superior vertebral segment relative to the one below it (Fig. 8.25). It is notable that coupled movements are inherent to the cylindrical shape of the spinal column, as well as the orientation of the apophyseal joints, variability in the shape of the thoracic spine, i.e., kyphosis, varying disc morphologies, presence of regional differences in the shapes of ribs, and ligamentous and other restraining soft tissues. Although the upper thoracic spine (T1 to T4) shows a segmental coupling behavior similar to the cervical spine, showing ipsilateral axial rotation during primary sidebending, no distinct coupling pattern has been shown for the lower thoracic spinal motion segments. Experimental evidence seems to indicate that the overall coupling behavior from T1 to T12,

**Table 8.4** Nordic (Kaltenborn) descriptions of coupled and non-coupled movement patterns in the thoracic spine (T3-T12)

| Movement | Movement classification |
|---|---|
| Flexion, sidebending, and rotation opposite | Non-coupled |
| Flexion, sidebending, and rotation same | Coupled |
| Neutral spine, sidebending, and rotation opposite | Non-coupled |
| Neutral spine, sidebending, and rotation same | Coupled |
| Extension, sidebending, and rotation opposite | Coupled |
| Extension, sidebending, and rotation same | Non-coupled |

Note: With the thoracic spine in its neutral position in standing, sidebending and rotation couple in the opposite direction

including the effect of the rib cage (which is more representative), exhibits a strong coupled ipsilateral axial rotation during primary sidebending as well as a small coupled contralateral sidebending during primary axial rotation when loaded with pure bending moments. However, in vivo studies have demonstrated considerable variability in coupling patterns among individuals.

In the clinical world, although several theories about the coupling of frontal and transverse plane movements in the thoracic spine have been described, two dominant schools of thought are discussed here: Nordic (i.e., F. Kaltenborn, a physiotherapist) and osteopathic. Kaltenborn's description of spinal coupling patterns (Table 8.4) was born primarily out of empirical evidence, i.e., his clinical experience. Most of his claims have been validated experimentally, yet disagreements between clinicians remain. Kaltenborn distinguished coupled and non-coupled movements because he uses specific tests for these patterns as part of his physical examination.

Let's explain what Kaltenborn means by coupled and non-coupled movement patterns entail. First, the non-coupled pattern. What do flexion, sidebending, and rotation opposite mean? If an individual flexes their trunk and holds that position, then chooses to sidebend their thorax to the left and then rotates their thorax to the right, they perform a non-coupled movement; that is, all three of the movements, i.e., flexion, left sidebending, and right rotation, are voluntarily chosen. Second, the coupled pattern. When we say that flexion, left sidebending, and left rotation are coupled movements, this is what happens: the individual flexes their trunk and holds that position, then chooses to sidebend their thorax to the left, and there occurs a simultaneous left rotation; that is, flexion and left sidebending is voluntarily chosen, but the third movement, the coupled axial rotation, is involuntary, i.e., accessory or automatic.

Kaltenborn's use of the phrase "non-coupled" sidebending and rotation implies that the motion occurs naturally; note that in this case the magnitude of the non-coupled pattern is more constrained by joint incongruence and intrinsic and extrinsic ligaments than are coupled movements. As just noted, non-coupled motions are voluntary, i.e., or may be part of a particular motor pattern, e.g., swinging a golf club, dance positions, etc. In other settings, manual medicine practitioners can also passively place an individual in positions where spinal segments are held in non-coupled positions as part of a treatment sequence, i.e., spinal manipulation. Reminder: the key characteristic of non-coupled positions of the spine as part of a posture or active movement pattern is that the joints involved are more constrained by the joint morphologies and/or periarticular soft tissues.

The osteopathic theory of spinal coupled movements originated with Dr. Harrison Fryette. He developed his ideas about intervertebral coupling based solely on his osteopathic practice: observation, palpation, and motion testing. Osteopathic physicians and other manual therapy practitioners, e.g., physical therapists, athletic trainers, etc., have used Fryette's examination and examination-based spine mobilization/manipulations methods based on his movement theories with great success. Similar to Kaltenborn, controlled experiments have provided only some support for Fryette's theories. Fryette summarized his theories as three principles of spinal motion:

1. When the spine is in neutral, sidebending to one side is accompanied by horizontal rotation to the opposite side. Fryette claimed that this law was observed in type I somatic dysfunction, where more than one vertebra was out of alignment and could not be returned to neutral by flexion or extension of the spine. The involved group of vertebrae demonstrated a coupled relationship between sidebending and rotation.
2. When the spine is in a flexed or extended position, i.e., non-neutral, sidebending to one side is accompanied by rotation to the same side. This law was observed in what Fryette defined as type II somatic dysfunction, where only one vertebral segment was restricted in motion and became worse on flexion or extension.
3. When motion is introduced in one plane, it modifies, i.e., reduces, motion in the other two planes. This is actually what is encountered with non-coupled, i.e., Kaltenborn's, movements. The third principle sums up the other two laws by stating that dysfunction in one plane negatively affects mobility in all other planes of motion.

Fryette did not describe non-coupled movements; however, spinal segment manipulations based on Fryette's laws have always used the idea of non-coupled postures as a method to stabilize spinal segments adjacent to the segments they attempt to mobilize during high-velocity, short-amplitude manipulations. Recall that according to Kaltenborn, placing a vertebral unit in a non-coupled position constrains excessive movements.

Trained practitioners in manual (or manipulative) therapy of the spine usually adopt one or the other spine biomechanical theoretical paradigm, i.e., Nordic or osteopathic, as a basis for their examination and treatment choices. This is so because each paradigm may dictate a different examination and treatment approach. Although we do not discuss a treatment approach for Scott, we develop hypotheses regarding his physical examination findings based on the *osteopathic* model; not because that is the only correct model, but simply because it simplifies our problem-solving process. Therefore, our second assumption, in this case, is that Scott's physical examination findings can be explained using an osteopathic model of vertebral coupling patterns. We come back to this after we detail the mechanism of injury.

In brief, there are two phases that are used to describe the biomechanics of splitting wood: (1) wind-up and (2) acceleration. With regard to the thoracic spine and rib cage, we claim that these phases included the following movement patterns: extension with right axial rotation and left sidebending (wind-up) followed by flexion, left rotation, and left sidebending (acceleration). In the wind-up phase, we have to acknowledge/account for the length and weight of the ax, added to the length and weight of both upper extremities since the totality must be lifted to place the ax-head in an optimal overhead position. In other words, the mass of the ax and upper extremities are external forces that must be moved against gravity to position the ax. Since, in this case, we are not dealing with any injuries to upper extremity joints or soft tissues, we just acknowledge that several upper extremity muscles (trapezius, serratus anterior, deltoid, rotator cuff, elbow extensors, elbow flexors, etc.), working in synergy, were used to accomplish this task successfully. Notice that the mechanism of injury, in this case, is distinguished from several previous cases because the potentially damaging forces and moments do not appear to be external to the body, e.g., like falling on an outstretched hand, but are self-induced. Therefore, in what follows, remember that when a movement is described, internal muscle-tendon actuator forces (i.e., tension) and internal joint moments are implied; for example, when the thoracic spine is extending and rotating, it does so due to actions (shortening) of the spinal extensors and rotators that produce extensor and axial rotation moments.

Let's first examine movements of the thoracic spine and rib cage during the wind-up. We only describe the osteokinematic movements and leave it to the reader to consult arthrokinematics descriptions of the thoracic spine and ribs in Chap. 3. Since the spine is extending and rotating to the right, it is likely that there were synergistic unilateral activations of the right erector spinae, coupled with shortening of the left obliquus externus abdominis and right obliquus internal abdominis (Fig. 8.26). Additionally, during this time period, it seems intuitive that raising the arms and ax and

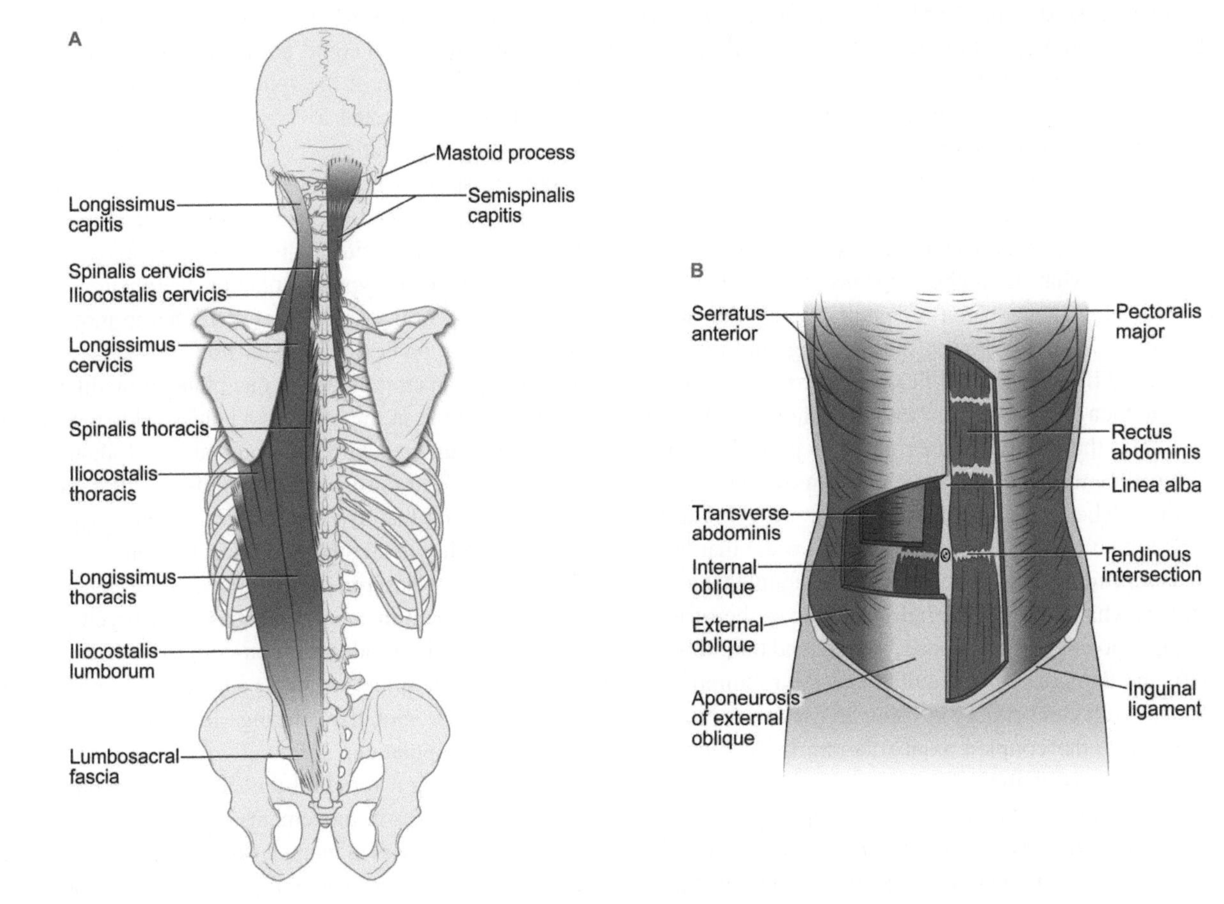


**Fig. 8.26** (**A**) Posterior view of erector spinae and (**B**) anterior view of abdominal muscles

turning the trunk to the right must have been controlled to some degree by the contralateral, i.e., respective right- and left-sided, muscles of the torso, as well as the one- and two-joint muscles, e.g., rotatores and multifidi, which were acting in some combination of isometry and/or eccentrically. Note that generally the primary movements of the ribs are associated with respiration, which we discussed in detail, and the muscles that attach to the ribs are not typically considered synergists for movements of the spine; that is, when the thoracic spine extends and rotates, the ribs simply move with the spine. However, the muscles identified as those that assist with inhalation and exhalation (e.g., intercostals, etc. – see Supplementary 8.1 for this chapter for images) and have various attachments to the ribs were likely responding, via muscle spindle activation, to vertebral rotations, and perhaps assisting with axial rotation and/or at the least providing chest wall stabilization.

At this point, we want to introduce conceptual ideas regarding the proposed actions of the erector spinae. Bogduk presented a model of the functions of the iliocostalis, longissimus, and multifidi in more detail than typical presentations by decomposing the muscle force vectors into vertical and horizontal components. He demonstrated this concept for the lumbar region, which may be influenced by the anterior convexity in that region, so it is not clear if his analysis can be applied to the thoracic spine, but we should at least consider it. For the iliocostalis lumborum pars lumborum and longissimus thoracis par lumborum, the vertical component for both muscles induces spinal extension when the muscles act bilaterally, and sidebending if either muscle acts unilaterally, with only the iliocostalis lumborum contributing to axial rotation. However, Bogduk showed qualitatively that the horizontal component in the sagittal plane is greater than the horizontal component in the frontal plane, suggesting that there is greater potential for these muscles to produce, when acting bilaterally, posterior translation; that is, lateral translation that could occur during sidebending or axial rotation if the muscles acted unilaterally is substantially less (Fig. 8.27). Abstractly, these results may have implications for how spinal segments may be stabilized during particular activities.

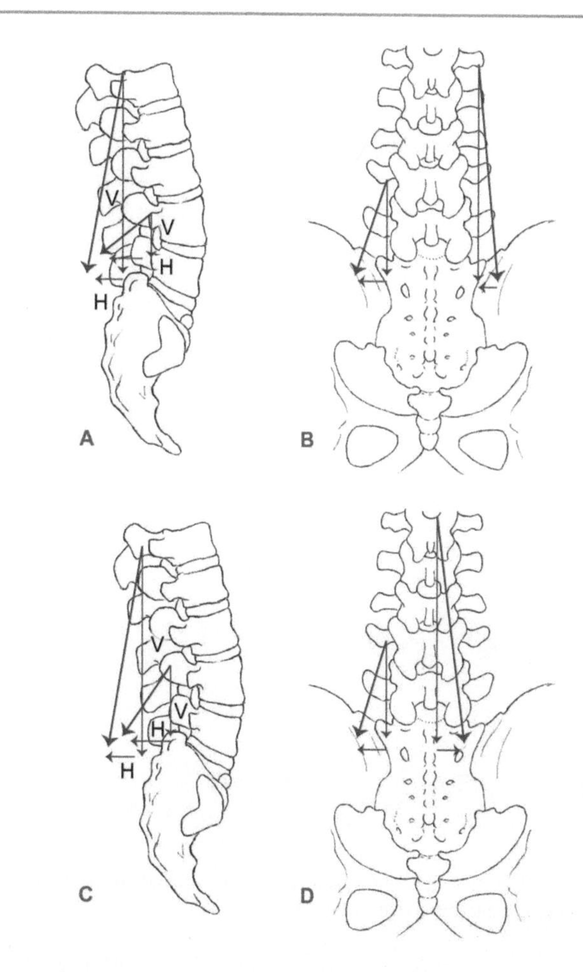


**Fig. 8.27** Schematic of (1) iliocostalis lumborum pars lumborum in (**A**) sagittal and (**B**) frontal view and (2) longissimus thoracis par lumborum in (**C**) sagittal and (**D**) frontal views. Note that the magnitude of the horizontal component seen in a sagittal view is larger than the magnitude of the horizontal component seen in a frontal view. (Modified with permission from Bogduk (2012))

What about the thoracic left-sidebending that was coupled with the extended/right rotated spine? This movement may be induced by concentric actions of several left-sided muscles, including the spinalis, erector spinae, semispinalis, multifidi, rotatores, and intertransversarius (Figs. 8.26A and 8.27), although Bogduk suggests that the smaller segmental muscles do not have the moment arms to rotate vertebral segments. He proposed that since these muscles have high densities of muscle spindles, their primary role is proprioceptive in nature. Based on Fryette's third law, we know that the magnitude of this left sidebending movement is likely reduced compared to the extension and right rotation. We also understand that, based on osteopathic theory, that contralateral sidebending is not the preferred or "natural" movement of the spine, and may induce excessive stresses on multiple joints, e.g., facet, costovertebral, etc., and their periarticular structures (Fig. 8.28).

Bogduk also examined the decomposition of lumbar multifidi force vectors into vertical and horizontal components (Fig. 8.29). In the frontal view, it is clear that the horizontal component is much smaller than the vertical component, suggesting that the multifidi role in single- or multi-segment sidebending and rotation may be quite limited. From the sagittal view, we can see that the bilateral action of the multifidi extends lumbar segments. Bogduk suggested that perhaps the primary role of the multifidi is to restrain the flexion moment during axial rotation of the trunk (since the primary muscles which rotate the trunk are strong flexors, e.g., internal and external obliques). As we note earlier, we need to be careful about extrapolating this analysis of the lumbar multifidi potential actions to segments in the thoracic spine.

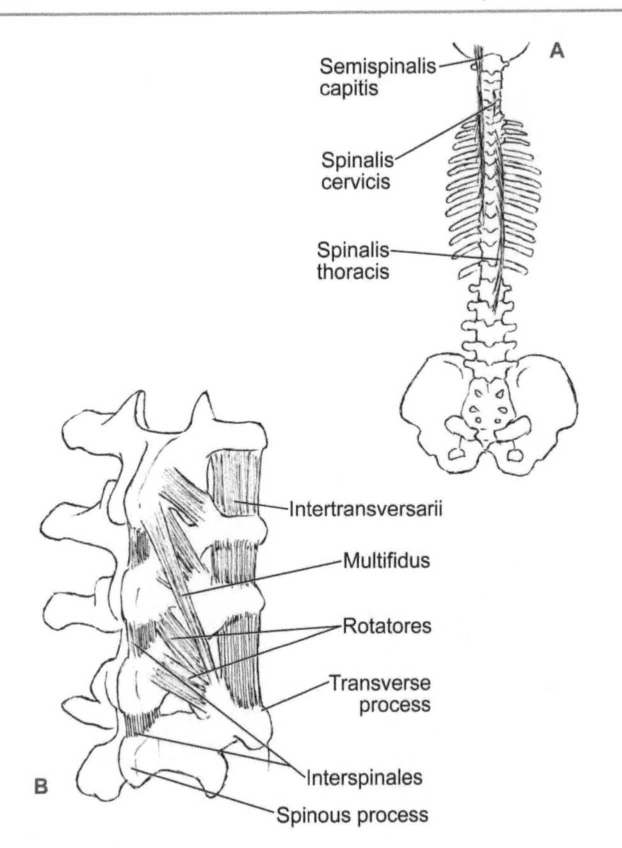


**Fig. 8.28** Posterior view of (**A**) semispinalis group and (**B**) deep muscles forming the transversospinal group

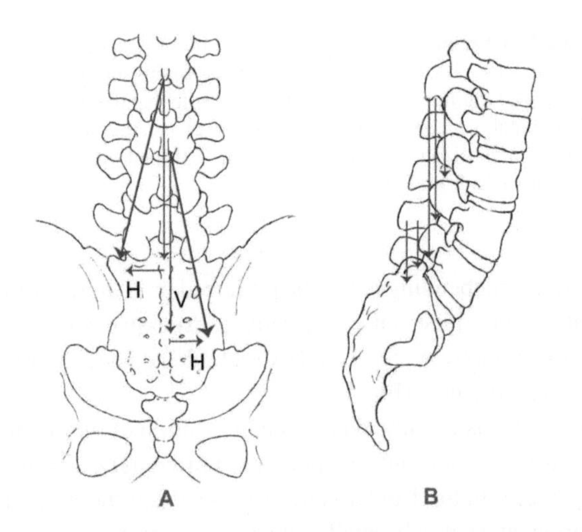


**Fig. 8.29** Schematic of decomposition of the multifidi force vector into vertical and horizontal components from (**A**) frontal and (**B**) sagittal views, noting that the horizontal components are of a smaller magnitude suggesting a lesser role of multifidi in sidebending and/or axial rotation. (Modified with permission from Bogduk (2012))

Now, the forces used to raise the ax and move the thorax were such that controlled positioning of the respective joints was made to optimize the effectiveness of the swinging ax. To accomplish this control, the agonists acted concentrically, while the contralateral, i.e., antagonistic, muscles acted eccentrically, with both groups acting at relatively slow shortening and lengthening velocities, respectively. Since the preparatory phase involved controlled concentric and eccentric muscle actions, we can conclude that the tensile stresses on the contractile and non-contractile elements of each muscle were relatively small during the preparatory phase of swinging the ax. Therefore, although minor muscle strains were possible, we can hypothesize that they were not likely, which is corroborated by Scott's physical examination findings.

As we noted earlier, the thoracic cage is unique in that it houses a great number of potential degrees of freedom of movement, yet its range of motion is moderately constrained in each plane, i.e., the chest wall sacrifices mobility for stability. Thus, there are many joints, and associated soft tissues, that can share in the distribution and mitigation of the internal forces generated by muscle actions on joints and their stabilizing structures. During the wind-up phase, it is likely that these forces were relatively small, yet could induce minor strain injuries. Let's examine the nature and location of these internal stresses. On the right side of the thoracic spine, the stresses were likely distributed in this fashion:

1. Muscle forces make the largest contribution to increased compression and shear stress on the facet and interbody joints (mitigated primarily by hyaline- and fibro-cartilage).
2. Thoracic spine extension induces facet closure, i.e., approximation, adding compression stress to those joint structures.
3. Right axial rotation induces rotation moments at the interbody, costovertebral, and sternocostal joints and torsion (which creates tension in the restraining soft tissues, e.g., inner and outer annulus of IVD and joint capsules, and shear stress to the rib shaft and chondral cartilages).

On the left side, or concave side of the bent spine, the facet joints must distribute compression and shear stresses. Additionally, since the left side of the intervertebral disc was compressed, displacement of the nucleus to the right likely increased tension on the inner annular fibers. One can imagine that one or more of the structures associated with the joints listed could have sustained a mild strain injury in the early preparatory phase of splitting wood and contributed to Scott's clinical presentation (Review Figs. 8.5, 8.6, and 8.7 as you imagine the movements and structures during the wind-up phase of splitting wood).

During the acceleration phase, the rate of thoracic rotation was significantly increased, suggesting that the internal forces and moments increased as well. During acceleration of the ax and trunk in the direction of its object, the thoracic spine moved from extension, right rotation, and left sidebending into flexion, left rotation, and left sidebending, a much more natural movement pattern of the spine, according to osteopathic theory. A large left rotation moment imposed on the chest wall was produced primarily by the right obliquus externus, left obliquus internus, and left-sided erector spinae, aided perhaps by several smaller synergists, e.g., rotatores, etc. It is notable that the shortening velocities of the prime movers were relatively rapid, while the muscles antagonistic to the prime movers were working eccentrically at a relatively high rate in order to provide control. The potential for strains to the muscles working eccentrically is greater, because of the greater potential for generation of tension (i.e., force-velocity phenomenon) with this type of muscle action. Besides the potential for strain injuries to muscle-tendon actuators, it is also possible that the gamma neuromotor system was over-facilitated most likely related to the muscles that were lengthened rapidly, i.e., those on the right side of the spinal column. When that happens, muscle spindle signals to the spinal cord reflexively initiate contraction of the same muscles. A common outcome of this apparent paradoxical predicament is a chronic state of tonic muscle firing long after the particular movement sequence has ended. Because the rotatores and multifidi are densely populated with muscle spindles, we hypothesize that it is these muscles that may contribute to a vertebral joint impairment similar to what Scott exhibited. Likewise, although the rib intercostals may contribute slightly to the rotations of these bones during vertebral movements, there were likely several spindle-activated muscles that could have contributed to post-injury impairment of rib mobility. For a list of the trunk muscles and their primary and secondary actions, see the table in the supplementary material for this chapter.

Joint bone-on-bone forces increase under the conditions of greater generation of muscle tension with eccentric muscle actions. Therefore, the facet, interbody, and costovertebral joints, i.e., hyaline- and fibro-cartilaginous structures, would have greater compression and shear stresses to distribute/mitigate on both the right and left sides of the vertebral column. In addition, the shear stress to the annulus of the IVD, shafts of the ribs, and chondrocartilages would also have increased. However, there would have been an asymmetry of increased tensile stresses on ligamentous constraints, e.g., joint capsules, intertransverse ligaments, etc., with greater stress on the convex side of the lateral bending curve.

Of course, in order to split the wood, this accelerating force needed to be maintained until the wood was struck and split. We define this force as impulsive because of its magnitude and mitigation over a short time period. This impulsive force would generally add to the increased joint compression and shear loads, but likely be distributed reasonably well through the many joints of the arms, and culminating in the vertebral column.

Based on Scott's medical and injury history, the self-imposed injuries he sustained are not likely related to repetitive overuse (as we saw with Brad, i.e., Chap. 4), but can be situated with one specific episode. Fortunately, we had an opportunity to evaluate Scott soon after the incident. Given our evaluation of the mechanism of injury and hypothetical listing of possible structural injuries, let's relook at Scott's physical examination data and suggest a possible relationship between these two data sets:

> In the frontal posterior view, with the patient standing, we noted a mild c-curve in the mid-thoracic (T5 to T8) spine (convex right) (Fig. 8.3B). When Scott was asked if he was aware of the scoliotic curve in the mid-back region, he denied any knowledge of it. At the end range of the standing forward flexion test, palpation of the transverse processes of T5 to T8 revealed a posterior prominence on the left side. Trunk extension was not tested in standing. With the testing of trunk hyperextension when Scott was placed in the prone position palpation of the transverse processes revealed no asymmetry between the right and left sides, i.e. the spine was straight. Trunk sidebending (tested in standing) and rotation (tested while Scott was seated) were limited bilaterally but were more limited toward the right compared to motion testing toward the left. With over-pressure at the end of the range of motion of right rotation Scott's symptoms were somewhat reproduced and a muscle guarding end-feel was notable. He was tender to palpation over the spinous processes from T7 to T10, and over the rib angles and costochondral junction of the corresponding ribs on the right side. On testing of gross respiratory function, we found mildly decreased rib cage movement with inhalation and exhalation, bilaterally, but more so on the right side.

A c-curve is another name for scoliosis. Scoliosis is defined as a sideways curvature of the spine. It is a common finding in the general population and can be categorized as structural or functional. Structural scoliosis is a fixed, i.e., permanent plastic changes in vertebral and chest wall elements, and often progressive curvature of the spine that can be c- or s-shaped. Its manifestations do not change with different postural, i.e., natural pose or with the trunk flexed or extended, positions. In structural scoliosis, the associated rib cage deformity is most obvious in fully flexed posture, i.e., rib hump on the convex side (Fig. 8.30). As seen in the figure, the ribs move and become distorted as the vertebral bodies rotate multi-dimensionally and deform. Structural scoliosis is an interesting topic and would take its own chapter to thoroughly explore, so we leave it to the curious reader to explore this further on their own.

Scott had never previously been told by his parents or health care providers that he had scoliosis, so we hypothesize that he presented with functional scoliosis. Functional scoliosis presents as a spinal curvature that is not fixed and is usually associated with another condition such as a leg length discrepancy. For example, the spine curvature associated with a leg length discrepancy is an adaptation by the spine to compensate for the asymmetry created by the leg length difference. The cardinal sign in a physical examination that a spinal curve is functional is that it changes its appearance with a change in posture, which is what we saw during Scott's examination. In Scott's case, his scoliotic curve was apparent in standing and appeared to be exacerbated when we placed him in a fully flexed posture, but his curvature

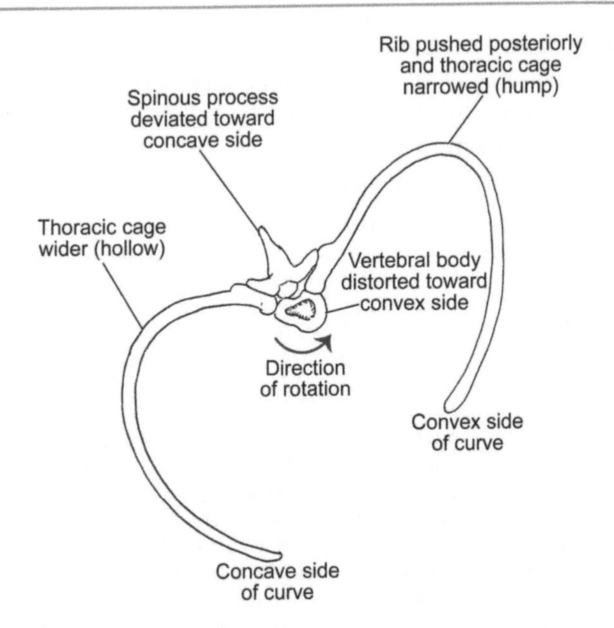


**Fig. 8.30** Schematic of antero-superior vies of a scoliotic segment illustrating the thoracic vertebral deformation and rotation that forces the rib cage to deform as well

disappeared when we asked him to hyperextend his back (Fig. 8.4). Scott did not have a leg length difference, so we hypothesized that his scoliosis was related to impaired movements within the thoracic spine and rib cage. Let us explain.

When we tested the active range of motion of Scott's trunk, we found that he had a mild limitation of right sidebending and rotation, compared to left sidebending and rotation. Right sidebending and rotation could be limited by (1) inability of the facets on the right side to close or for the facets on the left side to open, OR (2) inability of the ribs on the right to rotate posteriorly or the ribs on the left to rotate anteriorly, OR (3) some combination of both. With palpation of the transverse processes, we can make some assumptions about the direction vertebra are rotated and perhaps held in that position. In Scott's case, when he was placed in a flexed position, we found that the transverse processes of T5 to T8 were more posterior on the left side suggesting that those vertebral segments were rotated to the left; and based on Fryette's principles, the segments were also sidebent to the left (this would be labeled as a type II somatic dysfunction). Why would those vertebrae be rotated left? Recall, that with flexion of the spine, the facet joints should glide symmetrically in an antero-superior direction, i.e., an equal amount on both the right and left sides. We often use the phrase "during flexion the facets open," meaning that when they glide, as described, the intervertebral foramen opening increases, or the foramen "opens." However, when Scott flexed it appeared that the facets of vertebral segments T5 to T8 opened normally on the right side but did not on the left side. Mechanically, when facets do not move as they should, they essentially become the axis of rotation about which the vertebra rotates. Thus, we concluded that when Scott was placed in a flexed position, the segments that appeared, based on palpation, asymmetrical were rotated (and sidebent) to the left.

In order to verify our hypothesis of facet opening dysfunctions, we tested spinal extension. Normally during extension, the facets should glide symmetrically in a postero-inferior direction, to an equal degree on both the right and left sides. With facet gliding as described, the intervertebral foramen diameter is reduced, and we often refer to this as facet closure. If proper facet closure occurs, we expect to find symmetry in the positioning of the transverse processes during palpation. Since we found that spine hyperextension straightened Scott's spine, and noted transverse process symmetry, we concluded that the facet joints had closed properly.

An obvious question remains. What caused this condition? There is no one simple answer. So, all we can do is re-examine the list of hypotheses we develop with regard to the tissues that might have been strained as a result of the mechanism of injury, and then try to prioritize the list. We leave the bulk of this process to the reader but provide a cliff-note version of the process. As stated earlier, it is not likely that strain injuries were related to the sequence of events during the wind-up phase. On the other hand, we cannot rule that out entirely because there are occasions when joint dysfunctions are created even under normal loading conditions. However, it is more likely that any strain injury that occurred was likely related to the acceleration phase when dynamic forces were greater. Therefore, the most likely scenario is that Scott sustained mild strain injuries to the facet capsules and outer annulus of the IVD, as well as to one or more of the small, deep muscles of the thorax, e.g., multifidi. Strain to the structures listed would cause an inflammatory response, which includes mild bleeding and effusion, that would elicit arthrogenic pain and a muscle spasm (or guarding) response. This muscle response is typically protective, but also constrains joint, i.e., facet, motion.

But there is something else going on because Scott also had tenderness to palpation over the rib angles, posteriorly, and at the costochondral junctions anteriorly. Additionally, upon motion testing of the ribs by asking Scott to forcibly inhale and exhale, we found that he had reduced elevation and depression of the ribs on the right side, compared to the left. Based on these findings, we hypothesize that Scott also had some combination of costovertebral and sternocostal joint impairment. As before, we wonder what specific tissues might be involved. Since, as discussed earlier, respiratory muscles do not contribute significantly to thoracic rotation, we might rule out a muscle strain of specific respiratory muscles. On the other hand, it has been shown that intercostals are electrically active when the thorax is rotated and held, so it is possible Scott sustained minor internal and/or external intercostal muscle strains on the right side that was responsible for the restricted elevation and depression of the involved ribs. It was also possible that during the acceleration phase of the wood-splitting act, the facet capsules of the costocorporeal and costotransverse joints, as well as the costochondral cartilages and radiate ligaments anteriorly, were strained, which then could have facilitated segmental, e.g., rotatory or multifidi, muscle guarding.

## 8.7 Chapter Summary

Scott was a middle-aged, overweight individual with a comorbidity of COPD related to a long smoking history. He was otherwise reasonably healthy and active. He reported to his physical therapy appointment with symptoms of 3-day duration likely related to splitting wood. After developing some preliminary ideas about the etiology of Scott's complaints, we first examine the nature of COPD, its etiology, pathophysiology, and pathomechanics, and speculate how this progressive disease may or may not be related to his present status, i.e., musculoskeletal problems. Consider that both his medical diagnosis (of COPD) and the preliminary ideas about his musculoskeletal condition represent two complex systems.

In the past two decades or so, a theory of complex systems encompassing physics, biology, and social science has evolved. This theory is a theory of generalized time-varying interactions between elements that are characterized by states. These interactions typically take place on networks that connect the elements, which may cause the states of the elements to alter over time. The essence of a complex system is that the interacting networks may change and rearrange as a consequence of changes in the states of the elements. Thus, complex systems are systems whose states change as a result of interactions and whose interactions change concurrently as a result of the states. Complex systems show a spectrum of behavior that is adaptive and evolutionary, path-dependent and emergent, can produce and destroy diversity, are prone to collapse, and yet are resilient. The theory of complex systems tries to understand these properties based on foundations, i.e., building blocks, and on the interactions between those building blocks that take place on networks. We cannot do justice to the theory of complexity with this thumbnail sketch, so we encourage the reader who is curious to explore complexity further. In this case, we demonstrate a process of deconstructing the respiratory system and related musculoskeletal system, i.e., the thoracic wall, in an attempt to discern possible interacting parts.

We also perform a qualitative analysis of the biomechanics of splitting wood. We recall from Chap. 4, where we examine the biomechanics of pitching, that deconstructing the whole into its constituent parts is helpful in making something complex, simpler. In our deconstruction of the act of splitting wood, we define the internal forces and moments, where they were applied, and list the tissues subjected to those forces and moments. In that process, we wonder about the possible relationship between the pathomechanics of COPD and the mechanics of injury, as well as the most likely tissues impacted by the wind-up and acceleration phases of the activity.

Because of the complexity of the systems we examine and the nature of the likely mechanism of injury, we cannot make definitive statements about cause and effect. (And, by the way, this is typical of most cases presented to healthcare practitioners.) However, we can conclude that one of the advanced manifestations of COPD, i.e., the barrel chest, probably predisposed Scott to musculoskeletal injuries to the thoracic cage because of the limitations in mobility imposed by the plastic changes in the spine and rib cage associated with the barrel chest. In our final analysis, we also conclude that Scott's symptoms were related to impaired mobility of several facet joints in the mid-thoracic region, and dysfunctional respiratory mechanics associated with impaired mobility of the costovertebral and sternocostal joints.

We want to remind the reader that throughout this case you are often referred to our work in previous chapters. In other words, you are asked to draw on previously established anatomical and biomechanical generalizations and apply them to the specifics of the current case. What can you generalize or abstract from this case in anticipation of our next problem(s)?

## Bibliography


Baraldo S, Turato G, Beghé B et al (2002) Pathology of chronic obstructive pulmonary disease and asthma. In: Aliverti A, Brusasco V, Macklem PT, Pedotti A (eds) Mechanics of breathing, physiology, diagnosis and treatment. Springer, Milano, pp 183–193

Baraldo S, Turato G, Saetta M (2012) Pathophysiology of the small airways in chronic obstructive pulmonary disease. Respiration 84:89–97. https://doi.org/10.1159/000341382

Bogduk N (2012) The lumbar muscles and their fasciae. In: Bogduk N (ed) Clinical and radiological anatomy of the lumbar spine, 5th edn. Churchill Livingstone, Elsevier, Edinburgh

Boissonnault WG (1999) Prevalence of comorbid conditions, surgeries, and medication use in a physical therapy outpatient population: a multicentered study. J Orthop Sports Phys Ther 29:506–525

Bonini M, Usmani OS (2015) The role of small airways in the pathophysiology of asthma and chronic obstructive pulmonary disease. Ther Adv Respir Dis 9(6):281–293. https://doi.org/10.1177/1753465815588064

Brusasco V (2002) Structure-to-function relationships in chronic obstructive pulmonary disease and asthma. In: Aliverti A, Brusasco V, Macklem PT, Pedotti A (eds) Mechanics of breathing, physiology, diagnosis and treatment. Springer, Milano, pp 194–200

Butler JE, Hudson AL, Gandevia SC (2014) The neural control of human inspiratory muscles. Prog Brain Res 209:295–308. https://doi.org/10.1016/B978-0-444-63274.00015-1

Celli BRV (2002) Pathophysiology of chronic obstructive pulmonary disease. In: Aliverti A, Brusasco V, Macklem PT, Pedotti A (eds) Mechanics of breathing, physiology, diagnosis and treatment. Springer, Milano, pp 218–231

Coronado RA, Alappattu MJ, Hart DL et al (2011) Total number and severity of comorbidities do not differ based on anatomical region of musculoskeletal pain. J Orthop Sports Phys Ther 41(7):477–485. https://doi.org/10.2519/jospt.2011.3686

Cosio M, Ghezzo H, Hogg JC et al (1977) The relations between structural changes in small airways and pulmonary-function tests. N Engl J Med 296(23):1277–1281

DeStefano L (2017) Normal vertebral motion. In: De Stefano K (ed) Greenman's principles of manual medicine, 5th edn. Wolters Kluwer, Philadelphia

Factors affecting pulmonary ventilation: airway Resistance (2020, Aug 13) Retrieved 21 Aug 2021. https://med.libretexts.org/@go/page/7991

Hogg JC (2004) Pathophysiology of airflow limitation in chronic obstructive pulmonary disease. Lancet 364:709–721

Hudson AL, Butler JE, Gandevia SC et al (2010) Interplay between the inspiratory and postural functions of the human parasternal intercostal muscles. J Neurophysiol 103:1622–1629. https://doi.org/10.1152/jn.00887.2009

Hudson AL, Butler JE, Gandevia SC et al (2011) Role of the diaphragm in trunk rotation in humans. J Neurophysiol 106:1622–1628. https://doi.org/10.1152/jn.00155.2011

Kaltenborn FM (2018) Spinal movement. In: Kaltenborn FM (ed) Manual mobilization of the joints, volume II, the spine, 7th edn. Orthopedic Physical Therapy Products

Kaminsky DA (2012) What does airway resistance tell us about lung function? Respir Care 57(1):85–96. https://doi.org/10.4187/respcare.01411

Kapandji AI (2019) The thoracic spine and the thorax. In: Kapandji AI (ed) The physiology of the joints, volume 3, the spinal column, pelvic girdle and head, 7th edn. Handspring Publishing Limited, Scotland

Liebsch C, Wilke H-J (2018) Basic biomechanics of the thoracic spine and rib cage. In: Galbusera F, Wilke H-J (eds) Basic biomechanics of the spine, basic concepts, spinal disorders and treatments. Elservier, Academic Press, London, UK, pp 35–50. https://doi.org/10.1016/B978-0-12-812851-0.00003-3

Macklem PT (1998) The physiology of small airways. Am J Respir Crit Care Med 157:5181–5183

MacNee W (2006) ABC of chronic obstructive pulmonary disease, pathology, pathogenesis and pathophysiology. Br Med J 332:1202–1204

Neumann DA (2017) Kinesiology of mastication and ventilation. In: Neumann DA (ed) Kinesiology of the musculoskeletal system, foundations for rehabilitation, 3rd edn. Elsevier, St, Louis, MO

Nordon M, Frankel VH (2022) Basic biomechanics of the musculoskeletal system, 5th edn. Wolters Kluwer, Philadelphia

Rimmer KP, Ford GT, Whitelaw WA (1995) Interaction between postural and respiratory control of human intercostal muscles. J Appl Physiol 79(5):1556–1561

Sizer PS, Brismée J-M, Cook C (2007) Coupling behavior of the thoracic spine: a systematic review of the literature. J Manipulative Physiol Ther 30:390–399. https://doi.org/10.1016/j.jmpt.2007.04.009

Stewart JI, Criner GJ (2013) The small airways in chronic obstructive pulmonary disease, pathology and effects on disease progression and survival. Curr Opin Pulm Med 19(2):109–115

Thurner S, Hanel R, Klimek P (2018) Introduction to the theory of complex systems. Oxford University Press, Oxford, UK

Tu J, Inthavong K, Ahmadi G (2013a) The human respiratory system. In: Computational fluid and particle dynamics in the human respiratory system. Springer Science+Business Media Dordrecht, Dordrecht, pp 19–44. https://doi.org/10.1007/978-94-007-4488-2_2

Tu J, Inthavong K, Ahmadi G (2013b) Fundamentals of fluid dynamics. In: Computational fluid and particle dynamics in the human respiratory system. Springer Science+Business Media Dordrecht, Dordrecht, pp 101–138. https://doi.org/10.1007/978-94-007-4488-2_2

White AA, Panjabi MM (1990) Physical properties and functional biomechanics of the spine. In: White AA (ed) Panjabi clinical biomechanics of the spine, 3rd edn. J.B. Lippincott Company, Philadelphia

Whitelaw WA, Ford GT, Rimmer KP et al (1992) Intercostal muscles are used during rotation of the thorax in humans. J Appl Physiol 72(5):1940–1944

# 9 Lumbopelvic Region: Chronic Low Back Pain

## 9.1 Introduction

This case invites us to examine the lumbar spine and its relationship to the sacrum and pelvis. Like the shoulder case, a medical diagnosis for this case was provided as well; thus our task is to clarify it from a biomechanical perspective. One thing we do in regard to this case is examine lumbar and sacral kinematics and revisit the biomechanics of bone and the intervertebral disc. Additionally, as part of our analysis of the probable mechanism of injury, we search the sports medicine/biomechanics literature to see what it reveals about football-related stress fractures (review Chap. 3 for our first discussion of stress fractures). As in previous cases, this one has layers of complexity due to the roles played by the lumbar spine, pelvis, and lower extremities. Finally, we solve statics problems whose solutions may provide some insight into the chronicity of the pain.

Therefore, for this case, we explore and explain the (1) original mechanism of injury, (2) biomechanical rationale for clinical special tests, (3) stress fracture mechanics, (4) pathomechanics of the lumbopelvic region associated with the medical diagnosis (i.e., abnormal kinematics and kinetics), (5) biomechanical and functional implications of the solutions to lumbar statics problems, and (6) potential sources of mechanical pain generators, i.e., hypotheses.

## 9.2 Case Presentation

Jim Bishop was a 35-year-old small business owner who repaired small motors and generators. Jim was referred to our care with a diagnosis of chronic central low back pain. Jim recalled having this pain, more or less, since his college football playing days; he was a linebacker. Over the years he learned to cope with the pain, but prior to his visit with us, it seemed to be more bothersome, so he sought medical advice.

His symptoms included a constant, deep ache, which was worse with prolonged standing and at the end of the day. Jim was fairly active in regular physical exercise, e.g., weight- and cross-aerobic training, and hunting and fishing. His overall health was otherwise particularly good.

Jim presented with a meso-endomorphic, athletic-looking, body type. Observation of Jim's gait revealed no observable deviations. On physical examination, we noted a moderate increase in his lumbar lordosis/anterior pelvic tilt, with a "step-off" at L5/S1. Jim had a mild right lumbar scoliosis (c-curve convex right). We found the following with active motion testing of the trunk:

1. Flexion mildly limited with abnormal lumbopelvic rhythm, i.e., limited lumbar lordotic curve reversal with mild deviation to the left during the flexion phase, and with deviation to the right during the return extension phase.
2. Extension markedly limited with moderate pain at L5/S1.
3. Sidebending and axial rotation range of motion mildly limited bilaterally, but symmetrical without reproduction of symptoms.

Two special tests for the lumbar spine were performed. Jim had sharp central pain (L5/S1) with the quadrant test bilaterally. With the prone instability test, Jim had moderate muscle guarding with legs relaxed (hanging over the end of the examination table), and when the test was repeated with the hips extended and legs raised to be parallel with the tabletop. The lower extremity range of motion was normal, but he had moderate hamstring tightness bilaterally. Jim's right leg was shorter by 15 mm. Palpation revealed increased "resting tone" (often referred to as muscle spasms) in the erector spinae muscles and point tenderness over the posterior superior iliac spines bilaterally, and the L4 and L5 spinous processes. Neurological and vascular tests, i.e., upper and lower extremity


G. J. Alderink, B. M. Ashby, *Clinical Kinesiology and Biomechanics*, https://doi.org/10.1007/978-3-031-25322-5_9

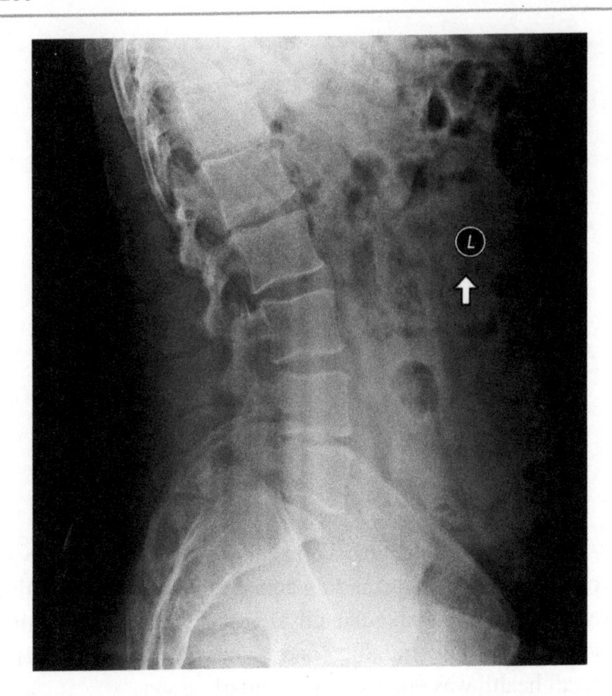


**Fig. 9.1** Lateral (L) radiograph showing apparent isthmic spondylolisthesis at L5/S1 measuring 9 mm with flexion and 7 mm with extension. There is a 2 mm retrolisthesis of L4 on L5. Disc space narrowing is most pronounced at L4/L5 and L5/S1

pulses, were normal. Radiographic images of the lumbar spine that were ordered by his primary care physician showed a fracture of the pars interarticularis of the L5 segment; the radiologist's interpretation was spondylolytic spondylolisthesis with forward slippage (Fig. 9.1).

## 9.3 Key Facts, Learning Issues, and Hypotheses

Jim's job entailed repairing small motors, suggesting that his work likely involved prolonged standing over a workbench or table, repetitive pushing, pulling and manipulating various tools, and raising and lowering motors either on the workbench or to-and-from the floor. We were interested in the details of Jim's work situation, including the average weight of the motors he was repairing, and asked him to demonstrate the variety of postures he assumed throughout the workday.

Based on the posture examination, we know that he had a mild leg length difference, possibly related to his scoliosis, and an accentuated anterior pelvic tilt/lumbar lordosis. Early hypotheses regarding his chronic pain, and possible pain generators, appear to be related to his static working postures, the length of time he resides in those postures, and related external forces and moments placed on his lower back. The fact that he maintained a regular exercise program suggested that his trunk (i.e., core) and lower extremity

**Table 9.1** Neurological tests of the lumbosacral plexus

| Nerve root | Myotome (motor) | Dermatome (sensory) | Reflex |
|---|---|---|---|
| L2 | Hip flexors, knee extensors | Anterior thigh | None (partial patellar) |
| L3 | Knee extensors | Lower anterior thigh | Patellar |
| L4 | Tibialis anterior | Medial calf/ medial foot | Patellar |
| L5 | Extensor hallucis longus | Lateral calf/ dorsal aspect foot | None (tibialis posterior) |
| S1 | Peroneus longus/ brevis, gastrocnemius | Lateral foot | Achilles |
| S2-S4 | Intrinsic foot muscles; bladder | Anus | None (superficial anal reflex) |

strength (and endurance) would likely be better than average, and something that predicted a good prognosis with proper treatment (note: we do not discuss treatment intervention in this chapter).

With an active range of motion testing of trunk flexion, we observed reduced lumbar curve reversal (i.e., from the neutral lordotic or extended position), a moderate reduction in the total range of motion, and abnormal coupling of lumbar and pelvic movements, i.e., lumbopelvic rhythm. (We need to define lumbopelvic rhythm and think about its pathomechanics and possible relationship to the spondylolytic spondylolisthesis.) Trunk extension reproduced Jim's pain, which may have contributed to the loss of range of motion that we observed. Reviewing the osteo- and arthrokinematics of lumbar segments may help explain these signs and symptoms related to his medical diagnosis.

Based on the medical diagnosis and the findings associated with lumbar extension testing, two additional special tests were performed. The quadrant test, which traditionally attempts to provocate symptoms related to degenerative disc disease (DDD), may also provocate symptoms in someone with a spondylolytic condition, similar to Jim's. Whereas the prone stability test specifically targets possible hypermobility/instability in the lower lumbar region secondary to DDD and other destabilizing conditions (like Jim's). We demonstrate how these tests are performed and discuss the external forces and moments that these tests induce on specific biological structures. Typically, lumbar DDD may lead to nerve palsy and/or inflammation secondary to constant or intermittent impingement (or compression) on the nerve roots exiting the intervertebral foramen. However, since the neurological tests that were performed, which included myotomes (specific muscles associated with specific nerve roots), dermatomes (sensory tests, e.g., light touch, etc., over the skin of specific regions of the lower extremities), and reflexes (Table 9.1), were normal, we hypothesized that Jim did not

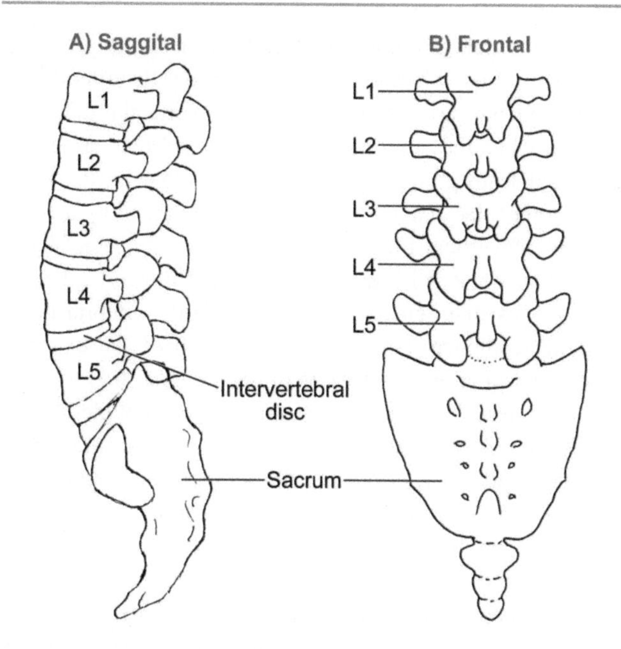


**Fig. 9.2** Schematic of (**A**) lateral (sagittal) and (**B**) posterior (frontal) views of the lumbosacral spine. In the lateral view note the angulation of the interbody joint between L5 and the sacral base

have any neurological impairments (a good thing) and that his signs and symptoms were more likely the results of other mechanical stresses.

## 9.4 Lumbopelvic Structure and Function

The primary function of this region of interest is to distribute and mitigate the large compression and shear stresses that are imposed by the forces of gravity and work-related activities created by the superincumbent weight of the head, arms, and trunk, as well as the ground reaction forces that are distributed from the feet through the pelvis. In examining the specific morphologies of bony structures in this region and their supporting soft tissues, we are reminded of the common adage, "function dictates structure."

### 9.4.1 Structure

The lumbar and sacral region of the spine consists of five separate (usually) lumbar vertebrae, named according to their location in the intact column, e.g., L1, L2, etc., and the sacrum (Fig. 9.2). In a sagittal view, the normal standing resting position ("neutral") of the lumbar curve is described as lordotic, i.e., convex anterior, which places the lumbar vertebral segments in a relatively extended position. Three angles are typically used to describe the lumbosacral spine: (1) the angle formed by the superior aspect of the L1 vertebra and the sacrum (a measure of lumbar lordosis), (2) the sacral angle formed by the base

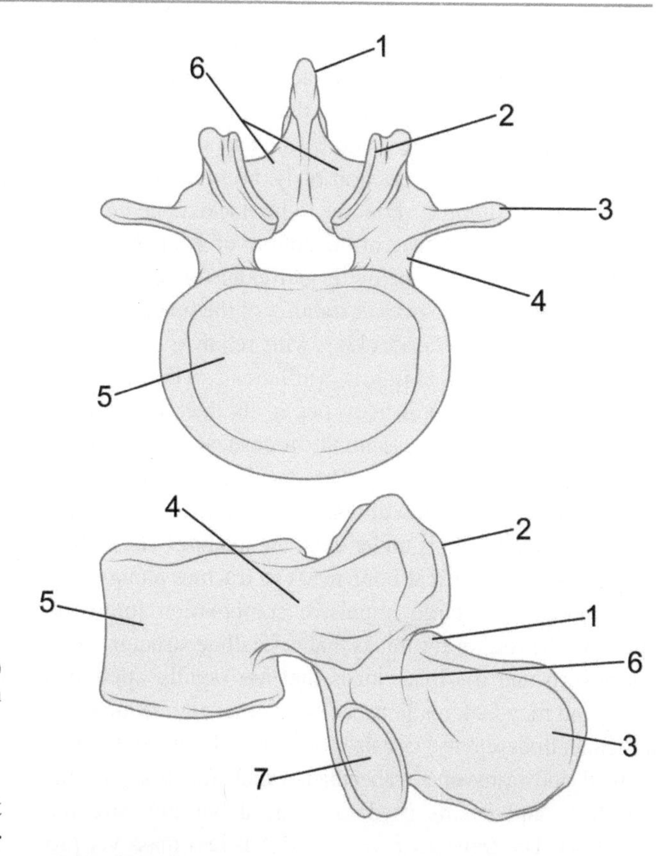


**Fig. 9.3** Schematic of key parts of a typical lumbar vertebra

of the sacrum and the horizontal plane (i.e., normally about 40°), and (3) the angle between the inferior aspect of L5 and the base of the sacrum (i.e., normally about 16°). Radiographs are required to document and quantify the magnitude of these angles.

In Jim's case, both the sacral angle, which was approximately 63°, and the angle between L5 and the base of the sacrum were greater than normal. It is likely that these abnormal angles predisposed Jim to his back injury. We return to this topic and explore it in some detail later in the text.

The lumbar vertebrae are irregular bones with a number of special parts (Fig. 9.3). The anterior part of the vertebra is a box-shaped bone referred to as the vertebral body. The top and bottom of the surface of the vertebral body are smooth and perforated by tiny holes and ringed by a smoother outer perimeter with fewer perforations, which represents the fused ring apophysis. The posterior surface of the body is obscured for inspection by the posterior elements but is marked by one or more holes known as nutrient foramina, which transmit the nutrient arteries of the vertebral body and the basivertebral veins. The size of lumbar vertebral bodies is larger than their cervical and thoracic counterparts, which is a result of their primary function, that is, reacting to and distributing/mitigating large compression and shear forces.

The flatness of the vertebral body, both the top and bottom, allows the lumbar vertebrae to be stacked one on top of the other. The congruence of the bottom of a superior vertebra relative to the one below it and axial (or longitudinal) loads contribute to the overall stability of the functional lumbar unit. However, the stacking with reliance only on the static axial load for stability is insufficient, so the functional strength of the vertebral body lies in its internal structure. The vertebral body is essentially a cortical shell that surrounds a cancellous cavity. The disadvantages of a body made up of entirely cortical bone are (1) it is too heavy, and (2) it more readily fails under dynamic loading. The crystalline structure of cortical bone tends to fracture along cleavage planes when large impulsive compression forces are applied. The reason for this is that crystalline structures cannot absorb and dissipate loads that are rapidly applied to them, and they lack resilience. The internal architecture, i.e., the cancellous cavity, consists of thin rods of bone called vertical and transverse trabeculae, which function just like the struts and beams used to frame a building structure (Fig. 9.4). The trabecular framework has less mass yet provides the vertebral body with weight-bearing strength and resilience. Thus, when axial or longitudinal dynamic loads are applied, they are first mitigated by the vertical trabeculae, but then when they deform, i.e., bow, the vertical struts are restrained from going beyond some threshold by the horizontal trabeculae. Thus, large loads are sustained by a combination of vertical pressure and transverse tension in the trabeculae; that is, it is the transfer of load from vertical pressure to transverse tension that provides the vertebral body with the resilience it needs to manage all of the external forces associated with activities of daily living. It is notable that this internal structure is porous, which provides channels for the blood supply and venous drainage of the vertebral body, and an accessory site for hemopoiesis (creation of red blood cells). The presence of blood in the cancellous cavity of the vertebral body appears like a sponge (i.e., this region is therefore sometimes referred to as vertebral spongiosa), which actually adds to the viscoelastic material properties of the vertebral body. Although we can see that the internal structure of the vertebral body makes this structure ideally suited to sustain large static and impulsive axial (or longitudinal) loads, it fails to provide adequate horizontal stability.

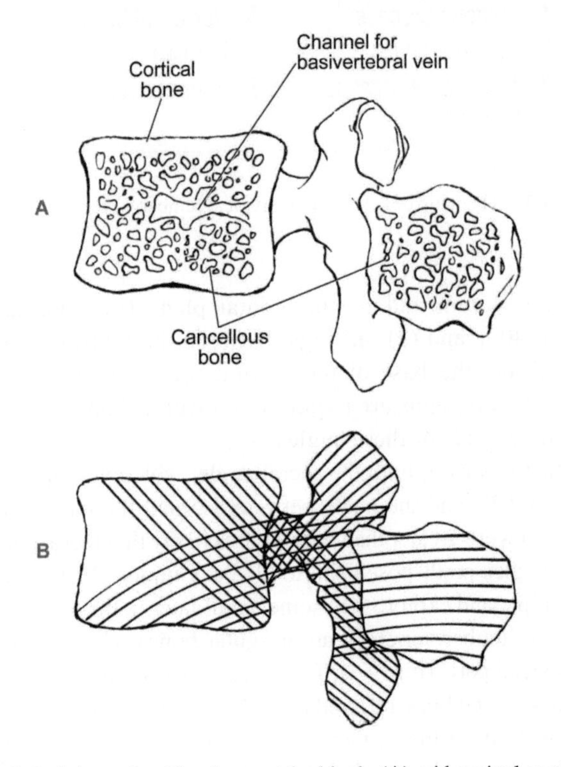


**Fig. 9.4** Schematic of lumbar vertebral body (**A**) mid-sagittal section demarking the outer cortical and inner cancellous (or trabecular) bone and (**B**) lateral sagittal section showing the trabeculae passing through the pedicle into the articular processes. Note that the region regions where the trabecular struts cross are those subjected to multidimensional forces, thus areas of greater strength

Let us now examine a few other important features of the lumbar vertebral segment. Projecting from the back of the vertebral body are two strong pillars of bone, called pedicles (Fig. 9.3). These projections form the legs of the neural arch, so named because it surrounds the spinal cord. The pedicles are the link between the posterior elements and the vertebral bodies. Projecting from each pedicle towards the midline are paired laminae, which fuse in the midline of the vertebrae to form a roof over, or wall around, the neural arch. The laminae are extended and enlarged bilaterally into inferior and superior articular processes. Thus, each vertebra has four articular processes where the medial surface of the superior articular processes and lateral surface of the inferior articular processes are covered by hyaline cartilage, an area referred to as the articular facet. In a functional vertebral unit, the inferior articular processes of the superior segment articulate with the superior articular processes of the inferior segment to form the zygapophyseal (or apophyseal or facet) joints. As one of the posterior elements, the facet joints in the lumbar spine constrain excessive vertebral rotations and forward translations. The articular processes also channel forces towards the pedicles, which then transmit the benefit of these forces to the vertebral bodies. Projecting posteriorly from the junction of the paired laminae are the spinous processes and projecting laterally from the junction of the pedicle and lamina are the transverse processes. The paired lamina and articular, spinous, and transverse processes are generally referred to as the posterior elements, each with unique functions.

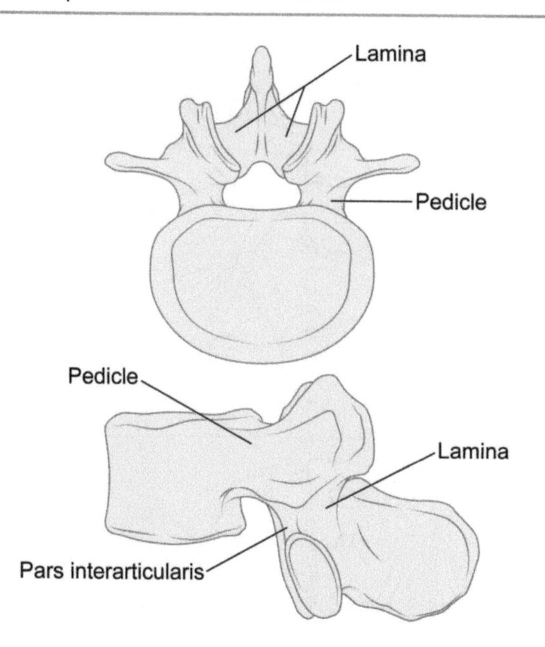


**Fig. 9.5** Schematic demonstrating the relationship among the pedicle, lamina, and pars interarticularis

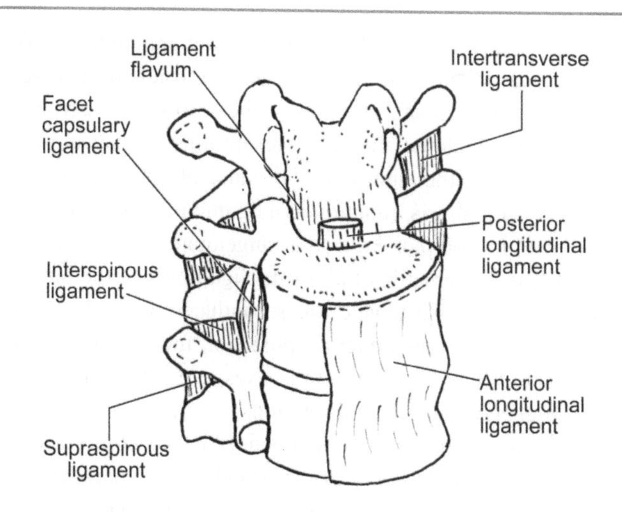


**Fig. 9.6** Schematic of a typical lumbar functional unit illustrating the primary ligaments that provide stability

The spinous and transverse (along with accessory and mamillary, not shown in Fig. 9.3) processes serve as attachment sites for muscles. In fact, every muscle that acts directly on a lumbar vertebra is attached to one or more of the posterior elements, except the crura of the diaphragm and parts of the psoas muscles. In particular, the spinous and transverse processes, because of their length, serve as effective levers for their muscle-tendon partners. Thus, the substantial tensile forces that are generated by muscle-tendon actuators are transmitted to the lamina and pedicles via the posterior elements.

The lamina may have a more substantial mechanical function beyond serving as a roof over the neural elements of the vertebral canal. Of the posterior elements, the laminae are centrally placed so that the various forces that act on the spinous and inferior articular processes are ultimately transmitted to the laminae. The section of the lamina that intervenes between the superior and inferior articular processes on each side is called the pars interarticularis (Fig. 9.5). This region entails the volume from the lateral border of the lamina to its upper border. Since the pars interarticularis lies at the junction of the vertically oriented lamina and horizontally projecting pedicle, it is subjected to considerable bending moments during various normal daily and sports activities.

Fortunately, the cortical bone in the pars interarticularis is generally thicker than in other regions of the lamina (note: the cortical bone here is likely thicker due to the magnitude and frequency of the applied external forces). However, recall that external bending forces induce complex internal bone stresses, which poses a particular challenge to this region of a lumbar vertebra. In fact, the pars interarticularis is a common fracture site secondary to repetitive, excessive, and/or impulsive forces.

The etiology of Jim's diagnosis of spondylolytic spondylolisthesis was likely a stress fracture of the pars interarticularis that was sustained during his adolescence, probably related to his abnormal lumbosacral angle and football activities. Exploration of details surrounding this hypothesis is forthcoming.

The pedicles transmit both tension forces and bending moments. For example, if a vertebral body glides anteriorly, the inferior articular processes of the vertebra lock against the superior articular processes of the next lower vertebra and resist excessive translation. This resistance is transmitted to the vertebral body as tension along the pedicles. Moreover, bending moments induced by muscle action, i.e., by the muscles with attachments to the posterior elements, are also transmitted to the vertebral body via the pedicles. It is no surprise that since the pedicles are stiff, cylindrical bones with relatively thick walls, i.e., they are hollow, they effectively resist bending in any direction. Thus, when a pedicle is bent downwards, the upper wall is placed under tension while its lower wall is compressed. Likewise, if a pedicle is bent medially, its outer wall is tensed, and its inner wall is compressed. From an engineering standpoint, the pedicle responds to forces similar to a beam that resists deformation with its peripheral surfaces, while the forces in the center reduce to zero.

We have discussed the role of the various bony features of a lumbar vertebral segment suggesting that longitudinal stability is inherent to some degree, but not sufficient. Let us take another look at a lumbar functional unit (a superior segment, intervening disc, and inferior segment) and the ligaments that provide translational, shear, and torsional stability (Fig. 9.6). Previously, we discussed the essential features of the intervertebral disc at length (see Chap. 7), including the separate roles played by the inner and outer annual fibers. We learned that in addition to their collective roles in the disc's hydraulic mechanism, the inner fibers form an envelope around the nucleus, therefore resembling a capsule-like structure, while the outer fibers, because of their attachment to the ring apophysis, function as a ligament. It is no surprise that, given its complex layered arrangement, the outer annulus is the most versatile of all of the ligaments of the human body in that it resists tensile stresses induced by lumbar rotations and translations (anterior, posterior, medio-lateral, and axial) about, and along, multiple axes, e.g., flexion, extension, sidebending, and axial rotations.

The anterior longitudinal ligament (ALL) is comprised of short, deep (spanning only one interbody joint), and superficial (spanning two to five interbody joints) layers that run from the cervical spine to the sacrum. The ALL is more developed in the lumbar spine and constrains excessive extension and anterior bowing of the lumbar spine. Bogduk noted that, in the region of the first three lumbar vertebrae, fascial connections from the ALL to the crura of the diaphragm suggest that parts of the ALL may also serve as a tendon.

Similar to the ALL, the posterior longitudinal ligament (PLL) spans the entire spinal column and has deep and superficial fibers. The difference lies in its coverage of the posterior aspect of the lumbar segments; that is, it forms a narrow band over the backs of the vertebral bodies but is broader over the IVDs, which would appear to be an extra safeguard for those structures. The PLL's fibers mesh with the outer annulus, as well as insert onto the posterior margins of the vertebral bodies. The PLL constrains excessive lumbar flexion and can serve as a hammock for minor annular herniations (bulges).

The primary ligaments of the posterior elements include the facet capsules, supra- (SL) and interspinous (IL) ligaments, and ligamentum flavum (LF). The three additional ligaments that could be included as part of the posterior element group, namely, the intertransverse, transforaminal, and mamillo-accessory ligaments, do not provide the typical constraint function of ordinary ligaments and are considered false ligaments. The ligamentum flavum is a short, thick ligament that connects consecutive laminae of two adjoining lumbar segments. The LF is unique among all lumbar ligaments because of the histological dominance of elastin (80%) over collagen (20%). Since this ligament is immediately posterior to the vertebral canal, biomechanists speculate that the most important function of the LF is to help protect neural elements as the lumbar spine returns from flexion postures or movements. Some have also suggested that the elastic energy stored in the deformed LF following flexion contributes to the lumbar extensor moment that initiates a return from flexed positions. The interspinous ligaments connect adjacent spinous processes. Intuition would suggest that the IL would constrain separation of the spinous processes; however, that is not the case. Therefore, the interspinous ligaments provide very little to no control of excessive lumbar flexion. The supraspinous ligament lies in the midline, bridges the interspinous spaces, and consists of three layers: superficial, middle, and deep. The histological nature of all three layers finds intertwining tendinous fibers of the dorsal layer of the thoracolumbar fascia and aponeurosis of the longissimus thoracis, suggesting that the SL may function more like an extended tendon and less like a ligament.

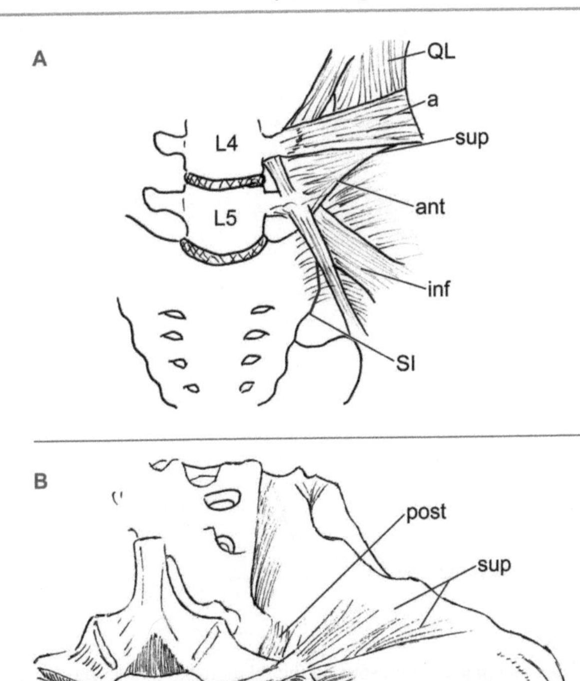


**Fig. 9.7** Schematic of (**A**) anterior and (**B**) posterior views of the iliolumbar ligaments. QL = quadratus lumborum, a = anterior layer of the thoracolumbar fascia, ant = anterior band, inf = inferior band, post = posterior band, sup = superior band, ver = vertical band, SI = sacroiliac band

The iliolumbar ligaments (ILL) connect the transverse processes of L5 to the ilium (there may also be slips of tissue coming off the transverse process of L4). It is reported that there are several parts to this ligament, but all of them transverse the distance from L5 to an area on the anteromedial surface of the ilium (Fig. 9.7). The description of this ligament brings an interesting history because there has been disagreement about its existence, with claims that the complex was more likely the anterior fascia of the quadratus lumborum because it lacked the features of a multi-dimensional

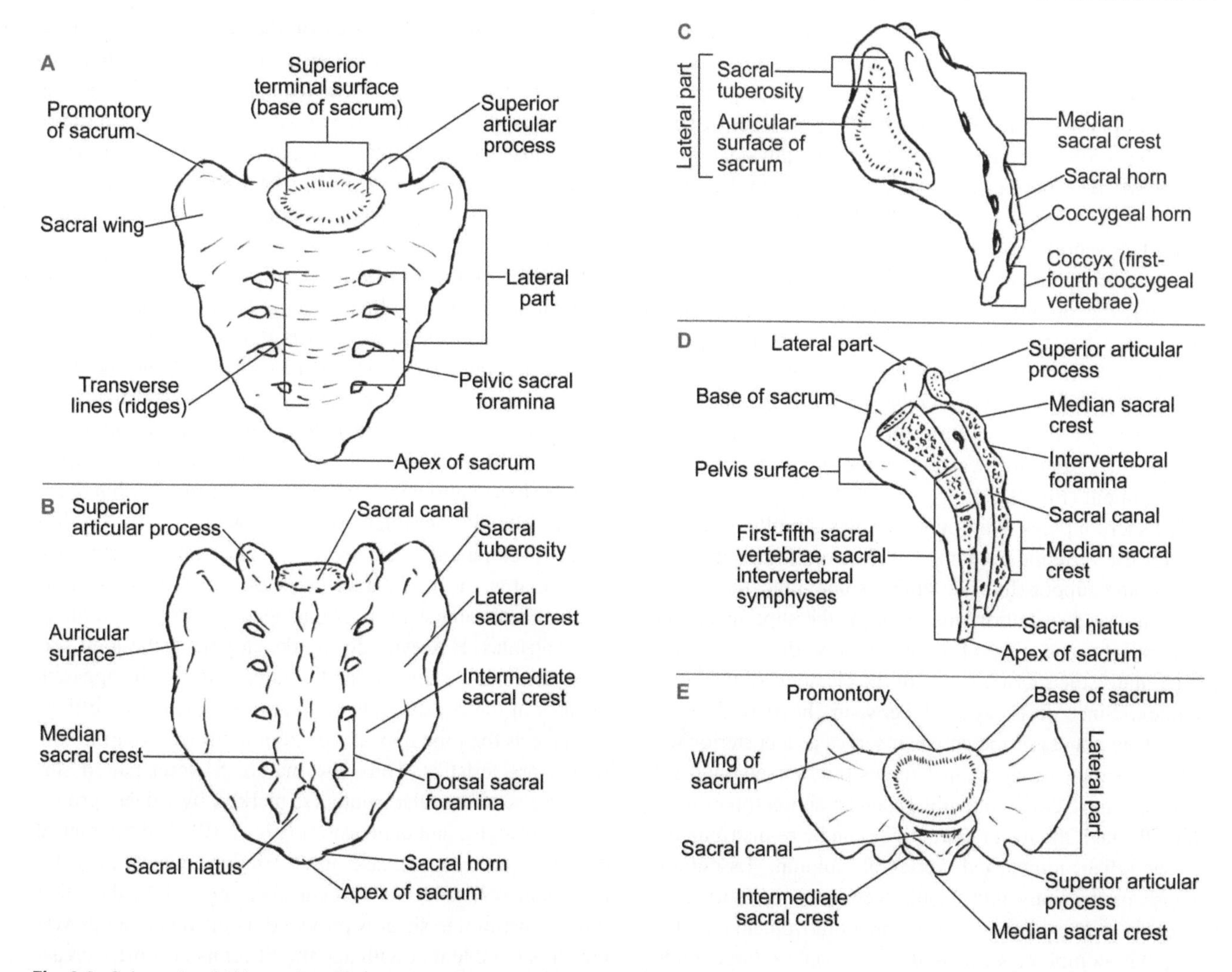


**Fig. 9.8** Schematic of (**A**) anterior, (**B**) posterior, (**C**) lateral, (**D**) sagittal longitudinal cross-section, and (**E**) superior views of the sacrum

collagen-laced structure consistent with a true ligament. Others have demonstrated that in neonates and children the mass of tissue in the region of the ILL was represented by a bundle of muscle, which develops into a ligament by the third decade. Regardless of the controversy regarding its origin, biomechanists agree that the ILL forms a strong bond between L5 and the ilium and is primarily responsible for constraining anterior sliding of the L5 segment relative to the sacrum and controls excessive flexion, extension, and side-bending of L5. Moreover, with the lumbar spine fixed, the sacroiliac band of the elements of the ILL restrains forward (nutation) and backward (counternutation) of the sacrum suggesting this ligament also provides mechanical stability to the sacroiliac joints. Based on the likely functions of the ILL, training of the quadratus lumborum may be critical in Jim's rehabilitation.

Let us now examine the characteristics of the sacrum and sacroiliac joints. The sacrum is a large bone that consists of five fused vertebrae (Fig. 9.8). The sacrum has a triangular shape with a broad upper end and tapered caudal end. Its volar surface is concave and relatively smooth, while the somewhat undulating dorsal surface is rougher. Anterior and posterior sacral foramina perforate the volar and dorsal surfaces that provide conduits for sacral ventral primary rami and dorsal primary rami, respectively, as well as sacral blood vessels. Anteriorly there are rectangular regions that resemble embedded vertebral bodies, with transverse ridges that represent vestiges of intervertebral discs. Transverse processes lie lateral to the foramina. Posteriorly, the fused segments show prominent spinous processes along the midline; S1 is typically most palpable. The line of spinous processes is defined as the median sacral crest. The sacral hiatus is an opening created by the failure of the laminae of S5 to meet in the midline and form a spinous process. The rim around the sacral hiatus represents the tubercles of the S5 vertebra that form the sacral horn or cornua. Both the sacral hiatus and cornua are easily palpable and are useful landmarks to assist in the diagnosis of sacral rotations. The sacral base's superior surface matches the inferior surface of L5, while laterally the broad transverse processes of S1 resemble wings (and are referred to as the ala of the sacrum). Posteriorly, the sacral base houses a pair of superior articular processes that articulate with the inferior articular processes of L5 to form the L5/S1 or lumbosacral (zygapophysial) joints (Fig. 9.8A, B). There is a neural arch, similar to the neural arch found in lumbar segments that helps to form a sacral canal. The sacral canal traverses the entire length of the sacrum and opens at the sacral hiatus (Fig. 9.8D, E). Laterally, the sacrum shows two distinct areas: a smooth surface that has the shape of an ear, called the auricular surface, and a more irregular surface appearance posterior to that. The two-armed auricular surface articulates with the innominate bones to form the sacroiliac joints, while the more irregular surface houses the interosseous sacroiliac ligament (Fig. 9.8C).

The sacrum plays two important roles. Axially, it forms a base for the vertebral column, therefore serving as an extension of, and support for, the lumbar spine; thus, the sacrum assists in the distribution/mitigation of the superincumbent body weight and related longitudinally directed forces. Horizontally, the sacrum is one of the elements of the pelvic complex. Since it is wedged between the two ilia (i.e., innominate bones) of the pelvis, it serves as a posterior wall and re-distributes longitudinal forces transversely, and into the lower limbs; concomitantly, ground reaction forces transmitted through the lower extremities can be re-distributed to the sacroiliac joints and vertebral column. Essentially, because of the bony morphology/geometry and soft tissue restraints, and six degrees of freedom of movement (small as it is), this complex is set as it is to manage large multidimensional forces. So, how is this managed?

The function of the sacrum is not accomplished in isolation, but in how it relates to the other parts of the pelvis. The pelvic complex is often referred to as the pelvic ring. In addition to the sacrum, the elements of this ring include a pair of sacroiliac joints, three bones of the hemipelvis (ilium or innominate, pubic, and ischium), and the pubic symphysis (Fig. 9.9). The keystone of the ring appears to be the sacrum, which is wedged between the two ilia, forming paired sacroiliac joints, while the pubic symphysis, anteriorly, provides additional structural integrity to the ring (note: the pelvic ring complex is very much analogous to the thoracic cage… interesting).

Because of the magnitude of forces absorbed and re-distributed by the sacroiliac (SI) joints, it is imperative that these joints are stable to, as Bogduk suggests, "lock the sacrum into the pelvis." The word lock should not be taken literally because if that were the case, we would likely see more fractures within the pelvic ring. But evolution has devised its own solution. The SI joint partners have reciprocally L-shaped auricular surfaces that, during childhood, are smooth, allowing for relatively free movement. The SI joints in the developmental years resemble most other diarthrodial joints; that is, the articular surfaces are covered with hyaline cartilage and joint stability is managed by strong, compliant joint capsules. However, during development, the joint surfaces become irregular and roughened, with apparent changes in the constituents that comprise the joint cartilage interface as the joint appears to adopt modified synarthrodial joint characteristics. Thus, the mature SI joint paired surfaces possess irregular contours, marked by ridges, prominences, troughs, and depressions (Fig. 9.10). The roughened articular surfaces increase joint friction, which provides additional resistance to vertical shearing, while the other major constraint to shear is provided by the ligamentous system. It is notable that with age the SI joints become increasingly less mobile, taking on characteristics of degenerative joints, and can become a source of low back pain.

Thus, the sacrum, in its articulation with the paired ilia, is held firmly in place by a type of self-locking mechanism.

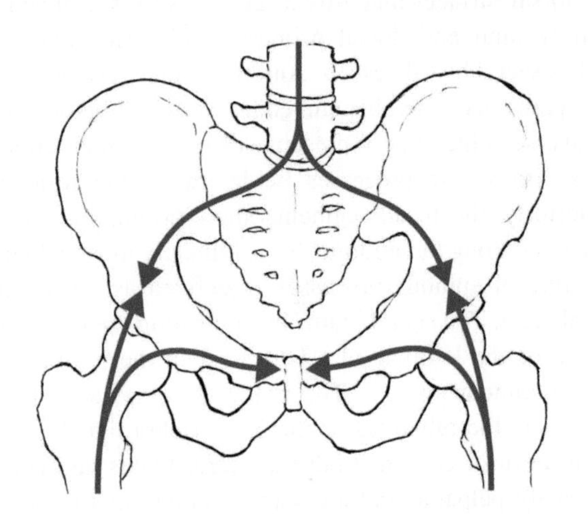

**Fig. 9.9** Elements of the pelvic ring with theoretical re-distribution of axially and horizontally directed forces from above and below

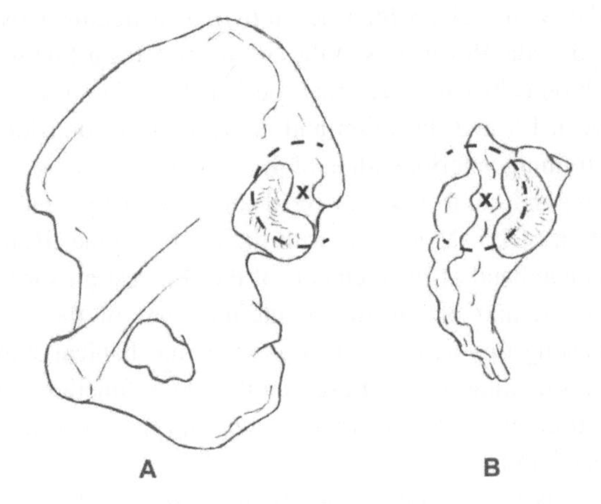


**Fig. 9.10** Exposed sacroiliac joint illustrating (**A**) iliac and (**B**) sacral auricular surfaces. Note the "x" locates the second sacral segment, which approximates a "joint" center for sacral rotation in the sagittal plane

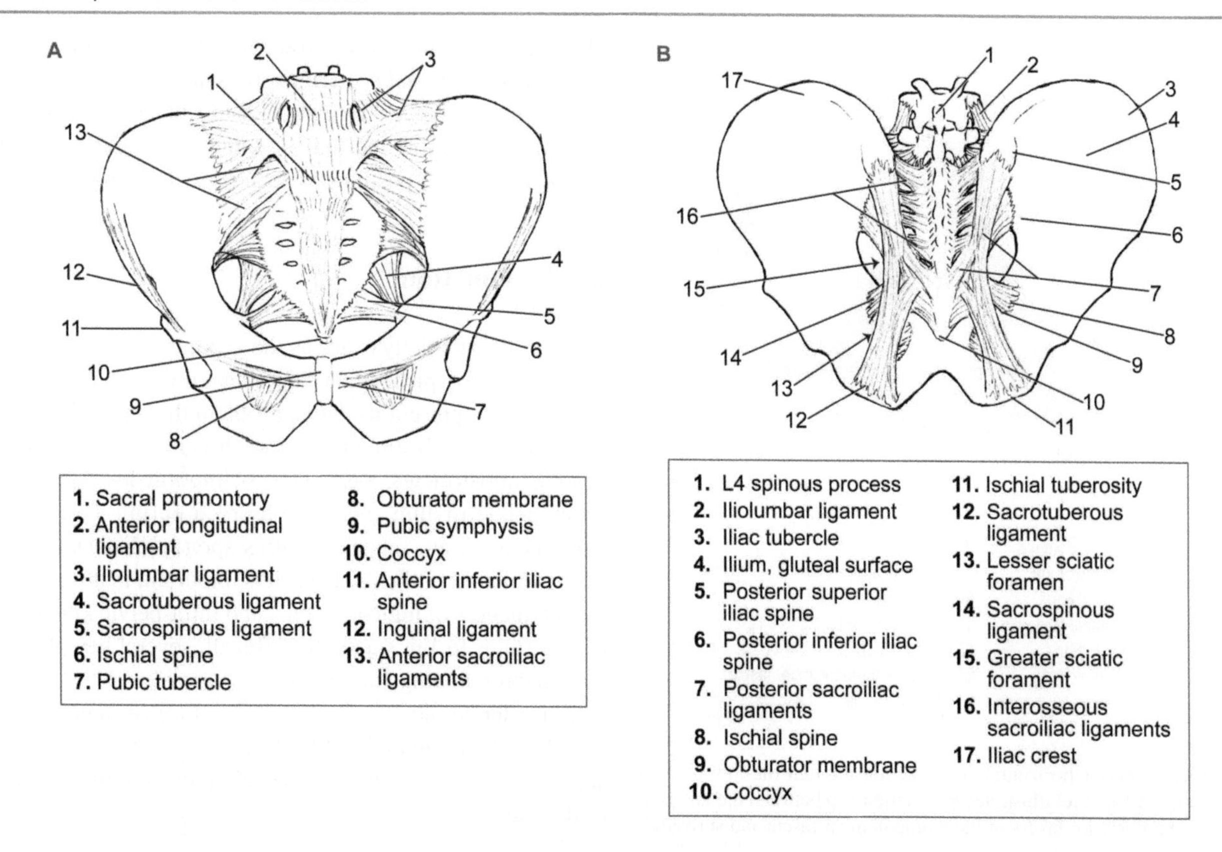


**Fig. 9.11** (**A**) Anterior and (**B**) posterior views of the sacroiliac ligaments. The density of ligaments in this region is an indication of the degree to which it is subjected to large, repetitive and complex forces

Vleeming suggested that the self-locking of the SI joints consisted of form- and force-closure. Form closure relates to the "keystone-like" shape of the sacrum. The sacrum is set obliquely between the ilia such that in a neutral standing stance, its volar surface leans anteriorly so that under axially directed loads the sacral base tends to tilt forward (i.e., nutate) and downward, which is opposed by the wedge-shaped sacrum. This potential forward tilt causes the wider end of the S1 segment to move inferiorly, tending to separate the ilia, while, at the same time, the wider end of the S3 segment moves upward also tending to separate the ilia. Under normal conditions, separation of the paired ilia does not occur because as the sacrum nutates, the ilia press against the sacrum and the engaged (and irregular) joint surfaces prevent the sacrum from sliding caudally. Form closure is necessary, but not sufficient. Thus, during dynamic activities, a force-closure mechanism is essential to assure SI stability. The enhanced stability induced by force-closure comes from the ligaments (from their restraint functions and passive elastic moment potential) associated with the pelvic ring, the thoracolumbar fascia, and muscle activation. It is hypothesized that the relationship among these elements significantly increases SI joint frictional constraints, which augments the friction created by form closure.

There are several dorsal sacroiliac ligaments. Of the many SI ligaments that contribute to force closure, the interosseous ligament has been tabbed as the most important (Fig. 9.11). This thick ligament lies deep

in the narrow recess between the sacrum and ilium posterior to the cavity of the joint. It is obvious by its location that the interosseous ligament's primary function is to reinforce the interlocking of the sacrum and ilia. The posterior sacroiliac ligament lies posterior to the interosseous ligament. Its superior superficial band is often identified as the short posterior sacroiliac ligament. The long version of the posterior sacroiliac ligament runs from the intermediate and lateral crests of the sacrum to the posterior superior iliac spine and inner lip of the iliac crest. While the short posterior sacroiliac ligament augments the function of the interosseous ligament and works to prevent diastasis of the ilium, the longer posterior SI ligament, with its more vertical orientation, checks sacral counternutation, i.e., a posterior tilt of the sacral base. The large, thick sacrotuberous ligament spans the gap between the ischial tuberosity and the lateral margin of the sacrum, transverse tubercles of the lower sacrum, and posterior superior iliac spine; it is notable that this ligament blends with the long posterior SI ligament cranially and tendon of

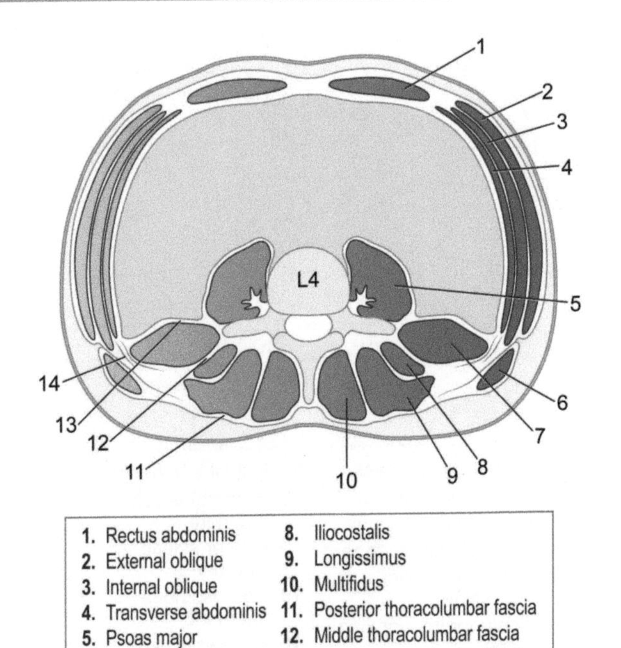


**Fig. 9.12** Superior horizontal cross-section through the low back at the fourth lumbar level illustrates the relationship between the anterior, middle, and posterior layers of the thoracolumbar fascia and surrounding muscle

the biceps femoris caudally. The sacrospinous ligament attachments run deep to the sacrotuberous ligament from the lateral margin of the distal sacrum and coccyx to the ischial spine. These two more distant posterior SI ligaments likely restrain sacral nutation.

Most ascribe the primary role of the thoracolumbar fascia (TLF) as the enhancement of the mechanical stability of the lumbar spine as a result of its relationship to the ventral (abdominal) and dorsal (extensor) muscles of the lower trunk (Fig. 9.12). However, as noted above, the thoracolumbar fascia may also play a significant role in sacroiliac stability via the force-closure mechanism and relationship with several dorsal spinal muscles, e.g., quadratus lumborum, latissimus dorsi, multifidi, etc., and pelvic girdle muscles, e.g., gluteus maximus. The anterior and middle layers of the TLF engulf the quadratus lumborum muscles, with medial attachments to lumbar transverse processes and lateral ones to the iliac crests. The posterior layer of the TLF covers the latissimus dorsi (superficially) and surfaces of the erector spinae and multifidi. The posterior layer, with its attachments to lumbar spinous processes, the sacrum, and ilia, also strongly suggests a supporting role of the TLF in sacroiliac stability. At the lateral margins of the thoracolumbar fascia, where the anterior and middle layers come together as the lateral raphe, we see a direct continuity of the TLF with the transverse abdominis, and to a lesser extent the internal oblique muscles. Thus, with these arrangements, i.e., connections between anterior and posterior soft tissues, we can imagine a significant mechanical role for TLF to enhance core stability of both the SI joint and lumbar spine, particularly with the flexed and rotated postures associated with lifting.

### 9.4.2 Function

In Chap. 3, we briefly examine the general osteo- and arthrokinematics of typical spinal functional units. This case does not appear to present us with a problem that primarily stems from kinematic dysfunction per se. It is more likely that the motion impairments, i.e., range of motion loss, abnormal lumbopelvic rhythm, etc., we observed during the physical examination were related to Jim's spondylolisthesis, instability, and mild scoliosis. Thus, it is important that we look at the uniqueness of lumbar, pelvic, and intra-pelvic kinematics as it pertains to Jim's activities. Recall, several key physical examination findings in this case:

A moderate increase in the lumbar lordosis/anterior pelvic tilt, with a "step-off" at L5/S1, with a mild right lumbar scoliosis (C-curve convex right). With active motion testing of the trunk:

1. Flexion mildly limited with abnormal lumbopelvic rhythm, i.e., limited lumbar lordotic curve reversal with mild deviation to the left during the flexion phase, and with deviation to the right during the return extension phase.
2. Extension markedly limited with moderate pain at L5/S1.
3. Sidebending and axial rotation range of motion was mildly limited bilaterally, but symmetrical without reproduction of symptoms.

Note: Let us interlude to define "step-off." Typically, with an observation of the lumbar spine posture from a sagittal perspective, we notice a mild lordosis from the base of the sacrum to the T12/L1 junction; that is, there is a smooth and continuous spinal curve that is convexity anteriorly. If we palpated and marked the spinous processes from S1 to T12, it would be like we highlighted discrete points that, if joined, comprise the continuous spinal curve that we observed. However, if, as in Jim's case, the L5 segment has been displaced anteriorly (secondary to spondylolisthesis), its spinous process, when marked, would be an outlier because it would not contribute to the smooth spinal curve. In other words, when we move from the spinous process of S1 to the spinous process of L5, it would feel as if our palpating hand had stepped forward or too far anteriorly, i.e., "stepped-off."

We begin with a description of normal lumbopelvic rhythm. Most of the activities we do involve bending forward (flexion) and backward (extension), often with some

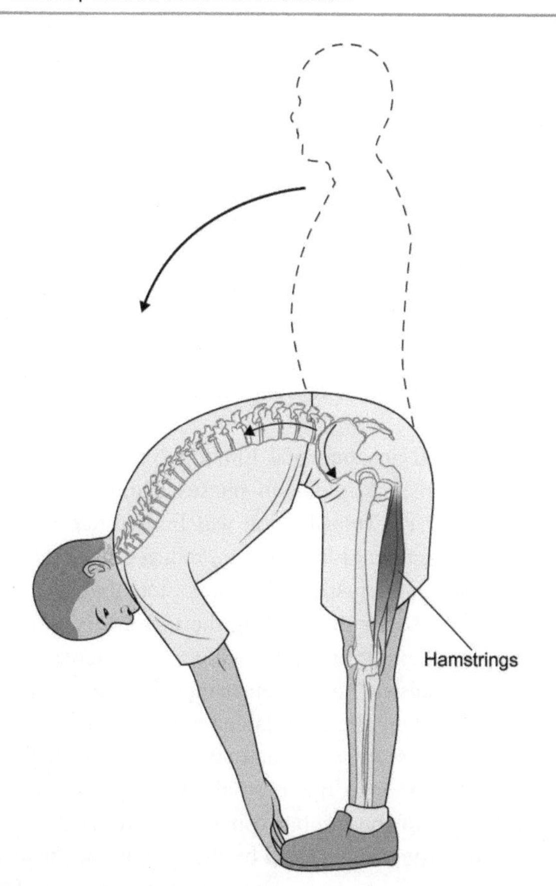


Fig. 9.13 Normal lumbopelvic rhythm, with knees extended showing partial reversal of the lumbar lordosis and anterior pelvic rotation. (Note: normal hamstring flexibility, i.e., length, is required to complete this movement)

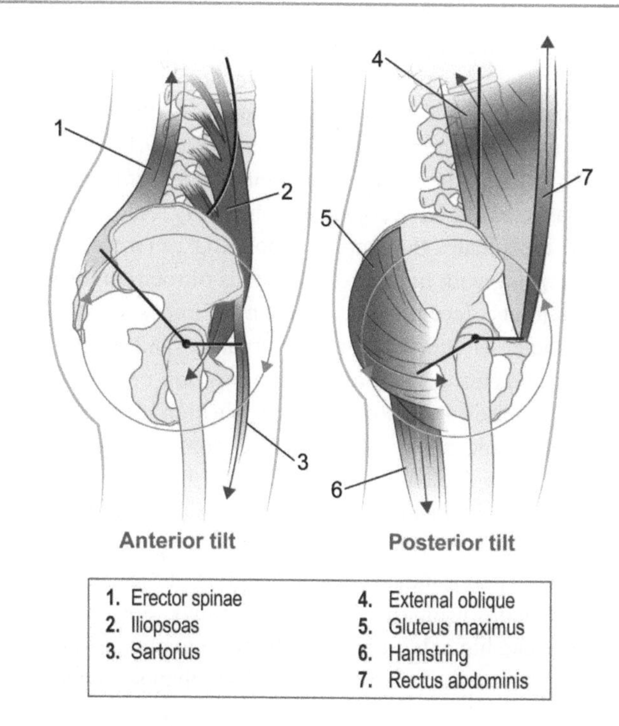


Fig. 9.14 Schematic of coupled pelvic and lumbar movements anterior pelvic tilt with lumbar extension and posterior pelvic tilt with lumbar flexion. (Note: the black lines represent the moment arms of the muscles that create a force couple about the axis of rotation through the hip joint center)

degree of turning (rotation). Flexion from a standing position requires the coordination of hip joint and lumbar spine flexion, coupled with an anterior tilt (rotation) of the pelvis (Fig. 9.13) (Note: in general, pelvic tilt or rotation is a forward (or backward) rotation about a horizontal axis in the sagittal plane). During an anterior tilt of the pelvis, the anterior superior iliac spines move inferior, while the posterior superior iliac spines move superior; the reverse sequence occurs with a posterior tilt of the pelvis. From a sagittal perspective, normal lumbopelvic rhythm requires approximately 45° of lumbar flexion and 60° of hip flexion, i.e., coupled flexion at the hip joint proper with anterior pelvic rotation. The initial flexion phase is initiated by concentric action of the trunk flexors but quickly followed by cessation of abdominal flexor activity and the onset of eccentric action of the hip and spinal extensors until the end range of lumbar flexion is achieved. If one statically holds the fully flexed forward bent posture at the end range, the spinal extensors become electrically silent, and the external flexor moment (produced by the weight of the head, arms, and trunk and their moment arms) is counteracted by the posterior annulus of the IVD, facet joint capsules, posterior longitudinal, supraspinous and interspinous ligaments, iliolumbar ligaments, thoracolumbar fascia, and series- and parallel components of the dorsal spinal muscles. The reason we describe this as a "rhythm" is that this movement occurs in a controlled and coordinated kinematic sequence with lumbar flexion occurring approximately during the first 25% of the movement, and anterior rotation of the pelvis occurring during the last 25% of the bend; note the hips are flexing as the pelvis rotates forward on fixed femurs. When we observe this test from a posterior frontal view, we would normally see the trunk move forward and down along a straight path. Normal rhythm during the return (extension) phase is just a simple reversal, that is, an early phase dominated by posterior rotation of the pelvis coupled with hip extension (with activation of the gluteus maximus and hamstrings, in synergy with trunk extensors), followed by lumbar spine extension (with significant activation of trunk extensors) and restoration of the lumbar lordosis.

The movement strategy, e.g., coupled lumbar flexion and anterior pelvic rotation, described above illustrates the most commonly used strategy for bending, lifting, and reaching activities. The strategy of combining movements of the spine and pelvis appears to maximize the range of motion. However, it is possible to implement a different strategy that utilizes primarily pelvic control, i.e., anterior or posterior tilt, but which creates apparent paradoxical lumbar movements, that is, in comparison to the movement pattern just described above (Fig. 9.14). The sequence of movements we now describe are not ones typically observed as part of normal daily activities but are presented as an example of an alternative illustration of the coupling dynamics of the hip joint, pelvis, and lumbar spine. As illustrated (Fig. 9.14), we see that activation of the erector spinae and hip flexors (primarily iliopsoas) create a force couple that anteriorly rotates the pelvis and accentuates the lumbar lordosis, i.e., increases lumbar extension, with the approximate axis of rotation through both hip joints. Accentuation of the lumbar lordosis decreases the diameter of the intervertebral foramen and places the lumbar segments in a more extreme position of extension, which increases the closed-packed position of the facet joints, thus increasing joint surface compression loads. Concomitantly, increased compression of the posterior IVD causes the posterior annulus to buckle, forcing the nucleus pulposis anteriorly which induces annular bulging. Posterior tilting of the pelvis induced by the coupling of the abdominal muscles, e.g., primarily rectus abdominis and external obliques, and hip extensors, e.g., gluteus maximus and hamstrings, reverses the effect on the lumbar spinal segments.

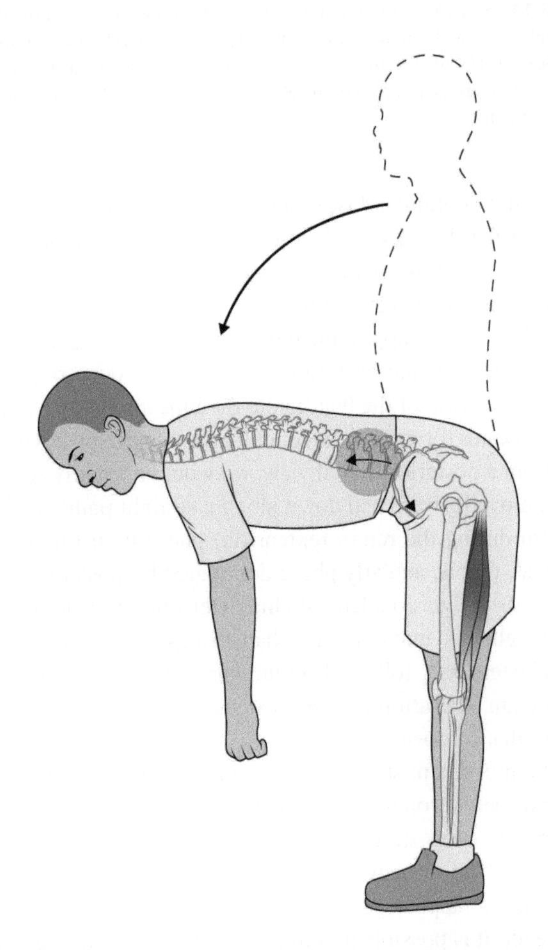

**Fig. 9.15** Illustration of an abnormal lumbopelvic rhythm where lumbar flexion is limited and anterior rotation of the pelvis is dominant. It is notable that the lumbar lordosis has not reversed but is held in a neutral, i.e., not flexed or extended, position

The notable difference between these two movement strategies, i.e., (1) bending forward that combines lumbar flexion, anterior pelvic rotation, and hip flexion and (2) activating the lumbar extensors and hip flexors to anteriorly tilt the pelvis, is that flexing the trunk in a cranial-to-caudal sequence (Fig. 9.13) is the result of a movement synergy that is only initiated by voluntary muscle action and later controlled involuntarily by eccentric actions of the erector spinae and hip extensors, whereas the second strategy (Fig. 9.14A) must be entirely controlled by voluntarily activating the erector spinae and hip flexors to anteriorly rotate the pelvis.

Jim displayed an abnormal lumbopelvic rhythm during both the flexion and extension phases, similar to what is shown in Fig. 9.15. Total flexion was limited due to inadequate lumbar segmental flexion, i.e., reduced reversal of the lordotic curve, and hamstring tightness (note that asking Jim to keep his knees extended during the forward bending test kept him from flexing his knees which would have allowed him to relieve tension on his hamstrings). As he reached the end of the movement, he complained of a "stretch" feeling in the back of his legs. Jim deviated to the left during the flexion phase, which was also not normal. We hypothesized that reduced flexion of the lumbar spine was compensatory and related to the instability created by the spondylolisthesis and that the left deviation was related to his scoliosis. These hypotheses make more sense after we complete a more detailed study of the arthrokinematics of the lumbopelvic complex.

Let us move on and examine more specific segmental lumbar movements. Axial compression of the lumbar spine causes a movement along the axis of the vertebral column that occurs during weight-bearing in the upright posture, or secondary to strong activation of the dorsal spinal muscles (although co-activation of both the ventral and dorsal muscles induces the same movement). These longitudinal loads induce approximation of the interbody joints, increasing compression stress within vertebral body, IVD, and vertebral endplates (see Chap. 7 for a review of the biomechanics of the IVD and vertebral endplate). Although the interbody joints are principally designed to distribute/mitigate most of the compression loading, the zygapophyseal (facet) joints may carry up to 20% of the vertical load. However, because of the general orientation of the lumbar facets, when the lumbar spine is in its neutral (slightly extended) posture, the first thing that occurs during axial compression is that the lumbar facets glide past one another, inducing tensile stresses on the facet capsules. However, if the lumbar lordosis is increased, and the axial compression is prolonged, e.g., prolonged standing, the tips of the inferior articular processes are driven into the superior articular processes and lamina of the segment below, particularly borne by the lower joints, i.e., L3/L4, L4/L5, and L5/S1. Although the facet joints are covered by hyaline cartilage, they were not designed for large and prolonged compression loading, so over time, these joints may develop degenerative disease. Finally, we need to consider at least two other defenses against axial compression loads when the lumbar lordosis is accentuated, the anterior annulus and anterior longitudinal ligament (ALL). With axial compression when the lumbar lordosis is accentuated, increased tensile stress is realized on all anterior soft tissue constraints, with the ALL being the principal element. Under normal conditions, the elastic tension in the ALL from axial loading can be stored and later used to restore a normal lumbar curve. However, more extreme and repetitive axial loading may induce plastic changes in the ALL and a reduction in its ability to provide needed stability.

Jim was noted to have an increased lumbar lordosis, with radiographic evidence of a significant increase in the pelvic angle. Jim also spent much of his work time holding small motors and manipulating tools in prolonged standing, and he stated that his central low back pain was made worse when standing. As a result of these signs and symptoms, we hypothesized that the repetitive and frequent axial movements and compression loads on Jim's lumbar spine were related to standing activities that may have strained, or induced intermittent irritating strains, to one or more structures in the lumbar region, e.g., vertebral body, IVD, vertebral endplates, facet joint surfaces, lamina, facet capsules, and anterior longitudinal ligament.

A distraction of the interbody and facet joints is another movement that occurs in the spine. However, activities of daily living involve muscle actions that usually induce joint compression, not a distraction, so let's move on. Remember that joint motion is dictated by bony partner orientation, joint geometry, soft tissue constraints, and muscle action. Since we have already discussed the movements and forces related to the interbody joint (see Chap. 7), our attention now shifts to the kinematics of the facet joints. Before we do that, let's take one more look at the lumbosacral junction.

The L5/S1 junction is different from the other spinal units in the lumbar region because of the angulation described earlier, i.e., (1) approximate 16° angle between the lower surface of L5 and the sacral base, and (2) the approximate 40° sacral angle. Intuition suggests that these angles have kinematic and kinetic implications. Biomechanical analysis has shown that these angles predispose the L5 segment to "slip" forward (anteriorly) even under normal resting standing positions. Soft tissues can restrain this forward slipping, which places tensile stresses on the anterior longitudinal ligament, anterior annulus fibrosus, and iliolumbar ligaments, with secondary support provided by the wide and frontally oriented L5/S1 apophyseal joints, and horizontal vectors of the iliocostalis lumborum pars lumborum and longissimus thoracis pars lumborum (see Chap. 8). Let us look at this situation more closely. The superincumbent body

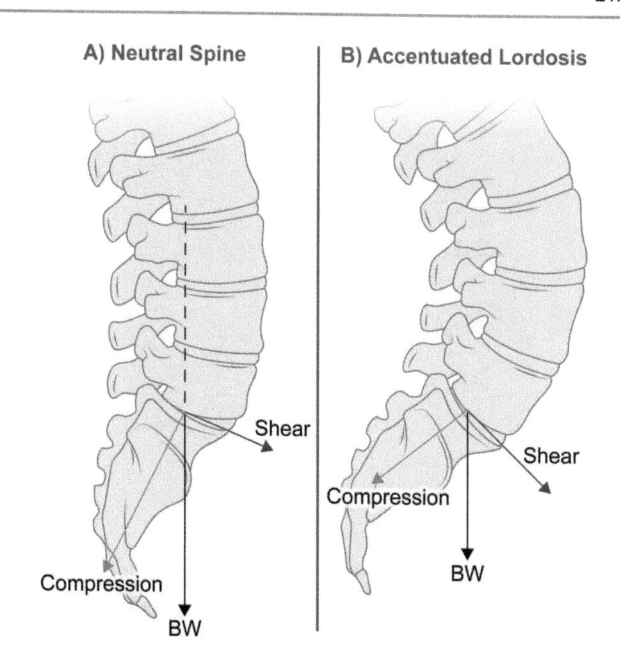


**Fig. 9.16** Sagittal views illustrating the compression and shear force components of the superincumbent body weight (dotted line in (**A**) at L5/S1) with (**A**) a normal lordosis and (**B**) an increased lordosis. It is notable that shear forces increase at L5/S1 with an accentuated lumbar lordosis and anterior sacrum

weight (BW) creates an axially directed force (this is the vertical force due to gravity acting on the head, arms, and trunk) on the sacral promontory that can be decomposed into shear and compressive components (Fig. 9.16). With the normal sacral angle, the shear load at L5/S1 is estimated to be approximately 60% of total BW. However, notice that when the lumbar lordosis/sacral angle is increased, the interbody compressive load is reduced, while the shear load is increased (approaching approximately 80% of total BW, depending on the sacral angle). The increase in shear force with an increase in the sacral angle is illustrated in the statics problems at the end of the chapter. An increased shear load implies that L5 is more disposed to anterior slippage, increasing compressive loads on the L5/S1 facet hyaline surfaces, and tensile loads on the soft tissue structures listed above. It is likely that the increased sacral angle which induces forward slippage of L5 facilitates L5/S1 mechanoreceptors at the interbody and facet articulations and alerts the central proprioceptive systems. One of those related systems includes secondary activation of the gamma motor system and activation of the spinal extensors to increase static muscle tonus. A greater magnitude of spinal muscle unit recruitment concomitantly increases muscle forces, with subsequent increased facet joint compression. Additionally, chronic muscle activity above normal resting levels could lead to muscle fatigue, eventual strain, and/or chronic inflammation.

It is clear that Jim's lumbosacral junction is exposed to external axial and shear forces that his local tissues are

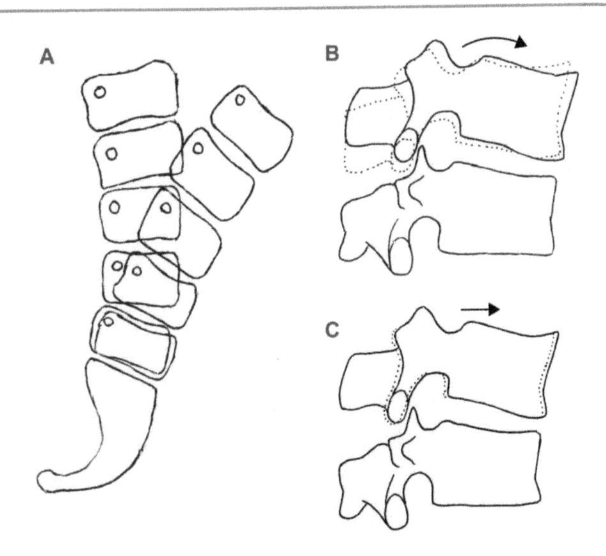


Fig. 9.17 Sagittal perspective illustrating anterior rotation (A, B) and translation (C) of lumbar vertebral segments during forward flexion

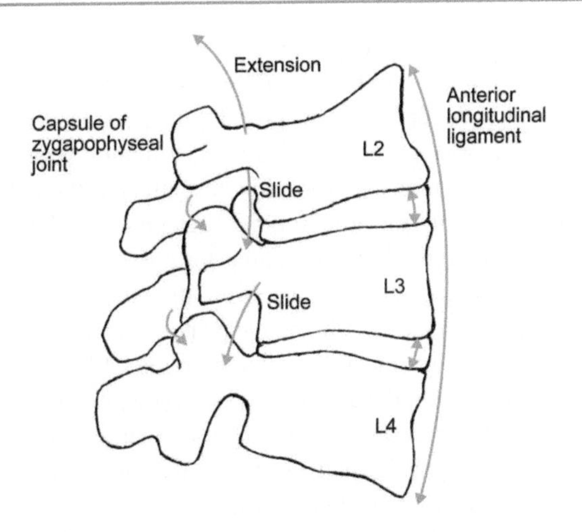


Fig. 9.19 Arthrokinematics, i.e., postero-inferior slide at the facet joints, of a normal lumbar segment during extension. Note that tension in the anterior longitudinal ligament and anterior annulus constrain excessive extension. Posteriorly the facet joint capsules slacken

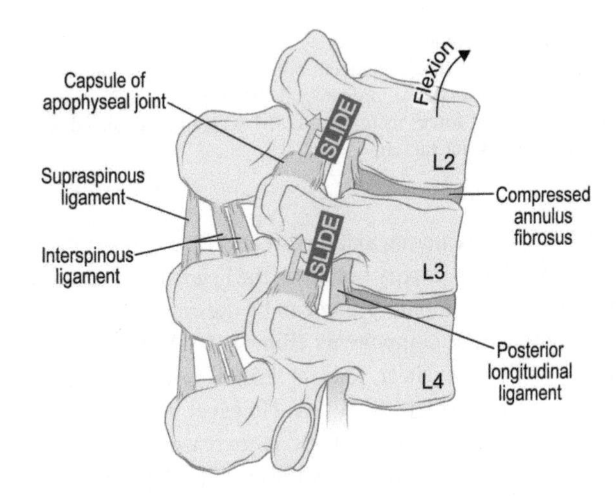


Fig. 9.18 Arthrokinematics of a normal lumbar segment during flexion as the lumbar segment translates anteriorly, there is a supero-anterior glide of the inferior facet of the superior segment on the superior facet of the inferior segment. (Note the posterior ligaments that constrain excessive lumbar flexion)

paying a price for. It is likely that Jim's increased lumbar lordosis is related to the measured increased sacral angle (which probably was genetically or developmentally determined), with perhaps both conditions predisposing him to a spondylolytic condition.

From L1 to L4 the facet surfaces are oriented nearly vertically, which largely favors sagittal plane motion, i.e., flexion and extension. Biomechanical research has verified this, demonstrating, on average for the entire lumbar spine, 45° to 55° of flexion and 15° to 25° of extension in a healthy adult spine. With the L4/L5 segment as somewhat transitional, the L5/S1 interbody junction and its facet orientation are quite different. The facet surfaces of the L5/S1 facet joints are oriented more frontally than sagitally, suggesting more freedom of movement in the frontal and transverse planes. Mobility at this junction is critical because it must compensate for the relative immobility of the sacrum caudally, as well as redistribute and mitigate the loads from above and below. Based on the anatomical and biomechanical complexity of the lumbosacral junction, it is not surprising that this is also the region that sees the majority of injuries.

During flexion, the lumbar spinal region moves forward and the lordotic curve begins a reversal, i.e., it straightens; we note that at the end range of flexion the lower lumbar spine does not fully reverse its lordosis but appears flattened (Fig. 9.17A). Full flexion is achieved as each vertebra rotates and translates anteriorly from its relatively extended position (Fig. 9.17B, C). As each vertebra rocks forward over its intervertebral disc, its inferior articular processes glide superior and slightly backward, which opens a gap between each inferior articular facet and the superior articular facet at the zygapophyseal joint and increases the diameter of the intervertebral foramen (Fig. 9.18). As the motion continues to its end range, the translation paradoxically closes the gap between the facets in the zygapophyseal joints and is halted by impaction of the inferior on the superior articular facets. The zygapophyseal joints play a role in maintaining the stability of the lumbar spine in flexion, assisted by tension in the facet capsules, posterior annulus, and several other ligaments, e.g., posterior longitudinal, supraspinous, interspinous ligaments, and ligamentum flavum. In the fully flexed lumbar spine, the total compression load on a given zygapophyseal joint is reduced, but the force per area, i.e., stress, is actually increased because surface areas of contact are reduced. Concomitantly, at any phase of lumbar flexion, the lumbar extensor muscle forces needed to maintain the position significantly increase facet bone-on-bone compression.

Interbody and zygapophyseal movements during lumbar extension are essentially a reversal of the kinematics of flexion (Fig. 9.19). As the lumbar segments extend one over the other, the superior element tips backward, or rotates posteriorly, and translates posteriorly. Gliding at the facet joints involves a downward movement of the inferior articular processes and spinous processes. Segmental extension is somewhat constrained by the anterior annulus and anterior longitudinal ligament, but mostly by impaction of the spinous processes. Impaction of the spinous processes induces buckling, with possible impingement of the interspinous ligaments, which could be a source of pain generation over time. End range extension in the lumbar spine might be considered a close-packed position for the zygapophyseal joints because the contact area between the joint surfaces is maximized. Intuition suggests that with maximal joint surface contact, the compression stress would be better distributed or mitigated, but that is not the case because the overall magnitude of compression is large at the end range of extension, particularly in weight-bearing positions. Repetitive prolonged standing with the lumbar spine fully extended induces plastic joint surface, i.e., articular cartilage, deformations over time that could become symptomatic.

Axial rotation in the lumbar spine is limited to a total of approximately 5° to 7° to each side, i.e., right rotation and left rotation. These movements induce a torsion, with resulting shear strains, of the intervertebral disc and impaction of the zygapophyseal joints. Because of facet orientation in the upper lumbar region, rotation is extremely limited, and impaction of the facets occurs before any significant deformation of the IVD occurs. For example, during right axial rotation, the rotation initially occurs about an axis in the vertebral body. As the posterior elements rotate, the left inferior articular process of the superior segment impacts the superior articular process of the inferior segment. Gapping and posterior distraction occur at the opposite zygapophyseal joint (Fig. 9.20). However, when maximum rotation, i.e., approximately 3°, has been reached in a particular functional lumbar unit, the axis of rotation is shifted from the vertebral body to the side of the approximated facets, which induces a lateral shear on the IVD. It has been shown that this limit (≥3° rotation) is associated with micro-failure, i.e., a yield point has been passed, of the IVD. It should be noted that although the frontal plane orientation of the L5/S1 facets should offer more axial rotation of the L5 segment than the upper segments, that hypothesis has not been substantiated by research findings.

For a normal lumbar functional unit, the zygapophyseal joints, and to some extent the posterior ligaments, e.g., supra- and interspinous ligaments and contralateral face capsules, protect the IVD from excessive torsion. Some have suggested that axial rotation with flexion creates larger tensile strains on the IVD making it more susceptible to nuclear herniation. However, this is only true if rotation at any one segment exceeds 3°. Otherwise, with a normal IVD, the annulus is well protected. In summary, during rotation, the impacted zygapophyseal joints are strained by compression, the IVD by torsion and shear, and the capsule of the opposite zygapophyseal joint by tension.

The lumbar spine allows a total of about 20° of sidebending in either direction. Coupled rotation may be more pronounced when sidebending is the primary movement, adding some complexity to overall kinematic patterns. The arthrokinematics of pure sidebending, however, are straightforward (Fig. 9.21). For example, during right sidebending the

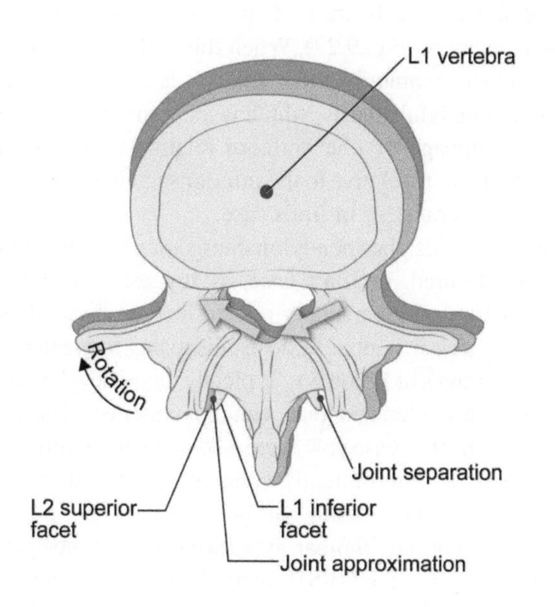


**Fig. 9.20** Arthrokinematics of a normal lumbar segment during right axial rotation. Note impaction (red dot) of the facets contralateral to the direction of segment rotation and gapping (red dot) of the facets on the right side

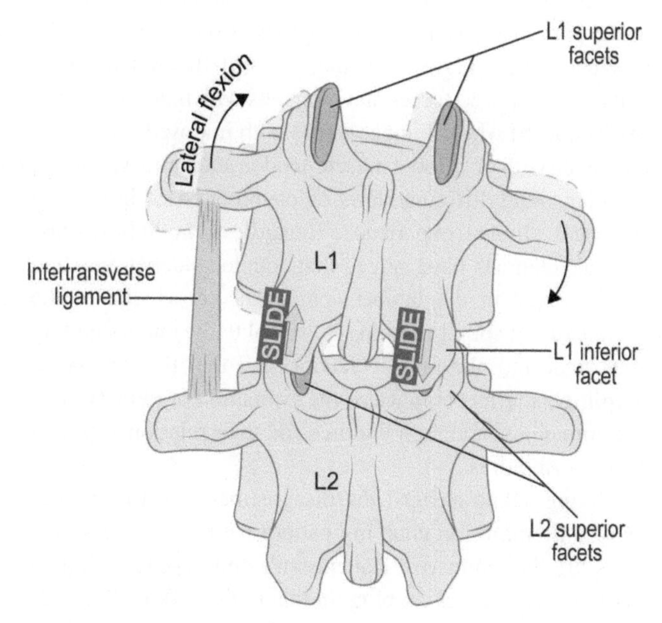


**Fig. 9.21** Arthrokinematics of a normal lumbar segment during right sidebending. During right sidebending, the facets on the right side "close," i.e., narrowing the intervertebral foramen, while the facets on the left side "open." Note tension in the contralateral, i.e., left intertransverse ligament

superior segment tilts to the right, with little to no concomitant translation. The left inferior facet of the superior segment glides superiorly, while its counterpart glides inferiorly. An alternative way of imagining facet movements during right sidebending can be thought of as "opening" (or flexion) of the facets on the left side, and "closing" (or extension) of the facets on the right side. With this in mind, the facet capsules on the left are placed under tension, while the facet joint surfaces on the right are placed under compression, and vice versa during left sidebending.

In Chap. 8, we described the coupling of sidebending and rotation in the thoracic spine, as suggested by the clinical theories provided by Kaltenborn and Fryette. Both clinicians believed that the coupling patterns differed by postural attitudes of the lumbar region, i.e., neutral, flexed, or extended. We also introduce Kaltenborn's theory regarding "coupled" and "non-coupled" mechanics. In Scott's particular case, we base our evaluation of the examination findings and development of hypotheses regarding thoracic pathokinematics on Fryette's principles. However, in that case, we leave the reader with the caveat that there is little agreement among interested biomechanists and clinicians regarding normal coupling patterns in the thoracic spine.

Before we go any further, let us be clear that there is also no consensus regarding coupled sidebending/rotation patterns for lumbar segments. In the biomechanics world, the lack of consensus has to do with the variety of experimental methods that have been used: (1) in vivo using optical motion capture, (2) in vivo using inertial motion sensors, (3) in vivo using bi-planar radiography/stereophotogrammetry, (4) in vivo using electromagnetic motion analyzers, (5) in vivo using percutaneous pins with optoelectronic tracking, (6) in vitro using cadaver specimens of (a) functional spinal units with imposed forces and moments with material testing devices or (b) whole lumbar spines with imposed forces and moments, with or without muscle-tendon actuators. However, clinicians hold differing views on coupling based largely on empirical clinical experience. Regardless, both basic and clinical scientists have agreed that coupled sidebending and rotation exist in the lumbar spine, that knowledge of the complexity of spinal coupling is critical to the understanding of normal and abnormal behavior, and that the variety of coupling patterns complicates the evaluation of radiographic and clinical examination findings and their relationship with patient symptoms.

Having acknowledged the disagreements in the biomechanics literature on coupling patterns in the lumbar spine, let us briefly summarize Kaltenborn's and Fryette's clinical theories on vertebral coupling in that region (Table 9.2). We think it important to reiterate the difference between coupling that appears to occur more naturally, i.e., part of a motor pattern learned over time, as implied by Kaltenborn's idea of coupled motion that is voluntarily (or passively) induced. An example of a more natural coupling movement probably occurs when one reaches over and bends to pick up something from the ground. Whereas a coupled movement that is forced, i.e., not part of a routine motor pattern, for example, may be one where an individual reaches up over their shoulder when in a flexed trunk position. Although Kaltenborn and Fryette may have differed on rotation coupled patterns, they agreed that interbody translations that occur with flexion and extension, i.e., anterior and posterior, respectively, are an inherent characteristic of spinal movements.

**Table 9.2** Proposed lumbar motion coupling: Kaltenborn and Fryette

| Clinicians | Proposed coupling |
|---|---|
| Kaltenborn | Contralateral rotation-sidebending in extension[a] (coupled)<br>Ipsilateral rotation-sidebending in extension (non-coupled)<br>Ipsilateral rotation-sidebending in flexion[a] (coupled)<br>Contralateral rotation-sidebending in flexion (non-coupled) |
| Fryette | Contralateral rotation coupled to sidebending in spine neutral and extension<br>Ipsilateral rotation coupled to sidebending in flexion |

[a]The phrases "in extension" and "in flexion" imply a situation when the lumbar spine is extended or flexed, respectively

Having concluded our review of normal lumbar kinematics, it is possible to offer hypotheses in an attempt to explain most of Jim's physical examination findings. Let's digress a bit into the details of spondylolytic spondylolisthesis before we do that and provide an operational definition. Spondylolysis is a fracture of the pars interarticularis of a lumbar segment (Fig. 9.22). When this occurs in the lumbar spine, most commonly there follows a progression to an anterior spondylolisthesis, which is the term that describes a forward slippage of one vertebra relative to another (the superior segment relative to the inferior segment), e.g., L5 on S1 (or the sacrum), as in Jim's case.

In some cases, the spondylolisthesis may be congenital or may be acquired. In Jim's case, as suggested earlier, it is likely that a developmentally related increased sacral angle induced a greater lumbar lordosis, disposing to greater anterior shear forces at L5 on S1. Typically, bony developmental abnormalities, like this one, can go unnoticed for years, although in the case of increased lumbar lordosis, an individual may intermittently complain of central low back pain, i.e., "the pain comes and goes." Over time, however, this disposition, i.e., lumbar hyperextension, creates greater shearing forces at the L5/S1 junction and bending moments at the pars interarticularis that eventually lead to a stress reaction and fracture (Fig. 9.23) (review Chap. 3 and the section on general stress fracture mechanics).

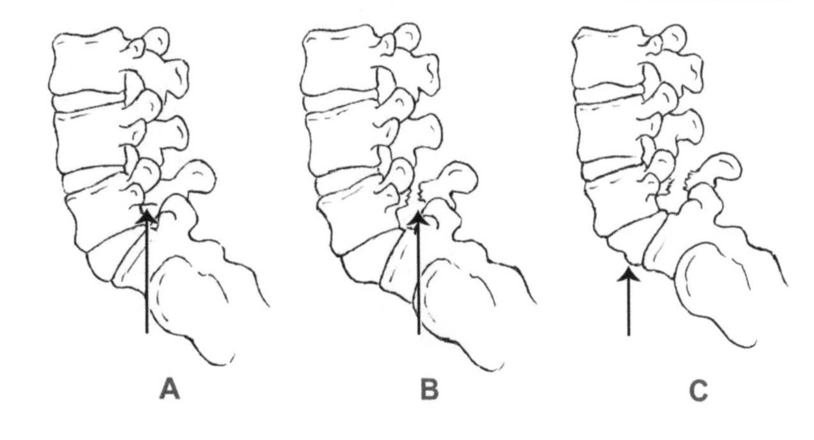


**Fig. 9.22** Schematic comparing (**A**) normal spine to one with (**B**) spondylolysis at L5 and to one with (**C**) spondylolytic spondylolisthesis

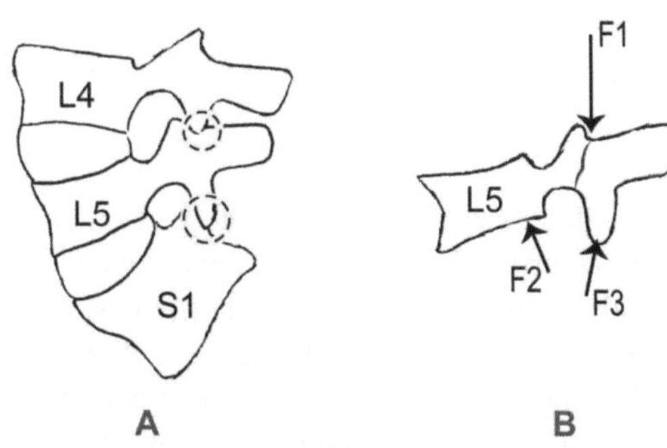


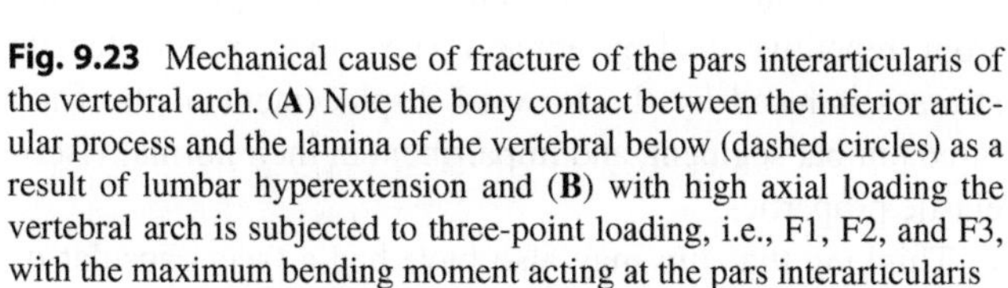

**Fig. 9.23** Mechanical cause of fracture of the pars interarticularis of the vertebral arch. (**A**) Note the bony contact between the inferior articular process and the lamina of the vertebral below (dashed circles) as a result of lumbar hyperextension and (**B**) with high axial loading the vertebral arch is subjected to three-point loading, i.e., F1, F2, and F3, with the maximum bending moment acting at the pars interarticularis

**Fig. 9.24** Image of typical encounter of offensive and defensive football players that demonstrate the lumbosacral extension injury mechanism of spondylolysis

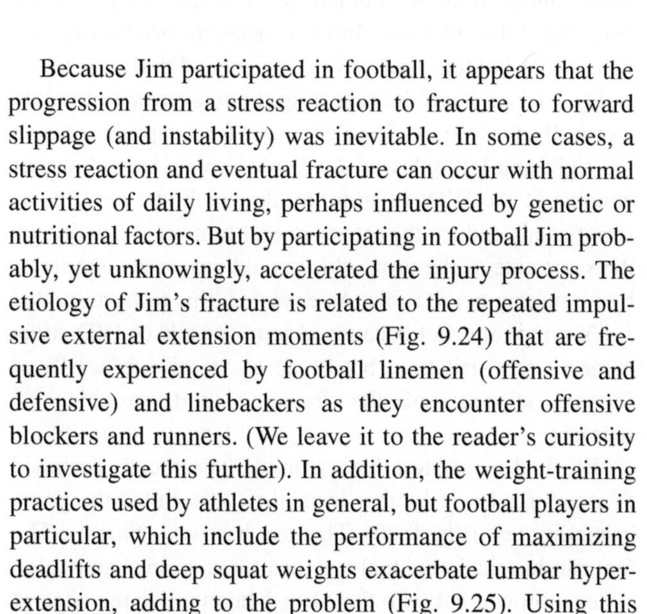

Because Jim participated in football, it appears that the progression from a stress reaction to fracture to forward slippage (and instability) was inevitable. In some cases, a stress reaction and eventual fracture can occur with normal activities of daily living, perhaps influenced by genetic or nutritional factors. But by participating in football Jim probably, yet unknowingly, accelerated the injury process. The etiology of Jim's fracture is related to the repeated impulsive external extension moments (Fig. 9.24) that are frequently experienced by football linemen (offensive and defensive) and linebackers as they encounter offensive blockers and runners. (We leave it to the reader's curiosity to investigate this further). In addition, the weight-training practices used by athletes in general, but football players in particular, which include the performance of maximizing deadlifts and deep squat weights exacerbate lumbar hyperextension, adding to the problem (Fig. 9.25). Using this information let's offer possible explanations for Jim's movement impairments.

When Jim was asked, during his physical examination, to bend as far forward as he could, he demonstrated an abnormal lumbopelvic rhythm characterized by limited flexion without reversal of his lumbar lordosis, i.e., he held his lumbar spine somewhat extended. Recall that normal flexion of a lumbar segment is inherently coupled with an anterior translation of the superior segment relative to the segment below it. Since Jim's L5 segment was already in an untenable position, i.e., anteriorly displaced, and unstable, forward flexion would have exacerbated the problem, thus an adaptive movement pattern was required. Typically, in situations like this, sensory signals from the pathological site to the central nervous system would initiate a series of adaptive motor control strategies, based on sensory input, in an attempt to create a more stable movement pattern, e.g., activating muscle in an attempt to stabilize the lumbar spine, which is what we observed. In other words, the abnormal lumbopelvic pattern of movement was a normal kinematic and kinetic adaptation to the unstable L5/S1 movement unit.

When Jim was asked to bend backward in standing, i.e., lumbar extension, his symptoms were also reproduced (Fig. 9.26). Normally, during extension, a lumbar segment

**Fig. 9.25** Schematics illustrating (**A**) deadlift and (**B**) deep squat. Note the exaggerated lumbar extension used during these exercises

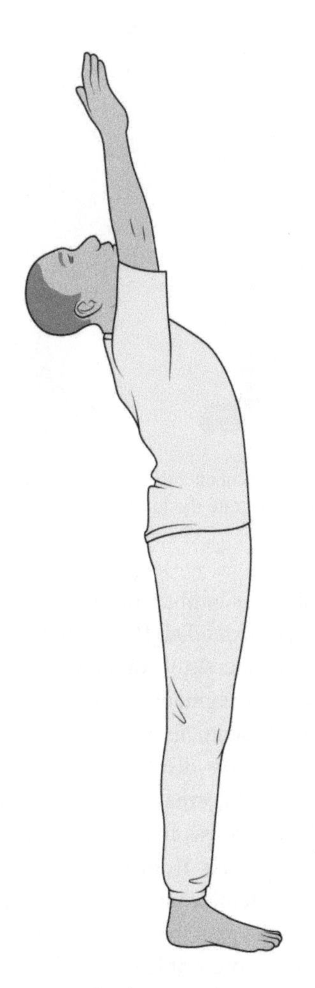

**Fig. 9.26** Standing test for lumbar extension

tilts (or rotates) and glides posteriorly. However, with spondylolytic spondylolisthesis as severe as Jim's, the L5 segment was wedged so far anterior that a posterior glide was not possible. Part of the problem is related to plastic deformations in the ligaments that normally restrain anterior translation of vertebral segments, e.g., anterior and posterior longitudinal ligaments, inter- and supraspinous ligaments, annulus of the IVD, ligamentum flavum, and iliolumbar ligaments. The plastic deformation, i.e., overlengthening, of those ligaments that we hypothesized took place resulted from repetitive and prolonged (i.e., over years) tensile stresses induced by the forward displacement of the fractured lumbar segment, and impairment of their normal viscoelastic properties.

Thinking that Jim may also have had a facet capsular or IVD annular strain, the lumbar quadrant test was performed (Fig. 9.27). For this test, Jim was asked to bend backward, then while holding that position, sidebend and turn to the right as far as possible, repeating the maneuver to the left. Generally, this test maximally stresses facet capsules on both the right and left sides of the lumbar spine, as well as the annulus of the IVD. In addition, the combined movements significantly reduce the intervertebral foraminal opening on the same side as the turn, which could provocate symptoms by impinging nerve roots. As expected, this test provocated Jim's symptoms, not because of joint capsule or IVD strain or nerve root irritation, but because extension exacerbated forward slippage of the fractured, and unstable, L5 segment.

The final special test that was used is specifically designed to assess for lower lumbar instability. It is called the prone instability test (Fig. 9.28). This test has two elements. The first part has the individual rest their torso on an examination table as shown, but with their legs hanging over the edge of the table (not shown), which should put the spinal and hip extensors in a relatively relaxed or resting state. To stabilize the upper body position, the individual grasps the edge of the table with their hands. Once the therapist and individual are properly positioned, the therapist applies a ventral force, i.e., an anterior-to-posterior force, over the spinous process of the L5 segment, which moves the already anteriorly subluxed segment forward even further, provocating symptoms. For part II, the individual raises both legs so that the feet are off the floor and the legs are nearly parallel to the table. Once the new position is secured, the therapist again applies a ventral force over L5. It is hypothesized that raising the legs activates the spinal extensors, which should provide dynamic stabilization, i.e., either hold the unstable segment or posteriorly translate it, and reduce or control symptoms. However, when the second part of the prone instability test was performed, Jim's symptoms were again aggravated, confirming that his primary signs and symptoms were related to the radiographic diagnosis.

**Fig. 9.27** Standing quadrant test of the lumbar spine that couples side-bending and rotation to the same side during extension

A positive prone instability test in both test positions suggested a moderately severe instability at L5/S1. Fortunately, Jim did not have neurological symptoms, meaning that IVD degeneration had not progressed to the point where the spinal cord or nerve roots were compromised. However, based on our previous understanding of the mechanics of key lumbar extensors, i.e., longissimus thoracis pars lumborum, iliocostalis lumborum pars lumborum, and multifidi (see the section in Chap. 8 where we reviewed the actions of the erector spinae and multifidi, as described by Bogduk), we wonder why might have these muscles appeared to fail during part II of the prone instability test? The answer may be addressed if we examine the decomposition of the erector spinae (ES) muscles for individuals with an increased sacral angle (Fig. 9.29). Recall, Bogduk suggested that when the longissimus thoracis pars lumborum and iliocostalis lumborum pars lumborum muscles are activated their horizontal force components induce posterior translation of the lumbar segments crossed by those muscles. However, Bogduk claimed that due to the shape of the lumbar lordosis and sacral angulation, the horizontal force component of the muscles that cross the L5/S1 spinal unit paradoxically induces an anterior translation. This appears to occur when, during maximum concentric action of the back extensor muscles, the upper lumbar vertebrae are drawn backward and forced downward under compression, with a companion

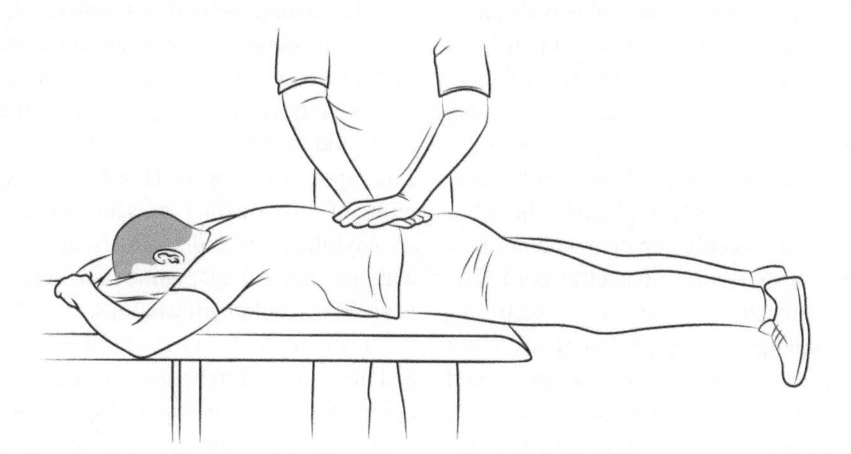

**Fig. 9.28** Prone stability test for lumbar stability. Bilateral prone leg lifts held isometrically, which activates the spinal extensors should dynamically stabilize the L5 segment against the anterior pressure applied to that segment. A positive test is indicated if L5 is felt to translate forward, and symptoms are reproduced

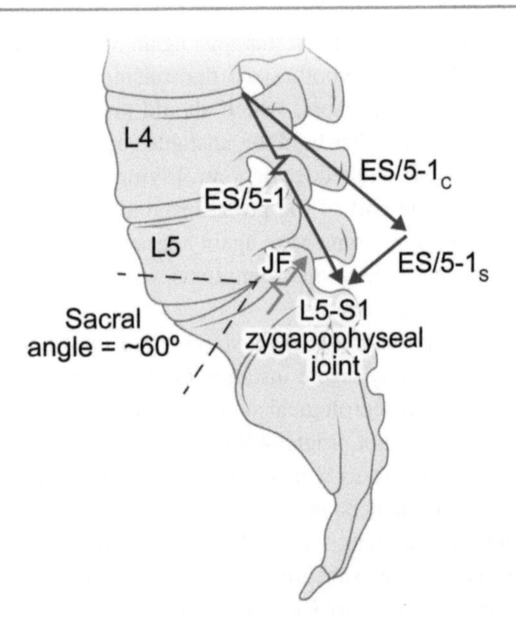


**Fig. 9.29** Schematic of decomposition of the erector spinae (ES) muscles in the lower lumbar segments in erect standing. Notice that the horizontal ES force component (ES/5-$1_s$) tends to push L5 in an anterior direction. (Note: ES/5-$1_c$ = compression component; ES/5-$1_s$ = shear component of the force produced by the ES muscles)

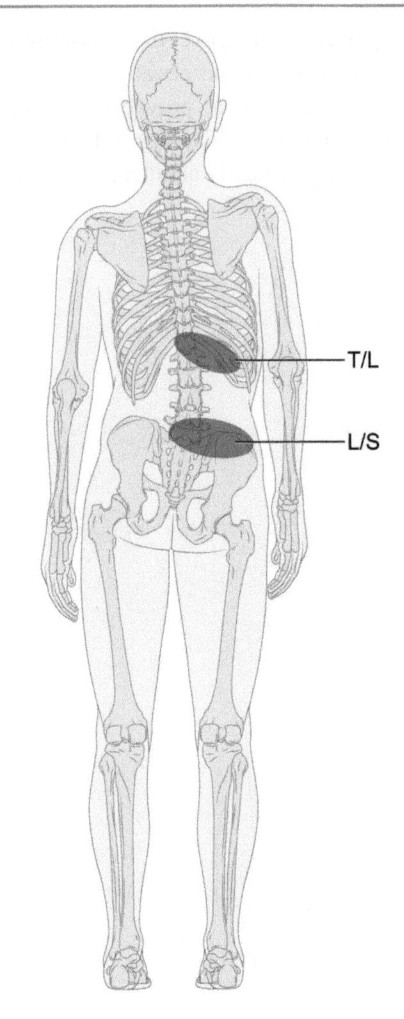


**Fig. 9.30** Posterior view of leg length discrepancy, i.e., right leg short relationship to right C-curve scoliosis and regions of possible pain. T/L thoraco-lumbar, L/S lumbosacral

force that pushes L5 ventrally across the upper slope of the sacrum. This finding may further explain why the active trunk extension and prone instability tests were positive. This is unfortunate since the primary conservative rehabilitative strategy for treating spondylolisthesis is the implementation of an exercise program targeting the transversus abdominis and spinal extensor muscles. The theory of this rehabilitative approach is that muscle could provide dynamic stabilization that can augment the static stabilizers, e.g., ligaments, thoracolumbar fascia, etc., in order to control excessive movements of unstable lumbar segments.

Although the origin of Jim's symptoms was clearly related to spondylolisthesis, he had a right lumbar scoliosis that may have complicated matters (Fig. 9.30). We believed that Jim's scoliosis was functional (see Chap. 8 for the definition of structural and functional scoliosis) and related to his leg length discrepancy (which, by the way, may either be structural or functional). We are able to make that claim because the scoliosis was not evident when he was placed in flexed or extended postures; that is, it was only apparent when he was standing with his spine in its natural or neutral pose. The clinical literature has demonstrated a number of pain syndromes related to the coupling of leg length discrepancy and functional scoliosis, particularly in the thoraco-lumbar and lumbosacral transitional regions. We can imagine that since Jim works on his feet for long periods of time, his scoliotic curve imposes prolonged, and perhaps excessive, compression and shear stress at the left zygapophyseal joints, with prolonged tensile stresses on the outer annulus, zygapophyseal joint capsules, and other lateral soft tissue structures on the contralateral side. In addition, the c-curve alters spinal muscle lengths, and perhaps contractile efficiency, with muscles on the concave side of the curve shortened and muscles on the convex side of the curve lengthened. It is unclear how these changes affect tensile forces at the interbody and facet joints and interact with the pathomechanics associated with Jim's spondylolisthesis. However, it is possible that the coupling of two distinct musculoskeletal abnormalities, i.e., spondylolisthesis and scoliosis, each with their own unique pathomechanical adaptations can only compound the other's adaptations and manifestations.

This completes our study of normal lumbar kinematics, and the relationship between spondylolisthesis and the adaptations/alterations in kinematics and kinetics of that spinal region. We conclude our examination of the function of the lumbosacral spine by examining the normal kinematics and kinetics of the sacroiliac joints. The results of in vitro biomechanical research on the mechanics of the sacroiliac joints

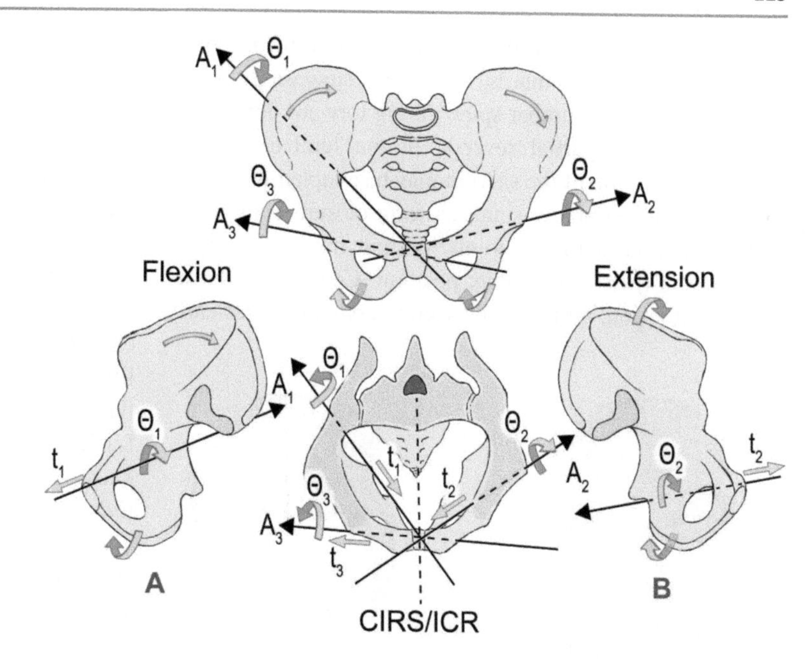


**Fig. 9.31** Schematic of the pelvic ring showing movements within the pelvis during right hip flexion and left hip extension. (**A**) right, (**B**) left. $A_1$ axis of rotation of the right iliac bone relative to the sacrum, $A_2$ axis of rotation left iliac bone relative to the sacrum, $A_3$ axis of rotation of right ilia to left ilia; $t_1$, $t_2$, and $t_3$ translations along $A_1$, $A_2$, and $A_3$ axes, respectively; $\theta_1$, $\theta_2$, $\theta_3$ rotations about $A_1$, $A_2$, and $A_3$ axes, respectively. ICR = instantaneous centers of rotation in the sagittal plane, i.e., the intersection of $A_1$, $A_2$, and $A_3$ with the median sagittal plane. (Modified with permission from Lavignolle et al. (1983))

are dubious, at best, for several reasons: (1) generalizations based on the use of small samples of cadaver specimens cannot be applied to the movements that occur in vivo; (2) applying external forces and moments on an isolated SI joint changes all of the mechanics since the pelvic ring has been interrupted, and (3) fixing or constraining the pelvic ring in order to apply pure moments to the sacrum does not account for the important roles of the public symphysis or movements of the ilia. Yet, shifting the focus to in vivo research using planar or bi-planar radiographs or three-dimensional photostereogrammetric methods and implanted devices, e.g., bone pins, tantalum spheres, etc., fixed to unique bony landmarks, or using optical motion capture and surface markers has presented its own unique set of challenges. What has been claimed, with reasonable agreement among biomechanists, is that the complex movements at the SI joints are critical for the effective distribution of large compressive and shear stresses from above and below, but that the magnitude of those movements (e.g., rotations and translations) are very limited, irregular, and likely cannot be accurately represented by our usual application of a Cartesian coordinate system. For example, in their in vivo study, Lavignolle et al. demonstrated that the axes of movement of the SI joints passed obliquely across the pelvis during reciprocal flexion of the right hip and extension of the left hip (Fig. 9.31). Given these results, it appears that during flexion and extension of the lower limbs, i.e., similar to what occurs during walking, the sacral and/or ilia movements are complex; that is, during flexion of the hip, the ipsilateral ilium glides posterior and downward relative to the sacrum, compressing the SI joint, while during hip extension the ilium glides anterior and flares away from the sacrum. Although Lavignolle et al. reported, on average, rotation and translation magnitudes up to 12° and 6 mm, respectively, later work, reported by others, based on more accurate measurement systems, demonstrated rotations of no more than 4° and translations ranging from 1 to 2 mm.

Prior to their in vivo study, Lavignolle et al.'s preliminary research employed one fresh mortem pelvis to examine innominate rotations and translations. They fixed the sacrum and applied linear forces and moments to the right and left iliac bones, measuring displacements, i.e., translation, and rotation, with comparators and an optical system, respectively. Based on their measures, they determined that each iliac bone presented with six degrees of freedom, i.e., three translations and three rotations. Earlier (see Chap. 3), we discussed the importance of recognizing the availability of six degrees of freedom (DoF) of movement for all synovial joints, regardless of limitations in the magnitude of the rotations and translations that took place. However, we also recognize two commonly used terms to describe sacral motion that appear to be most clinically relevant: *nutation* and *counternutation* (Fig. 9.32). *Nutation* (meaning to nod forward), sometimes referred to as anterior nutation, is defined as a relative anterior tilt (or flexion) of the sacral base (or promontory) relative to the ilium, with the sacral apex moving postero-superior. As the sacrum *nutates*, there may be a simultaneous inferior translatory movement, while the wings of the ilia approximate (move closer together) and the ischial tuberosities move apart. *Counternutation* (or posterior nutation) is movement in the opposite direction or relative posterior tilt (or extension) of the sacral base, as the sacral apex moves antero-inferiorly. With sacral *counternutation*, the wings of the ilia separate while the ischial tuberosities are drawn closer together.

If we consider the sacrum as a "sixth" lumbar segment, *nutation* is often associated with flexion of the lumbar region;

this rotation can be described as the sacrum moving on the relatively fixed ilium; that is, it has been suggested that the sacrum follows the lumbar spine during forward flexion and contributes to a partial reversal of the lumbar lordosis. Inferior translation of the sacrum may be coupled with *nutation*, but it is extremely small as form closure compresses and approximates the SI joints, putting them in a close-packed, and their most stable, position. Conversely, during lumbar extension, although the sacrum *counternutates* early in the range, the sacrum's contribution is halted due to bony and soft tissue constraints as the lumbar segments continue to rotate and translate posteriorly. It has been suggested that the *counternutated* sacrum may render the SI joints relatively less stable, were it not for force-closure, i.e., ligamentous constraint coupled with the dynamics of muscle control.

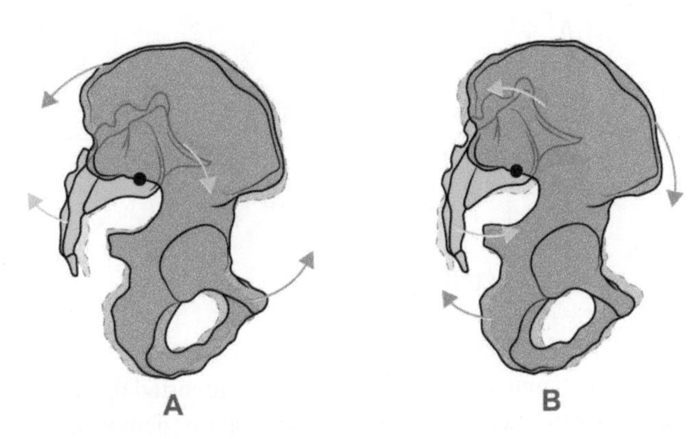


**Fig. 9.32** Kinematics of sacroiliac joints. (**A**) nutation and (**B**) counternutation These movements occur about a sagittal plane axis (small black dot) centered near the second sacral spinous process. Note the relative rotation between the sacrum and ilia, i.e., counter-directed arrows; that is, in (**A**) as the sacrum nutates, the innominate appears to rotate posteriorly, i.e., posterior tilt. The dotted lines indicate the position of the ilia after the movement is completed

According to the osteopathic mechanical model of the pelvis, the innominate bone as an extension of the lower extremity linkage, i.e., connected to the foot-shank-thigh linkage, contributes to SI rotation, which is associated with flexion and extension of the hip joint (as demonstrated by Lavignolle et al.). For example, during walking, it is well known that the pelvis, as a unit, in coordination with the hip joint, moves three-dimensionally (Fig. 9.33). Briefly, the pelvis simultaneously cycles through an anterior/posterior tilt, lateral (i.e., downward and upward) tilting on the right and left sides, and external (i.e., retraction)/internal (i.e., protraction) rotations during different subphases of the gait cycle. Under the assumption that the pelvis is a rigid body, these three-dimensional movements are relatively easy to measure using an optical motion capture system (see Appendix J). However, optical motion capture systems cannot capture intra-pelvic (sacroiliac or iliosacral) movements. Despite the lack of sensitivity to intra-pelvic movements, it has been suggested that during the walking cycle the ilia rotate (in the sagittal plane relative to the sacrum) in distinct patterns in conjunction with the hip joint. Let's see (Fig. 9.33) how this

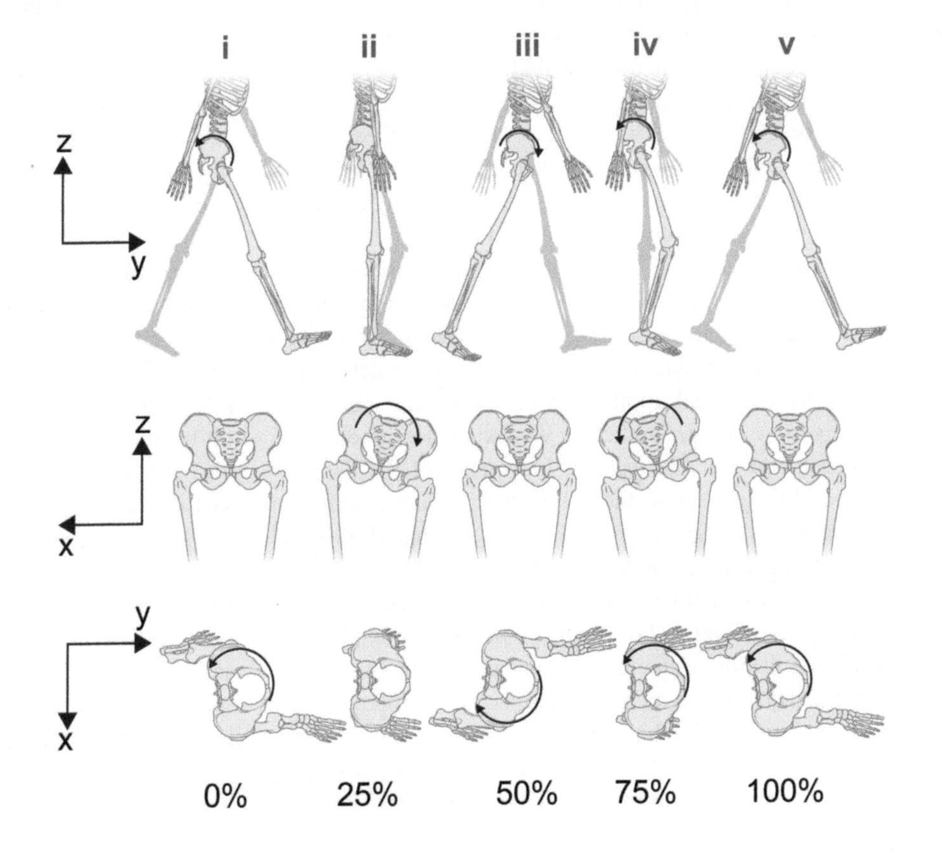


**Fig. 9.33** Schematic depiction of pelvic sagittal (top), coronal (middle) and transverse (bottom) plane rotations about the x-, y-, and z-axes respectively, during a gait cycle. Note that phase **i** corresponds with 0% of the gait cycle (initial contact), phase **ii** with midstance, phase **iii** with terminal stance, phase **iv** with initial swing, and phase **v** with second initial contact, i.e., 100% of a gait cycle

might work for the reference *right* limb beginning a new gait cycle which begins at initial contact (0%) and ends with second initial contact (100%): (1) at initial contact (0% to 2% of the gait cycle), the hip is flexed which induces posterior rotation of the ilium on the same side; (2) during loading response (2% to 12% of the gait cycle), the ground reaction force, through the lower extremity with the hip still flexed, induces a slight increase in posterior rotation of the ilium; (3) from loading response to terminal stance (12% to 50% of the gait cycle), the ilium rotates anteriorly as the hip extends; (4) during pre-swing (approximately 50% to 62% of the gait cycle), the ilium again rotates posteriorly as the hip flexes; and (5) during initial and terminal swing (62% to 100% of the gait cycle), the ilium posteriorly rotates to a maximum, as the hip flexes to a maximum of approximately 40°. What about the contralateral (left) limb, i.e., what is the coupled motion between the left hip and ilium? Essentially, it has been suggested that the left ilium rotates in the opposite direction to its counterpart, e.g., during initial contact through the loading response as the right ilium rotates posteriorly the left extending hip is coupled with anterior ilium rotation, etc. (you complete the left cycle based on what we shared about the right hip-ilium movement pattern).

Neither the sacrum nor ilia are fixed when we move so the sacrum and ilia are moving simultaneously. Using the osteopathic mechanical model of SI function allows us to further explore possible applications of the research that has suggested that the complexity of sacroiliac motion is better described by movements about/along multiple axes that are not strictly Cartesian. To this point, we have defined two types of sacral rotations, i.e., *nutation* and *counternutation*, which occur about a Cartesian-like transverse or a mediolateral axis. Extending our application of the osteopathic model of sacroiliac motion, yet acknowledging that experimental data corroborating the clinical model is lacking, we note that, in addition to sacral rotation about a transverse axis, sacral rotations about right and left oblique axes may also occur (Fig. 9.34).

The right oblique axis runs from the left inferior lateral angle to the right sacral base and the left oblique axis from the right inferior lateral angle to the left sacral base. Movements about the oblique axes have been labeled as torsions, which assume a coupling of rotation and sidebending to opposite sides. For example, during a forward rotation about the right oblique axis (i.e., right anterior nutation or forward torsion), the ventral surface of the sacrum moves to face to the right, while its superior surface declines to the left (i.e., left sidebending). Conversely, a forward torsion about the left oblique axis induces the ventral surface of the sacrum to face left, as its superior surface declines to the right (i.e., right sidebending). The osteopathic literature suggests that these torsional movements are essential elements of the gait cycle in that they assist in the mitigation of compression and

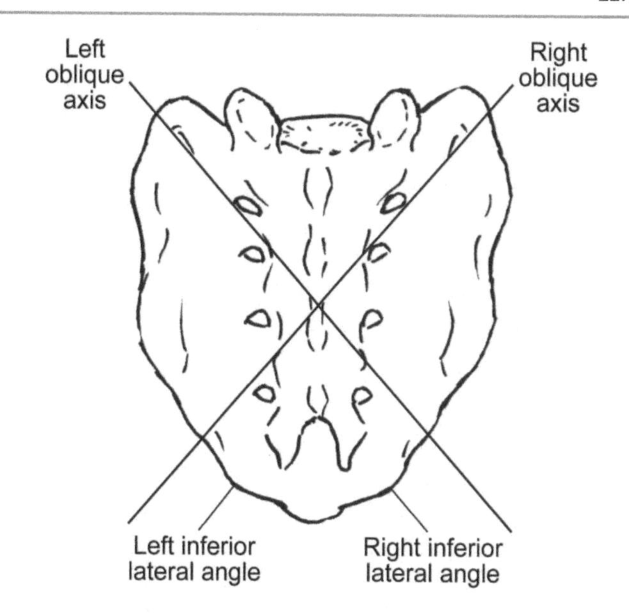


**Fig. 9.34** Right and left obliques axes for sacral rotations sometimes described as sacral torsions

shear stresses at the SI joints that are induced by large ground reaction forces.

Let's see how that might work. During the normal walking cycle, it has been suggested that the sacrum rotates anteriorly in an oscillating fashion around alternating oblique axes (Fig. 9.35). For example, at right midstance, as the hip begins to extend, (i.e., A3 and A4 in the figure), the sacral base on the left moves into anterior nutation, i.e., right rotation about the right oblique axis, as it is carried forward by the advancing left ilia; as it rotates right, it sidebends to the left. Concomitantly, the lumbar spine segments rotate to the left and sidebend to the right, an illustration of the concept of dynamic kinematic coupling. In a succeeding phase of the gait cycle, as the hip transitions from its hyperextended position in terminal stance (i.e., B5 and B6 in the figure) to flexion at initial swing (i.e., B7 and B8 in the figure), the ipsilateral ilium posteriorly rotates carrying the sacrum into a left anterior nutation, i.e., left rotation, about the left oblique axis with simultaneous right sidebending; and, in concordance with Fryette's law of neutral spine mechanics, the

lumbar spine sidebends to the left and rotates right. Due to the soft-tissue artifact and lack of fidelity to extremely small motions when using optical motion capture, the claims about these hypothetical SI rotations during the gait cycle based on clinical observation and palpation, as well as application of a theoretical framework, have yet to be demonstrated experimentally. However, these ideas provide a model that seems to be useful as a tool for addressing gait problems that might be induced by abnormal lumbosacral kinematics.

Our focus on only the posterior elements of the lower body, i.e., anatomy and mechanics of the lumbosacral spine,

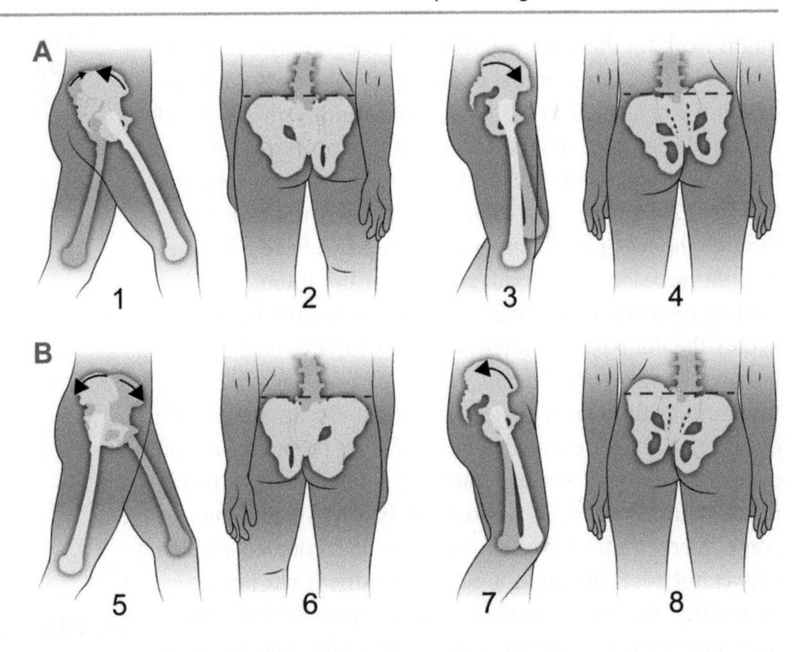


**Fig. 9.35** Relationship between hip flexion/extension, iliosacral posterior/anterior rotation, and left/right sacroiliac rotation during a gait cycle; (**A**) loading response to midstance, (**B**) terminal stance to initial swing of the gait cycle. (Modified with permission from Greenman (1990))

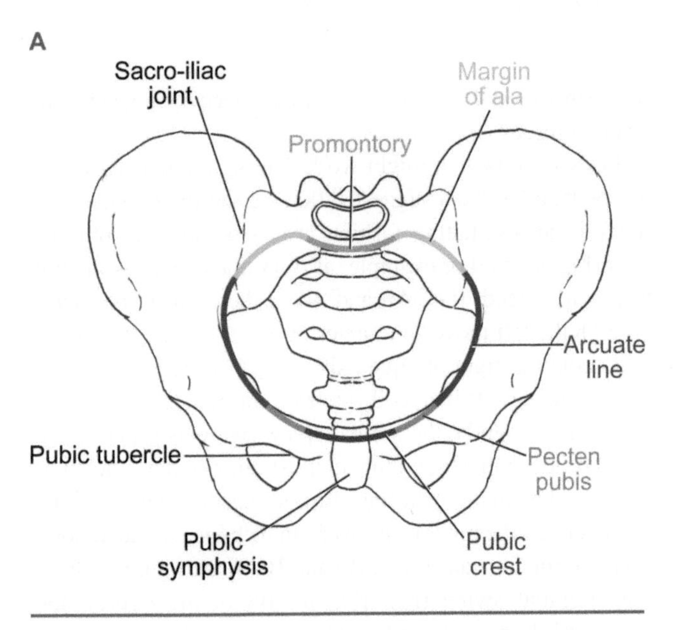


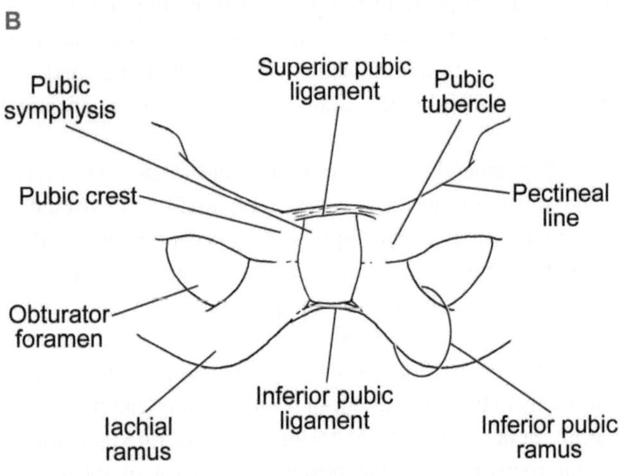


**Fig. 9.36** (**A**) Perspective of posterior and anterior elements of the pelvic ring and (**B**) more detailed view of pubic symphysis with stabilizing structures

up to this point can be easily justified: Jim's symptoms were located posteriorly. However, in keeping with the concept of complex systems and the importance of remaining tuned into the interrelationship among the elements that make up those systems, we have to consider an essential element of the anterior pelvic ring, the pubic symphysis (see Figs. 9.9 and 9.11). The pubic symphysis forms a strong midline union between the pubic bones of the pelvis that augments stability yet provides enough mobility to accommodate movements of the hip joints, as well as ilia rotations.

We begin with an examination of essential details of the anatomy of the anterior ring, pubic bones, and connecting ligaments (Fig. 9.36). Although not clearly evident in the figures, the articular surfaces of the public bones are oval in shape, slightly convex, and oriented obliquely in the sagittal plane, running postero-inferiorly in a craniocaudal direction. Although the articular surfaces of this joint are covered in hyaline cartilage, it is classified as an amphiarthrosis (or secondary cartilaginous) joint. The pubic bones are secured by a fibrocartilaginous interpubic disc (sometimes referred to as the interosseous ligament of the public symphysis) with fibers that run transversely and obliquely (Fig. 9.37). It is notable that the material and mechanical features of the interpubic and intervertebral discs are thought to be similar, suggesting that the interpubic disc's inherent viscoelastic properties are used to distribute the compression and shear stresses that originate cranially and caudally. A unique feature located in the center of the disc is a narrow, slit-like, oval-shaped cavity or cleft (Fig. 9.37). The function of this cleft has yet to be defined.

There are four ligaments that stabilize the pubic symphysis. The superior pubic ligament bridges the superior margins of the joint and is attached to the pubic crest as far lateral as the pubic tubercles. An inferior pubic (also referred to as the arcuate or subpubic) ligament forms an

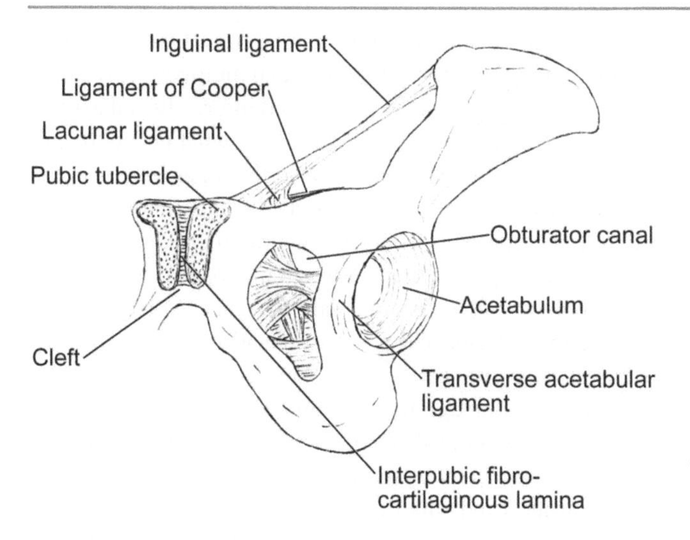


**Fig. 9.37** Cross-section of pubic symphysis demonstrating its relationship to the interpubic fibrocartilaginous lamina (or disc)

arc spanning the inferior pubic rami. Some have suggested that this ligament is stiffer than its superior counterpart. The very thick anterior pubic ligament has a collagen matrix that runs transversely and obliquely and merges with the periosteum of the pubic bones laterally. Biomechanical testing suggests that this ligament is second only to the interpubic disc in terms of tensile strength. It is notable that tendinous insertions from the adductor longus and brevis, and gracilis, blend with the anterior pubic ligament; additional connections to the ischiocavernosus muscles and corpora cavernosa have also been identified (Fig. 9.38). Having fascia connections to the tendons of several muscles suggests, perhaps, a more complex role for the anterior ligament, or alternatively, more expansive functions of the muscles that have these connections. For example, a common strain injury in sports often involves one or more of the groin (adductor) muscles. It is not uncommon for such injuries to be resistive to interventions typically prescribed, i.e., recovery times can be prolonged. Because of the association of the adductor longus and gracilis tendons with the pubic symphysis, it is possible that delayed recovery may be related to a strain of the anterior pubic ligament or some type of pubic symphysis joint dysfunction Moreover, an inspection of Fig. 9.38 suggests that an apparent "simple" adductor longus or gracilis strain could affect the function of many related muscles and joints. Finally, the posterior pubic ligament is a thin fibrous membrane continuous with the periosteum posteriorly, whose role in symphysis stability is somewhat obscure.

Biomechanical research related to the material properties of supporting ligaments and pubic symphysis kinematics is sparse. Most investigators agree that the pubic symphysis is subjected to a variety of different forces related to simple activities of daily living. The forces include tension primarily at the inferior aspect of the joint, compression of the superior region when standing and sitting, and compression and shearing with unequal weight-bearing while standing or walking. In an investigation reported by Beck et al., mechanical testing of the material properties, using four groups of adult cadavers (male, nulliparous females, non-pregnant multiparous females, and primigravidae in the last trimester of pregnancy) showed that the anterior pubic ligament had the greatest ultimate strength at failure when placed under tensile loads, followed by the inferior and then superior ligaments. Ligaments were strongest in men, slightly stronger in nulliparous compared with multiparous women, and weakest in primigravidae in the last trimester of pregnancy.

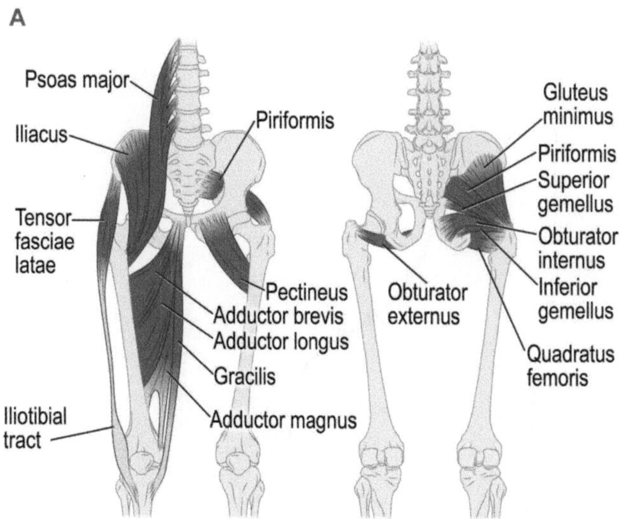


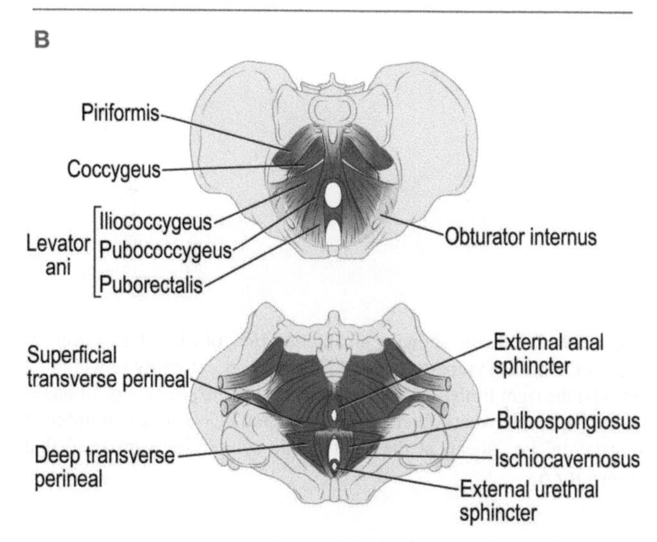


**Fig. 9.38** Relationships between the pubic symphysis and (**A**) superficial and deep anterior, posterior, and medial muscles of the thigh and (**B**) pelvic floor muscles

Although assessing the biomechanics of the pubic symphysis using cadaveric specimens has provided some insight into function, we previously have identified the limitations that in vitro research imposes on the clinical

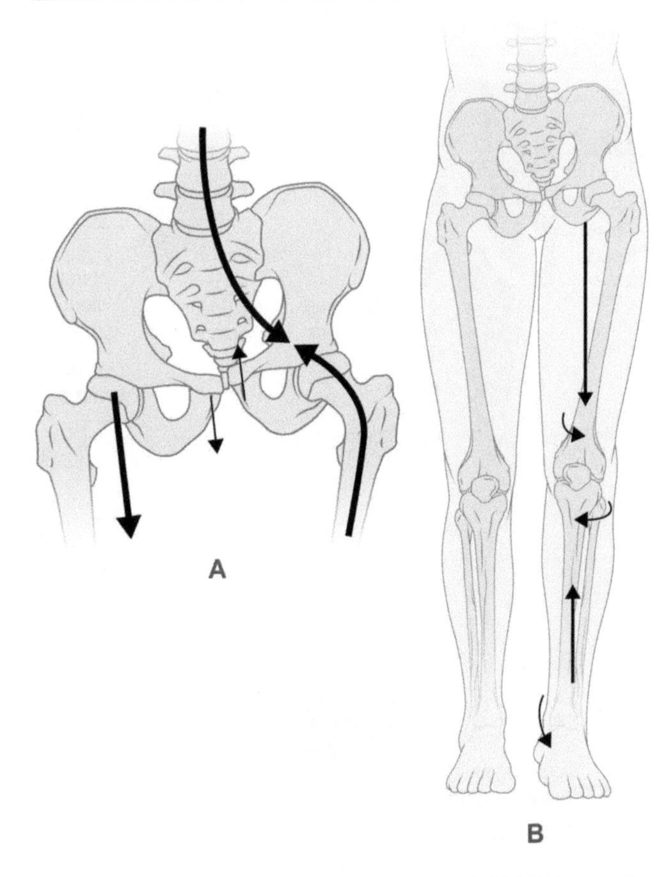


**Fig. 9.39** Schematic showing an anterior view of (**A**) inferior translation at the pubic symphysis with the right lower extremity unweighted and (**B**) the right limb being unweighted during the pre-swing subphase of the gait cycle where forces from above and below results in inferior translation, i.e., shear, at the right public symphysis, similar to that shown in (**A**)

application of experimental results. Measuring in vivo movements of the pubic symphysis obviously cannot be done using skin markers and optical motion capture. Therefore, researchers have one of two options at their disposal, the use of (1) optical motion capture systems with the application of marked steel pins inserted into the superior pubic rami on either side of the symphysis, or (2) bi-planar radiography. Both methods impose potential risks to participants. The advantage of using motion capture systems is that movements of the lower extremities and trunk are recorded in continuity, as opposed to the use of radiographs that can only record discrete joint positions, e.g., a starting joint position and at the end of the joint motion. And, of course, radiation exposure with the use of bi-planar radiography secondary to multiple trials likely leads to greater risk for research participants than motion capture. Research on pubic symphysis mobility to date is generally sparse and limited by small sample sizes and variable testing methods. Given the limitations, however, it is clear that the pubic symphysis allows both rotational and translational movements (see Becker et al. 2010, systematic review). Recorded translations ranged from 1 to 2 mm, with greater magnitudes for nulliparous and multiparous women. The largest translations were directed inferiorly and associated with single-leg standing, similar to what occurs in gait when one leg is weight-bearing, and the other limb is in the swing phase (Fig. 9.39). Research has also demonstrated sagittal and coronal plane rotations ranging from 1° to 3°. Some have suggested that, although the rotations that occur at the pubic symphysis are small, they necessarily accommodate the anterior and posterior rotations of the ilia that occur during the gait cycle. We need to interpret the extremely small magnitude of pubic symphysis movements with caution without knowing the sensitivity and reliability of the measurement systems. On the other hand, as we noted earlier, even small movements at joints can be critically important considering that normal function relies on dynamic kinematic coupling produced by all interrelated elements within and between systems.

As noted previously, Jim did not report any symptoms in the abdominal or groin region. Thus, although he did report central low back pain, we hypothesized that sacroiliac and pubic symphysis mechanics were normal. That is not to say that the supporting soft tissues at the pubic symphysis and sacroiliac joints were not being stressed in abnormal ways. Only that the origin of his symptoms did not likely originate in those tissues, but in the tissues immediately related to his spondylolisthesis. On the other hand, we are reminded that our hypotheses must be contingent.

## 9.5 Lumbar Muscles

We have alluded to the conundrum that the rehabilitation expert may encounter when considering a conservative intervention for a case of moderately advanced spondylolisthesis, similar to Jim's. We had earlier hypothesized that the normal passive constraints, i.e., the geometry of zygapophyseal joints, anterior longitudinal ligament, annulus of the IVD, iliolumbar ligaments, etc., that typically control the predisposition of L5 to translate anteriorly may no longer be able to do so for reasons that are related: (1) altered anatomy secondary to the spondylolisthesis and (2) over-lengthening, i.e., plastic deformation, of the soft tissues, due to the application of repetitive and prolonged tensile forces causing tissue creep. Our intuition suggests that the anatomical solution to the inadequacy or loss of passive constraints is to train the muscles of the region to provide dynamic stability. In Chap. 8, we discuss the actions of several thoracic and lumbar muscles related to the mechanism of injury, i.e., chopping wood. Normally, we would refer the reader to consult that information. However, given the potential importance of considering dynamic lumbar stabilization as a treatment option, we think it relevant in this case to re-examine the lumbar components

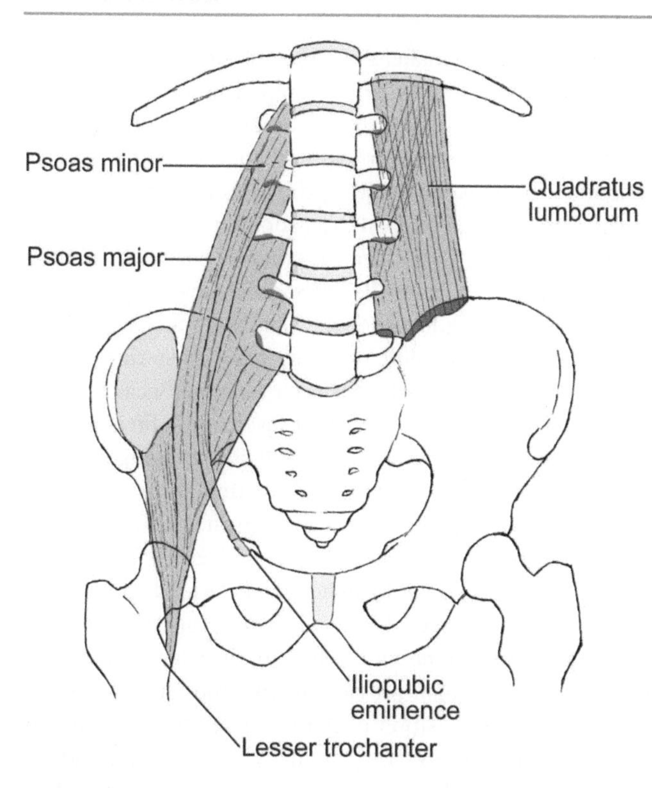


**Fig. 9.40** Anterior view of the psoas and quadratus lumborum muscles. The oblique and vertical lines in the quadratus lumborum are attempting to depict its three-dimensional layering

of those muscles. We draw heavily on the work of Bogduk yet focus more on the mechanics of muscle action and not on the details of the anatomy (the curious reader can do so).

Based on function, Bogduk suggested dividing the muscles of the lumbar region into three groups:

1. Psoas major, covering the anterolateral region
2. Intertransversarii laterales and quadratus lumborum because they connect and cover the transverse processes anteriorly
3. Lumbar back muscles, which cover the posterior elements

**Psoas Major** The psoas major has diverse attachments to the lumbar spine including, at each segmental level from T12/L1 to L4/L5, insertions to the medial three-quarters of the anterior surfaces of the transverse processes, intervertebral disc, and to the margins of the vertebral bodies adjacent to the IVD (Fig. 9.40). It is a long muscle that wraps over the pelvic brim to insert onto the lesser trochanter of the femur. Although intuition suggests that the psoas major could draw the lumbar spine and pelvis forward, i.e., increasing the anterior convexity, its fibers lie very close to the axis of rotation and therefore do not have the leverage to do so; therefore it acts primarily as a flexor of the hip. On the other hand,

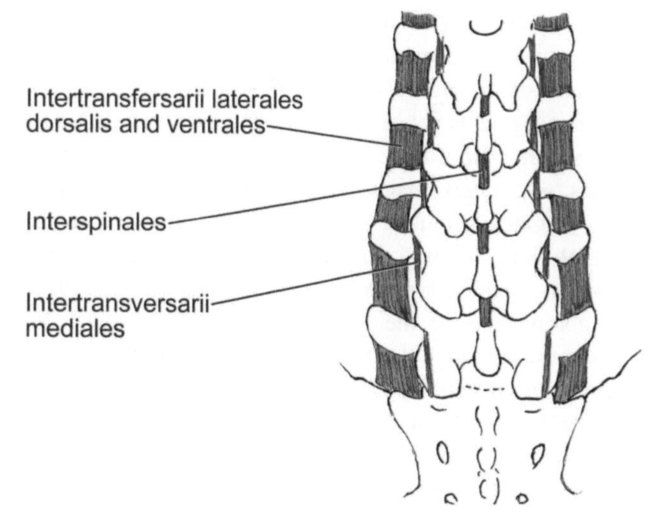


**Fig. 9.41** Short intersegmental muscles in the lumbar region

because of its relatively large cross-section, the tension the psoas major generates during isometric, concentric, or eccentric actions has the potential to create large interbody joint compression forces. This capacity has an advantage and a disadvantage. The advantage, in Jim's case, is that compression of the interbody joint at L5/S1 could provide needed stability. The disadvantage is that large compression forces on the IVD, over time, could contribute to the development of degenerative disc disease (DDD). We explore this topic later.

**Intertransversarii laterales and quadratus lumborum** The intertransversarii laterales have two elements, ventrales and dorsalis, neither of which are known to make significant, if any, mechanical contributions to the spine (Fig. 9.41). The quadratus lumborum (QL) comprises a complex array of oblique and longitudinal fibers that connect selected lumbar transverse processes, ilium, and 12th rib. Detailed dissection demonstrates that this muscle consists of four types of fascicles: iliocostal fibers that connect the ilium and the 12th rib; iliolumbar lumbar fibers that connect the ilium and lumbar transverse processes; lumbosacral fibers that connect lumbar transverse processes with the 12th rib; with the fourth type connecting the ilium and the body of T12; and arranged in posterior, middle, and anterior layers (Fig. 9.40). Historically, anatomists have assigned this muscle to the task of fixing the 12th rib during respiration, thereby providing a base from which the lower thoracic fibers of the diaphragm can act. It is likely that the primary function of the QL is to anchor the lumbar transverse processes and the 12th rib to the ilium to facilitate lumbar lateral flexion (unilateral activation) and extension (bilateral activation). In either case, it is estimated that the QL makes small contributions to these lumbar movements.

**Lumbar back muscles** These include all of the dorsal muscles that have attachments to the lumbar segments and act directly on them, and muscles that do not have attachments to lumbar segments yet can move the lumbar spine. Bogduk organized them into three groups: (1) short intersegmental muscles, e.g., interspinales and intertransversarii mediales; (2) polysegmental muscles that attach to lumbar segments, e.g., multifidus, lumbar components of the longissimus and iliocostalis; and (3) long polysegmental muscles, e.g., thoracic components of the longissimus and iliocostalis. As opposed to traditional descriptions of the origins and insertions of these muscles, Bogduk suggests that we consider the posterior muscles from above downward because it appears to be more consistent with their pattern of nerve supply, clarifies the identity of the regionality of certain muscles, and their relationship to the erector spinae aponeurosis, and mechanical disposition of the various muscles.

**Interspinalis and intertransversarii mediales** The interspinalis muscles run from spinous process to spinous process lateral to the interspinous ligaments and appear to contribute to the posterior rotation of the superior segment on the one below it. The intertransversarii mediales lie lateral to the axis of lateral flexion and posterior to the axis of posterior rotation, but the fibers lie so close to the respective axes of rotation effectively eliminating their moment-producing function. Neither the interspinalis nor intertransversarii mediales effectively rotate lumbar segments, fine-tune vertebral positioning, nor can they provide significant dynamic stability. However, because of the preponderance of intrafusal muscle fibers, i.e., muscle spindles, it is highly likely that they play a leading role in proprioception (Fig. 9.41). In fact, all unisegmental muscles of the vertebral column have up to six times the density of muscle spindles found in the longer polysegmental muscles.

**Multifidus** These are the largest and most medial of the lumbar extensors, comprised of repeating fascicles that run from the laminae and spinous processes proximally to consistent attachments caudally. The shortest fascicles, or laminar fibers, span two segments running from lamina to lamina, with the exception of the L5 fascicle, which attaches caudally to the dorsal aspect of the proximal sacrum. The bulk of multifidi mass is composed of five overlapping groups of larger fascicles that radiate from the lumbar spinous processes to caudal attachments on mamillary processes, iliac crest, and the sacrum (Fig. 9.42). It is notable that some of the deeper fibers of multifidus fascicles attach to the facet joint capsules next to the mammillary processes, allowing the muscle to prevent impingement of the capsule during segmental extension.

The oblique caudolateral orientation of the multifidus can be mechanically resolved into component vectors: vertical and horizontal (Fig. 9.43). Based on the length of each vector, it is clear that the vertical component has the greater magnitude, suggesting that the primary action of the multifidus is to produce an internal extensor moment across lumbar interbody joints, i.e., posterior rotation of a superior lumbar segment on the one below it. The sagittal view of the multifidus line of action indicates that when these muscles shorten, they produce nearly a pure rotation; that is, posterior translation is not coupled with the posterior rotation. Because of the polysegmental nature of the larger fascicles, indirect actions on interposed vertebra cause a type of bowstringing effect of the lumbar segments. Therefore, the action of the multifidus may also accentuate the lumbar lordosis, inducing increased IVD compression posteriorly, while increasing tension on the anterior longitudinal ligament and anterior annulus.

Given what we have discussed to this point, we can conclude that although the multifidi, with their large density of muscle spindles, could provide timely proprioceptive information about the position of the lower lumbar segments to the central nervous system, any activity that would induce isolated action of these muscles would likely increase forward slippage of L5. There are those who believe that the multifidi are the principal muscles to target in a lumbar stabilization rehabilitation program. Based on what we know now, targeting the multifidi may actually be contraindicated.

The direction of the horizontal component of the multifidus force vector suggests that when it pulls the spinous process laterally, the lumbar segment sidebends ipsilaterally and rotates contralaterally. However, under normal circumstances impaction of the contralateral zygapophyseal joints limits axial rotation, unless a large shear force is also applied to the IVD. With attachments to spinous processes, however, the multifidi do not have the leverage to exert large shear forces, thus making it unlikely that these muscles are effective at rotating lumbar segments.

Since it appears that the multifidi do not make isolated and large contributions to lumbar rotation, what role might they play during activities that involve axial rotation? It turns out that rotation of the lumbar region occurs secondary to the rotation of the thorax, which is initiated by activation of the abdominal oblique muscles, e.g., left rotation is induced by concentric action of the left internal and right external abdominal obliques. Generally, muscle forces produced by their fibers that are oriented multi-dimensionally can be modeled mechanically as vectors with multiple lines of action (e.g., as illustrated in Fig. 9.43). However, when a muscle with a multidimensional structure shortens (or lengthens), those vectors do not operate independently. For example, when the abdominal oblique muscles, which have lines of action for both horizontal rotation and flexion, are activated, the thorax turns in one direction and bends for-

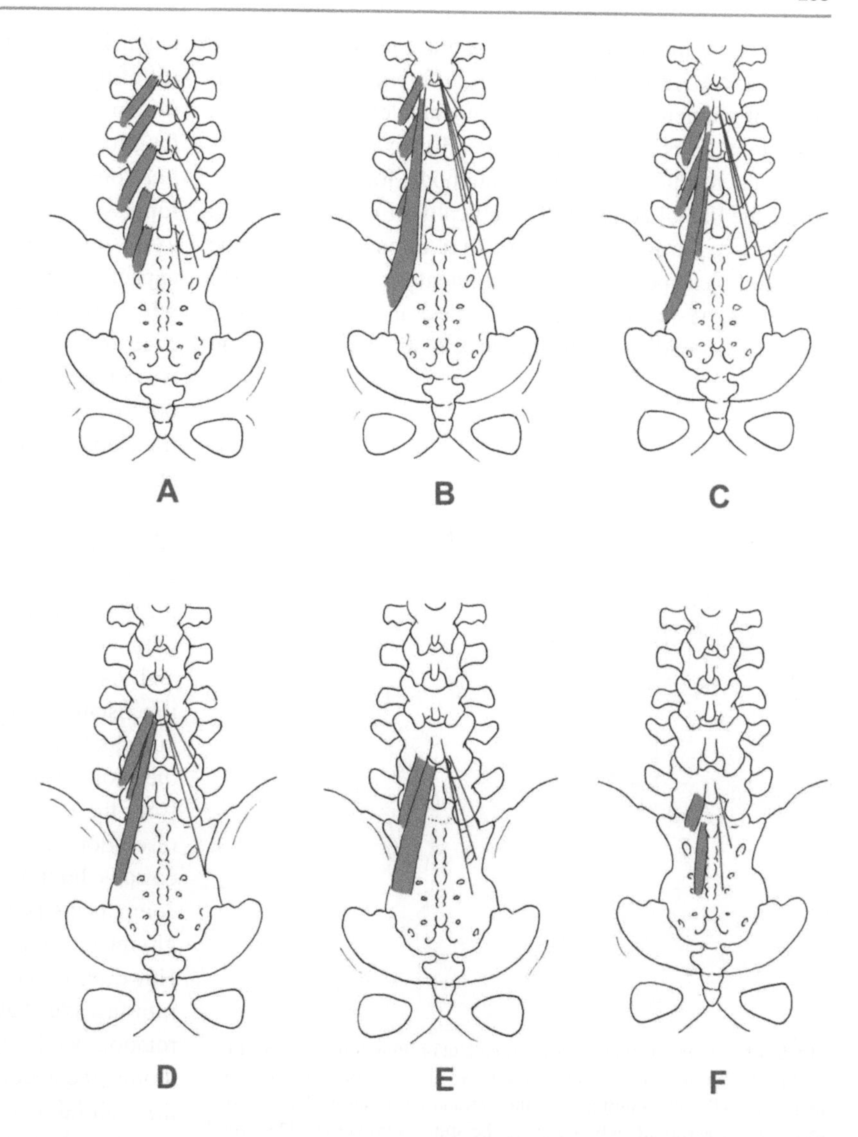


**Fig. 9.42** Schematic of (**A**) the short laminar fibers of multifidus, (**B–F**) the more extended fascicles of the multifidus spanning L1 to L5. (Note: the straight lines represent a line of action for specific muscle fascicles. Modified with permission from Bogduk (2012))

ward simultaneously. Of course, when we turn in one direction or another, we sometimes want to combine this movement with forward bending, but at other times with a backward bend. Therefore, we need a mechanism to provide that kind of control. It turns out that the action of the multifidi likely provides that mechanism so that when the abdominal oblique muscles create trunk rotation the unwanted action of flexion can be inhibited or controlled by simultaneous activation of the multifidi.

**Lumbar erector spinae** Most anatomists include three muscles in this category: spinalis, longissimus, and iliocostalis. Based on his dissections, Bogduk suggested that the erector spinae (ES) consists of two muscles, the longissimus thoracis and iliocostalis lumborum, with each of those having two elements. Thus, we have four parts identified as longissimus thoracis pars lumborum, iliocostalis lumborum pars lumborum, longissimus thoracis pars thoracis, and iliocostalis lumborum pars thoracis. Because of the relevance to Jim's diagnosis, the muscles we focus on include the longissimus thoracis pars lumborum and iliocostalis lumborum pars lumborum. We encourage the reader to pursue further details on ES anatomy and function by consulting the reference provided.

**Longissimus thoracis pars lumborum** This muscle is composed of five fascicles, each with attachments from the accessory process and adjacent medial end of the dorsal surface of the transverse processes of each lumbar segment (Fig. 9.44A). The fascicle of L5 is the deepest, with each cranial fascicle more superficial. Fascicles from L1 to L4 converge caudally to form tendons and the lumbar intermuscular aponeurosis, which represents a common tendon, or the aponeurosis, of lumbar fibers of the longissimus.

As with the multifidus, the fascicles of the lumbar longissimus can be resolved into vertical and horizontal vectors.

The relative magnitudes of the vector components change from L1 to L5, resulting in relative differences in the magnitudes of moments of force at a segmental level (Fig. 9.44B). Additionally, bilateral and unilateral activation of the longissimus results in different interbody joint movements. Since the line of action of the vertical vector lies lateral and posterior to the approximate axis of rotation located near the center of the superior vertebral segment for sidebending and extension, respectively, the fascicles of the longissimus induce sidebending when the muscle acts unilaterally and extension when the muscle acts bilaterally. The lesser capacity of the fascicles of the longissimus, compared to multifidi, to induce extensor moments lies in the fact that the length of their moment arms is smaller. The direction of the horizontal vectors suggests that when the fascicles of the longissimus act bilaterally, lumbar backward rotation, i.e., extension, is coupled with posterior translation, with this capacity greatest at the lowest lumbar levels. Conversely, when the lumbar spine flexes, the longissimus may act eccentrically to control anterior translation. Just a reminder: we noted earlier that at the lowest lumbar level, the horizontal force vector of the ES paradoxically induces anterior translation, particularly with spines that are hyperlordotic and/or have increased sacral angles.

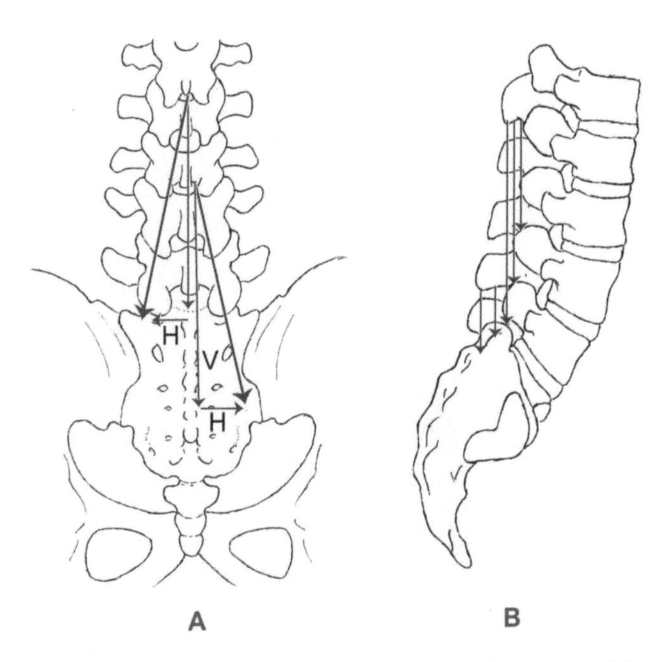


**Fig. 9.43** (**A**) In a frontal view the resultant oblique force vector of the multifidus can be resolved into vertical (V) and horizontal (H) components, and (**B**) in a sagittal view the vertical orientation of multifidus force vector acting at right angles to the spinous processes. (Modified with permission from Bogduk (2012)))

**Iliocostalis lumborum pars lumborum** This muscular system consists of four fascicles from L1 to L4 that run from the tips of transverse processes to an area extending laterally onto the middle layer of the thoracolumbar fascia (Figs. 9.12 and 9.45). Many have suggested that an L5 fascicle of the iliocostalis lumborum had existed in neonates and children but was replaced by collagen to become the posterior band of the iliolumbar ligament, while the anterior band originated from the quadratus lumborum. As shown in Fig. 9.45, the resolution of the more laterally placed fascicles of the iliocostalis lumborum into its vertical and horizontal vector components is nearly identical to those created for the longissimus, suggesting that their actions are also similar. However, because of their laterality, the iliocostalis lumborum fascicles have greater moment arms for inducing axial rotation, while still secondary to the primary rotators of the thorax, the abdominal obliques; the iliocostalis likely assists the multifidi in controlling the flexor moment of the rotating upper trunk predisposed to flexing.

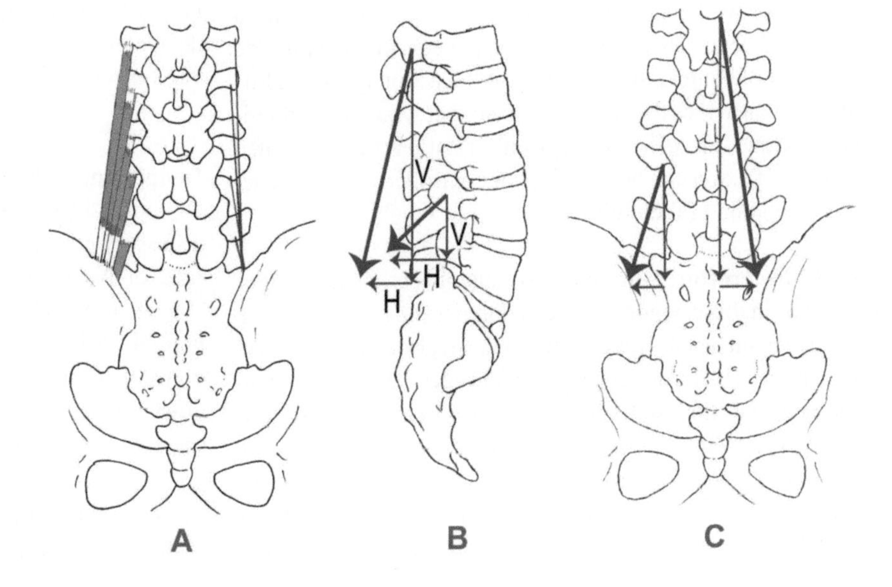


**Fig. 9.44** (**A**) Fascicles of longissimus thoracis pars lumborum and (**B, C**) sagittal and posterior frontal perspective of vertical (V) and horizontal (H) force vector components of the fascicles of the longissimus thoracis pars lumborum. (Note: the horizontal force vector appears to have an ineffective moment arm to produce axial rotation. Modified with permission from Bogduk (2012))

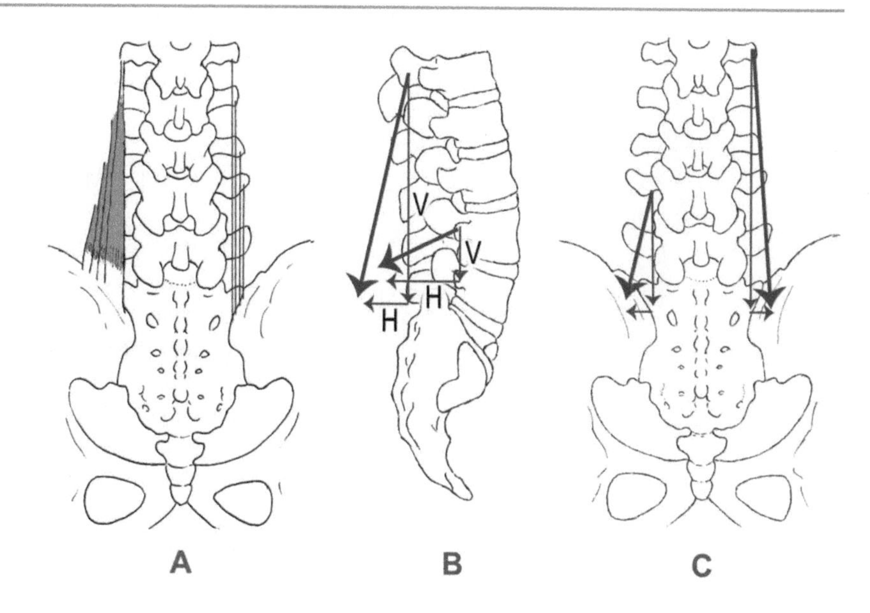


**Fig. 9.45** (A) Fascicles of iliocostalis lumborum pars lumborum and (B, C) sagittal and posterior frontal perspective of vertical (V) and horizontal (H) force vector components of the fascicles of the iliocostalis lumborum pars lumborum. (Modified with permission from Bogduk (2012))

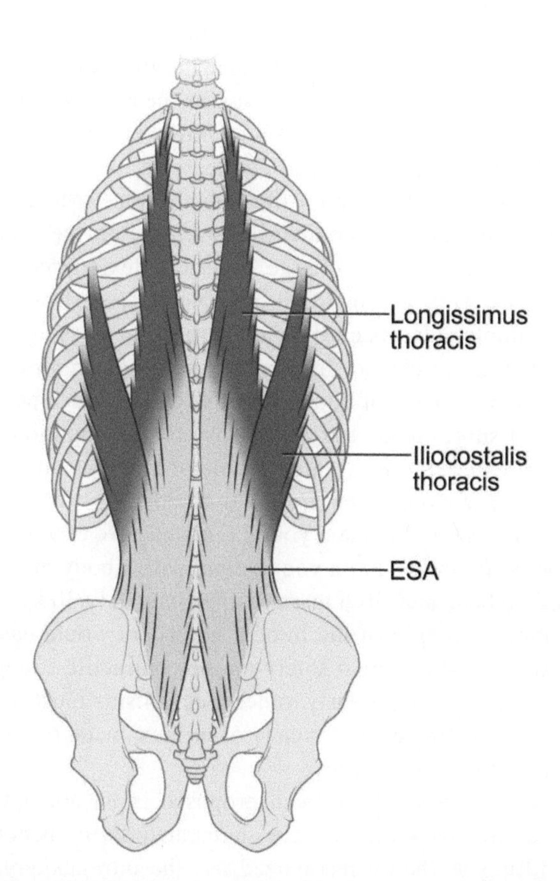


**Fig. 9.46** Erector spinae aponeurosis (ESA) is formed by the caudal tendons of the longissimus thoracis pars thoracis (medial fascicles) and iliocostalis lumborum pars thoracis (lateral fascicles). The darkened areas represent the muscle bellies and their span is shown on the right. (Modified with permission from Bogduk (2012))

It appears that enhancing the actions and effectiveness of both the longissimus thoracis pars lumborum and iliocostalis lumborum pars lumborum through a rehabilitation program tailored to Jim's case might be appropriate. This is so because of the potential mechanical effect of the horizontal vector component that induces posterior translation of the lower lumbar segments when these muscles are activated. However, we should remember two things: (1) when a muscle acts, it does not isolate one vector component over the other, so great care must be taken to avoid lumbar hyperextension, and (2) the lumbar fibers of ES exert posterior shear forces on the vertebrae to which they are attached, but anterior shear forces on vertebrae below these; the thoracic fibers of lumbar ES induce posterior shear forces on upper lumbar segments but anterior shear forces on L4 and L5. These facts pose a challenge to the rehabilitation professional managing someone with a problem like Jim's.

**Longissimus thoracis pars thoracis and iliocostalis lumborum pars thoracis** Fascicles from these muscles arise from transverse processes and ribs, forming tendons that find purchase into the erector spinae aponeurosis, and eventually the sacrum and posterior ilia (Fig. 9.46). When acting bilaterally both muscles extend the thorax and exert a bowstring effect that exaggerates the lumbar lordosis. Acting unilaterally the fascicles induce ipsilateral thoracic, and indirectly, lumbar sidebending.

**Thoracolumbar fascia** We briefly described the thoracolumbar fascia in a different context (see Fig. 9.12), but now give it the attention it deserves because of its role to enhance lumbar stability, particularly during work activities in flexed postures. It is notable that the posterior layer of the thoracolumbar fascia consists of a superficial (with collagen fibers running caudomedially) and a deep (with collagen fibers running caudolaterally) laminae (Fig. 9.47).

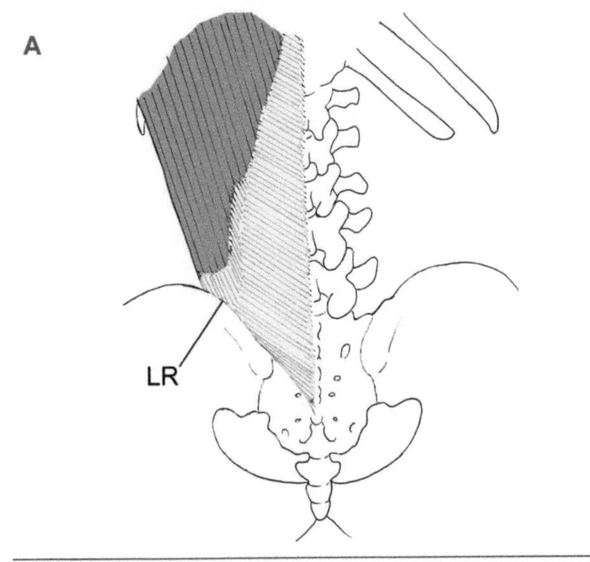


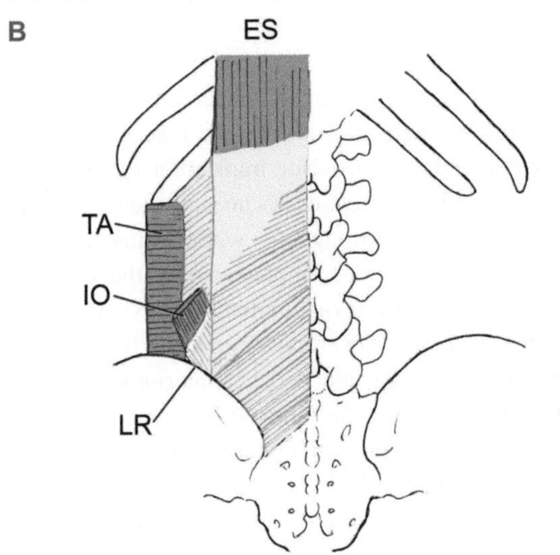


**Fig. 9.47** (**A**) Superficial and (**B**) deep lamina layers of the thoracolumbar fascia The superficial layer has extensive connections to the latissimus dorsi laterally and the spinous process of the thoracolumbar spine, iliac crest, and midline of the sacrum. The deeper lamina shows collagen fibers from the midline to the lateral raphe (LR) and posterior superior iliac spine, with aponeurotic connections to the erector spinae (ES) superiorly, and transversus abdominis (TA) and posterior fibers of the internal oblique (IO) laterally. (Modified with permission from Bogduk (2012))

Together, the superficial and deep laminae of the posterior thoracolumbar fascia form a retinacular-like structure over the dorsal muscles of the lower back; that is, this fascia covering prevents the dorsal displacement of the lower back muscles. Additionally, the collagenous bands that course from the spinous processes of L3, L4, and L5 to the ilia provide ligament-like structures that may control excessive anterior translation of those segments. Finally, the posterior layers with their fascia connections from the spinous processes of the lumbar spine to the transversus abdominis seem to provide a significant mechanical link between the ventral and dorsal muscles of the lower trunk.

### 9.5.1 Mechanics of Lifting

We have alluded to the role that the thoracolumbar fascia may play in the mechanical stability of the lumbar spine, in general, as well as during lifting. Let it be said at the outset that the thoracolumbar fascia is only one of many players in this act. The act of lifting is a matter of balancing, sometimes very large, external flexion moments (and sometimes flexion/rotation moments), which involves many elements of the hip joint and lumbosacral regions. Getting insight into lifting mechanisms is important for this case since the majority of Jim's work entails lifting and/or working in partially flexed postures daily and for extended periods of time. The summary which follows comes largely from Bogduk.

To start, we acknowledge the ability of the mass of muscles about the hip to easily manage the external flexion moment at the hip joint, as well as the role those muscles play in controlling the pelvis; however, though there may be some indirect fascial connections, the hip extensors do not act directly on the lumbar spine. Moreover, relying on only the contractile strength of the spinal extensors to balance a large external flexion moment incurred during a bent lift is insufficient (see Bogduk). For example, it has been shown that the total mass of muscles that act to extend the lumbar spine has an absolute maximum strength of about 4000 N, which, when acting via their short moment arms (relative to the long moment arms of external forces), are capable of inducing an internal extensor moment no greater than 200 Nm (for males under the age of 30 years; young females have 60% of the capacity of males). For a young man, with a body mass of 70 kg, to bend and lift a mass of approximately 30 kg, the maximum strength of the lumbar extensors would easily be exceeded if he relied solely on the contractile strength of his muscles. So, clearly, other mechanisms must be in play when lifting because many of us have made multiple similar lifts without injury.

Several theories have been proposed in an attempt to explain how back muscles are augmented during bending and lifting. A theory recognized as "the intra-abdominal balloon mechanism" suggested that a Valsalva maneuver prior to a lift increases the intra-abdominal pressure, serving to protect lumbar extensor muscles. However, later research showed that intra-abdominal pressure alone did not correlate well with the size of the load lifted, nor by increases in intra-discal pressure. Moreover, strengthening the abdominal muscles did not increase intra-abdominal pressure during lifting. Additional criticisms of the balloon mechanism theory suggested that (1) the pressure needed to counter a larger flexion moment would exceed the maximum hoop tension stress of the abdominal muscles and would obstruct flow within the abdominal aorta, and (2) action of the abdominal muscles to create intra-abdominal pressure would, itself, induce a flexor moment negating any anti-flexion value.

Two alternative theories account for a primary role of the thoracolumbar fascia. In one of these theories, lateral tension developed in the posterior layer of the thoracolumbar fascia via the abdominal muscles, e.g., transversus abdominis and internal obliques, could pull on lumbar spinous processes and produce an internal extension moment of lumbar segments. With this theory, the abdominal muscles not only braced, but extended, the lumbar spine during heavy lifts, and an increase in intra-abdominal pressure was merely a by-product. Gaps in this proposal include (1) the posterior layer of the thoracolumbar fascia was fully developed in only the lower lumbar segments; (2) irregular fascial connections to the abdominal muscles included only the internal oblique; and (3) although the transversus abdominis has consistent attachments to the thoracolumbar fascia, its contribution to the lumbar extensor moment appears to be small, i.e., approximately 6 Nm. A second theory involving the thoracolumbar fascia, referred to as the hydraulic amplifier effect, asserted that since the thoracolumbar fascia surrounded the back muscles like a retinaculum, it could brace and multiply the lumbar extensor power. Subsequent modeling, yet to be validated, has shown that this mechanism enhances the strength of the back muscles significantly, i.e., increased by up to 30%. More research may improve the application and understanding of this alternative theory. Note: enlisting the concept of the hydraulic amplifier effect in Jim's rehabilitation should be considered.

Another model of lifting does not involve the thoracolumbar fascia per se, but the posterior ligamentous system, namely, the capsules of the zygapophyseal joints, the interspinous and supraspinous ligaments, posterior longitudinal ligament, ligamentum flavum, and posterior layer of the thoracolumbar fascia (particularly the fan-shaped fibers running from the spinous processes of L3 to L5 and the ilium). This theory proposes that extension of the lumbar spine was not required to lift heavy loads; that is, the lumbar spine would remain flexed, relying on the viscoelastic properties of the posterior ligamentous system, and the activation energy for the lift would be provided by the hip extensors. As the hip extensors rotated the pelvis backward, the lumbar spine would be passively raised while it remained flexed until the flexion moment arm was decreased to the point where lumbar extensors could be activated to restore the lordotic curve. This model showed promise until further research demonstrated that the extensor moment produced by the entire posterior ligamentous system did not exceed 71 Nm, well short of what was necessary to augment the muscular system. One thing that was not included in this model was the contribution that the passive elastic properties of the muscle contractile, series elastic and parallel elastic elements could make to total muscle force.

It is likely that there is some unique combination of all of the elements identified in this abbreviated review that will eventually explain how we manage to bend and lift large loads safely. In Jim's case, the rehabilitation expert could examine Jim's lifting mechanics and may be able to make recommendations to offset the large compression and shear stresses imposed on his lower lumbar spine. However, based on the permanent structural damage that is apparent in Jim's lumbosacral region, a good prognosis is guarded.

## 9.6 Degenerative Disc Disease

Degeneration of the intervertebral disc is a common occurrence, beginning as early as the second decade, and is present with detectable signs on magnetic resonance imaging studies in approximately 40% of adults by their third or fourth decade. Disc degeneration appears to be part of the normal aging of the disc, with its etiology often assigned to a "wear and tear" phenomenon yet remains symptomless for many years. The development of advanced disc degeneration associated with pain and functional limitations in the younger population, however, is not part of normal aging and is assigned the name degenerative disc disease (DDD). It is unclear what triggers a more rapid advancement of degeneration, but it is likely related to some combination of genetic, systemic (biological), discogenic nutrition, loading history, and injury. Adams et al. (2013) have defined DDD as:

1. "The process of disc degeneration is an aberrant, cell-mediated response to progressive structural failure;
2. A degenerated disc is one with structural failure combined with accelerated or advanced signs of aging;
3. Early degenerative changes should refer to accelerated age-related changes in a structurally intact disc; and
4. Degenerative disc disease should be applied to a degenerated disc which also is painful."

These points are consistent with a previous statement on disc degeneration that claimed: "mechanical damage which results in a pattern of morphological and histological changes, and a sluggish adaptation to gravity loading followed by obstructed healing" differentiates DDD from the normal aging process.

To fully understand DDD we must obviously have a thorough and ready knowledge of normal intervertebral disc and vertebral endplate anatomy, function, and biomechanical

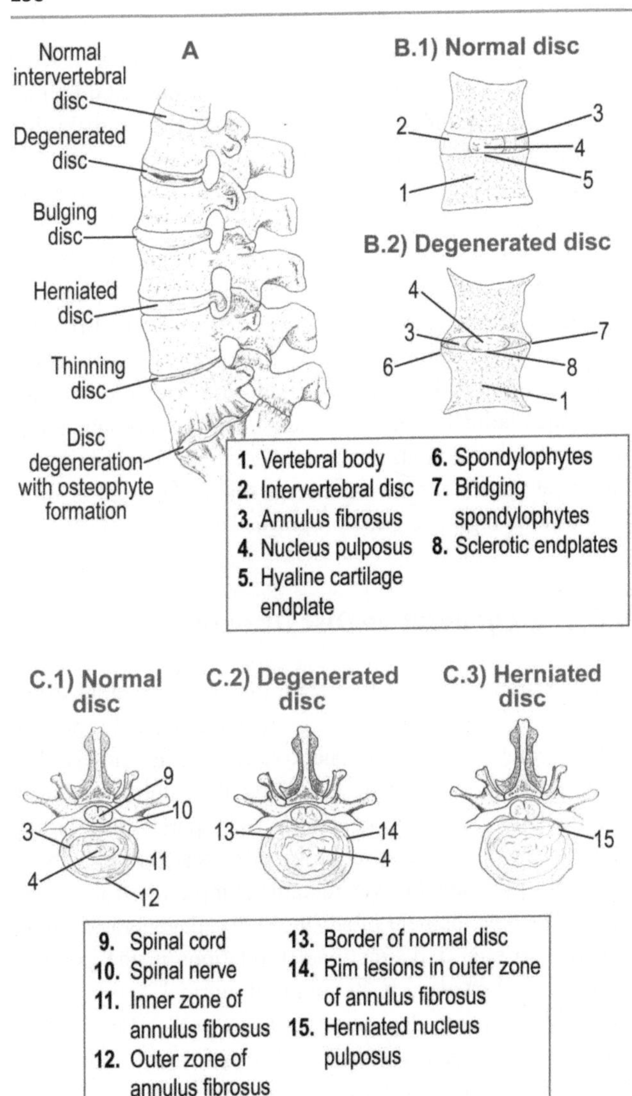


**Fig. 9.48** Typical characteristics of disc degeneration are shown in (**A**) in toto, (**B**) sagittal cross-sections, and (**C**) superior cross-sectional views

properties (see Chap. 7 for this review). The typical structural changes found in DDD include circumferential and radial tears in the annulus, inward buckling of the inner annulus, increased radial bulging of the outer annulus, reduced disc height (secondary to loss of the normal proteoglycan-collagen matrix), endplate defects, and vertical bulging of the endplates into the adjacent vertebral bodies (Fig. 9.48). As these structural failures emerge, there is a biological response and attempt to repair, regenerate, and remodel, but the defining feature of DDD is that the failures are permanent.

Early attempts at healing may begin with neovascularization, where vascularized granulation tissue, stimulated by chondrocytes and fibroblasts, appears in the damaged region, e.g., an inner radial strain of the annulus fibrosis. These early attempts at healing are associated with the invasion of afferent neurons transmitting pain (nociceptors) in the same area (where normally they do not exist). The inflammatory process brings mast cells, macrophages, enzymes, cytokines, etc. to the injured area as well, which promotes a whole series of events that stimulates the growth and sensitivity of nociceptors and may enhance or prolong the inflammation. Moreover, detection of minute changes in the ability of the IVD to distribute compression and tensile stresses, thereby altering nuclear response to changes in osmotic and hydrostatic pressure, can induce a mass response by the chondrocytes and fibroblasts to produce a new extracellular matrix, e.g., proteoglycans, collagen, etc. Often this process results in a functionally inferior matrix less able to manage external loads, thus accelerating the degenerative process. Additionally, the abnormal osmotic and hydrostatic pressures associated with IVD degeneration impair the normal nutritional processes, further inhibiting a normal response to micro-injuries. Finally, damage to one part of a disc changes/increases load-bearing to adjacent tissues, e.g., zygapophyseal joints, etc., inducing micro-injury and the initiation of additional degenerative processes. In summary, what may start out as a normal response to an injury, cascades into an unregulated (or perhaps over-regulated) balance between degeneration and tissue healing and remodeling, made worse by repeated microtrauma associated with normal activities of daily living.

We hypothesized that Jim's central low back pain may also be related to the development of DDD. The extent of his DDD cannot be determined by his clinical examination alone and would require imaging studies that may include a discogram and/or magnetic resonance imaging. But Jim has been fortunate in a couple of ways: he appears to be reasonably fit, i.e., he has functional core strength, and he has not developed any neurological impairment, e.g., pressure on lumbar nerve roots that induces nerve palsy and muscle weakness.

The most obvious cause of Jim's disc degeneration might be the stress fracture of the pars interarticularis and subsequent spondylolisthesis. However, the complex, repetitive array of impulsive forces imposed on Jim's low back during his football-playing days, combined with his increased sacral angle, likely also initiated a degenerative process of the lower lumbar discs. Then the fracture itself, before the eventual slippage of L5, created an abnormal framework for managing external loads by both the IVD and zygapophyseal joints. Finally, as the restraining ligaments of the interbody joints became overstretched over time and L5 progressively slipped ventrally, the increased interbody shear stress overmatched the ability of the IVDs to distribute even normal-sized loads.

## 9.7 Forces at the Lumbosacral Junction in Flexed Postures as Determined by Static Equilibrium Analysis

We want to determine the erector spinae (ES) muscle force and total L5/S1 interbody joint forces (and approximate the L5/S1 compression and shear forces) somewhat simulating what Jim might be doing at his worktable. He works standing, but his trunk is inclined approximately 30° from the vertical. Thus, we solve two static equilibrium problems: (1) forward bent without holding a weight and (2) forward bent while holding a 30-pound motor (as Jim might do at work). Since it was noted that Jim had a significantly increased sacral angle, we solve both problems to compare the shear loads for two sacral angles: a more normal 40° and an increased angle demonstrated by Jim of 63°.

As we have done with previous static equilibrium problems, let's list the assumptions related to solving this set of problems:

1. All posterior spinal muscles are represented by the erector spinae (ES).
   (a) An oversimplification since even the erector spinae are made up of three muscles, i.e., spinalis, iliocostalis, and longissimus (and there are distinct regions of these muscles, as well as fascicles that span from one or two vertebral segments to many vertebral segments). Additionally, there are several ES synergists, e.g., multifidi, latissimus dorsi, quadratus lumborum, etc., as well as passive elastic elements, e.g., PLL, interspinous ligament, supraspinous ligament, annulus of the intervertebral disc, thoracolumbar fascia, etc., that may contribute.
   (b) Although it may be true that, as a group, the ES has a moment arm of 0.05 m, the moment arms of the many other spinal extensors are quite variable depending on their depth from the skin surface. Furthermore, the given moment arm is with the trunk flexed 30°, and we know that the magnitude of the moment arm changes as the trunk orientation changes, i.e., changes in joint angle.
   (c) There may be antagonistic muscle action, i.e., coactions, when standing flexed and holding an object but these forces are neglected.
2. The trunk is modeled as a rigid body hinged to the sacrum (pelvis) at L5/S1.
   (a) This is not true. The spine is a lumped segment consisting of 24 segments, each of which can move with six degrees of freedom.
   (b) L5/S1 is a crude estimate for the location of the axis of rotation for trunk motion relative to the sacrum (pelvis).
3. Rigid body mechanics, i.e., there is no deformation of bone, ligaments, etc.
   (a) This is not true, even for a static standing position. For example, we know that when the spine is flexed, there is an increase in internal compression and shear at the intervertebral joint deforming the intervertebral disc; that is, because of the viscoelastic nature of the disc, it undergoes both creep and stress relaxation over some time period.
4. A static equilibrium solution is a good estimate of an inverse dynamic solution.
   (a) Perhaps only a reasonable estimate.

The solution begins with listing what we know:

- Body weight of $W = 215$ lb $= 956.3$ N.
- Weight of head/arms/trunk (HAT) minus the pelvis, $W_{HAT} = (0.678 - 0.142)W$[1] $= 513$ N
- Weight of motor, $W_{mot} = 30$ lb $= 133$ N
- Moment arms of ES muscle force ($F_{ES}$), $W_{HAT}$, and $W_{mot}$ are $d_{ES} = 0.05$ m, $d_{HAT} = 0.15$ m, and $d_{mot} = 0.30$ m, respectively
- Trunk flexed 30° relative to the vertical
- ES oriented 8° relative to the long axis of the trunk
- Coordinate axes oriented with the plane of the lumbosacral joint (the $x$-axis is perpendicular to the plane and the $y$-axis is parallel to the plane)

Note: the estimate of the head, arms, and trunk weight were taken from LeVeau (1977).

We need to define the "signs" of forces and moments. Forces that point in the $+x$ and $+y$ directions and counterclockwise moments are all positive. The free-body diagram is a simplified sketch that depicts the relationship between the segments of interest and the environment, as well as the body-fixed reference system (see Fig. 9.49). In this case the "free body" represents all of Jim's body superior to the L5/S1 joint. All relevant, known and unknown, external forces (e.g., gravity, muscle, friction, ground reaction, bone-on-bone or joint reaction, etc.) acting on the system are shown. Note that some forces internal to the body become "external" forces acting on the body when it is "freed" from the inferior part of the body for this analysis. A force is a vector with both magnitude and direction and so is represented by an arrow that indicates its point of application, line of action, and positive sense (direction). For this example, we represent both problems (with and without the motor) with one free-body diagram since the only difference between the two conditions is the weight.

[1]The mass fraction for the HAT is 0.678 and the mass fraction of the pelvis is 0.142 (see Table A1 in Appendix A).

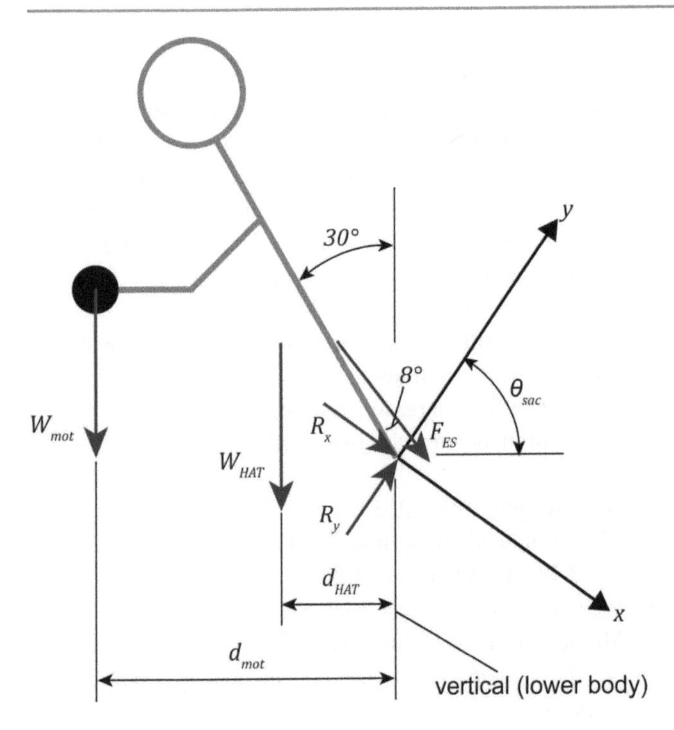


**Fig. 9.49** Free body diagram depicting external and internal forces during forward bending to 30°

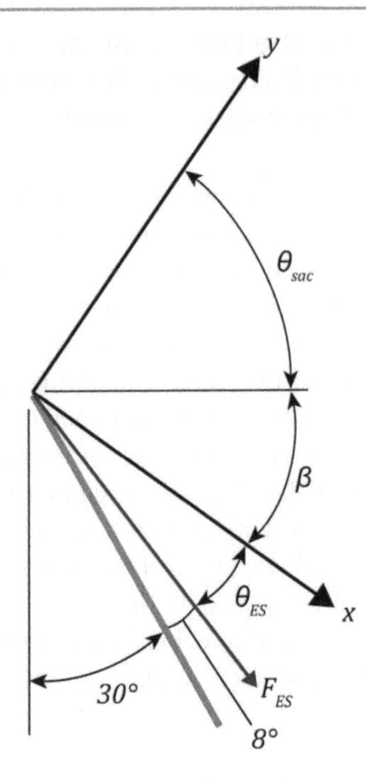


**Fig. 9.50** Determining angle of $F_{ES}$ with +$x$-axis

Now, we write the static equilibrium equations for the problem with Jim bent forward without holding a weight. The equilibrium conditions are: $\sum F_x = 0$, $\sum F_y = 0$, and $\sum M = 0$. We use the moment equation first to determine the muscle force, $F_{ES}$. Recall that a moment = force times the shortest perpendicular distance from the action line of the force to the point of reference, and that we define a counterclockwise moment as positive.

Therefore:

$$\sum M = W_{HAT}d_{HAT} - F_{ES}d_{ES} = 0$$

Note that the moment due to the weight of the HAT is positive because it tends to induce a counterclockwise rotation about the L5/S1 joint and the moment due to the force in the ES is negative because it tends to induce a clockwise rotation about the joint.

Solving for $F_{ES}$:

$$F_{ES} = \frac{W_{HAT}d_{HAT}}{d_{ES}} = \frac{(513\,\text{N})(0.15\,\text{m})}{(0.05\,\text{m})} = 1538\,\text{N}$$

The resultant force of the erector spinae is equal to 1538 N, oriented 8° relative to long axis of the trunk. Since earlier we report that the maximum strength of the lumbar extensors approximates 4000 N, it is clear that the forward bent position is well-controlled with little chance of an acute strain (tension) injury to the back extensor muscle-tendon actuators.

Now, we set up the equations to solve for the reaction forces at the L5/S1 interbody joint in the $x$-direction. To do that, we need to figure out the angles between the forces and the $x$-axis. The angle between the weights and the +$x$-axis is $\theta_{sac}$. Figure 9.50 can be helpful in determining the angle that $F_{ES}$ makes with the +$x$-axis, $\theta_{ES}$.

Understanding that angle between the $x$- and $y$- axes is 90°:

$$\beta = 90° - \theta_{sac}$$

Also, the angle between the horizontal and vertical directions is 90°:

$$90° = 30° + 8° + \theta_{ES} + \beta = 30° + 8° + \theta_{ES} + (90° - \theta_{sac})$$

Solving for $\theta_{ES}$:

$$\theta_{ES} = \theta_{sac} - 38°$$

Now, summing all of the forces in the $x$-direction:

$$\sum F_x = W_{HAT}\cos\theta_{sac} + F_{ES}\cos(\theta_{ES} - 38°) + R_x = 0$$

Solving for $R_x$:

$$R_x = -W_{HAT}\cos\theta_{sac} - F_{ES}\cos(\theta_{ES} - 38°)$$

For a 40° sacral angle:

$$R_x = -(513\,\text{N})\cos 40° - (1538\,\text{N})\cos(40° - 38°) = -1929\,\text{N}$$

The negative sign in the answer for $R_x$ simply means that the actual direction for $R_x$ is opposite to that drawn in the free-body diagram, i.e., −1929 N in the positive $x$-direction is equivalent to +1929 N in the negative $x$-direction. The direction of $R_x$ indicates that it is a compression force at L5/S1, approximately 2.0 times Jim's body weight. Experimental research has demonstrated that under pure compression loading the vertebral endplate fractures before the IVD ruptures. The maximum strength of the endplate approximates 10,000 N, suggesting that forward bending without a load would not likely damage a healthy endplate.

Next, solving for the joint force in the $y$-direction:

$$\Sigma F_y = -W_{\text{HAT}} \sin\theta_{\text{sac}} - F_{\text{ES}} \sin\left(\theta_{\text{ES}} - 38^\circ\right) + R_y = 0$$

Note that the $y$-component for $F_{\text{ES}}$ is negative because it is below the $+x$ axis. Solving for $R_y$:

$$R_y = W_{\text{HAT}} \sin\theta_{\text{sac}} + F_{\text{ES}} \sin\left(\theta_{\text{ES}} - 38^\circ\right)$$

For a 40° sacral angle:

$$R_y = (513\,\text{N})\sin 40^\circ + (1538\,\text{N})\sin\left(40^\circ - 38^\circ\right) = 383\,\text{N}$$

Thus, we find a 383 N shear force between the L5 body and the L5/S1 IVD, a relatively small force for someone with a normal lordosis and sacral angle.

Finally, we compute the resultant ($R$) interbody joint force:

$$R = \sqrt{R_x^2 + R_y^2} = 1967\,\text{N}$$

This is not much larger than the compression force, meaning that the shear force contributes little to the total force.

The internal erector spinae (ES) moment of 76.9 Nm ($F_{\text{ES}}d_{\text{ES}}$ = (1538 N)(0.05 m)) is well below the maximum measured moment the ES are capable of producing, i.e., 200 Nm (see Bogduk), before failure. Therefore, the margin of safety, (how much the load can be increased without failure) is $\frac{200}{76.9} - 1 = 1.60$, i.e., the moment could be increased 160% without exceeding the maximum moment the ES can produce. Under the condition of this statics problem, there is minimal risk of sustaining a first-degree injury to the ES. It is likely that the margin of safety for the ES is even greater, given that there are many spinal extensors working in concert with the ES.

Let's repeat these steps for the forward bend while holding a 30-lb motor. First, we solve for the ES muscle force using a moment equation with respect to the lumbosacral joint:

$$\Sigma M = W_{\text{HAT}} d_{\text{HAT}} + W_{\text{mot}} d_{\text{mot}} - F_{\text{ES}} d_{\text{ES}} = 0$$

Solving for $F_{\text{ES}}$:

$$F_{\text{ES}} = \frac{W_{\text{HAT}} d_{\text{HAT}} + W_{\text{mot}} d_{\text{mot}}}{d_{\text{ES}}} = \frac{(513\,\text{N})(0.15\,\text{m}) + (133\,\text{N})(0.30\,\text{m})}{(0.05\,\text{m})} = 2338\,\text{N}$$

Thus, when holding a 30-lb motor an ES force of 2338 N induces a 117 Nm moment, which reduces the factor of safety to $\frac{200}{117} - 1 = 0.71$, i.e., the moment could be increased 71% without exceeding the maximum moment the ES can produce. This means that the chance of a significant strain injury to the ES is small, but with repeated and prolonged activity like this, chronic minor strains are more likely.

Now, we set up the equations to solve for the reaction forces at the L5/S1 interbody joint in the $x$-direction. Summing the forces in the $x$-direction:

$$\Sigma F_x = W_{\text{mot}} \cos\theta_{\text{sac}} + W_{\text{HAT}} \cos\theta_{\text{sac}} + F_{\text{ES}} \cos\left(\theta_{\text{ES}} - 38^\circ\right) + R_x = 0$$

Solving for $R_x$:

$$R_x = -W_{\text{mot}} \cos\theta_{\text{sac}} - W_{\text{HAT}} \cos\theta_{\text{sac}} - F_{\text{ES}} \cos\left(\theta_{\text{ES}} - 38^\circ\right)$$

For a 40° sacral angle:

$$R_x = -(133\,\text{N})\cos 40^\circ - (513\,\text{N})\cos 40^\circ - (2338\,\text{N})\cos\left(40^\circ - 38^\circ\right) = -2831\,\text{N}$$

Once again, the negative sign in the answer for $R_x$ simply means that the actual direction for $R_x$ is opposite to that drawn in the free-body diagram. Thus, the 2831 N of L5/S1 compression force is significantly increased while holding the 30-lb weight to ~3.0 times bodyweight.

Now, summing the forces in the $y$-direction to obtain the shear force:

$$\Sigma F_y = -W_{\text{mot}} \sin\theta_{\text{sac}} - W_{\text{HAT}} \sin\theta_{\text{sac}} - F_{\text{ES}} \sin\left(\theta_{\text{ES}} - 38^\circ\right) + R_y = 0$$

Solving for $R_y$:

$$R_y = W_{\text{mot}} \sin\theta_{\text{sac}} + W_{\text{HAT}} \sin\theta_{\text{sac}} + F_{\text{ES}} \sin\left(\theta_{\text{ES}} - 38^\circ\right)$$

For a 40° sacral angle:

$$R_y = (513\,\text{N})\sin 40^\circ + (513\,\text{N})\sin 40^\circ + (2338\,\text{N})\sin\left(40^\circ - 38^\circ\right) = 497\,\text{N}$$

We see that holding the 30-lb weight increases the shear load by about 30%.

Finally, obtaining the resultant force for the lumbosacral joint:

$$R = \sqrt{R_x^2 + R_y^2} = 2875\,\text{N}$$

We have just solved for the loads at the lumbosacral joint during forward bending with, and without, lifting an external load while assuming a normal sacral angle of 40°. We are curious how the loads at L5/S1 are affected by Jim's increased sacral angle. So let's repeat these analyses with a 63° sacral angle. Increasing the sacral angle would not change the forces in the ES of 1538 N without lifting the motor and 2338 N with lifting the motor. The total resultant forces at the joint of 1967 N and 2875 N, respectively, would be the same as well. However, with a different sacral angle, the $x$- and $y$-components of the resultant force would differ.

For the case without lifting the motor and assuming a 63° sacral angle, the $x$- and $y$-components of the joint reaction force would be:

$$R_x = -(513\,\text{N})\cos 63° - (1538\,\text{N})\cos(63° - 38°) = -1626\,\text{N}$$

$$R_y = (513\,\text{N})\sin 63° + (1538\,\text{N})\sin(63° - 38°) = 1107\,\text{N}$$

The calculated compressive force is 16% lower, but the shear force is 189% higher for the 63° sacral angle as compared to the normal sacral angle.

When lifting the 30-lb motor and assuming a 63° sacral angle, the $x$- and $y$-components of the joint reaction force would be:

$$R_x = -(133\,\text{N})\cos 63° - (513\,\text{N})\cos 63° - (2338\,\text{N})\cos(63° - 38°) = -2413\,\text{N}$$

$$R_y = (513\,\text{N})\sin 63° + (513\,\text{N})\sin 63° + (2338\,\text{N})\sin(63° - 38°) = 1564\,\text{N}$$

The calculated compressive force is 15% lower, but the shear force is 215% higher, or more than triple.

Thus, we see that the shear forces are increased considerably when the sacral angle is increased. In Jim's case, these increased forces, over time, will likely contribute to degenerative changes at the interbody and apophyseal joints. Particularly concerning is how the shear loads could likely exacerbate forward slippage of the L5 vertebral segment.

## 9.8 Chapter Summary

In this case we were confronted with a relatively young, yet active and fit, man with chronic low back pain secondary to the medical diagnosis of spondylolytic spondylolisthesis. We hypothesized that he acquired this injury perhaps as a result of a genetically engineered increased sacral angle/lumbar lordosis, which predisposed Jim to a subsequent stress fracture to the pars interarticularis. We believe that playing football likely also contributed significantly to his pars fracture and instability. A leg-length difference-induced mild scoliosis has probably contributed to Jim's central low back pain as well. At the point that we encountered Jim, he did not have any signs or symptoms of neurological impairment, i.e., sensory, reflex, or motor changes, which is fortunate, but may be short-lived given the progressive nature of spondylolisthesis and related degenerative disc disease.

In reframing the medical problem biomechanically in an attempt to explain the pathomechanics, we examine normal anatomy and biomechanics of the lumber spine and pelvic ring, including the dynamic coupling of the kinematics of the lumbar spine, sacroiliac and pubic symphysis joints, the mechanism of injury and stress fracture mechanics, lifting mechanics, and a related degenerative lumbar condition. Part of our process is to activate previous knowledge (from other cases already presented) that can be generalized, and then used, to frame and explain some of the issues we identify in this case. Solving a statics problem related to posture and lifting helps us to quantify some of the abstract biomechanical concepts presented. Finally, this is yet another example in which we are forced to recognize the interaction of the elements of complex systems and acknowledging the limitations of reductive methods of analysis.

## Bibliography

Adams MA, Bogduk N, Burton K et al (2013) The biomechanics of back pain, 3rd edn. Churchill Livingstone Elsevier, Edinburgh

Aihara T, Takahashi K, Yamagata M et al (2000a) Does the iliolumbar ligament prevent anterior displacement of the fifth lumbar vertebra with defects of the pars? J Bone Joint Surg (British) 82-B:846–850

Aihara T, Takahashi K, Yamagata M et al (2000b) Biomechanical function of the iliolumbar ligament in L5 spondylosis. J Orthop Sci 5:238–242

Becker I, Woodley SJ, Stringer MD (2010) The adult human pubic symphysis: a systematic review. J Anat 217:475–487. https://doi.org/10.1111/j.1469-7580.2010.01300.x

Bogduk N (2012) Clinical and radiological anatomy of the lumbar spine, 5th edn. Churchill Livingstone Elsevier, Edinburgh

Brinckmann P, Frobin W, Leivseth G et al (2016) Orthopaedic biomechanics, 2nd edn. Thieme, Stuttgart

DeStefano L (2017) Greenman's principles of manual medicine, 5th edn. Wolters Kluwer, Philadelphia

Goode A, Hegedus EJ, Sizer P Jr et al (2008) Three-dimensional movements of the sacroiliac joint: a systematic review of the literature and assessment of clinical utility. J Man Manip Ther 16(1):25–38

Greenman PE (1990) Clinical aspects of sacroiliac function in walking. J Man Med 5:15–130

Hammer N, Scholze M, Kibsgård T et al (2019) Physiological *in vitro* sacroiliac joint motion: a study on three-dimensional posterior pelvic ring kinematics. J Anat 234:346–358. https://doi.org/10.1111/joa.12924

Huijbregts P (2004) Lumbar spine coupled motions: a literature review with clinical implications. Orthop Div Rev

Ikata T, Miyake R, Katoh S et al (1996) Pathogenesis of sports-related spondylolisthesis in adolescents, radiographic and magnetic resonance imaging study. Am J Sports Med 24(1):94–98

Kaltenborn FM (2018) Manual mobilization of the joints, Volume II, the spine, 7th edn. Orthopedic Physical Therapy Products

Kapandji AI (2019) The physiology of the joints, Volume 3, The spinal column, pelvic girdle and head, 7th edn. Handspring Publishing Limited, East Lothian

Kirnaz S, Capadona C, Lintz M et al (2021) Pathomechanism and biomechanics of degenerative disc disease: Features of healthy and degenerated discs. Int J Spine Surg 15(s1):10–25. https://doi.org/10.14444/8052

Lavignolle B, Vital JM, Senegas J et al (1983) An approach to the functional anatomy of the sacroiliac joints in vivo. Anat Clin 5:169–176

Legaspi O, Edmond SL (2007) Does the evidence support the existence of lumbar spine coupled motion? A critical review of the literature. J Orthop Sports Phys Ther 37(4):169–178. https://doi.org/10.2519/jospt.2007.2300

Leong JCY, Luk KDK, Chow DHK et al (1987) The biomechanical functions of the iliolumbar ligament in maintaining stability of the lumbosacral junction. Spine 12(7):669–674

LeVeau B (1977) Williams and Lissner: Biomechanics of human motion, 2nd edn. W. B. Saunders Company, Philadelphia

Neumann DA (2017) Kinesiology of the musculoskeletal system, foundations for rehabilitation, 3rd edn. Elsevier, St. Louis

Nordon M, Frankel VH (2022) Basic biomechanics of the musculoskeletal system, 5th edn. Wolters Kluwer, Philadelphia

Odeh K, Wu W, Taylor B et al (2021) In-vitro 3D analysis of sacroiliac joint kinematics, primary and coupled motions. Spine 46(8):E467–E473. https://doi.org/10.1097/BRS.000000000000.3841

Panjabi MM (1992a) The stabilizing system of the spine. Part I. Function, dysfunction, adaptation, and enhancement. J Spinal Disord 5(4):383–389

Panjabi MM (1992b) The stabilizing system of the spine. Part II. Neutral zone and instability hypothesis. Clin Spine Disord 5(4):390–397

Pool-Goudzwaard AL, Kleinrensink GJ, Snijders CJ et al (2001) The sacroiliac part of the iliolumbar ligament. J Anat 199:457–463

Pool-Goudzwaard A, Hoek van Dijke G, Mulder P et al (2003) The iliolumbar ligament: Its influence on stability of the sacroiliac joint. Clin Biomech 18:99–105

van Rijsbergen MM, Barthelemy VMP, Vranchen ACT et al (2017) Moderately degenerated lumbar motion segments: Are they truly unstable? Biomech Model Mechanobiol 16:537–547. https://doi.org/10.1007/s10237-016-0835-9

Vazirian M, Van Dillen L, Bazrgari B (2016) Lumbopelvic rhythm during trunk motion in the sagittal plane: a review of the kinematic measurement methods and characterization approaches. Phys Ther Rehabil 3(article 5). https://doi.org/10.7243/2055-2386-3-5

Vleeming A, Scheunke M (2019) Idiopathic pelvic girdle pain as it relates to the sacroiliac joint, form and force closure of the sacroiliac joints. PM&R 11:S24–S31. https://doi.org/10.1002/pmrj.12205

Vleeming A, Pool-Goudzwaard AL, Stoechart R et al (1995) The posterior layer of the thoracolumbar fascia, its function in load transfer from spine to legs. Spine 20(7):753–758

# 10 Hip Joint Complex: A Degenerative Joint

## 10.1 Introduction

The hip joint complex, along with the pelvic ring, is the result of unique evolutionary adaptations to bipedal locomotion induced by the need to mitigate and distribute large forces associated with the superincumbent body weight and ground reaction forces. The complex structural and functional relationship between the femur and pelvis is similar to that of the humerus and scapula, except that stability is favored over mobility. Despite this difference, the hip joint's six degrees of freedom are necessary and sufficient to manage the variety of movements imposed during activities of daily living. Yet, because of the heavy duties that are imposed, the hip is susceptible to a number of different insults resulting in injury, with degenerative hip joint disease counted as one of the most common of all lower extremity pathologies.

This case is different than those preceding it because it came with a definitive medical diagnosis, which did not appear to be induced by a specific mechanism of injury. Therefore, we use biomechanics to develop hypotheses in regard to potential etiologies of degenerative disease specific to the hip and to describe in some detail the physical and gait manifestations of hip osteoarthritis. We previously examine the general causes of degeneration of synovial joints, i.e., osteoarthritis (see Chaps. 3 and 4), and do not need to do that in this case. However, we do review the muscular anatomy of the pelvic and hip complex, and its role in joint stability and mobility, as well as muscles' contribution to the very large joint compression and shear stresses associated with standing and walking.

Since we note the importance of the hip joint to daily activities, we need to inquire about the characteristics of bipedal locomotion. Therefore, with this case (and the two cases that follow), we examine the "simple" task of walking and introduce three-dimensional instrumented motion (gait) analysis (IGA) as an additional clinical and biomechanical tool. Instrumented gait measurements/analyses are seen as tools used to enhance our understanding of how lower extremity joint degenerative disease and injury can be manifested during walking. For this case, we review and compare to a control data set, Ralph's temporospatial gait parameters, trunk, pelvis, and lower extremity joint kinematics and kinetics to describe the unique manifestation of his hip disease during gait. The reader can find foundational information on observational and instrumented gait analysis in Appendix J.

## 10.2 Case Presentation

Ralph Wieber was a 50-year-old man (height 184.7 cm; mass 81.4 kg; BMI 23.2 kg/m$^2$), who worked as the Vice President of Finance of a non-profit organization, was referred to the Motion Analysis Center for a gait analysis. The Motion Analysis Center was established by a local rehabilitation hospital. Its primary purpose is to provide instrumented gait analysis to assist in the medical, surgical, and rehabilitative care of individuals with acute and chronic orthopedic and neuromuscular pathologies.

Ralph had no significant medical or surgical history and claimed a relatively healthy lifestyle. He reported having left hip symptoms for about 13 years and right hip symptoms for 5 years. He claimed that both hips ached daily, the left greater than the right. Left hip pain was centered around the hip joint, into the groin region, and lateral thigh down to the knee, whereas right hip pain was restricted only to the hip joint region. He complained of bilateral hip joint stiffness in the morning. The ache was usually worse one day after activities, such as yard work, bending, squatting, climbing, golfing, and the like. He occasionally had sharp pains but could not pinpoint a location. Often, he felt a "catch" or "locking" feeling in his hips after prolonged sitting.

During his high school years, Ralph played basketball and golf. While in college Ralph participated in intramural


G. J. Alderink, B. M. Ashby, *Clinical Kinesiology and Biomechanics*, https://doi.org/10.1007/978-3-031-25322-5_10

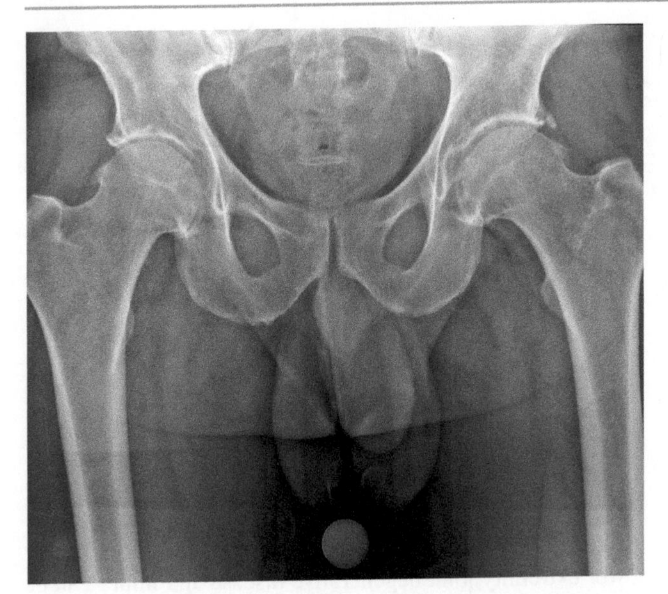

**Fig. 10.1** Radiograph of the typical presentation of bilateral moderate to severe osteoarthritis in both hips, noting moderately decreased joint space and relatively large marginal osteophytes, i.e., bone spurs. Relatively good overall alignment and good overall bone quality

basketball, football, and volleyball. Following his college years and up until 3 years ago, he played pick-up basketball three times a week. He denied a history of major trauma to his hips but stated that he always had poor flexibility in his legs. At the age of 21, he "hyperextended" his left knee while playing football. The knee injury was treated with a course of immobilization and seemed to resolve spontaneously (no formal rehabilitation). At the age of 22 years, Ralph sustained a left ankle dislocation, which was treated primarily only with immobilization; his treatment had a good outcome and Ralph was able to continue an active lifestyle. Approximately 14 years ago he sustained a left hip contusion while playing basketball. As a result, he could not walk for 1 to 2 days and needed to use crutches for about 5 days. At the time of his gait analysis, he was still able to play golf (using a cart) at least once a week without significant difficulty.

Radiographs were taken 14 years earlier and revealed moderate-to-severe degenerative changes in both hips with moderately decreased joint space and relatively large marginal osteophytes (Fig. 10.1). There was relatively good alignment with no substantial leg length discrepancy and good overall bone quality.

Based on physical examination Ralph had several remarkable findings (see Table 10.1 for details). With resisted muscle testing, he demonstrated normal motor function, but bilateral weakness of the hip flexor, extensor (including hamstrings), and abductor muscles. A squat test was used as a general screen for lower extremity range of motion and strength (see Magee). Ralph's attempt was limited by 50% and reproduced pain in both hips.

Passive range of motion testing revealed (in degrees):

| | Right | Left | Normal |
|---|---|---|---|
| Hip flexion | 90 | 80 | 120–135 |
| Thomas test | Lacked 5 | Lacked 5 | 0 |
| Hip abduction | 10 | 10 | 30–50 |
| Hip internal rotation | 0 | 0 | 30–45 |
| Hip external rotation | 30 | 40 | 30–45 |

Note: Femoral antetorsion, as clinically estimated using Ryder's (or Craig's) test, was not measured due to Ralph's inability to internally rotate his hip joints. Based on observation of his foot progression angles in standing and walking, estimations of tibial torsion using the thigh-foot angle or transmalleolar measures were not indicated

Observation of Ralph's posture in standing revealed that the right knee joint was held in mild flexion, a moderate forward head, mildly increased thoracic kyphosis and lumbar lordosis, but no scoliosis and a level pelvis. Leg length was measured (with Ralph in the supine position on the examination table) from the tip of the anterior superior iliac spine to the inferior aspect of the medial malleolus and revealed a 1 cm difference between limbs (left 103, right 102); this leg length difference may explain the mild right knee flexion that was observed during the posture examination. Weight- and not-weight bearing examination of shank and foot mechanical alignment was within normal limits, except for the presence of bilateral, but flexible, mild forefoot varus and related plantar flexed first ray.

Observational gait analysis was performed with the aid of two standard video cameras, providing side (sagittal plane) and ventral/dorsal (frontal plane) views simultaneously. With observational gait analysis, the examiner looks for potential relationships between the physical examination findings of neuromuscular and joint impairments and primary, secondary, compensatory sagittal, frontal, and transverse plane kinematic gait deviations. A problem-solving approach is used to develop hypotheses regarding the likely causes of gait abnormalities. We present these findings later when we examine gait data from the instrumented portion of the gait study.

## 10.3 Key Facts, Learning Issues, and Hypotheses

Before we move forward, let's pause and take some time to exercise what we have learned at this point in the textbook about how to frame, or re-frame, the presentation of a clinical problem. We then use a line of inquiry that begins with what you think you know (drawing on previous experiences) that leads to the development of ideas, i.e., hypotheses. Finally, taking these hypotheses that might begin to explain this particular scenario by tickling your curiosity and help to identify learning issues, i.e., what else do I need/want to know?

**Table 10.1** Physical examination data

| HIP | Normal ROM (degrees) | L | R | L | R | Weight bearing | L | R |
|---|---|---|---|---|---|---|---|---|
| Flexion | 120–135 | 80 | 90 | 4/5 | 4/5 | HF varus*/valgus | WNL | WNL |
| Extension/Mod. Thomas Test: | | | | | | *Coleman block | | |
| Knee extended | 0 | 5 | 5 | 5/5 | 5/5 | MF planus#/cavus | WNL | WNL |
| Knee flexed 90° | 0 | 5 | 5 | 5/5 | 5/5 | #reverses w/toe standing | | |
| Abduction (hip ext.) | 30–50 | 10 | 10 | 4/5 | 4/5 | FF Ab/adductus | WNL | WNL |
| Adduction | *** | | | 5/5 | 5/5 | Hallux valgus | WNL | WNL |
| Ober Test (IT band) | Neg | POS | POS | | | | | |
| Internal rotation | 30–45 | 0 | 0 | | | *Non-wt. bearing* | *L* | *R* |
| External rotation | 30–45 | 30 | 40 | | | HF varus/valgus | WNL | WNL |
| Femoral anteversion | Adult: 12–15 | NA | NA | | | Flexible? | | |
| | | | | | | MF planus/cavus | WNL | WNL |
| *KNEE* | | *L* | *R* | *L* | *R* | Flexible? | | |
| Flexion (supine) | 135–160 | WNL | WNL | | | FF varus/valgus | Mild varus | Mod varus |
| Flexion (prone) | ~135 | WNL | WNL | 4/5 | 4/5 | Flexible? | Yes | Yes |
| Extension (passive) | 0 | WNL | WNL | | | FF Ab/adductus | | |
| Extension (active): | | | | | | Flexible? | | |
| Strength (sitting) | *** | | | 5/5 | 5/5 | PT 1st ray | Mild PF | Mild PF |
| Active lag (supine) | 0 | WNL | WNL | | | Callus pattern: | Neg | Neg |
| Popliteal angle (from vert.) | | | | | | *Spasticity* | *L* | *R* |
| Traditional (spastic/end) | 0–30 | WNL | WNL | | | Hip flexors | NT | NT |
| Neutral pelvis (end) | 0–20 | NT | NT | | | Hip adductors | NT | NT |
| Transmalleolar axis | ~20 ER | NT | NT | | | Hip internal rotators | NT | NT |
| Thigh-foot angle | ~15 ER | NT | NT | | | Hip external rotators | NT | NT |
| *Ankle/foot* | | *L* | *R* | *L* | *R* | Hamstrings | NT | NT |
| Dorsiflexion: | | | | | | Vasti | NT | NT |
| Knee ext'd (spastic/end) | ~10–15 | WNL | WNL | | | Rect. fem./Duncan-Ely | NT | NT |
| Knee flexed (end) | 15–25 | WNL | WNL | 5/5 | 5/5 | Ankle clonus | NT | NT |
| Plantarflexion | 45–50 | Mild ↓ | Mild ↓ | 5/5 | 5/5 | Gastrocnemius | NT | NT |
| Tarsal inversion (supination) | 35–50 | WNL | WNL | 5/5 | 5/5 | Soleus | NT | NT |
| Tarsal eversion (pronation) | 35–50 | WNL | WNL | 5/5 | 5/5 | Tibialis anterior | NT | NT |
| 1st MTP extension | 70–90 | WNL | WNL | | | Tibialis posterior | NT | NT |
| *Trunk strength* | | *Posture* | | | | Peroneals | NT | NT |
| Lower abdominals (leg lift) | NT | * Mild right knee flexion; moderate forward head; | | | | Toe flexors (FDL,FHL) | NT | NT |
| Upper abdominals (crunch) | NT | mild ↑ thoracic kyphosis | | | | Confusion test | NT | NT |
| Trunk extensors (prone) | NT | and lumbar lordosis | | | | *Balance* | | |
| *Notes/other* | | | | | | Single leg stance time | NT | NT |
| * Genu valgus (NWB): WNL | | | | | | *Leg length* | 102 cm | 103 cm |
| * Squat test: moderate ↓ secondary to hip pain (dull quality) | | | | | | *Motor selectivity scale:* | | |
| * Knee stress tests WNL | | | | | | *(No manual resistance/gravity only)* | | |
| | | | | | | *U* = unable to produce requested motion | | |
| | | | | | | *S* = able to produce requested motion in *synergy* | | |
| | | | | | | *I* = able to produce requested motion in *isolation* | | |

| *Abbreviations:* | *Spasticity scale (Modified Ashworth):* | *Strength scale (manual muscle test):* |
|---|---|---|
| *NA* = not applicable | 0 = no increase in tone (normal) | 0 = no palpable muscle activity |
| *NT* = not tested | 1 = slight ↑ in resistance: catch and release | 1 = palpable muscle tension or flicker; no movement |
| *UTA* = unable to assess | 1+ = slight ↑ in resistance: catch and min. resistance | 2 = full ROM in horizontal plane; 0 against gravity |
| *P/PL* = painful/pain-limited | 2 = more marked ↑ in resistance thru most ROM | 3 = full ROM against gravity only; no resistance |
| *WNL* = within normal limits | 3 = considerable ↑ in resistance; PROM difficult | 4 = full ROM against gravity and mod. resist. |
| *WFL* = within function limits | 4 = rigid | 5 = full ROM against gravity and max. resist. |

Note: *IT* iliotibial band; *FDL* flexor digitorum longus; *FHL* flexor hallucis longus; *MTP* metatarsophalangeal; *Neg* negative; *ext'd* extended; *HF* hindfoot; *MF* midfoot; *FF* forefoot

Based on the radiographic images, supported by historical and physical examination data, it is established that Ralph has advanced degenerative joint disease in both hips. We know from the literature that symptomatic hip osteoarthritis (OA) has a prevalence of 9.2% among adults aged 45 years and older, with men having a higher prevalence of hip OA before 50 years. We also know that Ralph's cluster of signs and symptoms, hip range of motion, and strength deficits are consistent with hip OA, as reported in the literature. The question that these findings pose is how might the physical examination findings relate to his walking pattern? Making cause/effect conclusions without sufficient data can lead to inaccurate judgments, which may lead to incorrect decisions about appropriate medical and rehabilitative management. Given these thoughts, a major learning issue associated with this case is to explore the biomechanics of normal walking and evaluate Ralph's gait relative to what is normal.

Although the radiologist's analysis provided evidence of the extent of Ralph's hip OA and made some general comments on his leg length and bony alignment, it did not feature analysis of other morphological aspects of the acetabulum and proximal femur that might have predisposed Ralph to early degenerative changes. For example, a shallow acetabulum, i.e., socket portion of the joint, defined as hip dysplasia, can be genetic or acquired. A mild dysplastic acetabulum, either inherited or acquired developmentally, may not induce joint instability, nor be symptomatic, thus escaping detection for years. Yet, a mildly dysplastic acetabulum allows excessive translation of the femoral head during normal activities inducing shear stresses that, over time, may lead to the degeneration of the hyaline cartilage on both sides of the joint. It may be that plain radiographs are not suitable to examine for these special bony features. Regardless, we take the opportunity to examine a number of these unique, but normal, morphological features of the acetabulum and proximal femur, and then ask if these elements are not normal, what might the abnormalities look like and how could they alter bone and joint forces that might have predisposed Ralph to early hip OA?

We were not offered any information about Ralph's family history (i.e., genetic predisposition), which may have provided clues as to why he developed hip OA at such a young age. We know that Ralph was not obese and that he led a healthy lifestyle, e.g., nutrition and sleep habits, and was not a smoker, avoiding typical risk factors for hip OA. Thus, we hypothesized that the etiology of his hip disease was not idiopathic and/or systemically induced, but more likely mechanically related. Ralph was quite active physically into his late 40s, participating in golf and basketball. Whereas chronic low back pain is commonly associated with golf, hip OA is not. However, activities like soccer, handball, hockey, track and field, long-distance running, and basketball, most of which are associated with a high frequency of cutting, large volumes of running, and multi-directional movements, have been associated with the development of early-onset hip OA. Research has shown that male elite-level athletes involved in high-impact sports, such as soccer, handball, hockey, and track and field, are at an increased risk for the development of hip OA; however, there is little agreement on whether the increased risk is associated with long-distance running. Because Ralph played basketball, a sport associated with impact loading of the lower extremity joints during jumping and cutting, it is possible that this activity along with undetected hip joint irregularities contributed to early degenerative changes. We later review how impact loading coupled with abnormal lower extremity alignment could overload the hip joint.

## 10.4 Hip Complex Structure and Function

### 10.4.1 Bony Structure

At the outset, remember that the hip joint is part of a complex, with many inter-related elements, that includes the lumbar spine, sacroiliac (SI) joint, and the pubic symphysis. Recall also (Chap. 9) that the osteopathic biomechanical model of the lower half of the body contends that the innominate rotations/translations that occur at the SI joints and pubic symphysis are simply extensions of the multi-direction movements at the hip joint that are produced during locomotion activities.

The paired innominate bones, formed by three bones, ilium, ischium, and pubis, are constrained anteriorly at the pubic symphysis and posteriorly at the paired sacroiliac joints (Fig. 10.2A, B). These constraints form the pelvic ring that allows enough mobility to shield the pelvis from fractures yet restrains the entire system through form closure to re-distribute large compression and shear stresses that are produced from superincumbent and ground reaction loads. The many pelvic bony protuberances, ridges, and contours located anteriorly, posteriorly, and laterally serve as anchors for the muscle-tendon actuators from proximal and distal segments that contribute to force closure of the SI joints and serve as a muscular floor of the pelvic diaphragm.

The most striking feature of the external surface of the pelvis is where the three bones come together to form the acetabulum, the cup-shaped socket of the hip joint (Fig. 10.3). The rim of the acetabulum forms an incomplete circle, leaving an acetabular notch inferiorly, which is spanned by the transverse acetabular ligament. The acetabular fossa is a depression in the floor of the acetabulum that normally does not contact the femoral head. This area lacks articular cartilage, allowing space for fat, synovium, vessels, and the ligamentum teres; the ligamentum teres appears to be a vestigial structure that provides little mechanical constraint or arterial

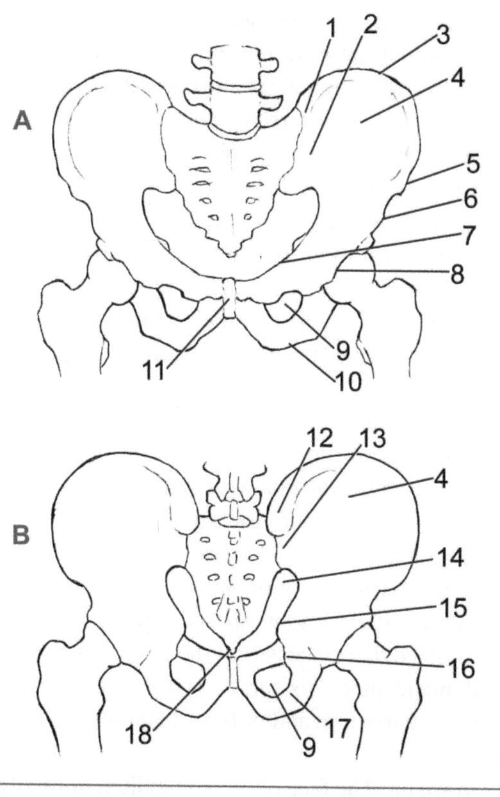


| | |
|---|---|
| **1.** Iliac tuberosity | **10.** Ischial ramus |
| **2.** Auricular surface | **11.** Interpubic disc |
| **3.** Iliiac crest | **12.** Posterior-superior iliac spine |
| **4.** Ilium | **13.** Posterior-inferior iliac spine |
| **5.** Anterior-superior iliac spine | **14.** Greater sciatic notch |
| **6.** Anterior-inferior iliac spine | **15.** Ischial spine |
| **7.** Superiour pubic ramus | **16.** Lesser sciatic notch |
| **8.** Acetabulum | **17.** Ischial tuberosity |
| **9.** Obturator foramen | **18.** Coccyx |

**Fig. 10.2** (**A**) Anterior and (**B**) posterior aspects of the pelvis and right proximal femur. In (**A**), note the exposed auricular surface of the left ilium and in (**B**) note the attachments of the long posterior sacroiliac, sacrospinous, and sacrotuberous ligaments

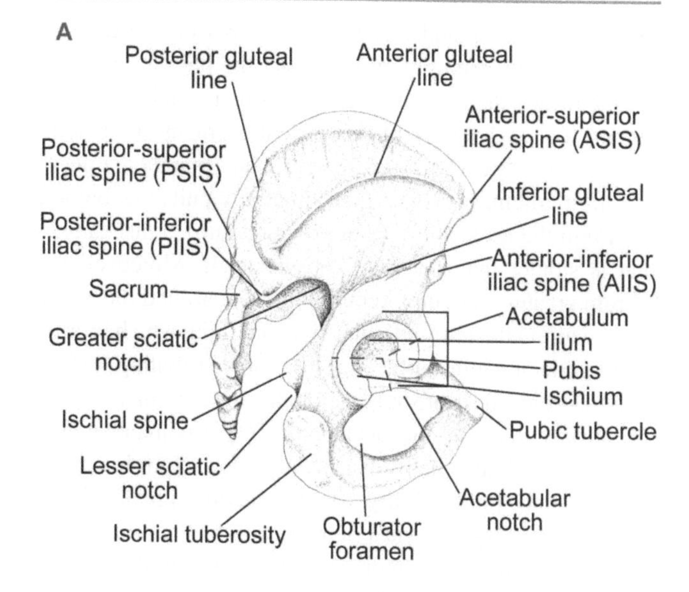


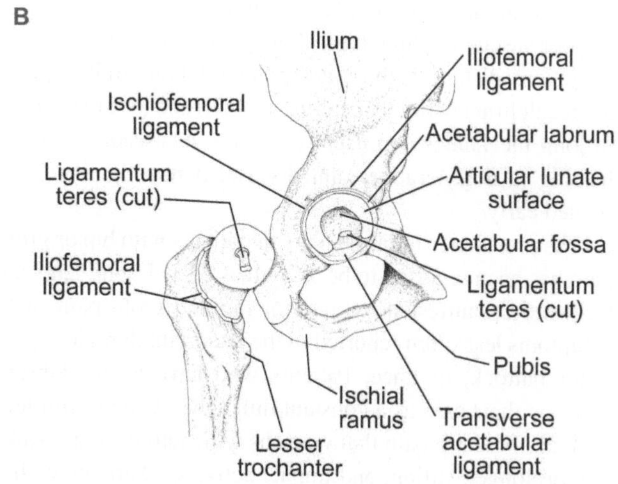


**Fig. 10.3** (**A**) Lateral view of the right innominate illustrating the contribution of each pelvic bone to the formation of the acetabulum and (**B**) exposed hip joint showing lunar-shaped articular cartilage, acetabular labrum, and intrinsic and extrinsic ligaments

supply. Under circumstances when external forces at the hip joint are lesser, the femoral head contacts the acetabulum only along the crescent-shaped lunate surface, normally covered by articular cartilage, which is the thickest anterosuperiorly. Intuition tells us that the area of the hip socket which houses the thickest cartilage must be the region that mitigates and distributes the largest compression and shear stresses induced during activities of daily living. During walking at a self-selected pace, for example, hip contact forces have been estimated to range from approximately 13% of body weight (swing phase) up to 300% of body weight (midstance). During certain phases of activities, i.e., like the midstance phase of walking, that impose large bone-on-bone forces, the acetabular articular cartilage, true to its inherent viscoelastic nature deforms, widening slightly at the acetabular notch, thus re-distributing peak stresses. Although the acetabulum has substantial depth, its gap inferiorly and insufficient contact with the entire femoral head requires additional structures to optimize stress distribution and joint stability.

The labrum of the acetabulum is a strong fibrocartilaginous tissue that rings the outer surface of the socket, thus providing an additional articulating and stabilizing structure. The labrum merges with the transverse acetabular ligament inferiorly, and internally with the lunar-shaped articular cartilage (forming a labro-chondral junction). The acetabular labrum deepens the hip socket, creating a type of seal around the femoral head, thus significantly enhancing joint stability by resisting lateral and vertical motion, as well as excessive translation. Interestingly, it is not just the physical connections between the labrum, articular cartilage, and periarticular soft tissues that prevent excessive movements at the joint surfaces. The "seal" appears to create and maintain a negative intra-articular pressure that essentially pulls the femoral head into the acetabulum (like a suction cup), a mechanism that is stouter than the joint capsule itself. Moreover, the labrum creates a circumferential fluid seal around the entire joint securing the intra-articular synovial fluid, which is critical for reducing joint surface friction and distributing surface compression and shear stresses. Finally, the labrum increases the surface area of the acetabulum (approximately 28% greater), which also helps distribute the load. Injuries to the labrum reduce joint stability, which is readily signaled to the central nervous system via a rich array of the hip joint mechanoreceptors and nociceptors that are housed in the labrum and surrounding periarticular soft tissues. Unfortunately, due to poor vascularization, typical of fibrocartilage, particularly in regions distant to the joint capsule, labral strain injuries are recalcitrant to timely healing (if at all) and lead to changes in joint mechanics that usually initiates a cascade of events leading to degenerative joint disease, if not identified and treated early.

The prevalence of labral tears in patients with hip or groin pain has been reported to be as high as 25%. Labral tears are frequently manifested by anterior hip and groin pain, with symptoms less often reported in the lateral thigh region, posterior buttock, or knee. Patients who have a torn labrum describe their pain as a constant dull ache, with intermittent episodes of sharp pain that worsens with activity, e.g., walking, prolonged sitting, and impact activities. Patients with a labral tear also report a variety of mechanical symptoms, including clicking, locking, or catching and giving way; patients may also complain of feelings of hip joint instability. When asked about the onset, the majority of patients cannot recall a specific incident, and therefore labral tears and impairments commonly go undiagnosed over an extended period of time. In one report, approximately 55% of patients with mechanical hip pain of unknown etiology were found to have a labral tear upon further examination.

The importance of the pathology of the acetabulum labrum is its association with degenerative changes. One study reported that 73% of patients with arthroscopically diagnosed labral fraying or tearing had chondral damage and that the chondral damage was more severe in patients with labral lesions. This research also reported that in 94% of the patients, articular cartilage damage occurred in the same zone of the acetabulum as the labral lesions. These findings suggest that the relative risk of significant chondral erosion approximately doubles in the presence of a labral lesion.

Ralph's radiographic studies, which showed evidence of advanced hip OA, were completed 14 years ago. Because the acetabular labrum is composed of fibrocartilage, a plain radiograph would not have revealed a labral injury, even if one was suspected. Magnetic resonance arthrography (MRA) is a more sensitive measure for detecting a torn or frayed labrum. When Ralph's radiographs were taken, the standard of care in the differential diagnosis of hip pain often did not include labral tears. Only with the advent, and more frequent use, of diagnostic arthroscopy and MRA in years following was a labral tear routinely considered. Of course, if advanced joint destruction is notable on a radiograph, as was the case for Ralph, it is well past the time when recognition of a labral tear matters.

When Ralph was seen at the Motion Analysis Center, he reported a cluster of hip-related symptoms similar to what has been reported by patients with labrum tears, as noted. It is likely that he had many of the same symptoms even prior to his radiographic studies 14 years prior, and although irritating, did not significantly impair his activities, and therefore went unreported for years. Based on what we know now, we hypothesized that since part of the labrum is richly innervated with nociceptors, his current symptoms were probably related as much to a torn and/or frayed labrum as a degenerative joint.

Let's move on. The face of the acetabulum normally projects outward, with a slight antero-inferior oblique orientation. Examination of acetabular geometry and alignment using a variety of imaging techniques is commonly used to ascertain congenital and developmental abnormalities. These images are usually obtained in individuals, much younger than Ralph, who present with abnormal positioning and/or movements of the hip joint, e.g., infants, or have complaints of hip pain of unknown etiology. For example, as we alluded to earlier, a common malformation of the acetabulum is one where the socket is shallower than normal, i.e., dysplastic, leaving the femoral head poorly covered. The natural history of hip dysplasia includes joint instability, abnormal joint stress distribution, and, ultimately, progressive degeneration. It is possible that Ralph had a mild form of hip dysplasia that was asymptomatic in his early years and was not detected as part of his radiographic analysis. Often ultrasonography or magnetic resonance imaging (MRI) is the methods of choice in detecting and measuring the shape and structure of the acetabulum and its orientation relative to the femoral head. Two particular measurement metrics that are often used to measure acetabular alignment include the center-edge and acetabular anteversion angles (Fig. 10.4).

The center-edge (CE) angle (or angle of Wiberg) can be measured using standard anterior-posterior radiographs. The angle normally ranges from 25° to 35° in the adult and indicates the degree of femoral head coverage (Fig. 10.4A). As suggested by the figure, a decreased CE angle indicates less acetabular coverage of the femoral head. Conversely, an increased CE angle indicates greater femoral head coverage

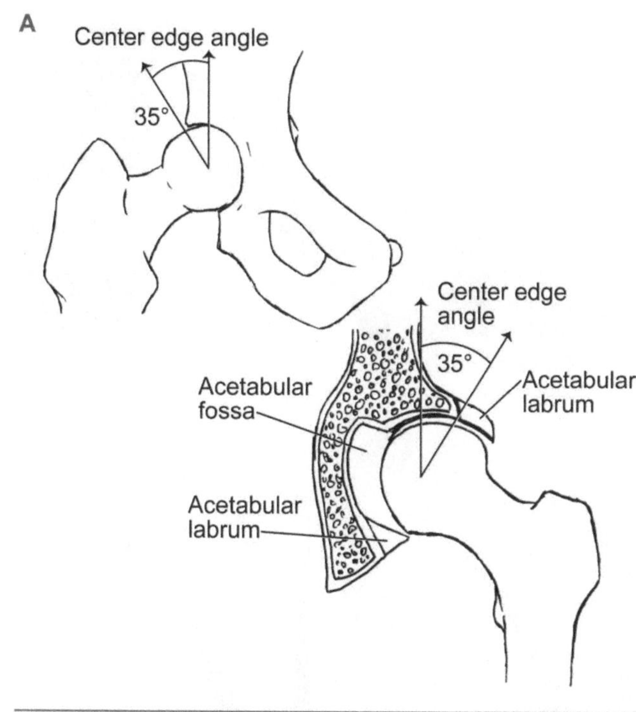


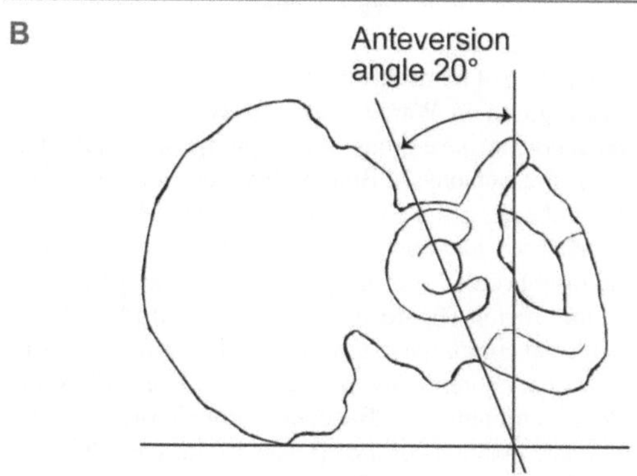


**Fig. 10.4** (**A**) Lateral perspectives illustrating center-edge (CE) angle that measures the fixed orientation of the acetabulum within the frontal plane, relative to the pelvis, and (**B**) superior view showing the acetabular anteversion angle that measures the fixed orientation of the acetabulum, within the horizontal plane, relative to the pelvis

than normal. In either case, one mechanical implication of an abnormal CE angle includes an altered surface area of contact between the acetabulum and the femoral head. For example, if the CE is significantly decreased, in addition to a reduction in the surface area of contact between the joint partners, there is also a predisposition to hip joint instability, i.e., femoral subluxation or dislocation. Given the large bone-on-bone forces at the hip joint that are induced during the single-limb stance phase of gait, for example, a reduced surface area of contact would significantly decrease the joint's ability to distribute large joint surface compression and shear stress (recall that stress = force/area). Over time, of course, an inadequate ability to distribute large compression and shear stresses could initiate a cascade of events leading to hip OA. Although an increased CE angle would increase the joint surface area of contact, it has been associated with labral impingement, a condition that can result in fraying and tearing, ultimately leading to hip joint degeneration as well.

Using computed tomography (CT) scanning, which more readily images the horizontal plane of the pelvis, the acetabular anteversion angle can be measured (Fig. 10.4B). This angle normally approximates 20° and measures the extent to which the acetabulum faces anteriorly. The normal angulation of the acetabulum leaves part of the femoral head exposed anteriorly, a type of predisposition toward anterior subluxation. Additionally, it appears that acetabular and femoral anteversion and antetorsion (discussed later) are related so that if both angles are increased, the potential for dislocation becomes greater. The anterior capsule and iliopsoas tendon, as well as acetabular labrum (by deepening the socket), provide some restraint to excessive anterior translation. Additionally, it appears that there may also be a functional adaptation during gait to compensate for the static instability produced by these torsional deformities.

It is also possible that congenital or developmental abnormalities influence the creation of a retroverted acetabulum. In this case, the acetabular anteversion angle is significantly decreased (to zero or a negative angulation), resulting in a retroverted acetabulum, i.e., the acetabulum faces laterally. As with the case of an abnormal CE angle, abnormalities in acetabular anteversion change joint surface contact locations and areas of contact, leading to a lesser ability of the articular cartilage to optimally distribute large joint stresses.

Because Ralph sought medical attention for hip pain during his middle years, it was not likely that his physician considered measuring the center-edge or acetabular anteversion angles. If they were measured and found to be abnormal, there would have been no treatment recommendation considering Ralph's age, and the fact that the abnormal angles were the result of fixed torsional aberrations of the pelvis. On the other hand, if Ralph developed such abnormalities during his growing years (that is the only possibility because Ralph did not report any major traumatic accidents, e.g., fractures, as part of his history), we speculate that the regular stressors of life's activities, as well as the high impact loading of joints associated with basketball, may have contributed to early degenerative changes in his hip joint.

Let's examine the acetabulum's bony partner: the femur. The femur has four distinct regions: head, neck, shaft, and distal femoral condyles (which are studied more closely as part of the knee case) (Fig. 10.5). The femoral head projects antero-medially to articulate with the acetabulum; that is, the proximal femur, including the head and neck, also has an antetorsion orientation (more on that later). The femoral head forms two-thirds of a sphere, where the hyaline cartilage covering is thickest on the medial-central surface surrounding the fovea, and thinnest toward the periphery. With

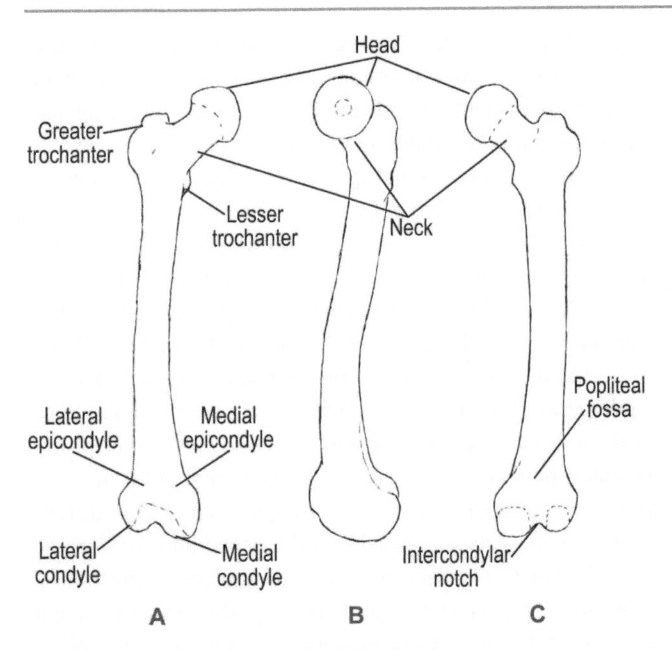


**Fig. 10.5** (**A**) Anterior, (**B**) medial, and (**C**) posterior surfaces of the right femur

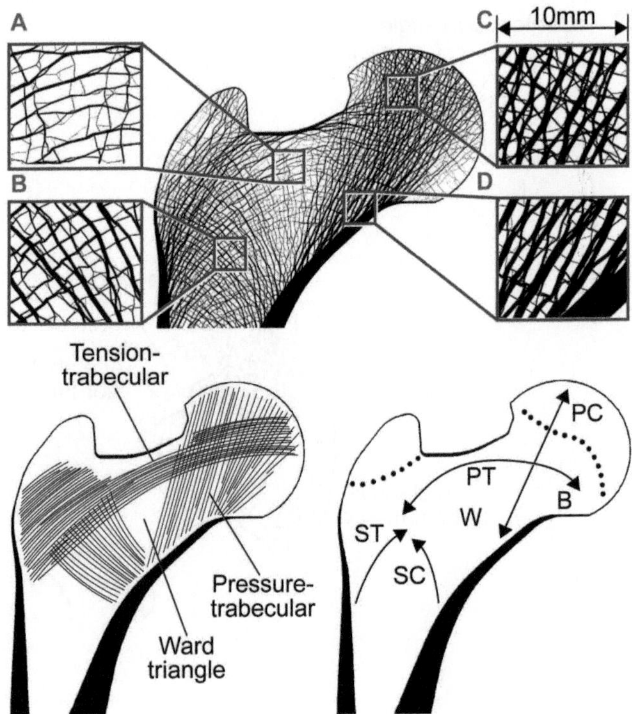


**Fig. 10.6** Top: simulation of human proximal femoral trabecular bone. (**A**) principal tensile group in Ward's triangle, (**B**) intertrochanteric arches, (**C**) principal compressive group in the epiphysis, (**D**) principal compressive group in the metaphysis. Bottom: two figures that show the arrangement of the trabecular framework of the proximal femur. Note: the Ward triangle has been identified as an area of weakness, as suggested by the void of trabecular overlap and high incidence of femoral neck fractures in this region (figure on the left), the cross-sectional areas: PC = the principal compressive group; PT = the principal tensile group; SC = the secondary compressive group; ST = the secondary tensile group; W = Ward's triangle; B = Babcock's triangle (figure on the right). The approximate positions of the epiphyseal lines are shown as large dots. (Modified with permission from Jang and Kim (2008))

lesser axial loads, load-bearing is concentrated at the periphery of the lunate surface of the femoral head, whereas load-bearing under greater compression shifts to the center of the lunate surface, as well as to the anterior and posterior horns. For our normal daily activities, most of the load is carried by the anterior and medial lunate surfaces of the articular cartilage on the femoral head, suggesting that hip joint loads associated with run-of-the-mill activities are moderately large. As noted earlier, the ligamentum teres (also known as the ligament to the head of the femur) may have had two important roles for the neonate: hip joint stability and a source of blood to the head of the femur. However, though there is conflicting evidence regarding any role the ligamentum teres might play in stabilizing the adult hip joint, the mechanoreceptors of this ligament may suggest that it is an important structure for relaying proprioception information to central processors, e.g., the spinal cord.

The neck of the femur connects the head to the shaft and displaces the proximal shaft of the femur laterally. More distally, the femoral shaft curves medially so that the femoral condyles are closer to the midline, which effectively makes the distal femur the proximal contribution to the normal genu valgus angulation of the lower extremities. The shaft of the femur has a mild convexity anteriorly, giving this long bone a slight "bow" shape. As a result, compressive loads from above and below create a bending moment, i.e., eccentric loading, along the shaft, with tensile stress distributed posteriorly and compression stress distributed anteriorly.

When we discussed the constituents and material properties of bone earlier in the text (Chap. 3), it became clear that because of its primary functions, i.e., support, protection, and movement, it had to be able to distribute/mitigate multiple stresses, e.g., from compression, tension, torsion, shear, and bending. Briefly, let's review how this plays out in the proximal femur. Like all bone, this region of the femur is composed of cortical (compact) and cancellous (trabecular) bone, but with its own unique organization of cancellous bone (Fig. 10.6). Recall that cortical bone has a dense constituency making it quite stiff, providing it with an ability to withstand large loads before failure. This type of bone is particularly thick in the outer shell of the lower neck and the entire shaft of the femur, where shear and torsional stresses are large. Cancellous bone, in contrast, is characterized by its porosity and elasticity, features ideally suited to distribute large compressive repetitive loads, thus providing a degree of protection to the articular surfaces. It is also characterized by a multi-dimensional trabecular lattice, whose pattern is dictated by the lines of action and type of applied load. For

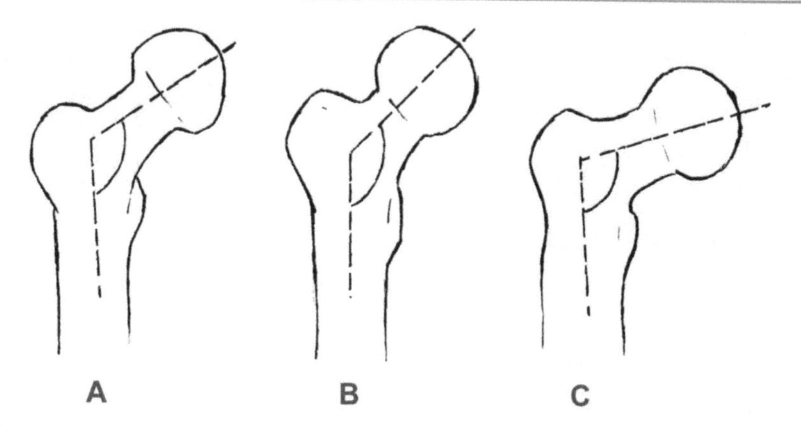


**Fig. 10.7** (**A**) Normal angle of inclination (~125°), (**B**) increased angle of inclination (>125°) or coxa valga, and (**C**) reduced angle of inclination (<125°) or coxa vara

example, the nearly vertical medial lattice (labeled "pressure-trabecular" in the figure) is the bony response to axial compression loads applied to the femoral head. Whereas the lattice that runs lateral to medial (sometimes referred to as the arcuate network) appears to be related to the acetabular-femoral head joint compression that results from muscle actions, e.g., primarily gluteus medius, gluteus minimus, and tensor fascia latae; the label "tension-trabecular" is a reminder that the tension generated by the muscle-tendon actuator induces compression on the joint that the muscles span. The third lattice pattern running medial to lateral (not labeled) likely represents the bony response to the tension created by the large hip muscle, e.g., gluteus medius, that inserts on the greater trochanter. We see in these trabecular arrangements perfect examples of the application of Wolff's law.

We know that the ultimate shape, configuration, and orientation of the developing proximal femur is determined by genetic disposition, differential growth of the ossification centers, i.e., growth plates, forces related to muscle actions, circulation, nutrition, and pathology. For example, abnormal growth and development related to cerebral palsy, a neurological disorder acquired during the birthing process, has been shown to significantly affect the shape and orientation of the proximal femur as the child exercises impaired motor, and related movement patterns, associated with locomotion. As expected, the abnormal orientation of the proximal femur (head and neck primarily) has implications for hip joint stability, as well as how joint stresses are managed. Let's examine two prominent angulations of the proximal femur that dictate its shape and function.

The angle of inclination is the angle made in the frontal plane between the femoral neck and the medial femoral shaft (Fig. 10.7). At birth, this angle ranges from 165° to 170°. There is a gradual decrease in this angle during development, influenced by motor, i.e., muscle, and loading patterns across the femoral neck associated with all forms of locomotion, e.g., crawling, walking, etc., which eventually results in the average normal adult value of approximately 125°. As expected, a normal angle of inclination is considered optimal because it creates a reasonable balance and distribution of the large compression and shear stresses at the joint surface and across the femoral neck and is associated with a bending moment at the femoral neck that is best tolerated. Moreover, the length tensions and internal moment arms of the muscle-tendon actuators that act on the hip joint are such that their function is optimized under conditions provided by a normal angle of inclination.

Coxa valga is an abnormal increase in the angle of inclination and is associated with significant changes in the external and internal forces and moments acting across the femoral neck. Coxa valga reduces the bending moment and shear stress across the femoral neck, while it increases the functional length of the hip abductor muscles, which are perhaps acceptable outcomes. Conversely, this condition reduces the moment arm of the hip abductors and may predispose the femoral head to superior subluxation. Although the body attempts to adapt to this bony pathology, this orientation alters how joint surface compression and shear stresses are distributed, perhaps predisposing to early degenerative changes.

Coxa vara is an abnormal decrease in the angle of inclination. As with coxa valga, this abnormal angle of inclination alters the distribution of joint surface compression and shear stresses and leads to other changes that have both advantages and disadvantages. On the positive side, coxa vara increases the moment arm of the hip abductor muscles and may promote a more stable joint configuration. However, it increases the bending moment across the femoral neck, adding significant shear stress to the same region, and decreases the functional length of the hip abductor muscles. This condition could also contribute to the onset of hip OA.

The angle of declination, more commonly known as femoral torsion (or femoral version), defines the orientation

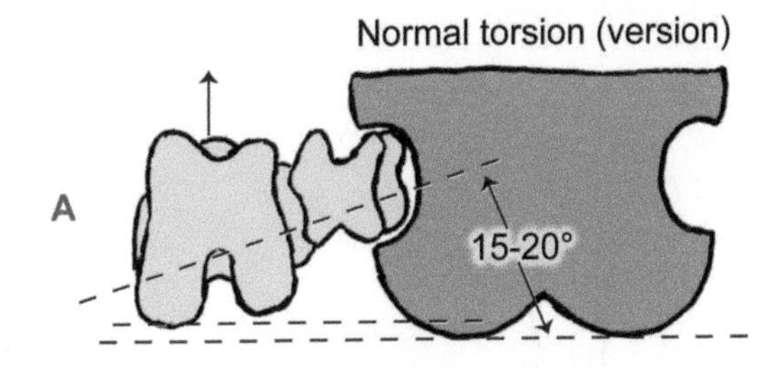


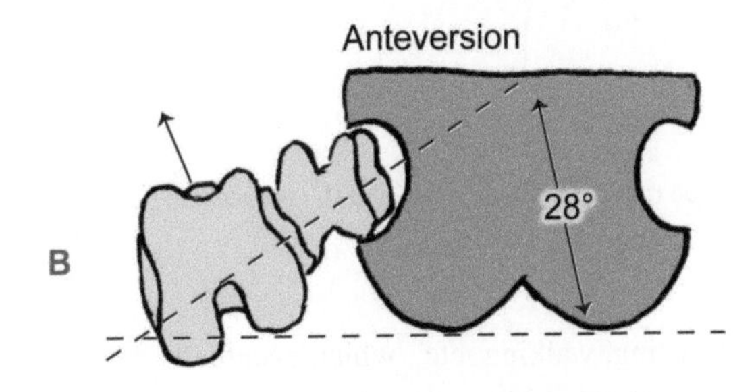


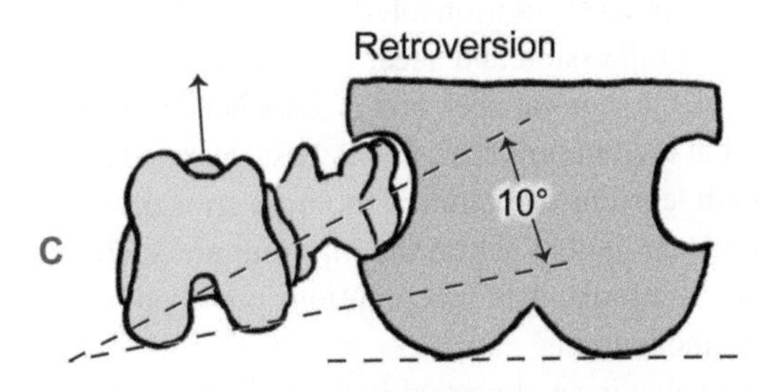


**Fig. 10.8** Schematics illustrating (**A**) normal femoral torsion (version), (**B**) anteversion, and (**C**) retroversion. Note that with anteversion there is an increase in the angle of the head and neck of the femur relative to the frontal plane of the body, and though the femur has a normal 15° to 20° of torsion, the femur is abnormally positioned in the acetabulum with the net effect of putting the limb in an externally rotated position. Conversely, retroversion induces a decrease in the angle of the head and neck of the normal femur relative to the frontal plane of the body, with the effect of putting the limb in internal rotation (Kirtley 2006)

between the femoral head and neck of the femur relative to the frontal plane of the body. As viewed from above, the normal orientation reveals that the femoral head and neck lie anterior to a medial-lateral axis through the femoral condyles (Fig. 10.8A). In the normal adult, the head and neck of the femur are angulated approximately 15° to 20° relative to the frontal plane of the body (i.e., version), which is identical to the angle that the head and neck of the femur make with the femoral condyles (i.e., torsion) when the normal femur is correctly positioned into the acetabulum.

It is possible to have a normal version, yet have a condition described as anteversion, which is an increase in the angle of the head and neck of the femur relative to the frontal plane of the body (Fig. 10.8B); that is, the femur is normal, but abnormally positioned in the acetabulum. The net effect of anteversion is an externally rotated leg. Retroversion is

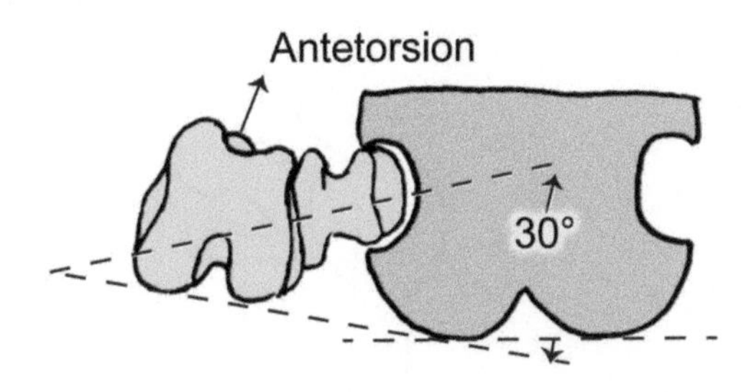


**Fig. 10.9** Schematic of antetorsion of the head and neck of the femur (Kirtley 2006)

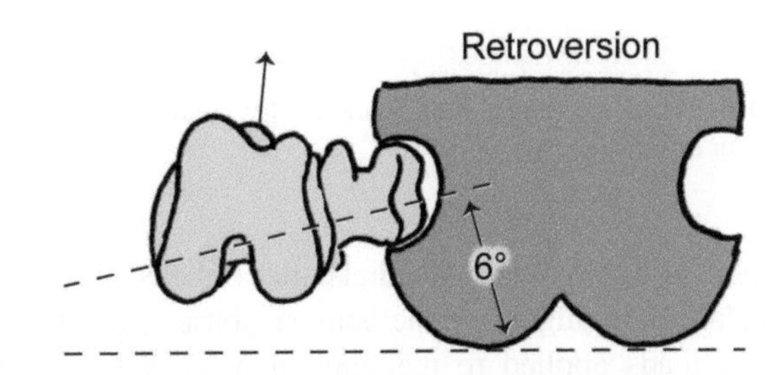


**Fig. 10.10** Schematic of retrotorsion of the head and neck of the femur (Kirtley 2006)

just the converse. That is, there is a normal femoral version, but a decrease in the angle of the head and neck of the femur relative to the frontal plane of the body, with the net effect of an internally rotated leg (Fig. 10.8C). We can imagine that anteversion or retroversion could be caused, for example, by muscles that may be spastic or tight, thus "holding" the lower limb in an abnormal position.

In the preceding figures, the angular relationship between the head and neck of the femur and the femoral condyles (torsion) is normal, and the illustrations depict three different positions of a normal femur relative to the frontal plane of the pelvis. However, it is possible that the angle of the head and neck of the femur relative to the pelvis may be normal but the torsion (a fixed twist) within the bone is abnormal. Antetorsion is an increase in the angle of the head and neck of the femur relative to the femoral condyles (Fig. 10.9). In the example, an angle of 30° is shown. Note that the angle that the head and neck of the femur make with the frontal plane of the body is normal, but the antetorsion brings the limb into an internally rotated position. Conversely, a femur that has retrotorsion has a decrease in the angle of the head and neck of the femur relative to the femoral condyles; in the illustration, we see an angle of approximately 6° (Fig. 10.10). In this example, the angle that the head and neck of the femur make with the frontal plane of the body is normal, but the retrotorsion induces an externally rotated position of the thigh. In Fig. 10.11 we see the superposition of normal torsion, antetorsion, and retrotorsion. Before we proceed, a reminder of two things: (1) in the normal adult the torsion angle equals the version angle, i.e., approximately 15° to 20°, and (2) when referring to version angles, the head and

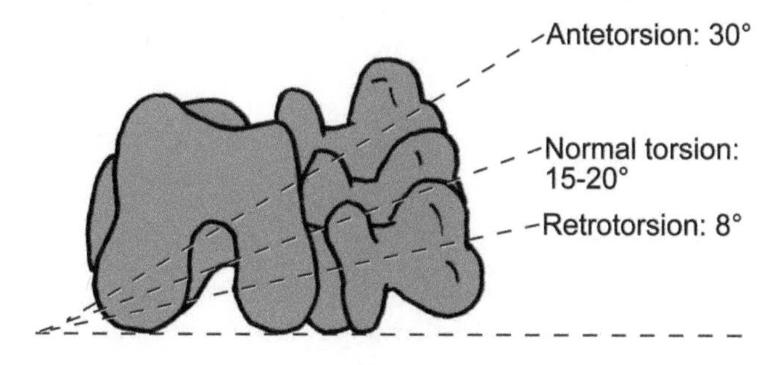


**Fig. 10.11** Superposition of normal and abnormal femoral torsions (Kirtley 2006)

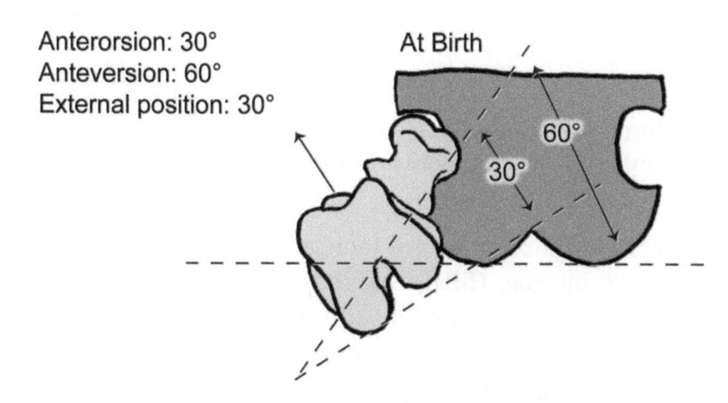


**Fig. 10.12** Schematic of anteversion (60°) and antetorsion (30°) that is typically seen at birth in a child without neuropathology (Kirtley 2006)

neck are measured relative to the frontal plane of the body, whereas when referring to torsion angles, the head and neck of the femur are measured relative to the condyles of the femur.

Now at birth, the torsion within the femur and the position of the femur relative to the frontal plane of the body are very different from adult values (Fig. 10.12). Measurement of the newborn's limb position would find approximately 30° of external rotation. The limb is externally rotated because the increase in anteversion (60°) is 30° greater than the increase in antetorsion (30°). During typical development (i.e., birth to adulthood), however, anteversion decreases from 60° to approximately 15° to 20°. The net effect of this decrease is a position change of approximately 45° resulting in a less externally rotated limb. At the same time, antetorsion undergoes structural changes decreasing from 30° to approximately 15°, for a difference of 15°, with the net effect of less internal deviation of the limb. These changes usually take place within the first 6 years of development, although delays up to adolescence have been reported.

The normal femoral torsional angle optimizes joint congruity and alignment while minimizing joint surface contact stresses. As was the case with the angle of inclination, newborns with a normal neurological system have a greater degree of femoral anteversion and antetorsion that gradually decreases as part of their normal development. Excessive

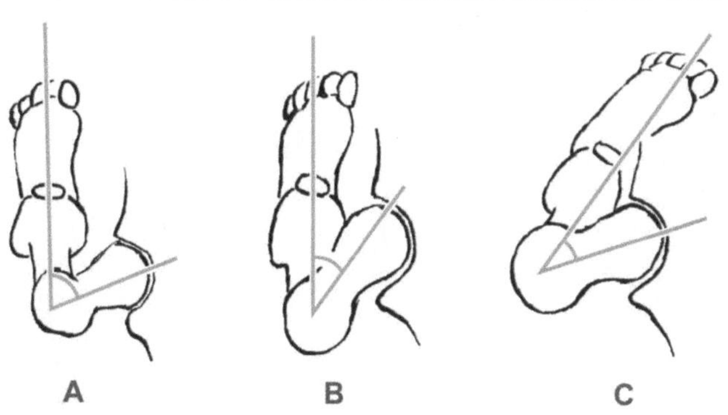


**Fig. 10.13** Schematic comparing (**A**) normal torsion to (**B**) antetorsion and (**C**) typical compensation for femoral antetorsion during gait

antetorsion is defined as femoral torsion angles that range from 20° to 40°. Excessive antetorsion affects the biomechanics of the hip joint, as moment arms and the line of action of muscles around the joint are altered. Greater antetorsion shortens the hip abductor moment arm, which changes the mechanics of the hip joint resulting in greater hip contact forces during gait. In addition, this condition increases the risk factors for osteoarthritis, slipped capital femoral epiphysis, and anterior hip subluxation (or dislocation).

Excessive femoral antetorsion is commonly seen in children with cerebral palsy (CP) and is associated with the finding of significantly increased hip internal rotation range of motion compared to the hip external range of motion when performing a physical examination. Functionally, excessive antetorsion is typically manifested by excessive internal rotation at the hip joint during walking; a compensation that moves the femoral head into better alignment with the acetabulum and appears to increase the moment arm of the hip abductor muscles, thereby restoring hip joint stability (Fig. 10.13). In some children with CP excessive antetorsion may improve, as may the gait pattern, over time with rehabilitation and gait training. However, in most cases, the abnormality does not improve which not only induces changes in muscle and ligament lengths about the hip joint but also creates abnormal loads distally inducing changes in tibial torsion; over time the abnormal loads can induce excessive external tibial torsion. Recalcitrance to normalization increases the likelihood of hip dislocation, ongoing joint incongruity, and abnormal joint surface stress, ultimately leading to hyaline cartilage disruption, labral fraying and tears, and hip OA.

Retrotorsion induces malalignment of the femoral head and neck relative to the femoral condyles, but femoral head coverage within the acetabulum appears to be adequate (Fig. 10.10), suggesting that this condition is more stable. On the other hand, retrotorsion may induce greater shear forces on the femoral neck-head junction and may be associated with progressive increases in patellofemoral contact pressures.

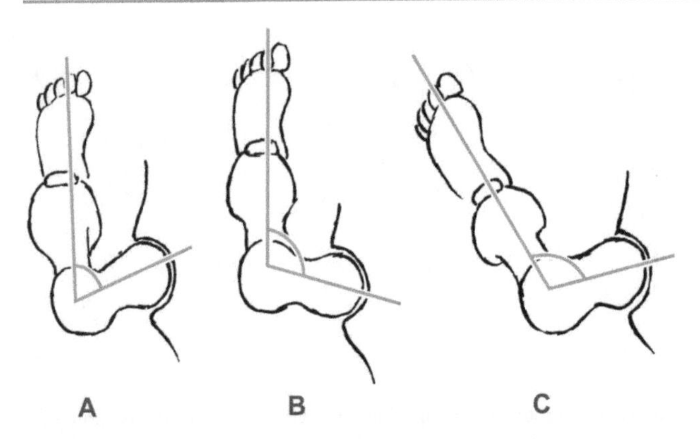


**Fig. 10.14** Schematic comparing (**A**) normal antetorsion to (**B**) retrotorsion and (**C**) typical compensation for retrotorsion during gait

Femoral retrotorsion is suspected, based on a physical examination, when there is a significant increase in hip external rotation range of motion compared to the hip internal range of motion. Individuals with retroversion are easily identified because of the excessive toed-out appearance during standing and walking (although there are other conditions that may also produce a toed-out gait, e.g., excessive external tibial torsion) (Fig. 10.14). Retrotorsion of the femoral head and neck is seen in children with cerebral palsy as well, although it is less common than excessive antetorsion. Because of the potential long-term sequelae, e.g., hip OA, for both increased femoral antetorsion and retrotorsion, surgical intervention, e.g., derotation osteotomies, is often the treatment of choice for children with cerebral palsy when these abnormalities do not seem to be normalizing.

You might wonder why we discussed the angles of inclination and declination. For sure, Ralph did not describe any development delays as a child, and he certainly did not have a neuropathic disorder, such as cerebral palsy. Yet, it is possible that he had mild abnormalities in the angles of inclination or declination, or both, that were not manifested in any of his childhood doctor visits or evident in how he walked, ran, or played, and did not become symptomatic. Based on the discussion above, we might speculate that if he did have mild abnormities in proximal femoral orientations, the incongruencies in joint surface alignment combined with the forces of activities of daily living, and sports participation, could have initiated minor strain injuries to joint hyaline cartilage, or the acetabular labrum, which became chronic, eventually leading to frank hip OA. There is no physical examination measurement available to screen for abnormal angles of inclination; they can only be ascertained using plain radiographs. Given Ralph's advanced joint degeneration when he was radiologically examined several years earlier, it would have made no sense to measure them at that time. Femoral torsion abnormalities can be picked up by screening tests as part of the physical examination test, e.g., Ryder's (or Craig's) test, but a definitive diagnosis of that

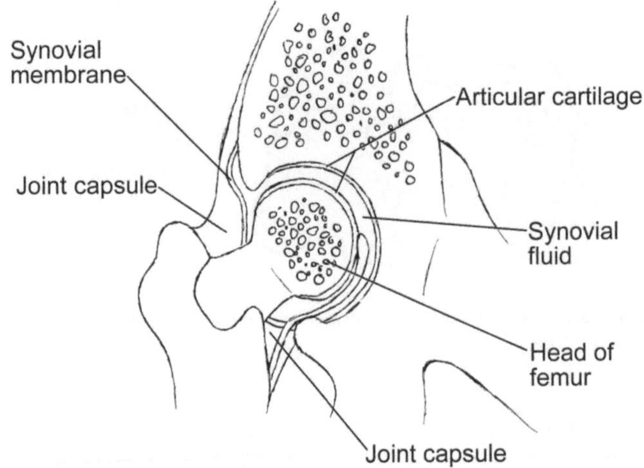


**Fig. 10.15** Cross-sectional view of the right hip joint illustrating the relationship between the synovium and joint capsule

condition requires computed tomography (CT) or magnetic resonance imaging (MRI). Again, it would have made no sense to go to the expense of performing a CT scan to quantify a femoral torsion abnormality for someone with Ralph's degenerative hip joint. Additionally, as noted earlier, Ryder's test, used to screen for excessive femoral antetorsion, was not performed as part of Ralph's physical examination because he lacked the internal rotation mobility at the hip needed to render a valid measurement. In summary, Ralph's early hip OA could have been initiated by abnormalities in orientations of the acetabulum and proximal femur, but we will never know, and at this point in his care it is irrelevant. On the other hand, this process of inquiry was most relevant.

### 10.4.2 Soft Tissue Structures

There are additional hip joint periarticular soft tissues that need to be studied. The hip joint is inherently stable because of the large acetabular socket and intimate labrum, negative intracapsular pressure, and the congruence of the femoral head with the acetabulum. Similar to the sacroiliac joints, the hip joint has both form- and force-closure; its force-closure is induced by the intrinsic (joint capsule) and extrinsic ligaments, as well as the muscles that cross the joint.

Let's first look at the ligamentous complex. There is an intimate relationship between the synovial membrane that lines the internal surface of the hip joint and its capsule (Fig. 10.15). The three major ligaments at the hip, iliofemoral, pubofemoral, and ischiofemoral, blend closely with the external surface of the capsule such that many refer to the composite of these structures as simply the capsular ligaments (Fig. 10.16). Parts of the anterior capsule are reinforced by attachments of the iliocapsularis, gluteus minimus,

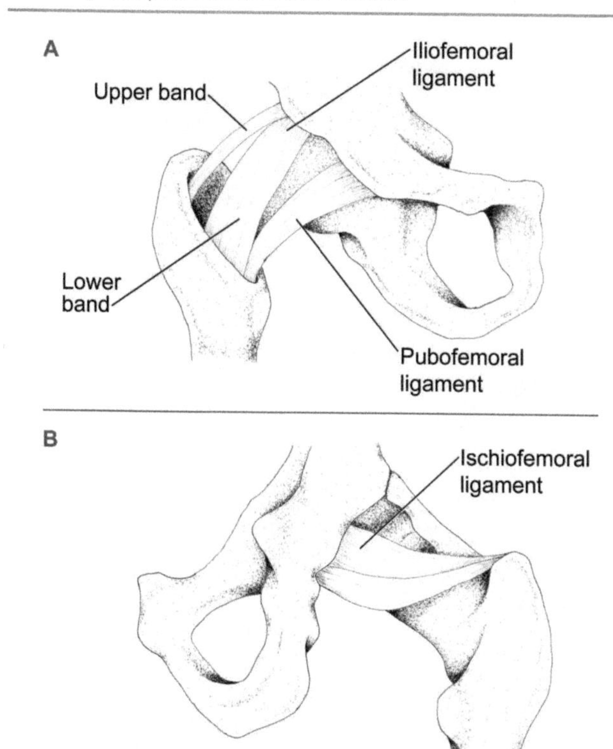


**Fig. 10.16** (**A**) Anterior and (**B**) posterior views of hip capsular ligaments

and head of the rectus femoris muscles. In general, each separate ligament is named relative to its specific proximal attachment to the acetabulum. The iliofemoral ligament (or Y-ligament of Bigelow) is so-named because it resembles an inverted-Y when viewed anteriorly. This ligament is the thickest and strongest of the three ligaments such that it and the superimposed passive tension of the iliopsoas tendon alone provide sufficient restraint to hip hyperextension. A combination of hip hyperextension and external rotation places maximum tension on the iliofemoral ligament and anterior capsule.

Both the pubofemoral and ischiofemoral ligaments lack the thickness and stiffness of the iliofemoral ligament yet blend with, and augment, the joint capsule in adjacent regions of the joint periphery. The pubofemoral ligament blends with the medial portion of the iliofemoral ligament and becomes taut with hip extension, abduction, and, to a lesser extent, external rotation. The ischiofemoral ligament spans, in a spiral-like fashion, the space from the postero-inferior acetabulum and ischium to the greater trochanter of the femur. It becomes most taut when the hip is simultaneously internally rotated and abducted (approximately 10 to 20°).

Based on the functional restrictions imposed by the capsular ligaments, as just described, we might surmise that the close-packed position of the hip joint is the combined positions of end-range hip extension plus mild abduction plus internal rotation. Recall (Chap. 3), we define the close-packed position of a joint as one in which the joint surfaces are maximally congruent, combined with the maximum tautness of the intrinsic and extrinsic joint ligaments. The close-packed position can be considered the position of maximum joint stability, i.e., or the position where joint accessory movements are minimal. What is unique about the hip joint is that full hip extension combined with internal rotation and abduction does not place the respective joint surfaces, i.e., acetabular cup and femoral head, in a position of maximum congruence. The hip position that maximizes joint congruency is actually 90° of hip flexion plus moderate abduction plus external rotation, a position that paradoxically untwists the capsular ligaments. We might wonder why that is. One possibility is that this position is one that is associated with common birthing positions. Let's leave that where it is for now, but if you are interested in taking it forward we encourage you to follow your curiosity.

We include ligaments as structures that are part of force-closure systems for good reason. When joints move, internal stresses within ligaments change, i.e., increase or decrease. When internal stresses increase, the ligaments approximate the joint bony partners, which can make the joint more stable. At the same time, this approximation increases bone-on-bone forces (sometimes referred to as joint reaction forces), with concomitant increases in the internal joint cartilage compression and shear stress. Actually, perhaps the phrase "force-closure" doesn't convey the whole story. Because, in fact, since the majority of the joint ligaments are eccentrically located relative to the joint, they induce joint passive-elastic moments that then change, i.e., increase, decrease, and/or re-distribute, joint surface stress. Similarly, then, when a muscle-tendon actuator is activated, its action creates an internal joint moment, which typically increases joint surface forces. From this point forward, we consider that the actions of ligaments and muscles, whether they are passive or active, contribute to the force-closure of a joint. This assumption is made in light of the fact that the primary role of the muscle-tendon actuator is to control or produce joint motion. With this background, let's briefly review the functional anatomy, i.e., muscles, surrounding the hip joint.

### 10.4.3 Muscle Function

The muscles of the hip region are innervated by nerve roots that originate from the lumbar (T12 to L4) and sacral (L4 to S4) plexuses. The nerves that form these plexuses arise from the ventral rami of the spinal cord, with those from the lumbar region innervating muscles of the anterior and medial thigh, and those from the sacral region innervating muscles of the posterior and lateral hip, posterior thigh, and entire lower limb (Fig. 10.17). Originating from the lumbar plexus, both the femoral and obturator nerves are formed by the L2

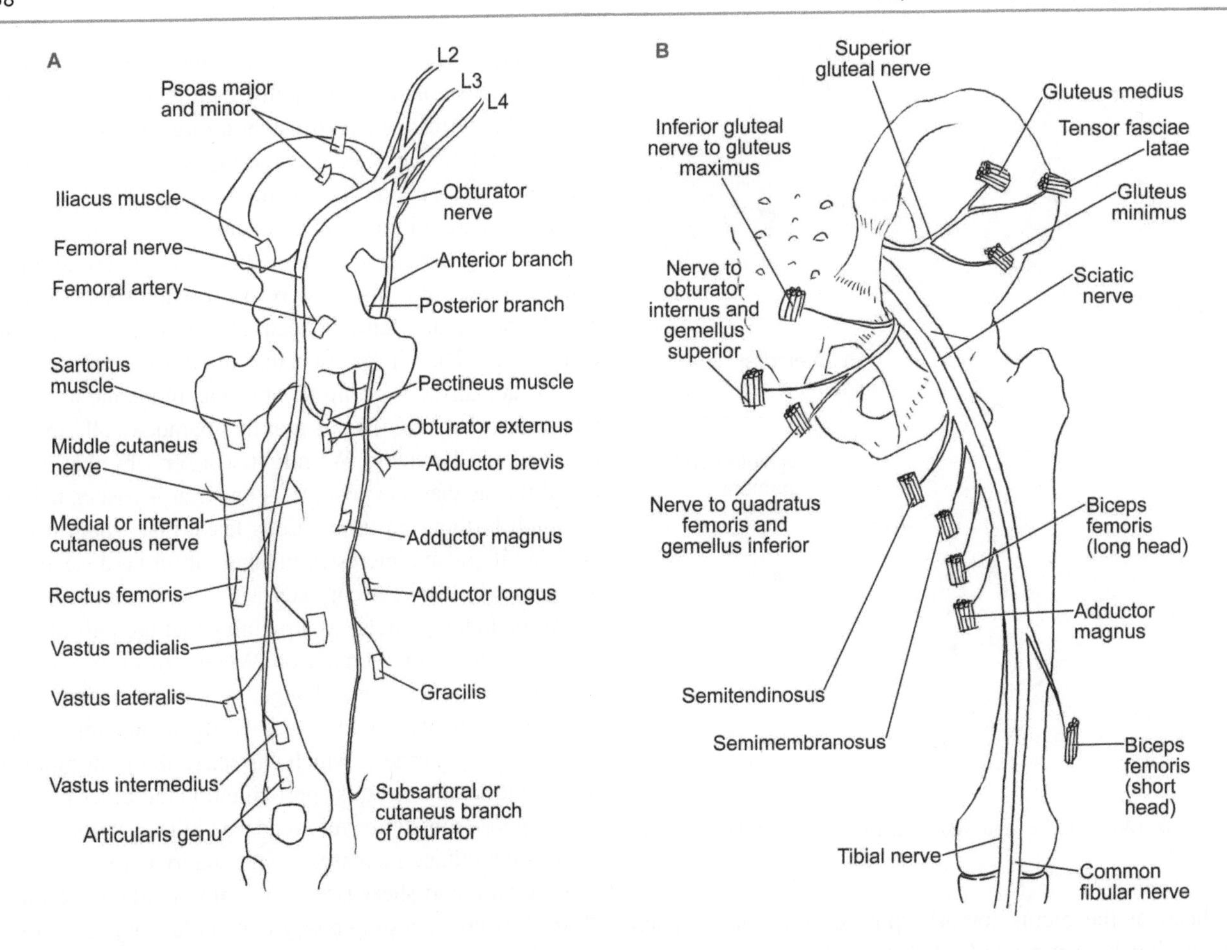


**Fig. 10.17** Cranial to caudal distribution of muscle innervation for (**A**) the femoral and the obturator and (**B**) the sciatic nerves that originate from the lumbar and sacral plexuses

to L4 nerve roots. Nerves originating from the sacral plexus include the superior and inferior gluteal, and sciatic nerves, as well as nerves named simply by the muscles they innervate, e.g., nerve to the piriformis, etc. In general, sensory innervation to and from hip periarticular structures, e.g., joint capsule, ligaments, and parts of the acetabular labrum, is carried by the nerve roots that supply neighboring muscles. One interesting fact is that the connective tissues of the antero-medial aspects of the hip and medial knee both receive sensory fibers from the obturator nerve (i.e., the L3 nerve root), which helps to explain symptoms that are commonly referred to the knee that originates from the hip joint, e.g., hip OA.

Since this case involves a degenerative hip that likely originated from mechanical, not neurological, causes, our muscular anatomical review is brief (Table 10.2), focusing on the key muscular functions associated with walking. Before we go any further, however, we want the reader to consider that when discussing muscle actions across any of the joints of the lower extremity, particularly related to walking, we sometimes describe these actions, and the motions produced or controlled by them, as being dependent on whether the muscle "origin" is relatively fixed, or whether the muscle "insertion" is relatively fixed. Let's look at an example involving the hip joint. If the lower limb is not fixed, i.e., the entire lower limb is free to move, activation of the iliopsoas elevates, i.e., flexes, the thigh relative to the pelvis (Neumann refers to this as "femoral-on-pelvic" hip flexion), a movement that is almost always associated with actions of the hip flexors. Conversely, if one were standing with both feet fixed to the ground, simultaneous activation of bilateral hip flexors and lumbar erector spinae, i.e., a type of force couple, anteriorly rotates the pelvis, i.e., called anterior pelvic tilt, relative to the fixed femurs (a kinematic movement Neumann refers to as "pelvic-on-femoral" hip flexion) (Fig. 10.18). Note, although differentiating distal-on-proximal (i.e., femoral-on-pelvis) and proximal-on-distal (i.e., pelvic-on-femoral) may provide an interesting perspective, we believe that in both cases the hip is flexing. From the perspective of the joint, distal-on-proximal motion is the same as proximal-on-distal motion. Therefore, going forward we restrict our presentation to classical references of

**Table 10.2** Primary and secondary muscle actions at the hip joint

| | Flexors | Extensors | Abductors | Adductors | External rotators | Internal rotators |
|---|---|---|---|---|---|---|
| Primary | Iliopsoas<br>Sartorius<br>Tensor fascia latae<br>Adductor longus<br>Pectineus | Gluteus maximus<br>Biceps femoris (long head)<br>Semitendinosus<br>Semimembranosus<br>Adductor magnus (posterior head) | Gluteus medius<br>Gluteus minimus<br>Tensor fascia latae | Pectineus<br>Adductor longus<br>Gracilis<br>Adductor brevis<br>Adductor magnus | Gluteus maximus<br>Piriformis<br>Obturator internus<br>Gemellus superior<br>Gemellus inferior<br>Quadratus femoris | N/A |
| Secondary | Adductor brevis<br>Gracilis<br>Gluteus minimus (anterior fibers) | Gluteus medius (middle and posterior fibers)<br>Adductor magnus (anterior head) | Piriformis<br>Sartorius<br>Rectus femoris<br>Gluteus maximus (anterior/superior fibers) | Biceps femoris (long head)<br>Gluteus maximus (inferior/posterior fibers)<br>Quadratus femoris<br>Obturator externus | Gluteus medius and minimus (posterior fibers) Obturator externus<br>Sartorius<br>Biceps Femoris (long head) | Gluteus minimus and medius (anterior fibers)<br>Tensor fascia latae<br>Adductor longus<br>Adductor brevis<br>Pectineus |

Modified with permission from Neumann (2017)
Note: These actions assume muscle shortening with the body in the anatomical position

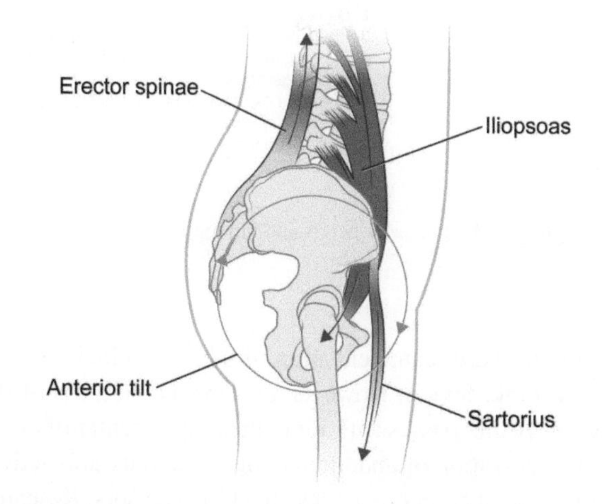


**Fig. 10.18** Example of hip flexion with the femur fixed, which results in anterior pelvic tilt, demonstrating a force couple induced by the hip flexors and lumbar extensors, which rotate the pelvis anteriorly about an axis fixed in the humeral head. Note that this action also accentuates the lumbar lordosis

joint rotations. We refer the reader to Neumann's text, which provides an encyclopedic review of the muscular functions across the hip joint.

The primary hip flexors include the iliopsoas, sartorius, tensor fascia latae, rectus femoris, adductor longus, and pectineus (Fig. 10.19). Of these, the iliopsoas plays a predominant role in all activities requiring flexion of the hip joint. The iliopsoas has two very unique features: (1) its moment arm is greatest when either the hip is fully extended or when the hip is flexed 90°, making it an efficient flexor at important points in the range of motion, and (2) when activated bilaterally, the iliopsoas provides frontal plane stability to the lumbar spine. It is notable that the region where the iliopsoas traverses the anterior pelvic brim and hip capsule, femoral head, and iliopubic eminence is a site where the muscle/tendon can become irritated, creating symptoms related to what has been referred to as "snapping hip syndrome." Two other primary hip flexors, the tensor fascia latae (and related iliotibial band) and rectus femoris, are unique in that they cross both the hip and knee joints, suggesting that the control of hip and knee kinematics during normal daily activities involves complex motor patterns. For example, since the tensor fascia latae secondarily abducts and internally rotates the hip joint, multi-dimensional control is likely a necessary requirement during many activities of daily living.

It is possible that the "catching" symptom that Ralph reported was related to irritation of the iliopsoas over the brim, as well as the loss of flexibility (or extensibility) of the iliopsoas, rectus femoris, and iliotibial band, as noted on physical examination. Ralph's physical examination also revealed a mild increase in his lumbar lordosis, with an associated mild anterior pelvic tilt. This phenomenon is related to the hip flexion we described earlier, which in Ralph's case, was not transient, but likely a chronic fixture related to the reduced length of the anterior hip capsule and hip flexor muscles (Fig. 10.20). Over time, the lumbar erector spinae can also become shortened, with concomitant abdominal muscle weakening. This condition increases the compressive loads on the lumbar zygapophyseal and posterior interbody joints while increasing the shear stress at the L5/S1 junction. Fortunately, at the time we saw Ralph he did not report low back pain symptoms. On the other hand, a hip flexion contracture can increase hip joint compression and shear stresses during standing activities, which may have initiated Ralph's hip OA, or perpetuated its progression (likely the latter). In many cases, we are faced with the "chicken or egg" dilemma in terms of cause/effect. At the time we saw Ralph, the cause was irrelevant. However, it is a good thing to wonder how things began for future cases.

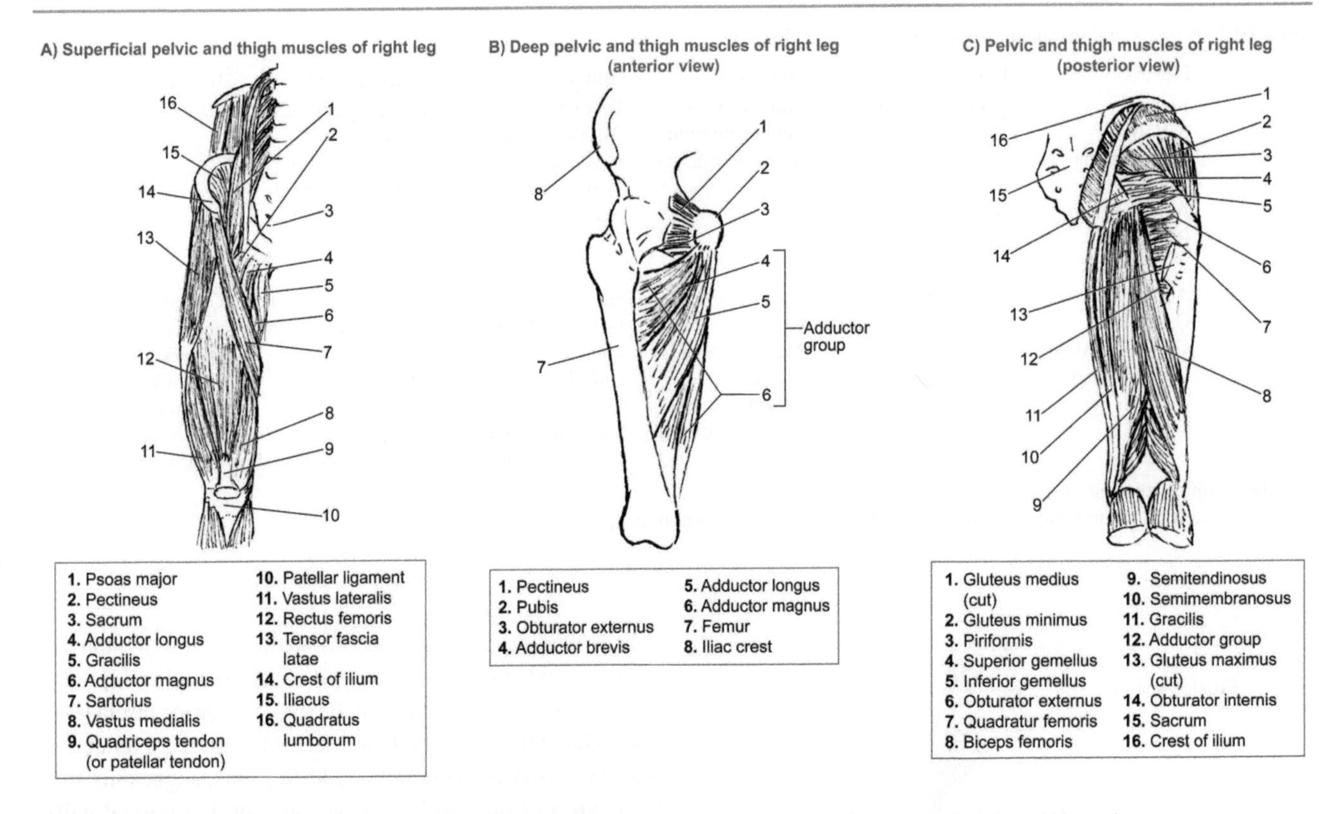


**Fig. 10.19** Muscles of the (**A**) superficial pelvis and thigh, (**B**) deep pelvis and thigh, and (**C**) posterior pelvis and hip region

The primary hip extensors include the gluteus maximus, long head of the biceps femoris, semitendinosus, semimembranosus, and posterior head of the adductor magnus; augmentation of a hip extensor moment is provided by anterior fibers of the adductor magnus and middle and posterior fibers of the gluteus medius (Fig. 10.19). For sporting activities that push the hip joint past 70° flexion, most of the hip adductors (except the pectineus) assist with hip extension, e.g., sprinting, climbing, etc. Two common hip extension movements include the performance of a posterior pelvic tilt (Fig. 10.21) and controlling a forward lean (Fig. 10.22). Posterior tilting the pelvis while standing or sitting is made possible by a force couple involving the hip extensors and abdominal flexors (i.e., primarily the external oblique, assisted by the rectus abdominus). This movement is not one commonly performed during daily activities but is often used in rehabilitation programs for individuals with low back dysfunction. In standing, the posterior pelvic tilt range of motion is constrained by the hip flexors and anterior hip joint capsule, and/or motor control of the individual; that is, performing a pelvic tilt is more difficult to control in the seated posture.

Control of a forward lean is something we do every day (Fig. 10.22). As an aside, the phases of the forward lean are described previously, in a different context, when we examine the mechanics of lumbopelvic rhythm (see Chap. 9). With the forward lean, activation of the abdominal muscles initiates trunk flexion. However, as soon as the mass of the head/arms/trunk passes anterior to the body's center of mass, there is a cessation of abdominal muscle activity and activation of the lumbar erector spinae and hip extensors. Research has shown that the moment arms of the hamstring muscles make them more effective hip extensors than the gluteus maximus in both phases of a forward lean. If the individual maintains a knee extended pose as the forward lean progresses, increased passive tension in the hamstring-tendon actuators provides additional control while sparing some metabolic cost.

This concludes our review of the hip flexor and extensor muscles, so let's now briefly look at the role of the hip flexors and extensors during a normal gait. During gait, the femur runs an excursion, relative to the pelvis, from approximately 20° flexion at initial contact to 10° extension at terminal stance to 40° flexion in mid-swing (Fig. 10.23; also see Appendix J). Therefore, it is clear that there is a cycling of hip flexor and extensor muscle activity when the limb is fixed, i.e., at initial contact through pre-swing, and when it is not fixed, i.e., from initial to terminal swing. In general, the hip flexor and extensor muscles serve to stabilize the trunk and stance limb, as well as to accelerate the body forward. To help visualize the relationship among the lines of force of the various muscles acting largely in the sagittal plane relative to

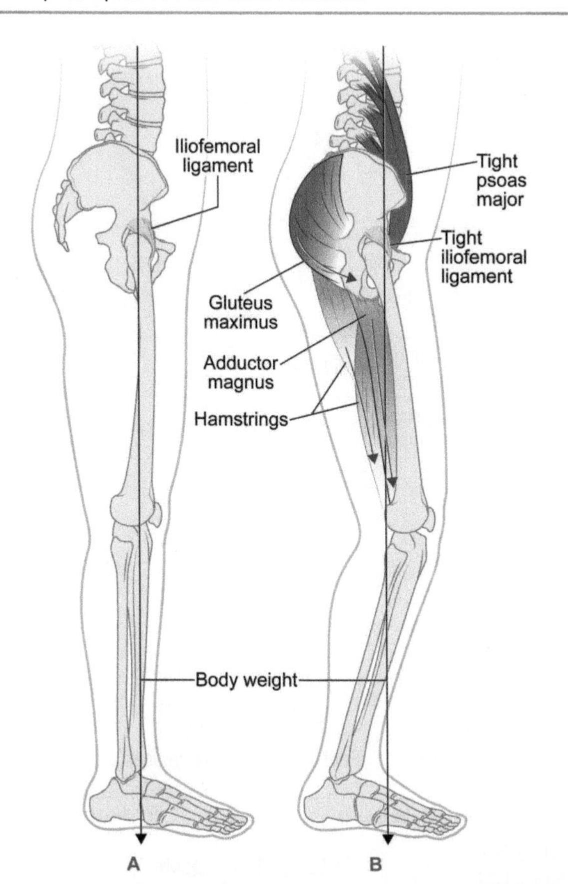


**Fig. 10.20** (**A**) Ideal lumbar posture and lumbar lordosis associated with a normal length of the psoas and lumbar erector spinae and (**B**) altered biomechanical alignment of stance associated with tightness of the anterior hip capsule and iliopsoas. Under normal conditions, alignment of the acetabulum and femoral head is such that compression and shear stresses are optimally distributed protecting the hyaline cartilage and subchondral bone. Moreover, resting muscle tonus is balanced between hip flexors and extensors in a way that the metabolic cost to maintain balance is minimized. With plastic shortening of the anterior hip capsule and hip flexor muscles, more energy, i.e., activation of hip extensors, is needed to control a "quiet" bipedal stance. Increased isometric muscle activation simultaneously increases joint compression stress, which is poorly distributed because of the misalignment of the acetabulum and femoral head secondary to a chronic hip flexion pose

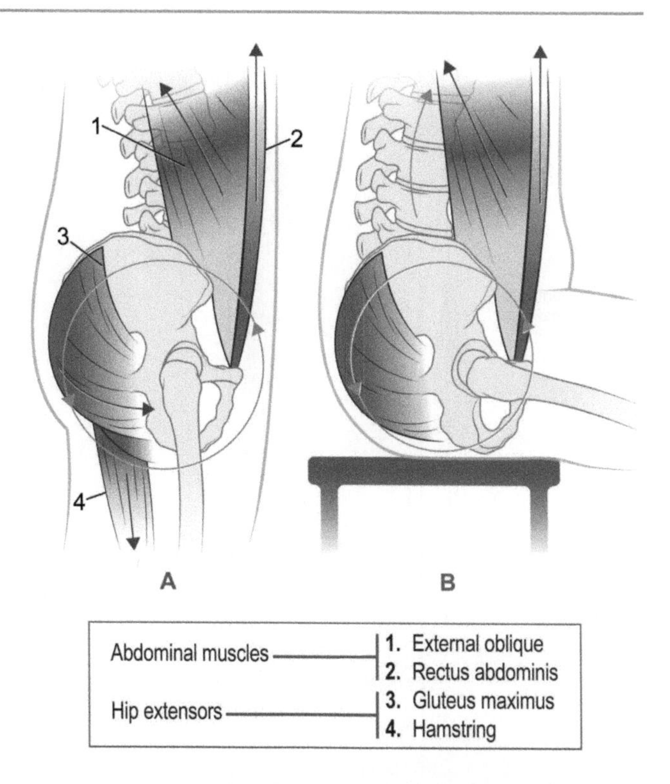


**Fig. 10.21** Schematic illustrating control of a posterior pelvic tilt induced by hip extension, in (**A**) bipedal stance and (**B**) while seated. Note that with a posterior pelvic tilt, the lumbar lordosis is reduced

the hip joint center, we refer the reader to Fig. 10.24. This figure may provide some insight into the complexity of muscles crossing the hip joint that are responsible for sagittal plane motion. However, we want to caution the reader that the action lines in the figure do not represent the force potential of particular muscles, but the approximate orientation of the muscles' force. Furthermore, the lines of force and lengths of the moment arms depicted in the sketch apply to only one point in the range of motion of the femur relative to the pelvis, that is, the anatomical, or neutral, position. We remind the reader that as the segments change their pose, i.e., orientation relative to its bony partner, the action lines representing the direction of the muscle tension and the muscle moment arm constantly change, meaning that the muscles' efficiency and function also change.

Let's now examine the hip muscles that act in the frontal plane (Figs. 10.19 and 10.25). The primary hip adductors include the adductor brevis, longus, and magnus; pectineus; and gracilis. The hip adductors can potentially produce internal hip joint moments in all three dimensions, particularly during sporting activities. For example, during a sprint when the hip is fully extended, the adductors have a flexor moment arm that can significantly augment the flexor moment needed to hyperflex the thigh; conversely, when the hip is maximally flexed, the adductors have an extensor moment arm that can augment hip extension. However, during straight-line walking at a self-selected pace, the hip adductors do not adduct the hip or pelvis but are active as hip extensors, e.g., adductor magnus, during loading response, and hip flexors, e.g., adductor longus and gracilis, during pre-swing and initial swing.

The primary hip abductors, particularly the posterior fibers of the gluteus medius, play a vital role as the hip adducts during first and second double support and single-limb support during walking. Of the three hip abductors (gluteus medius, gluteus minimus, and tensor fascia latae), the gluteus medius makes up the largest cross-sectional area and has a moment arm that advantages it over the other two muscles.

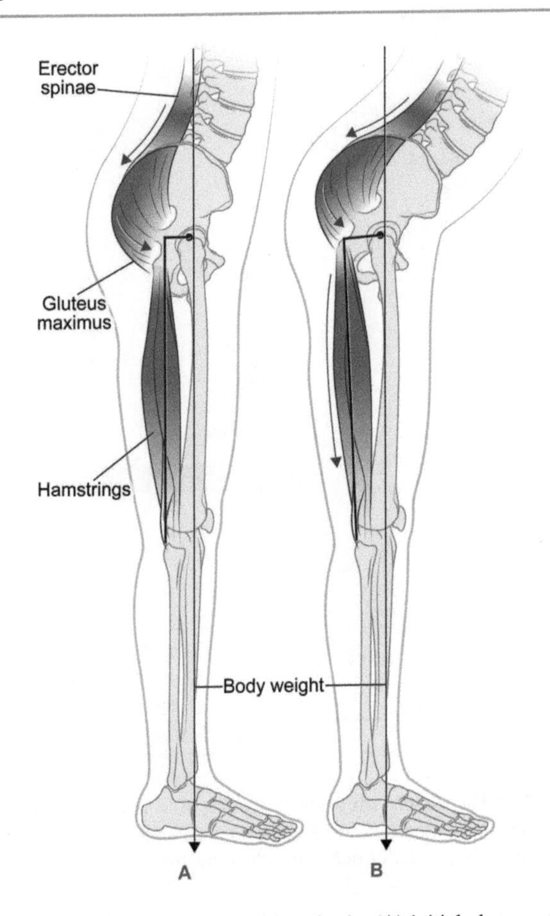


Fig. 10.22 Control of a forward lean in the (A) initial phase and (B) latter phase of the movement. The lumbar erector spinae, gluteus maximus, and hamstring muscles act eccentrically to control a flexor moment about an axis fixed in the femoral heads. Note that the moment arm of the hamstring muscles (small black arrowhead) increases with increased trunk flexion, allowing them to control a large flexion moment with somewhat reduced motor unit recruitment

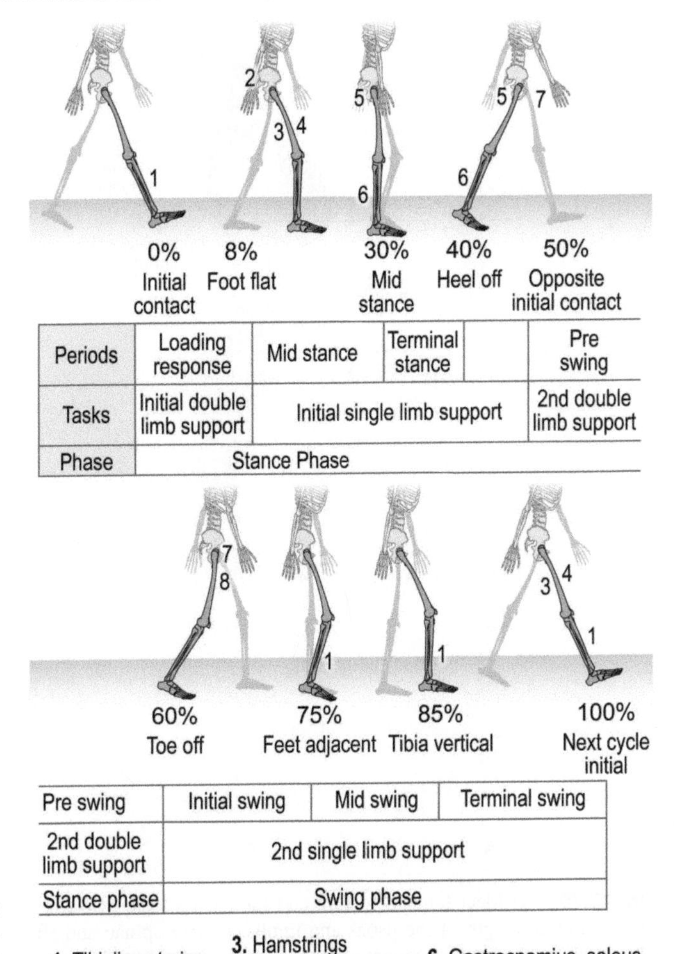


Fig. 10.23 Schematic illustrating major phases and subphases of the gait cycle, sagittal plane hip, knee, and ankle kinematics, and key muscles for the reference limb, i.e., right

Thus, beginning at loading response and continuing through terminal stance the hip abductors of the stance limb control adduction of the pelvis relative to the femur. It is during this time that the contralateral limb is first unloaded and then lifted off the ground during the swing phase as part of advancing the body forward. Although the gluteus minimus and tensor fascia latae contribute to the hip abductor moment during this crucial time period of gait, the trunk and pelvis (on the swing leg side) would drop too far, potentially causing a fall, if the gluteus medius were weak (Fig. 10.26). In most cases where significant gluteus weakness exists (on the left side as shown in the figure), we do not see a positive Trendelenburg on the right, but a compensated Trendelenburg in which the individual leans their trunk toward (or over) the weak side, i.e., to the left. Of note is that all hip abductor muscles can also induce hip internal and external rotation, so that in order to produce a pure hip abduction moment all three muscles need to be activated together.

Ralph demonstrated mild weakness of bilateral hip flexors and abductors. Following observation of his gait, we hypothesized that there would be reduced hip flexion in the swing phase, but that was not the case. On the other hand, the combination of hip pain, secondary to osteoarthritis, and hip abductor weakness induced a mild compensated Trendelenburg to one side. Given the advanced stage of Ralph's hip OA, we were surprised that his compensated Trendelenburg sign was not more pronounced.

There are no muscles at the hip that are considered primary internal rotators; however, several muscles, such as anterior fibers of gluteus medius and minimus, tensor fascia latae, adductor longus, and brevis and pectineus, include hip internal rotation as a secondary action (Figs. 10.19 and 10.27). Recall that the image in Fig. 10.27 depicting the lines of action of both external and internal hip rotators is relative to the anatomical position, i.e., the limb is in neutral rotation and extended at the hip and knee. Thus, given the location of the action lines of the internal rotators relative to the hip joint center (blue circle), which runs in a supero-inferior direction,

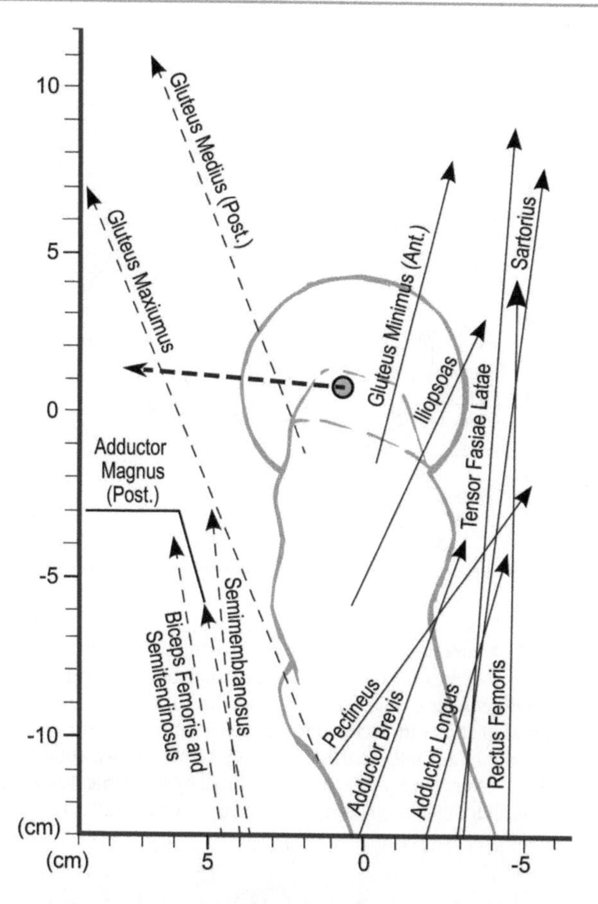


**Fig. 10.24** A lateral view of the right proximal femur depicting the action lines of forces for muscles anterior and posterior to the hip joint center (green circle). The flexors are indicated by solid lines and extensors by dashed lines. Note: the actual scale is indicated by the vertical and horizontal axes. (Modified with permission from Neumann (2017))

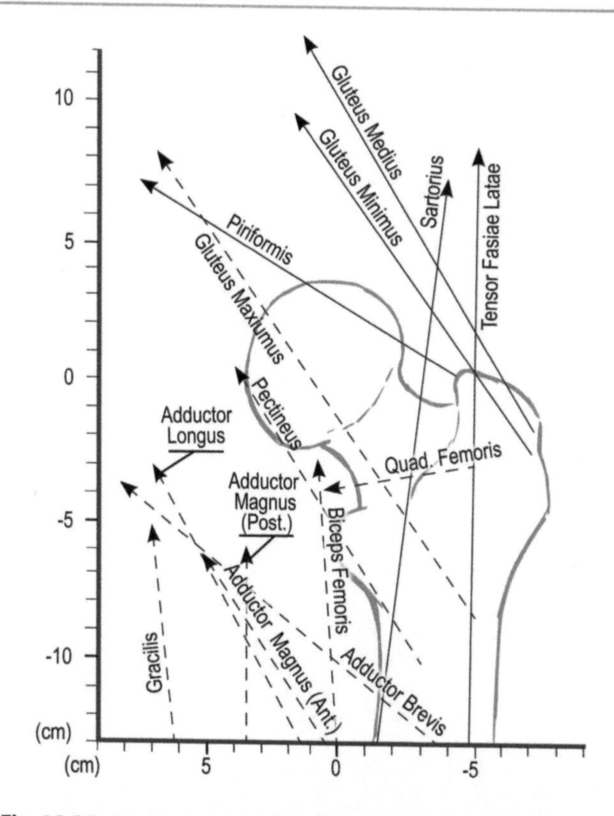


**Fig. 10.25** A posterior view of the right proximal femur depicting the action lines of forces for the primary and secondary hip abductor and adductor muscles relative to an imaginary hip joint center (geometric center of the humeral head). The abductor muscles' lines of force are indicated by solid lines and adductor lines of force by dashed lines. Notice the relatively large moment arms of the primary adductors, e.g., gracilis, adductor longus. Note: the actual scale is indicated by the vertical and horizontal axes. (Modified with permission from Neumann (2017))

we can imagine that the magnitude of their respective moment arms is relatively small, i.e., perhaps ranging from 1 to 3 cm. As a result, if we performed a resistive muscle test of these muscles, collectively, with the femur in the anatomical position, i.e., neutral rotation, we would expect a less-than-robust torque output, compared to testing the hip abductors as a group, for example. However, research has shown that when the hip is flexed close to 90°, the action lines of these muscles change from nearly parallel to nearly perpendicular to the longitudinal axis of the femur. With this change, the internal joint moment potential of these muscles significantly increases. In fact, the moment arms of several hip external rotators, including the piriformis, posterior fibers of the gluteus minimus, and anterior fibers of the gluteus maximus, change such that these muscles become internal rotators when the hip is flexed beyond 60°.

The primary hip external rotators include gluteus maximus, piriformis, obturator internus, superior and inferior gemelli, and quadratus femoris. Based on anatomical position orientation, the posterior fibers of the gluteus medius and minimus, obturator externus, sartorius, and long head of the biceps are considered secondary external hip rotators. As a group, in addition to their primary action, the external rotators effectively compress the head of the femur into the acetabulum, thus enhancing joint stability.

Although the hip external rotators appear to contribute to hip rotation during activities that require abrupt changes in the direction of body movements, e.g., running and cutting in basketball, these muscles do not appear to make significant contributions to normal walking. The internal rotators, on the other hand, appear to contribute to the transverse plane rotation of the pelvis during walking. Before we get into the details of our mechanical analysis, let's be clear that in using a reductionist approach, we fix the femur and assume that the right limb is the reference. The analysis of gait usually begins with initial contact of the reference limb, i.e., the right limb in this example. Thus, after making initial contact, the right hemipelvis is protracted, i.e., the anterior face of the right ilia rotates toward the left (sometimes also called internally rotated), while the left pelvis is retracted (or sometimes referred as externally rotated) (Fig. 10.28). We can see that at this point, we could describe the kinematics in one of two ways: (1) the

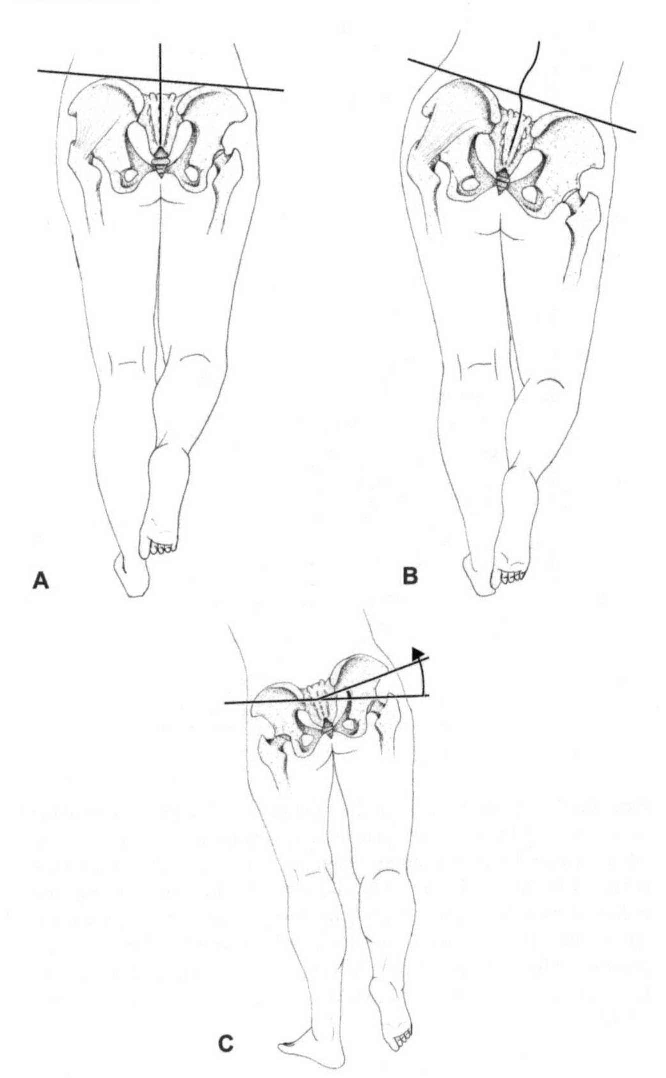


Fig. 10.26 Schematic comparing (A) normal, (B) abnormal hip adduction during left stance phase (right swing phase) of gait, termed uncompensated Trendelenburg, and (C) compensated Trendelenburg for left hip pathology, i.e., pain and/or gluteus medius weakness

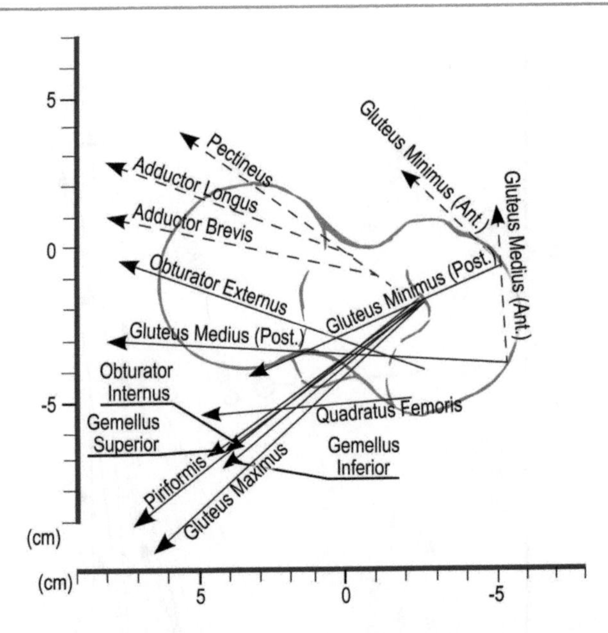


Fig. 10.27 A superior view of the right proximal femur depicting the action lines of forces for the hip muscles, relative to an imaginary hip joint center (geometric center of the femoral head) that produce internal and external rotation (not pictured tensor fascia latae and sartorius). The external rotators are indicated by solid lines and internal rotators by dashed lines. Note: the actual scale is indicated by the vertical and horizontal axes. (Modified with permission from Neumann (2017))

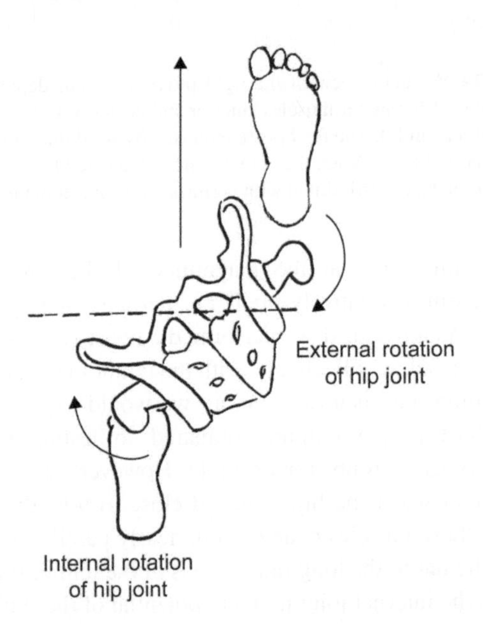


Fig. 10.28 Schematic showing the counterclockwise rotation of the pelvis about an inferior-superior axis related to the external rotation of the right hip and internal rotation of the left hip. This movement is apparent during right foot contact and loading response and left terminal stance of the gait cycle

right hip is externally rotated, or (2) the pelvis is protracted (or internally rotated as described above). On the contralateral side, just before toe-off, the hip could be described as internally rotated. As the individual advances forward, i.e., pushes off the left limb, it is hypothesized that the right hip internal rotators, i.e., anterior fibers of the gluteus medius and minimus, adductor longus, and tensor fascia latae, are activated to induce protraction, i.e., internal rotation, of the left hemipelvis, as the hip and knee flex during pre-swing, into the initial- and mid-swing subphases (Fig. 10.29). In reality, both the femur and pelvis are rotating simultaneously.

In three-dimensional instrumented gait analysis (see Appendix J), the pelvis and femur are modeled as separate segments. Typically, pelvic kinematics are determined by measuring the pose of the pelvis relative to the laboratory coordinate system, whereas hip joint angles are determined by measuring the pose of the femur relative to that of the pelvis. Therefore, we report the kinematics of both the pelvis and hip simultaneously (Fig. 10.30). During walking there

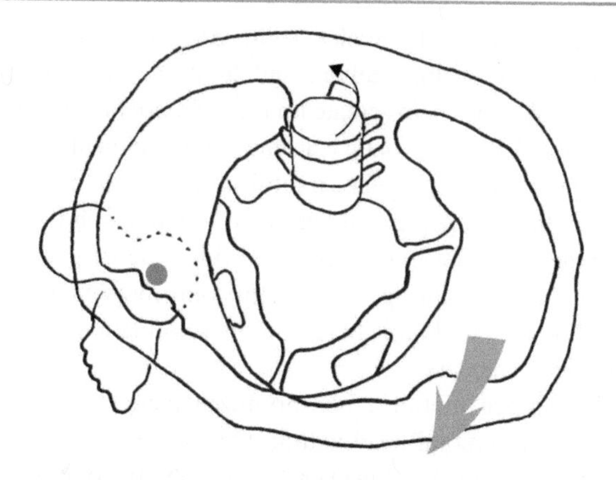

**Fig. 10.29** Schematic showing protraction of the left side of the pelvis during the left swing phase of walking. Note: from a cranial perspective, clockwise rotation of the pelvis (large green arrow) and counterclockwise rotation of the lumbar spine (small black arrow)

appears to be a pattern of pelvic protraction/hip internal rotation and pelvic retraction/hip external rotation that is cyclic.

### 10.4.4 Function: Osteo- and Arthrokinematics

One might argue that examining the kinematics of a joint should precede the study of the kinetics, i.e., muscle structure and function. However, in problem-based learning we find that the sequence by which we order our study of specific topics is not as important as finding the relevant information; thus, we tend to examine the information in a sequence that follows our curiosity or pursuing a particular question.

A succinct illustration of hip rotations in the three cardinal planes is shown in Fig. 10.31. As defined earlier for joints in the upper extremity, we assign similar tags to motion at the hip joint: sagittal plane motion includes flexion/extension; frontal plane motion includes abduction/adduction; horizontal (or transverse) plane motion includes internal/external rotation. Osteokinematically, all motions occur about an axis fixed in the geometric center of the femoral head: (1) for flexion/extension the axis runs medio-laterally; (2) for abduction/adduction the axis runs antero-posteriorly; and (3) for internal/external rotation the axis runs longitudinally, i.e., supero-inferiorly. Note: because of the angle of inclination and anterior bowing of the femoral shaft, the longitudinal axis lies outside of the femur, which has implications for the actions of some of the hip muscles.

Using the anatomical position of the lower limb as our reference, the following ranges of motion (in degrees) for a normal healthy adult are:

| Movement | Normal ROM |
|---|---|
| Flexion | 120–135 |
| Extension | 10–20 |
| Abduction | 30–50 |
| Adduction | 15–25 |
| Internal rotation | 30–45 |
| External rotation | 30–45 |

Note: *ROM* range of motion

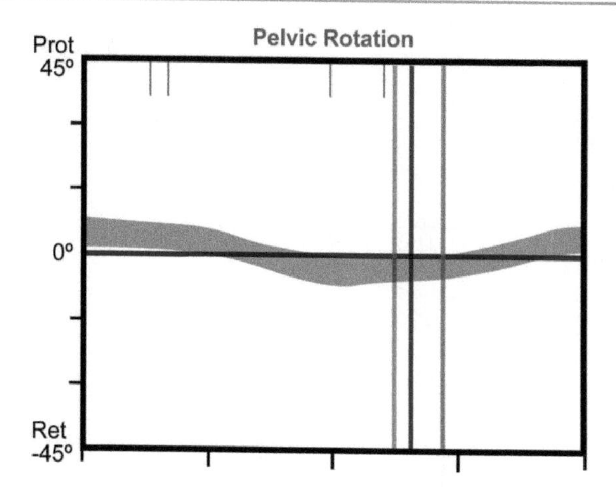


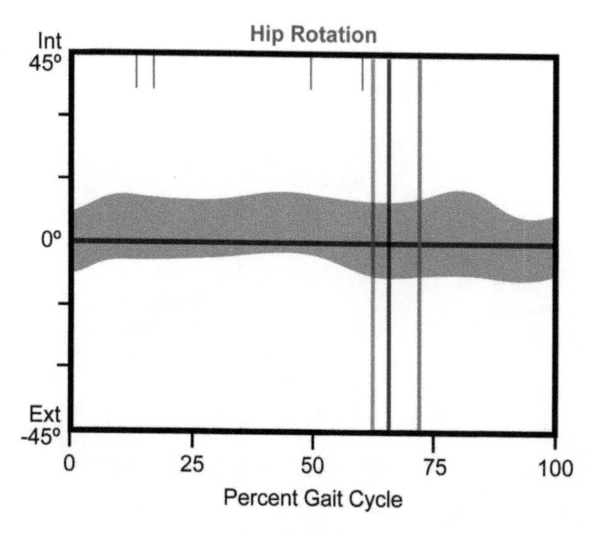


**Fig. 10.30** Kinematic plots during walking of the transverse plane rotation of the pelvic segment and hip joint. Notice that the pelvis cycles from protraction (internal rotation) to retraction (external rotation) and back to protraction. The hip joint rotation range of motion is small and seems to fluctuate from external to internal back to external from first initial contact to second initial contact. Note: the horizontal axis defines the gait cycle from 0% (far left) to 100% (far right). The vertical axis indicates the magnitude of joint motion in degrees, with a horizontal line representing 0°, which divides the motion descriptors, e.g., hip flexion has positive values and hip extension has negative values. The vertical line at about 60% of the GC represents toe-off; with stance phase to the left and swing phase to the right of the toe-off line. The dashed red line = right limb and the solid blue line = left limb. The gray band represents ±1 standard deviation from the control group mean. Top figure: Pro = protraction; Ret = retraction; bottom figure Int = internal rotation; Ext = external rotation. (Control data used with permission from the Motion Analysis Center, Shriners Children's, Chicago, IL)

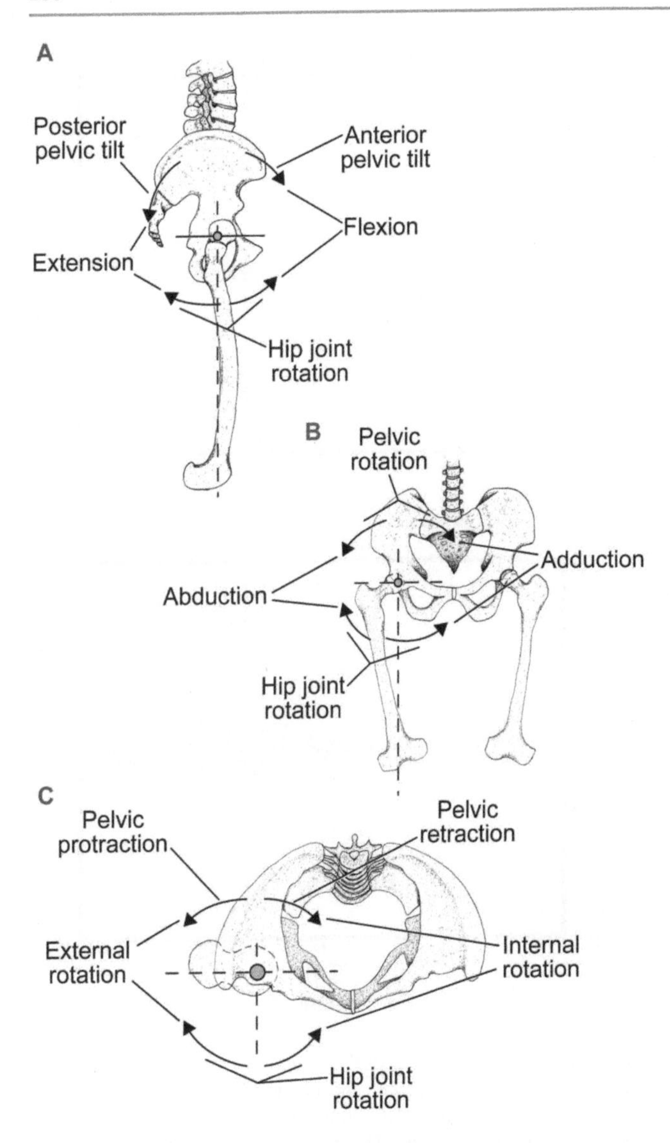


**Fig. 10.31** Illustrations of the osteokinematic movements of hip rotations in the (**A**) sagittal, (**B**) frontal, and (**C**) horizontal planes

We need to note that because of the bi-articular nature of several hip joint muscles, e.g., iliopsoas, rectus femoris, etc., and dynamic coupling, there are joint movements within and outside the pelvic ring during hip rotation, particularly as the hip reaches its end range in any one of the planes of movement. We have highlighted some of these previously, but let's now complete the list: (1) with an anterior tilt of the pelvis, the lumbar lordosis is increased; (2) with a posterior tilt of the pelvis, the lumbar lordosis is decreased; (3) with pelvic abduction/adduction, a lumbar convexity is induced toward the pelvis on the low side; (4) with internal rotation (protraction or right rotation of the pelvis about a fixed right femur), the lumbar spine rotates contralaterally, i.e., to the left; and (5) with external rotation (retraction or left rotation of the pelvis about a fixed right femur), the lumbar spine rotates contralaterally, i.e., to the right. Although our list refers to what the pelvis and lumbar spine are doing (not the hip per se), we encourage the reader to imagine the complex relationship between hip, pelvic, and lumbar spine rotations. Drawing sketches helps create images of these movements in your mind that fosters deep learning and retainment. We also leave it to the reader to make a sub-list of the movements that occur at the zygapophyseal and interbody joints in the lumbar spine, as well as the kinematics within the intervertebral disc that occurs during various hip joint movements. You should be able to do this based on the information provided in previous chapters.

The surface motions, i.e., arthrokinematic translations, between the femoral head and acetabular socket follow the convex/concave rule. However, because the femoral head is snugly contained within the bony and labrum-lined acetabulum, the magnitude of those translations is very small (~1 to 2 mm). Remember, however small the joint translations may be, they are still part of a system that grants, and requires, six degrees of freedom, all of which is important to ensure normal pain-free motion. In fact, one of the hypotheses for the etiology of secondary hip OA is related to loss of normal joint surface motion. Let's briefly review this hypothesis: with a loss, or change, of normal surface motion, the weight-bearing area of contact between the femoral head and acetabulum is altered to the point where the large compression and shear stresses may not be distributed optimally. Over time, the maldistribution of joint forces initiates a strain injury to hyaline cartilage, followed by a cascade of events, i.e., cycles of microinjury, from which the joint structures cannot recover and/or remodel.

Our activities of daily living involve all varieties of movements, including locomotion. We have provided a description of hip osteokinematics; now let us describe hip joint arthrokinematics. Recall that in the case of the hip motions, the convex surface of the head of the femur glides relative to the concave surface of the acetabulum, assuming: an anatomical position of the hip joint:

| Movement/roll | Direction of glide |
|---|---|
| Flexion (antero-superior roll) | Spin with posterior glide |
| Extension (postero-inferior roll) | Spin with anterior glide (see Fig. 10.32) |
| Abduction (supero-lateral roll) | Infero-medial glide |
| Adduction (infero-medial roll) | Supero-lateral glide |
| Internal rotation (anterior roll) | Posterior glide |
| External rotation (posterior roll) | Anterior glide |

Let's see how our understanding of one application of hip arthrokinematics is manifested during walking by looking at just two key subphases of gait. Immediately following initial contact, as the reference limb (right in this example) is being loaded during the first double support, the hip abductor muscles are activated ipsilaterally (on the right side) in order to control the pelvic drop on the left side (Fig. 10.33). Normally the pelvis adducts relative to the femur approximately 3° to

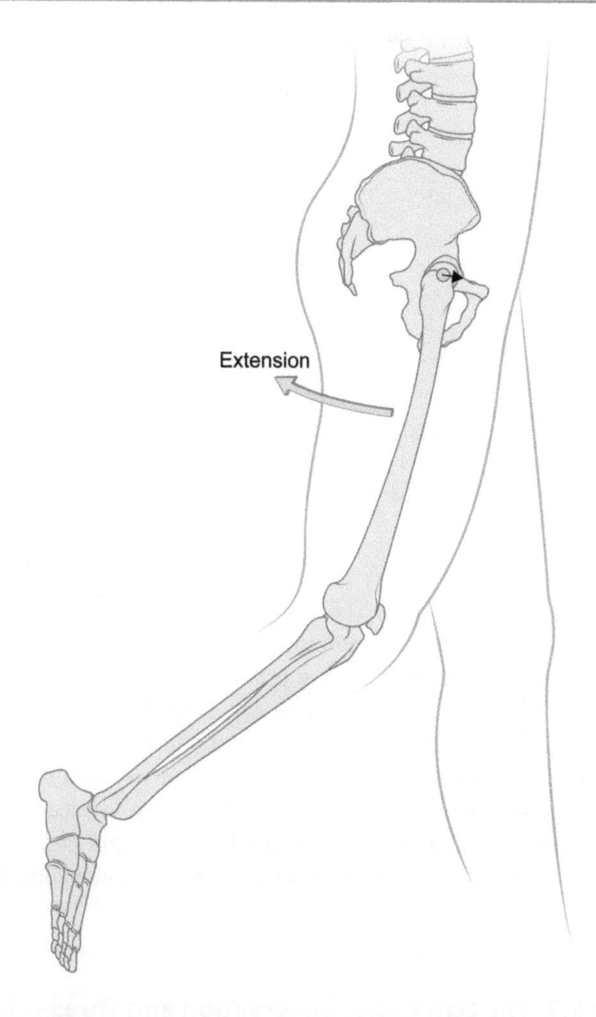


Fig. 10.32 Schematic illustrating osteo- and arthrokinematics during standing hip extension. Note femoral extension, i.e., posterior movement (or roll) of the femur, and anterior glide at the joint surface

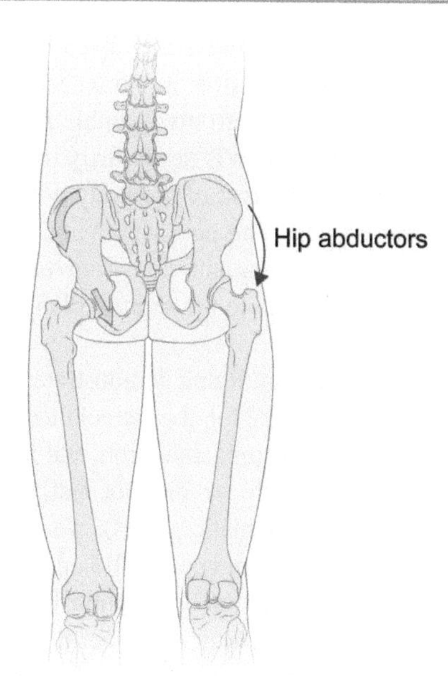


Fig. 10.33 Schematic of a posterior view of mild right hip adduction induced by a mild drop of the pelvis on the left side during loading response of the gait cycle. Note as the pelvis drops down on the left side (large arrow), there is a concomitant inferior glide (small arrow) of the pelvic acetabulum

5°. Since osteokinematics entails inferior movement of the pelvis, there is also an inferior glide of the acetabulum relative to the femoral head.

Our second example is drawn from the terminal stance subphase of gait (Fig. 10.34). At this point in the gait cycle, the individual is nearing the end of the single support period and just beginning the second double support period. Based on observation, the hip joint of the reference limb (right side) is fully extended approximately 20°. In reality (i.e., using instrumented gait measures), this appearance is a combination of approximately 10° of hip extension and 10° pelvic retraction (external rotation). We see then that the pelvis and femur are moving simultaneously; that is, with hip extension, the femur rotates (i.e., rolls) posteriorly which induces an anterior glide of the femoral head on the acetabulum, while the pelvis is rotating (or rolling) anteriorly. This combination of osteokinematic movements is associated with anterior glides of both the femoral head and

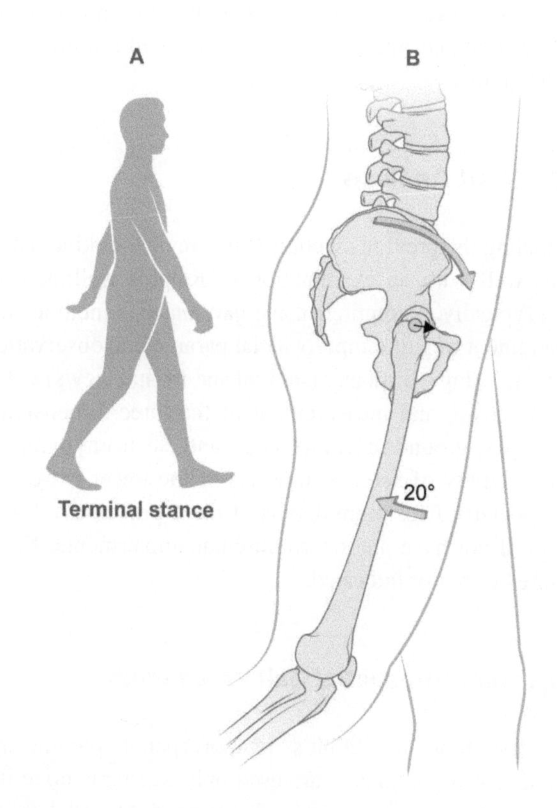


Fig. 10.34 (A) Terminal stance of gait with right hip extended and (B) schematic illustrating the relative movement of the pelvis and femur during this hip extension

acetabulum. As a point of interest, the hip extensors are not responsible for this hip extension, as evidenced by the electrical silence found with electromyographic recordings. In terminal stance, hip extension is necessarily induced by the forward momentum of the rest of the body with the stance limb fixed to the ground. The passive tension of the psoas major and iliofemoral ligament constrain excessive hip extension, thus contributing to the sagittal plane stability of the stance limb.

On physical examination, Ralph demonstrated limitations in passive range of motion for the osteokinematic movements of hip flexion, extension, abduction, and internal rotation (Table 10.1). Although we did not test for the joint surface motions (arthrokinematic glides), we assumed that they would be limited in all directions. It is possible that he developed a restricted range of hip motion beginning at a young age, or as he said, "I always had poor flexibility in my legs," which could have precipitated early degenerative processes (as we suggest earlier). However, we hypothesize that these motion losses were related to the changes in his joint, periarticular structures, and tight muscles, e.g., limited hip extension due to tightness in the anterior joint capsule, iliopsoas, and rectus femoris, associated with the hip OA that was diagnosed 14 years earlier. His loss of hip flexion and abduction would not likely create primary or secondary gait deviations. However, we hypothesize that the limitations of hip extension and internal rotation induce mild to moderate gait deviations.

## 10.5 Gait Analysis

Concluding the previous section with a review of kinematics segues well with an examination of Ralph's walking patterns. Typically, a comprehensive gait analysis includes the measurement of gait temporospatial parameters, observation of the gait using a bi-planar (sagittal and frontal views) video camera set up, and measurement of the three-dimensional joint angles, ground reaction forces, and electromyographic (EMG) activity of selected muscles of the lower extremity (see Appendix J for more details of these processes). Since Ralph did not have any neuromuscular impairments, EMG measures were not indicated.

### 10.5.1 Temporospatial Gait Parameters

Let's first look at Ralph's temporospatial parameters. Generally, gait parameters are used only to characterize the extent of symmetry/asymmetry between right and left, as well as for differences between the pathological and normal gait for several key variables. Examination of temporospatial parameters also allows us to develop an overall impression of the severity of the gait. The gait parameters may allow the clinical scientist to generate broad hypotheses regarding an abnormal gait, but they cannot provide insights that would help in the differentiation of primary, secondary, and compensatory gait deviations. The most important indicators we typically examine include walking velocity, step and stride length, cadence, and step width. In Ralph's case, we see that, although he had a small difference between right and left step length, all of his gait parameters were within the normal ranges (Table 10.3).

**Table 10.3** Temporospatial parameters

| *Gait parameter* | *Left* | *Right* | *Normal* |
|---|---|---|---|
| Stance (%gc) | 60.6 ± 1.0 | 61.1 ± 1.1 | 59.5 ± 1.5 |
| 1st double support (%gc) | 8.2 ± 1.1 | 11.2 ± 2.3 | 9.5 ± 1.5 |
| Single support (%gc) | 39.1 ± 1.8 | 39.5 ± 0.8 | 40.5 ± 4.5 |
| Step length (cm) | 66.0 ± 1.0 | 71.6 ± 2.5 | 70.5 |
| Stride length (cm) | 138 ± 3.0 | 138 ± 3.0 | 98% age normal |
| *Gait parameter* | | | |
| Step width (cm) | 4.3 ± 1.1 | | 7.5 ± 2.5 |
| Cadence (steps/min) | 110 ± 1.0 | | 97% age normal |
| Walking velocity (cm/s) | 126 ± 2.6 | | 92% age normal |

Note: Data are presented as mean ± SD; *% gc* percentage gait cycle; *cm* centimeters

### 10.5.2 The Gait Cycle; Observation and Three-Dimensional Gait Analysis and Problem-Solving Approach

Before we get into the details of Ralph's data, we need to review a few basic concepts related to the normal gait. The analysis of gait can become confusing very quickly without a standard approach to examination and evaluation of the many variables based on a broad framework of normal gait. Therefore, let's take a step back and construct a broader view of bipedal locomotion by defining and describing the prerequisites for a normal gait. According to Perry and Gage, several essential functions need to be accomplished while walking: (1) stability during stance, (2) foot clearance in swing, (3) pre-positioning of the foot in swing, (4) adequate step length, (5) shock absorption, (6) propulsion or forward progression of the body, and (7) minimization of energy cost. Briefly, upright stability is maintained by proper alignment of the limbs, ligamentous integrity, and appropriate and distinct motor patterns. Moving the body forward is accomplished by a series of events that enable each limb segment to continually move forward. This begins with (1) initiation of the first step from a bipedal stance; (2) a shift in the center of pressure; (3) relaxation of the soleus muscles bilaterally and

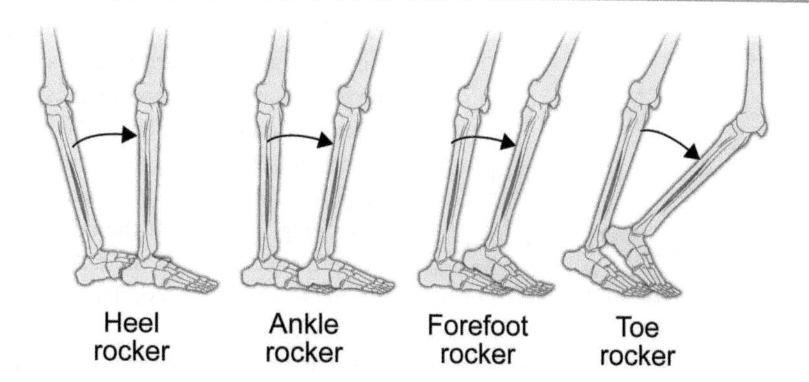


**Fig. 10.35** The four foot rockers that have been identified as part of a typical gait cycle. (Modified with permission from Richie (2021))

activation of the tibialis anterior, for example, which may assist in pulling the shank forward; (4) eventually leading to a toe-off and lifting of the limb into swing; and (5) followed by contacting the ground. Progression of the body over the stance limb depends on the stability of the trunk, pelvis, and hip; knee flexion; and actions at the supporting foot/ankle.

Assuming normal neuromuscular control at the hip and ankle during the stance phase, four rockers have been identified that appear to make important contributions to the progression of the body: (1) the heel rocker: at initial contact, a point on the heel acts as a fulcrum to allow controlled plantarflexion that puts the foot in a plantigrade position (foot flat); (2) ankle rocker: the ankle joint center becomes the axis about which the shank passively moves forward over the fixed foot; this is controlled by eccentric action, and elasticity, of the gastro-soleus complex; (3) forefoot rocker: as the body center of pressure reaches the metatarsal heads, the heel rises and the rounded contour of the metatarsals serves as the rocker, assisting in passive acceleration of the body; and (4) toe rocker: at pre-swing the medial forefoot and great toe serve as the base for final push-off to maintain or foster forward progression (Fig. 10.35).

Propulsion is aided by a knee flexion in pre-swing, hip flexion of the swing limb, and a controlled forward "fall" of body weight. Normal walking velocity, however, also uses other propulsion forces generated by the timely action of the hip and knee extensors and ankle plantar flexors. Shock absorption, i.e., mitigation of ground forces, occurs during the loading response and the period of first double support, managed by ankle plantarflexion (eccentric action by the anterior tibial muscles), knee flexion (eccentric action by the quadriceps), and hip adduction (eccentric action by the hip abductors). Minimizing energy cost has to do with the efficiency with which the work of walking can be accomplished. Work is the scalar product of the force vector and the displacement vector over which the force acts. When the force and displacement are aligned, then work is simply the product of the two. With walking, then, we are concerned with the duration of the muscle actions and their control during the gait cycle. Minimizing energy costs during the walking cycle is generally achieved through control of the body's center of mass via selective motor (or muscular) control. Research has suggested several small mechanical changes of the pelvis, and hip, knee, and ankle joints during the gait cycle that minimize the rise and fall, as well as the mediolateral movement, of the body's center of mass. Selective muscle control implies the intermittent activity of key muscles, acting either concentrically or eccentrically, during different phases of the gait cycle, as well as the utilization of the passive elastic properties of the muscle-tendon actuators and ligaments, packaged as locomotion synergies controlled by the central nervous system, i.e., the motor cortex and spinal cord (location of central pattern generators).

Over the years a variety of terms have been used to describe the gait cycle. We have used the system developed by the Rancho Los Amigos Gait Analysis Committee (Ranchos). Ranchos divides the gait cycle (GC) or stride into functional divisions (Figs. 10.23 and 10.36). A stride, the functional term for the gait cycle, is defined as the interval between two sequential initial contacts with the same foot, e.g., right heel contact to right heel contact again. The stride is divided into periods, stance and swing, which are delineated by foot contact. The three functional tasks, weight acceptance, single limb support, and swing limb advancement, are further divided into phases, which more directly identify the functional significance of different joint movements and body postures. The phases of gait also provide a means for examining the dynamic coupling of the individual joints and body segments into the overall patterns of total body and limb function. What we do going forward is provide important details about each functional task, and the corresponding phases under normal conditions. Knowing (or referring to) these details is paramount to conducting an accurate and insightful gait analysis.

The following discussion applies to the normal, healthy individual. The functional task of weight acceptance (WA) is one of the most challenging because three functional demands must be managed simultaneously: (1) shock

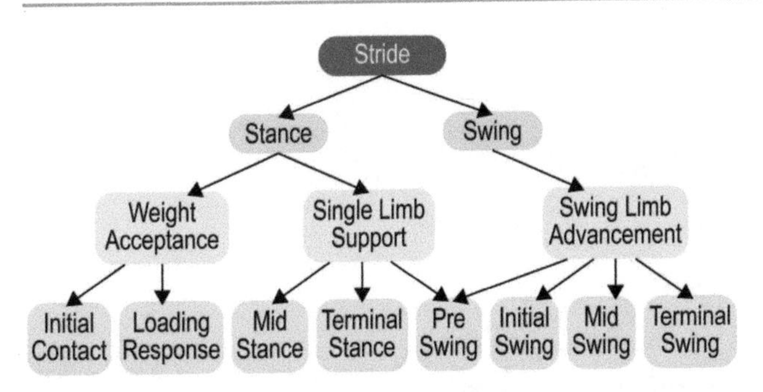


**Fig. 10.36** Divisions and phases of the gait (stride) cycle, as described by Ranchos Los Amigos Gait Analysis Committee. Note that pre-swing is transitional and is included under both the single limb support and swing limb advancement functional tasks. (Perry and Burnfield 2010)

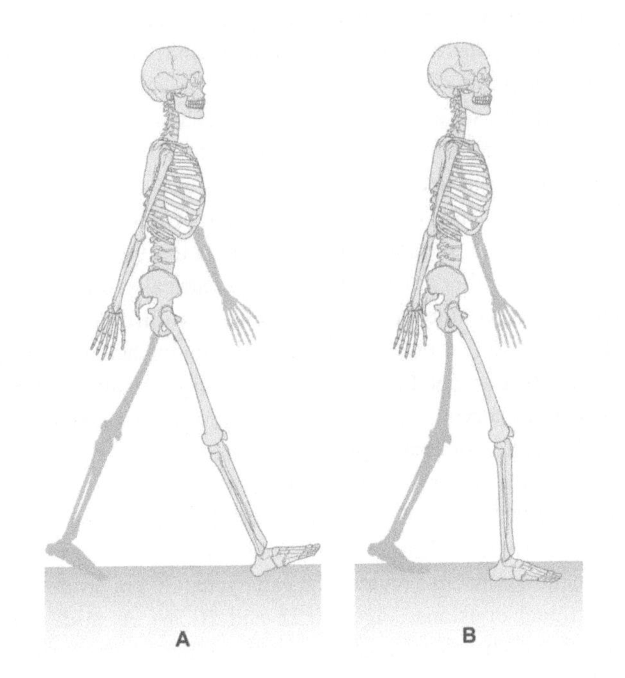


**Fig. 10.37** (**A**) Initial contact and (**B**) loading response phases that comprise weight acceptance. Note: the right limb is the reference limb and the other (clear) limb is at the beginning of pre-swing

absorption or control of large ground forces, (2) initial limb stability, and (3) maintaining forward progression. Two gait phases are involved: initial contact (IC) and loading response (LR) (Fig. 10.37). Initial contact spans the interval from 0% to 2% of the gait cycle (GC) and includes the instant the foot contacts the floor, and the body's immediate reaction to the transfer of body weight to the stance limb. Initial contact sees the hip flexed, knee extended, and ankle dorsiflexed (to neutral), with floor contact made by the heel; the key muscles in this phase include the hip extensors (gluteus maximus), quadriceps, and pre-tibial muscles, e.g., tibialis anterior. The objectives (or critical events) of this phase include initiation with the heel rocker and impact deceleration. The critical events of the loading response include controlled knee flexion, limb and postural stability, and preservation of forward progression, which are managed by the gluteus maximus and medius, quadriceps, and pre-tibial muscles. Loading response (2–12% of the GC) immediately follows foot contact, is the second phase in the initial double support period, and continues until the opposite limb is lifted into swing. During LR the body weight is transferred onto the reference limb, as the ankle plantarflexes (via the heel rocker) to achieve a foot flat position, while the knee simultaneously flexes.

It has been claimed that the quadriceps' eccentric action during LR "absorbs the shock" of the ground reaction force. During this time period, the quadriceps action may dissipate energy as it acts eccentrically and, in some way, dampens the ground force by (1) minimizing the impact velocity of the foot at IC and (2) softening deceleration of the body's COM by modest amounts of flexion. As we have repeatedly shown with the solutions to our statics problems, muscles that cross joints (whether their actions are concentric or eccentric) make the largest contributions to joint bone-on-bone (or joint reaction) forces. Thus, paradoxically, the eccentric action of the quadriceps (and be mindful that the hamstring muscles are working eccentrically to control hip flexion during the same time period) actually induces large knee joint stresses, which approximates, or exceeds, 2.5 times body weight. It is the responsibility of the articular cartilage and menisci, at the knee joint, to serve as a spring-and-damper system to mediate, i.e., "absorb shock for," these joint forces.

Single limb support (SLS) for the stance foot is initiated when the contralateral foot is lifted to swing forward. The tasks of SLS are complete once the contralateral foot contacts the floor again. During the interval of SLS one limb has total responsibility for supporting the body weight in the sagittal and frontal planes, while forward progression continues. This functional task is comprised of the midstance and terminal stance phases, which are differentiated primarily by their mechanisms for fostering forward progression and controlling limb and trunk stability (Fig. 10.38). Midstance (12–31% of the GC) is the first half of the SLS interval and begins once the other foot is lifted, continuing until the bodyweight is aligned over the forefoot of the reference limb. Stance limb advancement over the fixed foot is achieved as the ankle dorsiflexes (i.e., ankle or 2nd rocker), while the hip and knee extend. Key muscles that are active during midstance include the gluteus medius and ankle plantar flexors (i.e., acting eccentrically to control tibial advancement). The terminal stance (TS) interval runs from 31% to 50% of the GC, completing the second half of SLS. Terminal stance begins with heel rise, with the reference limb advancing over the forefoot rocker while the hip and knee achieve full extension, putting the limb in a trailing position. The posterior tibial muscles are active and control tibial advancement, but muscles that cross the hip and knee remain relatively electrically silent,

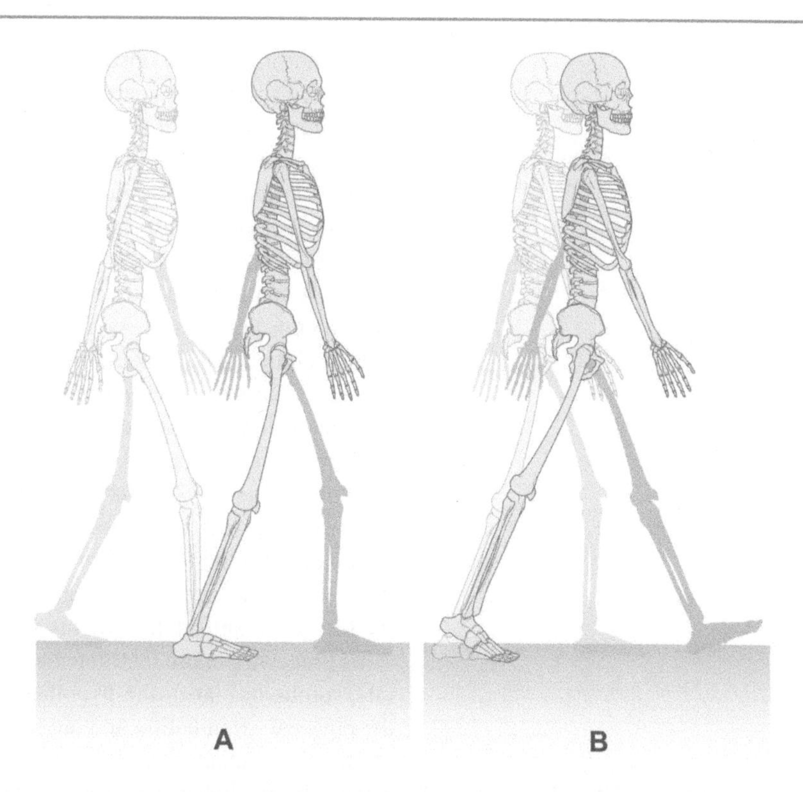


**Fig. 10.38** Single limb support includes (**A**) midstance where the opposite limb is advancing through the mid-swing phase and (**B**) terminal stance where the contralateral limb is completing terminal swing. Note: the right limb is the reference limb

save for the rectus femoris controlling hip extension; note, the EMG evidence indicates that there is variability in the activation of the rectus femoris during the period among individuals. Terminal stance ends when the other foot contacts the floor (period of second double support).

Swing limb advancement begins with preparatory posturing of the limbs and trunk in late stance phase and is comprised of four phases: pre-swing, initial swing, mid-swing, and terminal swing (Fig. 10.39).

Pre-swing (50–62% of the GC) is the second period of double stance and the final phase of the stance phase. It begins with initial contact of the opposite limb and ends with ipsilateral toe-off (secondary to concentric action of the plantar flexors); thus, there is a transition to greater weight-bearing on the contralateral limb as the reference limb is primed to accelerate the body forward and position the limb for swing. The critical events in pre-swing are accomplished as the result of ankle plantarflexion, rapid knee flexion (from 0° to 40°; it is notable that the hamstrings are electrically silent), and hip flexion. From 62% to 75% of the GC, initial swing is comprised of increased hip (secondary to iliopsoas activation) and knee flexion (which is not associated with hamstring activation) to assist in limb advancement and foot clearance, respectively; pretibial muscles are activated, but ankle dorsiflexion to neutral is not yet reached. Mid-swing spans the GC interval from 75% to 87%. The critical events of limb advancement and foot clearance during mid-swing are met as the swing limb is advanced past the center of mass as a result of hip flexion (secondary to activation of the hip flexors) and the ankle dorsiflexion (due to activation of the pretibial muscles) to neutral, respectively. Gravity appears to extend the knee passively. Terminal swing (87–100% of the GC) sees reduced hip flexion (hamstrings), maintenance of a neutral ankle (pretibial muscles), and full knee extension (hamstrings to control knee extension). Knee extension (quadriceps in anticipation of foot contact) completes limb advancement, while the neutral ankle prepares the limb for stance. With that brief review complete, we next outline the steps to the gait problem-solving approach that was used for the analysis of Ralph's gait.

Prior to accessing and examining the details of Ralph's three-dimensional kinematic and kinetic data, we use a three-step problem-solving algorithm combining data from the physical examination and observation of the gait. *Step 1.* Based on Ralph's physical examination, we identified several impairments that we hypothesized could lead to gait impairments (Table 10.1). These included limitations in hip range of flexion, extension, abduction, and internal rotation, mild weakness of the hip flexors and extensors (hamstrings), as well as hip abductors. *Step 2.* With what we know about the critical events of the gait cycle, we hypothesized that the subphases of mid- and terminal stance and pre-swing may be affected secondary to the limitation in hip extension and weak hip abductors; as well, the terminal swing may be

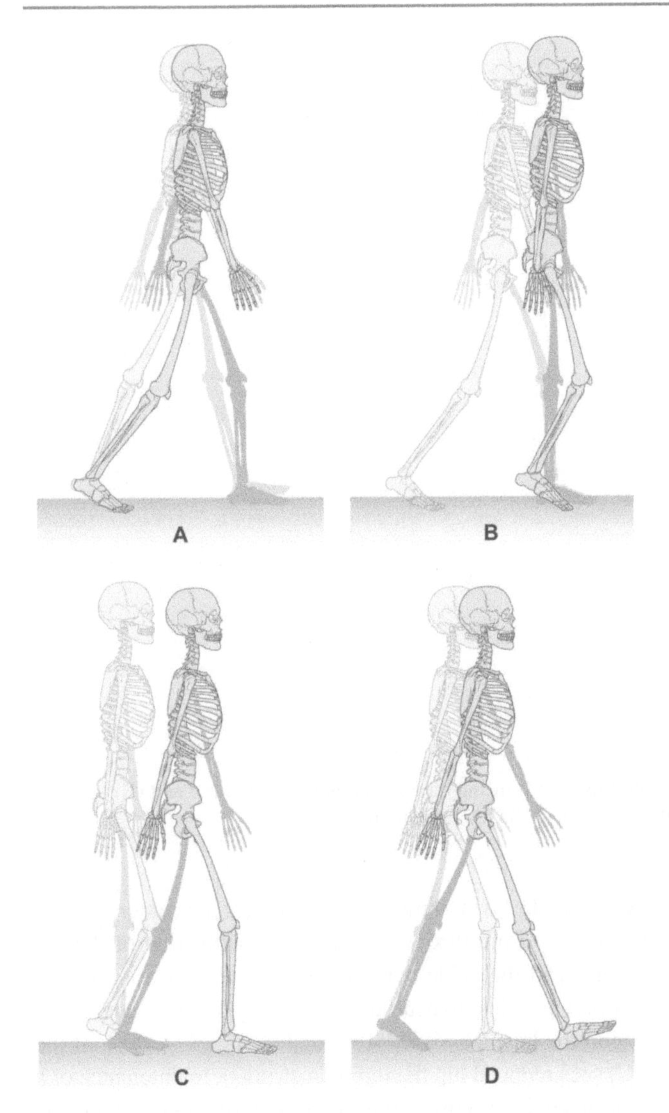


**Fig. 10.39** Swing limb advancement comprised of (**A**) pre-swing, where the opposite limb is in loading response; (**B**) initial swing, with the opposite limb in early mid-stance; (**C**) mid-swing and the opposite limb in later mid-stance; and (**D**) terminal swing with the opposite limb in terminal stance. Note: the right limb is the reference limb

affected due to weak hamstring muscles. *Step 3*. The gait deviations we identified using video observations were fairly mild but included (1) in weight acceptance and single support, excessive lateral trunk lean secondary to hip pain and hip abductor weakness; (2) in weight acceptance, excessive right knee flexion secondary to a 1-cm leg length difference; and (3) in single limb support, inadequate hip extension in terminal stance due to tightness hip flexors. Based on Ralph's radiographic findings of marked hip OA we expected to see greater gait deviations. Yet, for individuals who remain active, as Ralph did, it is possible that the gait can remain largely functional.

The primary purpose of the observational gait analysis is to get a sense of the major kinematic deviations in the sagittal, frontal, and transverse planes. Although previous research has demonstrated that observational gait analysis is moderately valid and reliable, particularly when making judgments about joint motions in the sagittal plane, instrumented gait measurements continue to be the gold standard for the analysis of abnormal walking. Instrumented gait analysis application to persons with neurological disorders, such as post-stroke or cerebral palsy, is invaluable in that the information it can provide aids not only in providing additional information but also in its ability (in the hands of highly trained clinicians) to differentiate between primary, secondary, and compensatory gait deviations. In other words, for complex gait abnormalities, instrumented gait measures can provide a biomechanical explanation for the deviations, something that physical examination in combination with observational analysis alone cannot do. In Ralph's case, we were not expected to use instrumented gait analysis to help differentiate the orthopedic diagnosis; we were asked to clarify the gait manifestation of his medical diagnosis. At the same time, we can use instrumented gait analysis to corroborate, refute, or clarify the hypotheses we develop based on the physical examination and observational gait findings.

Before we examine and analyze the kinematic data, let's briefly discuss what we mean by primary, secondary, and compensatory gait deviations. Normal human gait uses repetitive reciprocal limb motions in order to advance forward progression while simultaneously maintaining stance stability. This is accomplished by tightly regulated patterns of muscle activation, which generate joint moments and powers. Some of the muscle work is done to accelerate the body forward, and some are used to provide vertical support against the force of gravity. Pathologies, e.g., muscle weakness, nerve palsy, bone deformities, etc., and ensuing impairments, e.g., deficient joint mobility, abnormal motor control, etc., and/or pain interfere with these tightly regulated patterns. As a result, we observe/measure alterations, or gait deviations, in the normal joint kinematics and kinetics of gait. Primary deviations are directly due to the pathology. For example, a hip flexor contracture restricts normal hip extension during terminal stance. Secondary deviations include passive secondary effects that follow as a physical effect to the primary deviation. For example, because a hip flexor contracture limits hip extension, the step length on the side of the contracture is reduced. Another type of secondary deviation, which by some is termed *compensation*, is defined as an active (voluntary) deviation in order to actively offset primary deviations and secondary physical effects (Table 10.4). Related to the primary deviation we provide, a common compensation for a hip flexor contracture and limited hip extension is excessive pelvic retraction on the same side. Increased pelvic retraction can functionally introduce a pseudo-hip extension, effectively lengthening the limb, and normalizing step length. What makes differentiating gait deviations complex is that both secondary and compensatory

**Table 10.4** Common compensatory mechanisms that appear to be independent of primary pathologies

| Biomechanical constraints secondary to the primary pathology | Compensatory mechanisms |
|---|---|
| Hip extensor weakness | Posterior trunk extension |
| Hip abductor weakness | Compensated Trendelenburg (Duchenne limp) |
| Knee extensor weakness<br>Quadriceps avoidance | Hip extensors for knee extensors<br>Center of mass anterior to knee joint by:<br>increased activity of plantarflexion/knee extension couple<br>Hip flexion<br>Anterior pelvic tilt<br>Knee hyperextension |
| Ankle plantar flexor weakness | Eccentric work of hip flexors for progression in stance<br>Hip and knee extensors in stance<br>Hip flexors (pulling) in preswing<br>Hip extensor torque strategy in late stance (loading of flexor tissue)<br>Internal rotation of trunk and pelvis on contralateral side<br>Larger symmetrical hip power generation |
| General leg weakness | Hyperactivity ankle plantar flexors<br>Co-contraction around knee<br>Prolonged activity of contralateral hip abductors (weight acceptance) |
| Reduced foot clearance | Pelvic up tilt on unaffected side<br>Pelvic hike<br>Circumduction, hip abduction, hip external rotation<br>Hip flexion and/or knee flexion<br>Increased plantarflexion on unaffected side (vaulting) |
| Limited hip extension | Lumbar lordosis increased<br>Knee flexion to allow the pelvis to progress forward<br>Increased pelvic retraction on ipsilateral side |

Schmid et al. (2013)

deviations may be coupled with deviations across joints to different segments, e.g., deviation at the hip may produce deviations at the trunk, pelvis, knee, or ankle, as well as across planes, e.g., we saw that limited hip extension secondary to hip flexor contracture induced a compensatory transverse plane rotation of the pelvis. Thus, the reason it is important to be intentional about working through the complexity of these deviations and differentiate gait deviations is because it matters when it comes time to develop a treatment strategy. We want to identify and direct our intervention to primary deviations, and not the secondary deviations. Addressing treatment toward a secondary deviation would not correct the gait abnormalities and thus functional limitations.

Let's now examine Ralph's joint kinematics data. Normally, when we perform the kinematic analysis, we want to examine the graphs of the trunk, pelvis, hip, knee, and ankle/foot that are all on the same page (for the simple reason that gait deviations can cross joints and planes and we want to be intentional about looking for coupled patterns). However, for the purposes of this case, we divide the kinematics into more manageable sections. For this first section let's examine the trunk and pelvis (Fig. 10.40A).

By convention, trunk and pelvic segment kinematics are determined by calculating the respective segment orientation relative to the laboratory coordinate system. In the frontal plane, it is clear that there was excessive trunk lateral lean, i.e., a compensated Trendelenburg or compensatory gait deviation, to both the right and the left sides, which we attributed to hip pain and abductor weakness. We also noticed some sort of coupling between the frontal plane movements of the trunk and pelvis, i.e., during swing the left hemipelvis was up when it should have been down, and during stance, the right hemipelvis was down when it should have been up. We believe that this was also secondary to the hip abductor weakness. Excessive anterior pelvic tilt (an example of a primary gait deviation), with its double-bump pattern, is indicative of hip flexor tightness, particularly of the rectus femoris.

Calculation of joint angles for the hip, knee, and ankle joints is done by determining the orientation of the distal segment relative to the proximal segment, similar to how clinicians measure a joint angle in one Cartesian plane using a goniometer. For example, hip joint angles are determined in three dimensions by the use of Euler angles derived by the position of a segmental coordinate system attached to the thigh relative to a segmental coordinate system attached to the pelvis (see Appendix J for a review of Euler angle calculations). Ralph demonstrated a significant reduction in hip extension during the stance phase (a primary gait deviation), which we attributed to the tightness of the hip anterior capsule and hip flexors. Recall that physical examination revealed a complete lack of hip internal rotation bilaterally. Although the kinematic graphs do not clearly manifest the loss of hip rotation, there appears to be a trend of insufficient hip internal rotation bilaterally during the early stance phase. The significance of these findings is unclear since hip internal rotation is not a major event in a normal gait. The excessive pelvic transverse plane rotation, i.e., protraction and retraction, is a compensation (secondary gait deviation) to increase the step/stride length that was reduced by inadequate hip extension in terminal stance (Fig. 10.40B).

The kinematic data for the knee and ankle joints were, for the most part, unremarkable. We note a reduction in right knee extension (or increased right knee flexion), and likely related increased right ankle dorsiflexion, during stance, which we attribute to a combination of reduced hip extension

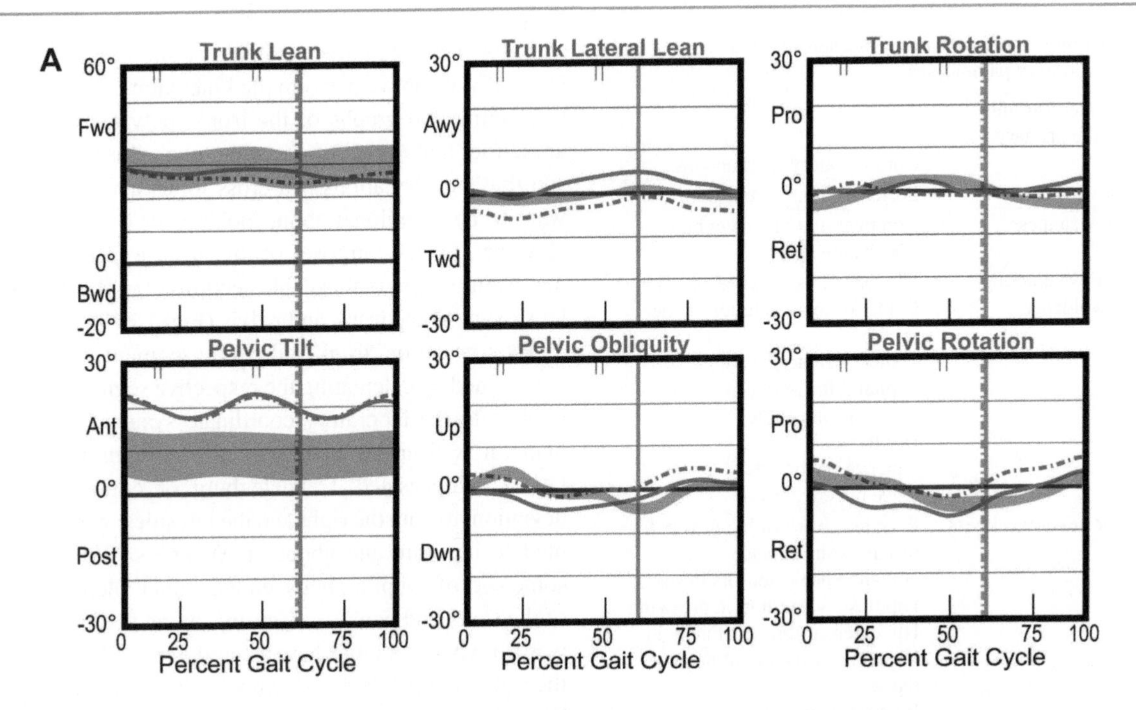


**Fig. 10.40A** Segmental angles of the trunk and pelvis during barefoot walking. Sagittal, frontal, and transverse plane data are organized by columns going from left to right. Data are plotted relative to the percent gait cycle (from 0% far left to 100% far right) along the horizontal axis. The vertical axis indicates the magnitude of joint motion in degrees, with a horizontal line representing 0°, which divides the motion descriptors, e.g., trunk forward lean (Fwd) has positive values and trunk backward lean (Bwd) has negative values. The vertical line at about 60% of the GC represents toe-off; with stance phase to the left and swing phase to the right of the toe-off line. The dashed red line = right limb and the solid blue line = left limb. The gray band represents ± 1 standard deviation from the control group mean. Twd = toward reference limb; Awy = away from reference limb; Pro = protraction; Retr = retraction; Up = reference side of pelvis is up; Dwn = reference side of pelvis is down; Ant = anterior; Post = posterior

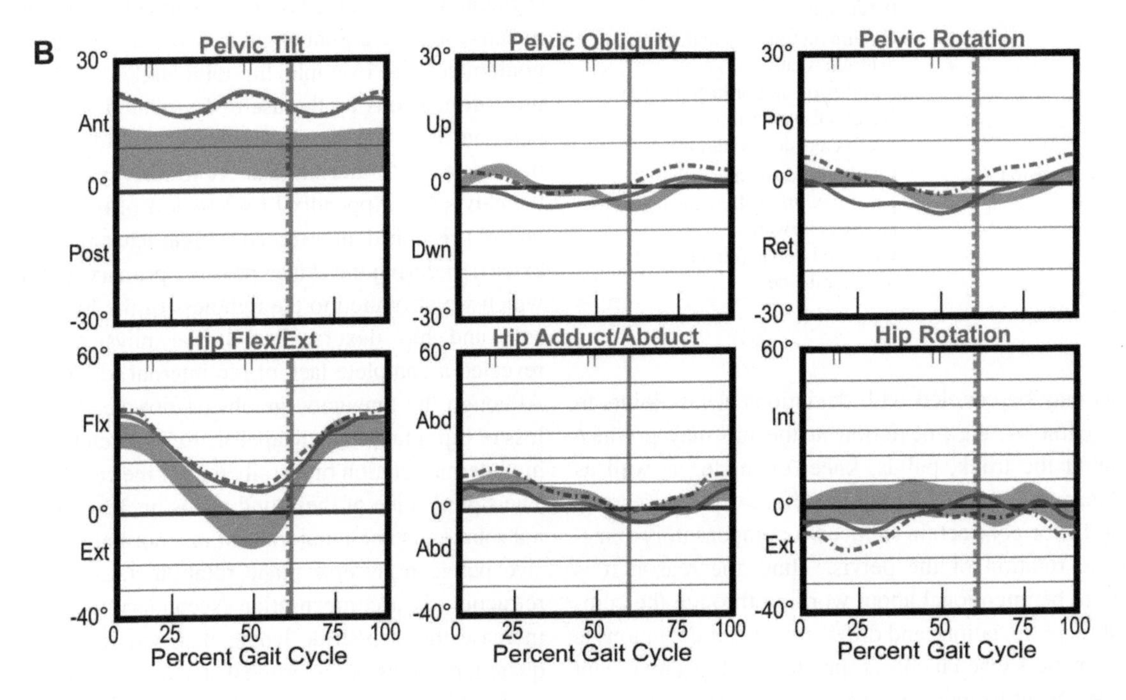


**Fig. 10.40B** Segment and joint angles of the pelvis and hip during barefoot walking. Sagittal, frontal, and transverse plane data are organized by columns going from left to right. Data are plotted relative to the percent gait cycle (from 0% far left to 100% far right) along the horizontal axis. The vertical axis indicates the magnitude of joint motion in degrees, with a horizontal line representing 0°, which divides the motion descriptors, e.g., hip flexion (Flx) has positive values and hip extension (Ext) has negative values. The vertical line at about 60% of the GC represents toe-off; with stance phase to the left and swing phase to the right of the toe-off line. The dashed red line = right limb and the solid blue line = left limb. The gray band represents ± 1 standard deviation from the control group mean. Ant = anterior; Post = posterior; Up = reference side of pelvis is up; Dwn = reference side of pelvis is down; Pro = protraction; Retr = retraction; Add = adduction; Abd = abduction; Int = internal; Ext = external

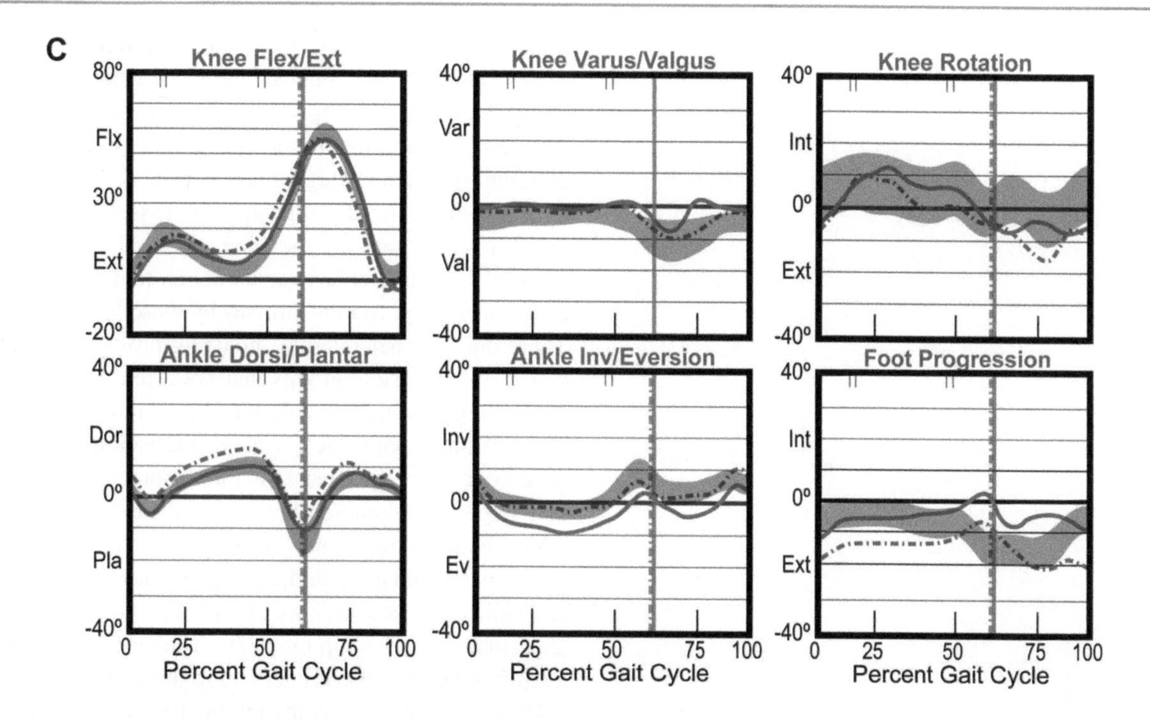


**Fig. 10.40C** Joint angles of the knee and ankle during barefoot walking. Sagittal, frontal, and transverse plane data are organized by columns going from left to right. Data are plotted relative to the percent gait cycle (from 0% far left to 100% far right) along the horizontal axis. The vertical axis indicates the magnitude of joint motion in degrees, with a horizontal line representing 0°, which divides the motion descriptors, e.g., knee flexion (Flx) has positive values and knee extension (Ext) has negative values. The vertical line at about 60% of the GC represents toe-off; with stance phase to the left and swing phase to the right of the toe-off line. The dashed red line = right limb and the solid blue line = left limb. The gray band represents ± 1 standard deviation from the control group mean. Var = varus; Val = valgus; Dor = dorsiflexion; Plan = plantarflexion; Inv = inversion; Ev = eversion; Int = internal; Ext = external

and the leg length inequality (right leg was longer by 1 cm) (Fig. 10.40C).

Let's digress at this point and briefly examine the relationship between leg length inequality and hip OA. Leg length inequalities (LLI) or leg length discrepancies (LLD) are common with a prevalence of approximately 60% for differences of 5 mm or more, yet greater than 99% of the general population with LLI less than 3 cm are classified as mild. Three categories of LLI have been identified based on the magnitude of the discrepancy: mild (differences <3 cm), moderate (differences ≥3 and ≤6 cm), and severe (differences >6 cm). Many hold the view that LLI of less than 2 cm is not clinically significant, while others dispute that claim suggesting that even mild LLI may be deleterious if an individual participates regularly in locomotor activities involving repetitive impulsive loading. Mild LLI has been associated with three orthopedic disorders, stress fractures, low back pain, and osteoarthritis, although there is no consensus in the literature about the significance of these associations. Despite the lack of agreement about the role a LLI may play in the development of hip OA, we should consider how even mild LLI alters the biomechanics at the hip joint. An LLI induces a frontal plane pelvic obliquity with the pelvis higher on the long leg side (Fig. 10.41). This un-leveling alters the distribution of compression and shear stresses at the sacroiliac, lumbar zygapophyseal, and vertebral interbody joints. In addition, an LLI may impose bilaterally unequal stresses within the joints of the hip, knee, and ankle. It is hypothesized that the tilted pelvis shifts the line of action of the center of mass away from the hip joint on the side of the longer limb. As a result, the greater muscle activity necessary to compensate for the shift may increase the magnitude of the femoro-acetabular joint forces (both compressive and shear). Additionally, the lateral pelvic tilt (or pelvic obliquity) likely changes (reduces) the contacting area of the articulating hip joint surfaces. During locomotion lower extremity joint stresses are further increased by muscle activity required to move and control segments, by inertial forces developed by the moving segments, and by the impulsive ground reaction forces that are distributed from the foot to the ankle, knee, and hip. Some have speculated that the shorter leg sustains greater impulsive forces (compared to normal) at foot contact, while others have suggested that the longer limb sustains excessive loads while weight-bearing.

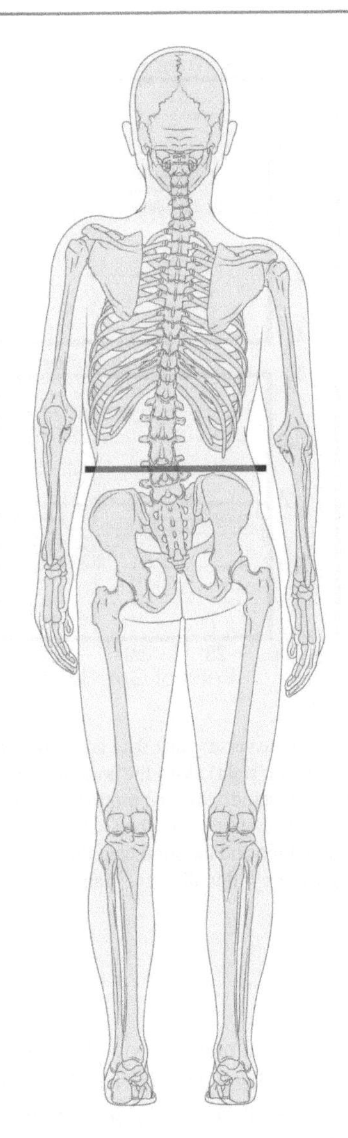

**Fig. 10.41** Posterior view of the compensatory abnormal alignment of the lower extremities, pelvis, and lumbar spine associated with a leg length inequality, i.e., right long leg

Ralph's physical examination revealed a 1 cm LLI, with the right leg longer. This LLI would be classified as mild, so it is likely that, even if it had been discovered in his early years, it would not have been medically addressed. On the other hand, theory suggests that the increased and abnormally distributed large hip joint loads associated with normal daily activities, as well as recreational sports, could have initiated micro-damage to the hip joint cartilage and acetabular labrum eventually contributing to the early onset of Ralph's hip OA.

Instrumented gait analysis is invaluable for the assessment of complex gait impairments (which Ralph did not have) because it can provide information that one cannot observe. Analyzing net internal joint moments and powers that can be determined with the aid of instrumented measurement systems, i.e., optical motion cameras and floor-embedded force platforms, is an invaluable asset. We know that muscles are the prime movers of joints. Moreover, although we acknowledge that a portion of joint forces and moments are induced by ligaments and associated periarticular structures, most of the biomechanical models we use clinically are not sophisticated enough to delineate the precise contribution to joint stresses by these "passive" tissues. Therefore, the net internal moments provide indirect evidence of the muscle groups that are active in producing or controlling joint motions. Reduced joint moments may indicate dysfunctional muscle synergies related to injury and/or weakness, whereas excessive joint moments may suggest that large joint bone-on-bone forces are being generated, which could induce joint degeneration. Based on Ralph's physical examination findings of hip flexor and abductor, and hamstring weakness we hypothesized that we might see abnormalities in his hip moments. The only abnormality we identified was a reduced internal hip flexor moment at terminal stance, bilaterally. Although his hip flexors were mildly weak, we did not attribute this reduced hip flexor moment at terminal stance to this weakness, but to inadequate hip extension related to the anterior hip capsule and hip flexor tightness. We can be confident that this was true because the hip flexors typically do not demonstrate EMG activity during terminal stance, but function as passive restraints to excessive hip extension. Despite the hip abductor weakness, and compensated Trendelenburg cited earlier, Ralph's hip abductor moments were within normal limits (Fig. 10.42).

Net internal joint power is a scalar product of joint angular velocity and net internal joint moment. Like joint moments, joint powers can provide indirect evidence of the actions of muscle groups. In instrumented clinical gait analysis, we are primarily interested in only the sagittal plane net joint powers since these are associated with the most important muscle groups during gait. The sign of the net power, i.e., in its graphical representation, is an indication of what kind of muscle actions are utilized, e.g., concentric versus eccentric. Power generation is associated with concentric muscle actions while power absorption is associated with eccentric muscle actions. Examining joint power along with joint kinematics, the net joint moments, and the EMG activity of key muscles, if that is available, can enhance our confidence in the interpretation of the power graphs. As indicated earlier, we chose not to collect EMG data in Ralph's case because his medical diagnosis did not warrant it. Since power is a function of angular velocity and joint moments, we graph joint angles, moments, and powers together (Fig. 10.43). We do not graph the joint angular velocities, but angular velocity can be inferred from joint angle graphs if one has a bit of a calculus background

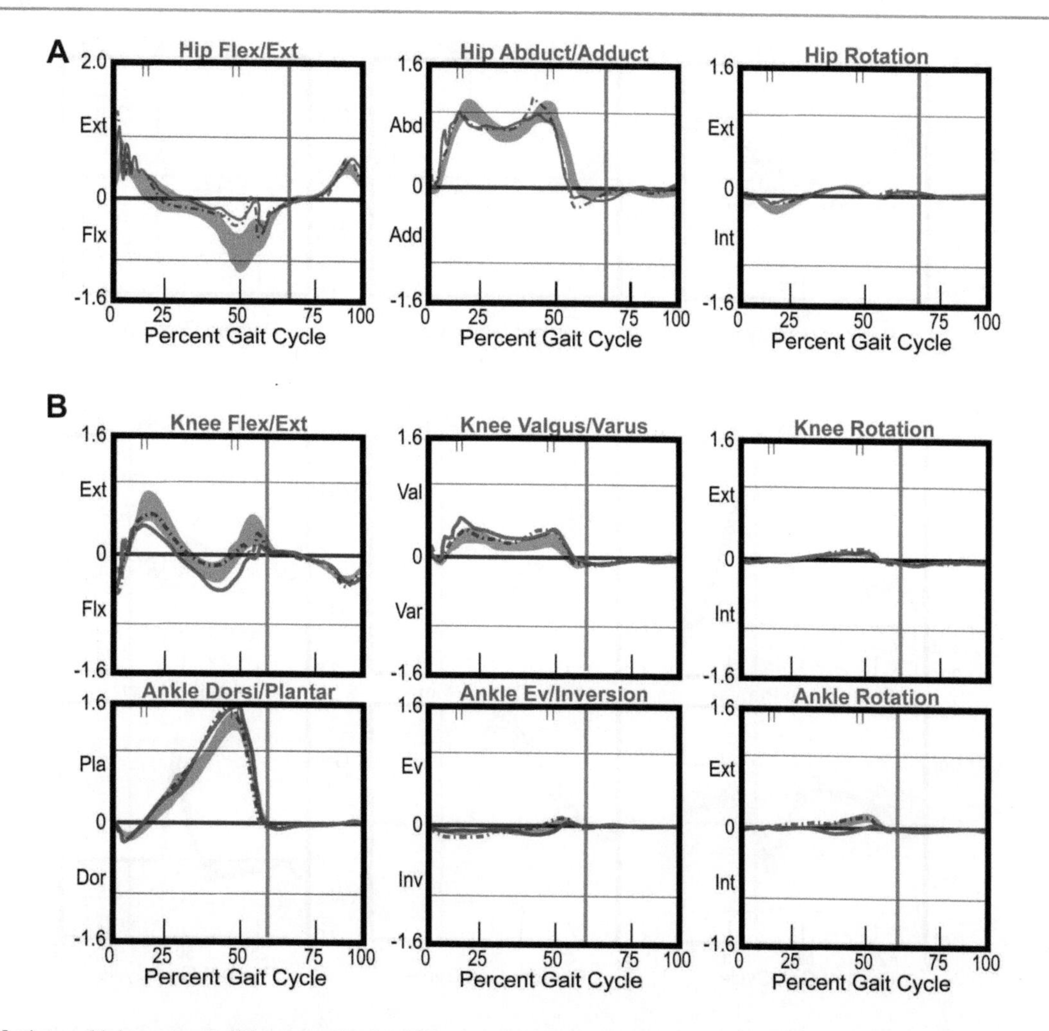


**Fig. 10.42** Net internal joint moments (Nm normalized to body mass) at (**A**) the hip and (**B**) the knee and ankle during barefoot walking. Sagittal, frontal, and transverse plane data are organized by columns going from left to right. Data are plotted relative to the percent gait cycle (from 0% far left to 100% far right) along the horizontal axis. The vertical axis indicates the magnitude of the net joint moment, expressed in Nm/kg (newton meters per kilogram), with a horizontal line representing a zero net moment, which divides the moment descriptors, e.g., hip extensor moment (Ext) is positive and hip flexor moment (Flx) is negative. The vertical line at about 60% of the GC represents toe-off; with stance phase to the left and swing phase to the right of the toe-off line. The red dashed line = right limb and the blue solid line = left limb. The gray band represents ±1 standard deviation from the control group mean. Flx = flexor; Ext = Extensor; Add = adductor; Abd = abductor; Int = internal rotator; Ext = external rotator; Pln = plantar flexor; Dor = dorsiflexor; Var = varus; Val = valgus; Ev = evertor; Inv = invertor

(review Appendix D to see how calculus may be applied to the joint angle graph to estimate the angular velocity). The only abnormalities we saw in Ralph's power data set involving the hip joint: (1) abnormal power generation in loading response and early stance and (2) reduced power generation in pre-swing. The abnormal hip power generation can be explained by the excessive hip flexion, during the same gait interval, which is reduced by concentric action of the hip extensors (note internal hip extensor moment). Power generation that is misplaced, i.e., occurs during an interval of the gait cycle that is not normal, is inefficient, and leads to excessive energy cost. The evidence of reduced hip power generation in pre-swing was expected based on Ralph's inadequate hip extension in terminal stance and reduced net internal flexor moment.

Note: in the determination of internal net joint moments and powers you should have noticed that normalized the magnitude relative to body mass, i.e., Nm/kg (moments) and Watts/kg (power). This is necessary since it allows us to compare individuals.

In summary, the three-dimensional kinematic and kinetic data derived from the instrumented gait study corroborated our preliminary conclusions about Ralph's gait abnormalities based on physical examination and observations; as well, it clarified and added information that provided a more robust characterization of Ralph's gait modified by severe

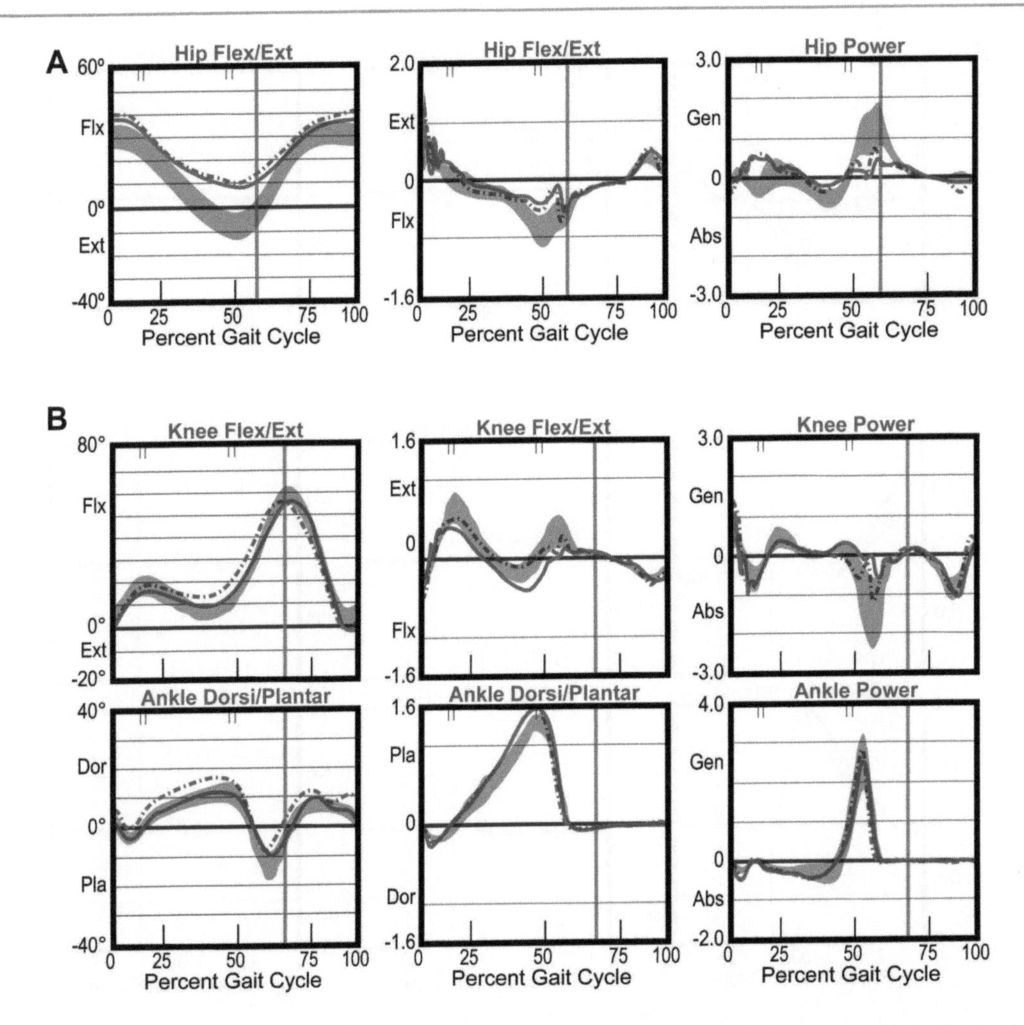


**Fig. 10.43** Sagittal plane joint angles, and net internal joint moments and powers (Watts normalized to body mass) for (**A**) the hip and (**B**) knee and ankle during barefoot walking for the ankle, knee, and hip joints. Data are plotted relative to the percent gait cycle (from 0% far left to 100% far right) along the horizontal axis. We have described the vertical axis for joint angles and moments. For the power curves, the vertical axis indicates the magnitude of the net joint power, expressed in Watts/kg, with a horizontal line representing zero net power, which divides the power descriptors, e.g., ankle power generation (Gen) is positive and ankle power absorption (Abs) is negative. The vertical line at about 60% of the GC represents toe-off; with stance phase to the left and swing phase to the right of the toe-off line. The red dashed line = right limb and the blue solid line = left limb. The gray band represents ±1 standard deviation from the control group mean. Dor = dorsiflexion or dorsiflexor; Pln = plantarflexion or plantar flexor; Flx = flexion or flexor; Ext = extension or extensor; W/kg = watts per kilogram

hip OA. As noted in an earlier conclusion, we were surprised at the mildness of the gait abnormalities given that Ralph appeared to carry severe hip OA for at least 14 years.

## 10.6 Hip Joint Loads: A Static Equilibrium Estimate During Single Limb Stance

Osteoarthritis (OA) is the most common joint disorder in the United States. Symptomatic hip osteoarthritis is not as common as knee or hand OA yet has a prevalence among adults aged 45 years and older at 9.2%, with 27% showing radiographic signs of the disease; prevalence is slightly higher among women. Men have a higher prevalence of hip OA before age 50, whereas women have a higher prevalence thereafter. A hallmark of hip OA is pain along the lateral upper thigh, with referral to the groin and medial knee areas. As illustrated in this case, common impairments associated with hip OA include (1) restricted hip range of motion, particularly hip extension, abduction, and internal rotation, due to joint capsule and hip flexor tightness; (2) hip muscle weakness generally affecting the hip extensors and abductors; and (3) gait impairments highlighted by a compensated Trendelenburg due to hip pain and hip abductor weakness, and reduced step and stride length secondary to restrictions in hip extension at terminal stance.

In the previous static problems that we have solved, we were interested in estimating joint bone-on-bone forces related to physical examination procedures or activities of daily living, as well as estimating the force produced by a key muscle acting across the joint of interest. Note, that when solving two-dimensional statics equilibrium problems, we are limited in the number of unknowns that can be determined, and thus can estimate the force of only one muscle. The internal hip abductor moment produced by the hip abductor muscles, with the gluteus medius being primary, is essential for controlling the frontal plane position of the pelvis during unipedal stance (balancing statically on one leg), or during the single-limb support interval of gait. As we discussed earlier, the combination of hip pain and/or hip abductor weakness commonly associated with hip OA results in the situation, during single limb support, where the superincumbent body weight cannot be controlled as the swing limb is advanced, causing the pelvis to drop on the side opposite the stance limb. We referred to this as a Trendelenburg sign. Therefore, the force produced by the hip abductor muscles is a very important part of assuring a normal gait. Generalizing from the solutions to our previous static problems, we can imagine, then, that the hip abductor muscle force must be quite large in order to balance the weight of the head, arms, and trunk during gait, which also suggests that the hip joint bone-on-bone force would be large. The purpose of this section is to estimate the hip joint bone-on-bone force and the force of the gluteus medius needed to balance the superincumbent body weight in single-limb standing.

Let's first list the assumptions associated with this problem:

1. Static equilibrium simulation, i.e., standing still and balancing on one leg, is a reasonable estimate of what happens during the single-limb support during walking.
2. Rigid body mechanics is assumed so that relevant tissues, e.g., bone, cartilage, etc., do not deform, i.e., strain under external forces.
3. Friction between the femoral head and acetabulum is absent.
   (a) This is likely *not* true given the advanced stage of hip OA.
4. Only one muscle is acting, e.g., gluteus medius (GMed), to counter the external hip adductor moment.
   (a) This is an oversimplification since the gluteus minimus and tensor fascia latae make moderate contributions to the internal abductor moment during walking; likewise, co-action of agonists and antagonists may contribute to internal joint forces.
5. The periarticular tissues, e.g., intrinsic and extrinsic ligaments, do not contribute to internal hip moments.
   (a) This is not true yet the contribution to joint moments from periarticular tissues is small.

The solution begins with listing what we know:

- Body weight, $W_b$ = 179 pounds (lb) = 796 Newtons (N) equals the vertical ground reaction force (GRF).
- Lower extremity weight, $W_{LE} = (0.161)\ W_b = 128.2$ N.
- GMed moment arm with respect to the hip joint center, $d_{GMed} = 0.056$ m (Henderson et al.)
- The angle of the GMed force is $\theta = 70°$ with respect to the horizontal
- Lateral distance from the center of pressure (COP) to the hip joint center, $d_{GRF} = 0.11$ m
- Lateral distance from the lower extremity center of mass (COM) to the hip joint center, $d_{LE} = 0.032$ m.

Note: the estimates for the lower extremity dimensions (p. 102) and for lower extremity weight (p. 211 from Clauser) were taken from LeVeau B (1977) *Williams and Lissner: Biomechanics of Human Motion* (2nd ed). W. B. Saunders Company, Philadelphia.

Figure 10.44A illustrates the static standing situation, where we see that the body is balanced over the right foot. Let's imagine that a force plate positioned under the foot records the two reaction force components ($F_{GRF_x}$ and $F_{GRF_y}$) where $x$ represents the medial-lateral direction with positive to the right and $y$ represents the vertical direction with positive up as shown. Referring to the free body diagram in Fig. 10.44A, we can calculate $F_{GRF_x}$, $F_{GRF_y}$, and the distance $d$ between the line of action of $F_y$ and $W_b$ using static equilibrium:

$$\Sigma F_x = F_{GRF_x} = 0$$

$$\Sigma F_y = F_{GRF_y} - W_b = 0 \rightarrow F_{GRF_y} = W_b = 128.2\,\text{N}$$

$$\Sigma M_{COP} = -W_b d = 0 \rightarrow d = 0$$

Note that in the moment equation, the moment for the weight is negative, because it would induce a clockwise moment about the center of pressure (COP), and counterclockwise is positive. Thus, the vertical component of the ground reaction force is equal to the total body weight $W_b$, and acts along the same line of action of $W_b$, but in the opposite direction.

With that exercise complete, let's find a solution to the main problem. We start with a free body diagram of just the right lower extremity as shown in Fig. 10.44B. All relevant external forces acting on the system along with the relevant dimension to be able to calculate the moments of the forces are shown.

We solve the problem with Ralph standing and balancing on one leg with the hip and knee extended and the ankle dorsiflexed to neutral.

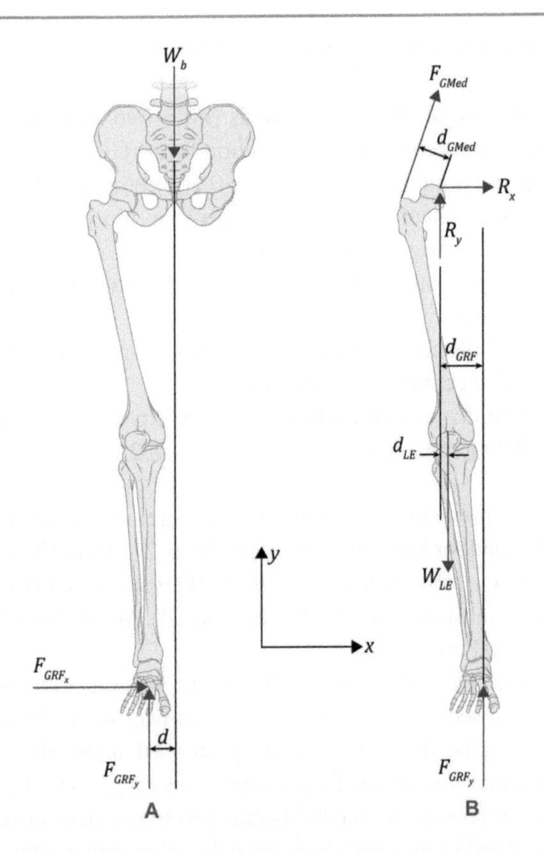


**Fig. 10.44** (**A**) Free body diagram of the whole body standing on one foot used to calculate the ground reaction force components, $F_{\mathrm{GRF}_x}$ and $F_{\mathrm{GRF}_y}$, and (**B**) free body diagram of the lower extremity used to calculate the gluteus medius force, $F_{\mathrm{GMed}}$, and the hip joint contact force components, $R_x$ and $R_y$

The equilibrium conditions are: $\sum F_x = 0$, $\sum F_y = 0$ and $\sum M_{\mathrm{hip}} = 0$. Let's first set up the moment equation. Recall, that a moment = force times the shortest perpendicular distance from the action line of the force to the point of reference, and that we defined a counterclockwise moment as positive:

$$\sum M_{\mathrm{hip}} = -F_{\mathrm{GMed}} d_{\mathrm{GMed}} - W_{\mathrm{LE}} d_{\mathrm{LE}} + F_{\mathrm{GRF}_y} d_{\mathrm{GRF}} = 0$$

Note that $F_{\mathrm{GMed}}$ and $W_{\mathrm{LE}}$ forces both induce negative (i.e., clockwise) moments and $F_y$ induces a positive (i.e., counterclockwise) force. Solving for $F_{\mathrm{GMed}}$:

$$F_{\mathrm{GMed}} = \frac{-W_{\mathrm{LE}} d_{\mathrm{LE}} + F_y d_{\mathrm{COP}}}{d_{\mathrm{GMed}}} =$$

$$\frac{-(128.2\,\mathrm{N})(0.032\,\mathrm{m}) + (796.2\,\mathrm{N})(0.11\,\mathrm{m})}{0.056\,\mathrm{m}} = 1491\,\mathrm{N}$$

Now, let's resolve the $F_{\mathrm{GMed}}$ into its $x$- and $y$-components, which is needed for the force equations:

$$F_{\mathrm{GMed}x} = F_{\mathrm{GMed}} \cos 70° = 510\,\mathrm{N}$$

$$F_{\mathrm{GMed}y} = F_{\mathrm{GMed}} \sin 70° = 1401\,\mathrm{N}.$$

Now, let's calculate the joint contact force components at the hip. First looking at forces in the $x$-direction:

$$\sum F_x = F_{\mathrm{GMed}x} + R_x = 0$$

$$R_x = -F_{\mathrm{GMed}x} = -510\,\mathrm{N}$$

The negative sign simply means that the hip joint contact force in the $x$-direction is 510 N to the left, opposite of what is shown in the free body diagram. The force component represents both compression and shear forces at the interface of the femoral head and acetabulum

Then looking at forces in the $y$-direction:

$$\sum F_y = F_{\mathrm{GMed}y} + F_{\mathrm{GRF}_y} - W_{\mathrm{LE}} + R_y = 0$$

Solving for $R_y$:

$$R_y = -F_{\mathrm{GMed}_y} - F_{\mathrm{GRF}y} + W_{\mathrm{LE}} = -2069\,\mathrm{N}$$

Once again, the negative sign means the hip joint contact force in the $y$-direction is 2069 N, acting down. This represents a combination of both compression and shear force at the joint surface.

Finally, solving for the resultant bone-on-bone force:

$$R = \sqrt{R_x^2 + R_y^2} = 2131\,\mathrm{N}$$

This is approximately 2.7 times body weight. $F_{\mathrm{GMed}}$ clearly makes the largest contribution to the joint bone-on-bone force. Why might the abductor muscle force be so large? This occurs because, in general, muscle moment arms are much smaller than the moment arms of external forces. In this problem, the moment arm of the hip abductors is approximately half the length of the moment arm of the ground reaction force.

Now, let's put these numbers into perspective and abstract something that may be useful going forward. For every step we take, the acetabulum and femoral head are compressed by conflation of the ground reaction and hip abductor muscle forces. Under the best of conditions, these large forces are distributed to a small area of articular cartilage on the head of the femur that is matched between the ball and socket, as well as by the subchondral (trabecular) bone in the proximal femur. Remember that our solution was derived using static assumptions. However, since walking is not static, the velocities and accelerations associated with walking induce larger abductor muscle forces, and therefore greater joint forces. Thus, the viscoclastic properties of the articular and fibrocartilage, as well as the subchondral bone, are vital to distribute/mitigate the large bone-on-bone forces, and dissipate stored energy, with each step. Let's see if this is true.

The hip joint load that is induced during gait can be determined using a biomechanical model that accounts for a per-

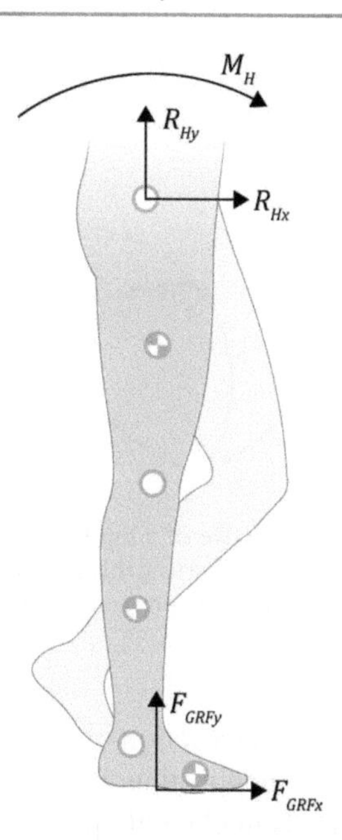


**Fig. 10.45** Determination of the hip joint loads from gait analysis. $\boldsymbol{R}_H$ (with $\boldsymbol{R}_{Hx}$ and $\boldsymbol{R}_{Hy}$ components) = proximal reaction force; $\boldsymbol{M}_H$ = reaction moment; $\boldsymbol{F}_{GRF}$ (with $\boldsymbol{F}_{GRFx}$ and $\boldsymbol{F}_{GRFy}$ components) = distal reaction force acting on the body. Note: bolded letters denote a vector representation. (Modified with permission form Brinckmann et al. (2016))

son's anthropometry (dimensions and mass distribution of body segments), and the kinematic (segmental accelerations, angular velocities, angular accelerations) and kinetic (ground reaction forces) data that are measured. Instrumented gait analysis provides important information on hip joint loading beyond what can be obtained by looking at more simple models such as the stance phase of slow walking. To make these calculations the standing leg is regarded as a free body (Fig. 10.45). Segment (foot, lower leg, and upper leg) masses, locations of the centers of mass, and segment mass moments of inertia with respect to the centers of mass must be known. The time course of the linear accelerations of the centers of mass of the segments, the angular accelerations of the segments, and the ground reaction forces are measured during walking. Using inverse dynamics, starting at the foot, the intersegmental forces and net moments at the ankle, knee, and hip joints are calculated. To determine the hip joint load, a model of the joint, which accurately describes its geometry and the lines of actions and moment arms of the muscles that cross the joint, is established and used. Because there is more than one muscle acting across the hip at any given interval during the gait cycle, additional simplifying assumptions are needed to find a determinate solution. Using inverse dynamics, pioneering research in the late 1970s estimated maximum values of hip joint loading during the stance phase of gait, i.e., from initial contact to toe-off. Because of the alternating decelerations and accelerations of the body mass, peak loads were found to occur after initial contact and during push-off and were found to rise with increasing gait velocity. Peak hip loads ranged from three times bodyweight to seven times body weight during slow and fast walking after heel strike and ranged from four to seven times bodyweight at push-off. In more recent years, direct measures of hip joint loading during the gait cycle have been reported following the employment of instrumented hip prostheses. There appears to be some agreement in the literature that model estimates and instrumented hip joint prosthetic measures of the magnitudes and timing of hip joint loads during gait are similar.

To measure hip joint loads in vivo artificial hip joint replacements have been equipped with force transducers. Thus, during normal daily activities the transducer signals can be transmitted telemetrically to instruments outside the body, allowing the hip joint load to be determined directly, free from the theoretical assumptions used in modeling, e.g., muscle forces are distributed in a manner that minimizes the sum of the muscle stresses squared. Moreover, contributions to the joint load from the spring-like properties of muscles, tendons, joint capsules, and ligaments are assumed to be included in the measured forces. With instrumented hip joint prostheses, joint loads may be more accurately obtained compared to loads determined by musculoskeletal modeling methods. Instrumented hip joint loads appear to be invaluable for the development of artificial joint replacements by providing insight into bone growth and remodeling around implants and in providing guidelines for postoperative rehabilitation. The value of direct measures of hip loads is validated when we look at the peak load during bipedal stance (Table 10.5). The data in Table 10.5 was generated from several individuals whose degenerative hip disease was treated with an instrumented hip joint replacement and shows that the magnitude of hip joint loads change with different modes of bipedal stance. It is notable that hip joint forces measured by an instrumented hip joint replacement also then account for the changes in passive and active muscle tension, as well activation of the viscoelastic properties of hip articular and periarticular structures. Note, although passive and active tension of muscle and contributions of periarticular structures are accounted for, we still do not know their unique specific contributions.

Recall from our discussion of the biomechanics of deformable bodies that relative to the joint structure(s), e.g., hyaline cartilage, etc., that must distribute bone-on-bone forces, the magnitude of joint pressures (stress) is more relevant than the absolute joint load. Joint stress depends on both the direction and magnitude of the force transmitted by the joint, as well as the shape and fit (congruency) of the articulating bones. Having information about joint stresses is about more than academic curiosity because it also provides insight into the causes of secondary hip OA, and may assist in the design and construction of joint replacements. In a

**Table 10.5** Peak hip joint loads during normal activities of daily living

| Activity | Load (% body weight) |
| --- | --- |
| Bipedal stance (equal weight bearing) | 70 |
| Walking (4 km/h) | 250 |
| Walking (fast) | 350 |
| Jogging | 500 |
| Stumbling (unintentional) | 800 |
| Ascending stairs | 250 |
| Descending stairs | 300 |
| Lifting the pelvis lying supine | 300 |
| Lifting the leg against resistance lying supine | 250 |
| Bicycling | 200 |
| Walking using a cane | 190 |

Modified with permission from Brinckmann et al. (2016)

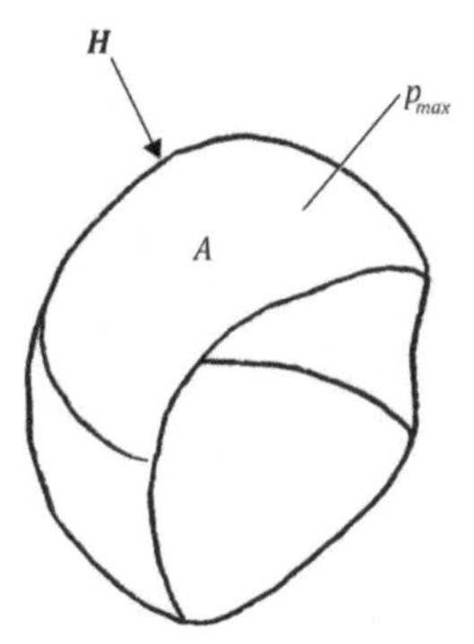


**Fig. 10.46** Anterior perspective of a hip joint model, where part $A$ is the only part of the articulating surface that transmits pressure, $H$ = load vector, and $p_{max}$ point of maximum pressure. (Modified with permission From Brinckmann et al. (2016))

first, i.e., simplistic, approach to estimate the magnitude of the pressure on the surface of the hip joint the mean pressure, $p_{mean}$, is determined as,

$$p_{mean} = \frac{H}{A},$$

where $H$ represents the magnitude of the force on the hip joint and $A$ denotes the projected area of the joint, i.e., the area as seen when looking along the direction of the force vector onto the joint (Fig. 10.46). Simplifying, we assume that the (1) femoral head is fully covered by the acetabulum; (2) area of the femoral head is similar to that of a circle, i.e., $A = \pi r^2$, with radius = 2.5 cm; (3) the individual has a mass of 60 kg (~600 N); and (4) load on the hip joint is three times body weight during single limb stance (typical for walking). Using those assumptions, we determine a mean pressure of:

$$p_{mean} = \frac{3(600\,\text{N})}{\pi\left(0.025\,\text{m}\right)^2} = 917{,}195\,\text{N}/\text{m}^2 = 0.917\,\text{MPa},$$

where MPa = megapascals ($10^6$ pascals). Your response to this number is likely underwhelming because of its abstractness. To provide some perspective, a stress 0.917 MPa is well within the elastic range of normal cartilage in general, and thus much smaller in magnitude than the ultimate failure stress. On the other hand, we need to acknowledge that determining a mean pressure does not accurately represent the actual distribution of stresses on the femoral head.

More realistic information on the pressure distribution within the hip joint requires the creation of a model based on measurements of the shape of the articulating surfaces and viscoelastic properties of the articular cartilage. Earlier in this chapter we acknowledge that there was incomplete coverage of the femoral head secondary to the irregularly shaped rim of the acetabulum. Because of this, pressure is transmitted to only part of the surface of the acetabular rim and related labrum. As a result, pressure distribution is not uniform and depends on the location of the rim of the acetabulum relative to the application of reaction joint forces (Fig. 10.47). In this example, only part A of the articulating surface distributes pressure, and depending on the location of the rim of the socket with respect to the point of application of the vector $\boldsymbol{H}$, the point of maximum pressure $p_{max}$ is located eccentrically, i.e., not centrally located but somewhere between the rim and the point of force application.

Using hip joint force data previously published, Yoshida et al. measured hip pressures relative to specific regions of the acetabulum during activities of daily living and delineated the magnitude and specific location of the peak pressures on the acetabulum (Fig. 10.47). They demonstrated that high pressures were encountered with changes in position of the femur and when the contact area within the acetabulum was reduced. Conversely, when the articulating contact areas were larger, joint surface pressure was reduced relative to hip joint force magnitude (Table 10.6).

Discussing the magnitude and distribution of hip joint forces relative to the unique anatomy of the hip joint and normal daily activities suggest their potential contribution to the development of secondary hip OA. Throughout this chapter, we present several hypotheses that may explain the development of Ralph's hip joint disease at an early age. Unfortunately, there are disagreements in the literature regarding potential causes of secondary hip OA. Let's briefly highlight some of the issues: (1) one study found first signs of hip articular cartilage degeneration in regions of the joint surface that are not typically subjected to large stresses, e.g., on the medio-caudal and lateral surfaces of the femoral head during erect stance, which suggested that perhaps areas which receive too little pressure may trigger degeneration; (2) evidence of increased bone density on the lateral aspect of the acetabulum, as well as cartilage degeneration in the same area, may indicate that excessive pressure on certain parts of the joint surface instigate hip OA; (3) damping of repetitive impulsive forces by subchondral bone, not joint

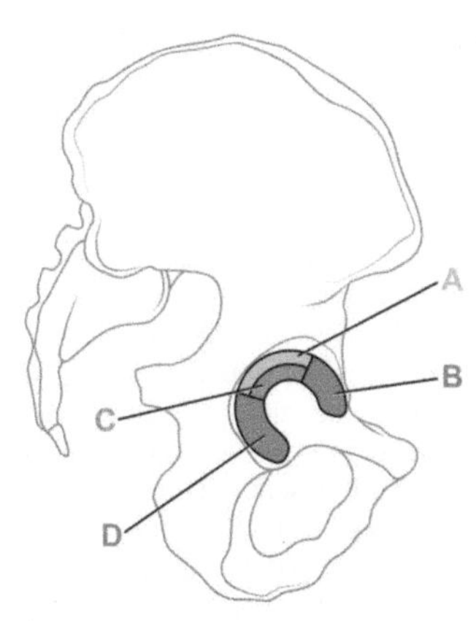


**Fig. 10.47** Three-dimensional regions of the acetabulum that are loaded differently during activities of daily living. A = lateral roof; B = anterior horn (of labrum); C = medial roof; D = posterior horn (of labrum). (Modified with permission from Yoshida et al. (2006))

**Table 10.6** Peak pressures and regions where they occur during activities of daily living

| Activity | Peak pressure (MPa) | Location peak pressure | % of joint surface used |
|---|---|---|---|
| Fast walking | 3.28 | A | 78.7 |
| Normal walking | 3.26 | A | 76.3 |
| Slow walking | 2.87 | A | 81.2 |
| Rising from a chair | 8.97 | D | 19.7 |
| Sitting down | 9.36 | D | 17.6 |
| Squatting | 3.65 | D | 51.6 |
| Ascending stairs | 5.71 | A | 52.1 |
| Descending stairs | 3.77 | A | 80.6 |

Yoshida et al. (2006)
Note: body weight assumed 700 N; A = lateral roof of acetabulum; D = posterior horn of acetabulum

cartilage, damages the subchondral bone but not the cartilage; microdamage to subchondral bone with subsequent healing and remodeling makes the subchondral bone stiffer, which then subjects articular cartilage to greater repetitive mechanical loads that initiate a degenerative process; (4) a simple imbalance between mechanical repetitive- or overloading and the repair processes of joints; (5) a small acetabulum, an abnormally extended acetabulum, or subtle deviations from normal joint geometry such as in asymmetries in the sphericity of the femoral head and local irregularities in cartilage thickness may precipitate early degeneration; (6) persons with physically demanding occupations may be at risk suggesting that high loads may cause the disease; (7) obesity; and (8) variations in the application of the load vector and related point of maximum pressure secondary to injury, e.g., fracture, loss of motion, etc., that changes the ability of the articular cartilage to adapt and redistribute high joint forces. All of these hypotheses must be considered as candidates because they may provide insight into prevention and intervention strategies.

How might the bone-on-bone joint forces change with hip osteoarthritis? The magnitude of the compression and shear stresses associated with activity does not change unless an individual alters their lifestyle, but because of the damage to articular cartilage, and perhaps the labrum, how those forces would be mitigated and distributed would be impaired. With OA, the collagen-PG-water complex is damaged resulting in increased permeability and exudation of interstitial fluid; and the viscoelastic properties of the articular cartilage would be significantly impaired as well. These impairments could result in increased loading of subchondral bone, which could undergo its own series of adaptions. The inability of the joint cartilage to mitigate the very large loads during gait results in pain, and subsequent gait deviations/compensations. Finally, the joint changes in the osteoarthritic hip also affect other periarticular structures, e.g., nociceptors and mechanoreceptors, which, combined with joint pain, can produce arthrogenic muscle inhibition. This neurological inhibition is thought to induce abnormal resting tonus and activation of the stabilizing muscles that cross the hip joint, e.g., gluteus medius, resulting in atrophy and weakness, which feeds into a negative feedback loop and cascade of degeneration. Some might claim that weak hip abductors cause an increase in the hip joint forces. But, in fact, as we just demonstrated, it is the muscle forces that make the largest contribution to joint forces, so if the hip abductors have become atrophic, they are unable to generate the force, and moment, needed to stabilize the pelvis during walking. Our nervous system has a ready compensation, which we already described, called a compensated Trendelenburg, the hallmark of which, lateral leaning over the impaired hip, actually reduces the hip joint forces because it reduces the moment arm of the external force, i.e., the ground reaction force.

## 10.7 Chapter Summary

The hip joint is similar to the sacroiliac joints in that both complexes are foundational to the movements and stability of the joints they link from above and below. As such, the hip joints are subjected to large and multi-dimensional internal stresses. Fortunately, the acetabular-labral-femoral head geometry and joint capsular-ligamentous complex have evolved to manage large external forces, while, at the same time, allowing a moderately liberal joint range of motion. Moreover, actions of a large redundant set of muscles crossing the joint contribute to the six degrees of freedom of movement at the joint, as well as add significantly to the joint's dynamic stability. Paradoxically, the internal muscle moments that are critical for both mobility and stability also create large joint bone-on-bone forces during weight-bearing activities, e.g., walking, running, jumping, etc., that can contribute to the initiation, and perpetuation, of joint degeneration.

The primary purpose of this case is to see how biomechanics is used to understand/explain the manifestation of hip OA, one of the most common orthopedic conditions in the industrial world. Secondarily, this case calls on the reader to draw on prior experience/knowledge/generalizations, i.e., from previous cases in this text, to both initiate and sustain an inquiry. This case reminds us of the nature of dynamic coupling of human joint function as we explore the concepts of femoral-on-pelvic and pelvic-on-femoral kinematics, as well as the kinematic relationship between the pelvic ring and lumbar spine. We explore several hypotheses in an attempt to explain the development of secondary hip osteoarthritis, that is, arthritis that appears to have a mechanical etiology. Most of those hypotheses force us to review the biomechanics of deformable bodies, which is presented in an earlier chapter.

What is new to this case is the introduction of the biomechanics of gait, and how gait analysis can be used for cases involving orthopedic disease, like hip OA. It is clear from this case that the combination of a detailed physical examination and observational and instrumented gait measures provides additional, and critical, information on the functional manifestation of Ralph's hip disease. This chapter provides foundational information on the phases of normal gait, combined with a more technical review of instrumented gait analysis in Appendix J, that are applied in a specific problem-solving approach to identify and analyze Ralph's gait deviations. The reader should draw on these principles for the knee and ankle/foot cases that follow.

Finally, we use another static equilibrium problem set-up to quantify the abductor mechanism of the gait cycle. This provides an additional opportunity to practice examination of problem-solving assumptions, in general, but also those unique to the hip joint; setting up a free body diagram with its associated definitions; accounting for the known (external) and unknown (internal) forces and moments; following pre-scribed steps for setting equations; and solving and checking solutions. For the problem in this case, we see again, that because muscle moment arms are much smaller than the moment arms of the external forces, the marked joint bone-on-bone forces are largely induced by the muscle forces acting at that joint. Our solution shows that the resultant hip joint reaction force is greater than twice body weight during the single limb support interval of normal walking, and that this force is increased threefold with running or stair walking. We then discuss the implications of these large joint forces in the instance of hip OA.

## Bibliography


Arnold AS, Askkawa DJ, Delp SL (2000) Do the hamstrings and adductors contribute to excessive internal rotation of the hip in persons with cerebral palsy? Gait Posture 11:181–190

Athanasiou KA, Agarwal A, Dzida FJ (1994) Comparative study of the intrinsic mechanical properties of the human acetabular and femoral head cartilage. J Orthop Res 12:340–349

Brinckmann R, Frobin W, Hierholzer E (1981) Stress on the articular surface of the hip joint in health adults and persons with idiopathic osteoarthritis of the hip joint. J Biomech 14(3):149–156

Brinckmann R, Frobin W, Leivseth G et al (2016) Mechanical aspects of the hip joint. In: Brinckmann R, Frobin W, Leivseth G et al (eds) Orthopedic biomechanics, 2nd edn. Thieme Publishers, Stuttgart

Cusick BD, Stuberg WA (1992) Assessment of lower-extremity alignment in the transverse plane: implications for management of children with neuromotor dysfunction. Phys Ther 72(1):3–15. https://doi.org/10.1093/ptj/72.1.3

Dahlke TM, Jabs WL (1997) Utilizing the Ryder's and thigh-foot angle tests to establish normal values of femoral anteversion and tibiofibular torsion in children aged 5 through 10 years. Masters Thesis, Grand Valley State University.

Delp SC, Hess WE, Hungerford DS et al (1999) Variation of rotation moment arms with hip flexion. J Biomech 32:493–501

Gage JR, Schwartz MH, Koop SE, Novacheck T (eds) (2009) The identification and treatment of gait problems in cerebral palsy, 2nd edn. Mac Keith Press

Groh MM, Herrera J (2009) A comprehensive review of hip labral tears. Curr Rev Musculoskelet Med 2:105–117. https://doi.org/10.1007/s12178-009-9052-9

Henderson ER, Marulanda GA, Cheong D et al (2011) Hip abductor moment arm – A mathematical analysis for proximal femoral replacement. J Orthop Surg Res 6:6. https://www.josr-online.com/content/6/1/6

Jang IG, Kim IY (2008) Computational study of Wolff's law with trabecular architecture in human proximal femur using topology optimization. J Biomech 41:2353–2361. https://doi.org/10.1016/j.jbiomech.2008.05.037

Kim C, Nevittt M, Guermazi A et al (2018) Leg length discrepancy and hip osteoarthritis in the multicenter osteoarthritis study and the osteoarthritis initiative. Arthritis Rheumatol 70(10):1572–1576. https://doi.org/10.1002/art.40537

Kirtley C (2006) Clinical gait analysis, theory and clinical practice. Elsevier Churchill Livingstone, Edinburgh. and www.clinicalgaitanalysis.com - frequently asked questions/femoral anteversion & tibial (malleolar) torsion

Lepasio MJ, Sultan AA, Piuzzi NS et al (2018) Hip osteoarthritis: a primer. Perm J 22:17–084. https://doi.org/10.7812/TPP/17-084

LeVeau B (1977) Williams and Lissner: biomechanics of human motion, 2nd edn. W. B. Saunders Company, Philadelphia

Liu RW, Streit JJ, Weinberg DS et al (2018) No relationship between mild limb length discrepancy and spine, hip or knee degenerative disease in a large cadaveric collection. Orthop Traumatol Surg Res 104:603–607. https://doi.org/10.1016/j.otsr.2017.11.025

Magee DJ (2013) Orthopedic physical assessment, 6th edn. Elsevier Saunders, St. Louis

Martin RL, Enseki KR, Draovitch et al (2006) Acetabular labral tears of the hip: examination and diagnostic challenges. J Orthop Sports Phys Ther 36(7):503–515. https://doi.org/10.2519/jospt.2006.2135

McCaw ST, Bates BT (1991) Biomechanical implications of mild leg length inequality. Br J Sports Med 25(1):10–13

Murray KJ, Axari MF (2015) Leg length discrepancy and osteoarthritis in the knee, hip and lumbar spine. J Can Chirop Assoc 59(3):226–237

Neumann DA (2017) Hip. In: Neumann DA (ed) Kinesiology of the musculoskeletal system, foundations for rehabilitation, 3rd edn. Elsevier, St. Louis

Nordin M, Frankel VH (2022) Basic biomechanics of the musculoskeletal system, 5th edn. Wolters Kluwer, Philadelphia

Perry J, Burnfield JM (2010) Phases of gait. In: Perry J, Burnfield JM (eds) Gait analysis, normal and pathological function, 2nd edn. SLACK Incorporated, Thorofare

Richie DH (2021) Human walking: the gait cycle. In: Richie DH (ed) Pathomechanics of common foot disorders (eBook). Springer Nature Switzerland AG, pp 45–62. https://doi.org/10.1007/978-3-030-54201-6

Scorcelletti M, Reeves ND, Rittweger J et al (2020) Femoral anteversion: significance and measurement. J Anat 237:811–826. https://doi.org/10.1111/joa.13249

Silverling S, O'Sullivan E, Garofalo M et al (2012) Hip osteoarthritis and the active patient: will I run again? Curr Rev Musculoskelet Med 5:24–31. https://doi.org/10.1007/s12178-011-9102-y

Schmid S, Schweizer K, Romkes J et al (2013) Secondary gait deviations in patients with and without neurological involvement: a systematic review. Gait Posture 37:480–493. https://doi.org/10.1016/j.gaitpost.2012.09.006

Stansfield BW, Nicol AC, Paul JP et al (2003) Direct comparison of calculated hip joint contact forces with those measured using instrumented implants. An evaluation of a three-dimensional mathematical model of the lower limb. J Biomech 36(7):929–936. https://doi.org/10.1016/S0021-9290(03)00072-1

Taylor JB, Wright AA, Dischiavi SL et al (2017) Activity demands during multi-directional team sports: a systematic review. Sports Med 47:2533–2551. https://doi.org/10.1007/s40279-017-0772-5

Vigdorchik JM, Nepple JJ, Eftekhary N et al (2016) What is the association of elite sporting activities with the development of hip osteoarthritis? Am J Sports Med 45(4):961–964. https://doi.org/10.1177/0363546516656359

Yoshida H, Faust A, Wilchens J et al (2006) Three-dimensional dynamic hip contact area and pressure distribution during activities of daily living. J Biomech 39:1996–2004. https://doi.org/10.1016/j.jbiomech.2005.06.026

Zeng W-N, Wang F-Y, Chen C et al (2016) Investigation of association between hip morphology and prevalence of osteoarthritis. Sci Rep 6:23477. https://doi.org/10.1018/srep23477

Zang Y, Jordan JM (2010) Epidemiology of osteoarthritis. Clin Geriatr Med 26(3):355–369. https://doi.org/10.1016/j.cger.2010.03.001

# Knee Joint Complex: Failed Lateral Meniscus Allograft

# 11

## 11.1 Introduction

Interposed between the ankle and hip joints, the knee complex consists of three joints: tibiofemoral and patellofemoral, and proximal tibiofibular. The tibiofemoral and patellofemoral joints sustain and transmit large multidimensional loads and facilitate movements within the lower extremity to accommodate an unlimited array of activities. The nearby third joint, the proximal tibiofibular joint, is not directly involved with the knee joint proper, yet its function is crucial to the operation of the knee and the entire lower extremity. Because the knee joint is interposed between two very large bony levers, i.e., the femur and tibia, and crossed by large (in terms of cross-sectional area) one- and two-joint muscles, all of its structures are challenged to provide stability, as well as mobility, as it moves through a large range of motion during normal daily activities. Neither the tibiofemoral nor patellofemoral joints demonstrate joint configurations, i.e., congruent bony partners, that provide inherent stability, suggesting that its periarticular soft tissue constraints, e.g., joint capsule, ligaments, fasciae, menisci, and muscles, are paramount for providing stability. Because of their configurations, both the tibio- and patellofemoral joints are particularly susceptible to injury, often leading to early advanced osteoarthritis and eventual knee joint replacement.

## 11.2 Case Presentation

Cathy Grieve (height: 182 cm, mass: 97.7 kg) was a 35-year-old full-time homemaker with right knee pathology and a long history of knee impairment and pain, who was referred for a clinical instrumented gait analysis. She was otherwise healthy.

Cathy was an elite high school basketball and track athlete. She originally injured her right knee at the age of 16 years as she rose from the floor to stand. The injury occurred in the early spring following basketball season but before the competitive track season. The injury was diagnosed arthroscopically as a partial tear of the lateral meniscus and was treated by arthroscopic repair. Following surgical treatment, Cathy used a knee immobilizer and crutches, with weight-bearing as tolerated, for about 2 weeks. She reported that she did not receive any formal education or rehabilitation for her knee and returned to compete in track 6–8 weeks following surgery without incident. She finished her high school athletic career, using a meniscus brace during all practice and game situations, without any further knee complaints.

Cathy accepted a scholarship to play basketball at a Division I University. She recalled that the training, compared to high school, was much more intense. She thought that the second right knee injury toward the end of the basketball season in her freshman year was related to "being worn down." Cathy recalled reinjuring the right knee (again the lateral meniscus) one morning while getting out of bed. Again, the diagnosis and treatment were made arthroscopically, and the residual lateral meniscus was debrided and repaired. She wore a knee immobilizer and used crutches for approximately 2 weeks, as before, but received no formal rehabilitation. As a result of this second injury, Cathy decided to discontinue her collegiate basketball career but remained otherwise quite active in recreational sports.

Six years after the original injury, Cathy reinjured her right knee while getting up from the floor. Prior to that, she was quite active, recreationally participating in softball, bike riding, and rollerblading. Arthroscopic debridement and repair were performed again, followed by 2 weeks of crutch walking, with a gradual return to normal activity; again, no formal rehabilitation was prescribed.

Over a 9-year period following her last re-injury, Cathy significantly curtailed her recreational activities, and many of her normal daily activities were also impaired, secondary to right knee pain. Since her injury, at age 20, she had

**Supplementary Information** The online version contains supplementary material available at https://doi.org/10.1007/978-3-031-25322-5_11.


G. J. Alderink, B. M. Ashby, *Clinical Kinesiology and Biomechanics*, https://doi.org/10.1007/978-3-031-25322-5_11

undergone six additional knee procedures (four arthroscopic procedures and two arthrotomies, including two meniscal allograft transplantations). At age 32 years, Cathy received her first meniscal allograft procedure (which was relatively novel at the time). Following surgery, she was non-weight-bearing (NWB) for 8 weeks and was guided in postoperative care by her physician but received no formal rehabilitation. Three years later, it was apparent that the first meniscal allograft had failed, and her surgeon recommended and performed a second allograft procedure. Following surgery, Cathy followed the postoperative instructions of walking with crutches, weight-bearing as tolerated, while wearing a knee immobilizer. She used crutches for approximately 2–3 weeks, followed instructions for doing daily range of motion and isometric muscle exercises, but received no formal rehabilitation.

At the time we saw Cathy, it was apparent that the second meniscus allograft procedure had also failed. She was referred for an instrumented gait analysis, seeking possible biomechanical reasons for the failed allograft procedure and continued knee pain. Radiographs were not provided.

A comprehensive physical examination revealed several significant orthopedic impairments (Table 11.1). A brief look at Cathy's walking revealed an antalgic (painful and reduced right stance time) gait pattern, but despite this, she could walk on her heels and toes. (Heel-toe walking is a functional test for anterior and posterior tibial muscle strength, respectively.) Posture examination for spinal alignment was normal, but Cathy held the right knee slightly flexed, likely related to pain with weight-bearing and mild swelling (knee joint line circumference measures: left 38.5 cm, right 42.5 cm). The standing squat test (tests for general lower extremity range of motion and strength; see Magee) was reduced by approximately 50% due to right knee pain. There was noticeable atrophy of the right calf and quadriceps musculature and moderate reduction in right quadriceps muscle definition with a quadriceps muscle setting task.

Passive range of motion testing revealed (in degrees):

| | *Right* | *Left* | *Normal* |
|---|---|---|---|
| Knee flexion supine | 90° | WNL | 135°–160° |
| Knee flexion prone | 90° | WNL | ~135° |
| Knee extension | −10° | WNL | 0° |
| Knee extension (active) | 10° | WNL | 0° |

Note: *WNL* within normal limits. A −10° of extension denotes knee hyperextension, which for Cathy was not normal. It is not uncommon for individuals to exhibit knee hyperextension, but if the limbs are asymmetrical, we consider the difference abnormal. The 10° recorded for active knee extension meant that Cathy could not fully extend her right knee, i.e., the knee was flexed 10°. This denotes, what is referred to as, an "active lag" of knee extension

Resistive muscle tests (i.e., manual muscle tests) also revealed several significant deficits (see examination form (Table 11.1) for operational definitions for the manual grading of muscle strength):

| | *Right* | *Left* |
|---|---|---|
| Hip flexors | 4/5 | 5/5 |
| Hip extensors | 3+/5 | 4+/5 |
| Hip abductors | 3+/5 | 4+/5 |
| Hip adductors | 5/5 | 5/5 |
| Knee flexors | 4/5 | 5/5 |
| Knee extensors | 5/5 (pain) | 5/5 |

Note: The hip internal and external rotators were not tested; resisted testing of the right ankle invertors reproduced right knee pain

Stress testing of the anterior and posterior cruciate and medial and lateral collateral ligaments revealed only mild laxity and pain with the varus stress test on the right knee. There are several provocation tests for meniscal injuries, but only the McMurray's was performed, which was positive with pain (no clicking) and guarding when the right knee was tested.

## 11.3 Key Facts, Learning Issues, and Hypotheses

Cathy was referred for an instrumented gait analysis with the following clinical question: can gait analysis provide a biomechanical explanation for her knee pain and gait dysfunctions? On the surface, this seemed like a reasonable question. However, given the long history of repeated injuries to her right knee and long-term failure of the surgical interventions, it was unlikely that the three-dimensional gait analysis would provide insight into the mechanism of injury. Rather, the gait test may provide biomechanical insight into the manifestation of her right knee problem. Although we lacked radiographic evidence, we hypothesized that because of Cathy's repeated injuries and failed treatments over many years, she was most likely experiencing the effects of advanced degenerative knee osteoarthritis. Therefore, we need to examine the literature on the epidemiology of knee osteoarthritis related to meniscal injuries. Since Cathy was only 16 at the time of her first injury, we are particularly interested in what is known about the natural history of meniscal injuries in young athletes. In previous chapters in this text, we examine, in detail, the constituents and material properties of hyaline and fibrocartilage, as well as the cascade of events leading to degenerative joint disease. In this case, we look at the specifics of normal meniscal structure and function, as well as the pathomechanics of the knee following meniscal injuries. The reader is encouraged to review the details of degenerative joint disease discussed in previous chapters, e.g., Chap. 3.

It seemed unusual that Cathy reported having received no formal rehabilitation following injury and arthroscopic repair, especially during her brief time as an intercollegiate

**Table 11.1** Physical examination data

| | | Range of motion | | Strength | |
|---|---|---|---|---|---|
| *Hip* | *Normal ROM* | *L* | *R* | *L* | *R* |
| Flexion | 120–135 | WNL | WNL | 5/5 | 4/5 |
| Extension/mod. Thomas test: | | | | | |
| Knee extended | 0 | WNL | WNL | 4+/5 | 3+/5 |
| Knee flexed 90° | 0 | WNL | WNL | 4+/5 | 3+/5 |
| Abduction (hip ext.) | 30–50 | WNL | WNL | 4+/5 | 3+/5 |
| Adduction | *** | | | 5/5 | 5/5 |
| Ober test (IT band) | Neg | NT | NT | | |
| Internal rotation | 30–45 | WNL | WNL | | |
| External rotation | 30–45 | WNL | WNL | | |
| Femoral anteversion | Adult: 12–15 | WNL | WNL | | |
| *Knee* | | *L* | *R* | *L* | *R* |
| Flexion (supine) | 135–160 | WNL | 90 | | |
| Flexion (prone) | ~135 | WNL | 90 | 5/5 | 4/5 |
| Extension (passive) | 0 | WNL | −10 | | |
| Extension (active): | | | | | |
| Strength (sitting) | *** | | | 5/5 | 5/5, pain |
| Active lag (supine) | 0 | WNL | 10 | | |
| Popliteal angle (from vert.): | | | | | |
| Traditional (spastic/end) | 0–30 | NT | NT | | |
| Neutral pelvis (end) | 0–20 | NT | NT | | |
| Transmalleolar axis | ~20 ER | WNL | WNL | | |
| Thigh-foot angle | ~15 ER | WNL | WNL | | |
| *Ankle/foot* | | *L* | *R* | *L* | *R* |
| Dorsiflexion: | | | | | |
| Knee ext'd (spastic/end) | ~10–15 | WNL | WNL | | |
| Knee flexed (end) | 15–25 | WNL | WNL | 5/5 | 5/5 |
| Plantarflexion | 45–50 | WNL | Mild ↓ | 5/5 | 5/5 |
| Tarsal inversion (supination) | 35–50 | WNL | Mild ↓ | 5/5 | 4/5, pain |
| Tarsal eversion (pronation) | 35–50 | WNL | Mild ↓ | 5/5 | 5/5 |
| 1st MTP extension | 70–90 | WNL | WNL | | |
| *Trunk strength* | | *Posture* | | | |
| Lower abdominals (leg lift) | NT | *Mild right knee flexed; note two surgical scars: | | | |
| Upper abdominals (crunch) | NT | Right lateral and anterolateral knee | | | |
| Trunk extensors (prone) | NT | | | | |

| Foot posture | | |
|---|---|---|
| *Weight-bearing* | *L* | *R* |
| HF varus*/valgus | WNL | WNL |
| *Coleman block | | |
| MF planus#/cavus | WNL | WNL |
| #Reverses w/ toe standing | | |
| FF ab/adductus | WNL | WNL |
| Hallux valgus | WNL | WNL |
| *Non-weight-bearing* | *L* | *R* |
| HF varus/valgus | NT | NT |
| Flexible? | | |
| MF planus/cavus | NT | NT |
| Flexible? | | |
| FF varus/valgus | NT | NT |
| Flexible? | | |
| FF ab/adductus | NT | NT |
| Flexible? | | |
| PT first ray | NT | NT |
| Callus pattern: | | |
| *Spasticity* | *L* | *R* |
| Hip flexors | NA | NA |
| Hip adductors | NA | NA |
| Hip internal rotators | NA | NA |
| Hip external rotators | NA | NA |
| Hamstrings | NA | NA |
| Vasti | NA | NA |
| Rect. Fem./Duncan-Ely | NA | NA |
| Ankle clonus | NA | NA |
| Gastrocnemius | NA | NA |
| Soleus | NA | NA |
| Tibialis anterior | NA | NA |
| Tibialis posterior | NA | NA |
| Peroneals | NA | NA |
| Toe flexors (FDL, FHL) | NA | NA |
| Confusion test | NA | NA |
| *Balance* | | |
| Single-leg stance time | >10 s | >10 s |
| *Leg length* | 92 cm | 92 cm |
| *Circumference measures* | *L* | *R* |
| *Joint line (cm)* | 38.5 | 42.5 |
| *15 cm sup joint line (cm)* | 51 | 50.5 |
| *10 cm inf joint line (cm)* | 42.5 | 42 |

*Notes/others*

*Genu valgus (NWB): WNL
*Atrophic R calf and quadriceps, R quad set noted moderate ↓ muscle definition
*Heel/toe walking WNL
*Squat 50% ROM with R knee pain
*Mild laxity/pain during R knee varus stress test; R (+) McMurray – Pain, no click

| *Abbreviations* | *Spasticity scale (modified Ashworth)* | *Strength scale (manual muscle test)* |
|---|---|---|
| *NA* = not applicable | 0 = no increase in tone (normal) | 0 = no palpable muscle activity |
| *NT* = not tested | 1 = slight ↑ in resistance: Catch & release | 1 = palpable muscle tension or flicker, no movement |
| *UTA* = unable to assess | 1+ = slight ↑ in resistance: Catch & min. Resistance | 2 = full ROM in horizontal plane, 0 against gravity |
| *P/PL* = painful/pain limited | 2 = more marked ↑ in resistance thru most ROM | 3 = full ROM against gravity only, no resistance |
| *WNL* = within normal limits | 3 = considerable ↑ in resistance, PROM difficult | 4 = full ROM against gravity & mod. Resist. |
| *WFL* = within function limits | 4 = rigid | 5 = full ROM against gravity & max. Resist. |

Note: *IT* iliotibial band, *FDL* flexor digitorum longus, *FHL* flexor hallucis longus, *MTP* metatarsophalangeal, *neg* negative, *ext'd* extended, *HF* hindfoot, *MF* midfoot, *FF* forefoot athlete at a Division I University. It was not surprising that she received little care as a high school athlete since many secondary schools typically do not have full-time athletic trainers. The fact that, according to Cathy, she received no formal treatment or education following surgical intervention may have been a factor in her ongoing problems. Because rehabilitation is such an important component of comprehensive care following a meniscal repair and allograft transplantation, we examine a typical rehabilitation protocol for a post-meniscus-allograft procedure. As well, we attempt to define a biomechanical rationale related to the rehabilitation protocol relative to phases of healing, modeling, and maturation of the substitute meniscus.

Cathy had several lower extremity range of motion and strength deficits. We discuss the possible relationships between these impairments and her knee pathology and how the impairments can be used to make predictions about specific gait deviations. For example, although the significant loss of passive knee flexion can be explained by advanced knee joint degeneration, this loss of motion would not likely impact her ability to produce normal knee flexion during the swing phase (since only approximately 70° of knee flexion is needed). Additionally, moderate weakness of the right-side hip extensors and abductors may challenge the critical events of sagittal and frontal plane stance stability. This case is another opportunity to practice and apply what we know about normal gait to develop hypotheses regarding Cathy's abnormal gait. If you recall from the hip case (Chap. 10), when we approach a gait analysis, we should examine not only the dysfunctional joint but must also consider the dynamic coupling of the ankle, knee, hip, pelvis, and trunk on both the ipsi- and contralateral extremities during the walking cycle.

Several special tests were selected as part of Cathy's physical examination. These tests focus on the knee joint and included stress tests of the major knee ligaments, as well as a provocative test for a suspected meniscal tear. As in Chap. 3, for this case, we provide a biomechanical rationale for the special tests that were used in the physical examination. Thus, some of the learning issues in this case are to examine the anatomy of the knee ligaments, their relationship to knee kinematics, and their specific function and to use that information to explain Cathy's findings. Likewise, since this case is primarily about recurrent injuries to the lateral meniscus, we explore details of the structure and function of the medial and lateral meniscus.

Cathy had pain with resisted muscle testing of the quadriceps, yet surprisingly she demonstrated "normal" strength. This finding would typically suggest a minor strain injury to the knee extensor muscles (recall that we cannot isolate the strength of each part of the quadriceps group) and/or disturbance of the extensor mechanism, i.e., patellofemoral joint, and/or pain secondary to knee (tibiofemoral) osteoarthritis. Although we did not believe Cathy had a patella-femoral problem, we examine the patella-femoral joint anatomy and biomechanics, i.e., stability and extensor mechanism, to help in our biomechanical differential diagnosis.

Unfortunately, we do not have biplanar video of her walking as part of this case, so we apply our problem-solving approach in gait analysis using only data from the instrumented measures. For this case, we also collected muscle activity (using surface electromyography (EMG) sensors) while she was walking from several lower extremity muscles despite the fact that she demonstrated normal selectivity when we performed resisted muscle testing. Because of her long-standing meniscus problem, which we think included mild instability of the knee (based on a positive tibiofemoral varus stress test), we hypothesized that abnormal coactivity of key muscles may be evident during the gait cycle (see Appendix J for a review of electromyography methodology and data interpretation).

Stress testing of the knee's major ligaments is routine when an individual reports knee pathology, particularly, as in Cathy's case, when there is meniscus involvement. As we discuss in more detail later, the menisci make important contributions to knee stability. Thus, when they are compromised, it is likely that one or more periarticular soft tissue constraints, i.e., ligaments, adapt/compensate and experience unusual or excessive forces, eventually leading to plastic deformations, i.e., overstretching, that render those constraints compromised. Table 11.2 summarizes the functions

**Table 11.2** Major supporting ligaments at the knee, their function, and clinical tests of integrity

| Structure | Function | Clinical test |
|---|---|---|
| Medial collateral ligament (and posteromedial capsule) | Resists valgus (abduction).<br>Resists knee extension.<br>Resists extremes of axial rotation (especially knee external rotation). | Valgus stress test with the knee flexed 20°–30° and with the knee fully extended. |
| Lateral collateral ligament | Resists varus (adduction).<br>Resists knee extension.<br>Resists extremes of axial rotation. | Varus stress test with the knee flexed 20°–30° and with the knee fully extended. |
| Anterior cruciate ligament | Most fibers resist knee extension, i.e., excessive anterior translation of the tibia or posterior translation of the femur.<br>Resists extremes of knee valgus, varus, and axial rotation. | Anterior drawer test.<br>Lachman's test. |
| Posterior cruciate ligament | Most fibers resist knee flexion, i.e., excessive posterior translation of the tibia or anterior translation of the femur.<br>Resists extremes of valgus, varus, and axial rotation. | Posterior drawer test.<br>Reverse Lachman's test.<br>Posterior sag sign. |

Modified with permission from Neumann (2017)

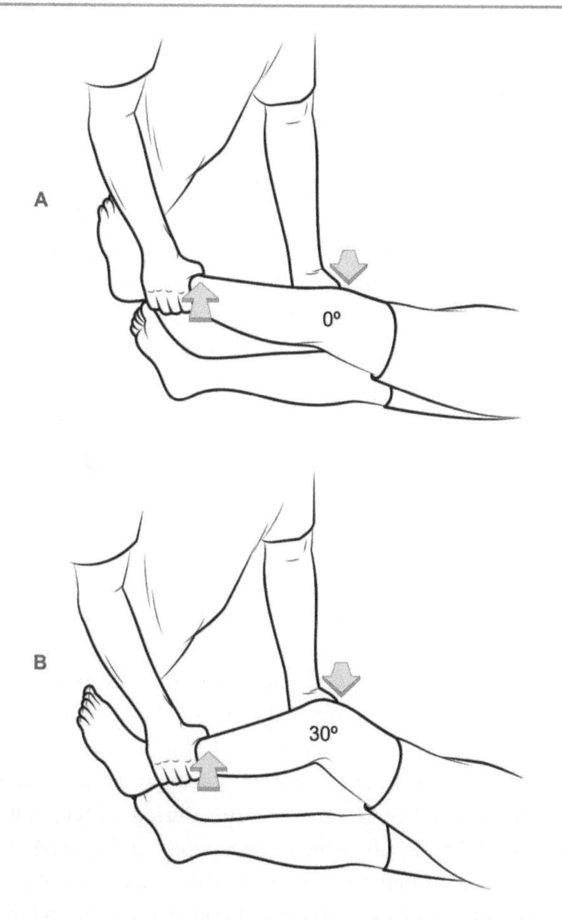


**Fig. 11.1** Varus stress test with knee (**A**) extended and (**B**) flexed 30°. The operator stabilizes the thigh with the left hand while attempting to adduct the tibia relative to the thigh by pushing with the right hand. (Note: a valgus stress test was also performed (not shown))

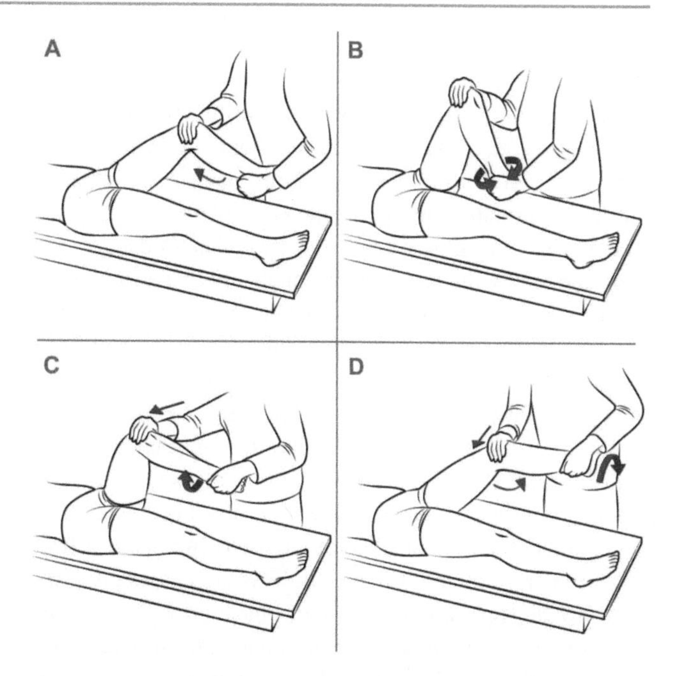


**Fig. 11.2** McMurray's test was conducted sequentially to provocate both the medial and lateral menisci. (**A**) Passive movement of the knee through a full range of flexion and extension, applying overpressure at either extreme of motion while palpating the medial and lateral joint lines for tenderness and/or clicking. (**B**) With the knee flexed, passive internal and external rotation of the tibia, applying overpressure at either extreme of motion while palpating the medial and lateral joint lines for tenderness and/or clicking. (**C**, **D**) With the knee flexed and externally rotated, applying a valgus stress on the knee as the knee is extended. Note: the final two movements are repeated starting with the knee flexed and internally rotated, applying a varus stress on the knee as the knee is extended

of the major ligaments that act on the tibiofemoral joint. We performed stress tests on each major knee ligament as part of Cathy's examination. When we applied the varus stress test to her right knee, Cathy complained of pain, and we detected mild laxity (Fig. 11.1). With a varus stress test, the external force applied to the tibia attempts to adduct it relative to the femur, which places tensile forces on the lateral collateral ligament (really on all the soft tissue constraints laterally) and compression forces on the medial knee joint compartment. Pain and mild laxity with this test suggested a second-degree strain injury, i.e., a plastic deformation, to the lateral collateral ligament (and lateral joint capsule). We hypothesize that these findings were perhaps due to the chronic nature of Cathy's lateral meniscus pathology, which created joint hypermobility, eventually damaging other primary constraints. It is likely that lateral knee joint hypermobility also affected the medial knee joint compartment. We had previously suggested that Cathy had likely developed advanced degenerative joint disease affecting both the lateral and medial knee compartments. Thus, the compression force induced on a degenerated medial knee joint compartment, as part of the varus stress test, could also have contributed to her pain response.

Many different special tests have been described to test for meniscal injury, sometimes referrred to as an internal derangement. We chose only the McMurray test because it seems to test the multiple functions of the menisci in a series of maneuvers (Fig. 11.2). McMurray's test is performed in a sequence that goes from least to most provocative, and each sequence stresses both menisci. A positive test, in one or more of the tests in the sequence, is indicated by reproduction of pain, a palpable or audible click near either the medial or lateral joint line, and/or muscle guarding, which sometimes makes it difficult to complete the test. If the operator perceives a positive test with the simplest passive movement, e.g., flexion/extension, other aspects of the test may not need to be completed. Briefly, the passive flexion/extension movements test the mobility of the menisci as the tibia rolls/glides relative to the femur, and passive overpressure at the extremes of flexion and extension applies added compression to the anterior and posterior horns of the menisci. Likewise, internal and external rotations test the ability of the menisci to glide, as well as change their shape, i.e., deform, as the tibia rotates relative to the femur. Axial rotation of the knee induces torsional and tensile stress to each meniscus. When movements in the sagittal (flexion/extension) and transverse (internal/external rotation) planes are combined with valgus or varus stress to the knee joint, both menisci are stressed multidimensionally, mimicking the kinds of stresses the knee experiences during normal daily activities (although the loads applied by the operator are much less than those experienced during walking, for instance). Because of the dynamic coupling of the functions of the menisci and knee ligaments that accompany the sequence of actions during the most provocative part of McMurray's test, we remind the reader that these tests do not isolate the menisci. In other words, Cathy's positive McMurray test likely indicated pathology of the right knee articular cartilage in both medial and lateral compartments, lateral meniscus (what was left of it), and lateral collateral ligament. This concludes an early look into the application of biomechanical principles as part of the clinical examination. Let us review the structure and function of the knee before we do anything else.

## 11.4 Knee Structure and Function

### 11.4.1 Structure

Kapandji takes an evolutionary/developmental approach in his exposition of the long bones that make up the tibiofemoral joint. By first viewing the general architecture of the lower limb segments, we can imagine factors that influence the orientation of the articular surfaces of the knee (Fig. 11.3). The knee joint is traditionally described as a mobile trochoginglymus (pivotal hinge) joint, suggesting that, at most, it has two degrees of freedom, which primarily promote knee flexion/extension and internal/external rotation. As suggested by Wolff's law, the femur and tibia are molded by functional movements, actions of large muscles, and external forces. First, then, we see that to accommodate knee flexion greater than 90°, the distal shaft of the femur is hollowed or indented just above the condyles (Fig. 11.3C), which prevents the impact of the tibia onto the femur at the end range of flexion. This is offset by an anterior bowing (or displacement) of the shaft as the condyles become bent posteriorly. As an offset, the tibial shaft is thinned posteriorly and reinforced anteriorly, inducing the tibia plateau to bend posteriorly (Fig. 11.4B). The bends in the femur can be explained by Euler's laws governing the behavior of columns loaded eccentrically. Thus, depending on whether a column is jointed at both ends or fixed below and mobile above, the axial loads have a differential effect. Both cases apply to the femur. Therefore, when the femur is mobile proximally and relatively fixed distally, e.g., in the stance phase of gait, an axial load creates two bends, with the more proximal bend taking up two-thirds of the column (Fig. 11.3A). In case the column is fixed at both ends, the bend occurs along the entire length of the femur (Fig. 11.3C). An axial load applied to a column that is fixed at both ends, e.g., the tibia during stance, induces a bend in the midshaft, which corresponds to the bend in the coronal plane, as depicted in Fig. 11.4B. In the sagittal plane, axial loads produce (1) a posterior bend proximally and in the midshaft and (2) an approximate 5° posterior slope of the tibial plateau. The opposite concavities of the femur, i.e., anterior, and the tibia, i.e., posterior, allow space for the large

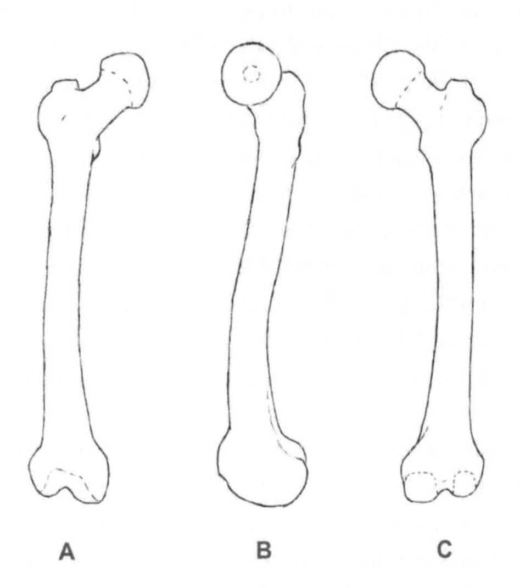


**Fig. 11.3** (**A**) Anterior, (**B**) medial, and (**C**) posterior aspects of the right femur. In (**B**), note the slight anterior convexity in its midshaft and the slight posterior bend of the femoral condyles

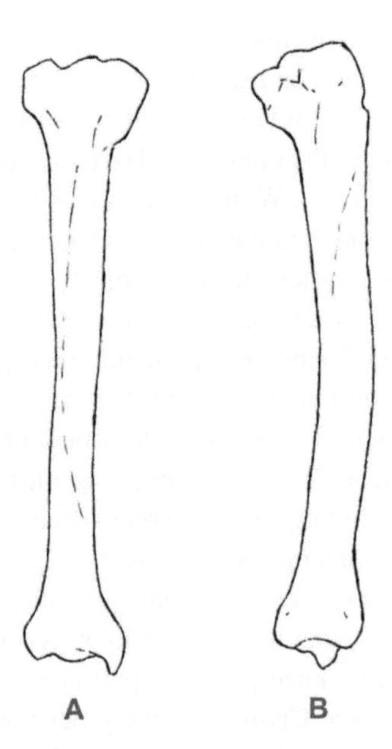


**Fig. 11.4** (**A**) Anterior and (**B**) lateral aspects of the right tibia. In (**A**), note the mild coronal plane bend and concave medial and, in (**B**), the slope of the tibial plateau and posterior concavity

posterior thigh and tibial muscles as they approximate during extreme knee flexion.

The sagittal and coronal plane bends in the femur and tibia certainly influence the alignment of the distal femur and proximal tibia at their respective joint surfaces, which adds complexity to how the knee must move to navigate daily activities. However, Wolff's law also suggests that the transverse plane torsions that exist in both the proximal femur and tibia influence the lower extremity alignment, knee, and ankle joint surface contacts and movements. In Chap. 10, we discuss both acetabular anteversion and the normal femoral antetorsion that positions the head and neck approximately 15–20° anteriorly relative to the femoral condyles. Our emphasis when examining the hip joint is how this normal femoral torsion optimally matched the acetabular anteversion, which likely maximized the surface area of contact between the femoral head and the acetabulum for weight-bearing positions. In the context of the knee joint, the anterior femoral offset has repercussions distally: (1) the need for a conjunct external rotation of the tibial relative to the femur during terminal knee extension (discussed in more detail later); (2) the development of external tibial torsion, which normally ranges from 15° to 20°; and (3) an offset of the ankle joint axis that appears to accommodate tibial torsion, i.e., the ankle joint axis (talocrural joint or what is commonly called the ankle mortise) being oriented obliquely, medial to lateral, posteroinferiorly (Fig. 11.5). These interacting bony bends and torsions create a mild genu valgus, i.e., abduction of the tibia relative to the femur, and an approximate 15° external foot progression during walking.

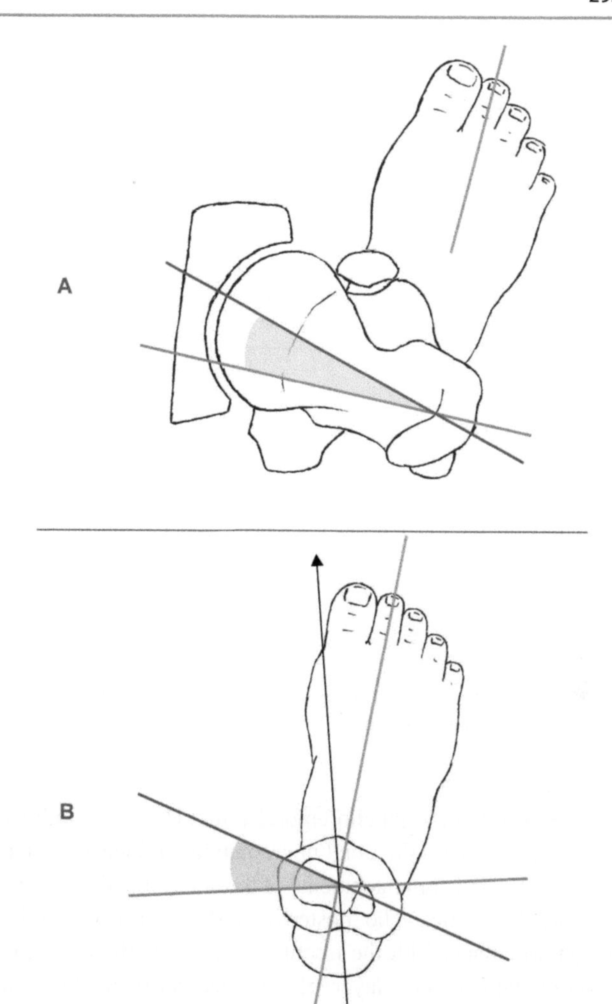


**Fig. 11.5** (**A**) Relative orientation of the proximal femur, i.e., antetorsion position relative to femoral condyles, and (**B**) relative orientation of the knee, i.e., tibial plateau to the transmalleolar axis of the ankle joint. Note the approximate 15° external "rotation" appearance of the foot, which is the result of the proximal limb countertorsions. Also note that external rotation of the foot contributes to external foot progression observed in typical gait

In Chap. 10, we speculate that proximal limb malalignments secondary to abnormal angles of inclination and declination, as well as acetabular anteversion, could contribute to the initiation of early hip joint degeneration. We now show that there also exists a complex coupling of countertorsions involving the femur, tibia, and ankle joints. Therefore, it is possible that mechanical malalignments distal to the hip joint could also induce bony and soft tissue, as well as joint movement, compensations involving the hip, knee, and/or ankle that could predispose an individual to a number of acute or chronic strain injuries. In Cathy's case, it seems strange that merely rising from the floor, which was her original mechanism of injury, could produce forces large enough to create a third-degree lateral meniscus injury. However, if over many years her menisci were not functioning normally because of some preexisting lower limb malalignment, it is possible that failure of the lateral meniscus was one of "fatigue" and not due to excessively large forces. Recall that the original clinical question posed to the gait analysis center staff was related to finding a potential biomechanical cause of Cathy's injury (or continued series of injuries). Knowing the complex nature of the alignment of the lower extremity, the examiner elected to measure genu valgus, femoral antetorsion, and tibial torsion, as well as look for the foot progression angles in a quiet stance and determined that they were normal (see Table 11.1). Thus, it appears that the mechanism of injury was most likely related to the complex kinematics of the lower limb while rising from the floor, which, by the way, is one of the most common mechanisms of meniscal injury.

The bones that comprise the three joints around the knee are shown in Fig. 11.6. At the distal end of the femur, we find the convex-shaped medial and lateral femoral condyles that articulate with the shallow gutters of the superior tibial condylar surfaces, i.e., tibial plateaus. The medial and lateral epicondyles project laterally from the femoral condyles and serve as sites of attachment for the collateral ligaments. The femoral condyles meet anteriorly to form the trochlear (intercondylar) groove, which houses the posterior surface of the

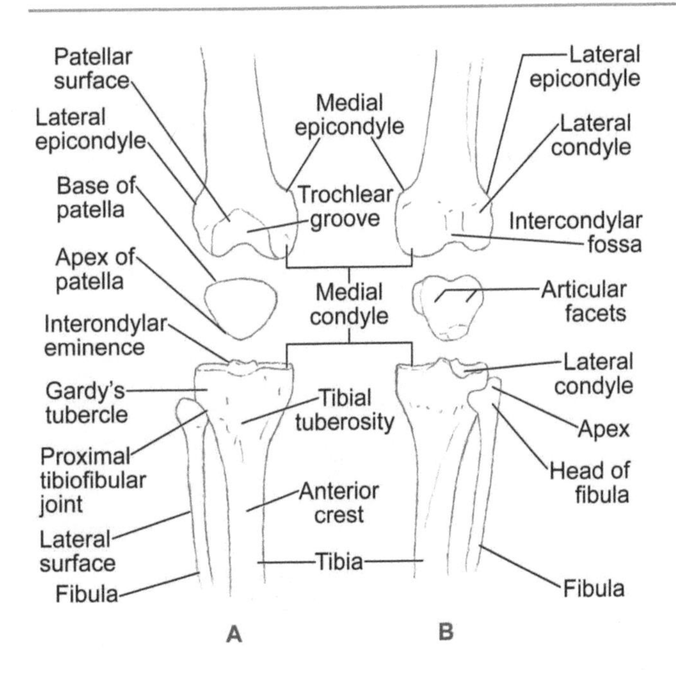


**Fig. 11.6** (**A**) Anterior and (**B**) posterior coronal views of the bones, which comprise the knee joints

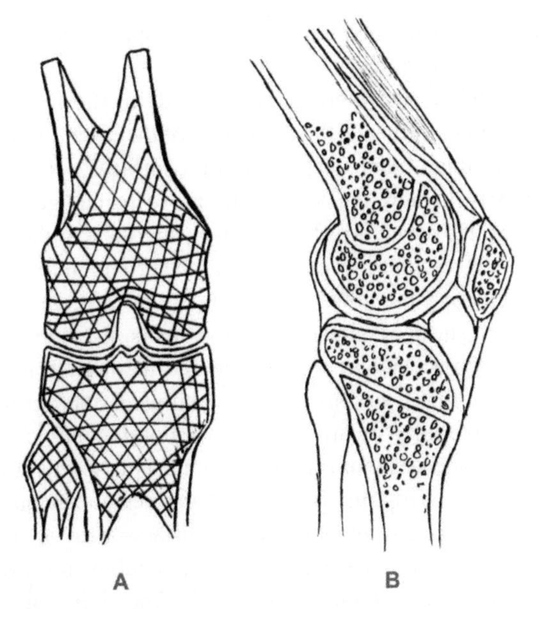


**Fig. 11.7** (**A**) Trabecular patterns in the distal femur, proximal tibia, and fibular head laid along the lines of functional stresses and (**B**) lateral view of knee joint trabeculae relative to tibiofemoral and patellofemoral joint surfaces

patella, forming the patellofemoral joint. The intercondylar fossa is paired with the tibial intercondylar eminence, which serves as the site for attachments of the anterior and posterior cruciate ligaments. The posterior surface of the patella, which articulates with the trochlear groove of the femur, is covered with a thick layer of articular cartilage that has evolved to distribute the large compression forces at that joint created by the activation of the quadriceps during functional activities. A vertical ridge on the posterior surface of the patella divides the medial and lateral facets, which articulate in the trochlear groove; a third "odd" facet on the patella is found along the medial border of the medial facet.

Although the proximal tibiofibular joint does not participate directly in knee function, the head of the fibula is the site of attachment for the lateral collateral ligament and biceps femoris. Note that although the fibula provides some lateral support to the tibia, the tibia is the major load-bearing bone as it transfers approximately 80% of the load between the ground/foot and knee.

Let us now look more closely at the bony architecture of the distal femur and proximal tibia and the shapes of their articulating surfaces. With most joints, it is typical that the articulating portions of the bones do not exactly match; that is, the ends of the bones are not perfectly congruent. This is particularly true at the tibiofemoral joint. Because of this incongruency, contact between these rigid bodies can occur only at points or along lines, suggesting that only a fraction of the contact area is available to mitigate and/or redistribute axial compression, i.e., pressure. For both the femur and tibia, the cortical bone in the midsection is relatively thick, but beneath the articulating cartilage, it is very thin. Therefore, close to the joint, both the femur and tibia are filled with trabecular (cancellous) bone (Fig. 11.7). Nature has decided that the combination of thin cortical bone and a thicker region of trabecular bone beneath the joint surface is more resilient than only a thick layer of cortical bone. It is notable that cortical bone is stiffer than cancellous bone, but the viscoelastic nature of the cancellous bone allows it to absorb more energy and effectively dampen multidimensional joint loads. At the same time, this arrangement increases the area by which pressure can be distributed, thus reducing stress on the joint surface (i.e., articular cartilage and menisci). Generally, the greater ratio of cancellous to cortical bone near joint surfaces is not unique to the knee joint. However, introducing this idea now is particularly relevant, given the fact that the respective articulating surfaces at the tibiofemoral joint present such a mismatch. In other words, it is likely that, given Cathy's hypothetical advanced knee joint osteoarthritis, the subchondral cancellous bone has also undergone significant adaptation in order for the joint to manage peak compression forces.

Examining the asymmetry of the shapes of the femoral and tibial condyles provides insight into the challenges imposed on the tibiofemoral joint, reinforcing the importance of the subchondral bone. The inferior surface of the femoral condyles is convex, with greater anteroposterior than mediolateral dimensions, and their anteroposterior axes are not parallel. Moreover, the medial femoral condyle is longer anteroposteriorly and is narrower than the lateral femoral condyle (Fig. 11.8). Because of the differences between

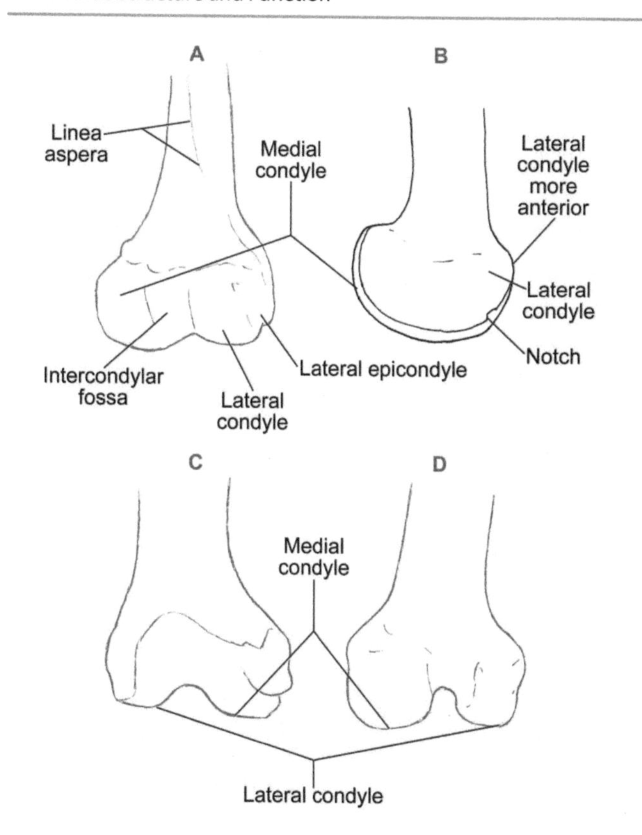


**Fig. 11.8** (**A**) Oblique, (**B**) lateral, (**C**) anterior, and (**D**) posterior views of femoral condyles

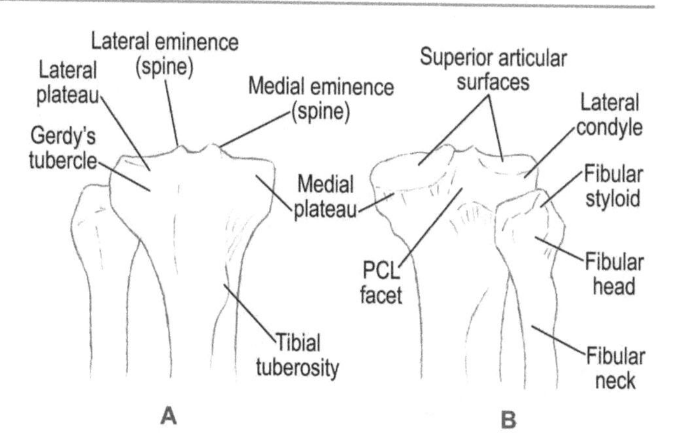


**Fig. 11.9** (**A**) Anterior and (**B**) posterior views of tibial plateaus, which only partially capture the unique architecture of the lateral tibial surface

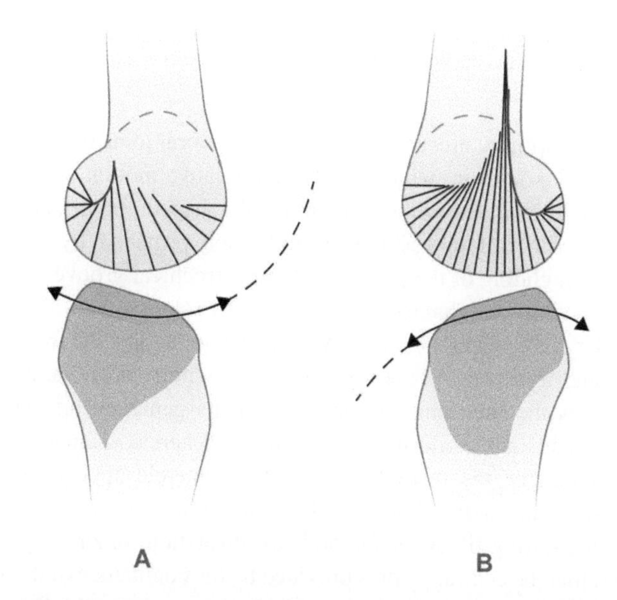


**Fig. 11.10** Schematic illustrating tibiofemoral joint surface geometries for (**A**) medial femoral condyle (top) and medial tibial plateau (bottom) and (**B**) lateral femoral condyle (top) and lateral tibial plateau (bottom). Kapandji suggested that the geometric spiral of Archimedes could be used to illustrate differences in the radii of curvature and centers of rotation between the medial and lateral femoral condyles and medial and lateral tibial condyles

medial and lateral sides, we can imagine that the tibiofemoral joint arthrokinematics may not conform exactly to the convex-concave rule, as defined in Chap. 3. But there is more. For example, when we view the knee articulations from an anterior perspective (Fig. 11.7A), it appears that the medial and lateral femoral condyle convexities match their respective tibial condyles well. However, this sketch does not capture the detailed nature of the tibial plateaus. In reality, the anteroposterior profiles of the tibial articular surfaces differ from one another: the medial articular surface is concave, while the lateral articular surface is convex. Furthermore, whereas the medial articular surface of the tibial plateau is biconcave, the lateral surface has more of a saddle shape, i.e., concave transversely and convex sagitally (Fig. 11.9).

Taking vertical sagittal sections at two different levels of the medial and lateral knee compartments in a sample of fresh bone, Kapandji discussed how the femoral and tibial condyle asymmetries may theoretically affect joint centers of motion (Fig. 11.10). Kapandji first illustrated the asymmetries in morphologies of the medial and lateral femoral and medial and lateral tibial condyles. He then quantified the incongruencies within the tibiofemoral joint compartments. Because the convexity of the medial femoral condyle better matches its concave partner, Kapandji concluded that the medial compartment is more congruent than the lateral compartment, suggesting that the medial femoral condyle would be slightly more stable within the tibial medial plateau. On the whole, however, the radii of curvature of the femoral condylar and tibial articular surfaces are unequal, pointing to incongruencies that render the tibiofemoral joint relatively unstable. Fortunately, the knee menisci effectively improve tibiofemoral joint congruency, thus increasing the surface area of contact between the proximal and distal bony partners. We return to a more detailed discussion of the knee menisci after we briefly examine the knee geometry that fosters transverse plane rotation.

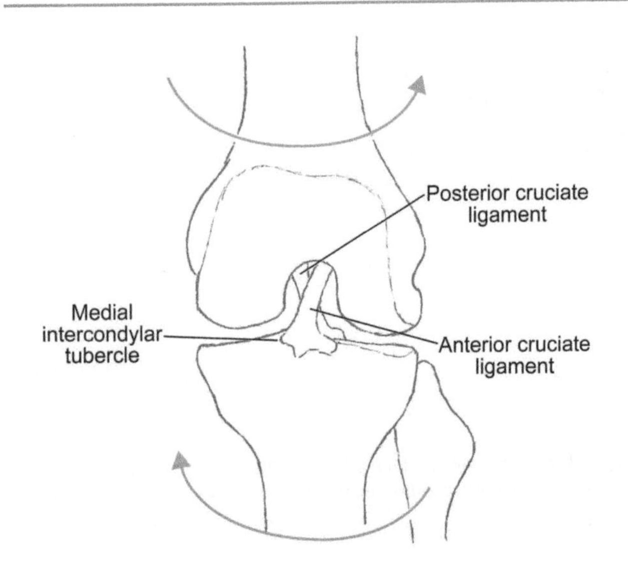


**Fig. 11.11** During axial rotation, the tibia (or femur) appears to pivot around either the medial intercondylar tubercle or cruciate ligaments

The primary movements at the tibiofemoral joint are flexion and extension, which involve the relative motion of the femoral condyles to their "reciprocally shaped" tibial partners, with full powerful knee extension provided by the smooth gliding of the patella within the trochlear groove via the extensor mechanism. However, axial rotation, as a secondary movement, is vital in order to accommodate the frontal and transverse plane torsions of the femur and tibia, as well as the multidimensional movements generated at the foot and ankle during locomotion. Whereas Kapandji claimed that the medial intercondylar tubercle appears to serve as the point about which slight sliding and pivoting occurs during tibial (or femoral) axial rotation, others suggest that the central pivot is provided by the confluence of the cruciate ligaments at the intercondylar eminence (Fig. 11.11). Functionally, these slight differences in perspective may not be that important.

#### 11.4.1.1 Soft Tissue Constraints

We have established the importance of the knee's role in transmitting distal forces and moments proximally while providing a large degree of sagittal mobility. To match functional needs, however, tibiofemoral stability is sacrificed to a large degree. Since the distal femur and proximal tibia are insufficiently congruent to provide bony stability, the knee is largely dependent on a complex system of passive, e.g., ligaments, and active, e.g., muscles, soft tissues for stability. The capsule of the knee encloses the medial and lateral compartments of the tibiofemoral joint, as well as the patellofemoral joint (Fig. 11.12). The capsule is reinforced by many ligaments, fasciae, and muscles (Table 11.3). The importance of providing significant details of the joint capsule and all its supporting structures

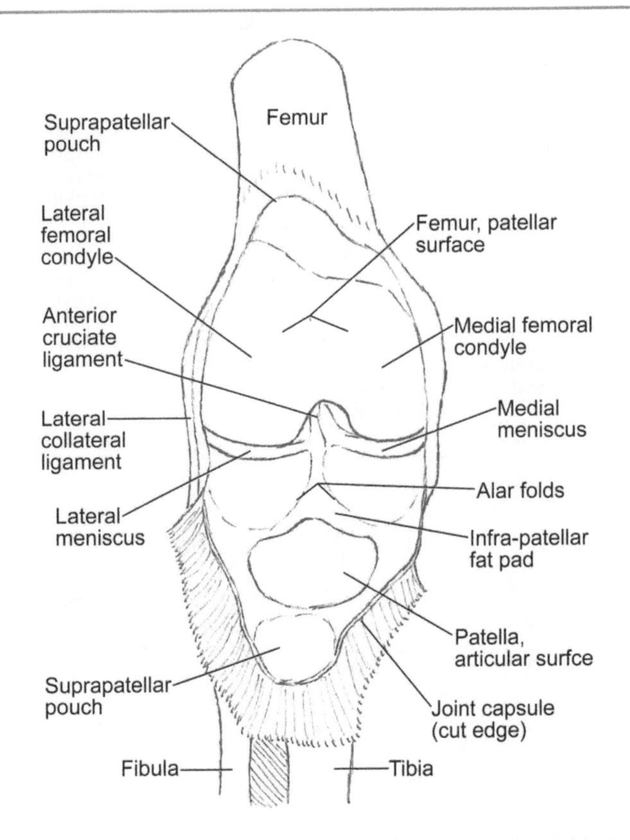


**Fig. 11.12** Knee joint capsule and its proximity to other periarticular joint soft tissues

**Table 11.3** Reinforcing structures of the knee joint capsule

| Location of reinforcement | Passive restraint | Dynamic restraint |
|---|---|---|
| Anterior | Patellar tendon<br>Patellar retinacula (medial and lateral) | Quadriceps |
| Lateral | Lateral collateral (fibular) ligament<br>Lateral patellar retinaculum<br>Iliotibial band | Biceps femoris<br>Popliteus tendon<br>Gastrocnemius (lateral head) |
| Posterior | Oblique popliteal ligament<br>Arcuate ligament | Popliteus<br>Gastrocnemius<br>Hamstrings (especially tendon of semimembranosus) |
| Posterolateral | Arcuate ligament<br>Lateral (fibular) collateral ligament<br>Popliteofibular ligament[a] | Popliteus tendon |
| Medial | Medial patellar retinaculum<br>Medial collateral ligament<br>Posterior oblique ligament[b] | Fasciae from semimembranosus tendon<br>Sartorius, gracilis, and semitendinosus tendons |

[a]Also referred to as medial patellofemoral ligament
[b]Also referred to as posterior-medial capsule

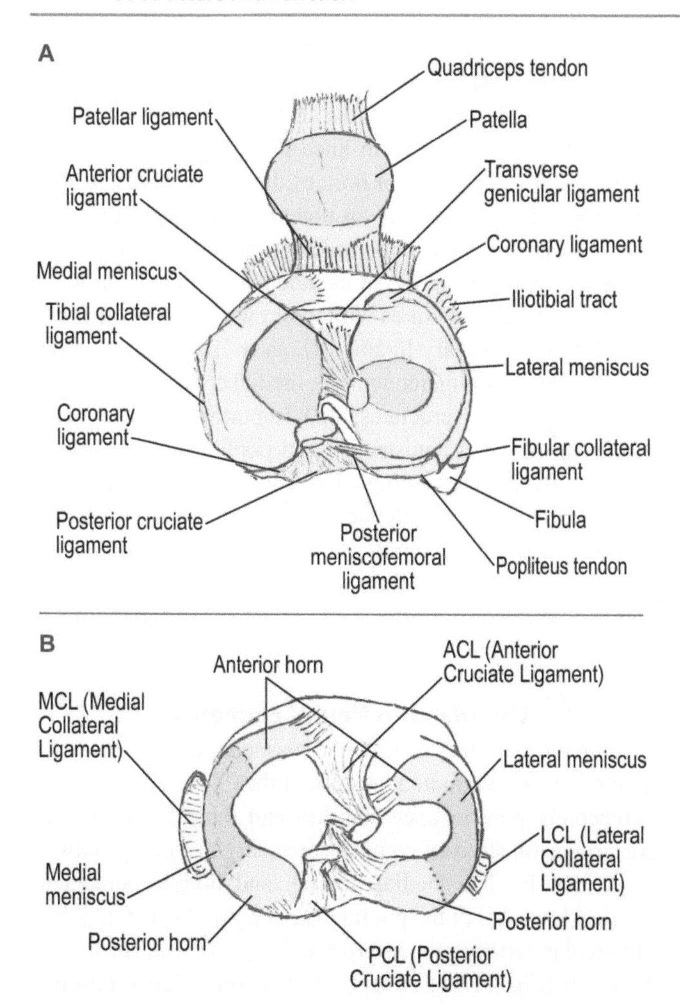


**Fig. 11.13** (**A**) Superior view of the right tibia illustrating the medial and lateral menisci and associated ligaments and (**B**) attachments of the menisci anterior and posterior horns relative to the cruciate ligaments

is that one, or more, of these likely experiences unusual and excessive tensile loads induced by a significant meniscus injury, similar to what Cathy experienced. See Supplementary 11.1 for additional detailed images of knee joint capsular, ligamentous, and muscular anatomy.

### 11.4.1.2 Menisci

The wedge-shaped medial and lateral menisci (or semi-lunar fibrocartilages) that fit between the mismatched femoral and tibial condyles significantly increase the contact areas between the femur and tibia (Fig. 11.13). However, even with intact menisci, the contact areas still do not extend over the entire tibial plateau (Fig. 11.14), and the pressure-loaded area has been shown to shift locations at different points throughout the knee range of motion. For example, as flexion increases, the contact area shifts from anterior to posterior. It is notable that if menisci were not present, the contact area would be reduced by a factor of two to three. Most have concluded that partial or full loss of a meniscus is associated

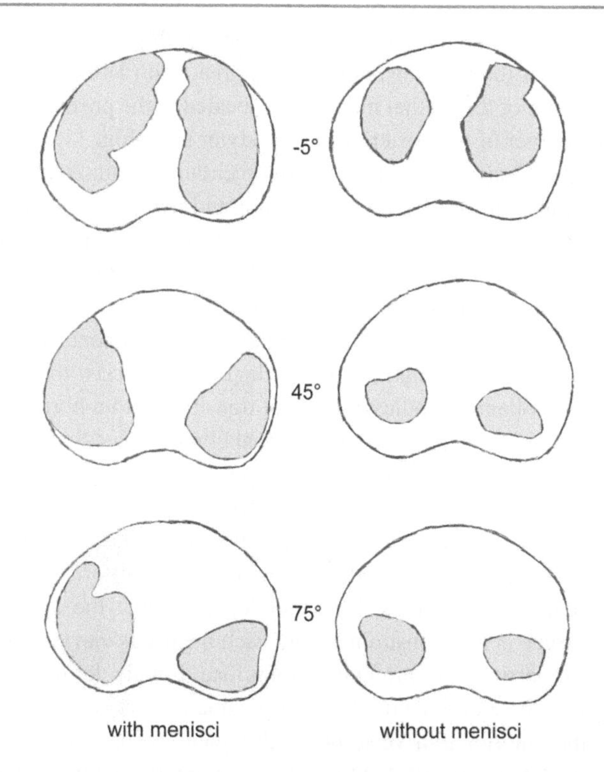


**Fig. 11.14** Size and location changes in contact area between the femur and tibia with knee flexion. (Note: −5 extension = hyperextension, the significant reduction in the size of the contact area when menisci are not present. (Modified with permission from Brinckmann et al. (2016))

with the situation where the large axial loads at the knee cannot be mitigated/redistributed well, ultimately leading to the premature development of degenerative changes in the tibiofemoral joint, as we believe happened to Cathy.

The architecture of the menisci creates a concavity over the tibial condyles that much more readily matches the convexities of the femoral condyles yet cannot fully accommodate the asymmetrical shapes of the condyles that contribute to the complexity of joint kinematics. The menisci are incomplete rings at the level of the intercondylar tubercles, with the medial being more crescent shaped and the lateral meniscus resembling an incomplete circle, each with anterior and posterior horns. As clearly shown in Fig. 11.13A, the menisci have functionally important attachments that anchor them but do not impede their movement. For example, the knee joint capsule (also referred to as coronary or meniscotibial ligaments) secures the peripheral aspects of the menisci to the tibia. From earlier presentations, we know that, in general, joint capsules contribute to joint surface kinematics, so one might wonder how the knee joint capsule might assist meniscal biomechanics.

In the anterior intercondylar area, (1) the anterior horn of the lateral meniscus lies just in front of the lateral intercondylar tubercle, (2) the posterior horn of the lateral meniscus lies behind the lateral intercondylar tubercle, (3) the anterior horn of the medial meniscus is located in the anteromedial corner of the anterior intercondylar area, and (4) the posterior horn of the medial meniscus is located in the posteromedial corner of the posterior intercondylar area (Fig. 11.13B). For the menisci to be able to develop circumferential tension and transmit load between the femur and tibia, they need to be held firmly in these locations. The anterior and posterior horns (or roots) are ligamentous-like structures that exhibit high tensile stiffness and strength, yet are flexible. The roots insert into the tibial eminence similar to how ligaments insert into the bone through a fibrocartilaginous enthesis, that is, where collagen bundles at the root transition into a fibrocartilaginous zone, followed by calcified fibrocartilage that connects to the underlying cortical bone. It is notable that the anterior and posterior roots of the medial and lateral menisci vary in their collagen orientation at their insertions, resulting in altered material properties, likely related to differences in their mechanical roles; for example, recall that the lateral meniscus is less constrained, and each meniscus carries different portions of tibiofemoral joint load. Finally, let us recognize the anterior intermeniscal ligament (also referred to as the anterior transverse or meniscomenisco or transverse geniculate ligament) that links the more anterior fibers of the anterior horns. It is proposed that this ligament restrains anterior subluxation and excessive posterior translation when the menisci are under load, as well as acts as a tie between the menisci, controlling their relative positioning on the tibial plateau during axial rotation. Also, the presence of fascial attachments from the patella, via strands from the infrapatellar fat pad (not shown), to the anterior transverse ligament suggests a biomechanical connection between the quadriceps femoris and menisci. Not evident in Fig. 11.13A are the fascial links, i.e., patellar retinacula, from the lateral borders of the patella to the peripheral borders of both menisci.

The medial meniscus is securely anchored by the posterior fibers of the medial collateral ligament (MCL). The high stiffness value of the MCL suggests a significant role in resisting tension and in controlling excessive mobility of the medial meniscus. On the other hand, the lateral meniscus is separated from the lateral collateral ligament by the popliteus tendon, which sends fascial connections to the posterior border of the meniscus to form the posterolateral corner of the knee. Despite its fibrous connection to the popliteus tendon, the lateral meniscus has significantly more mobility than its medial counterpart. The semimembranosus has a fascial expansion to the posterior border of the medial meniscus to form the posteromedial corner of the knee. Finally, fibers of the posterior cruciate ligament attached to the posterior horn of the lateral meniscus form the menisco-femoral ligaments (MFLs); where the ligament of Humphrey (aMFL) runs anterior to the posterior cruciate ligament, and the ligament of Wrisberg (pMFL) runs posterior to the posterior cruciate ligament. The high tensile moduli of these ligaments suggest that they play important functional roles within the joint: (1) the pMFL is tight in the extended knee and is lax with the knee flexed, and (2) the aMFL is lax in the extended knee but becomes taut with knee flexion. Both pMFL and aMFL control the posterior horn of the lateral meniscus during knee flexion. Finally, let us note that fibers from the anterior cruciate ligament into the anterior horn of the medial meniscus may have a role in its mechanics. In summary, we see the meniscal structures with their own attachments to the tibia (via the coronary ligaments), as well as fibrous extensions to important ligaments, e.g., medial collateral and anterior and posterior cruciates, and important muscles, e.g., quadriceps, medial hamstrings, and popliteus, suggesting a complex dynamic coupling of passive and dynamic structures, which fosters both the stability and mobility of the menisci and tibiofemoral joint. The interrelationships described suggest that injury to either key tibiofemoral ligaments or the menisci affects the internal mechanics of the knee in hard-to-predict ways.

#### 11.4.1.3 Vascular and Neural Elements

Examination of the neural and vascular anatomy of the menisci is also warranted because of the role menisci mechanoreceptors play in knee stability and the relative healing constraints on strained menisci secondary to a compromised blood supply. The medial, lateral, and middle geniculate arteries, branches of the popliteal artery, provide the majority of vascularization to the inferior and superior aspects of each meniscus (Fig. 11.15). A capillary network originating from branches of these arteries originates within the synovial and capsular tissues of the knee along the meniscal peripheries. The peripheral 10–30% of the medial meniscus border and 10–25% of the lateral meniscus are referred to as the "red zone" because they are relatively well vascularized. The inner two-thirds of the menisci are avascular and are referred to as the "white zone" and receive nourishment primarily from synovial fluid via diffusion or mechanical pumping, i.e., joint motion. The prognosis for healing following a meniscal strain injury is directly related to the robustness of, and the location relative to, its vascular supply.

According to Cathy's history, her physician elected to repair the meniscus strain, suggesting that the injury was located in a well-vascularized peripheral region. It appears that this was attempted several times over many years. However, a surgical repair is only as good as the nutrition to the healing meniscus and adherence to the healing constraints of the soft tissue via a well-structured rehabilitation program. Cathy's surgical history also included several instances of arthroscopic debridement, meaning that some damaged tissue was irreparable and removed. It has been suggested that lateral meniscectomies are more dangerous than medial meniscectomies because articular cartilage lesions tend to deteriorate much faster. The faster degeneration

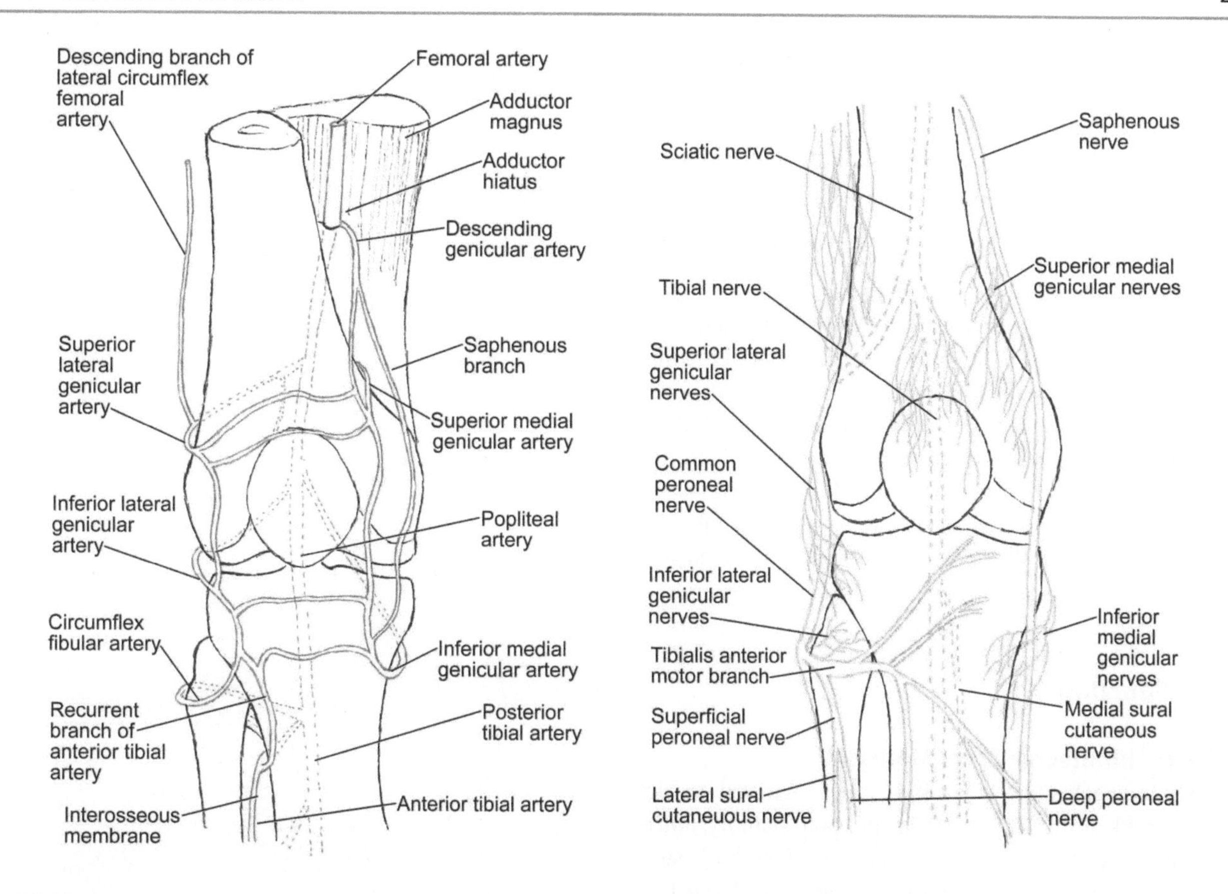


**Fig. 11.15** Anterior view of the (**A**) confluence of geniculate arteries and (**B**) neural innervations supplying the knee joint

on the lateral side is thought to be related to the fact that the fraction of load transmitted through the lateral meniscus is greater, particularly at its posterior horn. In the case we are studying, it is likely that poor vascularization in the regions where the repair was attempted and insufficient postoperative rehabilitation and the removal of damaged meniscal tissue likely precipitated early tibiofemoral joint degeneration.

The knee is innervated by the posterior branch of the posterior tibial nerve and terminal branches of the obturator and femoral nerves. Additionally, the femoral nerve sends branches to the vastus medialis, intermedius, and lateralis muscles, while the obturator nerve extends a genicular branch from its posterior division. Both the tibial and common peroneal nerves give off branches as the superior medial genicular, inferior medial genicular, and middle genicular and the superior lateral genicular, inferior lateral genicular, and recurrent genicular, respectively. The joint capsule receives innervation from the recurrent peroneal branch of the common peroneal nerve. These branches pierce the capsule and accompany the vascular supply to the peripheral regions of the menisci, as well as the anterior and posterior horns, where most of the nerve fibers for mechanoreceptors and nociceptors (pain transmission) are concentrated. The middle third of the meniscus is less densely innervated. Mechanoreceptors, commonly identified as Ruffini endings, Pacinian corpuscles, and Golgi tendon organs within the menisci and other periarticular structures, function as transducers converting compression and tension stress into nerve impulses that provide afferent information to the spinal cord, cerebellum, and other central control structures. The low threshold and slowly adapting Ruffini mechanoreceptors respond to changes in joint deformation and pressure, while the Pacinian corpuscles (also low threshold) respond rapidly to changes in tension. Golgi tendon organs are high-threshold responders to tension and are inhibitory to active muscle. It appears that mechanoreception is most facilitated in the meniscal horns when the end range of knee flexion and extension is reached. The neurosensory signals from joint mechanoreceptors induce automatic, as well as conscious, perceptions that are vital to normal knee function and tissue homeostasis. Although mechanoreceptors are essential to control the functional aspects of the knee joint, nociception is as important.

The repeated injury and subsequent debridement of the lateral meniscus noted in this case undoubtedly affected the sensory apparatus and normal homeostasis of the periarticular and intraarticular structures. And we cannot assume that attempts to repair a peripheral meniscus tear can restore the normal proprioceptive function of the menisci. Therefore, the absence of normal mechanoreception would prevent the central nervous system from getting the information it needed to assist in adaptations to manage abnormally high stresses and extremes of motion. That inability would likely inhibit normal joint proprioception, perhaps contributing to events leading to early degenerative processes, as we have already speculated.

Cathy's physical examination revealed the weakness of the right-sided hip flexors, hip extensors, and abductors bilaterally. Based on our review of the neural anatomy around the knee joint, we hypothesized that it is likely that Cathy's weakness proximally was related to afferent feedback directly from the femoral, obturator, and sciatic nerves, which may have indirectly affected the superior and inferior gluteal nerves. These afferent signals could be responsible for inducing adaptations in the activation of the hip muscles, as well as arthrogenic (i.e., knee pain) inhibition of muscle activation.

## 11.4.2 Function

### 11.4.2.1 Biomechanics of Menisci

Multiple biomechanical roles have been ascribed to the menisci: shock absorption, load transmission, stability, nutrition, joint lubrication, and proprioception. Many have suggested that the primary function of the menisci is to act as shock absorbers from the impulsive loading that accompanies bipedal locomotive activities. Yet the viscoelastic properties between intact and torn menisci have been shown to be similar, and the strain rate dependence of Young's modulus under higher rates of loading increased only moderately, suggesting that the shock-absorbing function of the menisci may not be primary. Moreover, other studies have demonstrated that the dynamic modulus of elasticity in hyaline cartilage is approximately ten times greater than in the menisci, suggesting that shock absorption and energy dissipation are best left to articular cartilage. Although dissipating large loads imposed at higher rates is certainly important, we suggest that the complexity of the meniscus apparatus entails the interdependency of multiple functions. To the reader, do not conclude that our order of presentation that follows suggests a particular priority of importance. It really does not matter where we begin our discussion of the functions of the menisci. So let us first look at the role menisci play during the sagittal and transverse plane movements of the tibiofemoral joint. Examining meniscal kinematics requires us to briefly introduce tibiofemoral arthrokinematics, and in doing so, one aspect of the interdependency of meniscal functions becomes clear.

### 11.4.2.2 Meniscal Kinematics and Architecture

It is clear that during flexion and extension, the contact point between the femur and tibia is not stationary, and one role of the menisci is to optimize congruency between the articulating bones at the joint surface. During flexion, it has been shown that the contact point between the femur and tibia moves posteriorly, while during extension, the contact point moves anteriorly. The menisci follow the movements of the femur relative to the tibia. Therefore, starting from the position of extension, the menisci, already in an anterior location, move unequally in a posterior direction (Fig. 11.16). By unequal we mean that the lateral meniscus recedes twice as far as the medial meniscus, on average 12 mm versus 6 mm, respectively. Because the menisci are anchored at their anterior and posterior horns, internal shear loads result in menisci deformation simultaneous with their translations. Thus, in addition to gliding a greater distance, the lateral meniscus also deforms to a greater degree. We encourage the reader to review Chap. 3, where we detail the viscoelastic properties of fibrocartilage since it is those properties that are challenged when the tibiofemoral joint is loaded and moved. Just briefly, we understand that fibrocartilage has very similar material properties as articular cartilage and becomes stiffer under higher rates of loading. The major difference between

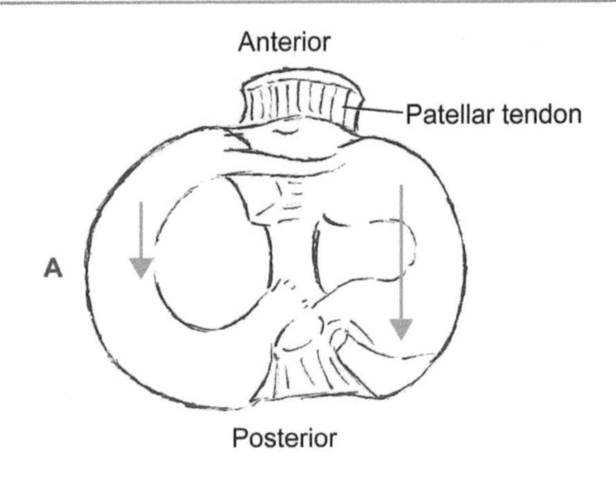


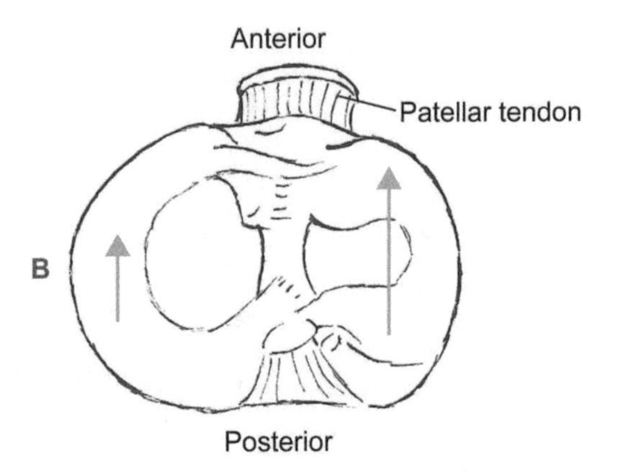


**Fig. 11.16** Superior view of the sliding of the medial and lateral meniscus during (**A**) flexion and (**B**) extension. (Note: the larger arrow indicates the greater movement of the lateral meniscus, which suggests that axial rotation is also an important aspect of tibiofemoral flexion/extension)

the typical hyaline cartilage role at the joint surface is that it generally responds, i.e., mitigates, primarily to axial loads, whereas the menisci must respond to the combination of axial and shear loads; that is, loading of the menisci is more complex. Intuition suggests that perhaps the greater mobility and deformation of the lateral meniscus may more readily predispose it to strain injury. On the other hand, being less constrained may be an advantage.

It is notable that the differences in kinematics between the two menisci are related to differences in the radius of curvature of the femoral condyles, as well as asymmetry in the morphology of the medial and lateral tibial condyles. During extension, the condyles present their greatest radii of curvature on top of the tibial articular surfaces, forcing the menisci to be tightly interposed between the two articulating bones. This tight interposition at full tibiofemoral extension is enhanced by the tautness of the collateral and cruciate ligaments. In any case, full tibiofemoral extension is the most stable position of the joint (i.e., it is the close-packed position), thus also promoting the optimal transmission of compression loads. With knee flexion, on the other hand, the condyles present their smallest radii of curvature, causing the menisci to have only partial contact with the condyles. At the same time, the collateral and cruciate ligaments are less taut during flexion movements, making the tibiofemoral complex less stable; i.e., the joint operates in a more loose-packed configuration.

The movement of the menisci during functional activities is the result of two general mechanisms: passive and active. In the first case, the menisci are simply pushed by the femoral condyles. The production of synovial fluid under normal circumstances maintains joint lubrication, which creates near-friction-free articular surfaces, allowing for easy gliding of the menisci. The active mechanism of meniscal gliding is related to the fascia connections between the menisci and the muscles we described earlier (Fig. 11.13A). During extension, the menisci are pulled anteriorly via the patellar retinacula, which become taut as the patella glides superiorly, also pulling on the transverse ligament. The posterior horn of the lateral meniscus receives an additional pull through tension received via the meniscofemoral ligament. With flexion, the medial meniscus is drawn posteriorly by its posterior fascial attachment to the semimembranosus, while the anterior horn paradoxically is pulled forward by the anterior cruciate ligament; the lateral meniscus is pulled posteriorly through fascial connections to the popliteus muscle.

As with tibiofemoral knee flexion and extension, the menisci follow the femoral condyles passively during axial rotation; however, as the patella moves relative to the tibia during axial rotation, the tension in the patellar retinacula may pull one of the menisci anteriorly. Axial rotation, i.e., internal and external rotation, at the tibiofemoral joint is very restricted when the joint is extended but is increasingly free

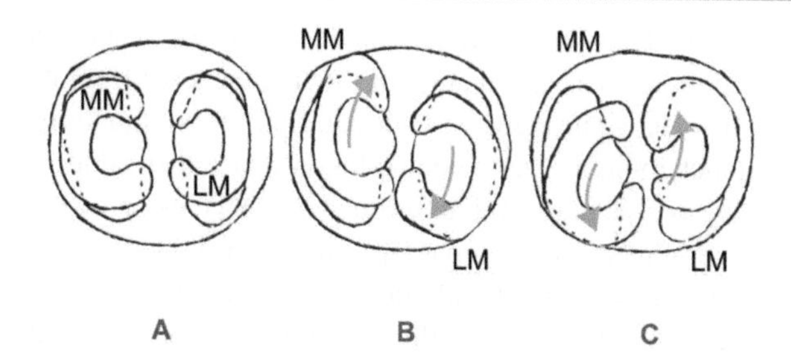


**Fig. 11.17** Menisci position: (**A**) in a neutral position, (**B**) during external femoral rotation, and (**C**) during internal femoral rotation. (Note: the dotted lines outlining the menisci relative to the solid lines represent their deformation)

as the joint flexes, with a maximum range at 90° knee flexion. Whether the foot is fixed to the ground or moving unconstrained in an open chain with the tibia free to internally or externally rotate relative to the femur, the menisci follow the path of the femoral condyles. Nevertheless, most meniscal injuries occur when the foot is fixed and the tibiofemoral joint is placed in (1) combinations of flexion-abduction-external rotation (most commonly), (2) combinations of flexion-abduction, or (3) any combination of flexion-axial rotation. With the knee in a neutral position, i.e., extended and neutral rotation, the medial and lateral menisci are resting slightly anterior but well centered over their corresponding tibial condyles (Fig. 11.17A). With external rotation (femur relative to the tibia), the lateral femoral condyle moves posteriorly as the medial condyle moves anteriorly (note: in relative terms, femoral external rotation is the "same as" tibial internal rotation). In that case, the medial meniscus is pulled anteriorly, while the lateral meniscus is pushed posteriorly (Fig. 11.17B). Internal rotation of the femur induces anterior advancement (anterior glide) of the lateral meniscus and a recession (posterior glide) of the medial meniscus (Fig. 11.17C). Because of differences in constraints, the lateral meniscus glides approximately twice as far as the medial meniscus. During axial rotation, as with flexion/extension, the menisci deform around their points of constraint, i.e., horns.

It seems clear that the kinematics and deformation of the menisci have a clear role in the arthrokinematics of the tibiofemoral joint while concomitantly promoting low-friction movements between the two articulating partners and transferring compression and shear stresses between the femur and tibia. And because of the viscoelastic properties of these fibrocartilaginous structures, the impulse loading at the tibiofemoral joint, for example, related to skipping, running, and jumping, can be well managed over thousands of repetitions as long as the health of the menisci is maintained.

So let us speculate about the likely mechanism of injury when Cathy originally injured her lateral meniscus. It is clear from our description that the sequence of events during multidimensional movements at the tibiofemoral joint is complex. Thus, as Cathy attempted to rise from the floor, which requires a combination of moving the knee from flexion to extension, there was, in addition, a combination of conjunct rotations, i.e., axial rotation and either adduction (varus) or abduction (valgus), which led to a tension strain of the lateral meniscus as it perhaps failed to follow the condylar movements on the tibial plateau. Interestingly, laboratory studies have shown that although fibrocartilage has greater tensile strength than hyaline cartilage, it has much less tensile strength than a tendon.

We have alluded to the protective role that menisci play as load-bearing (sharing) structures (revisit Fig. 11.14). For example, in a research project that examined the contribution of menisci to load bearing, one experiment showed that removing greater than 20% of the meniscus drastically increased the stresses imposed on the articular cartilage of the tibiofemoral joint. Additionally, with a 65% meniscectomy, there was a 225% increase in maximal shear stress in the articular cartilage. There is abundant evidence linking the progression of the early degeneration of the tibiofemoral joint following partial or whole meniscectomy, suggesting the important role that menisci play in preserving the articular cartilage. We conclude this section on biomechanical function by looking at the load transmission ability of the menisci. Although we previously examine the ultrastructure and biochemistry of fibrocartilage in general (Chap. 3), we need to see how meniscal fibrocartilage may be different (from the acetabulum labrum, for example) before we describe load transmission mechanics.

The meniscal fibrocartilage is comprised of a dense extracellular matrix (ECM) composed of approximately 75% water, 20% type I collagen fibers, and 5% noncollagenous substances, e.g., proteoglycans, matrix glycoproteins, and elastin. Meniscal cells within the matrix synthesize and maintain the ECM, which, as you may recall, largely determines the material properties of the tissue. The cells of the menisci are referred to as fibrochondrocytes because they appear to have characteristics of a mixture of fibroblasts and chondrocytes. In normal meniscus, tissue fluid represents 65–70% of the total weight, with most of the water retained by the hydrophilic proteoglycan aggregates. The water content of the meniscus is greater in the posterior region than in the central or anterior regions, suggesting that this region may have greater exposure to axial loads. The interaction between water and the matrix macromolecule network (i.e., collagen + proteoglycan aggregate) creates large hydrostatic pressures, which are used to overcome the drag of frictional resistance of forcing fluid through the meniscal tissue, endowing the tissue with its unique viscoelastic properties. Collagen is primarily responsible for the tensile strength of menisci, with the predominance of type I distinguishing the fibrocartilage of menisci from hyaline (articular) cartilage; the type II collagen found in articular cartilage is associated

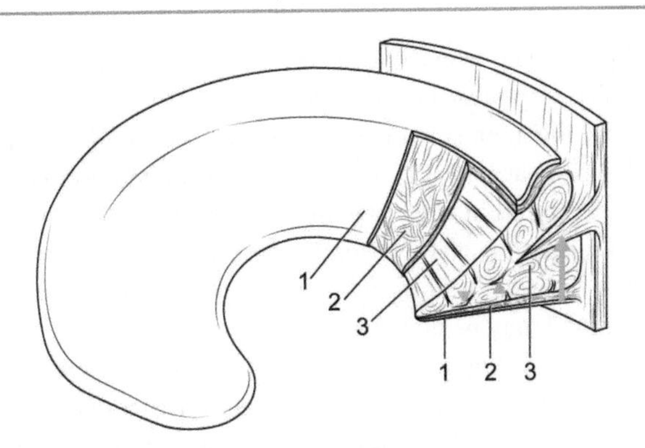


**Fig. 11.18** Schematic of meniscus microstructure, 1) superficial network, 2) lamellar layer, and 3) central main layer. Arrowheads indicate the region where radial collagen fibrils are interwoven with the circular fibril bundles. Note in section 3 the arrow where the external circumference loose connective tissue from the joint capsule penetrates radially between the circular fibril bundles. Modified with permission from Peterson W, Tillman B (1998) Collagenous fibril texture of the human knee joint menisci. Anat Embryol 197:317-324

with a greater propensity for water and the proteoglycan aggregate, a necessity for better distributing axial loads. The presence of proteoglycan aggregates provides for the capacity of menisci to resist compression loads. However, the mass of proteoglycans in the meniscus is one-eighth that of articular cartilage, suggesting, again, that the meniscus's role in mediating impulsive loads (acting as a shock absorber) is not as great as the ability of the articular cartilage.

The presence of the menisci makes a significant contribution to a relatively uniform stress distribution over much of the tibiofemoral joint contact area. It has been suggested that this distribution reduces the lateral movement of fluid in the articular cartilage and its flow into the joint space, allowing a load to be transferred through the cartilage mainly by means of fluid hydrostatic pressure. That is, since the fluid does not flow away from the area, it continues to transmit load, i.e., mediate compression, rather than transferring load to the ECM of the articular cartilage. The mechanism of transmission can be explained and understood by first looking at the microstructure of meniscal tissue (Fig. 11.18).

There appear to be three distinct layers: a superficial fibril layer network, a lamellar layer, and a central main region. The superficial and lamellar layers are present in both the tibial and femoral surfaces of the meniscal tissue. The superficial fibril layer has a multidimensional orientation very similar to that found in articular cartilage. It is believed that this layer functions as an interface with the articular cartilage to maintain the low-friction contact between bony and meniscal surfaces. Additionally, the larger concentration of elastin in this layer suggests a role in the elastic recovery required after experiencing shear forces associated with joint contact. That is, as the femoral condyles contact the menisci, applying forces that tend to induce strains biaxially along the surface of the tissue, the interconnected mesh of elastin in the superficial layer may apply a restorative force, aiding in the recovery of the collagenous framework. The second or lamellar layer contains fiber bundles that intersect at right angles apart from the anterior and superior aspects, where they are arranged radially (these fibers are sometimes referred to as "tie-fibers"), and which appear to keep the circumferential fibers in the main region from separating. In the main portion, the aggregate form appears as fibril bundles with the fibrils highly interwoven to form collagen fibers that are oriented circumferentially. In the "red-white" zone, the architecture appears to change with a less uniform organization, where collagen types I and II are coexpressed. Toward the less vascularized zone (the "white zone"), collagen type II predominates, leading to smaller fiber bundles and a structure more similar to articular cartilage. Therefore, since the architecture of the menisci is heterogenous regionally, i.e., peripherally or centrally, it appears that the menisci have also evolved their mechanical properties into an elaborate system of load transmission. Let us examine those material properties of meniscal tissue.

Experimental research examining the compressive and tensile material properties of the meniscus faces many methodological challenges, which can explain some of the discrepancies reported in the literature. In general, however, because the meniscal ECM is similar to what is found in articular cartilage, we know that the menisci must mediate compression loads. However, it has been shown that the hydraulic permeability of meniscal tissue is an order of magnitude lower than that of articular cartilage, suggesting that its viscoelastic response to large loads and higher rates of loading is less robust as well. On the other hand, menisci appear to be approximately 1000 times stiffer in tension than in compression. The important characteristic of this property, i.e., greater stiffness under tensile loads, is the ability of the meniscal tissue to deform, i.e., readily change its shape, in response to the variable geometry of the femoral condyles as the joint is loaded. Since the adage "function dictates structure" is a biological reality, we have to conclude that because the meniscus appears to be stronger under tensile loading, it must mean the menisci play a predominant role under tensile loads. This seems counterintuitive because, functionally, the tibiofemoral joint is loaded axially. What can explain this apparent paradox? Let us look more closely at how axial forces are transformed into tensile loads.

When bearing load, the tibiofemoral joint is subjected to axial or compression forces. These compression forces are distributed over an articulating contact area, resulting in contact stresses or pressures. These stresses are proportional to the load and inversely proportional to the contact area. Without menisci, the contact area between the articulating bony surfaces is much smaller, given the relative incongruency between the femoral and tibial condyles. With intact and healthy menisci, the load transmission is optimized, i.e., the contact pressures are minimized, due to increased joint surface con-

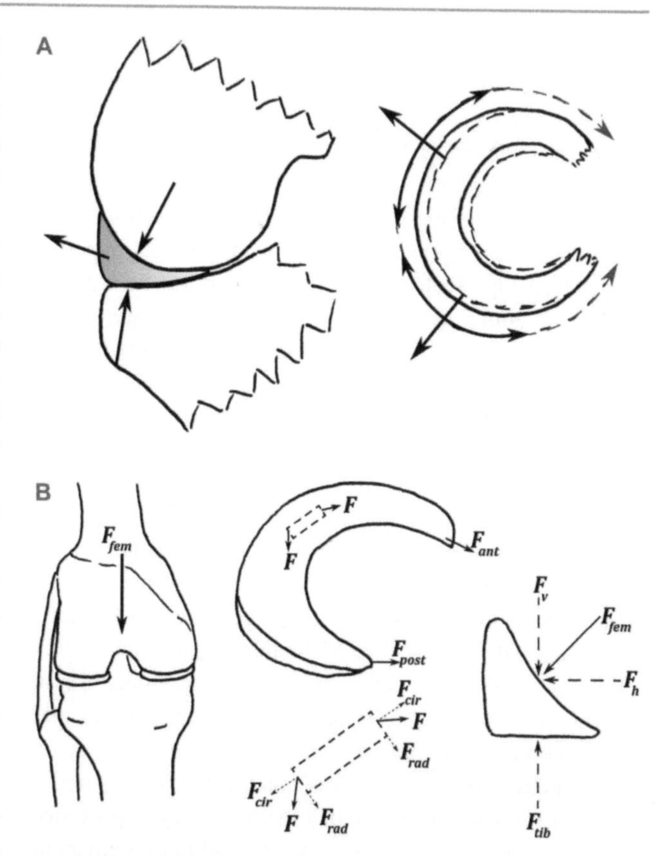


**Fig. 11.19** (**A**) General representation of tibiofemoral joint loading demonstrating radial deformation, which induces circumference hoop stress in the deeper layers of the meniscal tissue. (**B**) Free body diagram of forces ($F_{fem}$) acting on the meniscus during axial loading, which occurs during everyday activities. The vertical force of the tibia ($F_{tib}$) counteracts the downward vertical force of the femur ($F_{fem}$). The horizontal force exerted radially outward is countered by anchoring forces of the attachments at the anterior and posterior horns of the meniscus. (Note: $F_{ant}$ and $F_{post}$ = forces at anterior and posterior horns; $F_{cir}$ = tensile hoop force; $F_v$ and $F_h$ = vertical and horizontal forces, respectively, induced by $F_{fem}$; $F_{rad}$ = radial reaction force balances $F_h$. ((B) Modified with permission from Makris et al. (2011))

gruency. As the femoral condyles bear down onto the menisci, the meniscal wedged cross-section induces the tissue to extrude radially (i.e., "out away from the joint line") while also compressing the tissues between the two bones. The radial extrusion deforms the menisci, causing their circumference to increase. Because of the constraints imposed by the anterior and posterior horns, the increased circumference induces large tensile forces in the main portion of the three-layered meniscal microstructure. This circumferential tension is referred to as hoop stress (Fig. 11.19). Since the menisci become stiff circumferentially, the large internal tensile stresses create a snug contact between the menisci and femur. However, not all of the axial forces are transformed into hoop stress. Since the menisci are also being compressed between the femur and tibial condyles, these compression forces are transmitted from one bone to the other, compressing the intervening cartilage layers in the process. On the other hand, this remaining compression has a lesser magnitude (due to a partial transformation of it into hoop stress), which is readily distributed/mitigated by the articular cartilage.

This mechanism of load bearing occurs throughout the range of tibiofemoral joint flexion because the primary attachments of the menisci to the tibia are the insertional ligaments at their horns. At full extension or mild hyperextension (see Fig. 11.14), the femoral condyles have larger radii of curvatures so that they contact the entire area of the menisci. As the knee flexes, the lesser radii of curvature of the posterior aspect of the femoral condyles create reduced areas of contact that move posteriorly onto the posterior horns of both menisci. Because the lateral meniscus is more mobile and the lateral and tibial condyles are less congruent between the intervening fibrocartilaginous tissue, the lateral meniscus is displaced further posteriorly than the medial meniscus. These differences in the mobility of the menisci dictate that at the extreme of end-range knee flexion, e.g., deep squat, reduced mobility of the medial meniscus puts it in a position where it is subjected to very large compression stress without being able to increase its circumference and develop hoop stresses. Recall that the ultimate failure strain under compression for menisci is much less than it is under tension. It is no wonder, then, that the frequency of posterior horn medial meniscal tears is high. The ability of menisci to resist large compression loads by increasing their circumference and developing hoop stresses explains the variable severity and frequency of meniscal strain injury types (Fig. 11.20).

For example, a longitudinal strain does not have a major effect on meniscal function since it does not disrupt the continuity of the circumferentially oriented fibers that bear a load. In contrast, a radial strain disrupts the continuity of the circumferential fibers, reducing the number of fibers available to mediate large tensile stresses. This situation leads to increased articular cartilage-to-cartilage contact stresses and perhaps earlier degenerative changes.

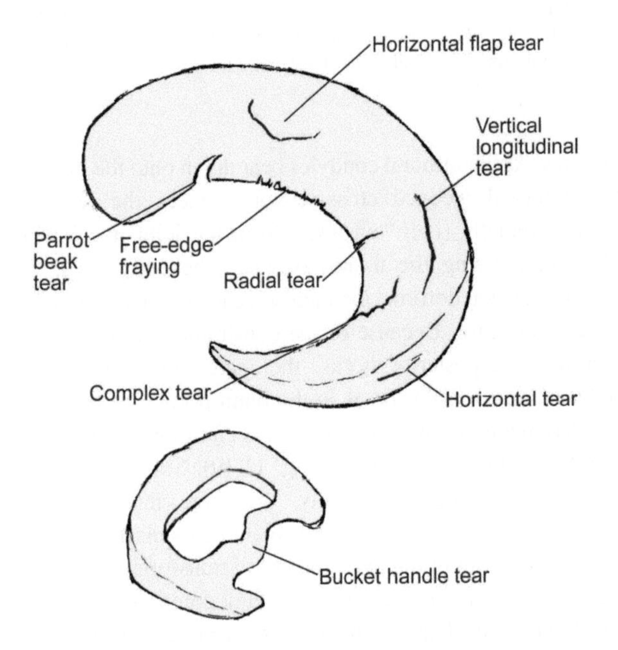


**Fig. 11.20** Illustration of the variety of meniscal strain injuries

In summary, the internal mechanisms within the menisci are perfectly structured to interface with multiple tissues to resist combinations of compression and tensile stresses repeatedly without fatigue failure. At the same time, there is a necessity for a low shear modulus to allow the surfaces of the menisci to adapt readily to the changing geometry of the bony partners during the movements of flexion and extension and for the conjunct axial rotations that accompany flexion and extension. The menisco-microstructure integration of radially oriented collagen tie-fibers and sheets of collagen helps keep the structure together as it transitions from a ligament-like tissue in its outer regions to more cartilage-like at its inner circumference. The changing density of proteoglycans likewise reflects the change in the composition, architecture, and material properties needed to resist the changing stress state in the tissue, from great circumferential tension and lower axial compression in the outer region moving to higher compression and lower circumferential tension toward the inner region. Finally, we cannot ignore the meniscal roots, i.e., horns, which simultaneously permit rigid body motion of the menisci and a tensile load-bearing function – a truly intricate and remarkable system!

#### 11.4.2.3 Muscle

We are going to conclude the section on knee structure with a brief examination of key muscles associated with the knee (and ankle and hip since there are several important two-joint muscles to consider). Earlier we allude to the motor and sensory control of the knee by identifying the femoral, obturator, and sciatic (tibial portion) nerves in our review of the neuroanatomy related to the menisci. All three of these nerves provide ample afferent information, from the knee joint capsule to the surrounding ligaments to internal structures. We remind the reader that the femoral and obturator nerves also have neural connections to the hip region, suggesting that knee pathology may partly be manifested by altered function there, e.g., impaired muscle activation. Table 11.4 summarizes basic information regarding the muscles involved, directly and indirectly, with knee function.

In problem-based learning (PBL), we focus on the information we immediately need to understand and explain the case at hand. A criticism of the PBL approach claims that students do not get "all of the information that is available." Our retort is that "all of the information" is many times not needed, and even if it was made available, most of it would have been presented out of context and would likely be lost, i.e., not remembered. In this case, the injury clearly directly involved the internal mechanisms of the tibiofemoral joint,

**Table 11.4** Innervations and actions of hip, knee, and ankle muscles

| Muscle | Actions | Innervation | Plexus |
|---|---|---|---|
| Gluteus medius | Hip extension, abduction, and external rotation | Superior gluteal nerve | Sacral |
| Gluteus minimus | Hip flexion, abduction, external and internal rotation | Superior gluteal nerve | Sacral |
| Tensor fascia latae | Hip flexion, abduction, and internal rotation | Superior gluteal nerve | Sacral |
| Sartorius | Hip flexion, abduction, and external rotation<br>Knee flexion and internal rotation | Femoral | Lumbar |
| Gracilis | Hip flexion and adduction<br>Knee flexion and internal rotation | Obturator | Lumbar |
| Rectus femoris | Hip flexion and abduction<br>Knee extension | Femoral | Lumbar |
| Vasti (medialis, intermedius, and lateralis) | Knee extension | Femoral | Lumbar |
| Popliteus | Knee flexion and internal rotation | Tibial | Sacral |
| Semimembranosus | Hip extension<br>Knee flexion and internal rotation | Sciatic (tibial) | Sacral |
| Semitendinosus | Hip extension<br>Knee flexion and internal rotation | Sciatic (tibial) | Sacral |
| Biceps femoris (short) | Knee flexion and external rotation | Sciatic (fibular) | Sacral |
| Biceps femoris (long) | Hip extension<br>Knee flexion and external rotation | Sciatic (tibial portion) | Sacral |
| Gastrocnemius | Knee flexion<br>Ankle plantar flexion | Tibial | Sacral |
| Plantaris | Knee flexion<br>Ankle plantar flexion | Tibial | Sacral |

Modified with permission from Neumann (2017)

with no concomitant strain injuries to muscles acting about the knee joint. Thus, we do not make an exhaustive review of all the knee-related musculature. We have already acknowledged that the weakness in several of Cathy's hip muscles was likely related to altered function and gait compensations and/or to knee arthrogenic neural inhibition. We have previously acknowledged the possible roles of the quadriceps and patellar retinacula relative to the dynamic control of the medial and lateral menisci, as well as the semimembranosus contribution to the kinematics and stabilization of the posterior horn of the medial meniscus.

Some research has suggested that simulated radial strains of the lateral meniscus increase contact pressure and reduce contact area in the patellofemoral joint. These findings need to be corroborated, but at the least, this information suggests that we take a close look at the role of the quadriceps and knee extensor mechanism. Finally, we detail a bit more the relationship between the lateral meniscus and popliteus muscle.

The quadriceps femoris is comprised of four muscles: vastus medialis (obliquus + longus), vastus intermedius, vastus lateralis, and rectus femoris; only the rectus femoris also has action at the hip joint. The quadriceps form a large cross-sectional area and contribute to a knee extensor moment that is about two-thirds greater than the knee flexor moment produced by the hamstring muscles. The four muscles converge to forge a tendon that inserts into the base and sides of the patella; fascial continuity across the patella forms the patellar tendon, which inserts into the tibial tuberosity (Fig. 11.21). Deep portions of the vastus medialis and lateralis attach to the joint capsule and menisci via lateral patellar retinacular fibers. The oblique portion of the vastus medialis is named

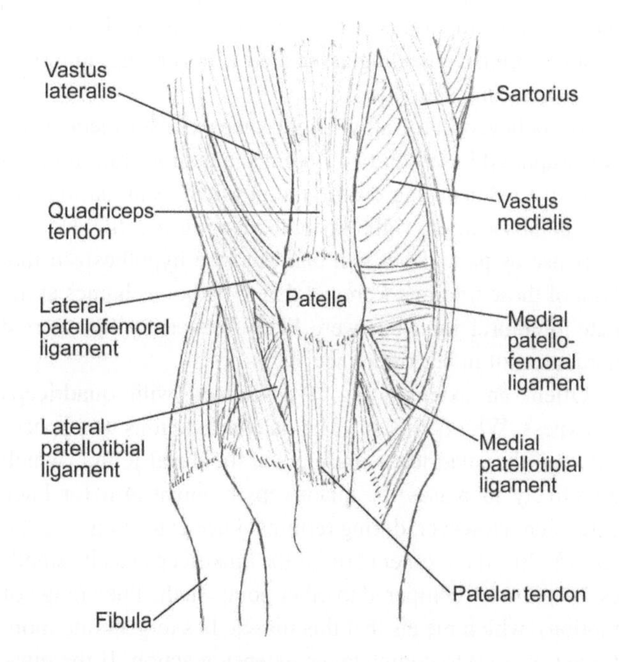


**Fig. 11.21** Schematic illustrating the relationship of the structures making up the extensor mechanism at the knee, i.e., the tibiofemoral and patellofemoral joints

such because the fibers of this deeper portion of the medialis approach the patella at approximately 50° medial to the quadriceps tendon. Although the cross-sectional area of the obliquus is small, it appears to play an important role in stabilizing the patella medially during tibiofemoral flexion and extension. Deep to the vastus intermedius (not shown), the articularis genu runs from the distal femur to the anterior capsule, suggesting that its function may be to pull the capsule and synovial membrane proximally during knee extension. In total, these tissues make up the knee extensor mechanism, a complex system with interrelated joints, i.e., tibiofemoral and patellofemoral; ligaments, i.e., collaterals and cruciates; and menisci, which are subjected to very large and repetitive forces related to normal and sporting activities. For example, during normal gait, the quadriceps group acts eccentrically during the loading response of the gait cycle to control knee movement from its extended position (at initial contact) to about 15° to 20° of flexion. The loading response of the gait cycle sees large compression forces on the tibiofemoral and patellofemoral joints that approximate 2.5 times the body weight; these forces are increased to 3 to 3.5 times the body weight when ascending and descending stairs, respectively. The eccentric action of the quadriceps functions like a spring and damper system during the loading response. This particular function of the quadriceps becomes increasingly important with activities associated with the greater impulsive loading associated with running, jumping, and descending stairs.

We believed that Cathy's right knee extensor mechanism was impaired based on two findings: a demonstration of (1) an active extensor lag during the physical examination and (2) an absence of right knee flexion during the loading response as part of her gait analysis. We hypothesized that both of these findings were not due to pathomechanics at the patellofemoral joint but were likely related to the internal derangement in her right knee.

Often, an extensor lag is associated with quadriceps weakness. Why is this so? The patella functions as a spacer between the quadriceps muscle and the distal femur, which effectively increases the quadriceps moment arm for knee extension. However, during terminal knee extension, i.e., the last 15–20°, the moment arm of the quadriceps has its smallest magnitude (compared to other points in the knee range of motion), which means that this muscle has to generate more force (tension) to complete the extension action. If the muscle has significant weakness due to nerve palsy or disuse atrophy, an individual may not be able to perform a terminal extension when lifting the shank against gravity. When we performed a resisted quadriceps strength test on Cathy, she demonstrated normal strength with the knee set at 45° flexion, suggesting that her quadriceps were being activated normally. Thus, we concluded that her inability to terminally extend the knee was likely related to intracapsular swelling (which was evident on physical examination), arthrogenic inhibition of the quadriceps due to abnormal meniscus mechanics, increased intraarticular pressure (due to swelling), and joint pain.

Besides eccentric control of knee flexion during the loading response, what other roles do the quadriceps play during the gait cycle? Prior to controlling knee extension in the loading response, the vasti begin their activation in terminal swing (80–95% of the gait cycle (GC)) (see the gait section of Appendix J if you need assistance visualizing the gait cycle). The level of effort of the vasti gradually diminishes to approximately 12% of the GC at the end of loading, then ceases by 20% of the GC (i.e., midstance). Surprisingly, the rectus femoris (RF) is not active during the loading response (perhaps the RF is electrically silent during the loading response because its two-joint functions inhibit it as a knee extensor, while the hip remains flexed) but only demonstrates variable and low-intensity activity between late preswing (57% GC) and early initial swing (65% GC). It is thought that the RF during this period controls rapid knee flexion, while it assists in flexing the hip.

Table 11.4 outlines the accepted concentric actions of the knee flexors during knee flexion. Let us look at their activity during a gait cycle. Two of them, the popliteus and short head of the biceps femoris (BFSH), provide direct knee flexion. The BFSH is primarily active in the initial and midswing subphases (65–82% GC), while the popliteus shows no consistent pattern of activity. The popliteus shows its greatest effort generally at the initiation of preswing (50% GC), a period when the knee moves from extension into rapid knee flexion. It is likely that its role is to "unlock" the knee by internally rotating the tibia as it initiates flexion. Another period of moderate effort of the popliteus begins in terminal swing (eccentric action to assist in controlling knee extension and external tibial rotation) and continues through the loading response (internally rotating the tibia to again "unlock" knee extension).

The three hamstring muscles (semimembranosus, semitendinosus, and long head of the biceps femoris) act primarily as hip extensors at initial contact into loading to control the external hip flexor moment. However, these muscles are better known as knee flexors. The semimembranosus and long head of the biceps femoris initiate activity during midswing and are joined in terminal swing by the semitendinosus. All three muscles show the highest intensity during the terminal swing as they control knee extension and prepare to control the hip flexor moment at initial contact.

Besides the rectus femoris, two other two-joint muscles, the gracilis and sartorius, become active in preswing to assist in both hip and knee flexion. The duration of sartorius activity concludes in the initial swing, whereas the gracilis is active through the loading response. Whereas the gastrocnemius is listed as a knee flexor and ankle plantar flexor, during gait, it controls knee flexion, that is, excessive tibial advancement over a fixed foot from mid- to terminal stance, asserting fairly high EMG intensity until the middle of the terminal stance at 40% GC. The eccentric actions of the gastrocnemius and soleus provide a damping effect as they function like springs storing elastic energy, which is subsequently used to initiate knee flexion and plantar flexion at the ankle at the onset of preswing; this occurs even as its EMG intensity decreases.

In the coronal plane, an external abductor moment develops in the knee on the stance side in response to the rapid unloading and pelvic drop on the opposite side. Tensing of the iliotibial (IT) band provides a lateral counterforce to stabilize the knee. This appears to persist to some degree into midstance (12–31% GC). Because of the position of its tendon within the tibiofemoral joint, the popliteus may also provide medial stability, i.e., resist the external tibial abductor moment during the first double-limb support.

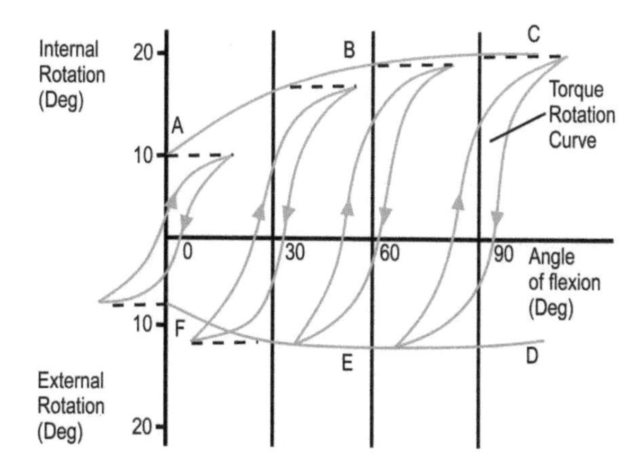


**Fig. 11.22** Torque-rotation laxity curve for tibiofemoral axial rotation at 0°, 30°, 60°, and 90° knee flexion. Moving from the extreme external to extreme internal back to extreme external rotation creates a hysteresis-like loop. Note that the total range of tibiofemoral axial rotation increases as the knee flexes up to 90°. (Modified with permission from Walker et al. (2022))

#### 11.4.2.4 Osteo- and Arthrokinematics of the Tibiofemoral and Patellofemoral Joints

Traditionally, the tibiofemoral joint has been described as a mobile trocho-ginglymus (pivotal hinge) joint possessing two degrees of freedom: flexion/extension and internal/external rotation. Yet frontal plane movement at the tibiofemoral joint exists when performed passively with the knee in slight flexion, yielding a total of approximately 6–7° abduction (valgus)/adduction (varus). With increasing knee extension, tibiofemoral abduction/adduction decreases because of ligamentous constraints; in general, there is less tibial abduction due to restraint provided by the medial collateral ligament.

The largest movements of the tibiofemoral joint occur in the sagittal plane. Movement ranges from approximately 5–10° hyperextension to 130–150° flexion. End-range flexion is met when the posterior femoral condyles impact the posterior horns of the menisci. Typical ranges of flexion during normal activities include walking 0–67°, ascending stairs 0–83°, descending stairs 0–90°, sitting down into a chair 0–93°, tying a shoe 0–106°, and squatting/lifting an object 0–117°. In general, increasing the speed of locomotion, i.e., walking, walking transition to running, and running, results in the need for more knee flexion.

Nordin and Frankel described knee motion in the transverse plane as laxity and graphically illustrated the continuum of laxity related to internal and external tibial rotation (Fig. 11.22). Each laxity curve is shaped like a hysteresis loop (see Appendix H for the definition of a hysteresis loop), reflecting a conflation of the viscoelastic properties, i.e., stiffness, time dependency, etc., of the ligaments (including the joint capsule), menisci, and other soft tissues that restrain excessive knee movements. When the knee is extended, the knee's rotational laxity is minimal, so it's apparent that minimal internal and external rotation is available. However, as the knee flexes, the axial rotation in either direction increases to a maximum at about 90° of knee flexion. Thus, with increasing flexion, the rotational stiffness, i.e., the slope of the elastic portion of the hysteresis loop, decreases. With the knee in full extension, i.e., close-packed position, rotational laxity is restricted by tibiofemoral joint congruency, menisci, and collateral and cruciate ligament tautness. When the knee is flexed, the tibiofemoral joint is less congruent, and the ligaments become relatively lax, thus allowing greater rotational mobility. Maximum internal and external rotation can be reached, and maintained, after approximately 40° of knee flexion. A total of approximately 45° of axial rotation is possible, with tibial external rotation exceeding internal rotation by a ratio of 2:1.

Before we look at the details of the joint surface motion at the tibiofemoral and patellofemoral joints, we remind the reader, although this should be obvious given what we have discussed to this point, that the tibiofemoral joint has six degrees of freedom (Fig. 11.23). We define rotations about the mediolateral axis as flexion/extension, rotations about the anteroposterior axis as abduction (valgus)/adduction (varus), and rotations about the superoinferior axis as internal and external rotation. These definitions are made under the assumption that the distal segment is moving relative to the proximal segment. In our arthrokinematic descriptions, then, we use the terms anterior/posterior glides, superior/inferior glides, and medial/lateral glides (with the reference being a right or left limb).

Let us now delve into tibiofemoral and patellofemoral arthrokinematics. With tibiofemoral extension, since the "concave" tibial condyles move relative to the convex femoral

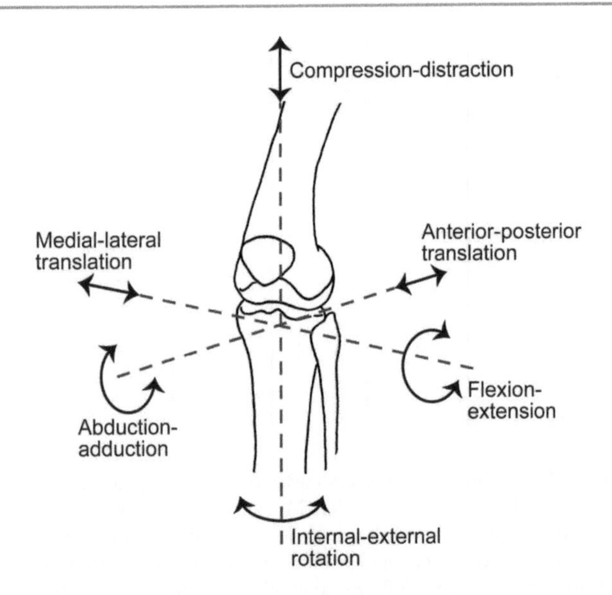


**Fig. 11.23** Schematic illustrating all six degrees of freedom that take place at the tibiofemoral joint. With a Cartesian coordinate axis centered between the tibial and femur, we can imagine and name rotations and translations around mediolateral, anteroposterior, and superoinferior axes

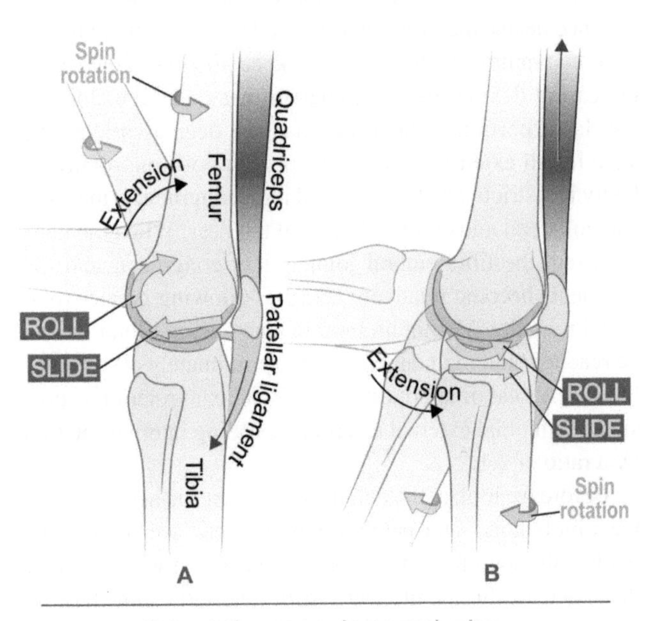


**Fig. 11.24** Schematic illustrating tibiofemoral arthrokinematics during (**A**) the latter part of the loading response of gait and (**B**) the terminal swing of gait. (Note that there is a concomitant external rotation (i.e., spin) of the tibial relative to the femur during terminal extension)

condyles, the anterior roll of the tibia is accompanied by an anterior glide (Fig. 11.24). The relative movements of the femoral condyles, assisted by tension in the medial and lateral patellar retinacula, simultaneously push/pull the menisci anteriorly. Because the medial femoral condyle extends further anteriorly than the lateral femoral condyle, the lateral

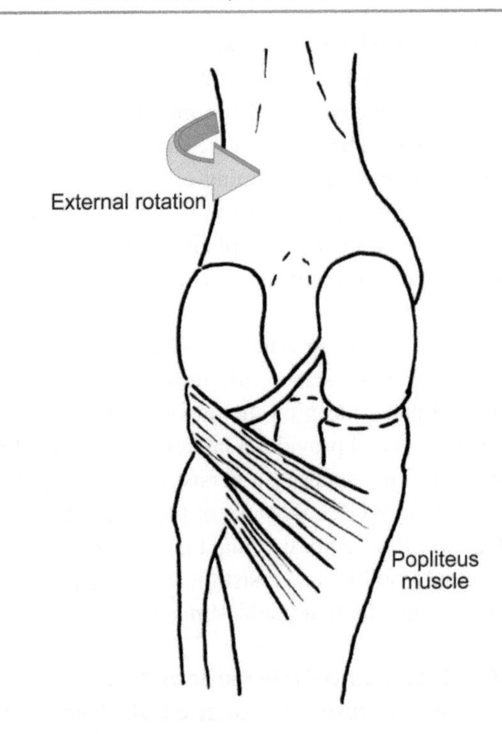


**Fig. 11.25** Posterior view illustrating the external rotation of the femur induced by a shortening action of the popliteus as the first step to unlocking the extended knee

compartment of the tibia continues to move (even after the medial femoral condyles have stopped moving) during terminal extension, inducing a conjunct external rotation of the tibia. Two additional anatomical factors that contribute to what is referred to as the screw-home mechanism include the tautness of the cruciate and collateral ligaments and a slight lateral pull of the quadriceps (related to the normal genu valgus alignment). External rotation of the tibia occurs along and about an axis slightly medial and posterior to the intercondylar eminence; some refer to the rotation of the tibia during terminal extension (or during the initiation of flexion) as a spin rotation. The end range of extension puts the tibiofemoral joint in a close-packed position, that is, the position of greatest stability, where the joint achieves maximum surface congruity and many of the periarticular structures, e.g., capsule, cruciate ligaments, etc., become taut.

Arthrokinematics during knee flexion is simply the reversal of what takes place during extension. To initiate knee flexion, the tibia rotates internally or is "unlocked" through the action of the popliteus muscle (Fig. 11.25). Because of its position relative to the structures within the knee, the popliteus can also externally rotate the femur if the distal segment, i.e., the foot, is fixed to initiate knee flexion.

In presenting an alternative explanation of tibiofemoral joint surface motion during knee flexion, Otis offered evidence that suggested that the femur translated posteriorly (not anteriorly) during its posterior roll. With this alternative explanation, it is possible that the apparent posterior

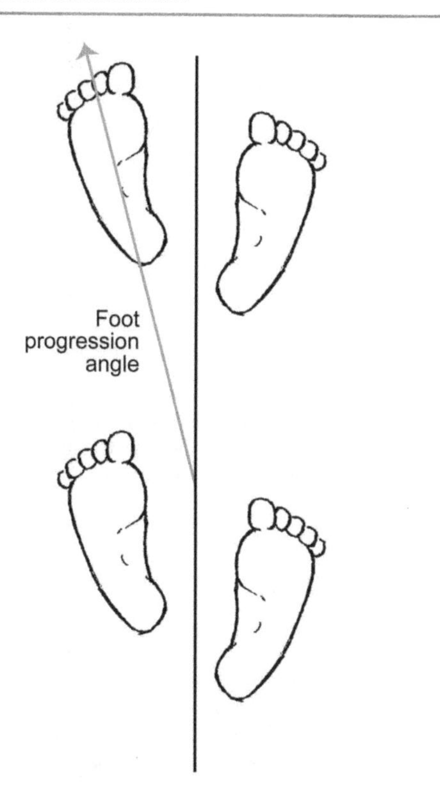


**Fig. 11.26** The normal external foot progression angle is determined by the angle spanning the long axis of the foot and the straight-line direction of walking; i.e., the longitudinal axis of the laboratory coordinate system

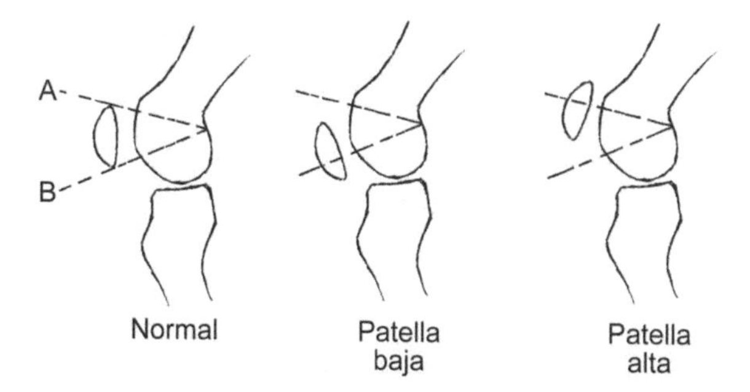


**Fig. 11.27** Schematic showing differences between the normal and abnormal proximal-distal patellar alignments

translation occurred closer to the lateral compartment, given that the lateral femoral condyle is more mobile (as described earlier). The above differences in the interpretation of data regarding the arthrokinematics of knee flexion must always be considered relative to the differences in experimental methods and whether or not reported research has been examined for accuracy and precision.

Before we describe patellofemoral mechanics, we need to briefly relook at the constitution of a normal mechanical alignment of the lower extremity. Recall that normal transverse plane alignment begins proximally with mild torsion (15–20°) of the femoral head/neck. This out-of-plane torsion is compensated by a 10–15° external torsion in the proximal tibia. The combination of these torsions induces, when standing or walking, an approximate 15–20° external foot progression angle (Fig. 11.26); the foot progression angle is the angle between the longitudinal axis of the foot, defined by a line through the mid axis of the second metatarsal to the heel (see fuller explanation in Appendix J) and a straight line representing the direction of walking. The foot progression angle is usually maintained well during gait when the joints of the foot are stable and well aligned. With this normal alignment, the patella is somewhat loose packed, residing more laterally in the trochlear groove of the femur when the knee is extended, but more well seated in the trochlear groove, i.e., more constrained or stable, as the knee is flexed.

However, from an anterior view (a view from in front), the patella normally presents itself as if we are looking at the face of a clock straight on, neither turned inward ("squinting" = patella resting medially) nor turned outward ("frog-eyed" = patella resting laterally). A squinting patella is often a sign of abnormal femoral antetorsion, while a frog-eyed patella is a sign of femoral retroversion. A sagittal (i.e., side) view can also provide information about whether the patella is sitting too far superior (patella alta) or too far inferior (patella baja or patella infera) (Fig. 11.27). Any one of these resting patellar abnormalities changes patellar kinematics and can induce changes in how the patellofemoral reaction force compression loads are distributed, often leading to a common patellofemoral painful condition, e.g., chondromalacia.

Other measures of patellar positioning, e.g., patellar tilt, sulcus angle, and congruence angle, require computed tomography or magnetic resonance imaging (MRI) images. If there are one or more deviations from the normal lower-limb alignments, as presented, it is likely that the patellofemoral kinematics is abnormal as well.

Although the nature of Cathy's knee signs and symptoms did not suggest a patellofemoral problem, the physical examination included observations of the lower extremity alignments just described. Her lower extremity alignment in all planes was considered normal. It is unlikely, again given her meniscal injuries, that her physician made imaging studies of the patella profile. The motion analysis center staff was largely convinced that her knee pain was not patellofemoral in origin but generated by the chronicity of the loss of normal lateral meniscus function and early tibiofemoral joint degeneration.

The patellofemoral joint is a synovial joint, yet unlike most of the joints we have encountered, there is no significant rolling action between the patella and femur. The patella is housed in the trochlear groove of the femur, stabilized by the congruity of the joint partners, the quadriceps femoris, and other related passive restraints. The articular cartilage on the underside of the patella counts as the thickest in the human body, an evolving story that is consistent with the large patellofemoral reaction forces induced by knee flexion activities.

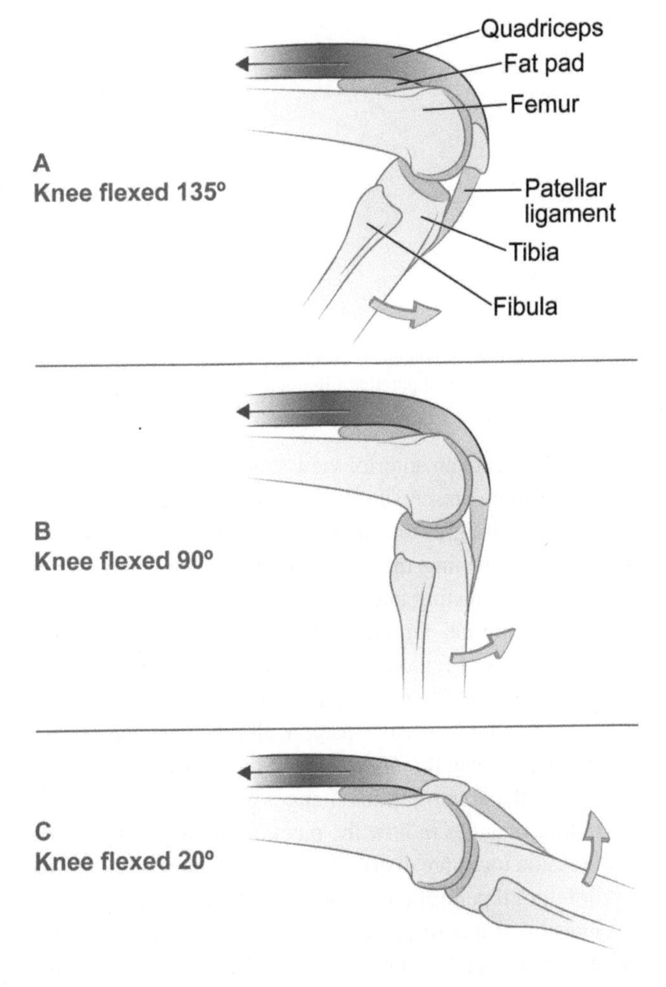


**Fig. 11.28** Patellofemoral kinematics during tibiofemoral extension. As the tibia moves in a superior direction, so glides the patella. The contact point on the patella moves from its superior pole when the knee is flexed (**A**) 135°, (**B**) 90°, then to its inferior pole (**C**) near full extension

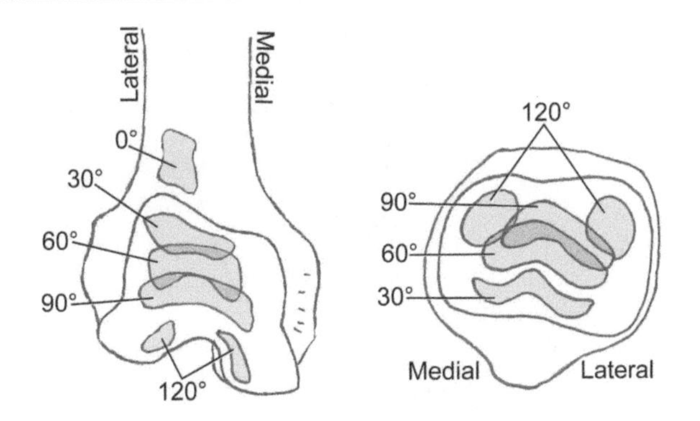


**Fig. 11.29** Illustration of the changes in contact locations between the undersurface of the patella and the trochlear groove and the surface areas as the knee progresses from full extension to 120° flexion

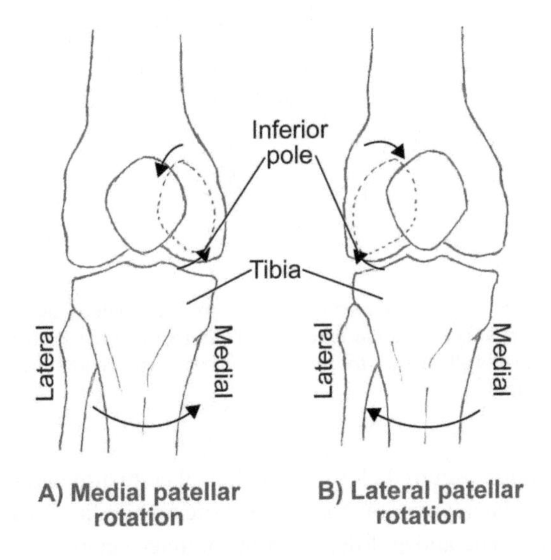


**Fig. 11.30** Mediolateral movements of the patella with (**A**) medial and (**B**) lateral tibial rotation

During tibiofemoral flexion without the foot fixed on the ground, the patella glides inferiorly, relative to the fixed trochlear groove of the femur (Fig. 11.28). Similarly, for flexion during stair descent, when the foot is fixed on a step, the femur moves under the relatively fixed patella, which is stabilized by eccentric action of the quadriceps and its extension, the patellar tendon. As the tibiofemoral joint flexes from an extended position, the contact areas between the distal femur and patella change in location and size (Fig. 11.29). Notice that at full extension, the patella is resting laterally and outside of the trochlea and that at 120° flexion, the contact area is located near the superior (proximal) pole of the patella. It appears that the surface area of contact is largest between 60° and 90° knee flexion, an important finding given that the largest knee extension moments are generated during that interval of knee flexion.

We now know that tibiofemoral flexion and extension are kinematically coupled with the axial rotations of the tibia. We can imagine that this axial rotation (either internal or external) must also affect patellar positioning. Thus, when the tibiofemoral joint is flexed, the patellar ligament (tendon) runs a slight oblique course inferolaterally. From that position and with the femur fixed, lateral (external) rotation of the tibia moves the tibial tubercle laterally, pulling the inferior pole of the patella in the same direction; in this instance, the patellar ligament's inferolateral course is exaggerated. Conversely, medial (internal) rotation of the tibia pulls the inferior pole of the patella medially, thus reversing the oblique course of the patellar ligament (Fig. 11.30). Being mindful of accessory motions of the patella during knee flexion and extension is important when examining knee osteo- and arthrokinematics.

In the past three decades or so, advances in radiological imaging processes and biomechanical modeling and the increasing prevalence of patellofemoral dysfunctions have motivated additional research on patellofemoral kinematics. This research has identified multidimensional accessory

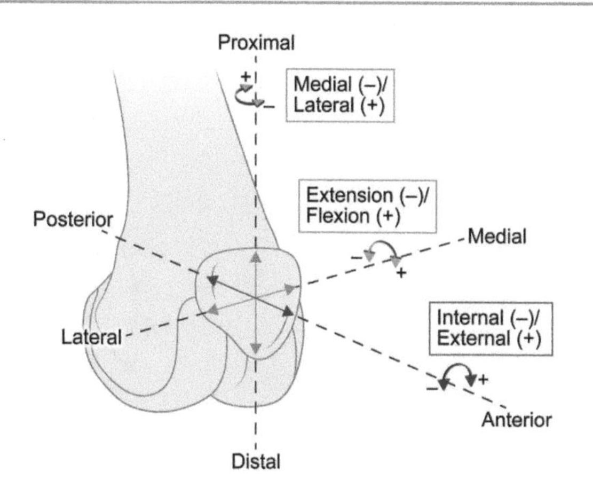


**Fig. 11.31** Six-degree-of-freedom patellar kinematics, some of which are accessory

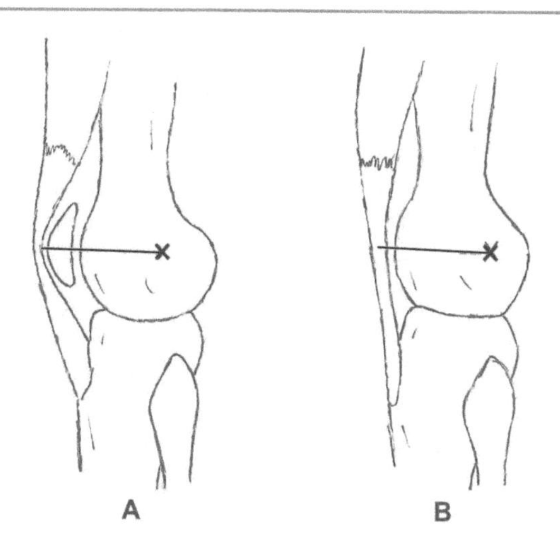


**Fig. 11.32** Noted difference in magnitude of the extensor moment arm (**A**) with and (**B**) without the patella. (Note that "x" marks the approximate center of rotation for the tibiofemoral joint)

patellar movements that appear to pattern in particular ways with the more global patellar movements (Fig. 11.31). For example, the most consistent pattern involves small medial and lateral translations of the patella, coupled with tibiofemoral knee flexion and extension, respectively; this is likely related to the coupled axial rotations of the tibia that occur during knee sagittal plane movements, particularly in terminal knee extension and when initiating knee flexion. The likely factors responsible for these accessory movements include (1) a mismatch between the patella and trochlear geometry throughout the range of flexion/extension; (2) normal genu valgus, which biases toward a lateral pull of the patella by the quadriceps; (3) normal axial rotation at the tibiofemoral joint during terminal knee extension and at the initiation of knee flexion; and (4) changing tensions in the periarticular soft tissues, e.g., patellar retinacular, etc. However, the accuracy and precision of the measures of these accessory patellar movements have yet to be established. Moreover, the amount and variability of these movements among individuals are significant, considering, for example, the number of combinations of lower extremity malalignments, e.g., excessive femoral torsion, increased genu valgus, etc., that are feasible, even in a normal healthy population, which can influence the resting pose of the patella. Thus, it is difficult to draw general conclusions about these accessory patellar movement patterns that might be related to particular patellofemoral pathologies. Clearly, more research is needed.

Let us conclude this section with an exploration of the primary role of the patellofemoral joint. Simply put, the patella functions like a pulley or spacer between the quadriceps and femur while serving to increase the mechanical advantage (i.e., increase the moment arm of the quadriceps) of the knee extensor mechanism (Fig. 11.32). As discussed in Chap. 3, we know that (1) the orientation of the action lines of muscles that cross a particular joint changes continuously as the joint partners move relative to the other; (2) as a result of #1, the muscle length-tension changes; and (3) again, as a result of #1, the internal moment arms of each muscle constantly change. At the tibiofemoral joint, the knee extensor moment arm is most influenced by the shape and position of the patella and the shape of the distal femur and migrating tibiofemoral joint axis during sagittal plane rotations, i.e., flexion/extension. Research to date has demonstrated that the knee extensor moment arm is greatest when the knee is flexed between 60° and 90°, with a dramatic reduction in quadriceps femoris extensor moment production during the last 30° of knee extension.

Many normal and sporting activities put the knee in flexed positions where the external flexor moment is potentially large, e.g., knee flexor moment during the loading response of gait. These situations challenge both the concentric and eccentric actions of the quadriceps, resulting in the generation of large extensor moments. Recall that, in many cases, external forces, e.g., gravity or other applied forces, have greater moment arms than the muscles that cross the joints (even if a particular muscle lever system may increase an internal moment arm), dictating the generation of relatively large muscle-tendon forces. With the use of vector addition, it is shown in Fig. 11.33 that the resultant patellofemoral bone-on-bone force (PF) has a greater magnitude when the knee is flexed than when the knee is extended. In this example, we see that when the knee is fully extended (in quiet standing), the moment arm of the quadriceps, enhanced by the presence of the patella, results in a relatively small patellofemoral joint force (PF), whereas with knee flexion (beginning to squat, for example), the required knee extension moment increases. This increased extension moment is achieved through a combination of an increase in the moment arm of the quadriceps (with the help of the patella), along

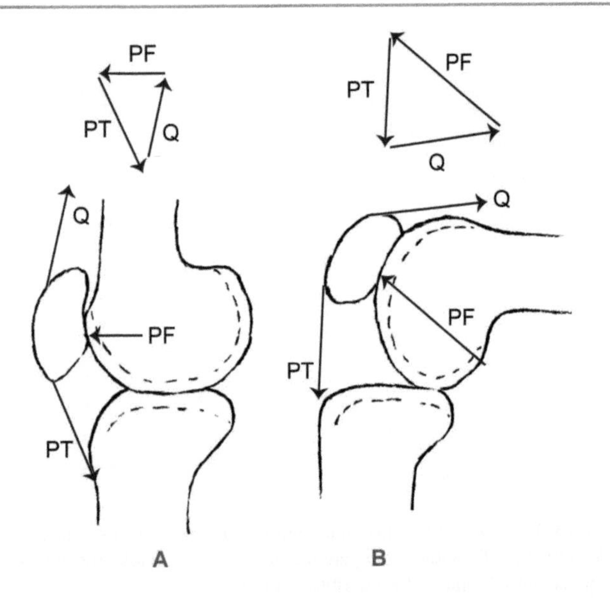


**Fig. 11.33** Schematic of the patellofemoral joint, patellar tendon (ligament), and quadriceps force vectors with (**A**) knee extended and (**B**) knee flexed. An examination of the respective vector addition results suggests an increase in the PF and PT forces with the knee flexed. PF = patellofemoral reaction force, PT = force in patellar tendon, Q = quadriceps force

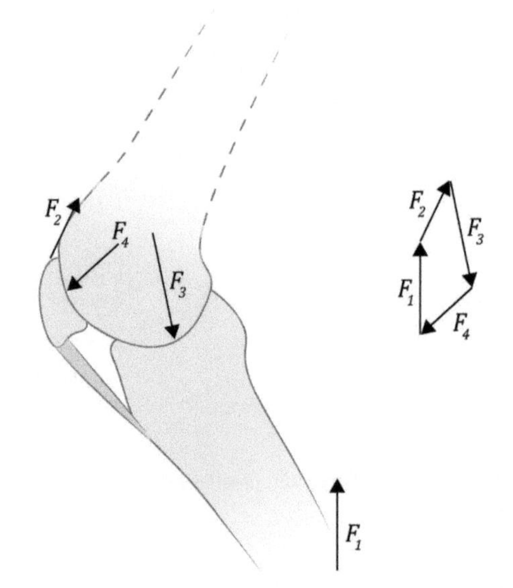


**Fig. 11.34** Schematic illustrating the distribution of the ground reaction force ($F_1$) among the quadriceps tendon ($F_2$), tibiofemoral joint ($F_3$), and patellofemoral forces ($F_4$). The vector addition diagram demonstrates how $F_3$ and $F_4$ work together to balance the forces of $F_1$ and $F_2$. Note: we assume that the ground reaction force is vertically aligned. (Modified with permission from Brinckmann et al. (2016))

with modestly larger quadriceps (Q) and patellar tendon (PT) forces. Knee flexion also changes the relative angle between the Q and PT forces. To achieve static equilibrium (see vector diagram), the patellofemoral force must increase considerably more than the Q and PT forces do.

Since the knee complex has two primary joints, tibiofemoral and patellofemoral, it would seem that these two joints may share/distribute the compression and shear forces that are induced by the quadriceps and ground forces as a result of activities when a knee extension moment is required. In a schematic of a static equilibrium situation (Fig. 11.34), we see a simple vector representation of the primary forces acting on the tibiofemoral and patellofemoral joints (neglecting the weight of the leg and foot). Fig. 11.34 suggests that there is a complex coupling of reaction forces at the patellofemoral ($F_4$) and tibiofemoral ($F_3$) joints. These reaction forces must together balance the forces from the ground ($F_1$) and the quadriceps femoris muscles ($F_2$). And despite the fact that the patella helps create a larger functional moment arm to the quadriceps femoris (i.e., potentially reducing the amount of force produced by the quadriceps femoris), it has been shown that the magnitude of patellofemoral compression forces associated with daily activities can be surprisingly high (Table 11.5).

We have already discussed the anatomical and biomechanical mechanisms at the tibiofemoral joint that manage large compression loads. At the patellofemoral joint, thick articular cartilage is combined with a relatively large surface area when the knee is flexed, reaching its maximum

**Table 11.5** Patellofemoral joint reaction (bone-on-bone) forces related to exercise and daily activities

| Activity | Reaction force |
|---|---|
| Walking | 0.5 × BW |
| Straight leg raise | 0.5 × BW |
| Cycling | 1.2 × BW |
| Rising from a chair using arms | ~3.0 × BW |
| Stairs (up or down) | 3.3 × BW at 60° knee flexion |
| Jogging | 6.0 × BW |
| Squat rise | 6.0 × BW |
| Squat descent | 7.6 × BW at 140° knee flexion |
| Jumping | 12 × BW |

Note: *BW* = body weight

contact area between 60° and 90° flexion. However, during knee flexion, increased patellofemoral compression forces somewhat overmatch the relative increase in the contact area, which significantly increases patellofemoral stress (recall that stress = force/area and that it is the internal joint stresses that challenge articular cartilage). On the other hand, having the greatest surface contact between joint partners at knee angles where the largest joint forces are generated offers the best protection possible, as long as tibiofemoral and patellofemoral joint congruity is optimal and the mechanical alignment of the lower extremities is reasonably normal. Normal mechanical alignment of the lower extremity is critical to the normal tracking of the patella during knee flexion/extension; that is, proper positioning and gliding of the patella within the trochlear groove during normal activities optimize contact surface congruity

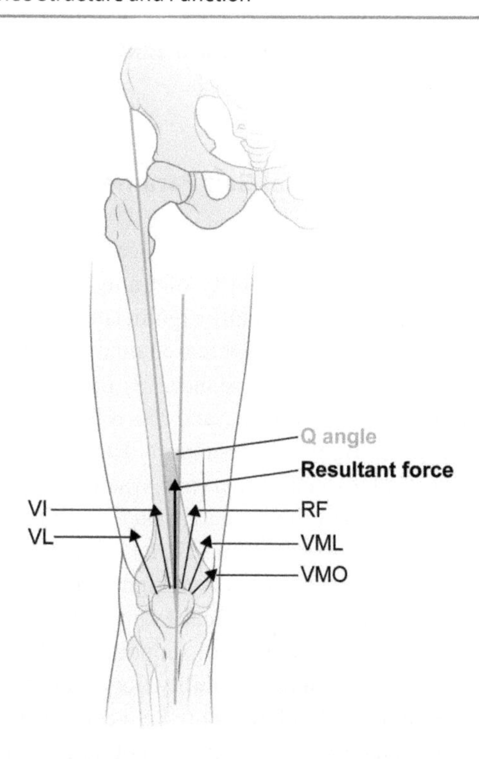


**Fig. 11.35** The resultant quadriceps force as well as separate lines of action of its four components. (Note the formation of a Q angle determined by the angle between the mechanical axis of the femur and the axial axis of the thigh. VL = vastus lateralis, VI = vastus intermedius, RF = rectus femoris, VML = vastus medialis longus, VMO = vastus medialis obliquus)

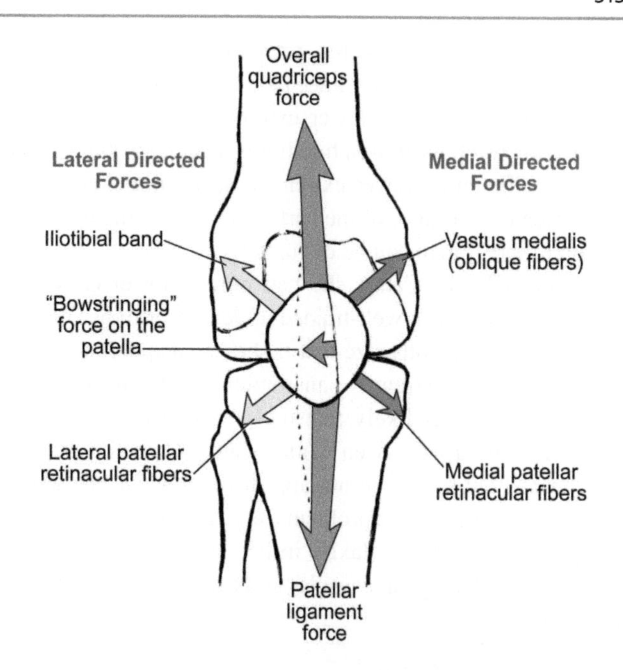


**Fig. 11.36** Schematic of some of the competing soft tissue forces that influence patellofemoral tracking. (Note that the combined effect of the lateral pull of the quadriceps femoris (vastus lateralis) and the lateral offset of the patellar tendon (ligament) induce a "bowstringing" effect (dotted line))

between the patella and trochlea, thus protecting against degenerative joint changes. On the other hand, abnormal patellar tracking appears to put individuals at risk of developing patellofemoral pain syndromes, one of the most common groups of knee injuries.

Besides good congruency between the patella and trochlea, normal genu valgus and action of the quadriceps femoris may have the most influence on optimal patellar tracking (Fig. 11.35). Genu valgus creates a lateral angle (Q angle) between the resultant quadriceps force and a line running from the midpoint of the patella to the tibial tubercle. This lateralized resultant force may be partly created by the larger cross-sectional area of the vastus lateralis. Thus, when the quadriceps femoris shortens and pulls the patella superiorly, there are also lateral and posterior forces influencing patellar positioning. The posterior force on the patellar surface contributes to the patellofemoral joint compression forces but also serves to stabilize the patella into the trochlear groove, at least until the terminal extension is reached. The lateral force vector is potentially destabilizing, particularly if the Q angle is increased, usually secondary to excessive genu valgus, and/or medial soft-tissue restraints are impaired.

Besides genu valgus, and the predisposition of the vastus lateralis to pull the patella off track, there is a balance of other competing anatomical and physiological factors that control patellar tracking (Fig. 11.36 and Table 11.6). We

**Table 11.6** Factors related to abnormal (lateral) tracking of the patella

| General causes | Anatomical and physiological factors |
|---|---|
| Dysplasia | Dysplastic lateral facet of the trochlear groove of the femur |
| Patellar displacement | Patella alta (superior riding patella) |
| Laxity of periarticular tissue | Medial patellar retinacular fibers, i.e., medial patellofemoral ligament<br>Attrition of the medial collateral ligament<br>Laxity and/or reduced height of the medial longitudinal arch of the foot (related to increased pronation of the subtalar joint) |
| Stiffness/tightness of periarticular connective tissue and/or muscle | Lateral patellar retinacular fibers or iliotibial band<br>Hip internal rotator or adductor muscles |
| Abnormal mechanical alignment of the lower extremity | Coxa varus<br>Femoral antetorsion<br>Excessive tibial external torsion<br>Tibial internal torsion, excessive tibial internal rotation during walking related to excessive subtalar pronation<br>Increased Q angle<br>Excessive genu valgum and/or in combination with abduction and external tibial rotation |
| Muscle weakness and/or disrupted motor control | Hip external rotators and abductors<br>Vastus medialis obliquus<br>Tibialis posterior (related to excessive subtalar pronation) |

Modified with permission from Neumann (2017)

allude to most of these factors previously when we describe a physical examination for knee pathology, which included a detailed observation of lower extremity mechanical alignment. This brief discussion highlights, again, a complex system, i.e., the entire lower extremity, with interrelated parts that when one or more of the parts "fails," the efficiency and health of the system can be disrupted.

Fortunately for Cathy, she had normal lower extremity alignment and was a well-trained athlete prior to her meniscus injury. We hypothesize that if she had acquired any preexisting lower extremity anatomical and physiological abnormalities, it is likely that the meniscal injury, and its sequelae, would have been worse. Cathy's knee injury, bad as it turned out to be, did not appear to induce a cascade of several other orthopedic problems, e.g., abnormal tracking of the patella, besides the laxity in her lateral collateral ligament (and perhaps joint capsule) and early tibiofemoral joint degeneration.

## 11.5 Gait Analysis

Recall that the original clinical question, in this case, was related to finding a biomechanical explanation for Cathy's knee problem based on gait analysis. On the surface, the assumption that her abnormal gait (which was obvious) was the cause of a continuing knee problem is consistent with a reductive thinking model. In most cases like this, the events need to be inverted; that is, her abnormal gait is most likely a complex neural response and the manifestation of the chronic nature and progression of the knee pathology. At this point in the chapter, we have clearly detailed the complexity of the knee joint and the importance of considering the related parts and being careful not to fall into a cause/effect trap.

Unfortunately, we did not have data from a standard biplanar video of Cathy's walking to begin our problem-solving approach to gait analysis. However, because of the chronic nature of the injury and proximal muscle weakness, surface electromyographic (EMG) data were collected on key lower extremity muscles. For this case, then, we use a combination of physical examination, three-dimensional kinematic and kinetic analysis, and EMG data in our gait analysis problem-solving approach.

### 11.5.1 Temporospatial Gait Parameters

As before, let us first examine the gait temporospatial parameters to frame the severity of Cathy's gait problem (Table 11.7). The first clinically significant group of findings includes the (1) relative increased time spent in the left stance compared to the right stance, (2) increased time in the right first double support, and (3) relative decreased time spent in the right single support. These asymmetries suggest that Cathy's central nervous system has "chosen" to spend less time on bearing weight on the right side. This was likely related to pain, functional or neurologically imposed weakness, and/or relative right knee instability, factors determined by the physical examination. Increased step width creates a wider base of support, which is often a compensation for dynamic instability. Research has shown a strong proportional relationship between stride length and walking velocity. Therefore, it is not surprising to see reduced stride length, walking velocity, and cadence, given their relationship. This combination of findings is typically found in individuals with painful and unstable joints. A slower walk is created by reducing the step and stride lengths and is safer. Moreover, a reduced walking velocity reduces the ground reaction and, thus, the joint reaction forces.

### 11.5.2 Problem-Solving Approach: Physical Examination, Joint Kinematics and Kinetics, and Electromyography

Recall the following significant impairments in range of motion and muscle strength that were found during Cathy's physical examination:

1. Mild weakness of the left hip extensors and abductors.
2. Moderate weakness of the right hip extensors (hamstrings) and abductors.

**Table 11.7** Temporospatial parameters

| Gait parameter | Left | Right | Normal |
|---|---|---|---|
| Stance (%gc) | 66.3 ± 1.8 | 61.7 ± 1.2 | 59.5 ± 1.5 |
| 1st double support (%gc) | 10.9 ± 0.8 | 13.7 ± 0.5 | 9.5 ± 1.5 |
| Single support (%gc) | 37.7 ± 1.3 | 34.9 ± 0.4 | 40.5 ± 4.5 |
| Step length (cm) | 60.7 ± 3.0 | 57.5 ± 4.0 | 70.5 |
| Stride length (cm) | 118 ± 4.9 | 118 ± 4.9 | 84% age normal |
| Step width (cm) | 10.8 ± 0.7 | | 7.5 ± 2.5 |
| Cadence (steps/min) | 82.1 ± 1.2 | | 73% age normal |
| Walking velocity (cm/s) | 81.0 ± 3.2 | | 60% age normal |

Note: Data are presented as mean ± SD. *%gc* percentage gait cycle, *cm* centimeters

3. Moderate limitation in right knee flexion (perhaps related to internal derangement and likely joint degenerative changes).
4. Mild right knee hyperextension, with associated mild pain and laxity of the right lateral collateral ligament.
5. Right 10° active extensor lag (likely related to internal derangement and arthrogenic pain secondary to degenerative joint disease).

With these findings, we hypothesized that a number of gait-critical events would be adversely affected (see the gait analysis section in Chap. 10 for a review). Because of the mild weakness of the left hip extensors and abductors, we predicted impairments in the functional tasks of weight acceptance and single-limb support. Normally, the hip extensors work eccentrically to control the external trunk/hip flexor moment from initial contact through the loading response, while the hip abductors are activated beginning in first double support through midstance to control the external hip adductor moment, i.e., contralateral pelvic drop. Since Cathy was not using an assistive device to help control these external moments, we predicted that the trunk may demonstrate compensatory backward or lateral leaning movements.

Moderate weakness of the right hip extensors and abductors would also have affected the loading response and midstance subphases during the weight acceptance and swing limb support tasks of walking. Because right hip muscle weakness was greater, we expected to see greater trunk compensations. Right knee hypertension coupled with mild laxity of the lateral collateral ligament suggested an unstable tibiofemoral joint, which likely also had an effect during weight acceptance and single-limb support. Knee stability is important immediately after initial contact when a large impulsive force is created by ground forces. Normally, the knee is in a close-packed position at initial contact, which makes the tibiofemoral joint surfaces maximally congruent and the collateral and cruciate ligaments (as well as the joint capsule) taut. As the knee flexes, it moves into a more open-packed position, making it rely more on the knee extensors for stability. Although Cathy's knee extensor strength, as tested manually, appeared to be within functional limits, we predicted that the mild knee instability from the loss of the lateral meniscus, the failed meniscal allograph procedures, the relatively lax lateral collateral ligament, and pain (related to the degenerative condition of the knee) would impair the normal loading response knee flexion. We also hypothesized that the EMG activation patterns (normal onset and cessation) of the quadriceps and hamstrings might be altered. Although Cathy had sufficient knee flexion (90°), we wondered if her motion impairment would have an effect on right swing limb advancement. Finally, we predicted that the active knee extensor lag may impair her ability to achieve full knee extension in the terminal swing phase.

Similar to the gait testing protocol presented in the hip case, we tested Cathy walking barefoot at her self-selected pace. She did not use an assistive device. We collected kinematic, kinetic (force plate), and EMG data simultaneously until we obtained several clean force plate strikes for both the right and left limbs. Let us first examine joint kinematics (Fig. 11.37A). A reminder that a clinically significant gait deviation is determined if the individual's data fall outside of the gray data band. Beginning with the top row, we see a mild increase in a forward trunk lean throughout the gait cycle, as well as a mildly excessive lateral trunk lean bilaterally during early stance. The lateral trunk lean has been defined as a compensated Trendelenburg and is likely related to hip abductor weakness. The pelvis had mild increases in anterior tilt and transverse plane rotation. Increasing pelvic protraction and retraction is often a compensation for shortened step lengths.

At the hip, we notice a mild limitation in hip extension during the terminal stance bilaterally (Fig. 11.37B). This is surprising in that her hip extension range of motion was normal. The reduction in hip extension could be related to a cautious gait due to pain and knee instability and reduced step length and could explain the mild increase in pelvic rotation. There was a moderate increase in right hip internal rotation and left hip external rotation throughout the gait cycle. These deviations cannot be explained by abnormal femoral torsion since her femoral torsional profile, i.e., femoral antetorsion, was determined to be normal as part of the physical examination. The increased right hip internal rotation may be related to the moderate weakness of the hip abductor muscles on the same side.

The most notable primary gait deviations were found at the knee (Fig. 11.37C). At initial contact on the right side, the knee was hyperextended compared to the left. This is a reminder that we recognized right knee hyperextension during Cathy's physical examination and attributed it to hypermobility/instability of the tibiofemoral joint secondary to the loss of the lateral meniscus and to mild laxity of the lateral collateral ligament (and perhaps associated joint capsule). During walking, the right knee hyperextension may also be related to impaired proprioception and an effort to reduce the loading response knee flexion, which places a high demand on the quadriceps muscle. We did not think that her quadriceps had a strength deficiency, but they were perhaps inhibited by arthrogenic pain secondary to the degenerative changes in her knee. Moreover, if she did not use the quadriceps in a normal manner during the loading response phase, the large axial load normally placed on the tibiofemoral joint would be significantly reduced. Recall that normally during loading response, the quadriceps acts eccentrically to control the external flexor moment, which contributes to a joint bone-on-bone force at least 2.5 times the body weight.

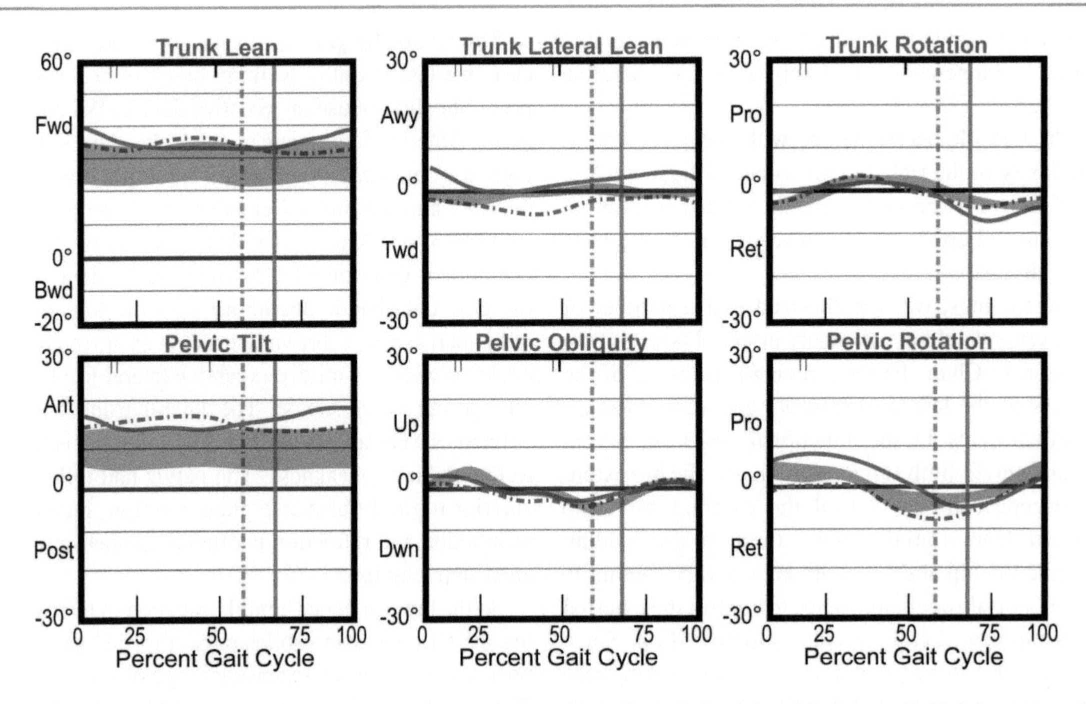


**Fig. 11.37A** Segmental angles of the trunk and pelvis during barefoot walking. Sagittal, frontal, and transverse plane data are organized by columns going from left to right. Data are plotted relative to the percent gait cycle (from 0% far left to 100% far right) along the horizontal axis. The vertical axis indicates the magnitude of joint motion in degrees, with a horizontal line representing 0°, which divides the motion descriptors; e.g., trunk forward lean (Fwd) has positive values, and trunk backward lean (Bwd) has negative values. The vertical line at about 60% of the GC represents toe-off, with stance phase to the left and swing phase to the right of the toe-off line. The dashed red line = right limb, and the solid blue line = left limb. The gray band represents ±1 standard deviation from the control group mean. Awy = away from the reference limb, Twd = toward the reference limb, Pro = protraction, Retr = retraction, Ant = anterior, Post = posterior, Up = reference side of pelvis is up, Dwn = reference side of pelvis is down

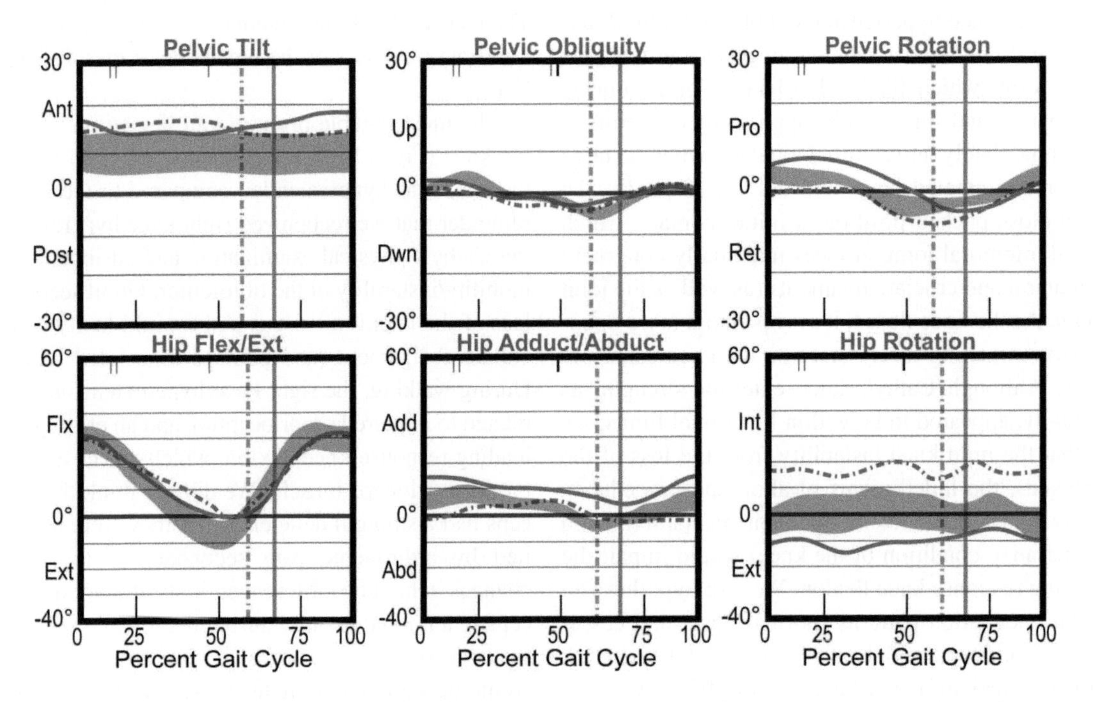


**Fig. 11.37B** Segment and joint angles of the pelvis and hip during barefoot walking. Sagittal, frontal, and transverse plane data are organized by columns going from left to right. Data are plotted relative to the percent gait cycle (from 0% far left to 100% far right) along the horizontal axis. The vertical axis indicates the magnitude of joint motion in degrees, with a horizontal line representing 0°, which divides the motion descriptors; e.g., hip flexion (Flx) has positive values, and hip extension (Ext) has negative values. The vertical line at about 60% of the GC represents toe-off, with stance phase to the left and swing phase to the right of the toe-off line. The dashed red line = right limb, and the solid blue line = left limb. The gray band represents ±1 standard deviation from the control group mean. Up = reference side of pelvis is up, Dwn = reference side of pelvis is down, Ant = anterior, Post = posterior, Pro = protraction, Retr = retraction, Add = adduction, Abd = abduction, Int = internal, Ext = external

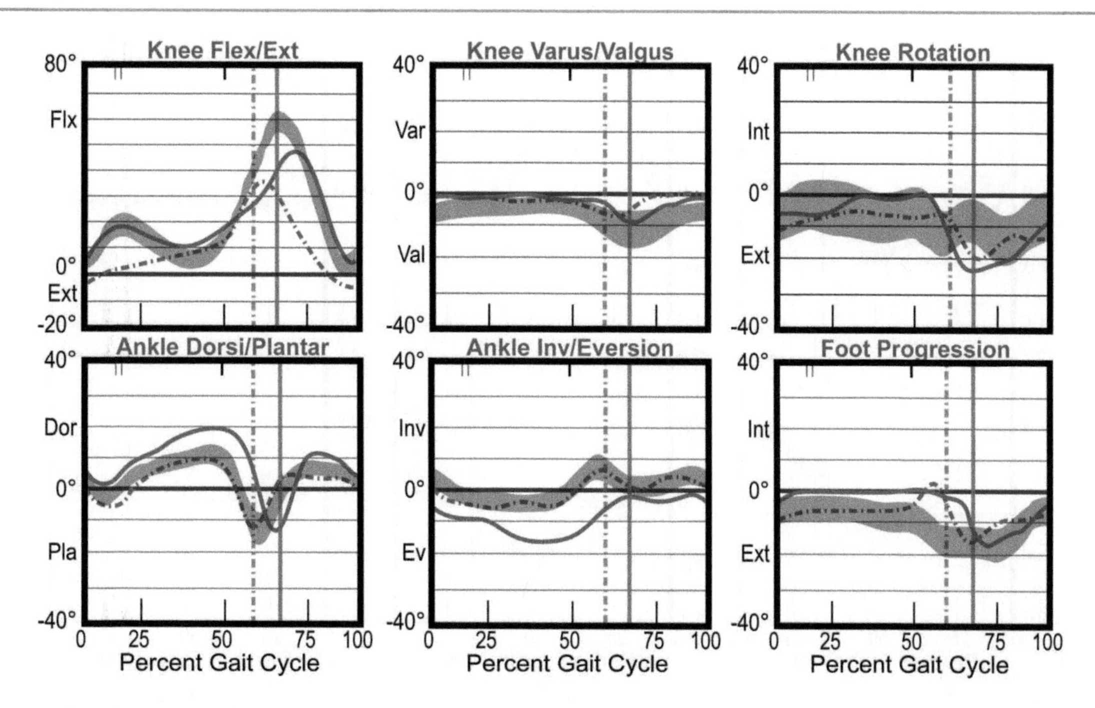


**Fig. 11.37C** Joint angles of the knee and ankle during barefoot walking. Sagittal, frontal, and transverse plane data are organized by columns going from left to right. Data are plotted relative to the percent gait cycle (from 0% far left to 100% far right) along the horizontal axis. The vertical axis indicates the magnitude of joint motion in degrees, with a horizontal line representing 0°, which divides the motion descriptors; e.g., knee flexion (Flx) has positive values, and knee extension (Ext) has negative values. The vertical line at about 60% of the GC represents toe-off, with stance phase to the left and swing phase to the right of the toe-off line. The dashed red line = right limb, and the solid blue line = left limb. The gray band represents ±1 standard deviation from the control group mean. Var = varus, Val = valgus, Dorsi = dorsiflexion, Plan = plantar flexion, Inv = inversion, Ev = eversion, Int = internal, Ext = external

Both the right and left knees demonstrated reduced knee flexion from preswing through midswing. The knee should rapidly flex from 0° to 40° during preswing as part of the acceleration of the stance limb produced by the hip flexors and ankle plantar flexors, which, by the way, also assist with toe clearance. Cathy's reduced rapid hip flexion was likely due to her cautious and reduced gait velocity. Although Cathy had enough knee flexion, i.e., 90° of passive knee flexion, as part of her physical examination, normal midswing knee flexion was significantly reduced bilaterally. On the right, this may have been related to impaired motor control, and it is likely that reduced gait velocity affected both limbs.

Abnormal left ankle kinematics is difficult to explain relative to the other primary and secondary deviations, particularly since the ankle range of motion tests and non-weight-bearing and weight-bearing ankle/foot alignment were normal. It is possible that the increased left ankle eversion was related to excessive hip external rotation. In any case, we do not think that these findings are particularly remarkable (Fig. 11.37C).

Examination of Figs. 11.38A, 11.38B, and 11.38C simultaneously allows us to evaluate the net internal joint moments and powers, respectively, in context. We note a decreased right hip net internal flexor moment in the terminal stance; that, in combination with reduced hip extension and mildly weak hip flexors, was likely related to the decreased net power generation from the terminal stance to preswing. It is possible that reduced walking velocity was also a factor since right and left hip power generation was reduced.

We were surprised to find an increased left internal knee extensor moment from mid- to terminal stance, given that the left knee kinematics was reasonably normal during the same time period. However, this could be explained by Cathy's tendency to walk with a forward trunk lean and excessive left ankle dorsiflexion. The reduced left knee net power absorption in preswing can be explained by Cathy's reduced and slow rate of knee flexion, likely related to decreased gait velocity.

There are two related findings at the right knee: absent net internal extensor moments during the loading response through the terminal stance and absent power generation and absorption during the same time period. We had anticipated this, given Cathy's internal derangement, instability, and arthrogenic knee pain. The absent knee internal valgus moments bilaterally cannot be explained by knee kinematics but could be the result of the coaction of muscles around both knees, which may be activated to enhance limb stability.

Bilaterally increased net internal inversion moments were evident at the ankle. The left can be explained by the exces-

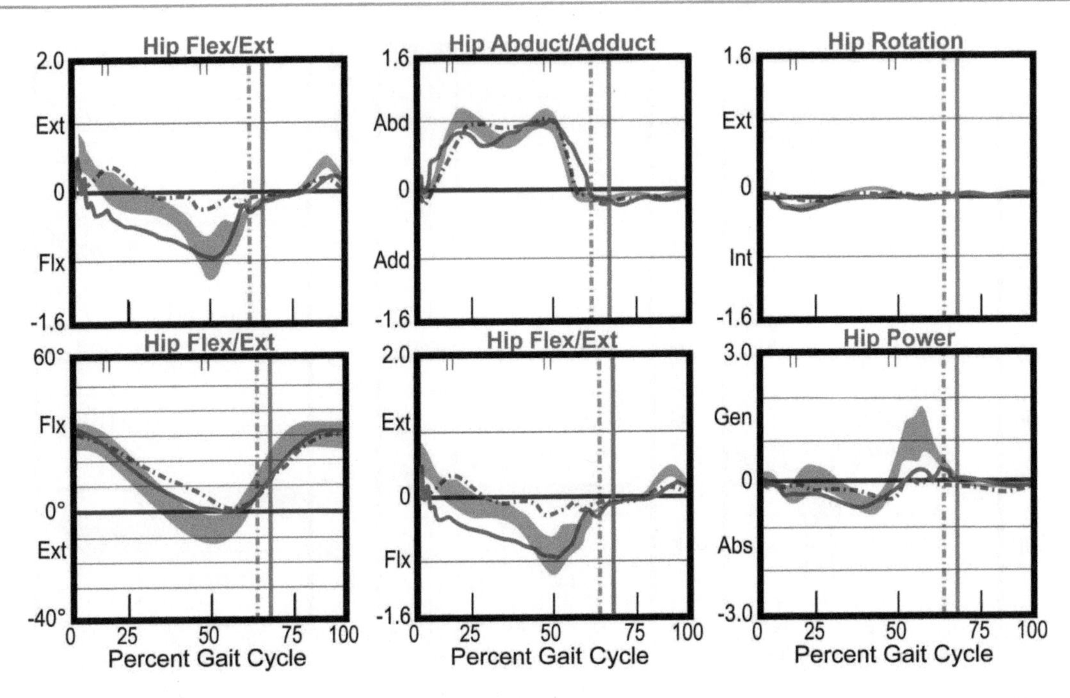


**Fig. 11.38A** *Top row*: Net internal joint moments for the hip during barefoot walking. Sagittal, frontal, and transverse plane kinematic data are organized by columns going from left to right. Data are plotted relative to the percent gait cycle (from 0% far left to 100% far right) along the horizontal axis. The vertical axis indicates the magnitude of the net joint moment, expressed in Nm/kg (newton meter per kilogram), with a horizontal line representing a zero net moment, which divides the moment descriptors; e.g., the hip extensor moment is positive, and the hip flexor moment is negative. The vertical line at about 60% of the GC represents toe-off, with stance phase to the left and swing phase to the right of the toe-off line. The red dashed line = right limb, and the blue solid line = left limb. The gray band represents ±1 standard deviation from the control group mean. Flx = flexor, Ext = extensor, Add = adductor, Abd = abductor, Int = internal rotator, Ex = external rotator. *Bottom row*: Sagittal plane joint angles and net internal joint moments and powers (watts normalized to body mass) during barefoot walking for the hip joint. Data are plotted relative to the percent gait cycle (from 0% far left to 100% far right) along the horizontal axis. We have described the vertical axis for joint angles and moments. For the power curves, the vertical axis indicates the magnitude of the net joint power with a horizontal line representing zero net power, which divides the power descriptors; e.g., hip power generation (Gen) is positive, and hip power absorption (Abs) is negative. The vertical line at about 60% of the GC represents toe-off, with stance phase to the left and swing phase to the right of the toe-off line. The red dashed line = right limb, and the blue solid line = left limb. The gray band represents ±1 standard deviation from the control group mean. Flx = flexion or flexor, Ext = extension or extensor, W/kg = watts per kilogram

sive ankle eversion during stance, but it was difficult to resolve a possible cause for the excessive inversion moment on the right side. There appeared to be a slight increase in ankle net power absorption in midstance bilaterally (left greater than right), which may be an effort by the posterior tibial muscles to stabilize the knee.

Electromyographic analysis (see Appendix J for an explanation of EMG and its application in gait analysis) provides a window into the motor control of the limbs during the gait cycle. The primary interest in clinical gait analysis is in the evaluation of the onset and cessation of the EMG signal, compared to a normal control data set. Thus, we typically do not evaluate the magnitude of the raw signal. The EMG signal is quite complex, and although the magnitude and density of the raw EMG account for the number of motor unit action potentials, they cannot provide any insight into the strength of a muscle. Since it has been shown that the motor control of muscles during a gait cycle may be dependent on neurological development, a clinical gait analysis center would have two different control data sets, one for children and one for adults.

Although Cathy's physical examination informed the provider that she had normal muscle selectivity, i.e., the ability to voluntarily recruit muscle synergies, during the resisted muscle tests, some of the activity patterns of key lower extremity muscles bilaterally during the gait cycle demonstrated abnormal timing (Fig. 11.39). Although the rectus femoris EMG showed a "noisy" baseline, it appeared that there was a prolonged activity of the rectus femoris during the stance phase, with early onset at terminal swing (more evident on the left side, with right-sided data less clear because of a noisy baseline). We saw a very similar pattern for the vastus lateralis bilaterally. Prolonged activity of these two components of the quadriceps may be compensatory to enhance stance stability or may provide evidence of abnormal motor control related to knee pain. It is difficult to get separate EMG signals from the semimembranosus and semitendinosus using surface electrodes. Thus, we describe the

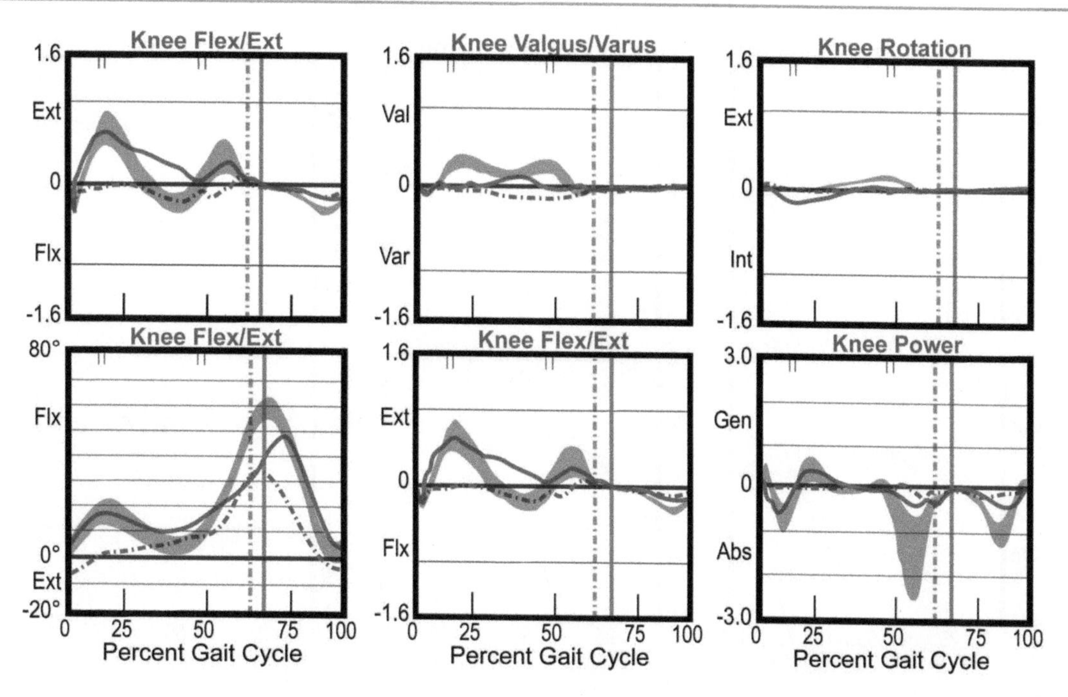


**Fig. 11.38B** *Top row*: Net internal joint moments and powers for the knee during barefoot walking. Sagittal, frontal, and transverse plane kinematic data are organized by columns going from left to right. Data are plotted relative to the percent gait cycle (from 0% far left to 100% far right) along the horizontal axis. The vertical axis indicates the magnitude of the net joint moment, expressed in Nm/kg (newton meter per kilogram), with a horizontal line representing a zero net moment, which divides the moment descriptors; e.g., the hip extensor moment is positive, and the hip flexor moment is negative. The vertical line at about 60% of the GC represents toe-off, with stance phase to the left and swing phase to the right of the toe-off line. The red dashed line = right limb, and the blue solid line = left limb. The gray band represents ±1 standard deviation from the control group mean. Flx = flexor, Ext = extensor, Val = valgus, Var = varus, Int = internal rotator, Ex = external rotator. *Bottom row*: Sagittal plane joint angles and net internal joint moments and powers (watts normalized to body mass) during barefoot walking for the knee joint. Data are plotted relative to the percent gait cycle (from 0% far left to 100% far right) along the horizontal axis. We have described the vertical axis for joint angles and moments. For the power curves, the vertical axis indicates the magnitude of the net joint power with a horizontal line representing zero net power, which divides the power descriptors; e.g., knee power generation (Gen) is positive, and knee power absorption (Abs) is negative. The vertical line at about 60% of the GC represents toe-off, with stance phase to the left and swing phase to the right of the toe-off line. The red dashed line = right limb, and the blue solid line = left limb. The gray band represents ±1 standard deviation from the control group mean. Flx = flexion or flexor, Ext = extension or extensor, W/kg = watts per kilogram

EMG signal as emanating from the medial hamstrings (signals from both the semimembranosus and semitendinosus). Cathy's hamstring activity was nearly continuous for the entire gait cycle (the right more so than the left). Normally, the medial hamstrings are only active during early stance as hip extensors to control the external hip flexor moment and during terminal swing as knee flexors to control the knee extensor moment. It is likely that the continuous hamstring activity reflected another effort to provide stability for the right limb during the stance phase, yet some overactivity may reflect aberrant motor control. Gastrocnemius EMG appeared normal. However, bilateral tibialis anterior activity appeared to be prolonged during stance, with the activity of the left muscle nearly continuous.

Fig. 11.40 provides a comprehensive demonstration of EMG activation and the relationship among lower extremity muscles that control motion at the hip, knee, and ankle during the stance and swing phases of gait. Note, these data were not the control EMG data set used by the Motion Analysis Center, which studied Cathy's gait. However, reviewing this figure, relative to the limited number of muscle EMGs that were collected in this case, may provide the reader with additional insight into Cathy's gait kinematics and kinetics. In summary, it appears that, for the most part, the abnormal timing of the quadriceps, hamstrings, and tibialis anterior can be explained by an effort by the central nervous system to provide additional stability during the stance phase of gait, a time when Cathy's dynamic balance may be challenged by right knee pain and instability.

## 11.6 Meniscal Allograft

Arthroscopic meniscal repairs and meniscectomies have been the standard of care for meniscal lesions for decades, yet the natural history of partial or total meniscectomies ends in degenerative joint disease or osteoarthritis (OA). Furthermore, several follow-up studies have reported poorer

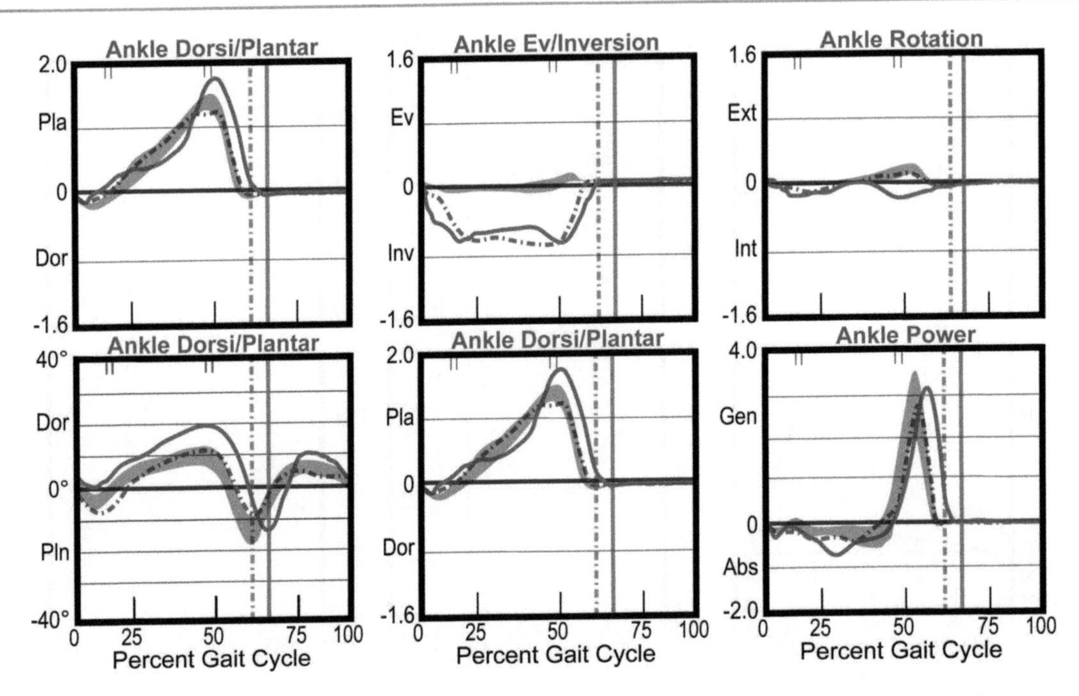


**Fig. 11.38C** *Top row*: Net internal joint moments and powers for the ankle during barefoot walking. Sagittal, frontal, and transverse plane kinematic data are organized by columns going from left to right. Data are plotted relative to the percent gait cycle (from 0% far left to 100% far right) along the horizontal axis. The vertical axis indicates the magnitude of the net joint moment, expressed in Nm/kg (newton meter per kilogram), with a horizontal line representing a zero net moment, which divides the moment descriptors; e.g., ankle plantar flexor moment (Pln) is positive, and ankle dorsiflexor moment (Dor) is negative. The vertical line at about 60% of the GC represents toe-off, with stance phase to the left and swing phase to the right of the toe-off line. The red dashed line = right limb, and the blue solid line = left limb. The gray band represents ±1 standard deviation from the control group mean. Pln = plantar flexor, Dor = dorsiflexor, Ev = evertor, Inv = invertor, Int = internal rotator, Ex = external rotator. *Bottom row*: Sagittal plane joint angles and net internal joint moments and powers (watts normalized to body mass) during barefoot walking for the ankle joint. Data are plotted relative to the percent gait cycle (from 0% far left to 100% far right) along the horizontal axis. We have described the vertical axis for joint angles and moments. For the power curves, the vertical axis indicates the magnitude of the net joint power with a horizontal line representing zero net power, which divides the power descriptors; e.g., ankle power generation (Gen) is positive, and ankle power absorption (Abs) is negative. The vertical line at about 60% of the GC represents toe-off, with stance phase to the left and swing phase to the right of the toe-off line. The red dashed line = right limb, and the blue solid line = left limb. The gray band represents ±1 standard deviation from the control group mean. Pln = plantar flexion or plantar flexor, Dor = dorsiflexion or dorsiflexor, W/kg = watts per kilogram

outcomes following lateral meniscectomies. One study, in an effort to explain this, used a finite element model to compare a medial to a lateral meniscectomy. Their results showed that under axial femoral compression loads, the peak contact stress and maximum shear stress in the articular cartilage increased 200% more after a lateral meniscectomy. Failed subtotal or total meniscectomies are often followed by meniscal allograft transplantation (MAT), which appears to provide clinically acceptable chondroprotective effects. MAT is a well-established surgical procedure, demonstrating satisfactory short- and long-term results in terms of survival, clinical outcomes, return to sport, and efficacy in preventing or delaying knee OA. Although Cathy appeared to be a strong candidate for meniscal allograft transplantation, she experienced two failed attempts. Regardless of her poor outcome, we need to study this intervention and postoperative rehabilitation from a biomechanical perspective.

Meniscal allograft transplantation is defined as the surgical placement of a human donor meniscus into a meniscectomized knee. The meniscus is well suited to undergo transplantation because it does not typically elicit a large immune response. Further, it can function, to some degree, even when devoid of living cells; that is, being largely acellular, it has been claimed that most of the function of the meniscus is derived from its structure rather than cells. *In vitro* biomechanical studies following transplantation have demonstrated that the substitute adequately improves contact axial forces. The use of an allograft is indicated when a regular repair or conservative procedure is not possible. The indications when considering suitable candidates include (1) 50 years of age or younger, (2) prior meniscectomy, (3) pain localized to the meniscal-deficient compartment of the knee, (4) no radiological evidence of advanced joint degeneration, (5) no or only minimal bone exposed on tibiofemoral surfaces, and (6) normal axial alignment. Prophylactic meniscus transplantations are not considered in asymptomatic patients because predictable long-term success rates are not available.

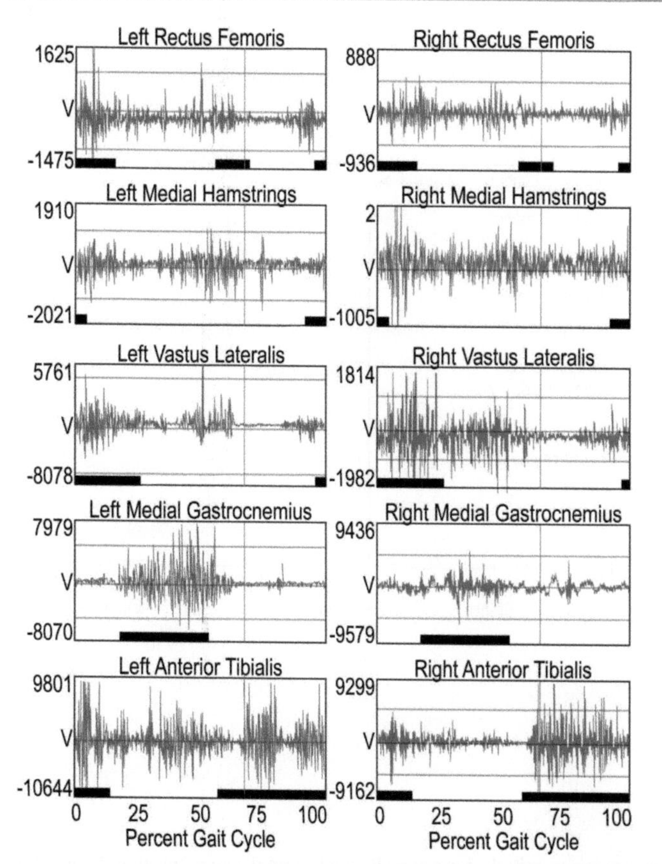


**Fig. 11.39** Left and right surface EMG raw signals during a representative gait cycle. The vertical axis labels the magnitude of the electrical signal (millivolts), and the horizontal axis represents the gait cycle, from the initial contact (0% GC) to the second initial contact (100% GC) of the same limb. The colored (blue = left, red = right) vertical lines represent toe-off. The black horizontal bars along the horizontal axis represent the onset and cessation of EMG recordings from a normal healthy adult control data set

It appears that Cathy had been an appropriate candidate for meniscal allograft transplantation, although we were not privy to a radiographic history that might have been available prior to her first transplantation surgery. It is possible that secondary to the many arthroscopic procedures that she experienced in her early years, she had more than minimal bone exposed in the lateral compartment of her knee.

Recall that one important function of the meniscus is to increase the contact area during weight-bearing. The menisci have a semi-lunar shape, are thicker at the periphery, and taper centrally. They are secured to the tibia at their anterior and posterior horns and effectively deepen the tibial side joint concavity and increase congruity with the femoral condyles. Without the menisci, high contact pressures are generated by the point-to-point contact between the femoral and tibial condyles. With the menisci, the effective contact area is increased, sharing the load over the entire tibial surface. Research has demonstrated repeatedly the direct relationship between the amount of meniscus resected and the amount of contact pressure generated.

It is not the case, however, that more meniscus is better. Thus, the allograft must be properly sized for optimum reduction of contact pressures; an oversized graft has been shown to increase contact pressures, while an undersized graft induces increased hoop stresses, which can result in meniscus rupture. Moreover, there are important differences in morphology between the medial and lateral menisci that need to be considered. The medial meniscus is more C-shaped (covering 64% of the medial tibial plateau and bearing approximately 50% of the axial load), while the lateral meniscus is more O-shaped (covering 84% of the lateral tibial plateau and bearing approximately 70% of the load). Other differences between the medial and lateral compartments are notable. For example, the medial tibial plateau is concave, which may promote more congruity with the articulating medial femoral condyle, while the lateral tibial plateau is convex, which means that in the absence of a lateral meniscus, there is almost a point-to-point contact between the joint surfaces. In sum, when considering MAT, appropriate sizing, matching the respective meniscus to the femoral and tibial condyles, and stable fixation, including ligamentous and bony attachments, are crucial factors for restoring the physiological load distribution and optimizing surgical and clinical outcomes.

Another important function of the menisci is load transmission among multiple heterogeneous joint structures, e.g., articular cartilage, subchondral bone, etc. With weight-bearing, the axial load is partially borne by the menisci, which normally extrude to some degree. As described previously, normal menisci resist extrusion by creating circumferential hoop stress, then transmit the remainder of the compression forces to the underlying articular cartilage. To distribute the hoop stress, the menisci must have an intact circumferential collagenous framework, as well as secure attachments to their anterior and posterior horns. These biomechanical factors come into play when considering the type of fixation for the allograft. Several techniques have been described for allograft transplantation, which appear to fit into three categories: open versus arthroscopically assisted, suture fixation versus bony fixation, and bone plug versus bone bridge (Fig. 11.41). In the early years, open techniques were used, with proponents claiming that this method more readily allowed secure peripheral fixation. However, arthroscopically assisted transplantation surgery has become the method of choice because of reduced surgical morbidity, avoidance of collateral ligament injury, and the potential for implementation of early rehabilitation.

Biomechanical studies have demonstrated that bony fixation is superior to suture alone. This is the case for two reasons: (1) the allograft is completely whole with its native bony

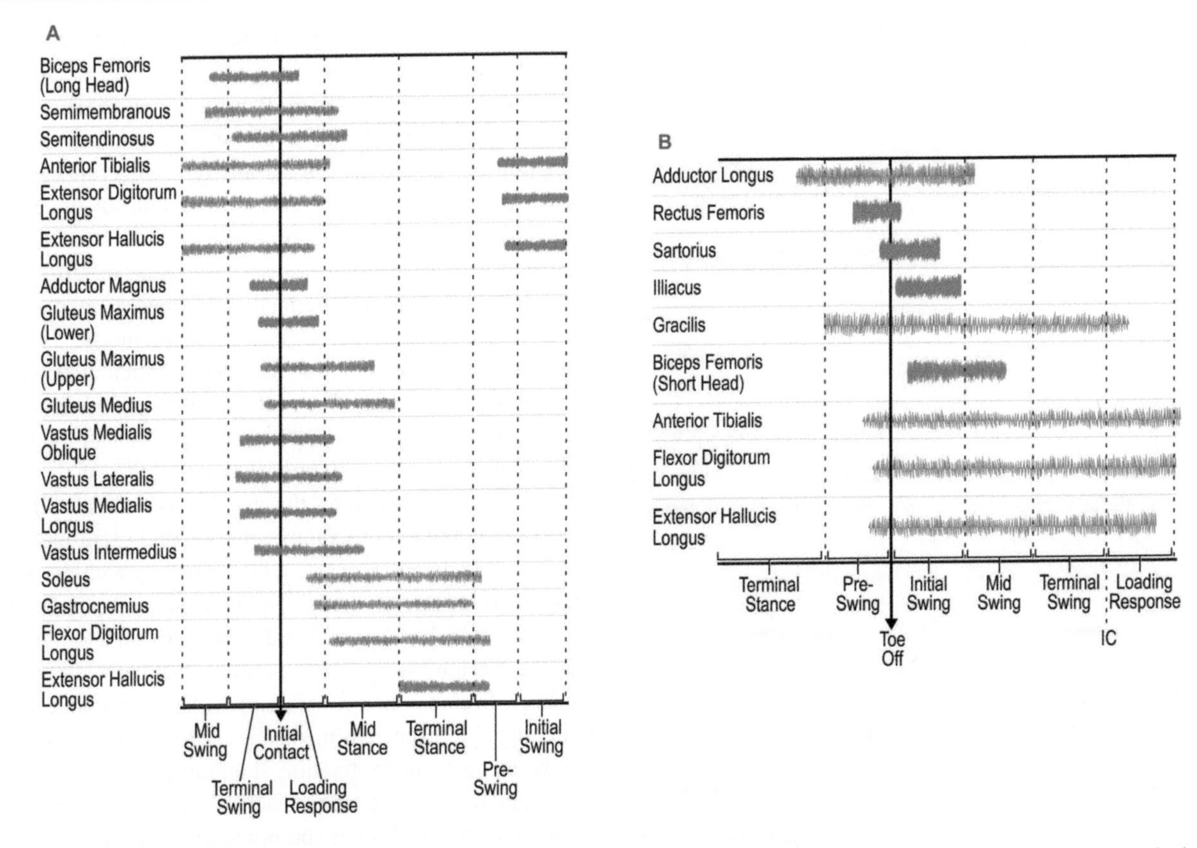


**Fig. 11.40** Onset and cessation of lower extremity muscles during (**A**) midswing to initial swing and (**B**) the terminal stance through the loading response phases of the gait cycle. (Modified with permission from Perry and Burnfield (2010))

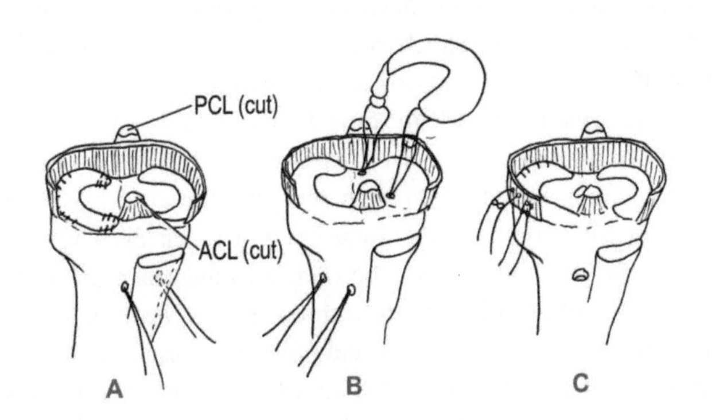


**Fig. 11.41** Schematic representations of meniscal allograft transplantation fixation methods: (**A**) suture, (**B**) bone plug, and (**C**) bone bridge. (Note the proximity of the PCL = posterior cruciate ligament and ACL = anterior cruciate ligament)

attachments, and (2) following transplantation, the inherent superior vascularization of bone reduces degeneration and enhances the revascularization of the allograft. For example, fixation without sufficient anchorage (i.e., use of bone plugs) has been shown to lead to biomechanical results comparable to a meniscectomy. Bone plugs use the allograft menisci still attached to the anterior and posterior horns, which are then brought down and placed into separate tibial tunnels. This has the advantage of preserving the tibial eminence. By contrast, a trough is created in the tibia between the anterior and posterior horns of the resected meniscus using the bone bridge technique. With this alternative bone technique, the allograft is prepared to leave the meniscus attached to an entire bridge of bone between the meniscal horns; the bridge slides directly into the prepared trough. This technique is more forgiving than the bone plug technique and preserves the native distance between the horns, but it sacrifices the tibial eminence. However, there do not appear to be any differences between the two bone fixation methods with regard to functional outcomes.

As we recount several times, menisci have several important functions that are, in fact, interrelated: contribute to tibiofemoral kinematics while enhancing joint stability, maximize contact area to distribute peak pressures, effectively maintain joint surface friction, and transmit compression and shear loads. We noted earlier in this account that the primary goal of meniscal allograft transplantation was chondroprotection, which can be achieved by restoring more normal contact areas and pressures, as well as optimizing load transmission. What does outcomes research tell us about MAT procedures? Almost all short-term (2–3 years) follow-up studies demonstrate favorable results, particularly with regard to pain relief. One notable group reported a 58% failure in their cohort after 30 months but attributed this to, perhaps, a poor selection of candidates; preoperatively, they had high-grade tibiofemoral arthrosis. At 2–5 years postsurgery, patient satisfaction was reduced, with failures reported to range from 10% to 15%. MAT outcomes beyond 10 years are inconclusive; generally, approximately 75–90% of the patients have experienced good results, but secondary MRI investigations at that point have revealed graft shrinkage, articular cartilage and meniscus degeneration, and extrusion. Reported complications related to MAT include disease transmission and immune reaction, which are rare. The most commonly reported complication is tearing (approximate 8% tear rate) of the transplanted meniscus.

What do we know about how well MAT restores biomechanical function? This information is limited and mostly speaks to chondroprotection factors. First, the clinical outcomes just reported indirectly address that question quite well. However, some research using finite element modeling and *in vitro* methods suggest that biomechanical function following MAT is only partially restored. For example, joint contact pressures, which were increased following meniscectomy, were significantly reduced following MAT, suggesting that the primary goal of MAT, i.e., chondroprotection, was mostly met. There appears to be a paucity of experimental research examining the effect of MAT on knee instability. However, *in vitro* research has demonstrated that MAT may partially reduce anteroposterior laxity, with a recent *in vivo* clinical study showing that a lateral MAT was able to restore anteroposterior stability, particularly during a pivot-shift maneuver (a common provocative knee ligamentous stress test).

The success of meniscal allograft transplantation depends on both biological and biomechanical integration. While biological integration creates a connection to the host tissue, leading in the long-term, to normal cell activity that reestablishes a matrix composition typical of the tissue being replaced, biomechanical integration is needed for a good functional outcome. That is, the graft must experience and manage the stresses and strains similar in character and magnitude typical to those of the replaced tissue. Thus, skilled placement of the meniscus allograft is only as good as a well-designed and managed rehabilitation program. The postoperative rehabilitation program must strictly adhere to the healing constraints imposed by the inflammatory, revascularization, and remodeling phases of graft healing (see this discussion in Chap. 6) while imposing compression and shear forces that are similar in nature to normal function but graded in order to protect the graft site. Failure to strictly follow protocol guidelines would likely result in a suboptimal outcome or complete failure.

Perhaps that is what happened to Cathy. According to Cathy, although she was provided some instruction in home care activities, i.e., managing pain, how to control joint swelling, and crutch use, she did not participate in a formal rehabilitation program, which could have provided structured and skilled supervision.

Briefly, three phases of graft integration are distinguished: inflammatory response with immediate degeneration of the graft cells, tissue maturation with revascularization (using bone plugs is a major advantage in the case of MAT, considering that normal vascularization of the meniscus is poor to begin with), and graft healing with tissue remodeling. The inflammatory phase is associated with cell death and graft degeneration, with attempts of the body to revascularize but with significant loss of strength and stiffness of the graft. This weakening continues into the revascularization phase when blood vessels begin to invade the new tissue. The process of revascularization is slower with allografts, keeping the meniscus graft weakened for up to 6–7 weeks. Graft remodeling and maturation comprise the third phase of graft integration when the vascularization process has progressed to the point where new cells invade and deposit a collagenase matrix that enhances the biomechanical properties of the tissue. This final process can persist for up to 12 months. It is notable that the allograft, even after several months of maturation, will never attain the biomechanical properties of the original meniscus.

Rehabilitation protocols based on the three phases of graft integration provide the following general guidelines, with the caveat that procedures to address chondral issues (microfracture, orthobiologic repair) may require additional modifications to weight-bearing and appropriate open- and closed-chain strengthening progressions:

- The patient is non-weight-bearing for the first 3–6 weeks postoperatively in a hinged, but locked, brace with the knee in full extension; at 6 weeks, the patient can begin weight-bearing but will continue to don the brace for locomotion and balance activities.
- Range of motion (ROM) expectations include terminal knee extensions (0° but not forced hyperextension for the first 6 weeks) in the first week and progression of flexion ROM to 45° by week 3, 90° by week 4, and full symmetrical ROM by weeks 10–12.
- Walking, elliptical, and stair master (only single-step depth) for cardiovascular exercise are allowed at week 12.
- Straight-line running and return to sports activities are not considered until week 24.

As we review the specific stages of the protocol, and each activity within a stage, it is important to consider that the protocol is designed to find the right balance between the protection of the graft and the application of multidimensional forces that lead to an optimal biological and biomechanical integration. Phase I entails the education of the patient, which includes a discussion of the surgical procedure and postsurgical expectations and brief instructions on preoperative exercises and immediate postoperative home management, e.g., the use of crutches, edema control (rest, ice, compression, and elevation), manual lymphatic drainage, ankle pumps, and patellar mobilizations.

Phase II (maximum protection) begins immediately after surgery and extends for approximately 3 weeks. It includes the home management of the preoperative instructions, with an emphasis on maintaining full knee extension when walking, patellar mobilization but no active knee flexion, and non-weight-bearing walking. This phase includes the acute and subacute inflammatory cycle, which is an important part of the biological integrations, and a time when the new graft undergoes degeneration; thus, the graft is especially susceptible to injury. Recall that the posterior horn of the medial meniscus is partially anchored by the semimembranosus, and the lateral meniscus has fascia links to the popliteus muscle. Thus, during this very early stage of healing, it makes biomechanical sense to refrain from active knee flexion; that is, the menisci may not have the strength/stiffness to manage the tensile stress imposed by the pull of the knee flexor muscles.

"Protected motion" is absolutely critical in Phase III, which runs from weeks 3 to 6; that is, restoring knee motion is important, but the activities that promote this must be closely supervised. Formal physical therapy is initiated and includes the initiation of weight-bearing to tolerance using the locked hinge brace, single-leg balance training in the locked brace (with protected weight-bearing, as with walking), progressive passive range of motion from 90° (by week 4) to 120° (by week 6) knee flexion, quadriceps setting exercises (submaximal to maximal, as tolerated), active knee extension (90° to full extension) without resistance, and straight-leg raises for hip flexion, abduction, and adduction (once a knee extensor lag is resolved). Progressive weight-bearing begins to challenge the contact areas and the ability of the meniscus to transmit the compression and shear loads to circumferential hoop stresses. Passive and active sagittal plane range of motion exercises engage graft translation, which is coupled with femoral and tibial rolling and gliding. Conjunct axial rotation, which naturally occurs with flexion/extension, adds a multidimensional element to graft stress; however, the initiation of knee flexion and the termination of knee extension must be guarded carefully. Exercise that includes both sagittal and transverse plane passive and active movements forces the new meniscus to experience necessary torsional deformations.

Phase IV spans the postoperative period from weeks 6 to 12. The individual progresses to full weight-bearing with the brace unlocked, allowing up to 90° flexion (as tolerated). Passive and active ROM continues until the full active range of motion is achieved; a bicycle, with minimal resistance, is used to enhance multiple repetition ROM exercises. A progressive increase in resistive antigravity exercise for the knee extensors and all hip muscles is instituted; partial squats to 60° are allowed, with care to control for tibial rotation and abduction. Finally, proprioceptive training in double- and single-leg standing using variable surfaces to add perturbation is initiated in an attempt to restore the normal neural feedback systems. It should be apparent that this phase is merely adding increased functional demands on the meniscus graft, as the individual can tolerate; the therapist closely monitors for signs, e.g., pain, swelling, prolonged soreness, etc., of graft and joint irritation and makes appropriate changes in the program, as needed. The individual is allowed to advance to Phase V if there is (1) full passive ROM, (2) a demonstration of normal gait on level ground, (3) a trace of or no joint effusion, and (4) manual resisted strength grades of good (4 out of 5 rating) for the quadriceps and hip abductors and hip external rotators.

Phase V is a more aggressive continuation of Phase IV activities, spanning the interval from postoperative weeks 12 to 24. More intensive squat exercises are added, as are elliptical and stair master for additional cardiovascular and muscle endurance training. Passing to Phase VI is approved if the individual demonstrates full, pain-free active and passive ROM; no joint effusion; and the normal performance of all daily activities. The final phase of rehabilitation begins at 6 months post-op. In consultation with the surgeon, and with the passing of strict testing criteria, the final phase includes a progression to advanced movement and impact types of activities, e.g., running, jumping, cutting, etc., and a gradual return to sports-specific skill training (for athletes). A return to full activities may require up to an additional 3 to 6 months of training.

A final note: the progression to functional activities is allowed only as pain, joint swelling, and proper biomechanics allow. During this particular phase of rehabilitation, the therapist closely monitors the proper control of dynamic abduction (valgus) stress at the knee with each task. And progression to single-limb-based tasks, e.g., deceleration and hopping, is not allowed until double-limb activities have been mastered (performed correctly and without pain or postexercise joint effusion). Activities requiring dynamic control of multidimensional movements at the knee, e.g., cutting and multiple plane lunges/jumps/hops, are not allowed until sagittal, frontal, and transverse (associated with single-plane movements) have been mastered. This multiple-staged program is the result of decades of research that has clarified the important role of the menisci for normal knee and lower-limb functioning, as well as years of clinical experience, including failures in the care of meniscal injuries.

## 11.7 Chapter Summary

Because the knee is interposed between the hip and ankle, it must do two things really well: (1) provide a large range of motion in one plane yet with the ability to make multidimensional adaptations and (2) provide a stable platform upon which movements can occur. Unfortunately, the task is nearly impossible, given the range of positions and activities we impose on our knees. This comment is supported by the fact that the knee has a higher rate of injury, and development of degenerative joint disease, than the hip and ankle. On the other hand, the knee joint does remarkably well. This case provided a great opportunity to study the unique anatomy and biomechanics of the knee joint in general and of the menisci in particular. Moreover, we present another example of a complex system and its interrelated parts.

Meniscal (and often related ligamentous) injuries are probably the most prevalent of all knee injuries. What we learn by studying Cathy's history is how vulnerable the menisci are even with seemingly innocuous movements, i.e., getting up from the floor to stand. The standard of care for an injury similar to Cathy's includes an accurate history, comprehensive physical examination, and arthroscopic diagnosis and treatment. According to Cathy, there was an attempt to repair her torn lateral meniscus, which is in keeping with the idea that the menisci are very important structures in their roles of distributing contact stresses, transmitting forces, and assisting in normal tibiofemoral kinematics. Unfortunately for Cathy, she experienced a series of reinjuries, which together initiated a premature progression of lateral compartment joint degeneration. When she was referred for gait analysis in an effort to understand her biomechanical deficiencies, it was clear that the knee degeneration had progressed to a point beyond simple debridement and repairs. Thus, the case also allows us to review the normal mechanics of articular and fibrocartilage, as well as the stages of joint degeneration (we actually do not do this in this chapter, but the curious reader, of course, could review this information, which is presented in appendices or previous chapters of the text).

This case provides the reader with biomechanical insights into how lower-limb mechanical alignments can affect normal and abnormal knee functioning, the clinical tests used for the examination of a knee injury, and the use of instrumented gait data to help explain primary and secondary gait deviations. Finally, although it seemed that Cathy did not have this experience, we review the biomechanical rationale for the stages of a formal meniscal allograft transplantation rehabilitation protocol. Recall that in the hand case, we presented a treatment protocol for postoperative care following a repair of the flexor pollicis longus tendon. With that presentation, we discover the biomechanical rationale to be based on the protection and application of controlled tension. In this case, we present a very similar type of general rehabilitation program based on the integration of biological and biomechanical factors but with different particulars related to meniscal function. Thus, in this effort, we continue to challenge the reader to look for and apply general biomechanical principles while, at the same time, gaining insights into the process and particulars of the case at hand that might be used in future cases.

## Bibliography


Andrews SHJ, Adesida AB, Abusara Z et al (2017) Current concepts on structure-function relationships in menisci. Connect Tissue Res 58(3–4):271–281. https://doi.org/10.1080/03008207.2017.1303489

Baroon J, Ravani B (2010) Three-dimensional generalizations of Reuleaux's and instant center methods based on line geometry. J Mech Robot 2:041011-1–041011-8. https://doi.org/10.1115/1.40001727

Blankevoort L, Huskies R, De Lange A (1990) Helical axes of passive knee joint motions. J Biomech 23(12):1219–1229

Brinckmann R, Frobin W, Leivseth G et al (2016) Mechanical aspects of the knee. In: Brinckmann R, Frobin W, Leivseth G et al (eds) Orthopedic biomechanics, 2nd edn. Thieme Publishers, Stuttgart

Chang PS, Brophy RH (2020) As goes the meniscus goes the knee. Early, intermediate, and late evidence for the detrimental effect of meniscal tears. Clin Sports Med 39:29–36. https://doi.org/10.1016/j.csm.2019.08.001

Chen H-N, Yang K, Dong Q-R et al (2014) Assessment of tibial rotation and meniscal movement using kinematic magnetic resonance imaging. J Orthop Surg Res 9:65. https://www.josr-online/content/9/1/65

England M (2009) The role of the meniscus in osteoarthritis genesis. Med Clin N Am 93:37–43. https://doi.org/10.1016/j.mcna.2008.08.005

Fox AJS, Bedi A, Rodeo SA (2012) The basic science of human knee menisci: structure, composition, and function. Sports Health 4(4):340–351. https://doi.org/10.1177/1941738111429419

Frankel VH, Burstein AH, Brooks DB (1971) Biomechanics of internal derangement of the knee, pathomechanics as determined by analysis of the instant centers of motion. J Bone Joint Surg 53-A(5):945–962

Gray JC (1999) Neural and vascular anatomy of the menisci of the human knee. J Orthop Sports Phys Ther 29(1):23–30

Guess TM, Razu S, Jahandar H et al (2015) Predicted loading on the menisci during gait: the effect of horn laxity. J Biomech 48:1490–1498. https://doi.org/10.1016/j.biomech.2015.01.047

Hashemi J, Chandrashekar N, Gill B et al (2008) The geometry of the tibial plateau and its influence on the biomechanics of the tibiofemoral joint. J Bone Joint Surg Am 90:2724–2734. https://doi.org/10.2016/JBJS.G.01358

Jarraya M, Roemer FW, Englund M et al (2017) Meniscus morphology: does tear type matter? A narrative review with focus on evidence for osteoarthritis research. Semin Arthritis Rheum 46:552–561. https://doi.org/10.1016/j.semarthrit.2016.11.005

Kapandji AI (2019) The knee. In: Kapandji AI (ed) The physiology of the joints, The lower limb, vol 2, 7th edn. Handspring Publishing Limited, Scotland

Kerin AJ, Wisnom MR, Adams MA (1998) The compressive strength of articular cartilage. Proc Inst Mech Eng H J Eng Med 212:273–280

Koh Y-G, Lee J-A, Kim Y-S (2019) Biomechanical influence of lateral meniscal allograft transplantation on knee joint mechanics during the gait cycle. J Orthop Res 14:300. https://doi.org/10.1186/s13018-019-1347-y

Koh J, Zaffagnini S, Kuroda R et al (eds) (2021) Orthopaedic biomechanics in sports medicine. Cham, Switzerland, ISOKOS, Springer

Kosel J, Giouroudi I, Scheffer C et al (2010) Anatomical study of the radius and center of curvature on the distal femoral condyle. J Biomech Eng 132:091002-1–091002-6. https://doi.org/10.1115/1.40002061

Lenhart RL, Brandon Scott CE, Smith CR et al (2017) Influence of patellar position on the knee extensor mechanism in normal and crouched walking. J Biomech 51:1–7. https://doi.org/10.1016/j.biomech.2016.11.052

Li G, Zhou C, Zhang Z et al (2022) Articulartion of the femoral condyle during knee flexion. J Biomech 131:110906. https://doi.org/10.1016/j.jbiomech.2021.110906

Lin Y, Zhang K, Li Q et al (2019) Innervation of nociceptors in intact human menisci along the longitudinal axis: semi-quantitative histological evaluation and clinical implications. BMC Musculoskelet Disord 20:338. https://doi.org/10.1186/s12891-019-2706-x

Lubowitz JH, Verdonk Peter CM, Reid JB III et al (2007) Meniscus allograft transplantation: a current concepts review. Knee Surg Sports Traumatol Arthrosc 15:476. https://doi.org/10.1017/s00167-006-0216-5

Luvzkiewicz P, Daszkiewicz K, Witkowski W et al (2015) Influence of meniscus shape in the cross sectional plane on the knee contact mechanics. J Biomech 48:1356–1363. https://doi.org/10.1016/j.biomech.2015.03.002

Macchiarola L, Di Paolo S, Grassi A et al (2021) In vivo kinematic analysis of lateral meniscal allograft transplantation with soft tissue fixation. Orthop J Sports Med 9(5):23259671211000459. https://doi.org/10.1177/23259671211000459

Magee DJ (2013) Orthopedic physical assessment, 6th edn. Elsevier Saunders, St. Louis

Makris EA, Hadidi P, Athanasiou KA (2011) The knee meniscus: structure-function, pathophysiology, current repair techniques, and prospects for regeneration. Biomaterials 32(30):7411–7431. https://doi.org/10.1016/j.biomaterials.2011.06.037

Mannel H, Marin F, Claes L (2004) Establishment of a knee-joint coordinate system from helical axes analysis – a kinematic approach without anatomical referencing. IEEE Trans Biomed Eng 51(8):1341–1347. https://doi.org/10.1109/TBME.2004.828051

Markes AR, Hodax JD, Benjamin C (2020) Meniscus form and function. Clin Sports Med 39:1–12. https://doi.org/10.1016/j.csm.2019.08.007

McDermott ID, Masouros SD, Amis AA (2008) Biomechanics of the menisci of the knee. Curr Orthop 22:193–201. https://doi.org/10.1016/j.cuor.2008.04.005

McNulty AL, Guilak F (2015) Mechanobiology of the meniscus. J Biomech 48:1469–1478. https://doi.org/10.1016/j.biomech.2015.02.008

Mine T, Kimura M, Sakka A et al (2000) Innervation of nociceptors in the menisci of the knee joint: an immunohistochemical study. Arch Orthop Trauma Surg 120:201–204

Mohamadi A, Momenzadeh K, Masoudi A et al (2021) Evolution of knowledge on meniscal biomechanics: a 40 year perspective. BMC Musculoskelet Disord 22:625. https://doi.org/10.1186/s12891-021-04492-2

Mosouros SD, McDermott LD, Amis AA et al (2008) Biomechanics of the meniscus-meniscal ligament construct of the knee. Knee Surg Sports Traumatol Arthrosc 16:1121–1132. https://doi.org/10.1007/s00167-008-0616-9

Neumann DA (2017) Knee. In: Neumann DA (ed) Kinesiology of the musculoskeletal system, foundations for rehabilitation, 3rd edn. Elsevier, St. Louis

Nilsson MK, Friis R, Michaelsen MS et al (2012) Classification of the height and flexibility of the medial longitudinal arch of the foot. J Foot Ankle Res 5:3. https://www.jfootankleres.com/content.5/1/3

Noyes FR, Heckmann TP, Barber-Westin SD (2012) Meniscus repair and transplantation, a comprehensive update. J Orthop Sports Phys Ther 42(3):274–290. https://doi.org/10.2519/jospt.2012.3588

Oatis CA (2017) Kinesiology, the mechanics and pathomechanics of human movement, 3rd edn. Wolters Kluwer, Philadelphia

Panjabi MM (1979) Centers and angles of rotation of body joints: a study of errors and optimization. J Biomech 12:911–920

Peña E, Calvo B, Martinez MA et al (2005) Finite element analysis of meniscal tears and meniscectomies on human knee biomechanics. Clin Biomech 20:498–507. https://doi.org/10.1016/j.clinbiomech.2005.01.009

Peña E, Calvo B, Martinez MA et al (2006) Why lateral meniscectomy is more dangerous than medical meniscectomy. A finite element study. J Orthop Res 24:1001–1020. https://doi.org/10.1002/jar.20037

Perry J, Burnfield JM (2010) Total limb function and bilateral synergistic relationships. In: Perry J, Burnfield JM (eds) Gait analysis, normal and pathological function, 2nd edn. SLACK Incorporated, Thorofare

Peterson W, Tillman B (1998) Collagenous fibril texture of the human knee joint menisci. Anat Embryol 197:317–324

Pytel A, Kiusalaas J (2010a) Virtual work and potential energy. In: Pytel A, Kiusalaas J (eds) Engineering mechanics statics, 3rd edn. Cengage Learning, Stamford

Pytel A, Kiusalaas J (2010b) Planar kinematics of rigid bodies. In: Pytel A, Kiusalaas J (eds) Engineering mechanics dynamics, 4th edn. Cengage Learning, Boston

Roos H, Laurén M, Adalberth T et al (1998) Knee osteoarthritis after meniscectomy, prevalence of radiographic changes after twenty-one years, compared with matched controls. Arthritis Rheum 41(4):687–693

Samitier G, Alentorn-Geli E, Taylor DC et al (2014) Meniscal allograft transplantation. Part 2: systematic review of transplant timing, outcomes, return to competition, associated procedures and prevention of osteoarthritis. Knee Surg Sports Traumatol Arthrosc 23:323. https://doi.org/10.1007/s00167-014-3344-3

Schrijvers JC, van den Noort JC, van der Esch M et al (2021) Neuromechanical assessment of knee joint instability during perturbed gait in patients with knee arthritis. J Biomech 118:110325. https://doi.org/10.1016/j.jbiomech.2021.110325

Seitz AM, Dürselen L (2019) Biomechanical considerations are crucial for the success of tendon and meniscal allograft integration – a systematic review. Knee Surg Sports Traumatol Arthrosc 27:1708–1716. https://doi.org/10.1007/s00167-018-5158-y

Sheehan FT (2007) The finite helical axis of the knee joint (a non-invasive in vivo study using fast-PC MRI). J Biomech 40:1038–1047. https://doi.org/10.1016/j.biomech.2006.04.006

Smith PN, Refshauge KM, Scarvell JM (2003) Development of the concepts of knee kinematics. Arch Phys Med Rehabil 84:1895–1902. https://doi.org/10.1016/S0003-9993(03)00281-8

Smith MV, Nepple JJ, Wright RW et al (2016) Knee osteoarthritis is associated with meniscus and anterior cruciate ligament surgery among elite college American football players. Sports Health 9(3):247–251. https://doi.org/10.1177/1941738116683146

Sohn DH, Toth AP (2008) Meniscus transplantation. J Knee Surg 21:163–172

Spiegelman JJ, Woo Savio L-Y (1987) A rigid-body method for finding centers of rotation and angular displacements of planar joint motion. J Biomech 20(7):715–721

Tamea CD, Henning CE (1981) Pathomechanics of the pivot shift maneuver, an instant center analysis. Am J Sports Med 9(1):31–37

Voloshin AS, Wosk J (1983) Shock absorption of meniscectomized and painful knees: a comparative in vivo study. J Biomech Eng 5:157–160

Walker PS, Arno S, Bell C et al (2015) Function of the medial meniscus in force transmission and stability. J Biomech 48:1383–1388. https://doi.org/10.1016/j.biomech.2015.02.055

Walker PS, Frankle VH, Nordin M (2022) Biomechanics of the knee. In: Nordin M, Frankle VH (eds) Basic biomechanics of the musculoskeletal system, 5th edn. Wolters Kluwer, Philadelphia

Wang H, Gee AO, Hutchinson ID et al (2014) Bone plug versus suture-only fixation of meniscal grafts: effect on joint contact mechanics during simulated gait. Am J Sports Med 42(7):1682–1689. https://doi.org/10.1177/0363546514530867

Wilson DR, Feikes JD, O'Connor JJ (1998) Ligaments and articular contact guide passive knee flexion. J Biomech 31: 1127–1136

Yamamoto T, Taneichi H, Seo Y et al (2021) MRI-based kinematics of the menisci through full knee range of motion. J Orthop Surg 29(2):1–8. https://doi.org/10.1177/23094990211017349

Zaffagnini S, Di Paolo S, Stefanelli F et al (2019) The biomechanical role of meniscal allograft transplantation and preliminary in-vivo kinematic evaluation. J Exp Orthop 6:27. https://doi.org/10.1186/s40634-019-196-2

# Ankle/Foot Complex: Recurrent Stress Fractures

# 12

## 12.1 Introduction

"When the feet hit the ground, everything changes" was the title of a continuing education course that was offered to a cadre of health professionals for decades. It was particularly attractive to physical therapists, athletic trainers, strength and conditioning, exercise and movement science experts, kinesiologists, and biomechanists who evaluated and treated individuals with a variety of common injuries that involved the ankle/foot, tibio- and patellofemoral joints, hip and pelvic complex, etc., whose source of dysfunction may have, in fact, originated in the foot. The underlying framework of the course, as the title implies, was that the evolutionary process had very creatively designed a system to optimize bipedal locomotion, i.e., walking and running, whose base was the ankle and foot. For example, the human foot tarsal and toe structure have changed drastically since the days of our tree-dwelling ancestors to provide the greater sagittal plane mobility needed for walking and running overground. At the same time, the bipedal foot has evolved to provide a stable base to absorb and transmit large forces exerted by the ground, be adaptable enough to manage uneven walking/running surfaces, and serve as an efficient interface to accelerate and decelerate the body relative to the ground. Similar to our discovery of other anatomical complexes, the ankle/foot complex does not operate in isolation (although we reduce the system to study its constituent parts in isolation). It is linked with the complex systems proximal to it, e.g., knee, hip, pelvis, etc. The course mentioned above was directed toward health care practitioners, with the idea that a radical understanding of the body's base complex, i.e., the ankle and foot, was parament before an accurate diagnosis, prognosis, and intervention strategy could be made with regard to several lower limb and spinal injuries, particularly ones related to locomotion activities.

If what we have just described makes sense to you, you might wonder why we waited until the end of the book to discuss a joint complex that, some may say, is the basis for understanding all of the joint complexes that ride on top of it. Should not this section be examined first rather than last? Sorry, we do not have a good answer to that question, except to claim that in the end, it does not matter where we start our investigations. What matters is that all of our studies should maximize a systematic process of inquiry. Moreover, we must gather relevant facts related to the problem at hand; secure knowledge of the most fundamental concepts of systems, e.g., biological, biomechanical, etc., that apply; and best discern how these systems interact. So the purpose of this case is to examine the anatomy and function of the ankle/foot complex in the context of a young athlete who had a history of repeated stress fractures. In this case, problems were restricted to the lower leg, specifically the foot, and did not, as in so many other instances, create dysfunctions proximally. The referring physician posed a very relevant clinical question: "Is there a biomechanical explanation related to the walking and running gait of Ted that might explain the cause of the repeated stress fractures?" Recall that in the previous two cases, i.e., hip and knee, where we used instrumented gait measurements as part of the physical examination process, we hypothesize that our gait findings would not likely provide probable causes of the impairments but would provide additional data that describe the manifestations of the joint diseases. In the present case, we hypothesize that, in fact, a combination of the history, physical examination, and gait data could provide clues for a plausible etiology of recurrent stress fractures.

## 12.2 Case Presentation

Ted Kennedy (Ted) was a 15-year-old elite high school distance runner (height: 175.3 cm; mass: 65.8 kg; body mass index (BMI): 21.4 kg/m$^2$) who presented to the Motion

**Supplementary Information** The online version contains supplementary material available at https://doi.org/10.1007/978-3-031-25322-5_12.


G. J. Alderink, B. M. Ashby, *Clinical Kinesiology and Biomechanics*, https://doi.org/10.1007/978-3-031-25322-5_12

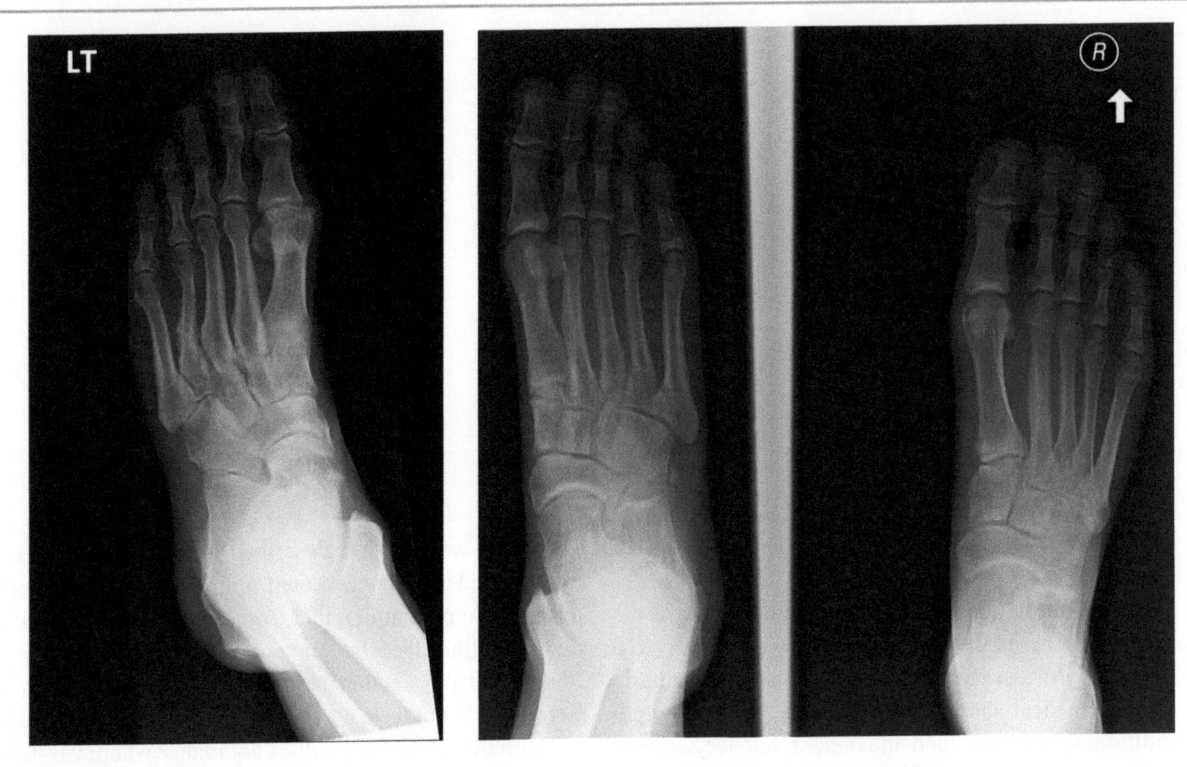


**Fig. 12.1** Healed fracture of the fourth metatarsal of the left foot and healed fracture of the second metatarsal of the right foot

Analysis Center with a history of metatarsal stress fractures. Symptom onset in the left foot began in the spring after a track meet in which Ted participated in three running events. His ability to practice and perform following that track meet was significantly impaired, yet he continued to train. He did not seek medical advice at that time. By fall, Ted recovered, and in preparation for cross-country, he made a major increase in training (frequency, duration, and intensity) and, during the same time period, a shoe change. He injured his right foot and sought medical advice. Radiographic imaging revealed a healed fourth metatarsal stress fracture in the left foot and a stress fracture of the second metatarsal in the right foot. He was immobilized for 4 weeks with a below-knee cast and recovered uneventfully. Ted was referred to the Motion Analysis Center by an orthopedic surgeon 3 months after recovery from the second metatarsal fracture for a biomechanical analysis of his walking and running gait (Fig. 12.1).

The findings of a comprehensive physical examination (Table 12.1) can be summarized as follows:

1. Moderate-marked tightness of the rectus femoris, hamstrings, tensor fascia latae/iliotibial (IT) band, and triceps surae, all bilaterally
2. Mild internal tibial torsion (left > right) with bilateral, symmetrical mild tibial varum
3. Weight-bearing foot position: mild midfoot cavus, mild/moderate adducted forefoot (L > R), first metatarsal/phalange mildly abducted, and Morton's toe bilaterally
4. Non-weight-bearing foot position: moderate midfoot cavus (L > R) and mild right/moderate left forefoot varus (flexible) with plantar flexed first ray (L > R)
5. Bilateral marked limitation subtalar inversion/eversion with moderate limitation of joint play at the transverse tarsal (midtarsal) and tarsometatarsal joints
6. Bilateral normal great toe extension range of motion
7. Multibeat ankle clonus bilaterally and mild left/moderate right posterior tibialis spasticity

**Note** Joint play tests consist of passive manipulations, i.e., translations, to the bones of the foot in dorsal and palmar directions, including, where possible, medial/lateral glides and long-axis rotations.

We were surprised when increased resting muscle tonus or spasticity, e.g., multibeat clonus, was elicited in several lower extremity muscles during the physical examination. These findings usually suggest an upper motor neuron (central nervous system) lesion, i.e., secondary to concussion, cerebral palsy, etc. When queried about any prior knowledge of a head injury; the identification, at birth, of mild neurological abnormalities by Ted's pediatrician; or early developmental delays related to a neurological condition, like cerebral palsy, Ted's father claimed that none of those possibilities had taken place. This information was important to gather because if a preexisting or current neurological condition was present, it would need to be accounted for relative to factors that might have contributed to Ted's stress fractures.

**Table 12.1** Physical examination data

| | | Range of motion | | Strength | |
|---|---|---|---|---|---|
| *Hip* | *Normal ROM* | *L* | *R* | *L* | *R* |
| Flexion | 120–135 | WNL | WNL | 5/5 | 5/5 |
| Extension/mod. Thomas test: | | | | | |
| Knee extended | 0 | WNL | WNL | 5/5 | 5/5 |
| Knee flexed 90° | 0 | 20 RF | 20 RF | 5/5 | 5/5 |
| Abduction (hip ext.) | 30–50 | WNL | WNL | 5/5 | 5/5 |
| Adduction | *** | | | 5/5 | 5/5 |
| Ober test (IT band) | Neg | Mod/mrk | Mod/mrk | | |
| Internal rotation | 30–45 | WNL | WNL | | |
| External rotation | 30–45 | WNL | WNL | | |
| Femoral anteversion | Adult: 12–15 | WNL | WNL | | |
| *Knee* | | *L* | *R* | *L* | *R* |
| Flexion (supine) | 135–160 | WNL | WNL | | |
| Flexion (prone) | ~135 | 120 | 120 | 5/5 | 5/5 |
| Extension (passive) | 0 | WNL | WNL | | |
| Extension (active): | | | | | |
| Strength (sitting) | *** | | | 5/5 | 5/5 |
| Active lag (supine) | 0 | WNL | WNL | | |
| Popliteal angle (from vert.): | | | | | |
| Traditional (spastic/end) | 0–30 | 40 | 50 | | |
| Neutral pelvis (end) | 0–20 | 30 | 40 | | |
| Transmalleolar axis | ~20 ER | 8 IR | 5 IR | | |
| Thigh-foot angle | ~15 ER | 10 IR | 5 IR | | |
| *Ankle/foot* | | *L* | *R* | *L* | *R* |
| Dorsiflexion: | | | | | |
| Knee ext'd (spastic/end) | ~10–15 | 0 | 0 | | |
| Knee flexed (end) | 15–25 | 5 | 5 | 5/5 | 5/5 |
| Plantar flexion | 45–50 | WNL | WNL | 5/5 | 5/5 |
| Tarsal inversion (supination) | 35–50 | 25 | 25 | 4+/5 | 5/5 |
| Tarsal eversion (pronation) | 35–50 | 20 | 20 | 5/5 | 5/5 |
| 1st MTP extension | 70–90 | WNL | WNL | | |
| *Trunk strength* | | *Posture* | | | |
| Lower abdominals (leg lift) | NT | *Mild/mod increase of anterior pelvic tilt and lumbar lordosis in standing position | | | |
| Upper abdominals (crunch) | NT | | | | |
| Trunk extensors (prone) | NT | | | | |
| *Notes/others* | | | | | |

| Foot posture | | |
|---|---|---|
| *Weight-bearing* | *L* | *R* |
| HF varus*/valgus | WNL | WNL |
| *Coleman block | | |
| MF planus#/cavus | Mild cavus | Mild cavus |
| #Reverses w/ toe standing | | |
| FF ab/adductus | Mild add | Mild add |
| Hallux valgus | WNL | WNL |
| *Non-weight-bearing* | *L* | *R* |
| HF varus/valgus | WNL | WNL |
| Flexible? | No | No |
| MF planus/cavus | Mod cavus | Mod cavus |
| Flexible? | No | No |
| FF varus/valgus | Mild varus | Mod varus |
| Flexible? | Yes | Yes |
| FF ab/adductus | Mild add | Mild add |
| Flexible? | Yes | Yes |
| PT 1st ray | Moderate | Mild |
| Callus pattern: | WNL | WNL |
| *Spasticity* | *L* | *R* |
| Hip flexors | 1 | 1 |
| Hip adductors | 1 | 1 |
| Hip internal rotators | NT | NT |
| Hip external rotators | NT | NT |
| Hamstrings | 1 | 1 |
| Vasti | NT | NT |
| Rect. fem./Duncan-Ely | 1 | 1 |
| Ankle clonus | 4 | 4 |
| Gastrocnemius | 1+ | 1+ |
| Soleus | 1+ | 1+ |
| Tibialis anterior | 0 | 0 |
| Tibialis posterior | 2 | 3 |
| Peroneals | NT | NT |
| Toe flexors (FDL, FHL) | NT | NT |
| Confusion test | WNL | WNL |
| *Balance* | | |
| Single-leg stance time | >10 s | >10 s |

(continued)

**Table 12.1** (continued)

| | Range of motion | Strength | Foot posture | | |
|---|---|---|---|---|---|
| *Hallux extensor and toe extensor strength I, 5/5 | | | *Leg length* | Equal | Equal |
| *Mild tibial varum and Morton's toe bilaterally | | | *Motor selectivity scale* | | |
| *Unable to perform full squats and to keep heels on the floor due to PF tightness | | | *(No manual resistance/gravity only)* | | |
| *Straight leg raise R50°, L55° (normal: 70–90°) | | | *U* = unable to produce requested motion | | |
| | | | *S* = able to produce requested motion in *synergy* | | |
| | | | *I* = able to produce requested motion in *isolation* | | |

| *Abbreviations* | *Spasticity scale (modified Ashworth)* | *Strength scale (manual muscle test)* |
|---|---|---|
| *NA* = not applicable | 0 = no increase in tone (normal) | 0 = no palpable muscle activity |
| *NT* = not tested | 1 = slight ↑ in resistance: catch & release | 1 = palpable muscle tension or flicker, no movement |
| *UTA* = unable to assess | 1+ = slight ↑ in resistance: catch & min. resistance | 2 = full ROM in horizontal plane, 0 against gravity |
| *P/PL* = painful/pain limited | 2 = more marked ↑ in resistance thru most ROM | 3 = full ROM against gravity only, no resistance |
| *WNL* = within normal limits | 3 = considerable ↑ in resistance, PROM difficult | 4 = full ROM against gravity and mod. resist. |
| *WFL* = within function limits | 4 = rigid | 5 = full ROM against gravity and max. resist. |

Note: *IT* iliotibial band, *FDL* flexor digitorum longus, *FHL* flexor hallucis longus, *MTP* metatarsophalangeal, *mrk* marked, *mod* moderate, *neg* negative, *ext'd* extended, *HF* hindfoot, *MF* midfoot, *FF* forefoot

Although Ted's father (and Ted) denied any neurological issues, the examining therapist suggested that they discuss our findings with the referring and their primary care physicians to further explore the possibility of a more serious disease (to ensure that the referring physician knew about our neurological findings, we included some concern to him in our final report). After the gait analysis, Ted was lost to follow-up, so no additional medical information was provided to the Motion Analysis Center staff.

Despite Ted's abnormal muscle tone, i.e., positively scored spasticity tests noted in the physical examination table, he demonstrated normal muscle selectivity with resisted muscle testing; that is, he was able to position and hold his extremities voluntarily during the resisted muscle tests. Although individual muscles cannot be isolated during resisted muscle tests, normal selectivity suggests that an individual can isolate muscle synergistic actions in contrast to performing joint actions with abnormal muscle patterns or synergies. In addition, Ted demonstrated normal muscle strength with all tests. These results suggested that Ted did not likely have an upper motor neural lesion. Ted was 15 years old, i.e., not fully mature physically, and demonstration of increased motor excitability is not uncommon for individuals who have not reached a mature physical and neurological stage of development.

Based on the reduced foot (tarsal) inversion and eversion passive range of motion and the associated firm/almost hard end feels, as well as the findings of the non-weight-bearing foot examination, the examiner performed additional passive motion tests on the tarsal bones. These included testing the gliding or translation mobility at the subtalar, talonavicular, calcaneocuboid, intercuneiform, and tarsometatarsal joints. As noted in the summary above, there was a significant reduction in mobility in several of the tarsal joints.

## 12.3 Key Facts, Learning Issues, Hypotheses

Although Ted did not have any obvious gait (walking) deviations based on visual inspection, Ted's unusual neurological findings were intriguing and suggested that he might have had a mild manifestation of spastic diplegic cerebral palsy (since both lower extremities were involved; his upper extremity reflexes were not tested). We need to briefly explore spastic diplegic cerebral palsy. In general, cerebral palsy (CP) describes a group of permanent disorders of movement and posture, causing activity limitation, which is attributed to nonprogressive disturbances occurring in the developing fetal or immature brain. The prevalence of CP for all live births ranges from 1.5 to 3 per 1000. Although the initial neuropathic lesion is nonprogressive, children with CP often develop a range of secondary conditions over time, e.g., extreme muscle tightness, muscle weakness, bony deformities, etc., which variably affect their functional abilities. Because, in many infants and children, abnormal neuromotor findings tend to resolve within the first few years, especially during the first 2–5 years of life, the reported prevalence of CP tends to be higher during infancy. (*Note: it is possible that Ted was one of those children.*) The motor disorders of CP are often accompanied by disturbances in sensation, perception, cognition, communication, behavior, and other pathologies, including epilepsy and secondary musculoskeletal problems. The persistence of primitive reflexes or primary motor patterns beyond the expected age is a key clinical characteristic of CP and prevents or delays the typical progression of motor development and the sequential acquisition of higher-level neuromotor skills. The neurologic impairments of the motor system in children with CP are characterized, in order of frequency, by spasticity (increased resting muscle tonus), dyskinesia (involuntary muscle movements), hypotonia (decreased resting muscle tonus), and ataxia (decreased coordination). Thirty-five percent of children with CP have spastic diplegia, the most common clinical phenotype. Spastic diplegia is induced by damage to the immature oligodendroglia between 20 and 34 weeks of gestation. In spastic diplegia, both the motor corticospinal and thalamocortical pathways are affected. Most children with spastic diplegia have a normal cognitive function and a good prognosis for independent ambulation. Note that there are several interesting learning issues here, e.g., oligodendroglia, corticospinal pathway, etc., but we leave them to the reader's curiosity to explore further since it is outside the scope of this case to pursue more details.

Spasticity (from Greek *spasmos* – "drawing, pulling") is a feature of altered muscle performance, secondary to central neuropathology, with a combination of paralysis, increased tendon reflex activity, and hypertonia. It is sometimes referred to as an unusual "tightness," "stiffness," or "pull" of muscles. Spasticity results from the imbalance of excitatory and inhibitory inputs to alpha motor neurons caused by damage to the spinal cord and/or central nervous system (as in CP). The damage induces a change in the balance of signals between the nervous system and the muscles, leading to increased muscle excitability. The stretch reflex, e.g., what is typically elicited during deep tendon reflex testing, is important in coordinating normal movements and is activated when a relaxed or shortened muscle is rapidly lengthened. The stretch reflex is encompassed within the gamma neuromotor system, wherein intrafusal muscle fibers, i.e., the muscle spindles, send signals about muscle length and the velocity of length changes to the spinal cord. It is damage to this sensory system that results in increased muscle activity (excitability). Thus, a defining feature of spasticity is that increased resistance to passive stretch is velocity-dependent. Spastic muscles typically lead to altered performance, e.g., during walking and running gait, and although the muscles appear stiff, they are also often weak and easily fatigable. Clonus (involuntary, rhythmic muscular contractions and relaxations) coexists with spasticity; it may be that clonus is an extended outcome of spasticity. Clonus appears to result from increased motor neuron excitability (i.e., decreased action potential threshold) and is common in muscles with long conduction delays, such as long reflex tracts found in distal muscle groups, e.g., posterior tibial muscles like the gastrocnemius.

Spastic diplegic cerebral palsy is a chronic neuromuscular condition of hypertonia and spasticity, manifested as an especially high and constant "tightness" or "stiffness," in the muscles of the lower extremities and pelvis, which affect the normal development of functional activities, such as gait. Muscle hyperresistance, i.e., increased tension in a passively stretched muscle, caused by neural impairments (spasticity), as well as by passive tissue resistance (stiffness), is the most prevalent problem in spastic diplegia. Stiffness is a mechanical resistance of the myotendinous tissue as it is passively lengthened. It has been suggested that muscle stiffness in CP already appears at an early age and that maladaptation to growth, not spasticity per se, plays a crucial role in developing muscle contractures.

In addition to his lower extremity spasticity (hypertonicity), Ted demonstrated a bilateral reduction in the length, i.e., contracture, of several two-joint muscles, e.g., rectus femoris, hamstrings, etc.

It is difficult to differentiate muscle contracture ("tight" muscles) from hypertonicity in individuals with CP. Although there is nothing physiologically aberrant about cerebral palsy, like metabolic imbalance similar to osteoporosis, that would predispose an individual to stress fractures, we hypothesized that Ted's hypertonicity (spasticity and clonus) and reduced muscle length were related and that this neuromuscular state might have resulted in adaptations to bone and joint mobility, predisposing Ted to injury. We need to explore these issues because if, in fact, Ted had a mild form of cerebral palsy, two things are true: (1) special instructions about training could be provided, yet (2) that condition may always be a limiting factor in Ted's ability to increase training beyond some threshold.

In the knee and hip cases, we compare and contrast normal and abnormal lower extremity mechanical alignment. Then we discuss the role that abnormal lower extremity alignment may play in either a predisposition toward or perpetuation of injuries to the hip or knee. For example, increased femoral antetorsion that results in a compensatory increase in external tibial torsion would create abnormal knee axial rotation, which could lead to abnormal tracking of the patella or perpetuate abnormal torsional stress on the tibiofemoral menisci. The growth, development, and maintenance of normal bony alignment are affected by several factors, including genetics, nutrition, drugs, hormones, and mechanical forces. The forces of gravity, and the response by the body to those forces in the form of muscle actions, are persistent in their effect on the size and shape of the body parts. Children's tissues are particularly plastic and perhaps more sensitive to the influencing factors cited above. For example, the deformities of clubfoot and tibial torsion may occur secondary to poor positioning in utero, during sleeping, or while sitting or secondary to abnormal muscle tonus. Ted clearly exhibited abnormal lower extremity alignment: mild tibial varum, internal tibial torsion, pes cavus, Morton's toe, and metatarsus adductus. Exploring concepts related to developmental biomechanics is relevant to this case and may provide insight into Ted's situation.

It appears that Ted's neurological and anatomical presentation may be related in some complex way and also are likely related to the reduced interjoint mobility in the foot. Therefore, as we have done with all of the previous cases, we examine the structure (osteology and arthrology) and function (osteo- and arthrokinematics) of the ankle/foot complex. Since Ted's injuries were related to running activities, we discuss the biomechanics of the ankle/foot during that mode of locomotion and contrast it to walking. Of course, since this all started with the clinical query about the potential causes of Ted's stress fractures, we have to reexamine the general concepts of stress fracture mechanics (see the section on stress fractures in Chap. 3 for additional information), as well as the particular factors in this case. Let us first review some fundamental concepts related to the mechanics of skeletal development. A study of developmental biomechanics is relevant in this case because although Ted's gait characteristics likely reached adult patterns by age 7 (as is typical), at age 15, he may not have reached skeletal maturity.

## 12.4 Developmental Biomechanics

Let us first acknowledge that several factors influence the growth, development, and maintenance of a healthy musculoskeletal system: genetics, nutrition, hormones, and daily activities. However, in the context of this case, we focus on how mechanical forces influence skeletal biology. The growth and maintenance principles presented in this section are based on the following research: (1) animal models where mechanical influences on the development of bone and muscle were studied in both normal and genetically engineered embryos, (2) examination of the forces that may induce bone deformity in fetuses, (3) tracing the effects of normal forces on infants and developing children, and (4) studying forces related to orthopedic problems in adults. It has become clear that the general shape, i.e., bony structure, of the human body depends on the proper development of skeletal structures (bone and muscle). Thus, any deficiency in development leads to abnormalities in appearance and function. Most agree that extrinsic factors have a major influence on the design of bone and other biological structures, e.g., articular cartilage, muscle, etc. Wolff's law of bone transformation suggests that every change in the form or function of a bone induces changes in the bone's internal and external architecture according to scripted mathematical laws. Some have suggested that this law applies only to the appositional growth of bone, yet it appears that the most general idea behind Wolff's law could be applied to endochondral growth, as well as to changes in soft tissues.

The type, duration, and magnitude of external loads influence the type of tissue or articulations being formed. All tissues respond in unique ways to tensile, compressive, shear, and torsional forces, as well as bending moments that ultimately contribute to their differentiation, maintenance, or, in some cases, deformation. For example, in general, chondrogenesis is associated with intermittent compression and shear loading, while osteogenesis is a response to continuous complex loading patterns, i.e., combinations of compression, shear, and torsional forces. Storey (as cited by LeVeau and Bernhardt 1984) proposed a continuum of types of loading patterns, including continuous and intermittent compression and tensile forces, which induce adaptations of connective tissues and joints:

1. Continuous compression in a constant direction causes bones to become connected by bars of cartilage, which act as shock absorbers and a "flexing system" (i.e., synchondrosis), e.g., sternocostal joint.
2. Intermittent compression and a range of movements between bones induce the formation and maintenance of articular cartilage.
3. Intermittent compression of a decreasing magnitude but an increasing amount of tension induces the formation of symphyses (i.e., the compressible fibrocartilaginous pad that connects two bones), e.g., pubic symphysis or the joints that connect the first pair of ribs to the sternum.
4. Intermittent tension and compression of equal magnitude, along with sliding (shear), create condylar cartilage, e.g., cartilage of the temporomandibular joint.
5. Increasing intermittent tension induces the formation of fibrous joints or sutures, i.e., "joints" in the skull.
6. Continuous tension in one direction creates thick collagenous structures, i.e., tendons, ligaments, and fascia.

Since the magnitude, duration, rate, and direction of loading influence the amount and direction of tissue growth, let us consider how specific types of loads on different tissues influence growth and development. The growth of long, and other-shaped, bones, i.e., endochondral growth, is dictated by how the epiphyseal growth plate is loaded so that growth can (1) increase longitudinally, (2) decrease longitudinally, (3) be deflected by shearing, and (4) as a result of continuous or intermittent twisting become torsional. Compression and/or tension forces that are applied perpendicular to the epiphyseal growth plate stimulate growth longitudinally. External compression forces, i.e., gravity, and forces parallel to the plane created by muscle actions, i.e., the muscle that crosses the joint that resides close to the growth plate, appear to induce a more rapid rate of growth than tension. An increase in compression or tension loading, continuous or intermittent, outside the normal magnitude inhibits chondrocyte mitosis and appears to retard longitudinal growth; that is, there may be premature closure of the growth plate. Increased load across the plate has also been shown to induce bone

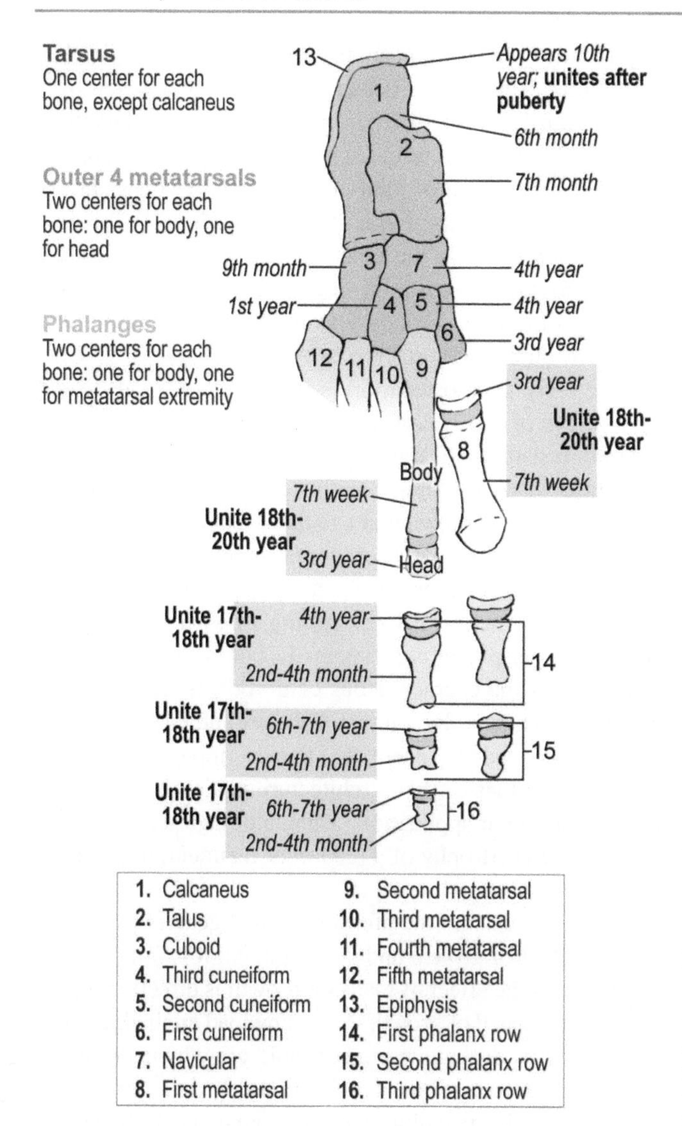


**Fig. 12.2** Growth plate sites within the foot, when they appear, and when they unite

resorption. Constant compression over a long duration can also cause bone atrophy, affecting the rate of long-bone growth.

There are several growth plate centers associated with the ankle/foot complex (Fig. 12.2). Since Ted exhibited several abnormal bone alignments, it is probable that they can be partially explained by a complex array of forces associated with the abnormal resting muscle tonus he exhibited, as well as how development forces may have been altered by his increased muscle tone.

A load applied to the epiphyseal plate that is not parallel to the direction of growth, i.e., longitudinally, and torsional loading across the epiphyseal growth plate is not normal and leads to deformity. A load that is applied, and maintained, parallel to the growth plate likely causes a lateral displacement of the epiphysis. Shear stress associated with a parallel direction of loading creates plastic changes in the chondral structure, effectively realigning the direction of growth and causing the growth plate to tilt. The epiphyseal growth plate is the least resistant to torsion; thus, torsional forces induce a rotational deflection of the growth columns. These torsional changes in the circumference of the epiphyseal plate induce new bone formation away from the epiphysis in a spiral pattern. In some cases, the torsional forces are normal. For example, a normally developing infant/toddler/child (ages 1 to 7 years), subject to the forces of gravity and muscle actions associated with locomotion activities, which create torsional stresses in the proximal tibial epiphysis, develops an external torsion of the tibia, ranging from approximately 15° to 30° (normal adult range). However, familial genetics and abnormal fetal positioning in utero, in addition to the normal forces associated with development, can lead to torsional forces on the epiphysis, which may result in tibial torsions exceeding 30°.

As noted in previous chapters, Wolff formulated several conclusions related to the appositional growth of bone. It is notable that bone is subjected to multidimensional loading under all types of forces: compression, shear, tension, and torsion, usually responding in typical ways. Bending moments induce a complex loading pattern, with the creation of tensile stress on the convex surface of the bend and compression stress on the concave surface of the bend. Generally, compression best stimulates appositional growth, and a lack of compression leads to a reduction in bone tissue. For example, greater weight-bearing forces increase the cross-sectional area (thickness) and density in the long bones of the lower extremity. Conversely, reduced weight-bearing or prolonged immobilization results in the loss of bone mass. Bending moments are more typically associated with remodeling the shape of a bone. Thus, when a prolonged bending moment is applied, there is a net loss of bone on the convex surface (due to osteoclastic activity) and a concomitant gain of bone on the concave surface (due to osteoblastic activity). For example, it must be the case that axial forces applied to the proximal femur create bending stresses in the shaft since it displays a slight anterior convexity, which is normal. On the other hand, it must also be the case that axial forces in some combinations of abnormal magnitude and/or duration and/or direction must have been applied to the proximal tibia to induce bowing, which we often define as tibial varum, a condition noted in Ted.

To this point, the principles of loading that have been presented assume an application to either the growth plate or cortical bone. However, we would be remiss if we did not include a brief review of trabecular bone (Fig. 12.3). Trabecular (cancellous or spongy) bone is a hierarchical, spongy, and porous material composed of hard and soft tissue components, located within the epiphyses and metaphyses of long bones and in vertebral bodies. Its macrostructure is composed of trabecular struts and plates, which form a

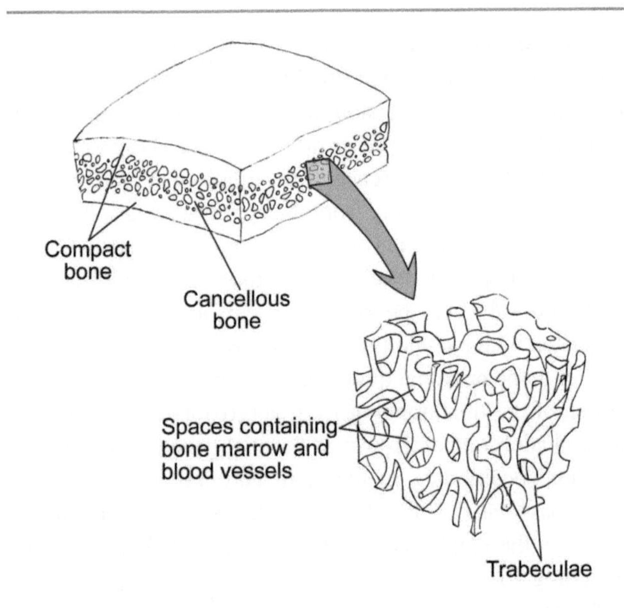


**Fig. 12.3** Spatial relationship between compact (cortical) and cancellous (trabecular) bones. Note the multidimensional struts (or rods) in trabecular bone

stiff yet ductile (more readily deformable without failure) structure that provides the framework for the soft, highly cellular bone marrow filling the intertrabecular spaces. Microstructurally, the trabecular architecture, primarily minerals and collagen, is organized to optimize load transfer. In the extremities, the trabecular bone transfers mechanical loads, i.e., primarily compression, from the articular surface to the cortical bone. In contrast, in vertebral bodies, the trabeculae compose the main load-bearing structure. The reader is reminded that in addition to transferring mechanical loads related to gravity in the upper extremities and gravity and weight-bearing in the spine and lower extremities, trabecular bone is also responsive to large forces applied to bones and across joints induced by muscle actions. As with cortical bone, without the continuity of daily external and internal loads, trabecular bone is resorbed, losing density and stiffness.

Two other types of tissues need to be considered: cartilage and fibrous tissue. The health of cartilage (hyaline as well as fibrocartilage) is maintained by a wide range of applied forces. The environment of articular cartilage is best managed under conditions of intermittent compression (and shear) loads. This is necessary since, generally, the vascularization of articular cartilage is poor, and intermittent loading fosters the movement and transfer of nutrients (and waste products). Conversely, if the joint compression loads imposed by gravitation and muscular forces are constant and/or excessive, cartilage undergoes plastic changes, initiating a cascade of events leading to joint degeneration. Recall that in four of the cases discussed to this point in the text, i.e., shoulder, elbow, hip, and knee, cartilage degeneration was a primary problem. In each of the previous cases we study, the cartilage degenerative process appears to be unique: (1) the glenohumeral joint degeneration (Chap. 3) was likely initiated by a large impulsive load to the joint surfaces secondary to a proximal humeral fracture, (2) radiocapitellar osteochondritis dissecans (a form of joint degeneration in young throwers) (Chap. 4) was induced by impulsive, repetitive compression loads at the elbow of an adolescent baseball player, (3) marked hip joint degeneration (Chap. 10) may have been initiated secondary to reduced hip mobility and repetitive impulsive joint forces related to sports, and (4) marked early knee osteoarthritis (Chap. 11) was instigated by the loss of the lateral knee meniscus and its ability to assist in resisting and transmitting the normal compression and shear loads associated with activities of daily living. Research has also shown that a reduction or absence of compression leads to the degeneration/atrophy of both articular and fibrocartilage. Fibrocartilage is unique because it has material properties mostly resembling articular cartilage, but because of peripheral attachments to bones, ligaments, and fascia, it has the characteristics of fibrous tissues like ligaments and tendons. In contrast to bones and cartilage, ligaments and tendons adapt to tensile forces. They are self-aligning; that is, their architecture is dictated along the lines of the force application. Thus, although tensile forces induce the hypertrophy of a tendon or ligament, they do not result in a change in its shape.

Before we discuss the aforementioned general principles in the context of growth and development, it is important to mention that the effect of loads on growth is directly proportional to the speed of growth. Thus, any load applied, even of short duration, during a period of rapid growth may result in the plastic deformation of a bone. Early fetal growth is extremely rapid, reaching a peak around the fifth month. Thus, for example, abnormal positioning in utero is a common cause of bone deformities typically seen in children with cerebral palsy, e.g., excessive femoral antetorsion. Later in the fetal period, although it may be exposed to increased external forces such as increased size, decreased amniotic fluid, and decreased movement, a slowing rate of growth rate and diminished plasticity may allow the fetus to better accommodate deforming forces. Varying rates of growth in early infancy and childhood also influence the development of greater or lesser (both magnitude and permanency) bone deformation. The presence of neurological diseases, like cerebral palsy, certainly negatively influences the development of genetically determined bony deformities.

Let us now examine the relationship between gravitational and muscle forces and normal and abnormal development of the weight-bearing areas of the body. We encourage the reader to consult and review previous chapters for the details of the anatomy presented in the following narrative. The general structure of the spinal elements, i.e., vertebra, provides for both static and dynamic stability. Each vertebra

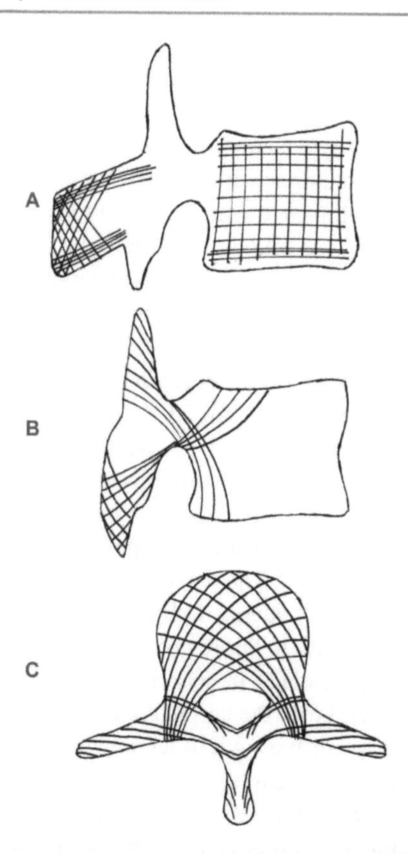


**Fig. 12.4** The internal structure of a lumbar vertebra segment. (**A**) Midsagittal section illustrating vertical and horizontal struts associated with compression and shear stress, as well as the trabeculae of the spinous process associated with tensile stress, (**B**) lateral sagittal section trabeculae associated with complex forces induced on the pedicles and articular processes, and (**C**) transverse section illustrating a trabecular strut pattern over the entire vertebral body, laminae, and transverse processes (Bogduk 2012)

consists of a body, which is the primary load-bearing region, and posterior elements, i.e., pedicles, laminae, transverse and spinous processes, and articular facets, which have weight-bearing responsibilities as well as serve as attachment sites for fibrous tissues, e.g., ligaments, muscles, etc. The size of the vertebral elements increases from cranial to caudal based on the magnitude of the superincumbent weight that must be borne. Axial loads on the spine contribute to the primary and secondary trabecular systems in the vertebrae: (1) primary, a vertically oriented system of struts that sustain the superincumbent body weight, and (2) secondary, an obliquely oriented structure that develops in response to external torsional and shear forces, as well as tensile forces produced by muscle action (Fig. 12.4). The sagittal plane curvature of the spine, i.e., cervical lordosis, thoracic kyphosis, lumbar lordosis, and sacral tilt, is a function of predictable developmental changes. At birth, the spine has posterior convexity, but as the infant begins to raise the head against gravity, the lordosis in the cervical spine increases. Residual tightness of the hip flexors (persisting from fetal flexion) coupled with the antigravity work of the infant in prone position, followed by crawling and high kneeling, induces the beginning of a lumbar lordosis; weak abdominal muscles initially cannot balance the lordotic tendencies. The sacral base forms an approximately 45° angle with the horizontal plane, which attempts to equalize the compressive and shear forces between the last lumbar and first sacral segments. As the infant rises and begins bipedal locomotion, the spine develops compensatory curves, i.e., thoracic kyphosis, which allow for the static and dynamic control (stability) of the body's center of mass.

Scoliosis (lateral curvature) and abnormal thoracic kyphosis (increased posterior convexity), or some combination of both, are common spinal deformities. Although scoliosis can be congenital, the majority of the cases develop after the embryonic period and may be idiopathic (unknown cause) or due to muscular strength imbalances, abnormal muscle tone, persistent abnormal positioning (postures), or abnormal spinal weight-bearing associated with various paralytic or neurological diseases. With scoliosis, abnormal forces cause plastic deformation in all of the bony and soft tissue elements in the spine, including the ribs. Ligaments and muscles shorten and thicken on the concave side and stretch and relax on the convex side of the spinal curve. With increased and persistent compression on structures on the concave side, e.g., articular facets, intervertebral disc, etc., a hyaline cartilage degeneration cascade may be initiated at the articular facet joint, while an imbalance of compression and tension within the intervertebral disc can induce internal derangement, nuclear migration, and annular bulging. Concomitant increased and persistent tension on the convex side alters the material properties of the facet joint capsules and lateral ligaments and the length-tension and function of spinal muscles. If these changes occur before the epiphyses fuse, bony deformation may also result. Compression on the concave side inhibits growth, while decreased loading on the convex side stimulates growth, and when combined, they induce vertebral wedging. A combination of side bending and rotation curves causes distortion of the vertebral elements, incongruity of the articular facets, as well as rib deformation and its posterior and anterior constraints, e.g., costovertebral joints, etc. These deformations create a self-perpetuating system that can progress, rapidly in some cases, and needs to be monitored closely.

In the case of a moderate or marked increase in the thoracic kyphosis typically induced by abnormal external and internal forces, a very similar cascade of events alters the anterior/posterior balance of the spine. Posterior structures, e.g., posterior annulus, facet joint capsules, ligaments, spinal muscles, etc., are placed under excessive and persistent tensile forces, which lengthen and weaken them; conversely, anterior structures shorten and eventually reduce thoracic spine extension. Fibrous replacement of the annulus fibrosis anteriorly leads to loss of vertebral mobility, as well as the creation of anterior wedging of the intervertebral disc and the vertebral body. Excessive compression anteriorly may also predispose a postmenopausal female to vertebral compression fractures. Finally, in order to control the body's center of mass, compensations in the cervical and lordotic regions increase their lordotic curves, ultimately leading to additional degenerative processes in those regions.

Another pathological condition we want to highlight that most typically develops at the lumbosacral junction, i.e., L5/S1, is spondylolysis (defect without a forward slippage) or spondylolisthesis (fracture and a forward slippage of the anterior position of L5 on S1). These conditions are caused by a defect/fracture of the pars interarticularis, often due to congenital weakness or repeated microtrauma induced by lumbar hyperextension. Because of the normal lumbar lordosis and sacral angle, compressive forces are transmitted primarily through the neural arch rather than the vertebral body. As a result, then, of a macrotraumatic event, e.g., a motor vehicle accident or repeated, impulsive lumbar hyperextensions, the L5 isthmus is pinched between the L4 articular facet and the upward projecting sacral processes (creating a bending moment), fracturing the weakest point, the pars interarticularis. Often, this injury is initiated as a stress reaction, but the cycle of repeated loading without sufficient time for tissue recovery from the stress eventually leads to fracture. Besides a congenital defect in the region, many factors can predispose this area to injury, including increased sacral angle and increased lumbar lordosis secondary to muscle imbalances in the spine and pelvic region or muscular tone or tightness of the hip flexors (primarily iliopsoas and rectus femoris) and lower erector spinae (see the case in Chap. 9 for a review of the spondylolytic condition).

You may wonder why we detoured to examine the development of scoliosis, for example. We did so because of the possibility that Ted's development may have been abnormal (as suggested by his muscle hypertonicity), and it is important that we consider other possibly relevant issues. This often happens in a problem-based learning (PBL) environment in order to clarify uncertainties in knowledge and/or understanding.

Fortunately, Ted did not report any back pain, nor did his physical examination reveal scoliosis or hyperkyphosis. On the other hand, he had moderate/marked bilateral tightness of the rectus femoris bilaterally, with a related mild/moderate increase in lumbar lordosis and anterior pelvic tilt. These impairments could have restricted hip extension in the terminal stance phase of his running gait, perhaps inducing compensations in the frontal and transverse planes at the knee and ankle/foot complexes. Thus, we suggest the hypothesis that impairment in hip extension led to compensations in knee and ankle/foot kinematics and kinetics, which may have predisposed Ted to stress fractures.

Although intuition suggests that we examine the developmental biomechanics of the paired sacroiliac (SI) joints, there is no need in the context of this case since Ted had no indication of SI dysfunction, nor are SI problems typically encountered in children with cerebral palsy. Therefore, we move on to the hip joint (see Chap. 10 to assist with your visualization of what follows relevant to the pelvis and hip). The adult hip complex includes an acetabulum, which faces outward, forward, and downward, and the head of the femur, with an angle of inclination (i.e., neck-shaft angle) approximating 125° and an angle of declination (femoral antetorsion) approximating 15° to 20°. With the help of the acetabular labrum, the head of the femur sits relatively deeply into its "socket," supported by strong ligaments and muscles. These bony shapes and joint configurations are the results of major mechanical forces applied in utero and during infant development. The pelvis is formed from three primary ossification centers: ischium, ilium, and pubis. Endochondral growth in this region allows the acetabulum to enlarge circumferentially in concert with the growth of the femoral head. At birth, the acetabulum has its most shallow configuration, predisposing it to femoral subluxation or dislocation. The normal angle of the acetabular roof with the horizontal is 30° at birth, which decreases to 20° by 3 years of age, remaining at this orientation through maturity. Unless the femoral head is malpositioned perinatally (from 5 months prior to and 1 month following birth), the acetabulum and femoral head develop congruently, which is essential for the proper development of both. The most important loads that guide this development are body weight and muscle tension. Any alterations in the compressive loads and/or incongruity of the joint structure lead to bony deformity, most commonly acetabular dysplasia (coxa plana).

Proximal femoral development is guided by three growth zones: a longitudinal growth plate, a trochanteric growth plate, and a femoral neck isthmus. These zones develop at linear rates at birth, but as the child ages, the longitudinal growth plate grows at a more rapid rate than the trochanteric plate, thus shaping the angles of inclination and declination. At birth, the angle of inclination approximates 150°, while the angle of declination approximates 40°. During child development under normal conditions, a combination of compression and tensile loads applied to the proximal femur reduces these angles to adult values (see above). Normal compression and tensile loads are also responsible for forming two distinct trabecular systems in the proximal femur: (1) trabecular columns aligned with typical weight-bearing compression forces and (2) lying perpendicular to the primary system, a secondary system that crosses the neck region and greater trochanter, which develops in response to compressions and tensions created by muscle actions. Bending moments experienced in the femoral neck are checked by cross-trabecular patterns.

Children with cerebral palsy are at risk of developing one or more hip deformities secondary to abnormal positioning during sleeping and sitting, muscular weakness and imbalances (e.g., in iliopsoas, hip adductors, and hamstrings), spasticity, delayed weight-bearing, and the application of weight-bearing forces with the hip joint malpositioned during gait. Thus, this population is prone to the development of coxa valga, excessive femoral antetorsion, and coxa plana. Coxa plana in children and adults with CP predisposes to hip subluxation or dislocation, with recurrent hip hypermobility/instability eventually leading to early joint degeneration. Likewise, coxa valga and increased femoral antetorsion alter the normal contact areas between the acetabulum and femoral head, which alters normal contact pressures and premature articular cartilage degeneration.

Coxa valga can only be determined with certainty using imaging studies, whereas hip dysplasia can be detected in the early neonatal period using clinical screening tests, such as the Ortolani test, Barlow maneuvers, and the Galeazzi sign (see Noordin et al. 2010). If those tests are positive, imaging studies would be ordered to verify clinical suspicions. Ted's father denied any history of childhood hip problems. Radiographic measures of coxa valga are made only if there are symptoms suggestive of hip disease. Since Ted reported no hip-related signs and symptoms during his physician visit, there was no indication to perform imaging studies beyond those of the ankle and foot. On the other hand, there is a routine clinical measure (Ryder's test) that is used to estimate femoral antetorsion. When the Ryder test was performed at the Motion Analysis Center, it was determined that Ted's femoral antetorsion was normal bilaterally. Therefore, based on the physical examination, there was no reason to think that the stress fractures might be related to hip pathomechanics related to a transverse plane abnormality.

The knee complex undergoes several alignment changes during development. In the adult, the shaft of the femur relative to the tibia forms a Q angle of approximately 170° secondary to a femoral obliquity (details about the Q angle can be revisited in Chap. 11). At birth, the Q angle is 180°, which progresses to 160° by age 3; the adult value is attained by approximately 6 years of age. A combination of weight-bearing and upper thigh muscle forces associated with locomotor activities leads to the normal sequence of development of the Q angle, from relative varus to moderate valgus and finally to normal valgus positioning. The tibial shaft also undergoes torsional changes during normal development as a result of the epiphyseal plate's sensitivity to torsional forces. There is an absence of tibial torsion in the newborn, which then progresses to the normal adult range of 10° to 20° external torsion by late childhood. Internal tibial torsion is not normal and may be present at birth secondary to the positioning of the limbs in utero. It is usually first noticed when a parent reports an intoeing (or "pigeon-toed" position) of the feet when the child stands and walks. In most cases and under normal neurological circumstances, internal tibial torsion gradually reverts to the normal external torsion by the time the child is 4–5 years of age.

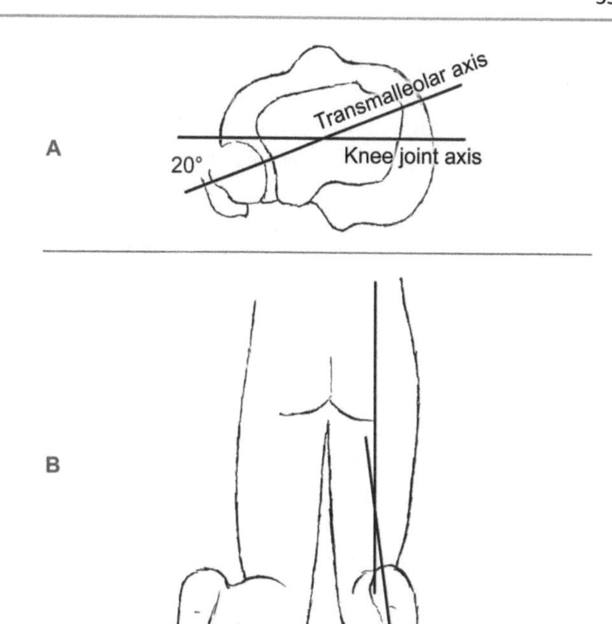


**Fig. 12.5** Schematics illustrating (**A**) transmalleolar and (**B**) thigh-foot angle estimates of tibial torsion. For the transmalleolar measure, the patient lies prone with the thigh resting on the table and the knee flexed to 90°; the rater aligns one arm of the goniometer parallel to the imaginary ankle joint axis, which runs from the medial to the lateral malleolus, and the other arm of the goniometer parallel to the imaginary knee joint axis, which runs from the medial to the lateral femoral condyles, then measures the angle between the two goniometric arms. For the thigh-foot angle measure, the patient lies prone with the thigh resting on the table and the knee flexed to 90°; holding the ankle and subtalar joints in neutral position, the rater places the axis of the goniometer over the center of the plantar surface of the calcaneus, aligns one arm of the goniometer parallel to the mid-posterior thigh with the other arm parallel to the long axis of the sole of the foot (parallel to the second ray = second phalange + metatarsal + midfoot), then reads the angle between the two goniometric arms

Ted demonstrated a mild pigeon-toed posture in standing and during gait (left greater than right), and since his femoral antetorsion was normal, we suspected he might have internal tibial torsion. The typical clinical measures for tibial torsion that were made part of Ted's physical examination, i.e., transmalleolar and thigh-foot angle methods (Fig. 12.5), also indicated mild right and moderate left internal tibial torsion.

It is likely that the magnitude of Ted's internal torsion was such that his pediatrician elected not to address it when he was a child. If the intoeing did not interfere with Ted's childhood activities, e.g., catching his toes that induced tripping and falling, then it seems reasonable to have not intervened.

However, it is possible that the internal tibial torsion was related to several other findings of abnormal alignment in Ted's physical examination, e.g., adducted forefoot, which may have been predisposed to abnormal stress distribution in the bones and joints of his feet.

The proximal tibia undergoes several other developmental adaptations in alignment: (1) induced by hamstring forces, retrotorsion is a posterior deflection of the upper tibia; (2) retroflexion is a posterior bend of the tibial shaft, which is produced by the tension generated by the gastrocnemius and soleus; and (3) retroversion is a posterior slant of the tibial plateau, reaching the adult value of approximately 5° by 19 years after undergoing a series of angular changes from birth to 10 years of age.

The trabecular systems in the distal femur and proximal tibia, aligned perpendicular to the joint surface, are formed by weight-bearing and muscle forces (Figs. 12.6 and 12.7). Secondary trabeculae develop in the distal femur, proximal tibia, and patella in reaction to tensile forces produced by muscle actions. With regard to the small angulations in the proximal tibia as well as the trabecular patterns, we can imagine how, with altered resting and dynamic muscular tensions associated with abnormal motor control in children with cerebral palsy, the normal architecture could easily be changed over time; also, then, how these changes may disrupt joint kinematics and kinetics proximally and distally.

As the child with diplegic cerebral palsy develops, albeit delayed depending on its severity, there are large and persistent forces (i.e., gravitation as well as muscular) that can create several deformities about the knee. Genu recurvatum (i.e., hyperextended knee) is common and can be induced congenitally or postnatally (Fig. 12.8). Malposition in utero when the feet are locked in the axilla, or under the mandible, due to breach positioning causes the knee to develop in extension instead of flexion. Recurvatum in utero results in contracture of the quadriceps and anterior capsule; development of a small, abnormal, or absent patella or lengthened or absent anterior cruciate ligament; and anterior dislocation of the hamstring tendons. Persistence of tibial retroversion, hamstring weakness, quadriceps weakness, and tightness and/or overactivity of the triceps surae can create genu recurvatum postnatally. Untreated knee recurvatum in an ambulatory child leads to plastic changes and insufficiency of the posterior knee capsule, as well as the cruciate and collateral

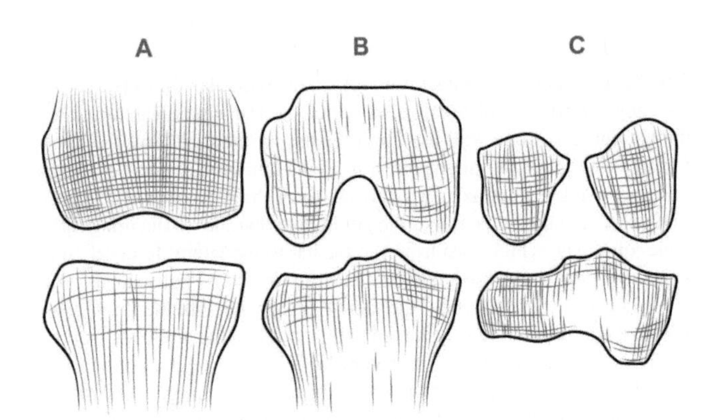


**Fig. 12.6** Frontal sections of the distal femur and proximal tibia, i.e., knee-joint (**A**) anterior, (**B**) middle, and (**C**) posterior cuts. Note that the predominant vertical trabecular struts and secondary horizontal trabeculae are closer to the joint surfaces (Takechi 1977)

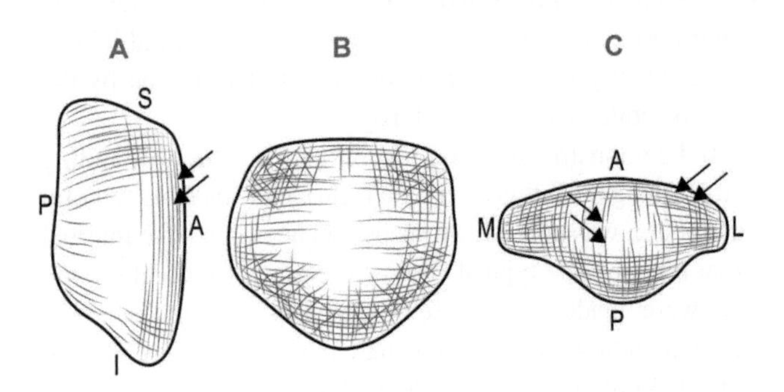


**Fig. 12.7** Sections of the patella: (**A**) parasagittal (note predominant vertical trabecular struts anteriorly), (**B**) frontal (note the horizontal, vertical, and oblique orientation of trabeculae), and (**C**) horizontal (note the predominant orientation of trabeculae from medial to lateral and anterior to posterior). P = posterior, A = anterior, S = superior, I = inferior, M = medial, L = lateral (Takechi 1977)

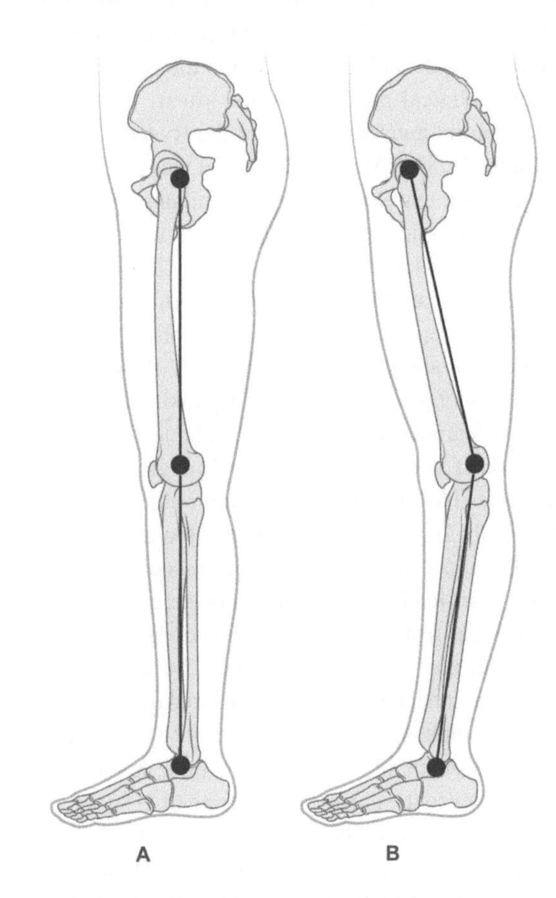


**Fig. 12.8** Sagittal view comparing (**A**) normal and (**B**) hyperextended (genu recurvatum) knee positions

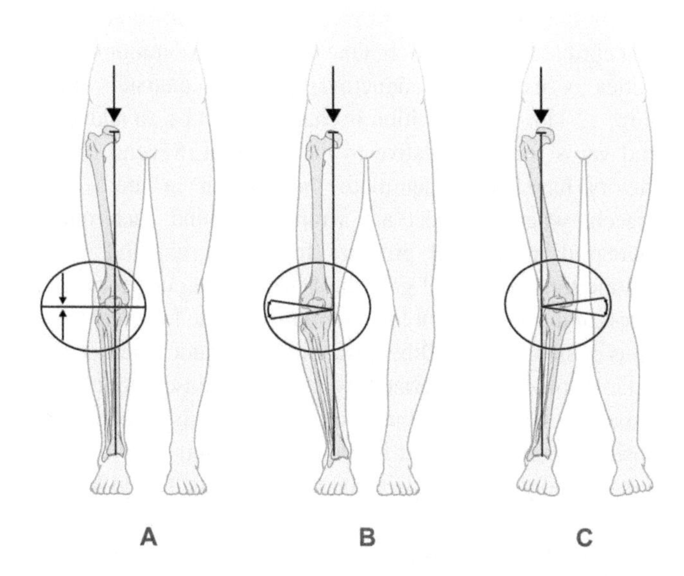


**Fig. 12.9** Frontal view of (**A**) normal, (**B**) genu varum, and (**C**) genu valgum lower limb alignments. Note that with abnormal alignment, the mechanical axis (arrows), i.e., the line of action of weight-bearing forces, is shifted, predisposing the tibiofemoral and patellofemoral joints to unbalanced bone-on-bone forces

ligaments, and may be associated with the abnormal tracking or lateral subluxation of the patella.

Excessive genu valgum and varum are caused by several potential agents: familial tendencies, obesity, epiphyseal growth plate deformity secondary to abnormal forces or metabolic deficiencies, overlengthened collateral ligaments, and muscular imbalances or abnormal muscle tone, i.e., spasticity (Fig. 12.9). In children with cerebral palsy, excessive genu valgum is often seen paired with excessive femoral antetorsion and external tibial torsion, and hypermobility of the hind- and midfoot, i.e., excessive pronation. This combination of deformities is most likely related to gravitation and muscle forces that come into play as the child begins locomotive activities. Some have claimed that genu valgum can also be induced by prolonged prone positioning and/or a sitting posture referred to as "W sitting," where the hips are internally rotated and the limbs are flexed with the tibiae externally rotated and placed in a valgus position. The persistence of genu valgum results in medial knee laxity and rotatory instability secondary to plastic changes in the cruciate and collateral ligaments, tightness of the tensor fascia latae and iliotibial band, and associated increased Q angle, deficient vastus medialis obliquus, and lateral tracking (and possible subluxation) of the patella. As discussed earlier (Chap. 11), abnormal patellar tracking is often associated with chondromalacia (breakdown of the articular cartilage on the underside of the patella) and early patellofemoral joint degeneration. It is likely that compensations for genu valgum may, or may not, be observed proximally or distally.

Genu varum may develop in response to persistent prone sleeping with the hips and knees flexed and tibiae internally rotated. Also, early weight-bearing, often in combination with abnormal nutrition, before structures can tolerate gravitational and muscle forces is implicated as a common cause. Often, genu varum is paired with the absence of normal external tibial torsion and is associated with distal compensations, including hind- and midfoot hypermobility, e.g., excessive pronation, talar adduction, and an intoeing gait.

Given the knee's position in the lower limb, it should be obvious that the combination of abnormal development and the forces associated with locomotion can create significant compensations proximally and distally. These compensations in combination with the need/desire to walk, run, and play and the associated high magnitude of forces can lead to injury in children at young ages. If not frank injury, these abnormalities could lead to early degenerative processes as the child progresses into adolescence and adulthood.

Ted's physical examination revealed mild tibial varum bilaterally. Although he demonstrated moderate/marked tightness of the tensor fascia latae/iliotibial band bilaterally, femoral antetorsion, genu valgum (the Q angle), and patellar tracking were normal. In addition, an observational evaluation noted good quadricep muscle definition medially (vastus medialis obliquus) and laterally, and Ted denied any anterior (patellofemoral) knee symptoms. It is likely that Ted's normal alignments proximally allowed small compensations for the large weight-bearing, and muscle forces, associated with running. Distally, however, we did not observe typical compensations at the hind- and midfoot complex, i.e., pronation and talar adduction. As a result, we hypothesize that Ted was unable to adequately compensate in the ankle/foot region for the combination of mild/moderate internal tibial torsion and tibial varum. In particular, we note that he did not have the hind- and midfoot mobility to do so.

In utero, the ankle/foot complex is subjected to a number of complex changes. Initially, the foot is directed medially and cranially, with the soles facing each other and the feet aligned with the legs; that is, there is no ankle angulation (called talipes equinus). In the seventh month, changes in the talus and calcaneus alter foot position. The talus widens as the head begins to torque, and the entire bone shifts laterally to align with the axis of the foot, changes that continue postnatally to counterbalance external tibial rotation effects. At the same time, the calcaneus grows broader and longer posteriorly, with a gradual reduction in the tuber calcanei angle relative to the tibia. Concomitantly, lower extremity flexion persistently increases over time, moving the feet to a dorsiflexed position held by pressure against the uterine wall. If normal intrauterine forces are maintained, then at birth, the foot is in the normal calcaneus, i.e., dorsiflexed position.

The foot continues to change substantially following birth, particularly beginning around the time the child initiates ambulatory activities (about 9–12 months). Before that, because the foot's soft tissues, i.e., ligaments, fasciae, and muscles, are poorly developed, the postnatal foot has not yet developed its longitudinal arches and is hypermobile. When a child initiates its bipedal stance, its wide base and external and pronated (i.e., flat) foot position assure some degree of stance stability. However, with repetition, i.e., standing and trying to walk, as well as over time, the passive and active soft tissues get stiffer and stronger, respectively, and the transverse and longitudinal arches begin to develop. Thus, by age 2 years, the hindfoot should be neutrally positioned, with a medial longitudinal arch well established and a normal external foot position in bipedal weight-bearing.

Trabecular systems in the ankle and foot develop based on the orientation of external forces (compression, torsion, and shear) and bending moments, i.e., induced by ground and muscle tensile forces (Fig. 12.10). The crossing pattern in the distal shank, termed interosseous trabeculae, is in response to multidirectional compression, tensile, and torsional loads. The calcaneus shows one pattern that runs longitudinally, cranial to caudal, forming in response to compression from weight-bearing and forces from gastrocnemius/soleus actions across the talocrural joint, and another pattern that runs proximal to distal in response to tension forces within the medial longitudinal arch.

The vertical trabecular columns in the astragal (more commonly referred to as the talus) are primarily derived from weight-bearing and muscular compression forces, while the talar longitudinal trabeculae are in response to bending forces across its arch. The trabeculae in the remaining tarsals and metatarsals are arranged along the lines of the compression forces (induced by ligaments and muscles) needed to sustain the longitudinal and transverse arches of the foot.

Eschewing genetics as a factor, our focus as we examine foot deformities (not necessarily related to cerebral palsy) centers on the influence of mechanical forces. Because the foot is the first complex to meet large ground forces, it is susceptible to deformation. One of the most common deformities is metatarsus adductovarus (or metatarsus varus) (Fig. 12.11). This condition is characterized by an adducted and varus forefoot relative to the hindfoot. Several causal factors have been identified: malposition in utero, i.e., breech, where the feet are wrapped around each other; increased mechanical pressure from maternal abdominal muscle tone; abnormal growth of the medial cuneiform and arrest of the normal ontological rotation of the foot; and large fetus or small uterus. Other possibilities include a neuromuscular disorder characterized by increased muscle tonus and imbalances, a paralysis that could contribute to the abnormal positioning of fetal lower limbs, and postnatal sleeping postures, usually stomach sleeping, that cause the feet to be in a turned-in, i.e., varus, position. A child who presents with this condition is seen to walk pigeon-toed (i.e., intoeing), with weight-bearing primarily on the outside of their foot and over the metatarsal heads; they may or may not make a heel-first contact, depending on the severity of the adductovarus angulation and the rigidity of the condition. Because of one's first impression of metatarsus adductovarus as an intoeing gait, it is important to consider that this gait pattern is also associated with increased femoral antetorsion and/or internal tibial torsion, so measures of those conditions must also be made.

Another common foot deformity, one very similar in appearance to metatarsus varus, is talipes equinovarus or clubfoot (Fig. 12.12). This condition is characterized by four abnormal foot positions: equinus (plantar flexion), hindfoot varus/supination, forefoot adductus, and cavus (increased height of the medial longitudinal arch). The causes of clubfoot may be idiopathic yet are multifactorial and often related to an isolated birth defect. The most common factors leading to this birth defect include malposition in utero, failure of lower limb rotation, muscular imbalances, and/or increased

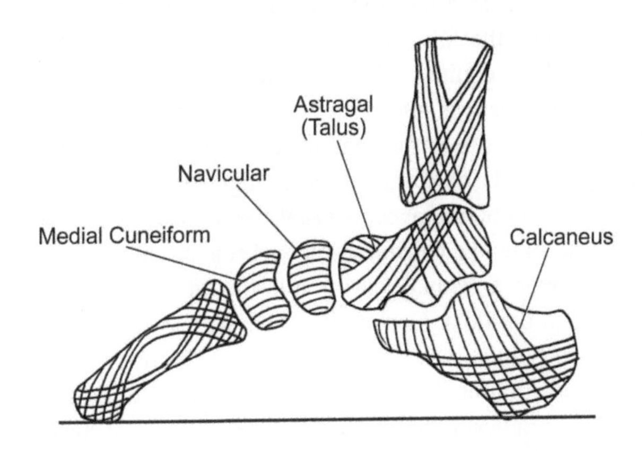


**Fig. 12.10** Schematic of trabecular patterns in the ankle and foot. Astragal is more commonly referred to as the talus. Note the complex trabecular strut patterns in the metatarsals

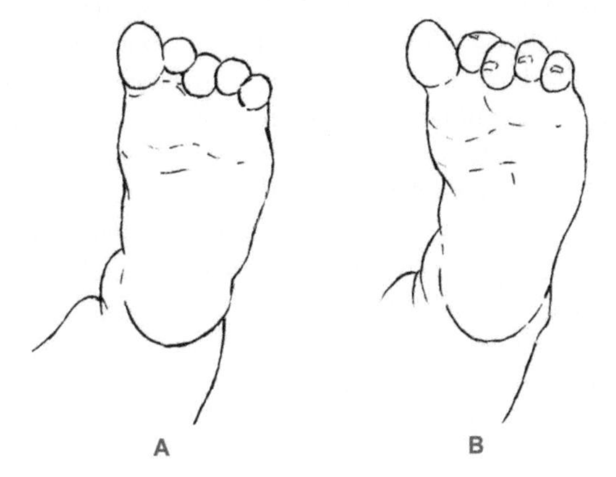


**Fig. 12.11** Comparison of (**A**) normal and (**B**) metatarsus adductovarus feet. In the metatarsus adductus foot, note the increased medial longitudinal arch and medial turn of the forefoot

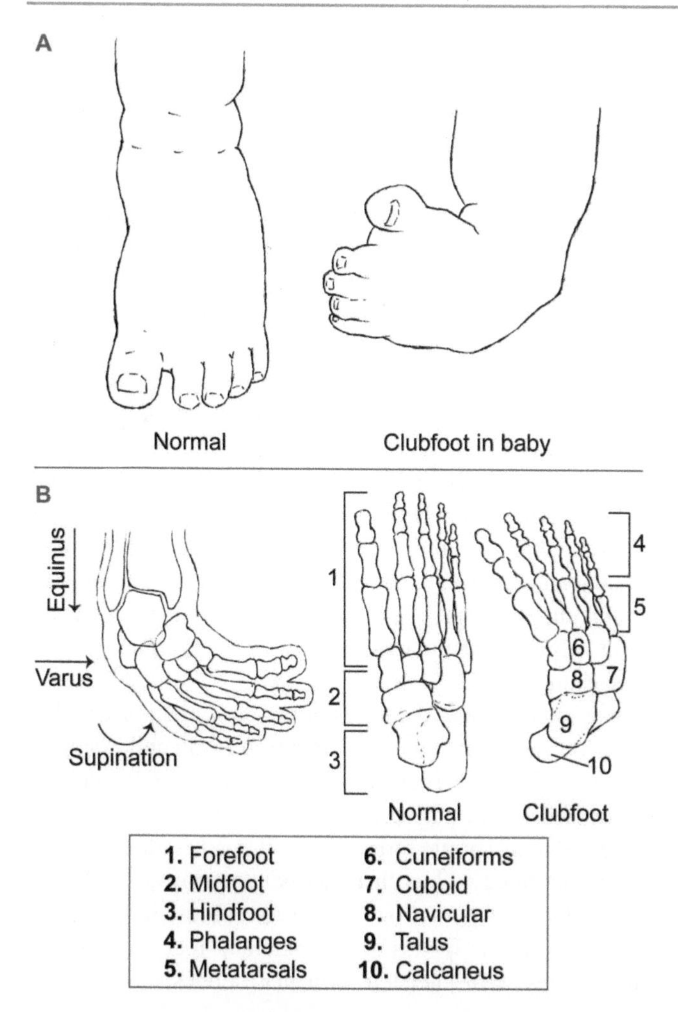


**Fig. 12.12** Comparison of normal and talipes equinovarus (clubfoot) foot: (**A**) gross ankle/foot configuration and (**B**) detailed demonstration of the malpositioned bones. With the clubfoot note the extreme supination/forefoot varus combined with plantar flexion

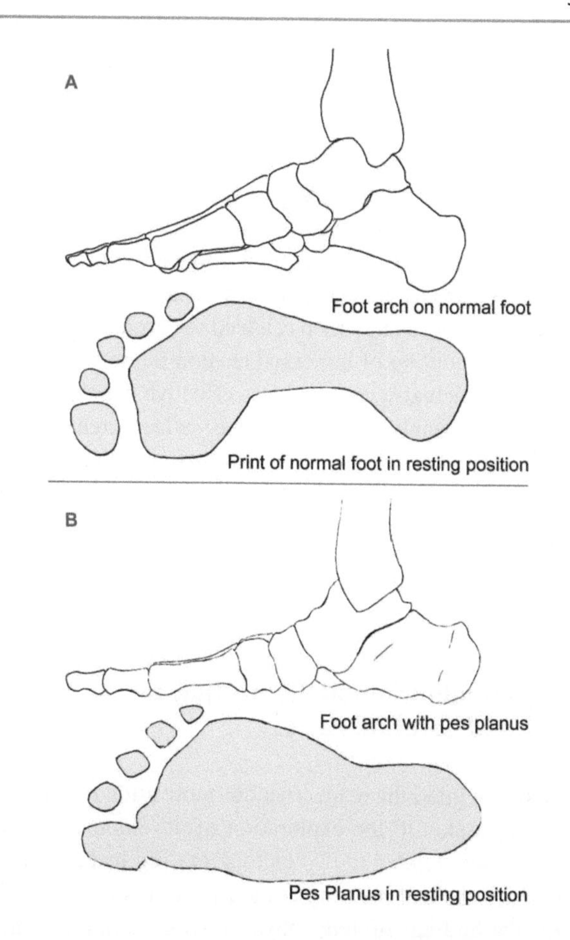


**Fig. 12.13** Skeletal and foot pressure prints of typical (**A**) normal and (**B**) pes planus (flat) feet. Note with pes planus the absence of a medial longitudinal arch

intrauterine pressure. A major deformation is the talipes equinus, where the navicular is medially displaced on the talus and the metatarsals are adducted. These bony deformities cause shortening of muscles, ligaments, and skin medially and posteriorly. The remainder of the tarsal bones rotates medially, developing abnormal shapes and poorly congruent articular surfaces. Compromised vascularity results in abnormal sequencing of ossification centers. The gait of a child with this condition is significantly impaired because of talocrural joint stiffness and reduced dynamic ankle range of motion, persistent intoeing, and significant gastroc-soleus weakness. The fixed equinus ankle position does not allow the child to dorsiflex the ankle during the swing phase in gait for toe clearance and prepositioning for the first ankle rocker at initial contact; in fact, in severe cases, the child may make foot contact laterally, causing stance instability. Additionally, the limited dorsiflexion available after the loading response does not allow for second ankle rocker action (see Fig. 12.19), which impedes forward progression. Failure of the second rocker, coupled with the marked weakness of the gastroc-soleus complex, results in the inability to control tibial advancement as well as a calcaneus gait, i.e., the inability to push off and accelerate the body in the transition from terminal stance to preswing. Persistent internal foot progression combined with the fixed equinus causes the toes to catch the floor or the contralateral limb.

Pes planus, or a fixed flat foot, is characterized by an absence of the medial longitudinal arch and is usually not diagnosed until the child reaches 16–28 months of age (Fig. 12.13). The causes of pes planus are recognized as the result of primary or secondary developmental abnormalities. In utero, shortening of the first metatarsal (metatarsus primus varus), muscular imbalances, and hypermobility syndromes are associated with primary causes. Secondary causes of pes planus include genu valgum and/or increased femoral antetorsion with a compensatory increase in external tibial torsion (often, the combination of increased femoral antetorsion and external tibial torsion is referred to as the "miserable malalignment syndrome"). The planus foot is associated with creep deformation of the plantar fascia (aponeurosis) and long plantar and spring ligaments induced by weight-bearing loads, which render the hind- and midfoot hypermobile. During the loading response of gait, excessive mobility of this region may be considered an advantage, given the large ground forces that need to be distributed. On the other hand, there are several disadvantages: (1) excessive mobility during the loading response requires greater reliance on muscles, e.g., gastroc-soleus and tibialis posterior, for stability and force transmission, which can lead to chronic strain injuries; (2) the coupling of increased tension in the plantar fascia (which is actually rendered less effective over time) and flexor hallucis longus (i.e., the windlass effect) creates a loss of the ability to transform the midfoot region into a more rigid lever for push-off; and (3) changes in the moment arm of the triceps surae makes it less effective as an ankle/foot invertor and plantar flexor, hampering the ability to push off during preswing.

### 12.4.1 Developmental Biomechanics and Cerebral Palsy

We want to remind the reader that our motivation for the preceding digression to the exploration of developmental biomechanics was related to two factors: (1) Ted had not likely reached full skeletal maturity at the time of his foot injuries, and (2) the findings of Ted's physical examination revealed bilateral lower extremity deformities. Now, we take another small detour and examine common foot deformities, and their natural history, in children with cerebral palsy. This is so because we suspected that Ted may have had a mild form of spastic diplegic cerebral palsy. By taking this small detour, we may gain some insight into the causes and sequelae of his foot deformities. Foot deformities in children with cerebral palsy (CP) are commonly the result of dynamic imbalances between the extrinsic muscles of the lower leg, which control ankle/foot alignment. These imbalances are typical of an abnormal neuromuscular complex with many interacting factors, including muscle spasticity, disrupted motor control (for gait and balance), soft tissue contractures, bony torsions, and/or joint instability. For example, the ankle plantar flexor muscles (most commonly gastrocnemius and tibialis posterior) are overactive (yet often weak), and the ankle dorsiflexor muscles are weak and ineffective, with variable imbalance patterns between the ankle/foot supinator and pronator muscle groups, which are manifested in locomotive activities.

Gait maturation in typically developing children begins when the child initiates walking through 7 years of age, during which time the movements, forces, muscle activity, and temporal parameters of gait evolve through predictable patterns of change. Between the ages of 4 and 7 years, adult gait patterns emerge with consistency, with some individual variation in the timing of maturation. The development of foot posture appears to be controlled by the plantar flexors, i.e., gastroc-soleus complex and tibialis posterior, and the peroneal muscles that provide mediolateral stability. Over time, foot intrinsic muscles and ligamentous constraints develop to fine-tune movements and enhance stability. The natural history of gait maturation in children with CP does not follow the typical patterns, especially related to the timing of motor maturation. However, there is evidence to show that children with CP tend to improve (depending on the type and severity of the disease) in their gait function until they are in the middle childhood years of ages 7–8. Children with CP who are more highly functioning tend to continue to improve well into adolescence. Their improvement is encouraged by neurologic maturation, where there is better independent motor control of the extrinsic and intrinsic muscles of the foot, which then assists in the development of the medial longitudinal arch, lateral peroneal arch, and distal transverse arch, all of which transmit large weight-bearing ground forces more evenly within the foot.

The factors cited above result in four common coupled ankle/foot segmental malalignment patterns in children with spastic CP. (Note: There are other forms of CP, such as ataxic, which will not be discussed here since we are only interested in its spastic form.) You will see that the word talipes is associated with each ankle/foot deformity, which will be described in what follows. This is so because talipes is the general term that applies to all ankle and foot deformities, whereas pes is the general term that applies only to foot deformities.

Equinus (Fig. 12.14) is characterized by excessive plantar flexion of the hindfoot relative to the tibia, which occurs secondary to overactivity of the plantar flexors; the mid- and forefoot tend to have normal alignment. The largest of the plantar flexor muscles is the gastrocnemius, implying that it creates the most influential deforming force. It is notable that the gastrocnemius, in crossing the knee joint, may also disrupt function and alignment at the knee. The spasticity of the plantar flexors tends to increase from the time the child

**Fig. 12.14** Talipes equinus or excessive hindfoot plantar flexion

begins walking, reaching its maximum severity between the ages of 5 and 6 years. Spasticity alters the contractile elements, as well as the length of the Achilles tendon, and, over time, may lead to weakness as well as contracture.

Equinoplanovalgus is a "collapsing" type of deformity and is characterized by equinus of the hindfoot, coupled with excessive pronation of the mid- and forefoot, and an imbalance between the ankle evertors and tibialis posterior (Fig. 12.15). Poor intrinsic muscle effort coupled with abnormal motor control induces hindfoot valgus, midfoot collapse into dorsiflexion, and forefoot abduction. This deformity leaves the hindfoot in an equinus position with a midfoot "break," i.e., hypermobility, which has the forefoot dorsiflexed relative to the hindfoot. On visual examination, the weight-bearing foot has a "rocker bottom" appearance and is pronated. The lateral column, i.e., calcaneus, cuboid, and fourth and fifth metatarsals, of the foot is functionally and/or structurally shorter than the medial column, i.e., talus, navicular, medial cuneiform, first metatarsal, and hallux. Often, ankle and hallux valgus develop over time in association with this deformity. Hallux valgus occurs because the combination of planovalgus with forefoot abduction, forefoot dorsiflexion, and external foot progression angle causes weight-bearing forces along the medial border of the hallux, and over time, these forces cause plastic changes in the lateral metatarsal-phalangeal constraints (i.e., joint capsule and collateral ligaments). The natural history of equinoplanovalgus often sees some improvement in 5- to 7-year-old children as muscle strength and motor control improve, but often, the rapid growth of adolescence sees further midfoot collapse, resulting in excessive external foot progression and lever arm dysfunction (altered effective lever arm of the gastroc-soleus complex) for push-off.

Equinocavovarus (or equinovarus) is more commonly seen in children with unilateral (spastic hemiplegia) CP. This initially develops as compensation for the equinus; that is, as the child walks with an equinus, the forefoot tends to adduct. It is characterized by equinus deformity of the hindfoot, coupled with supination of the midfoot, with variable alignment of the forefoot (Fig. 12.16). With this condition, the lateral column is functionally and/or structurally longer than the medial column. Compensatory ankle valgus may be associated with this deformity. In an 8- to 9-year-old child, the tibialis posterior muscle often develops hypertonicity over time, as well as contracture, which increases and fixes a forefoot varus. In an adolescent, the overactive tibialis posterior may also induce a fixed hindfoot joint contracture.

There are several forms of calcaneus deformities: calcaneus, calcaneovalgus, and calcaneovarus (Fig. 12.17). Calcaneovalgus appears to be the most common among children with spastic cerebral palsy. In addition to calcaneal valgus, this deformity is primarily characterized by excessive dorsiflexion. This is the least common deformity identified in children with spastic diplegic CP.

Typically, the common segmental malalignments found at the ankle and foot are flexible and can be corrected with manual/passive manipulation, i.e., passive range of motion testing, as part of a physical examination, particularly in children with less severe CP. However, as the child ages and grows, more plastic, i.e., permanent, changes occur in muscles (architectural alterations in the contractile elements and contracture), tendons (stiffening and shortening), and bones (fixed changes in size and shape). As a result, the malalignment patterns become more rigid and less (or not) correctable during passive range of motion testing.

As we note previously, we suspect that Ted might have had a mild form of spastic diplegic CP based on his physical examination. If he did and was never diagnosed, his natural history likely followed the pattern where, because of only mild spasticity and impaired motor control, the early walking period characterized by equinoplanovalgus feet naturally improved during his early childhood. Thus, by age 7 years, he acquired normal adult gait patterns. However, it appears that secondary to persistent mild spasticity and imbalance in both extrinsic and intrinsic muscles, perhaps guided by genetic predisposition (he appeared to have a form of clubfoot), Ted developed a talipescavovarus foot, with mild

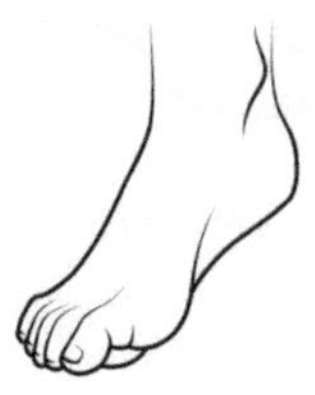

**Fig. 12.16** Talipes equinocavovarus

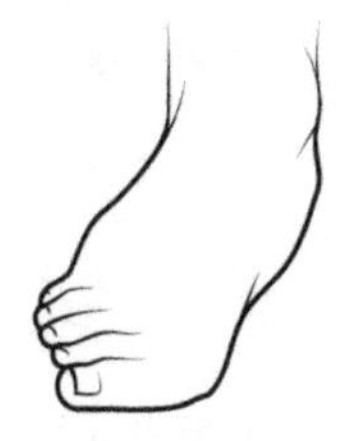

**Fig. 12.15** Talipes equinoplanovalgus with hallux valgus

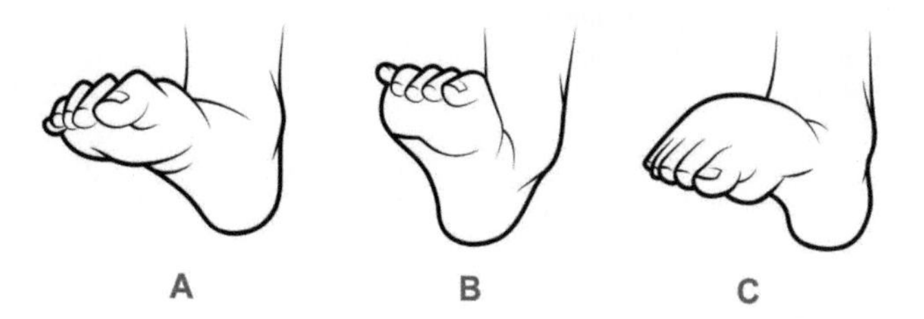


**Fig. 12.17** Three forms of talipes calcaneus: (**A**) calcaneus, (**B**) calcaneovalgus, and (**C**) calcaneovarus

forefoot adductus (Fig. 12.18). We assume that Ted acquired these deformities as he developed, but his malalignments were minor, which allowed him to play with his peers as a child and to successfully participate in sports. Despite his deformities, Ted eventually achieved success as a long-distance runner by the time he was seen at the Motion Analysis Center. However, because of his success as a competitive runner, i.e., lack of serious injury, it appears that over time, his foot deformities had become relatively fixed, i.e., not correctable with passive range of motion testing, likely now disposing him to injury.

For a typical child with spastic diplegia, the common ankle/foot deformities described above alter the gait during both the stance and swing phases. With the three most common deformities (excepting calcaneus deformities), a heel-first initial contact cannot occur, disrupting the first rocker (heel), which reduces the ability to dissipate the weight-bearing (i.e., ground) forces in the loading response. Equinus and equinocavovarus deformities disrupt the second (ankle) rocker by blocking ankle dorsiflexion, perhaps compromising midstance stability, whereas equinoplanovalgus maintains the mid- and forefoot segments in an "unlocked" (i.e., loose-packed) position, allowing greater dorsiflexion and creating midstance instability and excessive tensile loading of the soft tissues of the medial longitudinal arch. All three segmental malalignments may compromise the ability of the ankle plantar flexor muscles to generate adequate plantar flexor moments during the third (forefoot) rocker (Fig. 12.19; see Appendix J for a more detailed description). This is likely related to hindfoot malalignment, which shortens the length of the plantar flexor muscles, thus changing their length-tension and effectiveness in active tension generation. With equinoplanovalgus, the potential internal moment produced by the plantar flexors is reduced at the third rocker because their effective moment arms are reduced by malalignments of the mid- and forefoot. Excessive external tibial torsion, a malalignment problem often associated with equinoplanovalgus, contributes to increased external foot progression angle during gait, which further compromises the plantar flexor moment arm in terminal stance/preswing. Finally, all three deformities disallow normal ankle dorsiflexion in swing, which makes toe clearance and proper prepositioning of the foot for initial contact problematic.

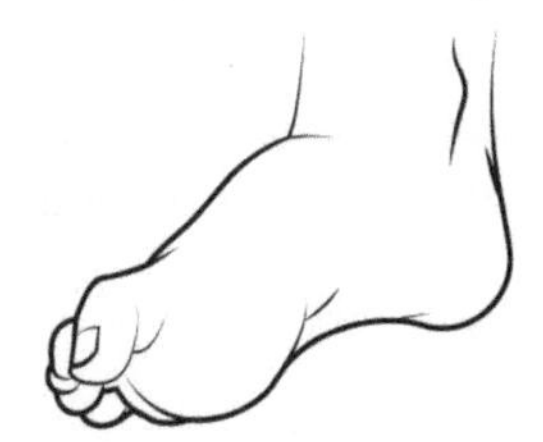

**Fig. 12.18** Talipescavovarus foot with mild forefoot adductus

## 12.5 Structure and Function of the Normal Ankle/Foot Complex

We give just a brief explanation of our method. In many instances, up to this point in this case discussion, we have used many terms without definition, which may be disconcerting to the reader. If we were meeting as a group (in PBL style) to discuss this case, we would naturally stop at any point when new words, phrases, and/or concepts were encountered; identify them as learning issues; and delegate individuals in the group to get the information needed in order to have a meaningful discussion. In many cases, individuals may seek, find, and present the new information immediately. At other times, it may be that a concept needs to be further reviewed and summarized and would not be presented until the next meeting. To some degree, we have digressed on particular topics but cannot do it as if we were working face-to-face. We encourage you, whenever you encounter a word or idea that has not been defined, to follow your inclination (or develop the inclination) and look for the information you need before you go forward. At the end of the chapter, we provide most of the definitions and explanations you need to understand the case. With PBL, however,

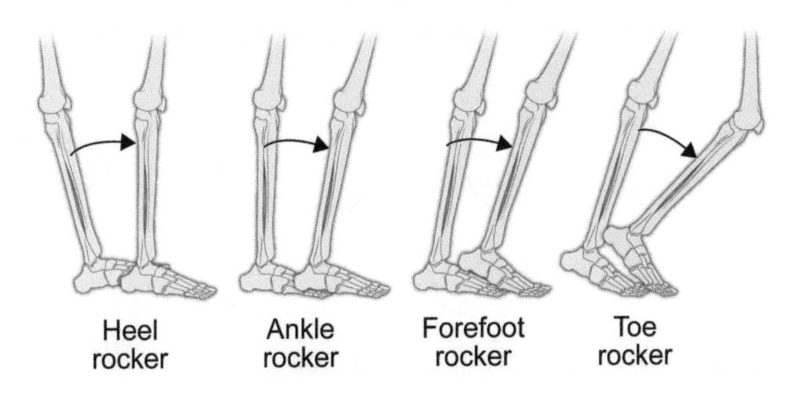


**Fig. 12.19** Four-foot rockers have been identified during a typical gait cycle. (Modified with permission from Richie (2021))

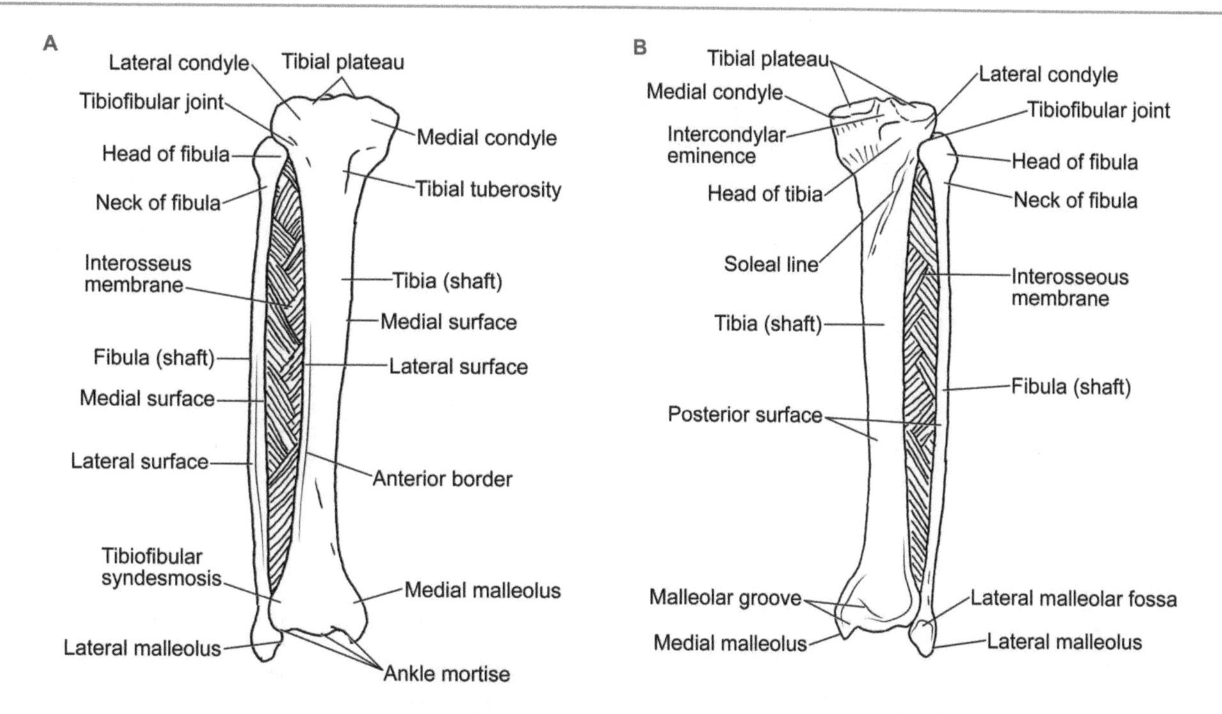


**Fig. 12.20** (**A**) anterior and (**B**) posterior views of key bony features of the lower leg, including the interosseous membrane

we want to remind the reader that we do not provide everything if it is not needed and leave it to you to get more information on your own.

## 12.5.1 Structure: Definitions of Regions and Kinematic Terms

The ankle/foot complex, which includes the proximal tibiofibular joint, may be the most integrated musculoskeletal system in the human body (Figs. 12.20 and 12.21). Before we venture into the anatomical details of this complex, it helps to establish an overview of the region and a few definitions. The proximal tibiofibular joint is often included in presentations of both the knee joint and ankle complexes. Despite its many fascial connections to tissues that act on the tibiofemoral joint, e.g., popliteus, the short head of biceps femoris, etc., the proximal tibiofibular joint does not directly contribute to knee kinematics. On the other hand, these same soft tissue connections likely influence the kinematics of the proximal tibiofemoral joint during tibial rotations and/or movements at the ankle joint.

The distal end of the fibula, i.e., lateral malleolus, and distal tibiofibular joint play a large role in ankle kinematics. Mechanically, the shaft of the fibula may provide some lateral support to the tibia, but it plays a very small part in transmitting the large ground reaction forces associated with locomotive activities, transferring only approximately 10–15% of the total load. The interosseous membrane (or middle tibiofibular ligament) extends between the interosseous crests of the tibia and fibula and separates the anterior and posterior muscles of the shank. This structure consists of a thin aponeurotic collagenous lamina composed of oblique fibers that run inferolaterally and superomedially, an architecture that suggests a mechanical function of controlling multidimensional forces between the two bones that comprise the shank. The primary effect of the interosseous membrane may be to stabilize the fibula and constrain its posterolateral bending, thus similar to an analogous structure in the forearm, allowing compression forces on the tibia to be shared with the fibula. The membrane is broader superiorly, with an upper aperture for the passage of vessels to the anterior aspect of the leg. The interosseous membrane is continuous inferiorly with the interosseous ligament of the tibiofibular syndesmosis.

The distal ends of the fibula and tibia, i.e., lateral and medial malleoli, respectively, form a large concave arch that articulates with the convex dome of the talus. This forms the ankle joint, more formally known as the talocrural joint. It is sometimes also referred to as the ankle mortise because its shape is very similar to a carpenter's mortise joint (Fig. 12.22). What is notable about the mortise is its near-perfect congruity and, with regard to the anatomical joint, a large surface contact area, i.e., articular cartilage matching of the lateral and medial malleoli and distal tibia with the talus. This feature is particularly significant because of the very large ground forces that must be distributed by and transmitted through this joint. It is likely that the rarity of talocrural

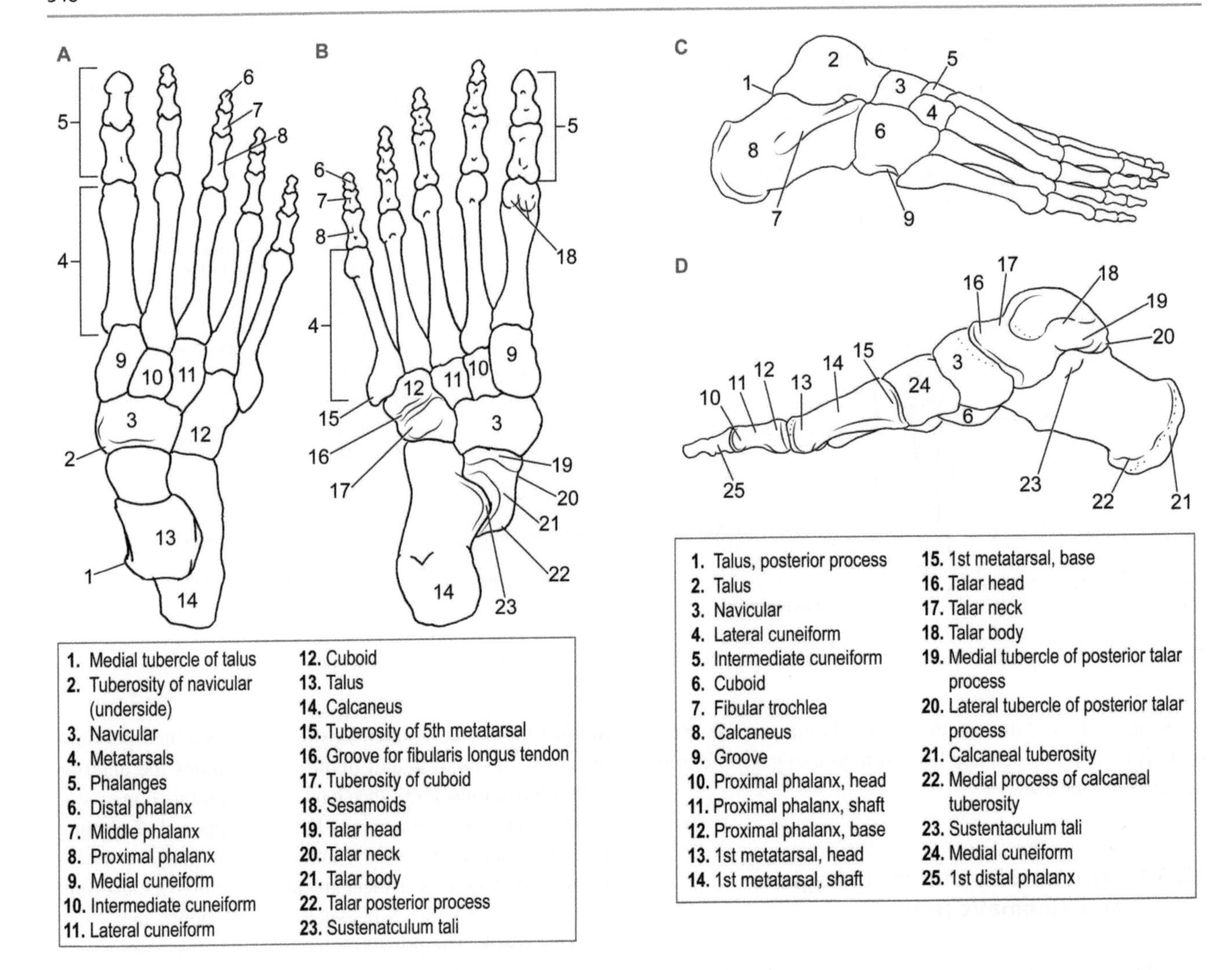


**Fig. 12.21** (**A**) Dorsal, (**B**) plantar, (**C**) lateral and (**D**) medial perspectives of foot anatomical structures

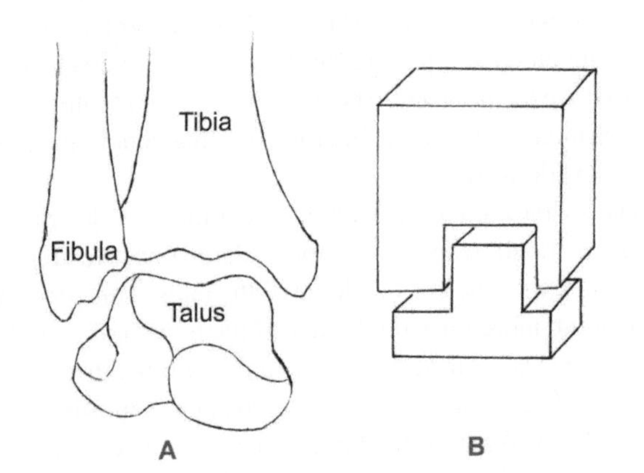


**Fig. 12.22** Schematic comparison of the (**A**) talocrural and (**B**) carpenter's mortise joints. If we imagine the ground forces produced during walking transmitted proximally, based on these figures, it is likely that the majority of the compression forces would be transmitted from the talus to the tibia and then proximally through the kinetic chain

joint degeneration is due largely to the mortise's configuration and its ability to distribute compression and shear stresses.

The ankle, then, refers to the articulation between the tibia, fibula, and talus, and the foot refers to the tarsal bones and joints distal to the ankle. The foot is generally divided into three regions, each with its own complex of bones and joints (Fig. 12.21). The rearfoot or hindfoot includes the talus, calcaneus, and subtalar joint. The midfoot consists of the navicular, cuboid, and cuneiforms, as well as the transverse tarsal joint, i.e., the talonavicular and calcaneocuboid joints. There are several other joints that comprise the midfoot region, named according to the bones that articulate with one another: distal intertarsal joint, cuneonavicular, cuboideonavicular, cuneocuboid, and intercuneiform. The forefoot consists of the metatarsals and phalanges, which include the tarsometatarsal, intermetatarsal, metatarsophalangeal, and interphalangeal joints. At times in our discussion, we use the term ray, which is defined as a metatarsal and its associated phalange.

With the altered orientation of the foot relative to the shank, the terms anterior and posterior are typically used to reference an area on the shank, which can be used interchangeably with the terms distal and proximal, respectively, to describe a region of the foot. The term dorsal refers to the top (superior) of the foot, and plantar refers to the bottom (inferior) of the foot. Later, when we detail the mechanics of the joints we are most interested in, i.e., talocrural, subtalar, and transverse tarsal, we introduce one unique naming convention associated with the ankle/foot complex. Because of the unique shapes of the bones that make up the talocrural, subtalar, and transverse tarsal joints, biomechanists and clinicians have devised terms to describe motion according to a standard right-handed Cartesian coordinate system, as well as a terminology more fitting to a helical axis system. Therefore, using the conventional kinematic naming scheme, dorsiflexion (flexion) and plantar flexion (extension) describe movements in the sagittal plane around a mediolateral axis; eversion and inversion describe movements in the frontal plane around an anteroposterior axis, and abduction/adduction describe movements in the horizontal (or transverse) plane around a superoinferior (or vertical) axis (Fig. 12.23). Remember that using the conventional axis system still allows for six degrees of freedom of movement: three rotations around each axis and three translations along each axis.

Due to the shapes of the bones that make up the joints in the ankle and foot, reality dictates that movements do not take place strictly in the cardinal planes. For example, the axis of the talocrural joint has been shown to be oblique in multiple planes, which appears to suggest that joint movement, such as dorsiflexion and plantar flexion, occurs in multiple planes simultaneously. This notion has been referred to as triplanar movement (Fig. 12.24). Note that the concept of triplanar movement is analogous to the notion of helical (or screw) rotations. Generally, the ankle joint center is found at the midpoint between the medial and lateral malleoli. Thus, because the lateral malleolus is more inferior and lies more proximal relative to the medial malleolus, the ankle joint axis runs in an inferoposterior direction from medial to lateral. Fig. 12.24 is specific to the talocrural joint, but analogous orientations are found at both the subtalar and transverse tarsal joints as well. Because of the oblique orientation of these axes, the terms pronation and supination are used to describe the composite motions that occur. Note that these definitions generally only apply when the movements of the foot occur relative to a proximal segment, i.e., the foot relative to the shank in an open kinematic chain. Pronation includes the

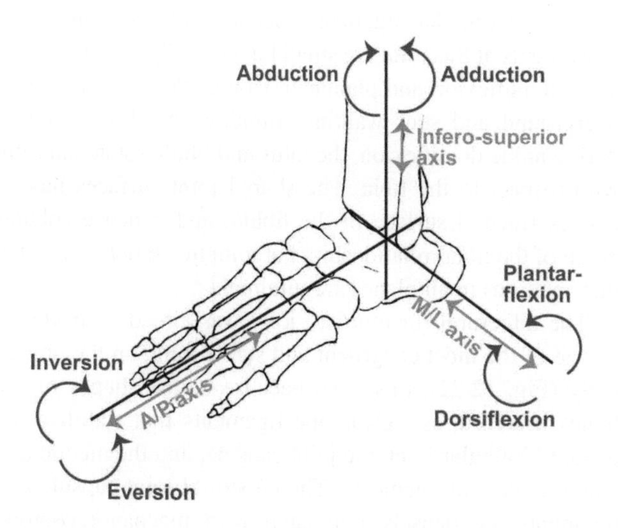


**Fig. 12.23** Schematic overview of a right-handed Cartesian coordinate system attached to the ankle joint center. This same system could be used to describe the movement at any of the joints making up the foot. Note: M/L = mediolateral, A/P = anteroposterior, I/S = inferosuperior

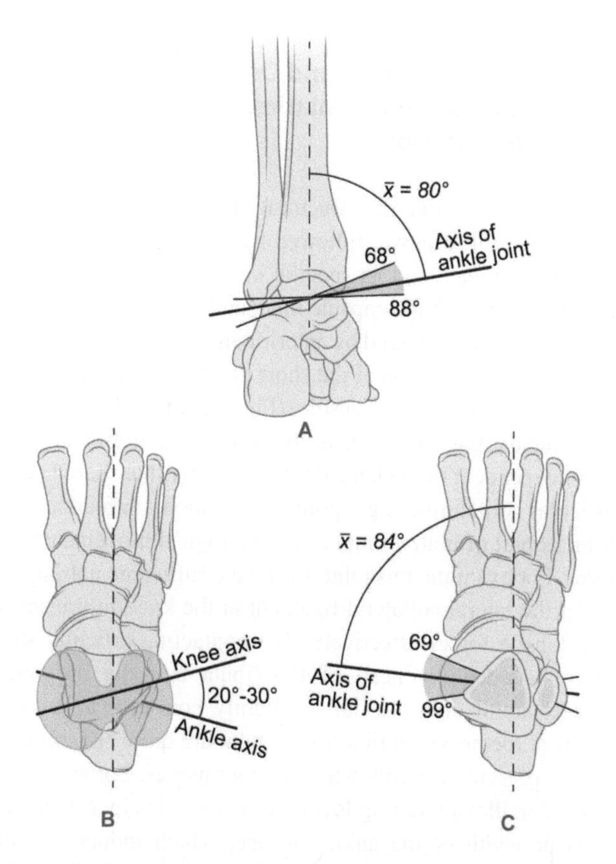


**Fig. 12.24** Schematic of talocrural triplanar axis. (**A**) Posterior view showing an angle between the ankle axis and the long axis of the tibia; on average, the ankle axis forms an angle of 80° relative to the long axis of the tibia. (**B**) Superior view showing the relationship between the knee, ankle, and foot axes; the 20° to 30° angle between the knee and ankle axes somewhat represents external tibial torsion. (**C**) Relationship between the talocrural axis and the longitudinal axis of the foot; on average, the ankle axis makes an angle of 84° relative to the longitudinal axis of the foot. Note: the dashed line in B and C represents the longitudinal axis of the foot. (Modified with permission from Musculoskeletalkey)

elements of eversion, abduction, and dorsiflexion, whereas supination is comprised of the movements of inversion, adduction, and plantar flexion. Therefore, using the concept of triplanar motion, if talocrural dorsiflexion is the primary motion, its axis orientation simultaneously dictates small degrees of eversion and abduction. The term triplanar can be confusing because it seems to imply that there are three rotations about three axes when in fact there is only one axis about which the joint movements occur. The idea of triplanar (or helical axis) movement patterns, where one segment rotates about and along an oblique axis relative to another segment, is truly the most concise way to think about joint movements. But, as noted previously, helical axes are not used clinically because it is difficult to assign clinically meaningful terms to that way of describing joint motion. The terms pronation and supination appear to be acceptable to clinicians because they can be imagined as three separate movements that have unambiguous clinical definitions.

### 12.5.2 Joint Structure and Osteo- and Arthrokinematics, Functional Application

Let us now examine the structural details of the joint complexes identified above, followed by a review of each joint, or joint complex, and its specific kinematics. The proximal tibiofibular joint has a capsule independent of the tibiofemoral joint and is reinforced by anterior and posterior ligaments, a portion of the tendon of the short head of the biceps femoris, and the popliteus muscle (Fig. 12.25). Although the movements that take place at the proximal tibiofibular joint are independent of the knee joint, it is likely that shared periarticular structures, e.g., popliteus tendon, influence this joint as part of both normal and pathological movement patterns. For example, this joint must have sufficient stability to assist the lateral collateral ligament at the knee in transmitting tensile forces effectively. The contacting articular surfaces between the head of the fibula and the proximal posterolateral tibia are flat or slightly convex. The movements at the proximal tibiofibular joint are quite constrained yet are present primarily when the foot moves. For example, ankle dorsiflexion during locomotion activities increases the anterior width of the ankle mortise, which induces small rotations and translations at both the proximal and distal tibiofibular joints.

The distal joint tibiofibular joint is equally as restrained in its movements, perhaps more so because of its intimate relationship with the ankle mortise and the much larger ground forces that are there encountered; in fact, the stability of the talocrural joint is highly dependent on the stability of the distal tibiofibular joint. Some have referred to this joint as a syndesmosis, a type of fibrous synarthrodial joint closely

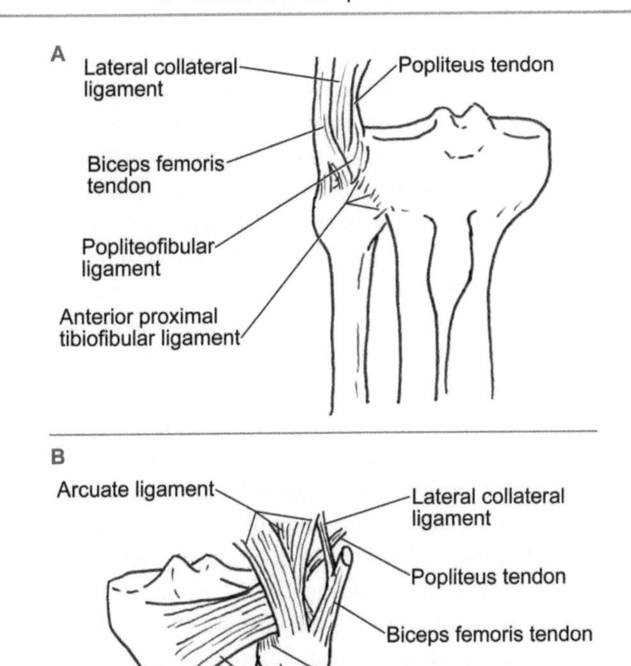


**Fig. 12.25** (**A**) Anterior and (**B**) posterior perspectives of the proximal tibiofibular joint and reinforcing soft tissue constraints

secured by the interosseous ligament, a ligament that is an extension of the interosseous membrane. Similar to its superior counterpart, the distal tibiofibular joint is controlled statically by anterior and posterior distal tibiofibular ligaments (Fig. 12.26), which are reinforced, to some degree, by the calcaneofibular ligament. Because of its constraints, movements at this joint are small but critically coupled to the ankle dorsiflexion and plantar flexion cycle associated with overground and stair walking, running, etc. For example, during ankle dorsiflexion, the talus and fibula rotate laterally with respect to the tibia. The shared joint surfaces have a convex (medial surface of the fibula) and concave (fibular notch of the tibia) relationship, but joint movements are such that rotations in all planes are minimal.

The talocrural (or mortise) joint was already introduced as one of the most congruent and stable joints in the human body (Fig. 12.22); this is so because of the shapes of the bones involved, as well as the ligaments that stabilize the distal tibiofibular joint, the joint capsule, and the medial and lateral collateral ligaments. The talocrural joint capsule and ligaments are densely populated with mechanoreceptors, whose sensory information is used to monitor/facilitate extrinsic muscle activation, i.e., the provision of dynamic stability. After the initial impulse of the ground forces is absorbed by the medial and lateral processes of the calcaneal

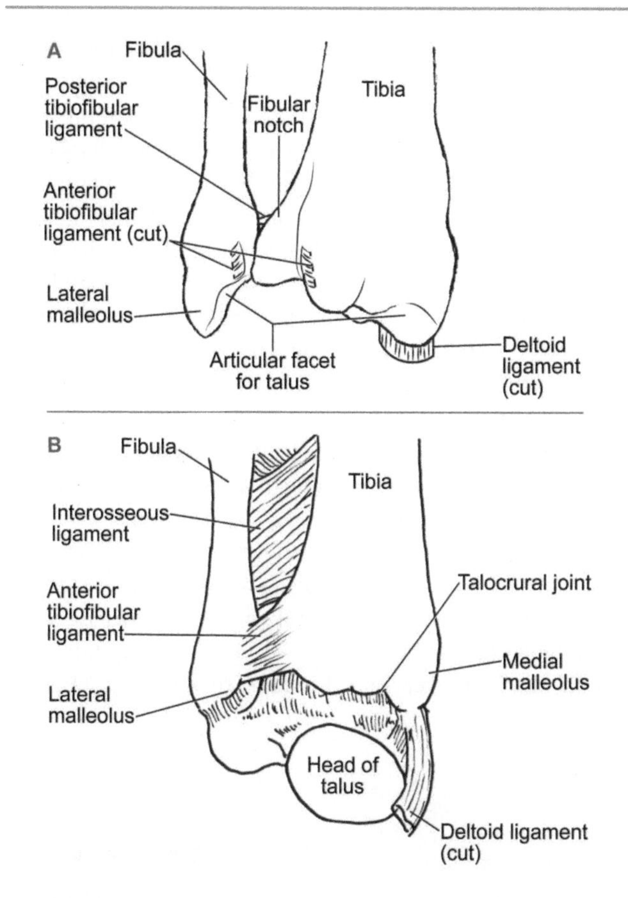


**Fig. 12.26** (**A**) Anterolateral and (**B**) anterior perspectives of the distal tibiofibular joint

tuberosity, a still very large force is propagated to the talocrural joint to be transmitted up the kinetic chain. Because of the asymmetry in the structure of the mortise contact surface, 90–95% of the weight-bearing forces pass through the medial talus and tibia, with the remaining forces transmitted through the lateral talus and fibula.

Like most other joints, the tissues that constrain excessive motion at the talocrural joint include an extensive joint capsule and primary and secondary ligaments named according to the bones or regions of their attachments (Fig. 12.27). What is somewhat unique about many of these ligaments is that they also constrain nearby joints, e.g., the calcaneofibular ligament. This review makes a few pertinent comments about the major ligaments and leaves it to the reader to consult Table 12.2 and other resources for more details. The medial collateral (also deltoid or triangular ligament) provides extensive coverage of the talocrural joint with its superficial and deep sets of bands. As a whole, the bands that make up the deltoid ligaments constrain excessive eversion at the talocrural, subtalar, and talonavicular joints, but the oblique orientation of some of the bands suggests that this ligament also constrains multidimensional movements, e.g., axial rotation, of the talocrural joint.

The lateral collateral ligaments include the anterior talofibular, calcaneofibular, and posterior talofibular ligaments. The anterior talofibular and calcaneofibular ligaments provide most of the restraint to excessive inversion throughout

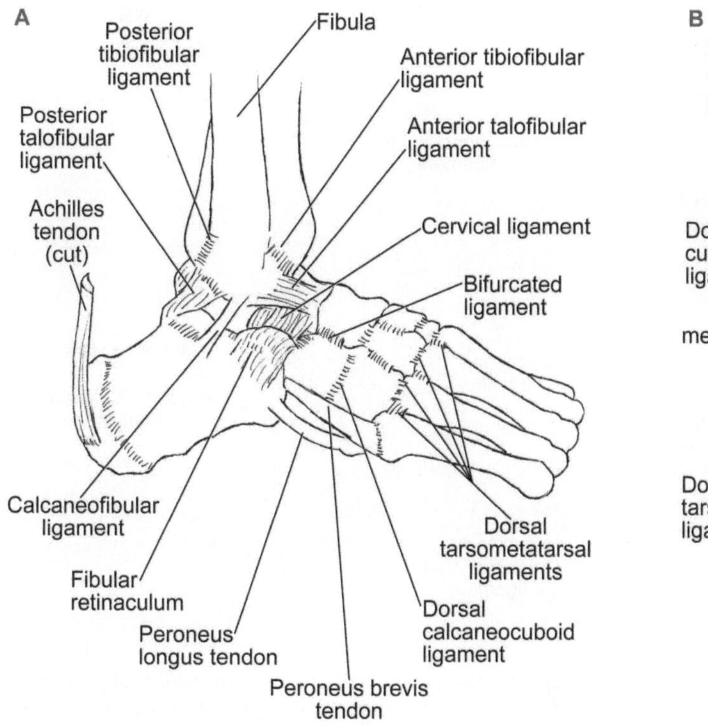


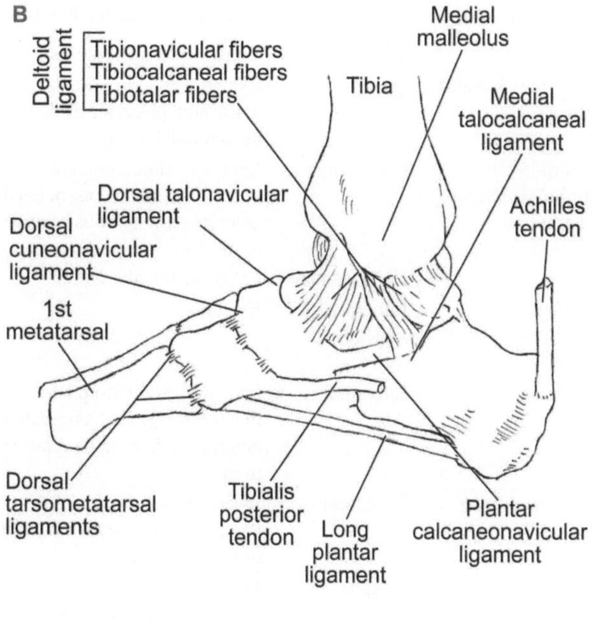


**Fig. 12.27** (**A**) Lateral and (**B**) medial perspectives of the ligamentous system of the ankle and foot. Note the relationship between ligamentous and selected muscle tendons

the range of talocrural dorsiflexion and plantar flexion; it is notable that the calcaneofibular ligament also secures the subtalar joint laterally. An inferior transverse ligament (not shown) appears to be a thickening that reinforces the posterior talofibular ligament.

Going forward, all presentations of joint/segment kinematics assume that the foot is not fixed, unless indicated otherwise. There are six degrees of freedom of movement at the talocrural joint, but osteokinematically, the primary motions at the talocrural joint reside in the sagittal plane. The neutral position of the foot, i.e., 0°, is when the bottom of the foot is perpendicular to the shank (or 90°). Maximum dorsiflexion is measured at approximately 15° to 25° (10° to 15° with the knee extended, with its range reduced secondary to the constraint of the gastrocnemius muscle), while maximum plantar flexion ranges from 40° to 50°. Approximately 20–30% of the dorsiflexion and plantar flexion range of motion comes from joints distal to the talocrural joint, e.g., subtalar, transverse tarsal, etc. Full dorsiflexion is considered the close-packed position of the talocrural joint. Recall that a close-packed position generally assumes that the bony joint partners are maximally congruent and selected periarticular joint constraints are taut, making this joint position the most stable. As suggested above, because the axis of the talocrural joint does not match a Cartesian axis convention (see Fig. 12.24), the foot everts and abducts a small amount during dorsiflexion and inverts and adducts slightly during plantar flexion.

Arthrokinematically, translations at the talocrural joint follow the convex/concave rule. Therefore, during dorsiflexion, as the foot moves anterosuperiorly (as does the talus), the talus glides posteriorly; conversely, plantar flexion includes a posterior roll and anterior glide of the talus (Fig. 12.28). As noted, end-range dorsiflexion is the close-packed position. It is notable that when this position is achieved, in addition to the tautness of several structures identified in Fig. 12.28, there is also increased tension in the distal tibiofibular and interosseous ligaments as the mortise separates anteriorly, secondary to the posterior talar glide; this wedging of the mortise occurs because the talus is wider anteriorly than posteriorly (Fig. 12.29). This event naturally occurs from the midstance to the terminal stance of the gait cycle. Since the foot is fixed during the stance phase, as the talus moves forward, the talocrural joint dorsiflexes. Locking of the ankle mortise is part of two events during this time period: ankle rocker and forefoot rocker (Fig. 12.19).

**Table 12.2** Talocrural joint ligaments and the movements they limit

| Ligaments | Joints | Constrained movements |
|---|---|---|
| Deltoid (tibiotalar fibers) | Talocrural | Eversion and dorsiflexion and associated posterior glide of talus relative to mortise |
| Deltoid (tibionavicular fibers) | Talocrural | Eversion, abduction, and plantar flexion and associated anterior glide of talus relative to mortise |
| | Talonavicular | Eversion and abduction |
| Deltoid (tibiocalcaneal fibers) | Talocrural, subtalar | Eversion |
| Anterior talofibular | Talocrural | Inversion, adduction, and plantar flexion and associated anterior glide of talus relative to mortise |
| Calcaneofibular | Talocrural | Inversion and dorsiflexion and associated posterior glide of talus relative to mortise |
| | Subtalar | Inversion |
| Posterior talofibular | Talocrural | Abduction, inversion, and dorsiflexion and associated posterior glide of talus relative to mortise |

Note: The movements described assume that the foot is not fixed, i.e., distal segment relative to the proximal segment. Modified with permission from Neumann (2017)

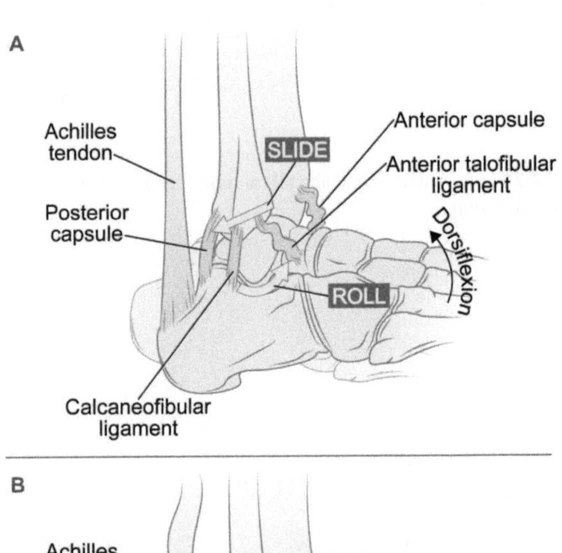


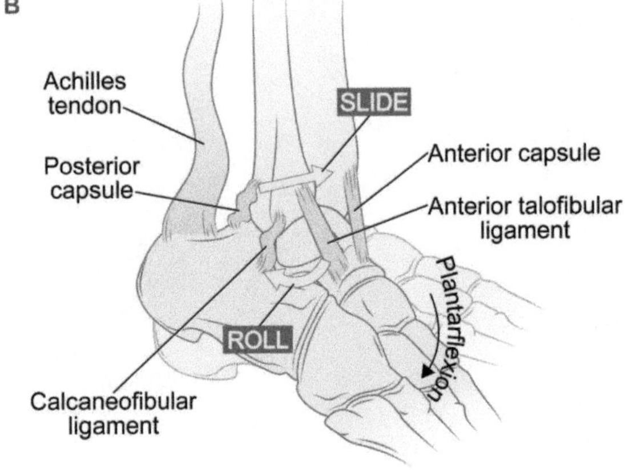


**Fig. 12.28** Osteo- and arthrokinematics at the talocrural joint during (**A**) dorsiflexion and (**B**) plantar flexion. Note that the ligaments that appear "crimped" suggest that they are slackened, and the ligaments that appear "straighter" suggest that they are taut, i.e., placed under tension

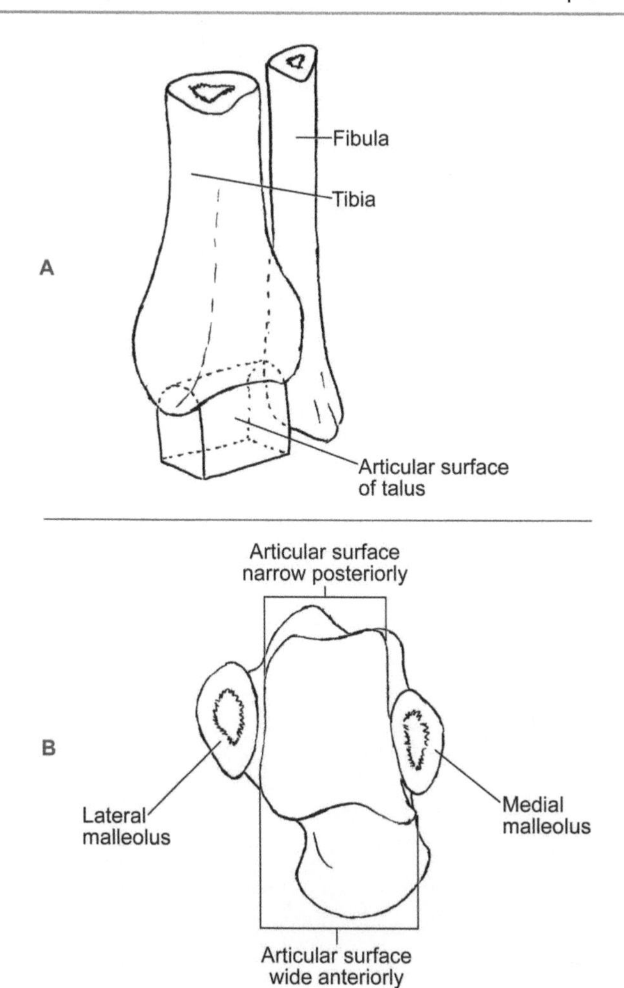


**Fig. 12.29** (**A**) Schematic of the mortise joint showing a talar model held within the mortise, i.e., lateral and medial malleoli, and (**B**) a superior view of the talus mounted between the malleoli, illustrating differences in width from the anterior to the posterior talar regions. It is clear that a posterior glide of the talus associated with ankle dorsiflexion forces the malleoli to separate

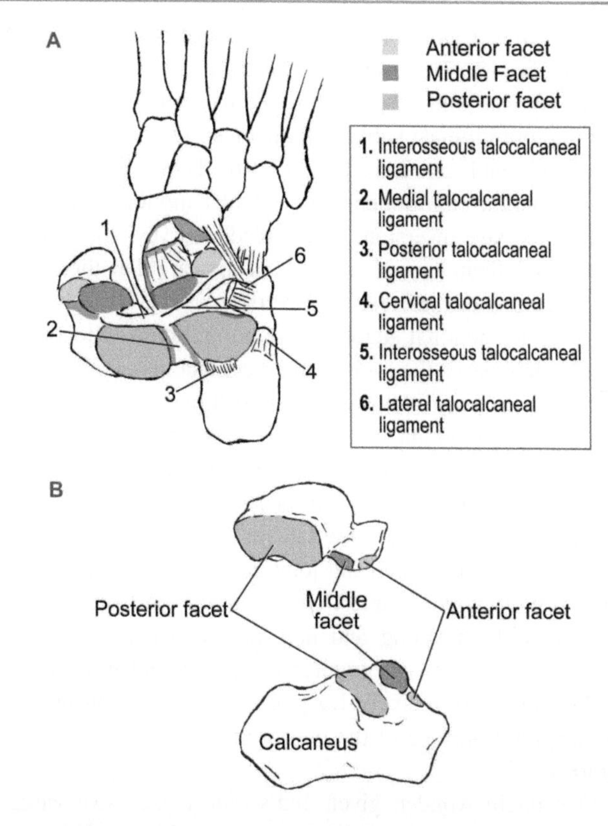


**Fig. 12.30** (**A**) Superior view of the subtalar joint illustrating the anterior middle and posterior facets of the talus and calcaneus and the ligaments that provide joint stability and (**B**) anterolateral view of the subtalar joint illustrating how the respective articulating facets of the talus and calcaneus match

Locking the mortise, i.e., putting the talocrural joint in a close-packed position, provides greater stability to the ankle during a period of late single-limb support. At the same time, the eccentric action and passive elastic properties of the posterior tibial muscles provide additional talocrural stability as they concomitantly prepare to shorten to assist in propulsion. Although we only explicitly described talocrural arthrokinematics with the foot free, it should be clear to the reader that one must be able to extrapolate this information in order to be able to also describe talocrural mechanics when the foot is fixed; this is so since almost all of the activities of daily living involve different forms of locomotion. We encourage the reader to chart out the mechanics at the talocrural joint when the tibia is moving over the fixed foot.

Next is the subtalar joint. As with the talocrural joint, perhaps the most important perspective of subtalar kinematics should come from the examination of how the shank/talus moves on a fixed calcaneus since this is the more functional approach. Thus, that is the tact we take in our review of subtalar kinematics and leave it to the reader to chart out subtalar osteo- and arthrokinematics when the distal segment (calcaneus) moves relative to the proximal (talus) segment when the foot is not fixed. To begin this investigation, we need first to closely examine the partner articular surfaces, i.e., the underside of the talus and the superior aspect of the calcaneus (Fig. 12.30). The talus has three facets: anterior, middle, and posterior, which interlock with corresponding facets on the superior surface of the calcaneus. The anterior and middle facets are small and flat on both sides of the joint, while the larger posterior talar concave facet rests snugly within the convex facet of the calcaneus. When we consider arthrokinematics at the subtalar joint, we should not dismiss the importance of the gliding that occurs anteromedially, but the convex-concave relationship at the more prominent facet posteriorly largely dictates this joint's helical action. Now, let us take a closer look at the soft tissue constraints that control subtalar movements.

There are separate joint capsules for the anteromedial and posterior subtalar articulations, with smaller supporting ligaments for the capsule of the larger posterior facets. The calcaneofibular and tibiocalcaneal fibers of the deltoid ligaments (Fig. 12.27) prevent excessive eversion and inversion at the subtalar joint, respectively, but may not be as effective as the more intimate subtalar ligaments: interosseous (talocalcaneal) and cervical (Fig. 12.30). These ligamentous bands restrain multidimensional subtalar motions but especially excessive inversion. As the reader surveys the medial and lateral ligaments of the foot, it appears that there are more ligaments that restrain inversion. This is likely the result of a functional adaptation to normal bipedal locomotion, which places the foot at initial contact in neutral dorsiflexion (it is not in a close-packed position and therefore is less stable) and slight inversion. Then as the ankle plantar flexes during the loading response, it is even less stable (talocrural plantar flexion creates a loose-packed position in the ankle/foot partly because the talus is less congruent within the mortise, which has distal joint implications) under very large weight-bearing forces. Given these events, the ankle/foot needs both strong and able passive (ligaments) and active (muscle) forces to counter the natural tendency for the ankle to invert excessively. Despite the safeguards, inversion ankle sprains are one of the most common lower extremity injuries.

One might wonder, given Ted's mildly excessive varus static foot position, why he did not report a history of inversion ankle sprains. It is possible that he did not report them because he was not asked. On the other hand, the robust strength of his ankle/foot muscles, neuromuscular efficiency for locomotor tasks, and relatively rigid cavovarus deformity likely shielded him from inversion sprains.

The bony morphology and orientation of the talus and calcaneus are key elements that dictate the oblique orientation of the subtalar joint axis and the primary and secondary osteokinematic movements (Fig. 12.31). Based on Fig. 12.31A, which illustrates the subtalar axis coursing from the posterolateral to the anteromedial aspect of the foot, it is clear that inversion and eversion are the primary motions at this joint. In normal adults, there is an approximate 2:1 ratio of inversion to eversion; generally, it is 20° inversion to 10° eversion. Staying consistent with the definition of pronation and supination, then, pronation is a combination of eversion, dorsiflexion, and abduction; conversely, supination combines inversion, plantar flexion, and adduction. However, because of the obliqueness specific to the subtalar axis, the primary components include inversion/eversion and abduction/adduction. Remember that although dorsi- and plantar flexion movements are minimal, their importance should not be ignored.

Although the amount of movement in the frontal plane is small compared to sagittal plane motion during walking, normal subtalar eversion/inversion is critical to the dissipation and transfer of large impulsive ground forces. We typically begin a gait analysis at initial contact (IC). At initial

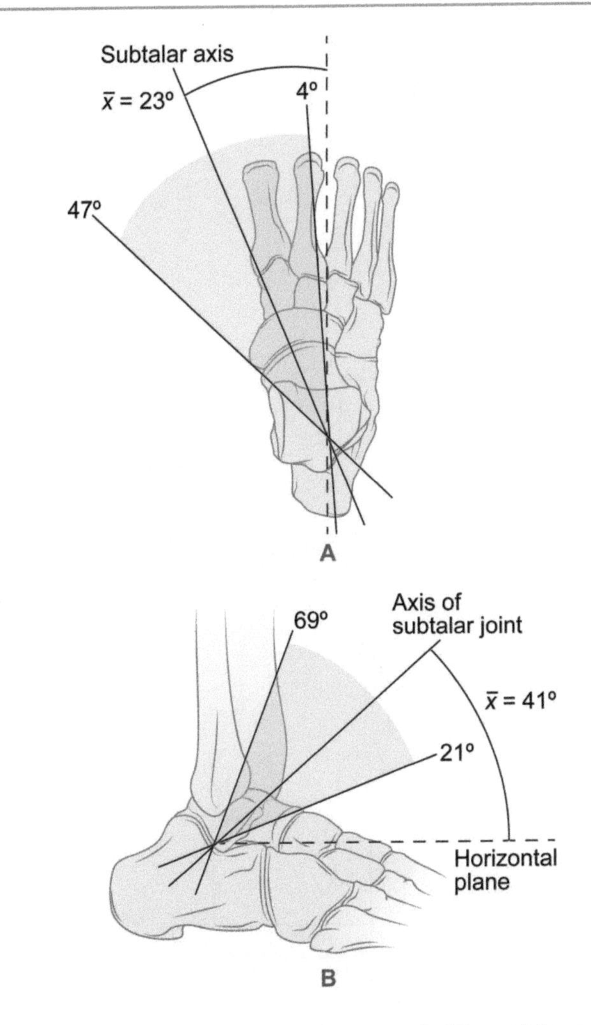


**Fig. 12.31** Schematic illustrating the proposed oblique triplanar axis of the subtalar joint (**A**) superiorly, where we see that the axis deviates approximately 23° medially, and (**B**) from a sagittal plane perspective, where we see that the axis deviates approximately 40° superiorly as it runs posterolaterally to anteromedially. The large variability in the magnitude of the orientation of the subtalar axis suggests a combination of significant variability in measurement precision and among individuals. (Modified with permission from Musculoskeletalkey)

contact, the talocrural joint is typically neutral (i.e., 0° dorsiflexion), whereas the subtalar joint is slightly inverted. Now, this seems contradictory since we claimed that with triplanar motion, dorsiflexion is combined with eversion and abduction. So how can it be that neutral dorsiflexion and inversion occur at the same time? Let us try to address this apparent paradox. Firstly, we described the spatial pose of two different joints: talocrural and subtalar. Secondly, recall that with complex systems, we find coupled motions, i.e., both osteo- and arthrokinematics, which may be different when segments are free or fixed. So, in fact, at initial contact, the talocrural joint may have a neutral pose three-dimensionally; at the same time, the subtalar joint is inverted, plantar flexed, and adducted.

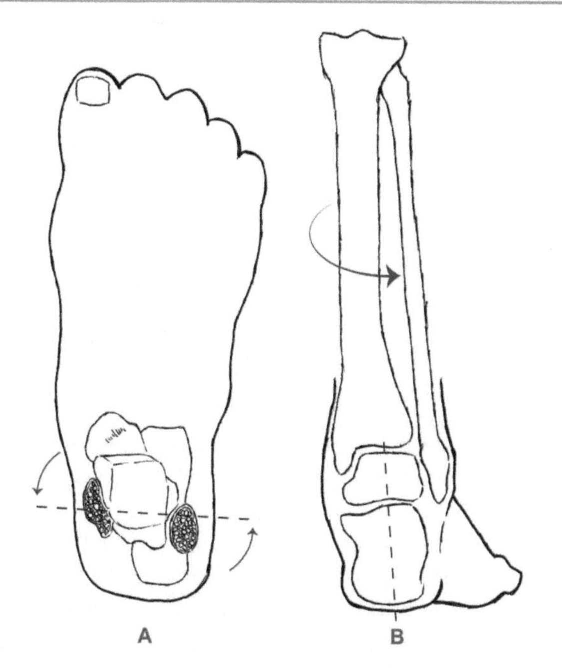


**Fig. 12.32** At loading response, (**A**) the posterolateral calcaneus is loaded, inducing eversion of the calcaneus and internal rotation of the talus, and (**B**) with concomitant internal rotation of the tibia

Immediately after IC, the foot, i.e., the posterolateral aspect of the calcaneus, is loaded (we call this the loading response (LR)), which induces eversion of the calcaneus (not pronation) (Fig. 12.32). Since the foot is fixed during LR, it appears (Fig. 12.32B) that the talus moves on the calcaneus in the direction of plantar flexion and adduction, which contradicts the definition of pronation (eversion, dorsiflexion, and abduction). How can we explain this paradox? It can be explained by what is actually happening at the subtalar joint surface based on the convex/concave rule. When the calcaneus everts, it moves (osteokinematically) laterally, which means it concomitantly must glide medially at its posterior facet. But a medial glide of the calcaneus is relative to one's perspective. Since the calcaneus is fixed, the glide that actually takes place is of the talus on the calcaneus; that is, the talus glides laterally at its concave posterior facet, which resembles subtalar abduction. Thus, we conclude that during LR, the subtalar joint pronates. Concomitantly the tibia internally rotates. This internal tibial rotation during the LR appears to be induced and (Fig. 12.32A) has been explained by, first noting the relationship between the talocrural and subtalar joints (i.e., mortise arrangement relative to the talus), and two plausible mechanisms: (1) internal rotation of the femur and shank after IC augments subtalar pronation or (2) subtalar pronation induces tibial internal rotation. This chicken-or-the-egg question has not been solved, but whichever way it may go, this mechanism is one of a series of interrelated mechanisms working within the lower extremity during locomotion activities to control large

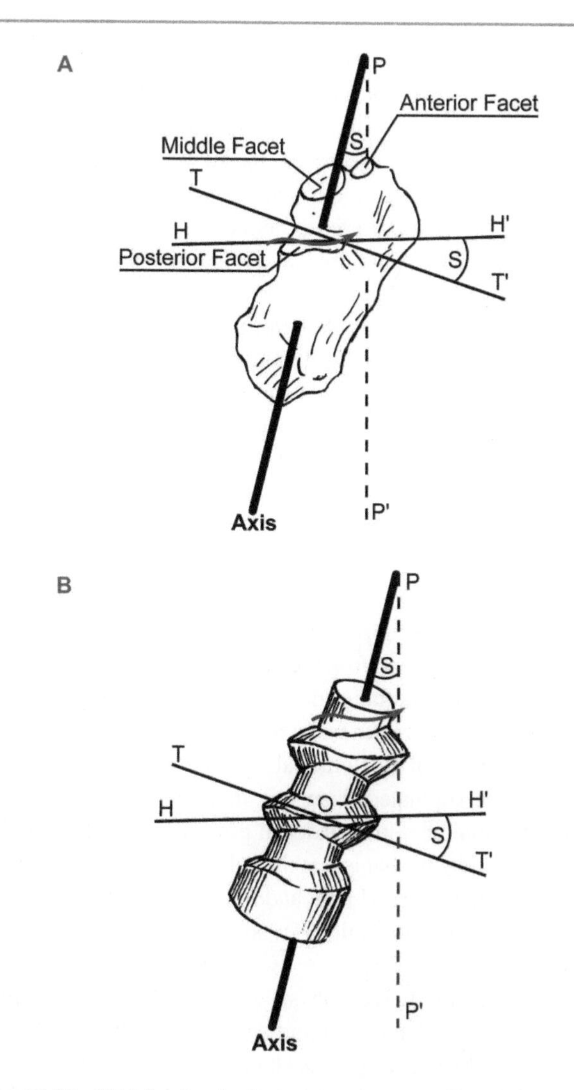


**Fig. 12.33** (**A**) Likening the facet arrangements on the superior aspect of the calcaneus of the right subtalar joint with a (**B**) right-handed screw. The red arrow in A and B represents the path of rotation following the screw. The horizontal plane in which motion is occurring is *HH'*; *TT'* is a plane perpendicular to the axis of the screw; *S* is the helix angle of the screw, which is determined by dropping a perpendicular (*PP'*) from the axis. As the calcaneus inverts (clockwise rotation), it simultaneously translates along the helical axis. (Modified with permission from Manter (1941))

weight-bearing forces. And this is, in general, another wonderful example of the interaction of subsystems within a complex system used to manage the environment and the forces of gravity.

**Note** We have successfully used a reductive method to explain, to the point of some understanding, the interaction between the talus and calcaneus during one subphase of the gait cycle. In reality, we should agree that, in fact, the talus and calcaneus are moving on each other simultaneously. This is also a point where we remind the reader how the subtalar facets resemble segments of a spiral of Archimedes similar to a screw (or helical) axis (Fig. 12.33). In this figure, the analogy is applied to the right foot that is not fixed, where the calcaneus translates anteriorly along the subtalar axis as it rotates clockwise, i.e., inverts, during subtalar supination. Similar to the threads of a screw, the subtalar joint is shown with a single degree of freedom in which the calcaneus moves in a triplanar pattern (as defined previously) of inversion/eversion and abduction/adduction as the more predominant motions (dorsi- and plantar flexion being so small that it can be ignored). Again, although triplanar motion seems to imply three separate axes of motion, the screw representation paints an image of movement patterns across all three planes simultaneously, and it should be noted that despite the appearance of the screw axis shown as a single degree of freedom, in reality, it is the most concise expression of six degrees of freedom of movement.

Let us pick up where we left off and finish our review of subtalar movements during gait. Soon after LR, as the tibia advances over the fixed foot (i.e., talocrural dorsiflexion), the subtalar joint sequentially moves from a pronated (everted calcaneus) to a neutral (in midstance) to a supinated (inverted calcaneus) position (in terminal stance); the subtalar joint then remains supinated from terminal stance through preswing. We challenge the reader to now chart out the details of the kinematic mechanism at the subtalar joint, and its relationship to the talocrural joint and rotations of the femur and tibia, during these transitions in single-limb support of the gait cycle.

Before we look at Ted's physical examination data relative to ankle/foot mechanical alignment and mobility and how that might influence his gait mechanics, we need to finish our review of the anatomical/mechanical complexes distal to the subtalar joint. The transverse tarsal joint (or Chopart's joint) has an intimate functional relationship to the subtalar joint and segments of the forefoot and appears to make a significant contribution to foot adaptations on even/uneven walking and running surfaces. This joint is not considered a true anatomical joint because it is not secured by one joint capsule. It is typically described as a compound functional joint since it is comprised of two separate synovial joints: talonavicular and calcaneocuboid. The talonavicular joint shares a joint capsule with the anterior and middle facets of the talus and their talar counterparts of the calcaneus; thus, these three articulations are often referred to as the talocalcaneonavicular joint. The talonavicular joint partners have reciprocally shaped saddle surfaces that create excellent joint congruency and, thus, some measure of inherent bony stability. Of course, more robust joint stability is enhanced by sets of ligaments. The spring (also referred to as the plantar calcaneonavicular) ligament is a thick, wide band of tissue composed of fibrocartilage that spans the gap between the sustentaculum talus of the calcaneus and the medioplantar surface of the navicular bone (Figs. 12.27B and 12.34). Its uniqueness lies in its dual

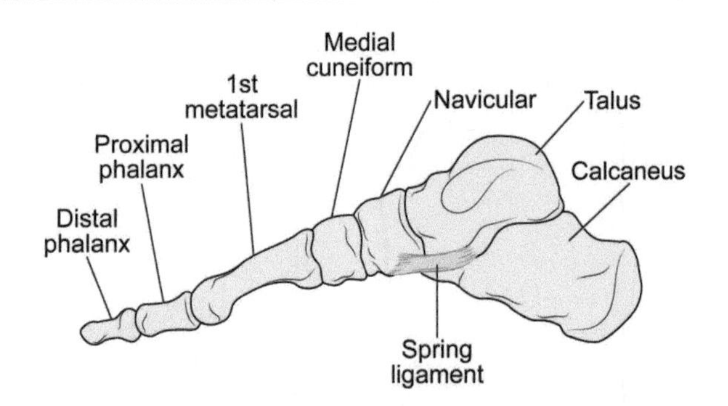


**Fig. 12.34** Medial view of foot illustrating the talonavicular joint. Note the sustentaculum talus, which supports the body of the talus and serves as the proximal attachment of the spring ligament, which courses to attach distally to the medioplantar surface of the navicular

function, serving (1) as a partial articular surface for the head of the talus, with responsibility for distributing/mitigating the compression and shear stresses related to weight-bearing, and, (2) like any other ligament, to constrain excessive joint movements, that is, resist the tensile stresses associated with the superincumbent body weight pushing the talus inferomedially during bi- and unipedal stance. Additional ligaments that provide stability to the talonavicular joint include (1) dorsal talonavicular, (2) bifurcated (for additional lateral support), and (3) tibionavicular and tibiospring fibers of the deltoid ligament (for additional medial support). Let us note that the tibialis posterior tendon also appears to provide significant dynamic support during walking. The bony and soft tissue structure of this medial longitudinal column is such that the longitudinal arch has functional spring-like characteristics that allow for the proper balance of deformation and stiffness required during the support and propulsion phases of walking and running.

The calcaneocuboid joint forms the lateral compartment of the transverse tarsal joint (Fig. 12.35). The articular surface of the calcaneus has a quadrilateral shape and a concavo-convex, undulating surface. The calcaneus is typically divided into superior and inferior parts, which have opposite orientations; the superior part is concave in the horizontal and vertical planes, while the inferior part is convex in the same planes. Its articulating partner, the cuboid, has reciprocal architectural characteristics. This matching creates a saddle-type joint arrangement that provides the calcaneocuboid joint and lateral longitudinal arch with more inherent bony stability than the medial longitudinal arch. The capsule of this joint is confined to the calcaneus and cuboid but is reinforced by the dorsal calcaneocuboid and bifurcated ligaments, as well as by the long and short plantar ligaments. The long plantar ligament is not specific to the calcaneocuboid joint, though it runs from the plantar surface of the calcaneus into the bases of the lateral three or four metatarsal bones. The short plantar ligament (plantar calcaneocuboid ligament) arises just distal to the long plantar ligament and

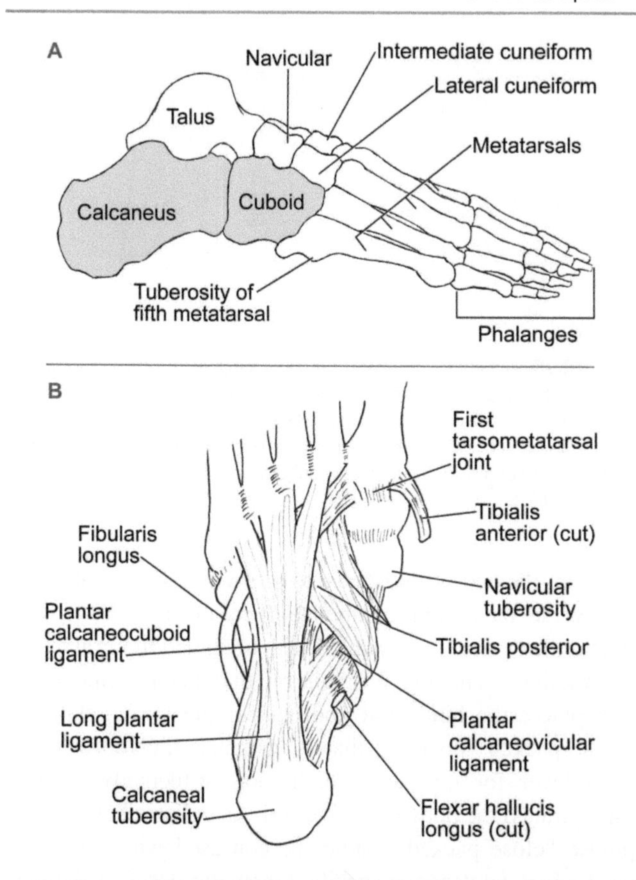


**Fig. 12.35** (**A**) lateral view of the right foot and calcaneocuboid joint and (**B**) selected ligaments and tendons of the plantar aspect of the left foot

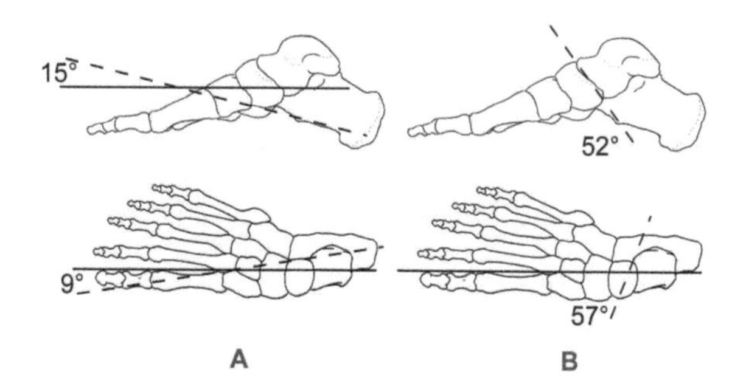


**Fig. 12.36** Proposed axes at the transverse tarsal joint: (**A**) longitudinal and (**B**) oblique. (Modified with permission from Manter (1941))

enters the plantar surface of the cuboid (Figs. 12.27A, B and 12.35). Both plantar ligaments provide additional support to the lateral longitudinal arch, but the longer plantar ligament plays an additional role in supporting the talonavicular joint and medial longitudinal arch.

The transverse tarsal joint, of course, includes an intermediate articulation, which between the navicular and cuboid is referred to as the cuboideonavicular joint. This joint has been labeled as a syndesmosis, constrained by dorsal, plantar, and interosseous cuboideonavicular ligaments.

Kinematically, transverse tarsal joint movements are strongly linked to those of the hindfoot. Yet this joint complex appears to exhibit more triplanar mobility than the subtalar joint; that is, the transverse tarsal joint is largely influenced by the ball-and-socket shape of the talonavicular joint, thus allowing large movements in all three cardinal planes. It is clear that the subtalar and transverse tarsal joints largely control the pronation and supination poses of the entire foot. We also see that the transverse tarsal joint is strongly coupled to triplanar movements distally, e.g., intercuneiform and cuneocuboid complex, Lisfranc's joints, etc. Unlike at the subtalar joint, where one can easily document, using a goniometer, the inversion/eversion range of motion, it is not readily feasible to goniometrically measure any aspect of transverse tarsal joint mobility. However, estimates have indicated that similar to the subtalar joint, there is substantially more inversion/supination than eversion/pronation.

Biomechanists agree that triplanar motion at the transverse tarsal joint mirrors that of the subtalar joint during gait, yet they have not reached a consensus on the definition of the axes that define the movements. Neumann cites Manter, who described two axes of rotation: oblique and longitudinal (Fig. 12.36). According to Manter, since the longitudinal axis is nearly parallel to an anteroposterior axis, the movements that occur around that axis have components of inversion/eversion. The oblique axis is pitched vertically, running posterolateral to anteromedial, suggesting that combined abduction/dorsiflexion and adduction/plantar flexion are strongly coupled. Although Manter provided images that conveniently illustrate how complex movement patterns can be depicted using cardinal plane definitions, research in the past two decades (with the help of advanced motion capture and imaging technologies that did not exist in Manter's time) has not validated his work. More recent research has suggested that we not attempt to define movements at the talonavicular and calcaneocuboid joints but instead refer to the mobility of the midtarsal joint for two reasons: (1) the ball-and-socket architecture of the talonavicular joint dominates its joint motion and (2) the relative immobility of the calcaneocuboid joint. Thus, researchers have concluded that midtarsal joint movements are triplanar around a single axis (see Fig. 12.33).

The arthrokinematics at the transverse tarsal joint have not been explicitly described, largely because the three joints in this complex, i.e., talonavicular, cuboideonavicular, and calcaneocuboid, appear to move primarily as a unit (Fig. 12.37). It is clear that rotations/translations at the talonavicular joint would likely be greater than those at the other two joints, but it is unlikely that the convex/concave rule can be applied, similar to what we have done at other joints in the lower extremity. What we do know is that the triplanar movements that exist at the transverse tarsal joint generally show a greater range when compared to those at the subtalar joint but follow, nearly exactly, the same pattern during the gait

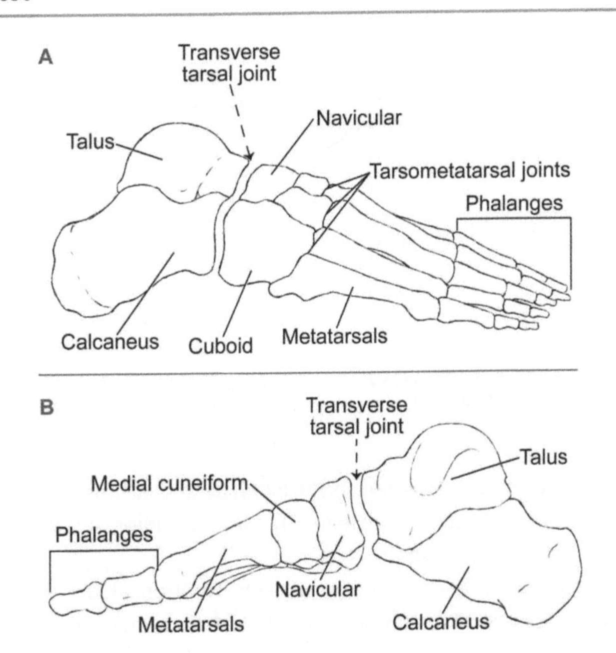


**Fig. 12.37** (**A**) Lateral and (**B**) medial views of the transverse tarsal joint complex

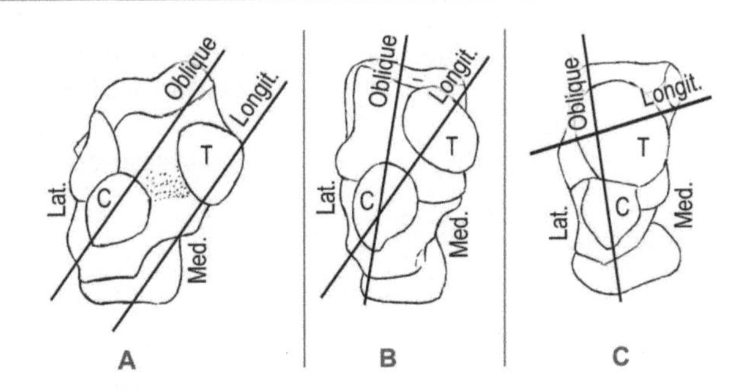


**Fig. 12.38** Schematic illustration of changes in the relationship between the longitudinal and oblique axes of the subtalar and transverse tarsal joints in the context of closed-chain, i.e., foot-fixed, (**A**) pronation – converged, (**B**) neutral – mildly diverged, and (**C**) supination – markedly diverged. Note that C = cuboid and T = talus (McPoil and Knecht 1985)

cycle. Thus, during the loading response and up to or slightly beyond midstance, the transverse tarsal joint everts, abducts, and dorsiflexes, i.e., pronates. This mechanism appears to place the hind- and midfoot complex in a relatively "open-packed" position, which theoretically could more handily dampen and transmit the large, impulsive ground forces effectively. And do not forget that these foot motions are coupled to tibial and femoral internal rotation, which are movements that are also part of the force distribution/dissipation system.

Recall, we earlier claim that end-range talocrural dorsiflexion was close packed, so the reader may be confused with the present comment that the combination of eversion, abduction, and dorsiflexion creates an open-packed position of the foot. We want to remind the reader that our previous claim was made in the context of isolated ankle dorsiflexion with the foot free, but the present comment is made in the context where the foot is relatively fixed. Moreover, during single-limb support, the talocrural joint moves from neutral dorsiflexion (IC) to plantar flexion (LR), back to neural dorsiflexion (midstance), and finally to approximately 10° dorsiflexion (which is not end range) at terminal stance, and these movements are coupled with subtalar and transverse tarsal joint movements. Our point is that biomechanists have agreed that this sequence of kinematic events under normal loading conditions optimizes the conditions for effectively managing ground forces. But this is not the end of the story because we have not yet examined the key role that passive and dynamic soft tissues play during the single-limb support phase of gait.

From the end of midstance to preswing, the subtalar and transverse tarsal joints supinate (i.e., invert, adduct, and plantar flex), a motion coupled with tibial and femoral external rotation. Generally, it is believed that this movement pattern places the hind- and midfoot complex in a relatively "close-packed" position; that is, it makes the foot a more rigid lever for muscles, which more effectively propels (accelerates) the body mass forward. As indicated above, the phrase "close packed" cannot be confused with how it is used when we describe specific kinematic poses of the isolated joint. As we move forward, it becomes clearer how a system of elements, i.e., joints, ligaments, and muscles, is coupled to optimize the system's most important function, i.e., gait.

Previous research has provided a possible explanation for one aspect of the "open-packed"/"close-packed" cycle during single-limb support, as just delineated. Recall that both subtalar and transverse tarsal joint movements have been described as rotations about an axis that has both longitudinal and oblique components. It has been suggested that when these joint complexes begin to pronate, the oblique and longitudinal axes of the talonavicular and calcaneocuboid joints converge, which appears to "unlock" the joints, allowing for greater freedom of movement, and/or perhaps put the joint structures, i.e., bones, joint contact areas, ligaments, etc., in optimal places for transferring ground forces (Fig. 12.38). Conversely, when, after midstance, the joint complexes begin to supinate, the respective foot axes may significantly diverge, creating a more constrained joint configuration (joint "locking"). Some have disputed the notion that the foot is "locked" in late stance by showing that as the hindfoot resupinates in terminal stance, the midfoot joints retain their mobility as they continue to move forward toward a position of extreme supination, and this occurs despite the fact that the relative joint complex axes diverge. We would argue that the reader avoid making dichotomous choices, i.e., joint axes

**Table 12.3** Structure and function of the more distal joints of the foot

| Joint | Key structural features | Kinematics |
|---|---|---|
| Distal intertarsal | | |
| Cuneonavicular | Stabilized by plantar and dorsal ligaments | Small gliding movements |
| Cuboideonavicular | Synarthrodial (fibrous) or sometimes synovial joints | Provides kinematic contact between lateral and medial longitudinal columns |
| Intercuneiform and cuneocuboid joint | Strengthened by plantar, dorsal, and interosseous ligaments; the intermediate cuneiform is the keystone of the transverse arch | Small gliding movements that allow the transverse arch to depress slightly under the superincumbent body weight |
| Tarsometatarsal (Lisfranc's joints) | The first tarsometatarsal has a joint capsule, all reinforced by plantar, dorsal, and interosseous ligaments | Variable mobility: greatest in the first, fourth, and fifth joints; least in the second and third joints |
| Intermetatarsal | Lateral four metatarsal bases are connected by plantar, dorsal, and interosseous ligaments; the first and second metatarsal bases do not form a true joint | Small gliding movements enhance tarsometatarsal movements; the absence of first-ray constraint increases its mobility |
| Metatarsophalangeal (MTP) | Stabilized by joint capsule and collateral and deep transverse metatarsal ligaments; a plantar plate reinforces the palmar surface and houses flexor tendons | Greater sagittal than frontal plane mobility; great toe MTP extension contributes to the toe (fourth) rocker in preswing |
| Interphalangeal (IP) | Stabilized by joint capsules, collateral ligaments, and plantar plates | Active motions limited to the sagittal plane |

convergence results in an "unlocking" of joint, etc., and be mindful that the methods used to examine these very complex movements are variable. Obviously, more research is needed, yet this should not prevent us from taking available information and balancing what is presently known to develop the most likely biomechanical and clinical hypotheses.

Let us continue our examination of the foot structure by reviewing the anatomy and biomechanics (related to walking) of the more distal joints (see Fig. 12.21). For this section, we rely on a table summary, highlighting only the most relevant items that relate to this case (Table 12.3). The distal intertarsal joints add to the range of pronation and supination across the midfoot and serve as a link to the forefoot while simultaneously enhancing overall stability across the foot as the keystone for the transverse arch. Movements at the cuboideonavicular and cuneonavicular joints are particularly crucial for the advancement of midfoot mobility; for example, the cuneonavicular joint has been shown to have as much as 5° to 10° of rotation in the transverse, frontal, and sagittal planes. The intercuneiform and cuneocuboid joint complex includes three articulations: two between the set of three cuneiforms and one between the lateral cuneiform and cuboid. It is this complex that comprises the bony contribution to the transverse arch of the foot. We examine the soft tissue elements of the transverse arch shortly.

The tarsometatarsal, or Lisfranc's, joints mark the boundary between the midfoot and forefoot. It is clear that the base of the first ray begins at the medial cuneiform, that the intermediate cuneiform marks the base of the second ray, and that the third ray begins with the lateral cuneiform. The bases of the fourth and fifth metatarsals articulate with the cuboid; thus, both the fourth and fifth rays claim the cuboid. Of all the tarsometatarsal joints, mobility at the second tarsometatarsal joint is most limited secondary to its apparent wedged position between the medial and lateral cuneiforms and stiff, constraining Lisfranc ligament. The less mobile second and third rays provide relative rigidity to the longitudinal column, which is particularly important during preswing when the ankle plantar flexes during push-off. The greater mobility of the first and lateral two rays allows the foot to accommodate uneven surfaces more easily, despite the fact that frontal and transverse plane motions are more limited in scope.

The first ray appears to have a special function from the loading response to midstance as the subtalar and transverse tarsal joints pronate. As the midfoot is slightly depressed by the superincumbent body weight, the first ray dorsiflexes about 5° to 10°; that is, as the height of the medial longitudinal arch is gradually reduced, the first column dorsiflexion helps transfer the stress of ground forces onto the foot. Conversely, during preswing, as the relatively "stiffer" foot is plantar flexed, the first tarsometatarsal joint also plantar flexes (approximately 10°), abducts, and everts; eversion of the first ray appears to stabilize the metatarsal in the sagittal plane. This mechanism is likely driven by concentric action of the fibularis (peroneus) longus, which appears to "shorten" the medial longitudinal column and raise the medial longitudinal arch, thereby increasing the stability of the mid- and forefoot when propulsion increases the tensile loading of the plantar soft tissues of the foot.

Bony, capsular, and ligamentous structural elements of the metatarsophalangeal (MTP) joints are very similar to those found in the hand. These joints are constrained by collateral ligaments and plantar plates, which are grooved to house the long digital flexor tendons. The plantar plates are particularly important in the toes. For example, during the toe rocker, i.e., when the great toe extends, the plates are pulled distally, which appears to protect the distal aspect of the articular surfaces. The deep plantar fascia provides additional support to the plantar plates and flexor tendon sheaths. The two sesamoid bones located within the tendon of the flexor hallucis brevis are constrained against the plantar plate of the first metatarsophalangeal joint, providing additional protection to the metatarsal head and increasing the angle of application of muscles to the great toe. At a deeper level, the transverse metatarsal ligament blends with the plantar plates of each metatarsophalangeal joint, adding to the stability of the anterior arch.

Although the MTP joints are reasonably mobile in the sagittal and frontal planes, abduction/adduction movements are restricted and are more accessory than voluntary. Kinematically, the convex/concave rule is applicable when describing arthrokinematic movements at the MTP joints. For example, during preswing when the first MTP joint extends the concave proximal phalangeal base rolls and glides dorsally relative to the convex metatarsal head. In mechanics, the instant center of rotation (also referred to as instantaneous velocity center, instantaneous center, or instant center) for a body undergoing planar (i.e., two-dimensional) movement is defined as the point that has zero velocity at a particular instant of time. In other words, the instant center is the point about which the body can be considered to be rotating about at that instant. Previous mechanical analyses of the first MTP joint using instant center analysis of the sagittal plane movements revealed the instant center clustered near the geometrical center of the metatarsal head (Fig. 12.39). Despite the close clustering of the instant centers of motion, some spacing demonstrates that gliding occurs throughout most of the range of motion, except at extreme extension (or dorsiflexion). Research has demonstrated that the first MTP joint compression contact area shifts dorsally with extension, with the joint reaction force peaking at full extension. The

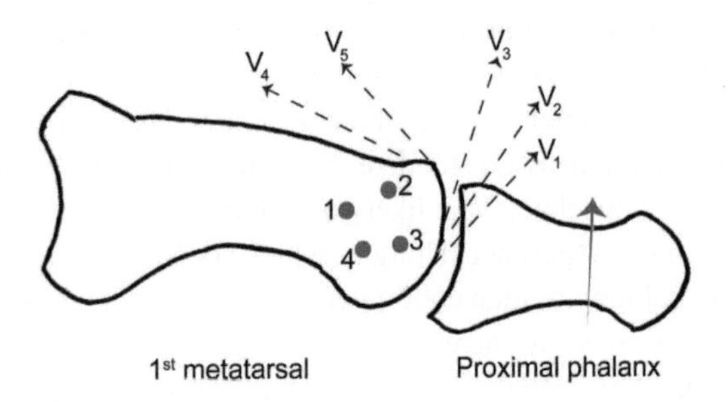


**Fig. 12.39** Instant centers of rotation at the first metatarsophalangeal joint in the sagittal plane with the proximal phalanx moving relative to the head of the metatarsal. Although the instant centers (points 1, 2, etc.) are relatively closely clustered, it is clear that gliding occurs throughout most of the range of motion, as indicated by the surface velocity vectors ($V_1$, $V_2$, etc.). (Modified with permission from Shereff et al. (1986))

joint forces at this joint have been shown to range from 80% to 100% of body weight during gait. During walking, passive first MTP extension to approximately 65° to 70° is critical for normal transition from stance to swing. The great toe provides stability to the medial aspect of the foot through the windlass mechanism of the plantar aponeurosis action of the fibularis (peroneus) longus. Two painful deformities of the first MTP, hallux rigidus (or limitus) and hallux valgus (with bunion), commonly disrupt the normal extension needed during walking.

In summary, the joints of the forefoot, like those of the hind- and midfoot, play vital roles during single-limb support. The distal joints compensate with a supinatory twist of the forefoot, as a unit, relative to hind- and midfoot pronation during the first half of stance, when weight-bearing and ground forces are introduced. Conversely, during the transition to late stance, as the hind- and midfoot supinate, the distal joints evert slightly in an effort to maintain contact with the ground as long as possible until toe-off. This response includes joint and soft tissue actions that raise the medial longitudinal arch and plantar flex the first ray, actions that add to the relative stiffness of the entire foot.

Now that we have acquired an extensive background into the structure and function of the foot, let us take a break and review selected physical examination findings that may offer clues to explain Ted's repeated stress fractures. Ted demonstrated mild tibial varum and a relatively rigid talipescavovarus foot with forefoot adductus, bilaterally. It was hypothesized that these deformities created nearly untenable options for the compensations or adaptations (i.e., hind- and midfoot pronation) his feet needed during the loading response of a running gait; it has been shown that the vertical ground reaction forces during running exceed 2.5 times the body weight, yet the total force magnitude within the foot is likely much greater secondary to muscle actions. Ted's inability to pronate sufficiently during loading likely exacerbated compression and shear stresses on all of the tarsal and metatarsal bones, as well as increased the tensile stresses on periarticular structures, predisposing him to significant injury. Fortunately, it appears that joint capsules, ligaments, and important fascial structures were not strained. Moreover, although Ted's cavovarus feet were too stiff to accommodate the large forces during the loading response, they were sufficiently stable to promote an efficient push-off and propulsion.

Two additional foot deformities were identified in both feet as part of the non-weight-bearing foot examination: (1) plantar flexed first ray and (2) Morton's toe. A plantar flexed first ray is not a primary deformity but a common compensation for forefoot varus. It is an attempt to provide a level forefoot plantar surface for single-limb support, which, if accomplished, would better distribute large ground compression forces. A plantar flexed first ray could potentiate a problem if it was rigid or fixed. In that case, a rigid first ray could not dorsiflex during the loading response and may sustain an injury, e.g., fracture. Physical examination revealed that Ted's plantar flexed first ray was mobile/flexible; that is, with manual manipulation, it could be dorsiflexed beyond neutral. Because of this mobility, it seems likely that the normal ground forces applied to the forefoot during running could easily dorsiflex the first ray in the stance phase without imposing damaging stresses. On the other hand, the abnormal position of the first ray in combination with the changes Ted made in training schedule/intensity, i.e., more training with less recovery time, and his running shoes may have been factors leading to a stress fracture.

Morton's toe is a condition where the second toe appears longer than the first. In most cases, this deformity has a genetic origin and results from the "dislodging" or "unwedging" of the second metatarsal base from between the medial and lateral cuneiform bones. The functional consequence of this decoupling is a change in the dynamics of forefoot mechanics: the second ray becomes more mobile, which may contribute to excessive and prolonged pronation of the foot during single-limb support, and/or it may shift the weight-bearing burden from the first to the second ray during late stance and preswing. We hypothesized that any increase in the mobility of the second ray secondary to Morton's deformity may have actually been an asset for Ted's otherwise very stiff foot, yet in this case, a second metatarsal stress fracture was sustained.

### 12.5.3 Plantar Vault, Longitudinal and Transverse Arches

To this point, we have reviewed the normal and abnormal bony structures and related joint kinematics in an effort to build a foundation that can help explain Ted's history. We learned that the ankle/foot mechanical system is complex, with normal function depending on the coupling of many interrelated elements. Now, we add to the picture as we define the mechanical arches of the foot and explore their relationship to the bone, soft tissue, and gait. Kapandji presents a somewhat elaborate overview of the concepts of vaults and arches, as they might be applied to the foot architecture, and puts forward a model of the plantar vault. A vault is characterized by supports at three points, which lie at the corners of an equilateral triangle (Fig. 12.40). Between each of the adjacent supports (AB, BC, and CA), there is an arch that constitutes the sides of the vault. The weight of the vault is applied at the keystone (arrow in Fig. 12.40B) and distributed by two buttresses to the support points B1 and B2. Because a vault structure is based on static mechanics, it was suggested that a foot structure, i.e., a combination of bone and soft tissues, might be readily compared to a truss or beam system. A classical truss model has a triangular structure that consists of two rafters (or struts), joined at the apex and kept together at its base by a tie beam. The struts are held in place by compression forces, and the tie rod is placed under tension (or traction) forces, which prevent the collapse of the triangle under load (Fig. 12.41). Thus, we can imagine that for the foot, there would be a single-axial truss with the main tie beam formed by the plantar ligaments and plantar muscles (extrinsic and intrinsic) as well as two secondary lateral tie beams corresponding to the medial and lateral longitudinal arches. For example, the superincumbent body

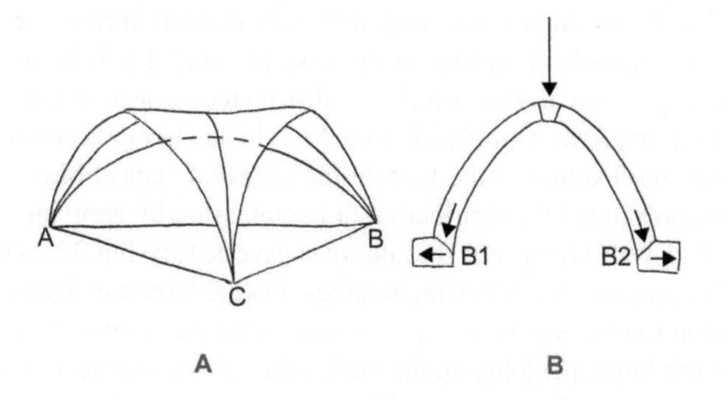


**Fig. 12.40** (**A**) Classic vault with three supports that constitute the sides of the vault and (**B**) the keystone (arrow) of the vault supported by buttresses B1 and B2 (Kapandji 2019)

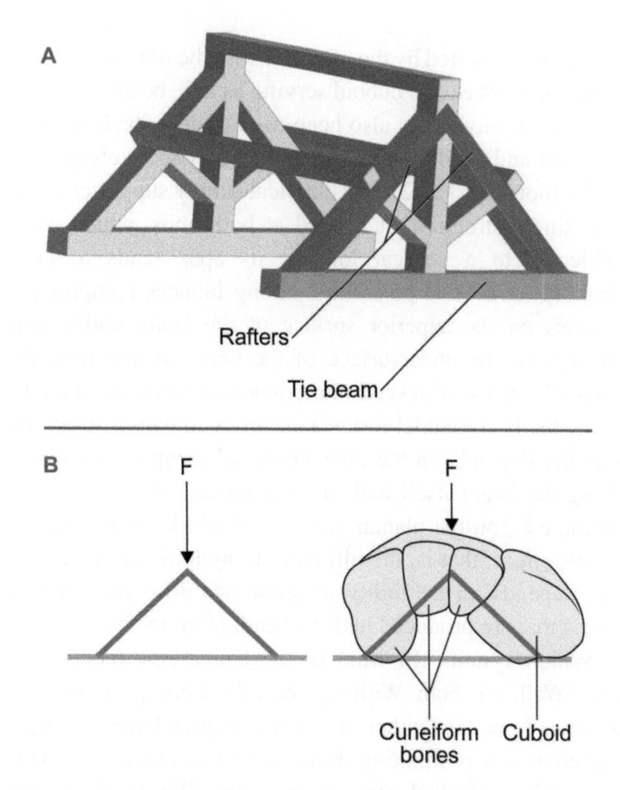


**Fig. 12.41** (**A**) Classic king post truss and (**B**) truss-like structure at the midtarsal joint, where the body weight (F) is supported by the link (beam) between the cuneiforms and the cuboid

A

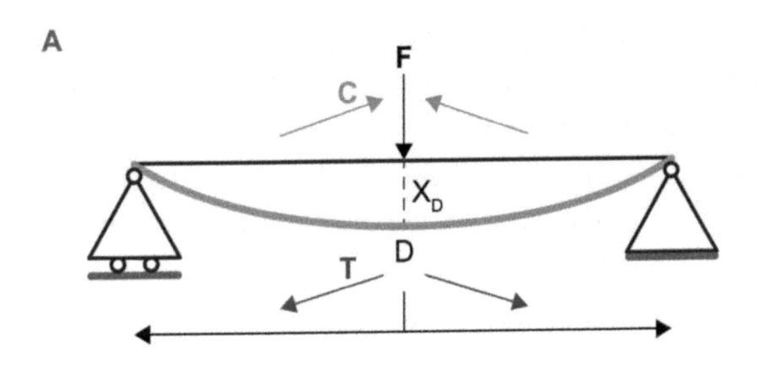


B

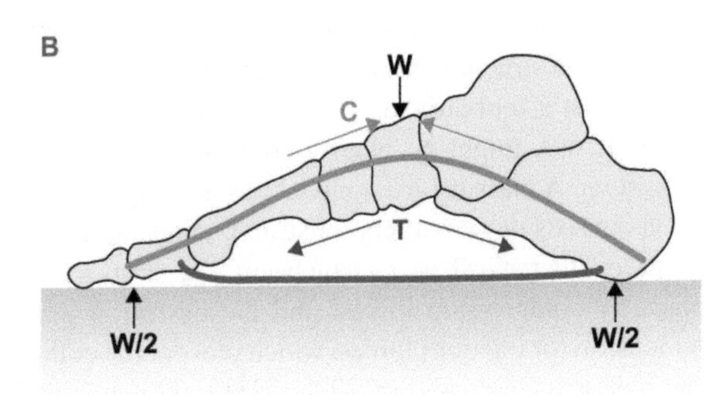


**Fig. 12.42** (**A**) Model of a simply supported beam that bends under force ***F***, deflecting (*D*) distance $X_D$, and (**B**) schematic of the foot illustrating the beam as connected foot bones and the longitudinal arch of the foot. The arch of the curved beam is supported by interconnecting joints and supporting soft tissues. The superincumbent body weight (***W***) induces compression (***C***) forces dorsally and tensile (***T***) forces along the planar surface of the foot

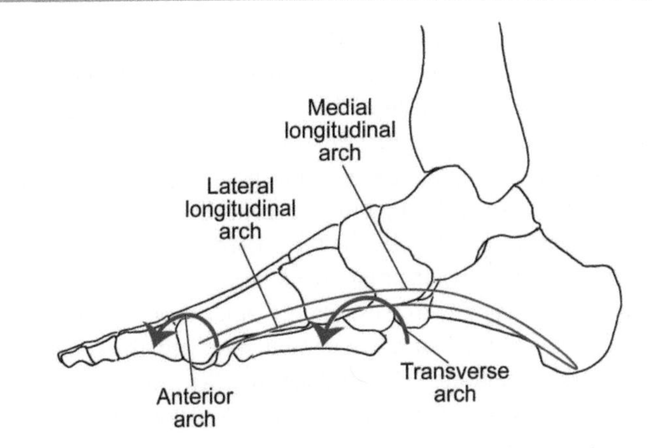


**Fig. 12.43** Plantar vault illustrating the general location of the anterior and transverse arches as well as the medial and lateral longitudinal arches. The mechanical coupling of these arches is critical to the maintenance of foot stability and the optimal distribution of ground reaction forces

weight is supported by the midfoot, with the link between the cuneiform bones and cuboid serving as a tie beam.

A beam model has also been used to describe how compression and tensile loads are managed by the arch systems of the foot (Fig. 12.42). The classical simply supported beam is a single structure, supported at both ends, which when subjected to a vertical load at its apex tends to bend (Fig. 12.42A). The potential bending induces compression stresses on the superior surface of the beam and tensile stresses on the undersurface of the beam. In the foot, the "beam" consists of several interconnected segments at joints. Thus, the beam model forms a curved arch whose structural stability depends on the intersegmental compression forces along the beam itself and the soft tissues, e.g., ligaments, fascia, etc., on the plantar surface, which develops internal tensile stress; that is, the ultimate strength of the beam system depends on its ability to resist two different types of forces that are produced under a bending moment.

Which system, the truss or the beam, best models the foot? Well, it is both. With the analysis of complex systems, we must reject an "either-or" choice approach dictated by a reductive way of thinking. Remember that models are typically only simplified versions of reality. We use them, but only with the idea that they are incomplete or contingent renditions of complex systems. Yet they are useful because when we combine the best insights from multiple models, we progress in our thinking/understanding. Therefore, we can think of the foot as a vault with multiple arches: anterior, transverse, medial longitudinal, and lateral longitudinal, which we can explain/understand using both truss and beam theories (Fig. 12.43).

The anterior arch is the shortest and lowest and spans the supports provided by the heads of the first and fifth metatarsals. A second transverse arch is created by the three cuneiforms and the cuboid and rests on the ground only at the cuboid, while the third transverse arch is subtended by the navicular and cuboid. The lateral longitudinal arch has intermediate length and height and stretches from the midpoint between the medial and lateral processes of the calcaneal tuberosity and the head of the fifth metatarsal. Finally, normally the longest and highest, and perhaps the most important functionally, the medial longitudinal arch stretches from the calcaneal tuberosity to the head of the first metatarsal.

The medial longitudinal arch is comprised of the following bones: first metatarsal, medial cuneiform, navicular, talus, and calcaneus; that is, these interconnected bones comprise the "beam" or, in a truss model, serve as the struts. The beam theory suggests that gravitational forces, i.e., superincumbent weight, in the case of bipedal gait induce compression forces, which are distributed in a near-linear fashion proximal to distal in the foot. In an attempt to simulate the loading of the foot during unipedal stance, Manter demonstrated the distribution of compression by applying a 60-lb thrust force over the talus of cadaveric feet (Fig. 12.44). He provides the following findings: (1) the largest compression forces were located proximally where the joint surfaces were large; (2) joints in the medial part of the foot received more stress than those positioned laterally, e.g., the talonavicular joint carried twice as much compression as the calcaneocuboid joint; (3) among the three joints of the

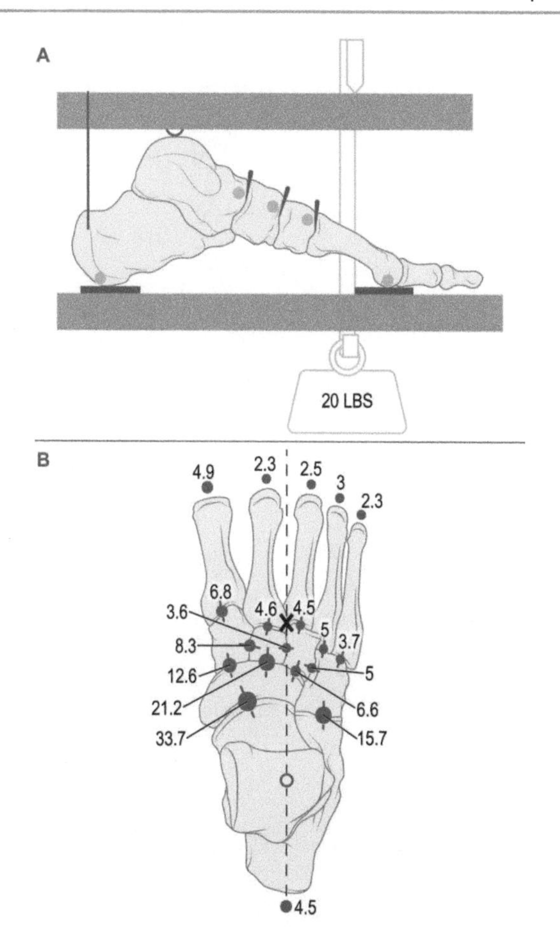


**Fig. 12.44** (**A**) Experimental setup where a 20-pound weight was suspended from a loading bar that rested on the upper surface of the talus, which was fastened to the calcaneus. Small balls and lead plates were inserted into joint spaces and at points of support under the foot to register compression forces. This apparatus induced a thrust of 60 pounds on the talus along a line between the second and third toes. (**B**) Schematic of the disarticulated foot showing the general pattern of compression distribution at joints and reactions of support under the foot. Numbers represent a force in pounds; the open circle is the point of application; X is the point of suspension of load; the dashed line indicates the position of the loading axis; approximate directions of joint compression forces are shown by lines through circles. (Modified with permission from Manter (1946))

cuneonavicular group, the second usually received the major share of compression; (4) the amount of compression in different zones appeared to be related to the intensity of bending movements in different parts of the foot; and (5) transverse compressions were registered between the cuboid and three cuneiforms, suggesting that these bones approximated, i.e., were pushed together closer, and there was a tendency for the transverse arch to flatten under the bending influence of the load. Although he did not report direct measures of stress within supporting soft tissues, he suggested that the compressions in the foot joints were directly related to supporting pressures. Manter did not use the beam or truss theory, but his work is often cited as foundational for our understanding of the important coupling of interconnected foot structures, ligaments, and foot architecture. As an aside, if we were to revisit Fig. 12.10, it would not be difficult to use the results of Manter's early work on the distribution of compression forces in the foot to provide a rationale for the distribution of trabecular struts in the ankle and foot.

Of course, the medial longitudinal arch is maintained not only by the stoutness of the talonavicular joint and other bony interconnections. Both beam and truss theories, as applied to the foot, dictate that the integrity of the passive and active soft tissues on the plantar side of the foot is also an essential element for the transmission of large impulsive ground forces and overall foot stability. In addition to joint capsules, other passive structures include the cuneometatarsal, naviculocuneiform, interosseous talocalcaneal, long and short plantar, and, especially, calcaneonavicular (spring) ligaments, as well as the plantar fascia. Most agree that the plantar fascia provides the primary support for the medial longitudinal arch. This structure has remarkable tensile strength, almost twice the strength of the foot's deltoid ligament.

The plantar fascia is made up of superficial and deep layers (or aponeurosis) (Fig. 12.45A). The deeper layer, being both thicker and stiffer, is extensive in its coverage of the plantar surface of the foot, with connections that extend from the calcaneal tuberosity to its distal attachments, which include the first layer of foot intrinsic muscles, the plantar plates of the metatarsophalangeal joints, and the fibrous sheaths of the flexor tendons of the digits and other fascial connections to the plantar aspects of the toes. The plantar fascia is also densely populated with nerve endings and mechanoreceptors capable of providing pain and proprioceptive information. Based on this anatomy, intuition suggests, then, that extension of the toes induces increased tension on the central fibers of the deep fascia, as well as the medial longitudinal arch. In fact, it has been shown that the stiffness of the deep plantar fascia provides primary passive support to the medial longitudinal arch during quiet stance and particularly during the action of the fourth rocker in preswing (Fig. 12.45B). When one rises from sitting to standing, the superincumbent weight that projects down onto the foot depresses the beam/truss of the foot, increasing the resting tension on the deep plantar fascia. This tissue deforms, within its elastic range, slightly as evidenced by a small reduction in the height of the medial longitudinal arch. This deformation is accompanied by minimal pronation of the hind- and midfoot. During the walking cycle, the height of the medial longitudinal arch and tension in the plantar fascia constantly change. During the swing phase, when the ankle is in neutral dorsiflexion with slight subtalar inversion, the medial longitudinal arch is also in its neutral position, with minimal tension on the plantar fascia. Immediately upon loading (LR), as the ankle plantar flexes and the subtalar and midtarsal joints pronate, the plantar fascia experiences

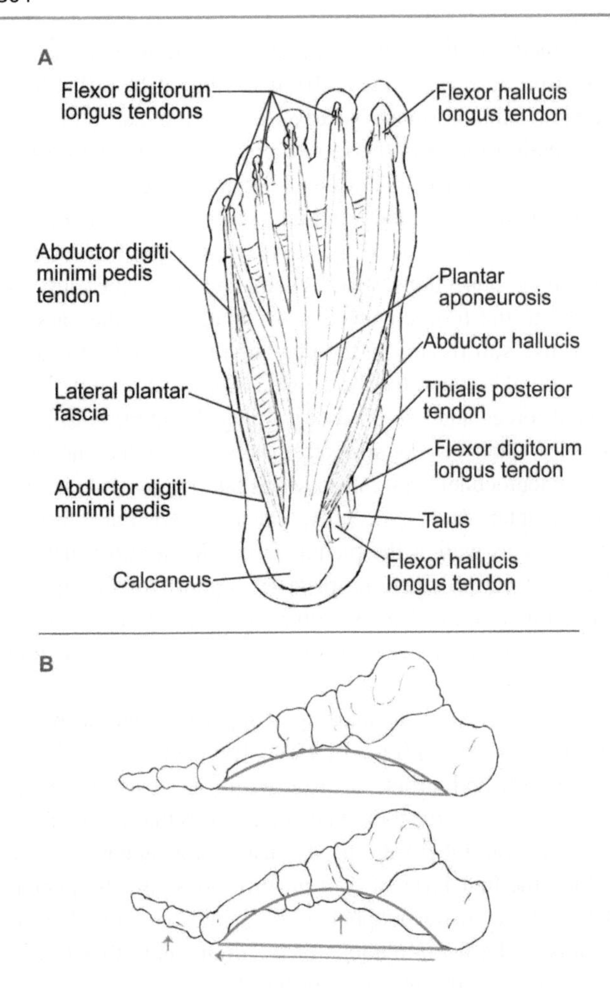


**Fig. 12.45** (**A**) Superficial and deep plantar fascia and relationship to selected intrinsic and extrinsic plantar muscles. (**B**) Schematic of the effect of the support of the medial aspect of the plantar vault by the plantar fascia in a normal bipedal stance (top) and the enhancement of the medial arch (windlass effect) when the hallux hyperextends as part of the fourth rocker during push-off (bottom)

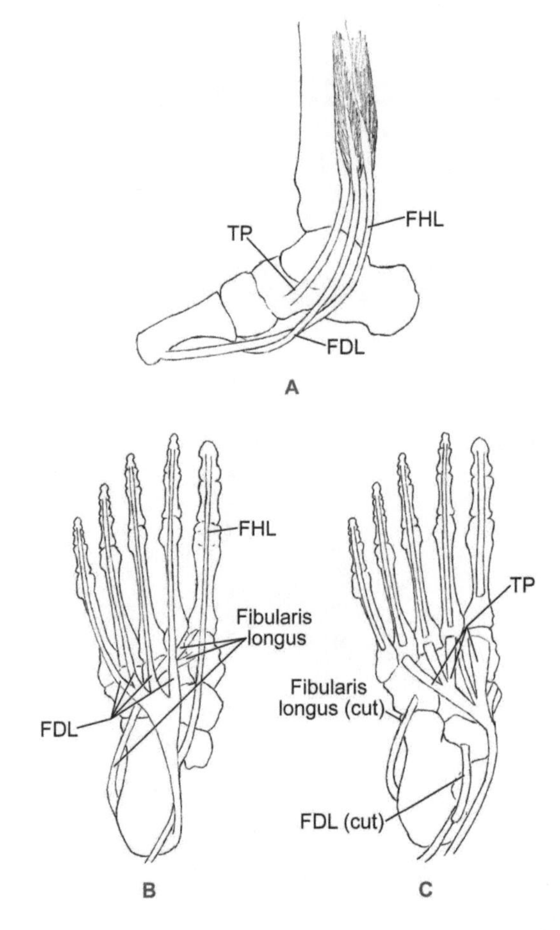


**Fig. 12.46** (**A**) Fixed pulley systems for the tibialis posterior (TP) and flexor digitorum longus (FDL, medial malleolus) and flexor hallucis longus (FHL, tubercles of talus and sustentaculum talus at medial calcaneus). (**B** and **C**) Plantar view of the TP, FDL, FHL, and fibularis longus (FL) tendons

impulsive tensile stress, deforming it, resulting in the lowering of the medial longitudinal arch as the foot achieves a foot-flat position. From the end of LR to approximately the midstance, as the hind- and midfoot remain pronated, the medial longitudinal arch remains at a lower height; however, the compensatory "forefoot twist," i.e., slight supination of the forefoot, which helps keep a plantigrade foot, maintains significant plantar fascia internal tensile stresses. At some optimal point in the transition from mid- to terminal stance, the hind- and midfoot joints supinate, and concomitant reactionary pronation twist of the forefoot begins to increase tension in the plantar fascia. This increased tension is exacerbated in preswing when the ankle plantar flexes and the toes extend just prior to toe-off. The significantly increased tensioning of the plantar fascia secondary to hyperextension of the hallux has been referred to as the "windlass effect" because of the apparent shortening of the distance between the metatarsal joints and the calcaneus, the elevation of the medial longitudinal arch, and the augmentation of hindfoot supination.

Augmentation of the control of large ground forces is dynamic and provided by intrinsic and extrinsic muscles. The tibialis posterior may be the most important, given its many insertions on the plantar aspect of the foot, i.e., navicular tuberosity, all cuneiforms, cuboid, metatarsal bases II–IV, and fascial connections to plantar ligaments (Fig. 12.46). This muscle begins its eccentric action during LR to control hind- and midfoot pronation, then continues activity, with the foot firmly plantigrade, by assisting the gastroc-soleus complex in controlling the advancement of the tibia, from mid- to terminal stance. Elastic loading of the parallel and series elastic elements of these posterior tibial muscles creates potential energy, which is subsequently converted to kinetic energy during their shortening actions in preswing. Thus, at the same time the tibialis posterior assists in propulsion, it stabilizes the medial and lateral longitudinal arches.

The flexor hallucis longus and flexor digitorum longus work together in supporting the medial arch. They do this by stabilizing the talus and calcaneus as they course that region on their way to their distal insertions. The tibialis posterior, flexor hallucis longus, and flexor digitorum longus share a unique route to their final destinations, but all provide major extrinsic support to the medial column (Fig. 12.46).

The fibularis longus is interesting in that it is considered a lateral compartment muscle and primarily acts to evert the subtalar joint and plantar flex the ankle. However, because of the course of its long tendon and insertion into the medial cuneiform and the base of the first metatarsal, this muscle accentuates the medial longitudinal arch by flexing the first metatarsal relative to the medial cuneiform and the medial cuneiform relative to the navicular; because of the path followed by its tendon, it also supports the lateral column and transverse arch (Fig. 12.47).

Key intrinsic muscle support of the medial column is provided by the abductor hallucis, by virtue of its expanse along the arch. When it shortens, it increases the curvature of the medial arch by bringing the distal and proximal ends closer (Fig. 12.48).

The lateral arch is more analogous to a beam than a truss system because its primary constituents include bones: calcaneus, cuboid, and fourth and fifth metatarsals. This column is not suspended, unlike its medial counterpart, and contacts the ground only through soft tissues, i.e., superficial and deep layers of the long plantar ligaments. The lateral arch is more rigid than the medial arch secondary to the congruity of its bony connections and the stiffness of the plantar ligaments. Its keystone is the anterior calcaneal process, which is buttressed proximally at the lateral process of the calcaneal tuberosity and distally by the fifth metatarsal head (Fig. 12.49). The three muscles that provide dynamic stability by compressing the joint connections include the fibularis brevis and longus (Fig. 12.47) and the abductor digiti minimi (Fig. 12.48).

We should briefly examine the trabecular patterns of the distal shank and bones making up the lateral column, as we did for the medial column (Fig. 12.50). There are two trabecular systems: (1) the posterior trabeculae arise from the anterior distal tibia, fanning inferoposteriorly below the shared posterior facet of the talus and calcaneus, and (2) the anterior trabeculae arise from the posterior distal tibia, spanning the head of the talus, where it meets the anterior calcaneal process, the cuboid, and fifth metatarsal. Additionally, the calcaneus holds two trabecular systems, superior and inferior arcuate. Concaved inferiorly, the superior system converges to the floor of the sinus tarsi in response primarily to compressive stresses. The inferior system is shown as

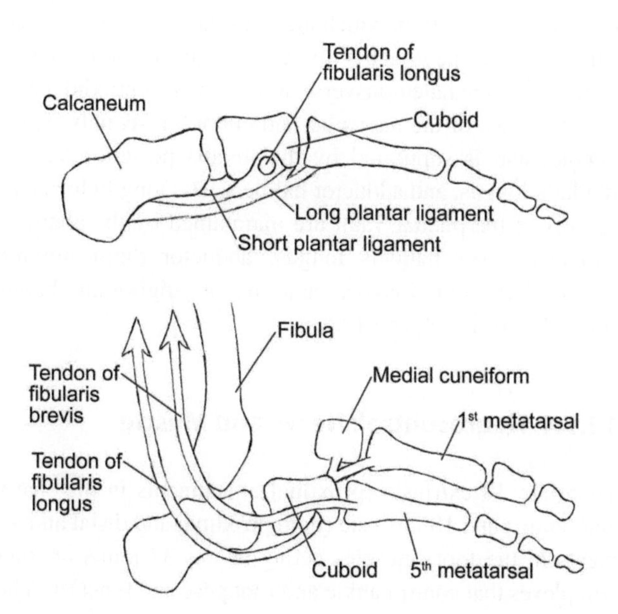


**Fig. 12.47** Two alternative lateral views illustrate the coursing of the tendons of the fibularis longus and brevis and their relationship to the long and short plantar ligaments

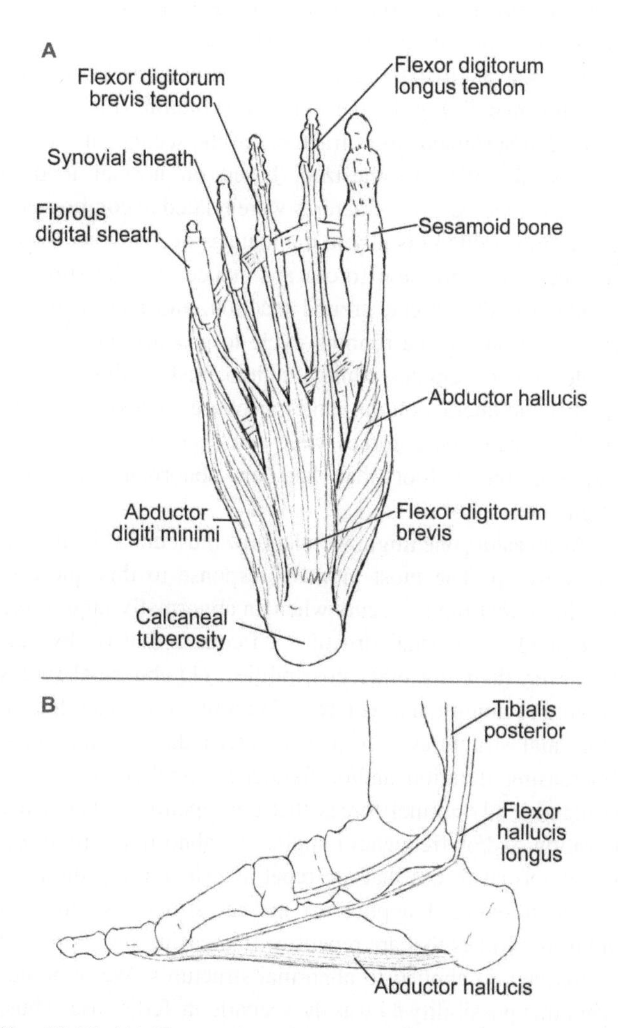


**Fig. 12.48** (**A**) Plantar view of the abductor hallucis in relation to selected superficial foot intrinsic muscles, i.e., the abductor digiti minimi and flexor digitorum brevis. (**B**) Medial view of the abductor hallucis muscle relative to the tibialis posterior and flexor hallucis longus tendons

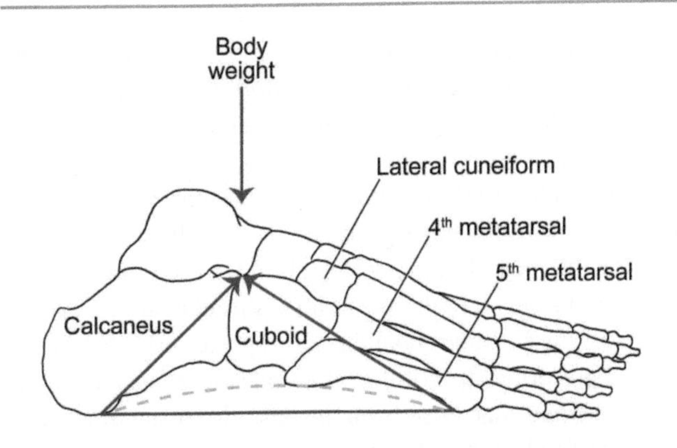


**Fig. 12.49** The lateral column/arch is comprised of the calcaneus, cuboid, and fourth and fifth metatarsals. The keystone of the arch is located where the most superodistal point of the calcaneus and the superoproximal point of the cuboid meet. The beam system is denoted by the triangle, whose apex meets at the keystone, and the base is formed by the long plantar ligaments (blue dashed line)

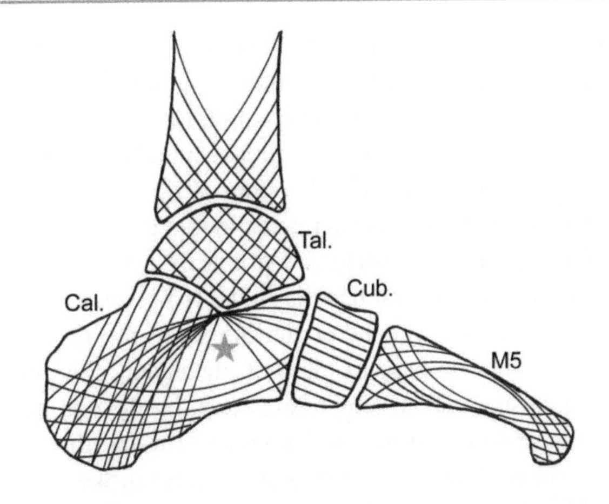


**Fig. 12.50** Lateral view of trabecular struts associated with the transmission of mechanical stresses within the lateral column. In this figure, we only see the fifth metatarsal but need to imagine the location of its medial partner, the fourth metatarsal, and its own trabeculae. A point of relative weakness, secondary to the absence of overlapping of trabecular struts, is marked by a star

concave superiorly, converging toward the proximal calcaneus, likely in response to tensile stresses.

It is relevant to explore the trabecular patterns, along with both the medial (Fig. 12.10) and lateral columns (Fig. 12.50), since Ted sustained stress fractures of the second and fourth metatarsals. We hypothesized, based on normal loading patterns, that the stress fractures were related to compression forces transmitted distally to the metatarsals. Now we know that these compression forces are induced by the superincumbent body weight, ground reaction, and muscle forces (activated during the running cycle to produce and control ankle/foot movements while simultaneously stabilizing the medial and lateral columns and transverse arches). Although in Ted's case these forces were not excessive, they were imposed on a foot that had an abnormal cavovarus alignment.

As an aside, one might wonder how musculoskeletal injuries develop. The most intuitive response to this question would be that injury occurs when an abnormally large force is applied to a normal structure. Of course, that is obvious. However, there are other possibilities: (1) abnormal forces applied to abnormal structures, (2) normal forces applied to abnormal structures, (3) normal forces that are repetitive (increasing duration and/or frequency) applied to normal structures, (4) normal forces that are repetitive (increasing duration and/or frequency) applied to abnormal structures, (5) abnormal forces that are repetitive (increasing duration and/or frequency) applied to normal structures, and (6) abnormal forces that are repetitive (increasing duration and/or frequency) applied to abnormal structures. We hypothesized that possibility #4 was the scenario in Ted's case. Thus, it was probably inevitable that the combination of changes in training (frequency, duration, and intensity); abnormal, somewhat rigid, lower extremity alignment (particularly in the foot); and normally large, impulsive ground forces associated with running would result in injury somewhere. In Ted's case, the "somewhere" was his metatarsals.

Let us conclude our study of the foot arch complex by examining the transverse arches. The anterior arch is formed by the metatarsal heads I through V, with the second metatarsal as the keystone. This arch has a low ceiling, i.e., curvature height, and is supported passively by the plantar metatarsal ligaments and dynamically by the adductor hallucis. The next more proximal arch consists of four bones, the three cuneiforms, and the cuboid. Normally, the medial cuneiform does not contact the ground, and the middle cuneiform forms the keystone; the middle cuneiform also forms the ridgeline of a longitudinal arch, which includes the second metatarsal. The tendon of the fibularis longus provides dynamic support for this intermediate transverse arch. The most proximal arch is at the level of the navicular and cuboid, rests only on the cuboid, and is supported by the tibialis posterior tendon, fibularis longus, and adductor hallucis. The longitudinal curvatures of the plantar vault are maintained by the abductor hallucis, flexor hallucis longus, abductor digiti minimi, flexor digitorum longus, and flexor digitorum brevis (Figs. 12.46, 12.48, and 12.51).

### 12.5.4 Neurocontrol: Nerve and Muscle

There are 12 extrinsic (proximal attachments in the shank and femur) and 18 intrinsic (both proximal and distal attachments in the foot) muscles acting across 32 joints or joint complexes that control ankle and foot pose and function. The extrinsic muscles are generally stronger, with a primary function related to the production and control of locomotion activities. These muscles can be organized into three compartments of the leg: anterior, lateral, and posterior. The intrinsic muscles generally provide static and dynamic stability of the foot arches during standing and locomotion activities, e.g., propulsion. There is a different motor nerve that innervates muscles within each compartment (Fig. 12.52). Each motor nerve is a branch of the sciatic nerve (sacral plexus, roots L4 to S3). The deep branch of the fibular nerve innervates muscles in the anterior compartment, the superficial branch innervates the fibularis longus and brevis (lateral compartment), and the tibial nerve, and its terminal branches, innervate extrinsic muscles in the posterior compartment and all of the intrinsic muscles (via medial and lateral plantar nerves). See the Supplementary 12.1 for a view of the extrinsic and intrinsic muscles.

The extrinsic muscles of the shank cross multiple joints and are therefore able to act across several planes of movement (Fig. 12.53). We are particularly interested in how they are activated during normal walking and running gait to ensure an energy-efficient transfer of muscle force that can accelerate and decelerate the body along a line of progression (Fig. 12.54). The four muscles of the anterior compartment

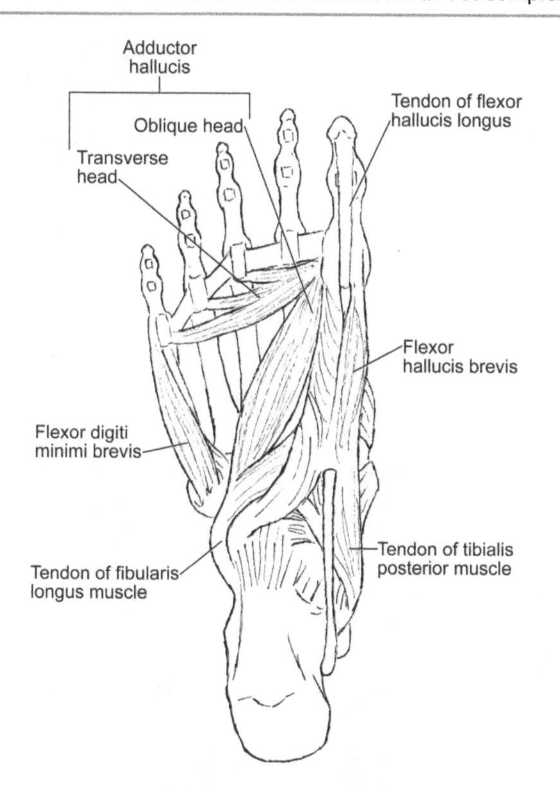


**Fig. 12.51** Plantar view of deep foot intrinsic muscles. Note the adductor hallucis and its potential to support the anterior arch

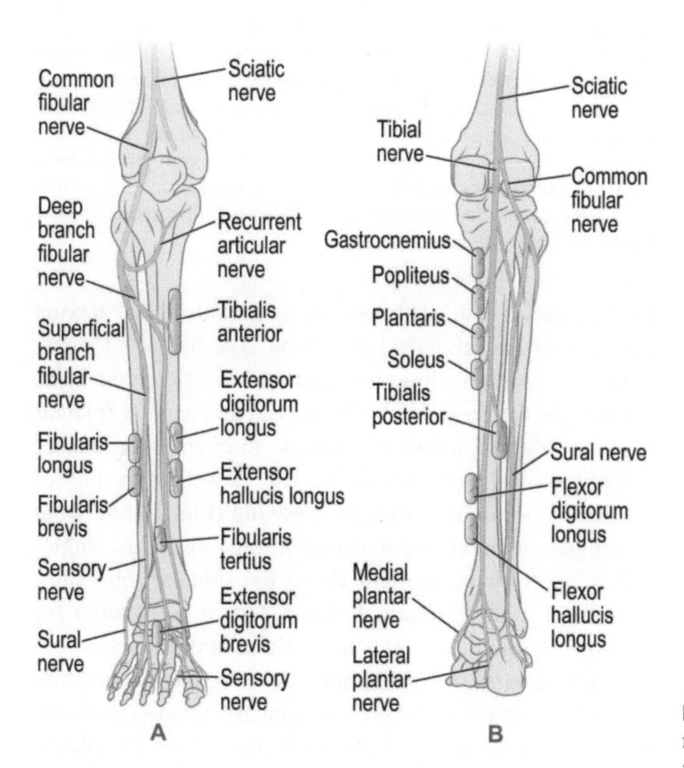


**Fig. 12.52** Proximal-to-distal path of muscle innervation for the (**A**) deep and superficial branches of the common fibular (peroneal) nerve and the (**B**) tibial nerve and its branches

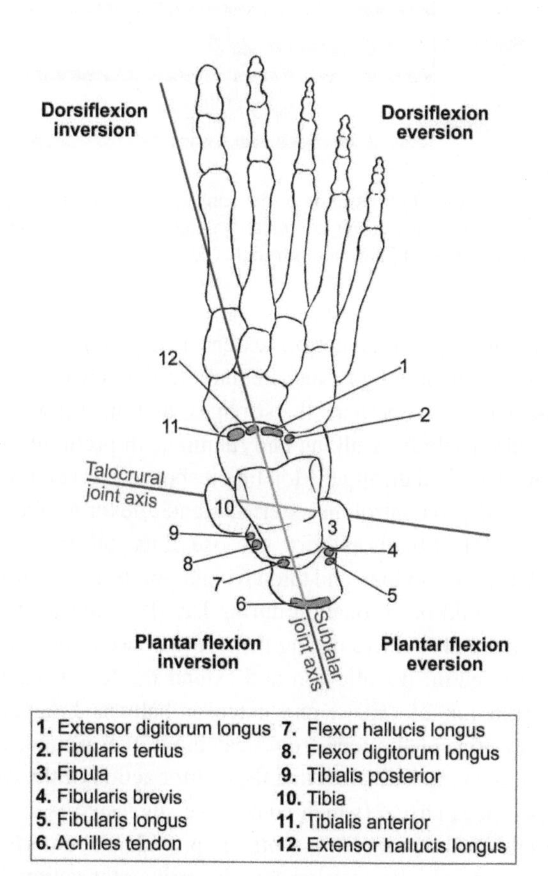


**Fig. 12.53** Note the imaginary axial line of actions for each extrinsic muscle relative to the talocrural and subtalar joint axes. It is clear that all of these muscles are capable of multiple actions. Note that the oblique mediolateral axis of the talocrural joint promotes primarily sagittal plane movements, while the oblique posteroanterior axis promotes primarily frontal plane movements

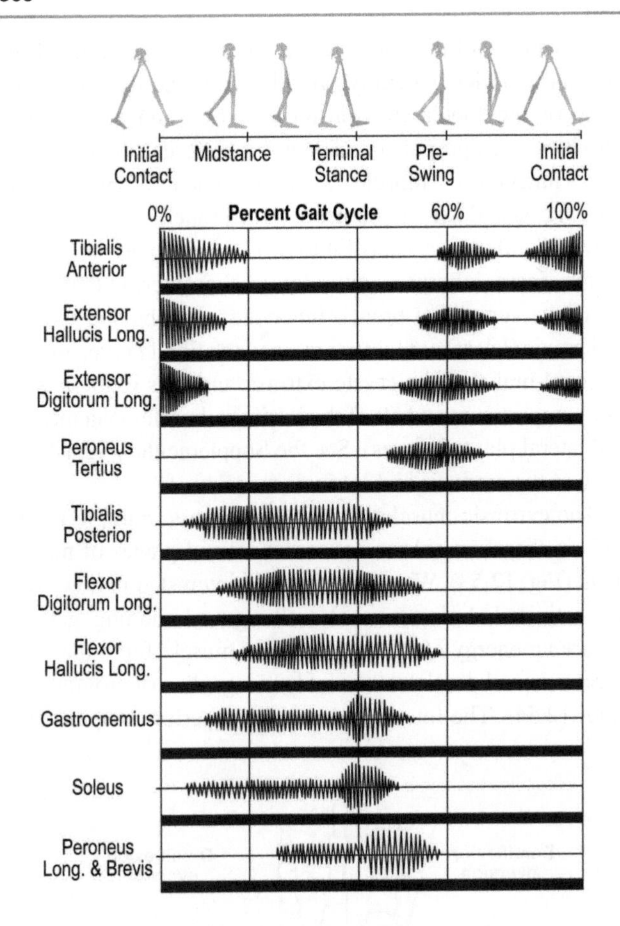


**Fig. 12.54** Onset and cessation of the activation (electromyography) of ankle/foot muscles during a typical walking gait cycle. (Modified with permission from Perry and Burnfield (2010))

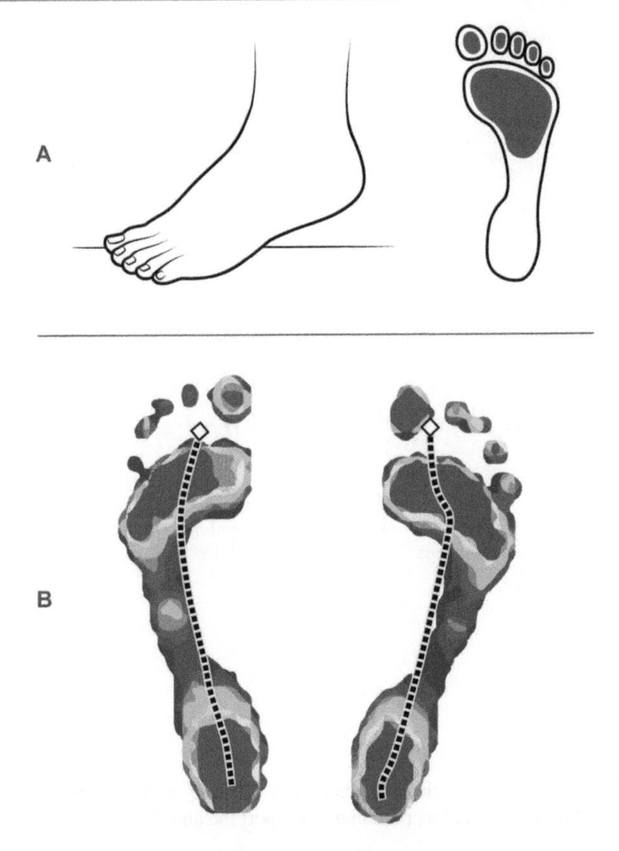


**Fig. 12.55** (**A**) Normal distribution of plantar pressures at the terminal stance and (**B**) averaged plantar pressure map distribution of normal plantar pressures from initial contact to toe-off during self-paced walking. (Used with permission from the Motion Analysis Laboratory at Mary Free Bed Rehabilitation Hospital, Grand Rapids, MI). The dashed line from the heel to the great toe represents the center of pressure path. The center of pressure is the path of the vertical ground reaction force during the stance phase of gait. Note: pressure = force/area; areas of low pressure are designated in various shades of blue, with progressively greater pressures represented by green, yellow, orange, and red (area of highest pressure). (See Appendix J for additional information on how plantar pressures are measured)

include the tibialis anterior, extensor digitorum longus, extensor hallucis longus, and fibularis tertius. Of these, the tibialis anterior is perhaps the strongest and most important pretibial muscle for walking and running. All pretibial muscles are activated during the loading response (LR) and work eccentrically to control the external plantar flexor moment at the ankle. The tibialis anterior, because of its ability to invert the talocrural, subtalar, and talonavicular joints, also controls hind- and midfoot pronation during LR. The anterior tibial muscles are also active during the swing phase to return the ankle to neutral dorsiflexion and extend the toes (extensor digitorum longus (EDL) and extensor hallucis longus) in order to prevent the toes from "catching" the ground, i.e., preventing a trip/fall. Note that the evertor action of the EDL and peroneus tertius (which some have suggested is simply the EDL's fifth tendon) is essential to provide a counterforce to the strong tibialis anterior dorsiflexor/invertor action during the swing phase of walking.

The fibularis longus and brevis are the only two muscles in the lateral compartment. These muscles are strong evertors of the foot and provide lateral static and dynamic stability at the talocrural joint, lateral column, and transverse arch. These muscles also contribute to talocrural plantar flexion and subtalar and midtarsal abduction. The fibularis longus' tendinous insertion allows it to create an eversion moment as distal as the forefoot. Its ability to evert and depress (plantar flex) the first ray allows this muscle to counter the medial pull of the tibialis anterior during LR, as well as the pull of the tibialis posterior muscle in preswing (Fig. 12.54). Both the longus and brevis are activated during most of the single-limb support phase, particularly as the subtalar joint supinates from mid- to terminal stance and in preparation for push-off. Beginning in midstance, these muscles act eccentrically to control the extent and rate of subtalar supination and fix the first ray to the ground (fibularis longus) to ensure the proper distribution of ground forces across the plantigrade foot (Fig. 12.55). Actions of both the fibularis longus and brevis are part (along with the gastroc-soleus and tibialis posterior) of the stretch-shortening cycle from the mid- through terminal stance, which prepares the ankle for push-off. During the same period, the ability of the fibularis muscles to create an evertor moment allows them to balance the strong invertor moment created by the tibialis posterior and gastroc-soleus complex. The coaction of these antagonists assures stabilization of the longitudinal and transverse arches in late stance and creates a relatively neutral talocrural joint going into the initial swing phase.

Plantar pressure measures (see Appendix J for a review of foot pressure measuring systems) are frequently used to study individuals with foot deformities because they provide indices of how the large ground forces during walking are distributed throughout the foot. Fig. 12.55B illustrates the typical pressure distribution for a normally aligned foot. Research has shown that many of the plantar pressure indices are significantly different in persons with a cavus foot, including (1) reduction in plantar surface weight-bearing area, (2) greater peak pressures and pressure-time integral (impulse) under the heel and lateral forefoot, and (3) reduced pressure-time integral, maximum force, force-time integral (a measure of the cumulative exposure to force over a time period), and contact area under the midfoot and hallux (Fig. 12.56). Plantar pressures were not measured as part of Ted's gait test. However, given what we knew about the deformities in his shank (mild tibial varum) and foot (talipes-cavovarus, with mild forefoot adductus and plantar flexed first ray) and what research has shown, we hypothesize that his plantar pressure distribution would be abnormal and is similar to the illustration in Fig. 12.56B. Moreover, intuition suggests that these abnormal pressure distributions likely contributed to Ted's stress fractures.

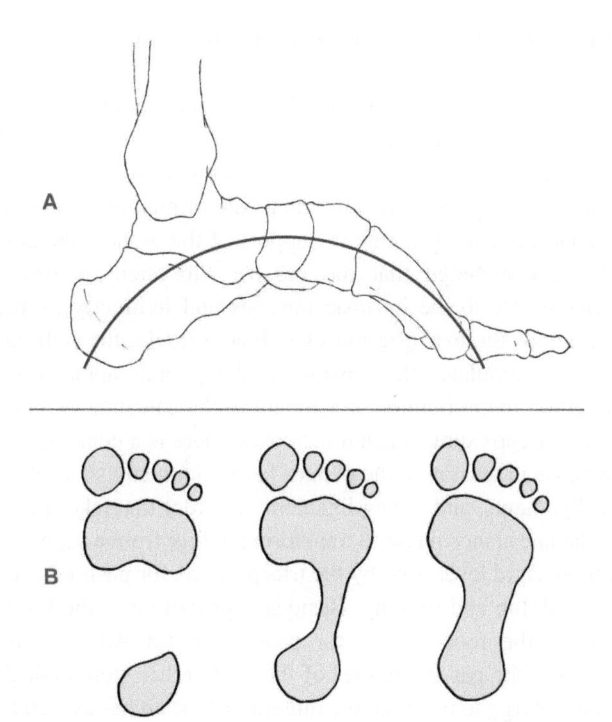


**Fig. 12.56** (**A**) Schematic of a cavus (high medial longitudinal arch) foot and (**B**) qualitative comparison of plantar pressure distribution in a marked cavus foot (far left), moderate cavus foot (middle), and normal foot (far right)

The posterior compartment muscles are divided into superficial (gastrocnemius, soleus, and plantaris) and deep (tibialis posterior, flexor digitorum longus, and flexor hallucis longus) groups. The gastrocnemius and soleus (often referred to as triceps surae) merge to form the Achilles tendon, which inserts into the posteromedial aspect of the calcaneus. These two muscles have a very large combined cross-sectional area and a relatively large moment arm and make the largest contribution to the preswing plantar flexor moment at the talocrural joint (approximately 80% of the total), easily dwarfing the combined moment potential of the other ankle plantar flexors. The tibialis posterior, flexor digitorum longus, and flexor hallucis longus lie deep to the soleus. Since each of their tendons course posterior to the medial malleolus, we can see that they invert (supinate), as well as plantar flex, the talocrural and subtalar joints (Figs. 12.46 and 12.53). Along with being the most effective supinator at the subtalar and transverse tarsal joints, the numerous insertions (every tarsal bone except the talus) of the tibialis posterior adduct the hind- and midfoot and create a supination "twist" that effectively "close packs" the midfoot during terminal stance.

Except for the plantaris, all of the plantar flexor/supinator muscles are active from the end of the loading response through preswing (Fig. 12.54). During the period from foot flat to terminal stance, these muscles, most notably the soleus, act eccentrically to control the external dorsiflexor moment (i.e., movement of the tibia over the relatively fixed talus). At the same time, the parallel and series elastic elements, e.g., Achilles tendon, etc., of the muscles are being loaded under tension, which creates potential elastic energy (this is the first part of a stretch-shortening cycle). At heel-off, the second phase of the stretch-shortening cycle is activated by the conversion of potential to kinetic energy within the passive tissues of the muscles by plantar flexor concentric action, which is used for push-off. During push-off, activity in the flexor hallucis longus, flexor digitorum longus, and foot intrinsic muscles holds the extending toes against the ground, which increases the surface area of contact between the foot and ground, minimizing contact pressures.

Although the deep posterior compartment muscles are all plantar flexors, their activation during the loading response of gait is used to control the subtalar and transverse tarsal joint pronation that occurs during that time period; that is, these posterior tibial muscles act eccentrically to manage the lowering of the medial longitudinal arch that accompanies subtalar pronation (Fig. 12.54). At the same time, these eccentric actions may better transfer the large impulsive ground forces. Based on the natural history of inflammatory conditions of distal anterior and posterior tibial tendons, i.e., sometimes referred to as shin splints, which, more often than not, involve the tibialis posterior tendon, biomechanists have suggested that the tibialis posterior takes the brunt of the load during the early part of the stance phase. As noted above, since the tibialis posterior is a prime invertor/plantar flexor, it is likewise involved in the stretch-shortening cycle, which takes place from foot flat to preswing; however, in the case of the tibialis posterior, its prime function, beginning around midstance, is to supinate the hind- and midfoot to ready the ankle/foot for push-off.

We need to pause here to make sure that the reader is clear about how we interpret the electromyographic (EMG) data illustrated in Fig. 12.54. Note that these were not data collected during Ted's gait study but were used to represent a control data set. In clinical gait analysis, the biomechanist is interested in the neurocontrol of muscle during the gait cycle, but usually only in the onset and cessation of the EMG signal (see EMG section in Appendix J). At times, the biomechanist attempts to quantify the EMG signal in an attempt to get an idea of motor unit recruitment. However, even in that instance, the interpreting biomechanist is still mostly interested in the timing of muscle actions. Finally, interpretation of the raw EMG signal similar to Fig. 12.54 does not allow us to make any conclusions about the strength (or quantification of muscle force or tension) of the muscles of interest or the type of muscle action, i.e., concentric or eccentric. We can make judgments about the type of muscle action only if we examine the joint kinematics, net joint power, and EMG simultaneously. If you recall in the hip case, we briefly discuss the types of muscle actions by looking at joint kinematics.

Although in this case Ted demonstrated hypertonus in several lower extremity muscles, he had normal selectivity and strength during resisted muscle testing. Furthermore, there was nothing in his history that suggested that he had sustained any muscle strains. Thus, it was assumed that Ted's intrinsic muscles were normal. However, for completeness, we need to briefly discuss a model that suggests that the intrinsic muscles provide the core stability for the foot by being the local stabilizers of the arches of the foot (Fig. 12.57 and Table 12.4). This model, based on Panjabi's work related to a stabilizing system of the spine, is rooted in the interdependence of the passive, active, and neural subsystems of the ankle/foot complex. The passive subsystem consists of bony and periarticular structures and provides for a balance between mobility and stability, whereas the active subsystem is comprised of muscle-tendon actuators and consists of two functional elements: local stabilizers and global movers.

As noted, the passive subsystem of the foot core consists of the bones, ligaments, and joint capsules, which maintain the longitudinal and transverse arches of the foot (Fig. 12.42).

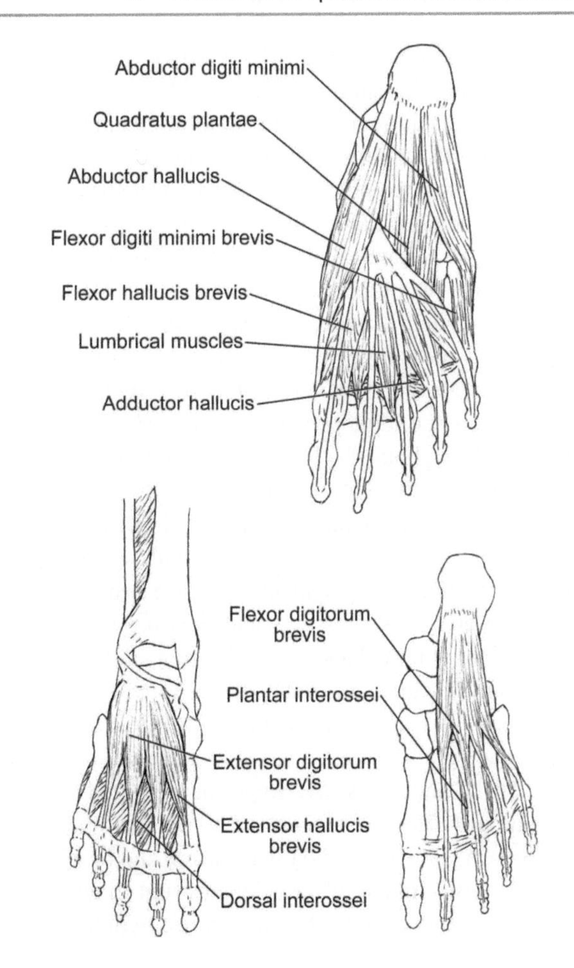


**Fig. 12.57** Dorsal and plantar views of foot intrinsic muscles

Kapandji's discussion of the plantar vault includes each of the arches and their adaptation to changes in the magnitude and location of ground and superincumbent weight forces. Earlier we highlight the important role of plantar aponeurosis and plantar ligaments in support of the arches. We now also acknowledge that the passive subsystem is linked dynamically to the intrinsic muscles and indirectly by the actions of the extrinsic muscles. For example, the Achilles tendon modulates the tension of the plantar aponeurosis based on their common attachment to the calcaneus, so that as the triceps surae tension increases, there is a concomitant increase in tension in the plantar fascia, long and short plantar ligaments, and spring ligament, an event that takes place in the late stance phase to transform the foot from a supple to a more rigid lever used by the triceps surae for push-off during walking and running. Besides contributing to the leverage of the foot needed for propulsion for walking and running, the passive tissues of the foot arches store enough strain energy to maintain the integrity of the arches as well as contribute to the increased energy efficiency noted in running (we will come back to this concept).

As part of the active subsystem, the plantar foot intrinsic muscles are considered the most important local stabilizers

**Table 12.4** Intrinsic muscle of the foot

| Muscle | Location | Actions |
|---|---|---|
| Extensor digitorum brevis | Dorsum of foot | Toe extension |
| Flexor digitorum brevis | Layer 1 | Flexion of the PIP and MTP joints of lesser toes |
| Abductor hallucis | Layer 1 | Abduction and flexion (assists) of the first MTP joint |
| Abductor digiti minimi | Layer 1 | Abduction and flexion (assists) of the fifth MTP joint |
| Quadratus plantae | Layer 2 | Medial stabilization of the common tendons of the FDL |
| Lumbricals | Layer 2 | Flexion of the MTP joints and extension of the IP joints of the lesser toes |
| Adductor hallucis | Layer 3 | Adduction and flexion (assists) of the first MTP joint |
| Flexor hallucis brevis | Layer 3 | Flexion of the first MTP joint |
| Flexor digiti minimi | Layer 3 | Flexion of the fifth MTP joint |
| Plantar interossei (three) | Layer 4 | Adduction of the MTP joints of the third, fourth, and fifth digits |
| Dorsal interossei (four) | Layer 4 | Abduction of the MTP joints of the second, third, and fourth digits |

Modified with permission from Neumann (2017)
*PIP* proximal interphalangeal, *MTP* metatarsophalangeal

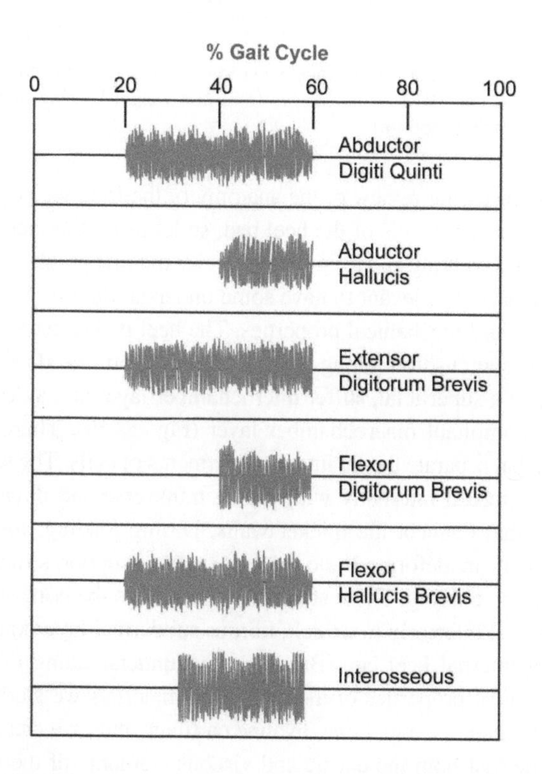


**Fig. 12.58** Intrinsic foot muscle electromyographic activity of a normal healthy individual while walking at a self-paced speed on a level surface. Note that this EMG profile represents sample control data, not data from Ted's gait study. (Modified with permission from Mann and Inman (1964))

because they originate and insert within the foot and because they are dynamically linked/coupled to the longitudinal and transverse arches. The two most superficial layers of foot intrinsic muscles appear to be aligned with the longitudinal arches, whereas the deep layers have their fibers, more or less, aligned with the proximal and distal transverse arches. Finally, although the moment-generating potential of the foot intrinsic muscles is small, these muscles appear to play a role in the fine adaptations of the feet, which are needed, depending on the terrain, particularly during the later phases of single-limb support during walking (Fig. 12.58). We do not know precisely the role of the foot intrinsic muscles, but intuition suggests that they play a similar role.

The neural subsystem consists of sensory receptors in the joint capsules, ligaments, muscles, and tendons involved with the passive and active subsystems. In general, these receptors monitor foot movements and forces and send afferent signals to the central nervous system (CNS). If those afferents exceed a given threshold, efferent signals are sent from the CNS to the appropriate muscles, i.e., both local stabilizers and global movers, to alter joint positions. It is well established that plantar sensation is critical to gait and balance. Based on the location of the intrinsic muscles, it seems that the primary role of their sensory systems is to provide immediate information about changes in foot posture.

The reader is reminded that although the intrinsic muscles did not appear to present a problem for Ted, both concentric and eccentric actions of these muscles, as part of their routine stabilizing and mobilizing functions, create multidimensional stresses, e.g., compression, tension, torsion, etc., onto all of the bones and joints they cross. Since Ted ran cross-country, in addition to track, we can imagine that the surfaces he ran on throughout the year posed significant challenges to the intrinsic muscles in his feet. Then it is plausible that multidimensional stresses induced by intrinsic muscle actions contributed in some complex manner to the development of stress fractures, given the abnormal alignment of his lower extremity and feet.

## 12.6 Stress Fracture: Epidemiology and Mechanism

Stress fractures are relatively uncommon injuries accounting for approximately 1–7% of all athletic injuries. In high school athletes, stress fractures occur at a rate of 1.54 per 100,000 athletes, whereas the stress fracture rate in collegiate athletes is 5.7 per 100,000. There is a higher incidence of stress fractures among military recruits, runners, and those involved in jumping sports, although any activity with repetitive loading can lead to a stress fracture. Any bone can sustain a stress fracture, but the tibia, tarsals, and metatarsals are most frequently affected in athletes of all genders and ages.

The onset of stress fractures is associated with extrinsic and intrinsic factors. Extrinsic factors include the type of activity, training regimens, type of equipment and footwear, training surfaces and techniques, nutrition, and improper rest/recovery. Intrinsic factors have to do with the individual's anatomy and biology, most commonly including an abnormal cavus or planus foot, leg length discrepancy, excessive forefoot varus tarsal coalitions, prominent posterior calcaneal process, tight posterior tibial muscles (including the Achilles tendon), poor bone density or vascular supply, and abnormal hormonal levels. The biomechanical effects of muscle fatigue have also been linked to the development of stress-related injuries.

We can imagine that there are complex interactions between these extrinsic and intrinsic factors relative to the development of stress fractures, which creates a challenge for the health care provider attempting to ascertain an etiology or mechanism of injury. For running-related injuries in general terms, the ground impact profile (ground reaction force), as indicated by the vertical ground reaction force at initial foot contact and initial loading, appears to be critical. In particular, the impact peak (the maximum force reached during the early period of contact) and the loading rate (the rate of force increase) are metrics of interest. It has been reported that the loading rate is significantly increased in runners who develop stress fractures.

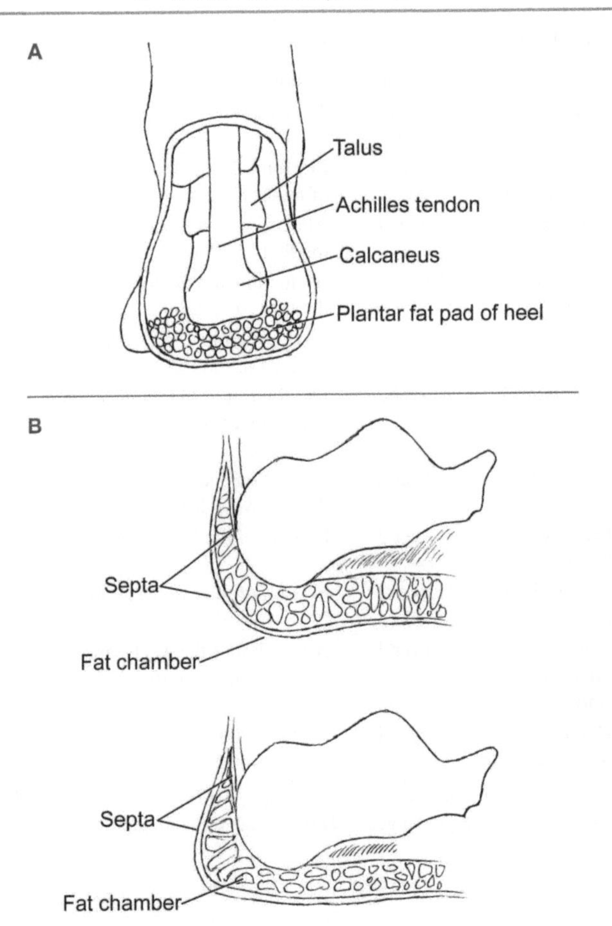


**Fig. 12.59** Schematic of the heel pad (**A**) relative to the bony anatomy of the foot and (**B**) how it changes from non-weight-bearing (top) to weight-bearing (bottom)

### 12.6.1 The Heel Pad

The transient peak at the beginning of the force curve during the loading response of walking and running depends on the deceleration of different body segments, i.e., the wobbling mass of soft tissue and the rigid mass of bones, as well as the deceleration of the bony segments, e.g., calcaneal (or heel) fat pad (hereafter referred to as the heel pad), in immediate contact with the ground surface. Thus, the impact profile can be influenced, i.e., attenuated, by several parameters, both active and passive. Active elements include muscle activity, and passive elements include bone, cartilage, and other soft tissues. The idea of impact attenuation implies the dissipation of the force over a longer period of time. Muscles can assist in this process through the anticipatory actions/adjustments of the central nervous system and by their eccentric actions when the feet actually contact the ground. The passive mechanisms include all elements in the interface between the skeleton and the ground, including the shoe sole, heel pad, bone, cartilage, etc., all of which work to attenuate the impact during the foot-ground contact. Each of these structures has different stiffnesses and energy returns, so the behavior of the interface is affected by how these many elements interact.

In our earlier review of the anatomy of the foot, we neglect to look at the details of the heel pad, so let us do that here. In non-shod walking and running, the heel is the first point of contact; thus, it is relevant to have some understanding of its anatomical and mechanical properties. The heel pad consists of a fatty tissue cushion, along with a structure comprised of two regions: a superficial, stiffer microchamber layer and a deeper more compliant macrochamber layer (Fig. 12.59). There are septa that separate the fat into compartments of cells. The septa are reinforced internally with elastic transverse and diagonal fibers that connect the thicker walls. During loading, the fat chambers are deformed, and the septa act as a support structure to prevent bulging. These structures are fixed to the bone superiorly and inferiorly to a thick, fibrous subdermal layer known as the internal heel cup. Based on our understanding of the mechanical properties of the biological materials we study in earlier chapters, e.g., bone, hyaline cartilage, etc., we can also assume that both the elastic and viscous elements of the heel pad are critical to its role in attenuating ground forces. It has been suggested that microchambers are more effective in resisting deformation than macrochambers.

*In vivo* and *in vitro*/*in situ* research has provided insights into the viscoelastic properties of the heel pad, but measuring the interaction between biology and footwear has been frustrated by the fact that the measurement tools, i.e., force platforms, can only provide information on external peak loads and loading rates. However, several findings may be important to our case here. For example, the active mechanism, i.e., muscle, has demonstrated its role in tuning the mechanical properties of the soft tissue; that is, muscle activity changes relative to shoe softness or hardness. A hard, compared to a soft, shoe induces small increases in impact peak, as well as loading rates, which appears to be related to changes in muscle activity, joint angles, and joint angular velocities. Moreover, increased compression and shear stresses were demonstrated in male and female simulated models running in hard shoes. These data suggest, then, that increases in the internal loading of bones and soft tissues can be expected when runners use harder shoes.

Ted did not provide information about his running shoes. Nor did he explain why he changed his running shoes and how the newer shoes were different. We know that shoe type, as an extrinsic factor in stress fracture risk, is important. Based on Ted's injury pattern, we hypothesize that his shoe change likely altered both the active and passive mechanisms relative to impulse force attenuation in complex ways that might have led to his stress fractures.

### 12.6.2 Risk of Stress Fractures, Common Sites of Fracture

Two stress fracture risk categories, low and high, are used to determine the relative risk of developing a stress reaction and, ultimately, a stress fracture. Low-risk stress fractures include those to the calcaneus, cuboid, cuneiforms, and lateral malleolus, whereas the risk of stress fractures is high for the navicular, talus, medial malleolus, metatarsals (primarily second, third, and fifth), and sesamoids. Fractures of the metatarsals are high risk, occur most frequently in the second and third metatarsals, and are common in runners, military recruits, ballet dancers, and basketball players; metatarsal stress fractures represent up to 20% of stress fractures in the athletic population. Reports of pain in the region of these bones are most often associated with increases in training.

Distal second metatarsal stress fractures are the most common. During walking and running, it has been demonstrated that the second metatarsal is exposed to the highest bending moment (compression + tension loading), as well as shear stress. This seems to be related to the fixed base and proximally hinged metatarsophalangeal joint that creates a bending moment at the proximal diaphysis during the stance phase of walking and running gait. It has been suggested that the relatively longer second metatarsal and excessively mobile first ray (Morton's foot) also contribute to increases in the reactive internal bone stresses.

Fractures of the fifth metatarsal are usually located at the diaphyseal-metaphyseal junction. Apparently, a repetitive adduction force with the ankle in plantar flexion induces increased tensile forces in the plantar fascia, which may be a factor in the development of a stress reaction in this bone, as well as ultimate fracture. In addition, it has been suggested that a varus hindfoot, cavus medial longitudinal arch, varus forefoot, and genu varum predispose individuals to fractures of the fifth metatarsal base. These deformities contribute to the inability to adequately pronate the hindfoot during early stance, thereby inhibiting the transfer of lateral foot forces to the medial side of the foot. The fourth metatarsal is susceptible to the same foot abnormalities and abnormal distribution of forces and perhaps more prone, secondary to its smaller cross-sectional area; that is, the redistribution and concentration of normal forces to the lateral column lead to increases in internal bone stress because its cross-sectional area is inherently less.

### 12.6.3 General Mechanisms of Injury: Stress Fracture

In Chap. 3, we conduct a thorough study of bone mechanics, which includes fracture and stress fracture (fatigue) mechanics. We do not repeat that here, but let us scratch your memory a bit. Bones are subjected to multidimensional forces related to the activities of daily living. The effects of those forces depend on the type of bone, i.e., cortical or trabecular, experiencing the stress. Wolff's law states that bones remodel in response to applied loads. External loads create microscopic changes that initiate a cascade of biochemical stimuli, which results in a remodeling process to alter the orientation of trabecular struts and the thickness of cortical bone. In general, cortical bone is most responsive to compression loads but more susceptible to bending moments, i.e., compression + tensile stress. Although the trabecular struts are generally organized along the line of application of external compression loads, they appear to be susceptible to compression forces. Fatigue-type stress fractures are typically seen in normal bones after excessive activity and, thus, could be placed into two force categories: tension or compression. Tension fatigue fractures are the more serious of the two because they are caused by a debonding of osteons, which induces transverse cracks, which can more quickly lead to a displaced fracture. With compression stress reactions, however, microfailure of bone is characterized by oblique cracks that isolate the cortical bone, leading to devascularization, which may initiate a fracture process. Compression fractures often appear more slowly, and thus may be more difficult to detect

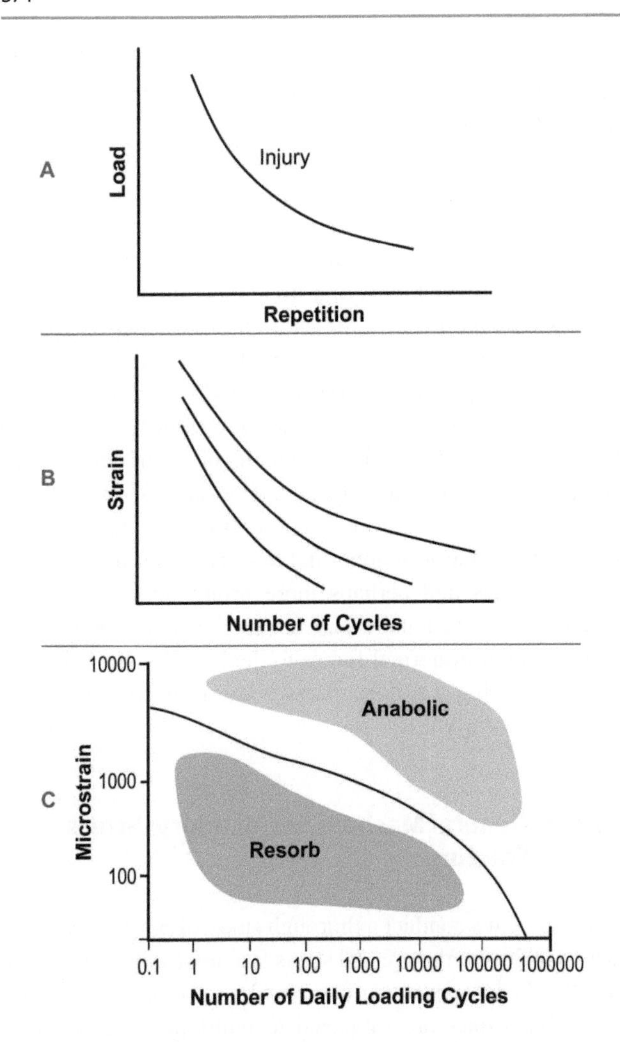


**Fig. 12.60** (**A**) Fatigue curve illustrating the relationship of injury to load and repetition. (**B**) Fatigue curve illustrating the coupling of the frequency of loading, i.e., an example of three loading cycles, and the coupling of strain and repetition on fatigue curves; the greater the strain (*y*-axis) and the greater the number of cycles (*x*-axis), the greater the bone deformation. (**C**) Illustration of the bone's nonlinear dependence on load intensity (microstrain) and frequency (cycle number) in the maintenance of bone mass and morphology. (Modified with permission from Seitz et al. (2022))

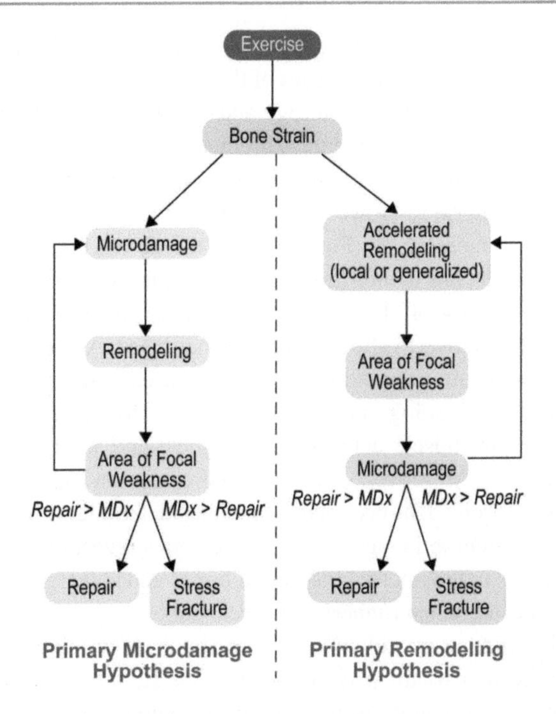


**Fig. 12.61** Representation of possible mechanisms for stress fracture development: primary microdamage (MDx) hypothesis and primary remodeling hypothesis. (Modified with permission from Bennell et al. (1999))

initially, and often heal on their own through normal remodeling processes (but only if the external loads are reduced in magnitude, applied less frequently, or more time is provided for recovery).

Fatigue fractures result from submaximal, repetitive loading, which induces an imbalance between bone formation and resorption. Athletes at the highest risk are those who abruptly increase the duration, intensity, and frequency of physical activity without adequate periods of rest. It appears that this vicious cycle of intense activity, inadequate recovery, and resumption of activity may cause increased osteoclastic activity, leading to an imbalance in bone resorption and bone formation/remodeling. The coupling of load and repetition for any material can be plotted on a fatigue curve (Fig. 12.60A). When bone is subjected to repetitive low loads, it may sustain microdamage. *In vitro* testing of bone has shown that microfractures typically develop rapidly when the load or deformation approaches its yield strength. Thus, fewer loading cycles are needed to produce a fatigue fracture if the load approaches the yield point of the bone with each repetition. The fatigue process is affected not only by the magnitude of the load and number of repetitions but also by the number of applications of the load within a given time (i.e., frequency of loading) (Fig. 12.60B). Fig. 12.60C illustrates the nonlinear relationship between load intensity and cycle number related to the maintenance (or not) of bone mass and morphology. It has been suggested that any point above the optimal ratio of load intensity to cycle number is anabolic (i.e., increased fracture risk), whereas any point below the curve triggers osteoclastic activity. In summary, it is clear that the coupling of load magnitude (normal and abnormal), cycles, and the number of daily loading cycles lead to bone fatigue and microfracture.

We understand that a stress fracture is a process and not an event and is the end product of bone fatigue as a result of repetitive loading. During this process, the microdamage of bone that results from repetitive loading accumulates as concomitant bone remodeling takes place. Both of these factors play an important part in the pathogenesis of stress fractures. Two models, not necessarily exclusive, have been offered to explain the development of stress fractures (Fig. 12.61).

Under the primary microdamage model, bone strain from repetitive loading initiates the development of microdamage (microcracks) at particular sites, i.e., weak links in the kinetic chain (or system). A remodeling response is initiated at the damaged site. Normal homeostatic mechanisms would usually create a balance between these two processes, eventually leading to the repair of the microdamage. However, a fatigue fracture occurs if the microdamage exceeds the capacity of the normal remodeling process, e.g., insufficient recovery time period, impairment of the local remodeling, or a combination of these factors.

The primary remodeling hypothesis suggests that accelerated bone remodeling is the initiating stimulus to stress fracture development. The impetus for accelerated bone remodeling could be genetic, nutritional, hormonal, exercise, or any combination of these factors. With this model, osteoclastic resorption precedes bone formation in the remodeling process, so that there is a lag time when the bone is in a weakened state. Microdamage then occurs in specific regions of weakness, perhaps resulting in a stress fracture (Fig. 12.60C). Although these two models highlight the general role that repetitive loading plays in microdamage to bone and the physiological processes of bone remodeling, they do not speculate on the specific nature of the forces and moments that might result in this microdamage.

Research has shown that bone is damaged from external forces and moments that create internal stresses as well as microstrains that exceed the bone's yield point. Interestingly, bone can fail in fatigue in as few as 1000–100,000 loading cycles at strain ranges less of than 1500 microstrains in tension and 2500 microstrains in compression. With muscle fatigue, strain magnitudes do not change, but strain rates appear to increase significantly. Thus, it has been suggested that excessive internal bone stresses and subsequent stress fractures occur because of muscle fatigue. Thus, fatigued muscle provides inadequate protection; i.e., they have a reduced stiffness in response to applied loads (Fig. 12.62).

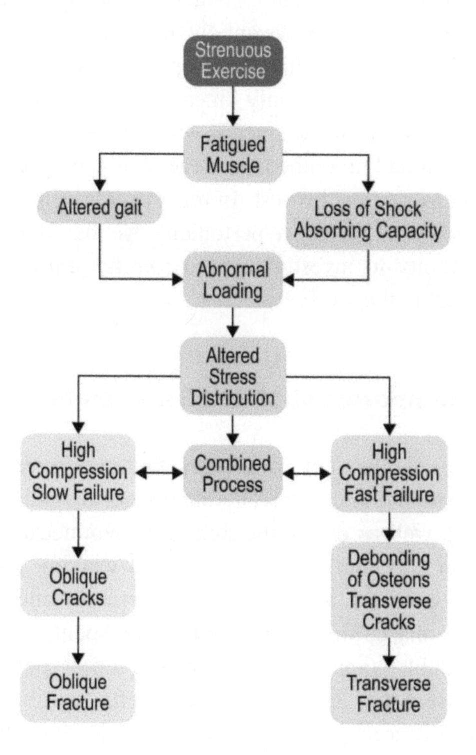


**Fig. 12.62** Overview of the relationship between muscle fatigue and stress fracture. (Modified with permission from Seitz et al. (2022))

Let us briefly examine the role that muscles may play in the mechanics of bone health. We can agree that the actions of muscles alter the stress distribution in bones. Of course, the primary role of the muscle-tendon actuator is to produce and/or control joint movement. At the same time, however, when they act across joints, muscle-tendon forces create large multidimensional forces on bones. For example, (1) tension forces are transmitted to the bone via their bony insertions; (2) muscle actions create large compression and shear stresses at the joint surface, which are transmitted to the subchondral bone; (3) during the acceleration phase of the throwing act, when the humerus is rapidly internally rotating, the humerus is subjected to large torsional stresses; and (4) during locomotion, bending moments (which induce external compression + tension forces) are applied to the femoral neck. It has been suggested that gluteus medius action produces compressive stress across the femoral neck, which neutralizes the potentially damaging tensile stress at the superior cortex of the neck, thus allowing the femoral neck to sustain higher loads than would otherwise be possible. The paradox here is that though the gluteus medius may create a "protective" compression force across the femoral neck, it induces large tensile stress at its insertion into the femur, i.e., the greater trochanter. This poses a conundrum: how can it be that compression forces produced by the gluteus medius are protective on the one hand when the large tensile forces it creates at its insertion are potentially injurious?

An alternative model suggests that as a molecular machine, muscles can protect joints and bones by absorbing mechanical impacts based on their ability to redistribute the energy of impact in time and space, e.g., the control of knee flexion during the loading response of gait through eccentric action of the quadriceps. For biological tissues in general, and for muscles in particular, the common mechanism for absorbing mechanical shock is achieved by their viscoelastic properties. The proponents of this hypothesis have suggested that most research on viscoelastic parameters as factors of muscle actions has focused on elastic properties, to the neglect of muscle viscosity. This hypothesis claims that since viscosity is a property that helps extend the impact time and absorb mechanical shock and is considered a damping factor preventing the muscle from generating damaging forces, it must be an important factor in protecting bones and joints. Research in support of this hypothesis claims that differences in muscle fiber viscosity mechanisms, i.e., active and passive, and differences in muscle responses i.e., slow or fast, to external mechanical loads, as well as agonist-antagonist muscle actions, provide some evidence for the importance of muscle viscosity in protecting the skeleton when absorbing large external forces.

There is no doubt that the viscoelastic properties of muscle, which include both passive (e.g., tendon) and active (e.g., muscle spindles) elements, are important for its function as a generator and controller of joint moments. However, its viscous nature does not change the fact that when muscles act, either as an agonist or as an agonist-antagonist pair, they produce large reactive (bone-on-bone) forces (primarily compression and shear) across joint surfaces and/or to the bones where their attachments lie. We remind the reader that we demonstrate this repeatedly through the solutions to statics problems, e.g., shoulder, elbow, hip, etc. Thus, for example, during the loading response of gait, the eccentric action of the tibialis posterior to control pronation across the subtalar and transverse tarsal joints is primarily controlling an external moment. However, it is not absorbing loads otherwise induced in bone; rather, it is creating large interjoint compression and shear stresses. In the process of lengthening, the viscoelastic properties, originating in both passive and active elements, contribute to the total force production of the tibialis posterior; i.e., recall the length-tension curves that include the roles of contractile and passive elements. In addition, based on the force-velocity relationship, we know that the impulsive lengthening of muscles inherently increases their tension-generating potential. Thus, the viscoelastic elements of the tibialis posterior allowing it to control pronation do not decrease the loading of bones but may actually increase the loads in some complex way. The viscosity of the tibialis posterior (and muscles in general) allows it to mitigate large impulsive ground forces, but it does not necessarily enhance the inherent viscous nature of bone and its ability to mitigate impulsive forces.

In summary, we can agree on the important role muscle plays in protecting bone from large or repeating forces. Hypotheses regarding the role that muscle actions play to counteract damaging compression and tensile stresses to bone and the role of muscle viscosity have been forwarded. And there is some evidence showing a relationship between muscle fatigue and metatarsal stress fractures. However, counterevidence for the relationship between muscle actions and the creation of large multidimensional forces at joint surfaces and within bones creates a paradoxical situation that requires additional research. We concluded that Ted's stress fractures were not primarily the result of muscle fatigue but the result of normal forces applied repetitively to lower limb and foot structures with abnormal alignment. Moreover, we suggested that although muscle fatigue was likely an issue related to changes in the intensity, duration, and frequency of training and inadequate periods of recovery, it was not the likely cause of Ted's stress fractures, but more likely, the extrinsic and intrinsic muscle forces, combined with forces related to gravity, contributed to microdamage to the foot bones, which ultimately led to stress fractures.

## 12.7 Gait Analysis

The clinical question in this case, and the reason for referral to the Motion Analysis Center, was, "Is there a biomechanical explanation related to the walking and running gait of Ted that might explain the cause of the repeated stress fractures?" This was a good, well-directed question, and our effort to answer it began with hypotheses that were developed based on the understanding of normal gait and data from the physical examination. Reflections throughout this chapter demonstrate part of the process.

In this case, we collected data of Ted walking and running in barefoot and shod conditions, respectively. Since at the time of data collection the Motion Analysis Center was extremely limited in conducting overground running trials, we collected kinematic data (using both standard and high-speed cameras) with Ted running on a treadmill; we had Ted run at three different speeds on the treadmill (10-minute-, 7:15-minute- and 6-minute-per-mile pace). By having Ted run at increasing speeds, we anticipated changes in the initial foot contact position, which might provide clues about foot function. Unfortunately, only kinematic data were collected during treadmill running since it was not instrumented, i.e., the treadmill had no embedded force plates. We did not collect EMG as part of this test. In retrospect, given the nature of Ted's lower extremity hypertonicity, we may have missed an opportunity to investigate what role his muscles might have played in the stress fractures.

### 12.7.1 Temporospatial Gait Parameters

By examining the gait parameters first, we were able to gauge whether there were any significant deviations from normal, as well as determine right/left asymmetries. A first impression was that Ted's self-selected walking pace was somewhat cautious, as indicated by increased time spent in the stance phase, slightly increased time spent in first and second double support, and decreased time spent in the swing phase. Since it is common for patients to initially present with a different gait than their normal one when tested in a laboratory, we did not interpret a mildly slower walking gait as an indication of pathology. Additionally, we noticed a

**Table 12.5** Temporospatial gait parameters

| Gait parameter | Left | Right | Normal |
|---|---|---|---|
| Stance (%gc) | 63.2 ± 1.0 | 64.8 ± 0.6 | 59.5 ± 1.5 |
| Swing (%gc) | 36.8 ± 1.0 | 35.2 ± 0.6 | 40.5 ± 1.5 |
| 1st double support (%gc) | 12.1 ± 0.9 | 11.9 ± 0.9 | 9.5 ± 1.5 |
| Single support (%gc) | 37.4 ± 1.0 | 38.5 ± 1.3 | 40.5 ± 1.5 |
| 2nd double support (%gc) | 13.9 ± 1.1 | 14.3 ± 1.3 | 9.5 ± 1.5 |
| Gait cycle time (sec) | 1.1 ± 0.01 | 1.1 ± 0.01 | 1.0 |
| Step length (cm) | 60.6 ± 1.4 | 64.5 ± 2.3 | 65 ± 4.3 |
| Stride length (cm) | 122.5 ± 1.5 | 124.7 ± 2.3 | 128 ± 8.5 |
| *Gait parameter* | | | |
| Step width (cm) | 10.8 ± 1.5 | | 7.5 ± 2.5 |
| Cadence (steps/min) | 108.1 ± 1.3 | | 114 ± 8.5 |
| Walking velocity (cm/s) | 111.2 ± 2.2 | | 122 ± 12.0 |

NOTE: Data are presented as mean ± SD
*%gc* percent gait cycle, *cm* centimeters

significant increase in step width. Normally, this suggests some degree of dynamic instability, which was not likely with Ted since he was a highly conditioned track athlete. Recall that Ted had mild internal tibial torsion and an adducted forefoot bilaterally. These mild deformities produce an intoeing (internal foot progression) gait pattern, so that a greater step width was likely an adaptation, i.e., compensation, to avoid catching his toes while walking. Otherwise, Ted's temporospatial walking parameters were within normal limits. The only biomechanical insight gained from gait parameters was the likely relationship between step width and his tibial and foot deformities (Table 12.5).

### 12.7.2 Problem-Solving Approach: Physical Examination, Planar Video Analysis, and Three-Dimensional Joint Kinematics and Kinetics

Recall that the principal examination findings included the following:

1. Moderate-marked tightness of rectus femoris, hamstrings, tensor fascia latae/iliotibial (IT) band, and triceps surae, all bilaterally
2. Mild internal tibial torsion (left > right) with bilateral, symmetrical mild tibial varum
3. Weight-bearing foot position: mild midfoot cavus, adducted forefoot (L > R); first metatarsal/phalange mildly abducted with Morton's toe bilaterally
4. Non-weight-bearing foot position: moderate midfoot cavus (L > R), mild right/moderate left forefoot varus (flexible) with plantar flexed first ray (L > R)
5. Bilateral marked limitation in subtalar inversion/eversion with moderate limitation of joint play at the transverse tarsal (midtarsal) and tarsometatarsal joints
6. Bilateral normal great toe extension range of motion
7. Multibeat ankle clonus bilaterally, mild left/moderate right posterior tibialis spasticity

**Note** Muscle "tightness" refers to insufficient muscle length; it is possible that Ted's resting hypertonus, i.e., spasticity, masked the ability to differentiate muscle length from muscle stiffness secondary to hypertonus.

Ted's foot deformity was classified, bilaterally, as a relatively rigid talipescavovarus foot (Fig. 12.18), with mild tibial varum (Fig. 12.9) and adducted forefoot. Based on these findings, we predict that several critical events related to the subphases of gait would be affected. Firstly, the adducted forefoot and internal tibial torsion are static bony deformities that do not change with movement and thus would be present throughout the gait cycle. These deformities would likely be manifested as a tendency toward an internal foot progression and/or increased internal rotation at the knee. Secondly, and more critically, controlled pronation during weight acceptance and single-limb support (i.e., from the loading response through midstance) would be significantly limited; that is, because of the rigidity of Ted's hind- and midfoot cavovarus deformities, the normal sequence from LR to the terminal stance of subtalar/midtarsal pronation to supination could not take place. We hypothesize that the foot would likely remain supinated throughout the gait cycle. Normally, the tibial varum tends to predispose the ankle/foot to a varus position, thus typically requiring increased subtalar and transverse tarsal joint pronation during the loading response. However, because of Ted's rigid foot, we anticipate that he would not be able to compensate for the increased varus at the subtalar joint and transverse tarsal joints but would need to adapt to a greater extent in other regions of the foot, i.e., abnormal stress distribution. These adaptations may or may not be captured by the instrumented gait data. Finally, two other possibilities remain: (1) because of Ted's reduced hip extension range of motion secondary to rectus femoris tightness, we expected to see an increased anterior pelvic tilt and reduced hip extension in terminal stance, and (2) there would be difficulty dorsiflexing the talocrural joint during swing limb advancement secondary to mild tightness and hypertonus of the triceps surae.

Using biplanar video recordings and observational gait analysis (Rancho Los Amigos), several major problems were identified; these findings were described as "major" simply because they were clearly evident. Throughout the gait cycle (i.e., weight acceptance, single-limb support, and swing limb advancement), the pelvis appeared to have a mild increase in anterior tilt. From weight acceptance through single-limb support, there was internal foot progression (left > right), which may have been related to dynamic hip internal rotation and internal tibial torsion; increased hip internal was not expected because Ted's femoral antetorsion measurements were normal. During weight acceptance (at initial contact), there was an appearance of abnormal toe extension (especially the great toe), suggesting the facilitation of the extensor digitorum longus and extensor hallucis longus to aid in talocrural dorsiflexion for toe and foot clearance; the apparent overactivity of these synergistic ankle dorsiflexors was likely due to triceps surae tightness/hypertonicity. Finally, as predicted, the ankle/foot appeared to remain inverted during the entire gait cycle. The reader needs to be reminded that, although the Rancho observational gait analysis system is standardized and highly regarded in the biomechanics community, its concurrent validity (using three-dimensional instrumented gait analysis as the gold standard) is moderate at best. Thus, it is possible that the kinematic findings (from the instrumented analysis) that follow may dispute some of what was observed.

#### 12.7.2.1 Biomechanical Foot Models

A single-segment foot model was used for the instrumented measurement and analysis of Ted's gait (see Appendix J on three-dimensional instrumented gait analysis for additional information on biomechanical models), which presents some limitations to the kinematic data presented and how they could be interpreted. Thus, it is necessary to take a brief detour to examine two commonly used biomechanical foot models so that the reader can understand one of the limitations of Ted's gait study. In the years from about 1980 to 1990, biomechanists were limited to the use of a single-segment foot model (i.e., analysis of the kinematics of the entire foot relative to the shank) because of limits in camera resolution and, therefore, the size and number of markers that could be placed on the foot. One such model was described by Soutas-Little et al., which was very similar to the model used in Ted's gait study (Fig. 12.63). With this model, six markers are placed on the lower leg and foot, three on the shank (two on the distal posterior tibia and one on the lateral malleolus), and three on the hindfoot (two on the posterior calcaneus and one on the lateral calcaneus). In order to define a rigid segment in three-dimensional space, at least three points must be clearly marked on the segment(s) (see Appendix J). With an established laboratory (global) coordinate system and the application of vector algebra operations, the locations of the markers on the segments are used to determine local (segmental) coordinate systems. Once segmental coordinate systems are established, joint kinematics (angles) are calculated in the usual way: motion of the distal segment relative to the proximal segment (see Appendix J). For clinical gait analysis, the ankle joint motions of primary interest include plantar flexion/dorsiflexion, inversion/eversion, and foot progression angle; note that, typically, lateral and medial rotation of the foot relative to the shank (i.e., talocrural axial rotation) is so limited, secondary to joint morphology, that it is not analyzed. With this

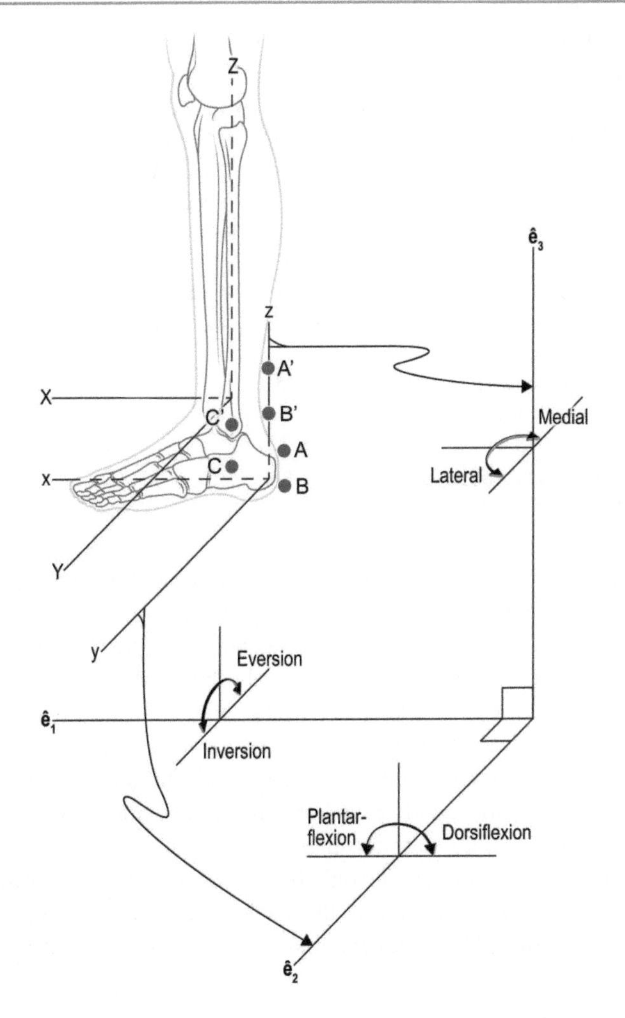


**Fig. 12.63** (**A**) Two marker sets: markers $A$, $B$, and $C$ are used to create a local (or segmental) coordinate system to define a foot segment, and markers $A'$, $B'$, and $C'$ are used to create a local coordinate system to define a shank segment. (**B**) Use of the joint coordinate system (JCS) to determine the three-dimensional pose/motion of the foot relative to the shank. Note: $e\wedge_1$ is a unit vector and virtual axis parallel to the $x$-axis of the shank and foot defining frontal plane motion (inversion/eversion), $e\wedge_2$ is a unit vector and virtual axis parallel to the $y$-axis of the shank and foot defining sagittal plane motion (plantar flexion/dorsiflexion), and $e\wedge_3$ is a unit vector and virtual axis parallel to the $z$-axis of the shank and foot defining transverse (or horizontal) plane motion (lateral/medial rotation). Note: medial or internal rotation is synonymous with adduction, and lateral or external rotation is synonymous with abduction. (Modified with permission from Soutas-Little et al. (1987))

explanation as background, the reader should be able to intuit that although determining three-dimensional foot motion using a single-segment foot model is possible, the inversion/eversion movements that are quantified cannot reveal the subtle frontal plane motions that occur at the subtalar, transverse tarsal, and forefoot segments. Nor can the small sagittal and transverse plane motions at the foot segments be expressed.

With advancements in optical motion capture technology beginning in the early 1990s, biomechanists were able to

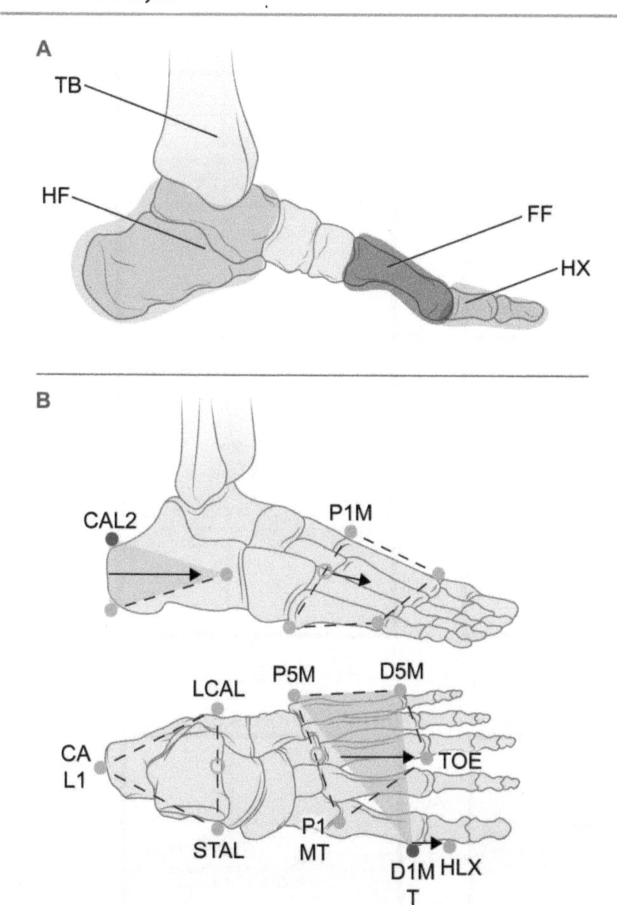


Fig. 12.64 The Oxford Foot Model (A) creates a three-segment foot, hindfoot, forefoot, and hallux, (B) with detailed marker placement. In A, TB = tibial, HF = hindfoot, FF = forefoot, and HX = hallux. In B, CAL 2 and CAL1 = placements of three markers on the posterior aspect of the calcaneus, LCAL = lateral calcaneus, STAL = sustentaculum talus, P5M and D5M = proximal (base) and distal (head) of the fifth metatarsal, P1M and D1M = proximal (base) and distal (head) of the first metatarsal, toe = marker placed between the second and third metatarsal heads, HLX = marker placed on the medial aspect of the phalanx of the great toe. (Modified with permission from Carson et al. (2001))

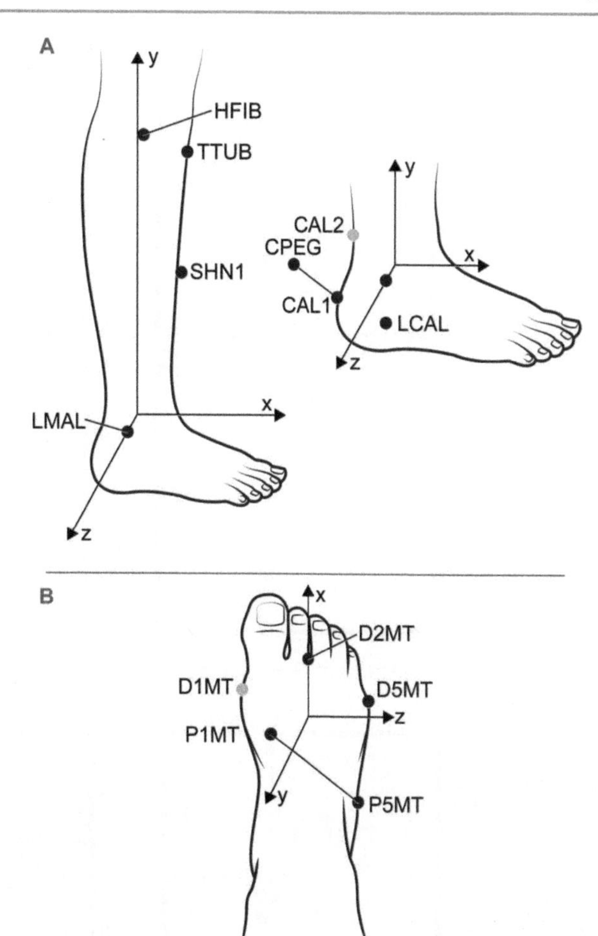


Fig. 12.65 Oxford Foot Model with (A) local (segmental) coordinate systems depicted on the tibia and hindfoot and (B) local coordinate systems depicted on the forefoot and hallux. In A, HFIB = head of the fibula, TTUB = tibial tubercle, SHN1 = anterior shank, LMAL = lateral malleolus, CPEG = calcaneal marker on a wand, CAL1 = posterior calcaneus, CAL2 = marker above CAL1 placed on the Achilles tendon. In B, other abbreviations are defined in Fig. 12.64. (Modified with permission from Carson et al. (2001)

develop multisegment biomechanical models of the foot. A multisegment foot model means that movements of subunits of the foot, e.g., the hind-, mid-, and forefoot segments, could be parceled out. Since then, many versions of these models have been created, tested, and validated. One such model was developed in the late 1990s at the Oxford Gait Laboratory in collaboration with Oxford University (Figs. 12.64 and 12.65). The Oxford Foot Model places a redundant (more than three) set of markers on the shank and foot to create a three-segment model: hindfoot, forefoot, and hallux. With these marker placements, local coordinate systems are created for each segment so that movements of the hindfoot relative to the tibia, forefoot relative to the hindfoot, forefoot relative to the tibia, and hallux relative to the forefoot can be determined using inverse kinematics. Note that with the Oxford foot model, only flexion/extension movements of the hallux can be expressed since the hallux holds only one marker (not the required three to determine a three-dimensional pose of a rigid segment). The obvious advantage of a multisegment over the single-segment foot model is that interfoot segmental movements can be determined and analyzed (Fig. 12.66); note that the data expressed in Fig. 12.66 were determined from a cohort of healthy children at the Oxford Gait Laboratory and were not part of the case study in this chapter. The development and use of multisegment foot models, in conjunction with plantar pressure measurement systems, is widespread and critically important for the biomechanical diagnosis and treatment of clubfoot deformities and foot deformities related to cerebral palsy. The use of a multisegment foot model has also been of value in the evaluation of athletes with lower extremity injuries, similar to those experienced by Ted. As we examine Ted's joint angles during walking

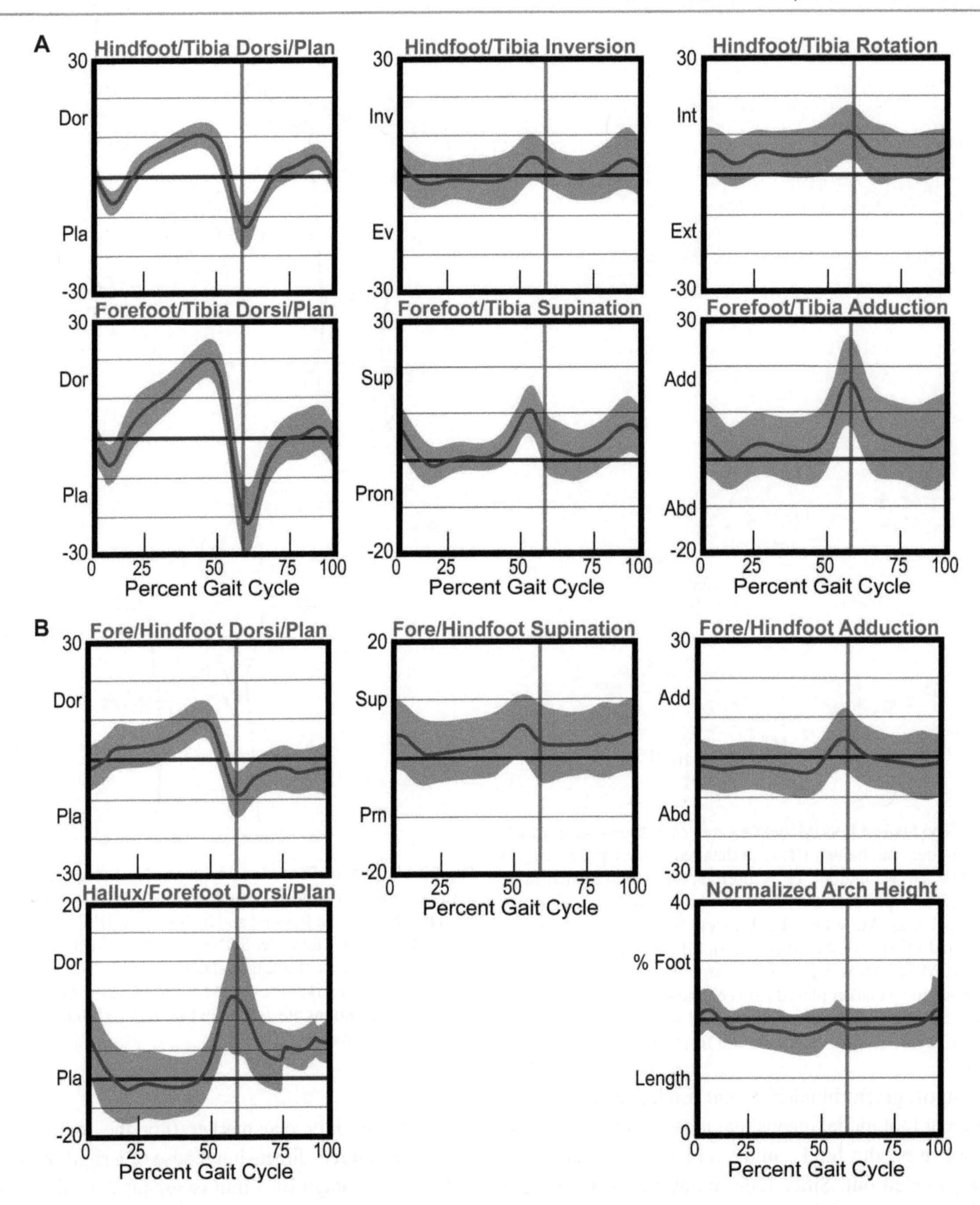


**Fig. 12.66** (**A**, **B**) Three-dimensional foot kinematic from a normal control pediatric data set. Data in the left column represent sagittal plane motion, middle column frontal plane motion, and right column transverse plane motion. (**A**) The top row represents the motion of the hindfoot relative to the tibia, and the second row represents forefoot movements relative to the tibia. (**B**) The top row represents forefoot movements relative to the hindfoot, and the second row includes motion of the hallux relative to the forefoot (only sagittal plane) as well as medial longitudinal arch height. Data are expressed in percent gait cycle with the far left of the horizontal axis 0% of the gait cycle, or initial contact, and the far right of the horizontal axis 100% of the gait cycle, or second initial contact; the data are expressed as a mean (dark solid line) plus/minus one standard deviation; the vertical gray line is the division between the stance and swing phases (toe-off). Dor = dorsiflexion, Pln = plantar flexion, Inv = inversion, Ev = eversion, Ext = cxternal, Int = internal, Sup = supination, Pron = pronation, Add = adduction, Abd = abduction. (Modified with permission from Stebbins et al. (2006))

and running, it is important to be careful in making definitive conclusions based on the limitations of the single-segment foot model that was used.

#### 12.7.2.2 Review of Ted's Kinematic and Kinetic Gait Data

Let us look at Ted's walking data first. Examination of the kinematic data for the self-paced walking demonstrated increased right pelvic protraction throughout the stance phase, which corresponds to increased left retraction (Fig. 12.67A). This could be related to his tight rectus femoris and an attempt to maintain symmetric step and stride lengths. However, Ted's hip rotation and extension throughout the entire gait cycle were normal. Ted demonstrated a mild increase in hip abduction from the end of the loading response through terminal stance, which was likely related to an increased step width. He demonstrated only mild deviations from normal in frontal and transverse plane rotations of the right and left knees. These knee kinematic deviations are difficult to reconcile, given the symmetry in knee physical examination findings, yet the mild increase in knee varus motion may be related to his tibial varum and/or an adaptation to impaired subtalar/transverse tarsal mobility. As expected, Ted demonstrated a mild internal foot progression angle on the left, which is consistent with the internal tibial torsion and forefoot adductus (Fig. 12.67B). It is likely that a mild increase in right knee external rotation masked a right internal foot progression angle. As we predicted, Ted was unable to evert his feet from weight acceptance through midstance, as evidenced by the ankle inversion (left > right). Based on this dynamic profile and his physical examination data, we hypothesized that Ted lacked the proper subtalar and transverse tarsal pronation needed during early and midstance to transfer large ground forces.

Now, we can examine Ted's running kinematic data (Fig. 12.68). Recall that we collected these data using a non-instrumented treadmill (i.e., not embedded with force plates) with Ted running at three different speeds. During running, Ted demonstrated the exaggerated anterior pelvic tilt we expected, but did not observe with walking. We hypothesize that increased anterior tilt of the pelvis became evident because of the increase in locomotor velocity, which likely increased the activation of the rectus femoris muscles (because of their attachments at the anterior superior iliac spines of the pelvis, they can pull the pelvis forward). Ted exhibited a moderate increase in hip abduction bilaterally from weight acceptance through terminal stance, again secondary to a widening step width related to internal foot progression (internal tibial torsion plus adducted forefoot) (Fig. 12.68A, C). Knee-joint kinematics were normal across all planes of movement bilaterally. Intuition would suggest that during running, the increase in ground forces would increase Ted's need to pronate (i.e., evert) during the loading response. In individuals with normal mobility in the hind- and midfoot joints, we generally see a slight increase in pronation at those joints during loading response, but we do not see that in Ted's data. Our data show just the opposite. That is, our frontal plane analysis of ankle motion shows inversion (i.e., a component of supination) from the loading response through midstance (Fig. 12.68B, D).

To better understand the kinetic data associated with walking, for this case, we encourage the reader to consult the section on kinetics in Appendix J. Ted's kinetic data during walking were within the normal limits of the control data set used by the Motion Analysis Center (Figs. 12.69 and 12.70A–C), with one exception. Ted demonstrated a mild increase in the net internal eversion moment at the ankle (Fig. 12.70C), which is likely related to the fact that the ankles remained inverted during most of the stance phase; thus, the ankle evertors (fibularis longus and brevis) were activated to provide dynamic stability.

In summary, based on the observational and instrumented kinematic and kinetic data, the most significant deviations that were identified included ankle inversion from the loading response (weight acceptance) through late midstance (single-limb support), mild left internal foot progression throughout stance, and increased toe extension during swing limb advancement. We conclude that excessive ankle inversion (or inability to evert the ankle) was secondary to the relatively rigid hind- and midfoot intersegment joints, perhaps related to preexisting genetic/developmental factors and/or chronic hypertonicity, i.e., spasticity, of the gastrosoleus complex and tibialis posterior. It is likely that increased toe extension during swing limb advancement was compensatory for the hypertonus of the posterior tibial muscles and necessary to achieve adequate talocrural dorsiflexion to clear Ted's toes in the initial swing phase. Finally, the internal foot progression was secondary to bilateral internal tibial torsion and forefoot adductus.

### 12.7.3 Walking Versus Running

The walking and running cycles are very similar. However, the running cycle does not have periods of double support but instead is characterized by a flight (or double float) phase, where both feet are airborne during the gait cycle, one at the beginning and one at the end of the swing phase (Fig. 12.71). As with walking, a running gait cycle is defined by two consecutive foot contact events on the same leg, with foot contact of the opposite limb occurring halfway through. Each leg has a stance phase when the foot is in contact with the ground and a swing phase when the foot is off the ground. With walking, the duration of stance decreases as walking velocity increases; thus, the periods of double support

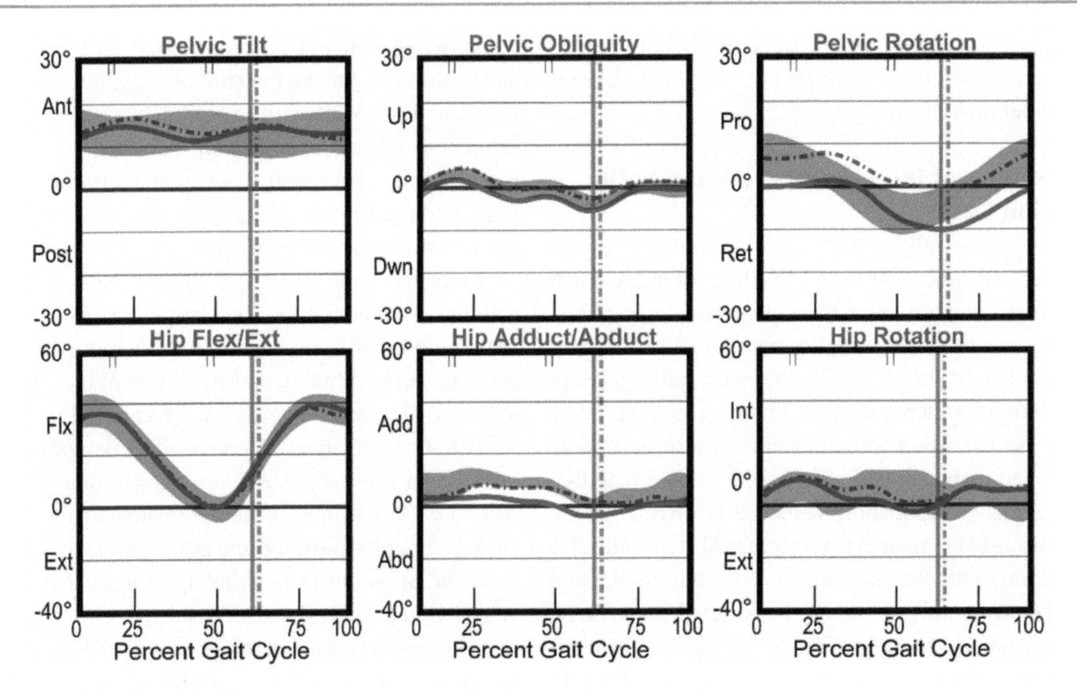


**Fig. 12.67A** Segment and joint angles of the pelvis and hip during barefoot walking. Sagittal, frontal, and transverse plane data are organized by columns going from left to right. Data are plotted relative to the percent gait cycle (from 0% far left to 100% far right) along the horizontal axis. The vertical axis indicates the magnitude of joint motion in degrees, with a horizontal line representing 0°, which divides the motion descriptors; e.g., hip flexion (Flx) has positive values, and hip extension (Ext) has negative values. The vertical line at about 60% of the GC represents toe-off, with stance phase to the left and swing phase to the right of the toe-off line. The dashed red line = right limb, and the solid blue line = left limb. The gray band represents ± 1 standard deviation from the control group mean. Ant = anterior, Post = posterior, Up = reference side of pelvis is up, Dwn = reference side of pelvis is down, Pro = protraction, Retr = retraction, Flx = flexion, Ext = extension, Add = adduction, Abd = abduction, Int = internal, Ext = external

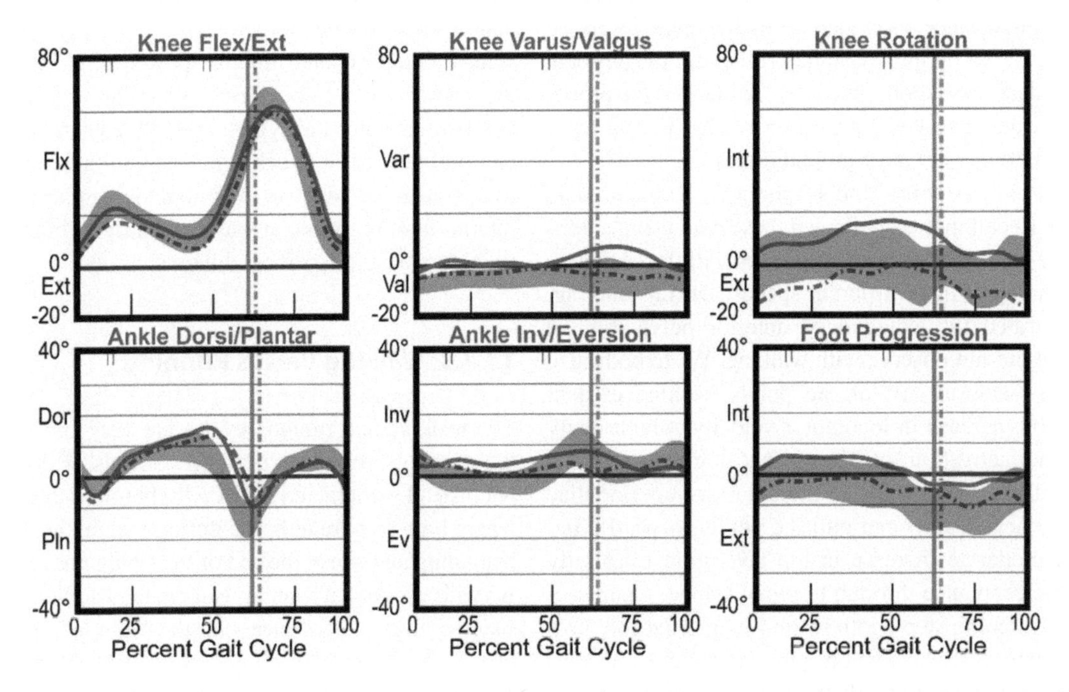


**Fig. 12.67B** Joint angles of the knee and ankle during barefoot walking. Sagittal, frontal, and transverse plane data are organized by columns going from left to right. Data are plotted relative to the percent gait cycle (from 0% far left to 100% far right) along the horizontal axis. The vertical axis indicates the magnitude of joint motion in degrees, with a horizontal line representing 0°, which divides the motion descriptors; e.g., knee flexion (Flx) has positive values, and knee extension (Ext) has negative values. The vertical line at about 60% of the GC represents toe-off, with stance phase to the left and swing phase to the right of the toe-off line. The dashed red line = right limb, and the solid blue line = left limb. The gray band represents ± 1 standard deviation from the control group mean. Var = varus, Val = valgus, Dor = dorsiflexion, Pln = plantar flexion, Inv = inversion, Ev = eversion, Int = internal, Ext = external

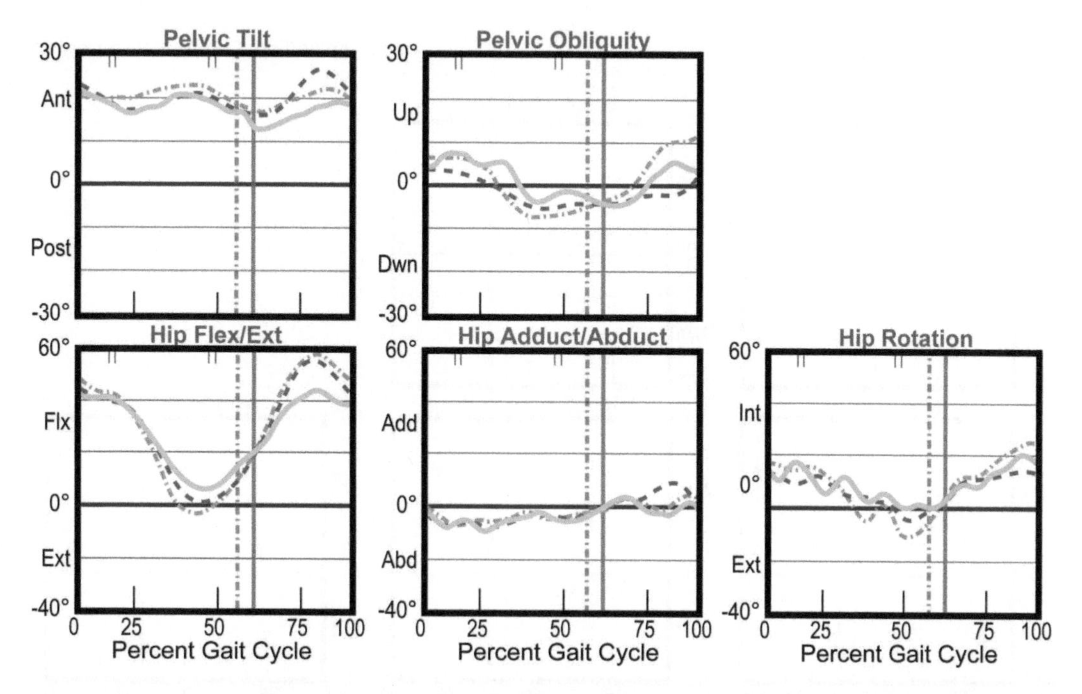


**Fig. 12.68A** Right segment and joint angles of the pelvis and hip during shod running on the treadmill. Sagittal, frontal, and transverse plane data are organized by columns going from left to right. Data are plotted relative to the percent gait cycle (from 0% far left to 100% far right) along the horizontal axis. The vertical axis indicates the magnitude of joint motion in degrees, with a horizontal line representing 0°, which divides the motion descriptors; e.g., hip flexion (Flx) has positive values, and hip extension (Ext) has negative values. The vertical line at about 55% of the GC represents toe-off, with stance phase to the left and swing phase to the right of the toe-off line. The solid line represents the 10-minute-per-mile pace; the dashed line represents the 7:15-minute-per-mile pace; the dash-dotted line represents the 6-minute-per-mile pace. Transverse plane rotation data of the pelvis were not available. Ant = anterior, Post = posterior, Up = reference side of pelvis is up, Dwn = reference side of pelvis is down, Add = adduction, Abd = abduction, Int = internal, Ext = external

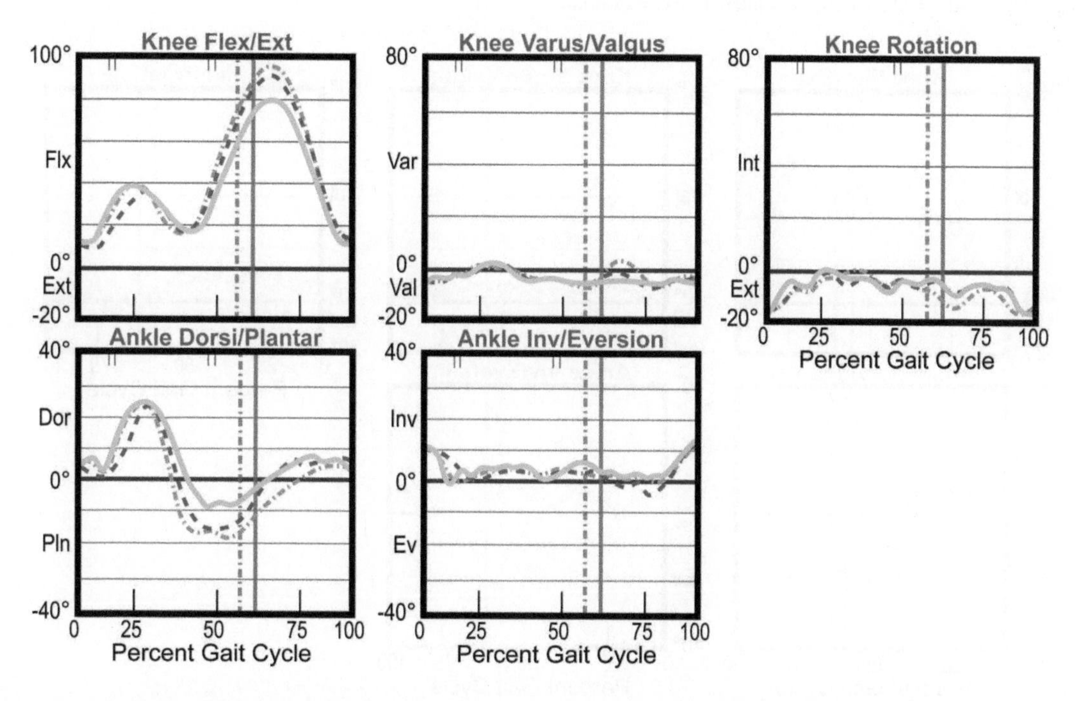


**Fig. 12.68B** Right joint angles of the knee and ankle during shot running on the treadmill. Sagittal, frontal, and transverse plane data are organized by columns going from left to right. Data are plotted relative to the percent gait cycle (from 0% far left to 100% far right) along the horizontal axis. The vertical axis indicates the magnitude of joint motion in degrees, with a horizontal line representing 0°, which divides the motion descriptors; e.g., knee flexion (Flx) has positive values, and knee extension (Ext) has negative values. The vertical line at about 55% of the GC represents toe-off, with stance phase to the left and swing phase to the right of the toe-off line. The solid line represents the 10-minute-per-mile pace; the dashed line represents the 7:15-minute-per-mile pace; the dash-dotted line represents the 6-minute-per-mile pace. Transverse plane rotation data of the ankle were not available. Var = varus, Val = valgus, Dor = dorsiflexion, Pln = plantar flexion, Inv = inversion, Ev = eversion, Int = internal, Ext = external

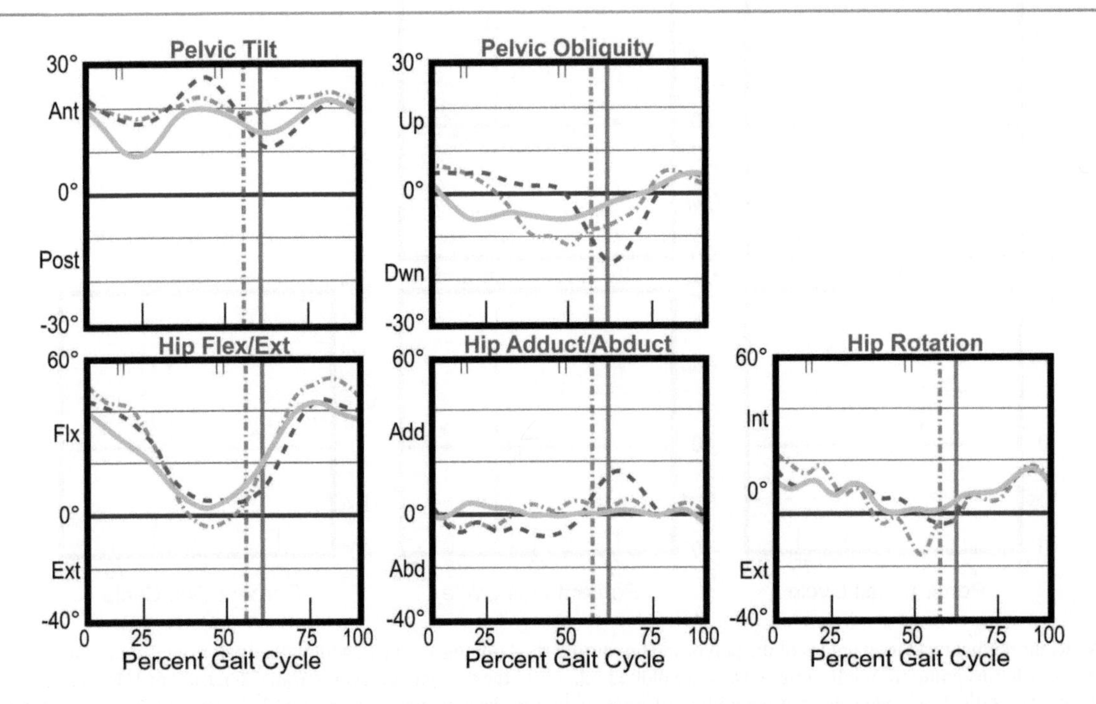


**Fig. 12.68C** Left segment and joint angles of the pelvis and hip during shod running on the treadmill. Sagittal, frontal, and transverse plane data are organized by columns going from left to right. Data are plotted relative to the percent gait cycle (from 0% far left to 100% far right) along the horizontal axis. The vertical axis indicates the magnitude of joint motion in degrees, with a horizontal line representing 0°, which divides the motion descriptors; e.g., hip flexion (Flx) has positive values, and hip extension (Ext) has negative values. The vertical line at about 55% of the GC represents toe-off, with stance phase to the left and swing phase to the right of the toe-off line. The solid line represents the 10-minute-per-mile pace; the dashed line represents the 7:15-minute-per-mile pace; the dash-dotted line represents the 6-minute-per-mile pace. Transverse plane rotation data of the pelvis were not available. Ant = anterior, Post = posterior, Up = reference side of pelvis is up, Dwn = reference side of pelvis is down, Add = adduction, Abd = abduction, Int = internal, Ext = external

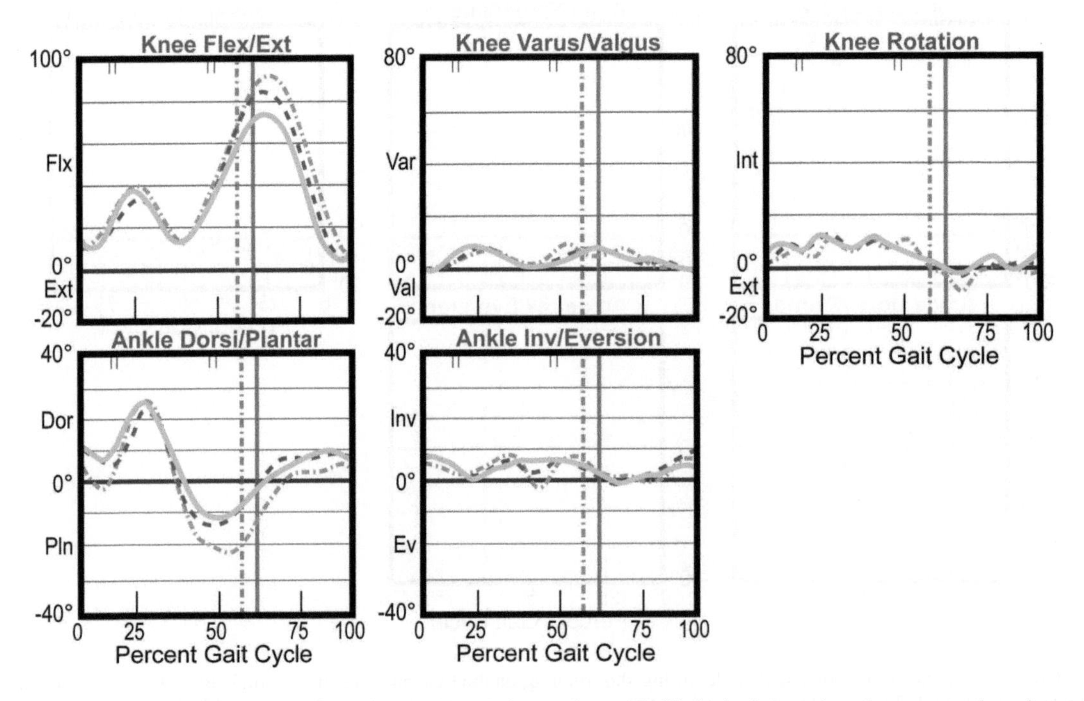


**Fig. 12.68D** Left joint angles of the knee and ankle during shot running on the treadmill. Sagittal, frontal, and transverse plane data are organized by columns going from left to right. Data are plotted relative to the percent gait cycle (from 0% far left to 100% far right) along the horizontal axis. The vertical axis indicates the magnitude of joint motion in degrees, with a horizontal line representing 0°, which divides the motion descriptors; e.g., knee flexion (Flx) has positive values, and knee extension (Ext) has negative values. The vertical line at about 55% of the GC represents toe-off, with stance phase to the left and swing phase to the right of the toe-off line. The solid line represents the 10-minute-per-mile pace; the dashed line represents the 7:15-minute-per-mile pace; the dash-dotted line represents the 6-minute-per-mile pace. Transverse plane rotation data of the ankle were not available. Var = varus, Val = valgus, Dor = dorsiflexion, Plan = plantar flexion, Inv = inversion, Ev = eversion, Int = internal, Ext = external

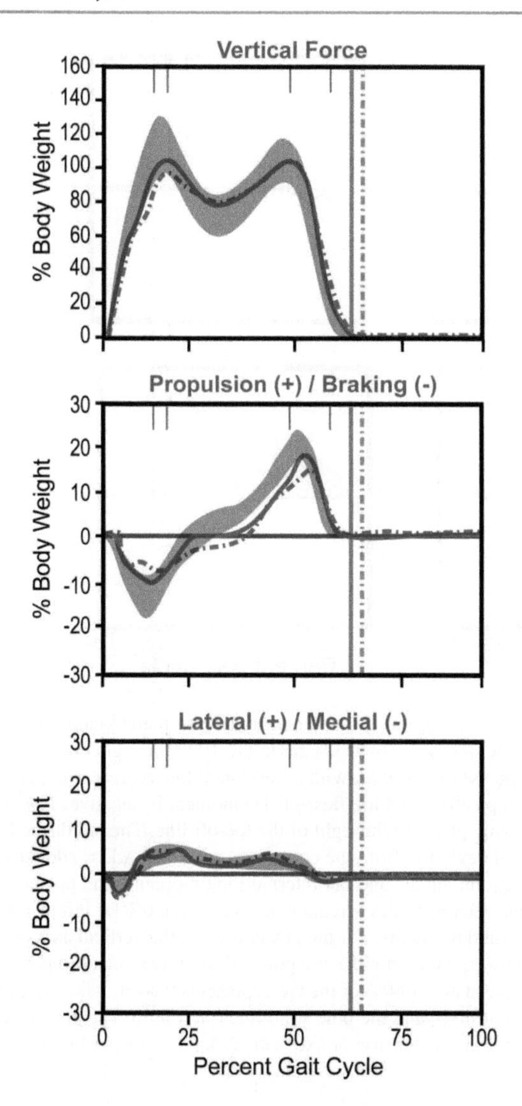


**Fig. 12.69** Vertical, propulsion/braking, and lateral/medial ground reaction forces (GRF) during the stance phase of walking. GRF magnitudes are normalized (expressed) to percent body weight. The vertical line at about 60% of the GC represents toe-off, with stance phase to the left and swing phase to the right of the toe-off line. The dashed red line = right limb, and the solid blue line = left limb. The gray band represents ±1 standard deviation from the control group mean. The solid line = left limb, the dashed line = right limb

shorten as well. With progressively increasing speeds of walking, the stance phase gives way to a flight phase. The flight phase is the signature characteristic of running, as is its efficiency, i.e., energy cost related to optimal step length and step frequency; that is, a transition from walking to running occurs around the speed at which running becomes more efficient or economical. Another difference between walking and running relates to initial foot contact. With walking, a heel-first initial contact (first ankle rocker) is a critical event to assist with forwarding progression, whereas in running, the initial foot contact may occur at the hind- (heel), mid- or forefoot. Variation in foot contact during running is typically related to running speed; at slower speeds, heel contact is most common, whereas faster-speed running typically induces a mid- or forefoot initial contact.

We can gain insight into the dynamics of locomotion by using measurements of the ground reaction forces (GRF) over time (also see Appendix J). There are notable differences between the vertical GRFs of walking and running (Fig. 12.72). During walking, the vertical reaction forces peak at approximately 1.2 times the body weight, whereas for running at moderate speeds, the vertical ground force typically exceeds two body weights. Fig. 12.72 makes it apparent that during heel-contact running, there is an initial spike associated with a steeper loading rate, i.e., the first slope of the ground force, which we have referred to as an impulsive force, i.e., a force or load that is induced over a short time period. Biomechanists have speculated that this impulsive load (sometimes very large) may be related to overuse injuries similar to what Ted experienced, particularly if combined with changes in training, running shoes, and lower-limb deformities.

Anterior/posterior and medial/lateral reaction forces are also recorded during instrumented measures of ground forces during walking (see Fig. 12.69) and running. Obviously, during running, the magnitude of these forces is also significantly greater (Fig. 12.73). The horizontal ground reaction force points backward (braking peak) during the first half of the stance phase, decelerating the center of mass. The magnitude of the peak braking impulse increases with an increase in running speed. The forefoot striker sees this braking impulse placed further in front of the body's center of mass. It is notable that the mechanics of the subtalar/transverse tarsal joint complexes are particularly critical to the transfer of the early-stance-phase braking forces when the runner makes a heel-first contact. Thus, in Ted's case, because of his rigid hind- and midfoot, the dissipation of these horizontal forces was likely impaired. On the other hand, if Ted were to run with a forefoot initial contact, his already relatively rigid subtalar/transverse tarsal joint complexes may have been able to well manage the impulsive loads. The second horizontal peak (propulsion) occurs during the end of stance, i.e., preswing; these forces could be easily managed by Ted because of his more rigid foot.

How a runner makes initial contact alters the vertical ground reaction force (Fig. 12.74). For example, the runner who utilizes a walking-type heel-to-toe running pattern (rearfoot striker (RFS)) demonstrates an early impact peak, then a second larger peak as the limb enters the loading response phase. As we noted above, although the first peak has a smaller amplitude, its force creates an impulse that challenges the viscoelastic properties of articular cartilage and subchondral bone primarily at the talocrural, subtalar, and transverse tarsal joints. Conversely, the runner who makes a forefoot initial contact (FFS) does not demonstrate this early impulse, although the location of the braking impulse is located more distally on the foot. Some have sug-

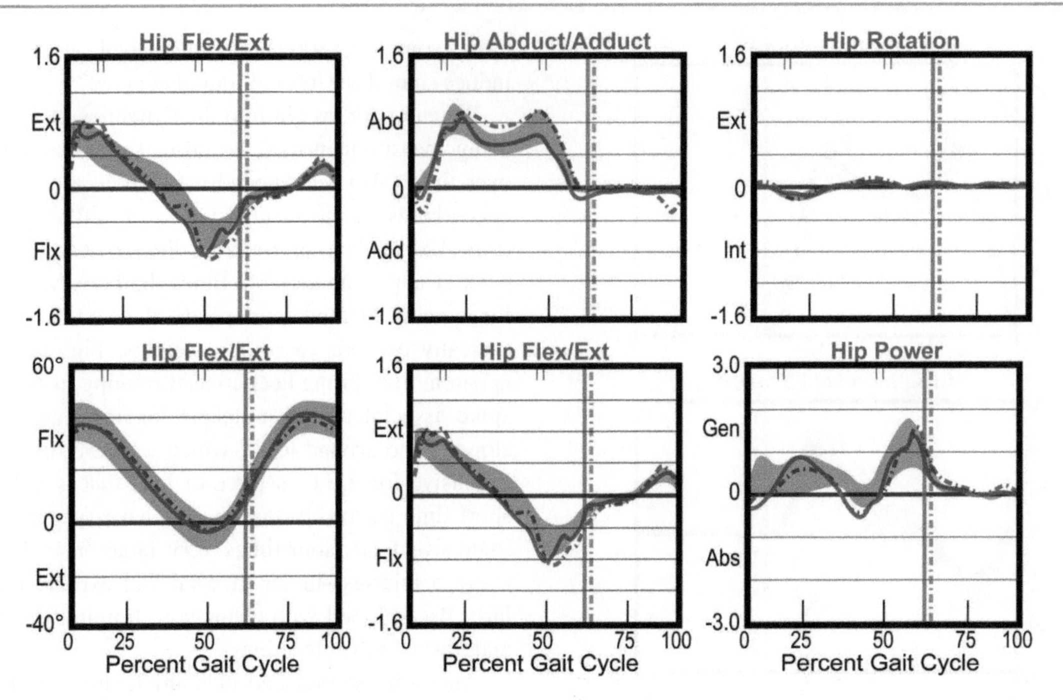


**Fig. 12.70A** *Top row*: Net internal joint moments for the hip during barefoot walking. Sagittal, frontal, and transverse plane kinematic data are organized by columns going from left to right. Data are plotted relative to the percent gait cycle (from 0% far left to 100% far right) along the horizontal axis. The vertical axis indicates the magnitude of the net joint moment, expressed in Nm/kg, with a horizontal line representing a zero net moment, which divides the moment descriptors; e.g., hip extensor (Ext) moment is positive, and hip flexor (Flx) moment is negative. The vertical line at about 60% of the GC represents toe-off, with stance phase to the left and swing phase to the right of the toe-off line. The red dashed line = right limb, and the blue solid line = left limb. The gray band represents ± 1 standard deviation from the control group mean. Add = adductor, Abd = abductor, Int = internal rotator, Ex = external rotator. *Bottom row*: Sagittal plane joint angles and net internal joint moments and powers (watts normalized to body mass) during barefoot walking for the hip joint. Data are plotted relative to the percent gait cycle (from 0% far left to 100% far right) along the horizontal axis. We have described the vertical axis for joint angles and moments. For the power curves, the vertical axis indicates the magnitude of the net joint power with a horizontal line representing zero net power, which divides the power descriptors; e.g., hip power generation (Gen) is positive, and hip power absorption (Abs) is negative. The vertical line at about 60% of the GC represents toe-off, with stance phase to the left and swing phase to the right of the toe-off line. The red dashed line = right limb, and the blue solid line = left limb. The gray band represents ± 1 standard deviation from the control group mean. Flx = flexion or flexor, Ext = extension or extensor, W/kg = watts per kilogram

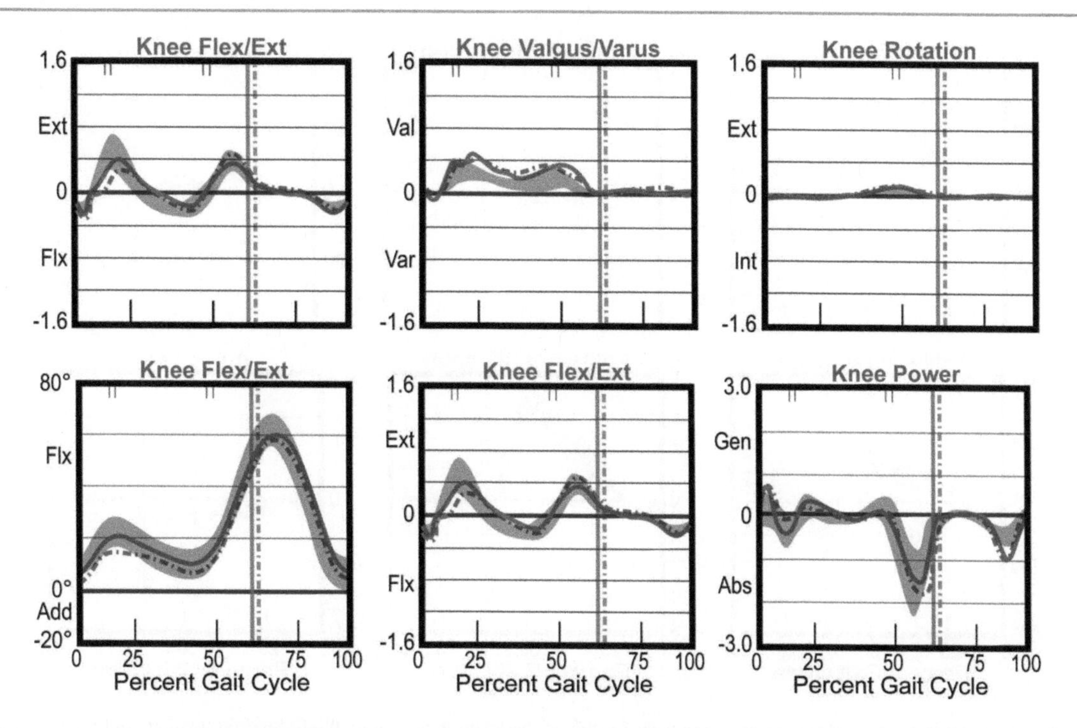


**Fig. 12.70B** *Top row*: Net internal joint moments and powers for the knee during barefoot walking. Sagittal, frontal, and transverse plane kinematic data are organized by columns going from left to right. Data are plotted relative to the percent gait cycle (from 0% far left to 100% far right) along the horizontal axis. The vertical axis indicates the magnitude of the net joint moment, expressed in Nm/kg, with a horizontal line representing a zero net moment, which divides the moment descriptors; e.g., hip extensor (Ext) moment is positive, and hip flexor (Flx) moment is negative. The vertical line at about 60% of the GC represents toe-off, with stance phase to the left and swing phase to the right of the toe-off line. The red dashed line = right limb, and the blue solid line = left limb. The gray band represents ± 1 standard deviation from the control group mean. Val = valgus, Var = varus, Int = internal rotator, Ex = external rotator. *Bottom row*: Sagittal plane joint angles and net internal joint moments and powers (watts normalized to body mass) during barefoot walking for the knee joint. Data are plotted relative to the percent gait cycle (from 0% far left to 100% far right) along the horizontal axis. We have described the vertical axis for joint angles and moments. For the power curves, the vertical axis indicates the magnitude of the net joint power with a horizontal line representing zero net power, which divides the power descriptors; e.g., knee power generation (Gen) is positive, and knee power absorption (Abs) is negative. The vertical line at about 60% of the GC represents toe-off, with stance phase to the left and swing phase to the right of the toe-off line. The red dashed line = right limb, and the blue solid line = left limb. The gray band represents ± 1 standard deviation from the control group mean. Flx = flexion or flexor, Ext = extension or extensor, W/kg = watts per kilogram

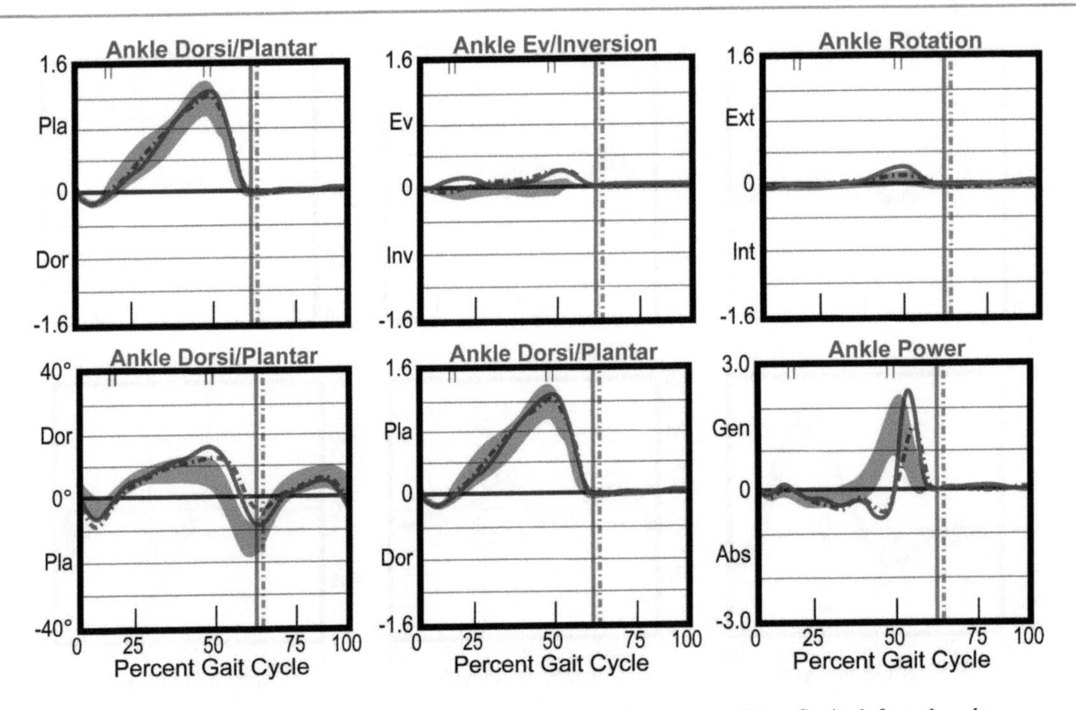


**Fig. 12.70C** *Top row*: Net internal joint moments and powers for the ankle during barefoot walking. Sagittal, frontal, and transverse plane kinematic data are organized by columns going from left to right. Data are plotted relative to the percent gait cycle (from 0% far left to 100% far right) along the horizontal axis. The vertical axis indicates the magnitude of the net joint moment, expressed in Nm/kg, with a horizontal line representing a zero net moment, which divides the moment descriptors; e.g., ankle plantar flexor (Pln) moment is positive, and ankle dorsiflexor (Dor) moment is negative. The vertical line at about 60% of the GC represents toe-off, with stance phase to the left and swing phase to the right of the toe-off line. The red dashed line = right limb, and the blue solid line = left limb. The gray band represents ± 1 standard deviation from the control group mean. Ev = evertor, Inv = invertor, Int = internal rotator, Ex = external rotator. *Bottom row*: Sagittal plane joint angles and net internal joint moments and powers (watts normalized to body mass) during barefoot walking for the ankle joint. Data are plotted relative to the percent gait cycle (from 0% far left to 100% far right) along the horizontal axis. We have described the vertical axis for joint angles and moments. For the power curves, the vertical axis indicates the magnitude of the net joint power with a horizontal line representing zero net power, which divides the power descriptors; e.g., ankle power generation (Gen) is positive, and ankle power absorption (Abs) is negative. The vertical line at about 60% of the GC represents toe-off, with stance phase to the left and swing phase to the right of the toe-off line. The red dashed line = right limb, and the blue solid line = left limb. The gray band represents ± 1 standard deviation from the control group mean. Pln = plantar flexion or plantar flexor, Dor = dorsiflexion or dorsiflexor, W/kg = watts per kilogram

gested that the forefoot striker makes a more natural ground first contact and may, therefore, be somewhat protected from large impulsive forces. On the other hand, using a subject-specific model, including accurate metatarsal geometry, it has been shown that the forefoot striker running barefoot demonstrates greater external loading, bending moments, compressive stresses, as well as greater second metatarsal stresses in early stance. In that study, there were no differences in peak pressures between the rearfoot and forefoot strikers. Thus, although it is clear that the peak magnitude of the ground forces is not different between the two types of running foot strikes (Fig. 12.74), in either case, the ankle/foot complex likely transfers these large forces, which are unique to individual bone geometry and joint constraints, using different mechanisms.

Additional insights into the differences between walking and running can be gained by examining the cycling of kinetic ($E_K$) and gravitational potential ($E_P$) energies. The forward speed (and, therefore, the kinetic energy due to forward speed) and gravitational potential energy are greatest during the flight phase of running (Fig. 12.75). Conversely, $E_K$ and $E_P$ are minimized during midstance; the forward velocity is the least when the center of mass (COM) is the lowest. As opposed to walking, the kinetic and potential energies during running reach their maxima and minima at the same time. Thus, with running, the benefit of the exchange of gravitational potential and kinetic energies (as accomplished during walking) is lost. Yet in running, potential energy is stored (in tendons, ligaments, and both extrinsic and intrinsic muscles) and released (through the use of passive elastic moments and muscle actions) similar to a mass and spring (Fig. 12.76). In the application of a mass-spring model to running, the spring compresses and the mass reaches its lowest point at midstance when the velocity of the mass center is at a minimum. The spring-like action in the bones and soft tissues in our legs actually makes running an

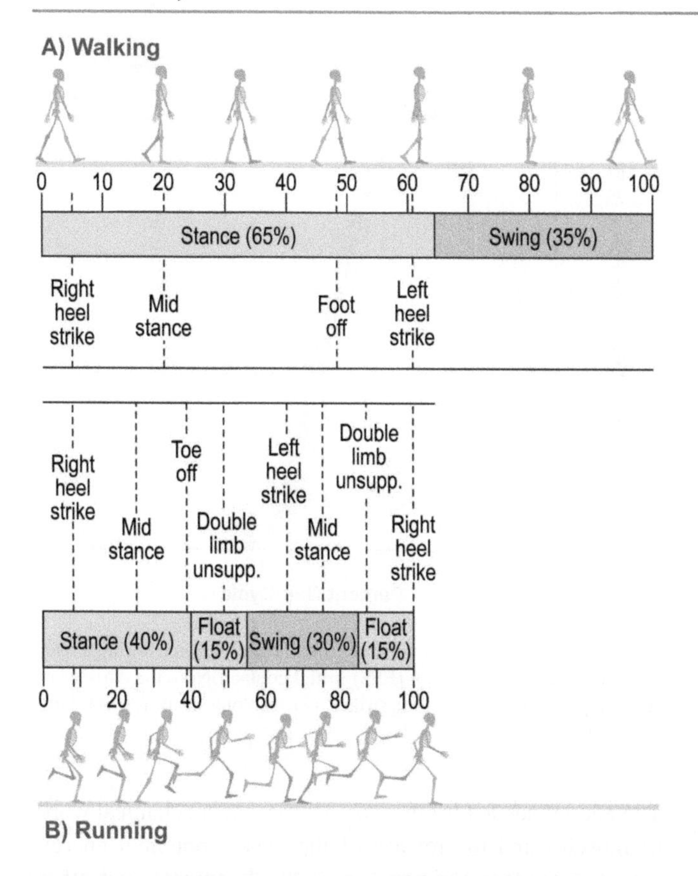


**Fig. 12.71** Comparison of (**A**) walking and (**B**) running gait cycles and their constituent events, e.g., foot contact, and phases, e.g., support. The percentages of time spent in stance and swing vary with running speed

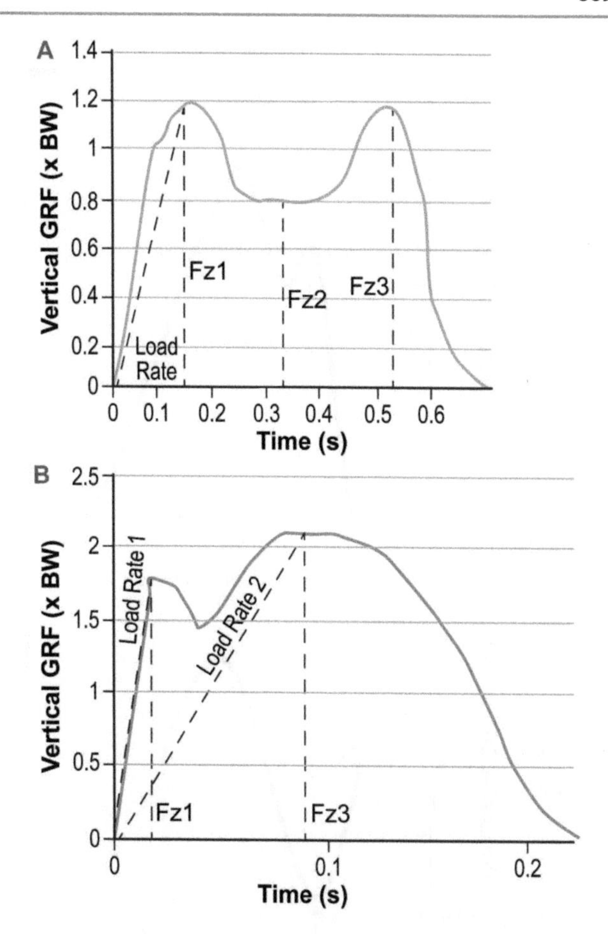


**Fig. 12.72** Vertical ground reaction force for (**A**) walking and (**B**) heel-contact running. It is notable that the loading rate and peak vertical forces are significantly greater during running. Fz1 = first vertical peak (loading), Fz2 = midstance peak during the midstance of walking, Fz3 = peak during propulsion

efficient form of locomotion. How does this happen, and what are the implications?

Earlier models for walking have often described it as a controlled fall or "inverted pendulum" gait, in which the body's COM moves from its zenith in midstance to its nadir during double support, actions very similar to a pendulum. Thus, the efficiency of walking is maintained by the effective interchange between gravitational potential and kinetic energies. On the other hand, more recent walking models have modified the simple inverted pendulum model by suggesting that walking can be better explained by a spring-loaded inverted pendulum. This modification can be explained by similar mechanisms, i.e., soft tissue storage of elastic energy, that appear to explain that the efficiencies of running are also present during walking to a lesser degree. For example, recall that during the stance phase in walking, as the tibia advances over the fixed foot, several posterior tibial muscles, e.g., gastroc-soleus complex, tibialis posterior, etc., control the tibia via their eccentric actions, with concomitant stretching of their tendons.

On the other hand, since forward kinetic and gravitational potential energies are in-phase during running, the efficiency of running is maintained primarily via the storage and return of elastic potential energy by the stretching of elastic structures, e.g., tendons, connective tissue that binds muscles, ligaments, and molecular springs within the muscles themselves. Although energy storage and return in tendons contribute to running efficiency, clinicians have proposed that the repetitive cycling of tendon stretch may be responsible for the many chronic overuse syndromes in runners, e.g., Achilles tendonitis, bone stress reactions, etc. On the other hand, for the plantar arch to be an effective spring during running, the dynamic coupling of the transverse tarsal, subtalar, and joints of the anterior tarsals must be such that the plantar ligaments are strained in an optimal fashion. If, as in Ted's case, the mechanics of these joints are not normal, compression and shear loads will not likely be distributed normally, thus creating an environment conducive to stress-related failure of bone.

During running, then, gravitational potential and forward kinetic energies peak during the flight phase. As the center of

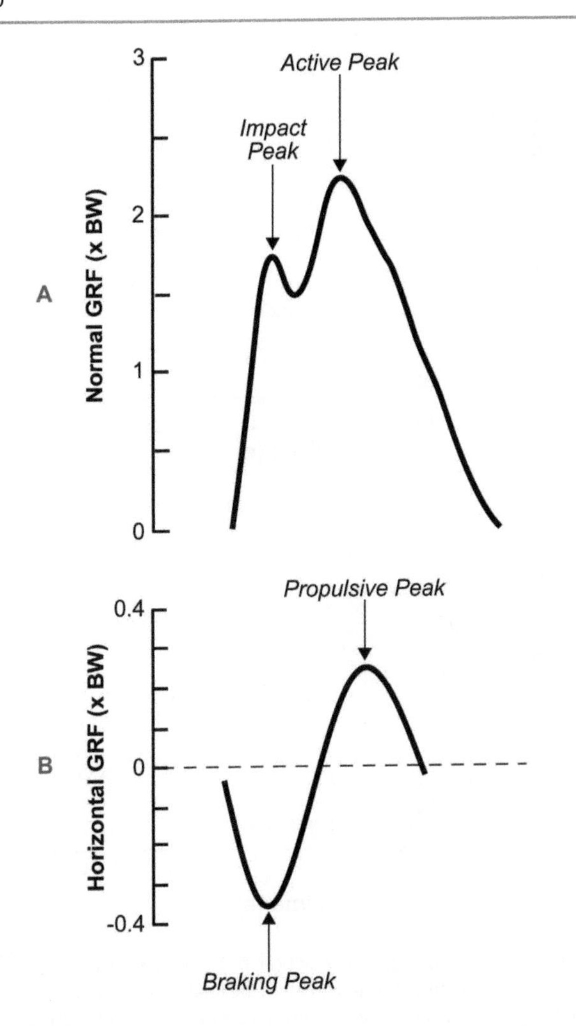


**Fig. 12.73** Typical (**A**) vertical and (**B**) horizontal (breaking/propulsion) ground reaction forces for running at a moderate speed. Note that the horizontal force magnitudes are much less than the vertical forces but substantially higher than in walking

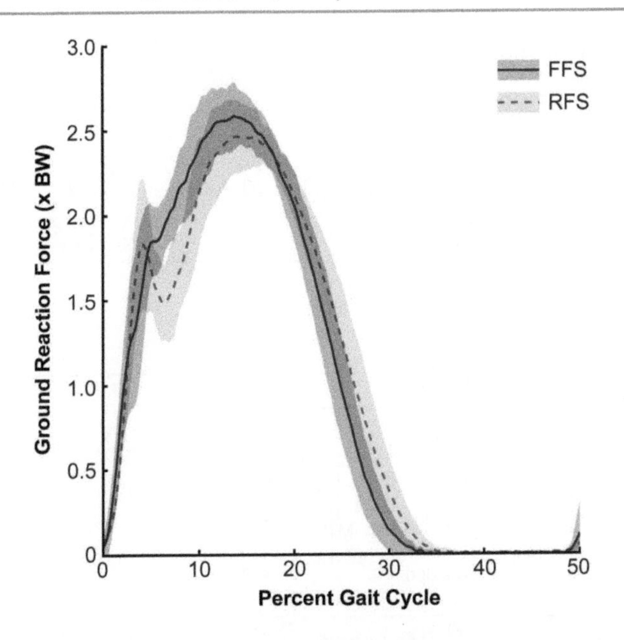


**Fig. 12.74** Comparison of vertical ground reaction force during a rearfoot (RFS) versus a forefoot (FFS) initial contact. Note the early transient impulse for the rearfoot striker. (From Yong et al. (2014), with permission)

mass is lowered during stance, the gravitational potential energy decreases, and as the foot contacts the ground, kinetic energy decreases. Much of the energy is converted into elastic potential energy and stored in the muscles, tendons, aponeuroses, and ligaments. Recall, our previous discussions on the viscoelastic properties of these tissues (see Chaps. 3 or 4 and Appendix H). For example, when a tendon is loaded under tension in the elastic region, it is temporally deformed, i.e., strained, until the load is released (Fig. 12.77). Upon release, the tendon recoils and the elastic energy stored can be used, along with muscle shortening, to control or move a joint; that is, when the energy is released by the tendon, it can contribute to the work of controlling or moving joints. We have already provided examples of this mechanism during the walking cycle, yet this is a major form of potential energy during running.

During the acceleration phase, i.e., when energy is generated, the center of mass accelerates upward and forward, and both the gravitational potential and kinetic energies increase. The energy needed for this comes from the mechanical work of muscles and the release of the elastic potential energy stored in tendons (primary), muscles, ligaments, and other related superficial and deep fasciae. So how does energy generated by distal segment muscles work to increase the energy of proximal segments? The answer is through the transfer of energy from one segment to another. This is accomplished through two mechanisms, sometimes referred to as "passive" and "active" energy (or power) transfers. Passive energy transfer occurs through the joint reaction forces. The power due to a force is the dot product of the force and the velocity of the point of application of that force:

$$P = \boldsymbol{F} \cdot \boldsymbol{v}$$

Reaction forces acting at a joint occur in equal-and-opposite, collinear pairs. For example, the force at the knee acting on the tibia, $\boldsymbol{F}_{k,\,\text{tibia}}$, is the negative of the force at the knee acting on the femur, $\boldsymbol{F}_{k,\,\text{tibia}}$:

$$\boldsymbol{F}_{k,\text{tibia}} = -\boldsymbol{F}_{k,\text{femur}}$$

The velocity of the knee joint, $\boldsymbol{v}_k$, is the same on both the proximal tibia and the distal femur. Thus, the power due to the reaction force acting at the knee on the tibia, $P_{k,\,\text{tibia}}$, is the negative of the power due to the reaction force acting at the knee on the femur, $P_{k,\,\text{femur}}$:

$$P_{k,\text{tibia}} = \boldsymbol{F}_{k,\text{tibia}} \cdot \boldsymbol{v}_k = -\boldsymbol{F}_{k,\text{femur}} \cdot \boldsymbol{v}_k = -P_{k,\text{femur}}$$

Note that the sum of these two powers is zero, meaning that the joint reaction force does not generate or absorb power.

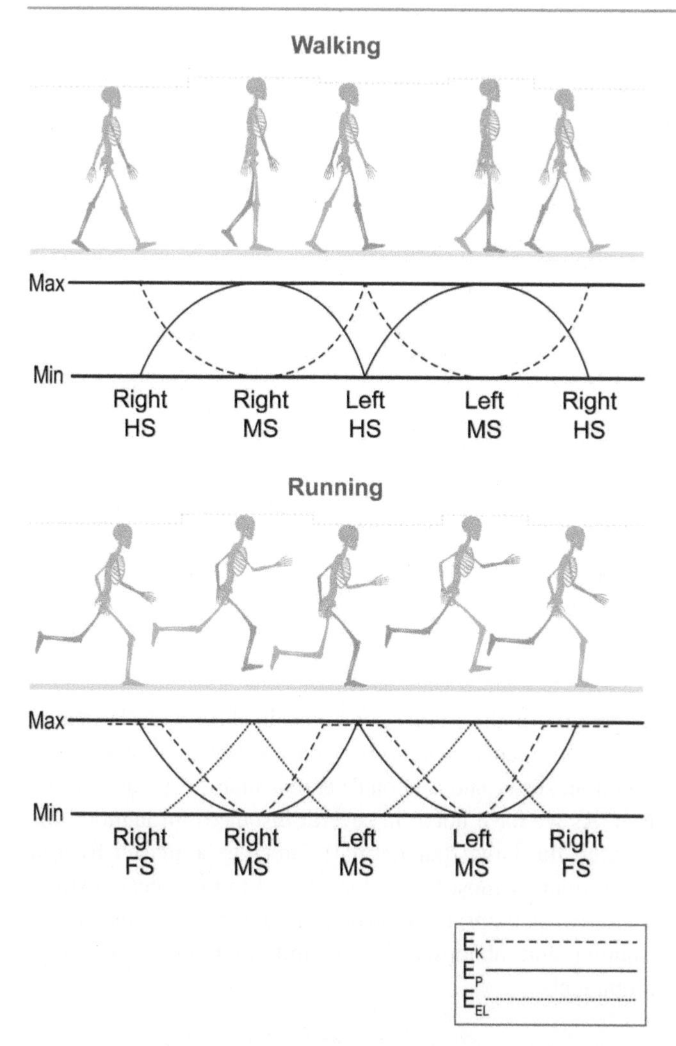


**Fig. 12.75** Phases of walking and running with corresponding cycles of mechanical energy relative to locomotion phases. It is notable that in the model for walking, i.e., termed "inverted pendulum," illustrated here, the center of mass vaults over a relatively extended leg during the stance phase, efficiently exchanging potential and kinetic energies out of phase with every step. Notice that for walking, $E_K$ and $E_P$ are out of phase, whereas for running, they are in phase. With running, the use of a mass-spring model, where the exchange of potential and kinetic energies is different, appears to show that greater locomotion speeds promote better metabolic efficiency. $E_K$ = kinetic energy, $E_P$ = potential energy, $E_{ES}$ = elastic strain energy, HS = heel strike, MS = midstance, TO = toe-off, FS = foot strike, Max. = maximum, Min. = minimum. (Modified with permission from Bramble and Lieberman (2004))

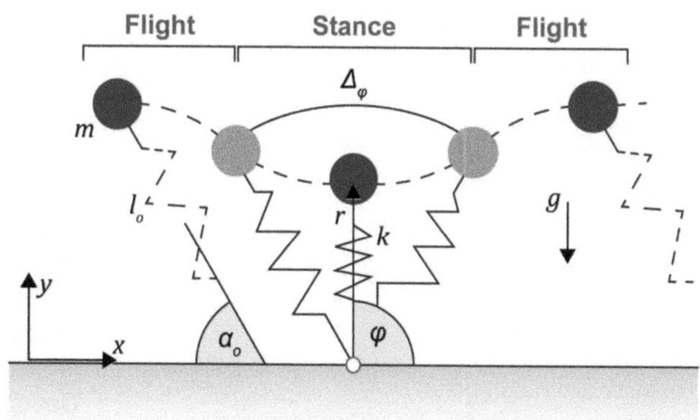


**Fig. 12.76** Schematic of a mass-spring model during phases of running. In this model, the mass of the body is lumped into a point mass, which sits on top of a massless linear spring representing the leg. $M$ = point mass, $l_0$ = rest length, $\alpha_0$ = leg angle of attack during flight, $g$ = gravitational acceleration, $k$ = spring stiffness, $r$ = radial and $\varphi$ = angular positions of the point mass, $\Delta\varphi$ = angle swept during stance. (Modified with permission from Geyer et al. (2005))

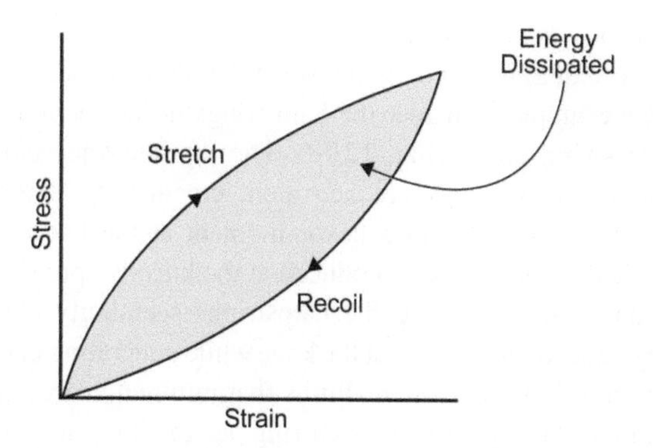


**Fig. 12.77** Typical stress/strain curve for a tendon loaded within its elastic limit, i.e., prior to the tissue's yield point, and hysteresis loop

However, the power due to the joint reaction force is negative on one segment and positive on the adjoining segment. For example, when $P_{k,\text{femur}}$ is positive, $P_{k,\text{tibia}}$ is negative, and energy is being taken away from the tibia and added to the femur. This type of energy transfer occurs continuously as we move.

Active energy transfer is achieved through muscle activity. When muscles act concentrically, i.e., shortening while producing force, energy is generated. When muscles act eccentrically, i.e., lengthening while producing force, energy is dissipated. When muscles act isometrically, i.e., producing force without changing length, no energy is generated or dissipated. However, in all three of these cases, energy can be transferred between the segments that the muscles are attached to, consistent with the same principles described above for passive energy transfer. Consider the isometric case, wherein no energy is generated. A muscle force acting at its origin is the same as the force acting at its insertion but opposite in direction. In addition, with the velocity of the origin point not being the same as the velocity of the insertion point, power could be negative on one segment and positive on the other, thus transferring energy from one segment to another, even when no energy is being generated or absorbed. When muscles act concentrically or eccentrically, energy transfer can still occur at the same time energy is being generated or dissipated. Monoarticular muscles can only transfer energy between adjacent segments. Biarticular muscles have the advantage of being able to transfer energy

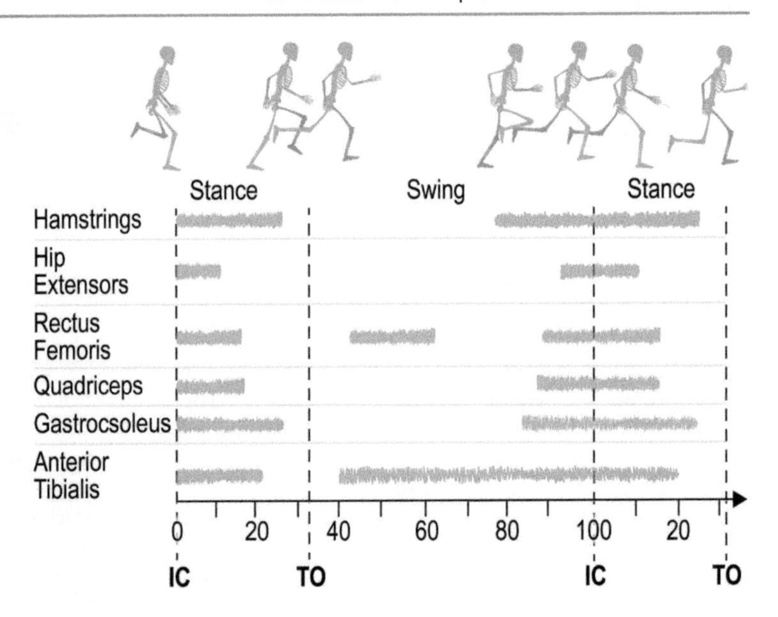


**Fig. 12.78** EMG of selected lower extremity muscles during a running gait cycle. The normal onset and cessation of muscle activity are represented by simulated raw EMG interference patterns. Note the interesting absence of muscle activation at the time of TO. TO = toe-off, IC = initial contact. (Modified with permission from Mann and Hagy (1980))

between segments that are not even adjacent; e.g., the rectus femoris can transfer energy between the tibia and pelvis.

For example, consider the hamstrings in the second half of the swing phase (Fig. 12.78). The hip and knee extend while the hamstrings are activated, creating an extensor moment at the hip and a flexor moment at the knee; it is notable that the moment produced at the knee is opposite its motion. Thus, in effect, the hamstrings seemingly absorb energy due to their action at the knee while generating energy at the hip. However, the reality is that minimal energy generation or dissipation is occurring as the muscle length change is minimal, but energy is being transferred from distal to proximal segments. Other examples of energy transfer between segments include the actions of the quadriceps, which include both one- and two-joint muscles and gastrocnemius muscles: they both transfer mechanical energy between the proximal and distal segments of the leg.

It is intuitive that the amplitudes of the joint kinematics, e.g., joint angles, and kinetics, e.g., net joint moments and powers, increase during running yet are dependent on the speed of running. However, the patterns of joint motions, joint kinetics, and muscle activity are very similar in walking and running. Given the differences in running dynamics, however, increases in ground forces and internal reactive stresses are significant. During running, the center of mass is accelerated by gravity and ground reaction forces. Ground reaction forces are heavily influenced by the velocities of impact of the feet, the center of mass velocities, and the muscle forces of the stance limb. Throughout most of the stance phase, the ankle plantar flexors (primarily soleus and gastrocnemius but also including the tibialis posterior and fibularis longus and brevis) are the main contributors to upward and forward accelerations of the center of mass, with the soleus as the largest contributor. We can imagine that the magnitude of the forces generated in these muscle-tendon actuators creates concomitant bone-on-bone forces (compression, shear, and torsional) across many structures in the foot. As we have noted in several instances throughout this chapter, the large (but normal) forces experienced by Ted while running most likely contributed to his repeated stress fractures, perhaps made worse as a result of changes in training and abnormal lower limb and ankle/foot bony alignments.

## 12.8 Chapter Summary

This case provides a wonderful opportunity to fashion an integrated model of the ankle/foot complex and discover its ability to provide a stable platform for functional activities, distribute/attenuate large ground forces, adapt to the terrain, and provide a rigid lever to assist in the acceleration of the body mass during locomotion activities. This case reinforces the importance of the ideas of redundant degrees of freedom with regard to joint kinematics and muscle synergies and the dynamic coupling of segments within a complex system, i.e., ankle/foot, as well as outside that system, i.e., shank, thigh, and pelvis.

Because of the discovery of abnormal muscle tone as part of Ted's physical examination, we explore the nature of spasticity and its possible link to cerebral palsy. Likewise, the shank and foot deformities Ted demonstrated directed us to examine developmental biomechanics and the natural history of foot deformities in children with cerebral palsy. We learn that abnormal muscle tone and the presence of foot deformities likely changed the ankle/foot system's ability to distribute/attenuate the large ground forces associated with running.

**Table 12.6** Summary of ankle/foot kinematics during normal walking

| Leg and rearfoot | Midfoot | Forefoot |
|---|---|---|
| *Weight acceptance: Initial contact to loading response (0–12% of the gait cycle)* | | |
| Tibial rotates from a position of external rotation at heel contact to internal rotation; Ankle plantar flexes approximately 10° (about longitudinal axis) Calcaneus everts at STJ Pronates | MTJ joint pronation (about oblique axis) with "unlocking" of cuboid and navicular | Forefoot contact by the end of loading response; Forefoot supinated relative to midfoot and rearfoot |
| *Single-limb support: Midstance to terminal stance (12–50% of the gait cycle)* | | |
| Anterior movement of tibia; Reversal of tibial rotation from maximum internal rotation to external rotation by terminal stance of this phase; STJ pronation maximum between 25% and 50% of stance; Calcaneus inverts to accompany STJ supination to the neutral position by the end of this period | MTJ pronation relative to rearfoot; Osseous "locking" of calcaneocuboid joint stabilizes forefoot against rearfoot by end of this period of the gait cycle | Plantar flexion of medial forefoot to maintain forefoot contact by the end of stance; Osseous locking of calcaneocuboid joint in combination with ligamentous tension stabilizes forefoot on rearfoot |
| *Swing limb advancement: Preswing (50–62% of the gait cycle)* | | |
| Tibial external rotation reaches maximum prior to toe-off; Ankle plantar flexion to 20° by end of pre-swing; STJ supination; Calcaneus inversion | Midtarsal joint supination (about oblique axis) accompanies STJ supination | Significant first ray plantar flexion at this phase may be described as "pronation" twist: of forefoot relative to midfoot |

Nawoczenski and Epler (1997)
*MTJ* midtarsal joint, *STJ* subtalar joint

We review the definition and nature of stress fractures and provide two hypothetical models that attempt to explain the general etiology or course of events leading to stress fractures. This process of investigation reinforces several of the hypotheses we develop with regard to the possible causes of Ted's recurrent stress fractures. We learn that stress fracture risk is related to extrinsic and intrinsic factors that interact in complex ways, leading to injury. Notions of the interaction between active and passive mechanisms relevant to the attenuation of ground forces are reviewed.

In order to understand Ted's problem related to running, we first apply our understanding of the anatomy and biomechanics of the ankle and foot to the events of normal walking. Although we focus on the ankle and foot, we learn that it influences and is influenced by the ground forces from below and the segments proximal to it. Along the way, we look for and propose relationships between Ted's deformities, impairments, and abnormal muscle tone with changes in his gait. The observational and instrumented measures of his gait corroborate most of our hypotheses. Then we examine the similarities between walking and running, and their differences, to cement our clinical and biomechanical conclusions.

What do we conclude? Let us first look at a summary of what should happen within the ankle/foot complex during walking (Table 12.6). Because of Teds' tibial varum and relatively rigid talipescavovarus foot, with a plantar flexed first ray and adducted forefoot, several abnormal events likely took place during walking and running. His tibial varum predisposed a varus foot position at initial contact, which normally would be accommodated by increased pronation during weight acceptance/loading response. His foot deformity could not make that accommodation, let alone pronate normally during weight acceptance into single-limb support. Thus, the large ground impulse could not be attenuated well by active and passive mechanisms; that is, large ground forces (which were not abnormally high) were likely redistributed in ways that loaded foot bones and joints in a repetitive manner, which led to a stress reaction and eventual fracture. For example, the abnormal distribution of compression forces within the lateral column likely led to the fourth metatarsal fracture. We note that although it has been reported that fourth metatarsal stress fractures are not categorized as high risk, i.e., not common, we believe that Ted's unique combination of foot abnormalities' inability to mitigate the repetitive impulsive loading associated with running provided a plausible explanation for Ted's fourth metatarsal fracture. Moreover, the abnormal distribution of ground and muscle forces, i.e., posterior tibialis and fibularis longus, along the medial column distal to the plantar flexed first ray likely led to the fracture of the second metatarsal. The greater peak and rate of loading associated with running only exacerbated Ted's force redistribution problems. We cannot forget the role that extrinsic factors of changes in training and running shoes played in the interaction with Ted's biology to create the environment leading to stress fractures. Finally, Ted's increased resting muscle tonus, and perhaps active muscle mechanisms, likely created joint and bone forces, which only increased internal stresses, predisposing his feet to stress fractures.

## Bibliography


Alonso-Vázquez A, Villarroy MA, Franco MA et al (2009) Kinematic assessment of paediatric forefoot varus. Gait Posture 29:214–219. https://doi.org/10.1016/j.gaitpost.2008.08.009

Araújo VL, Santo Thiago RT, Khuu A et al (2020) The effects of small and large varus alignment of the foot-ankle complex on lower limb kinematics and kinetics during walking: a cross-sectional study. Musculoskelet Sci Pract 47:102149. https://doi.org/10.1016/j.msksp.2020.102149

Benink RJ (1985) The constraint-mechanism of the human tarsus, a roentgenological experimental study. Acta Orthop Scand 56(sup215):1–135. https://doi.org/10.3109/17453678509154158

Bennell KL, Malcolm SA, Wark JD et al (1999) Models for the pathogenesis of stress fractures in athletes. Br J Sports Med 30:200–204

Blickhan R (1989) The spring-mass model for running and hopping. J Biomech 22(11–12):1217–1227

Bogduk N (2012) The lumbar vertebrae. In: Bogduk N (ed) Clinical and radiological anatomy of the lumbar spine, 5th edn. Elsevier, Churchill Livingstone, Edinburgh

Bramble DM, Lieberman DE (2004) Endurance running and the evolution of Homo. Nature 432:345–352

Breine B, Malcolm P, Van Caekenberghe I et al (2017) Initial foot contact and related kinematics affect impact loading rate in running. J Sports Sci 35(15):1556–1564. https://doi.org/10.1080/02640414.2016.1225970

Buldt AK, Levinger P, Murley GS et al (2015) Foot posture is associated with kinematics of the foot during gait: a comparison of normal, planus and cavus feet. Gait Posture 42:42–48. https://doi.org/10.1016/j.gaitpost.2015.03.004

Buldt AK, Allan JJ, Landorf KB et al (2018) The relationship between foot posture and plantar pressure during walking in adults: a systematic review. Gait Posture 62:56–67. https://doi.org/10.1016/j.gaitpost.2018.02.026

Burr DB, Martin RB, Schaffler MB et al (1985) Bone remodeling in response to *in vivo* fatigue microdamage. J Biomech 18(3):189–200

Cappellini G, Ivanenko YP, Poppele RE et al (2006) Motore pattens in human walking and running. J Neurophysiol 95:3426–3437. https://doi.org/10.1152/jn.00081.2006

Carson MC, Harrington ME, Thompson N et al (2001) Kinematic analysis of a multi-segment foot model for research and clinical applications: a repeatability analysis. J Biomech 34:1299–1307

Carter DR, Caler WE, Spengler DM et al (1981) Fatigue behavior of adult cortical bone: the influence of mean strain and strain range. Acta Orthop Scand 52:481–490

Chan CW, Rudins A (1994) Foot biomechanics during walking and running. Mayo Clin Proc 69:448–461

Davids J (2010) The foot and ankle in cerebral palsy. Orthop Clin North Am 41:579–593. https://doi.org/10.1016/j.oci.2010.06.002

De Wit B, De Clercq D, Aerts P (2000) Biomechanical analysis of the stance phase during barefoot and shot running. J Biomech 33(3):269–278

DeLeo AT, Dierks TA, Ferber R et al (2004) Lower extremity joint coupling during running: a current update. Clin Biomech 19:983–991. https://doi.org/10.1016/j.clinbiomech.2004.07.005

Deschamps K, Staes F, Roosen P et al (2011) Body of evidence supporting the clinical use of 3D multisegment foot models: a systematic review. Gait Posture 33:338–349. https://doi.org/10.1016/j.gaitpost.2010.12.018

Dierks TA, Davis I (2017) Discrete and continuous joint coupling relationships in uninjured recreational runners. Clin Biomech 22:581–591. https://doi.org/10.1016/j.clinbiomech.2007.01.012

Ellison MA, Kenny M, Fulford J et al (2020) Incorporating subject-specific geometry to compare metatarsal stress during running with different foot strike patterns. J Biomech 105:109792. https://doi.org/10.1016/j.jbiomech.2020.109792

Farley CT, González O (1996) Leg stiffness and stride frequency in human running. J Biomech 29(2):181–186

Fernández-Seguín LM, Diaz Mancha JA, Sánchez Rodríquez R et al (2014) Comparison of plantar pressures and contact areas between normal and cavus foot. Gait Posture 39:789–792. https://doi.org/10.1016/j.gaitpost.2013.10.018

Geyer H, Seyfarth A, Blickhan R (2005) Spring-mass running: simple approximate solution and application to gait stability. J Theor Biol 232(3):315–328. https://doi.org/10.1016/j.jtbi.2004.08.015

Geyer H, Seyfarth A, Blickhan R (2006) Compliant leg behavior explains basic dynamics of walking and running. Proc R Soc B 273:2861–2867. https://doi.org/10.1098/rspb.2006.3637

Gomes RBO, Souza TR, Paes BDC et al (2019) Foot pronation during walking is associated to the mechanical resistance of the midfoot complex. Gait Posture 70:20–23. https://doi.org/10.1016/j.gaitpost.2019.01.027

Greaser MC (2016) Foot and ankle stress fractures in athletes. Orthop Clin North Am 47:809–822. https://doi.org/10.1016/j.ocl.2016.05.016

Hamner SR, Seth A, Delp S (2010) Muscle contributions to propulsion and support during running. J Biomech 43:2709–2716. https://doi.org/10.1016/j.biomech.2010.06.025

Hicks JH (1954) The mechanics of the foot. II. The plantar aponeurosis and the arch. J Anat 88(1):25–30

Kapandji AI (2019) The plantar vault. In: Kapandji AI (ed) The physiology of the joints, the lower limb, 7th edn. Handspring Publishing Limited, Pencaitland

Kelly LA, Cresswell AG, Ricinais S et al (2014) Intrinsic foot muscles have the capacity to control deformation of the longitudinal arch. J R Soc Interface 11:20131188. https://doi.org/10.1098/rsif.2013.1188

Kelly LA, Lichtwark G, Cresswell AG (2015) Active regulation of longitudinal arch compression and recoil during walking and running. J R Soc Interface 12:20141076. https://doi.org/10.1098/rsif.2014.1076

Ker RF, Bennett MB, Bibby SR et al (1987) The spring in the arch of the human foot. Nature 325:147–149

Knorz S, Kluge F, Gelse K et al (2017) Three-dimensional biomechanical analysis of rearfoot and forefoot running. Orthop J Sports Med 5(7):2325967117719065. https://doi.org/10.1177/2325967117719065

Kondo M, Iwamoto Y, Kito N (2021) Relationship between forward propulsion and foot motion during gait in healthy young adults. J Biomech 121:110431. https://doi.org/10.1016/j.jbiomech.2021.110431

Leardini A, Caravaggi P, Theologis T et al (2019) Multi-segment foot models and their use in clinical populations. Gait Posture 69:50–59. https://doi.org/10.1016/j.gaitpost.2019.01.022

LeVeau BF, Bernhardt DB (1984) Developmental biomechanics, effect of forces on growth, development, and maintenance of the human body. Phys Ther 64(12):1874–1882

Lundberg A, Svensson OK (1993) The axes of rotation of the talocalcaneal and talonavicular joints. The Foot 3:65–70

Lundberg A, Svensson OK, Bylund C et al (1989) Kinematics of the ankle/foot complex-part 2: pronation and supination. Foot Ankle Int 9(5):248–253

Lundgren P, Nester C, Liu A et al (2008) Invasive in vivo measurement of rear-, mid- and forefoot motion during walking. Gait Posture 28:93–100. https://doi.org/10.1016/j.gaitpost.2007.10.009

Magalhães FA, Fonseca ST, Araújo VL et al (2021) Midfoot passive stiffness affects foot and ankle kinematics and kinetics during the propulsive phase of walking. J Biomech 119:110328. https://doi.org/10.1016/j.jbiomech.2021.110328

Mani H, Miyagishima S, Kozuka N et al (2021) Development of the relationships among dynamic balance control, inter-limb coordination, and torso coordination during gait in children ages 3–10 years. Front Hum Neurosci 15:740509. https://doi.org/10.3389/fnhum.2021.740509

Mann RA, Hagy J (1980) Biomechanics of walking, running, and sprinting. Am J Sports Med 8(5):345–350

Mann R, Inman VT (1964) Phasic activity of intrinsic muscles of the foot. J Bone Joint Surg 46A(3):469–481

Manter JT (1941) Movements of the subtalar and transverse tarsal joints. Anat Rec 80(4):397–410. https://doi.org/10.1002/ar.1090800402

Manter JT (1946) Distribution of compression forces in the joints of the human foot. Anat Rec 96(3):313–321. https://doi.org/10.1002/ar.1090960306

May AS, Keros ST (2017) The neurological exam in neonates and toddlers. In: Greenfield JP, Long CB (eds) Common neurosurgical conditions in the pediatric practice: recognition and management. Springer, New York, pp 11–26. https://doi.org/10.1017/978-1-4939-3807-0

Mayer SW, Joyner PW, Almekinders LC et al (2014) Stress fractures of the foot and ankle in athletes. Sports Health 6(6):481–491. https://doi.org/10.1177/1941738113486588

McClay I, Manal K (1999) Three-dimensional kinetic analysis of running: significance of secondary planes of motion. Med Sci Sports Exerc 31(11):1629–1673

McDonald KA, Stearne SM, Alderson JA et al (2016) The role of arch compression and metatarsophalangeal joint dynamics in modulating plantar fascia strain during running. PloS One 11(4):e0152602. https://doi.org/10.1371/journal.pone.0152602

McKeon PO, Hertel J, Bramble D et al (2015) The foot core system: a new paradigm for understanding intrinsic foot function. Br J Sports Med 49:290. https://doi.org/10.1136/bjsports-2013-092690

McPoil TG, Knecht HG (1985) Biomechanics of the foot in walking: functional approach. J Orthop Sports Phys Ther 7(2):69–72

Miller F (2020a) Motor control and muscle tone problems in cerebral palsy. In: Miller F, Bachrach S, Lennon N et al (eds) Cerebral palsy, 2nd edn. Springer, New York, pp 559–583. https://doi.org/10.1007/978-3-319-74558-9

Miller F (2020b) Foot deformities in children with cerebral palsy: an overview. In: Miller F, Bachrach S, Lennon N et al (eds) Cerebral palsy, 2nd edn. Springer, New York, pp 2211–2221. https://doi.org/10.1007/978-3-319-74558-9

Miller F, Church C (2020) Natural history of foot deformities in children with cerebral palsy. In: Miller F, Bachrach S, Lennon N et al (eds) Cerebral palsy, 2nd edn. Springer, New York, pp 2223–2232. https://doi.org/10.1007/978-3-319-74558-9

Monaghan GM, Lewis CL, Hsu W-H et al (2013) Forefoot angle determines duration and amplitude of pronation during walking. Gait Posture 38:8–13. https://doi.org/10.1016/j.gaitpost.2012.10.003

Mozafaripour E, Rajabi R, Minoonejad H (2018) Anatomical alignment of lower extremity in subjects with genu valgum and genu varum deformities. Phys Treat 8(1):27–36. https://doi.org/10.32598/ptj.8.1.27

Naemi R, Chockalingam N (2013) Mathematical models to assess foot-ground interaction: an overview. Med Sci Sports Exerc 45(8):1524–1533. https://doi.org/10.1249/MSS.0b013e3182be3a7

Nawoczenski DA, Epler ME (1997) Orthotics in functional rehabilitation of the lower limb. W.B. Saunders Company, Philadelphia

Nester CJ, Findlow AH (2006) Clinical and experimental models of the midtarsal joint. Proposed terms of reference and associated terminology. J Am Podiatr Med Assoc 96(1):24–31

Nester CJ, Findlow A, Bowker P (2001) Scientific approach to the axis of rotation at the midtarsal joint. J Am Podiatr Med Assoc 91(2):68–73

Neumann DA (2017) Ankle and foot. In: Neumann DA (ed) Kinesiology of the musculoskeletal system, foundations for rehabilitation, 3rd edn. Elsevier, St. Louis

Nigg BM, Liu W (1999) The effect of muscle stiffness and damping on simulated impact forces during running. J Biomech 32:849–856

Nilsson J, Thorstensson A (1989) Ground reaction forces at different speeds of human walking and running. Acta Physiol Scand 136:217–227

Noordin S, Umer M, Hafeez K et al (2010) Developmental dysplasia of the hip. Orthop Rev 2(2):e19. https://doi.org/10.4081/or.2010.e19

Novacheck TF (1998) The biomechanics of running. Gait Posture 7:77–95

Oatis CA (2017) Kinesiology, the mechanics and pathomechanics of human movement, 3rd edn. Wolters Kluwer, Philadelphia

Okita N, Meyers SA, Challis JH et al (2014) Midtarsal joint locking: new perspective on an old paradigm. J Orthop Res 32(1):110–115. https://doi.org/10.1002/jor.22477

Patel DR, Neelakantan M, Pandher K et al (2020) Cerebral palsy in children: a clinical overview. Transl Pediatr 9(Suppl 1):S125–S135. https://doi.org/10.21037/tp.2020.01.01

Periyasamy R, Anand S (2013) The effect of foot arch on plantar pressure distribution during standing. J Med Eng Technol 37(5):342–347. https://doi.org/10.3109/03091902.2013.810788

Perry J, Burnfield JM (2010) Total limb function and bilateral synergistic relationships. In: Perry J, Burnfield JM (eds) Gait analysis, normal and pathological function, 2nd edn. SLACK Incorporated, Thorofare

Phan C-B, Nguyen D-P, Kee KM et al (2018) Relative movement on the articular surfaces of the tibiotalar and subtalar joints during walking. Bone Joint Res 7(8):501–507. https://doi.org/10.1302/2046-3758.78.BJR-2018-0014.R1

Phan C-B, Shin G, Lee KM et al (2019) Skeletal kinematics of the midtarsal joint during walking: midtarsal joint locking revisited. J Biomech 95:109287. https://doi.org/10.1016/j.biomech.2019.07.031

Pohl MB, Messenger N, Buckley JG (2007) Forefoot, rearfoot and shank coupling: effect of variations in speed and mode of gait. Gait Posture 25:295–302. https://doi.org/10.1016/j.gaitpost.2006.04.012

Powell DW, Long B, Milner CE (2011) Frontal plane multi-segment foot kinematics in high- and low-arched females during dynamic loading tasks. Hum Mov Sci 30:105–114. https://doi.org/10.1016/j.humov.2011.08.015

Powell DW, Williams Blaise DS, Bulter RJ (2013) A comparison of two multisegment foot models in high- and low-arched athletes. J Am Podiatr Med Assoc 103(2):99–105. https://doi.org/10.7547/1030099

Powell DW, Blaise Williams DS 3rd, Windsor B et al (2014) Ankle work and dynamic joint stiffness in high- compared to low-arched athletes during a barefoot running task. Hum Mov Sci 34:147–156. https://doi.org/10.1016/j.humov.2014.007

Raychoudhury S, Hu D, Ren L (2014) Three-dimensional kinematics of the human metatarsophalangeal joint during level walking. Front Bioeng Biotechnol 2:73, 1–9. https://doi.org/10.3389/fbioe.2014.00073

Richie DH (2021) Human walking: the gait cycle. In: Richie DH (ed) Pathomechanics of common foot disorders (eBook). Springer Nature Switzerland AG, pp 45–62. https://doi.org/10.1007/978-3-030-54201-6

Roach KE, Wang B, Kapron AL et al (2016) In vivo kinematics of the tibiotalar and subtalar joints in asymptomatic subjects: a high-speed dual fluoroscopy study. J Biomech Eng 138:091006. https://doi.org/10.1115/1.4034263

Roberts TJ, Azizi E (2011) Flexible mechanisms: the diverse roles of biological springs in vertebrate movement. J Exp Biol 214:353–361. https://doi.org/10.1242/jeb.038588

Rose J, McGill KC (2005) Neuromuscular activation and motor-unit firing characteristics in cerebral palsy. Dev Med Child Neurol 47:329–336

Sarrafian SK (1987) Functional characteristics of the foot and plantar aponeurosis under tibiotalar loading. Foot Ankle Int 8(1):4–18

Sarvazyan A, Rudenko O, Aglyamov S et al (2014) Muscle as a molecular machine for protecting joints and bones by absorbing mechanical impacts. Med Hypotheses 83:6–10. https://doi.org/10.1016/j.mehy.2014.04.020

Sass P, Hassan G (2003) Lower extremity abnormalities in children. Am Fam Physician 68(3):461–468

Schaffler MB, Jepsen KJ (2000) Fatigue and repair in bone. Int J Fatigue 22:839–846

Seitz AM, Wilke H-J, Nordin M (2022) Biomechanics of bone. In: Nordin M, Frankel AH (eds) Basic biomechanics of the musculoskeletal system, 5th edn. Wolters Kluwer, Philadelphia

Seref-Ferlengez Z, Kennedy OD, Schaffler MB (2015) Bone microdamage, remodeling and bone fragility: how much damage is too much damage. Bonekey Rep 4:644, 1–7. https://doi.org/10.1038/bonekey.2015.11

Sheehan FT (2010) The instantaneous helical axis of the subtalar and talocrural joints: a non-invasive *in vivo* dynamic study. J Foot Ankle Res 3:13. https://www.jfootankleres.com/content/3/1/13

Shereff MJ, Bejjani FJ, Kummer FJ (1986) Kinematics of the first metatarsophalangeal joint. J Bone Joint Surg 68A(3):392–398

Shono H, Matsumoto Y, Kokubun T et al (2022) Determination of relationship between foot arch, hindfoot, and hallux motion using Oxford foot model: comparison between walking and running. Gait Posture 92:96–102. https://doi.org/10.1016/j.gaitpost.2021.10.043

Soutas-Little RW, Beavis GC, Verstraete MC et al (1987) Analysis of foot motion during running using a foot co-ordinate system. Med Sci Sports Exerc 19(3):285–293

Stacoff PW, Liu A, Nester C et al (2008) Functional units of the human foot. Gait Posture 28:434–441. https://doi.org/10.1016/j.gaitpost.2008.02.004

Stebbins J, Harrington M, Thompson N et al (2006) Repeatability of a model for measuring multi-segment foot kinematics in children. Gait Posture 23:401–410. https://doi.org/10.1016/j.gaitposture.2005.03.002

Stergiou N, Bates BT, James SL (1999) Asynchrony between subtalar and knee joint function during running. Med Sci Sports Exerc 31(11):1645–1655

Struzik A, Karamanidis K, Lorimer A et al (2021) Application of leg, vertical and joint stiffness in running performance: a literature review. Appl Bionics Biomech 2021:9914278. https://doi.org/10.1155/2021/9914278

Takabayashi T, Edama M, Nakamura E et al (2017) Coordination among the rearfoot, midfoot, and forefoot during walking. J Foot Ankle Res 10:42. https://doi.org/10.1186/s13047-017-0224-3

Takabayashi T, Edama M, Yokoyama E et al (2018) Quantifying coordination among the rearfoot, midfoot, and forefoot segments during running. Sports Biomech 17(1):18–32. https://doi.org/10.1080/14763141.2016.1271447

Takechi H (1977) Trabecular architecture of the knee joint. Acta Orthop Scand 48(6):673–681. https://doi.org/10.3109/17453677708994816

Thomas KA, Harris MB, Willis MC et al (1995) The effects of the interosseous membrane and partial fibulectomy on loading of the tibia: a biomechanical study. Orthopedics 18(4):373–383

Torburn L, Perry J, Gronley JK (1998) Assessment of rearfoot motion: passive positioning, one-legged standing, gait. Foot Ankle Int 19(10):688–693

Tweed JL, Campell JA, Thompson RJ et al (2008) The function of the midtarsal joint, a review of the literature. The Foot 18:106–112. https://doi.org/10.1016/j.foot.2008.01.002

Uchida TK, Delp SL (2020) Running. In: Uchida TK, Delp SL (eds) Biomechanics of movement, the science of sports, robotics, and rehabilitation. The MIT Press, Cambridge

Van der Krogt MM, Bar-On L, Kindt T et al (2016) Neuromusculolskeletal simulation of instrumented contracture and spasticity assessment in children with cerebral palsy. J Neuroeng Rehabil 13:64. https://doi.org/10.1186/s12984-016-0170-5

Van Gheluwe B, Dananberg HJ, Hagman F et al (2006) Effects of hallux limitus on plantar foot pressure and foot kinematics during walking. J Am Podiatr Med Assoc 96(5):428–436

Vera AM, Patel KA (2021) Stress fractures of the foot and ankle. Oper Tech Sports Med 00:150852. https://doi.org/10.1016/j.otsm.2021.150852

Voloshin AS (1994) Effect of fatigue on the attenuation capacity of human musculoskeletal system (abstract). J Biomech 27(6):708

Wakeling JM, Liphardt A-M, Nigg BM (2003) Muscle activity reduces soft-tissue resonance at heel-strike during walking. J Biomech 36:1761–1769. https://doi.org/10.1016/S0021-9290(03)00216-1

Wei Z, Li JX, Fu W et al (2020) Plantar load characteristics among runners with different strike patterns during preferred speed. J Exerc Sci Fit 18:89–93. https://doi.org/10.1016/j.esf.2020.01.003

Weist R, Eils E, Rosenbaum D (2004) The influence of muscle fatigue on electromyogram and plantar pressure patterns as an explanation for the incidence of metatarsal stress fractures. Am J Sports Med 32(8):1893–1898. https://doi.org/10.1177/0363546504265191

Wilder RP, Sethi S (2004) Overuse injuries: tendinopathies, stress fractures, compartment syndrome, and shin splints. Clin Sports Med 23:55–81. https://doi.org/10.1016/S0278-5919(03)00085-1

Wright IC, Neptune RR, van den Bogert AJ et al (1998) Passive regulation of impact forces in heel-toe running. Clin Biomech 13:521–531

Ye D, Sun X, Zhang C et al (2021) In vivo foot and ankle kinematics during activities measured by using dual fluoroscopic imaging system: a narrative review. Front Bioeng Biotechnol 9:693806. https://doi.org/10.3389/fbioe.2021.693806

Yong JR, Silder A, Delp SL (2014) Differences in muscle activity between natural forefoot and rearfoot strikers during running. J Biomech 47:3593–3597. https://doi.org/10.1016/j.jbiomech.2014.10.015

Yong JA, Dembia CL, Silder A et al (2020) Foot strike pattern during running alters muscle-tendon dynamics of the gastrocnemius and the soleus. Sci Rep 10:5872. https://doi.org/10.1038/s41598-020-62464-3

Zadpoor AA, Nikooyan AA (2010) Modeling muscle activity to study the effects of footwear on the impact forces and vibrations of the human body during running. J Biomech 43:186–193. https://doi.org/10.1016/j.biomech.2009.09.028

Zadpoor AA, Nikooyan AA (2011) The relationship between lower-extremity stress fractures and the ground reaction force: a systematic review. Clin Biomech 26:23–28. https://doi.org/10.1016/j.clinbiomech.2010.08.005

# Appendix A: Anthropometry

Anthropometry is a branch of anthropology that examines the physical measurements of the human body. In biomechanics, the study of muscle and joint forces, and net joint moments and powers, requires body measurements, including overall body height, mass, and center of mass, as well as the length, mass, center of mass, and mass moments of inertia of body segments. Historically, these measures were made using human cadavers, which were then used to estimate the physical properties of all body segments using an individual's mass and height. It is notable that using anthropometric measures based on cadaver studies is prone to errors related to the natural variation in anatomical proportions, age, and gender of the anatomical specimens. With the use of more advanced technology in recent decades, e.g., radiographic and magnetic resonance imaging, estimates of segment volumes and density have improved the accuracy of human anthropometric measures. Whereas with the more advanced methods of measurement an individual's characteristics relative to the natural variation in anatomical proportions related to age and gender can be accounted for, these methods are costlier and less efficient. Therefore, the work by Dempster, Drillis and Contini, and others, including measures of human body segment masses; positions of the center of mass; moments of inertia, i.e., the radius of gyration; and lengths based on total body mass and height, provides the framework for body segment estimates used by most biomechanists today.

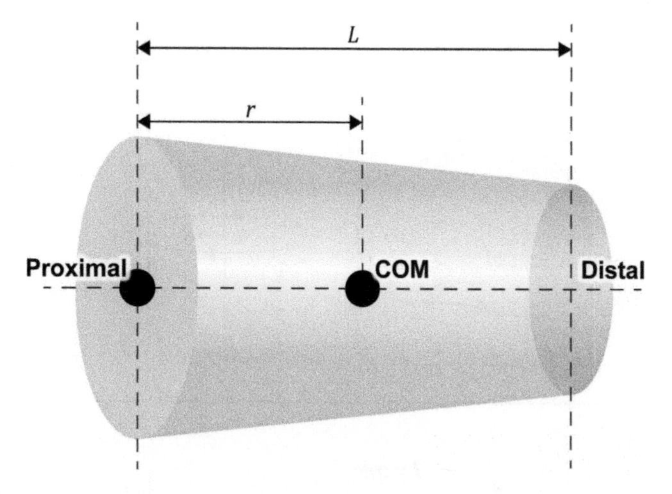


**Fig. A.1** Schematic of a rigid body segment, with mass $m$, including essential anthropometric measures relative to a point at the proximal end of the segment; $r$ = the distance of the center of mass from the proximal end; $L$ = length of the body segment; COM = the location of the segment's center of mass

The anthropometric measurements most commonly needed for what we do in biomechanics include the following for each segment: length ($L$), mass ($m$), and the distance of the center of mass (COM) from the proximal end ($r$) (Fig. A.1).

The location of the COM of each segment is needed in order to analyze the translational movement of a segment in space. As discussed in Chap. 2, Newton's second law, i.e., $\sum \boldsymbol{F} = m\boldsymbol{a}$, describes how the sum of the linear forces, $\boldsymbol{F}$, acting on the body is directly proportional to the resultant linear acceleration, $\boldsymbol{a}$, of the body's mass, $m$. The constant of proportionality in Newton's second law is mass, $m$, which represents resistance to changes in velocity. We can also apply Euler's equation for planar bodies, $\sum M = I\alpha$, which describes how the sum of the moments, $M$, with respect to the center of mass is directly proportional to the angular acceleration, $\alpha$, of the body. The constant of proportionality for Euler's equation is the mass moment of inertia, $I$, with respect to the mass center, which represents resistance to changes in angular velocity.

The mass moment of inertia of a single particle of mass $m$ with respect to a given axis is given by:

$$I = mr^2$$

where $r$ is the distance between the axis of rotation and the particle. For a body that contains many particles, this becomes:

$$I = \sum_i m_i r_i^2 = \int r^2 dm$$



G. J. Alderink, B. M. Ashby, *Clinical Kinesiology and Biomechanics*, https://doi.org/10.1007/978-3-031-25322-5

Calculating the summation or even the integral in the above equation can be difficult to do at times. Therefore, the concept of the *radius of gyration* is often used to assist in the calculation of the mass moment of inertia. In terms of the *radius of gyration*, the mass moment of inertia is expressed as:

$$I = mk^2$$

and the radius of gyration, $k$, can be determined as:

$$k = \sqrt{\frac{I}{m}}$$

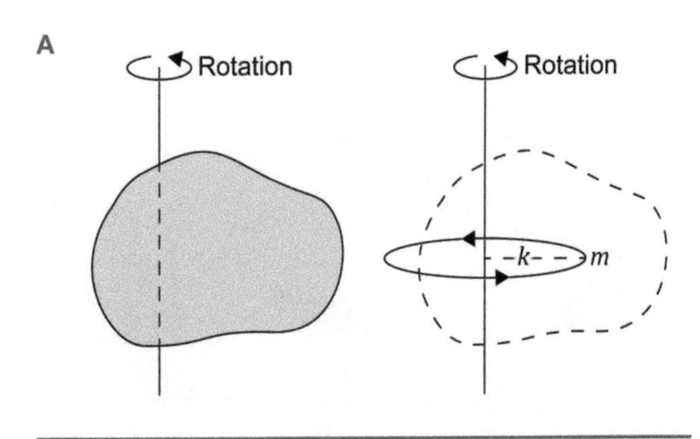


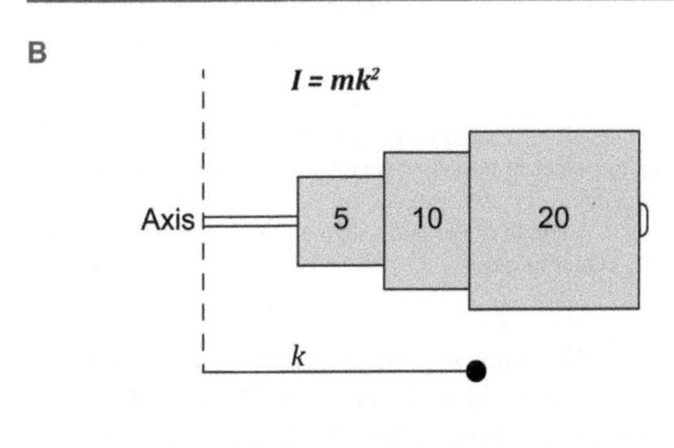


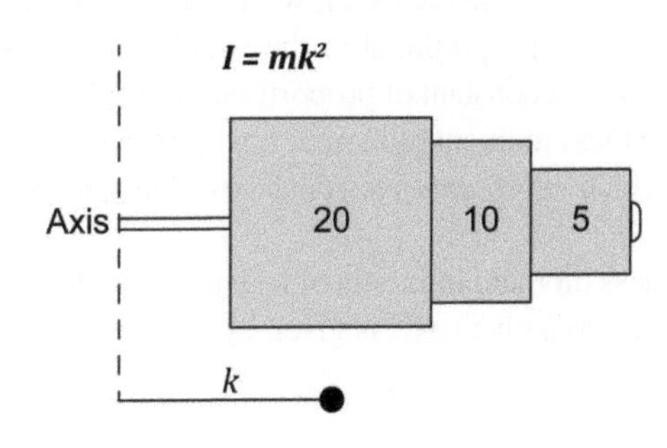


**Fig. A.2** (**A**) The rotation of a segment with distributed mass $m$ about an arbitrary axis of rotation with a *radius of gyration* $k$ and (**B**) an example of how the mass moment of inertia $I$ changes with changes in $k$ induced by changes in the distribution of mass. Although each system has equal masses, the upper figure has greater $I$ because its mass is distributed in a way that increases $k$. Note: the dashed lines represent the axis of rotation

So what does the radius of gyration physically represent? If you were to rearrange all of the particles into a ring of constant radius without changing the mass moment of inertia, the radius of that ring would be the *radius of gyration*. Since the *radius of gyration* represents a distance, its dimension is length, expressed in units such as feet or meters. Practically speaking, the *radius of gyration* is a way of representing the mass moment of inertia normalized by its mass (Fig. A.2). A notable feature of the *radius of gyration* is that it changes as the distribution of the segment's mass changes. Moreover, two bodies with the same shape and dimensions, but made of different materials, would have the same *radius of gyration* but have different mass moments of inertia.

Segment lengths between joints are one of the most basic physical measures needed for biomechanical analyses. Segment lengths and segment length proportions vary from individual to individual. Several researchers have summarized estimates of segment lengths that can be used when, for whatever reason, we cannot measure the segment lengths of an individual directly. Fig. A.3 provides an average set of segment lengths, expressed as percentages of body height, as summarized by Drillis and Cantini (1966). A simple example

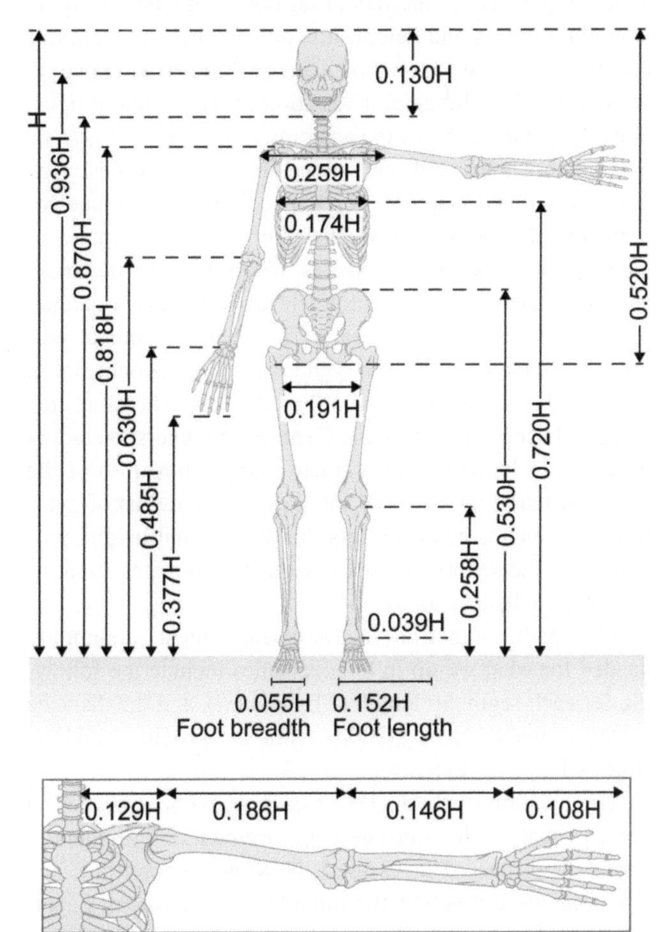


**Fig. A.3** Body segment lengths expressed relative to body height H. (Adapted from Drillis and Cantini (1966))

**Table A.1** Anthropometric data

| | Definition | Segment mass/ total body mass | Center of mass/*segment length* | | Radius of gyration/*segment length* |
|---|---|---|---|---|---|
| | | | Proximal | Distal | COM |
| Hand | Wrist axis/knuckle II middle finger | 0.006 | 0.506 | 0.494 | 0.297 |
| Forearm | Elbow axis/ulnar styloid | 0.016 | 0.430 | 0.570 | 0.303 |
| Upper arm | Glenohumeral axis/elbow axis | 0.028 | 0.436 | 0.564 | 0.322 |
| Forearm & hand | Elbow axis/ulnar styloid | 0.022 | 0.682 | 0.318 | 0.468 |
| Total arm | Glenohumeral joint/ulnar styloid | 0.050 | 0.530 | 0.470 | 0.368 |
| Foot | Lateral malleolus/head metatarsal II | 0.0145 | 0.50 | 0.50 | 0.475 |
| Leg | Femoral condyles/medial malleolus | 0.0465 | 0.433 | 0.567 | 0.302 |
| Thigh | Greater trochanter/femoral condyles | 0.100 | 0.433 | 0.567 | 0.323 |
| Foot & leg | Femoral condyles/medial malleolus | 0.061 | 0.606 | 0.394 | 0.416 |
| Total leg | Greater trochanter/medial malleolus | 0.161 | 0.447 | 0.553 | 0.326 |
| Head & neck | C7–T1 & 1st rib/ear canal | 0.081 | 1.000 | – | 0.495 |
| Thorax | C7–T1/T12–L1 & diaphragm | 0.216 | 0.82 | 0.18 | – |
| Pelvis | L4–L5/greater trochanter | 0.142 | 0.105 | 0.895 | – |
| Trunk | Greater trochanter/glenohumeral joint | 0.497 | 0.50[a] | 0.50 | – |
| Trunk/head/neck | Greater trochanter/glenohumeral joint | 0.578 | 0.66[a] | 0.34 | 0.503 |
| Head, arms, & trunk (HAT) | Greater trochanter/glenohumeral joint | 0.678 | 0.626[a] | 0.374 | 0.496 |
| HAT | Greater trochanter/midrib | 0.678 | 1.142[a] | – | 0.903 |

From Winter (2009a)
[a]Note: The proximal end is the greater trochanter; COM = center of mass

can show us how to use Fig. A.3. Let us assume an individual with a height of 1.8 m, then use the information from Fig. A.3 to estimate the length of a forearm segment:

Length of the forearm = (0.146)(1.8 m) = 0.263 m

Or with a bit more work, we find that the length of the thigh is 0.245 of the total body height. In this case, we have to subtract the height of the knee (0.285H) from the height of the hip (0.530H):

Length of the thigh = (0.530 – 0.285)(1.8 m) = 0.441 m.

Using Table A.1, we can determine many variables needed to find solutions for both static equilibrium and inverse dynamic biomechanical problems. The information in Table A.1 is used to determine the masses, centers of mass location, and mass moments of inertia for a variety of human body segments. All values in this table are normalized with the segment masses expressed as fractions of the total body mass, the center of mass locations from both the proximal and distal ends expressed as fractions of the segment length, and the radii of gyration with respect to the segment center of mass expressed as fractions of the segment length. Special care should be taken to understand how the reference segment lengths are defined in this table.

Building on the previous simple example where we determined the length of the forearm, let us now determine the location of its mass center relative to its proximal end. The distance of the center of mass of the forearm from its proximal end (i.e., the elbow) is 0.430 times the segment length:

COM of the forearm = (0.430)(0.263 m) = 0.113 m from the elbow.

In the next examples, let us imagine that from direct measures using an optical motion capture system, the following segment lengths were determined: foot = 0.195 m, leg = 0.435 m, thigh = 0.410 m, and HAT = 0.578 m (measured from the greater trochanter to the glenohumeral joint). With this information, we can determine the location of the mass centers relative to the proximal end of the segment:

COM of the foot = (0.50)(0.195 m) = 0.098 m from the malleolus marker
COM of the leg = (0.433)(0.435 m) = 0.188 m from the femoral condyle marker
COM of the thigh = (0.433)(0.410 m) = 0.178 m from the greater trochanter marker
COM of the HAT = (0.678)(0.578 m) = 0.392 m from the greater trochanter marker

Now, let us calculate the mass of the foot, leg (often referred to as the shank), thigh, and HAT (head-arms-trunk) segments. To do this, we need the total mass of the individual (80 kg). Therefore:

Mass of the foot = (0.0145)(80 kg) = 1.16 kg
Mass of the leg = (0.0465)(80 kg) = 3.72 kg
Mass of the thigh = (0.10)(80 kg) = 8.0 kg
Mass of the HAT = (0.678)(80 kg) = 54.2 kg

Note: see Appendix J for the description of an optical motion capture system and the use of anatomical markers.

Although the radius of gyration is not needed to solve static equilibrium problems, it is needed to solve dynamic problems involving rotations. We can determine the radius of gyration in a similar fashion to the calculation of the segment's mass center location, and then use that information to find the moment of inertia. Let us calculate the moment of inertia of the leg about its center of mass, once again assuming a total mass of 80 kg and an overall height of 1.8 m:

Mass of the leg (Table A.1) = 3.72 kg
Length of the leg (Fig. A.3) = (0.285 – 0.039)(1.8 m) = 0.443 m

The radius of gyration with respect to the center of mass divided by the segment length is 0.302. Thus, the radius of gyration and moment of inertia are calculated as:

$$k_{\text{leg}} = (0.302)(0.433\text{ m}) = 0.134\text{ m}.$$

$$I_{\text{leg}} = m_{\text{leg}} k_{\text{leg}}^2 = (3.72\,\text{kg})(0.134\,\text{m})^2 = 0.0665\,\text{kg}\cdot\text{m}^2.$$

# Appendix B: Language of Movement (Figs. B.1, B.2, and B.3)

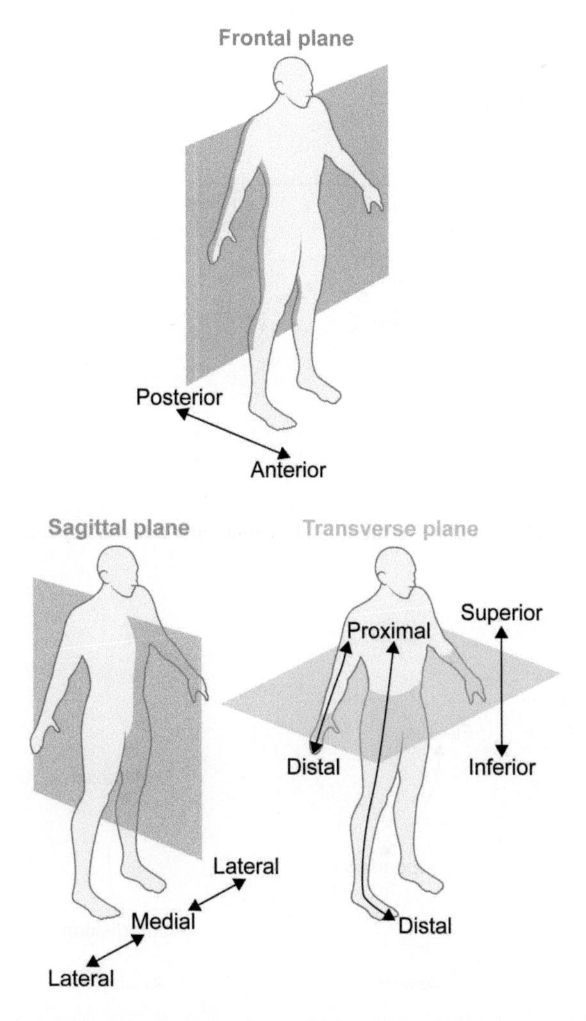


**Fig. B.1** Anatomical planes and directions employed in the study of human biomechanics. (Image copyright © David Delp, from Uchida and Delp (2020a). All rights reserved)



G. J. Alderink, B. M. Ashby, *Clinical Kinesiology and Biomechanics*, https://doi.org/10.1007/978-3-031-25322-5

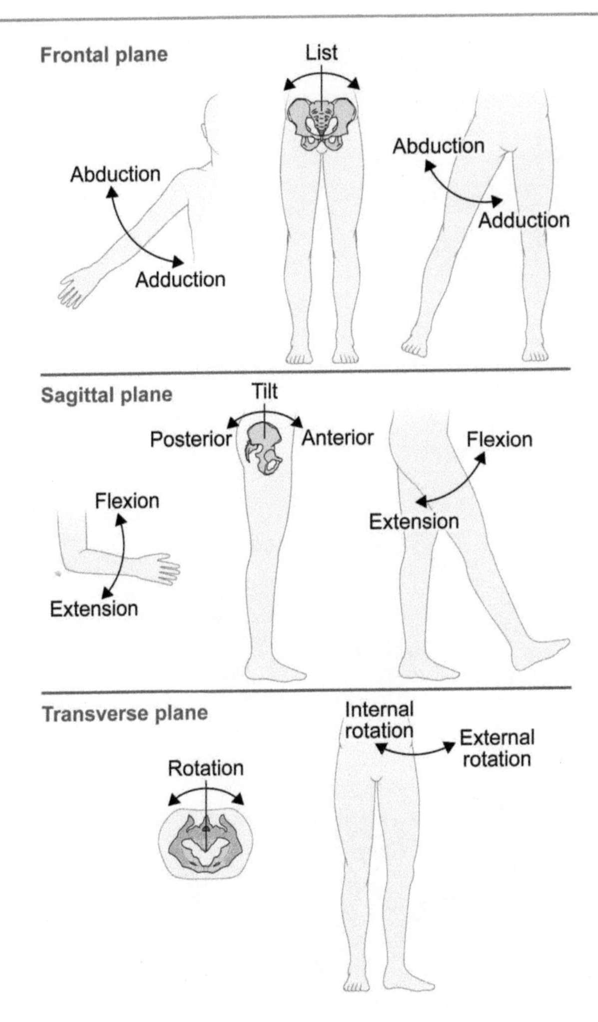


**Fig. B.2** Motions of the shoulder, elbow, pelvis, and hip in the frontal, sagittal, and transverse planes. (Image copyright © David Delp, from Uchida and Delp (2020a). All rights reserved)

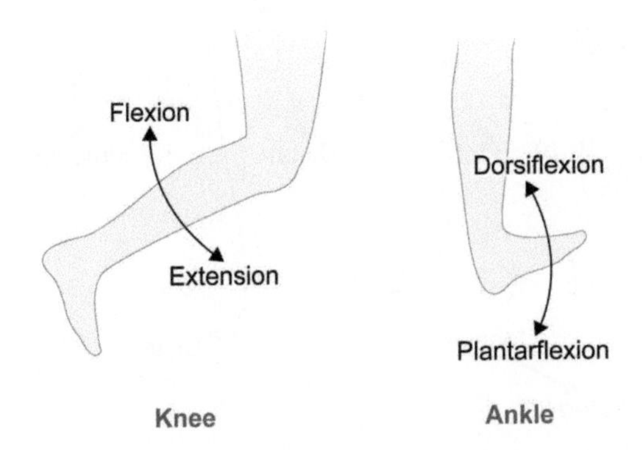


**Fig. B.3** Motions of the knee and ankle in the sagittal plane. (Image copyright © David Delp, from Uchida and Delp (2020a). All rights reserved)

# Appendix C: Euclidean Plane Geometry

## 1.1 Points, Lines, Angles, and Planes

Euclidian plane geometry refers to two-dimensional (2D) Euclidean geometry based on Euclid's postulates. For example, Euclid claimed that it is possible to draw a straight line between any two points, a straight line could be extended indefinitely, and within a circle, a straight-line segment could serve as a radius of the circle with one endpoint at its center. In geometry, we work with points, lines, and planes, where a point is dimensionless (no length or width), a line extends indefinitely in opposite directions, and a plane is a flat surface that also extends ad infinitum.

Some definitions related to points and lines are as follows:

- *Collinear points*: points that lie on the same line or line segment
- *Line segment*: a straight line with two endpoints
- *Ray*: a section of a straight line that contains a specific point
- *Intersection point*: a point where two lines intersect
- *Midpoint*: a point in the middle of a line segment
- *Parallel lines*: lines that, drawn on a 2D plane, may extend forever in either direction without ever intersecting
- *Perpendicular lines*: lines that intersect at exactly a 90° angle (i.e., right angles are considered orthogonal)
- *Concurrent lines*: lines that intersect at the same point
- *Skew lines*: lines that, drawn in a three-dimensional (3D) space, are neither parallel nor perpendicular and do not intersect

When lines intersect, they form angles. Below are definitions related to angles:

- *Adjacent angles*: have the same vertex (highest point or apex) and share a side
- *Complementary angles*: add up to 90°
- *Supplementary angles*: add up to 180°

The angles $\theta$ in Fig. C.1, which are formed by two intersecting straight lines, are equal; they are called *opposite angles*.

The angles $\theta$ in Fig. C.2 are formed by a straight line intersecting two parallel lines that are equal; they are called *alternate angles*.

The angles $\theta$ in Fig. C.3 are equal. Right angles (i.e., 90° or orthogonal angles) are formed by the line CC, which is perpendicular to BB, and DD, which is perpendicular to AA. Graphically, orthogonality is usually symbolized by small square boxes. The geometry in Fig. C.3 is often recognized in mechanics when analyzing motions on inclined surfaces, where AA represents the horizontal, DD represents the

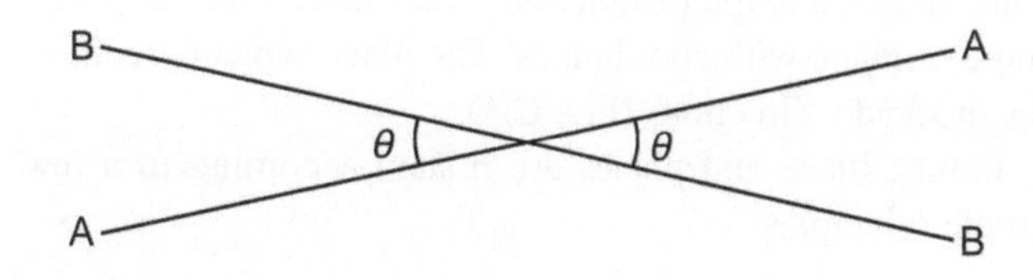


Fig. C.1 Example of opposite angles

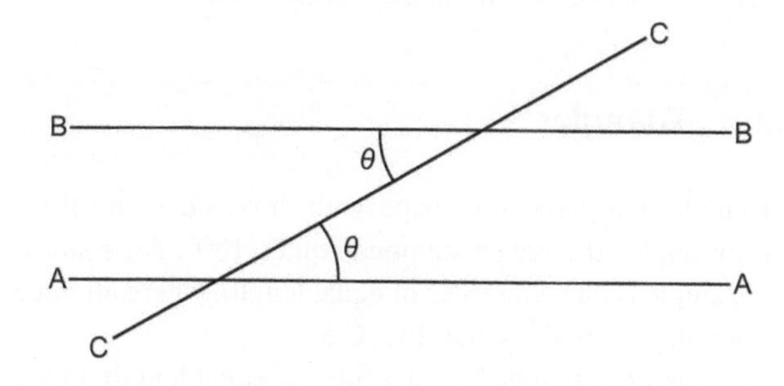


Fig. C.2 Example of alternate angles


G. J. Alderink, B. M. Ashby, *Clinical Kinesiology and Biomechanics*, https://doi.org/10.1007/978-3-031-25322-5

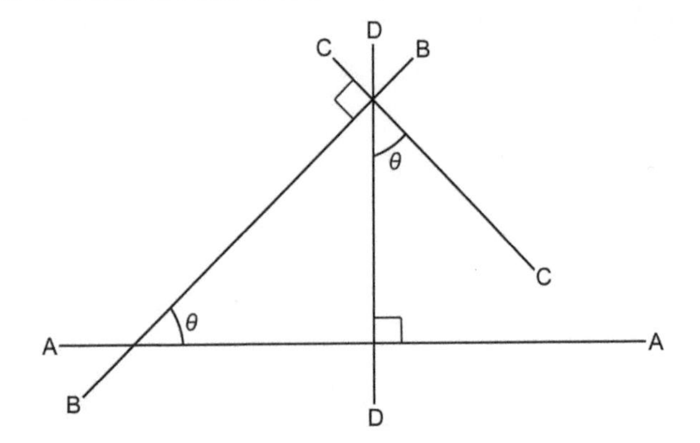


Fig. C.3 Perpendicular lines, e.g., BB and CC

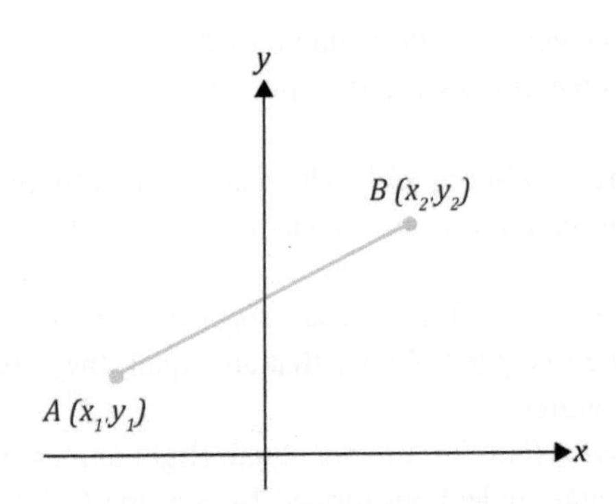


Fig. C.4 2D Cartesian coordinate plane

vertical, and BB represents the inclined surface that makes an angle $\theta$ with the horizontal. Note that *acute* angles measure < 90°, and *obtuse* angles measure > 90°.

A plane is a 2D space that extends indefinitely in all directions. In 2D, a graph of functions takes place on a Cartesian plane or plane with coordinates. The plane typically continues in $x$ and $y$ directions (Fig. C.4).

Points, lines, and planes are related according to a few simple principles:

1. A line is unique if it passes through two distinct points.
2. Three noncollinear points form a unique plane.
3. The intersection of two planes forms a line.

## 1.2 Triangles

A triangle is a geometric shape with three sides and three interior angles that when summed equals 180°. An *equilateral* triangle is one with sides of equal length, where all three interior angles are the same (Fig. C.5).

An *isosceles* triangle has two sides of equal length. In the isosceles triangle, the angles that are opposite the two equal sides are equal to each other (Fig. C.6).

An *irregular* triangle is one with all three sides of different lengths (Fig. C.7).

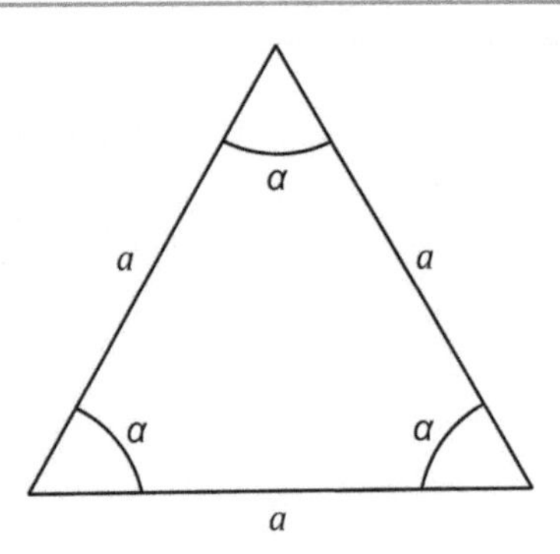


Fig. C.5 Example of an equilateral triangle

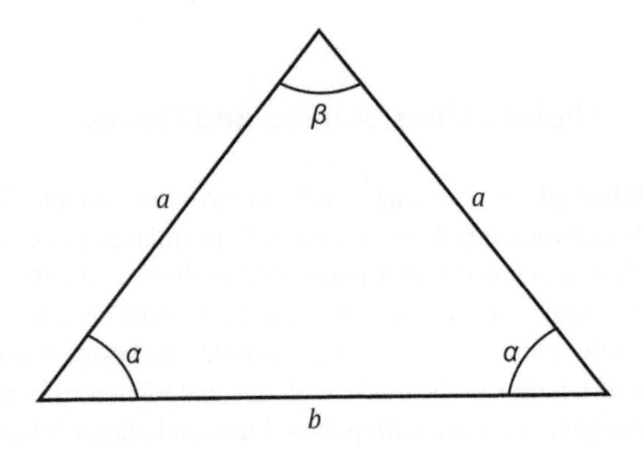


Fig. C.6 Example of an isosceles triangle

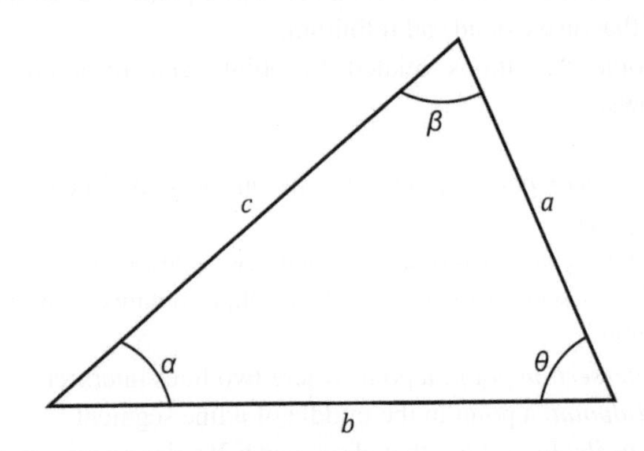


Fig. C.7 Example of an irregular triangle

## 1.3 Laws of Sines and Cosines

For any triangle, such as the one in Fig. C.7, the angles and sides of the triangle are related through the *law of sines*:

$$\frac{\sin\alpha}{a} = \frac{\sin\beta}{b} = \frac{\sin\theta}{c}.$$

For any triangle, such as the one in Fig. C.7, if two sides of the triangle ($a$ and $b$) and an angle between them ($\theta$) are known, the unknown third side ($c$) can be determined by applying the *law of cosines*:

$$c^2 = a^2 + b^2 - 2ab\cos\theta.$$

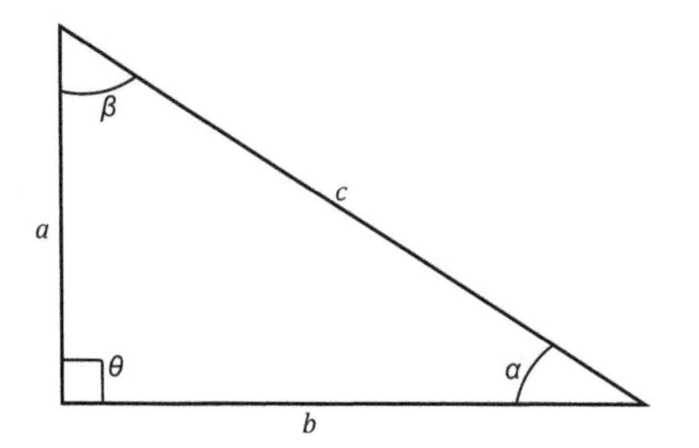


**Fig. C.8** Example of a right triangle

## 1.4 The Right Triangle

A right triangle has 90° as one of its angles (angle $\theta$ in Fig. C.8). The sum of the remaining angles is also equal to 90°:

$$\theta = \alpha + \beta = 90°.$$

The longest side of the triangle, side $c$, which is opposite to the right angle (angle $\theta$) is called the *hypotenuse*. With respect to angle $\alpha$, $b$ is the length of the *adjacent side* and $a$ is the length of the *side opposite*. The two sides are often referred to as the legs of the triangle.

## 1.5 Pythagorean Theorem

The Pythagorean theorem states that the square of the length of the hypotenuse of a right triangle equals the sum of the squares of the lengths of the other two sides:

$$c^2 = a^2 + b^2$$

or taking the square root of both sides of the equation:

$$c = \sqrt{a^2 + b^2}.$$

## 1.6 Sine, Cosine, and Tangent

Referring to the right triangle in Fig. C.8:

$$\sin\alpha = \frac{a}{c} \text{ and } \sin\beta = \frac{b}{c}$$

and

$$\cos\alpha = \frac{b}{c} \text{ and } \cos\beta = \frac{a}{c}.$$

Note that the sine of angle $\alpha$ equals the cosine of angle $\beta$, and the cosine of angle $\alpha$ equals the sine of angle $\beta$. Additionally:

$$a = c\sin\alpha$$
$$b = c\cos\alpha.$$

Notice that if we substitute the squares of $a$ and $b$ from above into $c^2 = a^2 + b^2$, we get:

$$c^2 = a^2 + b^2$$
$$c^2 = (c\cos\alpha)^2 + (c\sin\alpha)^2$$
$$c^2 = c^2\left(\cos^2\alpha + \sin^2\alpha\right)$$

Then dividing through by $c^2$ yields a useful trigonometric identity:

$$\cos^2\alpha + \sin^2\alpha = 1.$$

Referring to the right triangle in Fig. C.8, the tangent of an acute angle in a right triangle is the ratio of the lengths of the opposite and adjacent sides:

$$\tan\alpha = \frac{a}{b} \text{ or } \tan\alpha = \frac{\sin\alpha}{\cos\alpha}$$
$$\tan\beta = \frac{b}{a} \text{ or } \tan\beta = \frac{\sin\beta}{\cos\beta}.$$

When working with right triangles, if the length of at least two sides is known or if one of the acute angles and the length of one side is known, the remaining sides and angles can be determined using trigonometric identities.

## 1.7 Inverse Sine, Cosine, and Tangent

Often, we need to determine the magnitude of an angle. This can be accomplished if the sine, cosine, or tangent is known:

If $\sin\alpha = A$, then $\alpha = \sin^{-1} A$; $\cos\alpha = B$, then $\alpha = \cos^{-1} B$; and $\tan\alpha = C$, then $\alpha = \tan^{-1} C$.

The inverse of sine, cosine, and tangent is also known as arcsine, arccosine, and arctangent, respectively. Note that inverse trigonometric functions do not have unique solutions. For example, the sine of 30° and the sine of 150° are both 0.5. Similarly, the tangent of 45° and the tangent of -135° are both 1. Thus, the inverse tangent of 1 has more than one answer. Care must be taken when using inverse trigonometric functions to ensure that you select the angle that truly represents the correct answer.

## 1.8 Radians: Definition and Application of Its Use in Various Measurements of the Fingers

Angles can be measured in degrees or radians. There are $\pi$ radians in 180° (1 radian = 57.3°). Another geometric rule states that when a radius of a circle rotates through an angle of $\theta$, a point on that radius moves a distance along an arc on the perimeter of the circle $s$ equal to the radius times $\theta$ (expressed in radians) (Fig. C.9):

$$s = r\theta$$

Let us apply this rule of geometry to the measurement of tendon movement (excursion) relative to a measured angular movement at a joint. Note that when a lever moves around an axis, the distance moved by every point on the lever is proportional to its own distance from the axis. Imagine a rope (analogous to a tendon) that is wrapped around a pulley and a weight attached to the rope is released, resulting in a rotation $\theta$ (Fig. C.10). As the pulley rotates, it moves through an arc $s$; the excursion of the rope is equal to that arc of movement (equal to the radius of the pulley times $\theta$). For example, if a pulley with a radius $r = 5$ cm rotates through an angle of $\theta$ = 0.5 rad (28.6°), the excursion of the rope would be $s = (5\text{ cm})(0.5) = 2.5\text{ cm}$.

Let us illustrate this another way using a goniometer (a tool used by clinicians to measure joint range of motion) and a rope, i.e., tendon. We see that the lengthwise movement of the rope may be used to measure the moment arm of the joint (Fig. C.11). If the joint rotates an angle $\theta$, the length of the tendon that runs off of the pulley $s$ must be equal to its moment arm at the joint, i.e., the radius of the pulley (Fig. C.11A) times $\theta$. We could then calculate the moment arm of the tendon by dividing $s$ by $\theta$. With the model illustrated slightly differently in Fig. C.11B, we see a connecting link between a lever and a pulley wheel. If the wheel is fixed to the lever, a cord, e.g., tendon, that runs around the pulley wheel serves as an extension of the lever system, but with a constant leverage (or moment arm), as long as it runs off the convex wheel (like a tendon that runs along the surface of the head of a bone at a joint) (Fig. C.12). In Fig. C.12A, we see the recording of a 60° joint movement (roughly 1 radian) and related tendon excursion, which would be approximately 1 moment arm. In Fig. C.12B, we see a joint movement of 90° (roughly 1.5 rad) with a proportional tendon excursion of about 1.5 times the moment arm.

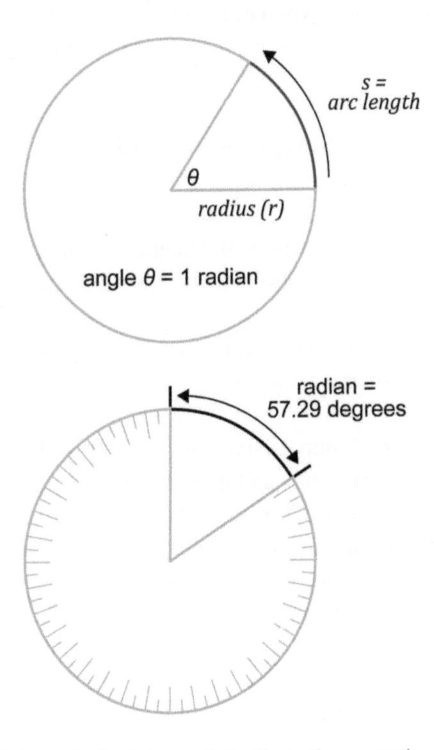


**Fig. C.9** Schematic depicting a circle illustrating a rotation $\theta$ about its center, where the radius, $r$, of the circle equals the arc length, $s$, produced by the rotation ($s = r\theta$)

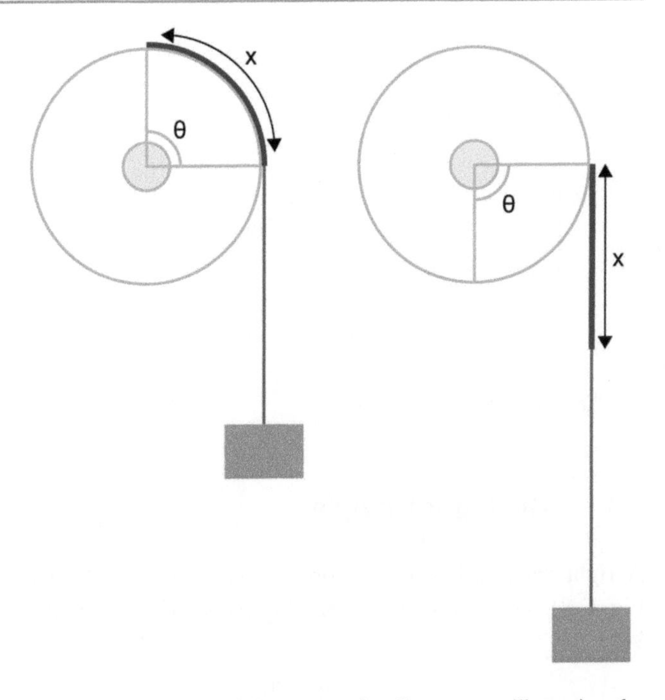


**Fig. C.10** Schematic of the rope and pulley system illustrating the excursion of the rope wrapped around the pulley

This simple model effectively demonstrates the way the lengthwise excursion of a tendon is used to estimate a tendon's moment arm. Brand et al. used this rope-pulley model to describe the digital tendon-pulley system, where for flexion of the fingers, i.e., digital flexion, the radius of the pulley represented the length of a tendon (associated with a muscle) moment arm and where the movement of the rope represented the tendon's excursion (Fig. C.13).

Brand et al. calculated that the metacarpophalangeal (MCP) joint rotating through its normal range of motion of 85° (1.48 rad) had a moment arm relative to the MCP joint of approximately 10 mm, with a concomitant excursion of the

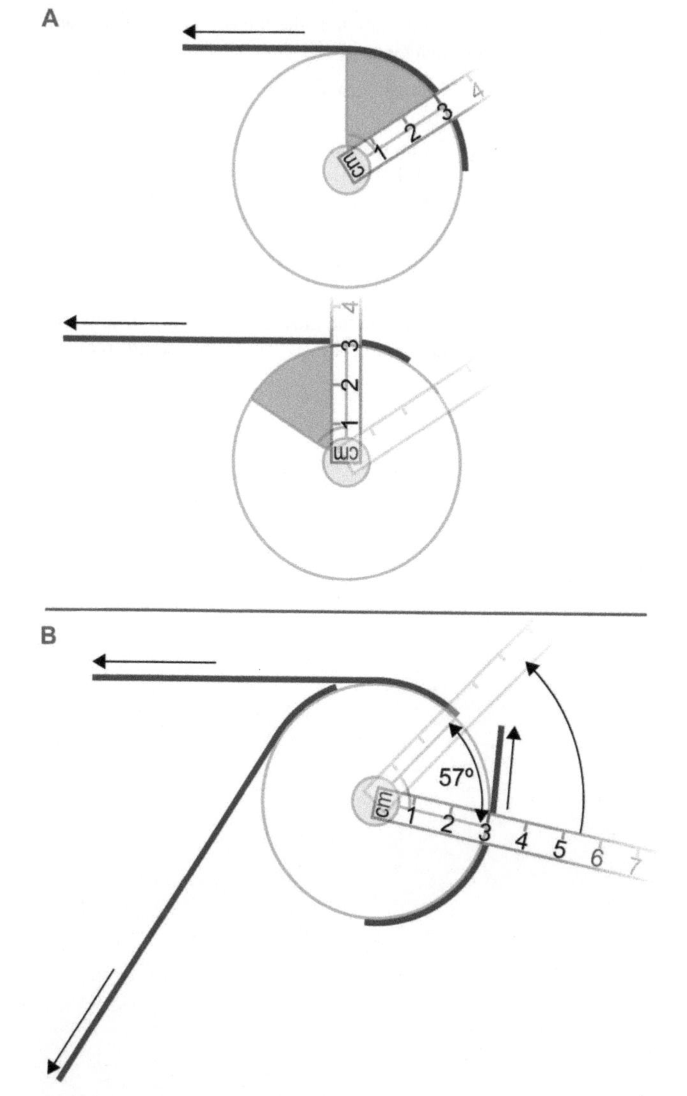


**Fig. C.11** (**A**) Lengthwise movement of a tendon is proportional to the length of the joint moment arm. (**B**) The relationship between tendon excursion and joint moment arm for each angular change. (Modified with permission from Brand et al. (1999))

flexor tendon of (10 mm) (1.48) = 14.8 mm. Brand et al. also provided an estimated joint range of motion, moment arm, and tendon excursion for the MCP and proximal (PIP) and distal (DIP) interphalangeal joints (Table C.1). Using his model, he projected the reduction in joint motion with increased moment arms of a digital flexor at the MCP, PIP, and DIP joints in the case of an injury, i.e., disruption, to an annular pulley (Table C.2).

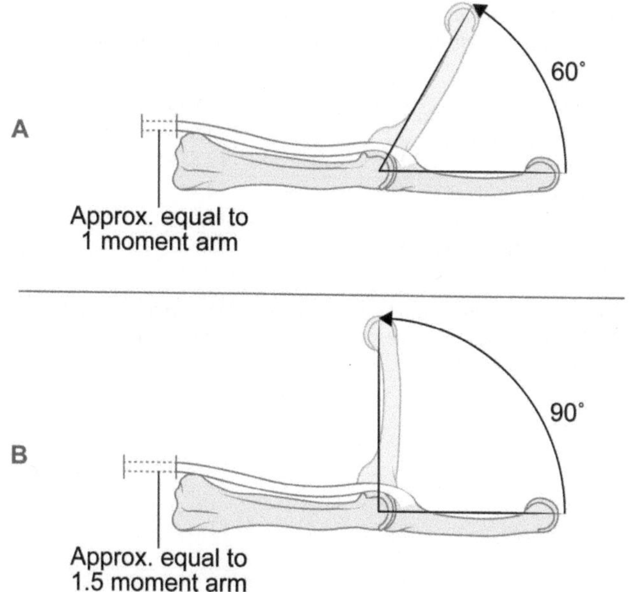


**Fig. C.12** Tendon movement (**A**) approximately equal to a 1.0 × moment arm and (**B**) approximately equal to 1.5 × moment arm. (Modified with permission from Brand et al. (1999))

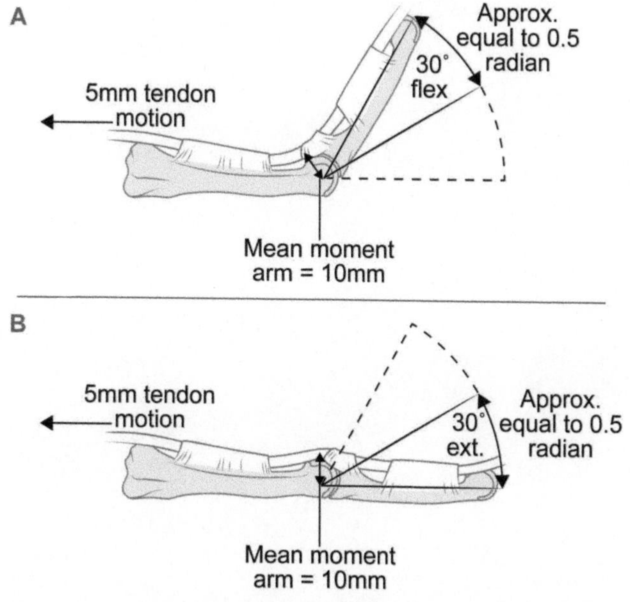


**Fig. C.13** The rope-pulley model and digital flexion/extension apparatus with (**A**) 30° flexion and (**B**) 30° extension illustrating little difference between the tendon excursions (and moment arms). (Modified with permission from Brand et al. (1999))

**Table C.1** Joint range of motion, moment arm, and tendon excursion for metacarpophalangeal and proximal and distal interphalangeal joints

| Joint (ROM) | Moment arm (MA) (mm) | Tendon excursion (mm) |
|---|---|---|
| Metacarpophalangeal (85°) | 10[a]<br>12 | 14.8[a]<br>17.8 |
| Proximal interphalangeal (110°) | 7.5[a]<br>9 | 14.4[a]<br>17.3 |
| Distal interphalangeal (65°) | 5[a] | 5.7[a] |

Notes: "a" denotes normal, where joint ROM divided by 1 radian (57.3°) multiplied by the moment arm yields excursion (joint ROM × MA = excursion)

**Table C.2** Moment arm and motion at the metacarpophalangeal and proximal and distal interphalangeal joints

| Joints | Moment arm (MA) (mm) | Lost joint motion (°) |
|---|---|---|
| Metacarpophalangeal | 2 mm increase (10–12) | 85 to 68 |
| Proximal interphalangeal | 1.5 mm increase (7.5–9) | 110 to 88 |
| Distal interphalangeal | 1 mm increase (5–6) | 65 to 52 |

Note: A relatively small increase in the muscle moment arm results in a significant loss of finger flexion range of motion (and reduced function)

# Appendix D: Calculus

## 1.1 Functions

Functions are fundamental to mathematics and, therefore, biomechanics. A function is used to denote the dependence or relationship of one quantity with another. Once a function relating two quantities is established, changes in one expression as a result of changes in the other may be predicted. Most physical and biological laws are expressed in the form of mathematical functions. For example, velocity is defined as the time rate of change of position. Therefore, if we can measure the change in the position of an object over time, we can express that change in terms of a function:

$$v = \frac{dx}{dt},$$

where $v$ is the velocity and $\frac{dx}{dt}$ is the derivative of position $x$ with respect to time $t$. The derivative is a fundamental function of calculus, which is defined later in this treatise.

The standard practice is to denote a function by a letter; e.g., $f$ or $g$ is commonly used. Therefore, if $f$ is a function, then the number which $f$ associates to a number $x$ is denoted by $f(x)$ (often referred to as "$f$ of $x$" or the value of $f$ at $x$). For example, $y = f(x)$ implies that $y$ is a function of $x$:

$$y = f(x) = x^2,$$

where $x$ in $y = f(x)$ is the "input" of the operation, while $y$ is the "output." We call the input of the function the *independent* variable and the output the *dependent* variable.

There are a variety of fundamental functions, each with unique characteristics that are frequently used in physical sciences. In many cases, these functions are combined to describe particular features of a physical entity.

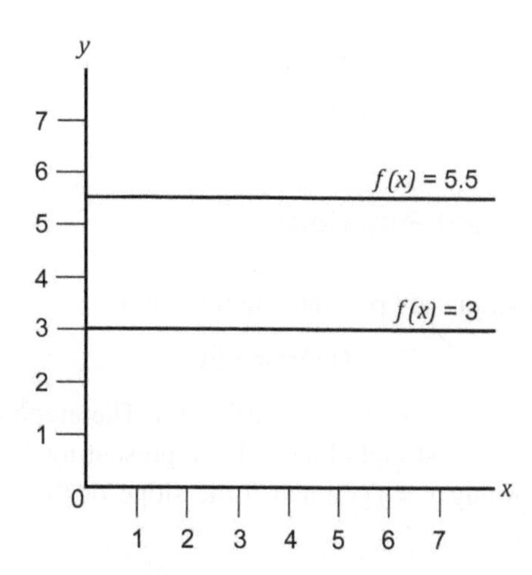


**Fig. D.1** Examples of constant functions

## 1.2 Constant Functions

Constant functions are usually represented in this form:

$$f(x) = c,$$

where $c$ symbolizes a constant number. Graphs of constant functions are depicted as horizontal lines (Fig. D.1).

## 1.3 Power Functions

Power functions are represented as:

$$f(x) = x^n,$$


G. J. Alderink, B. M. Ashby, *Clinical Kinesiology and Biomechanics*, https://doi.org/10.1007/978-3-031-25322-5

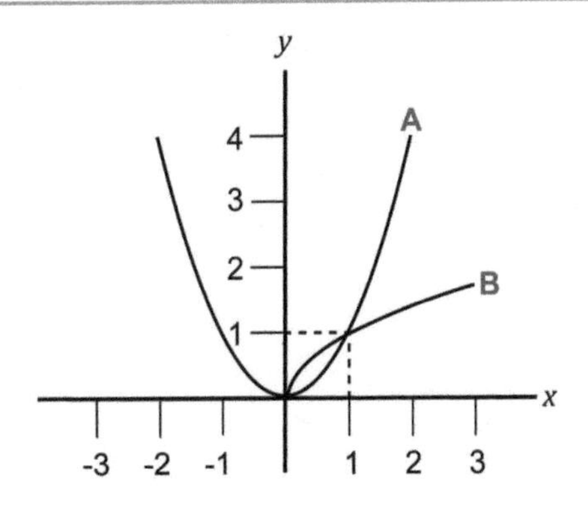


**Fig. D.2** Examples of power functions: (**A**) $f(x) = x^2$ and (**B**) $f(x) = \sqrt{x} = x^{1/2}$

where $n$ can be any real number. Examples of power functions can be seen in Fig. D.2.

## 1.4 Linear Functions

Linear functions are presented in this form:

$$f(x) = a + bx,$$

where $a$ and $b$ are constant coefficients. The graph of a linear function is a straight line with $a$ representing the $y$-intercept (assuming $y = f(x)$) and $b$ the slope of the line (Fig. D.3).

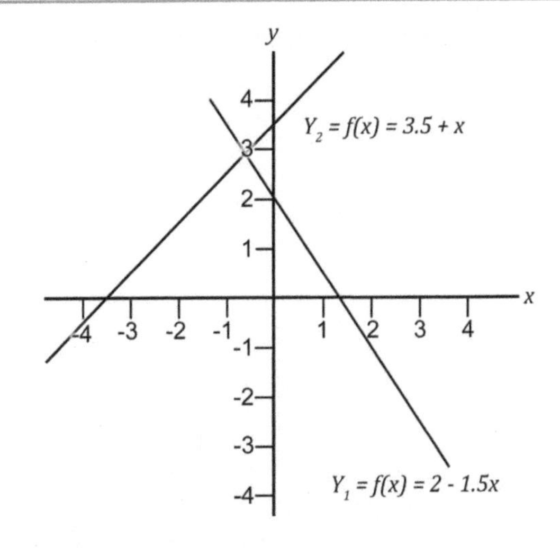


**Fig. D.3** Examples of graphs for two different linear functions, e.g., $Y_1 = f(x) = 2 - 1.5x$ and $Y_2 = f(x) = 3.5 + x$

## 1.5 Quadratic Functions

Quadratic functions are found in the form:

$$f(x) = a + bx + cx^2,$$

where $a$, $b$, and $c$ are positive or negative real or integer numbers. The coefficients $a$ and $b$ can be 0. A quadratic is characterized by the highest power equal to 2, e.g., $x^2$. A typical quadratic graph is shaped in the form of a parabola (Fig. D.4).

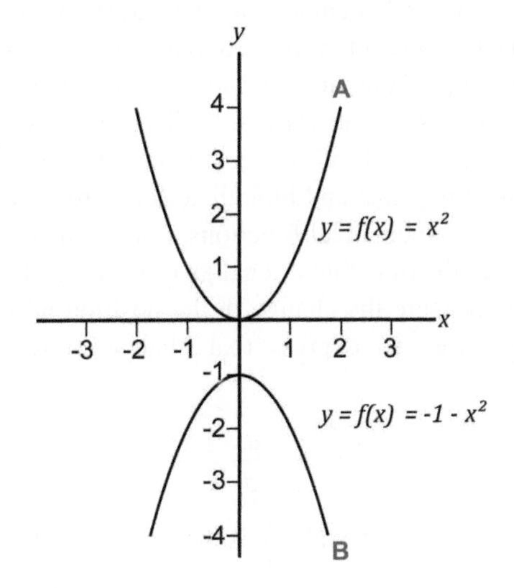


**Fig. D.4** Examples of quadratic functions: (**A**) $Y = f(x) = x^2$ and (**B**) $Y = f(x) = -1 - x^2$

## 1.6 Polynomial Functions

Polynomials take this form:

$$f(x) = a_0 + a_1x + a_2x^2 + a_3x^3 + \cdots + a_nx^n,$$

where the coefficients $a_0$, $a_1$, …, $a_n$ are real or integer, positive or negative, zero or non-zero numbers and $n$ is a positive integer corresponding to the highest power of $x$; that is, the general form has a power of $n$. Constant, linear, and quadratic functions are special forms of polynomial functions.

## 1.7 Trigonometric Functions

To say that $y$ is a trigonometric function of $x$ means that $y$ depends on another quantity $x$ through a trigonometric relationship, such as $y = \sin(x)$, $y = \cos(x)$, and $y = \tan(x)$ – the sine, cosine, and tangent functions, respectively. The argument of these functions, $x$, can be expressed either in degrees or radians. The conversion between radians and degrees is:

$$\text{Radians} = \frac{\pi}{180} \times \text{degrees}.$$

Trigonometric functions are periodic. For example, the graphs of $y = \sin(x)$ and $y = \cos(x)$ repeat every $2\pi$ radians or 360°; that is, the period of the sine and cosine function is $2\pi$ radians. Moreover, the graphed functions $f(x) = \sin(x)$ and $f(x) = \cos(x)$ assume values between −1 and +1 and therefore have an amplitude of 1 (Fig. D.5).

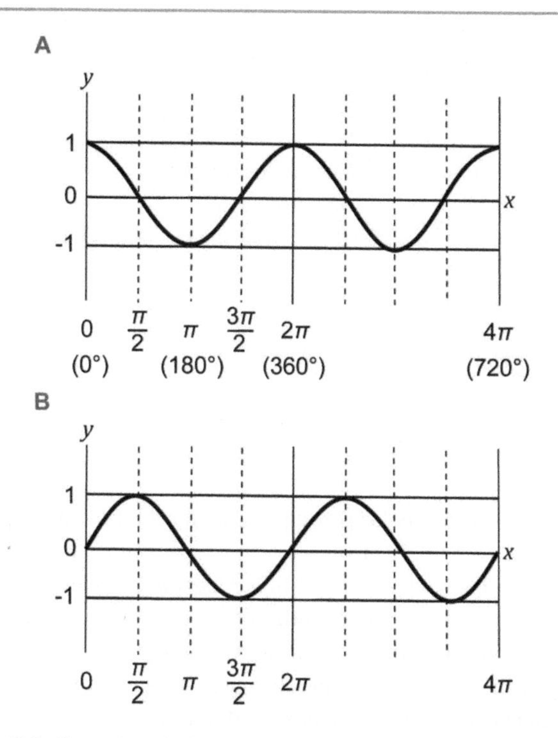


**Fig. D.5** Examples of trigonometric functions: (**A**) $Y = f(x) = \cos(x)$ and (**B**) $Y = f(x) = \sin(x)$

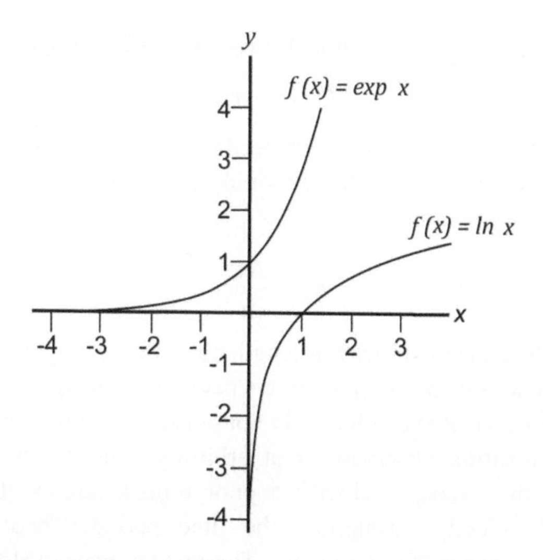


**Fig. D.6** Examples of exponential and logarithmic functions: $f(x) = e^x$ and $f(x) = \ln x$

## 1.8 Exponential and Logarithmic Functions

The general form of exponential functions is $b^x$, where $b$ is called the base. A commonly used base is $e$, an irrational number between 2.71 and 2.72, so that the function $e^x$ (or exp $x$) is referred to as the exponential function. The inverse of $e^x$ is called the natural logarithmic function, $\ln x$ ($\log_e x$); it can also be called logarithm with base $e$ (Fig. D.6). These functions have several useful properties (Table D.1).

**Table D.1** Properties of $e^x$ and $\ln x$

| |
|---|
| If $\ln x = y$, then $x = e^y$ |
| $\ln(e^x) = x$ and $e^{\ln(x)} = x$ |
| $\ln 1 = 0$ and $e^0 = 1$ |
| $e^x e^y = e^{x+y}$ |
| $\frac{e^x}{e^y} = e^{x-y}$ |
| $e^{-x} = \frac{1}{e^x}$ |
| $(e^x)^y = e^{xy}$ |
| $\ln(xy) = \ln x + \ln y$ |
| $\ln\left(\frac{x}{y}\right) = \ln x - \ln y$ |
| $\ln\left(\frac{1}{x}\right) = -\ln x$ |
| $\ln x^y = y \ln x$ |

Note: Taking the ln of a negative number results in an imaginary number. It is possible for $\ln x$ to be negative. $e^x$ is defined for all $x$ and is always positive

## 1.9 Derivatives

Differentiation is one of the most fundamental operations used in calculus and biomechanics. In simple mathematical terms, the derivative is the determination of the slope of the graph of a function. For example, the slope of the linear function $y = f(x) = 1 + 2x$ is determined by considering any two points along the line, e.g., $x_1 = 3$, $y_1 = 7$, $x_2 = 5$, and $y_2 = 11$:

$$\text{Slope} = \frac{y_2 - y_1}{x_2 - x_1} = \frac{11 - 7}{5 - 3} = 2$$

It turns out that with linear functions, the slope is also equal to the tangent of the angle between the line for the linear function and any horizontal line (Fig. D.7).

The slope of a linear function is always a constant. However, most functions in biomechanics are curvilinear, which means that the slope continuously changes. Differentiation of curves makes it more efficient to determine a slope. In dynamics, differentiation is used to determine important kinematic variables, such as velocity and acceleration. Before we discuss the relationship between position, velocity, and acceleration, a few additional basic concepts are needed.

Let us consider the function $y = f(x)$. If this function was differentiated with respect to $x$, it could be represented by one of several abbreviations: $f'$, $f'(x)$, $\frac{df}{dx}$, $y'$, or $\frac{dy}{dx}$. The derivative of a constant function, $f(x) = c$, is zero because it represents a horizontal line whose slope is 0, whereas the

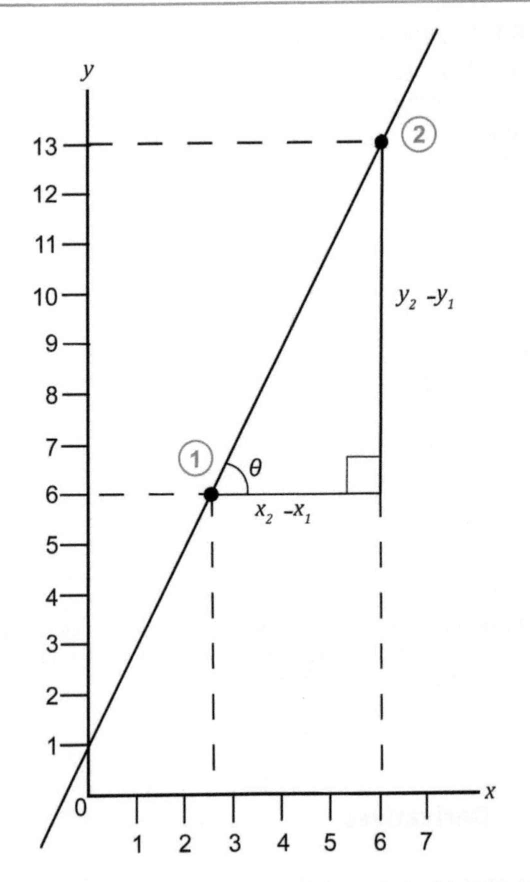


**Fig. D.7** The derivative of a function represents the slope. Note that the slope is $\tan\theta = \dfrac{y_2 - y_1}{x_2 - x_1}$

derivative of a linear function, $f(x) = x$, is 1 because it represents a straight line (not horizontal), which has a slope of 1. The slope of these simple functions can easily be determined by using the slope formula, but we can formalize the operation by taking the derivative; for example:

$$\frac{d}{dx}(c) = 0,$$

and

$$\frac{d}{dx}(x) = 1.$$

Here is how it works. To derive the derivative of a power function, $x^r$, multiply the function by its power and reduce the power by 1 (this is referred to as the *power rule*):

$$\frac{d}{dx}\left(x^r\right) = r \cdot x^{r-1}.$$

For example, using the *power rule*, consider the following functions and their derivatives:

(i) If $f(x) = x^4$, then $\dfrac{df}{dx} = 4x^3$

and

(ii) If $f(x) = x^{\frac{1}{2}}$, then $\dfrac{df}{dx} = \dfrac{1}{2} \cdot x^{-\frac{1}{2}} = \dfrac{1}{2\sqrt{x}}$.

Next, here are a few interesting examples:

(i) $\dfrac{d}{dx}(1) = \dfrac{d}{dx}\left(x^0\right) = 0 \cdot x^{-1} = 0$ since $x^0 = 1$ regardless of what $x$ is

(ii) $\dfrac{d}{dx}(x) = \dfrac{d}{dx}\left(x^1\right) = 1 \cdot x^0 = 1$

(iii) $\dfrac{d}{dx}\left[c \cdot f(x)\right] = c \cdot \dfrac{df}{dx} = c \cdot f'$

Note that in this last example, the constant $c$ can be factored out of the derivative.

Now, let us look at the rectilinear motion of a particle with a single degree of freedom and, using the derivative, examine the relationship between position, velocity, and acceleration. Suppose there is a particular change in the position of a particle during a specific time period; that is, at time $t_1$ a particle is at position $x_1$, and at time $t_2$, the particle is at position $x_2$. The change in position can be denoted by $\Delta x$:

$$\Delta x = x_2 - x_1.$$

Similarly, the change in time between the observations, $\Delta t$, is:

$$\Delta t = t_2 - t_1.$$

We now have the information needed to determine the *average* velocity ($v_{\text{ave}}$), or the rate of change of position, during the time period of interest:

$$v_{\text{ave}} = \frac{\Delta x}{\Delta t},$$

with the units of velocity as length per time, usually denoted in feet per second (ft/s) or meters per second (m/s).

However, if the velocity is not constant, using two discrete position observations at arbitrary time points may make the average velocity a poor representation of the actual velocity throughout the time period. Therefore, another approach is necessary. The *instantaneous* velocity, or the velocity at an instant in time, provides a better description of the actual velocity. We can determine the instantaneous velocity by averaging over a very small interval of time. Thus, the instantaneous velocity of the moving point at time $t_1$ is defined as the limit as the change in time approaches zero, or as the time interval $\Delta t$ becomes shorter and shorter. It turns out that the derivative is the ultimate reduction of the time interval. In mathematical (or calculus) terminology, the instantaneous velocity is defined as the limit of the average velocity as the time interval approaches zero:

$$v = \lim_{\Delta t \to 0} \frac{\Delta x}{\Delta t} = \frac{dx}{dt},$$

or the derivative of $x$ with respect to $t$. So we now see the practical use of differentiation in biomechanics. An alternative notation for velocity is to place a dot over position $x$ to indicate differentiation with respect to time:

$$\dot{x} = \frac{dx}{dt}.$$

Velocity is defined as the rate of change of position $x$ with respect to time, where position is considered a function of time $x(t)$ and velocity, the derivative of that function, is also a function of time $v(t)$. The velocity may be constant or may change over time, i.e., increasing or decreasing. Acceleration (or deceleration = a slowing down) measures the change in velocity with respect to time. We can define acceleration (deceleration) as the time rate of change in velocity and, as we did for change in position, define an average acceleration as the change in velocity divided by the corresponding change in time:

$$a_{\text{ave}} = \frac{\Delta v}{\Delta t}.$$

with the instantaneous acceleration as:

$$a = \lim_{\Delta t \to 0} \frac{\Delta v}{\Delta t} = \frac{dv}{dt}.$$

Thus, acceleration is the derivative of velocity with respect to time:

$$a = \frac{dv}{dt} = \dot{v},$$

or the second derivative of position with respect to time:

$$a = \frac{d^2 x}{d^2 t} = \ddot{x}.$$

Velocity, defined as the rate of change of position $x$, is the slope of the position versus the time curve at that instant. Likewise, acceleration as the rate of change of velocity $v$ is the slope of the velocity versus the time curve at that instant. Velocity is zero where there is a minimum or maximum in the position curve. Similarly, acceleration is zero at points on the velocity curve where there is a minimum or maximum. Velocity and acceleration are at their maximum when the slopes of the position and velocity curves, respectively, are the greatest.

## 1.10 The Integral

The other major fundamental operation of calculus seeks to determine the functions whose derivatives are known. This process reverses the operation of differentiation and is called integration. Mathematically, integration often involves a more difficult set of operations. Fortunately, those who came before created tables that contain commonly used integrals. In biomechanics, integration is most commonly used to determine a velocity function from acceleration or the position from the velocity function. Integration is also used for area and volume calculations, as well as work computations. For example, consider a typical strain rate-dependent stress-strain curves for tendons (Fig. D.8). If we dropped vertical lines from the ends of the stress-strain curves to the horizontal axis, the area under the respective curves would be closed. The strain energy is represented by the area under the stress-strain curve. If we could define a function for both the high and low strain rate ($\dot{\varepsilon}$) curves, we could then use integration to determine the strain energy related to different rates of loading.

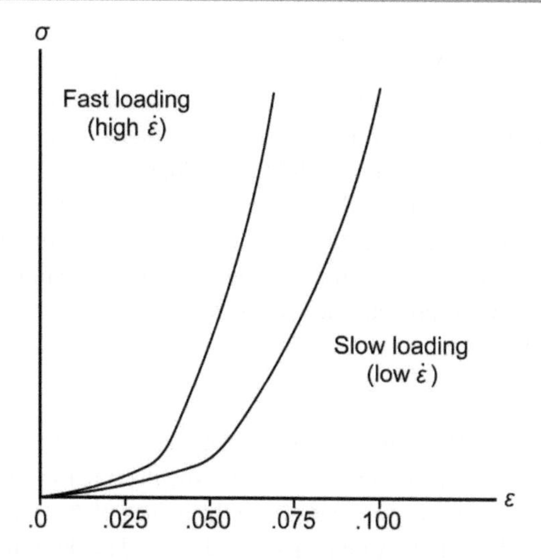


**Fig. D.8** Strain rate-dependent stress-strain curves for a tendon when loaded under tension, where $\dot{\varepsilon}$ is the rate of strain

The integral of the function $f(x) = y$ with respect to $x$ can be expressed in two ways:

$$\int f(x)dx \quad \text{(an indefinite integral)}$$

or

$$\int_a^b f(x)dx \quad \text{(a definite integral).}$$

The integral symbol $\int$ is similar to a summation, the function $f(x)$ to be integrated is called the integrand, $dx$ is an infinitesimally small increment in $x$, and $a$ and $b$ are called the lower and upper limits of integration, respectively.

For example, let $y = f(x) = 2x$ and $f_1(x) = x^2$. Note that the derivative of $f_1(x) = f(x)$, so that $f_1(x)$ can be the integral of $f(x) = 2x$. Now, consider the function $f_2(x) = x^2 + c_0$, where $c_0$ is a constant. The derivative of $f_2(x)$ is likewise equal to $f(x) = 2x$ because the derivative of a constant equals 0. We see that $f_1(x)$ is a special case of $f_2(x)$, for which $c_0 = 0$. Thus, the indefinite integral of $f(x) = 2x$ is:

$$\int(2x)dx = x^2 + c_0,$$

where $c_0$ is a constant of integration. It is notable that the indefinite integral of a function is not unique because there are different solutions for different values of $c_0$.

The definite integral of a function has a unique solution. This is so because the process of evaluating the definite integral effectively determines the value of the constant of integration. In the abstract case, consider:

$$\int_b^a f(x)dx.$$

To evaluate this definite integral, take the indefinite integral of the given function $f(x)$ to be $F(x)$ (neglecting the constant of integration), and then evaluate the values of $F(x)$ at $x = a$ and $x = b$; that is, evaluate $F(a)$ and $F(b)$. The definite integral of $f(x)$ between $a$ and $b$ is equal to $F(b)$ minus $F(a)$:

$$\int_b^a f(x)dx = F(b) - F(a).$$

Let us look at an example: let $y = f(x) = 2x$ and the integral be evaluated between $x = 1$ and $x = 3$. Recall that the indefinite integral of $f(x) = 2x$ is $f(x) = x^2$, neglecting the constant of integration:

(i) $\int_3^1 (2x)dx = \left[x^2\right]_1^3$

(ii) $= [(3^2) - (1^2)] = 8$

The physical meaning of the definite integral is that it represents the area bounded by the given function and the $x$-axis between the two limits of integration.

There are many properties of indefinite and definite integrals and several methods of integration. Interested readers can explore these on their own.

# Appendix E: Vector Algebra

## 1.1 Definitions

In biomechanics, we work with scalars and vectors. A scalar is simply a quantity with magnitude only, e.g., mass, energy, power, and temperature. The magnitude of a vector is a scalar quantity and is always positive. A vector has both magnitude and an associated direction, e.g., force, moment, velocity, and acceleration. For example, to describe a moment, one must state the magnitude of the moment and the direction to which it is applied, i.e., clockwise or counterclockwise.

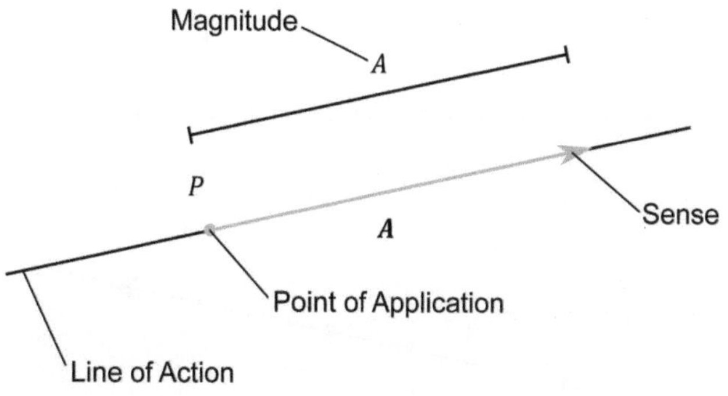

Magnitude $A$

$P$

$A$

Sense

Point of Application

Line of Action

**Fig. E.1** Example of representation of vector $\boldsymbol{A}$

## 1.2 Notation

Representations of scalars are unbolded identifiers, whereas representations of vectors are **bolded** in this textbook (there are other ways vectors are identified as well). Graphically, vectors are designated by an arrow (Fig. E.1). The orientation of the arrow indicates the line of action, the arrowhead denotes the direction (i.e., sense along its line of action), the base (or tail or tip) is the point of application, and the length is the magnitude of the vector; note that the length is proportional to the magnitude if one or more vectors are being used, but oftentimes the arrows are not drawn to scale. The magnitude of the vector must always be positive and is unbolded (since it is a scalar).

## 1.3 Cartesian Coordinate System

We define the location and changes in the location of objects in space relative to a reference frame or coordinate system. There are several coordinate systems to choose from, including Cartesian, polar, cylindrical, or spherical, but in biomechanics, the Cartesian coordinate (or rectangular) system is most commonly used.

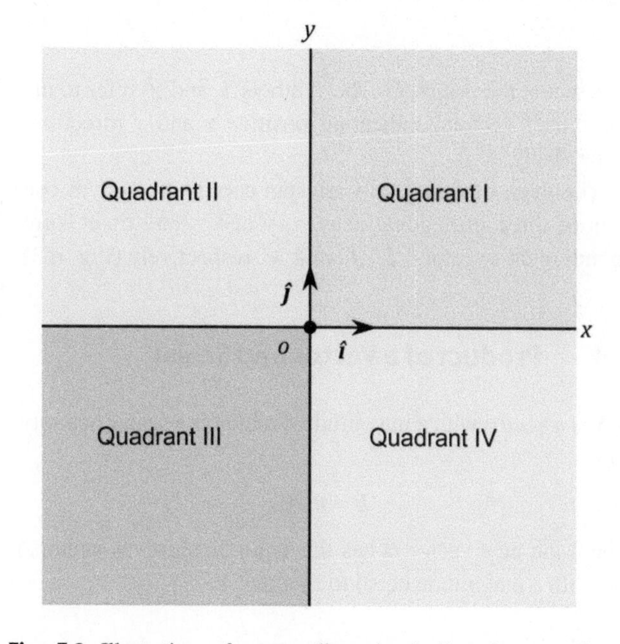


**Fig. E.2** Illustration of a two-dimensional Cartesian coordinate system

A two-dimensional Cartesian coordinate system consists of two orthogonal axes, $x$ and $y$, that divide a plane into four quadrants (Fig. E.2). The axes of the coordinate system


G. J. Alderink, B. M. Ashby, *Clinical Kinesiology and Biomechanics*, https://doi.org/10.1007/978-3-031-25322-5

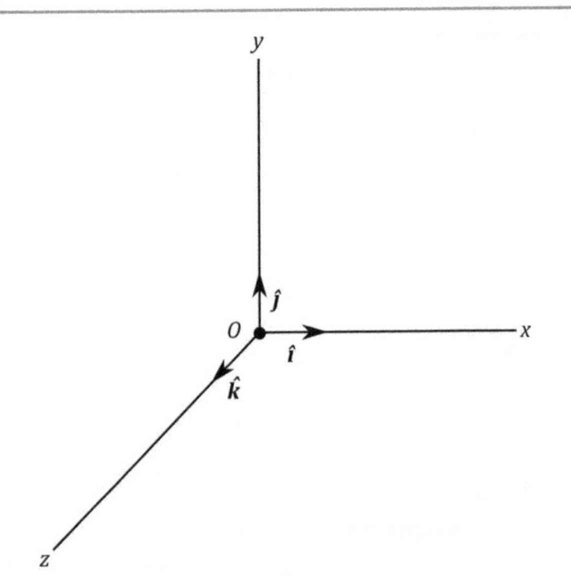


**Fig. E.3** Illustration of a three-dimensional Cartesian coordinate system

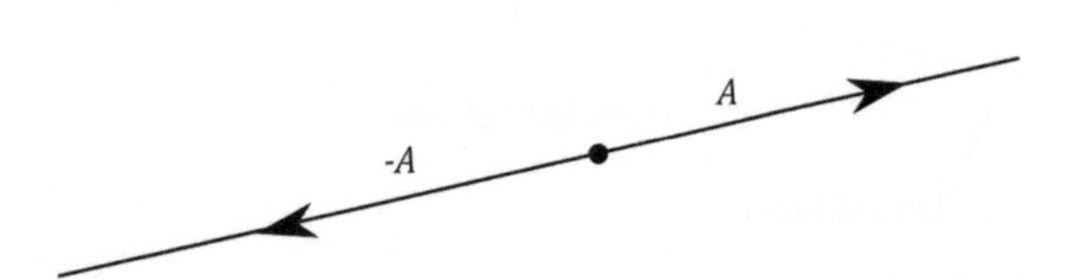


**Fig. E.4** A negative vector

intersect at the origin, $O$. The symbols $\hat{i}$ and $\hat{j}$ refer to unit coordinate vectors, indicating positive $x$ and $y$ directions, respectively.

The three-dimensional Cartesian coordinate system consists of three orthogonal axes, $x$, $y$ and $z$, and their corresponding unit vectors, $\hat{i}$ , $\hat{j}$ , and $\hat{k}$ , respectively (Fig. E.3).

## 1.4 Product of a Vector and Scalar

If $A$ is a vector with a magnitude $A$ and $m$ is a scalar quantity, then:

$$B = m\boldsymbol{A},$$

where the new vector $\boldsymbol{B}$ has the same direction as vector $\boldsymbol{A}$ but with a magnitude equal to $m$ times $\boldsymbol{A}$.

## 1.5 Negative Vectors

A negative vector, $-\boldsymbol{A}$, implies a change in the direction of vector $\boldsymbol{A}$ only. Thus, $\boldsymbol{A}$ and $-\boldsymbol{A}$ have magnitudes that are equal (Fig. E.4).

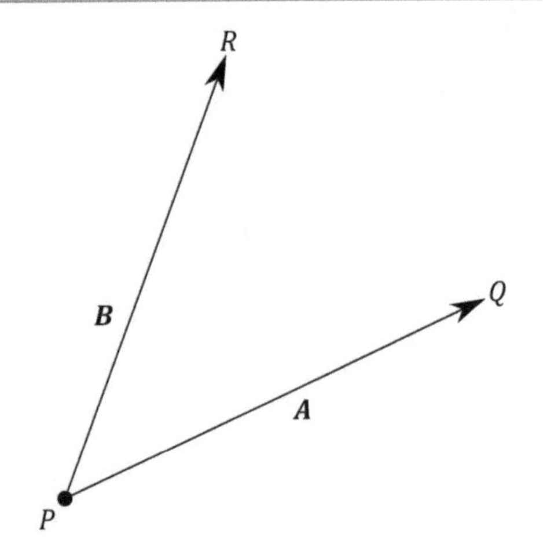


**Fig. E.5** Vectors $\boldsymbol{A}$ and $\boldsymbol{B}$

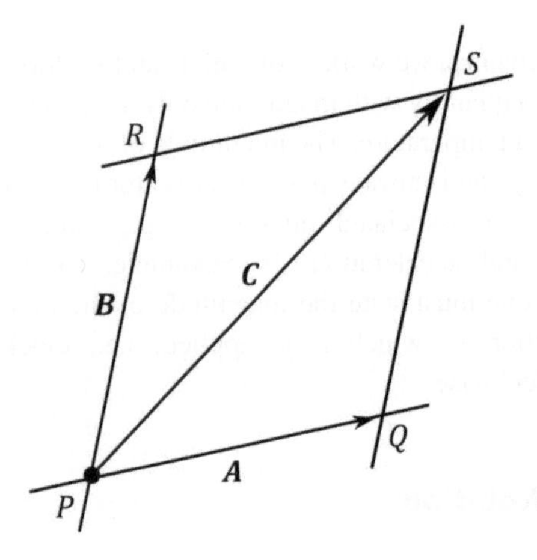


**Fig. E.6** Illustration of using vectors to construct a parallelogram to be used for vector addition

## 1.6 Vector Addition: Graphical Method

Two or more vectors can be added graphically using one of two methods: parallelogram or triangle (tail to tip). Note the two vectors, $\boldsymbol{A}$ and $\boldsymbol{B}$ (Fig. E.5). Vector $\boldsymbol{A}$ is pointing to $Q$, and vector $\boldsymbol{B}$ is pointing to $R$.

To draw a parallelogram, at the tip of one of the vectors, a line is drawn parallel to the other vector, then repeated for the second vector. Point $S$ corresponds to the point of intersection of the two parallel lines; the arrow drawn from point $P$ toward $S$ represents a vector that is equal to the sum of $\boldsymbol{A}$ and $\boldsymbol{B}$. This third vector is called the resultant vector, $\boldsymbol{C}$ (Fig. E.6), expressed as:

$$\boldsymbol{A} + \boldsymbol{B} = \boldsymbol{C}.$$

When using the tail-to-tip method, the tail of one of the vectors to be added is translated to coincide with the tip of the other vector. The resultant vector is simply a line drawn

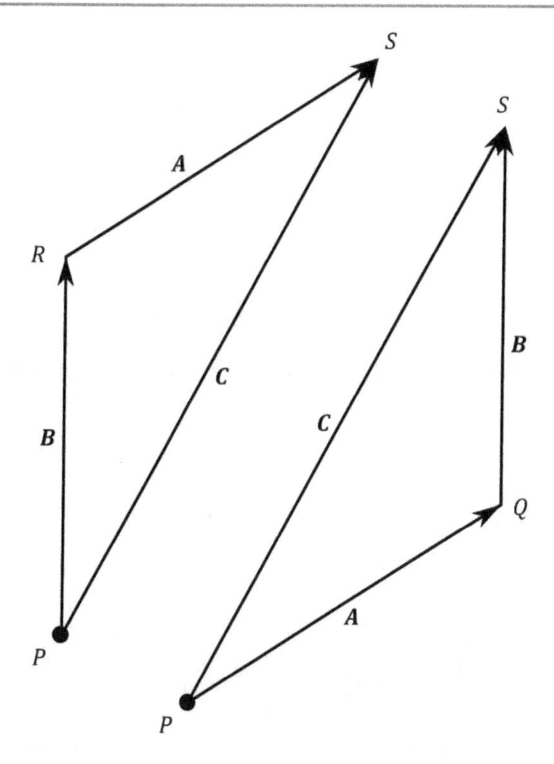


Fig. E.7 Illustration of the tip-to-tip method for adding vectors

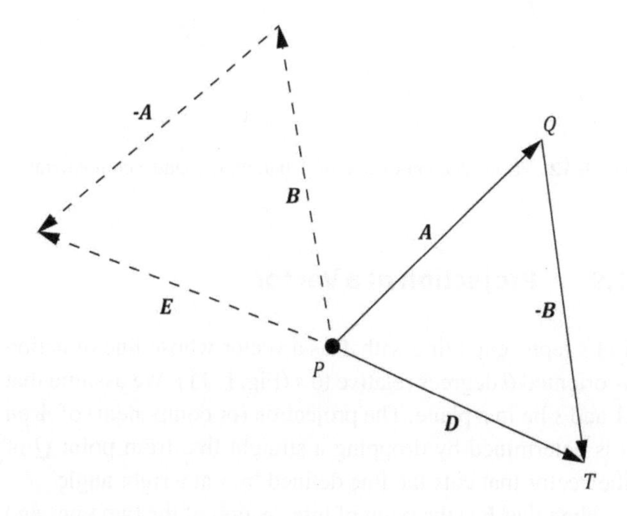


Fig. E.8 Illustration of the tip-to-tip method for subtracting vectors

from the tail of the first vector to the tip of the second vector (Fig. E.7).

Vector addition is commutative, that is:

$$\boldsymbol{A}+\boldsymbol{B}=\boldsymbol{B}+\boldsymbol{A}=\boldsymbol{C}.$$

## 1.7 Vector Subtraction

Recal that the negative of any vector is simply a vector with the same magnitude but pointing in the opposite direction (Fig. E.8). To subtract vector $\boldsymbol{B}$ from $\boldsymbol{A}$, $-\boldsymbol{B}$ can be added to $\boldsymbol{A}$ to determine a resultant.

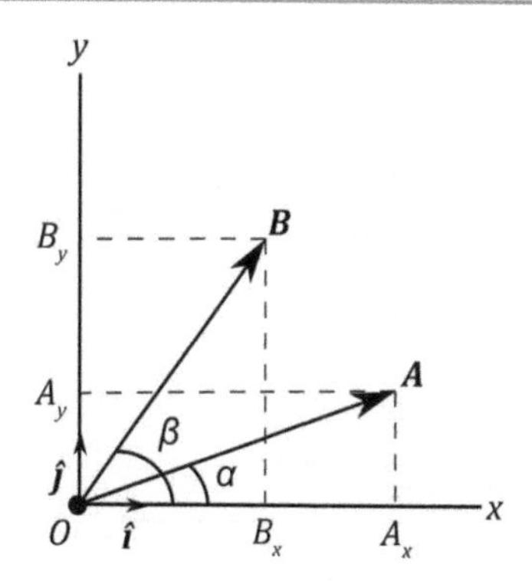


Fig. E.9 Illustration of the decomposition of vectors into trigonometric components for vector addition

For example,

$$\boldsymbol{A}-\boldsymbol{B}=\boldsymbol{D}$$
$$\boldsymbol{B}-\boldsymbol{A}=\boldsymbol{E}.$$

## 1.8 Trigonometric Addition (and Subtraction) of Vectors, Using Trigonometry to Resolve Vectors into Their Components

Consider vectors $\boldsymbol{A}$ and $\boldsymbol{B}$ in the $xy$-plane with magnitudes equal to $A$ and $B$, respectively (Fig. E.9).

Vectors $\boldsymbol{A}$ and $\boldsymbol{B}$ have orientations relative to the $x$-axis, $\alpha$ and $\beta$, respectively. We determine the scalar components of $\boldsymbol{A}$ and $\boldsymbol{B}$ using the properties of right triangles. For $\boldsymbol{A}$:

$$A_x = A\cos\alpha$$
$$A_y = A\sin\alpha.$$

Recall that $A_x$ and $A_y$ are the scalar components of $\boldsymbol{A}$ along the $x$ and $y$ axes, respectively. We can also make use of the unit vectors $\hat{\boldsymbol{i}}$ and $\hat{\boldsymbol{j}}$ that define the positive $x$ and $y$ directions to define the vector components of $\boldsymbol{A}$:

$$\boldsymbol{A}_x = A_x\hat{\boldsymbol{i}}$$
$$\boldsymbol{A}_y = A_y\hat{\boldsymbol{j}}.$$

So now, $\boldsymbol{A}$ can be expressed as:

$$\begin{aligned}\boldsymbol{A} &= \boldsymbol{A}_x + \boldsymbol{A}_y\\ &= A_x\hat{\boldsymbol{i}} + A_y\hat{\boldsymbol{j}}\\ &= A\cos\alpha\hat{\boldsymbol{i}} + A\sin\alpha\hat{\boldsymbol{j}}.\end{aligned}$$

Since $A_x$, $A_y$, and $A$ are legs of a right triangle, with $A$ being the hypotenuse:

$$A=\sqrt{A_x^2+A_y^2}.$$

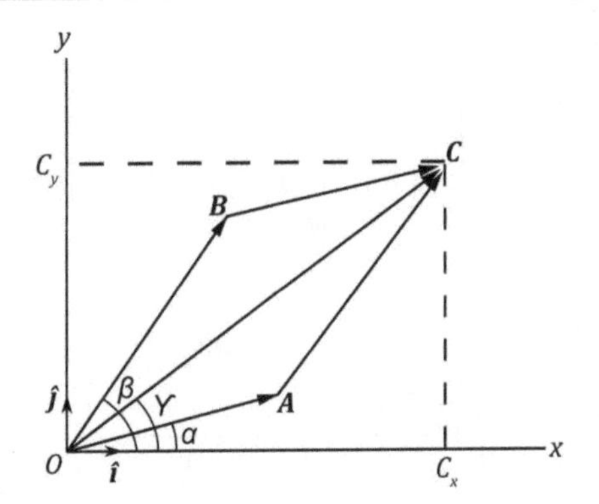


**Fig. E.10** Illustration of the vector summation of ***A*** and ***B*** to determine ***C***

Vector ***B*** can be expressed in a similar fashion:

$$\begin{aligned} \boldsymbol{B} &= \boldsymbol{B}_x + \boldsymbol{B}_y \\ &= B_x\hat{\boldsymbol{i}} + B_y\hat{\boldsymbol{j}} \\ &= B\cos\beta\hat{\boldsymbol{i}} + B\sin\beta\hat{\boldsymbol{j}}. \end{aligned}$$

Once the vectors are expressed in terms of their components, addition (or subtraction) is accomplished by simply adding like components:

$$\begin{aligned} \boldsymbol{A}+\boldsymbol{B} &= \left(A_x\hat{\boldsymbol{i}} + A_y\hat{\boldsymbol{j}}\right) + \left(B_x\hat{\boldsymbol{i}} + B_y\hat{\boldsymbol{j}}\right) \\ &= \left(A_x + B_x\right)\hat{\boldsymbol{i}} + \left(A_y + B_y\right)\hat{\boldsymbol{j}} \\ &= \left(A\cos\alpha + B\cos\beta\right)\hat{\boldsymbol{i}} + \left(A\sin\alpha + B\sin\beta\right)\hat{\boldsymbol{j}}. \end{aligned}$$

Let vector ***C*** be the vector sum of ***A*** and ***B*** (Fig. E.10):

$$\begin{aligned} \boldsymbol{C} &= \boldsymbol{A}+\boldsymbol{B} \\ &= \boldsymbol{C}_x + \boldsymbol{C}_y \\ &= C_x\hat{\boldsymbol{i}} + C_y\hat{\boldsymbol{j}}. \end{aligned}$$

Now, $C_x$ and $C_y$ can be determined:

$$\begin{aligned} C_x &= A_x + B_x = A\cos\alpha + B\cos\beta \\ C_y &= A_y + B_y = A\sin\alpha + B\sin\beta, \end{aligned}$$

as well as *C*:

$$C = \sqrt{C_x^2 + C_y^2}.$$

The resultant vector, ***C***, has an orientation, $\gamma$, relative to the *x*-axis, determined by using the inverse tangent function, $\tan^{-1}$ (arctangent or arctan):

$$\gamma = \tan^{-1}\left(\frac{C_y}{C_x}\right).$$

Subtraction looks like this:

$$\begin{aligned} \boldsymbol{A}-\boldsymbol{B} &= \left(A_x\hat{\boldsymbol{i}} + A_y\hat{\boldsymbol{j}}\right) - \left(B_x\hat{\boldsymbol{i}} + B_y\hat{\boldsymbol{j}}\right) \\ &= \left(A_x - B_x\right)\hat{\boldsymbol{i}} + \left(A_y - B_y\right)\hat{\boldsymbol{j}} \\ &= \left(A\cos\alpha - B\cos\beta\right)\hat{\boldsymbol{i}} + \left(A\sin\alpha - B\sin\beta\right)\hat{\boldsymbol{j}}. \end{aligned}$$

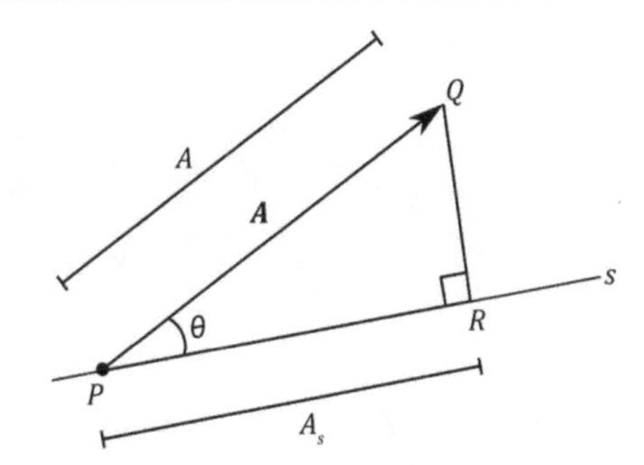


**Fig. E.11** Illustration of the projective of a vector in a given direction

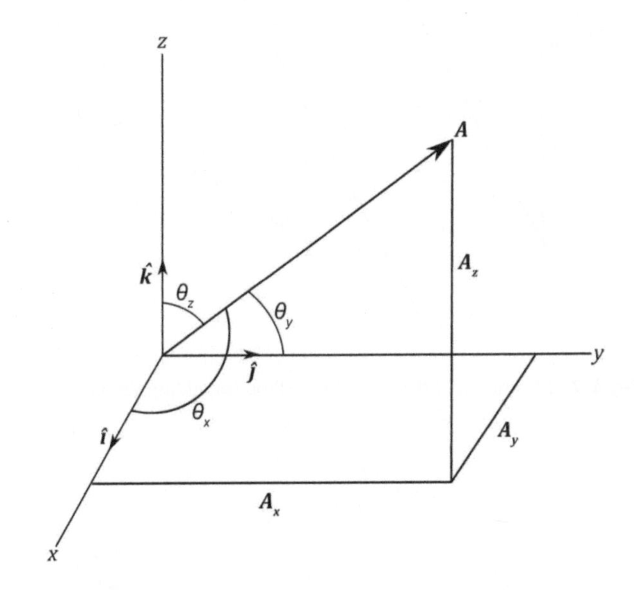


**Fig. E.12** Vector decomposition of ***A*** into its *x*, *y*, and *z* components

## 1.9 Projection of a Vector

Let *s* represent a line with ***A*** as a vector whose line of action is oriented $\theta$ degrees relative to *s* (Fig. E.11). We assume that ***A*** and *s* lie in a plane. The projection (or component) of ***A*** on *s* is determined by dropping a straight line from point *Q* of the vector that cuts the line defined by *s* at a right angle.

Note that *R* is the point of intersection of the two lines and $A_s$ is the length of the line segment between points *R* and *P*, then $A_s$ is the projection or scalar component of vector ***A*** on *s*. Points *P*, *Q*, and *R* define a right triangle, so:

$$A_s = A\cos\theta$$

## 1.10 Determining Vector Magnitude and Unit Vector

Recall that a vector is comprised of a magnitude and direction. Note the arbitrary vector ***A*** and its three component vectors $\boldsymbol{A}_x$, $\boldsymbol{A}_y$, and $\boldsymbol{A}_z$ oriented along the $\hat{\boldsymbol{i}}$, $\hat{\boldsymbol{j}}$, and $\hat{\boldsymbol{k}}$ directions (Fig. E.12).

This graphical representation can be expressed as:

$$\boldsymbol{A} = \boldsymbol{A}_x + \boldsymbol{A}_y + \boldsymbol{A}_z$$
$$\boldsymbol{A} = A_x\hat{\boldsymbol{i}} + A_y\hat{\boldsymbol{j}} + A_z\hat{\boldsymbol{k}}.$$

The relationship between the magnitudes of the vectors is given by:

$$|\mathbf{A}| = A = \sqrt{A_x^2 + A_y^2 + A_z^2}.$$

The unit vector $\hat{\boldsymbol{u}}$ that points in the same direction as $A$ can be determined by:

$$\hat{\boldsymbol{u}} = \frac{\boldsymbol{A}}{|\boldsymbol{A}|}.$$

The original vector $\boldsymbol{A}$ can be expressed as:

$$\boldsymbol{A} = A\hat{\boldsymbol{u}},$$

where the magnitude of $\hat{\boldsymbol{u}}$ is always equal to 1.

# Appendix F: Forces, Force Vectors, Composition, and Resolution

A force is simply defined as a push or a pull. Forces can deform a body, initiate motion, and change the direction of motion. In some cases, for example, when a force is balanced by other forces, it does not cause movement, e.g., pushing against a wall.

A force is a vector quantity with both magnitude and direction. This vector is illustrated by an arrow that indicates the point of application, line of action, and sense of direction (or orientation) with a length that indicates the magnitude (Fig. F.1). Note: if multiple force vectors of different magnitudes are present, each vector length must be proportional to the force magnitude.

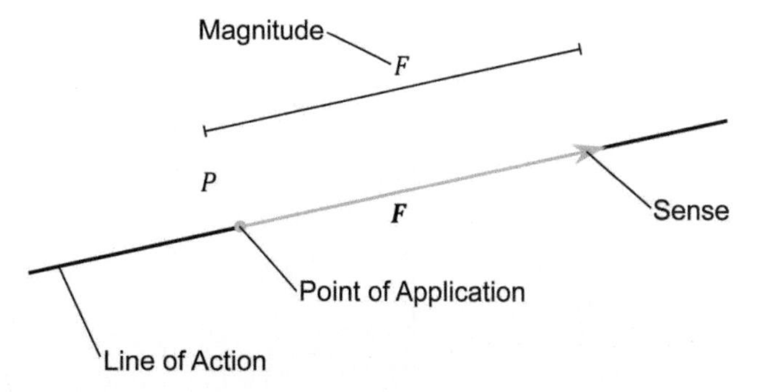


**Fig. F.1** Illustration of representation of a force vector, ***F***

Forces may be either external or internal. Examples of external forces include gravity, applying pressure to a body segment during resisted manual muscle testing (MMT), holding a weighted object, etc. When solving static equilibrium problems, force magnitudes may be known or unknown. Internal forces are those that reside inside the material and provide resistance to the deformation of biological tissues, e.g., bone, ligament, articular cartilage, etc. Internal forces also result from muscle actions, which, because of their unique nature, offer both active and passive resistance.

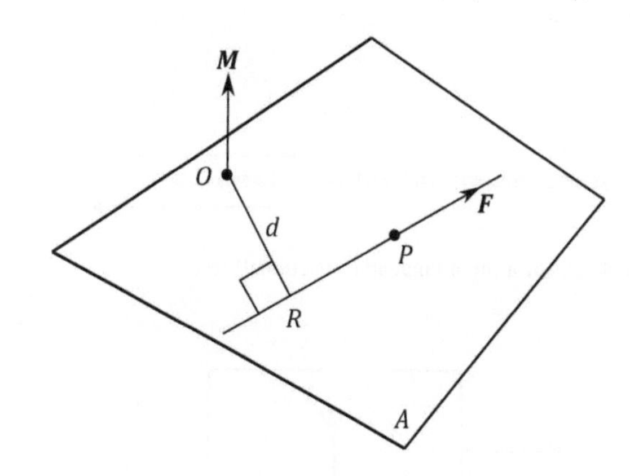


**Fig. F.2** Illustration of representation of a moment vector, ***M***

Force systems are recognized any time two or more forces act on a body. Although we recognize that all human activity involves three-dimensional (3D) movements, for the purposes of instruction, we limit ourselves to planar, i.e., two-dimensional (2D), systems. Therefore, when we refer to forces that are coplanar, we are thinking of two dimensions, and orienting force vectors, using a Cartesian coordinate (rectangular) system by considering the $x$ and $y$ components of external and internal forces. Planar problems can also be solved using a polar coordinate system, but we restrict ourselves to solving problems using a rectangular coordinate system. Note: the moment of force (moment or torque) calculated as part of static equilibrium problems is a vector that is perpendicular to the $xy$-plane or about the $z$-axis (Fig. F.2).

Our bodies, or the objects we manipulate, encounter several different kinds of forces. Normal forces are reaction forces that act perpendicularly to a body and enforce constraints between two bodies (Fig. F.3).

Tangential forces are reaction forces that act parallel to the contact surface (Fig. F.4). For example, tangential forces could be "seen" at a joint between the articular cartilage surfaces of two adjacent bones, e.g., the head of the femur and acetabulum. Tangential forces that act between surfaces are frictional forces, whereas a shear force, induced by a tangential force, is a force inside a body. A shear force applied over a given area, i.e., force/area, is defined as shear



G. J. Alderink, B. M. Ashby, *Clinical Kinesiology and Biomechanics*, https://doi.org/10.1007/978-3-031-25322-5

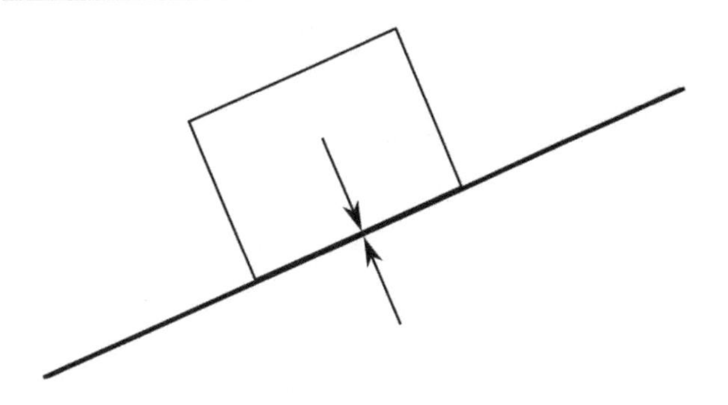

**Fig. F.3** Illustration of forces that are normal (perpendicular) to the surfaces in contact

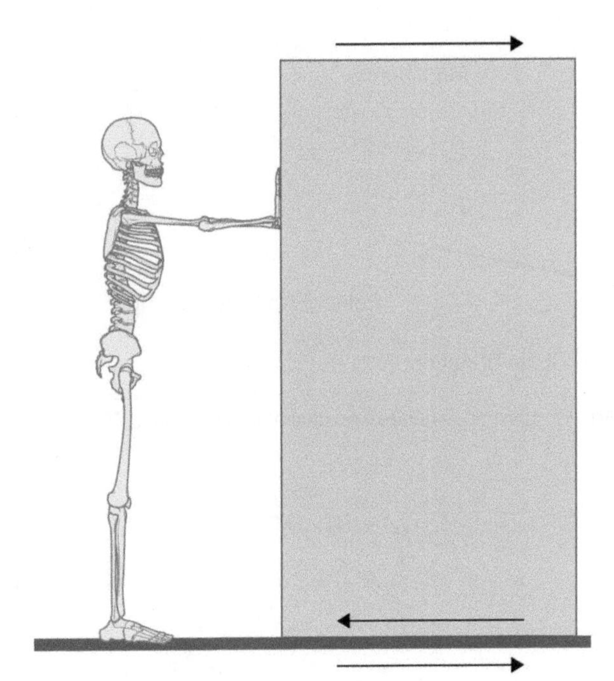

**Fig. F.4** Illustration of tangential (frictional) forces

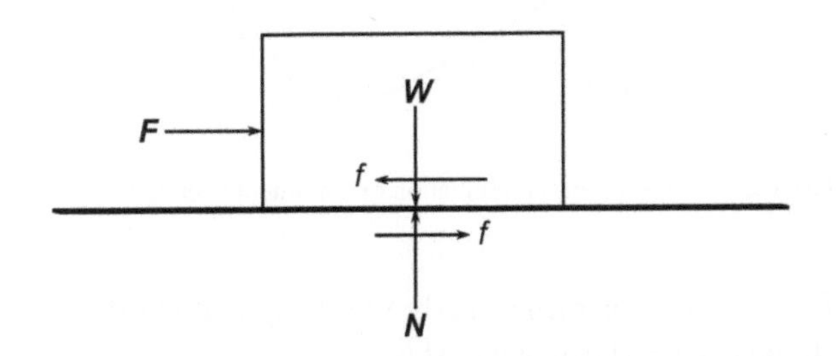


**Fig. F.5** Illustration showing that friction exists on surfaces when one surface slides or tends to slide relative to the other surface

stress. Note: tangential forces (friction) exist whenever one object moves in parallel to another object, but frictional forces may also exist in the absence of movement (Fig. F.5).

There are additional types of ways bodies or tissues are loaded (or experience a force). During tensile loading, equal and opposite loads are applied from the surface of a structure, and tensile stress and strain (i.e., deformation) result inside the structure. With compression loading, equal and opposite loads are applied toward the surface of the structure, and compressive stress and strain result inside the structure (Fig. F.6). With bending, loads are applied to a structure in a manner that causes it to bend about an axis (Fig. F.7). In this case, the structure is subjected to a combination of compression and tension; tensile stresses and strains are experienced on the convex surface of the bend, where as compressive stresses and strains are experienced on the concave surface of the bend. In torsion, a load is applied to a structure in a manner that causes it to twist about an axis, and a torque (moment) is produced within the structure (Fig. F.8). When a body is loaded with torsion, shear stresses are distributed over the entire body. With activities of daily living, the human body is typically subjected to different types of loads across multiple dimensions.

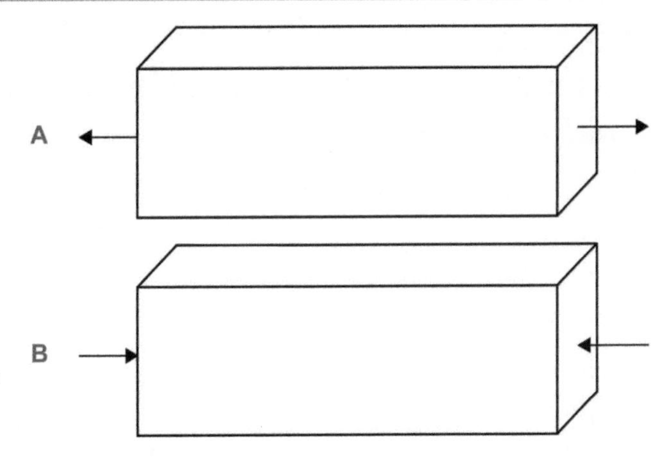


**Fig. F.6** Illustration of (**A**) tensile and (**B**) compressive forces

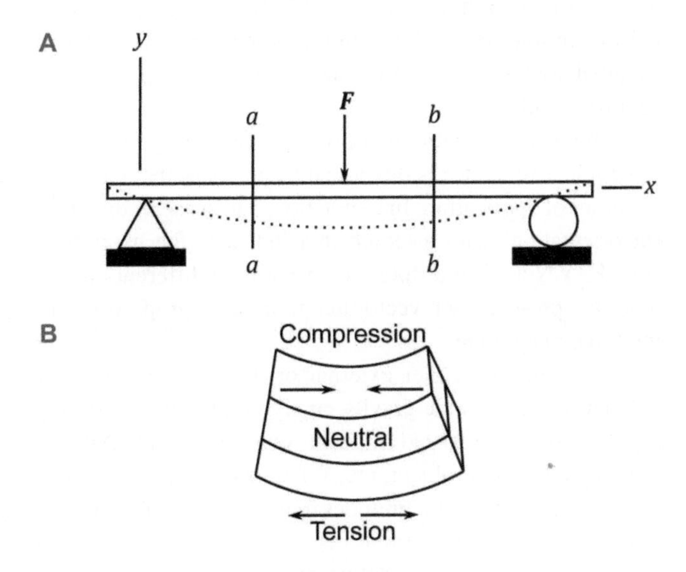


**Fig. F.7** Illustration of (**A**) a bending force and (**B**) the compressive and tensile stresses induced by the bending

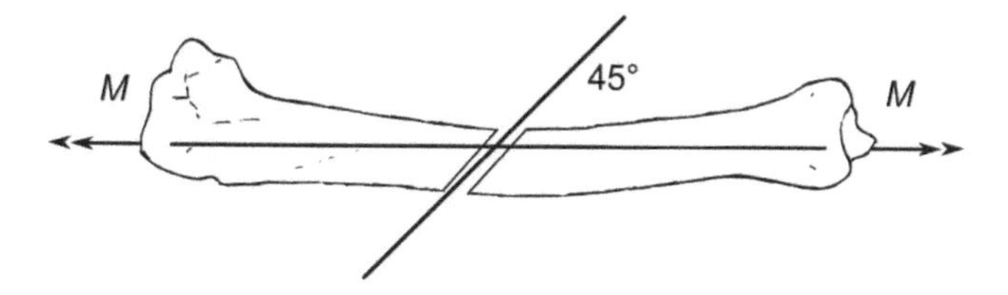


**Fig. F.8** Illustration of the potential for a spiral bone fracture induced by pure torsion, i.e., by moment $\boldsymbol{M}$

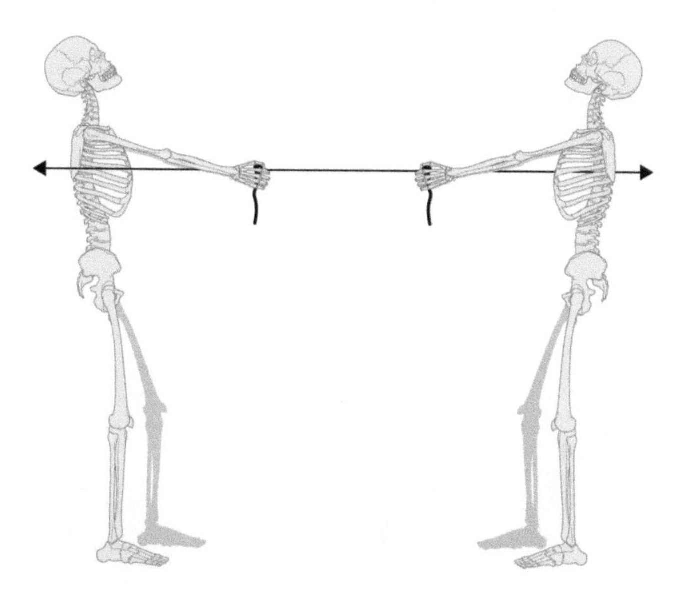

**Fig. F.9** Illustration of collinear forces

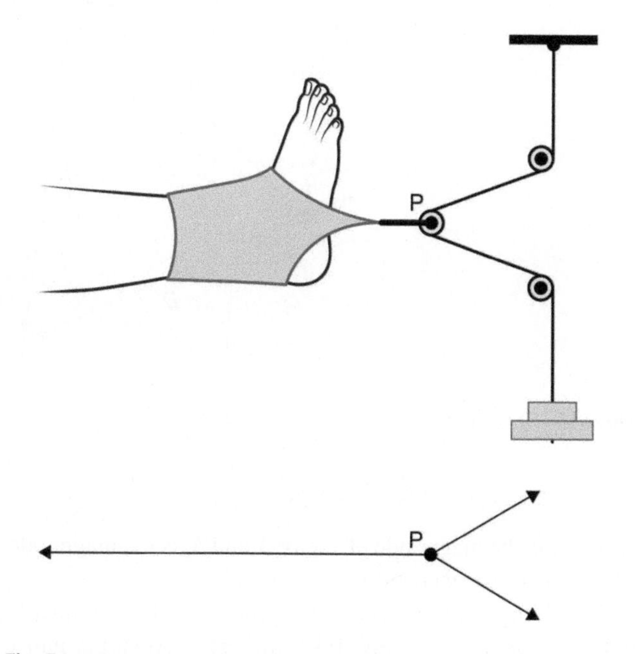


**Fig. F.10** Illustration of concurrent forces

When solving coplanar statics problems, forces may be described as colinear (Fig. F.9), concurrent (Fig. F.10), or parallel (Fig. F.11).

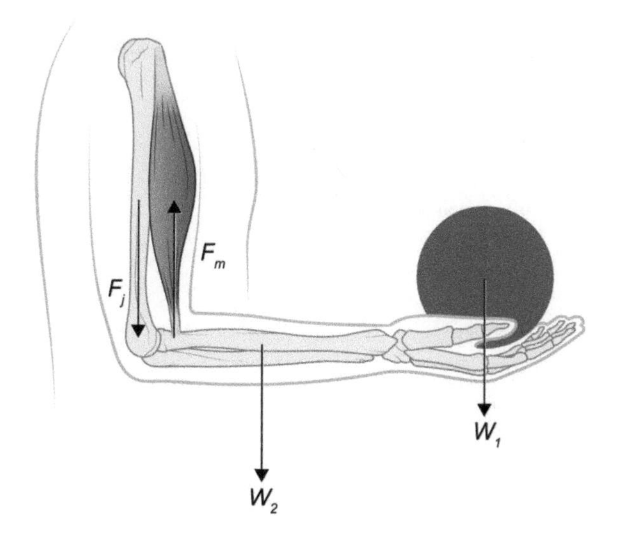


**Fig. F.11** Illustration of parallel forces

## 1.1 Composition of Forces

The composition of forces is used when we have two or more coplanar forces acting on an object, e.g., a parallel force system, and wish to find their combined effect as a single force, called a resultant force. There are several methods used to find the magnitude of a resultant force: (1) graphical, using a vector diagram; (2) algebraic, utilizing trigonometric functions; and (3) vector algebra, i.e., vector addition or subtraction.

Recall that a scalar has magnitude only, whereas a vector has both magnitude and direction. Since force is a vector both its magnitude and direction must be provided to fully describe its nature. The magnitude of a vector is a scalar quantity, which is always positive. Typically, we assign a positive value to a force vector if, in our Cartesian coordinate system, the vector points up (i.e., in the direction of the positive $y$-axis) or to the right (i.e., in the direction of the positive $x$-axis) and assign a negative value if the vector points down or to the left. In other words, for vector quantities, a negative sign implies a change in direction and has nothing to do with its magnitude. An example of positive and negative vectors is shown in Fig. F.12.

There are two ways of adding vectors graphically: parallelogram and triangle (or tip to tail). Consider (Fig. F.13) and imagine vectors $\boldsymbol{A}$ and $\boldsymbol{B}$ that represent forces.

We can see how to add these vectors using the parallelogram method in Fig. F.14 or the tip-to-tail method in Fig. F.15.

We can also "subtract" a vector, or add a negative vector, using graphical methods (Fig. F.16).

**Fig. F.12** Illustration of a negative vector

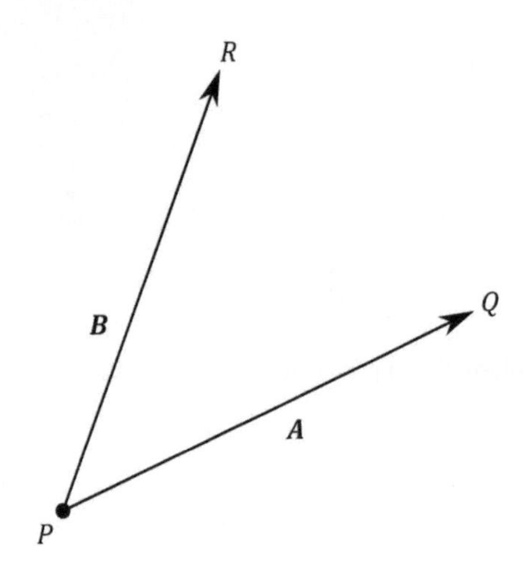


**Fig. F.13** Illustration of vectors ***A*** and ***B*** acting concurrently at point *P*

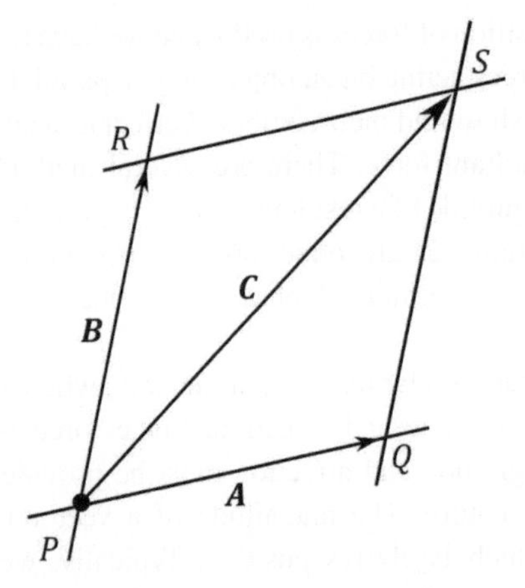


**Fig. F.14** Illustration of adding force vectors ***A*** and ***B*** using the parallelogram method

## 1.2 Resolution of Forces

To help us resolve vectors in a plane, we first illustrate how to project any vector onto another line in the plane. For example, Fig. F.17 shows vector ***A*** oriented by angle $\theta$ relative to line *s*, from point *P* to *Q*.

To determine the projection of ***A*** onto *s*, first drop a line from point *Q* to line *s* so the junction of *R* at *s* forms a 90° angle. So $A_s$ is the length of the line segment running from *P* to *R*, now referred to as the scalar component of ***A*** in the *s* direction or the projection of ***A*** to *s*. Since we created a right triangle, we can determine $A_s$ as:

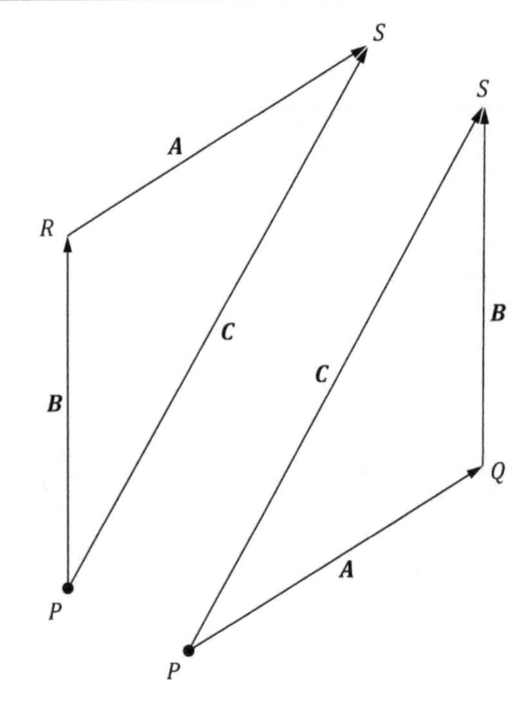


**Fig. F.15** Illustration of adding force vectors ***A*** and ***B*** using the tip-to-tail method

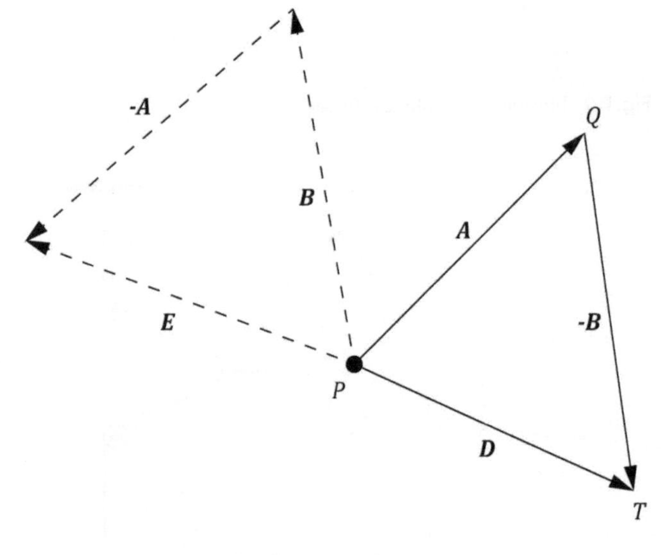


**Fig. F.16** Illustration of adding a negative vector

$$A_s = A\cos\theta,$$

where *A* is the magnitude of vector ***A*** and $A_s$ is the magnitude of vector ***A*** projected onto line *s*.

Resolving a force vector is the process of determining its planar components. In Fig. F.18, we resolve vector ***A*** into *x* and *y* components. Notice that we create two triangles that are similar (forming a rectangle or parallelogram), using the same steps for determining the projection of a vector.

Given that we know the magnitude of ***A*** and angle $\theta$, we can determine that

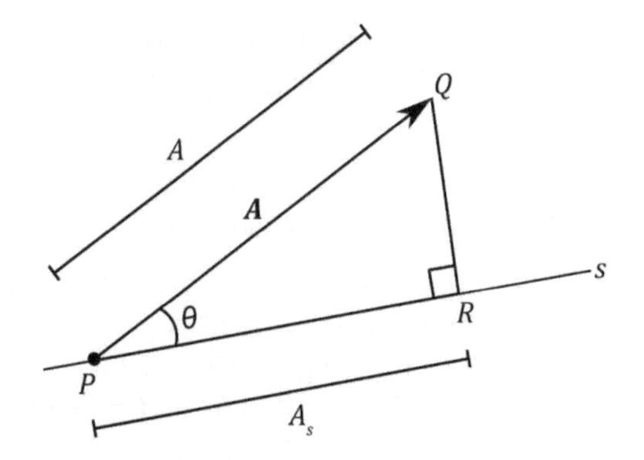


**Fig. F.17** Illustration of the projection of a vector in a given direction

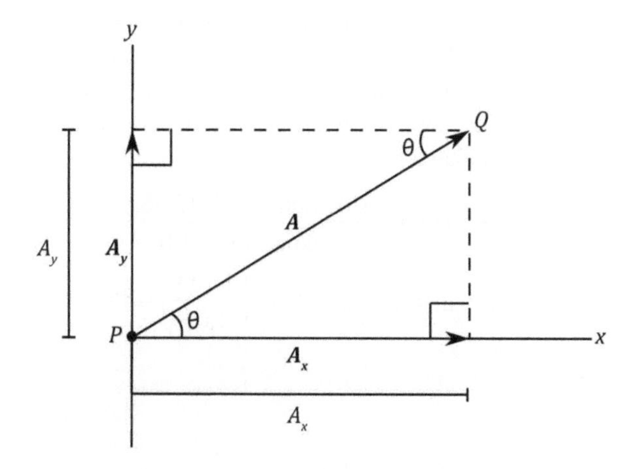


**Fig. F.18** Decomposition of vector $A$ into components along mutually perpendicular directions

$$A_x = A\cos\theta$$
$$A_y = A\sin\theta,$$

where $A_x$ and $A_y$ are the magnitudes of the components of $\boldsymbol{A}$ and $A$ is the magnitude of vector $\boldsymbol{A}$.

We use these simple algebraic and trigonometric functions to resolve individual forces into components in planar statics problems, then use principles of force composition, i.e., adding like components (see Appendix E), to determine the net force acting on an object or body segment.

**Examples**

1. Consider two forces, $\boldsymbol{F}_1$ and $\boldsymbol{F}_2$, applied to an object at point $O$ in the $xy$-plane (Fig. F.19).
   Assumptions/known variables:

   - Forces directed up/right are positive.
   - $F_1 = 20$ N oriented 30° above the positive $x$-axis.
   - $F_2 = 15$ N oriented 45° above the negative $x$-axis.

   Determine the magnitude of the resultant $\boldsymbol{F_R}$ and its orientation (relative to the $x$-axis).

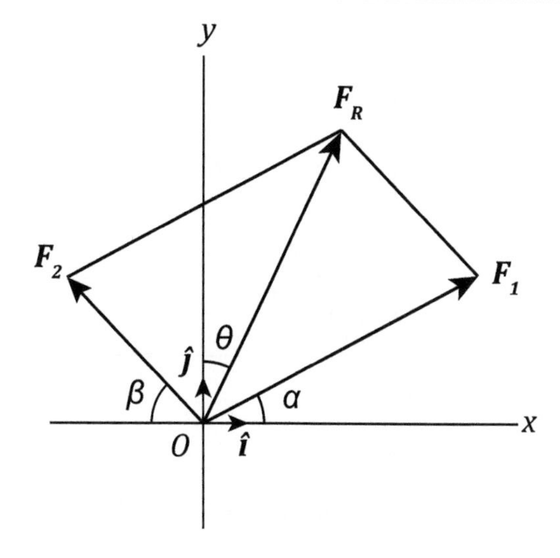


**Fig. F.19** Force vectors $\boldsymbol{F}_1$ and $\boldsymbol{F}_2$ sum to be the resultant force, $\boldsymbol{F_R}$

***Solution:*** 1. Resolve $\boldsymbol{F}_1$ and $\boldsymbol{F}_2$ into their respective $x$ and $y$ components:

- $F_{1x} = F_1 \cos 30° = 17.3$ N
- $F_{1y} = F_1 \sin 30° = 10$ N
- $F_{2x} = -F_2 \cos 45° = -10.6$ N
- $F_{2y} = F_2 \sin 45° = 10.6$ N

2. Determine the scalar magnitude of $\boldsymbol{F_R}$:

- $\sum F_x = F_{1x} + F_{2x} = 17.3$ N $+ (-10.6$ N$) = 6.7$ N
- $\sum F_y = F_{1y} + F_{2y} = 10$ N $+ 10.6$ N $= 20.6$ N
- $F_R = \sqrt{F_x^2 + F_y^2} = 21.7\text{N}$

3. Determine the orientation of $\boldsymbol{F_R}$:

- $\tan\theta = \dfrac{\text{side opposite}}{\text{side adjacent}} = \dfrac{20.6}{6.7};$

  therefore, $\theta = \tan^{-1}\left(\dfrac{20.6}{6.7}\right) = 72.0°$

2. Consider the planar forces shown in Fig. F.20.
   $\boldsymbol{F_R}$ represents the resultant of the sum of vectors $\boldsymbol{F}_1$ and $\boldsymbol{F}_2$.
   Assumptions/known variables:

   - Forces directed up/right are positive.
   - The magnitude of $\boldsymbol{F_R}$ is 25 N, and it is oriented 33° relative to the plus $x$-axis.
   - The scalar components of $\boldsymbol{F}_1$ are $F_{1x} = 22$ N and $F_{1y} = 26$ N.

   Determine the magnitude and direction of vectors $\boldsymbol{F}_1$ and $\boldsymbol{F}_2$.

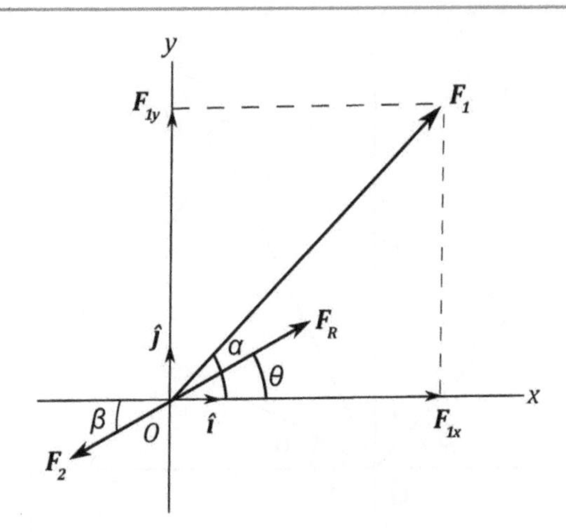


**Fig. F.20** Finding vector $\boldsymbol{F}_2$ given the components of $\boldsymbol{F}_1$ and the resultant vector $\boldsymbol{F}_R$

***Solution:*** 1. Calculate the magnitude of $\boldsymbol{F}_1$:

- $F_1 = \sqrt{F_{1x}^2 + F_{1y}^2} = 34.1\,\text{N}$

2. Determine the orientation, i.e., angle $\alpha$, that $F_1$ makes relative to the $x$-axis:

- $\tan\alpha = \dfrac{\text{side opposite}}{\text{side adjacent}} = \dfrac{26}{22}$;

  therefore, $\alpha = \tan^{-1}\left(\dfrac{26}{22}\right) = 49.8°$

3. Calculate the magnitude of $\boldsymbol{F}_2$:

- $F_{Rx}$ = (25 N) cos 33° = 20.97 N; $F_{Ry}$ = (25 N) sin 33° = 13.62 N
- $F_{2x} = F_{Rx} - F_{1x} = 20.97 - 22 = -1.03$ N; $F_{2y} = F_{Ry} - F_{1y} = 13.62 - 26 = -12.38$ N
- $F_2 = \sqrt{F_{2x}^2 + F_{2y}^2} = 12.43\,\text{N}$

4. Determine angle $\beta$ that $\boldsymbol{F}_2$ makes relative to the $x$-axis. Since both the $x$ and $y$ components of $\boldsymbol{F}_2$ are negative, $\boldsymbol{F}_2$ is in the third quadrant, as shown in Fig. F.20 above. Angle $\beta$, which is an angle below the negative $x$-axis, is then calculated as:

- $\tan\beta = \dfrac{\text{side opposite}}{\text{side adjacent}} = \dfrac{12.38}{1.03}$;

  therefore, $\beta = \tan^{-1}\left(\dfrac{12.38}{1.03}\right) = 85.2°$

# Appendix G: Moment (Torque) Vectors

When we solve problems assuming rigid body mechanics, any deformations of the body are neglected. When an unbalanced force is applied to a body, Newton's second law dictates that the body accelerates in the direction of the force with a magnitude that is proportional to that force. If the line of action of the force does not pass through the center of mass, the body is accelerated in rotation in addition to being accelerated in translation. The physical quantity that causes this rotation is called a "moment of force" or simply a "moment," for short. The moment of force is also frequently referred to as a "torque."

In two dimensions (2D), the magnitude of moment $\boldsymbol{M}$ is defined as the magnitude of force $\boldsymbol{F}$ times the shortest perpendicular distance $d$ (also called the moment or lever arm) from the action line of the applied force to the point of reference (Fig. G.1).

$$M = Fd. \tag{G.1}$$

Note: to determine the shortest perpendicular distance for the moment arm, we often need to extend the action line of the applied force and draw a line that forms a right angle between the action line and the point, i.e., most often the joint center, about which the moment is being calculated.

Since a moment is a vector, it has a magnitude and orientation or direction. In 2D, the direction of the moment is always perpendicular to the plane, e.g., in the $z$-direction, if the force and reference point lies in the $xy$-plane (Fig. G.2). The direction and sense of the moment vector can be determined using the right-hand rule; that is, when the fingers of the right hand curl in the direction where the applied force tends to rotate the body about point $O$, the right thumb points in the direction of the moment vector (Fig. G.3).

When describing the direction of a moment in 2D, we could state that it occurs about the $z$-axis, OR depending on the observer's frame of reference, claim that it is acting in a clockwise or counterclockwise direction. In either case, it is important that a frame of reference is defined. On the other hand, a reference frame is not required when using clinical terms, such as describing the bending of the elbow as a flexor moment.

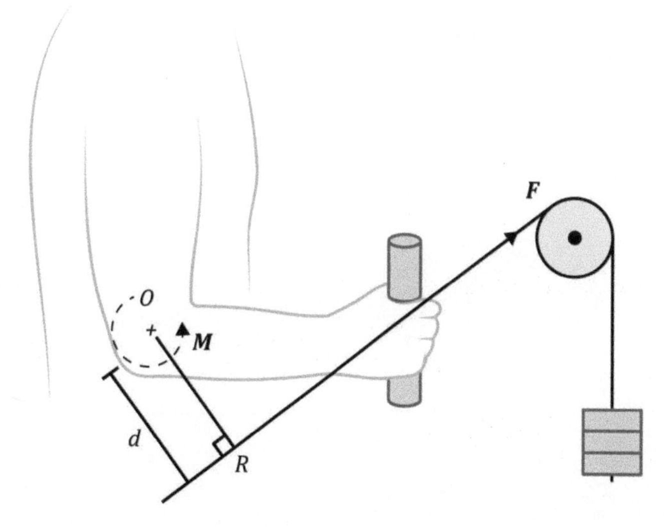

**Fig. G.1** Example of moment $\boldsymbol{M}$ about point $O$. Note that the applied force $\boldsymbol{F}$ has its action line extended, which intersects moment arm $d$ at point $R$

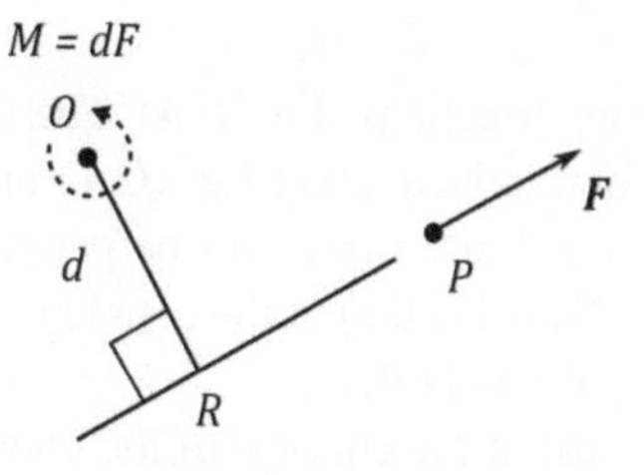

**Fig. G.2** A force with magnitude $F$ acting along a line of action that is a distance $d$ from point $O$ as shown results in a counterclockwise moment $\boldsymbol{M}$ with magnitude equal to d times $F$

There is an interesting special case of parallel forces acting on a body that seems to have some clinical relevance, which is the concept of a couple. A force couple or couple moment is present when two parallel forces with equal magnitudes are applied to an object along different lines of


G. J. Alderink, B. M. Ashby, *Clinical Kinesiology and Biomechanics*, https://doi.org/10.1007/978-3-031-25322-5

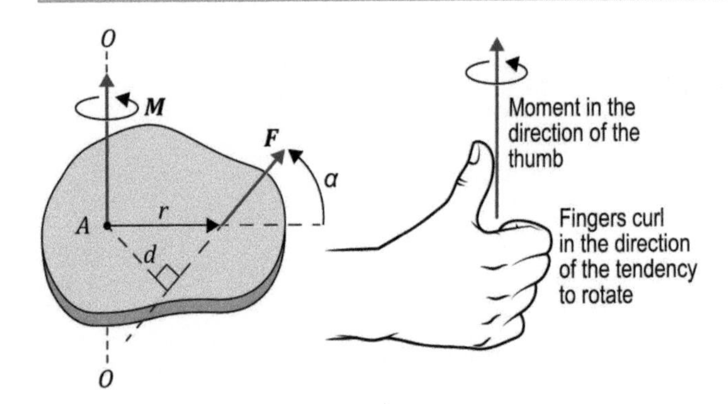


**Fig. G.3** Depiction of the right-hand rule: when a force $\boldsymbol{F}$ acts relative to point $A$ with a moment arm of magnitude $d$, a counterclockwise (when looking down at the plane that $\boldsymbol{r}$ and $\boldsymbol{F}$ are in) moment $\boldsymbol{M}$ is created. Note: $OO$ = axis of rotation; $\boldsymbol{M}$ = moment; $A$ is a point piercing the plane on axis $OO$; $\boldsymbol{r}$ = the vector from point $A$ to the tail of the force vector $\boldsymbol{F}$; $d$ = the moment arm of $\boldsymbol{F}$

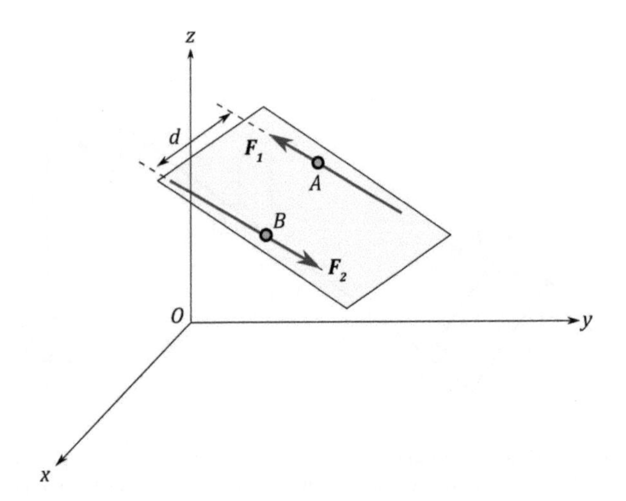


**Fig. G.4** A system of forces whose resultant force is equal to zero, but the resultant moment, i.e., the couple-moment, is not zero

action in opposite directions (Fig. G.4). The magnitude of a couple can be determined using Eq. (G.1) or is simply the magnitude of one force times the perpendicular distance between forces. The direction of the couple moment is determined by the right-hand rule.

There is interest in the concept of the couple moment in biomechanics. For example, a situation that approximates a couple moment is produced by the nearly equal and opposite muscle forces due to concentric actions of the upper and lower trapezius, assisted by the serratus anterior, to elevate and upwardly rotate the scapula during glenohumeral abduction and flexion (Fig. G.5).

Let us identify several "mechanical facts" about moments that are useful when solving static equilibrium problems:

1. One can slide a force vector along a line of action without changing the moment, as shown in Fig. G.6.

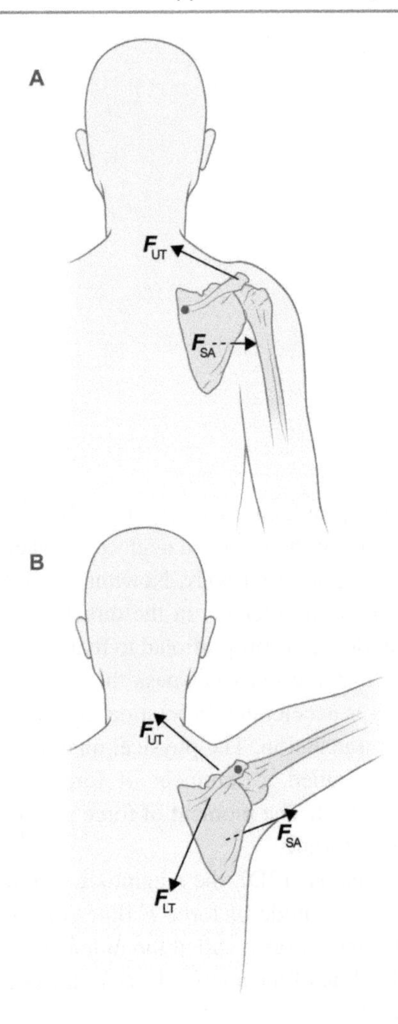


**Fig. G.5** (**A**) Upper trapezius-serratus anterior force couple to initiate upward rotation of the scapula during the initial 30° of arm elevation and (**B**) a force couple created by upper trapezius-lower trapezius-serratus anterior during the final phase of arm elevation. Note: UT = upper trapezius; LT = lower trapezius; SA = serratus anterior

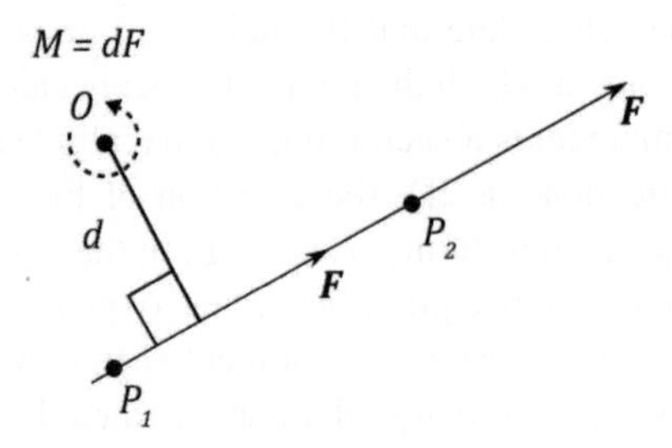


**Fig. G.6** Illustration showing that a moment is invariant as the force vector $\boldsymbol{F}$ is translated along its line of action

2. In Fig. G.7, we see that two moments have equal magnitudes, but in a different sense (i.e., rotate in opposite directions) if the respective forces acting on a body are equal in magnitude but acting in opposite directions along the same line of action.

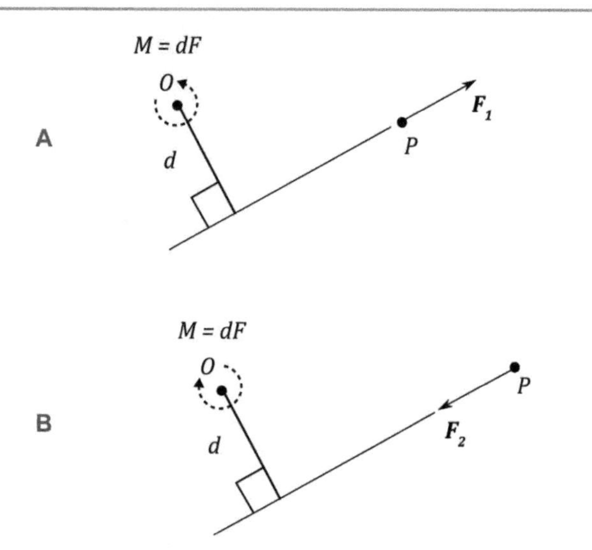


**Fig. G.7** Illustration of (**A**) counterclockwise and (**B**) clockwise moments that are equal in magnitude but opposite in direction dependent on the direction of ***F*** relative to *O*

3. The magnitude of a moment increases if (a) the moment arm of the applied force increases and/or (b) the magnitude of the force increases.
4. If the line of action of an applied force passes through the point about which the moment is being calculated, i.e., the joint center, the moment of force is zero because a moment arm does not exist.
5. A force applied to a body causes changes in the rotational velocity of the body either in a clockwise or counterclockwise direction, depending on the location of the reference point, i.e., the joint center, and the direction of the applied force relative to that point.

## Examples

1. Consider a patient sitting with a weighted boot attached to his/her foot performing knee extension exercises (Fig. G.8).

   Assumptions/known variables:

   - Forces directed up/right are positive.
   - Clockwise moments are positive.
   - Weight of the shank and foot, $W_1 = 60$ N; weight of the boot, $W_2 = 66.7$ N.
   - Distance from the center of gravity of the shank (point *A*) to the center of rotation (point *O*) of the knee joint, $a = 20$ cm (0.20 m).
   - Distance from the center of gravity of the weight boot (point *B*) to point *O*, $b = 50$ cm (0.50 m).

   Determine the net knee extensor moment when the knee is at position 1 (0° knee flexion), position 2 (30° knee flexion), and position 3 (90° knee flexion).

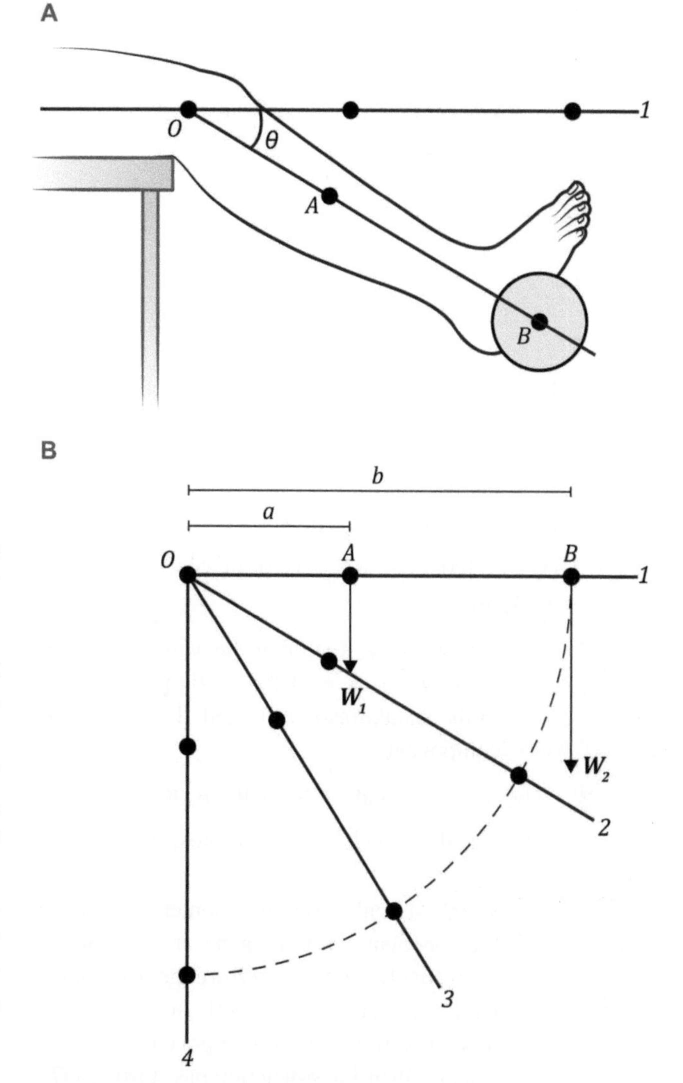


**Fig. G.8** Illustration of (**A**) the anatomical and kinesiological situation and (**B**) schematic setup and necessary variables

***Solution:*** 1. Position 1: note that *a* and *b* are the moment arms for $W_1$ and $W_2$, respectively:

$$\begin{aligned} M_{\text{net}} &= aW_1 + bW_2 \\ &= (0.20\,\text{m})(60\,\text{N}) + (0.05\,\text{m})(66.7\,\text{N}) \\ &= 45.4\,\text{Nm} \end{aligned}$$

Note that this 45.4 Nm clockwise external moment due to weights $W_1$ and $W_2$ must be balanced by a 45.4 Nm counterclockwise internal moment due to the knee extensor muscles acting on the shank.

2. Position 2: note $d_1$ (moment arm for $W_1$) $= a\cos\theta$ and $d_2$ (moment arm for $W_2$) $= b\cos\theta$ (Fig. G.9):

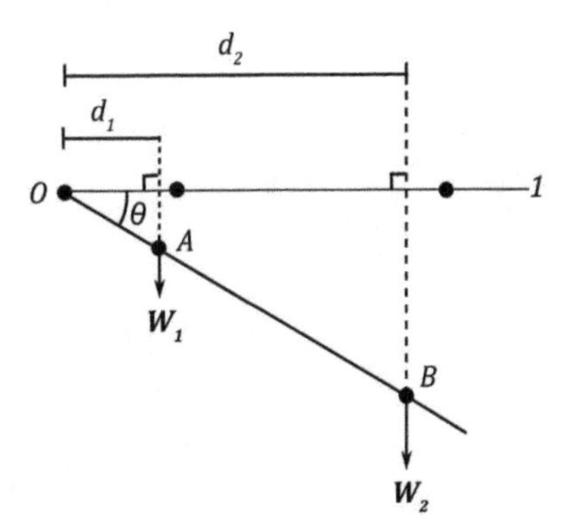


**Fig. G.9** Illustration of forces and moment arms as the knee joint angle, $\theta$, relative to the horizontal changes

$$
\begin{aligned}
M_{\text{net}} &= d_1 W_1 + d_2 W_2 \\
&= (a\cos\theta) W_1 + (b\cos\theta) W_2 \\
&= (0.20\,\text{m})\cos 30°(60\,\text{N}) + (0.05\,\text{m})\cos 30°(66.7\,\text{N}) \\
&= 39.3\,\text{Nm}
\end{aligned}
$$

Alternatively, assuming the $x$-axis is aligned with the shank with the $y$-axis perpendicular to the shank, resolve $W_1$ and $W_2$ into $x$ and $y$ components:

$$W_{1_x} = W_1 \sin 30° \quad \text{and} \quad W_{1_y} = W_1 \cos 30°$$
$$W_{2_x} = W_2 \sin 30° \quad \text{and} \quad W_{2_y} = W_2 \cos 30°$$

Then we only use the moments from the $y$-component with moment arms $a$ and $b$, respectively, since $a$ and $b$ are the perpendicular distances from $W_{1_y}$ and $W_{2y}$ to point $O$. The moments from the $x$ components are zero because their lines of action pass through $O$:

$$
\begin{aligned}
M_{\text{net}} &= aW_{1_y} + bW_{2_y} \\
&= a\,W_1 \cos 30° + b\,W_2 \cos 30° \\
&= (0.20\,\text{m})(60\,\text{N})\cos 30° + (0.05\,\text{m})(66.7\,\text{N})\cos 30° \\
&= 39.3\,\text{Nm}
\end{aligned}
$$

3. Position 4

Since both $W_1$ and $W_2$ have lines of action that pass through $O$, their moment arms are 0, resulting in a moment $M_{\text{net}} = 0$.

2. Consider the external forces applied to a total hip joint prosthesis (Fig. G.10).
   Assumptions/known variables:

- Counterclockwise moments are positive.
- The total body weight is 180 lb (801 N); in the single-limb stance during gait, the peak bone-on-bone force at the hip joint is estimated to be approximately 1.2 × body weight (BW) → 1.2 × 801 N = 961 N.

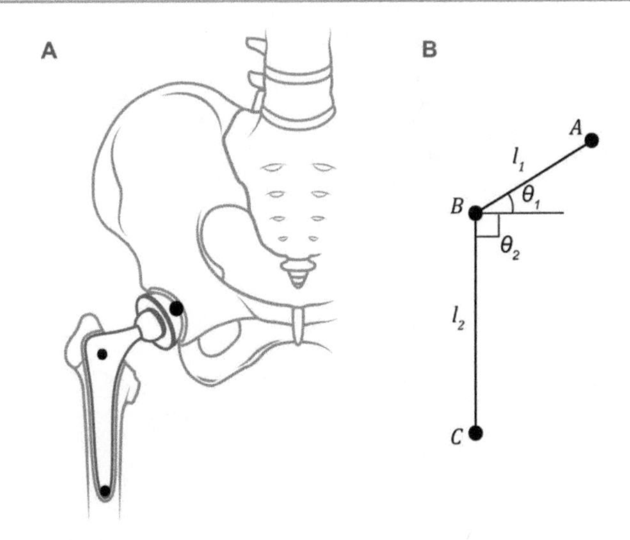


**Fig. G.10** Illustration of (**A**) total hip prosthesis and (**B**) related diagram with problem variables

- The geometric parameters of the prosthesis are as follows:
  - $l_1 = 50$ mm (0.05 m)
  - $l_2 = 100$ mm (0.10 m)
  - $\theta_1 = 45°$
  - $\theta_2 = 90°$

Estimate the moments generated during the single-limb stance by the force acting at the hip joint during gait generated about points $B$ and $C$, considering three different force lines of action (Fig. G.11).

***Solution:*** 1. For the condition in Fig. G.11A:

- The moment arm for force $F$ with respect to both $B$ and $C$ is:

$$d_1 = l_1 \cos\theta_1 = (0.05\,\text{m})\cos 45° = 0.0354\,\text{m}.$$

- The moments with respect to both $B$ and $C$ are:

$$M_B = M_C = -d_1 F = -(0.0354\,\text{m})(961\,\text{N}) = -34.0\,\text{Nm}.$$

- Note that the negative sign for the moment indicates that it is clockwise.

2. For the condition in Fig. G.11B:

- The moment arm for force $F$ with respect to $B$ is zero because its line of action passes through $B$. Therefore, $M_B = 0$.

- The moment arm for the force $F$ with respect to $C$ is:

$$d_2 = l_2 \cos\theta_1 = (0.10\,\text{m})\cos 45° = 0.0707\,\text{m}.$$

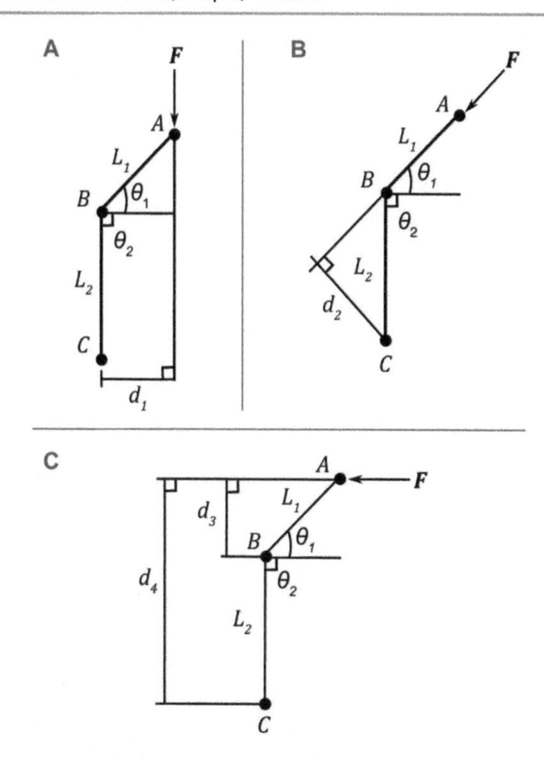


**Fig. G.11** Diagram of problem variables for forces applied along three different lines of action, i.e., (**A**), (**B**), and (**C**)

- Therefore, the moment with respect to $C$ is:

$$M_C = d_2F = (0.0707\,\text{m})(961\,\text{N}) = 68.0\,\text{Nm}.$$

- Note that the positive sign for the moment indicates that it is counterclockwise.

3. For the condition in Fig. G.11C:

- The moment arm for force $F$ with respect to $B$ is:

$$d_3 = l_1 \sin\theta_1 = (0.05\,\text{m})\sin 45° = 0.0354\,\text{m}.$$

- Therefore, the moment with respect to $B$ is:

$$M_B = d_3F = (0.0354\,\text{m})(961\,\text{N}) = 34.0\,\text{Nm}.$$

- The moment arm for force $F$ with respect to $C$ is:

$$d_4 = d_3 + l_2 = 0.0354\,\text{m} + 0.10\,\text{m} = 0.1354\,\text{m}.$$

- Therefore, the moment with respect to $C$ is:

$$M_C = d_4F = (0.1354\,\text{m})(961\,\text{N}) = 130\,\text{Nm}.$$

- Note that the positive signs for both moments indicate that they are both counterclockwise.

# Appendix H: Deformable Body Mechanics

We choose to simplify the models used in rigid body mechanics by assuming that deformations under applied loads are negligible. However, in the field of deformable body mechanics, we have to account for the magnitude of externally applied forces and moments, material properties of the segments of interest, the environment, and body deformation. *Homogeneous* materials demonstrate the same mechanical properties throughout the object or body; *heterogeneous* materials, on the other hand, demonstrate varying mechanical properties depending on the location within the material. Materials are *isotropic* if their properties are independent of the orientation or direction of measurement, whereas *anisotropic* materials exhibit properties that are dependent on load orientation. In general, biological structures are heterogeneous and anisotropic.

The human body is subjected to externally applied forces and moments during the course of daily activities. Consider the object in Fig. H.1, which demonstrates such forces.

Under both static and dynamic conditions, Newton's third law dictates that there are internal forces that balance the external loading. If an object is separated into two parts, as illustrated in Fig. H.2, these internal forces can be represented by resultant forces and moments acting on each part that are equal in magnitude and opposite in direction.

The coordinate directions for a sectioned three-dimensional (3D) object (Fig. H.3) can aid in describing the types of internal forces and moments. A force in the *x*-direction ($P_x$) is called an *axial* or *normal* force. A positive $P_x$ acts to elongate the object and is called a *tensile* force, whereas a negative $P_x$ acts to shorten the object and is called a *compressive* force. The force components $P_y$ and $P_z$ act to resist the sliding of one cut section relative to the other and are called *shear* forces. The *x*-component of moment $M_x$ represents the *torsional* (or twisting) moment, and $M_y$ and $M_z$ represent the bending moments.

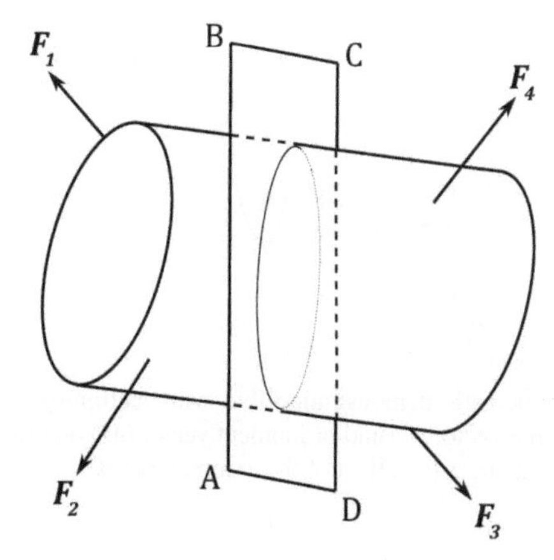


**Fig. H.1** Illustration of a deformable body subjected to externally applied forces

## 1.1 General Considerations

Mechanical properties of materials, either man-made or biological, are measured by the application of external forces and moments under experimental conditions (Fig. H.4 is a schematic of a typical setup for uniaxial tensile testing). Materials can be tested under a variety of conditions, with measured deformations dictated by the magnitude, direction, and duration of applied forces and moments, as well as the material properties of the object and the environmental conditions.

## 1.2 Load-Deformation: Stress-Strain

Laboratory experiments using consistent testing standards for measuring material deformation result in accurate and precise measurements of the mechanical properties of the


G. J. Alderink, B. M. Ashby, *Clinical Kinesiology and Biomechanics*, https://doi.org/10.1007/978-3-031-25322-5

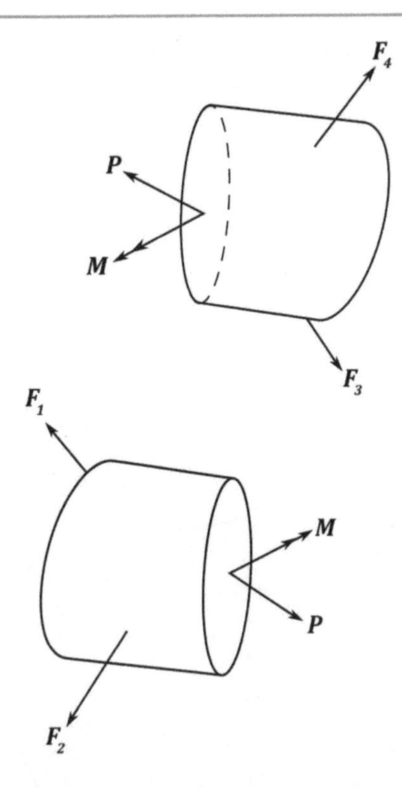


**Fig. H.2** Schematic demonstrating that with sectioning of the body, there is a force vector (***P***) and/or moment vector (***M***) on the cut section that counterbalances the effect of the external forces

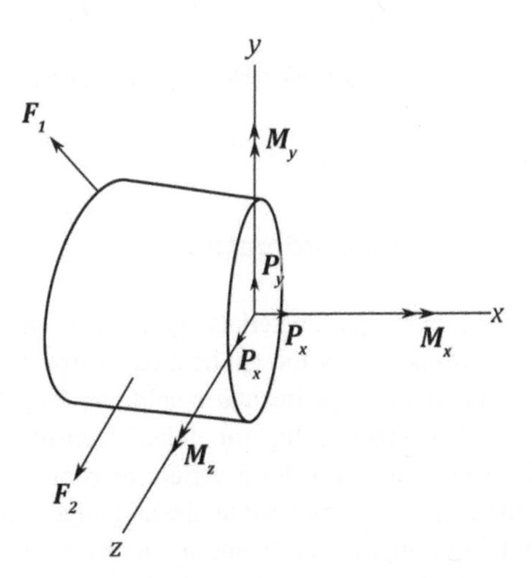


**Fig. H.3** For a three-dimensional body, the internal forces (***P***) and moments (***M***) are resolved into their components along three mutually perpendicular axes

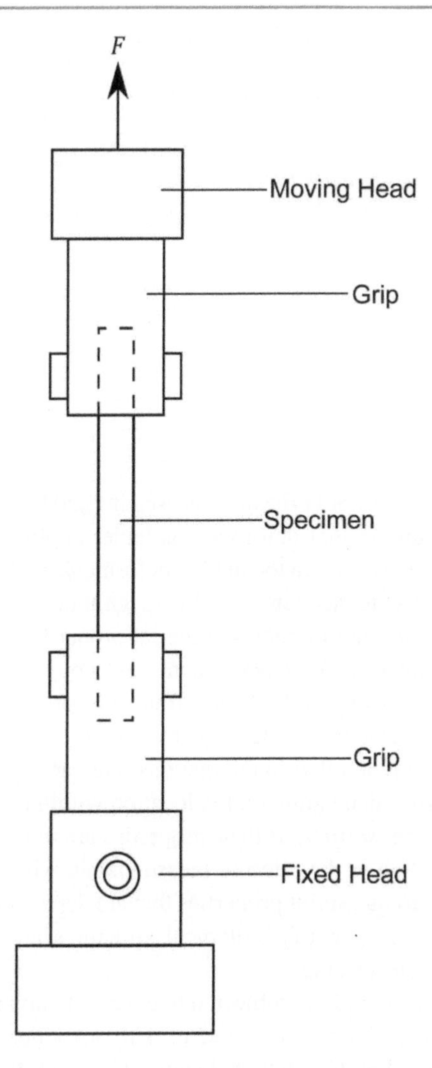


**Fig. H.4** Schematic of an experimental setup for conducting a uniaxial tension test. It consists of one fixed and moving head, with attachments to grip the specimen. The specimen is placed and firmly fixed, after which a tensile force of known magnitude is applied through the moving head. The dependent variable in this experiment is the measured elongation of the specimen. Material testing machines can also be used to measure the deformation of bodies under compression, shear, torsional, and bending

bodies of interest. A common expression of material testing is a load-deformation diagram (force is synonymous with load in this context). A load-deformation diagram illustrates the relationship between the amount of deformation relative to the magnitude of an applied load. Fig. H.5 illustrates tensile forces applied to three specimens made up of the same material but with different dimensions. In Fig. H.6, the three different curves represent the relationship between the applied force ($F$) and deformation ($\Delta L$) of three bars.

Fig. H.6 may lead to a mistaken conclusion that the three bars have different material properties despite the fact that they are made of the same material. To better illustrate the mechanical properties of materials relative to their geometries, the load and deformation are normalized. Load is normalized by dividing its value by the cross-sectional area of the material, resulting in a physical quantity known as *stress*, symbolized as $\sigma$ (sigma). And deformation is normalized by dividing the change in length, $\Delta L$, by the original length, $L$, resulting in a physical quantity known as *strain*, symbolized as $\varepsilon$ (epsilon). These are plotted to form a stress-strain diagram (Fig. H.7). Note that the three different force-length curves in Fig. H.6 collapse to the same stress ($\sigma$)-strain ($\varepsilon$) curve in Fig. H.6.

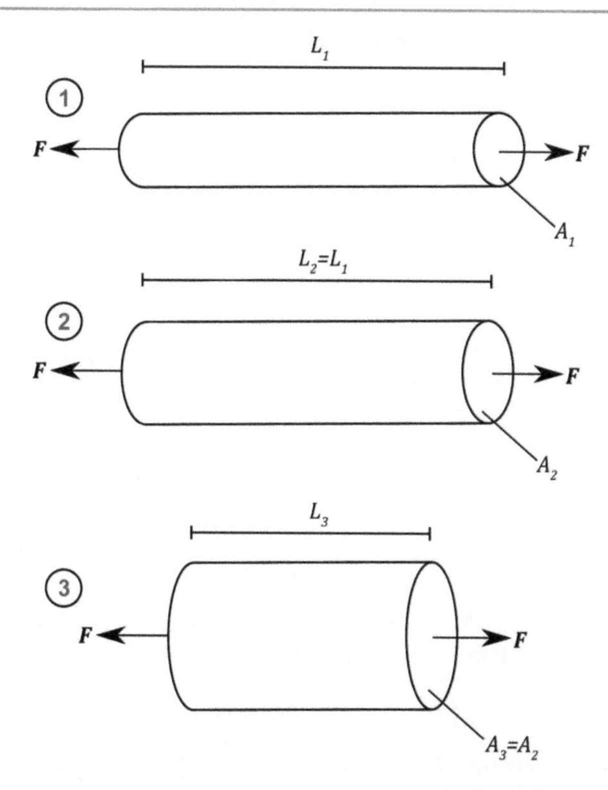


**Fig. H.5** Testing specimens made of the same material but with different lengths and cross-sectional areas. Specimens 1 and 2 have the same length but different cross-sectional areas, whereas specimens 2 and 3 have the same cross-sectional area but different lengths

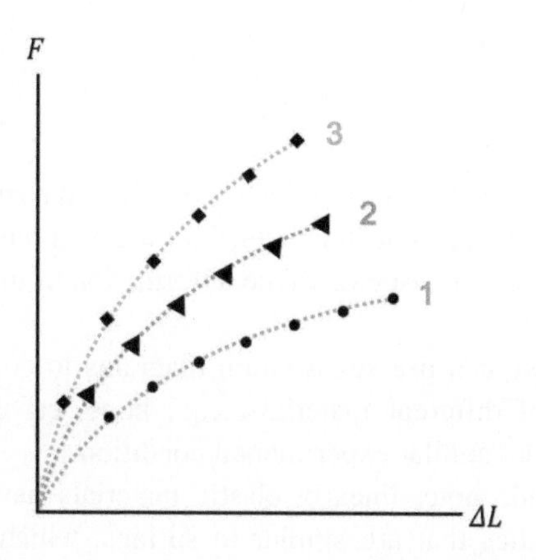


**Fig. H.6** Load-elongation curves (diagrams) resulting from a similarly applied tension force to each specimen (illustrated in Fig. H.5). Note that the specimen with the largest cross-sectional area elongates less (specimen 2 versus 1) and that, given similar cross-sectional areas, the longer specimen deforms more (specimen 2 versus 3). $F$ = load, $\Delta L$ = elongation or change in length

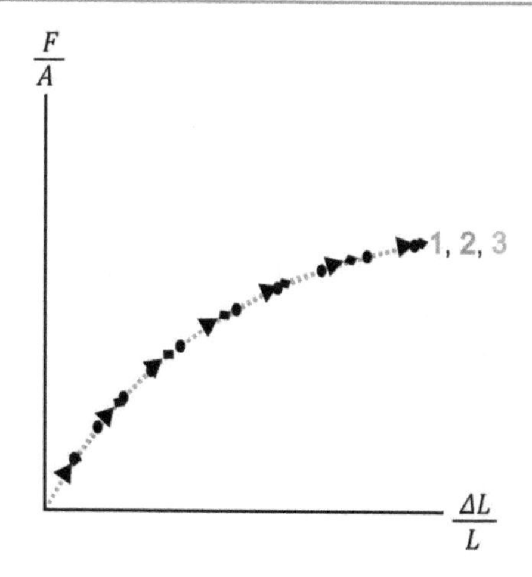


**Fig. H.7** The stress $\left(\frac{F}{A}\right)$ versus strain $\left(\frac{\Delta L}{L}\right)$ diagram shows that the normalization process produces curves for specimens 1, 2, and 3 (Fig. H.5), which are identical and unique for a particular material, independent of the geometries of the specimens. Using this representation allows for the comparison of the mechanical properties of different materials

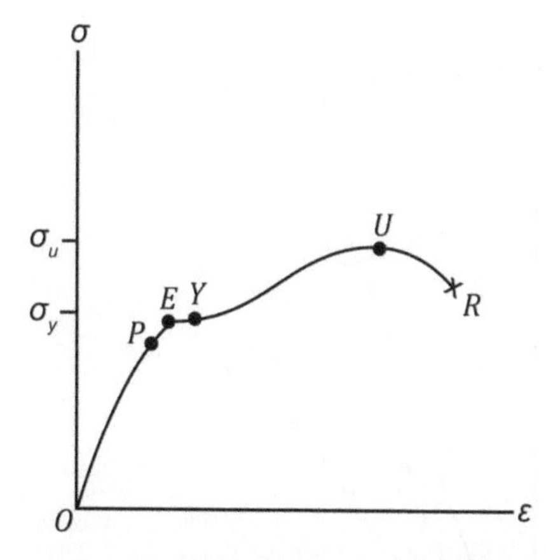


**Fig. H.8** A general representation of a stress-strain diagram for a material tested under axial loading conditions. Note: the line from $P$ to $O$ represents the elastic portion of the curve; $P$ = proportionality limit; $E$ = elastic limit; $Y$ = yield point; $U$ = greatest stress point; $R$ = point of rupture; $\sigma_u$ = ultimate stress; $\sigma_y$ = stress at the yield point

A typical stress-strain diagram for an engineering or biological material, e.g., steel, bone, etc., has several general features or points (Fig. H.8). Point $O$ is the origin, which corresponds to no load and deformation. The line from $O$ to $P$ represents the *elastic* portion of the $\sigma - \varepsilon$ diagram, where stress and strain are linearly proportional. Point $E$ represents the elastic limit of the material, that is, the greatest stress that can be applied without causing permanent deformation. It is notable that if the load is removed at point $E$, material deformation is fully recoverable. The *yield point*, $Y$, and its associated stress ($\sigma_y$) correspond to the yield strength of the material. At $\sigma_y$, considerable elongation (*yielding*) can occur without a proportional increase in load. Elongation after the yield point is referred to as *plastic*; it is permanent and is not recoverable if the load is removed. $U$ is the highest stress point, and stress$\sigma_u$ represents the *ultimate strength* of the material. Finally, point $R$ represents a *rupture* (*failure point*) of the material. Stress at $R$ is referred to as the *rupture strength* of the material. Although with the testing of biological materials, i.e., bone, ligament, etc., the shape of the $\sigma - \varepsilon$

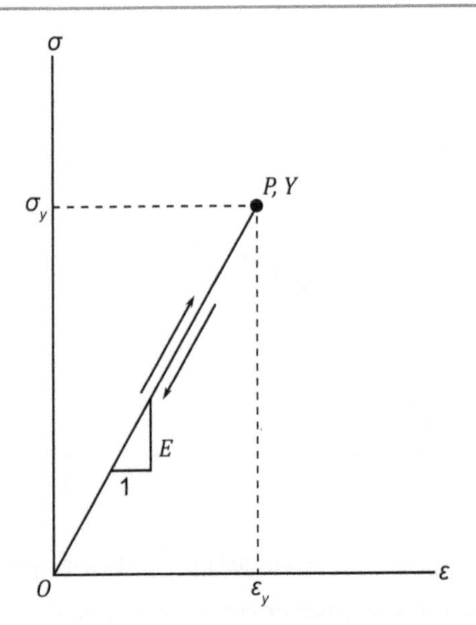


**Fig. H.9** Typical stress-strain diagram for a linear elastic material (↗ = loading; ↙ = unloading). Note: $E$ = the elastic or Young's modulus of the material since $\sigma_y$ (stress) and $\varepsilon_y$ (strain) are linearly proportional

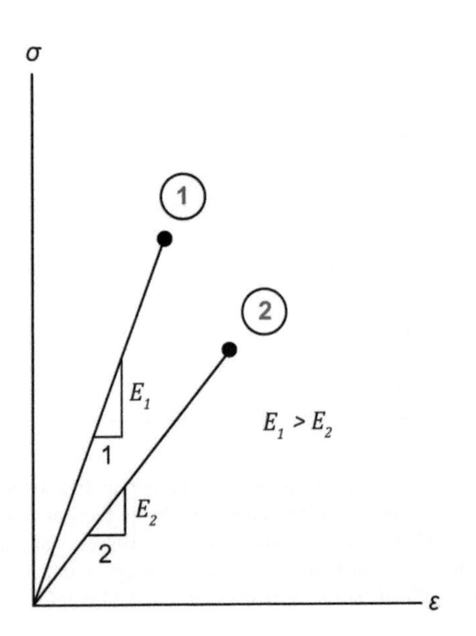


**Fig. H.10** Schematic showing that the greater slope of material 1 indicates that it is stiffer than material 2; i.e., $E_1 > E_2$

diagram may differ from Fig. H.8, the primary points of interest, i.e., *P*, *E*, *Y*, *U*, and *R*, are all utilized in the analysis of the body's material properties.

## 1.3 Elasticity, Plasticity, and Viscoelasticity

*Elasticity* is defined as the ability of a material to return to its original shape after the removal of the deforming load. Consider Fig. H.9, which illustrates the characteristics of an

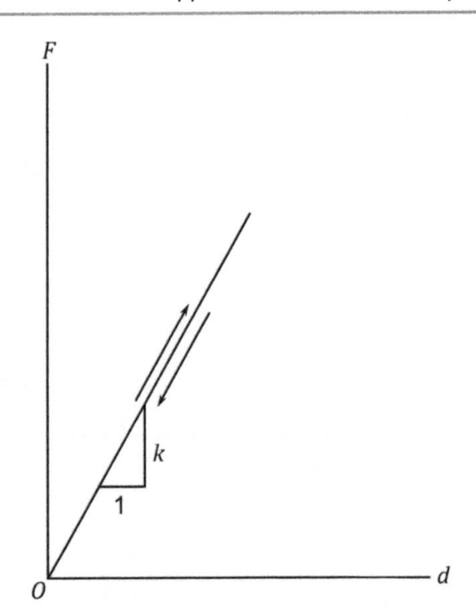


**Fig. H.11** Example of a load-elongation diagram for a linear spring where *k*, the spring constant, represents the stiffness of the spring

elastic material that is loaded to its yield point, *Y*, then unloaded.

Note the linear relationship between stress and strain. Stress is related to strain by a proportionality constant, *E*, referred to as the *elastic* or *Young's modulus*. *E* is expressed as:

$$E = \frac{\sigma}{\varepsilon} \tag{H.1}$$

where *E* is the slope of the linear portion of the stress-strain diagram and represents the *stiffness* of the material. Stiffness is a measure of resistance to the deformation of the material (Fig. H.10).

Thus, we can use stress-strain diagrams to compare the stiffness of different materials, e.g., bone, cartilage, etc., loaded under similar experimental conditions.

As a side note, linearly elastic materials have loading characteristics that are similar to springs, which was first identified by Robert Hooks. Like springs, elastic materials have the ability to store potential energy when they are loaded externally. During unloading, the release of the potential energy causes the material to resume its original shape. A linear spring that is loaded under tension elongates, with the applied load linearly proportional to the degree of elongation (Fig. H.11).

Thc constant of proportionality between the load and the deformation is called the *spring constant* or *stiffness* of thc spring, often denoted by the symbol *k*. For a linear spring with a constant *k*, the relationship between the applied load *F* and the magnitude of the elongation *d* is:

$$F = kd \tag{H.2}$$

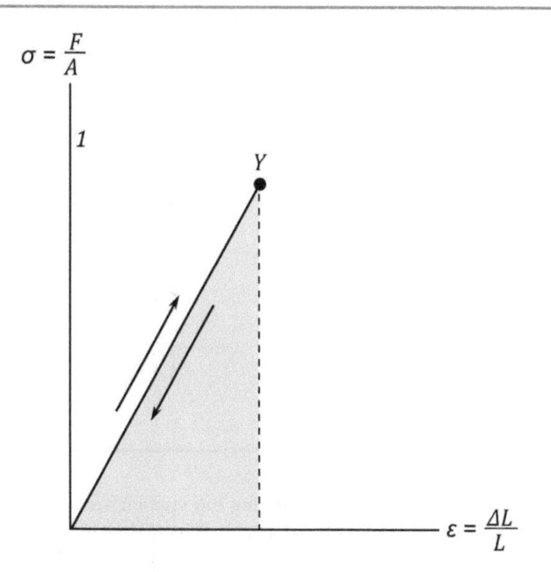


**Fig. H.12** Illustration of the internal work done (hatched area) and elastic strain energy per unit volume for a material placed under a tensile load. Note that stress multiplied by area is equal to force and that strain multiplied by length is displacement, so that the product of stress and strain is equal to the work done on a body per unit volume of the body, that is, the internal work done on the body by the externally applied forces. This *elastic strainenergy* is the release of energy that brings the body back to its original shape when it is unloaded

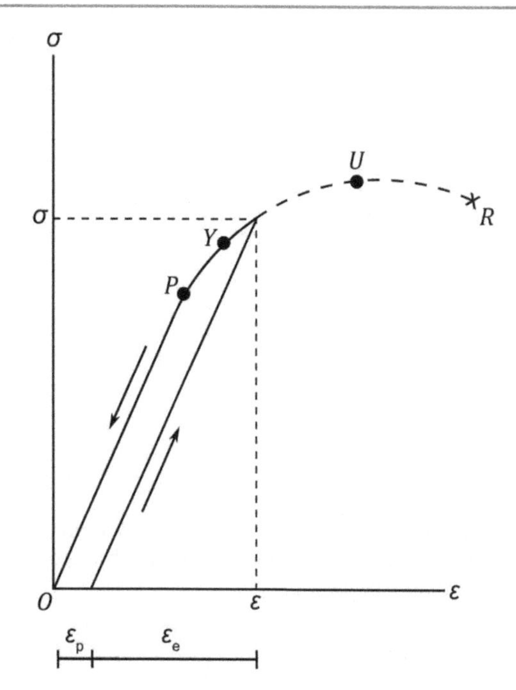


**Fig. H.13** Example of a material that is loaded beyond the yield point (*Y*), inducing plastic deformation. Note: $\varepsilon_p$ = plastic strain, i.e., an unrecoverable or permanent change in the length of the material. See Fig. H.8 for the definitions of *P*, *U*, and *R*

In comparing Eqs. (H.1) and (H.2), it is interesting to note the similarities in the stress in an elastic material and force applied to a spring, as well as the strain in an elastic material and the amount of deformation in a spring, where the elastic modulus, *E*, of an elastic material is analogous to the spring constant.

An important feature of springs is that when they are placed under tensile loading conditions, they store potential energy, or *elastic strainenergy*, which can be used to do work (Fig. H.12). The ability to store energy without permanent deformation is referred to as the *resilience* of the material.

The total amount of energy stored in the elastic material (per unit volume) is equal to the area under the stress-strain diagram in the elastic region (also called the *modulus of resilience*). In biomechanics research, tendons, for example, have been modeled as springs, which has provided important insights into our understanding of the internal forces related to the muscle-tendon actuator, as well as a tendon's contribution to total force production.

*Plasticity* defines a state of permanent deformation. Plastic changes are created when stress exceeds the elastic limit, i.e., yield point. In Fig. H.13, we see that the tensile load that was applied exceeded the yield point to some strain, $\varepsilon$. If the load is removed before *U*, or ultimate strength, is reached, the deformed body recovers with plastic deformation (see down arrow in the figure). Recovery ends at $\varepsilon_p$, referred to as *plastic strain*, which presents the magnitude of unrecoverable deformation.

Total strain, $\varepsilon$, under the conditions of this test is determined by:

$$\varepsilon = \varepsilon_e + \varepsilon_p$$

where $\varepsilon_e$ represents the amount of *elastic strain*, or the strain recovered upon unloading. Similar to elastic materials, the area associated with $(\varepsilon - \varepsilon_p)$ represents the strain energy or the potential energy that is capable of performing work. Unrecoverable energy due to permanent deformation is lost, usually in the form of heat.

Let us examine some properties of plastic deformation. For example, consider a comparison of different materials in Fig. H.14, where differences in the magnitude of plastic deformation before failure are demonstrated.

A more *ductile* material (1) demonstrates more plastic deformation before failure than a *brittle* one (2).

*Toughness* refers to the capacity of a material to sustain plastic deformation over time and represents the area under the stress-strain curve until failure (Fig. H.15).

In Fig. H.15, we see that material 1 has a greater capacity to sustain permanent deformation than material 2.

Most biological tissues exhibit viscoelastic properties. *Elasticity* is a solid material property, where the rate of loading does not affect the stress-strain relationship. Elastic materials deform in proportion to the load applied and completely recover their original shape when loads are removed. Viscosity is a fluid property and a measure of resistance to

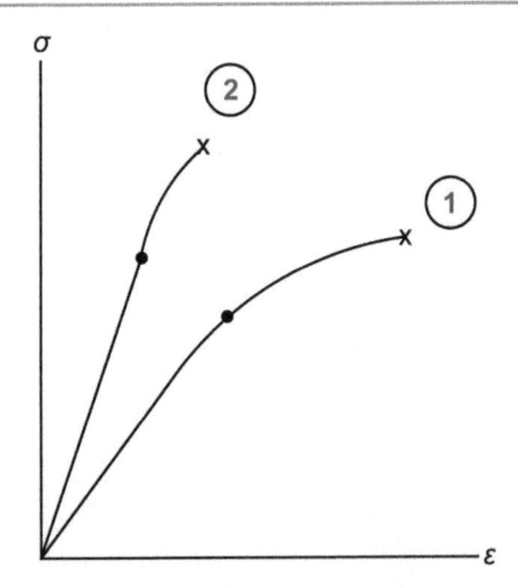


**Fig. H.14** Illustration demonstrating differences in ductility between two different materials: material 1 is more ductile because it deforms more before ultimate failure

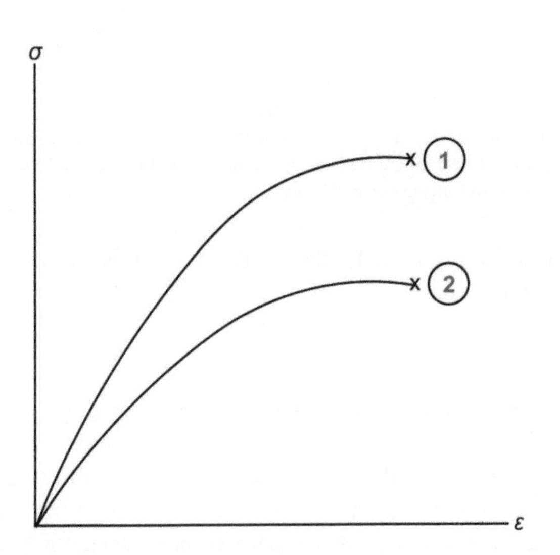


**Fig. H.15** Illustration demonstrating differences in toughness, i.e., the total area under the stress-strain diagram. Since the total area under material 1 is greater, it has greater toughness

deformation (or *flow*). The degree of resistance of a viscous material increases with an increase in the rate of loading. Biological materials possess both solid and fluid properties. The relationship between stress and strain in biological structures can be expressed as:

$$\sigma = \sigma(\varepsilon, \dot{\varepsilon})$$

where stress $\sigma$ is a function of strain$\varepsilon$ and the *strain rate* $\dot{\varepsilon} = \frac{d\varepsilon}{dt}$, with $t$ denoting time (see Appendix D to review the derivative). The viscoelastic stress function is demonstrated in Fig. H.16, where it is shown that material stiffness increases with increasing strain rate.

As noted previously, the spring has been used to model the material properties of solids. To model properties of

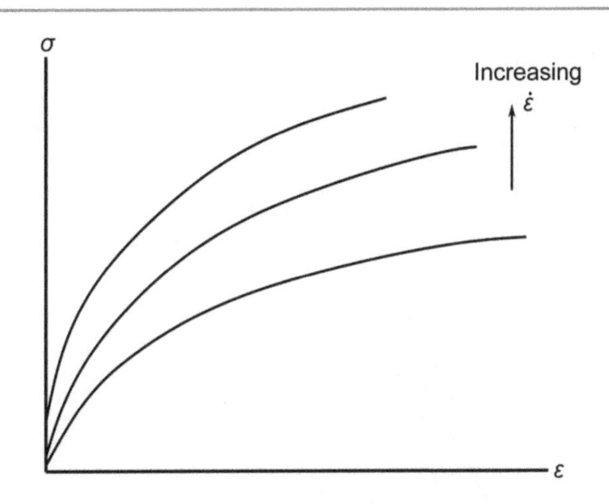


**Fig. H.16** Series of curves illustrating the viscoelastic behavior of a biologic material when the strain rate, $\dot{\varepsilon}$, is increased

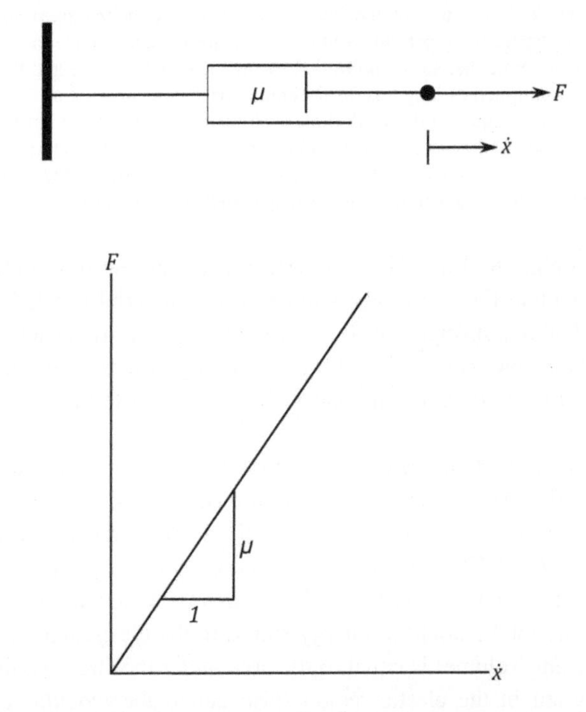


**Fig. H.17** Schematic of a force-displacement rate diagram illustrating the properties of a dashpot, where $\mu$ is the coefficient of friction and $\dot{x}$ is the rate of displacement

Newtonian fluids, engineers use the dashpot. The dashpot is a simple piston-cylinder, where the speed of the piston is dependent on the magnitude of the applied force and friction in the walls of the piston (Fig. H.17); note, however, that the friction in the walls of a piston is nearly negligible. Thus, the resistive force in a piston is induced by fluid being forced to flow through a small orifice. At faster speeds, the fluid experiences greater difficulty moving through the orifice, which increases the force of resistance. Thus, the speed of the piston depends on the force applied and the material properties of the fluid.

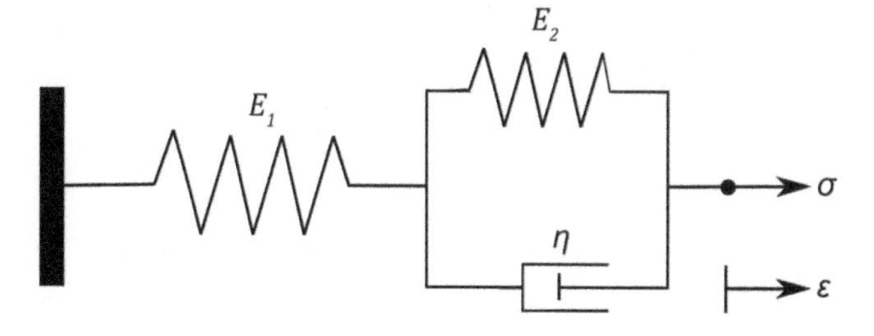


**Fig. H.18** Schematic of the Kelvin-Voight solid model illustrating a viscoelastic body connecting springs and a dashpot, where $E_1$ and $E_2$ represent spring constants and $\eta$, the coefficient of viscosity of the fluid

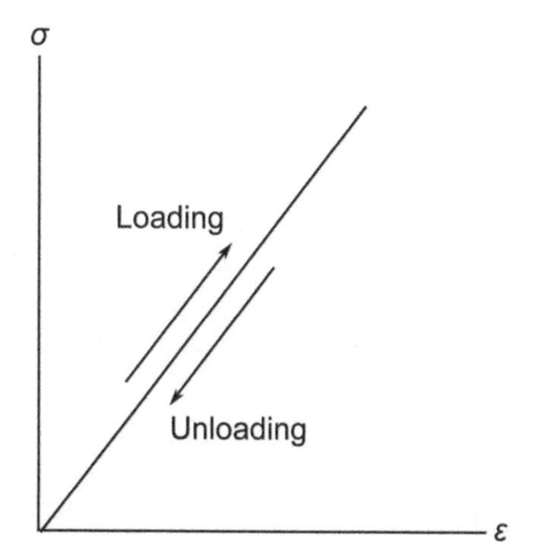


**Fig. H.19** Illustration demonstrating that for an elastic material, the loading and unloading paths coincide

For a linear dashpot, the magnitude of the applied load $F$ and the rate of displacement $\dot{x}$ are linearly proportional, where $\mu$ (*damping coefficient*) is the constant of proportionality:

$$F = \mu \dot{x}.$$

To model the viscoelastic nature of biological tissues, springs and dashpots can be linked in a variety of ways. In these simplified representations of reality, springs represent the solid elastic behavior and recoverable deformation, whereas dashpots represent the fluid behavior and permanent deformation. One such model used to study the material properties of cartilage and muscle, for example, is called the standard solid model. The standard solid model is a three-parameter ($E_1$, $E_2$, and $\eta$) model and is used to describe the viscoelastic nature or biological materials, e.g., cartilage, etc. This model combines spring and Kelvin-Voight solid and Maxwell fluid models connected in series (Fig. H.18).

The material function relating to stress, strain, and their rates is:

$$(E_1 + E_2)\sigma + \eta\dot{\sigma} = E_1 E_2 \epsilon + E_1 \eta \dot{\varepsilon}$$

where $\eta$, i.e., the *coefficient of viscosity* of the fluid, is the constant of proportionality between stress$\sigma$ and the strain rate $\dot{\varepsilon}$, $E_1$ and $E_2$ represent spring constants, and $\dot{\sigma} = \frac{d\sigma}{dt}$ is the stress rate.

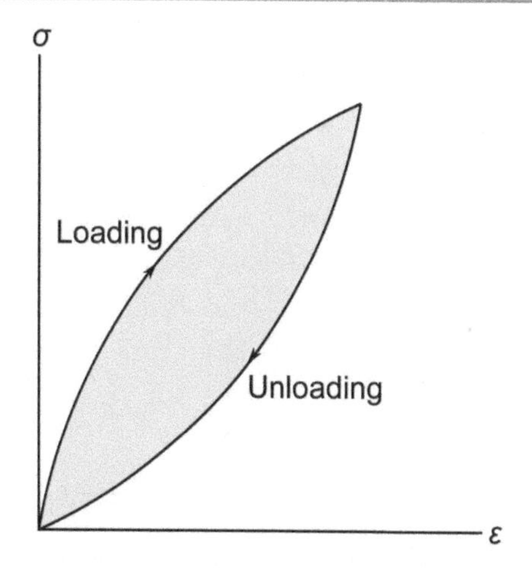


**Fig. H.20** Illustration of a hysteresis loop that is created when the viscoelastic body is loaded and unloaded in its elastic region

For an elastic solid, loading and unloading follow a similar path (Fig. H.19), so the deformation is fully recoverable and the energy stored (i.e., strain energy) can be used to perform work.

On the other hand, when viscoelastic materials are loaded and unloaded, recovery does not follow a linear progression (Fig. H.20). Consider the standard model (Fig. H.18) being loaded, then unloaded. We see that the loading/unloading cycle traces not a straight line but a loop called a *hysteresis-loop*. The area within this loop represents the energy lost as heat during the deformation and recovery phases.

Dissipation of heat in a viscoelastic material occurs whether or not the material undergoes plastic deformation. What is unique about a viscoelastic material is that the area contained within the hysteresis loop increases with an increase in the rate of loading. Note, however, that a solid elastic material, when loaded into the plastic region, can also exhibit hysteresis, indicating that heat was lost as a result of permanent deformation (Fig. H.21).

The difference between elastic solids and viscoelastic materials is most evident under experimental conditions when time-dependent behavior is tested. These tests are referred to as creep and stress relaxation experiments. A creep and recovery test is when constant stress is applied to a material for a period of time and then removed. Strain observations are documented over the course of time (Fig. H.22A).

If constant stress is applied and maintained over time, the elastic solid exhibits a strain that does not change (Fig. H.22B). If constant stress is applied to a viscoelastic solid (Fig. H.22C), the material demonstrates a creep (flow) response, i.e., strain or deformation. In other words, the gradual increase in strain from $t_0$ to $t_1$ under constant stress gradually decreases once the stress is removed. A viscoelas-

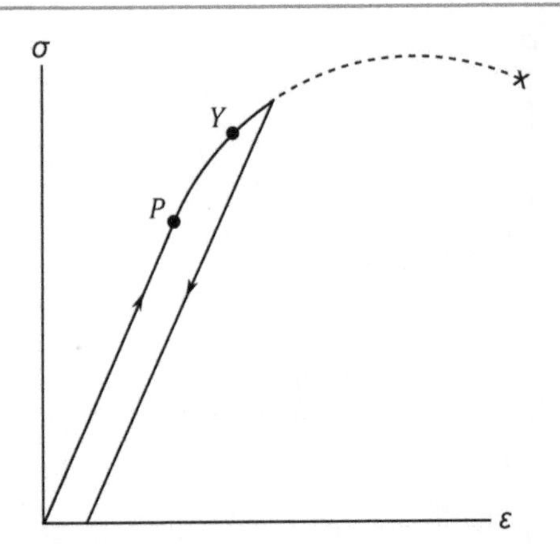


**Fig. H.21** Illustration of a hysteresis loop for an elastic material loaded past its yield ($Y$) point. In this case, the material does not return to its original shape; i.e., there is a permanent deformation of the material, and energy is lost as heat

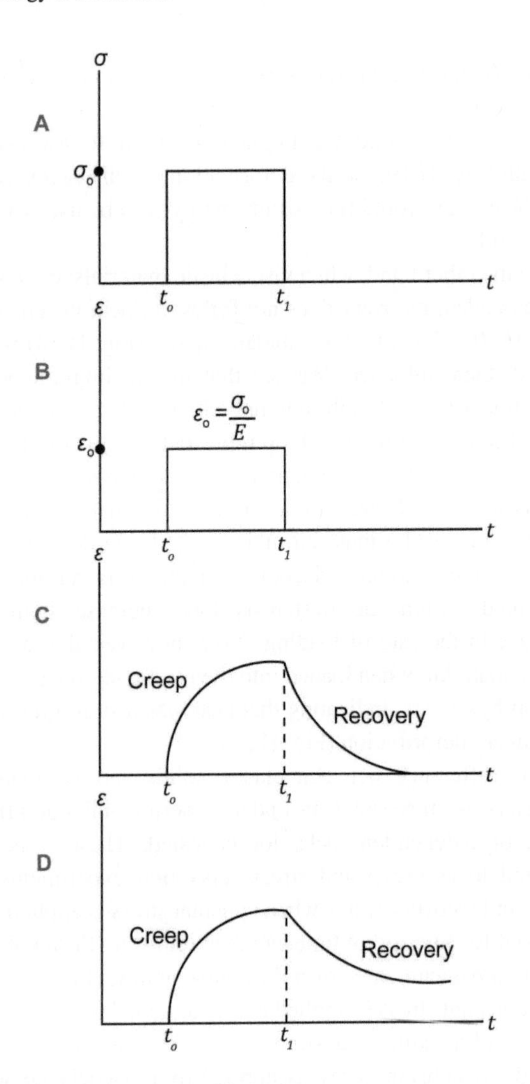


**Fig. H.22** Series of figures illustrating a creep-recovery experiment. Note: $\sigma_0$ = application of a constant stress at $t_0$; $\varepsilon_0 = \frac{\sigma_0}{E}$ or strain induced by the load at time $t_0$; $t_1$ = later time when the load is removed

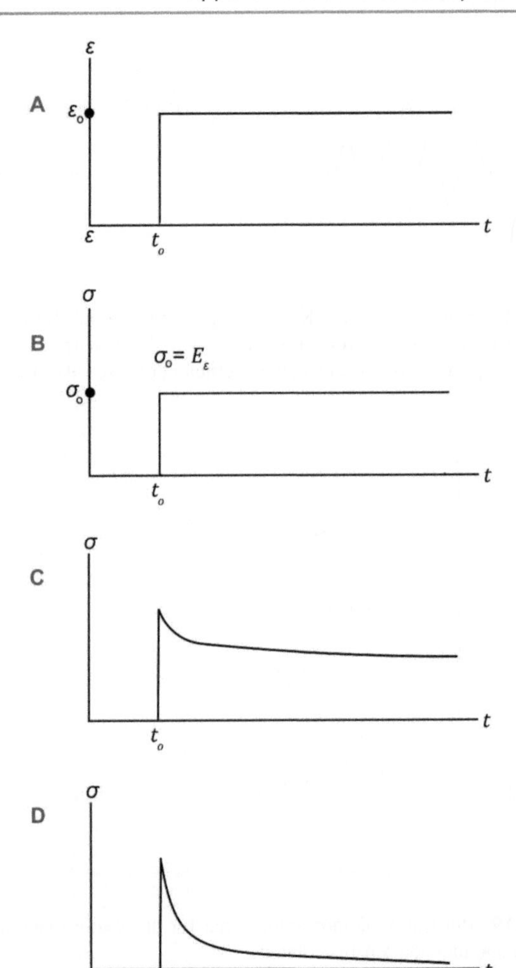


**Fig. H.23** Series of figures illustrating a stress-relaxation experiment. Note: $\varepsilon_0$ = application of a constant strain at $t_0$; $\sigma_0 = E\varepsilon$ or stress induced by the load at time $t_0$

tic fluid, however, gradually decreases after the stress is removed but never fully recovers, leaving a residual deformation (Fig. H.22D).

In a stress-relaxation experiment, the material is subjected to an instantaneous strain that is held constant, while material responses are observed (Fig. H.23A). An elastic solid shows no change in stress over time (Fig. H.23B). Viscoelastic materials demonstrate an initial spike in stress, followed by relaxation of stress over time. Stress recovery is only partial in a viscoelastic solid (Fig. H.23C), whereas stress disappears over time in a viscoelastic fluid (Fig. H.23D). As noted, most biological tissues exhibit both viscoelastic solid and fluid properties, which need to be accounted for in the study of injury mechanisms, recovery following injury, and rehabilitation.

# Appendix I: Static Equilibrium

Statics studies rigid bodies in equilibrium. With static equilibrium analyses, we assume that when forces are applied to the body, the deformations are negligible, so we ignore them. Equilibrium implies that a body is at rest, or if the body is moving, its velocity remains constant (Newton's first law). For our purposes, we solve equilibrium problems with the body at rest. In biomechanics, we are often interested in constraint, or reaction, forces, so we also invoke Newton's third law (law of action-reaction). What follows are general tips and guidelines for solving static problems:

1. Gather and organize the information that is given and/or derived from data provided in the problem statement.
2. Establish a convenient coordinate system, with $x$ and $y$ axes that are orthogonal, i.e., perpendicular, in the plane for each free-body diagram and resolve all forces into their components along these mutually perpendicular directions. A good selection of coordinate systems can simplify the analysis considerably.
3. *Free-body diagrams* are constructed to help identify the forces and moments acting on individual parts of the system and to ensure the correct use of the equations of statics. The parts constituting a system are isolated from their surroundings, and the corresponding forces and moments replace the effects of the surroundings. All known and unknown forces and moments are shown. A force is unknown if its magnitude or sense is not known. If the sense of a force is not known, then simply choose a direction for it. *Note: the sense of a force defines which way the force is acting along the line of action, whereas the direction of the force defines its line of action. If we do not know the direction, we break the force into components.* If this initial "guess" for the force's sense is incorrect, it appears in the solution as a negative force.
4. A system is said to be in translational equilibrium if the net force, i.e., the sum of all forces, is equal to 0:

$$\Sigma \boldsymbol{F} = 0.$$

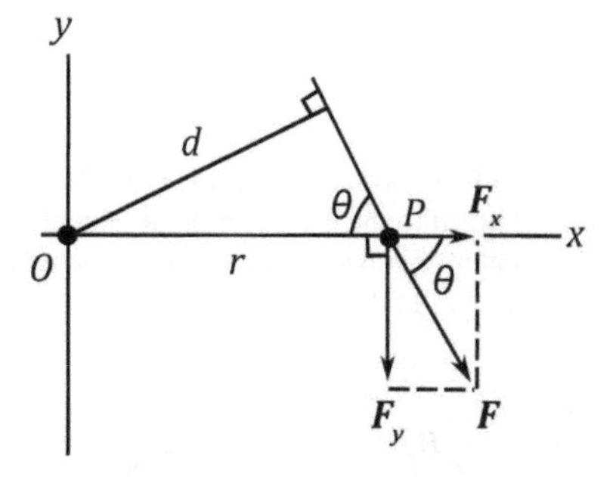


**Fig. I.1** Illustration of the components of force vector $\boldsymbol{F}$

And the system is in rotational equilibrium if the net moment, i.e., the sum of all moments, is equal to 0:

$$\Sigma \boldsymbol{M} = 0.$$

5. For each free-body diagram, apply the necessary translational and rotational equilibrium conditions. For two-dimensional (planar) problems, the number of equations available is three:

$$\Sigma F_x = 0;\ \Sigma F_y = 0;\ \Sigma M_O = 0$$

The moments are expressed about a $z$-axis passing through $O$ because the rotation induced by the forces occurs about a $z$-axis. Solve these equations simultaneously for the unknowns.

Note: resolving a force into its components can be used to simplify the calculation of moments.

Fig. I.1 provides an example of this:

where the applied $\boldsymbol{F}$ is resolved into its components, $F_x$ and $F_y$:

$$F_x = F\cos\theta$$

$$F_x = F\sin\theta$$

Note that the moment of force about $O$ due to $F_x$ is zero because $F_x$ has no moment arm. Since $r$ is the perpendicular distance from $F_y$ to $O$, $F_y$ produces a clockwise moment about $O$:


G. J. Alderink, B. M. Ashby, *Clinical Kinesiology and Biomechanics*, https://doi.org/10.1007/978-3-031-25322-5

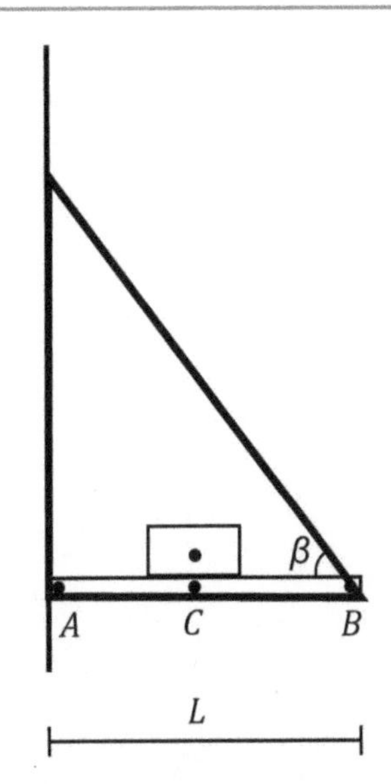


**Fig. I.2** Schematic of the horizontal beam setup. Note: $A$ = point at which the beam is hinged to the wall; $B$ = point of cable attachment to the beam; $C$ = the center of gravity of the beam; $L$ = length of the beam; $\beta$ = angle the cable makes to the horizontal

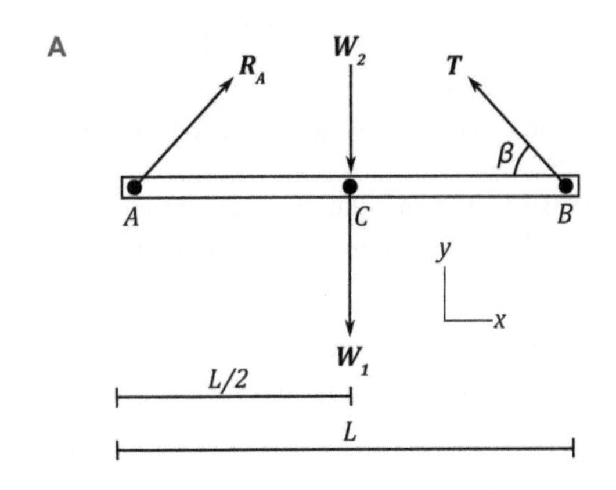


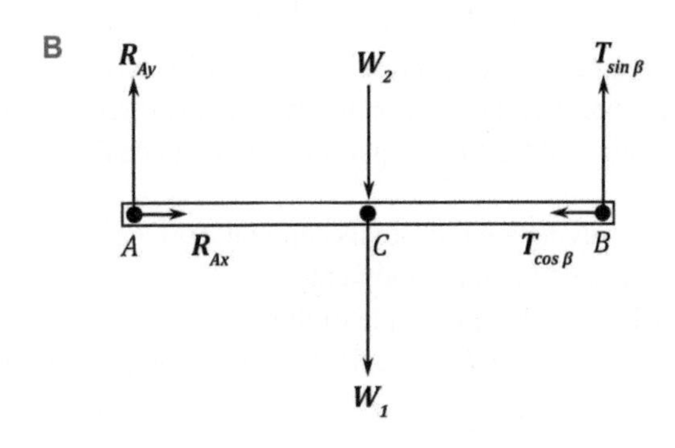


**Fig. I.3** (**A**) Free-body diagram and (**B**) forces acting on the beam. Note: $R_A$ = the reaction force; $T$ = tension in the cable; $W_1$ and $W_2$ = weight of and load on the beam, respectively

$$M = rF_y = rF\sin\theta$$

Alternatively, since $d = r\sin\theta$:

$$M = dF$$

6. Include the correct sense, i.e., "signs," and the units of forces and moments in the solution.
7. While applying the translational equilibrium conditions, pay special attention to the direction of the forces. Let a force be positive if it is acting in a positive $x$ (or $y$) direction, i.e., usually to the right (or up), and let a force be negative if it is acting in the negative $x$ (or $y$) direction.
8. When applying the rotational equilibrium condition, choose a proper (most appropriate or convenient) point about which all moments are to be calculated. The choice of location of the axis is arbitrary, but a good choice can simplify the computations. Additionally, pay attention to the direction of moments. Typically, assign a positive moment if the force induces counterclockwise rotation and a negative moment if it induces a clockwise rotation. This choice is arbitrary, but once you make your decision, be consistent in its application throughout the problem.
9. For the solution of a two-dimensional problem in statics, you may or may not need all three equations.
10. In general, the number of equations needed to solve a problem is equal to the number of unknowns. A system is called *statically determinate* if the number of unknowns is equal to the number of equations. If the number of unknowns exceeds the number of equations, then the system is called *statically indeterminate* or *underdetermined.* We do not pose and/or attempt to solve indeterminate problems in this book.

## Examples

1. Consider the horizontal beam that is connected to the wall at $A$ by a hinge and suspended by a cable attached to the beam at $B$ (Fig. I.2). The cable is attached to the wall, creating an angle $\beta$ (60°) relative to the horizontal at $B$. $W_1$ represents the weight of the beam, and $W_2$ represents the weight of the load on top of the beam. The center of mass of the beam is at point $C$, which is equidistant from points $A$ and $B$.

   Calculate tension $T$ in the cable and the reaction force at point $A$.

   Assumptions/known variables:

- Forces up/right are positive; counterclockwise moments are positive.
- $L = 4$ cm (0.04 m).
- $W_1 = 450$ N; $W_2 = 335$ N.

***Solution:*** Note in Fig. I.3 the free-body diagram that models the beam and associated forces. The hinge joint at $A$ constrains the translation movement of the beam in both $x$ and $y$ directions. The weight of the beam and external loads at point $C$ produces a tension force in the cable.

There are three unknowns ($R_{A_x}$, $R_{A_y}$, and $T$)and three equations. There are two unknowns for $\boldsymbol{R}_A$ because of the constraints at point $A$; i.e., the beam is hinged at $A$ preventing motion in the $x$ and $y$ directions. First, consider rotational equilibrium at $A$, which was chosen for convenience; i.e., it eliminates two of the three unknowns since the$x$- and $y$-components of $\boldsymbol{R}_A$ do not have moment arms relative to $A$. The $y$-component of $\boldsymbol{T}$ is $T \sin \beta$. Thus, summing the moments about $A$:

$$\sum M_A = 0: -\frac{L}{2}W_1 - \frac{L}{2}W_2 + LT\sin\beta = 0$$

Solving for $T$:

$$T = \frac{W_1 + W_2}{2\sin\beta}$$

$$T = \frac{450\,\text{N} + 335\,\text{N}}{2\sin 60°} = 453.2\,\text{N}$$

Next, consider the translational equilibrium of the beam ($\sum \boldsymbol{F} = 0$). First, in the $x$-direction:

$$\sum F_x = 0: R_{A_x} - T\cos\beta = 0$$

Solving for $R_{A_x}$:

$$R_{A_x} = T\cos\beta$$

$$R_{A_x} = (453.2\,\text{N})\cos 60° = 226.6\,\text{N}$$

Now, in the $y$-direction:

$$\sum F_y = 0: R_{Ay} - W_1 - W_2 + T\sin\beta = 0$$

Solving for $R_{A_y}$:

$$R_{A_y} = W_1 + W_2 - T\sin\beta$$

$$R_{A_y} = 450\,\text{N} + 335\,\text{N} - (453.2\,\text{N})\sin 60° = 392.5\,\text{N}$$

The magnitude of the reaction force at point $A$ is:

$$R_A = \sqrt{R_{A_x}{}^2 + R_{A_y}{}^2} = 453.2\,\text{N}$$

And the orientation, angle $\alpha$ relative to the horizontal, of $\boldsymbol{R}_A$ can be determined as:

$$\alpha = \tan^{-1}\left(\frac{R_{Ay}}{R_{Ax}}\right) = 60°$$

2. This example shows a flexed elbow grasping an object in the hand (Fig. I.4) and an example of a parallel force system. Note the free-body diagram illustrating this model, which assumes that the biceps brachii is the only muscle acting, with a vertical muscle force line of action (Fig. I.5).

   Point$O$ represents the elbow joint center, point $A$ is the point of application of the biceps, point $B$ is the center of mass of the forearm/hand, and point $C$ represents the center of gravity of the weight in the hand. The letters $a$, $b$, and $c$ represent the moment arms for the muscle force, weight of the forearm/hand, and weight of the object in the hand, respectively. $F_M$ and $F_J$ represent the magnitudes for the biceps and joint reaction forces (i.e., bone on bone), respectively. Note that since this is a parallel force system, all of the forces are parallel to the $y$-axis, with no forces parallel to the $x$-axis.

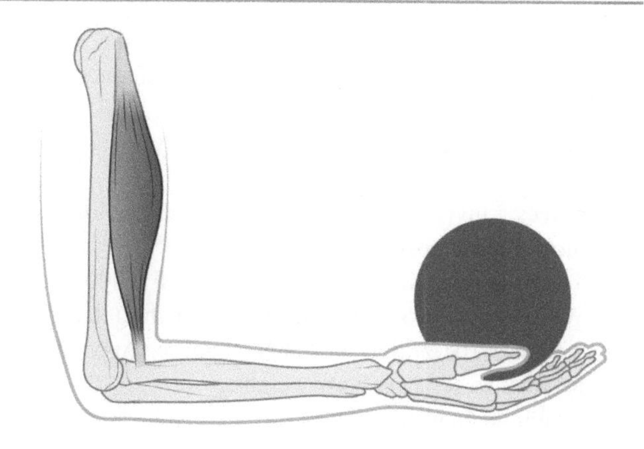

**Fig. I.4** Illustration of an individual holding an object in their hand with the elbow flexed 90°

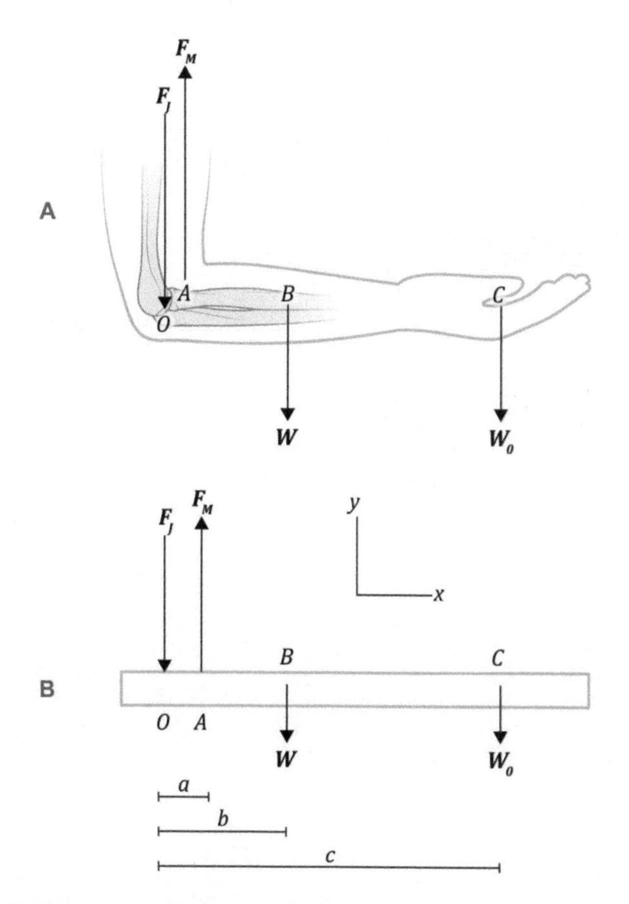


**Fig. I.5** (**A**) Free-body diagram and (**B**) forces acting on the forearm. Note: $\boldsymbol{F}_M$ = force vector of the muscle; $\boldsymbol{F}_J$ = joint reaction force vector; $\boldsymbol{W}$ = force vector of the weight of the forearm; $\boldsymbol{W}_O$ = force vector of the weight of the object in the hand

For this example, we wish to calculate the muscle force tension and bone-on-bone force at the humeroulnar joint.

Assumptions/known variables:

- Forces up/right are positive; counterclockwise moments are positive.
- Total body weight (BW) of an individual = 700 N.
- Weight of the forearm/hand (2.2% of the total body weight) = (0.022)BW = 15.4 N. Note: see Appendix A for relative segment body weights.
- Weight in hand $W_O = 90$ N.
- $a = 0.04$ m; $b = 0.18$ m; $c = 0.28$ m.

***Solution:*** Considering rotational equilibrium of the forearm/hand relative to the elbow joint center (point $O$):

$$\Sigma M_O = 0: -cW_O - bW + aF_M = 0$$

Solving for $F_M$:

$$F_M = \frac{cW_O + bW}{a}$$

$$F_M = \frac{(0.28\,\text{m})(90\,\text{N}) + (0.18\,\text{m})(15.4\,\text{N})}{0.04\,\text{m}} = 699.3\,\text{N}(\uparrow)$$

Translation equilibrium is expressed by $\sum F_x = 0$ and $\sum F_y = 0$. Since the $x$-components of all forces are 0, then the equation $\sum F_x = 0$ gives us nothing of use. The equilibrium equation in the $y$-direction, however, gives us:

$$\Sigma F_y = 0: -F_J + F_M - W - W_O = 0$$

Solving for $F_J$:

$$F_J = F_M - W - W_O$$

$$F_J = 699.3\,\text{N} - 15.4\,\text{N} - 90\,\text{N} = 593.9\,\text{N}(\downarrow).$$

Since the result for $F_J$ was positive, this means that the assumed direction in the free-body diagram (i.e., down) is correct.

Remarks: notice how large the bone-on-bone force is compared to the object's weight (almost eight times as large). Also, notice that the force at the elbow joint is approximately 85% of the total body weight, holding a weight of only 90 N. What might be a clinical implication of this finding?

# Appendix J: Three-Dimensional (3D) Instrumented Gait Analysis

## 1.1 Introduction

The analysis of human motion in general and walking in particular has been a topic of interest for more than a century. Significant advances in motion and force-sensing technology in the 1970s moved the equipment from research to clinical motion analysis laboratories. In this appendix, we briefly describe the theoretical and practical applications of the three major technologies used in clinical motion analysis: video-based motion capture (MOCAP), force platforms, and electromyography (EMG) (Fig. J.1). Although 3D motion can be determined using a two-camera system, the complexity of human gait requires the use of several cameras. Typically, for clinical motion analysis, the laboratory setup includes 10–16 cameras and 2 to 4 force platforms arranged linearly along the path of walking progression. The EMG data collection system may be wireless (as shown in Fig. J.1) or tethered to a power source via a thin cable.

## 1.2 Optical Motion Capture

### 1.2.1 Cameras

Kinematics is the study of motion without regard to the forces and moments that produce that motion. In biomechanics, video-based motion systems quantify body segment kinematics by taking successive images so that changes in the 3D position of each segment can be tracked relative to the laboratory (or global) coordinate system over time. We model the human body as a series of rigid segments (or links) connected by frictionless rotating joints, which enables us to compute joint angles between the segments.

As opposed to technologies that directly measure body segment kinematics, e.g., inertial motion sensors, video-based MOCAP is an indirect method for the determination of body segment motion. This is so because marker locations are used to deduce the 3D pose of body segments. The cameras used in MOCAP have advanced lens image resolution; capabilities that allow for the measurement of high-speed

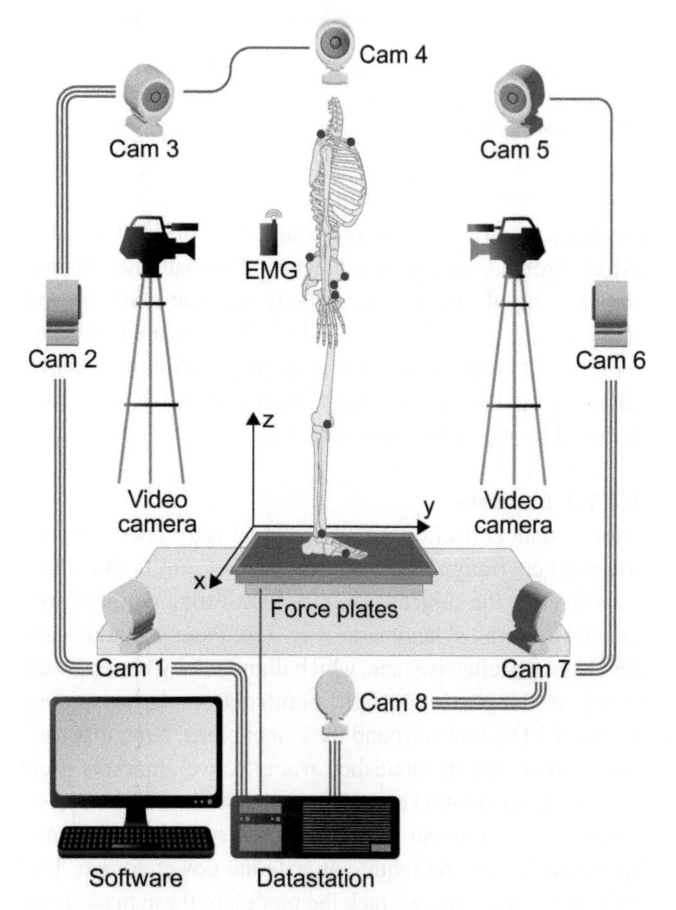


**Fig. J.1** Schematic representation illustrating a caricature of the synchronization of standard video, MOCAP, force platform, and electromyographic systems. Note the location of the laboratory (LCS or global (GCS)) coordinate system, with the origin located at the corner of the force platform. This location and orientation of the LCS, i.e., $x$ (mediolateral), $y$ (posteroanterior), and $z$ (inferosuperior), is just an example, since each laboratory may use different locations or orientations. The red-filled balls represent body segment markers



G. J. Alderink, B. M. Ashby, *Clinical Kinesiology and Biomechanics*, https://doi.org/10.1007/978-3-031-25322-5

**Fig. J.2** Image of a Vicon© motion capture camera. Note the light-emitting diodes (LEDs) that surround the camera lens. The LEDs emit light that induces body segment passive markers to reflect light back into the camera lens. Image of the Vantage camera with permission from Vicon Motion Systems, Ltd., UK

movements; advanced charge-coupled diodes (CCD) that use electronic shutter controls, which eliminate the blurring and skewing of images, i.e., of body segment markers; and active infrared light-emitting diodes (LEDs) around the camera lens that are pulsed at a set frequency to illuminate retroreflective markers (placed on the body), whose locations are recorded by the camera lens (Fig. J.2).

#### 1.2.1.1 Markers

Prior to data collection, small spherical retroreflective passive markers (ranging in size from 6 to 14 mm in diameter) are affixed to the subject's skin (or tight-fitting clothes) over specific anatomical landmarks (Fig. J.3). Passive markers are covered with reflective tape, which illuminates when exposed to the strobing infrared light emitted from light-emitting diodes (LEDs) that surround the camera lens. Note that passive markers are distinguished from "active" markers used by optoelectric motion capture systems. Active markers, also attached to anatomical landmarks, consist of small infrared lights that strobe and require an external power source. The MOCAP video cameras track the motion of these markers as individuals perform movements, e.g., walking, running, throwing, etc., within the data capture volume.

Marker locations are recorded as sequential images, which are used to determine 3D coordinates in the laboratory coordinate system (LCS) and to create the 3D pose of the underlying skeleton (or segments). A minimum of three noncollinear anatomical markers must be placed on a body segment to define it and determine its position and orientation in space. Many biomechanical models use a combination of anatomical markers and marker clusters (technical markers); the clusters consist of at least three noncollinear markers attached to a rigid plate affixed to a body segment, or clusters of markers are attached directly to the skin. Using marker clusters on a rigid plate reduces errors associated with intermarker movement on anatomical landmarks and enables the tracking of each segment independently (Fig. J.4).

The accuracy of most commercial passive marker-based systems has been validated. Root mean square errors are typically less than 2.0 mm for fully visible moving markers and 1.0 mm for stationary markers. Laboratory experiments have established the precision (reliability) of passive marker-based systems as well. Both the accuracy and precision of MOCAP are influenced by two sources of systematic and random extrinsic errors: accurate placement of markers and soft-tissue artifacts. Errors due to soft-tissue artifacts have been characterized as relative and absolute. Relative errors are associated with the movement between two or more markers that define a rigid body, whereas absolute errors occur when a marker moves relative to the anatomical landmark it is supposed to represent. Thus, despite the demonstrated accuracy and precision of MOCAP, the biomechanist must be cognizant of inherent errors that could affect the calculation of joint and segment kinematics and kinetics.

#### 1.2.1.2 Camera Calibration, Laboratory Coordinate System, Subject Anthropometric Measures, and Sources of Error

There are a number of biomechanical models used, e.g., the PIG marker set (Fig. J.3), but their fidelity for a clinical application depends on the accurate placement of anatomical markers and choices for how the segment pose and the location of joint centers are determined. Camera calibration and

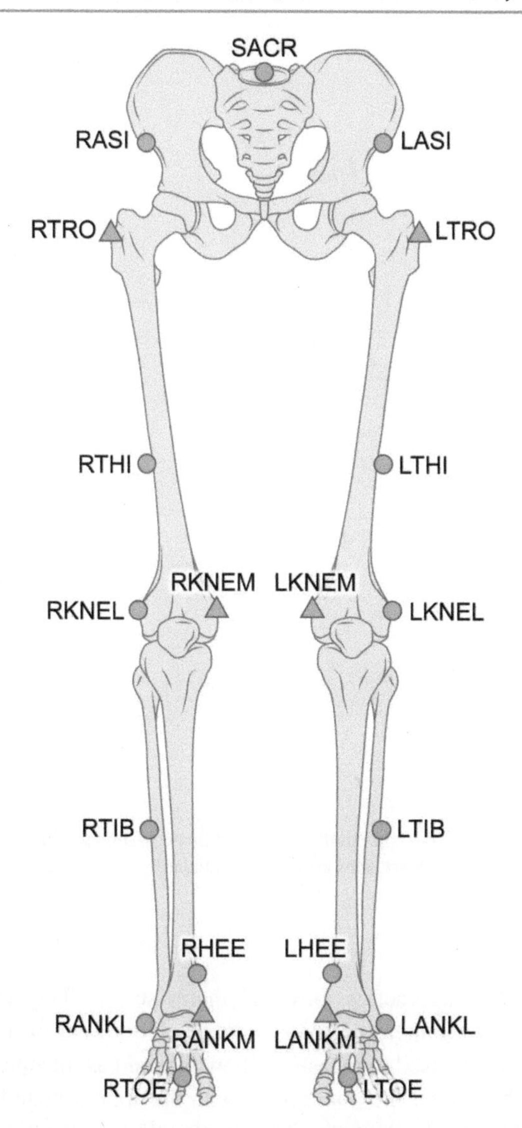


**Fig. J.3** Schematic of the location of markers using the lower-body Plug-in Gait (PIG) marker set used to model the pelvis and lower extremities. The Vicon PIG model (also referred to as Vicon Clinical Manager or the Conventional Gait Model) uses the direct kinematics (DK) computation method to calculate joint kinematics and outputs three rotations for the pelvis segment, hip, and knee joints and two rotations for the ankle joint. SACR = mid base of sacrum; RASI/LASI = right and left anterior superior iliac spines; RTRO/LTRO (optional) = right and left greater trochanters; RTHI/LTHI = right and left lateral mid-thighs; RKNEL/LKNEL = right and left lateral femoral condyles; RKNEM/LKNEM (optional) = right and left medial femoral condyles; RTIB/LTIB = right and left lateral mid-tibias; RANKL/LANKL = right and left lateral malleoli; RANKM/LANKM (optional) = right and left medial malleoli; RHEE/LHEE = right and left posterior mid-calcanea; RTOE/LTOE = between the right and left heads of the second and third metatarsals. Vicon Motion Systems, Ltd., UK

the establishment of the laboratory coordinate system prior to data collection are used to check the synchronization and accuracy of cameras by capturing the location of known marker locations. Another step that is required before the collection of dynamic, e.g., walking, data is the capture of marker location with respect to the LCS using a subject static

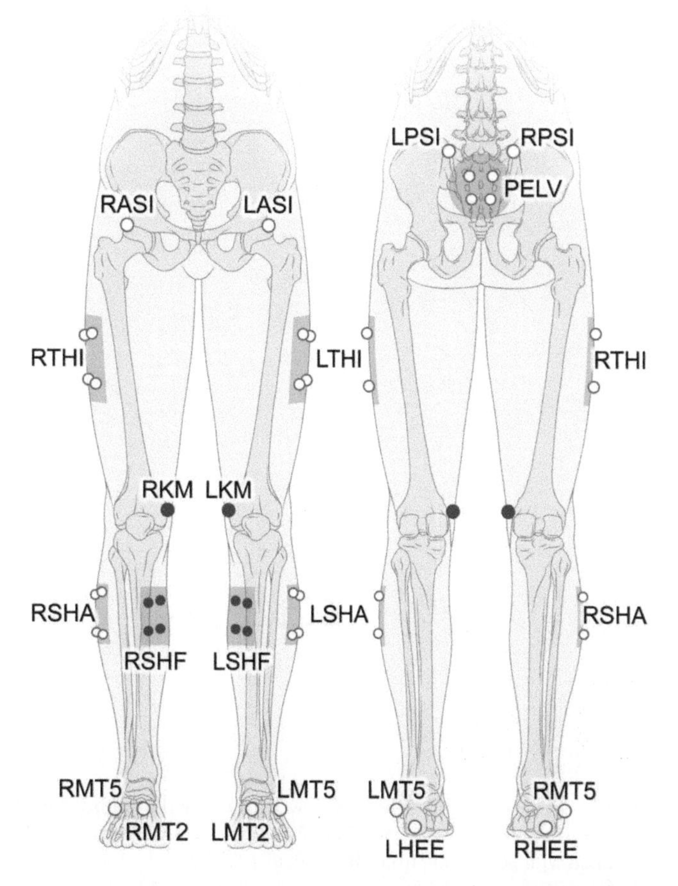


**Fig. J.4** Schematic illustrating a biomechanical model (Strathclyde Cluster Model) that uses a combination of technical markers and marker clusters. PELV = pelvis cluster; RASI/LASI = right and left anterior superior iliac spines; RPSI/LPSI = right and left posterior superior iliac spines; RTHI/LTHI = right and left thigh clusters; RKM/LKM = right and left medial femoral condyles; RSHA/LSHA = right and left lateral shank clusters; RSHF/LSHF = right and left anterior shanks; RMT5/LMT5 = right and left fifth metatarsal bases; RMT2/LMT2 = right and left second metatarsal bases; RHEE/LHEE = right and left posterior mid-calcanea. Note: open circles denote technical markers, and red-filled circles denote markers for calibration only

pose. Along with the input of anthropometric measures, e.g., height, mass, leg length, etc., into the calibration software, marker location data captured during the subject static pose determine the unique anatomical anthropometrics of the test individual, including the mass of body segments, location of segment centers of mass (COM), location of joint centers, and segmental coordinate systems (SCS) (Fig. J.5).

The errors in marker location due to soft-tissue artifacts and other sources are amplified when position data are differentiated to determine linear and angular velocities and accelerations (additional kinematic information is needed to later determine joint forces and moments). To minimize error, it is necessary to carefully perform several steps in preparing the system and subject. Errors cannot be completely eliminated, but a number of data reduction techniques, including signal filtering and algorithms, to enforce anatomical

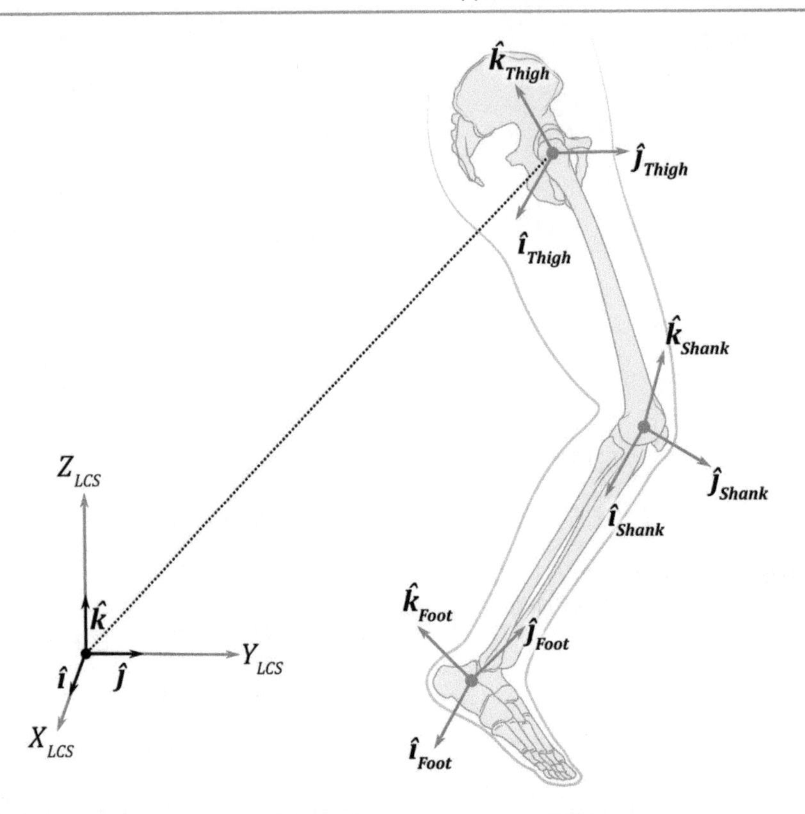


**Fig. J.5** Schematic of the pelvis and lower extremity illustrating the laboratory (LCS) and segmental (SCS) coordinate systems, with the SCS attached to the thigh, shank, and foot. Note that the LCS and SCS are aligned using the Cartesian coordinate standard, i.e., *x*, *y*, and *z* axes. (Modified with permission from Hamill et al. (2014))

constraints have been developed to minimize the effects of noise and improve the fidelity of kinematic data.

### 1.2.1.3 Determination of Three-Dimensional Marker Location, Coordinate Transformations, Segmental Coordinate Systems, and Determination of Joint Angles

During the measurement of the dynamic movement, e.g., gait, the position of each marker is captured and determined in each camera's local 2D image plane. The 2D locations combine with the information obtained during camera calibration to estimate the 3D location of each marker relative to the LCS. The next step in data reduction calls for label assignment (see labeled markers in Fig. J.3) to each marker location for each frame of data. Once the 3D location of markers is known, a series of vector algebra operations transform their location from the laboratory to segmental coordinate systems (Fig. J.5). In effect, this creates a virtual SCS for each segment in the chosen biomechanical model. For example, the markers illustrated in Fig. J.3 would be used to create seven segments: the pelvis and two thighs, shanks, and feet. The SCSs are aligned with the LCS and are used to calculate the 3D kinematics, e.g., segment and joint angles.

In clinical practice, physical therapists routinely use a goniometer to measure joint range of motion. For example, the principle used to measure elbow flexion has the operator align the axis of the goniometer with the approximate location of the elbow joint axis (i.e., running through the humeral epicondyles), the stationary arm parallel to the humerus (proximal segment), and the movable arm parallel to the forearm (distal segment). The degree of flexion of the forearm relative to the humerus is made by subtracting the value read at full elbow extension from the value read at full elbow flexion. This analogy is useful when visualizing, i.e., imagining, how 3D joint angles are determined.

Similar to the goniometric measure, the standard used in biomechanics dictates that we measure the change in the position of the distal SCS relative to the proximal SCS. In most clinical gait laboratories, the general computational approach to calculating segment kinematics is referred to as "direct" or "unconstrained" inverse kinematics because no constraints or limitations are imposed on limb dimensions or the joint motions of the skeletal model. The major limitation of unconstrained inverse kinematics is that it does not account for anatomical constraints and is significantly affected by soft-tissue artifacts. This limitation has downstream implications; that is, errors in the determination of net

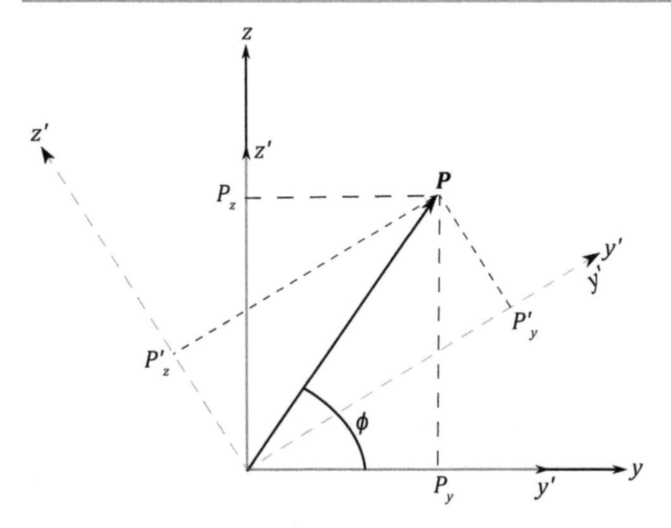


**Fig. J.6** Illustration of coordinate transformation from $xyz$ to $x'y'z'$. The rotation occurs about the $x$ and $x'$ axes (not shown). Vector $\boldsymbol{P}$ is represented in both $yz$ and $z'$ coordinate systems as $P_y$, $P_z$. If the coordinate system $yz$ is rotated to $y'z'$ through an angle $\phi$, a rotation matrix can be determined to transform vector $\boldsymbol{P}$ to $\boldsymbol{P}'$

internal joint forces and moments are also introduced. Fortunately, advanced mathematical methods are available that introduce anatomical constraints as part of the inverse kinematic calculations that control some of the errors associated with soft-tissue artifacts. The details of constrained inverse kinematics are beyond the scope of this appendix and so are not included.

Before reviewing how 3D joint angles are calculated, we provide the framework used to describe relative motion. Previously, we described the process of establishing LCS as part of preparing for MOCAP data collection. Recorded data are resolved into this fixed, right-handed coordinate system (Fig. J.5). The LCS is based on the Cartesian (or rectangular) system and is composed of three mutually perpendicular (orthogonal) axes, commonly labeled as $x$, $y$, and $z$. The Cartesian axes $x$, $y$, and $z$ are commonly expressed with the unit vectors (i.e., vectors with a magnitude of one), $\hat{i}$, $\hat{j}$, and $\hat{k}$, respectively.

Since we are interested in the movement of body segments, we need to be able to relate each body segment to the LCS. In Fig. J.5, we see that each segment, i.e., thigh, shank, and foot, is defined completely by its segmental coordinate system (SCS); as the segment moves, the SCS moves correspondingly. We designate the SCS in lowercase letters. Notice that the LCS and SCS are similar, and the orientation of the SCS with respect to the LCS defines the orientation of the body or segment in the LCS. The description of a body segment moving in a 3D space in different coordinate systems is related by means of a transformation (linear and rotational) between coordinate systems. This transformation allows for the conversion of coordinates expressed in one coordinate system to those expressed in another coordinate system. As noted above, we define a rigid segment by the location of at least three noncollinear points, i.e., marker locations, which enables the determination of a segmental coordinate system.

Joint angles describe the relative orientation of one SCS with respect to the SCS of an adjoining segment and are independent of the location of the origin of these coordinate systems. Several methods are used to parameterize the relative orientation of two coordinate systems, including the Cardan-Euler method, the joint coordinate system method, and the helical angle method. This review only describes the Cardan-Euler method. To help visualize the notion of a rotating coordinate system (i.e., coordinate transformation), let us look at a 2D rotation about a single axis, i.e., the $x$-axis (Fig. J.6).

Before the rotation, the coordinate, systems $xyz$ and $x'y'z'$ are superimposed, i.e., coincident; that is, they share an origin, and the $y$- and $y'$-axes are parallel. The SCS $x'y'z'$ is then rotated from the right horizontal SCS $xyz$ by an angle $\phi$, i.e., positive rotation about the $x$-axis. The rotation matrix $\boldsymbol{R}$ describes the transformation from the vector $\boldsymbol{P}$ expressed in $xyz$ to $\boldsymbol{P}'$ expressed in $x'y'z'$:

$$\boldsymbol{P}' = \boldsymbol{RP}$$

$$\begin{bmatrix} P_x' \\ P_y' \\ P_z' \end{bmatrix} = \begin{bmatrix} 1 & 0 & 0 \\ 0 & \cos\phi & \sin\phi \\ 0 & -\sin\phi & \cos\phi \end{bmatrix} \begin{bmatrix} P_x \\ P_y \\ P_z \end{bmatrix}$$

Expanding this results in:

$$\begin{aligned} P_x' &= P_x \\ P_y' &= P_y \cos\phi + P_z \sin\phi \\ P_z' &= -P_y \sin\phi + P_z \cos\phi \end{aligned}$$

Thus, vectors can be represented in different coordinate systems by means of a transformation matrix.

To determine 3D joint angles, we relate two SCSs (reference frames), e.g., shank relative to thigh, by three angles, called Euler angles (named after mathematician Leonhard Euler). Euler angle calculations are sequence-dependent (hierarchical), meaning angle magnitudes may change depending on the sequence by which they were calculated. If we label segmental coordinate axes as $x$, $y$, and $z$, there are 12 possible Euler rotation sequences to choose from: (1) six sequences described as symmetric, e.g., $x$-$y$-$x$, $x$-$z$-$x$, $y$-$z$-$y$, $y$-$x$-$y$, $z$-$x$-$z$, and $z$-$y$-$z$, and (2) six sequences (also referred to as Cardan or Tait-Bryan angles) described as asymmetric, e.g., $x$-$y$-$z$, $y$-$z$-$x$, $z$-$x$-$y$, $x$-$z$-$y$, $y$-$x$-$z$, and $z$-$y$-$x$. A rotation sequence is chosen primarily based on the nature of the research question and the movement analyzed. The Euler angle sequence used in gait analysis is most commonly defined by the Cardan, or $x$-$y$-$z$, sequence (Fig. J.7).

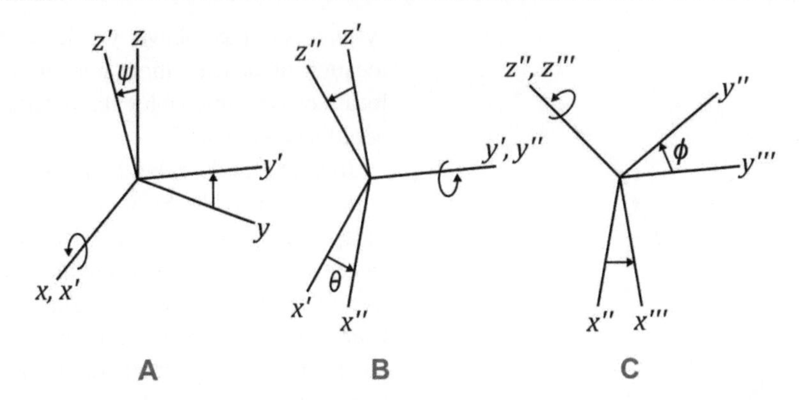


**Fig. J.7** Cardan ($x$-$y$-$z$) rotation sequence where the first rotation $\psi$ is taken about (A) the $x$-axis of the stationary coordinate system, (B) the second rotation $\theta$ about the $y'$-axis, and (C) the third rotation $\phi$ about the $z''$-axis

Imagine two axis systems superimposed, e.g., one attached to the foot (distal segment) and the other attached to the shank (proximal segment), labeled as $xyz$. To determine 3D joint angles at the ankle, we relate the orientations of these two segment coordinate systems by means of a 3 × 3 rotation matrix, which is repeated for the subsequent two rotations. The first rotation, angle $\psi$, of one segment relative to the other about the $x$-axis to produce the primed ($x'y'z'$) system is represented by the following rotation matrix:

$$R_x(\psi) = \begin{bmatrix} 1 & 0 & 0 \\ 0 & \cos\psi & \sin\psi \\ 0 & -\sin\psi & \cos\psi \end{bmatrix}$$

A second rotation, angle $\theta$, occurs about the $y'$ axis to create the double-primed ($x''y''z''$) system with the rotation matrix:

$$R_{y'}(\theta) = \begin{bmatrix} \cos\theta & 0 & -\sin\theta \\ 0 & 1 & 0 \\ \sin\theta & 0 & \cos\theta \end{bmatrix}$$

The third rotation taken about $z''$, angle $\phi$, creates the triple-primed axis system with the rotation matrix:

$$R_{z''}(\phi) = \begin{bmatrix} \cos\phi & \sin\phi & 0 \\ -\sin\phi & \cos\phi & 0 \\ 0 & 0 & 1 \end{bmatrix}$$

Note: we imagine the newly created triple primed system fixed to the foot. Matrix multiplication of the elementary rotation matrices, $R_x$, $R_{y'}$, and $R_{z''}$ results in the 3 × 3 transformation matrix $R$, which represents the 3D rotation from the shank to the foot. Note that the rotation matrix $R$ is computed by multiplying the three matrices in the specified order from right to left:

$$R = R_{z''}(\phi) R_{y'}(\theta) R_x(\psi)$$

$$R = \begin{bmatrix} c\phi & s\phi & 0 \\ -s\phi & c\phi & 0 \\ 0 & 0 & 1 \end{bmatrix} \begin{bmatrix} c\theta & 0 & -s\theta \\ 0 & 1 & 0 \\ s\theta & 0 & c\theta \end{bmatrix} \begin{bmatrix} 1 & 0 & 0 \\ 0 & c\psi & s\psi \\ 0 & -s\psi & c\psi \end{bmatrix}$$

$$R = \begin{bmatrix} c\phi c\theta & c\phi s\theta s\psi + s\gamma c\psi & s\phi s\psi - c\phi s\theta c\psi \\ -s\phi c\theta & c\psi c\phi - s\phi s\theta s\psi & s\phi s\theta c\psi + c\phi s\psi \\ s\theta & -c\theta s\psi & c\theta c\psi \end{bmatrix} = \begin{bmatrix} r_{11} & r_{12} & r_{13} \\ r_{21} & r_{22} & r_{23} \\ r_{31} & r_{32} & r_{33} \end{bmatrix}$$

where $s$ = sine and $c$ = cosine. The elements in the combined matrix $R$ represent the transformation from the SCS of the shank to the SCS of the foot. The Cardan angles are calculated directly using the inverse tangent function on selected elements, where $\psi$ represents rotations in the sagittal plane, i.e., clinical angle flexion/extension; $\theta$ represents rotations in the frontal plane, i.e., clinical angle abduction/adduction; and $\phi$ represents rotations in the transverse (or horizontal) plane, i.e., internal/external rotation:

$$\psi = \tan^{-1}\left(\frac{-r_{32}}{r_{33}}\right)$$

$$\theta = \tan^{-1}\left(\frac{r_{31}}{\sqrt{r_{11}^2 + r_{21}^2}}\right)$$

$$\phi = \tan^{-1}\left(\frac{-r_{21}}{r_{11}}\right).$$

Inverse trigonometric functions do not give unique results. There are always two correct answers between 0 and 360°. With the inverse sine or inverse cosine, there is no way to determine which of the two angles is correct. However, with the inverse tangent, you can know which is correct by tracking the numerator and denominator in the argument separately. The result is in the first quadrant (from 0° to 90°) if

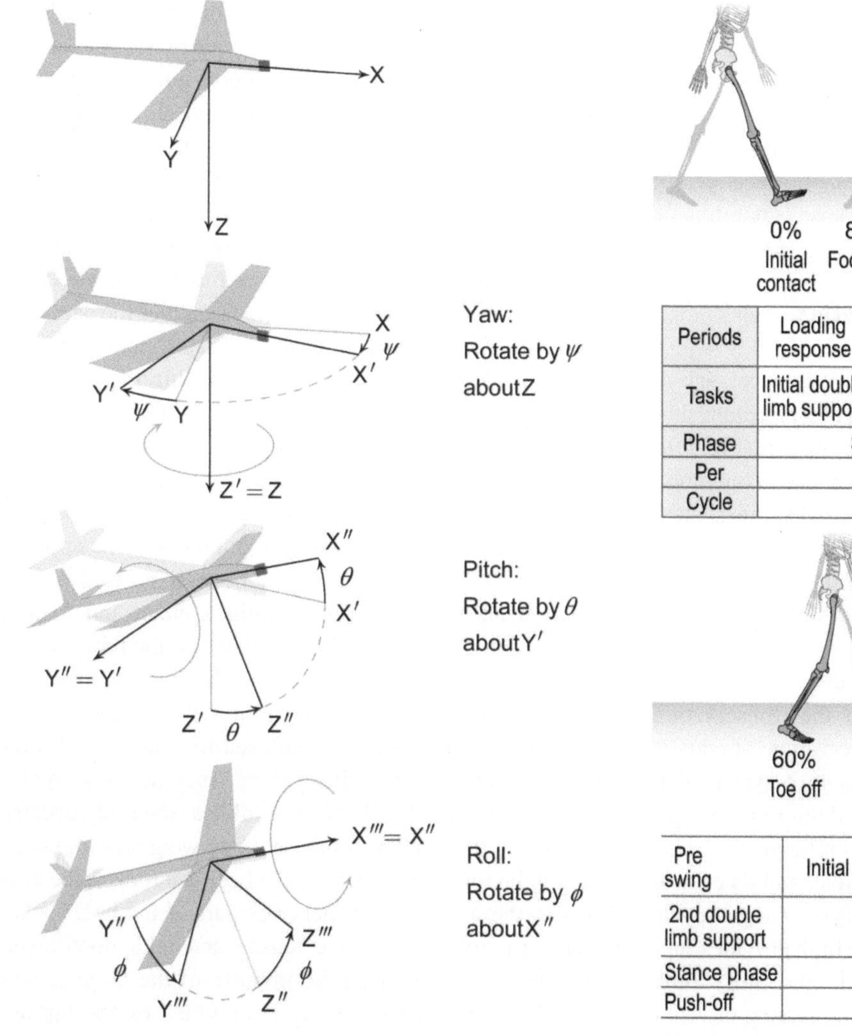


**Fig. J.8** Analogy between the three angles of rotation in Cardan Z-Y-X body-fixed rotation sequence and the yaw, pitch, and roll of an airplane. (Image copyright © David Delp, from Uchida and Delp (2020b). All rights reserved)

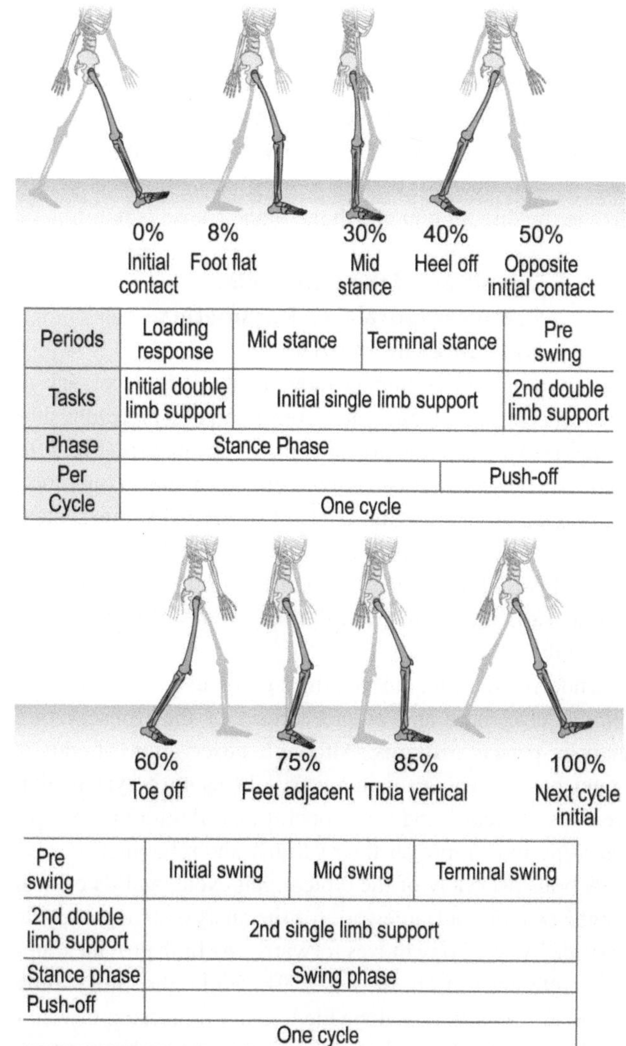

0% Initial contact
8% Foot flat
30% Mid stance
40% Heel off
50% Opposite initial contact
Periods | Loading response | Mid stance | Terminal stance | Pre swing
Tasks | Initial double limb support | Initial single limb support | 2nd double limb support
Phase | Stance Phase
Per | Push-off
Cycle | One cycle
60% Toe off
75% Feet adjacent
85% Tibia vertical
100% Next cycle initial
Pre swing | Initial swing | Mid swing | Terminal swing
2nd double limb support | 2nd single limb support
Stance phase | Swing phase
Push-off
One cycle


**Fig. J.9** Subdivisions of a gait cycle illustrating major phases of stance and swing, gait events (e.g., initial contact), periods (e.g., loading response, etc.), and tasks (periods of single and double support). Note that the right limb is the reference limb

both the numerator and denominator are positive, in the second quadrant (from 90° to 180°) if the numerator is positive and the denominator is negative, in the third quadrant (from 180° to 270°) if both the numerator and denominator are negative, and in the fourth quadrant (from 270° to 360°) if the numerator is negative and the denominator is positive. Note that assuming you select the positive root for the square root in the equation for $\theta$ above, this means that the result for $\theta$ is always in the first or fourth quadrants, i.e., from 90° to +90°.

To help with understanding the relationships between rotation matrices, let us imagine you are a passenger in an airplane and have defined a coordinate system whose $x$ axis points out the nose of the plane, whose $y$ axis points toward the right wing, and whose $z$ axis points down (Fig. J.8). We can recognize this "frame" as the laboratory coordinate system. The plane completes three maneuvers: yaw, pitch, and roll. First, as the plane initiates taxing down the runway, it turns to the right, rotating about the $z$ axis by an angle $\psi$. This turn, called a "yaw," results in a new set of basis vectors (i.e., unit vectors $\hat{i}$, $\hat{j}$, $\hat{k}$) for axes $\{x'y'z'\}$; note that $z' = z$ because the rotation took place about that axis. This new intermediate axis system $\{x'y'z'\}$ defines a new frame, $f_1$. And we got to this new coordinate system by using an elementary rotation matrix similar to $R_x$, shown previously.

Second, as the nose of the plane lifts (called "pitch") during take-off, there is a rotation $\theta$ about the $y'$ axis. This maneuver results in a new frame $f_2 = \{x'', y'', z''\}$, which was derived by using a second elementary rotation matrix (similar to $R_y$). Note that in this case, $y'' = y'$ because pitching the nose upward does not affect the direction the right wing is pointing.

Finally, the plane banks to one side, tipping the right wing down, i.e., rotating about the $x'$ axis, by an angle $\phi$. This maneuver, called a "roll," results in a third reference frame that we refer to as the segmental frame: $f_3 = \{x''', y''', z'''\}$, created by applying a third elementary rotation matrix. In this case, $x''' = x'$ because tipping a wing up or down does not change the direction in which the plane's nose is pointing.

#### 1.2.1.4 Definition of the Gait Cycle, Spatiotemporal Gait Parameters, and Joint Kinematics

Like many complex human movements, it is necessary to divide the gait cycle into major events and other relevant subdivisions to assist in the organization of an analysis (Fig. J.9). Although these divisional assignments, and related operation definitions, are not the direct result of MOCAP, this system provides a methodology that is critical to the analysis of normal and pathological gait by augmenting the kinematic and kinetic variables, which are derived from an instrumented gait study.

Understanding the normal timing of gait events is invaluable to clinicians and biomechanists in their observation analysis of gait patterns. Although observation of gait is somewhat subjective, gaining skill in using a system that provides standard and clear operational definitions of gait events/periods improves their validity and reliability.

A brief overview of the typical gait cycle, and its critical events, provides a framework for the analysis of a gait cycle or stride. As the body moves forward, one limb provides support (reference limb in Fig. J.9), while the other limb advances forward, i.e., along the line of progression of walking. During a portion of time during the gait cyclewhen both feet are in contact with the grounda safe transfer of weight from one limb to the other is achieved. A single sequence of these functions defines the gait cycle as an initial contact of the right heel to a second initial contact of the right heel, whereas a step refers to the timing between the initial contact of the right foot to the initial contact with the left foot. Each gait cycle is divided into two phases: stance and swing. Stance is the entire period when the foot is on the ground, while swing applies to the time when the foot is in the air. Swing begins as the foot is lifted from the floor, i.e., toe-off. Stance is subdivided into tasks of two instances of double support, one at the beginning and another at the end of stance, and one instance of single-limb support (only one foot on the ground). Single-limb support begins when the opposite foot is lifted for swing. The entire body weight must be supported by just one foot during single-limb support, which signals a time period of importance with regard to dynamic stability; a longer relative duration in single-limb support is an indication of greater single-limb stance stability. The precise duration of the gait cycle intervals depends on the individual's walking velocity, so that the change in stance and swing times becomes progressively greater as gait velocity increases; that is, as walking velocity increases, the single-stance time increases as double support intervals decrease. When the double-support intervals disappear, the individual has entered a running mode of locomotion.

Stance and swing are divided into three functional divisions: weight acceptance, single-limb support, and swing limb advancement, which are comprised of seven gait periods. The purpose of these divisions is that the analysis of an individual's walking pattern by periods enables the analyst to identify the functional significance of the kinematics more directly at each joint. Furthermore, dividing the gait cycle into periods provides an opportunity to examine the dynamic coupling of individual joint motion, i.e., both timing and joint angles, into patterns of total limb function. Let us briefly look at each functional division, related gait periods, and the critical events at each period; note that the approximate timing of each gait period can be seen in Fig. J.9.

Weight acceptance includes initial contact and loading response. Initial contact includes the instant the foot makes a heel-first contact with the floor and the body's immediate reaction to ground forces. The hip is flexed, the knee is extended, and the ankle is dorsiflexed to neutral, with mild hindfoot inversion. The critical events include a heel rocker and impulses at the hip, knee, and ankle as the body decelerates. During the loading response, body weight is supported more by the forward limb. With the heel as a rocker, the knee flexes as the ankle plantar flexes as part of the heel rocker. Eccentric action of the knee extensors and ankle dorsiflexors work together to dissipate the impulse of the large ground forces. Weight-bearing stability is provided by the hip and knee extensors as the heel rocker preserves walking progression.

Lifting the other foot for swing begins the phase of single-limb support. This interval continues until the opposite foot again makes contact. During single-limb support, the stance limb has total responsibility for supporting body weight; this is accomplished by actions of the hip extensor and abductor muscles. In its first half of single-limb support, the limb advances over the fixed foot by ankle dorsiflexion (ankle rocker) and hip and knee extension. The hip extensor and abductor muscles provide trunk and limb stability, while eccentric action of the posterior tibial muscles allows and controls forward progression. Terminal stance occupies the second half of single-limb support and is characterized by hip extension, terminal knee extension, and heel rise, with advancement over the forefoot (forefoot rocker). Trunk and limb stability and forward progression are critical events of this gait phase. The opposite limb is completing the terminal swing.

Swing limb advancement includes the following periods: preswing, initial swing, midswing, and terminal swing. Preswing typically occurs during the 50–60% interval of the

**Table J.1** Sample spatiotemporal table comparing right and left data from a patient to control (normal) data

| *Gait parameter* | *Left* | *Right* | *Normal* |
|---|---|---|---|
| Stance (%gc) | 63.2 ± 1.0% | 64.8 ± 0.6% | 59.5 ± 1.5 |
| Swing (%gc) | 36.8 ± 1.0% | 35.2 ± 0.6% | 40.5 ± 1.5 |
| First double support (%gc) | 12.1 ± 0.9% | 11.9 ± 0.9% | 9.5 ± 1.5 |
| Single support (%gc) | 37.4 ± 1.0% | 38.5 ± 1.3% | 40.5 ± 4.5 |
| Second double support (%gc) | 13.9 ± 1.1% | 14.3 ± 1.3% | 9.5 ± 1.5 |
| Gait cycle time (s) | 1.1 ± 0.01 | 1.1 ± 0.01 | 1.0 |
| Step length (cm) | 60.6 ± 1.4 | 64.5 ± 2.3 | 65(4.3) |
| Stride length (cm) | 122.5 ± 1.5 | 124.7 ± 2.3 | 128 ± 8.5 |
| Step width (cm) | 10.8 ± 1.5 | | 7.5(2.5) |
| *Gait parameter* | *Patient's data/barefoot* | | *Normal* |
| Cadence (steps/min) | 108.01 ± 1.3 | | 114(8.5) |
| Walking velocity (cm/s) | 111.2 ± 2.2 | | 122 ± 12.0 |

Note: Data are presented as mean ± SD; %gc = percent gait cycle, cm = centimeters. Control data used with permission from the Motion Analysis Laboratory, Mary Free Bed Rehabilitation Hospital, Grand Rapids, Michigan

gait cycle. This is the final phase of the second (terminal) double-limb support. It begins with the initial contact of the opposite limb and ends with toe-off. All motions and muscle actions during this interval relate to progression: the abrupt transfer of body weight rapidly unloads the limb, while the trailing limb contributes to progression with a forward push (acceleration), which also assists the limb for the swing periods. This interval is characterized by ankle plantar flexion to approximately 15°, the initiation of hip flexion, and rapid knee flexion to 40°. During the initial swing, foot clearance from the floor is critical as the limb is advanced with increasing hip and knee flexion; ankle dorsiflexion is incomplete. The opposite limb is in early midstance. Midswing sees the continued advancement of the limb as the foot is cleared by ankle dorsiflexion to neutral and by hip flexion. Terminal swing begins with a vertical lower leg and ends with the knee fully extended as the foot makes preparation for initial contact. The critical events of this period include the completion of limb advancement and preparing the limb for stance. This review of the gait cycle is just that, so we refer the reader to the text by Perry and Burnfield to get the depth of information you need to carry out a competent gait analysis and to satiate your curiosity.

Normal spatiotemporal (sometimes also referred to as temporo-spatial) parameters of gait provide valid and reliable quantitative data that are useful as baseline measures and for longitudinal comparisons when studying individuals with gait pathologies (Table J.1). However, the clinician or biomechanist should use caution when using gait parameters based on published research conducted at other institutions. For example, some institutions use MOCAP and force platforms to measure gait parameters (Fig. J.10), whereas electronic walkways, plantar pressure systems, foot switches,

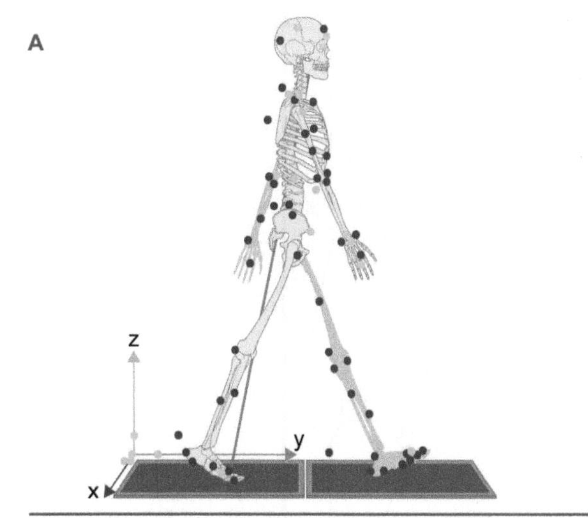


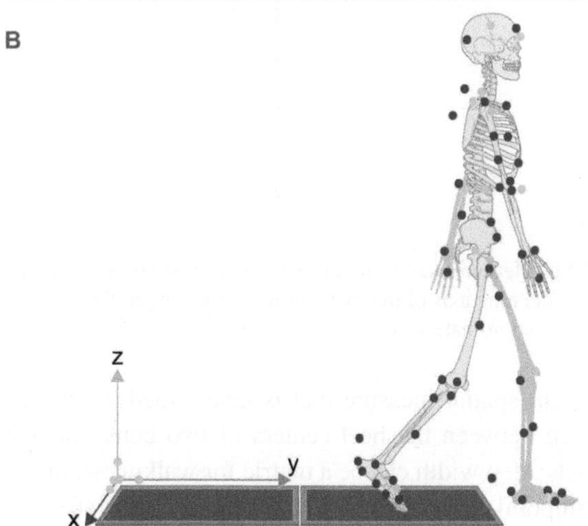


**Fig. J.10** Floor-embedded force platforms that record ground reaction forces at (**A**) initial contact and (**B**) toe-off are used in combination with heel marker locations to detect key gait events. Note the laboratory coordinate system, which was established during a calibration procedure preceding data collection, anchored in the corner of the first force platform. Also note the large blue vector under the right foot at initial contact/loading response, which represents the vertical ground reaction force

and inertial motion sensors have also been used. Unfortunately, the concurrent validity of different systems used to measure gait parameters is not fully established. Moreover, there are reported differences in gait parameter values when walking at self-selected speeds over ground versus on an instrumented treadmill. Therefore, it is important for the user to check the details of research reports for what instrumentation was used if they intend to compare their results to published data from other labs as part of their gait analysis.

The most basic spatial description of gait includes the length of the stride and the length of the step. Comparing right and left step lengths is a good way to assess one aspect of the symmetry of gait between limbs. Step width is another

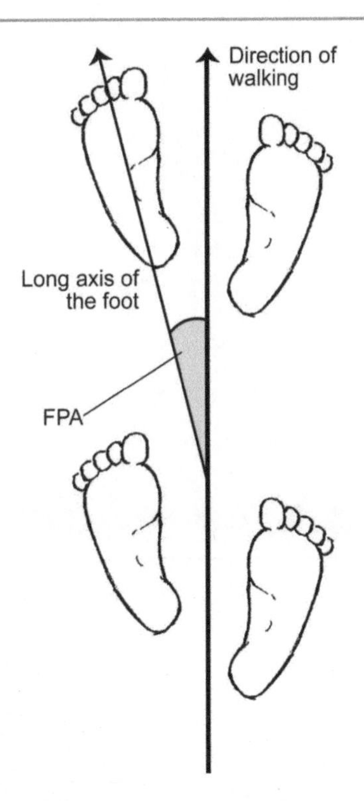


**Fig. J.11** Measurement of foot progression angle (FPA) as the planar angle of the long axis of the foot relative to the longitudinal axis of the laboratory coordinate system, i.e., the direction of walking

important spatial measure that is determined as the lateral distance between the heel centers of two consecutive foot contacts; step width can be a metric for walking stability.

Temporal descriptors of gait include cadence, stride time, and step time. Cadence (sometimes referred to as step rate) is typically expressed as the number of steps per minute. Stride time is the amount of time needed to complete one gait cycle, and step time is the time to complete a right or left step. Note that step time is the reciprocal of cadence. Walking speed combines both temporal and spatial measures and is strongly correlated with stride length. Walking speed and stride length are probably the most functional measures of walking ability. The units of walking speed are usually meters per second or miles per hour. Speed can be determined by measuring the time it takes to cover a given distance, the distance covered in a certain time, or by multiplying the step rate by step length.

In clinical gait analysis, a typical report includes figures that show the 3D kinematics for each segment, which were determined relative to a biomechanical model. As previously described, the standard for determining the three-dimensional joint angles is to measure the movement of the distal segment, i.e., its segmental coordinate system, relative to the proximal segment. For example, knee joint angles are determined by calculating the movement of the tibia relative to the femur. However, in the data below, we draw your attention to two exceptions to the standard. Firstly, pelvic motion is determined by measuring the movement of the pelvic coordinate system relative to the laboratory, i.e., fixed, coordinate system; in some cases, motion laboratories also determine movements of the trunk using the same convention, i.e., trunk relative to the laboratory. Secondly, in the transverse plane, we determine and analyze the foot progression angle in lieu of foot axial rotation. This is so because it has been determined by clinical gait biomechanists that the foot progression angle provides functional information that foot axial rotation cannot. For example, the foot progression angle can potentially provide information about compensatory rotations of the pelvis in the transverse plane; hip axial rotation compensations related to fixed proximal femoral torsions, e.g., antetorsion; and abnormal tibial torsion. To determine the foot progression angle, we measure the planar angle that the long axis of the foot makes with the longitudinal axis of the laboratory (Fig. J.11).

Fig. J.12 provides a sample of the 3D joint angles based on normal self-paced walking. In clinical gait analysis, it is incumbent on each motion analysis laboratory to collect a set of control data for the population(s) they serve, with separate sets for children and adults. In addition to the collection of these data, it is imperative that the biomechanical model used to determine 3D joint angles is validated (i.e., the joint angles are examined for their accuracy) and tested for intertrial and intersession precision, i.e., reliability. Likewise, the laboratory personnel who are responsible for marker placement should be involved with a quality assurance program that regularly checks the intra- and intertester reliability of kinematic data.

How are data from typically developing individuals (or healthy controls) used? After patient data are collected and processed, a representative gait cycle is selected for each limb, and those data are placed in graphs that overlay the normal bands. Generally, an abnormality in specific joint angles would be identified as those that lie outside a grey band (i.e., one standard deviation from the ensemble average of the control data set) over the gait cycle. Since the entire gait cycle is plotted, with specific time points or periods identified, clinicians and biomechanists can be quite precise in their identification of the kinematic abnormalities and where in the gait cycle they occur.

## 1.3 Gait Kinetics, Inverse Dynamics and Net Internal Joint Forces, Moments, and Powers

A dynamic analysis of human movement considers the motion of segments as well as the forces and moments responsible for and generated by the motion. Using a typical inverse dynamics approach and given a representative model of an individual, the individual's segment kinematics over

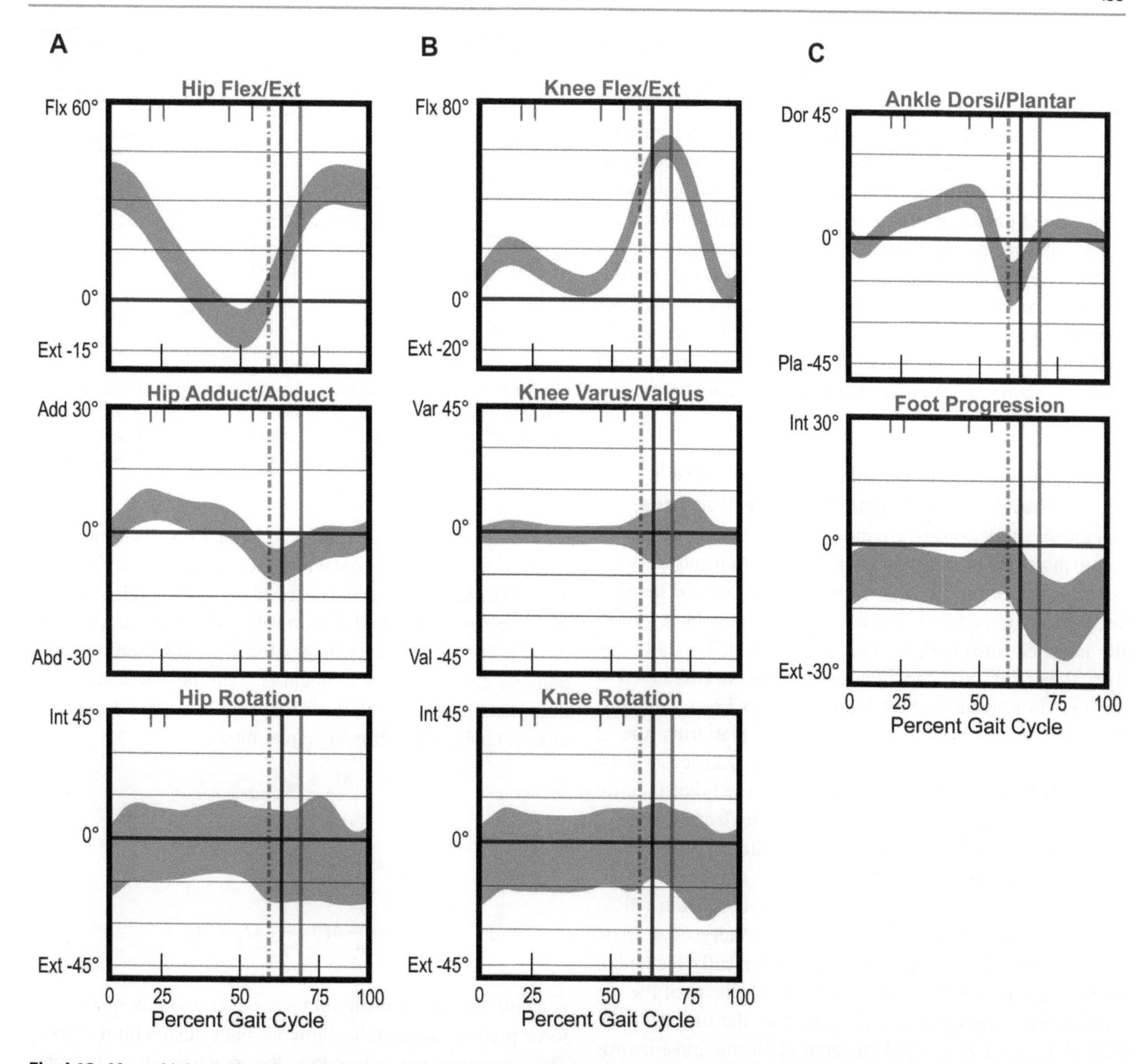


**Fig. J.12** Normal joint angles (degrees) for the (**A**) hip, (**B**) knee, and (**C**) ankle expressed in percent gait cycle (horizontal axis), proceeding from 0% to 100%, where 0% is first heel contact and 100% is the second heel contact. The gait cycle is separated into stance and swing, with the vertical dark grey line representing toe-off. Control joint angles are illustrated as grey bands, which represent the mean ± one standard deviation of the control data set. Note that flexion/extension and dorsiflexion/plantar flexion are movements in the sagittal plane, varus/valgus and adduction/abduction are movements in the frontal plane, and internal/external rotation are movements in the transverse plane. Flx = flexion; Ext = extension; Int = internal rotation; Ext = external rotation; Dor = dorsiflexion; Pla = plantar flexion; Add = adduction; Abd = abduction; Val = valgus; Var = varus. Typically, a kinematic control data set also includes motions of the trunk and pelvic segments (not shown). (Control data used with permission from the Motion Analysis Center, Shriners Children's, Chicago, IL)

time, and measurements of the external forces applied to the individual, we can determine the net joint forces, moments, and powers that were present to produce the given motion. Newton's second law lays the groundwork for dynamic solutions. It states that the rate of change of a particle's momentum is equal to and in the direction of an applied force. The most commonly expressed form of this law is $\boldsymbol{F} = m\boldsymbol{a}$, where $\boldsymbol{F}$ is the vector sum of all forces applied to a body, $m$ is the mass of the body, and $\boldsymbol{a}$ is the linear acceleration of its center of mass. Extending Newton's law is necessary in order to account for angular accelerations and moments of force, or torque. Thus, Euler's second law, which claims that the rate of change of a body's angular momentum about its center of mass is equal to the sum of all moments acting about this point, is used in conjunction with Newton's second law. For planar rotational problems, Euler's law is expressed as $M = I\alpha$, where $M$ is the sum of all moments with respect to the center of mass, $I$ is the mass moment of inertia with respect to the center of mass, and $\alpha$ is the body's angular acceleration. If the mass distribution and accelerations of each body segment are known, Newton's and Euler's laws can be used to determine the forces and moments acting on each of those segments. Angular velocities and accelerations, determined by differentiating segmental angles estimated from optical marker trajectories and inverse kinematic algorithms, and measured external forces, i.e., ground reaction forces (GRF), are necessary ingredients for the inverse dynamics problem. In summary, the application of the laws of mechanics to each body segment in a biomechanical model is used to determine the net internal forces and moments acting at each joint. Let us first look at how ground forces are measured.

### 1.3.1.1 Measuring External Forces: The Force Platform and Plantar Pressure Measures

The means of measuring ground reaction forces is necessary before discussing inverse dynamic computation and the analysis of gait. Ground reaction forces are measured using a force platform (or force plate), an instrumented plate generally installed flush with the ground (Fig. J.1). Typically, for clinical gait analysis, 2 to 4 force platforms are embedded in the floor in a linear fashion along a 10- to 12-m walkway. Presently, commercially available force platforms use 2 types of sensors: strain gauges and piezoelectric crystals. Strain-gauge models are the most widely used, but they do not have the range and sensitivity of piezoelectric models. This review is restricted to the characteristics of plates using strain gauge technology.

The force platform has its own reference system (PRS) with its axes often aligned with the laboratory coordinate system (LCS). The origin of the PRS is typically located at the center of the platform, slightly below the level of the top surface, with coordinates with respect to the origin of the LCS of $x_0$, $y_0$, and $z_0$. Modern force platforms have instrumented (strain gauges) columns or pedestals located near the corners of the plate. With this model, forces are accurate whenever the contact, i.e., deforming, force is applied within the area bounded by the pedestals. When someone engages the platform, the strain gauges deform and create a voltage signal that is proportional to the magnitude of the applied forces and moments. Most platforms used for research and applied clinical purposes report forces and moments along, and about, the three orthogonal axes of the LCS and PRS. Thus, the three force components from each of the four strain gauges give 12 independent measurements, which are combined to produce the resultant force vector and moment vector with respect to the force platform's origin. The components from these two vectors give six quantities: $F_x$ (mediolateral), $F_y$ (anteroposterior), $F_z$ (superoinferior) forces, and three moments, $M_x$, $M_y$, and $M_z$ (Fig. J.13).

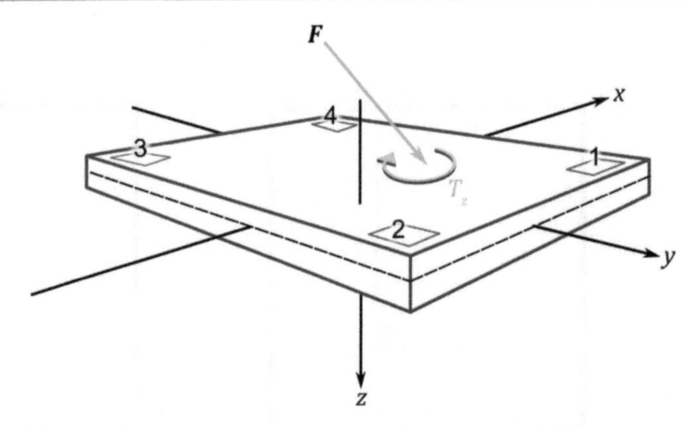


**Fig. J.13** Typical force platform with four corner pedestals containing strain gauges in each corner of the plate. Note the platform coordinate system (PRS); the vertical GRF, ***F***, intersecting the COP; and the free moment, $T_z$. The numbered boxes in the corners represent the approximate location of strain gauge pillars

The force vector is usually related to a distribution of forces over an area of contact on the surface of the platform. This distribution of forces is equivalent to the single resultant force vector acting at a location called the center of pressure (COP). The $x$ and $y$ coordinates for the COP location and a quantity called the free moment ($T_z$) can be calculated from the six components of the platform's resultant force and moment using the following equations:

$$x = -\frac{M_y + z_0 F_x}{F_z} + x_0$$

$$y = \frac{M_x + z_0 F_y}{F_z} + y_0$$

$$T_z = \frac{F_x}{F_z} M_x + \frac{F_y}{F_z} M_y + M_z$$

Note that the $z$ coordinate for the COP is 0, assuming that the force platform and LCS origin are both flush with the floor. The free moment represents the reaction to a twisting moment applied by the individual about a vertical axis. The free moments about the $x$ and $y$ axes are assumed to bezero because these can occur only if there is a direct connection between the shoe (or foot) and plate (as with glue). Typical COP progression and vertical, anteroposterior (i.e., braking-propulsion), and mediolateral GRFs during normal gait are illustrated in Figs. J.14 and J.15, respectively.

As noted previously, when discussing the spatial and temporal parameters of gait, GRFs are useful because they provide additional descriptive data that can be used to compare a pathological gait with a normal one. However, used alone, these data are primarily descriptive and rarely provide insight into primary gait deviations. Nevertheless, they are useful for comparing baseline and longitudinal data to get a sense of changes in gait following treatment. However, the disadvantage

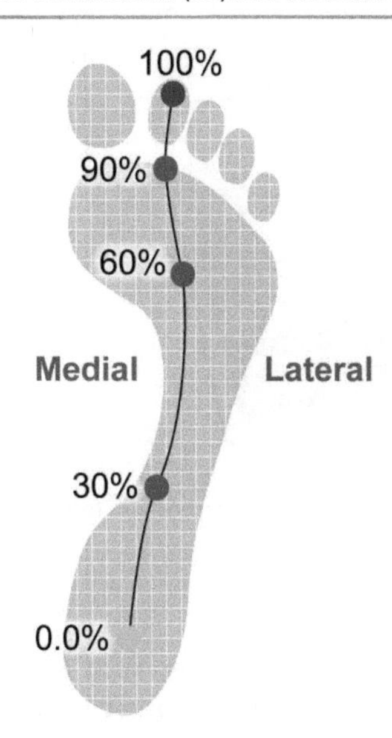


**Fig. J.14** Typical center of pressure (COP) progression during gait from initial heel contact to toe-off. Note that 0% indicates initial contact, and 100% indicates second initial contact

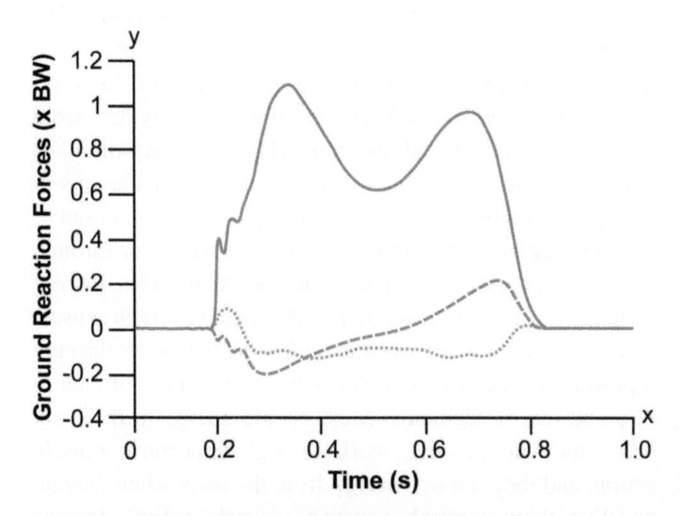


**Fig. J.15** Typical ground reaction forces (GRFs) during the gait stance phase. Note that GRF is expressed relative to body weight (BW). The solid line represents the vertical GRF, the dashed line the anteroposterior GRF, and the dotted line the mediolateral GRF

of COP and GRF data is that they are not able to provide insight into relationships between gait deviations within a limb or between limbs and so are not used in making treatment recommendations. Furthermore, since COP measured by a force plate is the weighted average of the distributed pressure under the feet, it cannot provide information about the pressure at any of the contact points under the feet. Therefore, to gain insight into pressures at all contact points, special pressure measurement systems have been developed, which are sometimes referred to as pedobarographic measures. There are several commercially available pressure measuring systems that offer either in-shoe devices or mat systems (Fig. J.16). These systems are particularly useful in the measurement and analysis of gait with children with cerebral palsy, club feet, or other congenital/acquired foot deformities who typically require extensive reconstructive foot surgery.

### 1.3.1.2 Kinetics: Inverse Dynamics

The inverse dynamic solution in gait analysis is based on two fundamental assumptions: (1) anatomical segments are rigid bodies, and (2) rigid bodies are formed into linked segments that represent the individual being tested. Note that the segments in the link model are assumed to have fixed anthropometric properties, i.e., mass, location of the center of mass, and moments of inertia, and are represented by a segmental coordinate system (SCS). Linked segments present a hierarchical model that presupposes two possible joint mechanisms: one that allows complete rotational and translational freedom of movement between the segments, i.e., six-degree-of-freedom (DoF) models, or one that specifies constraints at the joint to limit one or more of the rotations and/or translations, e.g., defining the glenohumeral joint as ball and socket with three DoF.

Before presenting a more detailed step sequence of the inverse dynamic approach, we should say a bit more about the complexity of 3D human movement analysis. A major obstacle to the inverse solution has to do with the indeterminacy of the solution. In a 3D inverse dynamics analysis (assuming a six DoF model), a rigid body has six equations of motion relating to the six DoF. But if we wished to compute the contributions of all the muscles toward a particular joint's motion, it is obvious there would be many more unknowns than equations. For example, to fully solve the ankle, knee, and hip sagittal plane kinetics during walking, we would have to account for 15 major muscles, along with the components for the reaction forces at each joint. This situation leads to an indeterminate solution because there are more internal forces that are unknown than equations. Therefore, to reduce the number of unknowns to six per body, the effect of all forces, e.g., from muscles, ligament, friction, etc., acting across a joint is replaced with a net joint moment and a resultant intersegmental force. It is important to note that these intersegmental forces do not represent bone-on-bone joint reaction forces. The bone-on-bone forces cannot be estimated without knowing the actual muscle forces.

The most common method used to calculate joint kinetics is the Newton-Euler formulation:

$$\sum \boldsymbol{F} = \frac{d}{dt}(m\boldsymbol{v}) = m\boldsymbol{a}$$

$$\sum \boldsymbol{M} = \frac{d}{dt}\boldsymbol{H} = \boldsymbol{I}\boldsymbol{\alpha} + \boldsymbol{\omega} \times \boldsymbol{I}\boldsymbol{\omega}$$

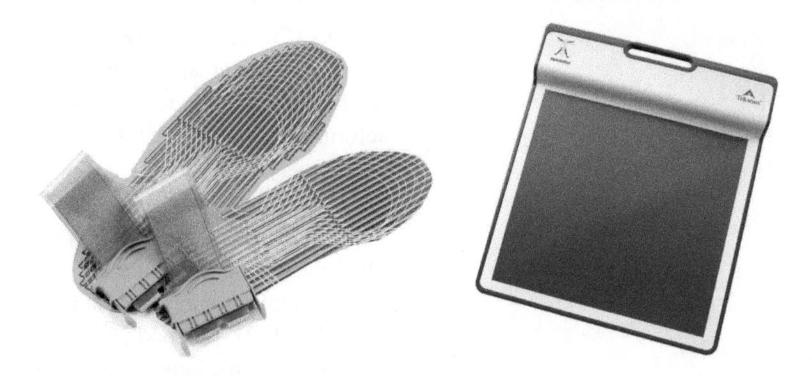

**Fig. J.16** Examples of in-shoe and mat pressure systems commonly used in clinical gait analysis. (With permission from Tekscan® (Tekscan, Inc., South Boston, MA))

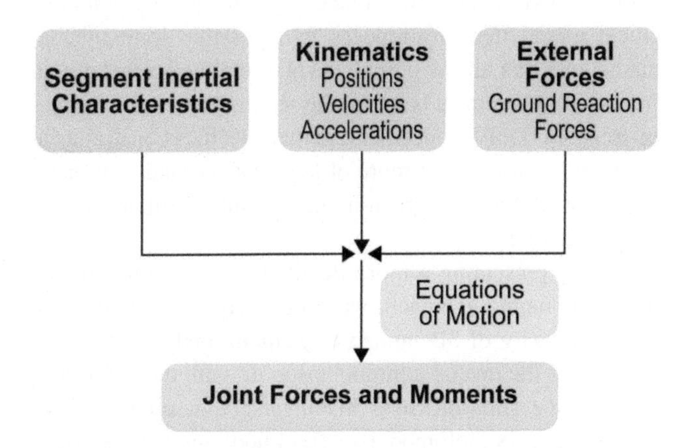


**Fig. J.17** Inputs required to solve the inverse dynamics problem in biomechanics

where $\sum \boldsymbol{F}$, the sum of all the forces acting on a rigid body, is equal to the rate of change of momentum of the body (Newton's equation) and where $\sum \boldsymbol{M}$, the sum of all moments acting on the body with respect to its center of mass, is equal to the rate of change of the angular momentum, $\boldsymbol{H}$, of the body with respect to the mass center (Euler's equation). Also, in these equations, $m$ represents the mass, $\boldsymbol{a}$ represents the acceleration, $\boldsymbol{I}$ represents the inertia tensor (matrix), $\boldsymbol{\alpha}$ represents the angular acceleration, and $\boldsymbol{\omega}$ represents the angular velocity. For planar (2D) problems, Euler's equation is simplified to:

$$\sum M = I\alpha$$

where $I$ represents the mass moment of inertia. Newton-Euler equations are applied separately to each segment (or body) in the system, providing six equations for 3D and three equations for 2D formulations. As inputs, the inverse dynamics solution requires segment anthropometric properties, such as segmental mass, and inertia properties and the location of the center of mass; segmental kinematics, such as mass center acceleration, angular velocity, and angular acceleration; and external loads, such as ground reaction forces and moments (Fig. J.17). The outputs of inverse dynamics are the net joint moments and intersegmental forces at the joints.

What does net joint power refer to? First, a brief detour is necessary to provide additional definitions. Mechanical energy and work have the same units (joules) but have different meanings. *Mechanical energy* is a measure of the state of a body at an instant in time as to its ability to do work, whereas *mechanical work* can measure the energy flow from one body to another, and time must elapse for that work to be done. The sum of all energy flows must equal the energy change of a segment. Energy can flow passively into or out of adjacent segments through joint contact surfaces or through soft tissues, such as ligaments, tendons, and muscles. Energy can also be actively generated or dissipated through muscle activity. Another way to look at the energy balance is through a power balance, which looks at the rate of flow of energy into and out of segments. Muscles add energy to the body when they do positive work through concentric muscle action, and they remove energy from the body when they do negative work through eccentric muscle action. Inverse dynamics does not allow us to calculate individual muscle forces, so muscle contributions are estimated through net joint moments, which are primarily a result of muscle activity. Assuming only agonist muscle activity (i.e., no cocontraction by antagonist muscles), when the muscle moment acts in the same direction as the angular velocity of the joint, energy is being generated through concentric muscle activity. When the muscle moment acts in the opposite direction to the movement of the joint, energy is being absorbed through eccentric muscle activity.

Joint power approximates the rate of doing work (positive or negative) by the muscles. The general equation for determining joint power $P$ is:

$$P = \tau \cdot \omega$$

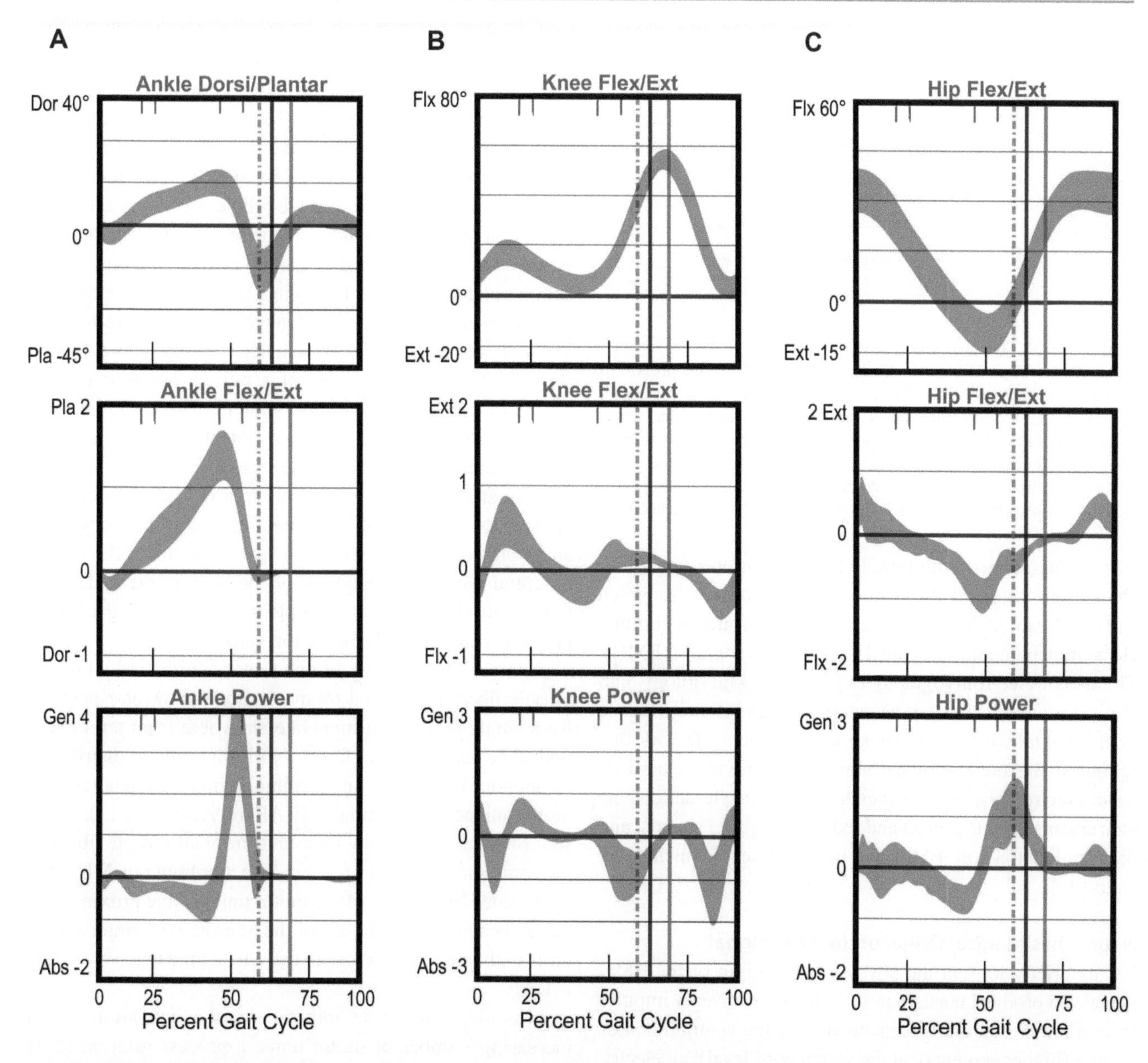


**Fig. J.18** Normal internal net joint moments (Nm/kg) and powers (W/kg) in the sagittal plane at the (**A**) ankle, (**B**) knee, and (**C**) hip joints in percent gait cycle (horizontal axis), proceeding from 0% to 100%, where 0% is heel contact and 100% is second heel contact. We typically combine the kinematic and kinetic data since net joint power is determined using joint angular velocity and net joint moments; that is, we can evaluate net joint power by qualitatively assessing the slope of the joint kinematic tracings and net joint moments. The moments and powers are normalized by each participant's total body mass, which allows for the comparison of individuals with different body masses. The gait cycle is separated into stance and swing, with the vertical grey line representing toe-off. Control joint motions, moments, and powers are illustrated as grey bands, which represent the mean ± one standard deviation. Joint angles, net joint moments, and net joint powers are provided in the top, middle, and bottom rows, respectively. Note: Flx/Ext = flexion/extension motion; Ext/Flx = extensor/flexor moments; Dor/Pln = dorsiflexion/plantar flexion motion; Pln/Dor = plantar flexor/dorsiflexor moments; and Gen/Abs = power generation/absorption. (Control data used with permission from the Motion Analysis Center, Shriners Children's, Chicago, IL)

where $\boldsymbol{\tau}$ is the net joint moment and $\boldsymbol{\omega}$ is the angular velocity at the joint. Note that this expression entails a dot product operation between two vectors, which results in a scalar quantity for power. By convention, concentric muscle actions are associated with power generation, while eccentric muscle actions are associated with power absorption. For clinical gait analysis, net joint powers are typically only determined at the ankle, knee, and hip in the sagittal plane.

Representative net joint moments and powers during gait representing a control (i.e., normal and healthy) data set are illustrated in Fig. J.18.

#### 1.3.1.3 Electromyography

When an end-plate potential is generated at a nerve-muscle synapse, it induces a muscle fiber action potential that propagates from the synapse to the ends of the muscle fiber. The

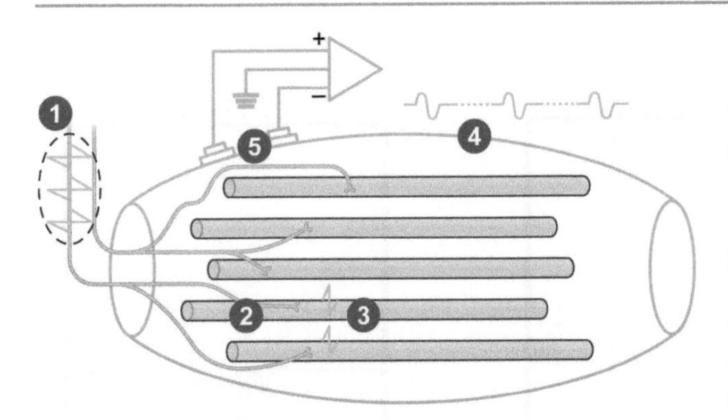


**Fig. J.19** (1) The motor neuron action potential initiates the process of muscle fiber activation. (2) The action potential arrives at the motor endplates. (3) By electrochemical processes, a muscle fiber action potential is initiated and moves along the muscle fiber. (4) The sum of all muscle fiber potentials activated by one motor neuron produces a motor unit action potential. (5) Recording of a motor unit action potential at the skin surface. (Modified with permission from Kamen (2014))

currents associated with muscle fiber action potential cause changes in potential differences within the muscle, which can be measured. The electrical signal associated with muscle action potentials is called an electromyogram (EMG). The use of electromyography in medicine, kinesiology, and clinical motion analysis is as much an art as science because of the complexity of the signal and the many factors that influence the fidelity of the signal. This review examines basic electrophysiology associated with muscle action, the instruments used to collect and reduce raw EMG signals, and the application of EMG in instrumented clinical gait analysis.

## Neurophysiological Origin of the EMG Signal

Let us briefly look at the physiological origin of the EMG signal. To produce tension, muscle fibers receive an impulse from a motor neuron. Once activated by the central nervous system (higher centers or at the spinal cord level), an electrical impulse propagates down the motor neuron to each motor endplate. At this junction, ionic events occur that result in the generation of a muscle fiber action potential. As an aside, note that even at rest, muscle fibers demonstrate a resting membrane potential that can be recorded, an important fact that needs to be considered when checking an EMG signal for comparing a normal resting signal versus noise, i.e., electrical signals that are not part of the desired EMG signal.

An action potential is the neural signal responsible for activating every segment of the muscle fiber so that each sarcomere contributes to the generation of muscular tension. This process begins with a change in the muscle fiber membrane's permeability to sodium ($Na^+$). Since $Na^+$ ions exist in a greater concentration outside the muscle fiber, any change in membrane permeability induces a movement of $Na^+$ across the membrane. This movement eventually leads to a reversal of the polarity of the membrane potential so that the outside of the muscle fiber becomes positive with respect to the surrounding extracellular matrix. As membrane permeability reverses, the permeability of the membrane to potassium ($K^+$) changes, causing it to leave the cell. The efflux of $K^+$ largely repolarizes the cell, restoring resting membrane potential. To make sure that the whole muscle is activated, the action potential produced in one region adjacent to a neuromuscular junction must spread to adjacent regions. As the action potential spreads, the membrane potential at each subsequent muscle fiber section changes from negative to positive and back to negative as each adjacent region is activated (Fig. J.19). Deeper portions of a muscle are activated by means of the transverse tubule system. The rates, i.e., frequency, of transmission of action potential determine some of the characteristics of EMG. For example, action potentials moving at slower rates contribute to low-frequency components in the surface EMG. It is notable that the histochemical and architectural features of a muscle fiber affect the amplitude and conduction velocity of the action potential. Fast-twitch muscle fibers, for example, tend to have larger action potentials and greater conduction velocities.

A motor unit consists of one motor neuron and all of the muscle fibers innervated by that motor unit. Motor neurons for a single motor unit innervate anywhere from ten to several thousand muscle fibers, with the fibers distributed throughout the muscle in a seemingly random manner. The motor unit action potential (MUAP) represents the aggregate sum of electrical activity measured from all muscle fibers in the same motor unit (Fig. J.20). The amplitude of MUAP is determined by the size of the motor unit and the proximity of the activated muscle fibers to the sensor; i.e., larger motor units with more fibers close to the sensor tend to have a larger MUAP.

Muscular tension is initiated by the activation of an increasing number of motor units, a process referred to as recruitment. Recruitment is typically hierarchical based on the size of the motor unit, a property known as Henneman's size principle. Therefore, smaller motor units, characterized by smaller-amplitude and long-duration force twitches, are recruited at lower contraction forces; i.e., they have a lower recruitment threshold. Meanwhile, larger motor units, characterized by larger-amplitude and shorter-duration force twitches, are recruited at higher contraction forces. The central nervous system also controls how frequently motor units are activated, called motor unit discharge or firing rate. Fast discharge rates produce increasing levels of muscle tension or force. The magnitude of a muscle force can also be increased when two or more motor units are activated simultaneously, a process called synchronization.

The force produced by a muscle is thus modulated by three processes: (1) variation in the number of active motor

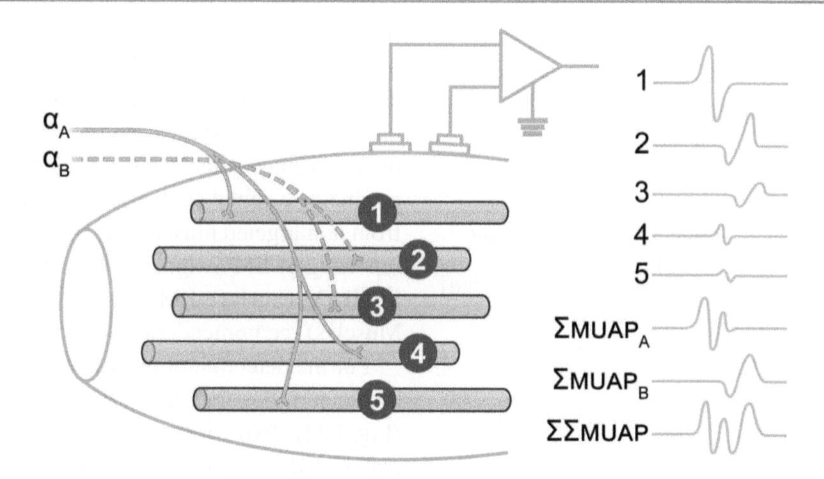


**Fig. J.20** The contribution that each muscle fiber's action potential (AP) makes to the EMG signal depends on the depth of the muscle fiber; e.g., fiber 5 makes the least contribution to the total EMG signal. The temporal characteristics of EMG depend on the electrode-motor endplate distance, terminal lengths, and diameters of the motor neurons. Two motor units are shown, with the EMG amplitude represented as the sum of the individual muscle fiber APs, i.e., $\sum$MUAP. The overall signal is the sum of both motor units, i.e., $\sum \sum$ MUAP. Note the schematic representation of surface EMG electrodes on the upper surface of the skin; $\alpha_A$ and $\alpha_B$ represent $\alpha$-motor neuron stimulation to motor units A and B, respectively. (Modified with permission from Kamen (2014))

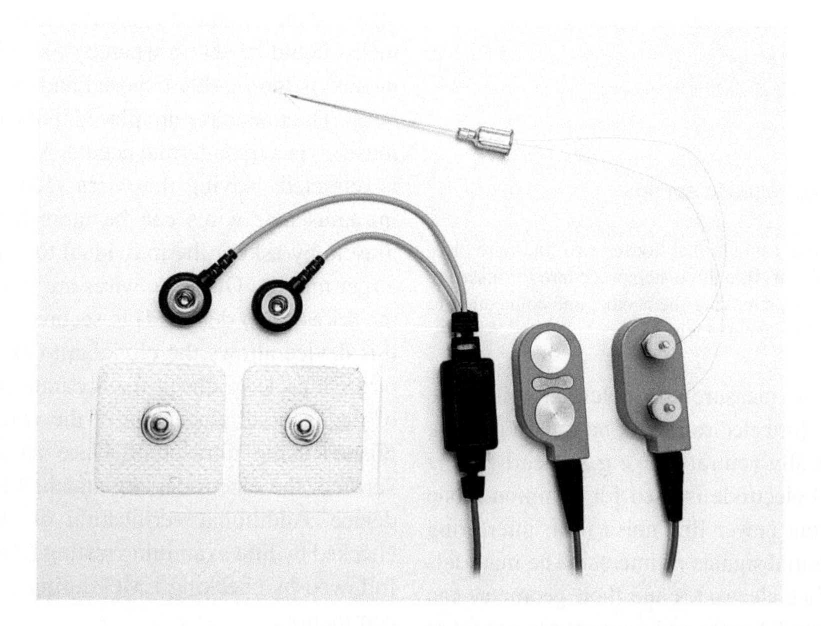

**Fig. J.21** Examples of find wire electrodes (top) and surface electrodes/amplifiers (bottom). With permission from Motion Lab Systems, Inc.

units, (2) variation in firing rate, and (3) motor unit synchronization. These phenomena translate to an increasing amplitude of a recorded EMG signal. In a very minimal muscle action where a single motor unit may be activated, a MUAP is recorded at the surface, followed by electrical silence. As the desired muscle force increases, other motor units are recruited, perhaps firing at increasing rates. Thus, the EMG signal is the aggregate sum of the electrical activity of all of the active motor units. The amplitude of the recorded EMG signal varies with the task demand, the specific muscles that are studied, and many other factors. It is clear that EMG amplitude increases as the intensity of the muscle action increases; however, the relationship between EMG amplitude and force magnitude is nonlinear. In addition, the coactivity of muscle antagonists may induce changes in agonistic muscle activity from which the EMG recordings are being made.

## Surface and Fine-Wire EMG

With kinesiological EMG, the electrical signals are recorded using either surface or fine-wire electrodes (Fig. J.21). In either case, two electrodes recording the signal are necessary

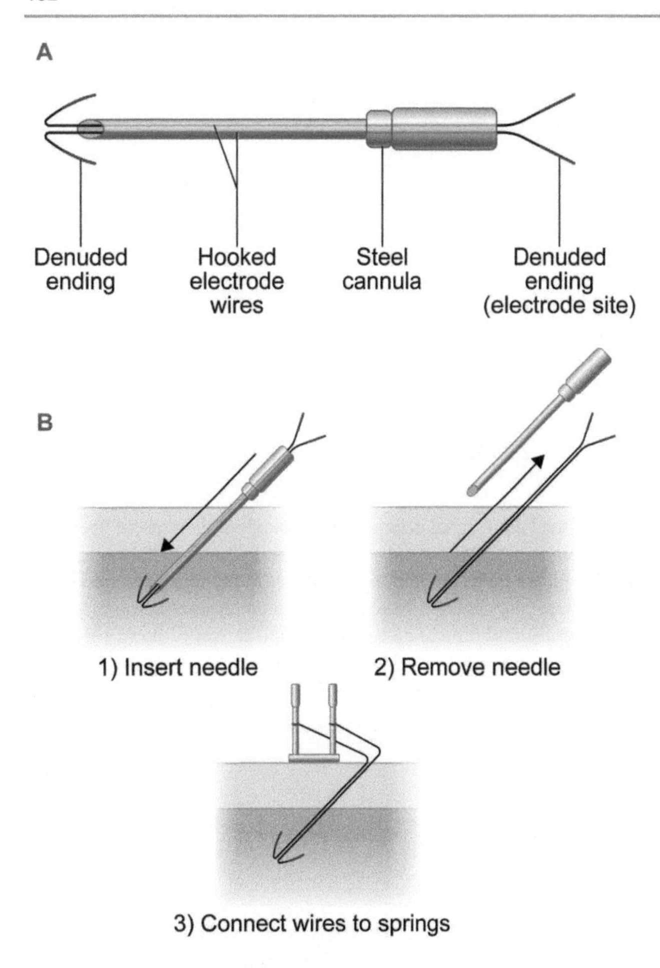


**Fig. J.22** (**A**) A hypodermic cannula that houses two fine-wire electrodes denuded at each end and (**B**) a three-step procedure for inserting the wires into a target muscle, removing the needle, and connecting to the EMG amplifier free of the denuded wires

because of the need to measure the potential difference between electrodes. A third electrode, i.e., neutral or ground, is placed at an electrically neutral site, e.g., a nearby bony landmark. The ground electrode is used for common mode rejection, i.e., to prevent power line noise from interfering with the small biopotential signals of interest. The materials used to construct surface electrodes and their geometry can affect the signal measured. Proper and accurate placement of surface electrodes is important as well. Because surface electrodes are placed on the skin overlying superficial muscles, it is important to rigorously clean the skin's surface before securing the electrode; likewise, the electrodes need to be cleaned before use. Electrodes are placed on the skin approximately over the mid-belly of the muscle, avoiding tendinous and motor point regions, making sure that the electrodes are placed parallel to muscle fiber orientation. This can be a challenge with muscles that are multipennate, e.g., deltoid or gastrocnemius. The geometry of the electrodes, i.e., the spacing of electrodes, is also important to consider. For example, a greater interelectrode distance records a larger volume of muscle activity, as well as greater EMG amplitude. In most cases, a biological amplifier has a built-in predetermined distance set between electrodes. If the interelectrode distance is too large, the EMG volume recorded may come from muscles near the target muscle, producing a signal contaminated with cross talk, i.e., a signal from nontargeted muscles. In 2000, standards for the use and reporting of EMG signals were established and published as SENIAM (Surface EMG for a Non-Invasive Assessment of Muscles) recommendations (Hermens et al. 2000)

For the detection of electrical signals from small muscles and/or muscles that are deep, fine-wire electrodes are used (Fig. J.22). Two wires are used, just as two surface electrodes are needed to measure the bipolar nature of the potential difference in electrical signals. These very thin wires, housed in a hypodermic needle, are insulated except for their tips, which have about a millimeter of insulation removed (denuded). It is notable that if a greater portion of the tip insulation is removed, a greater volume of muscle electrical activity is recorded. Preparation for fine wire insertion does not require a sterile environment; however, the clinician and all assistants should don sterile gloves, and a clean environment should be set up whereby the skin overlying the target muscle is thoroughly exposed and cleaned using an alcohol swab. The fine wires are placed in the mid-belly of the target muscle via a hypodermic needle. After placement, the needle is retracted, leaving the wires secured in the muscle. The intramuscular wires can be more firmly secured within a muscle by asking the individual to gently contract/relax the target muscle. Once the wires are in place, the other end of the wires (also denuded) is secured to a muscle stimulator; this device allows the clinician to electrically stimulate the target muscle to check for accurate placement. Verification of the accurate placement of the wires can also be accomplished using ultrasound. Once an accurate placement is verified, the electrodes are attached to the EMG recording device. Additional verification of electrode placement is checked by first examining resting EMG activity in real time, followed by checking EMG during voluntary active muscle contraction.

Several factors can influence or distort the EMG signal (Fig. J.23). If these factors are not carefully considered, the fidelity of the EMG signal could be compromised, perhaps invaliding the data and their interpretation. We know that electrode characteristics can affect the frequency and amplitude of the EMG signal. These characteristics include the type of electrode, electrode size, interelectrode distance, and electrode configuration (monopolar versus bipolar). Moreover, a poor skin-electrode interface, i.e., inadequate skin preparation, and the amount of subcutaneous tissue between the surface electrode and the underlying muscle can markedly reduce signal quality. The greater thickness of the

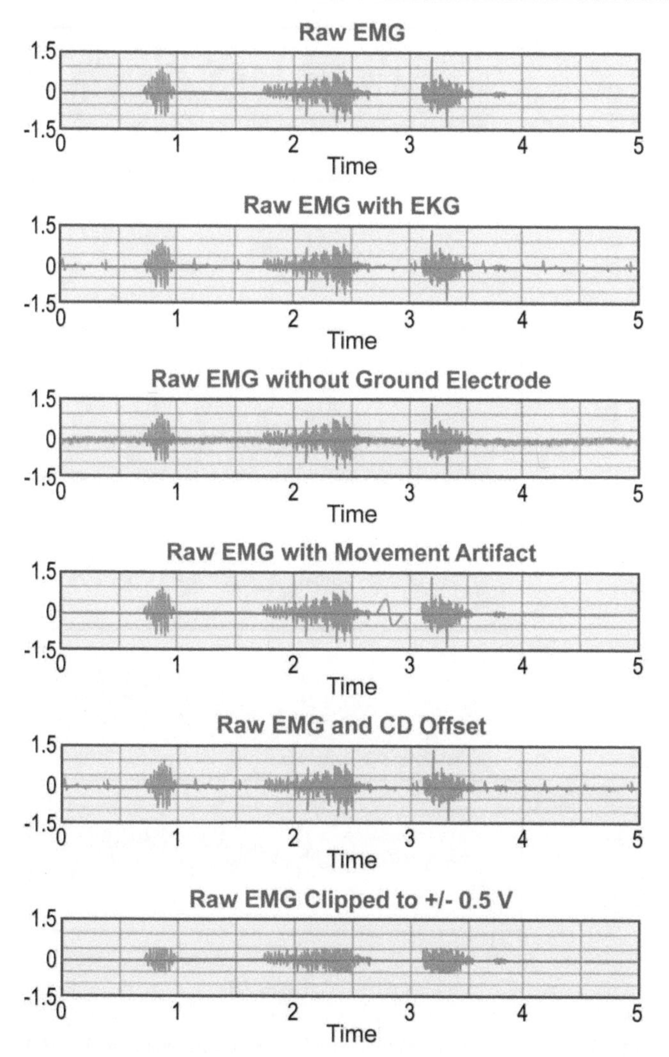


**Fig. J.23** Examples of distortions of the EMG signal, with the top signal as a true signal. (Modified with permission from Kamen (2014))

tissue between the surface electrode and the underlying muscle attenuates the higher frequency content of the EMG signal, while changes in muscle length during dynamic muscle actions can be dramatic, also affecting the frequency content of the EMG.

Surface electrodes are most suitable for recording the EMG of superficial and larger muscles. Attempting to record deeper muscles with surface electrodes, e.g., the tibialis posterior, will not likely provide a valid signal of the target muscle without being contaminated by crosstalk. Crosstalk is not just a problem when attempting to record deep muscle EMG using surface electrodes but is also a problem if the electrodes are recording from a very large volume. For example, surface electrode recording of one ankle peroneal muscle could include crosstalk from a neighboring peroneal muscle, the tibialis anterior, or the soleus. Additionally, crosstalk increases with the volume of the subcutaneous fat layer, e.g., gluteus maximus. Using fine-wire electrodes can effectively reduce crosstalk emanating from deep or small muscles. On the other hand, errors in the depth of insertion of the wires could also lead to crosstalk if insertion of the wires placed the wire tips are close to a neighboring deep muscle, e.g., flexor hallucis longus and tibialis posterior.

## EMG Data Acquisition, Reduction, and Interpretation

There are many technical aspects of an EMG amplifier that must be considered to ensure that the acquisition of signal accurately represents the summation of MUAPs and is free of noise or artifacts. An undistorted signal signifies that it has been amplified linearly over the range of the amplifier and recording system. Noise, either biological or man-made, consists of signals that are introduced from sources other than the target muscle. These could include an electrocardiogram (ECG) picked up by surface electrodes on the thoracic muscles and/or the "hum" from power lines, nearby machinery, or the EMG amplifier itself. Artifacts refer to false signals generated by electrodes (moving induced by body segment movements) or cabling. The technical aspects of the amplifier that need monitoring include amplifier gain, input impedance, common-mode rejection, and frequency response. Of these, we only briefly discuss frequency response. The reader may consult Winter (2009b) for additional technical information.

The collection frequency bandwidth of an EMG amplifier should be such as to amplify, without attenuation, all frequencies in the EMG. Quite simply, bandwidth refers to the difference between the upper cutoff frequency and the lower cutoff frequency. In order to select the proper data collection bandwidth, the spectrum of the EMG must be known. A power spectral analysis of EMG signals from normal muscles has determined that the spectrum of the EMG ranges from 5 to 2000 Hz. Fig. J.24 suggests that most of the EMG signal is concentrated in the band between 20 and 200 Hz for surface electrodes and 20–400 Hz for fine-wire electrodes, where movement artifacts are recognized in the 0–10 Hz frequency range. Based on the spectral analysis, the recommended data collection bandwidth range for surface EMG is 10–1000 Hz and 20–2000 Hz for fine-wire electrodes. Thus, typically for surface EMG, the signal is acquired using a high-pass filter set at about 20 Hz, with an aliasing low-pass filter set at about 1000 Hz; note that a low-pass filter retains (i.e., lets pass) frequency content below the cutoff frequency while attenuating higher frequency content.

Knowledge of the inherent frequency content of the EMG signal is important for choosing a sampling rate. Choosing a sampling rate is based on the Nyquist-Shannon sampling theorem. In brief, Nyquist-Shannon establishes a sufficient condition for a sampling rate that permits a discrete sequence of samples to capture all information from a continuous-time signal of finite bandwidth. The Nyquist limit requires that a sampling rate be at least twice that of the highest frequency

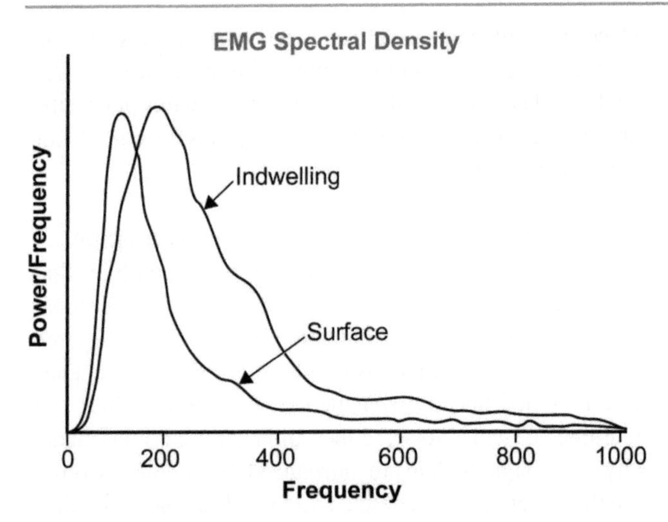


**Fig. J.24** Power frequency spectrum of EMG, as recorded using surface and fine-wire electrodes. Note the higher frequency content of fine wire recordings, likely related to the closer spacing between electrodes, as well as their closer proximity to active muscle fibers. (Modified with permission from Winter (2009b))

component in the signal. Thus, for the surface EMG signal, this typically requires a sampling rate of at least 1000 samples per second (i.e., Hz).

Unrelated to the Nyquist limit, but an important consideration, is the use of a notch filter. In the United States, 60 Hz signals related to overhead lights and other equipment can cause noise, which contaminates the EMG. Thus, in addition to a bandwidth filter, many biomechanists use a notch filter set at 60 Hz to eliminate this noise. Be aware, however, that based on the power frequency spectrum of the EMG (Fig. J.24), the use of a notch filter also eliminates real EMG signals.

A raw EMG signal consists of the superposition of electrical voltage associated with the propagation of action potentials along muscle fiber membranes. The waveform of MUAP depends on the orientation of surface sensors with respect to contracting fibers. Generally, MUAPs have a biphasic shape, i.e., positive and negative voltages. In human muscle, the amplitude of the action potential is dependent on the diameter of the muscle fibers, the distance between the muscle fibers and the detection site, and the filtering properties of the electrode. Moreover, the conduction velocity of the muscle fibers is dependent on the size of the motor unit; i.e., larger motor units produce greater conduction velocities. Thus, the raw EMG signal has a time-varying amplitude that fluctuates around zero and represents the amount of muscle activation, i.e., the number and size of motor units (Fig. J.25A).

The amplitude of the EMG signal varies over time during muscle action, as well as with the magnitude of the muscle force. Since raw EMG is biphasic, averaging the signal in an attempt to quantify it is not useful as it should always average out to be zero. As a result, several other data reduction methods have been developed to quantify the EMG signal.

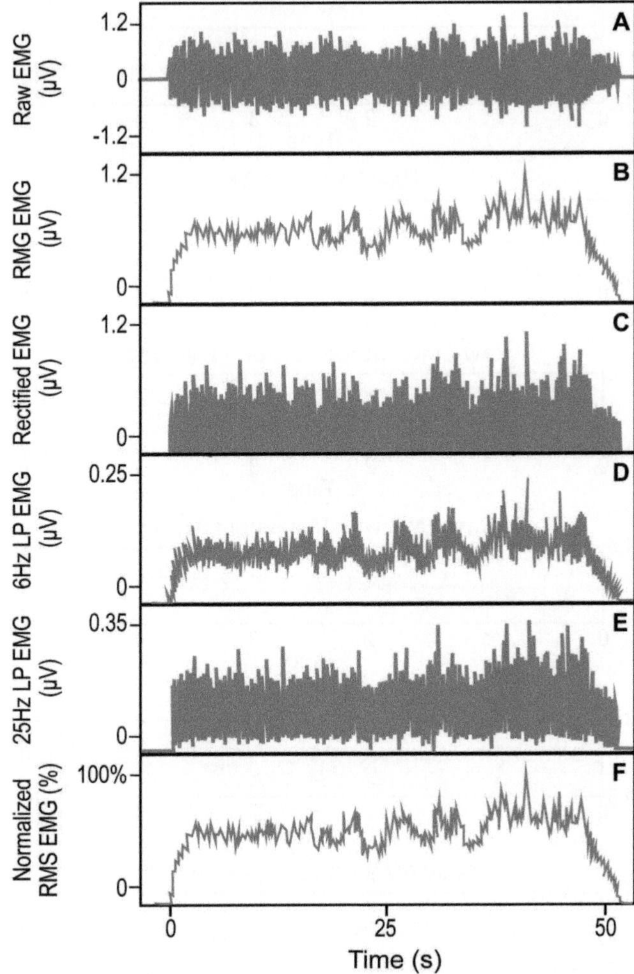


**Fig. J.25** (**A**) Raw EMG, (**B**) time-varying root mean square (RMS), (**C**) full-wave rectified EMG signal, (**D**) envelope calculated with a 6 Hz low-pass (LP) filter, (**E**) envelope calculated with a 25 Hz LP filter, and (**F**) time-varying RMS of the EMG signal normalized and displayed in percent (%). (Modified with permission from Contessa et al. (2018))

Rectification is a process that takes the absolute value of the EMG signal: (1) half-wave rectification passes the positive EMG values and blocks the negative values, whereas (2) full-wave rectification takes the absolute value of the EMG signal so that all negative values are transformed into positive voltage values (Fig. J.25C). Rectification is mainly used as a preprocessing step before applying another process, like averaging or computing the envelope of the signal. However, rectification can simply be used to present EMG data in a more usable format for determining the threshold for the onset or cessation of muscle action.

Calculation of the root-mean-square (RMS) value of the EMG signal is a common method for quantifying EMG signals (Fig. J.25B). It consists of a three-step calculation: (1) each data point in the signal is squared; (2) the average value over a specified window length or moving window, i.e., across a specified time interval, is determined; and (3) the square root of the value is then calculated for the moving window. Because RMS amplitude incorporates the squared values of the original EMG signal, it does not require full-wave rectification. RMS appears to better represent the amplitude of an EMG signal than does a raw or rectified signal and produces a waveform that is more easily analyzed.

Because the amplitude of a raw EMG signal is quite variable, low-pass filtering is commonly used to smooth the signal to provide a more "readable or interpretable" representation of its time-varying amplitude (Fig. J.25D, E). Using a full-wave rectified signal, the envelope procedure maintains the lower frequencies in the EMG signal while removing the higher frequencies; that is, removing some higher frequencies makes the waveform "smoother," i.e., with fewer high peaks. The area under the linear envelope appears to provide an estimate of the volume of EMG activity. From Fig. J.25 below, we can see that the degree of smoothing depends on the cutoff frequency of the low-pass filter. For example, there is more smoothing with the 6 Hz low-pass envelope, compared to the cutoff frequency at 25 Hz, which appears to promote a better visualization of peak values. The disadvantage of the lower cutoff is that it removes much of the frequency component of the EMG signal, possibly reducing signal amplitude.

We previously discussed the use of frequency spectrum analysis of surface and fine-wire EMG data to gain information about the frequency content of muscle action, information that is vital when choosing a sampling frequency. Another way to examine the frequency components of the EMG signal is through the use of the fast Fourier transform (FFT). This computation identifies the content of the EMG signal at each frequency. The frequency spectrum of the EMG signal depends primarily on the shape of its action potentials, which is related to the conduction velocity of muscle fibers. This technique is often used to quantify muscle fatigue. With muscle fatigue, there is a reduction in conduction velocity secondary to a shift toward the activation of lower-frequency motor units, i.e., smaller motor units, with a concomitant decrease in higher-frequency components. Thus, two parameters of the frequency spectrum of the EMG signal, i.e., median frequency and mean frequency, have been shown to decrease with muscle fatigue.

However, there are limitations in the use of the mean and median frequency analysis of muscles related to inherent variability in the system. For example, an increase in frequency does not necessarily indicate that more large motor units, i.e., fast twitch, are active. A greater frequency may indicate, instead, a greater firing rate of smaller motor units, i.e., slow twitch; the activation of muscle fibers with greater conduction velocities; decreased motor unit synchronization; or additional activation of synergist muscles. Conversely, a decrease in frequency may indicate not an increase in motor unit synchronization but a decrease in the total number of active motor units, a decrease in motor unit firing rate, or a reduction in contraction velocity. Finally, analysis of the EMG spectral frequency assumes that the EMG signal is stationary, i.e., linear. What does this infer? A stationary periodic signal suggests that there is a statistical similarity between the successive parts of a time series. It indicates that the mean and variance should not change as a function of time in the time series. In other words, linear methods interpret the structure of data through linear correlations, with the implication that the intrinsic dynamics of a system are governed by the notion that small changes in the system result in small effects. The multitude of possibilities for changes in frequency content, listed above, suggests that the EMG signal is nonstationary or nonlinear, a characteristic of dynamical or complex systems.

In fact, research has shown that the EMG signal is not stationary, nor does it have linear features. The EMG signal might better be characterized as an element of a dynamical system, one that is self-organizing and exhibits variability that is the result of nonlinear interactions with deterministic origins. A nonlinear signal cannot be analyzed using the same statistical tools used for linear systems. According to the dynamical system theory of movement, our central nervous system does not store patterns but rather stores a generative architecture that creates complex movements. Thus, the dynamical system approach focuses on how systems (1) maintain their current state, addressing questions of stability; (2) change or transition between states, addressing aspects of variability and adaptability; and (3) regulate complexity or fractal dynamics, characterized by interactions across different spatiotemporal scales, as well as invariance of processes across these scales. Nonlinear analysis of the variability, i.e., changes, in movement patterns provides a window into the neuromuscular workings of healthy, as well as neuropathic, individuals and, thus, insight into complex strategies for controlling movement and posture. In the past three decades, EMG research has begun to focus on techniques that more appropriately analyze the nonlinear characteristics of the EMG signal, including wavelet analysis, neural network classifications, recurrence quantification analysis, and fractal analyses. Obviously, detailing the nonlinear analysis of the EMG signal is well beyond the scope of this text, but we encourage the interested reader to seek other resources.

The most common approach to reliably (i.e., consistently) quantify the EMG amplitude of a muscle is to normalize the signal (Fig. J.25F). The magnitude of raw EMG represents two variables: the location of the electrode to the muscle's motor unit and the intensity of muscle action. However, many anatomical factors contribute to unreliable electrode placement, including the small size of muscle fibers, a varying mixture of slow (small motor units) and fast (large motor units) fiber types, a wide dispersion of motor units, fibrous planes separating muscle fiber bundles, and variations in the contour of individual muscles. These factors make it impossible to compare the raw EMG intensity of two similar muscles between individuals unless the signals are normalized. Normalization involves treating the data from each electrode as a ratio (%) of some reference value generated with the same electrode. For persons without neuropathology, the most convenient reference for the normalization process is the EMG signal from a maximum resisted manual muscle test using the peak EMG signal from a 4- or 5-s muscle action. The result is expressed as a percent of the reference value, i.e., % MMT or % MVIC (maximal voluntary isometric contraction). If the patient lacks normal motor control, two alternative normalization methods are used: (1) comparing the peak EMG of interest to the peak value obtained with the same electrode during the activity of interest, e.g., the gait cycle. However, the disadvantage of this technique is that peak EMG recordings of both weak and strong muscular activity are defined as 100%; or alternatively; (2) set and use a minimum threshold (submaximal muscle contraction) normalization value that can be assigned to represent the muscle's maximum voluntary contraction. Normalization is a widely accepted technique and a reasonably valid and reliable way of comparing activity among muscles and individuals and is a useful method for quantifying EMG amplitude in clinical gait analysis.

For clinical gait analysis, the intervals of EMG onset and cessation are of primary interest. These data can be determined directly from raw or full-wave rectified EMG by inspection. Of course, visual inspection is subjective as many recordings contain spikes (noise) or short bursts of extremely small amplitude signals. To circumvent these problems, it is important for the laboratory engineer to set minimum criteria for both the intensity and duration of the EMG signal. For example, computer analysis could be used to identify a minimal signal intensity equivalent to 5% of a maximum manual muscle test EMG, and a standard minimum duration of muscle action can be set to at least 5% of the gait cycle. In both instances, experienced practitioners with skill in discerning signal from noise are critical to making judgments about when a muscle is activated and when it is at "rest" and in determining valid EMG interpretations.

Abnormal motor control associated with neuropathology, muscle weakness, and both voluntary and reflex posturing/substitutions to accommodate pain or deformity leads to abnormal EMG patterns during walking. Both the timing and intensity of the EMG signal are altered in either particular phases and/or the entire gait cycle. Identifying the abnormal timing of muscle activity compared to normal is very important in the interpretation of pathological gait. The following definitions of timing errors of muscle activity are useful for gait interpretation:

- Premature – the action begins before the normal onset.
- Prolonged – the action continues beyond the normal cessation time.
- Continuous – EMG is uninterrupted for 90% or more of the gait cycle.
- Delayed – onset is later than normal.
- Curtailed – there is early termination of EMG.
- Absent – amplitude or duration of EMG is insufficient.
- Out of phase – the swing or stance EMG pattern is reversed

Examples of a few of these abnormalities can be seen in the images that follow. First,we provide a guide on how to interpret these data. The vertical axis expresses the magnitude of the EMG signal in volts; however, in this exercise, we do not evaluate the amplitude of the signal. The EMG signal is expressed as a percentage of a gait cycle (horizontal axis), where the beginning of the recording is at initial contact (0%) and the end of the recording is at the termination of the gait cycle (100% or second initial contact). The vertical line at about 60% of the gait cycle represents toe-off. The black horizontal bars below the raw EMG signal represent the onset (left end of the bar) and cessation (right end of the bar) of a normal EMG signal for the specific muscle of interest. For most of the example traces below, a "resting" or inactive muscle is represented by a baseline that is reasonably flat compared to the raw EMG signal. Note: each example presents data from 3 separate gait cycles, i.e., no*, one*, and two**.

- *Premature* – the action begins before the normal onset.
  - The diagnosis is cerebral palsy (CP) GMFCS II, toe initial contact pattern, and premature plantar flexion in midstance.

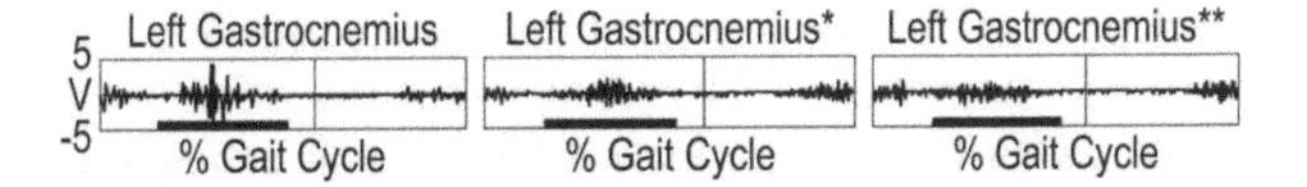


Note: this example expresses a very clean resting signal but an early onset of gastrocnemius activity, which actually begins in the terminal swing and is carried over into initial contact and the loading response.

With this, information you should be able to evaluate each of the tracings below, affirming how they have been categorized.

- *Prolonged* – the action continues beyond the normal cessation time.
  Example 1
  - The diagnosis is CP GMFCS III, mild crouch gait pattern.

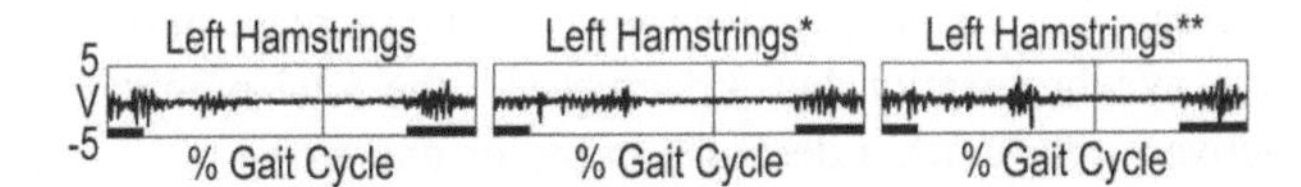

Example 2
– The diagnosis is CP GMFCS III, crouch gait pattern.

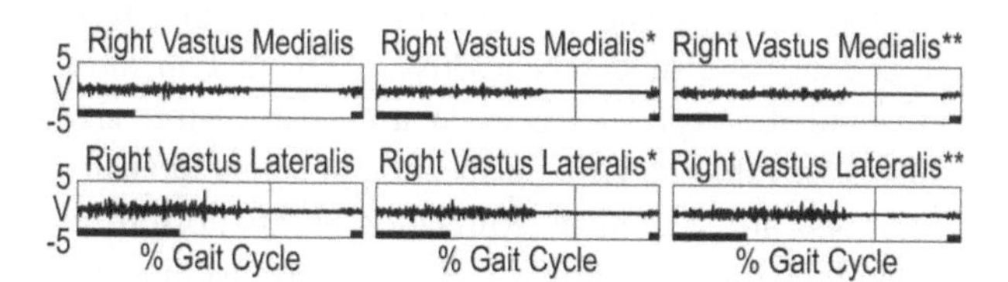


- *Continuous* – EMG is uninterrupted for 90% or more of the gait cycle.

Example 1

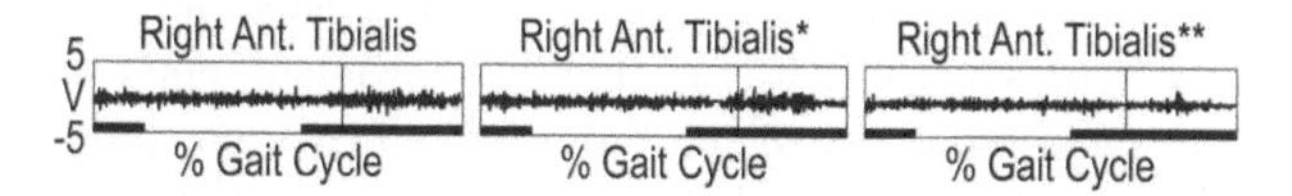


Example 2
– The diagnosis is CP GMFCS III, crouch gait pattern.

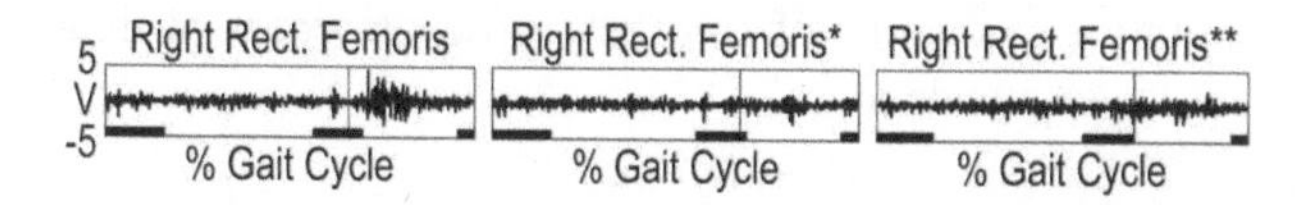


- *Delayed* – onset is later than normal.
  – The diagnosis is CP GMFCS II, toe-walking gait pattern.

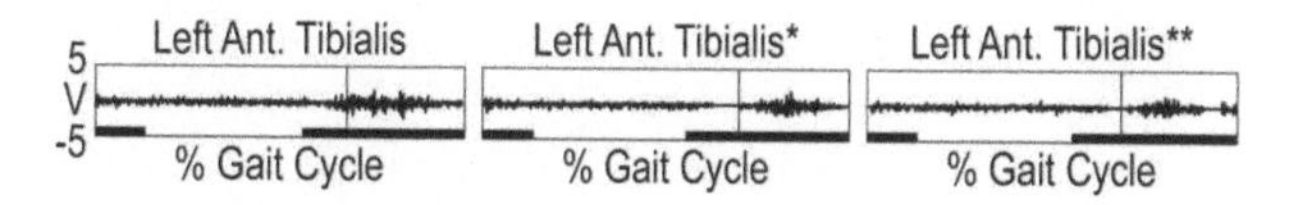


Note: in the example of a delayed signal, we see some noise in the signal, yet the EMG signal is clearly different.

- *Curtailed* – there is early termination of EMG.
  – The diagnosis is idiopathic toe walking, reflecting an increased equinus at initial contact and the absence of dorsiflexor moment.

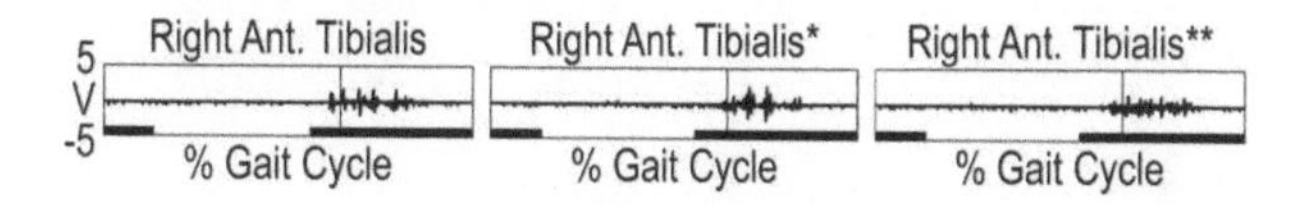


- *Absent* – the amplitude or duration of EMG is insufficient.
  Example 1
  – The diagnosis is Charcot-Marie Tooth (CMT) type 1, reflecting muscle weakness.

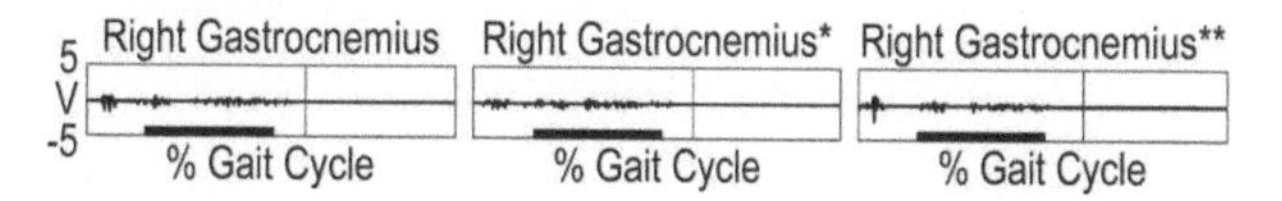


Example 2
– The diagnosis is CP GMFCS II, minimal quadriceps demand, slow walking velocity.

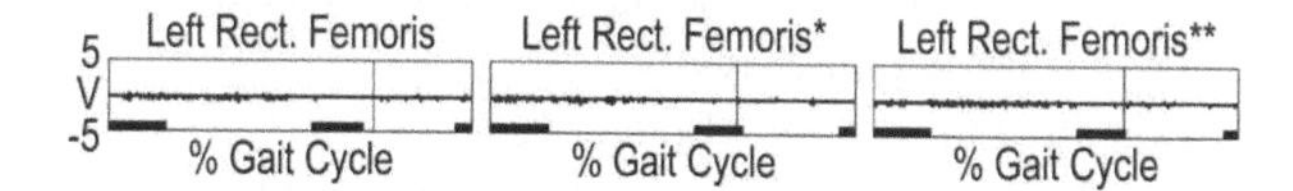


Note: in both of the examples in the "absent" category, we see signals that are not much different than a resting signal with some noise. A sufficient EMG signal should be comparable to the signals provided in the previous examples.

Note: the Gross Motor Functional Classification System (GMFCS) is a widely used description of the sitting and walking of, and the use of mobility devices for, children with neuropathology, e.g.,cerebral palsy. It can provide a clear description of a child's current motor function and a good idea of what equipment or mobility aids a child may need in the future. The GMFCS demarcates five levels of function:

**GMFCS Level I**

- Can walk indoors and outdoors and climb stairs without using hands for support
- Can perform usual activities, such as running and jumping
- Has decreased speed, balance, and coordination

**GMFCS Level II**

- Has the ability to walk indoors and outdoors and climb stairs with a railing
- Has difficulty with uneven surfaces and inclines or in crowds
- Has only minimal ability to run or jump

**GMFCS Level III**

- Walks with assistive mobility devices indoors and outdoors on level surfaces
- May be able to climb stairs using a railing
- May propel a manual wheelchair (may require assistance for long distances or uneven surfaces)

**GMFCS Level IV**

- Walking ability severely limited even with assistive devices
- Uses a wheelchair most of the time and may propel their own power wheelchair
- May participate in standing transfers

**GMFCS Level V**

- Has physical impairments that restrict the voluntary control of movement and the ability to maintain head and neck position against gravity
- Is impaired in all areas of motor function
- Cannot sit or stand independently, even with adaptive equipment
- Cannot independently walk, though may be able to use powered mobility

Identifying the abnormal intensity of the EMG signal is more challenging, particularly when only the raw signal is available. The operational definitions of pathological EMG intensity are as follows:

- Excessive – EMG value is greater than the normal band.
- Inadequate – EMG value is less than the normal band.
- Absent – EMG is insufficient to identify functional significance.

While these definitions are helpful, the evaluation of the signal remains difficult if only a raw EMG or full-wave rectified signal is available. As noted previously, although typically the density and magnitude of EMG are related to both the number and frequency of the firing of motor units, factors such as excessive subcutaneous tissue, changes in muscle length, and the type of electrode used can distort signal intensity. Therefore, a visual inspection of signal intensity is difficult and prone to misinterpretation unless postprocessing quantification of the raw signal is utilized. In clinical gait analysis, judgments about absent (or insufficient) signals can be valid and reliable, whereas judgments about signal intensity are not used.

In summary, evaluating EMG signals and synchronizing kinematic and kinetic data is effective for complex neuropathological gait patterns. Getting EMG signals that accurately represent muscle function is dependent upon the fidelity and reliability of the EMG instrumentation, proper preparation of the individual being tested, appropriate processing of the signals, and skillful evaluation of test measures. The integration of EMG, MOCAP, and inverse dynamics data maximizes the information available to assist in both the evaluation of gait and the development of treatment recommendations.

# Appendix K: Starting a Group in Problem-Based Learning

**Introductions** This may be the first step toward role definitions for each group member.

**Climate Setting** (outline goals and rules for group time; goals and rules might include time concerns, resources, schedules, and group member expectations; tutor also will define his/her role within the group)

- Express your ideas and thoughts freely.
- Comment on the ideas and opinions of others.
- Silence means assent (agreement).
- The role of the tutor: the tutor will ask students to reflect upon their own problem-solving behavior, and emphasize that acquiring knowledge is a means and end; knowledge is instrumental in the pursuit of competence in effectively managing problems.
- Your (student's) own role and responsibility.

### Starting with a Problem

- Setting group objectives (these are set as group dynamics are established and will evolve as groups learn more about their ability to collaborate and function)
- Presenting the problem or opening the case

### Assigning Tasks

- The reader
- The scribe

**Note:** These task assignments will change for each case

### Using the Board (Working Through a Problem Using a Whiteboard or Shared Electronic Document)

- Hypotheses: students' ideas for possible causes of the patient's problem. Students are encouraged to explain why they think these ideas are pertinent. Ideas are identified as students work through the patient's problem. Students usually identify these also as learning issues.
- Key facts: the growing synthesis of information gathered by the group through inquiry. This body of information is crucial for hypothesis generation. Data from the history and physical are recorded here, as well as results from diagnostics, the medical record, and other providers.
- Learning issues: students identify items/areas that need further study to enable them to understand the patient's problem. As students work through the problem, learning issues can be justified and more clearly identified.
- Actions: as students work through the problem, they begin to formulate a plan regarding further investigation and treatment modalities. Students learn about diagnostics by investigating how, when, and why tests are done. Additionally, students learn about therapeutics, appropriate consultations, etc.

### Reasoning Through the Problem

- Inquiry and analysis
- Problem syntheses
- Future actions (lab, treatments)
- Commitment to problem outcomes
- Learning issues shaping/assigning
- Resources identification
- Group follow-up

### Self-Directed Study

- Students seek resources (textbook, literature, internet, two-legged, etc.)

### Problem Follow-Up

- Resources: tutor and group members critique the types of resources used discussing quality, reliability, types of studies, etc.
- Reassess the problem, applying the information learned to discuss the hypotheses; this allows learning issues to be discussed, synthesized, and applied to help in the management of the patient



G. J. Alderink, B. M. Ashby, *Clinical Kinesiology and Biomechanics*, https://doi.org/10.1007/978-3-031-25322-5

- Review of hypothesis list and change it as appropriate
- Apply new learning to gain new facts from the patient's case
- Discuss and hand out references and notes from self-study
- Carry out laboratory and diagnostic procedures
- Identify new learning issues
- Final decisions about problem and management

**Conclusion**

- Summarize new knowledge and discuss generalizability
- Self-assessments and evaluations
- Peer, tutor, and group assessments and evaluations

# Appendix L: Behavioral Objectives: A Primer and Example

Many educators throughout the decades from the 1950s to the 1980s wrote about the importance of educational objectives, sometimes also referred to as outcomes, learning objectives, enabling objectives, performance objectives, and competencies. The early pioneers of this movement have provided a variety of definitions/characteristics:

"Intended change brought about in a learner" (Popham 1969).

"Explicit formulations of ways in which students are expected to be changed by the education process" (Bloom 1956).

"An objective is a description of a performance you want learners to be able to exhibit before you consider them competent. An objective describes an intended result of instruction, rather than the process of instruction itself" (Mager 1975).

"Properly constructed education objectives represent relatively specific statements about what students should be able to do following instruction" (Gallagher and Smith 1989).

According to Westberg and Jason (1993), some characteristics of effective educational objectives include the following:

- Consistent with the overall goals of the program
- Use of only one action verb and clearly stated
- Realistic and doable
- Appropriate for the learners' stages of development
- Appropriately comprehensive
- Worthy, complex outcomes
- Not treated as if they were etched in stone
- Not regarded as the only valuable outcomes

The purposes and functions of behavioral objectives include the following:

- To guide the teacher relative to the design of instruction
- To guide the teacher in evaluation/test design
- To guide the learner relative to the learning focus
- To guide the learner relative to self-assessment
- To assist in careful thinking about what is to be accomplished through instruction
- To assist in the relationship between the teacher and learner
- To enhance the possibility to create focused independent, self-directed learning
- To make teaching/learning more directed and organized
- To enable the teacher and learner to think carefully about what is important
- To assist the learner to make decisions regarding prioritizing
- To provide feedback to learners as objectives are accomplished

Bloom's taxonomy is a classification of the different objectives and skills that a teacher may utilize to create objectives for their learners. This taxonomy was proposed in 1956 by Benjamin Bloom, an educational psychologist at the University of Chicago. He suggested that the teacher structure learning objectives in the cognitive, affective, and psychomotor domains. The affective domain deals with the learner's emotions and how they are handled and includes the following major categories: receiving phenomena, responding to phenomena, valuing, organization, and internalizing values. The psychomotor domain, encompassing physical movement, coordination, and motor-skill usage, includes seven major categories: perception, set, guided response, mechanism, complex overt response, adaptation, and origination. For this primer, we will focus on objectives only in the cognitive domain.

There are six levels of learning in the cognitive domain:

1. *Remembering*: retrieving recognizing, and recalling relevant knowledge from long-term memory
2. *Understanding*: constructing meaning from oral, written, and graphic messages through interpreting, exemplifying classifying, summarizing, inferring, comparing, and explaining



G. J. Alderink, B. M. Ashby, *Clinical Kinesiology and Biomechanics*, https://doi.org/10.1007/978-3-031-25322-5

**Table L.1** Action verbs per category in the cognitive domain

| Bloom's level | Key verbs (keywords) |
|---|---|
| Create | Design, formulate, build, invent, create, compose, generate, derive, modify, develop |
| Evaluate | Choose, support, relate, determine, defend, judge, grade, compare, contrast, argue, justify, support, convince, select, evaluate |
| Analyze | Classify, break down, categorize, analyze, diagram, illustrate, criticize, simplify, associate |
| Apply | Calculate, predict, apply, solve, illustrate, use, demonstrate, determine, model perform, present |
| Understand | Describe, explain, paraphrase, restate, give original examples of, summarize, contrast, interpret, discuss |
| Remember | List, recite, outline, define, name, match, quote, recall, identify, label, recognize |

3. *Applying*: implementing or using a procedure for executing tasks
4. *Analyzing*: dividing material into constituent parts and determining how the parts relate to one another and to an overall structure or purpose through differentiating, organizing, and attributing
5. *Evaluating*: making judgments based on criteria and standards through checking and critiquing
6. *Creating*: synthesizing elements to form a coherent or functional whole and reorganizing elements into a new pattern or structure through generating, planning, or producing

Bloom's taxonomy is hierarchical, meaning that learning at higher levels is dependent on having attained prerequisite knowledge and skills at lower levels. However, it is not always necessary or prudent/efficacious to start with lower-order skills and step all the way through the entire taxonomy for each concept presented in a course.

A key to writing effective behavioral objectives using Bloom's taxonomy is the use of action verbs, i.e., "multilevel verbs", that apply to different activities for each domain of learning. Briefly, Bloom's original cognitive domain included six levels: knowledge, comprehension, application, analysis, synthesis, and evaluation. In the cognitive domain, the key verbs vary depending on the category (Table L.1).

The course objectives provided in a course syllabus are typically broad, covering major topics, e.g., functional anatomy, joint kinematics, etc., for a course in kinesiology/biomechanics. The primary purpose of assigning students to write case-specific objectives is to get them to be more actively engaged in their learning (a predominant goal in problem-based learning). Secondarily, the act of writing case-specific objectives assists students in meeting the course objectives. Writing lesson-specific (i.e., case-specific) objectives is not trivial, especially for students who, previously, likely rarely looked at syllabi objectives (perhaps a cynical perspective on my part), let alone had any prior experience in writing objectives. However, one of the advantages of the problem-based learning approach is its more holistic philosophical approach, which includes using the learning-by-doing strategy. On the other hand, students should not be left "out on a limb" with this assignment. Nor should students be expected to write case-specific objectives prior to beginning work on a case. Students can create their learning objectives as they work through a case, often with ongoing feedback from their tutor. After each case is completed, a summary of each case and students' lesson objectives can be collected, and tutors can provide formative feedback on both documents to help them direct their learning in future cases.

Case-specific objectives for the shoulder case can be provided as an introduction to the first case. These include objectives related to the application of problem-based learning, as well as objectives specific to the case. It is expected that students would use this model to help them construct objectives for each specific case that would follow over the course of the semester. In addition, general tips for writing functional learning objectives can be provided; for example, the objectives (1) should be short, concise, and easy to understand; (2) should be motivated by curiosity, what they need to do, and what they need to know in order to explain the clinical problem(s); (3) should be written using action verbs (see Table L.1) that indicate what they could expect to learn from each case, limiting to one action verb per objective; and (4) should be realistic, given the time constraints of the semester. When meeting to close each case, students can be provided with objectives written by the tutor, which could supplement their own work. Most students typically struggle with this assignment early on and generally make moderate progress in their objective-writing skills by the end of the semester.

Here are suggested objectives for the shoulder case (note: objectives 1–7 are generic and apply to all cases throughout the semester; the remaining objectives are germane to the first case and can serve as a template (although the reader is reminded that each case has its own unique aspects) for the remainder of the cases):

1. Apply concepts of group learning in the context of student-directed, tutor-assisted problem-based learning.
2. Apply concepts of learning by consensus.
3. Demonstrate professional behaviors in the context of group learning.
4. Demonstrate leadership and team skills.
5. Demonstrate a model of inquiry for clinical biomechanics.
6. Apply concepts of reflective and evidence-based practice in clinical biomechanics.
7. Apply analytical/logical/mathematical skills in the context of selected clinical biomechanicsstatics problems.
8. Describe relevant bony, muscular, ligamentous, and neural anatomy of the shoulder girdle complex and upper

quarter, i.e., cervical and upper thoracic spine and rib cage.
9. Describe the mechanism of injury.
10. Describe external and internal forces and moments across all relevant joints related to the mechanism of injury.
11. Describe normal mechanical properties of bones, the articular cartilage, other periarticular structures (i.e., triangular cartilage at the wrist, etc.), ligaments, and muscles and the role these properties play to attenuate external forces and moments related to the mechanism of injury.
12. Describe normal and pathological osteo- and arthrokinematics of all relevant joints of the shoulder girdle, cervical spine, etc.
13. Describe the pathomechanics as manifested by Mrs. Buckler's diagnosis of impingement syndrome.
14. Describe potential pathomechanics associated with osteoarthritis of the glenohumeral and acromioclavicular joints.
15. Explain the relationship between the impaired shoulder girdle force couples and impingement.
16. Explain possible fracture mechanics in the thoracic spine with osteoporosis.
17. Describe potential pathomechanics of shoulder girdler and thoracic cage function associated with healed stress fractures and abnormal postural alignment, i.e., Dowager's hump.
18. Clarify the medical diagnosis of impingement syndrome, i.e., identify possible mechanical etiologies and pain generators.
19. Explain the possible relationship between bilateral hand paresthesias and abnormal thoracic spine posture.
20. Solve shoulder statics equilibrium problem and discuss the clinical implications of solution(s).

Gordon Alderink, PT, PhD

# References

## Appendix A

Contini R (1972) Body segment parameters, part II. Artif Limbs 16:1–19

Dempster WT (1955) Space requirements of the seated operator. WADC-TR-159. Wright Patterson Air Force Base

Dempster WT, Gabel WC, Felts WJL (1959) The anthropometry of manual work space for the seated subjects. Am J Phys Anthropol 17:289–317

Drillis R, Cantini R (1966) Body segment parameters. Rep. 1163-03. Office of Vocational Rehabilitation, Department of Health, Education, and Welfare, New York

Richards J (2018) The comprehensive textbook of clinical biomechanics, 2nd edn. Elsevier Ltd

Winter DA (2009a) Biomechanics and motor control of human movement, 4th edn. John Wiley & Sons, Inc., Hoboken

## Appendix B

Uchida TK, Delp SL (2020a) Quantifying movement. In: Uchida TK, Delp SL (eds) Biomechanics of movement, the science of sports, robotics and rehabilitation. The MIT Press, Cambridge, MA

## Appendix C

Brand PW, Hollister AM, Agee JM (1999) Transmission. In: Brand PW, Hollister AM (eds) Clinical mechanics of the hand, 3rd edn. Mosby, Inc., St. Louis

Doyle JR (2001) Palmar and digital flexor tendon pulleys. Clin Orthop Relat Res 383:84–96

Özkaya N, Leger D, Goldsheyder D et al (2017a) Fundamentals of biomechanics, equilibrium, motion, and deformation, 4th edn. Springer Nature, Cham

## Appendix D

Özkaya N, Leger D, Goldsheyder D et al (2017b) Fundamentals of biomechanics, equilibrium, motion, and deformation, 4th edn. Springer, New York

## Appendix E

Özkaya N, Leger D, Goldsheyder D et al (2017c) Fundamentals of biomechanics, equilibrium, motion, and deformation, 4th edn. Springer, New York

## Appendix F

Özkaya N, Leger D, Goldsheyder D et al (2017d) Fundamentals of biomechanics, equilibrium, motion, and deformation, 4th edn. Springer, New York

## Appendix G

Özkaya N, Leger D, Goldsheyder D et al (2017e) Fundamentals of biomechanics, equilibrium, motion, and deformation, 4th edn. Springer, New York

## Appendix H

Özkaya N, Leger D, Goldsheyder D et al (2017f) Fundamentals of biomechanics, equilibrium, motion, and deformation, 4th edn. Springer, New York



G. J. Alderink, B. M. Ashby, *Clinical Kinesiology and Biomechanics*, https://doi.org/10.1007/978-3-031-25322-5

## Appendix I

Özkaya N, Leger D, Goldsheyder D et al (2017g) Fundamentals of biomechanics, equilibrium, motion, and deformation, 4th edn. Springer, New York

## Appendix J

Caldwell GE, Robertson GE, Whittlesey SN (2014) Forces and their measurements. In: Robertson DE, Caldwell GE, Hamill J, Kamen G, Whittlesey SN (eds) Research methods in biomechanics, 2nd edn. Human Kinetics, Champaign

Contessa P, De Luca CJ, Roy SH et al (2018) Electromyography. In: Richards J (ed) The comprehensive textbook of clinical biomechanics, 2nd edn. Elsevier Ltd

Davis RB, Õunpuu S, Tyburski D et al (1991) A gait analysis data collection and reduction technique. Hum Mov Sci 10(5):575–587. https://doi.org/10.1016/0167-9457(91)90046-Z

Hamill J, Selbie WS, Kepple TM (2014) Three-dimensional kinematics. In: Robertson DE, Caldwell GE, Hamill J, Kamen G, Whittlesey SN (eds) Research methods in biomechanics, 2nd edn. Human Kinetics, Champaign

Hermens HJ, Freriks B, Disselhorst-Klug C et al (2000) Development of recommendations for SEMG sensors and sensor placement procedures. J Electromyogr Kinesiol 10:361–374

Kamen G (2014) Electromyographic kinesiology. In: Robertson DE, Caldwell GE, Hamill J, Kamen G, Whittlesey SN (eds) Research methods in biomechanics, 2nd edn. Human Kinetics, Champaign

Meng L, Millar L, Childs C et al (2020) A strathclyde cluster model for gait kinematic measurement using functional methods: a study of the inter-assessor reliability analysis with comparison to anatomical models. Comput Methods Biomech Biomed Engin 23(12):844–853. https://doi.org/10.1080/10255842.2020.1768246

Neumann DA (2017) Kinesiology of the musculoskeletal system, foundations for rehabilitation, 3rd edn. Elsevier, Inc, St. Louis

Padmanabhan P, Puthusserypady S (2004) Non-linear analysis of EMG signals – a chaotic approach. In: Proceedings of the 26th Annual Conference of the IEEE EMBS, San Francisco, CA, USA. https://doi.org/10.1109/IEMBS.2004.1403231

Perry J, Burnfield JM (2010) Gait analysis, normal and pathological function, 2nd edn. SLACK, Inc., Thorofare

Selbie WS, Hamill J, Kepple TM (2014) Three-dimensional kinetics. In: Robertson DE, Caldwell GE, Hamill J, Kamen G, Whittlesey SN (eds) Research methods in biomechanics, 2nd edn. Human Kinetics, Champaign

Stergiou N (ed) (2016) Nonlinear analysis for human movement variability. CRC Press, Taylor & Francis Group LLC, Boca Raton

Uchida TK, Delp SL (2020b) Quantifying movement. In: Uchida TK, Delp SL (eds) Biomechanics of movement, the science of sports, robotics and rehabilitation. The MIT Press, Cambridge, MA

Van Emmerik Richard EA, Ducharme SW, Amado AC et al (2016) Comparing dynamical systems concepts and techniques for biomechanical analysis. J Sport Health Sci 5(1):3–13. https://doi.org/10.1016/j.jshs.2016.01.013

Winter DA (2009b) Kinesiological electromyography. In: Winter DA (ed) Biomechanics and motor control of human movement, 4th edn. John Wiley & Sons, Inc., Hoboken

## Appendix K

Barrows H (1996) What your tutor may never tell you. Southern Illinois University School of Medicine, Springfield

## Appendix L

Bloom BS (1956) Taxonomy of educational objectives, Handbook: the cognitive domain. David McKay, New York

Gallagher RE, Smith DU (1989) Formulation of teaching/learning objectives useful for the development of lessons, courses, and programs. J Cancer Educ 4(4):231–234

Mager RF (1975) Preparing objectives for instruction gall, 2nd edn. Fearon-Pitman Publishers, Belmont

Popham WJ (1969) Curriculum materials. Rev Educ Res 39(3):319–338. https://doi.org/10.3102/00346543039003319

Westberg J, Jason H (1993) Collaborative clinical education. Springer Publishing Company, New York

# Index




G. J. Alderink, B. M. Ashby, *Clinical Kinesiology and Biomechanics*, https://doi.org/10.1007/978-3-031-25322-5

Printed by Libri Plureos GmbH
in Hamburg, Germany